Photobiomodulation in the Brain

Low-Level Laser (Light) Therapy in Neurology and Neuroscience

Photobiomodulation in the Brain

Low-Level Laser (Light) Therapy in Neurology and Neuroscience

Edited by

Michael R. Hamblin
Wellman Center for Photomedicine, Massachusetts General Hospital, Boston, MA, United States
Department of Dermatology, Harvard Medical School, Boston, MA, United States

Ying-Ying Huang
Wellman Center for Photomedicine, Massachusetts General Hospital, Boston, MA, United States
Department of Dermatology, Harvard Medical School, Boston, MA, United States

ELSEVIER

ACADEMIC PRESS
An imprint of Elsevier

Academic Press is an imprint of Elsevier
125 London Wall, London EC2Y 5AS, United Kingdom
525 B Street, Suite 1650, San Diego, CA 92101, United States
50 Hampshire Street, 5th Floor, Cambridge, MA 02139, United States
The Boulevard, Langford Lane, Kidlington, Oxford OX5 1GB, United Kingdom

Notices
Knowledge and best practice in this field are constantly changing. As new research and experience broaden our understanding, changes in research methods, professional practices, or medical treatment may become necessary.

Practitioners and researchers must always rely on their own experience and knowledge in evaluating and using any information, methods, compounds, or experiments described herein. In using such information or methods they should be mindful of their own safety and the safety of others, including parties for whom they have a professional responsibility.

To the fullest extent of the law, neither the Publisher nor the authors, contributors, or editors, assume any liability for any injury and/or damage to persons or property as a matter of products liability, negligence or otherwise, or from any use or operation of any methods, products, instructions, or ideas contained in the material herein.

British Library Cataloguing-in-Publication Data
A catalogue record for this book is available from the British Library

Library of Congress Cataloging-in-Publication Data
A catalog record for this book is available from the Library of Congress

ISBN: 978-0-12-815305-5

For Information on all Academic Press publications
visit our website at https://www.elsevier.com/books-and-journals

Publisher: Nikki Levy
Acquisition Editor: Natalie Farra
Editorial Project Manager: Pat Gonzalez
Production Project Manager: Sruthi Satheesh
Cover Designer: Christian Bilbow

Typeset by MPS Limited, Chennai, India

Dedication

To the love of my life, my beautiful wife, Angela
Michael R. Hamblin

To Sophie and Ryan, you have always been great sources of inspiration, joy, and pride
Ying-Ying Huang

Contents

List of Contributors

Praveen R. Arany Department of Oral Biology and Biomedical Engineering, School of Dental Medicine, University at Buffalo, Buffalo, NY, United States

Jorge L. Arias INEUROPA (Instituto de Neurociencias del Principado de Asturias), Oviedo, Spain; Laboratory of Neuroscience, Department of Psychology, University of Oviedo, Asturias, Spain

Natalia Arias Department of Basic and Clinical Neuroscience, Institute of Psychiatry, Psychology and Neuroscience, King's College London, London, United Kingdom; INEUROPA (Instituto de Neurociencias del Principado de Asturias), Oviedo, Spain

Ronald Aung-Din Sarasota, FL, United States

Evan Austin Department of Dermatology, University of California at Davis, Sacramento, CA, United States; Dermatology Service, Sacramento VA Medical Center, Mather, CA, United States

Louis Banas Laser Innovations, Amherst, New York, United States

Matthew Bennett Patterson, CA, United States

Patrick Benson Bosch Institute, University of Sydney, Sydney, NSW, Australia; Discipline of Physiology, University of Sydney, Sydney, NSW, Australia

Rhett Bergeron Real Health Medical, Roswell, GA, United States

Marvin H. Berman Quietmind Foundation, Elkins Park, PA, United States

Yelena Bogdanova VA Boston Healthcare System, Boston, MA, United States; Department of Psychiatry, Boston University School of Medicine, Boston, MA, United States

Marco Antonio Caldieraro Universidade Federal do Rio Grande do Sul, Department of Psychiatry and Forensic Medicine, Porto Alegre, RS, Brazil; Hospital de Clínicas de Porto Alegre, Department of Psychiatry, Porto Alegre, RS, Brazil

Victoria Campbell RaVive Health, Inc., Operation Stand Tall Against TBI – A Non-Profit Organization, Calhan, CO, United States, National Association of Social Workers, Washington, D.C., United States, American Psychological Association, Washington, D.C., United States, Campbell Method for Treating TBI

James D. Carroll Thor Photomedicine Ltd., Chesham, United Kingdom

Paolo Cassano Depression Clinical and Research Program, Department of Psychiatry, Massachusetts General Hospital, Boston, MA, United States; Center for Anxiety and Traumatic Stress Disorders, Department of Psychiatry, Massachusetts General Hospital, Boston, MA, United States; Department of Psychiatry, Harvard Medical School, Boston, MA, United States

Marucia Chacur Laboratory of Functional Neuroanatomy of Pain, Department of Anatomy—ICB, University of São Paulo, São Paulo, Brazil

Agnes S. Chan Department of Psychology, The Chinese University of Hong Kong, Hong Kong, China; Chanwuyi Research Center for Neuropsychological Well-Being, The Chinese University of Hong Kong, Hong Kong, China

Suk-tak Chan Harvard Medical School, Massachusetts General Hospital, Boston, MA, United States; Department of Radiology, Athinoula A. Martinos Center for Biomedical Imaging, Massachusetts General Hospital, Boston, MA, United States

Linda Chao Departments of Radiology & Biomedical Imaging and Psychiatry, University of California, San Francisco, CA, United States

Francisco José Cidral-Filho Experimental Neuroscience Laboratory (LaNEx), University of Southern Santa Catarina, Palhocça, Santa Catarina, Brazil; Postgraduate Program in Health Sciences, University of Southern Santa Catarina, Santa Catarina, Brazil

Thomas J. Covey Division of Cognitive and Behavioral Neurosciences, Department of Neurology, Jacobs School of Medicine and Biomedical Sciences, University at Buffalo, Buffalo, NY, United States

Luis De Taboada Chief Technology Officer, LiteCure LLC, New Castle, DE, United States

Janis T. Eells Department of Biomedical Sciences, University of Wisconsin-Milwaukee, Milwaukee, WI, United States

Nabil El Massri Department of Anatomy, University of Sydney, Sydney, NSW, Australia

Andrea Fedoruk VA Boston Healthcare System, Boston, MA, United States

Manuel Fierro Department of Dermatology, University of California at Davis, Sacramento, CA, United States; Dermatology Service, Sacramento VA Medical Center, Mather, CA, United States

Saša R. Filipović Institute for Medical Research, University of Belgrade, Belgrade, Serbia

Sherry Fox BioCare Systems, Inc., Parker, CO, United States, Colorado BioScience Association, Denver, CO, United States, National Association of Laser Therapy, Baltimore, MD, United States, LumiWave NIR Therapy Device, Operation Stand Tall Against TBI – A Non-Profit Organization, Calhan, CO, United States

Juan Díaz González Área de Tecnología Electrónica de la Universidad de Oviedo, Gijon, Spain; Grupo de Electrónica para la Innovación Industrial, Gijon, Spain

Luke Gordon Bosch Institute, University of Sydney, Sydney, NSW, Australia; Discipline of Physiology, University of Sydney, Sydney, NSW, Australia

Rajiv Gupta Harvard Medical School, Massachusetts General Hospital, Boston, MA, United States; Department of Radiology, Neuroradiology Division, Massachusetts General Hospital, Boston, MA, United States

Michael R. Hamblin Wellman Center for Photomedicine, Massachusetts General Hospital, Boston, MA, United States; Department of Dermatology, Harvard Medical School, Boston, MA, United States; Harvard-MIT Division of Health Sciences and Technology, Cambridge, MA, United States

Catherine Hamilton Department of Anatomy, University of Sydney, Sydney, NSW, Australia

David Hamilton Department of Anatomy, University of Sydney, Sydney, NSW, Australia

Theodore A. Henderson Neuro-Laser Foundation, Centennial, CO, United States; The Synaptic Space, Centennial, CO, United States

Michael D. Ho VA Boston Healthcare System, Boston, MA, United States

Jason Huang Department of Neurosurgery, Baylor Scott & White Health, Dallas, TX, United States

Ying-Ying Huang Wellman Center for Photomedicine, Massachusetts General Hospital, Boston, MA, United States; Department of Dermatology, Harvard Medical School, Boston, MA, United States

Jared Jagdeo Department of Dermatology, University of California at Davis, Sacramento, CA, United States; Dermatology Service, Sacramento VA Medical Center, Mather, CA, United States; Department of Dermatology, State University of New York, Downstate Medical Center, Brooklyn, NY, United States

Daniel M. Johnstone Bosch Institute, University of Sydney, Sydney, NSW, Australia; Discipline of Physiology, University of Sydney, Sydney, NSW, Australia

Linda Ramball Jones Department of Physics and Astronomy, College of Charleston, Charleston, SC, United States

Ramanjot Kaur Department of Dermatology, University of California at Davis, Sacramento, CA, United States

Ivo I. Kerppers Laboratory of Neuroanatomy and Neurophysiology, University of Centro-Oeste, Guarapuava, Brazil

Boaz Kim Bosch Institute, University of Sydney, Sydney, NSW, Australia; Discipline of Physiology, University of Sydney, Sydney, NSW, Australia

Ji Yeon Kim Bosch Institute, University of Sydney, Sydney, NSW, Australia; Discipline of Physiology, University of Sydney, Sydney, NSW, Australia; School of Medicine, University of Queensland Centre for Clinical Research, Brisbane, QLD, Australia

Jeffrey A. Knight VA Boston Healthcare System, Boston, MA, United States; Department of Psychiatry, Boston University School of Medicine, Boston, MA, United States; National Center for PTSD—Behavioral Sciences Division, VA Boston Healthcare System, Boston, MA, United States

Andrey V. Kochetkov Federal State-Funded Educational Institution of Continuing Professional Education "Institute of Advanced Training of FMBA of Russia", Moscow, Russia

Ljubica M. Konstantinović Department of Rehabilitation, Faculty of Medicine, University of Belgrade, Belgrade, Serbia; Clinic for Rehabilitation "Dr Miroslav Zotović", Belgrade, Serbia

Bang-Bon Koo Bio-imaging Informatics Lab, Department of Anatomy and Neurobiology, Boston University School of Medicine, Boston, MA, United States

Maxine H. Krengel VA Boston Healthcare System, Boston, MA, United States; Department of Neurology, Boston University School of Medicine, Boston, MA, United States

Randy Lamartiniere Photo medicine Clinic, Baton Rouge, LA, United States

Paul A. Lapchak Neurocore LLC, Pomona, CA, United States

Tsz L. Lee Department of Psychology, The Chinese University of Hong Kong, Hong Kong, China

Yong Li Department of Neuroscience and Regenerative Medicine, Medical College of Georgia, Augusta University, Augusta, GA, United States

Huan Ling Liang Department of Cell Biology, Neurobiology and Anatomy, Medical College of Wisconsin, Milwaukee, WI, United States

Lew Lim Vielight Inc., Toronto, ON, Canada

George Louis Lindenfeld RESET Therapy Professional Training Institute LLC, Sarasota, FL, United States; RESET Therapy Professional Training Institute LLC, Hendersonville, NC, United States

Genane Loheswaran Vielight Inc., Toronto, ON, Canada

L. Longo Institute for Laser Medicine, International Academy for Laser Medicine and Surgery, Florence, Italy

Maria Gabriela Longo Department of Radiology, Neuroradiology Division, Massachusetts General Hospital, Boston, MA, United States

J.A. Lyons College of Health Sciences, University of Wisconsin-Milwaukee, Milwaukee, WI, United States

Javad Mahmoudi Neurosciences Research Center, Tabriz University of Medical Sciences, Tabriz, Iran

Ivan V. Maksimovich Clinic of Cardiovascular Diseases Named after Most Holy John Tobolsky, Moscow, Russia

Thomas Mang Department of Oral and Maxillofacial Surgery, School of Dental Medicine, University at Buffalo, Buffalo, NY, United States

Paula I. Martin VA Boston Healthcare System, Boston, MA, United States; Department of Neurology, Boston University School of Medicine, Boston, MA, United States

Daniel Fernandes Martins Experimental Neuroscience Laboratory (LaNEx), University of Southern Santa Catarina, Palhocça, Santa Catarina, Brazil; Postgraduate Program in Health Sciences, University of Southern Santa Catarina, Santa Catarina, Brazil

Daniel Oliveira Martins Laboratory of Functional Neuroanatomy of Pain, Department of Anatomy—ICB, University of São Paulo, São Paulo, Brazil

Melissa Meynadasy Division of Cognitive and Behavioral Neurosciences, Department of Neurology, Jacobs School of Medicine and Biomedical Sciences, University at Buffalo, Buffalo, NY, United States

John Mitrofanis Department of Anatomy, University of Sydney, Sydney, NSW, Australia

Larry D. Morries Neuro-Laser Foundation, Centennial, CO, United States

Sergey V. Moskvin The Federal State-Financed Institution "O.K. Skobelkin State Scientific Center of Laser Medicine under the Federal Medical Biological Agency" of Russia, Moscow, Russia

Margaret A. Naeser VA Boston Healthcare System, Boston, MA, United States; Department of Neurology, Boston University School of Medicine, Boston, MA, United States

Trent Nichols Quietmind Foundation, Elkins Park, PA, United States

Frank Nicklason Department of Anatomy, University of Sydney, Sydney, NSW, Australia; Geriatric Medicine, Royal Hobart Hospital, Hobart, TAS, Australia

Damir Nizamutdinov Department of Neurosurgery, Baylor Scott & White Health, Dallas, TX, United States

Amir Oron Department of Orthopedic Surgery, Kaplan Medical Center, Rehovot, Israel

Uri Oron Department of Zoology, George S. Wise Faculty of Life Sciences and Sagol School of Neuroscience, Tel Aviv University, Tel Aviv, Israel

Rodolfo Borges Parreira Salgado Institute of Integrative Health, Londrina, Brazil; Residency Program in Integrative Physical Therapy at UNIFIL University, Londrina, Brazil

Alberto Martín Pernía Área de Tecnología Electrónica de la Universidad de Oviedo, Gijon, Spain; Grupo de Electrónica para la Innovación Industrial, Gijon, Spain

Claudia Petrucco Bosch Institute, University of Sydney, Sydney, NSW, Australia; Discipline of Physiology, University of Sydney, Sydney, NSW, Australia

Nathali Cordeiro Pinto Physiotherapy, Bright Photomedicine Ltd., São Paulo, Brazil

Marcelo Victor Pires de Sousa Bright Photomedicine Ltd., São Paulo, Brazil

Eva-Maria Ratai Harvard Medical School, Massachusetts General Hospital, Boston, MA, United States; Department of Radiology, Athinoula A. Martinos Center for Biomedical Imaging, Massachusetts General Hospital, Boston, MA, United States; Department of Radiology, Neuroradiology Division, Massachusetts General Hospital, Boston, MA, United States

Carlo Rondinoni Institute of Radiology (INRAD), Universidade de São Paulo, São Paulo, Brazil

George Rozelle MindSpa Integrative Wellness Center, Sarasota, FL, United States

Saeed Sadigh-Eteghad Neurosciences Research Center, Tabriz University of Medical Sciences, Tabriz, Iran

Farzad Salehpour Neurosciences Research Center, Tabriz University of Medical Sciences, Tabriz, Iran; ProNeuroLIGHT LLC, Phoenix, AZ, United States

Afonso Shiguemi Inoue Salgado Salgado Institute of Integrative Health, Londrina, Brazil; Residency Program in Integrative Physical Therapy at UNIFIL University, Londrina, Brazil

Anita Saltmarche Saltmarche Health & Associates Inc., Orangeville, ON, Canada

David W. Shucard Division of Cognitive and Behavioral Neurosciences, Department of Neurology, Jacobs School of Medicine and Biomedical Sciences, University at Buffalo, Buffalo, NY, United States

William Stephan Buffalo, New York, United States

Jonathan Stone Bosch Institute, University of Sydney, Sydney, NSW, Australia; Discipline of Physiology, University of Sydney, Sydney, NSW, Australia

Luis De Taboada Chief Technology Officer, LiteCure LLC, New Castle, DE, United States

M.A. Tolentino College of Health Sciences, University of Wisconsin-Milwaukee, Milwaukee, WI, United States

Lorelei Tucker Department of Neuroscience and Regenerative Medicine, Medical College of Georgia, Augusta University, Augusta, GA, United States

Erica B. Wang Department of Dermatology, University of California at Davis, Sacramento, CA, United States; Dermatology Service, Sacramento VA Medical Center, Mather, CA, United States

Nicholas Alexander Wise Department of Physical Medicine and Rehabilitation, UNC Chapel Hill School of Medicine, Chapel Hill, NC, United States

Margaret Wong-Riley Department of Cell Biology, Neurobiology and Anatomy, Medical College of Wisconsin, Milwaukee, WI, United States

Mei X. Wu Wellman Center for Photomedicine, Massachusetts General Hospital, Boston, MA, United States; Department of Dermatology, Harvard Medical School, Boston, MA, United States; Harvard-MIT Division of Health Sciences and Technology, Cambridge, MA, United States

Luodan Yang Department of Neuroscience and Regenerative Medicine, Medical College of Georgia, Augusta University, Augusta, GA, United States

Michael K. Yeung Department of Psychology, The Chinese University of Hong Kong, Hong Kong, China; Department of Neurology and Neurosurgery, Montreal Neurological Institute, McGill University, Montreal, QC, Canada

Elisabeth Mateus Yoshimura Institute of Physics, Laboratory of Radiation Dosimetry and Medical Physics, University of São Paulo, São Paulo, Brazil

Quanguang Zhang Department of Neuroscience and Regenerative Medicine, Medical College of Georgia, Augusta University, Augusta, GA, United States

Reza Zomorrodi Temerty Centre for Therapeutic Brain Intervention, Centre for Addiction and Mental Health, Toronto, ON, Canada

Preface

Photobiomodulation (PBM) also known as low-level laser (or light) therapy has been known for over 50 years (since 1967), but it is only relatively recently that it has begun to make the transition into the mainstream. PBM describes the use of red or near-infrared light at levels that do not produce undue heating of the tissue to produce beneficial effects on the human body. The introduction of light-emitting diodes (LEDs) has made this approach more accessible than the previously used laser sources, as LEDs are safer, cheaper, and can easily be used at home. Another factor that has led to PBM becoming more widely accepted is the growing understanding of the mechanisms of action at a molecular and cellular level. The lack of a clear mechanism of action was a deterrent to many biomedical scientists who maintained a healthy level of skepticism.

Among the wide range of tissues, organs, diseases, and conditions that can be beneficially affected by PBM, the subject of this book is the brain. The brain is probably the single human organ that engenders the most concern, interest, and expenditure in the 21st century. Brain disorders that cause widespread morbidity, mortality, and loss of quality of life can be divided into four broad categories. Traumatic brain disorders include stroke, traumatic brain injury (TBI), global ischemia, and perinatal difficulties. Neurodegenerative diseases include Alzheimer's disease, Parkinson's disease, and a range of dementias. Psychiatric disorders include major depression, anxiety, addiction, and insomnia, among many others. Finally there are neurodevelopmental disorders (autism and ADHD) and the possibility of cognitive enhancement in healthy individuals. Many of these brain disorders are specifically addressed in the present volume.

The book is divided into three parts. The first part covers some basic considerations, dosimetry, and devices, and discusses the mechanisms of action at a cellular level and on the brain as a whole organ. The second part includes contributions from researchers who have carried out studies on a variety of animal models in their investigations of brain disorders, stroke, TBI, and Alzheimer's and Parkinson's diseases, to name a few. The third part concentrates on human studies, including controlled clinical trials, pilot trials, case series, and clinical experience. Disorders treated include TBI, stroke, Alzheimer's and Parkinson's diseases, depression, and others.

The book is expected to play a role in stimulating the further increase and acceptance of PBM for brain disorders, which has really started to take off in recent years. It will also act as a resource for researchers and physicians wishing to get a broad overview of the field and who are contemplating entering it themselves. The number of individuals considering obtaining a home-use PBM device is also steadily increasing and this book will act as an authoritative source of unbiased, well-researched, information, which is all the more necessary in the Internet age.

Part I

Basic considerations and in vitro

Chapter 1

Photobiomodulation therapy and the brain: an innovative tool for therapy and discovery

Praveen R. Arany
Department of Oral Biology and Biomedical Engineering, School of Dental Medicine, University at Buffalo, Buffalo, NY, United States

1.1 Introduction

Light has played a central role in human health in various forms such as its regulation of the diurnal circadian rhythm, enabling vision, or sunlight in vitamin D metabolism. There are several well-established studies indicating the key role of light in psychological health and the correlation of poorly lit, dark spaces with depression. This book is dedicated to outlining the evidence for the therapeutic applications of the low dose light treatment termed photobiomodulation (PBM) therapy (Anders et al., 2015). However, before attempting to discuss the putative therapeutic use of light on the human brain, the very nature of this unique organ is addressed. This brief chapter will address the basic characteristics and rationale for the use of PBM therapies for human neurocognition (Hennessy and Hamblin, 2017).

The formidable complexity of the human brain clearly extends well beyond simple ultrastructural architecture and functional anatomy. It is a surprising fact that, unlike many organs, the cell-tissue structural homogeneity does not correlate with the functional distinction; and intricate interconnectivity and higher level integration are responsible for routine functions. It is an often cited fact that among most organs, the brain appears to be one of the most studied and least understood. It would seem appropriate to compare the brain to a black box due to its internal mechanistic nature concealed during routine operation. The physiological components are best highlighted when the brain is rendered dysfunctional by some disorder (loss of function), which is a commonly employed research laboratory strategy such as the use of advanced transgenic techniques. In this context, an intervention capable of reestablishing these biological functions could offer potential insights into the pathophysiological roles. Hence, brain pathologies and interventions to remedy them may provide significant avenues to better understand routine brain functions.

1.1.1 Beyond the structure-function architecture of the human brain

To begin, it is quite common to use the physical structural organization of the brain as a starting point for any scientific exploration. The discrete anatomical landmarks and histological composition of the brain have been well documented. The anatomy of the human brain based on evolutionary and developmental origins is divided into the cerebrum, cerebellum, and brain stem with several specialized functional subunits. Gross examination of the brain reveals gray and white matter areas with several specialized neuronal cell types contained within them. Among them, neurons function as the primary information processing cells of the brain which are supported by specialized cells such as the glia, astrocyte, and endothelial cells among others. This concept has laid the foundation of modern neurosciences as well as neurosurgical manipulation that relies on these gross architectural characteristics (Fig. 1.1). Functional assessment of these tissues and their interconnectivity has led to allocation of primary functional units of the brain responsible for specific sensory integration and responses such as the motor and sensory cortex. This has been the premise of clinical neurology and psychology. The popular use of functional magnetic resonance imaging has opened new vistas in our

Photobiomodulation in the Brain. DOI: https://doi.org/10.1016/B978-0-12-815305-5.00001-4

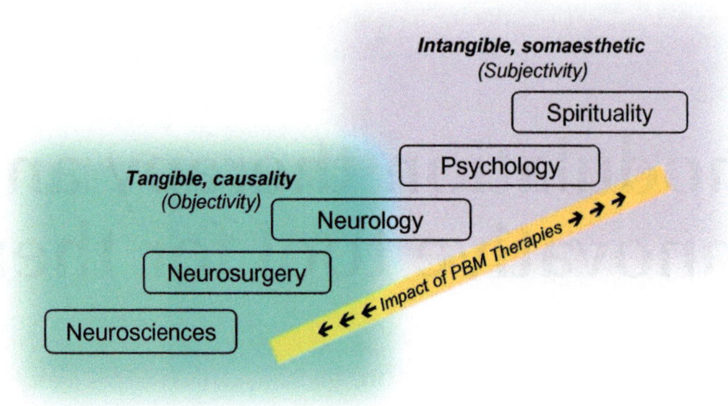

FIGURE 1.1 Spectrum of fields concerning the anatomical and pathophysiological aspects of the human brain involving both tangible and intangible aspects of functional determination. The breadth of potential assessment avenues are important to appreciate for thorough evaluation of various interventional therapies such as photobiomodulation therapy.

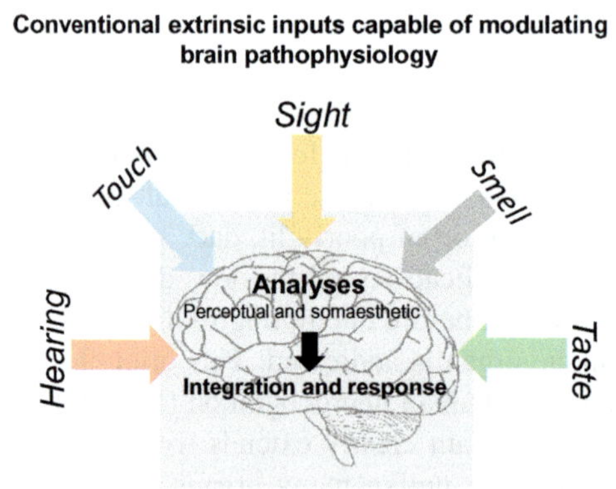

FIGURE 1.2 External sensory inputs that provide information to the brain for routine physiological functions. Their precise mechanisms and integrated neural pathways can provide potential avenues for a better understanding of biophotonic intervention during photobiomodulation therapy.

exploration of functional brain anatomy. These remarkable explorations have advanced our understanding of the physical structure, cellular constituents, and functional organization. Nonetheless, the brain's remarkable abilities to generate abstract thoughts, store and recall memories, and integrate sensory and motor responses remain to be explicitly defined. These seemingly simple questions appear to be intimately linked to perceptual integration within the brain and are influenced by routine daily activities such as the rejuvenating nature of sleep, physical exercise, daily dietary consumption, as well as the physical security and comfort of the living environment among many other factors. From the comfortable confines of the physical nature of neurosciences, these latter aspects of brain function extend into the study of the mind and behavior leading to psychology. A basic tenet of this field is the emphasis on an individual's interpretations of personal, assimilated experiences through formal (structured, instructional, or education) and informal (non-structured) interactions. Perhaps the epitome of intellectual exploration leads to existential questions about the purpose of life and inevitable final outcome of death. These questions appear to lead to the esoteric limits of spirituality, and eventually the characteristics of science and religion would seem to coalesce. Every biological field, including the study of pathophysiology of the human brain, appears to bridge this spectrum from rigorous causality to possible eventuality.

1.1.2 A bottom-up approach to brain neurosciences

To fully comprehend the ability of PBM therapy to modulate the human brain, it is imperative to first appreciate the nature and characteristics of brain function. The most interesting concepts of brain function appear to have extended from biology and medicine into the realms of engineering with the remarkable recent advancements in artificial intelligence (AI). The promise of AI regarding data acquisition, analysis, and interpretation has made tremendous strides (Fig. 1.2). It is worth emphasizing the critical importance of the first step involving reliance on external stimuli (sensors). Progress in artificial vision has received tremendous attention due to autonomous automation. Improvements in restoration of senses such as vision, touch, olfaction, taste, and hearing have benefited from artificial implantable prosthesis (Chuang et al., 2014; Fitzgerald et al., 2017; Kobayashi et al., 2010; Tisch, 2017; Lucarotti et al., 2013). The importance of exploring these synthetic approaches for replacing or improving conventional sensory organs could help outline the nature and pathways of propagation and interpretation of sensory stimuli within the brain. Besides the direct inputs from vision, these other sensory pathways may be potentially modulated by biophotonic stimulation providing nonconventional extrinsic avenues to modulate brain functions. The discovery of nonvisual photoreceptor cells in the eye, the intrinsic photoreceptive ganglion cells, is a good example (Melyan et al., 2005). Similar investigations of the visual system have attempted to carefully examine the roles of biophotonics and biological perception and functions. As alluded to previously, the direct role of visible (blue wavelength specifically) light in maintaining the pineal gland secretions of melatonin to maintain the circadian rhythm has been well established (Hattar et al., 2002). The predominantly deleterious effects of the more powerful, ionizing wavelengths (ultraviolet, X-ray, and gamma rays) that damage biomolecules are well established. Recent investigation into the biological perception of nonvisible (near-infrared) light raises interesting possibilities concerning the limits and pathways of perceptual capabilities of the human brain (Palczewska et al., 2014). Hence, there appears to be increasing overall recognition of the role of several light-sensitive, nonvisual phototransduction pathways in human health (Cronin and Johnsen, 2016; Van Gelder, 2008).

1.1.3 Modulating the "brain black box" with light

Interventions for brain disorders have ranged from innocuous, noninvasive extrinsic modification to extremely invasive, surgical resections of different lobes of the brain. Neurocognitive modulation techniques ranging from both environmental reengineering to behavioral modulation, such as mindfulness, have gained significant traction in recent years with rigorous feedback-based objective measurement of outcomes (Creswell, 2017; Hilton et al., 2017). The advent of functional imaging has greatly focused and directed some of these interventional studies into current techniques that can ablate epileptic foci and carry out deep brain stimulation (McGovern et al., 2016; Aum and Tierney, 2018; Miocinovic et al., 2013). Among these various approaches, there have been several intriguing observations concerning nonvisual roles of light in brain health (Fig. 1.3) (Hennessy and Hamblin, 2017). This book highlights specific information for the use of low dose light treatments or PBM therapy. There are several questions in this innovative field but two fundamental questions seem to dominate, namely, delivery of a clinically beneficial dose and basic biological mechanisms. First, several attempts have been made at careful dose modeling and physical assessment of light distribution following transcranial PBM treatment of the human brain (Yue and Humayun, 2015; Tedford et al., 2015). There is clear evidence that some, albeit miniscule, amounts of light are effectively transmitted to deeper parts of the brain following external PBM treatment to the head. There is surprising evidence that the human cells in the visual system are capable of detecting only a single photon (Tinsley et al., 2016; Ala-Laurila and Rieke, 2014). A remaining key

Big questions in brain pathophysiology	Potential insights with photobiomodulation therapies
Nature of memory–creation, storage and recall	Benefits noted in age-related dementia and Alzheimer disease
Complexity of abstract thoughts	Improved neurocognitive performance and executive functions
Rejuvenating nature of sleep	Improvements in disrupted sleep patterns following TBI
Daily nutritional energy requirements	Improved neurocognitive and skeletomuscular performances
Moods	Negative correlation of migraines and bright lights; Well-lit places correlates with less depression; Improvements in depression, PTSD and anxiety disorders
Coordinated and/fine motor skills	Improvement of motor tremors in Parkinson patients

FIGURE 1.3 Key questions in brain pathophysiology that could be potentially addressed with clinically observed benefits following photobiomodulation therapy in healthy and diseased individuals.

question is how high a PBM dose needs to be to drive a specific biological response; attempts are ongoing to integrate photon distribution with biological dose-response models (Arany, 2016). The implications of these PBM dose models are discussed in several chapters in this book. A second critical aspect of clinical translation of PBM therapy is the nature and responsiveness of specific biological targets and molecular mechanisms. Given the multiple tissue types and cell lineages involved in brain pathologies, several putative targets have been suggested. These include the vascular endothelial cells and perfusion-related blood supply; macrophages, mast cells neutrophils, and lymphocytes that modulate the inflammation and immune response; stem cells and tissue healing and regeneration among many others (Hamblin, 2016; Cassano et al., 2016). Significant advances have been made in optogenetics, generating some excitement for the possibility of unraveling functional neural pathways. Optogenetics relies on engineered, exogenous chromophores that can potentially be extended to current PBM investigation that relies on endogenous, naturally photoresponsive biological chromophores. This has been called "endogenous optogenetics" (Arany, 2016).

It is prudent to note that certain global biological responses are attractive PBM targets such as pain, inflammation, immune response, and wound healing. These basic Virchow signs are important across various neurodegenerative diseases that have all been shown to benefit from PBM therapy such as Parkinson disease, multiple sclerosis, Alzheimer disease, and concussion (traumatic brain injury). However, even though modulation of these pathophysiological responses could improve clinical symptoms, it may not directly address the underlying causal disease processes. Moreover, PBM therapy has been shown to be effective in disorders such as posttraumatic stress disorders, depression, addiction, and improved neurocognitive performance, even in healthy individuals, who do not have diagnosed alteration in these pathophysiological responses. Hence, it is reasonable to expect sustained, repeated treatments with varying dosimetry of PBM (including wavelength combinations, dose and pulsing regimens, coherence, and polarization among others) to generate rigorous, reproducible PBM therapeutic benefits. In conclusion, PBM therapy provides a noninvasive approach to modulate the human brain for both therapeutic benefits as well as serve as a discovery tool to better understand its complexity and critical normal functions.

References

Ala-Laurila, P., Rieke, F., 2014. Coincidence detection of single-photon responses in the inner retina at the sensitivity limit of vision. Curr. Biol. 24, 2888–2898. Available from: https://doi.org/10.1016/j.cub.2014.10.028.

Anders, J.J., Lanzafame, R.J., Arany, P.R., 2015. Low-level light/laser therapy versus photobiomodulation therapy. Photomed. Laser Surg. 33, 183–184. Available from: https://doi.org/10.1089/pho.2015.9848.

Arany, P.R., 2016. Craniofacial wound healing with photobiomodulation therapy: new insights and current challenges. J. Dent. Res. 95, 977–984. Available from: https://doi.org/10.1177/0022034516648939.

Aum, D.J., Tierney, T.S., 2018. Deep brain stimulation: foundations and future trends. Front. Biosci. (Landmark Ed.) 23, 162–182.

Cassano, P., Petrie, S.R., Hamblin, M.R., Henderson, T.A., Iosifescu, D.V., 2016. Review of transcranial photobiomodulation for major depressive disorder: targeting brain metabolism, inflammation, oxidative stress, and neurogenesis. Neurophotonics 3, 031404. Available from: https://doi.org/10.1117/1.NPh.3.3.031404.

Chuang, A.T., Margo, C.E., Greenberg, P.B., 2014. Retinal implants: a systematic review. Br. J. Ophthalmol. 98, 852–856. Available from: https://doi.org/10.1136/bjophthalmol-2013-303708.

Creswell, J.D., 2017. Mindfulness interventions. Annu. Rev. Psychol. 68, 491–516. Available from: https://doi.org/10.1146/annurev-psych-042716-051139.

Cronin, T.W., Johnsen, S., 2016. Extraocular, non-visual, and simple photoreceptors: an introduction to the symposium. Integr. Comp. Biol. 56, 758–763. Available from: https://doi.org/10.1093/icb/icw106.

Fitzgerald, J.E., Bui, E.T.H., Simon, N.M., Fenniri, H., 2017. Artificial nose technology: status and prospects in diagnostics. Trends Biotechnol. 35, 33–42. Available from: https://doi.org/10.1016/j.tibtech.2016.08.005.

Hamblin, M.R., 2016. Shining light on the head: photobiomodulation for brain disorders. BBA Clin. 6, 113–124. Available from: https://doi.org/10.1016/j.bbacli.2016.09.002.

Hattar, S., Liao, H.W., Takao, M., Berson, D.M., Yau, K.W., 2002. Melanopsin-containing retinal ganglion cells: architecture, projections, and intrinsic photosensitivity. Science 295, 1065–1070. Available from: https://doi.org/10.1126/science.1069609.

Hennessy, M., Hamblin, M.R., 2017. Photobiomodulation and the brain: a new paradigm. J. Opt. 19, 013003. Available from: https://doi.org/10.1088/2040-8986/19/1/013003.

Hilton, L., et al., 2017. Mindfulness meditation for chronic pain: systematic review and meta-analysis. Ann. Behav. Med. 51, 199–213. Available from: https://doi.org/10.1007/s12160-016-9844-2.

Kobayashi, Y., et al., 2010. Advanced taste sensors based on artificial lipids with global selectivity to basic taste qualities and high correlation to sensory scores. Sensors (Basel) 10, 3411–3443. Available from: https://doi.org/10.3390/s100403411.

Lucarotti, C., Oddo, C.M., Vitiello, N., Carrozza, M.C., 2013. Synthetic and bio-artificial tactile sensing: a review. Sensors (Basel) 13, 1435–1466. Available from: https://doi.org/10.3390/s130201435.

McGovern, R.A., Banks, G.P., McKhann 2nd, G.M., 2016. New techniques and progress in epilepsy surgery. Curr. Neurol. Neurosci. Rep. 16, 65. Available from: https://doi.org/10.1007/s11910-016-0661-6.

Melyan, Z., Tarttelin, E.E., Bellingham, J., Lucas, R.J., Hankins, M.W., 2005. Addition of human melanopsin renders mammalian cells photoresponsive. Nature 433, 741–745. Available from: https://doi.org/10.1038/nature03344.

Miocinovic, S., Somayajula, S., Chitnis, S., Vitek, J.L., 2013. History, applications, and mechanisms of deep brain stimulation. JAMA Neurol. 70, 163–171. Available from: https://doi.org/10.1001/2013.jamaneurol.45.

Palczewska, G., et al., 2014. Human infrared vision is triggered by two-photon chromophore isomerization. Proc. Natl. Acad. Sci. U.S.A. 111, E5445–E5454. Available from: https://doi.org/10.1073/pnas.1410162111.

Tedford, C.E., DeLapp, S., Jacques, S., Anders, J., 2015. Quantitative analysis of transcranial and intraparenchymal light penetration in human cadaver brain tissue. Lasers Surg. Med. 47, 312–322. Available from: https://doi.org/10.1002/lsm.22343.

Tinsley, J.N., et al., 2016. Direct detection of a single photon by humans. Nat. Commun. 7, 12172. Available from: https://doi.org/10.1038/ncomms12172.

Tisch, M., 2017. Implantable hearing devices. GMS Curr. Top Otorhinolaryngol. Head Neck Surg. 16. Available from: https://doi.org/10.3205/cto000145.

Van Gelder, R.N., 2008. Non-visual photoreception: sensing light without sight. Curr. Biol. 18, R38–R39. Available from: https://doi.org/10.1016/j.cub.2007.11.027.

Yue, L., Humayun, M.S., 2015. Monte Carlo analysis of the enhanced transcranial penetration using distributed near-infrared emitter array. J. Biomed. Opt. 20, 88001. Available from: https://doi.org/10.1117/1.JBO.20.8.088001.

Chapter 2

Theoretical neuroscience

Marcelo Victor Pires de Sousa[1], Marucia Chacur[2], Daniel Oliveira Martins[2] and Carlo Rondinoni[3]

[1]*Bright Photomedicine Ltd., São Paulo, Brazil,* [2]*Laboratory of Functional Neuroanatomy of Pain, Department of Anatomy—ICB, University of São Paulo, São Paulo, Brazil,* [3]*Institute of Radiology (INRAD), Universidade de São Paulo, São Paulo, Brazil*

2.1 Molecular and cellular neuroscience

The fundamental topics addressed in cellular and molecular neuroscience include the mechanisms of signal processing across all scales of living neural tissue—how signals are physiologically and electrochemically processed, and how neurotransmitters and electrical signals convey information to and from a neuron. Another major area of neuroscientific investigation is the embryonic development of the nervous system. These questions encompass the differentiation of neural stem cells, the organization of neuronal and glial cells, neuronal migration, axonal and dendritic development, trophic interactions, and synapse formation. Recently, models of computational neurogenetics have been created to better understand the development of brain function.

This chapter presents landmark discoveries in the field, addressing key questions in cellular and molecular neuroscience research, and also discusses a few prominent methods that can be applied to answer these questions. As one of the newest fields in neuroscience, the aim of cellular and molecular neuroscience is to explore how genes, signaling molecules, and cellular morphology interact together to form the nervous system.

2.1.1 History of neuroscience discovery over the decades

Some discoveries may be considered pivotal in the development of neuroscience. The 16th century saw the invention of the microscope, a simple apparatus that unveiled the never-before-seen organization of living and nonliving matter. The description of the first bacteria, muscular fibers and botanical specimens, among others, by Anton van Leeuwenhoek opened the possibility for a myriad of studies in the centuries to come. After that, the X-ray, which was discovered in 1850 by Wilhelm Conrad Roentgen, was the second great discovery regarding neuroscience research. In 1901 he received the Nobel Prize in Physics for his discovery. Angiography is a technique that can observe the internal body cavity, images of blood vessels, and different organs such as the heart. This technique was described by Egas Moniz; his innovation won him the 1949 Nobel Prize in Physiology and Medicine.

Several technologies were developed in the 20th century that enabled discoveries to be made in neuroscience. In 1932 the electron microscope was discovered, which allowed unprecedented magnification of cellular structures and a better visualization of the relationship between subcellular organelles and cell boundaries. The magnetic resonance technology developed in 1971, produced a noninvasive image whose detailed resolution no other device could reach. Currently there are advanced techniques such as ultra-high field functional magnetic resonance at 7 Tesla that can give hints about brain function at thc ncuronal level. This device opened opportunities for new discoveries and advances in imaging brain connectivity regarding the relationship between cortical layers in submillimeter resolution. All these discoveries and increasingly sophisticated techniques expand the field of knowledge and provide better tools to improve the quality of life in patients.

In the 1960s, the term neuroscience was introduced for the first time. It heralded the broadening of the vision of different disciplines and opened a new field for scientific research. Before that there were separate molecular techniques, anatomists and cell biologists who dominated the early history of neuroscience.

In the 1950s, an influx of physicists, chemists, and theoreticians swelled the ranks of biologists and started the molecular biology revolution, culminating in Watson and Crick's discovery of the double helix, the twisted-ladder

Photobiomodulation in the Brain. DOI: https://doi.org/10.1016/B978-0-12-815305-5.00002-6

structure of deoxyribonucleic acid (DNA). Watson and Crick explained how DNA (working through RNA) encodes the proteins that act as the functional units of cells (Watson and Crick, 1953). For the first time, neuroscientists were able to investigate the role specific genes and proteins played in the function of the nervous system (Crick, 1958).

In the 1960s, Eric Kandel discovered (using the marine mollusk Aplysia) the genes and proteins that make memory possible in neurons. He first analyzed the mechanism of memory, focusing on short-term memory (Kandel, 1976). Many insights emerged from this simple systems approach. His studies aimed to define the neural circuits that mediate behavior and synapses that are modified by learning and memory storage (Kandel, 2001).

In the 1990s, another important discovery in neural communication occurred due to the work of Thomas Sudhof. He discovered that calcium ions alter the shape of proteins that anchor neurotransmitter within vesicles, explaining how signals instruct vesicles to release their neurotransmitter cargoes with precision. In 2013 Sudhof together with Randy Schekman and James Rothman became Nobel Laureates for solving the mystery of how cells organize their neurotransmitter transport system (Balch et al., 1984; Kaiser and Schekman, 1990; Perin et al., 1990; Sollner et al., 1993).

The term neuroscience has been growing steadily more popular, and has been the object of major studies in the last few decades, mainly in academics. The interest of the community in this subject is also growing. Within social media, the term neuroscience has broadened its vision and disseminated its knowledge. Currently schools are teaching what neuroscience is and providing information on different scientific breakthroughs. The material ranges from scientific journals to magazines or even children's books.

Neuroscience deals with cognition, senses, receptors, motion, and emotions; therefore these concepts are of interest to the general public. Currently, education and knowledge dissemination in neuroscience plays an important role starting with children. Informal education has an important role to play in dissemination of this knowledge.

2.1.2 Molecular techniques in neuroscience research

Molecular techniques can be applied to better understand both the natural function of the central nervous system and its response to injury. By applying microarray techniques to a specific population of neurons, researchers can examine the differences in expression of genes in specific neurons. With these studies, scientists have proposed different functions and morphological characteristics for these neurons.

Specialized imaging techniques can be used to study the brain in detail. Using a confocal microscope, small regions of a specific brain structure can be analyzed. In addition, dramatic advances in imaging technology have allowed scientists to study smaller structures in greater detail. Regarding imaging tools, scientists can also use images to investigate molecular components in tissue from a specific area of the nervous system. Therefore the use of fluorescence microscopy in combination with immunohistochemistry assays, in which tissues are stained with fluorescent antibodies that mark the cellular localization of specific proteins, can be viewed using this technology.

Gene expression technology is used in neuroscience to analyze how proteins regulate the expression of specific genes (by identifying the DNA targets). For instance, the molecular mechanism that regulates ion transport across the cell membrane, resulting in the propagation of action potentials, can be studied using specific anesthetics (which block specific ion channels) resulting in the blocking of pain transmission signals to the brain. Alternatively, real-time or quantitative PCR, utilizes equipment that can indirectly measure the relative quantity of specific mRNA. This approach is extremely useful to detect individual gene expression.

Transgene technology is an important tool for the investigation of gene function. By this method, researchers can produce animals with their genomes altered by permanent or conditional removal of specific genes, known as knockouts, or with modified genes inserted into their genetic code, known as transgenics. Nervous system tissues from these animals can be analyzed in a variety of ways to determine how changes in gene expression impact cellular function. Currently there are efforts to develop technology to measure and manipulate the cognitive functioning of the brain, such as brain mapping systems. These methods integrate neuroscience technologies, designing and developing tools to detect and control the brain in animals and even human behavior. Furthermore, systems may encode intrinsic neural functions of the brain and decipher the brain's incomparable ability to understand complex phenomena.

Studies related to brain-machine interfaces are another recent topic of interest. One of the most important and well-developed applications is the study of mechanisms for rehabilitation of the motor system. Disabled patients, who suffer from some neurological injuries or motor loss can receive different implants of electrodes that are able to send neural messages to the brain (Hata et al., 1993). All these researches on biocompatible interfaces are opening new opportunities and new devices for studies on brain networks and neurodegenerative diseases. Moreover new structures and devices may be developed for processing neural circuits.

There are several other techniques that can applied to neuroscience. However, it is not possible to describe them all in one chapter. Here, some of the most frequently used techniques that are applied to basic science are described.

2.2 Translational research in neuroscience

Translational neuroscience applies findings from basic research translated into clinical practice. To accomplish this, translational research needs to overcome two important hurdles. Firstly, researchers need to test the ideas in studies first initiated on animals and then apply it in clinical trials, testing for significant differences which of course may depend on each individual human being. Second, translational research must deal with the human factors regarding behavioral and organizational inertia, infrastructure and resource constraints in the real-world hospital environment (Woolf, 2008). Translational neuroscience is a new and rapidly advancing area of biomedical or neuroscience research with massive therapeutic and commercial potential.

The advantages of translational research in medicine are enormous. One advantage involves the process of transferring knowledge from basic laboratory research to the discovery of new drugs for humans. Given that the use of new drugs in humans is only possible after the prospective drug has passed through a series of clinical trials (and moreover first preclinical studies are required), the whole process can unavoidably take years to complete. The second aspect of translational research refers to the application of the final product in the community. The focus in this domain is on how everyone will receive the products from the discovery originally made by researchers. To this end it should be considered how the ambulatory care service is improved. A systematic review was put forward to facilitate the implementation of these discoveries by applying a guideline into clinical practice, helping clinical doctors to effectively use this knowledge (Westfall et al., 2007; Mitchell, 2016).

There is great difficulty in translating basic research to clinical practice, especially in studies related to pain. Even though laboratory animals are considered to be similar to humans in terms of anatomy and physiology, these animals do not reproduce the cognitive and emotional factors that are typical of humans. There are some studies that try to describe how different animals compare to humans. Another relevant point in this type of study is the evaluation time. In animals, behavioral tests are usually performed over a period of 30–60 days, which does not really replicate what happens in humans. For these reasons many clinical trials do not reproduce animal studies, and do not work when applied in humans. This is often what happens in the development of new drugs. More than 90% of new drugs do not make it onto shelves.

There is still a long way to go to create new drugs that really meet the needs of patients. There are still many challenges to be overcome, however, in recent years, new diagnostic techniques have emerged to examine different animal behaviors. These diagnostic techniques include computer programs in which small differences could be observed. Additionally, functional magnetic resonance imaging (fMRI) has been growing, and increases the quality of visualization of brain function. All of this contributes to the growth of basic neuroscience research and greater translation into clinical application. Neuroscience tools are still needed to inform society. In the last decade, with the growth of the internet and information technology, these possible targets have been expanding. Making this approach real and shifting translational neuroscience out of the laboratory and into our communities is the way to go.

2.3 Approaches to simulations and computational neuroscience

It has long been known that systems simulations can improve learning and effective outcomes in health applications, and these simulation systems can be applied to either strategic, tactical, or operational levels of health institutions (Marshall, 2015). The ubiquity of simulation systems is such that the problem cannot be reduced to the level it should be, but in a much broader scope, why and how simulations should be health service priorities, and patient outcomes need to be refined (Brazil, 2017). Even though the first level of simulation-based medical systems is aimed at increasing medical learning competence in the educational laboratory, further steps are awaited to transfer this knowledge seamlessly into downstream patient care practice and improved patient and public health (McGaghie, 2011).

The gap between education/training and patient care practice has seen many initiatives to integrate basic research results into viable and configurable simulations. Firm basic theoretical and empirical foundations are essential to establish the minimal requirements for a thorough and sound parametrization of the simulation interface, so to avoid either oversimplification or unnecessary complications. Examples of simulations in computational neuroscience may include efforts in modeling the functioning of the neural system, the cognitive aspects of the human brain when faced with a simulated reality, or its interaction with a certain treatment modality. Here, some specific examples of each simulation domain are presented, aiming at pointing out what is common between them, offering a wider perspective of the applications in the field.

2.3.1 Neural function simulation

A large variety of applications in neural modeling is available in the specialized literature. These can be categorized into symbolic or network approaches. No matter the category, one can think of these simulations as an attempt to establish a multilevel framework or description of the neural system. Some attempts were made to tackle both symbolic and network approaches at the same time (Achler, 2014; Bonzon, 2017) but usually these two domains are kept separated for political and technical reasons (Bechtel and Abrahamsen, 1991). In cognitive psychology models, symbolic models aim at understanding segregation and integration of brain functions by linking the different levels of system abstraction to behavior through asynchronous communication (Deco et al., 2015; Willshaw et al., 1994). These efforts aim at approaching artificial intelligence and the mind using nonsymbolic coding, so as to achieve multiscale neural mechanisms with brain network modeling (Schirner et al., 2018)

Regarding the network approach. two of the most conspicuous endeavors are the Virtual Brain (Ritter et al., 2012; Proix et al., 2018; Woodman, 2014) and the Blue Brain Project (Markram, 2006; Markram et al., 2015; Eilemann et al., 2017). Taking two distinct approaches in terms of scaling and immediate applicability, both projects aim at building accurate models that incorporate a range of features of neuronal models and neural dynamics. Both projects aim at modeling the neural systems to the point that they respond to either internal modulations or to external stimuli. The Virtual Brain, a framework integrating system dynamics and whole-brain structural connectivity, allows the simulation of seizure propagation and the buildup of synthetic epileptogenic foci. The simulated seizures, modeled on a mesoscopic scale, rely on surface-based modeling approaches, in which a high-resolution cortical surface is equipped with a neural field and homogeneous short-range and diffusion magnetic resonance imaging (dMRI)-derived long-range connectivity. The system allows the simulation of seizure propagation and termination (Proix et al., 2018), and the study of simulated lesions (Falcon et al., 2015) (Figs. 2.1–2.3).

On the other side of the scaling continuum, the finely detailed cellular model proposed by the Blue Brain group is dedicated to simulating fine-grained calculations depending on the density of neuronal spines and the balance of synaptic enzymes. The ultra-detailed model integrates the visualization and ray-tracing capabilities, allowing the visualization of the 3D structure of 30,000 virtual neurons inside a single cortical column. Simulation runs based on previous data acquired from multipatch clamp set-ups for studies of the electrophysiological behavior of neural circuits and multielectrode arrays, allowing stimulation of and recording from brain slices. The results show that the organization of

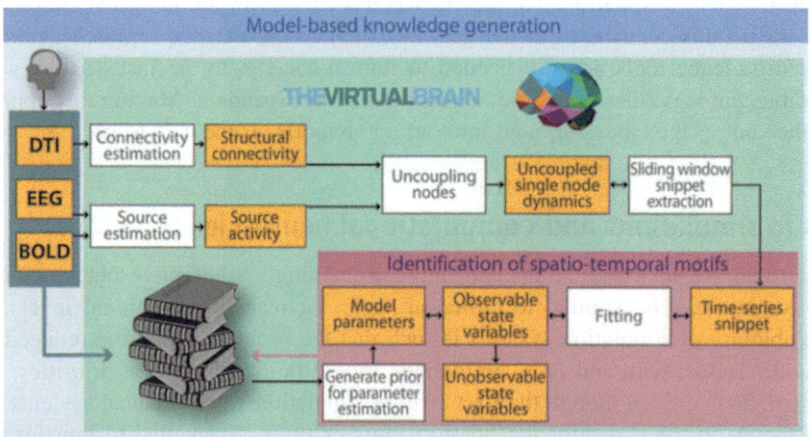

FIGURE 2.1 Model-based knowledge generation used in the Virtual Brain project. Empirical EEG and BOLD data are used to estimate electrical source activity parameter sets that enable the model-based replication of short empirical source activity. The internal model state variables are analyzed to infer knowledge about unobservable system states. Individual structural priors (fiber-tracts reconstructed by diffusion imaging) disentangle the influences between the system nodes, allowing the identification of spatiotemporal motifs. The simulation relies on a dictionary of initial parameter settings (priors) known to yield specific dynamic classes observed in empirical data. As the simulation is run in several times, the dictionary gets enriched for subsequent simulations, setting up an integrative method of induction and deduction for model optimization procedure. *Modified from Ritter, P., Schirner, M., McIntosh, A.R., Jirsa, V.K., 2012. The virtual brain integrates computational modeling and multimodal neuroimaging. Brain Connect. 3 (2), 121–145. doi:10.1089/brain.2012.0120.*

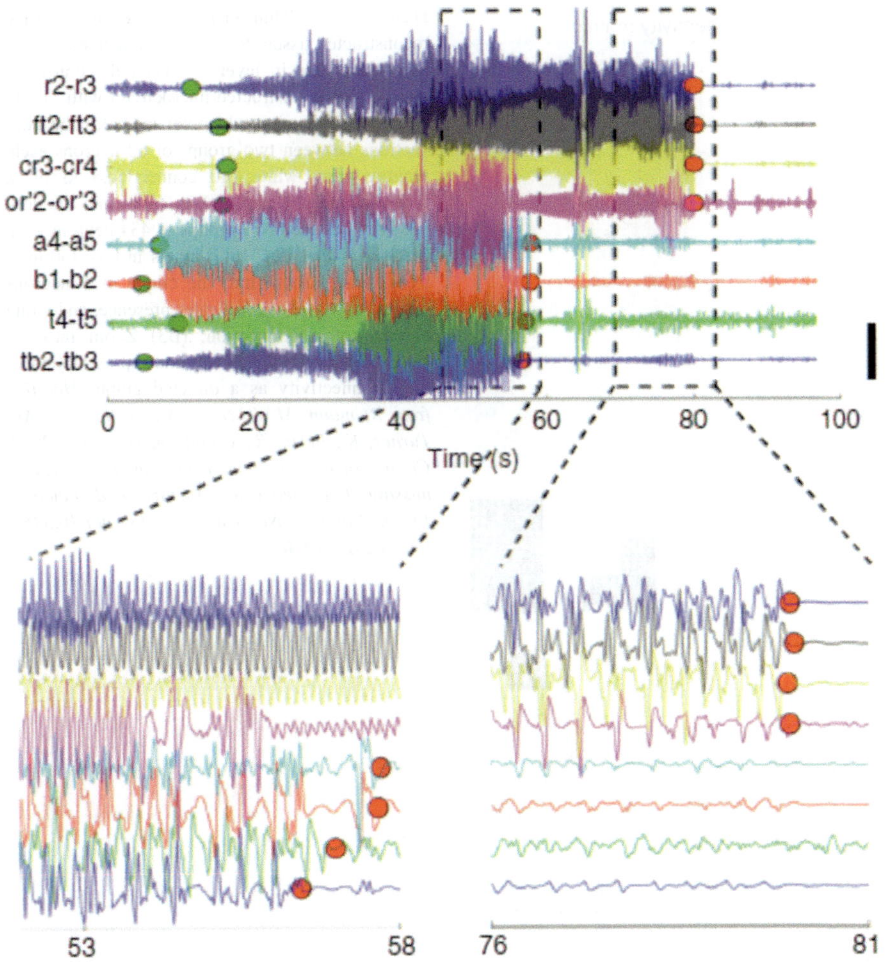

FIGURE 2.2 Example of a seizure recorded from a unifying neural field model. The top plot shows the full extent of a seizure, with green and red points indicating onset and offset of the seizure respectively. The plot magnifies the seizure termination for both clusters of recorded brain areas. Channels where the seizure ended simultaneously show coherent spike-and-wave activity. Colors denote the corresponding channels in the top and bottom plots. Scale bar: 1 mV. *Modified from Proix, T., Jirsa, V.K., Bartolomei, F., Guye, M., Truccolo, W., 2018. Predicting the spatiotemporal diversity of seizure propagation and termination in human focal epilepsy. Nat. Commun. 9, 1088. doi:10.1038/s41467-018-02973-y.*

subcolumnar brain signals may emerge from interactions spatial scales, configuring arrangements of local neuron and strange attractors or cliques of neurons (Reimann et al., 2017).

Given the different scales of the simulation, both approaches differ in the parameters they use to integrate and in the programming tools they employ. The Blue Brain project uses an IBM 65,536 core Blue Gene/Q supercomputer, whose simulations run away from public scrutiny, under the control of dedicated programmers. The Virtual Brain project, on the other hand, can be run by common public users on a local regular modern computer or on larger servers. Simulations can be tailored to the individual MRI and EEG data, opening the possibility of novel personalized strategies toward therapy and intervention (Jirsa et al., 2017). These two approaches have pros and cons, concerned with the subtleties of each approach and the scale of simulation (Kaiser, 2013).

2.4 Cognition and behavior

Aspects of perception and cognition have been studied for a long time in philosophy and later psychology. Currently, the analysis of cognitive psychology and cognitive science takes place in several scenarios including approaches such as, connectionist, dynamic, ecological, embodied, embedded, enactive, and extended. These scenarios differ in their conceptions of cognition and on the roles that the body and the environment play in these cognitive processes. Cognition can also be defined as how images in the brain are used to produce thought or behavior. Cognition includes perception, attention, episodic and semantic memories, associative learning, language, and executive control. These processes are coordinated by the individual to produce decisions, establish plans, and regulate behavior. Cognition also interacts with motivational and emotional processes and may involve social functions as well (Friedman et al., 2006).

The entire lifetime experience of the individual is another source of information; gained through learning and inferential processes, individuals can adjust their behavior and cognition to local requirements. And in some species,

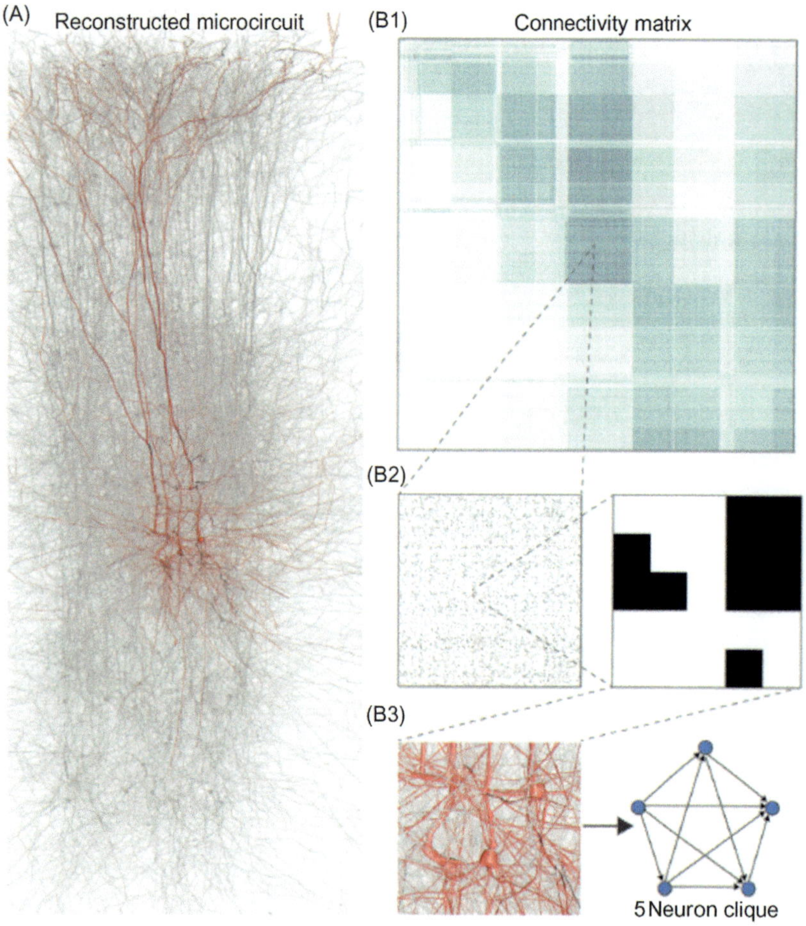

FIGURE 2.3 Thin (10 μm) slice of in silico reconstructed tissue. Red: A clique formed by five pyramidal cells in layer 5. (B1) Full connection matrix of a reconstructed microcircuit with 31,146 neurons. Each grayscale pixel indicates the connections between two groups of 62 neurons each, ranging from white (no connections) to black (≥8% connected pairs). (B2) Zoom into the connectivity between two groups of 434 neurons each in layer 5, that is, 7 by 7 pixels in (A), followed by a further zoom into the clique of 5 neurons shown in (A). Black indicates presence, and white absence of a connection. (B3) Zoom into the somata of the clique in (A) and representation of their connectivity as a directed graph. *Modified from Reimann, M.W., Nolte, M., Scolamiero, M., Turner, K., Perin, R., Chindemi, G., et al., 2017. Cliques of neurons bound into cavities provide a missing link between structure and function. Front. Comput. Neurosci. 11, 48. doi:10.3389/fncom.2017.00048.*

especially in humans, there are other potential sources of information: such as cultural traditions and social learning accumulated over generations (Henrich and McElreath, 2003). Individuals deal with their own worldviews by making use of information, with information being defined as a reduction of uncertainty about future events. There are several sources of such information. Given that we are a highly social species, the importance of such modeling for purposes of imitating, predicting, or understanding the behavior of others is potentially quite profound. The function of the mind is to guide action, and cognitive mechanisms such as perception and memory must be understood in terms of their ultimate contribution to appropriate behavior for a given situation. The individual agency is subordinated to the embodied cognition theory, in which triangulation must explain behavior between the world, body, and the brain (Shapiro, 2011).

One misleading approach to cognition is based on the premise that cognition depends constitutively on the presence of a living body, understood as an autonomous system operating in a complex open environment. The enactive approach is based on concepts such as autonomy, incarnation, creation of meaning in an environment and the activities it comprises, and the emergence of function and behavior arising from the interactions between the individual and his/her environment (Di Paolo et al., 2010). Ken Aizawa discusses an important question in the debate on embedded, enactive, and protracted cognition, which is "what is meant by cognition"? It might be questioned whether cognition should be a kind of behavior. Generally cognitive science has maintained that cognition is different from behavior. Behavior is considered to be the product of exogenous and endogenous factors or causes. Light waves, sound waves, aromatic chemical compounds, etc., are among some exogenous factors, while cognitive processes, along with motivation, attention, and so on are considered endogenous factors. Since cognitive processes are endogenous, it is a small step to adopt the view that the brain implements these processes (Maturana, 1980; Aizawa, 2017).

There is considerable evidence that behavior can be effectively modified through external interventions (Albarracin et al., 2005; Hobbs et al., 2013). However, evidence for the long-term sustainability of behavioral change in response to interventions is limited (Carpenter et al., 2013; Dombrowski et al., 2014). Moreover, what happens in the environment clearly influences cognitive processing, and cognitive processes are manifested not only in the brain, but also in the body and in the real-world. For example, light exerts a wide range of effects on the physiology and behavior of mammals. In addition to synchronizing circadian rhythms with the external environment, light has been shown to modulate autonomic and neuroendocrine responses, as well as regulate sleep and influence cognitive processes, such as attention, excitation, and performance (Fisk et al., 2018).

Moreover, the unique pattern of connective architecture among the billions of neurons making up the brain is initially formed by genetics, and then molded by experience over the entire lifetime (Kochunov et al., 2016; Yeh et al., 2016). In this regard, many groups are using neuroimaging techniques, particularly dMRI, to map this connective architecture in vivo (Le Bihan and Johansen-Berg, 2012). Another neuroimaging technique called molecular fMRI combines the specificity of cellular-level measurements with the noninvasive whole-brain coverage of fMRI, and has been used to associate integrative functions of the brain to mechanistically informative molecular and cellular variables. Establishing these relationships may be essential for understanding how low-level neurophysiology guides high-level behavior and cognition (Bartelle et al., 2016). Roberts showed using a task partial least squares analysis in which individual differences in cognitive flexibility were associated with the number of connections and differences in activity in several regions in the frontoparietal brain regions (Roberts et al., 2017).

Several studies have demonstrated that photobiomodulation or low-level laser therapy (LLLT) stimulates cognitive brain function, producing beneficial effects on prefrontal cortical functions related to sustained attention, working memory, and executive functioning, for instance using laser stimulation (Barrett and Gonzalez-Lima, 2013; Gonzalez-Lima and Barrett, 2014; Blanco et al., 2017). Similar to the results obtained from LLLT protocols, acute exercise-induced cognitive enhancement has been verified in neuroimaging studies (Yanagisawa et al., 2010; Li et al., 2014).

A growing number of human studies have reported the beneficial influences of acute (as well as chronic) exercise on cognitive functions. One study (in which the participants were tested before and after the treatments), using LLLT or acute exercise (EX) of high-intensity or combined treatment (LLLT + EX) showed that LLLT and EX treatments were similarly effective for cognitive enhancement, suggesting that both modalities augment prefrontal cognitive functions in a similar manner (Hwang et al., 2016). However, the optimal intensity of physical exercise for improved cognitive function might be related closely to exercise duration, exercise intensity, type of cognitive performance assessed, and participant fitness.

Neuroenhancement is a field that uses pharmacological or neuromodulatory interventions to improve cognitive abilities in normal humans (Clark and Parasuraman, 2014) A new and promising option for adjuvant neurodevelopment is LLLT using near-infrared transcranial lasers or LEDs. LLLT is noninvasive, therapeutically beneficial, and promotes a wide range of biological effects that regulate neuronal function in cell cultures, animal models, and clinical conditions (Eells et al., 2004). LLLT could be used as a noninvasive and effective approach to increase brain function, such as those related to cognitive and emotional dimensions. Recent efforts have been carried out to accomplish approaches to stimulate cognition and learning (Landriscina, 2013).

LLLT mechanisms involve the absorption of photons and the subsequent modulation of metabolic processes in a range of cell types, particularly including neurons (Anders et al., 2014). The primary molecular mechanism of action of LLLT seems to be photobiomodulation of mitochondrial cytochrome oxidase activity. Experimental studies showed that transcranial LLLT can increase cytochrome oxidase activity in the rat brain (Rojas et al., 2008), which can also improve the aerobic capacity of other tissues such as skeletal muscle (Hayworth et al., 2010).

LLLT also appears to have metabolic effects in the human brain and muscle tissue. LLLT could be used as a noninvasive and efficacious approach to increase brain functions such as those related to cognitive and emotional abilities (Barrett and Gonzalez-Lima, 2013). Naeser and collaborators showed improvement of cognitive functions in patients with a mild traumatic brain injury in a study using transcranial photobiomodulation (Naeser et al., 2016). Studies using photobiomodulation aimed at the right prefrontal cortex have been proven effective for increasing human cognitive and emotional functions (Hwang et al., 2016; Disner et al., 2016; Blanco et al., 2017).

Cognitive functions generally decline with age, and an experimental study showed that photobiomodulation (full body exposure) can improve working memory in middle-aged mice tested in a working memory test in a 3D maze (Michalikova et al., 2008). Another report in rats provided further evidence that LLLT modulates mood and may alleviate depression (Wu et al., 2012).

In a recent study, Vargas and collaborators showed for the first time that transcranial photobiomodulation could increase resting-state EEG alpha, beta, and gamma power; promoting more efficient prefrontal BOLD-fMRI activity and facilitating behavioral cognitive processing in middle-aged and older adults at risk for cognitive decline (Vargas et al., 2017). These data suggest a beneficial effect of transcranial photobiomodulation on the improvement of cognitive and emotional functions. Moreover, the mechanism by which this effect is induced, is presently unknown, but may involve several possible mechanisms including improved energy metabolism, promoting neuronal protection, modulation of antiapoptotic and proapoptotic mediators (Quirk et al., 2012).

Collectively, the data mentioned above highlight the use of photobiomodulation as a noninvasive and efficacious therapeutic tool to improve brain function, especially those related to cognitive and emotional dimensions. In terms of a potential therapeutic approach, photobiomodulation could be a healthy noninvasive and nonpharmacologic medical approach.

2.5 Neural treatment simulation

Besides using light to enhance cognitive performance, researchers may devise simulations to understand the interaction of light with neural systems and their fundamental constituents: the various biological tissues. While one traditional way of studying this is to use cadavers to study light scattering and transmission inside the biological tissue itself (Yue et al., 2015), the most common way of modeling the light transport is to simulate using Monte Carlo modeling with finite element analysis (Kienle and Hibst, 2006; Kirillin et al., 2010; Li et al., 2017). Tissues are described in terms of their optical properties, which, in turn, govern how light of different wavelengths interacts with different tissues. In the case of the human head, tissue anisotropy, water and blood content, and vessel densities are needed for a satisfactory volumetric model. Light is considered the stimulus, with specific wavelength, illumination profile, and source size (Figs. 2.4–2.6).

The interface needed for a specific use will affect the choice of the programming language as well as the supporting operating system (OS). The Fortran language and the Linux OS would be the choice for hardware programmers and

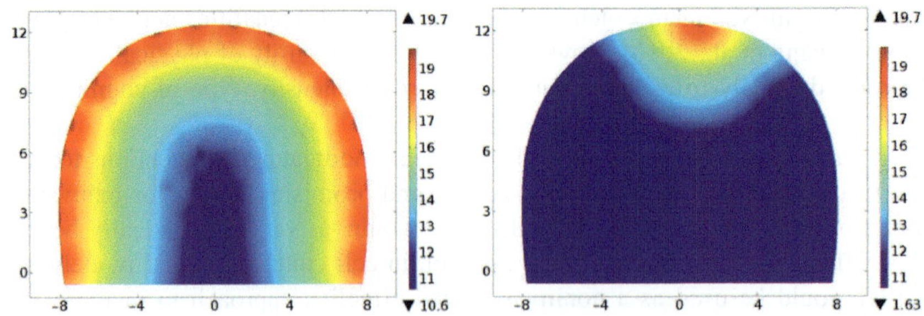

FIGURE 2.4 A slice view of log (fluence rate; cm^{-2}) in a frontal-view plane (x–z plane) at $y = 0$ cm. Left: simulation obtained with multiple light sources. Right: simulation obtained with only one light source. *Modified from Yue, L., Monge, M., Ozgur, M.H., Murphy, K., Louie, S., Miller, C.A., et al., 2015. Simulation and measurement of transcranial near infrared light penetration. Proc. SPIE 9321, Optical Interactions with Tissue and Cells XXVI, 93210S. doi: 10.1117/12.2077019.*

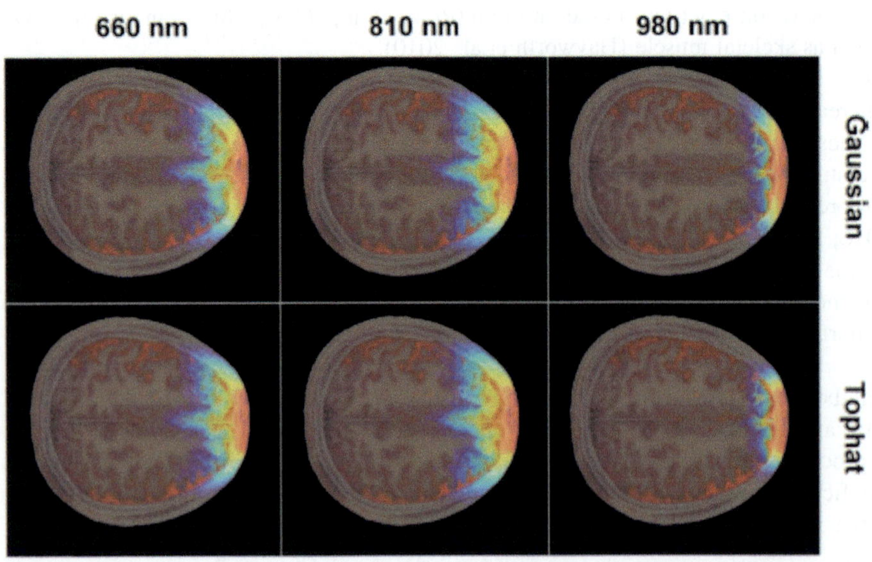

FIGURE 2.5 Light fluence distribution with the background of head geometry and 3D structure. The LLLT fluence distribution respective to Gaussian and Top-hat beams at different wavelengths (660, 810, and 980 nm). *Modified from Li, T., Xue, C., Wang, P., Li, Y., Wu, L., 2017. Photon penetration depth in human brain for light stimulation and treatment: a realistic Monte Carlo simulation study. J. Innov. Opt. Health Sci. 10 (5), 1743002. doi:10.1142/S1793545817430027.*

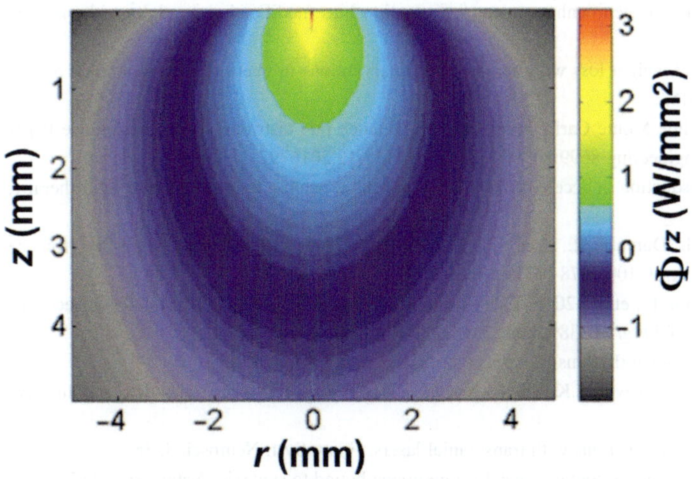

FIGURE 2.6 Gradient of fluence rate ϕ for 3×107 photon packets. z is the depth and r is the radial grid line in cylindrical coordinate system. The power density of the incident beam is 1 W/mm^2. *Modified from Doronin, A., Meglinski, I., 2013. Using peer-to-peer network for on-line Monte Carlo computation of fluence rate distribution. In: Conference Paper in Proceedings of SPIE—The International Society for Optical Engineering 8699:869909. doi:10.1117/12.2016797.*

low-level definition of processes. Interfaces using MATLAB, Microsoft Silverlight, ASP, NET in a Windows OS or a platform independent application (web-browser) would be adequate for a friendlier and more visually oriented user interface, aimed at the general user (Doronin and Meglinski, 2013).

Another aspect of the simulation that can be varied is the hardware specifications. Some platforms are based on the functioning of graphical processing units. The GPU-accelerated biophotonics integrates technical parameters and interactive aspects that influence the user experience (UX) directly. The goal of the whole setup is a practical way of simulating a certain aspect of a biological application. It is possible to create a system to simulate a therapeutic approach, a visualization of the light scattering volume, or an experimental setup. Some knowledge bases and information repositories are available for consultation in the field (Hellmers and Wriedt, 2009).

References

Achler, T., 2014. Symbolic neural networks for cognitive capacities. Biol. Inspired Cogn. Architect. 9, 71–81.

Aizawa, K., 2017. Cognition and behavior. Synthese 194 (11), 4269–4288.

Albarracin, D., Gillette, J.C., et al., 2005. A test of major assumptions about behavior change: a comprehensive look at the effects of passive and active HIV-prevention interventions since the beginning of the epidemic. Psychol. Bull. 131 (6), 856–897.

Anders, J.J., Moges, H., et al., 2014. In vitro and in vivo optimization of infrared laser treatment for injured peripheral nerves. Lasers Surg. Med. 46 (1), 34–45.

Balch, W.E., Dunphy, W.G., Braell, W.A., Rothman, J.E., 1984. Reconstitution of the transport of protein between successive compartments of the Golgi measured by the coupled incorporation of *N*-acetylglucosamine. Cell 39, 405–416.

Barrett, D.W., Gonzalez-Lima, F., 2013. Transcranial infrared laser stimulation produces beneficial cognitive and emotional effects in humans. Neuroscience 230, 13–23.

Bartelle, B.B., Barandov, A., et al., 2016. Molecular fMRI. J. Neurosci. 36 (15), 4139–4148.

Bechtel, W., Abrahamsen, A., 1991. Connectionism and the Mind. Blackwell Publishers.

Blanco, N.J., Saucedo, C.L., et al., 2017. Transcranial infrared laser stimulation improves rule-based, but not information-integration, category learning in humans. Neurobiol. Learn. Mem. 139, 69–75.

Blanco, N.J., Maddox, W.T., et al., 2017. Improving executive function using transcranial infrared laser stimulation. J. Neuropsychol. 11 (1), 14–25.

Bonzon, P., 2017. Towards neuro-inspired symbolic models of cognition: linking neural dynamics to behaviors through asynchronous communications. Cogn. Neurodyn. 11, 327–353. Available from: https://doi.org/10.1007/s11571-017-9435-3.

Brazil, V., 2017. Translational simulation: not 'where?' but 'why?' A functional view of in situ simulation. Adv. Simul. 2, 20. Available from: https://doi.org/10.1186/s41077-017-0052-3.

Carpenter, M.J., Jardin, B.F., et al., 2013. Clinical strategies to enhance the efficacy of nicotine replacement therapy for smoking cessation: a review of the literature. Drugs 73 (5), 407–426.

Clark, V.P., Parasuraman, R., 2014. Neuroenhancement: enhancing brain and mind in health and in disease. Neuroimage 85 (Pt 3), 889–894.

Crick, F.H., 1958. On protein synthesis. Symp. Soc. Exp. Biol. 12, 138–163.

Deco, G., Tononi, G., Boly, M., Kringelbach, M.L., 2015. Rethinking segregation and integration. Nat. Rev. Neurosci. 16 (7), 430–439. Available from: https://doi.org/10.1038/nrn3963.

Di Paolo, E.A., Rohde, M., De Jaegher, H., 2010. Horizons for the enactive mind: values, social interaction, and play. In: Stewart, J., Gapenne, O., Di Paolo, E.A. (Eds.), Enaction: Toward a New Paradigm for Cognitive Science. MAMIT Press, Cambridge, pp. 34–87.

Disner, S.G., Beevers, C.G., et al., 2016. Transcranial laser stimulation as neuroenhancement for attention bias modification in adults with elevated depression symptoms. Brain Stimul. 9 (5), 780–787.

Dombrowski, S.U., Knittle, K., et al., 2014. Long term maintenance of weight loss with non-surgical interventions in obese adults: systematic review and meta-analyses of randomised controlled trials. BMJ 348, g2646.

Doronin, A., Meglinski, I., 2013. Using peer-to-peer network for on-line Monte Carlo computation of fluence rate distribution. In: Conference Paper in Proceedings of SPIE—The International Society for Optical Engineering 8699:869909. DOI: 10.1117/12.2016797

Eells, J.T., Wong-Riley, M.T., et al., 2004. Mitochondrial signal transduction in accelerated wound and retinal healing by near-infrared light therapy. Mitochondrion 4 (5–6), 559–567.

Eilemann, S., Abdellah, M., Antille, N., Bilgili, A., Chevtchenko, G., Dumusc, R., et al., 2017. From Big Data to Big Displays High-Performance Visualization at Blue Brain. Springer. Available from: http://doi.org/10.1007/978-3-319-67630-2_47.

Falcon, M.I., Riley, J.D., Jirsa, V., McIntosh, A.R., Sheree, A.D., Chen, E., et al., 2015. The virtual brain: modeling biological correlates of recovery after chronic stroke. Front. Neurol. 6, 228. Available from: https://doi.org/10.3389/fneur.2015.00228.

Fisk, A.S., Tam, S.K.E., et al., 2018. Light and cognition: roles for circadian rhythms, sleep, and arousal. Front. Neurol. 9, 56.

Friedman, N.P., Miyake, A., Corley, R.P., Young, S.E., DeFries, J.C., Hewitt, J.K., 2006. Not all executive functions are related to intelligence. Psychol. Sci. 17 (2), 172–179.

Gonzalez-Lima, F., Barrett, D.W., 2014. Augmentation of cognitive brain functions with transcranial lasers. Front. Syst. Neurosci. 8, 36.

Hata, Y., Slaughter, C.A., Südhof, T.C., 1993. Synaptic vesicle fusion complex contains unc-18 homologue bound to syntaxin. Nature 366, 347–351.

Hayworth, C.R., Rojas, J.C., et al., 2010. In vivo low-level light therapy increases cytochrome oxidase in skeletal muscle. Photochem. Photobiol. 86 (3), 673–680.

Hellmers, J., Wriedt, T., 2009. New approaches for a light scattering Internet information portal and categorization schemes for light scattering software. J. Quant. Spectrosc. Radiat. Transf. 110 (14–16), 1511–1517. Available from: https://doi.org/10.1016/j.jqsrt.2009.01.023.

Henrich, J., McElreath, R., 2003. The evolution of cultural evolution. Evol. Anthropol.: Issues, News, Rev. 12 (3), 123–135.

Hobbs, N., Godfrey, A., et al., 2013. Are behavioral interventions effective in increasing physical activity at 12 to 36 months in adults aged 55 to 70 years? A systematic review and meta-analysis. BMC Med. 11, 75.

Hwang, J., Castelli, D.M., et al., 2016. Cognitive enhancement by transcranial laser stimulation and acute aerobic exercise. Lasers Med. Sci. 31 (6), 1151–1160.

Jirsa, V.K., Proix, T., Perdikis, D., Woodman, M.M., Wang, H., Gonzalez-Martinez, J., et al., 2017. The virtual epileptic patient: individualized whole-brain models of epilepsy spread. Neuroimage 145, 377–388.

Kaiser, C.A., Schekman, R., 1990. Distinct sets of SEC genes govern transport vesicle formation and fusion early in the secretory pathway. Cell 61, 723–733.

Kaiser, M., 2013. The potential of the human connectome as a biomarker of brain disease. Front. Hum. Neurosci. 7, 1–4. Available from: https://doi.org/10.3389/fnhum.2013.00484.

Kandel, E.R., 1976. Cellular Basis of Behavior: An Introduction to Behavioral Neurobiology. W.H. Freeman, San Francisco, CA.

Kandel, E.R., 2001. The molecular biology of memory storage: a dialogue between genes and synapses. Science 294, 1030–1038. Available from: https://doi.org/10.1126/science.1067020.

Kienle, A., Hibst, R., 2006. Light guiding in biological tissue due to scattering. Phys. Rev. Lett. 97 (1), 018104.

Kirillin, M., Meglinski, I., Sergeeva, E., Kuzmin, V.L., Myllyla, R., 2010. Simulation of optical coherence tomography images by Monte Carlo modeling based on polarization vector approach. Opt. Express 18 (21), 21714–21724.

Kochunov, P., Thompson, P.M., et al., 2016. The common genetic influence over processing speed and white matter microstructure: evidence from the Old Order Amish and Human Connectome Projects. Neuroimage 125, 189–197.

Landriscina, F., 2013. Simulation and Learning. A Model-Centered Approach. Springer, New York, NY. Available from: https://doi.org/10.1007/978-1-4614-1954-9.

Le Bihan, D., Johansen-Berg, H., 2012. Diffusion MRI at 25: exploring brain tissue structure and function. Neuroimage 61 (2), 324–341. Available from: https://doi.org/10.1016/j.neuroimage.2011.11.006.

Li, L., Men, W.W., et al., 2014. Acute aerobic exercise increases cortical activity during working memory: a functional MRI study in female college students. PLoS One 9 (6), e99222.

Li, T., Xue, C., Wang, P., Li, Y., Wu, L., 2017. Photon penetration depth in human brain for light stimulation and treatment: a realistic Monte Carlo simulation study. J. Innov. Opt. Health Sci. 10 (5), 1743002. Available from: https://doi.org/10.1142/S1793545817430027.

Markram, H., 2006. The blue brain project. Nat. Rev. Neurosci. 7, 153–160.

Markram, H., et al., 2015. Reconstruction and simulation of neocortical microcircuitry. Cell 163 (2), 456–492.

Marshall, D.A., 2015. Applying dynamic simulation modeling methods in health care delivery research—The SIMULATE Checklist: Report of the ISPOR simulation modeling emerging good practices task force. Value Health 18 (1), 5–16.

Maturana, H.V.F., 1980. Autopoiesis and Cognition: The Realization of the Living. Springer, Dordrecht.

McGaghie, W.C., 2011. Evaluating the impact of simulation on translational patient outcomes. Simul. Healthc. 6 (Suppl.), S42–S47. Available from: https://doi.org/10.1097/SIH.0b013e318222fde9.

Michalikova, S., Ennaceur, A., et al., 2008. Emotional responses and memory performance of middle-aged CD1 mice in a 3D maze: effects of low infrared light. Neurobiol. Learn. Mem. 89 (4), 480–488.

Mitchell, P., 2016. From Concept to Classroom. What Is Translational Research? Australian Council for Educational Research, Camberwell.

Naeser, M.A., Martin, P.I., et al., 2016. Transcranial, red/near-infrared light-emitting diode therapy to improve cognition in chronic traumatic brain injury. Photomed. Laser Surg. 34 (12), 610–626.

Perin, M.S., Fried, V.A., Mignery, G.A., Jahn, R., Südhof, T.C., 1990. Phospholipid binding by a synaptic vesicle protein homologous to the regulatory region of protein kinase C. Nature 345, 260–263.

Proix, T., Jirsa, V.K., Bartolomei, F., Guye, M., Truccolo, W., 2018. Predicting the spatiotemporal diversity of seizure propagation and termination in human focal epilepsy. Nat. Commun. 9, 1088. Available from: https://doi.org/10.1038/s41467-018-02973-y.

Quirk, B.J., Torbey, M., et al., 2012. Near-infrared photobiomodulation in an animal model of traumatic brain injury: improvements at the behavioral and biochemical levels. Photomed. Laser Surg. 30 (9), 523–529.

Reimann, M.W., Nolte, M., Scolamiero, M., Turner, K., Perin, R., Chindemi, G., et al., 2017. Cliques of neurons bound into cavities provide a missing link between structure and function. Front. Comput. Neurosci. 11, 48. Available from: https://doi.org/10.3389/fncom.2017.00048.

Ritter, P., Schirner, M., McIntosh, A.R., Jirsa, V.K., 2012. The virtual brain integrates computational modeling and multimodal neuroimaging. Brain Connect. 3 (2), 121–145. Available from: https://doi.org/10.1089/brain.2012.0120.

Roberts, R.P., Wiebels, K., et al., 2017. An fMRI investigation of the relationship between future imagination and cognitive flexibility. Neuropsychologia 95, 156–172.

Rojas, J.C., Lee, J., et al., 2008. Neuroprotective effects of near-infrared light in an in vivo model of mitochondrial optic neuropathy. J. Neurosci. 28 (50), 13511–13521.

Schirner, M., et al., 2018. Inferring multi-scale neural mechanisms with brain network modelling. eLife 7, e28927. Available from: https://doi.org/10.7554/eLife.28927.

Shapiro, L., 2011. Embodied cognition. Philos. Top. 39 (1).

Sollner, T., Whiteheart, W., Brunner, M., Erdjument-Bromage, H., Geromanos, S., Tempst, P., et al., 1993. SNAP receptor implicated in vesicle targeting and fusion. Nature 362, 318–324.

Vargas, E., Barrett, D.W., et al., 2017. Beneficial neurocognitive effects of transcranial laser in older adults. Lasers. Med. Sci. 32 (5), 1153–1162.

Westfall, J.M., Mold, J., Fagnan, L., 2007. Practice-based research—"blue highways" on the NIH roadmap. JAMA 297, 403–406.

Woolf, S.H., 2008. The meaning of translational research and why it matters. JAMA 299, 211–213.

Wu, X., Alberico, S.L., et al., 2012. Pulsed light irradiation improves behavioral outcome in a rat model of chronic mild stress. Lasers Surg. Med. 44 (3), 227–232.

Yanagisawa, H., Dan, I., et al., 2010. Acute moderate exercise elicits increased dorsolateral prefrontal activation and improves cognitive performance with Stroop test. Neuroimage 50 (4), 1702–1710.

Yeh, F.C., Vettel, J.M., et al., 2016. Quantifying differences and similarities in whole-brain white matter architecture using local connectome fingerprints. PLoS Comput. Biol. 12 (11), e1005203.

Yue, L., Monge, M., Ozgur, M.H., Murphy, K., Louie, S., Miller, C.A., et al., 2015. Simulation and measurement of transcranial near infrared light penetration. Proc. SPIE 9321, Optical Interactions with Tissue and Cells XXVI 93210S. Available from: https://doi.org/10.1117/12.2077019.

Watson, J.D., Crick, F.H.C., 1953. A structure for deoxyribose nucleic acid. Nature 171, 737–738.

Willshaw, D., Dennett, D., Partridge, D., 1994. Non-symbolic approaches to artificial intelligence and the mind [and discussion]. Phil. Trans.: Phys. Sci. Eng. 349 (1689), 87–102. <http://www.jstor.org/stable/54378>.

Woodman, M.M., 2014. Integrating neuroinformatics tools in TheVirtualBrain. Front. Neuroinform. 8, 36. Available from: https://doi.org/10.3389/fninf.2014.00036.

Chapter 3

Photobiomodulation of cultured primary neurons: role of cytochrome c oxidase

Margaret Wong-Riley and Huan Ling Liang

Department of Cell Biology, Neurobiology and Anatomy, Medical College of Wisconsin, Milwaukee, WI, United States

3.1 Introduction

Light close to and in the nearinfrared range (630–1000 nm; abbreviated as NIR in this chapter) has been used in the past few decades mainly to treat ischemic, infectious, and hypoxic wounds (Mester et al., 1971; Conlan et al., 1996). In recent years, however, it has also been applied to animal models and some human patients with stroke, traumatic brain injury, depression, and neurodegenerative disease, such as Parkinson disease and Alzheimer disease (reviewed in Salehpour et al., 2018). The most common mode of delivery is with lasers (Huang et al., 2012; also reviewed in many chapters in this book). With the advent of light-emitting diodes (LEDs), a prototype of which was developed by NASA for experiments in plant growth in space, NIR via LEDs has been employed with success in various types of applications, including wound healing in animals and humans, tissue growth, and animal models of methanol toxicity, stroke, Parkinson disease, and Alzheimer disease (Whelan et al., 2001; Duan et al., 2003; Eells et al., 2003; Choi et al., 2012; Quirk et al., 2012; Grillo et al., 2013).

At least three biological agents can absorb light in the NIR range: hemoglobin, myoglobin, and cytochrome c oxidase (Jöbsis, 1977; Cooper and Springett, 1997; Galkin et al., 1997; Karu,1999, 2010). Of these, hemoglobin has the strongest signal, but only cytochrome c oxidase is associated with energy generation within the mitochondria and thus, with a potential therapeutic effect. Using the noninvasive method of NIR spectroscopy, Jöbsis (1977) was able to monitor oxygen sufficiency for cytochrome c oxidase function, alterations in tissue blood volume, and hemoglobin-oxyhemoglobin ratios in the brain and heart. In a bloodfree rat liver preparation, Chance's group found that $\sim 50\%$ of NIR was absorbed by chromophores, such as cytochrome c oxidase, in the mitochondria (Beauvoit et al., 1994).

To more closely isolate the effect of NIR via LED (also referred to as photobiomodulation or PBM) on cytochrome c oxidase in biological samples without the confounding effects on hemoglobin-oxyhemoglobin and myoglobin, primary neurons from postnatal rodent brains can be grown in culture in the absence of blood vessels or muscles (Wong-Riley et al., 2001). These cultures are composed of mainly neurons from the cerebral cortex ($>95\%$) with only minor contributions from glia ($\leq 5\%$), and neurons stay healthy for up to 4 weeks (Zhang and Wong-Riley, 1999). Various chemicals and toxins can be applied to cultured neurons, NIR can be administered via an array of LED (25×10 cm) with a bandwidth of 25–30 nm, and results can be monitored at single neuronal or population levels by optical densitometry (Hevner et al., 1995; Zhang and Wong-Riley, 1999). Relatively little work has been done on the effect of PBM on primary neurons, presumably because the latter are delicate and technically challenging to culture and maintain in a healthy state. However, primary neurons are more physiological than cell lines, most of which are of cancerous origin. Thus, primary neurons are the preferred neuronal type to study in culture. This chapter will focus on the impact of photobiomodulaton on cultured primary neurons exposed to various toxins, and the mechanisms underlying the neuroprotective effect of PBM.

3.2 Cytochrome c oxidase: a biological mediator of photobiomodulation

Cytochrome c oxidase (cytochrome aa_3, ferrocytochrome c, oxygen oxidoreductase, EC 1.9.3.1) is the terminal enzyme or complex IV of the mitochondrial electron transport chain, without which oxidative metabolism cannot be carried to

Photobiomodulation in the Brain. DOI: https://doi.org/10.1016/B978-0-12-815305-5.00003-8

completion (Wikström et al., 1981). It has a long evolutionary history, portions of which evolved more than 1.5 billion years ago as part of the bacterial cell membrane and later as an integral transmembrane protein of the inner mitochondrial membrane. It catalyzes the oxidation of its substrate, cytochrome c, and the reduction of molecular oxygen to water. Complex IV, together with complexes I and III, actively pump protons from the mitochondrial matrix to the intermembranous space, setting up a proton as well as an electrical gradient. The backflow of protons down this gradient drives the synthesis of ATP from ADP and phosphate by complex V (Mitchell, 2011). Electron transport is, therefore, coupled to oxidative phosphorylation (Mitchell, 2011), and complexes I, III, and IV can be considered as energy-generating enzymes, while complex V is the energy-synthesizing enzyme.

The dinuclear Cu_A and Cu_B copper centers in cytochrome c oxidase are found to absorb light in the NIR (Gibson and Greenwood, 1965; Cooper and Springett, 1997; Karu, 1999). Based on the action spectra of HeLa DNA synthesis with various wavelengths of lasers, Karu (1999) proposed that 620, 680, 760, and 820 nm correspond to the absorption spectra of reduced Cu_A, oxidized Cu_B, reduced Cu_B, and oxidized Cu_A of cytochrome c oxidase, respectively.

3.3 Effect of photobiomodulaton on primary neurons exposed to tetrodotoxin

Tetrodotoxin (TTX) from the puffer fish is a potent inhibitor of neuronal activity, as it blocks the voltage-dependent sodium channels and, within a tested range, it blocks action potentials without interfering with other cellular functions, such as fast axoplasmic transport (Nakamura et al., 1965; Ochs and Hollingsworth, 1971). When primary neurons in culture are exposed to TTX, the blockage of action potentials leads to reduced levels of cytochrome c oxidase demonstrable histochemically and measurable with single neuron optical densitometry (Zhang and Wong-Riley, 1999). This approach enables the delineation of relative cytochrome c oxidase activity in individual neurons and, when carried to the electron microscopic level, resolves relative enzyme levels in individual mitochondrion (Wong-Riley, 1989).

Fig. 3.1A illustrates that primary neurons from postnatal visual cortical neurons have varying levels of cytochrome c oxidase activity that can be broadly classified into dark, moderate, and lightly-reactive metabolic cell types (Wong-Riley et al., 2001). Exposure to 0.4 μM TTX from day 11 to day 16 of culture, significantly reduced enzyme levels in all three metabolic cell types ($P < .001$) without altering their size, shape, or viability (Figs. 3.1B and 3.2A). LED emitted from GaAlAS arrays at 670 nm (bandwidth of 25–30 nm at 50% power) (Quantum Devices, Inc. Barnaveld, WI, USA) with power intensity of 50 mW/cm^2 and energy density of 4 J/cm^2 (when applied for 80 seconds) once a day for the last 5 of the 6 days in TTX significantly increased cytochrome c oxidase activity to control levels in all three metabolic cell types (Figs. 3.1C and 3.2A). Similar LED exposure in the absence of TTX significantly upregulated relative enzyme activity to levels above those of controls, again, in all three metabolic cell types (Figs. 3.1D and 3.2A). However, not all neurons responded to PBM to the same degree. For example, a single LED application to neurons

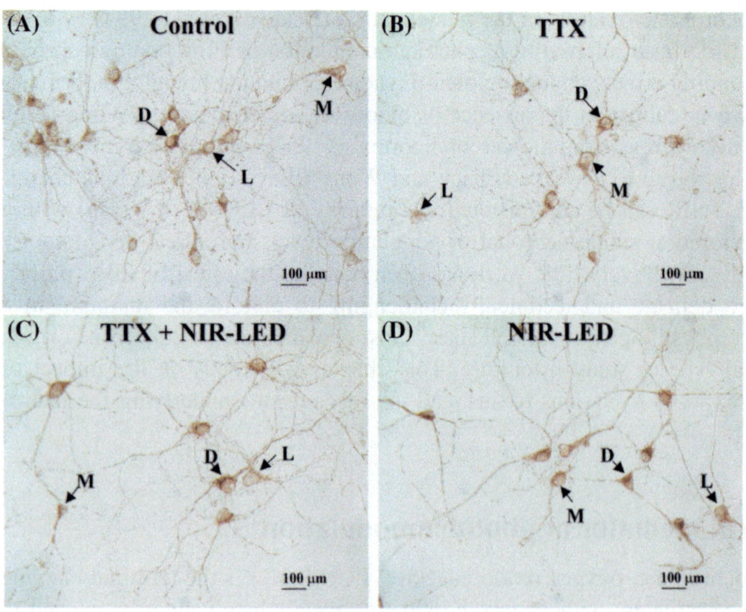

FIGURE 3.1 Examples of neurons taken from postnatal day 1 of the rat visual cortex and cultured for 16 days. Under both control (A) and experimental (B–D) conditions, neurons can be broadly classified into three metabolic cell types: darkly (D), moderately (M), and lightly (L) reactive for cytochrome c oxidase activity. (B) Neurons exposed to TTX on days 11-16 of culture had reduced enzyme levels in all three metabolic cell types. (C) In the presence of TTX, LED was administered once a day, from day 12 to 16 of culture. Enzyme level significantly increased in all three cell types. (D) In the absence of TTX, LED treatment for 5 days raised enzyme levels above that of controls in all three metabolic cell types. *Modified from Wong-Riley, M.T.T., Bai, X., Buchmann, E., Whelan, H.T., 2001. Light-emitting diode treatment reverses the effect of TTX on cytochrome oxidase in neurons. NeuroReport 12, 3033–3037.*

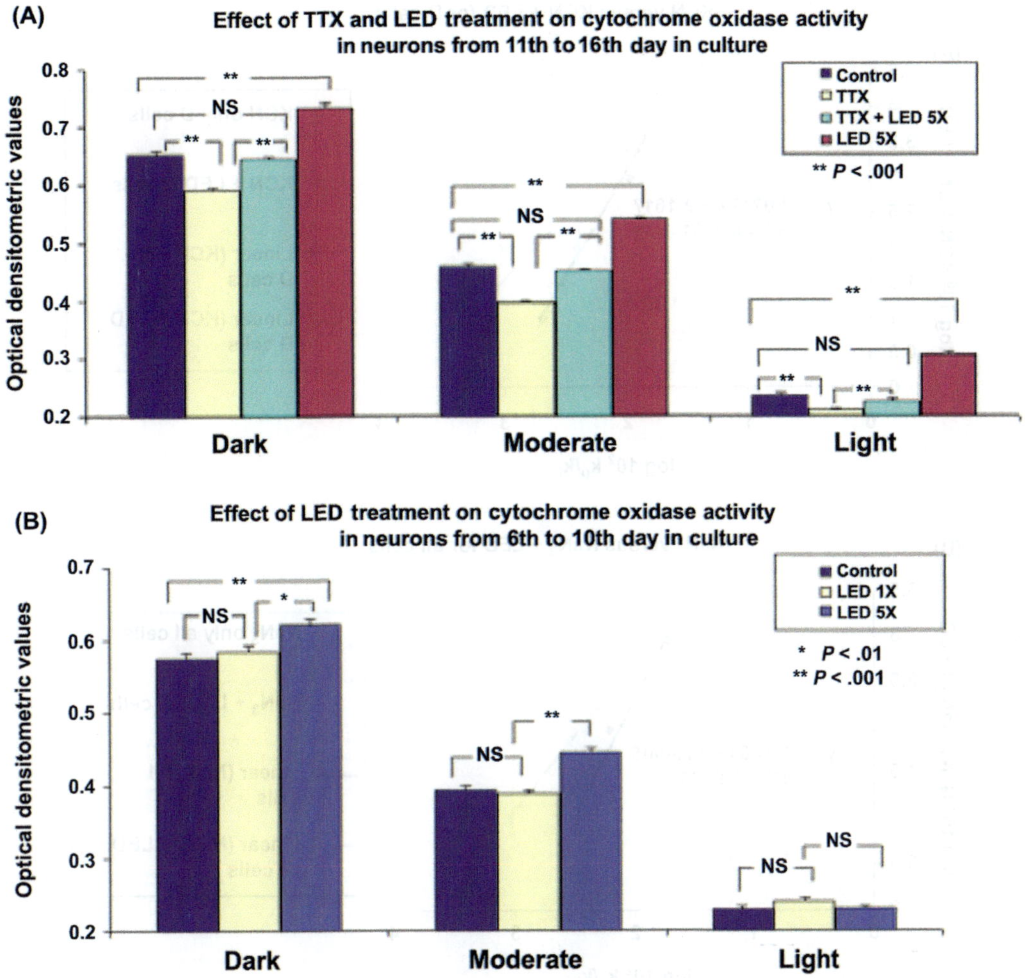

FIGURE 3.2 Optical densitometric measurements of cytochrome c oxidase reaction product in the three metabolic cell types from postnatal rat cortex. (A) TTX reduced enzyme activity in all three cell types, but photobiomodulation via LED once a day for 5 days restored them to control levels. Five times of LED increased enzyme levels of normal neurons significantly above that of controls. (B) LED one time on normal neurons had no effect, but five times over 5 days increased enzyme levels in dark and moderately reactive neurons. $^{**}P < .01$; NS: nonsignificant; when compared to controls. *Modified from Wong-Riley, M.T.T., Bai, X., Buchmann, E., Whelan, H.T., 2001. Light-emitting diode treatment reverses the effect of TTX on cytochrome oxidase in neurons. NeuroReport 12, 3033–3037.*

inactivated by TTX for 6 days increased enzyme activity (though not reaching control levels) only in darkly reactive neurons, but not in moderately or lightly reactive ones. On the other hand, a single LED administration to normal neurons in the absence of TTX did not yield any statistically significant change in enzyme levels in all metabolic cell types examined (Fig. 3.2B).

3.4 Equilibrium constants of azide and cyanide with cytochrome c oxidase

As cytochrome c oxidase can be inactivated by cyanide or azide (Wikström et al., 1981), the question is whether NIR has any effect on the equilibrium constants of these two toxins with the enzyme. Fig. 3.3 shows the results of such an experiment (Wong-Riley et al., 2005). The cyanide concentrations were 10, 100 μM, 1, and 10 mM (Fig. 3.3A), and those for azide were 10, 100 μM, and 1 mM (Fig. 3.3B). The "*a*" values in both cases were two to three times greater than the reported value of one (Wilson and Chance, 1967). In both cases, LED treatment reduced the "*a*" value of the constants, from 3.02 to 1.82 for KCN (Fig. 3.3A), and from 2.56 to 1.65 for NaN_3 (Fig. 3.3B). These findings strongly indicate that NIR induces an increase in the synthesis of cytochrome c oxidase in neurons.

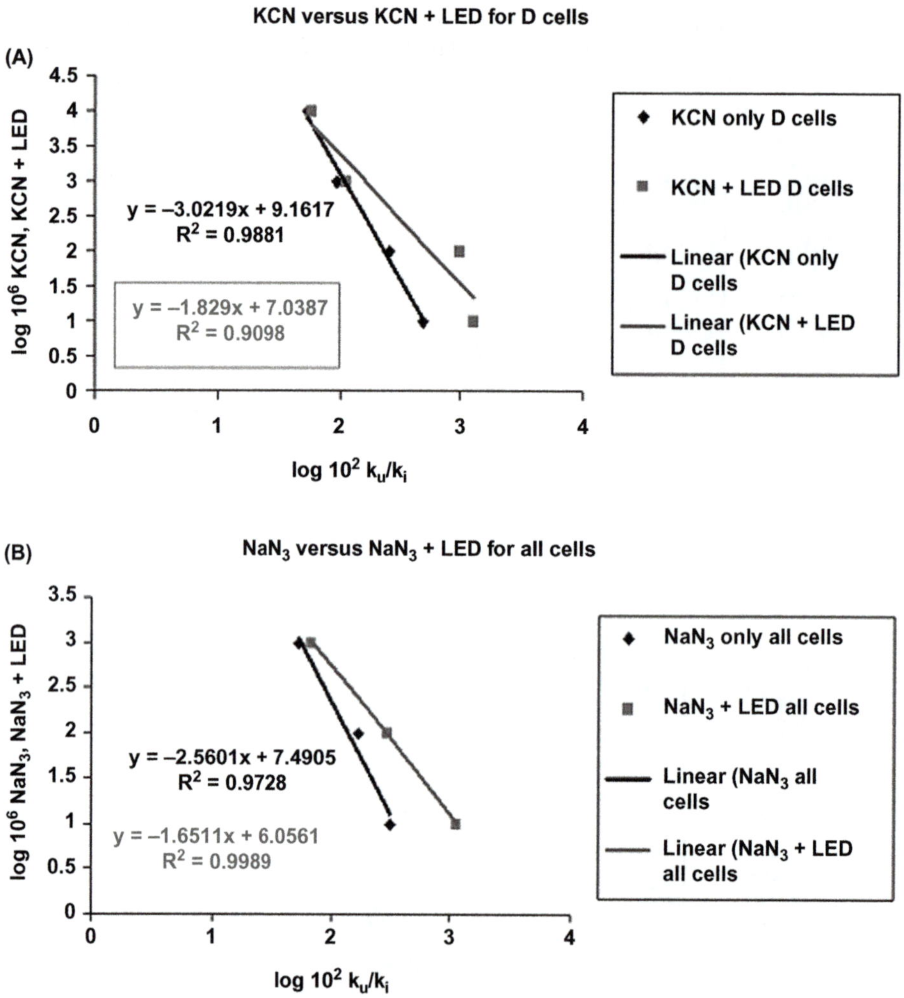

FIGURE 3.3 Equilibrium constants of cyanide (KCN) (A) and azide (NaN_3) (B) with cytochrome c oxidase. (A) The KCN concentrations plotted were 10, 100 μM, 1, and 10 mM, and the data were based on darkly-reactive (D) neurons. (B) The NaN_3 concentrations were 10, 100 μM, and 1 mM and the data were based on all three metabolic cell types. Note the reduction in the constants for both KCN and NaN_3 with LED treatment. *Adapted from Wong-Riley, M.T.T., Liang, H.L., Eells, J.T., Henry, M.M., Buchmann, E., Kane, M., et al., 2005. Photobiomodulation directly benefits primary neurons functionally inactivated by toxins: Role of cytochrome c oxidase. J. Biol. Chem. 280, 4761–4771.*

3.5 Effects of photobiomodulation at different wavelengths

If NIR's effect on cultured neurons is mediated mainly via its absorption by cytochrome c oxidase and the activation of ATP production, then the action spectra of these effects should correspond positively with the absorption spectrum of this enzyme. This question was tested in cultured visual cortical neurons exposed to the neurotoxin TTX and treated with various wavelengths of LED (Wong-Riley et al., 2005). In the presence of 0.4 μM TTX for 6 days, sister cultures of primary neurons received PBM via LED arrays at 670, 728, 770, 830, or 880 nm, respectively, once a day (all at 50 mW/cm^2 power intensity and 4 J/cm^2 energy density when exposed for 80 seconds), for the last 5 of the 6 days in TTX. Results indicate that not all wavelengths are of equal value (Fig. 3.4A). Whereas 830 and 670 nm were both effective in completely reversing the effects of TTX on enzyme levels, 830 nm was able to raise values even slightly, though not significantly, above that of controls (111.7%). 880 nm enabled 98.4% recovery, while 770 nm facilitated 67.2% recovery. The least effective one was 728 nm, which brought about only 27.5% reversal of enzyme activity.

Cellular ATP content revealed similar results (Fig. 3.4B): 6 days of TTX reduced ATP content to 82.5% of controls ($P < .001$). All above wavelengths tested, with the exception of 728 nm, were able to overcome the detrimental effect of TTX and return ATP content to levels not significantly different from controls. Both 670 and 830 nm were effective in raising ATP levels slightly above (110% and 106.8%, respectively), though not statistically different from that of

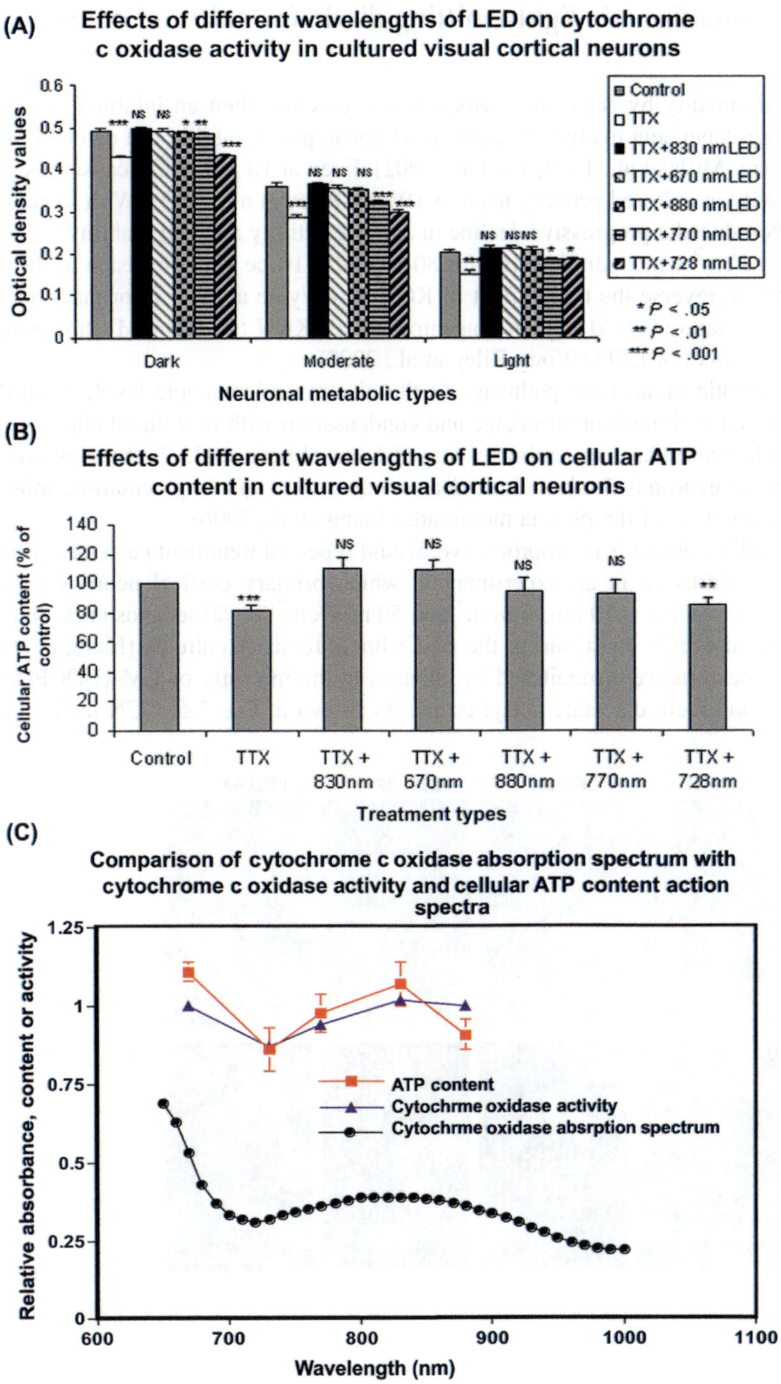

FIGURE 3.4 Effects of various wavelengths of LED on cytochrome c oxidase activity (A) and ATP content (B) in primary neurons exposed to TTX for 6 days. 670 and 830 nm were the most effective in restoring enzyme activity and ATP content to control levels. 880 and 770 nm were slightly less effective, and 728 nm was the least effective. (C) The action spectra of relative cytochrome c oxidase activity and relative ATP content in TTX-exposed and LED-treated neurons closely match the absorption spectrum of oxidized cytochrome c oxidase reported by Cooper and Springett (1997). Note that the least effective wavelength (728 nm) did not correspond to the absorption spectrum of the enzyme. *Modified from Wong-Riley, M.T.T., Liang, H.L., Eells, J.T., Henry, M.M., Buchmann, E., Kane, M., et al., 2005. Photobiomodulation directly benefits primary neurons functionally inactivated by toxins: Role of cytochrome c oxidase. J. Biol. Chem. 280, 4761–4771.*

controls, whereas 770 and 880 nm increased it close to controls (98.4% and 91.5%, respectively). The 728 nm wavelength, though in between those of 670 and 770 nm, was the least effective, and the resulting ATP content was 86.2% of controls, that is, not different from the value of TTX alone.

When the action spectra of cytochrome c oxidase activity and ATP content of neurons responding to varying wavelengths were plotted against known absorption spectrum of the enzyme (Cooper and Springett, 1997), the effective wavelengths (especially 670 and 830 nm) correlated positively with the reported absorption spectrum of oxidized cytochrome c oxidase. Significantly, the least effective wavelength, 728 nm, did not correspond to the absorption spectrum of the enzyme (Wong-Riley et al., 2005).

3.6 Optimal regimen of photobiomodulation via light-emitting diode for cultured neurons exposed to cyanide

If photobiomodulation stimulates ATP synthesis mainly by activating cytochrome c oxidase, then an inhibitor of this enzyme should lessen the effect of PBM in a dose-dependent manner. Cyanide is a known, potent inhibitor of cytochrome c oxidase (Wikström et al., 1981) and is cytotoxic (Mills et al., 1996; Li et al., 2002). Even at 10 μM, KCN causes a significant reduction in cytochrome c oxidase activity in cultured primary neurons (Wong-Riley et al., 2005). With increasing concentrations from 10 μM to 10 mM, KCN induced a progressive decline in enzyme activity and cell viability.

LED treatment (670 nm; energy density of 4 J/cm^2 when administered for 80 seconds), twice a day for each of the 5 days of exposure to 10–100 μM KCN, was able to reverse the toxic effect of KCN on enzyme activity in primary neurons ($P < .001$ for all three metabolic cell types examined). At higher concentrations of KCN (1–100 mM), however, increasing neuronal death negated any beneficial effect of LED (Wong-Riley et al., 2005).

Cell death can occur either through the apoptotic or necrotic pathway. At the electron microscopic level, 300 μM KCN was found to induce apoptotic neuronal death with nuclear shrinkage and condensation with or without chromatin aggregation and vacuole formation, but the nuclear and plasma membranes stayed intact (Liang et al., 2006). However, at 1 mM concentration, KCN brought on necrotic neuronal death with nuclear condensation and fragmentation, mitochondrial swelling, loss of organelles, and disintegration of the plasma membrane (Liang et al., 2006).

If LED treatment once or twice a day as described above is neuroprotective, would repeated treatment each day bring about further improvement? This question was addressed in an experiment in which primary cortical neurons were exposed to 300 μM KCN for 24 hours, and LED treatment (670 nm; 4 J/cm^2 and 50 mW/cm^2 for 80 seconds each time) was given once, twice, three, or four times spaced evenly apart during the first 8 hours to sister cultures (Liang et al., 2008). Reactive oxygen species (ROS) in these neurons were monitored by changes in the intensity of CM-H_2DCFDA (5-[and -6] chloromethyl-2′, 7-dichlorodihydrofluorescein diacetate acetyl ester). As shown in Fig. 3.5, KCN induced a

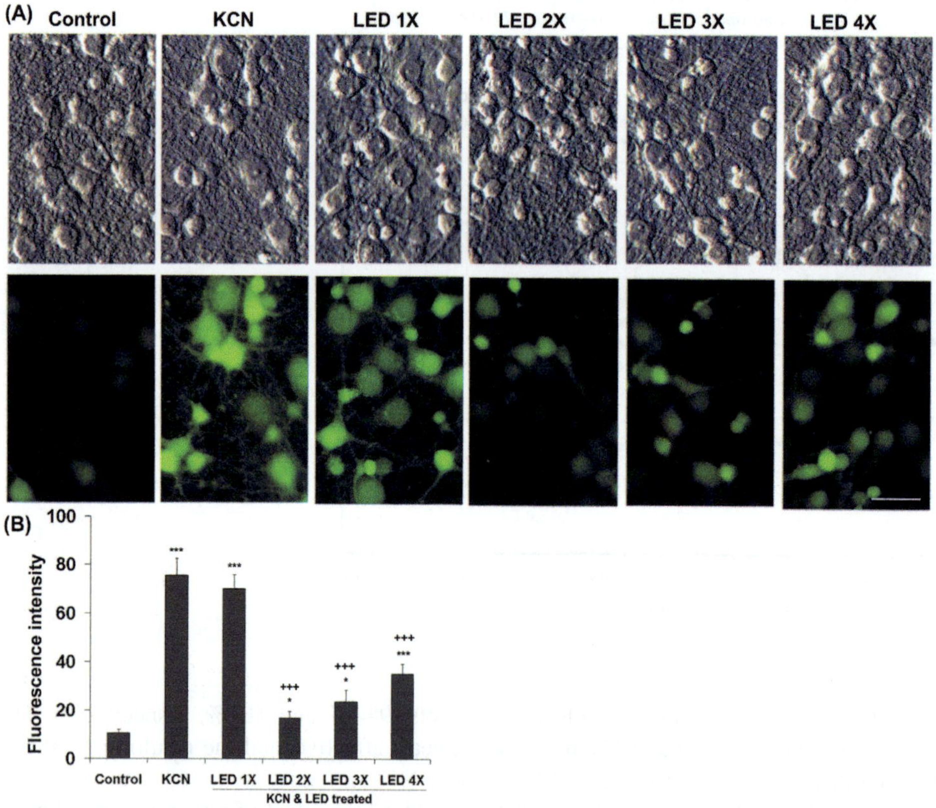

FIGURE 3.5 KCN-induced ROS production and the effect of various frequencies of LED treatment. (A) Top panels: phase contrast micrographs of cultured primary neurons. Lower panels: CM-H_2DCFDA fluorescence of ROS taken from the same fields as the corresponding top panels. Scale bar = 30 μm. (B) Fluorescence intensity of ROS was significantly enhanced by 24 h of KCN exposure. A single treatment with LED was not effective, but LED two times during the first 8-hour period was the most effective, with three and four times a day showing progressively less of an effect. All $*P$ values were compared to controls: $*P < .05$, $***P < .001$. $^{+++}P < .001$ when compared to KCN only. *Adapted from Liang, H.L., Whelan, H.T., Eells, J.T., Wong-Riley, M.T.T., 2008. Near-infrared light via light-emitting diode treatment is therapeutic against rotenone- and MPP$^+$-induced neurotoxicity. Neuroscience, 153, 963–974.*

significant increase in ROS formation within a 24-hour period. LED given once during that time was not effective in overcoming the effect of KCN. Remarkably, PBM given twice during the 24-hour period resulted in the lowest ROS formation, when compared with three or four times of treatment (Liang et al., 2008).

How would the various regimens of PBM affect cellular responses other than ROS production? Using paradigms similar to those described above (i.e., primary neurons were exposed to 300 μM KCN for 24 hours, and were treated with LED once, twice, three, or four times during the first 8 hours), it was found that two times per day of PBM was more effective than one, three, or four times a day in counteracting the toxic effect of KCN. Twice a day of PBM significantly increased cytochrome c oxidase activity, though not reaching control levels ($P < .05$ when compared with KCN alone; $P < .01$ when compared with controls) (Fig. 3.6A). ATP content actually exceeded, though not significantly different from, control levels ($P < .001$ when compared with KCN alone) (Fig. 3.6B). On the other hand, apoptotic neuronal numbers (shown with the Hoechst 33258 stain) were significantly reduced with PBM twice a day ($P < .001$ when compared with KCN alone) (Fig. 3.6C), and NO production as measured by DAF-2 DA fluorescence was also reduced

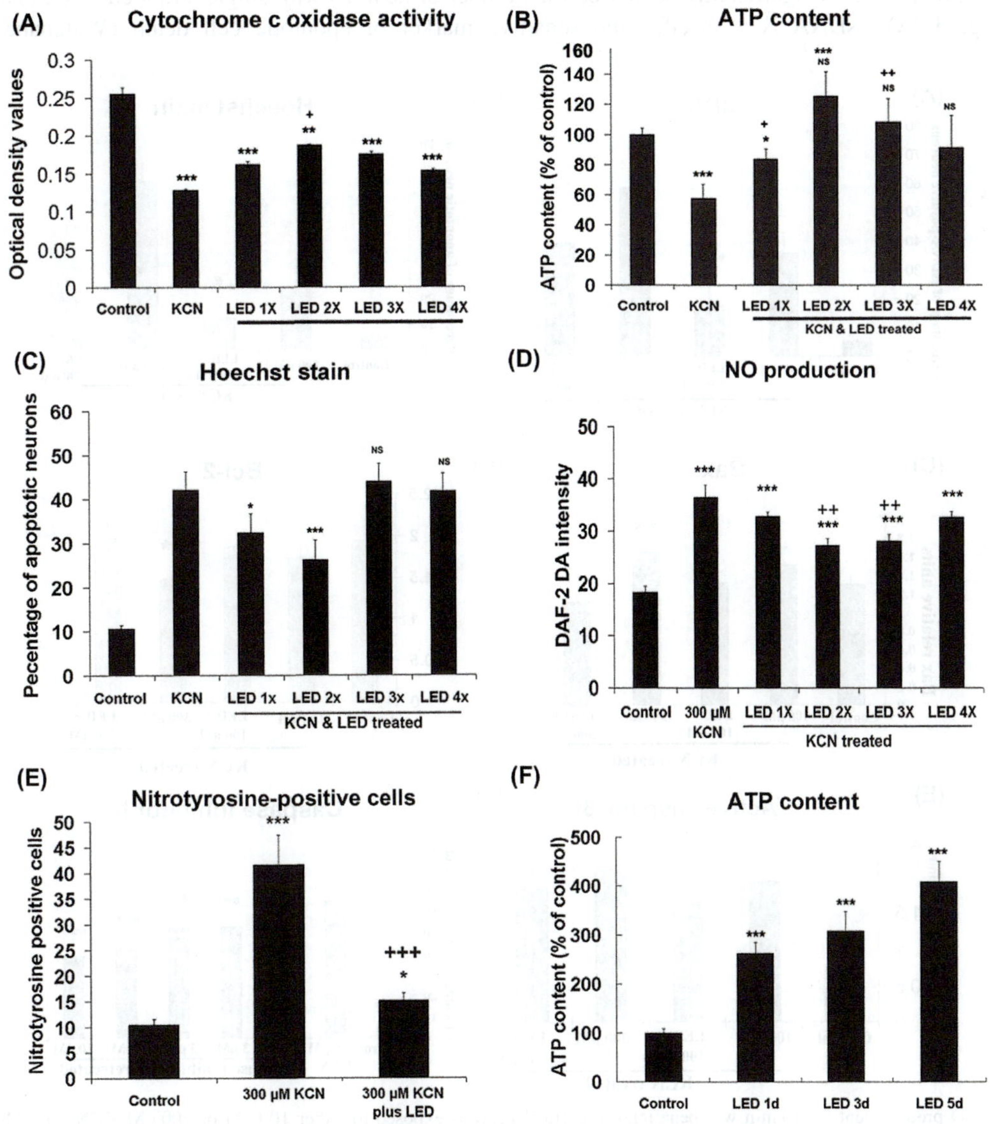

FIGURE 3.6 Effects of different frequencies of LED treatment (one, two, three, or four times during the first 8 h) on cortical neurons exposed to 300 μM KCN for 24 h. (A) Cytochrome c oxidase activity. (B) ATP content. (C) Hoechst stain for apoptotic cells; (D) DAF-2 DA for NO production; and (E) Nitrotyrosine expression. LED two times a day yielded the most beneficial result. (F) ATP content in cortical neurons after LED treatment two times a day for 1, 3, and 5 days. Energy production increased with treatment. All $^{*}P$ values were compared to controls. All ^{+}P values were compared to KCN alone. $^{*}P < .05$, $^{**}P < .01$, $^{***}P < .001$. $^{+}P < .05$, $^{++}P < .01$, $^{+++}P < .001$. NS, nonsignificant as compared to KCN alone. *Adapted from Liang, H.L., Whelan, H.T., Eells, J.T., Wong-Riley, M.T.T., 2008. Near-infrared light via light-emitting diode treatment is therapeutic against rotenone- and MPP*$^{+}$*-induced neurotoxicity. Neuroscience, 153, 963–974.*

in cultured cortical neurons ($P < .001$ when compared with controls; $P < .01$ when compared with KCN alone) (Fig. 3.6D). Twice a day PBM for 3 days significantly decreased the number of nitrotyrosine-positive cells as compared to KCN exposure alone ($P < .001$) (Fig. 3.6E). Twice a day of PBM also induced incremental cellular ATP content in normal cortical neurons by 2.63-, 3.08-, and 4.08-fold after 1, 3, and 5 days of treatment, respectively ($P < .001$ for all when compared to controls) (Fig. 3.6F).

Thus, under the paradigms tested, PBM twice a day appears to be the most effective strategy. It also indicates that the Goldilocks rule applies.

3.7 Photobiomodulation pretreatment has added benefits for neurons exposed to cyanide

Photobiomodulation is usually applied either concurrent with or after a toxic insult. Whether or not PBM pretreatment would yield added benefits, it has received relatively little attention. In a series of experiments on cultured primary neurons, Liang et al. (2006) found that LED pretreatment for 10 minutes (total energy density of 30 J/cm^2) before exposure to 100 or 300 μM KCN significantly decreased the number of neurons with single-stranded DNA (ssDNA) induced by KCN (Fig. 3.7A). ssDNA is a specific and sensitive marker of apoptotic cell death (Watanabe et al., 1999;

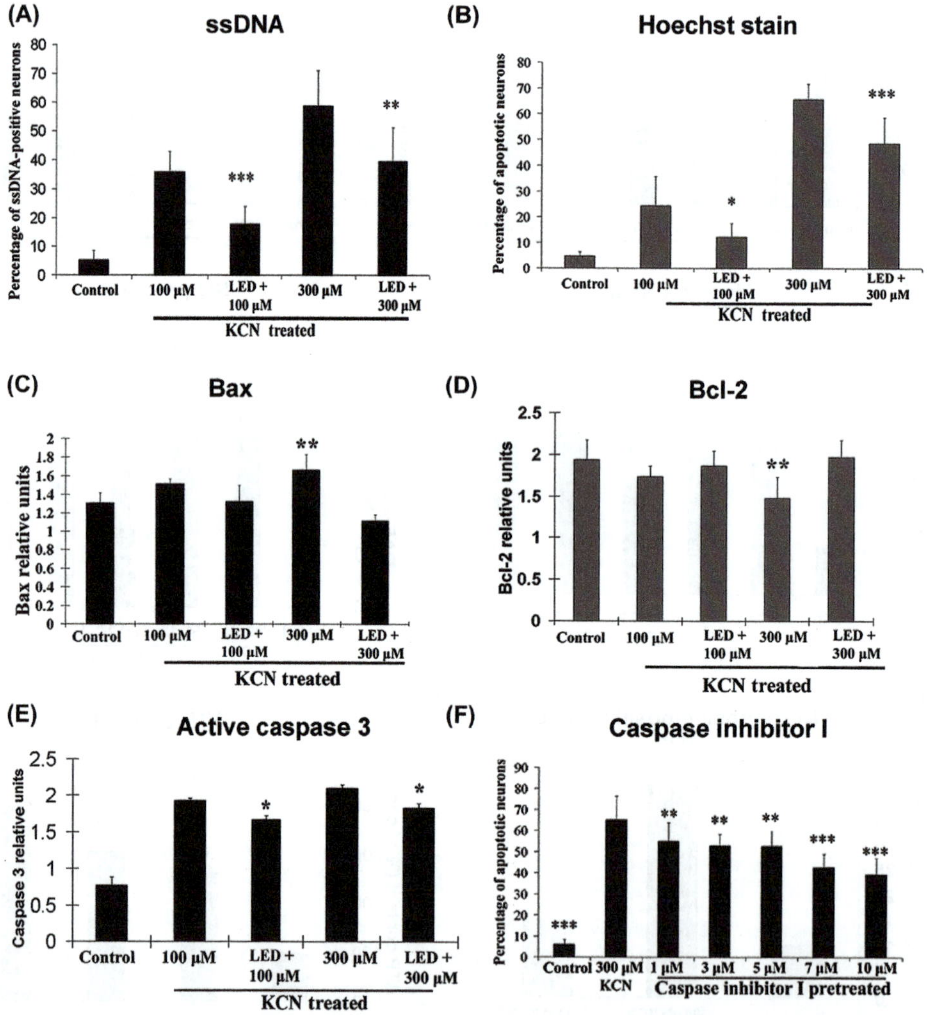

FIGURE 3.7 LED pretreatment for 10 min was beneficial to cortical neurons exposed to either 100 μM or 300 μM KCN for 28 h. (A) ssDNA for apoptotic cells; (B) Hoechst stain for apoptotic cells; (C) proapoptotic Bax; (D) antiapoptotic Bcl-2; and (E) caspase-3 as effector of apoptosis. (F) Increasing concentrations of caspase inhibitor I pretreatment progressively reduced the percentage of apoptotic neurons revealed by the Hoechst stain. *$P < .05$, **$P < .01$, ***$P < .001$. In (A), (B), (E), and (F): P values were compared to respective untreated KCN groups; in (C) and (D): P values were compared to controls. *Adapted from Liang, H.L., Whelan, H., Eells, J., Meng, H., Buchmann, E., Lerch-Gaggl, A., et al., 2006. Photobiomodulation partially rescues visual cortical neurons from cyanide-induced apoptosis. Neuroscience, 139, 639–649.*

Frankfurt and Krishan, 2001). Under normal conditions, only about 5.3% of cortical neurons in culture are ssDNA-positive. Exposure to 100 μM KCN increased this percentage to 36%, but PBM pretreatment reduced it to 17.9%, representing a 50.3% reduction ($P < .001$). Exposure to 300 μM KCN induced 58.9% of neurons to become ssDNA-positive, but PBM pretreatment decreased it to 39.6%, equivalent to a drop of 32.8% ($P < .01$). Comparable results were obtained with the Hoechst stain (Fig. 3.7B). PBM pretreatment also prevented the upregulation of proapoptotic Bax (Fig. 3.7C), and the downregulation of the antiapoptotic Bcl-2 (Fig. 3.7D) by 300 μM KCN. Moreover, it suppressed (though not to control levels) the upregulation of caspase-3, a potent effector of apoptosis, by KCN (Fig. 3.7E). The critical role of caspase in inducing apoptosis was verified when cultured neurons were pretreated for 30 minutes with caspase inhibitor I at 1, 3, 5, 7, and 10 μM concentrations before the cells were exposed to 300 μM KCN for 28 hours (Fig. 3.7F). The prevalence of apoptotic neurons was decreased in a dose-dependent manner, representing a 15.3%, 18.6%, 18.9%, 34.4%, and 39.6% reduction, respectively, by the various concentrations of caspase inhibitor I. All reductions were statistically significant ($P < .01 - .001$).

3.8 Therapeutic effect of photobiomodulation on primary neurons exposed to MPP^+ or rotenone

Parkinson disease is a debilitating, neurodegenerative disease characterized by the loss of dopaminergic neurons in the substantia nigra, pars compacta, of the midbrain, leading to tremor, rigidity, and bradykinesia (Hirsch et al., 1997; Parkinson, 2002). Two neurotoxins have been found to cause Parkinsonian-like symptoms in animal models, and both of them affect complex I of the electron transport chain in the mitochondria. The first is MPP^+ (1-methyl-4-phenylpyridinium ion), a metabolic product of MPTP (1-methyl-4-phenyl-1,2,3,6-tetrahydropyridine) (Langston et al., 1983; Schmidt and Ferger, 2001; Sherer et al., 2002). MPP^+ is taken up by dopaminergic and perhaps other neuronal types and accumulates in the mitochondria, where it inhibits complex I, increases ROS and reactive nitrogen species (RNS), and ultimately causes cell death (Metodiewa and Koska, 2000; Dawson and Dawson, 2003). The other toxin is rotenone, a broad-spectrum insecticide and pesticide found to inhibit the transfer of electrons from Fe-S centers to ubiquinone, thereby reducing cellular ATP production (Sherer et al., 2003; Hirata and Nagatsu, 2005). Both toxins induce oxidative stress, neuronal dysfunction, and eventual death (Cassarino et al., 1999; Fiskum et al., 2003; Watabe and Nakaki, 2004).

Would PBM via LED be protective against the detrimental effects of MPP^+ and rotenone? This was tested in both striatal and cortical neurons in culture (Liang et al., 2008). Neurons were exposed to varying concentrations of MPP^+ or rotenone for 48 hours. Sister cultures were treated with or without LED (670 nm; at 4 J/cm^2 and 50 mW/cm^2; 80 seconds each time) twice a day for each of the 2 days in the toxin, and additional normal cultures served as controls. Fig. 3.8A shows that occipital cortical neurons exposed to 100, 250, and 500 μM of MPP^+ had decreasing ATP content with increasing MPP^+ concentrations ($P < .001$ for all). However, LED treatment significantly increased the ATP content at all concentrations of MPP^+ ($P < .05 - .01$), though none of them reached control levels. Cortical neurons exposed to increasing concentrations of rotenone (50, 100, 200, and 400 nM) also exhibited decreasing ATP content that was reversed significantly in each case by LED treatment twice a day ($P < .05$ for all), though, again, they did not reach control levels (Fig. 3.8B). Striatal neurons (not shown) exhibited similar trends as cortical ones toward both MPP^+ and rotenone, although the loss in ATP content was not as severe as in cortical neurons, but nonetheless significant ($P < .01 - .001$). LED treatment twice a day, again, ameliorated the detrimental effect of MPP^+ and rotenone on striatal neurons, though to varying degrees depending on the concentrations of the toxins (Liang et al., 2008; Wong-Riley and Liang, 2017).

Exposure of cortical neurons to 250 or 500 μM of MPP^+ for 48 hours also led to apoptotic cell death, with nuclear condensation or decreased nuclear size, with or without nuclear fragmentation demonstrable with the Hoechst stain (Liang et al., 2008). LED treatment twice a day during the 48 hours of toxin exposure significantly reduced the number of apoptotic neurons by 36.4% and 42%, respectively, depending on the concentration of MPP^+ ($P < .001$) (Fig. 3.8C). Likewise, exposure to 50 or 100 nM of rotenone resulted in apoptotic death in 28% and 38.4% of cortical neurons, respectively (Fig. 3.8D). LED treatments twice a day reduced such incidence by 20.1% and 29.2%, respectively ($P < .01$) (Fig. 3.8D). Striatal neurons also underwent apoptosis in the presence of 250 or 500 μM of MPP^+, and LED treatment twice a day effectively decreased such numbers by 36% and 33.3%, respectively ($P < .001$) (Liang et al., 2008; Wong-Riley and Liang, 2017). Similarly, 50 or 100 nM of rotenone caused apoptotic cell death in 36.2% and 41.5% of striatal neurons, respectively, but LED twice a day lowered such numbers by 20.4% and 28.3%, respectively ($P < .01$) (Liang et al., 2008; Wong-Riley and Liang, 2017). Longer exposures of either striatal or cortical neurons to

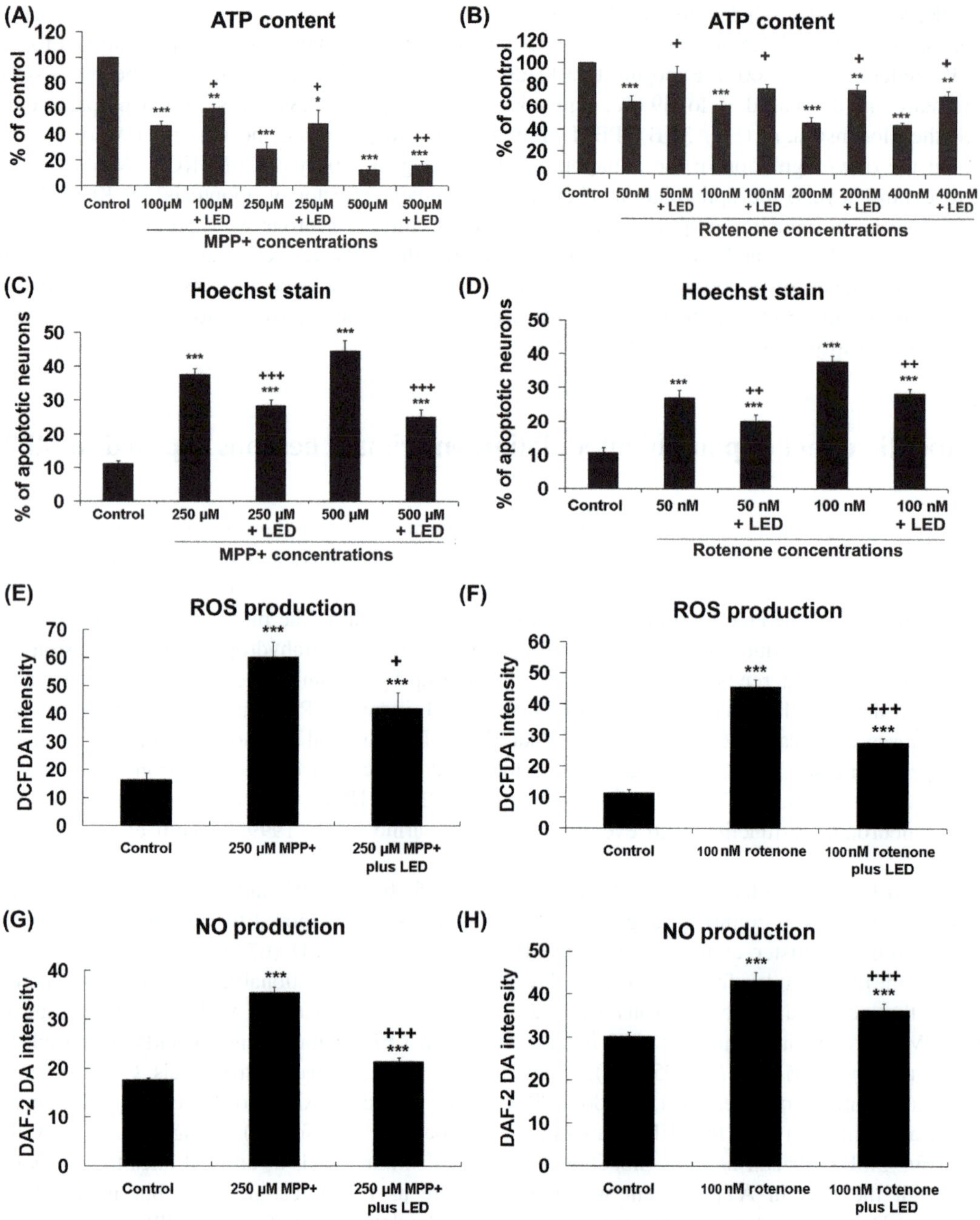

FIGURE 3.8 Effect of LED treatment twice a day on cortical neurons exposed to MPP^+ (A, C, E, and G) or rotenone (B, D, F, and H) at various concentrations. (A and B): ATP production. (C and D): Hoechst stain of apoptotic neurons. (E and F): ROS production. (G and H): NO production. All $*P$ values were compared to controls. $*P<.05$, $**P<.01$, $***P<.001$. All ^+P values were compared to respective toxins only. $^+P<.05$, $^{++}P<.01$, $^{+++}P<.001$. *Adapted from Liang, H.L., Whelan, H.T., Eells, J.T., Wong-Riley, M.T.T., 2008. Near-infrared light via light-emitting diode treatment is therapeutic against rotenone- and MPP^+-induced neurotoxicity. Neuroscience, 153, 963–974.*

MPP^+ or rotenone for 3–5 days resulted in further apoptotic cell death that was only partially, though significantly, reversed by LED treatment (Liang et al., 2008).

MPP^+ and rotenone also induced the production of ROS demonstrable with DCFDA. LED treatment twice a day, however, significantly reduced the DCFDA intensity induced by 250 μM MPP^+ from 60.3 to 42.0 in cortical neurons (Fig. 3.8E), and from 58.9 to 46.5 in striatal neurons (Liang et al., 2008) ($P<.05$ for both). LED treatment also decreased the DCFDA intensity induced by 100 nM rotenone from 47.5 to 29.8 in cortical neurons (Fig. 3.8F), and from 71.2 to 31.4 in striatal neurons (Liang et al., 2008) ($P<.001$ for both). However, PBM did not reverse the DCFDA intensity to control levels.

The effect of PBM on RNS production (shown with DAF-2 DA) brought on by MPP^+ or rotenone exposure was also investigated. Under the same condition as described above, LED treatment twice a day significantly reduced DAF-2 DA intensity induced by 250 μM MPP^+ or 100 nM rotenone in both cortical (Figs. 3.8G and H) and striatal neurons (Liang et al., 2008; Wong-Riley and Liang, 2017). Again, within the parameters tested, PBM did not revert DAF-2-DA intensities to control levels.

3.9 Pretreatment with photobiomodulation is beneficial for neurons exposed to MPP^+ or rotenone

If PBM is therapeutic once the neurons are exposed to MPP^+ or rotenone, will PBM pretreatment have added benefits? This question was addressed in a study of cultured striatal and visual cortical neurons (Ying et al., 2008). Sister cultures of each type of neurons were divided into five groups: (1) normal controls; (2) exposure to 250 μM MPP^+ for 48 hours; (3) pretreatment with 670 nm LED for 80 seconds (at 50 mM/cm^2 and 4 J/cm^2) twice a day for 2 days before exposing to MPP^+ for 48 hours; (4) exposure to MPP^+ for 48 hours, during which LED treatment was given twice a day for the 2 days in toxins; and (5) pretreatment for 2 days followed by neurotoxin exposure and LED treatment during exposure. Fig. 3.9A shows the results in cortical neurons. Forty-eight hours of MPP^+ induced 38.59% of cortical neurons to become apoptotic ($P < .001$). LED pretreatment for 2 days reduced the number of apoptotic neurons by 22.75% ($P < .001$). Without pretreatment, LED during the neurotoxin exposure decreased the prevalence of apoptotic neurons by 19.48% ($P < .001$). However, pretreatment plus treatment during toxin exposure brought added benefit and reduced the number of apoptotic neurons by 47.26% ($P < .001$). Similar results were obtained in striatal neurons (Ying et al., 2008; Wong-Riley and Liang, 2017). In all cases, PBM did not reduce the prevalence of apoptotic cells to control levels.

The effects of PBM pretreatment on neurons exposed to rotenone were also tested (Ying et al., 2008). Fig. 3.9B illustrates that 200 nM rotenone induced 42.96% of cortical neurons to undergo apoptosis ($P < .001$). LED treatment twice a day during the 2 days of toxin exposure reduced the number of apoptotic neurons by 41.43% ($P < .001$). However, PBM pretreatment for 2 days followed by PBM for 2 days during rotenone exposure resulted in 66.43%

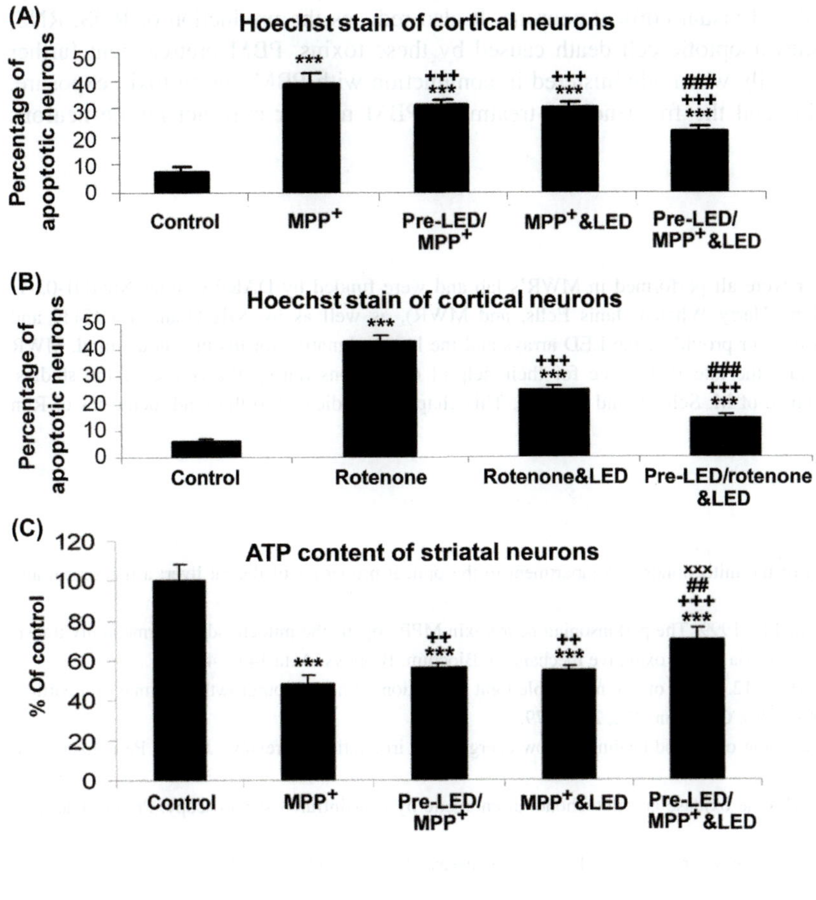

FIGURE 3.9 LED pretreatment had an added effect than LED alone on cortical (A and B) and striatal (C) neurons exposed to either MPP^+ (A and C) or rotenone (B). All **P* values were compared to controls. ***$P < .001$. All ^+P values were compared to respective toxins only. $^{++}P < .01$, $^{+++}P < .001$. All $^\#P$ values were compared to toxins plus LED without pretreatment. $^{\#\#}P < .01$, $^{\#\#\#}P < .001$. In C, $^{xxx}P < .001$ when compared to MPP^+ and LED. *Adapted from Ying, R., Liang, H.L., Whelan, H.T., Eells, J.T., Wong-Riley, M.T.T., 2008. Pretreatment with near-infrared light via light-emitting diode provides added benefit against rotenone- and MPP+-induced neurotoxicity. Brain Res. 1243, 167–173.*

decrease in the number of apoptotic neurons as compared to rotenone alone ($P < .001$), or a 42.68% reduction as compared to PBM without pretreatment ($P < .001$). Comparable results were found in striatal neurons (Ying et al., 2008; Wong-Riley and Liang, 2017). However, a lower percentage of striatal neurons underwent apoptosis as compared to cortical neurons (29.81% vs 42.96%). Likewise, LED treatment during toxin exposure resulted in a lower reduction of apoptotic neurons (33.65% vs 41.43%), and PBM pretreatment also caused a lower reduction of apoptosis in striatal as compared to cortical neurons (56.16% vs 66.43%) (Ying et al., 2008; Wong-Riley and Liang, 2017).

Besides reducing the prevalence of apoptosis in striatal and cortical neurons, does PBM pretreatment yield any benefit in ATP production? This question was tested in striatal neurons exposed to 250 μM MPP^+ for 48 hours (Ying et al., 2008). The toxin caused a severe downregulation of ATP content as compared to controls ($P < .001$) (Fig. 3.9C). Both LED treatment during toxin exposure and LED pretreatment without subsequent PBM in the presence of toxins increased ATP content above that of MPP^+ alone ($P < .01$). However, PBM pretreatment plus treatment during toxin exposure further increased ATP content above those of the other two treatment paradigms ($P < .01 - < .001$) (Fig. 3.9C). However, the values did not reach control levels.

Thus, both mitochondrial toxins, MPP^+ and rotenone, are detrimental to both striatal and cortical neurons in culture. The inhibition of complex I depletes the cells of ATP (and ATPdependent cellular processes) while increasing the production of free radicals and subsequent oxidative stress (Gerlach et al., 1991; Gandhi and Wood, 2005). PBM during toxin exposure proves to be neuroprotective, and PBM pretreatment has added benefit. Both of them are likely to boost the cells' energy production as well as trigger a cascade of events that alter gene expressions to upregulate energy generation and neuronal activity while downregulating cellular destructive genes induced by toxins (Eells et al., 2004; Wong-Riley and Liang, 2017).

3.10 Conclusions

Photobiomodulation with far red-to-NIR light via LEDs has a known therapeutic effect on wounds and a wide variety of ailments. This chapter reviews its application in cultured primary neurons that were exposed to various toxins. As little as 80 seconds of treatment with 670 nm LED (with energy density of 4 J/cm^2) and optimally twice a day is able to reverse the detrimental effect of nonlethal doses of TTX, KCN, MPP^+ or rotenone on cytochrome c oxidase activity, ATP production, and cell viability in striatal and visual cortical neurons. It also reduces the production of ROS, RNS, and nitrotyrosine, and significantly diminishes apoptotic cell death caused by these toxins. PBM pretreatment further enhances the therapeutic effect of LED, especially when administered in conjunction with PBM during toxin exposure. However, depending on the dosage of toxins and the frequency of treatment, PBM may or may not rescue neurons completely to control levels.

Acknowledgments

The experiments cited and illustrated in this chapter were all performed in MWR's lab and were funded by DARPA Grant N66001-03-1-8906, NIH/NICAM Grant R21-AT003002 (shared by Harry Whelan, Janis Eells, and MWR), as well as by NIH Grants EY05439 and EY018441 (to MWR). We thank Quantum Devices Inc. for providing the LED arrays and the late R. Ignatius for his ingenious mind. MWR also greatly appreciates Drs. H. Whelan, J. Eells, and the late B. Chance for their helpful discussions during the course of the studies. Additionally, we appreciate the able technical assistance of M. Scheidt and R. Ying. This chapter is dedicated to the fond memories of Ron Ignatius and Britton Chance.

References

Beauvoit, B., Kitai, T., Chance, B., 1994. Contribution of the mitochondrial compartment to the optical properties of the rat liver: a theoretical and practical approach. Biophys. J. 67, 2501–2510.

Cassarino, D.S., Parks, J.K., Parker Jr., W.D., Bennett Jr., J.P., 1999. The parkinsonian neurotoxin MPP^+ opens the mitochondrial permeability transition pore and releases cytochrome c in isolated mitochondria via an oxidative mechanism. Biochim. Biophys. Acta 1453, 49–62.

Choi, D.-H., Lee, K.-H., Kim, J.-H., Kim, M.Y., Lim, J.H., 2012. Effect of 710 nm visible light irradiation on neurite outgrowth in primary rat cortical neurons following ischemic insults. Biochim. Biophys. Res. Commun. 422, 274–279.

Conlan, M.J., Rapley, J.W., Cobb, C.M., 1996. Biostimulation of wound healing by low-energy laser irradiation. A review. J. Clin. Periodontol. 23, 492–496.

Cooper, C.E., Springett, R., 1997. Measurement of cytochrome oxidase and mitochondrial energetic by near-infrared spectroscopy. Philos. Trans. R. Soc. Lond. B Biol. Sci. 352, 669–676.

Dawson, T.M., Dawson, V.L., 2003. Molecular pathways of neurodegeneration in Parkinson's disease. Science 302, 819–822.

Duan, R., Zhu, L., Liu, T.C.-Y., Li, Y., Liu, J., Jiao, J., et al., 2003. Light emitting diode irradiation protect against the amyloid beta 25-35 induced apoptosis of PC12 cell in vitro. Lasers Surg. Med. 33, 199–203.

Eells, J.T., Henry, M.M., Summerfelt, P., Wong-Riley, M.T.T., Buchmann, E.V., Kane, M., et al., 2003. Therapeutic photobiomodulation for methanol-induced retinal toxicity. Proc. Natl. Acad. Sci. U.S.A. 100, 3439–3444.

Eells, J.T., Wong-Riley, M.T.T., VerHoeve, J., Henry, M., Buchman, E.V., Kane, M.P., et al., 2004. Mitochondrial signal transduction in accelerated wound and retinal healing by near-infrared light therapy. Mitochondrion 4, 559–568.

Fiskum, G., Starkov, A., Polster, B.M., Chinopoulos, C., 2003. Mitochondrial mechanisms of neural cell death and neuroprotective interventions in Parkinson's disease. Ann. N.Y. Acad. Sci 991, 111–119.

Frankfurt, O.S., Krishan, A., 2001. Identification of apoptotic cells by formamide-induced DNA denaturation in condensed chromatin. J. Histochem. Cytochem. 49, 369–378.

Galkin, O., Buchter, S., Tabirian, A., Schulte, A., 1997. Pressure effects on the proximal heme pocket in myoglobin probed by Raman and near-infrared absorption spectroscopy. Biophys. J. 73, 2752–2763.

Gandhi, S., Wood, N.W., 2005. Molecular pathogenesis of Parkinson's disease. Hum. Mol. Genet. 14, 2749–2755.

Gerlach, M., Riederer, P., Przuntek, H., Youdim, M.B., 1991. MPTP mechanisms of neurotoxicity and their implications for Parkinson's disease. Eur. J. Pharmacol. 12, 273–286.

Gibson, Q.H., Greenwood, C., 1965. Kinetic observations on the near infrared band of cytochrome c oxidase. J. Biol. Chem. 240, 2694–2698.

Grillo, S.L., Duggett, N.A., Ennaceur, A., Chazot, P.L., 2013. Non-invasive infra-red therapy (1072 nm) reduces β-amyloid protein levels in the brain of an Alzheimer's disease mouse model, TASTPM. J. Photochem. Photobiol B: Biol. 123, 13–22.

Hevner, R.F., Liu, S., Wong-Riley, M.T.T., 1995. A metabolic map of cytochrome oxidase in the rat brain: histochemical, densitometric and biochemical studies. Neurosci. 65, 313–342.

Hirata, Y., Nagatsu, T., 2005. Rotenone and CCCP inhibit tyrosine hydroxylation in rat striatal tissue slices. Toxicology 216, 9–14.

Hirsch, E.C., Faucheux, B., Damier, P., Mouatt-Prigent, A., Agid, Y., 1997. Neuronal vulnerability in Parkinson's diseae. J. Neural. Transm. Suppl. 50, 79–88.

Huang, Y.Y., Gupta, A., Vecchio, D., de Arce, V.J., Huang, S.F., Xuan, W., et al., 2012. Transcranial low level laser (light) therapy for traumatic brain injury. J. Biophotonics 5, 827–837.

Jöbsis, F.F., 1977. Noninvasive, infrared monitoring of cerebral and myocardial oxygen sufficiency and circulatory parameters. Science 198, 1264–1267.

Karu, T., 1999. Primary and secondary mechanisms of action of visible to near-IR radiation on cells. J. Photochem. Photobiol. B 49, 1–17.

Karu, T., 2010. Multiple roles of cytochrome c oxidase in mammalian cells under action of red and IR-A radiation. IUBMB Life 62, 607–610.

Langston, J.W., Ballard, P., Tetrud, J.W., Irwin, I., 1983. Chronic Parkinsonism in humans due to a product of meperidine-analog synthesis. Science 219, 979–980.

Li, L., Prabhakaran, K., Shou, Y., Borowitz, J.L., Isom, G.E., 2002. Oxidative stress and cyclooxygenase-2 induction mediate cyanide-induced apoptosis of cortical cells. Toxicol. Appl. Pharmacol. 185, 55–63.

Liang, H.L., Whelan, H., Eells, J., Meng, H., Buchmann, E., Lerch-Gaggl, A., et al., 2006. Photobiomodulation partially rescues visual cortical neurons from cyanide-induced apoptosis. Neuroscience 139, 639–649.

Liang, H.L., Whelan, H.T., Eells, J.T., Wong-Riley, M.T.T., 2008. Near-infrared light via light-emitting diode treatment is therapeutic against rotenone- and MPP^+-induced neurotoxicity. Neuroscience 153, 963–974.

Mester, E., Spiry, T., Szende, B., Tota, J.G., 1971. Effect of laser rays on wound healing. Am. J. Surg. 122, 532–535.

Metodiewa, D., Koska, C., 2000. Reactive oxygen species and reactive nitrogen species: relevance to cyto(neuro)toxic events and neurologic disorders. An overview. Neurotox. Res. 1, 197–233.

Mills, E.M., Gunasekar, P.G., Pavlakovic, G., Isom, G.E., 1996. Cyanide-induced apoptosis and oxidative stress in differentiated PC12 cells. J. Neurochem. 67, 1039–1046.

Mitchell, P., 2011. Chemiosmotic coupling in oxidative and photosynthetic phosphorylation. 1966. Biochim. Biophys. Acta 1807, 1507–1538.

Nakamura, Y., Nakajima, S., Grundfest, H., 1965. The action of tetrodotoxin on electrogenic components of squid giant axons. J. Gen. Physiol. 48, 985–996.

Ochs, S., Hollingsworth, D., 1971. Dependence of fast axoplasmic transport in nerve on oxidative metabolism. J. Neurochem. 18, 107–114.

Parkinson, J., 2002. An essay on the shaking palsy. 1817. J. Neuropsychiatry Clin. Neurosci. 14, 223–236.

Quirk, B.J., DeSmet, K.D., Henry, M., Buchmann, E., Wong-Riley, M., Eells, J.T., et al., 2012. Therapeutic effect of near infrared (NIR) light on Parkinson's disease models. Front. Biosci. 4, 818–823.

Salehpour, F., Mahmoudi, J., Kamari, F., Sadigh-Eteghad, S., Rasta, S.H., Hamblin, M.R., 2018. Brain photobiomodulation therapy: a narrative review. Mol. Neurobiol. Available from: https://doi.org/10.1007/s12035-017-0852-4 [Epub ahead of print].

Schmidt, N., Ferger, B., 2001. Neuroprotective effects of (+/-)-kavain in the MPTP mouse model of Parkinson's disease. Synapse 40, 47–54.

Sherer, T.B., Betarbet, R., Greenamyre, J.T., 2002. Environment, mitochondria, and Parkinson's disease. Neuroscientist 8, 192–197.

Sherer, T.B., Betarbet, R., Testa, C.M., Seo, B.B., Richardson, J.R., Kim, J.H., et al., 2003. Mechanism of toxicity in rotenone models of Parkinson's Disease. J. Neurosci. 23, 10756–10764.

Watabe, M., Nakaki, T., 2004. Rotenone induces apoptosis via activation of bad in human dopaminergic SH-SY5Y cells. J. Pharmacol. Exp. Ther. 311, 948–953.

Watanabe, I., Toyoda, M., Okuda, J., Tenjo, T., Tanaka, K., Yamamoto, T., et al., 1999. Detection of apoptotic cells in human colorectal cancer by two different in situ methods: antibody against single-stranded DNA and terminal deoxynucleotidyl transferase-mediated dUTP-biotin nick end-labeling (TUNEL) methods. Jpn. J. Cancer Res. 90, 188–193.

Whelan, H.T., Smits Jr., R.L., Buchmann, E.V., Whelan, N.T., Turner, S.G., Margolis, D.A., et al., 2001. Effect of NASA light-emitting diode irradiation on wound healing. J. Clin. Laser Med. Surg. 19, 305–314.

Wikström, M., Krab, K., Saraste, M., 1981. Cytochrome Oxidase. A Synthesis. Academic Press, New York.

Wilson, D.F., Chance, B., 1967. Azide inhibition of mitochondrial electron transport I. The aerobic steady state of succinate oxidation. Biochim. Biophys. Acta 131, 421–430.

Wong-Riley, M.T.T., 1989. Cytochrome oxidase: an endogenous metabolic marker for neuronal activity. Trends Neurosci. 12, 94–101.

Wong-Riley, M.T.T., Liang, H.L., 2017. Cytoprotective effect of low level light therapy (LLLT) using light-emitting diode (LED) on neurons. In: Hamblin, M.R., Agrawal, T., de Sousa, M. (Eds.), Handbook of Low-Level Laser Therapy. Pan Stanford Publishing, Singapore, pp. 185–206. (Chapter 11).

Wong-Riley, M.T.T., Bai, X., Buchmann, E., Whelan, H.T., 2001. Light-emitting diode treatment reverses the effect of TTX on cytochrome oxidase in neurons. NeuroReport 12, 3033–3037.

Wong-Riley, M.T.T., Liang, H.L., Eells, J.T., Henry, M.M., Buchmann, E., Kane, M., et al., 2005. Photobiomodulation directly benefits primary neurons functionally inactivated by toxins: Role of cytochrome c oxidase. J. Biol. Chem. 280, 4761–4771.

Ying, R., Liang, H.L., Whelan, H.T., Eells, J.T., Wong-Riley, M.T.T., 2008. Pretreatment with near-infrared light via light-emitting diode provides added benefit against rotenone- and MPP^+-induced neurotoxicity. Brain Res. 1243, 167–173.

Zhang, C., Wong-Riley, M., 1999. Expression and regulation of NMDA receptor subunit R1 and neuronal nitric oxide synthase in cortical neuronal cultures: Correlation with cytochrome oxidase. J. Neurocytol. 28, 525–539.

Chapter 4

Photobiomodulation on cultured cortical neurons

Ying-Ying Huang[1,2] and Michael R. Hamblin[1,2]

[1]*Wellman Center for Photomedicine, Massachusetts General Hospital, Boston, MA, United States,* [2]*Department of Dermatology, Harvard Medical School, Boston, MA, United States*

4.1 Introduction

Photobiomodulation (PBM) has gained considerable interest in recent years as a therapy for the treatment of a variety of diseases and injuries, by eliciting significant biological effects at a cellular level and at a tissular level. PBM is supported by studies of various animal models of different diseases and conditions, and by a large number of clinical trials, many of which were randomized controlled trials. Different conditions for which this therapy has shown positive outcome include pain relief (Peres, 2010; Fulop et al., 2010; Konstantinovic et al., 2010), inflammation (Xavier et al., 2010), and wound healing (Peplow et al., 2010; Lucas et al., 2002; Yasukawa et al., 2007). In the last few years, this therapy has been extended to the treatment of more severe conditions such as stroke, myocardial infarctions, neurodegenerative disorders, and traumatic brain injury (TBI; Hashmi et al., 2010). There are a growing number of reports demonstrating the promising outcome for PBM for diseases related to the central nervous system (CNS).

Some of the positive effects shown in neurons are the promotion of axonal growth and nerve regeneration in both rat spinal cord (Byrnes et al., 2005; Wu et al., 2009) and peripheral nerve injuries (Anders et al., 2004) after PBM. The efficacy of PBM in the nervous system has been further established in animal studies showing improved neurological and functional outcomes poststroke (Oron et al., 2006; Detaboada et al., 2006; Lapchak and De Taboada, 2010), post-TBI (Oron et al., 2007), and in a mouse model of Alzheimer's disease (De Taboada et al., 2011). PBM also improved emotional response and memory function of middle-aged CD-1 mice (Michalikova et al., 2008). The results of human clinical studies in patients with long term peripheral nerve injury (Lucas et al., 2002) and ischemic stroke (Lampl et al., 2007; Lapchak, 2010) have also been promising. Recently, a study was performed on two human chronic TBI patients which showed that transcranial LED could improve cognition (Naeser et al., 2011).

Although the data for therapeutic usefulness of PBM for neurological disorders is promising, the medical and research community at large is still skeptical about this therapy. The reason for these reservations is the lack of fundamental understanding into the basic mechanistic effects of PBM and how the cellular effects on neurons (that can be observed in cell culture studies) can be translated into improved brain function in vivo.

The proposed mechanism of PBM is largely assumed to involve stimulation of the cellular energy metabolism and energy production mediated by mitochondria acting as the primary cellular photoreceptor or target for absorbing the photons. Within mitochondria, cytochrome c oxidase (complex 4 of the respiratory chain) is considered to be the principal chromophore absorbing the incident light; although other cytochromes, porphyrins, and heme proteins could be involved. Light absorption further leads to increased activity of cytochrome c oxidase (Hu et al., 2007), release of nitric oxide (NO) (Karu et al., 2005), and an increase in ATP levels (Pastore et al., 1996; Passarella et al., 1984). Changes in intracellular signaling molecules such as calcium ions, reactive oxygen species (ROS), and redox-sensitive transcription factors like NF-kB are also thought to mediate the effects of light. Previous studies from our laboratory in mouse embryonic fibroblast cells (Chen et al., 2009) have shown that 810 nm laser induced ROS-mediated NF-kB activation.

Photobiomodulation in the Brain. DOI: https://doi.org/10.1016/B978-0-12-815305-5.00004-X

4.2 Dose response in cultured cortical neurons

PBM studies have demonstrated that light follows a biphasic response (Huang et al., 2009, 2011), along the lines of the frequently cited "Arndt-Schulz Law" (Lubart et al., 2006; Chow, 2006; Hawkins and Abrahamse, 2006a,b; Sommer et al., 2001). The Arndt-Schulz Law, formulated in 1888, stated that various poisons, if given in very low doses, had a stimulatory effect; small dosages were beneficial and large doses were harmful (Calabrese, 2016). Stated in a generic manner, the law states that small doses stimulate, moderate doses inhibit, and large doses kill.

This concept was developed into "hormesis," which has been extensively reviewed by Edward Calabrese (Agathokleous et al., 2018; Calabrese and Mattson, 2017; Calabrese et al., 2016). Such an axiom fits well for PBM; where, it has been clearly shown that a biphasic dose response exists. An insufficient dose of energy would result in no response (or an insignificant one); if the dose of light is increased past the necessary threshold, then biostimulation takes place. However, if too much energy is applied, then not only does biostimulation decrease, but bioinhibition may also occur if the light dose is increased even further. The biochemical mechanisms responsible for the biphasic dose response of PBM at the cellular level remain unknown, although reactive oxygen species have been implicated.

We decided to measure the cellular responses of primary cultured mouse cortical neurons to PBM (810 nm laser) (Sharma et al., 2011). A wide range of fluences spanning three orders of magnitude was tested to increase the chances of detecting the biphasic dose response.

Pregnant female C57BL/6 mice (aged 8–10 weeks) were sacrificed at 16 days postconception. The cortical lobes of 6–10 embryonic brains were separated from subcortical structures and dissociated mechanically. Cells were plated at a density of approximately 30,000 cells/well on poly-D-lysine–coated cell culture plates or sterile glass coverslips. The plating and maintenance media consisted of Neurobasal plus B27 supplement (NB27). This media formulation inhibits the outgrowth of glia resulting in a neuronal population which is >95% pure (Brewer, 1995). Experiments were performed on days 8–9 after plating the cells. The cells were irradiated in continuous wave mode using an 810 nm laser with a power density of 25 mW/cm^2 for different time periods to achieve energy densities of 0.03, 0.3, 3, 10, and 30 J/cm^2. Assays were conducted after the end of the longest irradiation period (20 minutes).

ROS production was measured using MitoSox Red, a fluorescent indicator of mitochondrial superoxide. Intracellular NO production was measured using DAF-FM (which is essentially nonfluorescent until it is nitrosylated by products of oxidation of NO) resulting in DAF-FM triazole which exhibits about a 160-fold greater fluorescence quantum yield. Intracellular calcium was measured by Fluo-4AM, which exhibits a large fluorescence intensity increase on binding to Ca^{2+}. Mitochondrial membrane potential (MMP) was measured by JC1, which is a cationic dye that accumulates in mitochondria. This accumulation is dependent on the membrane potential and indicated by a fluorescence emission shift from green (~525 nm) to red (~590 nm). Consequently, mitochondrial depolarization is indicated by a decrease in the red/green fluorescence intensity ratio. ATP was measured by the luminescence-based Cell-Titer Glo Assay.

The ROS levels showed an interesting triphasic pattern depicted in Fig. 4.1C. There was a significant initial increase with a fluence as low as 0.03 J/cm^2, a further increase at 0.3 J/cm^2,and a peak corresponding to a threefold increase over baseline that ROS was seen at 3 J/cm^2. When the fluence was further increased, there was a significant decrease (compared to 3 J/cm^2) in ROS observed at 10 J/cm^2. When the fluence was further increased to 30 J/cm^2 there was a second significant increase (compared to 10 J/cm^2) in ROS to a level higher (but not significantly) than that observed at 3 J/cm^2.

PBM gave an increase in intracellular DAF-FM fluorescence in the cortical neurons on irradiating with light as shown in the micrographs in Fig. 4.2A and B after a fluence of 0.3 J/cm^2, indicating an increase in nitric oxide levels postirradiation. The NO levels followed the same general triphasic pattern as the ROS levels but the increase was less prominent as compared to the ROS (Fig. 4.2C). There was a significant increase at 0.3 J/cm^2 compared to control, a significant decrease at 3 J/cm^2 compared to 0.3 J/cm^2, and a further increase at 30 J/cm^2 compared to 0.3 J/cm^2. The whole triphasic pattern appeared to be shifted toward lower fluences (the first peak was at 0.3 J/cm^2 instead of 3 J/cm^2 and the following trough at 3 J/cm^2 instead of 10 J/cm^2) compared to the pattern observed with ROS (compare Fig. 4.1C).

The changes in MMP were determined by measuring the red to green ratio of the fluorescence of JC1 dye. When the MMP is high, more dye accumulates in the mitochondrial membrane which leads to dye aggregation (J-aggregates) and a red shift in the fluorescence. By contrast, when MMP is low, the dye disaggregates and the fluorescence is green shifted (Reers et al., 1991). The MMP in neurons was significantly increased after 3 J/cm^2 of 810-nm light as seen in the micrographs in Fig. 4.3A and B. The dose response was biphasic in nature (Fig. 4.3C). There was a significant increase at 0.3 J/cm^2 reaching a peak of twice basal levels at 3 J/cm^2. Then there was a decrease at 10 J/cm^2, and outright depolarization (significantly lower than control) of MMP was observed at 30 J/cm^2.

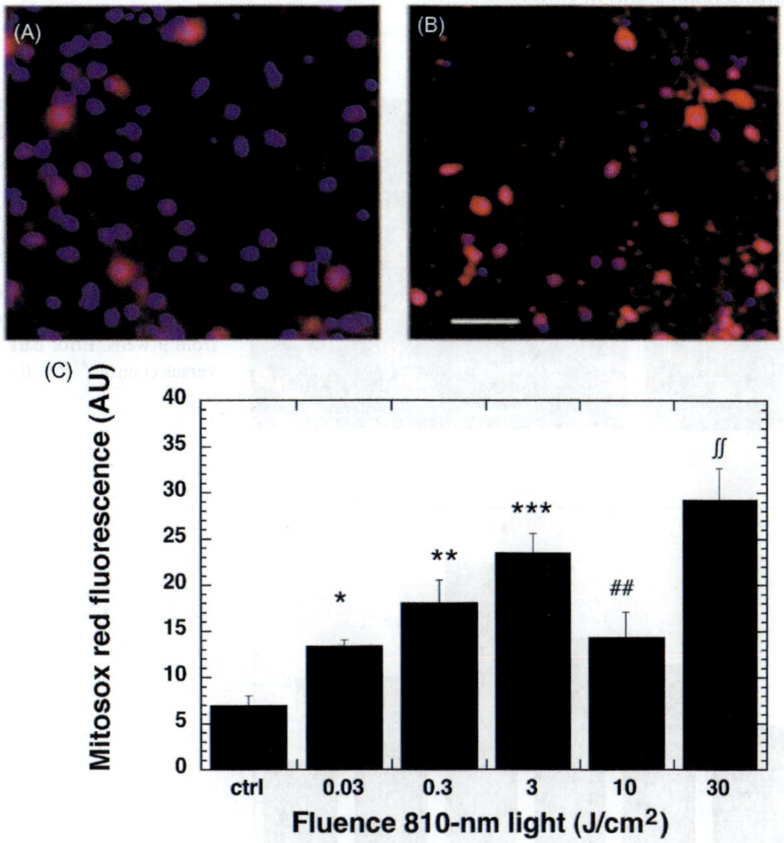

FIGURE 4.1 **Effect of 810-nm PBM on production of mitochondrial ROS in the cultured cortical neurons.** (A) Mitosox (*red*) and nuclear Hoechst (*blue*) fluorescence in control neurons. (B) Mitosox and nuclear Hoechst fluorescence in neurons treated with 3 J/cm^2 810-nm laser. Scale bar is 50 μm. (C) Quantification by fluorescence plate reader of mean Mitosox fluorescence values from 9 wells. Error bars are SD. $*P<.05$; $**P<.01$; $***P<.01$ versus control. $^{\#\#}P<.01$ versus 3 J/cm^2. $^{\int\int}P<.01$ versus 10 J/cm^2.

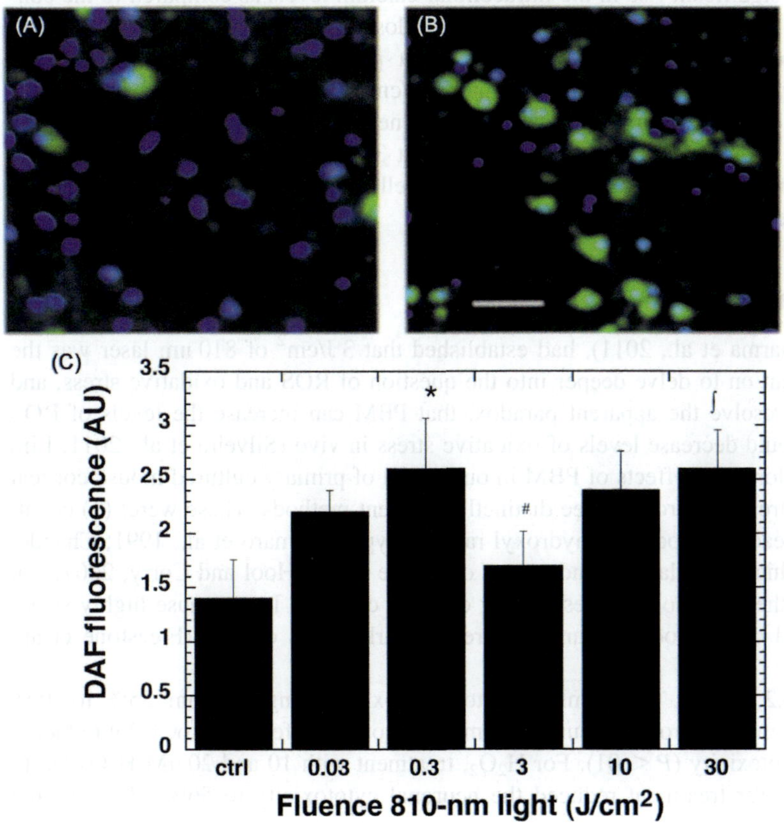

FIGURE 4.2 **Effect of 810-nm laser on production of intracellular NO in the cultured cortical neurons.** (A) DAF-DM (*green*) and nuclear Hoechst (*blue*) fluorescence in control neurons. (B) DAF-DM and nuclear Hoechst in neurons treated with 0.3 J/cm^2 810-nm laser. Scale bar is 50 μm. (C) Quantification by fluorescence plate reader of mean DAF-FM fluorescence values from 9 wells. Error bars are SD. $*P<.05$ versus control; $^{\#}P<.05$ versus 0.3 J/cm^2; $^{\int}P<.05$ versus 3 J/cm^2.

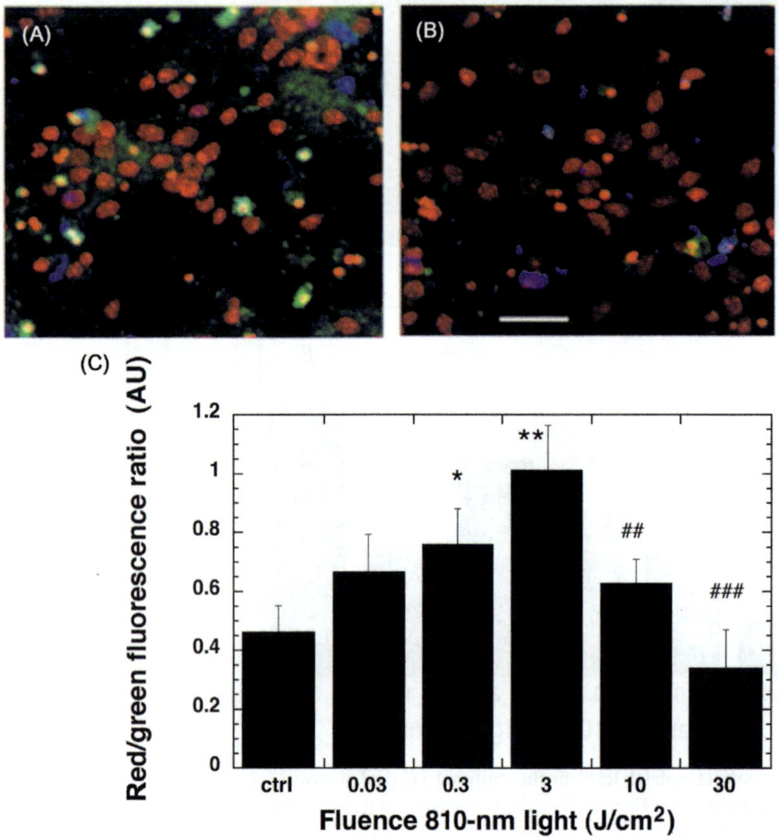

FIGURE 4.3 **Effect of 810-nm laser on mitochondrial membrane potential in the cultured cortical neurons.** (A) JC1 nonaggregated (*green*), JC1 aggregated (*red*), and nuclear Hoechst (*blue*) fluorescence in control neurons. (B) JC1 nonaggregated, JC1 aggregated, and nuclear Hoechst fluorescence in neurons treated with 3 J/cm^2 810-nm laser. Scale bar is 50 μm. (C) Quantification by fluorescence plate reader of the mean red/green fluorescence ratio values from 9 wells. Error bars are SD. $^{*}P<.05$; $^{**}P<.05$ versus control. $^{\#\#}P<.01$; $^{\#\#\#}P<.001$ versus 3 J/cm^2.

Neuronal cells treated with PBM showed a significant rise in the intracellular calcium levels as compared to the control as shown in the micrographs in Fig. 4.4A and B captured after 3 J/cm^2. The dose response of intracellular calcium was also biphasic as seen in Fig. 4.4C. The increase was significant when it peaked at the fluence of 3 J/cm^2. Interestingly, the levels decreased significantly in cells irradiated with 10 and 30 J/cm^2.

Fig. 4.5 shows the ATP content per mg cell protein in the mouse cortical neurons. ATP content was found to increase on treating the neurons with 810-nm light reaching a significant peak at 3 J/cm^2 as compared to the control cells. When the fluence was increased to 10 and 30 J/cm^2 the levels of the ATP fell and were equal to that of the control, and in the case of 30 J/cm^2 significantly less than the peak at 3 J/cm^2.

4.3 Oxidative stress in cultured cortical neurons

Since the previous study, described above (Sharma et al., 2011), had established that 3 J/cm^2 of 810 nm laser was the best dose to stimulate cortical neurons, we went on to delve deeper into the question of ROS and oxidative stress, and the cellular response to PBM. We wanted to resolve the apparent paradox, that PBM can increase the levels of ROS in vitro, while it was well-known that PBM could decrease levels of oxidative stress in vivo (Silveira et al., 2011; Lim et al., 2009, 2010; Rubio et al., 2009). We explored the effects of PBM in our model of primary cultured mouse cortical neurons that had been subjected to oxidative stress in vitro by three distinctly different methods. These were: (1) cobalt chloride ($CoCl_2$) which catalyzes the Fenton reaction producing hydroxyl radicals types (Tomaro et al., 1991; Chandel et al., 1998); (2) hydrogen peroxide (H_2O_2) which is a classical inducer of oxidative stress (Hool and Corry, 2007; Lin et al., 2004); and (3) rotenone which inhibits the mitochondrial respiratory chain at complex I, can cause highly selective degeneration of nigrostriatal neurons, and can reproduce many features of Parkinson's disease (Freestone et al., 2009; Betarbet et al., 2000).

For $CoCl_2$ treatment concentrations of 0.2, 0.5, 1, and 2 mM led to cytotoxicity ranging from 25% to 45% (Fig. 4.6A). PBM treatment with 3 J/cm^2 810-nm laser protects neurons from the cytotoxic effects of $CoCl_2$ by reducing the loss of viability to a maximum of 12% cytotoxicity ($P<.01$). For H_2O_2, treatment with 10 and 20 μM H_2O_2 led to cytotoxicity of 67% and 75%, respectively. Laser treatment reduced the neuronal cytotoxicity to 56% ($P<.01$) and

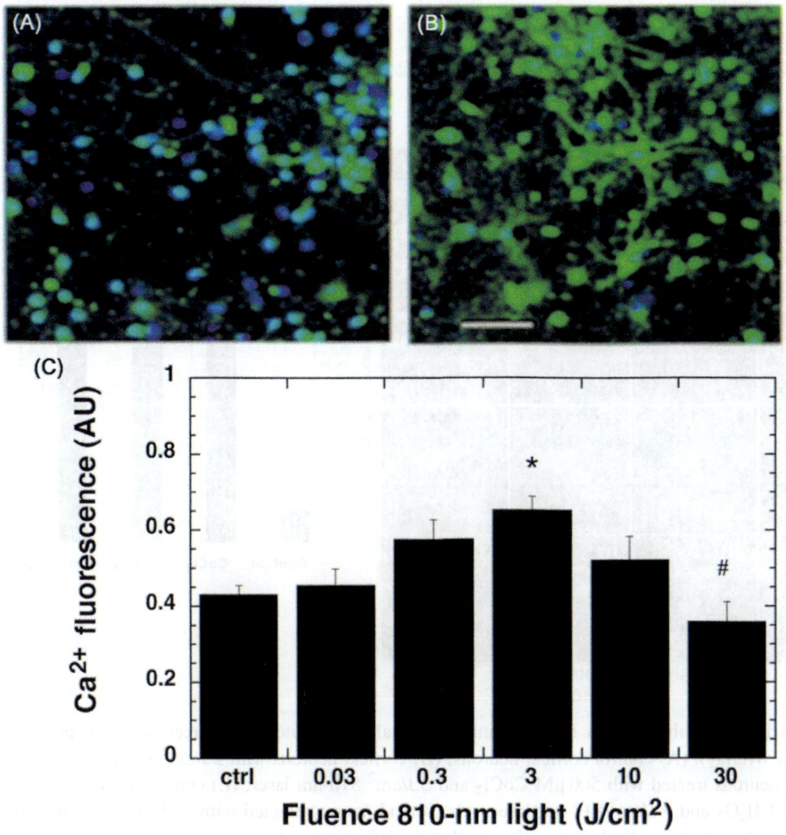

FIGURE 4.4 Effect of 810-nm laser on intracellular calcium in the cultured cortical neurons. (A) Fluo4 (*green*) and nuclear Hoechst (*blue*) fluorescence in control neurons. (B) Fluo4 (*green*) and nuclear Hoechst (*blue*) fluorescence in neurons treated with 3 J/cm^2 810-nm laser. Scale bar is 50 μm. (C) Quantification by fluorescence plate reader of the mean red/green fluorescence ratio values from 9 wells. Error bars are SD. $*P < .05$ versus control. $^{\#}P < .05$ versus 3 J/cm^2.

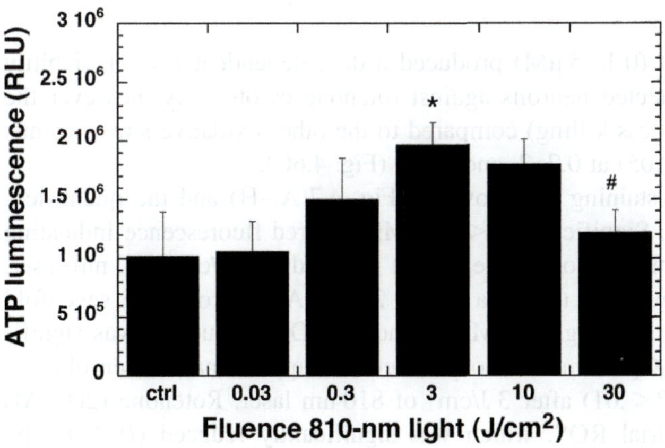

FIGURE 4.5 Effect of 810-nm laser on intracellular ATP in the cultured cortical neurons. Quantification by luminescence plate reader of the relative light unit values per mg cell protein from Cell-Titer Glo assay from 9 wells. Error bars are SD. $*P < 0.05$ versus control. $^{\#}P < .05$ versus 3 J/cm^2.

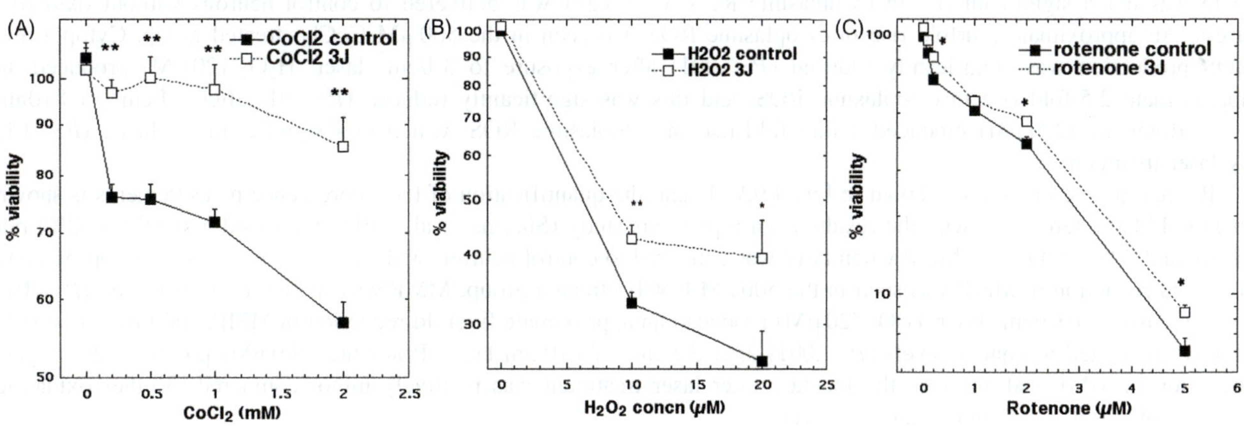

FIGURE 4.6 Cell viability of cortical neurons treated with increasing doses of (A) $CoCl_2$, (B) H_2O_2, (C) rotenone for 1 h with and without 3 J/cm^2 810-nm laser in first 2.5 min. $N = 6$ wells per group, error bars are SEM and $*P < .05$ and $**P < .01$.

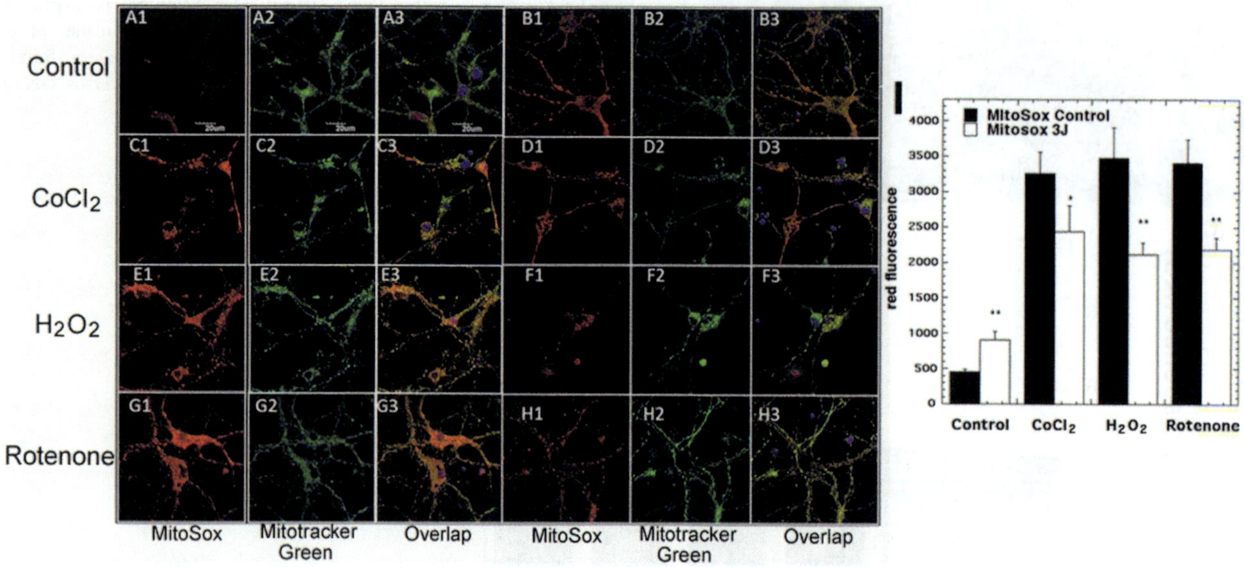

FIGURE 4.7 Representative images of cortical neurons labeled with MitoRox (*red*) for mitochondrial ROS, Mitotracker green for mitochondrial colocalization, and Hoescht (*blue*) for nuclei (seen in triple overlay). (A) control cortical neurons; (B) cortical neurons with 3 J/cm^2 810-nm laser; (C) cortical neurons treated with 500 μM $CoCl_2$; (D) cortical neurons treated with 500 μM $CoCl_2$ and 3 J/cm^2 810-nm laser; (E) cortical neurons treated with 20 μM H_2O_2; (F) cortical neurons treated with 20 μM H_2O_2 and 3 J/cm^2 810-nm laser. (G) cortical neurons treated with 200 nM rotenone; (H) cortical neurons treated with 200 nM rotenone and 3 J/cm^2 810-nm laser. Scale bar = 20 μm. (I) Quantification of fluorescence measurements from the groups shown. $N = 6$ fields per group, error bars are SEM and $*P < .05$ and $**P < .01$.

58% ($P < .05$), respectively (Fig. 4.6B). Rotenone treatment (0.1–5 μM) produced a dose-dependent loss of viability reaching cytotoxicity of 94% at 5 μM. Laser treatment protected neurons against rotenone cytotoxicity, however the protection in cell viability was much reduced (only 3%–5% less killing) compared to the other oxidative stress agents. Nevertheless, the protection was statistically significant ($P < .05$) at 0.2, 2, and 5 μM (Fig. 4.6C).

Representative images of the Mitosox and Mitotreacker staining are shown in Fig. 4.7(A–H) and the quantification of the fluorescence measurements is shown in Fig. 4.7I. Significant ($P < .01$) Mitosox red fluorescence indicating mitochondrial production of ROS was seen in control neurons (no oxidative stress) exposed to 3 J/cm^2 810 nm laser. This is in agreement with the results we have reported previously (Sharma et al., 2011). An approximate sevenfold rise in mitochondrial ROS was seen in the 500 μM $CoCl_2$–treated group. Mitochondrial ROS production was significantly reduced ($P < .05$) after exposure to 3 J/cm^2 laser. H_2O_2 (20 μM) also produced an approximate sevenfold rise in mitochondrial ROS, and this was significantly reduced ($P < .01$) after 3 J/cm^2 of 810 nm laser. Rotenone (200 nM) also produced an approximate sevenfold rise in mitochondrial ROS, which was significantly reduced ($P < .01$) by laser treatment.

Representative images of the CellRox fluorescence are shown in Fig. 4.8(A–H), and the quantification of the fluorescence measurements is shown in Fig. 4.8I. In contrast to the results that were found with mitochondrial ROS, there was not a significant rise in cytoplasmic ROS when laser was delivered to control neurons without oxidative stress. An approximate fourfold rise in cytoplasmic ROS was seen in the 500 μM $CoCl_2$–treated group. Cytoplasmic ROS production was significantly reduced ($P < .001$) after exposure to 3 J/cm^2 laser. H_2O_2 (20 μM) produced an approximate 2.5-fold rise in cytoplasmic ROS, and this was significantly reduced ($P < .01$) after 3 J/cm^2 of 810nm laser. Rotenone (200 nM) produced a threefold rise in cytoplasmic ROS, which was significantly reduced ($P < .01$) by laser treatment.

Representative images are shown in Fig. 4.9A–H and the quantification of the fluorescence measurements is shown in Fig. 4.9I. In agreement with the results from a previous study (Sharma et al., 2011) we found a small (~30%) but significant ($P < .01$) rise in MMP when laser was delivered to control neurons without oxidative stress. An approximate threefold reduction in MMP was seen in the 500 μM $CoCl_2$–treated group. MMP was significantly increased ($P < .01$) after exposure to 3 J/cm^2 laser. H_2O_2 (20 μM) produced an approximate fivefold reduction in MMP, and this was significantly attenuated to control levels ($P < .001$) after 3 J/cm^2 of 810 nm laser. Rotenone (200 nM) produced the largest reduction in MMP and although the increase after laser treatment was relatively minor compared to other oxidative stress agents, it was still significant ($P < .01$).

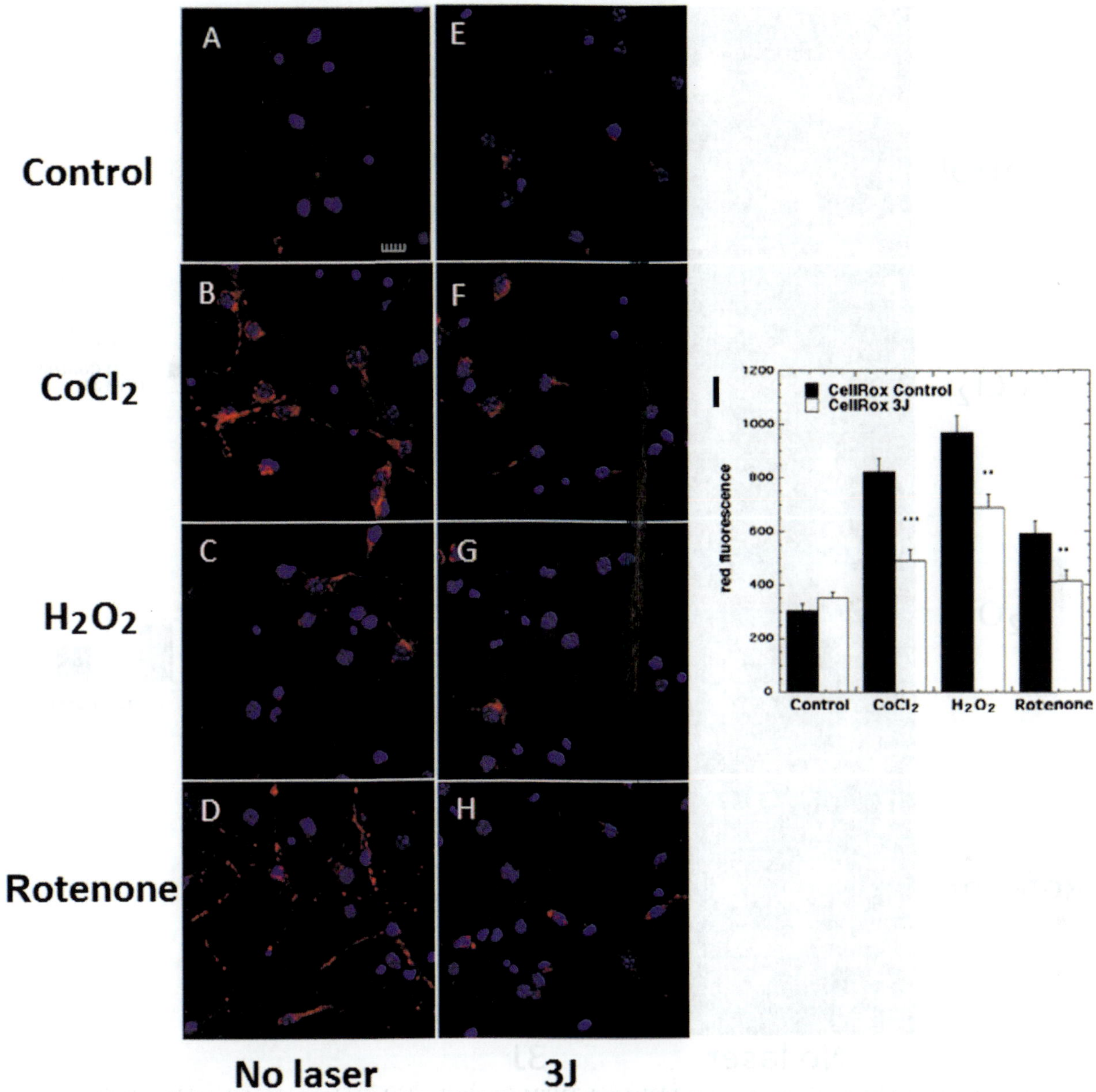

FIGURE 4.8 **Representative images of cortical neurons labeled with Cellrox (*deep red*) for cytoplasmic ROS, and Hoescht (*blue*) for nuclei.** (A) control cortical neurons; (B) cortical neurons with 3 J/cm^2 810-nm laser; (C) cortical neurons treated with 500 μM $CoCl_2$; (D) cortical neurons treated with 500 μM $CoCl_2$ and 3 J/cm^2 810-nm laser; (E) cortical neurons treated with 20 μM H_2O_2; (F) cortical neurons treated with 20 μM H_2O_2 and 3 J/cm^2 810-nm laser.;(G) cortical neurons treated with 200 nM rotenone; (H) cortical neurons treated with 200 nM rotenone and 3 J/cm^2 810-nm laser. Scale bar = 20 μm. (I) Quantification of fluorescence measurements from the groups. $N = 6$ fields per group, error bars are SEM and $*P < .05$; $**P < .01$ and $***P < .001$.

4.4 Excitotoxicity in cultured cortical neurons

Excitotoxicity is a pathological process by which neurons are damaged and killed by the excessive stimulation of receptors for the excitatory amino acid neurotransmitter glutamate (Glu) in the CNS. Excitotoxicity may be involved in CNS pathologies such as stroke, TBI, and spinal cord injury, and is also implicated in neurodegenerative diseases such as multiple sclerosis, Alzheimer's disease, amyotrophic lateral sclerosis, Parkinson's disease, and Huntington's disease.

We tested three different protocols to induce excitotoxicity in the neurons in vitro (Huang et al., 2014). These were (1) glutamate 30 μM, (2) *N*-methyl-D-aspartate (NMDA) 100 μM, and (3) kainic acid (KA) 50 μM; all incubated for 1 hour followed by a wash and exposure to real or sham PBM.

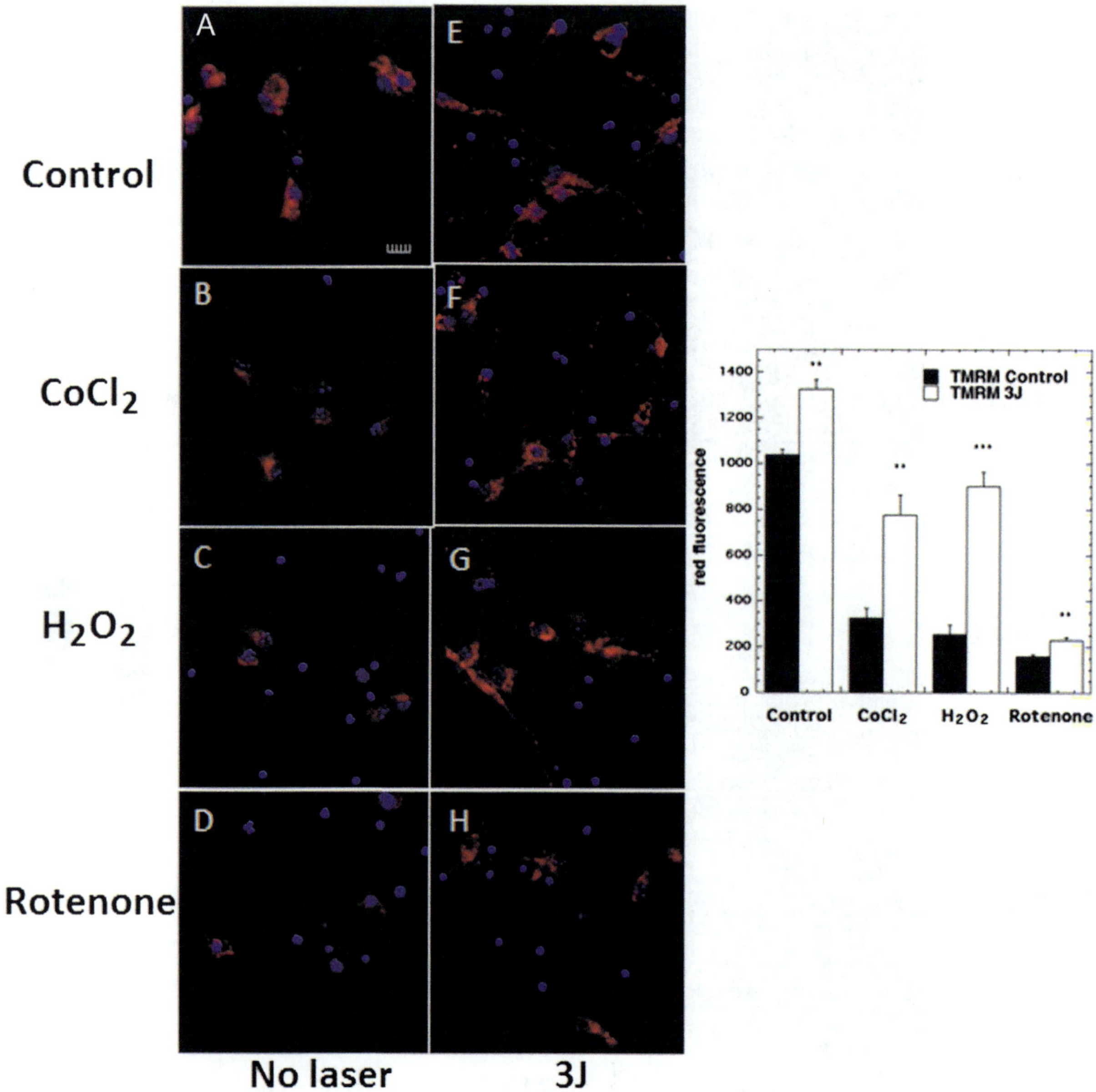

FIGURE 4.9 Representative images of cortical neurons labeled with TMRM for mitochondrial membrane potential, and Hoescht (*blue*) for nuclei. (A) control cortical neurons; (B) cortical neurons with 3 J/cm^2 810-nm laser; (C) cortical neurons treated with 500 μM $CoCl_2$; (D) cortical neurons treated with 500 μM $CoCl_2$ and 3 J/cm^2 810-nm laser; (E) cortical neurons treated with 20 μM H_2O_2; (F) cortical neurons treated with 20 μM H_2O_2 and 3 J/cm^2 810-nm laser; (G) cortical neurons treated with 200 nM rotenone; (H) cortical neurons treated with 200 nM rotenone and 3 J/cm^2 810-nm laser. Scale bar = 20 μm. (I) Quantification of fluorescence measurements from the groups. $N = 6$ fields per group, error bars are SEM and $^*P < .05$; $^{**}P < .01$ and $^{***}P < .001$.

One hour exposure to either of these excitotoxins followed by 24 h incubation produced reductions in cell viability. Glutamate (30 μM) led to 51% survival, NMDA (100 μM) led to 42% survival, and KA (50 μM) led to 50% survival. PBM treatment produced a modest but significant increase in survival to 65% for glutamate ($P < .05$), to 71% for NMDA ($P < .01$), and to 64% for kainate (Fig. 4.10A). Fig. 4.10B shows the ATP content per mg cell protein. ATP content was found to significantly increase (98% more, $P < .05$) when control neurons were treated with 810nm light. All three excitotoxins significantly reduced the ATP content of the neurons to between 10% and 23% of the control value ($P < .001$). PBM increased the ATP content of glutamate-treated neurons from 13% to 73% ($P < .001$), for NMDA-treated neurons from 23% to 80%, and for KA-treated neurons from 10% to 60%. In all three examples of excitotoxicity, the ATP content was at least tripled by PBM.

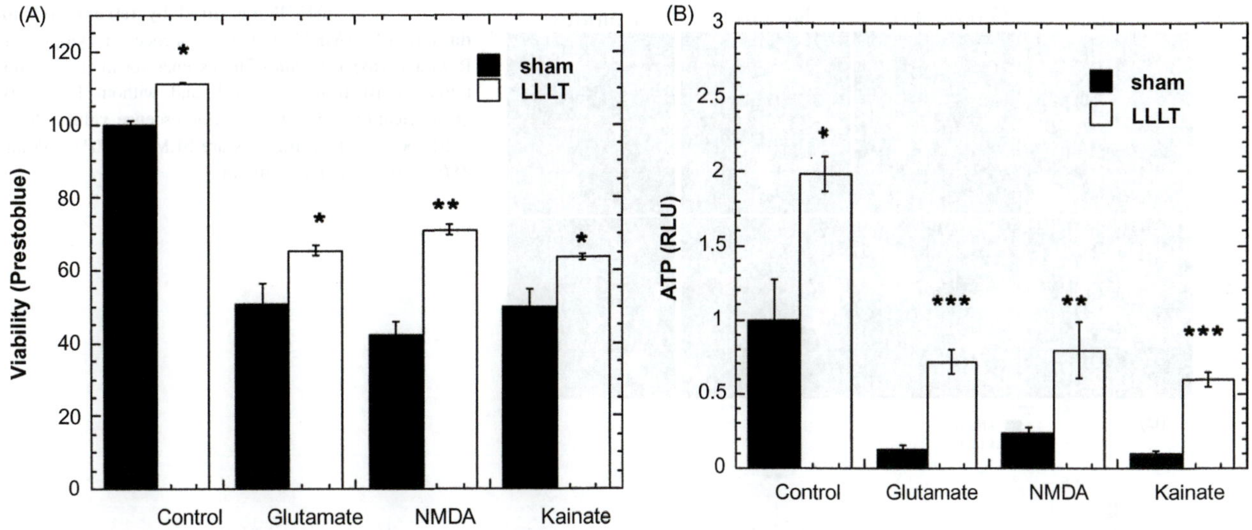

FIGURE 4.10 **Cell viability and ATP.** (A) Cell viability measured by Prestoblue and (B) ATP levels measured by Cell-Titer Glo of cortical neurons treated with excitotoxins for 1 h with and without 3 J/cm^2 810 nm laser in first 3 min. $N = 6$ wells per group, error bars are SEM and $*P < .05$; $**P < .01$ and $***P < .001$.

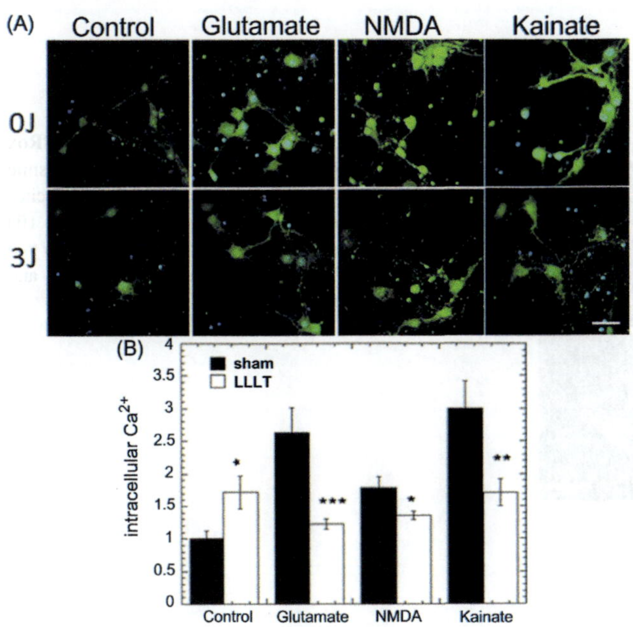

FIGURE 4.11 **Intracellular calcium measured by Fluo-4.** (A) Fluo-4 AM (*green*) fluorescence for calcium and Hoechst (*blue*) fluorescence for nuclei in excitotoxic cortical neurons with and without 3 J/cm^2 810 nm laser. B: Quantification of mean Fluo-4 fluorescence values. $N = 6$ fields per group. Error bars are SEM and $*P < .05$; $**P < .01$ and $***P < .001$. Scale bar = 20 μm.

Representative confocal microscope images are shown in Fig. 4.11A, and the quantification of the fluorescence measurements is shown in Fig. 4.11B. In agreement with our previously published data (Sharma et al., 2011), a 71% rise in intracellular Ca^{2+} was seen after PBM treatment of control neurons. All three excitotoxins significantly increased the intracellular Ca^{2+} content of the neurons by between 80% and 200% of the dark control value ($P < .001$). PBM reduced the Ca^{2+} content of glutamate-treated neurons by more than 50% ($P < 0.001$). PBM produced a smaller decrease in the Ca^{2+} content of NMDA-treated neurons of 25% ($P < .01$). The decrease in Ca^{2+} content in the kainate-treated neurons produced by PBM was almost as large (43%, $P < .01$) as that found with glutamate.

Representative images are shown in Fig. 4.12A, and the quantification of the fluorescence measurements is shown in Fig. 4.12B. Again in agreement with the results from a previous study (Sharma et al., 2011) we found a 29% rise in MMP ($P < .05$) when laser was delivered to control neurons without excitotoxicity. The three excitotoxins all produced drops in MMP ranging from 21% to 78% of control values. MMP in glutamate-treated neurons sharply increased by

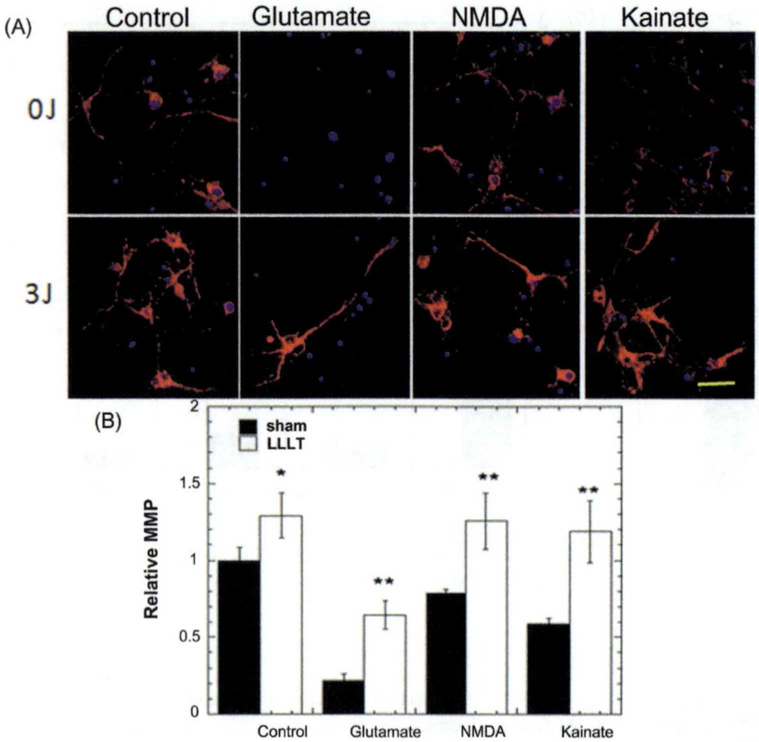

FIGURE 4.12 MMP measured by tetramethylrhodamine. (A) TMRM (*red*) fluorescence for cytoplasmic ROS and Hoechst (*blue*) fluorescence for nuclei in excitotoxic cortical neurons with and without PBM. (B) Quantification mean TMRM fluorescence values. $N = 6$ fields per group. Error bars are SEM and $*P < .05$ and $**P < .01$. Scale bar = 20 μm.

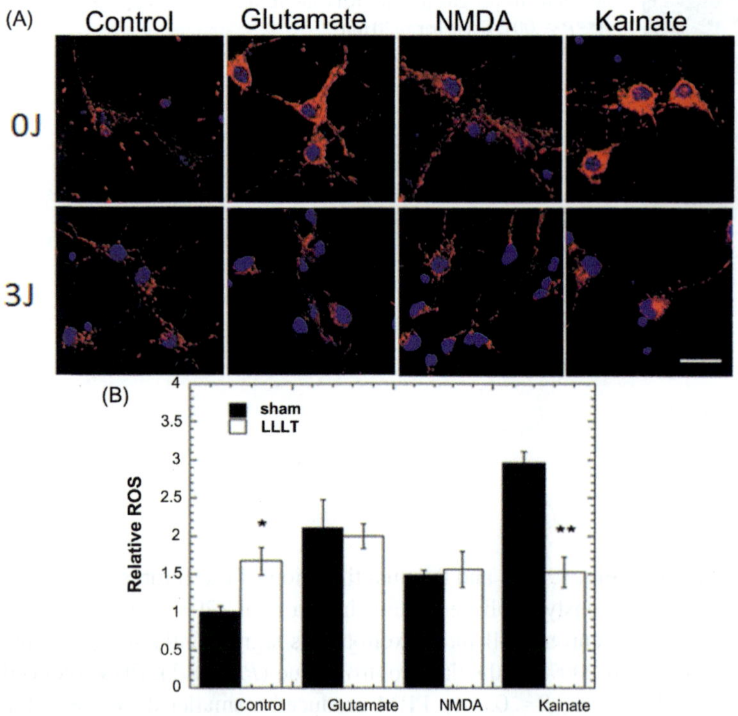

FIGURE 4.13 ROS production measured by CellRox red. (A) CellRox Red (*red*) fluorescence for cytoplasmic ROS and Hoechst (*blue*) fluorescence for nuclei in excitotoxic cortical neurons with and without PBM. (B) Quantification of mean CellRox fluorescence values. $N = 6$ fields per group. Error bars are SEM and $*P < .05$ and $**P < .01$. Scale bar = 20 μm.

200% ($P < .01$) after PBM. The MMP in NMDA-treated neurons increased by 60% ($P < .01$), and MMP in kainate-treated neurons doubled by PBM ($P < .01$).

Representative images are shown in Fig. 4.13A, and the quantification of the fluorescence measurements is shown in Fig. 4.13B. In agreement with our previous study (Sharma et al., 2011), we found a significant rise (66%, $P < .05$) in

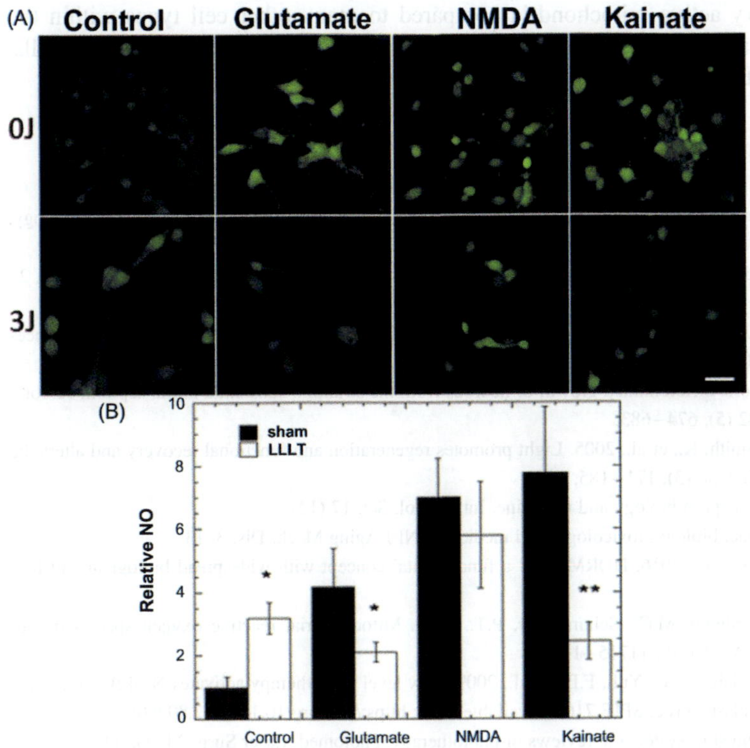

FIGURE 4.14 NO measured by DAF2-AM. (A) DAF2-FM (*green*) fluorescence for NO in excitotoxic cortical neurons with and without PBM. (B) Quantification of mean DAF-FM fluorescence values. $N = 6$ fields per group. Error bars are SEM and $*P < .05$ and $**P < .01$. Scale bar = 20 μm.

cytoplasmic ROS when PBM was delivered to control neurons without excitotoxins. The rise in ROS upon addition of excitotoxins ranged from 50% more (NMDA), 110% more (glutamate), and to 200% more (kainate). In this experiment, there was only a significant reduction in ROS (48% less, $P < .01$) produced by PBM in the case of kainate-treated neurons. The increased ROS produced by glutamate and by NMDA was not significantly affected by PBM.

Representative images are shown in Fig. 4.14A, and the quantification of the fluorescence measurements is shown in Fig. 4.14B. There was a rise in NO content of 220% in control neurons treated by PBM ($P < .05$) in agreement with our previous study (Sharma et al., 2011). All three excitotoxins produced a large increase in NO ranging from 318% for glutamate, 600% for NMDA, and to 677% for kainate ($P < .01$). PBM reduced the NO content of glutamate-treated neurons by 50% ($P < .05$). While there was a 20% reduction in NO for NMDA-treated neurons, it was not statistically significant. The NO in kainate-treated neurons was reduced most by 69% ($P < .01$) by PBM.

Conclusion

It is often said that PBM has much bigger effects on diseased, sick, damaged, or stressed cells, and comparatively few effects on normal, healthy, undamaged cells. Our data begin to shed some light on this observation. In normal cells at doses around the optimum value of 3 J/cm^2 (which incidentally is a fairly broad peak in the dose response curve), PBM will increase the MMP above baseline, accompanied by increases in ATP, intracellular calcium, accompanied by modest increases in ROS and NO. However, if the dose is increased beyond the optimum value (100 times higher in J/cm^2) then the MMP is decreased below baseline, the increases in ATP and calcium are lost, and a second peak of ROS and NO is observed. If the cells are subjected to damaging insults such as oxidative stress or excitotoxicity before PBM, then the picture is somewhat different. Because these toxic treatments lower the MMP (sometimes severely) and also lower the ATP levels and reduce cell viability, the effect of PBM is to raise MMP back toward baseline, and to reverse the decreases in ATP levels and cell viability. The effect of ROS depends on what treatment the cells were subjected to. In the case of oxidative stress, the MMP is lowered, and the mitochondria produce substantial amounts of ROS. Note the ROS measured in the oxidative stress cells are not due per se to the oxidizing agents added, but are produced by the damaged mitochondria. When the mitochondria are restored toward normality, the ROS levels come down. The levels of calcium are highly elevated by excitotoxicity due to the opening of ion channels, and PBM can lower these calcium levels possibly by increasing ATP levels back toward normal. The levels of ROS produced as a consequence of excitotoxicity is probably also due to the lowering of MMP by the high calcium levels.

Neurons are highly metabolic cells, with very active mitochondria compared to many other cell types within the human body. The beneficial activities of PBM may be more pronounced in neurons than they would be with other cells. Perhaps this explains why the brain does appear to respond particularly well to PBM.

References

Agathokleous, E., Kitao, M., Calabrese, E.J., 2018. Environmental hormesis and its fundamental biological basis: rewriting the history of toxicology. Environ. Res. 165, 274–278.

Anders, J.J., Geuna, S., Rochkind, S., 2004. Phototherapy promotes regeneration and functional recovery of injured peripheral nerve. Neurol. Res. 26 (2), 233–239.

Betarbet, R., Sherer, T.B., MacKenzie, G., Garcia-Osuna, M., Panov, A.V., Greenamyre, J.T., 2000. Chronic systemic pesticide exposure reproduces features of Parkinson's disease. Nat. Neurosci. 3 (12), 1301–1306.

Brewer, G.J., 1995. Serum-free B27/neurobasal medium supports differentiated growth of neurons from the striatum, substantia nigra, septum, cerebral cortex, cerebellum, and dentate gyrus. J. Neurosci. Res. 42 (5), 674–683.

Byrnes, K.R., Waynant, R.W., Ilev, I.K., Wu, X., Barna, L., Smith, K., et al., 2005. Light promotes regeneration and functional recovery and alters the immune response after spinal cord injury. Lasers Surg. Med. 36 (3), 171–185.

Calabrese, E.J., 2016. The emergence of the dose-response concept in biology and medicine. Int. J. Mol. Sci. 17 (12).

Calabrese, E.J., Mattson, M.P., 2017. How does hormesis impact biology, toxicology, and medicine? NPJ Aging Mech. Dis. 3, 13.

Calabrese, E.J., Dhawan, G., Kapoor, R., Iavicoli, I., Calabrese, V., 2016. HORMESIS: a fundamental concept with widespread biological and biomedical applications. Gerontology 62 (5), 530–535.

Chandel, N.S., Maltepe, E., Goldwasser, E., Mathieu, C.E., Simon, M.C., Schumacker, P.T., 1998. Mitochondrial reactive oxygen species trigger hypoxia-induced transcription. Proc. Natl. Acad. Sci. U.S.A. 95 (20), 11715–11720.

Chen, A.C.-H., Arany, P.R., Huang, Y.-Y., Tomkinson, E.M., Saleem, T., Yull, F.E., et al., 2009. Low level laser therapy activates NF-kB via generation of reactive oxygen species in mouse embryonic fibroblasts. Proc. SPIE 7165. Available from: https://doi.org/10.1117/12.809605.

Chow, R., 2006. Laser acupuncture studies should not be included in systematic reviews of phototherapy. Photomed. Laser Surg. 24 (1), 69.

De Taboada, L., Yu, J., El-Amouri, S., Gattoni-Celli, S., Richieri, S., McCarthy, T., et al., 2011. Transcranial laser therapy attenuates amyloid-beta peptide neuropathology in amyloid-beta protein precursor transgenic mice. J. Alzheimers Dis. 23 (3), 521–535.

Detaboada, L., Ilic, S., Leichliter-Martha, S., Oron, U., Oron, A., Streeter, J., 2006. Transcranial application of low-energy laser irradiation improves neurological deficits in rats following acute stroke. Lasers Surg. Med. 38 (1), 70–73.

Freestone, P.S., Chung, K.K., Guatteo, E., Mercuri, N.B., Nicholson, L.F., Lipski, J., 2009. Acute action of rotenone on nigral dopaminergic neurons--involvement of reactive oxygen species and disruption of Ca^{2+} homeostasis. Eur. J. Neurosci. 30 (10), 1849–1859.

Fulop, A.M., Dhimmer, S., Deluca, J.R., Johanson, D.D., Lenz, R.V., Patel, K.B., et al., 2010. A meta-analysis of the efficacy of laser phototherapy on pain relief. Clin. J. Pain 26 (8), 729–736.

Hashmi, J.T., Huang, Y.Y., Osmani, B.Z., Sharma, S.K., Naeser, M.A., Hamblin, M.R., 2010. Role of low-level laser therapy in neurorehabilitation. PM R 2 (12 Suppl. 2), S292–S305.

Hawkins, D., Abrahamse, H., 2006a. Effect of multiple exposures of low-level laser therapy on the cellular responses of wounded human skin fibroblasts. Photomed. Laser Surg. 24 (6), 705–714.

Hawkins, D.H., Abrahamse, H., 2006b. The role of laser fluence in cell viability, proliferation, and membrane integrity of wounded human skin fibroblasts following helium-neon laser irradiation. Lasers Surg. Med. 38 (1), 74–83.

Hool, L.C., Corry, B., 2007. Redox control of calcium channels: from mechanisms to therapeutic opportunities. Antioxid. Redox Signal. 9 (4), 409–435.

Hu, W.P., Wang, J.J., Yu, C.L., Lan, C.C., Chen, G.S., Yu, H.S., 2007. Helium-neon laser irradiation stimulates cell proliferation through photostimulatory effects in mitochondria. J. Invest. Dermatol. 127 (8), 2048–2057.

Huang, Y.Y., Chen, A.C., Carroll, J.D., Hamblin, M.R., 2009. Biphasic dose response in low level light therapy. Dose Response 7 (4), 358–383.

Huang, Y.Y., Sharma, S.K., Carroll, J.D., Hamblin, M.R., 2011. Biphasic dose response in low level light therapy – an update. Dose Response 9 (4), 602–618.

Huang, Y.Y., Nagata, K., Tedford, C.E., Hamblin, M.R., 2014. Low-level laser therapy (810 nm) protects primary cortical neurons against excitotoxicity in vitro. J. Biophotonics 7 (8), 656–664.

Karu, T.I., Pyatibrat, L.V., Afanasyeva, N.I., 2005. Cellular effects of low power laser therapy can be mediated by nitric oxide. Lasers Surg. Med. 36 (4), 307–314.

Konstantinovic, L.M., Cutovic, M.R., Milovanovic, A.N., Jovic, S.J., Dragin, A.S., Letic, M., et al., 2010. Low-level laser therapy for acute neck pain with radiculopathy: a double-blind placebo-controlled randomized study. Pain Med. 11 (8), 1169–1178.

Lampl, Y., Zivin, J.A., Fisher, M., Lew, R., Welin, L., Dahlof, B., et al., 2007. Infrared laser therapy for ischemic stroke: a new treatment strategy: results of the NeuroThera Effectiveness and Safety Trial-1 (NEST-1). Stroke 38 (6), 1843–1849.

Lapchak, P.A., 2010. Taking a light approach to treating acute ischemic stroke patients: transcranial near-infrared laser therapy translational science. Ann. Med. 42 (8), 576–586.

Lapchak, P.A., De Taboada, L., 2010. Transcranial near infrared laser treatment (NILT) increases cortical adenosine-5'-triphosphate (ATP) content following embolic strokes in rabbits. Brain Res. 1306, 100–105.

Lim, J., Ali, Z.M., Sanders, R.A., Snyder, A.C., Eells, J.T., Henshel, D.S., et al., 2009. Effects of low-level light therapy on hepatic antioxidant defense in acute and chronic diabetic rats. J. Biochem. Mol. Toxicol. 23 (1), 1–8.

Lim, J., Sanders, R.A., Snyder, A.C., Eells, J.T., Henshel, D.S., Watkins 3rd, J.B., 2010. Effects of low-level light therapy on streptozotocin-induced diabetic kidney. J. Photochem. Photobiol. B 99 (2), 105–110.

Lin, H.J., Wang, X., Shaffer, K.M., Sasaki, C.Y., Ma, W., 2004. Characterization of H_2O_2-induced acute apoptosis in cultured neural stem/progenitor cells. FEBS Lett. 570 (1-3), 102–106.

Lubart, R., Lavi, R., Friedmann, H., Rochkind, S., 2006. Photochemistry and photobiology of light absorption by living cells. Photomed. Laser Surg. 24 (2), 179–185.

Lucas, C., Criens-Poublon, L.J., Cockrell, C.T., de Haan, R.J., 2002. Wound healing in cell studies and animal model experiments by Low Level Laser Therapy; were clinical studies justified? a systematic review. Lasers Med. Sci. 17 (2), 110–134.

Michalikova, S., Ennaceur, A., van Rensburg, R., Chazot, P.L., 2008. Emotional responses and memory performance of middle-aged CD1 mice in a 3D maze: effects of low infrared light. Neurobiol. Learn. Mem. 89 (4), 480–488.

Naeser, M.A., Saltmarche, A., Krengel, M.H., Hamblin, M.R., Knight, J.A., 2011. Improved cognitive function after transcranial, light-emitting diode treatments in chronic, traumatic brain injury: two case reports. Photomed. Laser Surg. 29, 351–358.

Oron, A., Oron, U., Chen, J., Eilam, A., Zhang, C., Sadeh, M., et al., 2006. Low-level laser therapy applied transcranially to rats after induction of stroke significantly reduces long-term neurological deficits. Stroke 37 (10), 2620–2624.

Oron, A., Oron, U., Streeter, J., de Taboada, L., Alexandrovich, A., Trembovler, V., et al., 2007. Low-level laser therapy applied transcranially to mice following traumatic brain injury significantly reduces long-term neurological deficits. J. Neurotrauma 24 (4), 651–656.

Passarella, S., Casamassima, E., Molinari, S., Pastore, D., Quagliariello, E., Catalano, I.M., et al., 1984. Increase of proton electrochemical potential and ATP synthesis in rat liver mitochondria irradiated in vitro by helium-neon laser. FEBS Lett. 175 (1), 95–99.

Pastore, D., Di Martino, C., Bosco, G., Passarella, S., 1996. Stimulation of ATP synthesis via oxidative phosphorylation in wheat mitochondria irradiated with helium-neon laser. Biochem. Mol. Biol. Int. 39 (1), 149–157.

Peplow, P.V., Chung, T.Y., Baxter, G.D., 2010. Laser photobiomodulation of wound healing: a review of experimental studies in mouse and rat animal models. Photomed. Laser Surg. 28 (3), 291–325.

Peres, M.F., 2010. Low-level laser therapy for neck pain. Cephalalgia 30 (11), 1408.

Reers, M., Smith, T.W., Chen, L.B., 1991. J-aggregate formation of a carbocyanine as a quantitative fluorescent indicator of membrane potential. Biochemistry 30 (18), 4480–4486.

Rubio, C.R., Simes, J.C., Moya, M., Soriano, F., Palma, J.A., Campana, V., 2009. Inflammatory and oxidative stress markers in experimental crystalopathy: their modification by photostimulation. Photomed. Laser Surg. 27 (1), 79–84.

Sharma, S.K., Kharkwal, G.B., Sajo, M., Huang, Y.Y., De Taboada, L., McCarthy, T., et al., 2011. Dose response effects of 810 nm laser light on mouse primary cortical neurons. Lasers Surg. Med. 43 (8), 851–859.

Silveira, P.C., Silva, L.A., Freitas, T.P., Latini, A., Pinho, R.A., 2011. Effects of low-power laser irradiation (LPLI) at different wavelengths and doses on oxidative stress and fibrogenesis parameters in an animal model of wound healing. Lasers Med. Sci. 26 (1), 125–131.

Sommer, A.P., Pinheiro, A.L., Mester, A.R., Franke, R.P., Whelan, H.T., 2001. Biostimulatory windows in low-intensity laser activation: lasers, scanners, and NASA's light-emitting diode array system. J. Clin. Laser Med. Surg. 19 (1), 29–33.

Tomaro, M.L., Frydman, J., Frydman, R.B., 1991. Heme oxygenase induction by $CoCl_2$, Co-protoporphyrin IX, phenylhydrazine, and diamide: evidence for oxidative stress involvement. Arch. Biochem. Biophys. 286 (2), 610–617.

Wu, X., Dmitriev, A.E., Cardoso, M.J., Viers-Costello, A.G., Borke, R.C., Streeter, J., et al., 2009. 810 nm wavelength light: an effective therapy for transected or contused rat spinal cord. Lasers Surg. Med. 41 (1), 36–41.

Xavier, M., David, D.R., de Souza, R.A., Arrieiro, A.N., Miranda, H., Santana, E.T., et al., 2010. Anti-inflammatory effects of low-level light emitting diode therapy on *Achilles tendinitis* in rats. Lasers Surg. Med. 42 (6), 553–558.

Yasukawa, A., Hrui, H., Koyama, Y., Nagai, M., Takakuda, K., 2007. The effect of low reactive-level laser therapy (LLLT) with helium-neon laser on operative wound healing in a rat model. J. Vet. Med. Sci. 69 (8), 799–806.

Chapter 5

Safety and penetration of light into the brain

Erica B. Wang[1,2], Ramanjot Kaur[1], Manuel Fierro[1,2], Evan Austin[1,2], Linda Ramball Jones[3] and Jared Jagdeo[1,2,4]

[1]*Department of Dermatology, University of California at Davis, Sacramento, CA, United States,* [2]*Dermatology Service, Sacramento VA Medical Center, Mather, CA, United States,* [3]*Department of Physics and Astronomy, College of Charleston, Charleston, SC, United States,* [4]*Department of Dermatology, State University of New York, Downstate Medical Center, Brooklyn, NY, United States*

5.1 Introduction

Photobiomodulation was initially studied for stimulation of wound healing and pain relief but has transitioned to broader applications, including neurodegenerative disorders, neurotrauma, and neuropsychiatric conditions (Hashmi et al., 2010; Hamblin, 2016). Neurological conditions are rising global health concerns because of their high morbidity and increasing prevalence as the worldwide population ages (Hamblin, 2016; Henderson and Morries, 2017). Current treatments with pharmacological interventions and psychotherapy have limited efficacy or are associated with adverse effects (Hamblin, 2016; Henderson and Morries, 2017). Therefore, noninvasive therapies, such as transcranial light, have been investigated as potential treatment modalities. Emerging evidence has revealed the promising benefits of noninvasive transcranial light therapy in neuroprotection and repair (Hamblin, 2016). While not the focus of this chapter, transcranial photobiomodulation has potential clinical uses for a variety of acute and chronic neurological conditions, such as stroke, dementia, and traumatic brain injury.

5.2 Safety

Given that light absorption can lead to tissue heating (Joensen et al., 2011), the clinical safety of transcranial and intraparenchymal light on brain tissue may be a concern. Photobiomodulation of light therapy is wavelength-specific and not thermal or heat based (Mochizuki-Oda et al., 2002). Red light and near-infrared (NIR) light are nonionizing and, hence, present no ultraviolet (UV) light-associated risks (Lampl et al., 2007). Previous studies with transcranial near-infrared light therapy (NILT) found no treatment-related adverse effects on brain tissue structure and function (Ilic et al., 2006; Zivin et al., 2009; McCarthy et al., 2010; Naeser et al., 2011; Johnstone et al., 2015). Transcranial photobiomodulation is safe, based upon published literature, in animal and humans who are healthy or have a neurological condition (see Table 5.1). Additional clinical studies not mentioned in Table 5.1 have been performed for efficacy in neurological diseases, but did not state if there were any adverse events (AEs).

5.2.1 Animal studies

The thermal safety of NILT was studied in rabbits using an 808-nm laser placed on the scalp (Lapchak et al., 2004). After 10 minutes of irradiation with 25 mW/cm^2, the skin surface temperature beneath the laser probe increased up to 3°C, but the brain temperature increased by 0.8°C–1.8°C and normalized within 60 minutes after laser treatment (Lapchak et al., 2004). The heat threshold for neuronal damage is about 43°C (Yarmolenko et al., 2011). No short or long-term adverse effects were seen with this treatment regimen in a rabbit model (Lapchak et al., 2004). Similar temperature changes of 1°C–3°C were observed on the sheep and human skin surface with continuous wave NIR lasers (Henderson and Morries, 2015; Morries et al., 2015).

Photobiomodulation in the Brain. DOI: https://doi.org/10.1016/B978-0-12-815305-5.00005-1

TABLE 5.1 Adverse events noted in preclinical and clinical studies.

Author (year)	Model	Light parameters	Condition	Side effects
Oron et al. (2006)[a]	Rats	808 nm Ga-As diode laser, 7.5 mW/cm^2 at brain, 2 min, 0.9 J/cm^2, pulsed and CW	Stroke	Minimal heating of 1°C with 3.6 J/cm^2. No pathologic changes up to 3 months with 75 mW/cm^2 for 2 min
Meyer et al. (2016)[b]	Rabbit	808.5 nm NIR laser, 111 mW treatment of 2 min, 7.5 mW/cm^2, duty cycle 20%, 10 Hz	Stroke, embolic	No histological evidence of gross tissue necrosis or neuronal injury
Boonswang et al. (2012)[c]	Human ($n = 1$)	660 nm visible red and 850 nm NIR LED, 1400 mW, 2016 J, 2.95 J/cm^2 to 32 areas	Stroke, brain stem	Not stated
Xuan et al. (2013)[d]	Mice	810 nm laser, 25 mW/cm^2, 18 J/cm^2, spot diameter 1 cm. Single treatment, 3 daily treatments or 14 daily treatments	TBI	No difference in adverse events compared to control group
Khuman et al. (2012)[e]	Mice	800 nm laser, 60 J/cm^2, 500 mW/cm^2, beam size 1.32 cm^2, 2 min, transcranially over left and right parieto-temporal region	TBI	Increased brain temperature by 1.8°C with 60 J/cm^2. Temperature returned to baseline within 3–5 min
Henderson and Morries (2015)[f]	Human ($n = 1$)	810 and 980 nm NIR laser, 10–15 W, for 20 treatments	TBI, moderate	Not stated
Naeser et al. (2011)[g]	Human ($n = 2$)	LED cluster head (9 red 633 nm, 52 NIR 870 nm diodes); 500 mW; 22.2 mW/cm^2; 13.3 J/cm^2; CW	TBI	No negative side effects reported
Naeser et al. (2014)[h]	Human ($n = 11$)	LED cluster head (9 red 633 nm, 52 NIR 870 nm diodes); 500 mW; 22.2 mW/cm^2; 13 J/cm^2; CW; 10 min to 11 sites for 6 weeks	Chronic, mild TBI	No adverse events
Wu et al. (2012)[i]	Rat	810 nm laser, 3 mm diameter, power 350 mW, 100 Hz, 20% duty cycle, 2 min, 120 J/cm^2, 3 times a week for 3 weeks	Depression	No weight loss compared to control or antidepressant
Schiffer et al. (2009)[j]	Humans ($n = 10$)	810 nm NIR LED array, 250 mW/cm^2, 60 J/cm^2, 4 min at 2 sites	Depression and anxiety	No adverse events
Cassano et al. (2015)[k]	Humans ($n = 4$)	808 nm NIR laser, 700 mW/cm^2, 84 J/cm^2, 2.40 kJ, power 5 W, 2 min at 4 sites for 6 sessions	Depression	No adverse events related to NIR therapy
Naeser et al. (2012)[l]	Human ($n = 3$)	LED cluster head (9 red 633 nm, 52 NIR 870 nm diodes); 500 mW; 22.2 mW/cm^2; 146 Hz × 6 weeks	Aphasia	Not stated

CW, continuous wave; *Ga-As*, gallium arsenide; *LED*, light-emitting diode; *n*, number; *NIR*, near-infrared.

[a]*Oron, A., Oron, U., Chen, J., Eilam, A., Zhang, C., Sadeh, M., et al., 2006. Low-level laser therapy applied transcranially to rats after induction of stroke significantly reduces long-term neurological deficits. Stroke 37 (10), 2620–2624.*

[b]*Meyer, D.M., Chen, Y., Zivin, J.A., 2016. Dose-finding study of phototherapy on stroke outcome in a rabbit model of ischemic stroke. Neurosci. Lett. 630, 254–258.*

[c]*Boonswang, N.A., Chicchi, M., Lukachek, A., Curtiss, D., 2012. A new treatment protocol using photobiomodulation and muscle/bone/joint recovery techniques having a dramatic effect on a stroke patient's recovery: a new weapon for clinicians. BMJ Case Rep. 2012.*

[d]*Xuan, W., Vatansever, F., Huang, L., Wu, Q., Xuan, Y., Dai, T., et al., 2013. Transcranial low-level laser therapy improves neurological performance in traumatic brain injury in mice: effect of treatment repetition regimen. PLoS One 8 (1), e53454.*

[e]*Khuman, J., Zhang, J., Park, J., Carroll, J.D., Donahue, C., Whalen, M.J., 2012. Low-level laser light therapy improves cognitive deficits and inhibits microglial activation after controlled cortical impact in mice. J. Neurotrauma 29 (2), 408–417.*

[f]*Henderson, T.A., Morries, L.D., 2015. SPECT perfusion imaging demonstrates improvement of traumatic brain injury with transcranial near-infrared laser phototherapy. Adv. Mind Body Med. 29 (4), 27–33.*

[g]*Naeser, M.A., Saltmarche, A., Krengel, M.H., Hamblin, M.R., Knight, J.A., 2011. Improved cognitive function after transcranial, light-emitting diode treatments in chronic, traumatic brain injury: two case reports. Photomed. Laser Surg. 29 (5), 351–358.*

[h]*Naeser, M.A., Zafonte, R., Krengel, M.H., Martin, P.I., Frazier, J., Hamblin, M.R., et al., 2014. Significant improvements in cognitive performance post-transcranial, red/near-infrared light-emitting diode treatments in chronic, mild traumatic brain injury: open-protocol study. J. Neurotrauma 31 (11), 1008–1017.*

[i]*Wu, X., Alberico, S.L., Moges, H., De Taboada, L., Tedford, C.E., Anders, J.J., 2012. Pulsed light irradiation improves behavioral outcome in a rat model of chronic mild stress. Lasers Surg. Med. 44 (3), 227–232.*

[j]*Schiffer, F., Johnston, A.L., Ravichandran, C., Polcari, A., Teicher, M.H., Webb, R.H., et al., 2009. Psychological benefits 2 and 4 weeks after a single treatment with near infrared light to the forehead: a pilot study of 10 patients with major depression and anxiety. Behav. Brain Funct. 5, 46.*

[k]*Cassano, P., Cusin, C., Mischoulon, D., Hamblin, M.R., De Taboada, L., Pisoni, A., et al., 2015. Near-infrared transcranial radiation for major depressive disorder: proof of concept study. Psychiatry J. 2015, 352979.*

[l]*Naeser, M., Ho, M., Martin, P.E., Treglia, E.M., Krengel, M., Hamblin, M.R., et al. Improved language after scalp application of red/near-infrared light-emitting diodes: pilot study supporting a new, noninvasive treatment for chronic aphasia. Procedia Soc. Behav. Sci. 61, 138–139.*

Long- and short-term effects of 808 nm NILT on intact rat brain using varying power densities (maximum 750 mW/ cm^2), continuous wave, and pulse frequencies were also studied. There were no long-term differences between neurological tests and histopathological examination of laser-treated and control groups up to 30 and 70 days posttreatment (Ilic et al., 2006). However, adverse neurological damage was found in rats at the max settings in continuous wave NILT (Ilic et al., 2006). Additionally, no abnormalities in blood tests or brain and pituitary histopathology were noted between NIR continuous wave laser-treated (power 70 mW, irradiance 2230 mW/cm^2, and power density of 268 J/cm^2) and control healthy rats after daily treatment for 12 months (McCarthy et al., 2010). Similarly, NIR light applied intracranially in rats and mice caused no observable behavioral deficits or tissue necrosis, even at supratherapeutic doses for neuroprotection (Moro et al., 2014). Use of a NIR laser (808 nm, 111 mW/cm^2) for 2 minutes per session in rabbits after embolic stroke showed no detectable tissue damage (Meyer et al., 2016). As transcranial light therapy has a lack of adverse effects in animals, further safety studies in humans were warranted.

5.2.2 Clinical studies

Transcranial photobiomodulation has been demonstrated to be safe for use in neurological conditions in humans. It was initially investigated in small-scale clinical studies, which have shown no adverse effects (Naeser et al., 2014; Saltmarche et al., 2017; Vargas et al., 2017). No adverse effects were found in a pilot study using transcranial infrared laser stimulation (1064 nm, 250 mW/cm^2) for neurocognitive function augmentation in older adults (Vargas et al., 2017). Another pilot study of red/NIR light-emitting diode (LED) cluster (633 and 870 nm, 22.2 mW/cm^2) for chronic, mild traumatic brain injury showed no negative side effects (Naeser et al., 2014). In a case series of patients with mild to moderate dementia, no adverse reactions were demonstrated with NIR LED transcranial–intranasal photobiomodulation (810 nm, 41 mW/cm^2) weekly for 12 weeks (Saltmarche et al., 2017). However, larger randomized clinical trials were necessary to investigate the safety of this noninvasive intervention.

5.2.3 NeuroThera Effectiveness and Safety Trial clinical trials

Robust evidence of the clinical safety of NIR light comes from the three NeuroThera Effectiveness and Safety Trial (NEST) clinical trials. The NEST-1 double-blind, randomized controlled, Phase II trial initially established the safety of transcranial NILT for stroke (Lampl et al., 2007). The treatment group received approximately 1 J/cm^2 of 808 nm laser for 2 minutes at 20 predetermined scalp sites. There was no significant difference between the NILT group ($n = 79$) and sham (placebo) group ($n = 41$) in mortality rates, serious adverse events (SAEs) rates, worsening of underlying disease rates, cardiovascular SAEs rates, infection rates, or central nervous system SAEs.

In addition, the Phase III, double-blind NEST-2 study also verified that there was no significant difference in the safety outcomes between the NILT ($n = 331$) and the sham ($n = 327$) groups (Zivin et al., 2009). In comparing the mortality rates, SAE rates, and AE rates, there was no difference between the NILT and the sham groups. The NILT and sham groups had 58 (17.5%) and 57 (17.4%) deaths, 125 (37.8%) and 137 (41.8%) SAEs, and 92.7% and 93.6% of subjects with at least one AE, respectively (Zivin et al., 2009). Pooling of both NEST-1 and NEST-2 studies demonstrated no significant differences in mortality rates or SAEs between NILT and sham groups (Huisa et al., 2013). However, the mortality rates in the NEST-2 trial doubled the mortality rates of the NEST-1 trial, likely due to greater stroke severity and greater comorbidity in the NEST-2 subjects.

The larger NEST-3, Phase III clinical trial further proved that there were no clinically important safety concerns with the use of NILT compared to placebo (Hacke et al., 2014). With a total of 630 patients (316 in the NILT group and 314 in the sham group), 261 (82.6%) and 249 (79.3%) of patients reported an adverse effect, respectively. Additionally, 66 patients with NILT (20.9%) and 88 patients with sham therapy (28.0%) reported a serious adverse effect. While the NEST-3 trial failed to show efficacy in stroke outcomes possibly due to suboptimal NILT parameters, it solidified the safety profile of NILT and showed no clinical safety adverse outcomes attributable to NILT.

In summary, transcranial photobiomodulation has established safety and a notable lack of adverse effects in short- or long-term clinical studies. Accordingly, the Food and Drug Administration has approved the use of commercial LED arrays for NIR therapy.

5.3 Light penetration into the brain

Photobiomodulation for neurological diseases primarily requires light to reach target neurons in the brain. Alternatively, light may change brain function by modulating photoreceptors, such as opsins in the skin or eye, which yield secondary

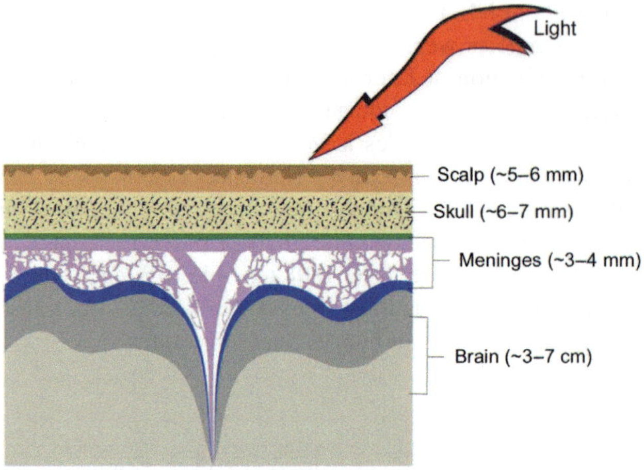

FIGURE 5.1 Layers of the skull with corresponding mean thickness measurements.

effects or trigger a cascade. Understanding light penetration into the brain is crucial to using transcranial photobiomodulation. In order to reach the depths of the brain, photons illuminated from a transcranial light source must penetrate the scalp, skull, meningeal layers, cerebrospinal fluid (CSF), blood, and brain parenchyma (see Fig. 5.1). In addition to overcoming these important barriers, light therapy must also maintain a sufficient fluence or dose at the depths of the target brain tissues to be effective. Thus, a critical concern is the ability of light to penetrate through the layers of the human cranium, including the scalp, skull, meningeal layers, and brain parenchyma.

Various studies have determined the feasibility of delivering sufficient light power to the targeted brain areas to induce the desired biological effect (see Table 5.2). Limitations in knowledge of photobiomodulation include the optimal wavelength, light source, doses, and pulsed or continuous waves (2010). The principal aim of this chapter is to discuss transcranial light penetration depth based on variables of tissue optical properties, wavelength, skull thickness, anatomical region or location, fluence, irradiance (i.e., power over unit area of the light source), coherence of the light source, pulsing, and tissue storage and processing. Within each section, the extent of light penetration will be reviewed in animal models, human studies, and computer simulation models (if available).

5.4 Mechanism of action

Photobiomodulation acts by photon absorption inducing a photochemical reaction to modulate biological processes in the cell. Although the precise cellular and molecular mechanisms underlying photobiomodulation are not yet fully understood, light in the 600–1200 nm wavelength range has significant photobiomodulation capability (Henderson and Morries, 2015). Different wavelengths have distinct chromophores, or tissue components that absorb light. The relevant segments of the electromagnetic spectrum for photobiomodulation include visible light and near-infrared regions. UV radiation ranges from 100 to 400 nm (Sankaran and Ehsani, 2014). The visible light spectrum covers violet/blue (400–470 nm), green (470–550 nm), yellow (550–590 nm), orange (590–630 nm), red (630–700 nm), and far-red light (700–750 nm) (Sankaran and Ehsani, 2014). NIR light extends from the visible region (750 nm) up to 1.2 μm (Sankaran and Ehsani, 2014).

Currently, the speculated primary mechanism of action for transcranial photobiomodulation centers around increased mitochondrial enzyme cytochrome c oxidase (CCO) activity (de Freitas and Hamblin, 2016). CCO has two copper centers and two heme-iron centers, which are oxidized when oxygen is reduced to water during respiration (Hashmi et al., 2010). Depending on its oxidation–reduction status, these metal centers have different light absorption peaks in the red (620 and 680 nm) and NIR (760 and 820 nm) spectral regions [2]. CCO is inhibited when nitric oxide (NO) produced by mitochondrial NO synthase displaces oxygen in damaged or hypoxic cells, resulting in down regulation of cellular respiration and subsequent decrease in ATP production (Hashmi et al., 2010). This inhibitory NO may be dissociated by photons of light absorbed by CCO (Hashmi et al., 2010). As NO is dissociated, the mitochondrial membrane potential is increased, more oxygen is consumed, more glucose is metabolized, and more ATP is produced (Hamblin, 2016).

Secondary mechanisms of action include greater blood and lymphatic flow from vasodilatory NO, activation of mitochondrial signaling pathways by reactive oxygen species and NO, opening of heat or light-sensitive ion channels,

TABLE 5.2 Summary of penetration depths in various models.

Study	Model	Wavelength (nm)	Power density (mW/cm^2)	Other settings	Penetration depth (mm)	Percentage of light remaining (%)
Lapchak et al. (2004)	Rabbit skull and brain	808	25	N/A	25–30	NS
Henderson and Morries (2015)	Ex vivo sheep head	810	NS	15 W	3000	2.90
Lapchak et al. (2015)	Mouse skull	800	700	N/A	0.44	40.10
	Rat skull				0.83	21.24
	Rabbit skull				2.11	11.36
	Human skull				7.19	4.18
Wan et al. (1981)	Human scalp and skull	NS	NS	8 J/cm^2	1000	2–3
					2000	0.2–0.3
Stolik et al. (2000)	Ex vivo human brain tissue	632.8	NS	N/A	0.92 ± 0.08	NS
		657	NS	N/A	1.38 ± 0.13	NS
		780	NS	N/A	3.46 ± 0.23	NS
		835	NS	N/A	2.52 ± 0.19	NS
Tedford et al. (2015)	Human scalp, skull, meninges, brain	808	1	1 W	40–50	NS
Jagdeo et al. (2012)	Human cadaver heads	633	67.5	N/A	10	0.7
		830	33	N/A	10	11.7
Haeussinger et al. (2011)	MCS	Near infrared	NS	N/A	23.6 ± 0.7	5

MCS, Monte Carlo simulation; *N/A*, not applicable; *NS*, not specified.

and activation of transcription factors. Additional photoacceptors other than CCO, such as water in light- or heat-sensitive ion channels, may play a role because wavelengths longer than red or NIR light have been reported to produce photobiomodulation effects (Hamblin, 2016).

5.5 Penetration depth

The minimum penetration depth that light must traverse before reaching the brain tissue can be estimated by the sum of the average thickness of the components of the human cranium. The average thickness of the human scalp and skull is 5–6 mm (Lapchak et al., 2015) and 6–7 mm (Parsons, 1929), respectively. The meninges with its associated CSF spaces (e.g., subdural space, subarachnoid space) add an additional 3–4 mm (Chung et al., 2012). These barriers translate to approximately 14–17 mm in depth. Based upon the specific neurological condition that requires intervention, light may need to penetrate an additional 3–7 cm to reach involved deeper brain structures (lobes/nuclei) (Henderson and Morries, 2015).

Table 5.2 details the light penetration depths reported in various studies (Wan et al., 1981; Stolik et al., 2000; Lapchak et al., 2004; Haeussinger et al., 2011; Jagdeo et al., 2012; Henderson and Morries, 2015; Lapchak et al., 2015; Tedford et al., 2015). Mild variability of light penetration depths exists in these studies, which may be due to differences in light parameters and model organisms.

5.6 Optical properties of tissue

5.6.1 Light–tissue interactions

Light propagation through biological tissues depends on light's natural behavior (Sankaran and Ehsani, 2014). Upon application of light to tissue, part of it is scattered, part is reflected, and part is absorbed (Sankaran and Ehsani, 2014). Scattering occurs when photons change their direction, and this phenomenon differs depending on the tissue's refraction index (Sankaran and Ehsani, 2014). An inherent property of biological tissue is scattering anisotropy factor, which indicates the scattering light in a forward or backward direction (Jacques, 2013). Anisotropy factor is dependent on the underlying tissue structure (Jacques, 2013). An increase in anisotropy factor approaching a value of 1 describes predominately forward-scattering, resulting in greater penetration depth (Jacques, 2013). The change between two adjacent tissues' refraction indices produces bouncing of light or light reflection (Sankaran and Ehsani, 2014). Following scattering and reflection, the remaining light is absorbed (Sankaran and Ehsani, 2014). Absorbed light energy can drive metabolic reactions, be re-released as light, or be transformed into heat (Sankaran and Ehsani, 2014).

Absorption, scattering, and reflection phenomena decrease or attenuate light intensity, which is associated with the attenuation coefficient (μ_{eff}) (Yaroslavsky et al., 2002). Smaller attenuation coefficients are associated with greater penetration depth and larger attenuation coefficients are associated with smaller penetration depth (Yaroslavsky et al., 2002). Furthermore, the Beer-Lambert law states that light intensity exponentially decays as a function of distance (Kocsis et al., 2006). As a result, increased penetration and delivery of light to the target tissue can theoretically be achieved by reducing absorption, scattering, and attenuation.

Light–tissue interactions depend not only on the incident light properties, but also on the optical properties of the irradiated tissue. Tissue optical properties are contingent upon each tissue morphology, composition, thickness, and histological substructures (e.g., cytoplasmic compounds, axonal structures) (Salomatina et al., 2006). Within tissues, photoacceptors or photoreceptor molecules absorb specific light wavelengths initiating a biological effect or regulation of a metabolic pathway (Sutherland, 2002). Conversion of the absorbed photon's energy excites the photoacceptor into a higher electronic state, leading to a physical or chemical molecular change that results in a biological response (Sutherland, 2002). Specialized photoacceptors include rhodopsin, flavoproteins, and porphyrins (see Table 5.3). Chromophores also absorb photons of light at certain wavelengths that result in distinct absorption spectra, but they are not part of specialized light-reception organs. The most common tissue chromophores are melanin in the skin and water and hemoglobin in the blood.

5.6.2 Melanin

The scalp, which includes skin, can affect transcranial light transmission. Within the skin, there are multiple layers that result in interfaces that can scatter light. Each layer of the skin, such as the epidermis and dermis, has different optical properties. The epidermis contains melanin, whereas the variably thick dermis is filled with hemoglobin-rich blood

TABLE 5.3 Endogenous tissue photoacceptors and chromophores and its absorption spectra.

Biological tissue molecule	Spectral ranges for maxima absorption
Photoacceptor	
Rhodopsin	Green (~510 nm)
Flavoproteins (e.g., NADH-dehydrogenase)	Violet to blue
Porphyrins	Blue
Cytochrome c oxidase	Red to near-infrared
Chromophore	
Melanin	Ultraviolet
Hemoglobin	Blue
Water	Mid-infrared

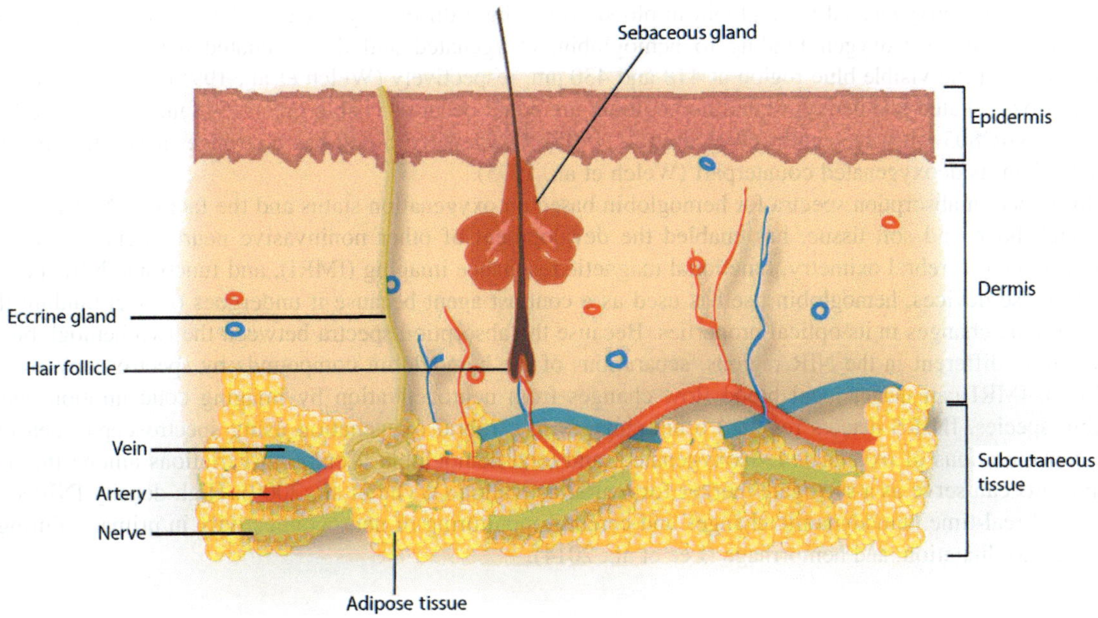

FIGURE 5.2 Skin layers.

vessels and has high scattering by collagen fibers (Anderson and Parrish, 1981) (see Fig. 5.2). In addition to these intrinsic absorption by the skin layers, the scalp in particular has other issues that can cause heavy, light attenuation. Hair can also provide light absorption and cause difficulty in attaching the light source to the scalp. Hair follicles also strongly absorb NIR light.

Melanin is a small pigment granules found in skin, hair, and iris. Melanin has high light absorbance in the UV range with a gradual decrease in absorption with increasing wavelength in the visible region ranging from 400 to 600 nm (Sandell and Zhu, 2011). Melanin appears black because it quantitatively absorbs blue and UV light more strongly than red light. The majority of NIR light energy was found to be absorbed in the initial 1 mm of skin (Esnouf et al., 2007).

5.6.3 Water

Water is the most abundant chromophore in soft tissues, comprising 65%, 73%, and 99% of volume in skin, adult brain tissue, and CSF, respectively (Forbes et al., 1953). Water's absorption is nonsignificant with wavelengths below 600 nm; it increases gradually with longer wavelengths between 600 and 900 nm, and increases rapidly with wavelengths beyond 900 nm (Pope and Fry, 1997). Water is an important strong infrared light absorber and has multiple absorption peaks in the mid-infrared wavelength region (1.2–7 μm) (Welch et al., 1995). For wavelengths above 1.4 μm, absorption by water dominates over light scattering in tissues and the absorption coefficient can be estimated by the water's percent concentration (Welch et al., 1995).

Since water is an important common chromophore, alterations in tissue water content or temperature can affect tissue optical properties. Variations of water concentration as seen in dehydration or edema can affect tissue optical properties (Barton, 2010). Ex vivo tissue samples with evaporative water loss showed decreased thickness, increased light transmission, increase in absorption, and constant scattering (Barton, 2010). Temperature changes can shift absorption peaks to a lower wavelength as water temperature increases (Barton, 2010). Therefore, the absorption of water has a strong influence on light penetration.

5.6.4 Hemoglobin

Blood attenuates NIR light transmission through biological tissue, including the brain (Donn et al., 1979; Jagdeo et al., 2012). Hemoglobin concentration in cerebral tissue is approximately 84 μm/dL. (Feng, 2013) Hemoglobin light absorption is based fundamentally on Beer-Lambert law and can be expressed as: $A = \varepsilon cL$, where absorption coefficient is ε, absorbance is A, L is optical path length of light, and c is concentration (Abitan et al., 2008). The larger the absorption coefficient a substance has, the higher absorption at that particular wavelength.

Oxygenated and deoxygenated hemoglobin in blood absorb light differently because there is a significant change in absorption spectrum after oxygen binding to hemoglobin. Oxygenated and deoxygenated hemoglobin have intense absorption peaks in the visible blue region at 418 and 430 nm, respectively (Welch et al., 1995). Isobestic points where absorption of oxygenated and deoxygenated hemoglobin are equal occur at visible green (548 nm), visible yellow (568, 587 nm), and NIR 805 nm wavelengths (Welch et al., 1995). Oxygenated blood has a stronger absorption band at NIR wavelengths than its deoxygenated counterpart (Welch et al., 1995).

The difference in absorption spectra for hemoglobin based on oxygenation status and the fact that NIR light can penetrate through bone and soft tissue, has enabled the development of other noninvasive neurodiagnostic applications, such as transcranial cerebral oximetry, functional magnetic resonance imaging (fMRI), and functional NIR spectroscopy (fNIRS). In these devices, hemoglobin itself is used as a contrast agent because it undergoes oxygen binding that produces measurable changes in its optical properties. Because the absorption spectra between the two hemoglobin species are significantly different in the NIR regions, separations of the hemoglobin compounds by spectroscopy are possible (Feng, 2013). fMRI can detect local blood flow changes from neuroactivation by imaging concentration changes of hemoglobin species. fNIRS detects regional brain changes in cerebral oxygenation using spectroscopic measurements of tissue concentrations of oxygenated, deoxygenated, and total hemoglobin. Temporal variations among the hemoglobin components can serve as a surrogate marker for neural activation (Wylie et al., 2009). High-density fNIRS can produce validated real-time hemodynamic changes and cerebral monitoring of neuroconnectivity in primates during events of ischemia, vasodilatation, and hemorrhage (Lee et al., 2014).

5.6.5 Optical window

Intense absorption by melanin and hemoglobin occurs below 600 nm while water in tissues absorbs energy at wavelengths greater than 1150 nm (Jacques, 2013). The minimized absorption and scattering of light by the principal tissue chromophores of melanin, hemoglobin, and water between red and NIR wavelengths (650–950 nm) effectively creates an "optical therapeutic window" for maximal light penetration in tissue (see Fig. 5.3) (Shi et al., 2016). A second (1100–1350 nm), third (1600–1870 nm), and fourth (2100–2350 nm) NIR optical windows have been reported

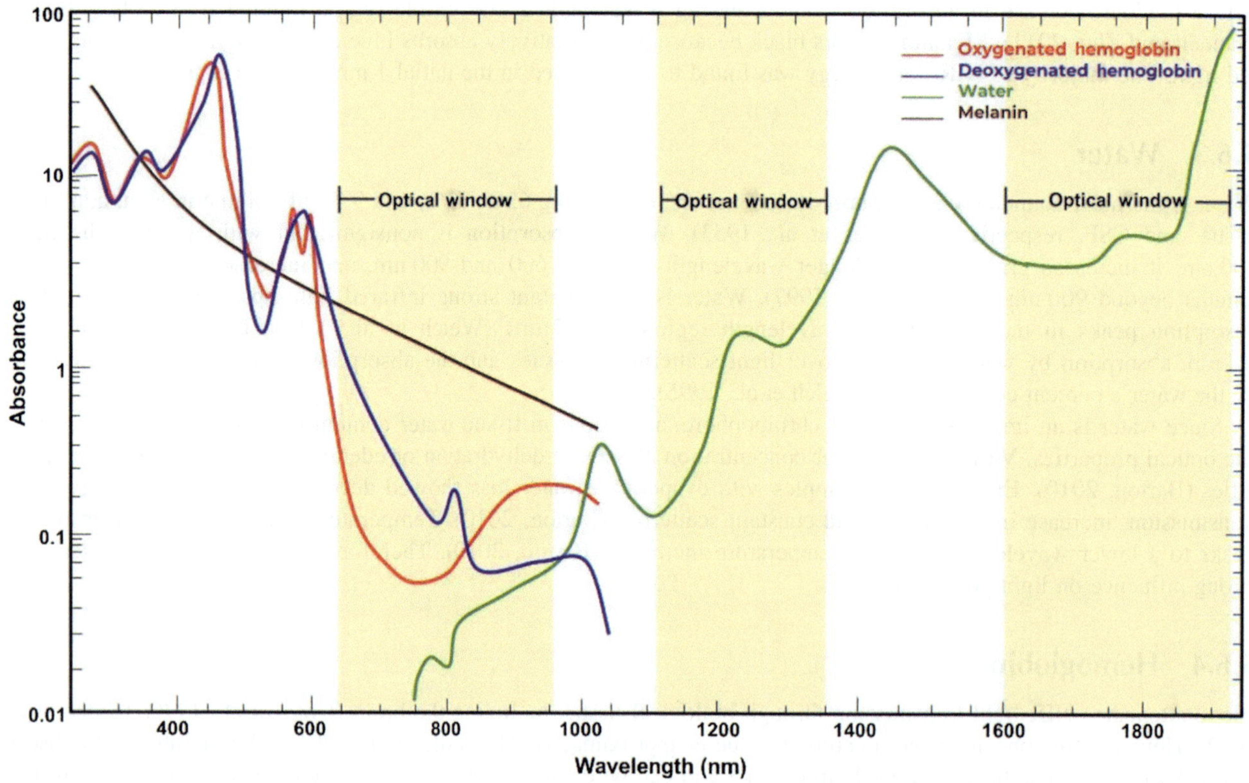

FIGURE 5.3 Absorption spectra of tissue chromophores. Tissue chromophores of oxygenated hemoglobin, deoxygenated hemoglobin, water, and melanin are represented. Note that molar absorption coefficient is in logarithmic scale.

(Sordillo et al., 2014; Shi et al., 2016). Of the four optical windows, the third optical window was optimal for deep imaging of rat brain tissue (Shi et al., 2016).

Optical windows are important because diagnostic imaging and therapeutic devices, such as optical coherence tomography, LEDs, and lasers, are centered on wavelengths within an optical window. Although wavelengths outside of the optical window, such as UV light, visible, blue, green, and yellow light, may have significant effects on in vitro cells, their use has not been investigated in transcranial photobiomodulation. Furthermore, the penetration depths of these wavelengths vary: less than 1 mm at 400 nm, 0.5–2 mm at 514 nm, 1–6 mm at 630 nm, and maximal at 700–900 nm (Simpson et al., 1998). Photobiomodulation in animals and humans almost exclusively uses red and NIR light in the optical window range.

5.7 Cerebrospinal fluid

The structure and thickness of the CSF influences the intensity of light (Okada and Delpy, 2003). The low absorption and low scattering properties of CSF strongly affects the light propagation through the cranium (Okada and Delpy, 2003). Generally, the thickness of CSF varies around the head because the brain can move to some extent within the skull. A Monte Carlo simulation (MCS) demonstrated that the CSF thickness did not independently affect the light penetration depth, as long as the combined thickness of the skull and CSF remains constant (Okada and Delpy, 2003).

5.7.1 Gray and white brain matter

Knowledge of the optical properties of the gray and white brain tissues may be fundamental to transcranial light penetration. There were substantial differences in light scattering properties in various regions of adult rat brain, including the brain stem, midbrain, and forebrain (Al-Juboori et al., 2013). This observation highlights the need for specific knowledge about the optical properties of brain areas and potentially a differential approach of light illumination across different brain areas.

Penetration depth in gray matter was greater than in white matter in both rat and human brains (Eggert and Blazek, 1987; Gottschalk, 1992; Roggan et al., 1994; Yaroslavsky et al., 2002; Abdo and Sahin, 2007). The mean penetration depth with 37% of NIR light remaining in rat gray and white brain matter was 0.41 ± 0.029 and 0.35 ± 0.026 mm, respectively (Abdo and Sahin, 2007). The reason for this difference in optical properties between white matter and gray matter is unclear. The increased higher absorption and scattering of white matter compared to gray matter may be secondary to tissue composition as gray matter contains numerous cell bodies and relatively few myelinated axons, whereas white matter is composed of dense myelinated axons and fewer cell bodies. The density of myelinated axons and myelination itself may contribute to the high scattering properties of white matter.

Optical properties of ex vivo human brain tissues were investigated from a variety of regions, including white matter, gray matter, and brain stem (Yaroslavsky et al., 2002). Within the brain stem, the scattering coefficients of the pons were lower than the scattering coefficients of the thalamus for all wavelengths tested (Yaroslavsky et al., 2002). Accordingly, in order of decreasing light penetration depth, light penetration depth was greatest for gray brain matter, pons, thalamus, and ultimately lowest in white brain matter (Yaroslavsky et al., 2002).

Because of the optical properties of brain and its surrounding tissues, the total dose of light will need to be increased to account for the loss of irradiance, as photons travel through various tissue layers to reach the targeted tissue. Moreover, neurological pathologies that transform brain tissue—for example, inflammation in meningitis, anoxia, brain edema, acute ischemia in strokes, or acute hemorrhage/hematoma – may considerably alter tissue cellularity, and thus, hamper light penetration (Eisenblatter, 2014).

5.8 Wavelength

Wavelength has a dual impact on transcranial light penetration because the scattering and absorption of light is highly wavelength-specific. Within the UV, visible, and NIR portions of the electromagnetic spectrum, longer wavelengths of light penetrate tissues to a greater depth because as wavelength increases, scattering decreases (Simpson et al., 1998). Furthermore, light absorption by different chromophores occurs at specific wavelengths based on the tissue's properties (Jobsis, 1977). In this section, we will discussthe animal and human studies that have investigated the wavelengths of light that are optimal in maximizing tissue penetration (Gottschalk, 1992; Roggan et al., 1994; Yaroslavsky et al., 2002; Jagdeo et al., 2012; Al-Juboori et al., 2013; Henderson and Morries, 2015; Pitzschke et al., 2015a,b; Tedford et al., 2015).

5.8.1 Animal studies

Due to NIR light's longer wavelength compared to red light, near infrared has superior penetration into calvaria and brain tissue. A study examined transcranial penetration of 670 nm red, 810 nm NIR, and 980 nm NIR wavelengths in ex vivo lamb skin, and lamb heads (Henderson and Morries, 2015). 810 nm NIR laser had significantly better penetration than 980 nm NIR laser, whereas penetration of 670 nm LED red light was negligible (~0.005%) in ex vivo lamb heads. Similar results of NIR light superiority were found in rabbit brains when comparing 635 nm red, 671 nm red, and 808 nm NIR (Pitzschke et al., 2015a,b). In this rabbit brain model, 808 nm was the best for light propagation into deep brain tissue. The maximal penetration through ex vivo rat brain tissues occurred at wavelengths of 940 nm when evaluating light from 360 to 1100 nm (Al-Juboori et al., 2013). Despite the premise that longer wavelengths penetrate further, others have consistently found that NIR light in the range of 808–810 nm had greater penetration than 980 nm NIR light in various animal tissue types (e.g., skin, connective tissue, muscle, nervous tissue, etc.), due to the influence of absorption of light by blood (Byrnes et al., 2005; Hudson et al., 2013).

5.8.2 Human studies

Since animal studies demonstrated that NIR wavelengths were optimal for light penetration, the superiority of NIR light was also investigated in ex vivo human studies. NIR light was ideal in ex vivo human studies, comparable to findings in animal studies. 810 nm NIR laser demonstrated significantly better transcranial penetration than 670 nm red and 980 nm NIR wavelengths in human skin [28,29]. In human cadaveric deep brain tissue, NIR light at 808 nm was superior compared to 635 and 671 nm red light [26,27]. In another study, the amount of 633 nm red light and 830 nm NIR light penetrating human cadaver skulls, soft tissue, and brain parenchyma were compared (Jagdeo et al., 2012). NIR light was able to penetrate both isolated human cadaveric skull and a 10-mm thickness of occipital skull with intact brain tissue (Jagdeo et al., 2012). Red light, in contrast, did not penetrate the same tissues to the same degree under similar conditions (Jagdeo et al., 2012).

Similarly, another study analyzed the penetration depths of 660 nm red, 808 nm NIR, and 940 nm NIR light using human cadaveric heads, including scalp, skull, meninges, and brain (Tedford et al., 2015). Of these three wavelengths, 808 nm light had the best penetration depth of 40–50 mm (Tedford et al., 2015). Within the human brain, light scattering created a gradient of energy (Tedford et al., 2015).

Although NIR light is superior at transcranial light penetration through the human skull, the maximal penetration of ex vivo human brain tissues occurred at the greatest wavelength of 1064 nm (Gottschalk, 1992; Roggan et al., 1994; Yaroslavsky et al., 2002). As wavelength increased from 360 to 1100 nm, there was better penetration because the scattering coefficient decreased, and the anisotropy factor increased while the absorption coefficient remained low (Yaroslavsky et al., 2002). Scattering coefficient decreases with longer wavelengths due to theories of Rayleigh scattering and Mie scattering (Shi et al., 2016).

Ideally, the wavelength that provides maximal penetration will also result in the optimal response of neurologic tissue. The action spectra describe the wavelength that results in prime physiological activity based on the absorption of light of the photoacceptor (Hartman, 1983). In regards to transcranial photobiomodulation, the action spectra of the targeted primary mitochondrial chromophore, CCO, is localized in the red-to-NIR region (670 and 830 nm) of the electromagnetic spectrum, which corresponds to wavelengths with high tissue penetration (see Fig. 5.4) (Karu, 2010).

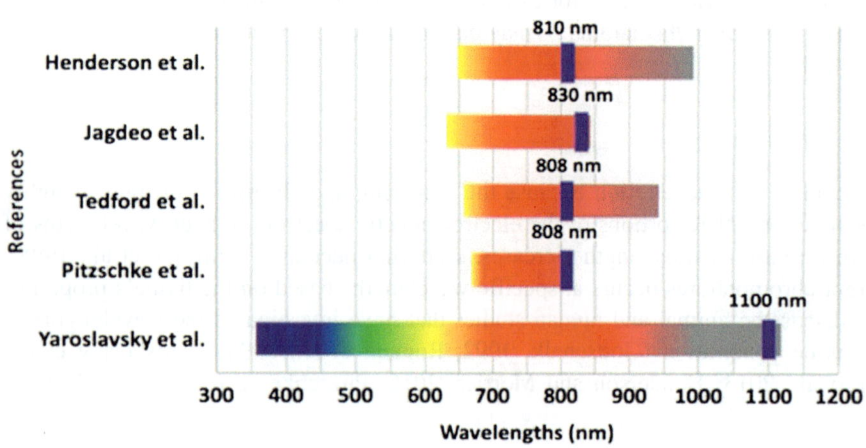

FIGURE 5.4 Optimal wavelength for transcranial light penetration across visible and infrared light spectrum. Wavelength spectrum based on reviewed studies. (Note: *blue bar* signifies optimal wavelength for penetration).

5.9 Skull anatomy

Each skull has a unique spatial distribution of thickness and density that varies by site, race, sex, and age. Frontal bones of white males were thicker than those of black males, whereas the parietal and occipital bones of white males were thinner (Adeloye et al., 1975). Skull thickness differences by sex is variable. As humans age, our skulls rapidly thicken in the first two decades of life, then it thickens gradually in the following three decades leading to a peak in the fifth and sixth decade. Thus, studies have investigated whether the light's location on the skull can impact the amount of light penetration as certain areas of the skull are thinner than others.

5.9.1 Animal studies

Enhanced penetration was observed in three other mammalian species with varying, albeit thinner, skull thicknesses: 40.10% of this light was transmitted in mouse skulls, 21.24% in rat skulls, and 11.36% in rabbit skulls (Lapchak et al., 2015). Hence, as skull thickness increased, there was a nonlinear decrease in light penetration. In contrast, no correlation was observed between NIR light transmission and skull density. Thus, translating light parameters from small animal studies to human clinical trials has limitations as differential penetration profiles among species exist due to the significant differences in skull thickness and penetration patterns (Lapchak and Boitano, 2016).

5.9.2 Human studies

Since light transmission is dependent on skull thickness which varies by site, the placement of the light source can impact light penetration. Lower irradiances by 10% were observed when light was placed on sagittally cut versus coronally sectioned human cadaveric skulls (Jagdeo et al., 2012). Furthermore, light penetration varied between the frontal, left parietal, and right parietal areas (Jagdeo et al., 2012). In two studies, there was extensive light attenuation with approximately 4.5% of NIR light penetrating the human skull, but this varied by region (Lapchak et al., 2015; Lapchak and Boitano, 2016). An 800 nm laser with irradiance of 700 mW/cm^2 transmitted 4.18% and 4.24% in human skulls at the bregma (7.19 mm thickness) and left parietal skull (5.91 mm thickness), respectively (see Fig. 5.5). Since human skull thickness varies depending upon the anatomical location, age and sex of the patient, different areas of the brain will likely received different dosages of light.

5.9.3 Monte Carlo modeling

Computerized MCS is a common technique for simulating light propagation using the trajectories of multiple photons under a variety of conditions, including different light sources and tissue types. Through a MCS of photon propagation in adult human head models, NIR light penetration was dependent on the skull thickness and, to a lesser extent, the

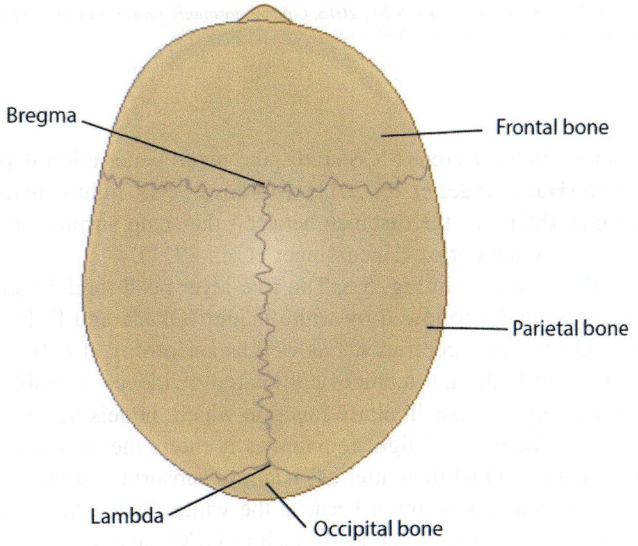

FIGURE 5.5 Superior view of calvaria. The positions of bregma and lambda are shown above.

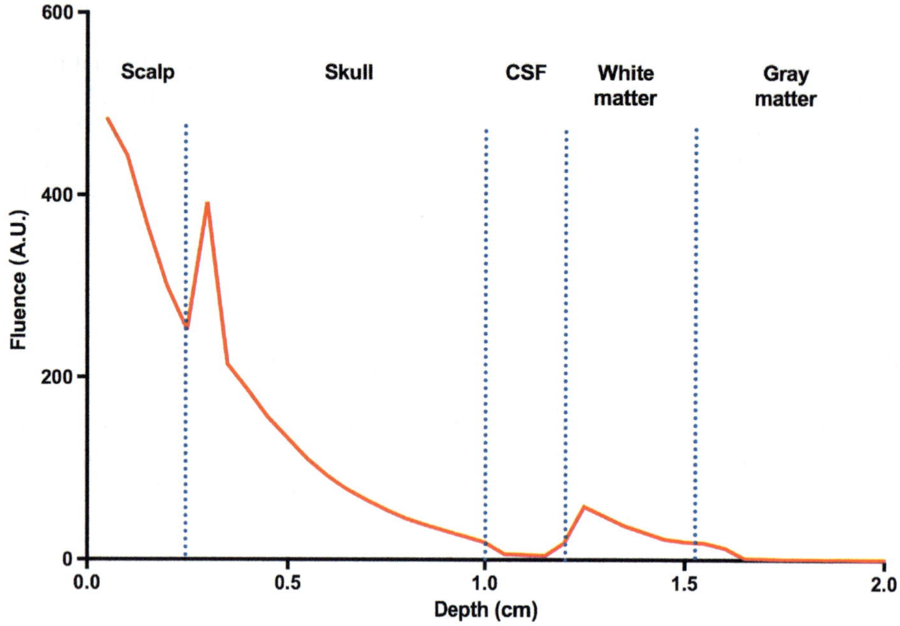

FIGURE 5.6 Monte Carlo simulation of 800-nm collimated light absorbed in a five-layer model of the human head including scalp, skull, CSF, white matter, and gray matter. Layer thickness and optical properties are given in Table 5.4. 5.3% of the incident light passes through the scalp, skull and CSF, into the gray matter (light fluence vs depth graph not shown).

TABLE 5.4 Thickness and optical coefficients for five-layer model of the human head.

800 nm	Thickness (cm)	Absorption (cm^{-1})	Scattering (cm^{-1})	Anisotropy
Scalp	0.3[a]	0.375[b]	100[b]	0.86[b]
Skull	0.7[a]	0.240[c]	184[c]	0.90[c]
CSF[#]	0.2[a]	0.090[a]	160[a]	0.90[a]
Gray matter	0.4[a]	0.350[c]	700[c]	0.965[c]
White matter	0.9[a]	0.050[c]	550[c]	0.85[c]

800-nm light is characterized by absorption and scattering coefficients as well as the anisotropy for each layer. #Arachnoid trabeculae scattering. Absorption is a combination of CSF and arachnoid trabeculae.[1]

[a]*Okada, E., Delpy, D.T., 2003. Near-infrared light propagation in an adult head model. II. Effect of superficial tissue thickness on the sensitivity of the nearinfrared spectroscopy signal. Appl. Opt. 42 (16), 2915–2922.*

[b]*Bashkatov, A.N., Genina, E.A., Kochubey, V.I., Tuchin, V.V., 2005. Optical properties of human skin, subcutaneous and mucous tissues in the wavelength range of 400 to 2000 nm. J. Phys. D: Appl. Phys. 38 (15), 2543–2555.*

[c]*Verleker, A.P., Shaffer, M., Fang, Q., Choi, M.R., Clare, S., Stantz, K.M., 2016. Optical dosimetry probes to validate Monte Carlo and empirical-method-based NIR dose planning in the brain. Appl. Opt. 55 (34), 9875–9888.*

CSF layer (Okada and Delpy, 2003). In a different MCS study, the mean penetration depth of the most distal reaching NIR light (5%) was 23.6 ± 0.7 mm (Haeussinger et al., 2011). Moreover, NIR light penetration depth was inversely correlated with scalp to cortex distance (SCD) or the distance between the scalp surface and brain cortex. SCD increased from right lateral to posterior medial frontal cortex (Haeussinger et al., 2011).

An example of a multilayer MCS is shown in Fig. 5.6. The five-layer head model includes a 3.0-mm scalp, 7.0-mm skull, 2.0-mm CSF, and 4.0-mm gray matter followed by white matter (Okada and Delpy, 2003). Each layer is defined by published optical absorption and scattering coefficients as well as anisotropy at 800 nm (see Table 5.4). The simulation is written from the model of Scott Prahl and includes collimated incidence, variable step size, index of refraction mismatch at the surface, and the Henyey-Greenstein phase function which models Mie scattering (Prahl, 1989).

According to this model, 5.3% of the incident light remains as it enters the gray matter. Multiple scattering events enhance the light flux in the gray matter and white matter as shown in subsurface peaks within each layer. The light fluence in the white matter results in minimal absorption because the white matter has a very low absorption coefficient (see Fig. 5.7). However, microscopic targets within the brain such as CCO have larger absorption coefficients.

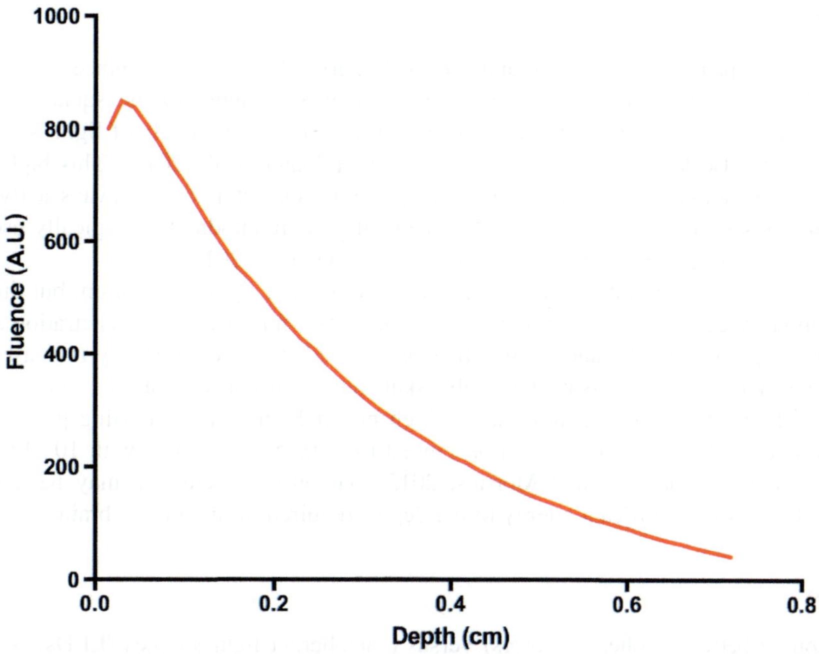

FIGURE 5.7 Monte Carlo simulation of 800-nm collimated light fluence through a human skull. 4.15% ± 0.18% of the incident light is transmitted through the skull. This is in excellent agreement to an experimental literature value of 4.18% for penetration of 800-nm light through a human skull in vitro.

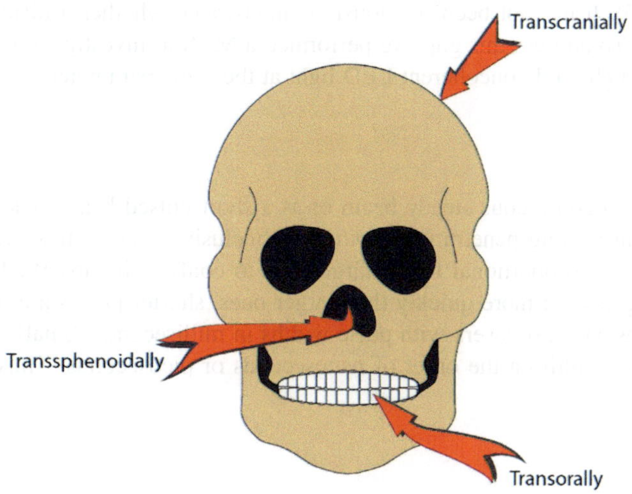

FIGURE 5.8 Potential routes of light delivery.

A study explored various potential light delivery routes: intracranially through the third ventricle, transsphenoidally through the sphenoid sinus, and transorally via the oral cavity (see Fig. 5.8) (Pitzschke et al., 2015a,b). Between the transcranial and transsphenoidal route, the best path for efficient light penetration into the brain was transsphenoidally (Pitzschke et al., 2015a,b). Therefore, one can conclude that light penetration is dependent on the light source's location on the skull.

Since skull thickness is a factor for light penetration and each individual has a unique skull thickness, one may speculate whether individualized LED or laser treatment parameters specific to an individual's skull anatomy would be beneficial, especially in an era of personalized medicine. Although this prospect may be enticing and ideal to base light parameters on an individual's skull anatomy, it could be also impractical as it may require measurements using imaging modalities that may expose patients to radiation.

5.10 Irradiance

Transcranial light therapy depends on irradiance and the total energy delivered. Irradiance, or rate of energy delivery per second per unit area (W/cm^2) during a single pulse, is inversely related to the square of the beam diameter. Photobiomodulation is known to have biphasic, dose-dependent effects as low levels of light is stimulatory while high levels of light is inhibitory, also known as the Arndt-Schulz law (Huang et al., 2011). This biphasic pattern has been explained by excessive generation of reactive oxygen species and NO, cytotoxic pathways activation, and decreased nF-kB activity at higher doses. (Huang et al., 2011) Transcranial photobiomodulation typically uses relatively low fluences of 0.04–50 J/cm^2 and low power output of 1–500 mW (Zhang et al., 2014).

Greater irradiance is thought to be important because it allows greater light penetration, but this remains to be confirmed. One study demonstrated a positive correlation between an 800-nm laser light penetration and irradiance on the skull surface of rabbits(Lapchak and Boitano, 2016). Increasing irradiance incrementally increased laser light penetration, but the percentage of laser light crossing the rabbit skull remained constant at 11% regardless of power density (Lapchak and Boitano, 2016). Likewise, in lamb heads, high power lasers had increasing penetration with increasing irradiance (Henderson and Morries, 2015). Penetration ranged from 0.45% to 2.90% with 10–15 W 810 nm NIR laser in 3 cm of lamb head tissue (Henderson and Morries, 2015). Therefore, irradiance may be a potentially important parameter that allows delivery of significant energy to the depths required in the human brain.

5.11 Coherence

The choice of light source between coherent (lasers) versus noncoherent light sources (LEDs) is currently a source of debate. The coherence of laser light was initially thought to be important for the photobiomodulatory effects, but the promising results of the noncoherent LEDs have challenged this belief. In a study, high power lasers had greater penetration depth and a 100-fold greater fluence delivery compared to low power NIR emitters in lamb heads; however, they had notably different power settings (Henderson and Morries, 2015). Low power NIR LEDs ranging from 50 mW to 0.2 W showed no penetration or detection at 2 mm of human skin, 2 mm of lamb skin, or 3 cm of lamb head (Henderson and Morries, 2015). It has not been previously demonstrated whether a difference between laser and LED exists at the same parameters. To address this gap, we performed a MCS to investigate whether light penetration differs between monochromatic laser light and noncoherent LED light at the same parameters.

5.12 Pulsing

Laser light can be delivered as a continuous steady beam or as a short pulsed beam. The debate on whether pulsing or continuous light is better for increasing penetration depth is inconclusive. Laser pulse width, or the time duration that laser energy delivers the beam, is proportional to the target size to confine the absorbed energy within the target (see Fig. 5.9). Because smaller targets cool more quickly than larger ones, shorter pulses are needed for smaller targets. For example, larger targets, such as hair, use lasers with pulse widths in milliseconds. Smaller targets like tattoo pigment in cells require lasers with a pulse width on the order of nanoseconds or picoseconds. Thus, the smallest targets, such as

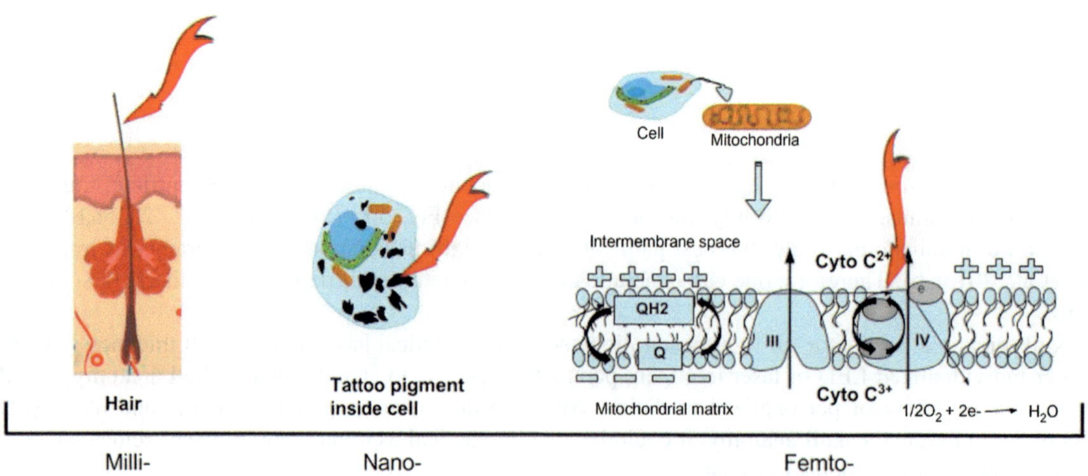

FIGURE 5.9 Pulse widths for different laser targets.

CCO in the mitochondria for photobiomodulation, would require femtosecond lasers. Therefore, we believe pulsing may not be a factor for transcranial photobiomodulation as the current pulse widths of the lasers and LEDs currently used are too long to target the mitochondrial unit. Furthermore, as lasers include shorter pulse widths, the cost of the device also increases; therefore, cost may be prohibitive to using femtosecond lasers for transcranial photobiomodulation.

Another hypothesis of why a difference in penetration depth between continuous wave and pulse-wave light exists is that at an identical average power output as a continuous-wave light system, the pulse wave mode may theoretically emit more photons deeper into the brain at pulsed peaks. However, this hypothesis has not been demonstrated in studies. Pulsed light yielded greater penetration through sheep skin, intact sheep head, and living human tissue, but this was not statistically significant and resulted in overall lower irradiance compared to continuous light (Henderson and Morries, 2015). Conversely, in fixed human cadaver heads, there was no difference between pulse- and continuous-wave laser light (Tedford et al., 2015). Additional ex vivo and in vivo human models are required for better delineation of advantages and disadvantages for pulse and continuous light.

5.13 Tissue storage and processing

Numerous studies that investigated penetration depth utilized ex vivo animal or human tissue that underwent chemical preparation, storage, and processing. These procedures may also affect the optical properties of tissues, consequently affecting penetration depth. This raises an important concern of whether ex vivo studies with tissue after preparation techniques can reliably be used to extrapolate the optical properties of living tissues to design in vivo transcranial applications.

To examine the impact of tissue preparation and storing processes on the tissue's optical properties, the effects of long-term (6 weeks) freezing versus formalin-fixation on the μ_{eff} of rabbit brains were observed (Yaroslavsky et al., 2002). Compared to in vivo tissue, tissue freezing decreased μ_{eff} by 15%–25% whereas formalin-fixation increased μ_{eff} by 5%–15% (Yaroslavsky et al., 2002). In contrast, the effects of two different methods of euthanization (exsanguination or potassium chloride injection) on tissue optical properties were negligible (Yaroslavsky et al., 2002). Regardless of processing, 808 nm NIR light had the lowest μ_{eff} compared to 671 and 635 nm red light (Yaroslavsky et al., 2002). Although hydration increased light penetration in animal skulls, there was no significant effect of hydration on light penetration in human skulls (Lapchak et al., 2015).

In summary, it is important to be aware of tissue preparation techniques used as it may affect translating results from animal models into human studies. Additionally, differences in tissue sample preparation techniques may explain inconsistencies when comparing data across studies.

5.14 Conclusion

As transcranial photobiomodulation for brain disorders emerges as a promising, safe, and effective treatment modality, it is vitally important to examine the extent that light penetrates the human cranium. Overcoming this significant challenge requires understanding light properties in the brain and considering the anatomy, optical properties, and characteristics associated with the human head.

Evidence supports using red-to-NIR wavelengths for the most therapeutic effects given the action spectra of CCO. However, NIR light has been used for better penetration due to its longer wavelength. NIR light has wide therapeutic applications beyond neurological diseases, not only in treating neurologic conditions but also in developing diagnostic imaging or neuromonitoring techniques. Noninvasive NIR spectroscopy-based imaging or neuromonitoring also requires light penetration through the scalp, skull, and CSF to measure changes in the brain. Furthermore, the clinical safety profile of transcranial NIR light is a distinct advantage over other invasive neuromodulation and monitoring methods.

Light penetration depth depends on wavelength, skull anatomy, tissue optical properties, irradiance, coherence, and pulsing. Leveraging these properties can aid researchers and clinicians in translating basic science findings into therapeutic interventions (bench-to-bedside), designing clinical trials, and developing standardized treatment regimens for patients. This is especially pertinent now because, as of yet, NILT has not been optimized or consistent in animal models or in humans. Some have postulated that the mid-point futility of the NEST-3 trial may have been due to dosing errors leading to limited effectiveness of light penetration, and not optimizing the use of NIR light prior to clinical trials, therefore, the continuous wave laser light could not reach the deeper stroke lesions (Morries et al., 2015; Lapchak and Boitano, 2016; Meyer et al., 2016). These optimal parameters consist of selecting the appropriate wavelengths and light settings that will maximize light penetration depth and reaching neurons through a transcranial route. Understanding light propagation in tissues is critical in maximizing efficacy of transcranial photobiomodulation by providing sufficient or adequate light penetration through the human cranium to activate mechanisms for neuromodulation.

References

Abdo, A., Sahin, M., 2007. NIR light penetration depth in the rat peripheral nerve and brain cortex. Conf. Proc. IEEE Eng. Med. Biol. Soc. 2007, pp. 1723–1725.

Abitan, H., Bohr, H., Buchhave, P., 2008. Correction to the Beer-Lambert-Bouguer law for optical absorption. Appl. Opt. 47 (29), 5354–5357.

Adeloye, A., Kattan, K.R., Silverman, F.N., 1975. Thickness of the normal skull in the American Blacks and Whites. Am. J. Phys. Anthropol. 43 (1), 23–30.

Al-Juboori, S.I., Dondzillo, A., Stubblefield, E.A., Felsen, G., Lei, T.C., Klug, A., 2013. Light scattering properties vary across different regions of the adult mouse brain. PLoS One 8 (7), e67626.

Anderson, R.R., Parrish, J.A., 1981. The optics of human skin. J. Invest. Dermatol. 77 (1), 13–19.

Barton, J., 2010. Dynamic changes in optical properties. In: Welch, A.J., van Gemert, M.J.C. (Eds.), Optical-Thermal Response of Laser-Irradiated Tissue. Springer, Dordrecht.

Byrnes, K.R., Waynant, R.W., Ilev, I.K., Wu, X., Barna, L., Smith, K., et al., 2005. Light promotes regeneration and functional recovery and alters the immune response after spinal cord injury. Lasers Surg. Med. 36 (3), 171–185.

Chung, H., Dai, T., Sharma, S.K., Huang, Y.Y., Carroll, J.D., Hamblin, M.R., 2012. The nuts and bolts of low-level laser (light) therapy. Ann. Biomed. Eng. 40 (2), 516–533.

de Freitas, L.F., Hamblin, M.R., 2016. Proposed mechanisms of photobiomodulation or low-level light therapy. IEEE J. Sel. Top. Quantum Electron. 22 (3).

Donn, S.M., Sharp, M.J., Kuhns, L.R., Uy, J.O., Knake, J.E., Duchinsky, B.J., 1979. Rapid detection of neonatal intracranial hemorrhage by transillumination. Pediatrics 64 (6), 843–847.

Eggert, H.R., Blazek, V., 1987. Optical properties of human brain tissue, meninges, and brain tumors in the spectral range of 200 to 900 nm. Neurosurgery 21 (4), 459–464.

Eisenblatter, M., 2014. In: Brahme, A. (Ed.), Comprehensive biomedical physics, vol. 1. Elsevier, p. 374.

Esnouf, A., Wright, P.A., Moore, J.C., Ahmed, S., 2007. Depth of penetration of an 850 nm wavelength low level laser in human skin. Acupunct. Electrother. Res. 32 (1–2), 81–86.

Feng, G., 2013. Medical background and optical properties of human tissue. In: Tian, J. (Ed.), Molecular Imaging: Fundamentals and Applications. Springer, Hangzhou.

Forbes, R.M., Cooper, A.R., Mitchell, H.H., 1953. The composition of the adult human body as determined by chemical analysis. J. Biol. Chem. 203 (1), 359–366.

Gottschalk, W., 1992. Ein Messverfahren zur Bestimmung der Optischen Parameter biologischer Gevebe in vitro. Dissertation 93 HA8984 Universitaet Fridriciana, Karlsruhe.

Hacke, W., Schellinger, P.D., Albers, G.W., Bornstein, N.M., Dahlof, B.L., Fulton, R., et al., 2014. Transcranial laser therapy in acute stroke treatment: results of neurothera effectiveness and safety trial 3, a phase III clinical end point device trial. Stroke 45 (11), 3187–3193.

Haeussinger, F.B., Heinzel, S., Hahn, T., Schecklmann, M., Ehlis, A.C., Fallgatter, A.J., 2011. Simulation of near-infrared light absorption considering individual head and prefrontal cortex anatomy: implications for optical neuroimaging. PLoS One 6 (10), e26377.

Hamblin, M.R., 2016. Shining light on the head: photobiomodulation for brain disorders. BBA Clin. 6, 113–124.

Hartman, K.M., 1983. Action spectroscopy. In: Hoppe, W., Lohmann, W., Marke, H., Ziegler, H. (Eds.), Biophysics. Springer-Verlag, Heidelberg.

Hashmi, J.T., Huang, Y.Y., Osmani, B.Z., Sharma, S.K., Naeser, M.A., Hamblin, M.R., 2010. Role of low-level laser therapy in neurorehabilitation. PM R 2 (12 Suppl. 2), S292–S305.

Henderson, T.A., Morries, L.D., 2015. Near-infrared photonic energy penetration: can infrared phototherapy effectively reach the human brain? Neuropsychiatr. Dis. Treat. 11, 2191–2208.

Henderson, T.A., Morries, L.D., 2017. Multi-watt near-infrared phototherapy for the treatment of comorbid depression: an open-label single-arm study. Front. Psychiatry 8, 187.

Huang, Y.Y., Sharma, S.K., Carroll, J., Hamblin, M.R., 2011. Biphasic dose response in low level light therapy – an update. Dose Response 9 (4), 602–618.

Hudson, D.E., Hudson, D.O., Wininger, J.M., Richardson, B.D., 2013. Penetration of laser light at 808 and 980 nm in bovine tissue samples. Photomed. Laser Surg. 31 (4), 163–168.

Huisa, B.N., Stemer, A.B., Walker, M.G., Rapp, K., Meyer, B.C., Zivin, J.A., 2013. Transcranial laser therapy for acute ischemic stroke: a pooled analysis of NEST-1 and NEST-2. Int. J. Stroke 8 (5), 315–320.

Ilic, S., Leichliter, S., Streeter, J., Oron, A., DeTaboada, L., Oron, U., 2006. Effects of power densities, continuous and pulse frequencies, and number of sessions of low-level laser therapy on intact rat brain. Photomed. Laser Surg. 24 (4), 458–466.

Jacques, S.L., 2013. Optical properties of biological tissues: a review. Phys. Med. Biol. 58 (11), R37–R61.

Jagdeo, J.R., Adams, L.E., Brody, N.I., Siegel, D.M., 2012. Transcranial red and near infrared light transmission in a cadaveric model. PLoS One 7 (10), e47460.

Jobsis, F.F., 1977. Noninvasive, infrared monitoring of cerebral and myocardial oxygen sufficiency and circulatory parameters. Science 198 (4323), 1264–1267.

Joensen, J., Demmink, J.H., Johnson, M.I., Iversen, V.V., Lopes-Martins, R.A., Bjordal, J.M., 2011. The thermal effects of therapeutic lasers with 810 and 904 nm wavelengths on human skin. Photomed. Laser Surg. 29 (3), 145–153.

Johnstone, D.M., Moro, C., Stone, J., Benabid, A.L., Mitrofanis, J., 2015. Turning on lights to stop neurodegeneration: the potential of near infrared light therapy in Alzheimer's and Parkinson's Disease. Front. Neurosci. 9, 500.

Karu, T., 2010. Mitochondrial mechanisms of photobiomodulation in context of new data about multiple roles of ATP. Photomed. Laser Surg. 28 (2), 159–160.

Kocsis, L., Herman, P., Eke, A., 2006. The modified Beer-Lambert law revisited. Phys. Med. Biol. 51 (5), N91–N98.

Lampl, Y., Zivin, J.A., Fisher, M., Lew, R., Welin, L., Dahlof, B., et al., 2007. Infrared laser therapy for ischemic stroke: a new treatment strategy: results of the NeuroThera Effectiveness and Safety Trial-1 (NEST-1). Stroke 38 (6), 1843–1849.

Lapchak, P.A., Boitano, P.D., 2016. Transcranial near-infrared laser therapy for stroke: how to recover from futility in the NEST-3 Clinical Trial. Acta Neurochir. Suppl. 121, 7–12.

Lapchak, P.A., Wei, J., Zivin, J.A., 2004. Transcranial infrared laser therapy improves clinical rating scores after embolic strokes in rabbits. Stroke 35 (8), 1985–1988.

Lapchak, P.A., Boitano, P.D., Butte, P.V., Fisher, D.J., Holscher, T., Ley, E.J., et al., 2015. Transcranial near-infrared laser transmission (NILT) profiles (800 nm): systematic comparison in four common research species. PLoS One 10 (6), e0127580.

Lee, D.C., Gevorgyan, T., Graber, H.L., Pfeil, D.S., Xu, Y., Mangla, S., et al., 2014. Feasibility of near-infrared spectroscopic tomography for intraoperative functional cerebral monitoring: a primate study. J. Thorac. Cardiovasc. Surg. 148 (6), 3204–3210. e3201–3202.

McCarthy, T.J., De Taboada, L., Hildebrandt, P.K., Ziemer, E.L., Richieri, S.P., Streeter, J., 2010. Long-term safety of single and multiple infrared transcranial laser treatments in Sprague-Dawley rats. Photomed. Laser Surg. 28 (5), 663–667.

Meyer, D.M., Chen, Y., Zivin, J.A., 2016. Dose-finding study of phototherapy on stroke outcome in a rabbit model of ischemic stroke. Neurosci. Lett. 630, 254–258.

Mochizuki-Oda, N., Kataoka, Y., Cui, Y., Yamada, H., Heya, M., Awazu, K., 2002. Effects of near-infra-red laser irradiation on adenosine triphosphate and adenosine diphosphate contents of rat brain tissue. Neurosci. Lett. 323 (3), 207–210.

Moro, C., Massri, N.E., Torres, N., Ratel, D., De Jaeger, X., Chabrol, C., et al., 2014. Photobiomodulation inside the brain: a novel method of applying near-infrared light intracranially and its impact on dopaminergic cell survival in MPTP-treated mice. J. Neurosurg. 120 (3), 670–683.

Morries, L.D., Cassano, P., Henderson, T.A., 2015. Treatments for traumatic brain injury with emphasis on transcranial near-infrared laser phototherapy. Neuropsychiatr. Dis. Treat. 11, 2159–2175.

Naeser, M.A., Saltmarche, A., Krengel, M.H., Hamblin, M.R., Knight, J.A., 2011. Improved cognitive function after transcranial, light-emitting diode treatments in chronic, traumatic brain injury: two case reports. Photomed. Laser Surg. 29 (5), 351–358.

Naeser, M.A., Zafonte, R., Krengel, M.H., Martin, P.I., Frazier, J., Hamblin, M.R., et al., 2014. Significant improvements in cognitive performance post-transcranial, red/near-infrared light-emitting diode treatments in chronic, mild traumatic brain injury: open-protocol study. J. Neurotrauma 31 (11), 1008–1017.

Okada, E., Delpy, D.T., 2003. Near-infrared light propagation in an adult head model. II. Effect of superficial tissue thickness on the sensitivity of the near-infrared spectroscopy signal. Appl. Opt. 42 (16), 2915–2922.

Parsons, F.G., 1929. The thickness of the living scalp. J. Anat. 63 (Pt 4), 427–429.

Pitzschke, A., Lovisa, B., Seydoux, O., Zellweger, M., Pfleiderer, M., Tardy, Y., et al., 2015a. Red and NIR light dosimetry in the human deep brain. Phys. Med. Biol. 60 (7), 2921–2937.

Pitzschke, A., Lovisa, B., Seydoux, O., Haenggi, M., Oertel, M.F., Zellweger, M., et al., 2015b. Optical properties of rabbit brain in the red and near-infrared: changes observed under in vivo, postmortem, frozen, and formalin-fixated conditions. J. Biomed. Opt. 20 (2), 25006.

Pope, R.M., Fry, E.S., 1997. Absorption spectrum (380-700 nm) of pure water. II. Integrating cavity measurements. Appl. Opt. 36 (33), 8710–8723.

Prahl, S.A., 1989. A Monte Carlo model of light propagation in tissue. Institutes for Advanced Optical Technologies. SPIE.

Roggan, A., Schroder, M.O., Muller, C.G., 1994. The determination of optical tissue properties with double integrating sphere technique and Monte Carlo simulationsSPIE Cell and Biotissue Optics .

Salomatina, E., Jiang, B., Novak, J., Yaroslavsky, A.N., 2006. Optical properties of normal and cancerous human skin in the visible and near-infrared spectral range. J. Biomed. Opt. 11 (6), 064026.

Saltmarche, A.E., Naeser, M.A., Ho, K.F., Hamblin, M.R., Lim, L., 2017. Significant improvement in cognition in mild to moderately severe dementia cases treated with transcranial plus intranasal photobiomodulation: case series report. Photomed. Laser Surg. 35 (8), 432–441.

Sandell, J.L., Zhu, T.C., 2011. A review of in-vivo optical properties of human tissues and its impact on PDT. J. Biophotonics 4 (11–12), 773–787.

Sankaran, S., Ehsani, R., 2014. Introduction to the electromagnetic spectrum. In: Manickavasagan, A., Jayasuriya, H. (Eds.), Imaging with Electromagnetic Spectrum: Applications in Food and Agriculture. Springer Berlin Heidelberg, Berlin, Heidelberg, pp. 1–15.

Shi, L., Sordillo, L.A., Rodriguez-Contreras, A., Alfano, R., 2016. Transmission in near-infrared optical windows for deep brain imaging. J. Biophotonics 9 (1–2), 38–43.

Simpson, C.R., Kohl, M., Essenpreis, M., Cope, M., 1998. Near-infrared optical properties of ex vivo human skin and subcutaneous tissues measured using the Monte Carlo inversion technique. Phys. Med. Biol. 43 (9), 2465–2478.

Sordillo, L.A., Pu, Y., Pratavieira, S., Budansky, Y., Alfano, R.R., 2014. Deep optical imaging of tissue using the second and third near-infrared spectral windows. J. Biomed. Opt. 19 (5), 056004.

Stolik, S., Delgado, J.A., Perez, A., Anasagasti, L., 2000. Measurement of the penetration depths of red and near infrared light in human "ex vivo" tissues. J. Photochem. Photobiol. B 57 (2–3), 90–93.

Sutherland, J.C., 2002. Biological effects of polychromatic light. Photochem. Photobiol. 76 (2), 164–170.

Tedford, C.E., DeLapp, S., Jacques, S., Anders, J., 2015. Quantitative analysis of transcranial and intraparenchymal light penetration in human cadaver brain tissue. Lasers Surg. Med. 47 (4), 312–322.

Vargas, E., Barrett, D.W., Saucedo, C.L., Huang, L.D., Abraham, J.A., Tanaka, H., et al., 2017. Beneficial neurocognitive effects of transcranial laser in older adults. Lasers Med. Sci. 32 (5), 1153–1162.

Wan, S., Parrish, J.A., Anderson, R.R., Madden, M., 1981. Transmittance of nonionizing radiation in human tissues. Photochem. Photobiol. 34 (6), 679–681.

Welch, A.J., van Gemert, M.J.C., Star, W.M., Wilson, B.C., 1995. Definitions and overview of tissue optics. In: Welch, A.J., Van Gemert, M.J.C. (Eds.), Optical-Thermal Response of Laser-Irradiated Tissue. Springer US, Boston, MA, pp. 15–46.

Wylie, G.R., Graber, H.L., Voelbel, G.T., Kohl, A.D., DeLuca, J., Pei, Y., et al., 2009. Using co-variations in the Hb signal to detect visual activation: a near infrared spectroscopic imaging study. Neuroimage 47 (2), 473–481.

Yarmolenko, P.S., Moon, E.J., Landon, C., Manzoor, A., Hochman, D.W., Viglianti, B.L., et al., 2011. Thresholds for thermal damage to normal tissues: an update. Int. J. Hyperthermia 27 (4), 320–343.

Yaroslavsky, A.N., Schulze, P.C., Yaroslavsky, I.V., Schober, R., Ulrich, F., Schwarzmaier, H.J., 2002. Optical properties of selected native and coagulated human brain tissues in vitro in the visible and near infrared spectral range. Phys. Med. Biol. 47 (12), 2059–2073.

Zhang, Q., Zhou, C., Hamblin, M.R., Wu, M.X., 2014. Low-level laser therapy effectively prevents secondary brain injury induced by immediate early responsive gene X-1 deficiency. J. Cereb. Blood Flow Metab. 34 (8), 1391–1401.

Zivin, J.A., Albers, G.W., Bornstein, N., Chippendale, T., Dahlof, B., Devlin, T., et al., 2009. Effectiveness and safety of transcranial laser therapy for acute ischemic stroke. Stroke 40 (4), 1359–1364.

Further reading

Huang, Y.Y., Hamblin, M.R., Taboada, L.D., 2010. Low-level laser therapy in stroke and central nervous system. In: Tuchin, V.V. (Ed.), Handbook of Photonics for Biomedical Science. CRC Press, Boca Raton.

Chapter 6

Near-infrared photonic energy penetration—principles and practice

Theodore A. Henderson and Larry D. Morries
Neuro-Laser Foundation, Centennial, CO, United States

6.1 Introduction

Near-infrared (NIR) light has a growing range of applications in medicine and shows particular promise as a nonpharmacological treatment for brain disorders. Much work has been done in small animal models to determine the optimal wavelengths, potential mechanisms of action, and molecular consequences of NIR phototherapy. However, a fundamental problem in NIR phototherapy has been how to "scale up" the therapy appropriately for clinical use in humans. Given that a mouse skull is 0.2 mm thick and the thickness of the entire mouse brain is 5 mm, scaling the treatment to be equally effective in a human with 10 mm of scalp and skull is problematic. Yet, the barriers with light penetration do not end at the calvaria. For example, Aulakh and colleagues (2016) examined this problem in an ex vivo pig head and found that not only was the scalp and skull a significant light barrier, but 90.8% of the remaining light energy was lost with every 5 mm of depth into the brain (Aulakh et al., 2016). They noted the use of more powerful laser emitters yielded deeper penetration. These NIR-tissue interactions will be explored in order to shed light on the barriers to clinically effective transcranial phototherapy. The behavior of laser light compared to that derived from a laser diode, as well as a light emitting diode (LED), will be explored to enhance understanding of the limits of current clinical techniques.

6.1.1 Understanding near-infrared light

The fundamental physical properties of light are relevant to its clinical use. Light is a form of electromagnetic radiation that has properties of both waves and particles. Light is characterized by its wavelength (distance between two peaks), frequency, and amplitude. Light is also characterized by its energy content. This energy is quantified as joules (J). The amount of energy delivered per unit time constitutes the power of light in watts (W = J/s). For medical applications, light is typically reported in terms of wavelength (nm), energy (J), irradiance or power density (W/cm^2), and radiant exposure [fluence or dose (J/cm^2)] (Rojas and Gonzalez-Lima, 2011; Jenkins and Carroll, 2011; Henderson and Morries, 2015b).

Coherence is a property of light waves in which waves of monochromatic light are aligned such that any point on the wave has the same amplitude and position as the equivalent point in an adjacent wave. Temporal coherence reflects the slight variations in the waveform over time. The more consistent the waveform is, the higher the temporal coherence. Monochromatic light typically has high temporal coherence. Spatial coherence is degraded by the divergence of the light from the point of emission (Karu, 2003). Laser has virtually no spatial divergence and creates a long narrow beam of coherent light. LEDs are not monochromatic, but emit light in a relatively narrow band on either side of the peak wavelength (Fig. 6.1). LEDs also have significant spatial divergence, and therefore a relatively wide volume of space is irradiated. The result is that noncoherent LED sources likely only provide coherent light in thin volumes, usually at surfaces (Karu, 2003). In contrast, laser generates a long narrow volume of coherent light which can penetrate deeper into tissues (Fig. 6.1).

LASER is an acronym for "light amplification by stimulated emission of radiation." Originally invented in 1960, LASER depends upon the use of atoms or molecules with electrons capable of being elevated into more excited states

Photobiomodulation in the Brain. DOI: https://doi.org/10.1016/B978-0-12-815305-5.00006-3

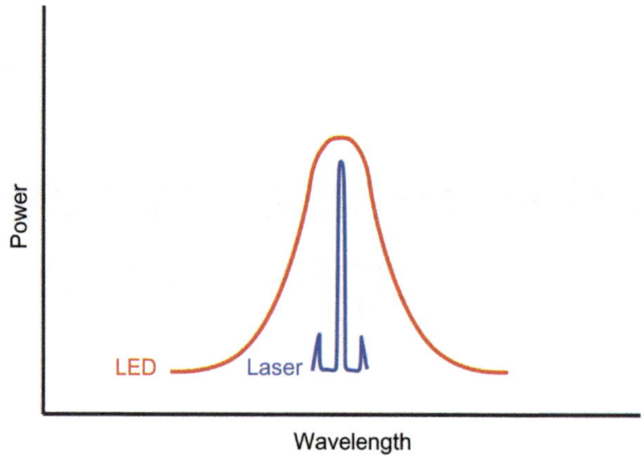

FIGURE 6.1 A comparison of emission spectra of light emitting diode (LED) emission relative to that of laser emission. LEDs provide a diverging conical beam of light wherein the energy and frequency of the light diverges with greater distance from the center of the beam. The full-width, half-maximum (FWHM) of an LED is in the range of 30 nm, while the FWHM of a laser is on the order of 2 nm.

by absorbing external energy. When electrons are in certain energy states in certain atoms, they can be induced to release energy in the form of a photon when dropping to a lower energy state. The resulting photon can collide with other atoms and cause the release of additional photons. The resulting photons have exactly the same energy and direction of travel as the impinging photon (due to the laws of conservation of energy and momentum) and, thus, one photon leads to two identical photons in an amplification process. In order to get a large population of electrons in the excited state capable of this amplification process, it is necessary to pump energy into the medium. With gas-based lasers, it is possible to excite electrons by applying an electric field. In the case of helium-neon (HeNe) lasers, the electric field excites the electrons in the helium, and the energy of the excited helium atoms is transferred to the neon atoms through collisions. The electrons in the neon atoms then undergo a transition to a lower energy state and release the coherent photons. Laser light has a very narrow spectral width—all of the photons are of the same energy and wavelength. Laser light has a narrow beam width which is mostly adjusted by the size of the emitter lens. Laser light is also highly coherent—the photons travel in identical phase with the waves of light.

A LED is capable of producing coherent laser light, but only under certain conditions. LEDs produce light mostly by the spontaneous emission of photons from solid state conductors. That is, when an electron drops from one energy level to another within an atom, a single photon of light is generated and LEDs take advantage of electron shifts in semiconductor materials (e.g., silicon). In nonlaser LEDs, the silicon is tainted with a small amount of another element (aluminum, indium, etc.) in a process referred to as "doping." Simply put, electrons are shunted from an atom with five valence electrons (N type) to an atom with three valence electrons (P type). The energy lost by an electron as it moves from an N-type atom to combine with a positive hole in a P-type atom dictates the wavelength of the light produced. LEDs created a conical beam of light which diverges further with distance from the source. Also, the emission spectrum of an LED is much broader than that of a laser. The spectral width is often 2–3 nm compared to the typical spectral width of 10^{-3} nm for a laser.

In a semiconductor laser diode, the typical example being a gallium-aluminum-arsenide (GaAlAs) diode, (an additional layer of gallium arsenide which is free of impurities) is sandwiched in between the P and N layers of the diode. This undoped layer is where the laser light is produced and is referred to as the active layer. Much larger amounts of energy are required to produce laser light compared to an LED. Indeed, laser cannot be produced until the diode is operating at or above 80% of its maximum current. A laser diode must be on the verge of burning out before laser light can be produced. At this level of current, the atoms of the P and N layers are highly energized. Electrons at higher energy states drop to lower states and release the energy as photons within the active layer. The photons bounce off the reflective walls of the active layer and in turn collide with atoms and produce more energized electrons. These, in turn, lose energy and release more photons. This process is referred to as "pumping." The photons, which are produced, all oscillate at a precise frequency. The initial output of photons is small, but it is reflected back onto the semiconductor material by the highly polished reflective ends of the active material. This reflected feedback generates more photons by stimulated emission. A laser diode operates at a much higher current, generally ten times that of a normal LED.

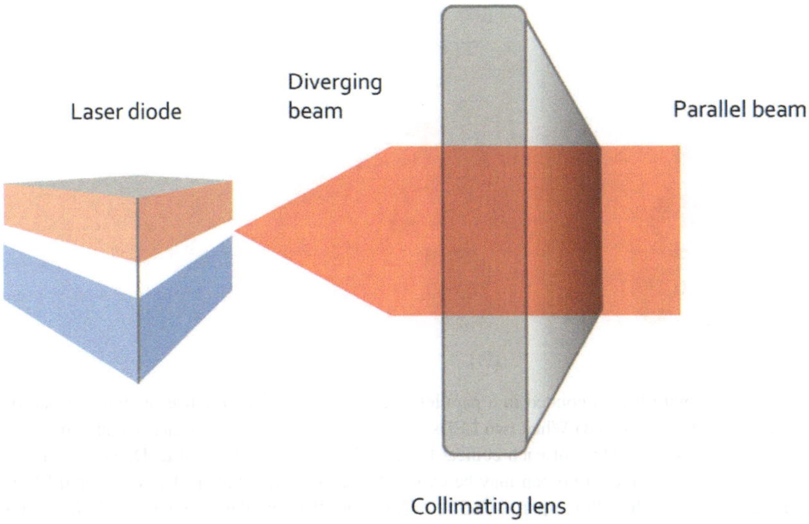

FIGURE 6.2 The light produced by a laser diode is divergent and elliptical and must be focused using a collimating lens.

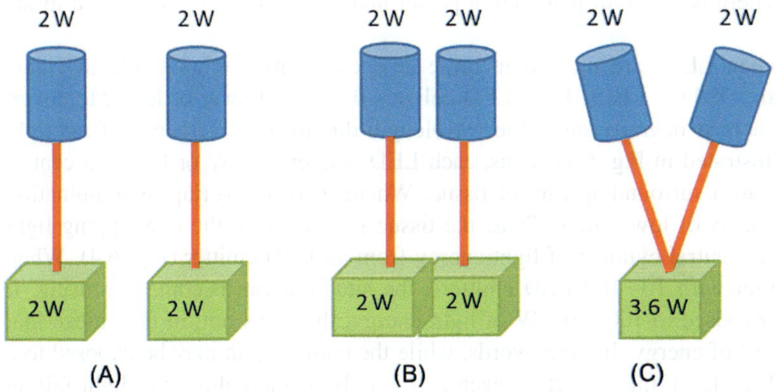

FIGURE 6.3 (A) Two lasers were aligned with the beams traveling in parallel and impinging on two light meters placed adjacent to one another. There is no additive effect. Each beam shines on its own cylindrical volume of tissue. (B) Even when the lasers were immediately adjacent to each other, there was no additive effect. (C) When the lasers were angled such that they were impinging on the same point on a single light meter, then an additive effect occurred, but it was not absolute or 100%.

The light beam produced by a laser diode is elliptical and must be focused into a parallel beam of light., see Fig. 6.2. This is accomplished with a collimating lens. Diverging light waves will begin to lose power and therefore the wavelength will drift. These waves at the edge of the diode emission are eliminated by the collimation process.

The output of light from LEDs and lasers is perhaps the most important distinction for the purposes of practical photobiomodulation. The infrared light must be able to reach the target tissue. Laser provides a coherent monochromatic cylindrical beam of light. The full-width, half-maximum (FWHM) of a laser is on the order of 2 nm. In contrast, LEDs provide a diverging conical beam of light wherein the energy and frequency of the light diverges with greater distance from the center of the beam. The FWHMs of LEDs are in the range of 30 nm. This issue becomes particularly important when looking at the effect of multiple sources of light, such as in LED panels. Some seem to believe that grouping multiple LEDs adjacent to one another produces a cumulative effect and delivers greater power to an area being treated. This is why many LED devices advertised for treating traumatic brain injury (TBI), described an array of many LEDs. In an effort to illustrate a presented logical flaw, we performed the following demonstrative experiment.

We examined the situation in which two light emitters are placed side by side as is the case in LED panels or LED treatment heads. When two lasers are aligned and the beams travel in parallel, there is no additive effect. A demonstration experiment is illustrated in Fig. 6.3. Each beam shines on its own cylindrical volume of tissue. Even if the lasers are immediately adjacent to each other, there is no additive effect. However, if the lasers are angled such that they are impinging on the same tissue, then an additive effect occurs—nonetheless, it is not 100%. Rather, we find that some decrease occurs: 1 + 1 does not equal 2 in this situation. Hence, with laser diodes grouped together in an array, there is not a cumulative effect. Each laser diode illuminates its own cylindrical column of tissue. Even when there is overlap due to angling of diodes, the result is not entirely additive. More importantly, the notion that 20 laser diodes each with a 0.5 W of power deliver 10 W of power to the brain is fallacious. Rather, what occurs is that each diode delivers

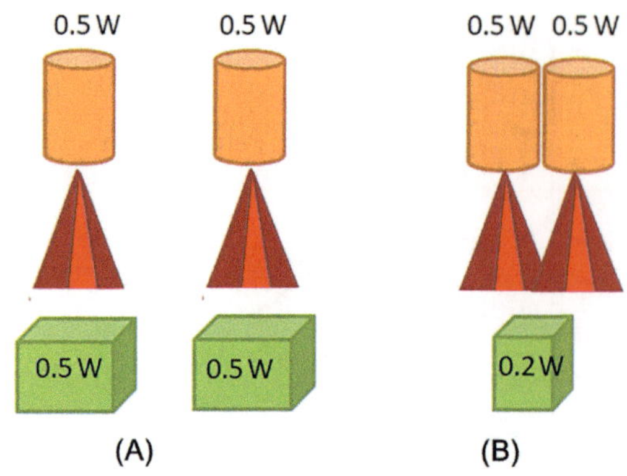

FIGURE 6.4 (A) Two light emitting diodes (LEDs) were aligned with beams pointed in a parallel direction. Each LED produces a conical beam of light with degradation of energy (shown in *dark red*) at the edges of the cone. (B) When two LEDs are placed adjacent to each other, the area of overlapping light energy consists of the overlap of the degraded light from the edges of each conical beam of light emitted by each LED. There is not a cumulative effect of increased light energy delivered to a tissue. While the entire organ may be exposed to a multiple number of low-powered LEDs (0.5 W or less), the impact on each cell is only that of the cylinder of light falling upon it (0.5 W or less), not the cumulative amount of light falling on the entire tissue over an area much larger than a single cell.

0.5 W of power to a given cylinder of tissue. There is 0.5 W of penetrating power and 0.5 W of energy delivered to the scalp (hair notwithstanding—see below).

The situation is even more pronounced when LEDs are utilized in these devices. Many commercially available devices have a panel of LEDs with both 810 and 980 nm LEDs. Each LED delivers 0.5 W of power or less. Moreover, an LED delivers a conical beam of light and there is degradation of the wavelength due to the divergence of the light energy at the edges of the conical beam (as illustrated in Fig. 6.1). Thus, each LED delivers 0.5 W or less to a central column of tissue and considerably less energy to a surrounding ring of tissue. Where there is overlap of illumination between two LEDs, the overlap occurs in the areas of low power. Thus, the tissue irradiated by the overlapping light energy is exposed to less energy than that of the central cylinder of light energy from an LED emitter (Fig. 6.4). When this model is scaled up to the size of an emitter with 10–100 LED emitters, the situation can become confusing. It would appear that the tissue or organ is being exposed to 100×0.5 W of light energy; this does not, in fact, mean that each cell of that organ is exposed to 100×0.5 W of energy. In other words, while the entire organ may be exposed to a multiple number of low-powered LEDs (0.5 W or less), the impact on each cell is only of the cylinder of light falling upon it (0.5 W or less), not the cumulative amount of light falling on the entire tissue over an area much larger than a single cell. This returns to one of the basic measures of NIR energy with which treatment is described. Energy delivered over time is the watt (W = J/s), power density is expressed as W/cm^2, and fluence is expressed as J/cm^2. Simply put, the area over which the light energy is delivered matters. While this concept is obvious to experts in the field, it seems to be misunderstood by countless commercial ventures and practitioners claiming cumulative benefits from LED arrays.

6.2 Light interactions with tissue

As a first step to explore the therapeutic application of NIR light, it is necessary to understand the physical interactions between tissue and light. The interaction of light with tissue results from both its properties of a wave and of a particle. These particles are, of course, photons. Photons can change direction without loss of energy. The photons impinging on a tissue can be reflected, refracted, scattered, absorbed, or transmitted through the tissue (Fig. 6.5).

6.2.1 Reflection and refraction

When light impinges on the surface, a portion of the energy is reflected (Steiner, 2011). The energy, which is transmitted through the surface, is refracted or bent toward a line perpendicular to the surface.

In NIR photobiomodulation this is a particularly important issue as NIR light must pass through multiple interfaces between different tissue types. At the border between two tissues, a phenomenon referred to as Fresnel reflection can occur. This occurs because of the difference between the refractive index (n) of each tissue. The amount of light that is reflected at a tissue interface represented by Reflectance (R) is equal to the square of the difference between the two refractive indices of the two tissues divided by the square of their sum (Jacques, 2013).

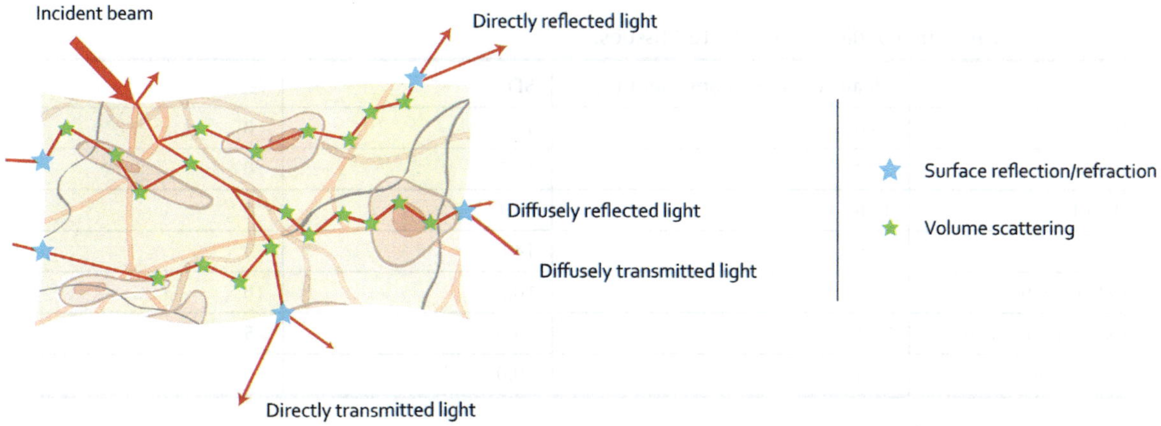

FIGURE 6.5 Light impinging on tissue is attenuated by reflection, refraction, scatter, and absorption. Considerable light is also scattered at interfaces between tissues.

TABLE 6.1 Refractive indices of cellular components.

Material	Refractive index (n)
Air	1.00029
Water	1.33
Extracellular fluid	1.35–1.36
Cytoplasm	1.36–1.375
Cell membrane	1.46
Cell nucleus	1.38–1.41
Mitochondria	1.38–1.41
Melanin	1.6–1.7

1. $R = \frac{(n_1 - n_2)^2}{(n_1 + n_2)^2}$

A simple way to think of refractive index is that it provides a numerical index of how much faster light would travel in a vacuum compared to the tissue. For example, the refractive index of water is 1.333. Thus, light travels 1.333 times faster in a vacuum than it does in water. For most tissues, the refractive index is in the range of 1.35–1.60. Doublik and colleagues (2013) provided refractive indices of several cellular components (Table 6.1). In the case of transcranial photobiomodulation, NIR light must pass through epidermis, dermis, subcutaneous fat, subcutaneous blood vessels, muscle (in some areas), aponeurosis, connective tissue, skull, dura mater, cerebrospinal fluid, and pia mater before reaching the surface of the cortex (Clemente, 1981). Each of these structures has different absorption and refraction properties, and each interface between different materials generates a separate Fresnel reflection (Jacques, 2013).

6.2.2 Scattering

Scattering can be defined as a change in direction of photon travel. Scattering can deflect a photon forward, backward, laterally, or some combination thereof. The greatest barrier to forward scattering is due to particles with a dimension equivalent or up to ten times the width of the wavelength of the incident light. Scattering increases the volume of tissue impacted by the light. Scattering is particularly likely at interfaces between different tissues. Scattering, together with reflection, refraction and absorption, contribute to shortening the distance to which light will travel through or penetrate into a tissue (Steiner, 2011; Lister et al., 2012; Wan et al., 1981).

TABLE 6.2 Scattering coefficient of selected tissues.

	Mean scattering coefficient (μ_s)	SD	N
Skin	46.0	13.7	8
Brain	24.2	11.7	8
Breast	16.8	8.1	8
Bone	22.9	14.6	3
Other soft tissues	18.9	10.2	18
Other fibrous tissues	27.1	5.0	5
Fatty tissue	18.4	9.0	6

Cellular organelles and fibrous tissue structures contribute to light scattering. Even proteinaceous cytoplasm has a higher refractive index compared to extracellular fluid. Collagen and elastin fibers also can scatter NIR light. Collagen can form individual fibrils, sheets, or bundles, which can have differing refractive indices.

The effects of a given tissue on scattering can be expressed as a scattering coefficient (μ_s). The scattering coefficient is equal to the fraction of light energy dispersed from a light beam per unit of distance, typically per cm. The larger the scattering coefficient, the larger the portion of light which is scattered over the distance of 1 cm of tissue. Scattering coefficient of selected tissues is shown in Table 6.2. For example, skin has a scattering coefficient of 46.0. Of course, this varies depending on where on the human body the skin is irradiated. The average skin thickness is 2 mm, but the scalp is approximately 8 mm thick. When photons pass through the skin, approximately half of the photons are scattered. Some are scattered backwards (back scatter), some are scattered laterally, and some are scattered in the forward direction (forward scatter). The scattering coefficient of bone is much lower at about 22.9. Fewer photons are scattered by bone compared to skin. However, the interface between scalp and skull is a source of additional scattering and reflectance. The brain has a scattering coefficient of 24.2, also causing less scatter than skin (Table 6.2).

Scattering has been studied in a wide variety of tissues and this work was recently reviewed by Jacques (2013). Scattering is often described by the terms "Rayleigh scattering and Mie scattering." Rayleigh scattering refers to scattering by small particles or mass density fluctuations much smaller than the wavelength of light, and Mie scattering refers to scattering by particles larger than the wavelength of light. Scattering in various tissues is plotted in Fig. 6.6.

Flowing blood presents a particularly important source of scatter. Both water and hemoglobin will absorb NIR light. The peak of water absorption is approximately 1000 nm, while that of hemoglobin is dependent upon its oxygenation status. Deoxyhemoglobin found in veins has absorption peaks at 550, 758, and 910 nm, while oxyhemoglobin has peaks at 418, 542, 577, and 925 nm (Douplik et al., 2013). Cellular components, such as white blood cells that contain nuclei, create additional scattering. The movement of blood also affects NIR light. Depending on the direction of blood flow, the frequency of the NIR light will increase or decrease due to a Doppler shift effect (Douplik et al., 2013).

6.2.3 Absorption

Most tissues have the capacity to absorb light energy. Usually this is mediated by a molecule absorbing a photon. Molecules containing metal ions have a strong capacity for absorbing photonic energy, but conjugated molecules such as porphyrins and flavins, DNA, and water also can absorb light. The absorption of energy can induce a change in the conformation and/or function of the molecule. One example of this is the effects of ultraviolet light on collagen and elastin leading to aging of the skin.

At the target tissue, absorption underlies the positive benefits that are derived from NIR photobiomodulation. The purported effects of NIR are illustrated in Fig. 6.7. Light in the wavelength range of 600–1200 nm has significant photobiomodulation capability (Karu and Kolyakov, 2005). Current data support the absorption of NIR photons by cytochrome c oxidase (COX) in the mitochondrial respiratory chain as the key initiating event in photobiomodulation (Rojas and Gonzalez-Lima, 2011; Henderson and Morries, 2015b; Karu and Kolyakov, 2005; Chung et al., 2012; Henderson, 2016; Salehpour et al., 2018). COX is a large transmembrane protein of the inner mitochondrial membrane. It contains two copper centers (Cu) and two heme-iron centers. These copper centers have different light absorption

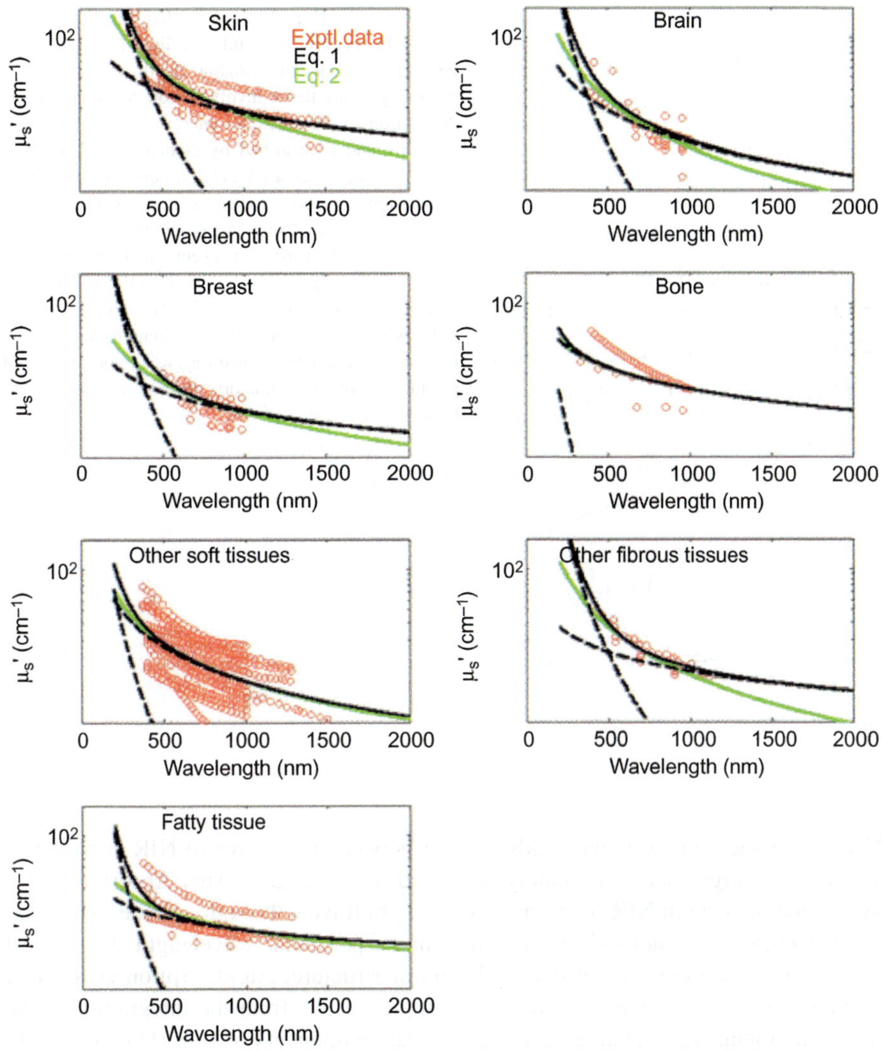

FIGURE 6.6 Scattering coefficient data summarized and reviewed in Jacques (2013). Seven different tissues are presented. Data from multiple sources that are summarized in the red circles. The lines represent fit lines derived from different equations for deriving the scatter coefficient. *From Jacques, S.L., 2013. Optical properties of biological tissues: a review. Phys. Med. Biol. 58 (11), R37–R61,*

peaks. Reduction of CuA occurs with 620 nm, oxidation of CuA occurs with 825 nm, reduction of CuB occurs with 760 nm, and oxidation of CuB occurs at 680 nm (Rojas and Gonzalez-Lima, 2011). These peaks correspond to the "optical window" associated with the biological effects of NIR. Irradiation of COX increases the activity of the entire electron transport chain producing more adenosine triphosphate (ATP). In addition, COX is autoinducible and its gene expression is activity-dependent, such that NIR irradiation may increase the amount of available COX over time (Wong-Riley et al., 2005). The effects of NIR have been studied in isolated mitochondrial preparations. Irradiation with 632 nm light results in increased proton electrochemical potential and increased ATP production (Passarella, 1989). COX activity increases with NIR irradiation, leading to activation of several electron transport chain components and increased oxygen consumption (Pastore et al., 2000; Yu et al., 1997).

During NIR photobiomodulation, absorption of red or NIR photons by COX and other less well-characterized mediators causes secondary molecular and cellular events, including activation of second messenger pathways, changes in nitric oxide (NO) levels, and growth factor production (Henderson and Morries, 2015b; Chung et al., 2012). When applied to the brain, NIR leads to reduction of excitotoxicity, the production of neurotrophic factors, the modulation of reactive oxygen species, the transcription of new gene products with protective or proproliferative properties, and the release of numerous growth factors for neurons and other cells (Chung et al., 2012; Henderson, 2016; Liang et al., 2008; Xuan et al., 2015; Leung et al., 2002; Chen et al., 2011; Frank et al., 2004; Lubart et al., 2005; Mirsky et al., 2002). NIR appears to initiate a cascade of subcellular events which can yield immediate, delayed, and persistent beneficial changes in the injured neurons or other cells.

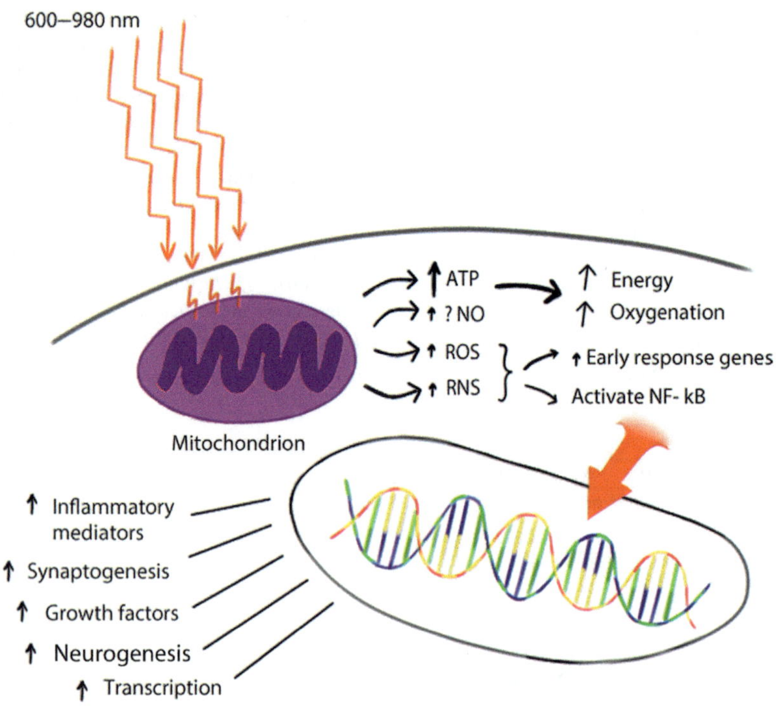

FIGURE 6.7 Hypothesized mechanism of action of NIR light therapy. Near infrared light (600–980 nm) penetrates tissue to variable depth depending on wavelength, the tissue involved, coherence, and time. A fraction of the photonic energy reaches the mitochondria and is absorbed by cytochrome C oxidase. This activates increased ATP production, increased production of reactive oxygen species (ROS), reactive nitrogen species (RNS), and possibly increases nitric oxide (NO). Downstream events include increased early response genes—cfos, cjun, and activation of nuclear factor kappa B (NF-kB), which in turn induces increased transcription of gene products leading to synaptogenesis, neurogenesis, and increased production of inflammatory mediators and growth factors.

On the other hand, absorption of NIR in the intervening tissue, fluids, and air between the source of NIR and the target tissue can markedly reduce the amount of energy that is ultimately delivered to the target tissue. Absorption also underlies tissue heating. Tissue heating is a major issue in NIR phototherapy. As light travels through a tissue, particular molecules or structures will absorb light energy. Each molecule or structure has a particular wavelength that is best absorbed. This can be defined as the absorption coefficient μ_a of that molecule or structure. An absorption coefficient can be defined for a tissue as well. The absorption coefficient of skin, for example, results from the integration of the absorption coefficients of melanocytes, hemoglobin, water ,and the other structural components of skin (Douplik et al., 2013). Without proper technique, NIR light can cause burning or drying of the skin.

6.2.4 Penetration

It is clear that penetration of NIR through tissues is determined by several factors: wavelength, energy, attenuation coefficient (composed of scattering, refraction, and absorption), area of irradiance, coherence, and pulsing. The attenuation coefficient of a tissue characterizes how easily it can be penetrated by a beam of light. A large attenuation coefficient indicates that the light beam is quickly attenuated or weakened as it passes through the tissue. In contrast, a small attenuation coefficient indicates that light passes through the tissue with little loss of energy. In general, longer wavelengths (up to 1000 nm) will penetrate deeper; however, the absorption of water begins to predominate above 1000 nm (Steiner, 2011; Jacques, 2013).

Increases in power density, in general, will lead to greater penetration. More photons will traverse the tissue. Aulakh and colleagues (2016) illustrated this point in porcine brain. The penetration of light into brain tissue from a 1.5 W NIR emitter was greater than that from a 0.5 W emitter. The area of surface irradiation also affects penetration due to scattering effects. Pulsing of NIR also increases the depth of penetration and the amount of energy delivered to any given point at the peak of a pulse. Yet, pulsing allows for troughs of energy output, such that the overall energy delivered to the tissue can be equivalent or even lower than that delivered by a continuous emission. We have previously shown that pulsing at 10 Hz reduced the dose of light energy to the surface by 50%, but virtually the same amount of energy was delivered to the depth of 3 cm (Henderson and Morries, 2015b). We also found that pulsing greatly enhanced penetration through skin. Pulsing is a property of lasers and laser diodes, which cannot easily be duplicated by LEDs.

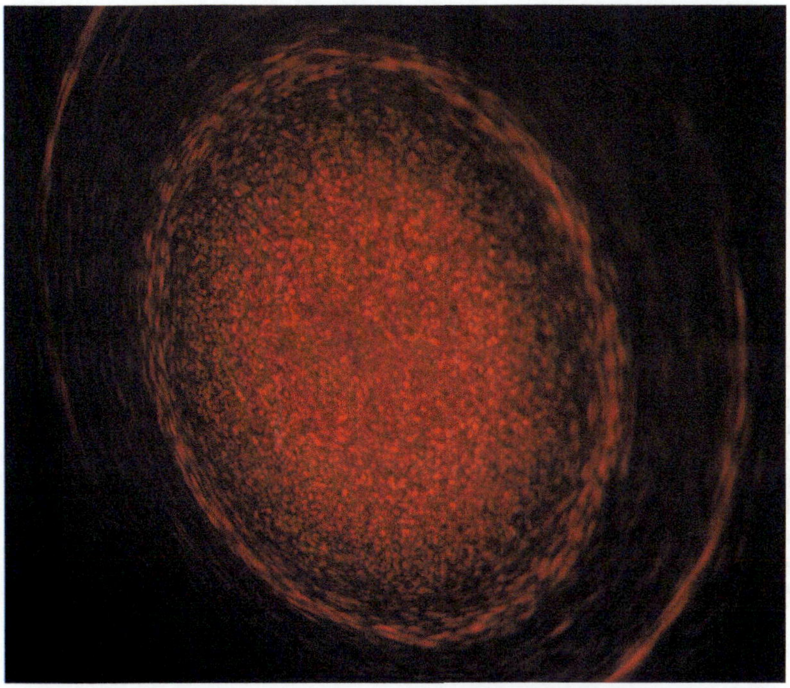

FIGURE 6.8 An 810 nm laser was projected onto a flat surface through one meter of air and was photographed with an infrared camera. The speckle of coherent light can be seen in the distribution of light with areas of increased intensity interspersed among areas of decreased intensity.

6.2.5 Speckling

A final aspect of the interaction of NIR light with tissue is the effect of scattering on coherent light. When coherent light enters a tissue, slight distortions in the timing and the shape of the waves occur. As a result, interference can occur between the waves. Polarization, the angle at which a wave is vibrating, also contributes to interference. On a single wave basis, interference results when the amplitude of the wave at a given point is different from that of an adjacent wave and of the population of coherent light waves. At the point of difference, the amplitudes can either cancel each other out $\{[+x]+[-x]\}$, be additive $\{[+x]+[+x]\}$, or any variation in between $\{[+x]+[-y]\}$. The result of these interactions is a field of randomly distributed points of increased and decreased light intensity, referred to as a speckle intensity pattern, as illustrated in Fig. 6.8. Speckling can have a significant impact on the effective penetration depth (Hode, 2005). As such, areas of high intensity will penetrate further or will have two to three orders of magnitude greater energy at a given depth (Hode, 2005).

The physical mechanisms that influence the penetration of light energy through materials are detailed here. Reflection, refraction, scattering, and absorption all combine to attenuate the energy of incident light as it passes through or fails to pass through a material. Differing wavelengths have differing coefficients for each of these parameters. Ultimately, the ability of light energy to penetrate an anatomical structure is mitigated by the wavelength, power density, coherence, and the characteristics of the tissue. We contend that the physical parameters described above account for the limited benefit of low-energy NIR light in clinical settings. Additionally, the very nature of light emission from an LED limits its effectiveness in penetrating tissues. Moreover, the edges of the cone of light energy emitted by an LED are likely to be of a different and lower energy than that found at the center of the cone of LED light emission. To illustrate these physical parameters in a simple fashion, we examined light penetration through ballistic gel with and without scattering additives. It is evident from Fig. 6.9, that even a low power (5 mW) laser is able to penetrate 8 cm of ballistic gel and continue penetrating the air beyond (Fig. 6.9A). In contrast, it is self-evident that the light emission of an equivalently powered LED penetrates less than 35% of the distance into a clear ballistic gel (Fig. 6.9B). Also, the conical nature of LED light emission is clearly evident. The extensive and diffuse scattering of the LED light emission can be appreciated. We then prepared ballistic gel with salt crystals and vigorous stirring to create a suspension of air bubbles and salt crystals within the gel. These created extensive scattering and absorption (Fig. 6.10). As can be seen in Fig. 6.9C, the 5 mW laser beam remained coherent through at least 70% of the gel. On the other hand, light energy from the LED is rapidly scattered and the light does not penetrate more that 25% into the gel (Fig. 6.9D).

The behavior of these two different forms of light in the ballistic gel illustrates an important distinction between coherent light and noncoherent light. Some have suggested that there is no difference in the scattering effects of tissue

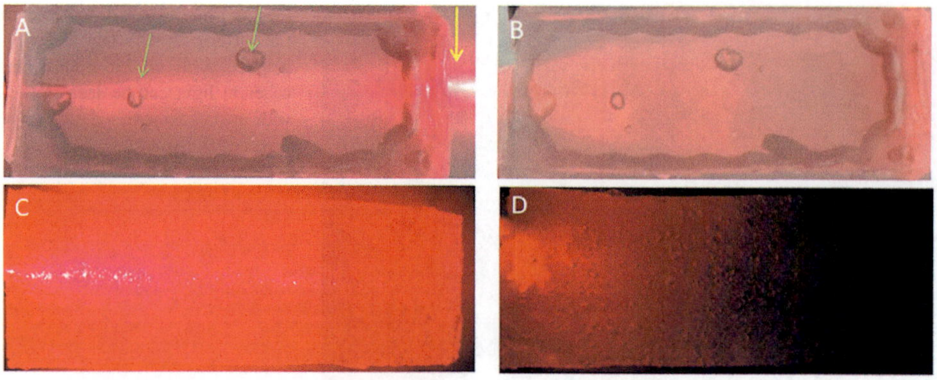

FIGURE 6.9 Illustration of the beam of laser light versus LED emitted light. (A) Path of light from a 620–670 nm 5 mW laser in ballistic gel (Humimic Medical, Fort Smith AZ, USA) with no impurities. Gel measures 8 cm in length. Light beam from laser is highly coherent and focused over at least 50% of the length and is still relatively focused upon exiting the gel (*yellow arrow*). Bubbles (*green arrows*) are not, in fact, within the gel but occur at the interface between the gel and the substrate. (B) Path of (approx.) 640–670 nm 5 mW LED through clear ballistic gel. Note the conical spread of the light and the scatter of the light with failure of penetration by approximately 65% of the length of the ballistic gel. (C) A model of attenuation (scatter and absorption) was created by preparing the ballistic gel with salt crystals (2% by weight). The block measured 8 cm in length. Path of the same laser through 8 cm of ballistic gel contained salt crystals and air bubbles. The laser beam remains coherent through at least 70% of the gel. Scatter reduces the depth of penetration and no laser light is seen emitted from the other end of the gel. (D) The path of light emitted from an LED through gel containing salt crystals and air bubbles. The light is highly scattered and penetration was no greater than 25%.

FIGURE 6.10 Magnified view of ballistic gel containing salt crystals and air bubbles to simulate tissue scatter and absorption.

on coherent and noncoherent light. The expectation is that laser light will be highly scattered and "light up the whole brain." However, observation of the laser in Fig. 6.8C reveals that coherent light does indeed have a greater penetrating ability into a complex scattering medium.

6.3 Infrared light—on a journey to the brain

6.3.1 Penetration of skin

Skin represents a substantial barrier to NIR light energy penetration. Skin is composed of layers—the major layers are the epidermis and dermis. The epidermis is composed of multiple surface layers which are keratinized to provide a protective layer. In contrast, the dermal layer consists of dense fibroelastic connective tissue which also contains glands and hair follicles (Fig. 6.11). Directly below the dermis is a subcutaneous layer of fat which varies widely in thickness.

There are five sublayers of the epidermis—stratum corneum, stratum lucidum, stratum granulosum, stratum spinosum, and stratum basale (Junqueira and Carneiro, 1982; Leeson and Leeson, 1979). Keratinocytes produced in the stratum basale migrate in a superficial direction and eventually are sloughed off. The epidermis varies in thickness for 50 microns to 150 microns depending on the location.

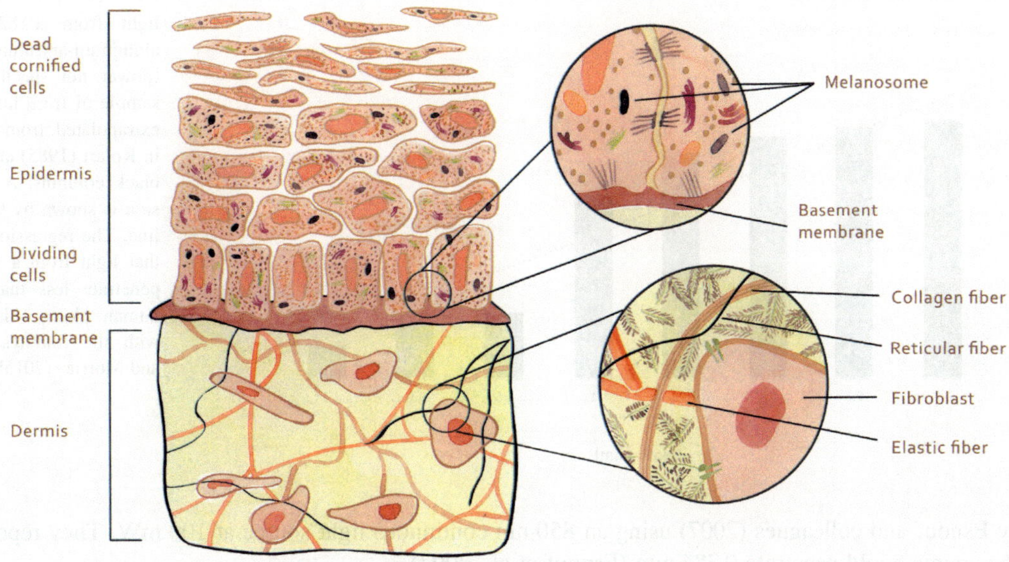

FIGURE 6.11 Diagram illustrating the composition of skin. There are five sublayers of the epidermis—stratum corneum, stratum lucidum, stratum granulosum, stratum spinosum, and stratum basale. Melanosomes, cell nuclei, collagen bundles, and other connective tissue fibers form substantial barriers to light penetration.

The dermis is composed of a variety of cells, connective tissues, and dermal structures. Fibroblasts are most plentiful, followed by mast cells, histiocytes, Langerhans cells, lymphocytes, and endothelial cells. All of these cells are contained within a matrix of collagen and elastin with a complex network of fibers. Dermal structures, such as hair follicles and glands, vary in their density depending upon location (Junqueira and Carneiro, 1982).

As is evident from this description, the skin presents a complex and heterogeneous structure through which infrared light must pass to reach deeper tissues. As has been discussed, the optical properties of a barrier are characterized by the tendency of each layer to cause light absorption and scattering, as well as the scattering and reflection that occurs at the interface of each layer. Adding to this is the highly variable contribution of hemoglobin and other blood-related chromophores, as well as the pigmentary chromophores of the individual patient. Nonetheless, certain general principles concerning the NIR absorption, scattering, and penetration in skin can be ascertained.

The epidermis is composed of living and nonliving layers of keratinocytes. The deep, living layers of epidermis contain most of the melanocytes and particles of melanin. Melanosomes are large melanin particles contained within melanocytes that produce significant forward scattering. Melanosomes produce diffuse scattering and that scattering is greater in those with greater pigmentation.

The complex fibrous network of the dermal layer with bundles of collagen fibers and intersecting elastin fibers, creates an interlacement of fibers and lamellae. Light energy is scattered by single fibers, intersections of fibers, and at layer interfaces.

The vascular structures in the dermis where blood, water, hemoglobin, bilirubin, and carotene are concentrated have significant NIR absorption. In addition, the effects of blood flow in the dermal layer are an incomplete studied aspect of this structure.

The subcutaneous fat layer consists of connective tissue, vascular structures, and adipocytes. The spherical droplets of lipids within adipocytes, which vary from 15 to 250 μm, create scattering of NIR light energy. In the scalp, this layer is thicker. Underneath this layer in the scalp is a dense fibrous layer referred to as the aponeurosis. The optical properties of the aponeurosis are not fully appreciated.

One of the earliest studies of NIR light penetration and human skin examined transmittance of 633 nm light through progressively thicker sections of human skin (Kolari, 1985). Transmittance through 0.4 mm of epidermis was 78% for 633 nm light, but with 2 mm thick skin, the energy had dropped to approximately 5% of the incident 633 nm light (Fig. 6.12). Bjordal and colleagues (2003) also concluded that 90% of the energy from 632 nm laser is lost in the skin. NIR light at 820 nm has slightly greater skin penetration. Approximately 89% of 820 nm light will penetrate 0.4 mm of epidermis and about 13.5% traverses 2 mm of skin, but 0% reaches a depth of 3 mm (Kolari and Airaksinen, 1993). Others found 80% of the energy from an 820 nm source is lost in the skin (Bjordal et al., 2003). Human skin also was

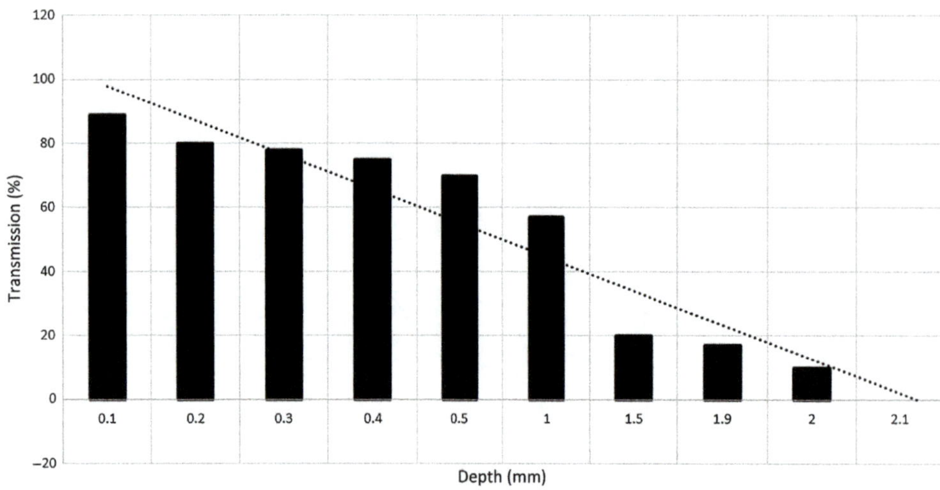

FIGURE 6.12 Penetration of light from a 820 nm gallium-aluminum-arsenium laser diode (power not specified) through a sample of fresh human skin. Data extrapolated from data presented in Kolari (1985) and shown in the black columns. A line of regression is shown by the black dotted line. The regression line indicates that light from a laser diode can penetrate less than 2.2 mm into human skin. This is consistent with the findings of Henderson and Morries (2015b).

examined by Esnouf and colleagues (2007) using an 850 nm continuous light source at 100 mW. They reported 34% of light from the source could penetrate 0.784 mm (Esnouf et al., 2007).

We examined NIR light penetration through unpreserved human skin and unpreserved sheep skin using several different NIR light sources (Henderson and Morries, 2015b). We showed considerably less penetration by low-level NIR light. We found energy from a 50 mW 810 nm LED did not penetrate 2 mm of human or sheep skin. No energy could be detected penetrating either human skin or sheep skin from a commercially available 200 mW LED (650 + 880 nm). In contrast, 9% of the energy from a 10 W 810 nm continuous mode infrared laser passed through 2 mm of skin (human or sheep). A 15 W 810 laser in continuous mode delivered 33% of its energy through 2 mm of skin. While longer wavelength is typically associated with greater penetration, this does not appear to be true for the complex anatomical structure of skin. Some have reported better transmission of 810 nm light compared to 980 nm light (Henderson and Morries, 2015b). For example, only 14% of the energy from a 15 W 980 nm laser in continuous mode penetrated 2 mm of sheep skin, while 33% of 810 nm light passed through the same skin sample. Using pulsed infrared light yielded much greater penetration. For example, 41% of the power density of an 810 nm infrared laser with a pulse frequency of 10 Hz penetrated 1.9 mm of human skin compared to only 11% of the continuous mode light of similar parameters. A 15 W 810 nm laser with a pulse frequency of 10 Hz delivered 69% of the energy through 1.9 mm of human skin compared to only 17% of the continuous mode light of similar parameters (Henderson and Morries, 2015b). This also illustrates the point that greater power yields greater penetration as Aulakh and colleagues (2016) demonstrated.

Simply put, it does not matter how long an LED is shone on a human head if the light energy cannot penetrate further than 3 mm. The LED-derived energy will not penetrate the thickness of the scalp. Very little infrared energy will be delivered to the brain and LED-derived energy will certainly not penetrate to depths of 3 cm or more. Some have suggested that energy from LEDs penetrate deeper with longer exposure times. This reflects a fundamental misunderstanding of the difference between energy delivered to the skin surface and energy reaching the target tissue, which is often several cm below the surface. If the line of regression shown by data provided by Kolari (1985) is correct (Fig. 6.11), then the energy of a laser diode is stopped at about 2.1 mm of human skin. Longer exposure times simply put more energy into the epidermis and dermis of the skin. They do not yield deeper penetration.

There is more. The skin is not the only barrier to light penetration.

6.3.2 Penetration of skull

To reach the brain, it is necessary to not only penetrate the overlying scalp, but the bone of the skull must be penetrated as well. Penetration through mouse skull (0.2 mm) (Choi et al., 2007) and overlying skin with 800–810 nm light from a 0.5 W LED emitter, ranges from 6.3% to 46% (Khuman et al., 2012). Fitzgerald and colleagues (2013) modeled light penetration into the human brain for a 670 and a 1064 nm LED light source emitting 28 mW/cm^2. Jagdeo and colleagues (2012) examined the penetration of low power NIR light through an isolated human skull from a cadaver and found that it was limited. Only 7.4% of the light energy from an 830 nm 0.5 W continuous LED device (Omnilux) penetrated the bone of human skull. Moreover, they found that no more than 0.5% of NIR at 830 nm penetrated approximately 10 mm of the frontal skull and overlying tissue in a cadaver model. Also, no detectable NIR energy from the

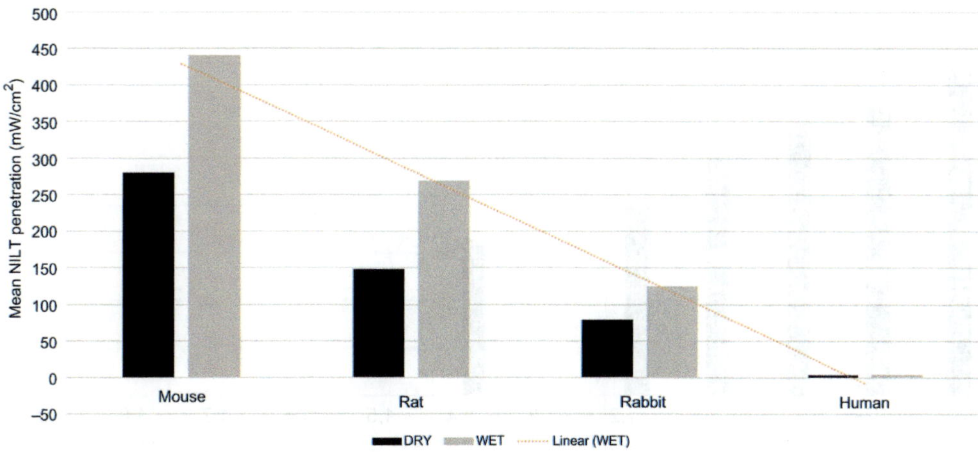

FIGURE 6.13 NIR light penetration through the skull of three animal species and human cadaver. Line of regression shown as dotted line. *Adapted from data provided in Lapchak, P.A., Boitano, P.D., Butte, P.V., Fisher, D.J., Hölscher, T., Ley, E.J., et al., 2015. Transcranial Near-Infrared Laser Transmission (NILT) profiles (800 nm): systematic comparison in four common research species. PLoS One 10 (6), e0127580.*

same light source penetrated the temporal bone and overlying tissue (Jagdeo et al., 2012). These studies of low-level light therapy (LLLT) and the computer model (Fitzgerald et al., 2013) indicate that photonic energy from low power LED emitters does not appear to deliver significant fluence to the depths required to treat the human brain.

Could this be a failing of only LEDs, or are low level laser light emitters similarly hampered? The clinical trials of LLLT using a laser diode device for the acute treatment of stroke delivered 808 nm NIR light at 70 mW and 268 J/cm^2 to the scalp (Lampl et al., 2007). The authors estimated that they delivered 10 mW/cm^2 or 1.2 J/cm^2 to the human cortical surface. This represents less than half of 1% of the light energy delivered to the scalp reaches the subjacent cortical surface. Lapchak and colleagues (2015) examined the penetration of NIR light through the skull of three animal species commonly used in NIR studies, as well as human skull samples (Fig. 6.13). They used a dual wavelength laser (800 and 970 nm) and applied 700 mW to the surface of the skull and measured the light penetration. For the mouse skull (0.44 mm thick), 40.1% of the light applied to the surface of the skull penetrated to the other side. In the case of the rat skull (0.83 mm thick), only 21.2% of the light applied penetrated the skull. In the case of rabbit skull (2.11 mm thick), the amount of NIR light penetrating the bone dropped to 11.4%. In the case of human skulls, two points were examined—the bregma and a section of the parietal bone. The thickness of the human skull varied from 5.9 to 7.2 mm. The amount of incident NIR light that penetrated the human skull was only 4.2%. Lapchak and colleagues (2015) calculated the incident power density to be 700 mW/cm^2 and the power density at the opposite surface of the human skull would have dropped to 29 mW/cm^2.

Ando and colleagues (2011) examined NIR light energy penetration of the mouse scalp and the combination of the mouse scalp and skull. Using a laser diode at 810 nm, they applied progressively higher power to the scalp surface from 50 to 500 mW and measured light energy transmitted to the other side of the tissue. They found approximately 15% of the energy applied to the scalp surface reached the other side of the scalp. Penetration through scalp + skull was considerably less with only 6% reaching the subjacent surface of the skull. Correlating these results with that of Lapchak and colleagues (2015) above, it is again apparent that the skin of the scalp represents a formidable barrier to NIR light energy, given that 40% of the NIR energy is capable of penetrating the mouse skull.

Lychagov and colleagues used a 1 W 810 nm laser to study penetration of NIR light through a human cadaver scalp and skull (Lychagov et al., 2006). They examined five areas—frontal bone at the forehead, temporal bone at right and left temple, occipital bone at occiput, and the vertex (Fig. 6.14). They confirmed that transmission of NIR energy decreased with increased sample thickness. They documented that the energy transmitted through the combination of scalp and skull was approximately one-third of that transmitted through the same thickness of skull.

6.3.3 Penetration of heterogeneous tissues

The question of penetrating the human scalp, skull, and brain has been explored in computer models. In a diffusion simulation computer model delivering 28 mW/cm^2 as described above, Fitzgerald and colleagues (2013) derived that at the center of the brain (10 mm skull + 46 mm brain = 56 mm), power density would be approximately 1.2×10^{-11} W/cm^2

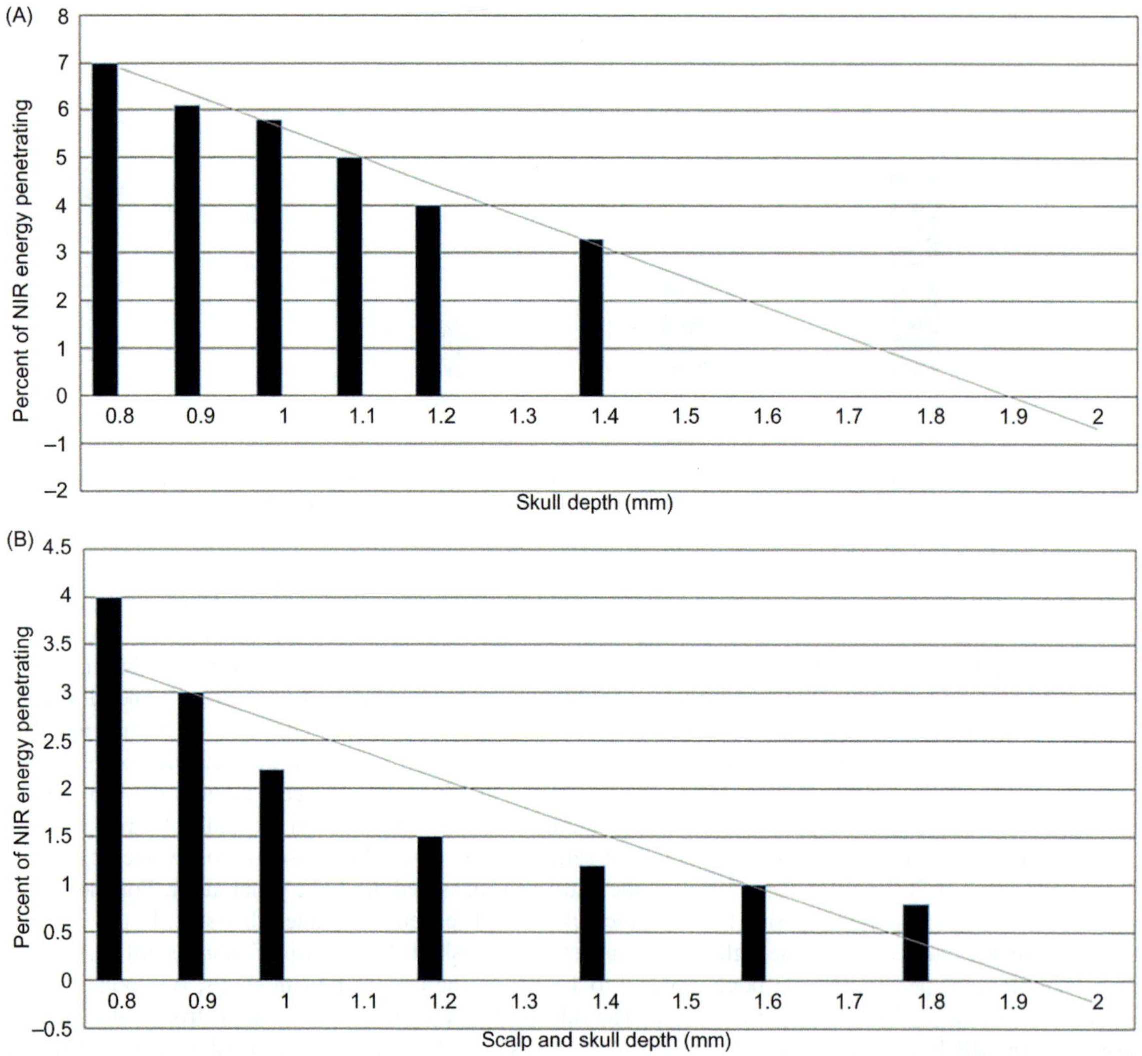

FIGURE 6.14 (A) Percent of 810 nm at 1 W delivered to the surface of the skull which penetrates to the internal surface of the skull. Line of regression shown as gray line and indicates light at this power cannot penetrate more than 2 mm. (B) Percent of 810 nm at 1 W delivered to the surface of the scalp which penetrates both the scalp and skull to the internal surface of skull. Line of regression shown as gray line and indicates light at this power cannot penetrate more than 2 mm. *Adapted from data provided in Lychagov, V.V., Tuchin, V.V., Vilensky, M.A., Reznik, B.N., Ichim, T., DeTaboada, L., 2006. Experimental study of NIR transmittance of the human skull. In: Tuchin, V.V. (Ed.), Complex Dynamics and Fluctuations in Biomedical Photonics III. Proc of SPIE V, vol. 6085, pp. 60850T1–60855T1.*

for 670 nm light and 1.4×10^{-7} W/cm^2 for 1064 nm light. This falls below the fluence range predicted to have a biological benefit. Lapchak (2010) reached similar conclusions about NIR light penetration into the human brain based upon the fate of the NEST trials. Therein, NIR laser light of 808 nm wavelength with infrared energy densities of 0.9 J/cm^2 was applied to the human scalp for a total of 40 minutes with applications at multiple sites during that time (Lampl et al., 2007; Zivin et al., 2009; Huisa et al., 2013). Lapchak (2010) asserted that the light energy failed to reach adequate depth with adequate energy to affect a clinically detectable change in patients with acute stroke.

Thicker segments of tissue have been studied in the laboratory by several groups. Khuman and colleagues (2012) used an 800 nm LED emitter at 0.5 W in a mouse model. Over 95% of the light was attenuated in passing through the structures of a mouse head (scalp, skull, and brain) which encompassed less than 10 mm. Byrnes and colleagues (2005) examined penetration of 810 nm continuous light from a 150 mW source through skin, muscle, bone, and spinal cord of the rat (approximately 24 mm). They found 6% of the energy penetrated these tissues. Similarly, Giacci and colleagues (2014) measured photonic energy transmission (670 and 830 nm) through an unspecified thickness of skin, muscle, and bone overlying the spinal column of the rat, but this distance was probably also approximately 24 mm. They found 6.6% of 670 nm light from a 500 mW emitter penetrated this distance. Penetration of 830 nm light was slightly greater at 11.3% (Giacci et al., 2014). Hudson and colleagues (2013) precisely measured photonic energy at varying depths of bovine muscle using a 1 W 808/980 nm emitter. They found 808 nm light had greater penetration, despite the premise that longer wavelengths travel further. The attenuation of 808 nm light at 3 cm was 99%.

NIR light transmission through living tissue is different than through postmortem tissue for a number of reasons. First, cross-linking of proteins is an early and progressive event in death. Second, changes in interstitial fluid occur within hours of death. Third, the perfusion of dermis and deeper tissues in vivo creates scattering and refraction of NIR light. Fourth, the flow of blood also disperses heat from the site of NIR application. Lastly, some authors have suggested that NO created at the site of irradiation and carried throughout the body in the blood is responsible for the beneficial effects of NIR phototherapy (Samoilova et al., 2004). This effect could account for the clinical benefits seen in TBI if NIR light does not penetrate to the depths of 3 cm or greater in living tissue.

Jagdeo and colleagues (2012) had previously modeled NIR penetration in living human tissue by serving as their own author/volunteers. For example, in a living human hand, Jagdeo and colleagues (2012) found only 0.01%–0.09% of 830 nm light from a 0.5 W emitter penetrated 25 mm. We have replicated some of their work using a higher power NIR light source (Henderson and Morries, 2015b). Hand, cheek, ear, and tricep tissues were used to model the complex mixture of skin, bone, connective tissues, and other tissues found in the pathway of NIR light en route to the brain. For example, applying 13.5 W of 810 nm laser to 20 mm of a thick segment of skin and subcutaneous flesh yielded only 0.005 W of light energy penetration over that distance (0.32%) (Henderson and Morries, 2015b). Pulsed NIR appeared to have greater penetration in living tissue, which stood in contrast to our findings in postmortem tissue. Only 0.6% of continuous wave NIR energy passed through 2.5–3.0 cm of the human hand, while 0.8% of pulsed NIR penetrated the same distance. Penetration also appeared to be greater in living tissues that contained bone (human hand) compared to a similar thickness of subcutaneous flesh (tricep tissue). Over 0.8% of the NIR energy passed through 25 mm of the human hand, while only 0.3% of NIR penetrated a similar depth of tissue with no bones (subcutaneous flesh) as described above. Also, cartilage (ear) appeared to allow greater energy transmission with 2.93% of the energy penetrating 5 mm of the human ear compared to a similar thickness of the hand, wherein only 0.44% of the light energy penetrated (Henderson and Morries, 2015b).

In the living human hand (25 mm), Jagdeo and colleagues (2012) found only 0.01%–0.09% of 830 nm light from a 500 mW emitter penetrated this distance. In contrast, Henderson and Morries (2015b) found that 0.6% of 810 nm light from a higher powered source (10 W) was able to penetrate 25 mm of skin, bone, and soft tissue. This represents at least a sixfold increase in penetration compared to previous work with low power emitters, and once again underscores that greater power yields greater penetration.

This human model was further studied with ourselves as volunteers. The hypothesis tested was one put forth by proponents of LED emitters as a method to treat human intracranial conditions. "Given enough time, even a low power LED emitter can deliver enough fluence transcranially for therapeutic benefit," would summarize this argument. Using a human hand with heterogeneous composition (including bone with a scattering coefficient close to that of brain), we studied the cumulative dose of NIR energy transmitted through 25 mm of tissue thickness over time. In Fig. 6.15, the amount of NIR energy transmitted through a human hand from an LED emitter with 500 mW at 830 nm is illustrated. As is evident from Fig. 6.15C, no NIR light energy is transmitted through 2.5 cm over 60 seconds.

Our studies of how NIR energy penetrates 3 cm of sheep skull, brain, and overlying soft tissue are a close model of the clinical practice of near-infrared light therapy (NILT) for TBI (Henderson and Morries, 2015b). Again, NIR energy from devices generating less than 1 W could not be detected at the depth of 3 cm. The energy from a 6 W LED system showed a 99.995% drop across 3 cm of tissue. In contrast, 0.14% of the energy from a 10 W 810/980 nm device penetrated 3 cm of tissue. At 15 W, an 810 nm emitter in continuous mode delivered 1.26% of the surface power density and 0.80% of the 980 nm device emissions reached the 3 cm depth through brain tissue. Using pulsed NIR emission settings, a lower overall power density was delivered to the surface; however, similar penetration was achieved.

Prior articles on NILT for the treatment of TBI or stroke in humans have focused on getting photonic energy through the skull to the cortex surface which traverses a distance of about 6–10 mm; however, this model is flawed in that the distance to the areas of damage within the brain may be far greater. In other words, the cortex immediately subjacent to the skull at the vertex may be 10 mm from the surface, but the NIR light energy may need to penetrate 3–7 cm to reach areas of damage. Much of the cortical surface is actually lining the walls and floors of sulci, rather than lying immediately subjacent to the skull. In addition, review of the neuroimaging literature on TBI has revealed that the most common areas injured in TBI are the orbitofrontal cortex (at the ventral surface of the frontal lobe) and the anterior and medial temporal lobes (Raji et al., 2014). It is not anatomically possible to position an NIR light emitter immediately exterior to the skull overlying these areas. Indeed, the orbitofrontal cortex positioned immediately above the eyes can only be reached from the forehead by angling the light emitter. Similarly, the temporal lobes are separated

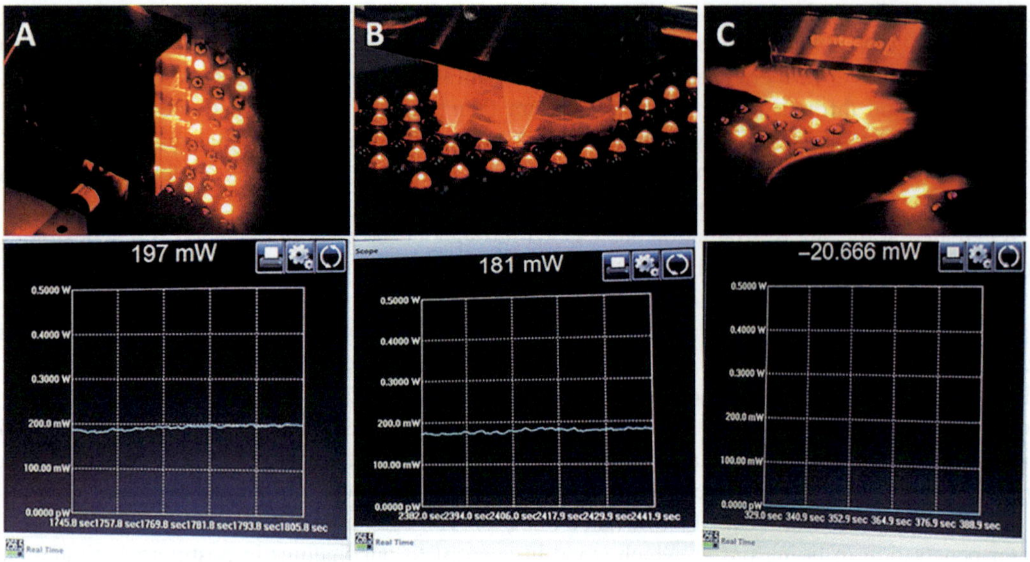

FIGURE 6.15 (A) An LED emitter at 830 nm with 500 mW output is placed in direct contact with a light meter. The meter registers 197 mW, considerably less than 500 nW. (B) A cardboard cylinder measuring 2.5 cm in depth is placed between the LED emitter and the light meter. The meter registers 181 mW is transmitted through 2.5 cm of air. (C) A human hand measuring 2.5 cm in thickness is placed in between the LED emitter and the light meter. The meter registers no light energy being transmitted through the human hand.

from the surface by epidermis, dermis, subcutaneous fat, subcutaneous blood vessels, accessory head of the temporalis muscle, connective tissue, temporalis muscle, skull, and dura mater (Clemente, 1981). Each of these structures has different absorption and refraction properties, and each interface between different materials also creates a barrier to transmission of photonic energy (Strangman et al., 2013). Blood flowing in the subcutaneous vessels creates a unique barrier to transmission (Douplik et al., 2013). In summary, effectively targeting the areas most commonly injured in TBI with sufficient photonic energy to initiate reparative processes represents a significant challenge in NILT.

6.3.4 A hairy problem

Naeser and colleagues described an open-label series of eleven subjects with TBI treated with low-power NILT (Naeser et al., 2014). The treatment was conducted with a device with three LED cluster heads (MedX Health Model 1100, Toronto, ON, Canada). The parameters used for the treatment were the following: NIR wavelength 870 and 633 nm (red light), irradiance 22.2 mW/cm^2, fluence 13 J/cm^2, approximate time 10 minutes per site (Naeser et al., 2014). The sites for irradiation on the skull were at the midline, and bilaterally on frontal, parietal and temporal areas. As illustrated in their article (Naeser et al., 2014) and in prior work (Naeser et al., 2011), the LED heads were applied over the hair of the subject, except at treatment sites on the forehead. However, hair is a potent barrier to NIR light.

Examining any number of websites of companies (e.g., Cerescan, Emerson, Vielight, BrainThor) and publications by such companies (Hipskind et al, 2018), claiming to have -an LED_based treatment device for TBI or any other neurological condition reveals images of patients sporting a device placed over their hair (Fig. 6.16). These images put in question these advertised claims by said companies and illustrate their poor understanding of NIR light energy. Hair acts as a barrier to NIR light.

To illustrate this point, we obtained hair samples and observed the amount of NIR energy that could penetrate a 2 mm mat of hair. For example, in Fig. 6.16B, the amount of energy from a 10 W 810 nm laser penetrating 2 mm of air was recorded. A 2 mm thick collection of blond hair was then interposed between the laser emitter and the laser meter (Fig. 6.16C). As can be seen in Fig. 6.16D, roughly 98% of the NIR energy was absorbed or reflected by the hair. Indeed, in continuous mode, the hair sample heated and released an odor of burnt hair.

As this demonstration illustrates, hair represents a profound barrier to the penetration of NIR energy. When LED-based therapy devices are placed over hair, undoubtedly very little NIR energy reaches the scalp, let alone the brain. If 98% of the energy from a 0.5 W LED is absorbed by hair, 80%−90% is absorbed by 2 mm of skin (Bjordal et al., 2003; Kolari and Airaksinen, 1993), and 96% of incident energy is attenuated by skull (Lapchak et al., 2015), then the claims of neurophysiological benefits of LED-based devices become highly questionable.

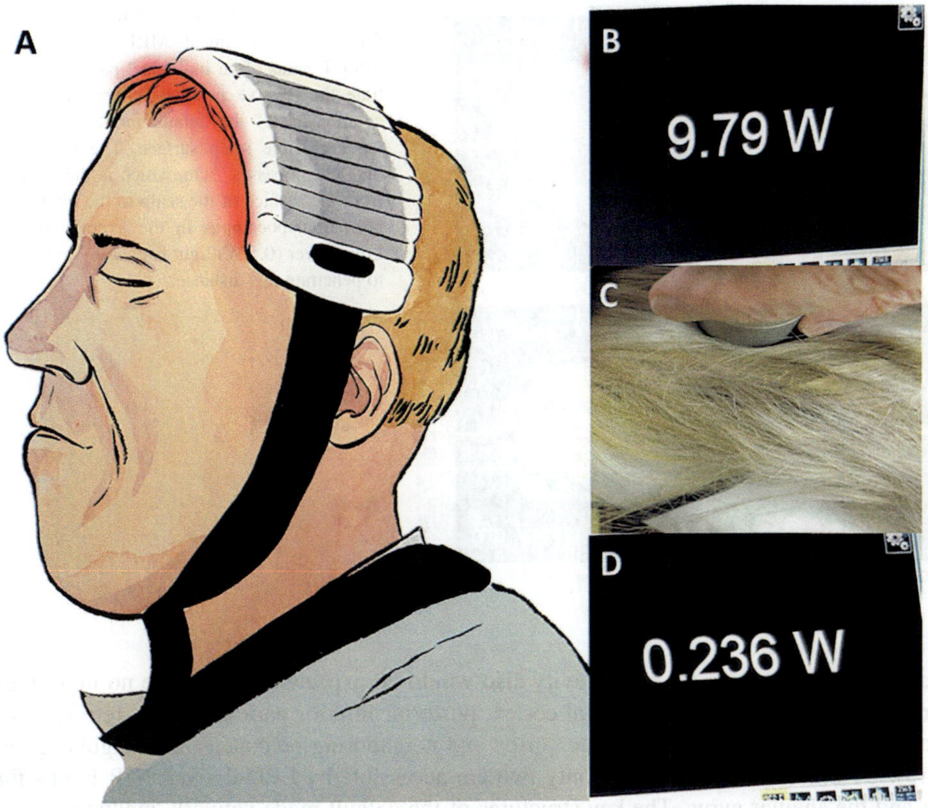

FIGURE 6.16 Numerous companies advertise LED devices that are placed on the head purportedly as a treatment for TBI and other neurological or psychiatric disorders (A). Typically, the devices are shown placed on top of the person's hair. Since, NIR is heavily attenuated by hair this undoubtedly limits or eliminates any potential benefit from NIR irradiation. (B) Image of light meter display when a 10 W laser at 810 nm is placed 2 mm from a light meter. The meter registers 9.79 W penetrating this distance of air. (C) A portion of hair approximately 2 mm in thickness is placed in between the NIR emitter and the light meter. (D) Image of meter display—only 0.236 W of NIR energy from the 10 W emitter can penetrate the hair to register on the meter.

6.3.5 Effectively treating the brain

The penetration data presented herein challenge the presumption that only LLLT protocols spanning weeks or months could possibly be clinically effective or that photonic energy delivered by LLLT even reaches the brain. For example, using a 0.5 W 830 nm LED, Jagdeo and colleagues (2012) estimated the power density reaching the cerebral cortex based on their model was 3 mW/cm^2, which equates to a fluence of 0.0064 J/cm^2–1/140th of the minimum thought to be necessary for ideal photobiomodulation (Karu, 2003). Anders and colleagues (Anders, 2015) have noted NIR penetrates 4 cm into human cadaver skin, skull, and brain using a 5 W laser. From our data, we estimate that in our clinical applications of high-powered NIR lasers (Morries et al., 2015), we are delivering 1.65–3.7 J/cm^2 to a depth of 3 cm. This is 100-fold greater fluence than that delivered by an LLLT system, but within the range of fluence shown to have beneficial biological effects (0.9–36 J/cm^2) in laboratory studies (Henderson and Morries, 2015b; Chung et al., 2012; Ando et al., 2011; Yip et al., 2011). Evidence from multiple sources detailed above indicate that none of the photonic energy from any LED-based device reaches the depth of 3 cm.

Our clinical experience supports and informs this position. Patients treated with 20 treatments, each lasting approximately 20–30 minutes, have experienced significant and sometimes dramatic improvements (Morries et al., 2015). Neurophysiological changes consistent with the clinical improvements can be documented by perfusion single photon emission computed tomography (SPECT) scans (Henderson and Morries, 2015a). Moreover, these improvements have been persistent over months and years. This contrasts sharply with reported findings so far using LLLT to treat patients wherein benefits begin to fade as soon as the treatment regimen is halted (Naeser et al., 2011, 2014; Cassano et al., 2015).

6.4 Alternative hypotheses to direct near-infrared light energy effects

Given the profound limitations on penetration of NIR energy from LEDs and low-power (less than 5 W) lasers, one must question the assumptions about the mechanism of clinical responses which have been reported in a small number of clinical studies. For example, a recent news article quoted Naeser stating that LED-derived NIR energy delivered to the scalp and the intranasal cavity was treating the default mode network (Richmond, 2018) and, thus, was treating posttraumatic stress disorder (PTSD). The challenge is that most of the elements of the default mode network are deeper

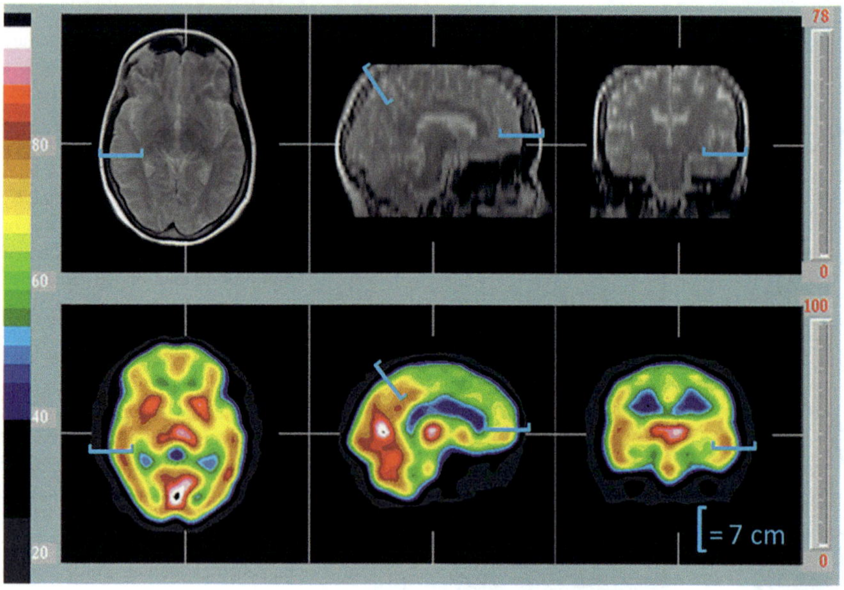

FIGURE 6.17 A portion of the display used for fusing anatomical MRI and perfusion SPECT scans of an individual patient to illustrate the relative position of key anatomical components of the default mode network to the accessible scalp surface. The bracket in blue encompasses 7 cm, which is the distance from the surface of the scalp to the hippocampus/parahippocampus in the average person. Low power (0.5 W) light is unlikely to be able to penetrate this distance.

than 3 cm from the scalp surface. Illumination from the nasal cavity also would be expected to penetrate no more than 2 cm. The default mode network is made up of medial prefrontal cortex, posterior inferior parietal cortex, retrosplenial cortex, hippocampus, parahippocampal gyrus, posterior cingulate cortex and neighboring precuneus, and angular gyrus (Buckner, 2012; Raji et al., 2015). Of these anatomical areas, only two are accessible by LED-derived NIR light—the posterior inferior parietal cortex and the angular gyrus. The key structures of the default mode network, medial prefrontal cortex, posterior cingulate cortex, hippocampus, and parahippocampus are too deep to be reached by LED-derived NIR light based on all of the evidence presented heretofore. This is illustrated in Fig. 6.17. Herein, selected images from an anatomical magnetic resonance imaging (MRI) scan and a SPECT scan of the same patient are computer-aligned. The bracket in blue encompasses 7 cm, which is the distance from the surface of the scalp to the hippocampus/parahippocampus in the average person. In the sagittal images, it is evident that there is at least 3.5 cm from the surface of the scalp to the posterior cingulate gyrus, although the precuneous is more superficial. Similarly, the areas of the medial frontal cortex which are involved in the default mode network lie at least 4–5 cm from the skin of the forehead. The authors of the aforementioned study of the default mode network have indicated they are delivering 26 J/cm^2 at the skin surface using 500 mW LEDs (Richmond, 2018). However, there is no information on what amount of this light energy is reaching 4–7 cm into the brain. Evidence presented herein would strongly suggest that no detectable energy would be expected to be found at these depths in the brain.

Yet, despite these limitations in terms of the penetrating ability of low-power NIR light energy, clinical benefits have been reported. For example, Naeser and colleagues (2014) reported improvement in cognitive testing (Stroop Test, California Verbal Learning Test)—results of almost one standard deviation after a course of 18 treatments using an LED-based device in an open-label trial. They also reported decreases symptoms of PTSD in some, but not all, patients. In a case report on two patients with TBI, Naeser and colleagues (2011) reported NIR (870 + 633 nm) was administered with an instrument made of three separate LED cluster heads. LED cluster heads were positioned sequentially over the bilateral forehead, temples, parietal region, high frontal region, and the vertex. An LED cluster was also placed on the foot. Notably, the patients had only transient benefit from this protocol. If the patients stopped treatment, then symptoms returned within two weeks. Also, the authors had no neuroimaging to localize the lesion to which they were trying to direct NIR energy. Moreover, the application of NIR energy to the foot has no direct effect on the brain. The authors (Naeser et al., 2011) suggest a corollary to acupuncture points and further hypothesize that the NIR energy is having an effect on blood flow in the frontal lobe. Cassano and colleagues (2015) described an open-label trial of four patients treated for depression with an LED-based device. Hamilton Depression Scale scores decreased from 19.8 ± 4.4 to 13 ± 5.35. A large, NIH-funded clinical trial of transcranial NILT for acute stroke in 120 humans (NEST-1) showed clinical improvement (Lampl et al., 2007); however, a subsequent Phase III clinical trial of 660 patients (NEST-2) failed to show benefit at an interim futility analysis (Zivin et al., 2009). Lapchak (Lapchak, 2010) later hypothesized that the success of the Phase II clinical trial was due to the selective inclusion of patients with middle cerebral artery strokes yielding strokes in the lateral aspects of the frontal and temporal lobes. He further hypothesized that the failure of the Phase III clinical trial was due to the inclusion of patients with deep strokes.

How could NIR energy be influencing brain function if it is not directly illuminating injured neurons?

One potential explanation for the changes in patients with a brain injury or other neurological abnormalities who are treated with low-power NIR light is that the, mostly transient, changes are the result of a systemic effect. Braverman and colleagues (1989) demonstrated systemic effects of localized laser irradiation (632 and 904 nm) on distant skin wounds. They treated one of two full-thickness skin wounds in a rabbit model. The nonirradiated lesion showed accelerated healing compared to lesions on untreated animals (Braverman et al., 1989). Rochkind and colleagues (1989) demonstrated similar systemic effects with bilateral wounds in which only one side was treated, but both sides showed accelerated healing. Moreover, they demonstrated that nerve crush injury also responded to systemic effects. By applying irradiation to the right sciatic nerve with HeNe laser (632 nm) after bilateral nerve crush injury, highly significant increases in action potentials were found in both the right sciatic nerve and the untreated left sciatic nerve, compared to nonirradiated controls (Rochkind et al., 1989). Rodrigo and colleagues (2009) examined three wounds in which only one wound was treated with either 830 or 632 nm light at 50 mW. Curiously, the wound furthest from the point of light treatment showed the greatest histological evidence of healing.

Support for a systemic effect of low-level NIR light therapy in humans can be derived from the work of Naeser and colleagues (2011). In this description of two cases treated for many months with low-level NIR derived from LED clusters, NIR was consistently applied to an acupuncture point on the foot. Also, the positioning of the LED clusters in an open-label trial was attributed to acupuncture points (Naeser et al., 2011). The authors hypothesized on a benefit to cerebral perfusion by this technique. In support of this, Quah-Smith and colleagues (2013) showed that laser acupuncture to points on the trunk and limb of patients with depression led to demonstrable perfusion changes in the default mode network as seen by functional MRI. The mechanism of action for acupuncture is poorly understood and beyond the scope of this work. Nonetheless, acupuncture mechanisms would not have a ***direct effect*** on neurons beneath the LED cluster. Rather, as Quah-Smith and colleagues (2013) showed a systemic effect from extracranial acupuncture points had demonstrable effects on intracranial structures.

Low-level NIR light may induce metabolic modulators, which have been shown to modulate the extent of excitotoxic secondary brain injury (Dong et al., 2015). NIR light may induce nitric oxide levels (Henderson and Morries, 2015b; Chung et al., 2012), although nitric oxide is an extremely short-lived molecule (Thomas et al., 2001). Nitric oxide synthase activity could possibly be induced, resulting in overall increases in nitric oxide levels. NIR light has been shown to induce certain growth factors, which may have systemic effects. NIR light can also have effects on inflammatory cytokines. These systemically acting agents, either separately or in some combination, may positively affect brain function. These systemic attributes to NIR light therapy warrant further evaluation.

One area in which inroads have been made, is the understanding of the effects of NIR light on the induction of anti-inflammatory cytokines (Konchurova and Gorbunov, 2012). Irradiation of a small area of skin on human volunteers results in elevation of growth factors (Samoilova et al., 2004). Plasma from treated subjects had growth promoting activity in cell culture. Moreover, a single treatment with NIR light, as well as four daily treatments, elevated levels of circulating antiinflammatory cytokines (Zhevago et al., 2005). Interleuken10 (IL-10) increased by approximately three-fold, while tumor growth factor beta-1 (TGFbeta-1) increased 150%. In contrast, pro-inflammatory cytokine levels were decreased. Tumor necrosis factor alpha (TNF-alpha) decreased 34-fold. Interleukin-6 (IL-6) decreased 12-fold (Zhevago et al., 2005). Similar results were obtained in a mouse model (Fukuda et al., 2013).

6.5 Conclusion

NIR photobiomodulation represents a remarkable and powerful tool for influencing energetics, inflammation, perfusion, tissue repair mechanisms, and neuromodulation. The challenge of translating laboratory data with small animal models to the situation in the patient has been presented. At this point, the actual mechanisms underlying the clinical improvement of symptoms of TBI or other neurological conditions in humans as a result of NIR photobiomodulation remain unclear. This review of the data on NIR energy attenuation as it passes through tissue suggests that the clinical benefit ***may not be the result of the direct effect*** of NIR energy on neurons. Rather, systemic effects may induce neurological changes that yield favorable changes in the patient's symptoms. In contrast, our work with multiwatt NILT may activate the same molecular mechanisms within the neurons of the human brain which have been demonstrated in animal studies, because NIR energy can demonstrably reach those neurons. If multiwatt NILT is, indeed, having a ***direct effect*** on the neurons of human brain, then this may explain why patients see marked and persistent clinical improvement in symptoms.

The issue is a matter of scale. Low-power laser therapy may lack sufficient energy to penetrate the extent of overlying tissue in order to reach the brain. Devices that utilize the LED, rather than a laser diode, have even less likelihood of efficacy. These LED devices, particularly when placed over the patient's hair, may be nothing more than a placebo device. This raises an important concern to the field as a whole. Claims made by those who do not understand NIR physics and light-tissue interactions, can create confusion, doubt, and distrust concerning NIR photobiomodulation. The field would do well to police itself. Claims should be supported by data and subjected to scrutiny based on an understanding of light-tissue interactions.

Acknowledgments

The authors would like to acknowledge the technical assistance of Ms. Stefa Grabchenko and Ms. Jeannie Rylands. The authors also acknowledge the contribution of Ms. Taylor Tuteur in the artistic creation of Figs. 6.1, 6.2, 6.5, 6.7, 6.11, and 6.16. The authors acknowledge Ms. Valentina Henderson for her assistance in translating Russian literature.

References

Anders, J.J., Personal Communication, January 13, 2015.

Ando, T., Xuan, W., Xu, T., Dai, T., Sharma, S.K., Kharkwal, G.B., et al., 2011. Comparison of therapeutic effects between pulsed and continuous wave 810-nm wavelength laser irradiation for traumatic brain injury in mice. PLoS One 6 (10), e26212.

Aulakh, K., Zakaib, S., Willmore, W.G., Ye, W.N., 2016. Transcranial light-tissue interaction analysis. In: Proc. SPIE 9706, Optical Interactions with Tissue and Cells XXVII, 97061B. https://doi.org/10.1117/12.2212831.

Bjordal, J.M., Couppé, C., Chow, R.T., Tunér, J., Ljunggren, E.A., 2003. A systematic review of low level laser therapy with location-specific doses for pain from chronic joint disorders. Aust. J. Physiother. 49 (2), 107–116.

Braverman, B., McCarthy, R.J., Ivankovich, A.D., Forde, D.E., Overfield, M., Bapna, M.S., 1989. Effect of helium-neon and infrared laser irradiation on wound healing in rabbits. Lasers Surg. Med. 9 (1), 50–58.

Buckner, R.L., 2012. The serendipitous discovery of the brain's default network. Neuroimage 62 (2), 1137–1145.

Byrnes, K.R., Waynant, R.W., Ilev, I.K., Wu, X., Barna, L., Smith, K., et al., 2005. Light promotes regeneration and functional recovery and alters the immune response after spinal cord injury. Lasers Surg. Med. 36 (3), 171–185.

Cassano, P., Cusin, C., Mischoulon, D., Hamblin, M.R., De Taboada, L., Pisoni, A., et al., 2015. Near-infrared transcranial radiation for major depressive disorder: proof of concept study. Psychiatry J. 2015, 352979.

Chen, A.C., Arany, P.R., Huang, Y.Y., Tomkinson, E.M., Sharma, S.K., Kharkwal, G.B., et al., 2011. Low-level laser therapy activates NF-kB via generation of reactive oxygen species in mouse embryonic fibroblasts. PLoS One 6 (7), e22453.

Choi, J.J., Pernot, M., Brown, T.R., Small, S.A., Konofagou, E.E., 2007. Spatio-temporal analysis of molecular delivery through the blood-brain barrier using focused ultrasound. Phys. Med. Biol. 52 (18), 5509–5530.

Chung, H., Dai, T., Sharma, S.K., Huang, Y.Y., Carroll, J.D., Hamblin, M.R., 2012. The nuts and bolts of low-level laser (light) therapy. Ann. Biomed. Eng. 40 (2), 516–533.

Clemente, C., 1981. Anatomy – A Regional Atlas of the Human Body, second ed. Urban & Schwarzenburg Inc. (Part VII).

Dong, T., Zhang, Q., Hamblin, M.R., Wu, M.X., 2015. Low-level light in combination with metabolic modulators for effective therapy of injured brain. J. Cereb. Blood Flow Metab. 35 (9), 1435–1444.

Douplik, A., Saiko, G., Schelkanova, I., 2013. The response of tissue to laser light. In: Jelinkova, H. (Ed.), Lasers for Medical Applications, Diagnostics, Therapy, and Surgery. Woodhead Publishing, Sawston, pp. 47–109.

Esnouf, A., Wright, P.A., Moore, J.C., Ahmed, S., 2007. Depth of penetration of an 850nm wavelength low level laser in human skin. Acupunct. Electrother. Res. 32 (1–2), 81–86.

Fitzgerald, M., Hodgetts, S., Van Den Heuvel, C., Natoli, R., Hart, N.S., Valter, K., et al., 2013. Red/near-infrared irradiation therapy for treatment of central nervous system injuries and disorders. Rev. Neurosci. 24 (2), 205–226.

Frank, S., Oliver, L., Lebreton-De, C.C., Moreau, C., Lecabellec, M.T., Michel, L., et al., 2004. Infrared radiation affects the mitochondrial pathway of apoptosis in human fibroblasts. J. Invest. Dermatol. 123 (5), 823–831.

Fukuda, T.Y., Tanji, M.M., Silva, S.R., Sato, M.N., Plapler, H., 2013. Infrared low-level diode laser on inflammatory process modulation in mice: pro- and anti-inflammatory cytokines. Lasers Med. Sci. 28 (5), 1305–1313.

Giacci, M.K., Wheeler, L., Lovett, S., Dishington, E., Majda, B., Bartlett, C.A., et al., 2014. Differential effects of 670 and 830 nm red near infrared irradiation therapy: a comparative study of optic nerve injury, retinal degeneration, traumatic brain and spinal cord injury. PLoS One 9 (8), e104565.

Henderson, T.A., 2016. Multi-watt near infrared light therapy as a neuroregenerative treatment for TBI. Neural Regen. Res. 11 (4), 563–565.

Henderson, T.A., Morries, L.D., 2015a. SPECT perfusion imaging demonstrates improvement of TBI with transcranial near infrared laser phototherapy. Adv. Mind Body Med. 29 (4), 27–33.

Henderson, T.A., Morries, L.D., 2015b. Near-infrared photonic energy penetration: can infrared phototherapy effectively reach the human brain? Neuropsychiatr. Dis. Treat. 11, 2191–2208.

Hipskind, S.G., Grover Jr, F.L., Fort, T.R., Helffenstein, D., Burke, T.J., Quint, S.A., Bussiere, G., Stone, M., Hurtado, T., 2018. Pulsed Transcranial Red/Near-Infrared Light Therapy Using Light-Emitting Diodes Improves Cerebral Blood Flow and Cognitive Function in Veterans with Chronic Traumatic Brain Injury: A Case Series. Photomed Laser Surg 37 (2), 77–84. Nov 28.

Hode, L., 2005. The importance of the coherency. Photomed. Laser Surg. 23 (4), 431–434.

Hudson, D.E., Hudson, D.O., Wininger, J.M., Richardson, B., 2013. Penetration of laser light at 808 nm and 980 nm in bovine tissue samples. Photomed. Laser Surg. 31 (4), 163–168 (2013).

Huisa, B.N., Stemer, A.B., Walker, M.G., Rapp, K., Meyer, B.C., Zivin, J.A., et al., 2013. Transcranial laser therapy for acute ischemic stroke: a pooled analysis of NEST-1 and NEST-2. Int. J. Stroke 8 (5), 315–320.

Jacques, S.L., 2013. Optical properties of biological tissues: a review. Phys. Med. Biol. 58 (11), R37–R61.

Jagdeo, J.R., Adams, L.E., Brody, N.I., Siegel, D.M., 2012. Transcranial red and near infrared light transmission in a cadaveric model. PLoS One 7 (10), e47460.

Jenkins, P.A., Carroll, J.D., 2011. How to report low-level laser therapy (LLLT)/photomedicine dose and beam parameters in clinical and laboratory studies. Photomed. Laser Surg. 29 (12), 785–787.

Junqueira, L.C., Carneiro, J., 1982. Basic Histology, fourth ed. Lange Medical Publications, Los Altos, pp. 70–72.

Karu, T.I., 2003. Cellular mechanism of low power laser therapy: new questions. In: Simunovic, F. (Ed.), Lasers in Medicine and Dentistry, vol. 3. Z. Vitgraf, Rijeka, pp. 79–100.

Karu, T.I., Kolyakov, S.F., 2005. Exact action spectra for cellular responses relevant to phototherapy. Photomed. Laser Surg. 23 (4), 355–361.

Khuman, J., Zhang, J., Park, J., Carroll, J.D., Donahue, C., Whalen, M.J., 2012. Low-level laser light therapy improves cognitive deficits and inhibits microglial activation after controlled cortical impact in mice. J. Neurotrauma 29 (2), 408–417.

Kolari, P.J., 1985. Penetration of unfocused laser light into the skin. Arch. Dermatol. Res. 277 (4), 342–344.

Kolari, P.J., Airaksinen, O., 1993. Poor penetration of infra-red and helium neon low power laser light into the dermal tissue. Acupunct. Electrother. Res. 18 (1), 17–21.

Konchurova, T.B., Gorbunov, A.K., 2012. Laser Therapy and Prophylaxis. Practical Medicine, Practical Medicine, Moscow. Russian – translated, pp. 41–76. (Chapter 2).

Lampl, Y., Zivin, J.A., Fisher, M., Lew, R., Welin, L., Dahlof, B., et al., 2007. Infrared laser therapy for ischemic stroke: a new treatment strategy: results of the NeuroThera Effectiveness and Safety Trial-1 (NEST-1). Stroke 38 (6), 1843–1849.

Lapchak, P.A., 2010. Taking a light approach to treating acute ischemic stroke patients: transcranial near-infrared laser therapy translational science. Ann. Med. 42 (8), 576–586.

Lapchak, P.A., Boitano, P.D., Butte, P.V., Fisher, D.J., Hölscher, T., Ley, E.J., et al., 2015. Transcranial Near-Infrared Laser Transmission (NILT) profiles (800 nm): systematic comparison in four common research species. PLoS One 10 (6), e0127580.

Leeson, T.S., Leeson, R.C., 1979. A Brief Atlas of Histology. W.B. Saunders Company, Philadelphia, pp. 26–27.

Leung, M.C., Lo, S.C., Siu, F.K., So, K.F., 2002. Treatment of experimentally induced transient cerebral ischemia with low energy laser inhibits nitric oxide synthase activity and up-regulates the expression of transforming growth factor-beta 1. Lasers Surg. Med. 31 (4), 283–288.

Liang, H.L., Whelan, H.T., Eells, J.T., Wong-Riley, M.T., 2008. Near-infrared light via light-emitting diode treatment is therapeutic against rotenone- and 1-methyl-4-phenylpyridinium ion-induced neurotoxicity. Neuroscience 153 (4), 963–974.

Lister, T., Wright, P.A., Chappell, P.H., 2012. Optical properties of human skin. J. Biomed. Opt. 17 (9), 90901.

Lubart, R., Eichler, M., Lavi, R., Friedman, H., Shainberg, A., 2005. Low-energy laser irradiation promotes cellular redox activity. Photomed. Laser Surg. 23 (1), 3–9.

Lychagov, V.V., Tuchin, V.V., Vilensky, M.A., Reznik, B.N., Ichim, T., DeTaboada, L., 2006. Experimental study of NIR transmittance of the human skull. In: Tuchin, V.V. (Ed.), Complex Dynamics and Fluctuations in Biomedical Photonics III. Proc of SPIE V, vol. 6085. pp. 60850T1–60855T1.

Mirsky, N., Krispel, Y., Shoshany, Y., Maltz, L., Oron, U., 2002. Promotion of angiogenesis by low energy laser irradiation. Antioxid. Redox Signal. 4 (5), 785–790.

Morries, L.D., Cassano, P., Henderson, T.A., 2015. Treatments for traumatic brain injury with emphasis on transcranial near infrared laser phototherapy. Neuropsychiatr. Dis. Treat. 11, 2159–2175.

Naeser, M.A., Saltmarche, A., Krengel, M.A., Hamblin, M.R., Knight, J.A., 2011. Improved cognitive function after transcranial, light-emitting diode treatments in chronic, traumatic brain injury: two case reports. Photomed. Laser Surg. 29 (5), 351–358.

Naeser, M.A., Zafonte, R., Krengel, M.H., Martin, P.I., Frazier, J., Hamblin, M.R., et al., 2014. Significant improvements in cognitive performance post-transcranial, red/near-infrared light-emitting diode treatments in chronic, mild traumatic brain injury: open-protocol study. J. Neurotrauma 31 (11), 1008–1017.

Passarella, S., 1989. He-Ne laser irradiation of isolated mitochondria. J. Photochem. Photobiol. B 3 (4), 642–643.

Pastore, D., Greco, M., Passarella, S., 2000. Specific helium-neon laser sensitivity of the purified cytochrome c oxidase. Int. J. Radiat. Biol. 76 (6), 863–870.

Quah-Smith, I., Suo, C., Williams, M.A., Sachdev, P.S., 2013. The antidepressant effect of laser acupuncture: a comparison of the resting brain's default mode network in healthy and depressed subjects during functional magnetic resonance imaging. Med. Acupunct. 25 (2), 124–133.

Raji, C.A., Tarzwell, R., Pavel, D., Schneider, H., Uszler, M., Thornton, J., et al., 2014. Clinical utility of SPECT neuroimaging in the diagnosis and treatment of traumatic brain injury: a systematic review. PLoS One 9 (3), e91088.

Raji, C.A., Willeumier, K., Taylor, D., Tarzwell, R., Newberg, A., Henderson, T.A., et al., 2015. Functional neuroimaging with default mode network regions distinguishes PTSD from TBI in a military veteran population. Brain Imaging Behav. 9 (3), 527–534.

Richmond, L.M., 2018. LED therapy offers new hope to veterans with TBI, PTSD. Psychiatric News. Published Online: March 1, 2018. https://doi.org/10.1176/appi.pn.2018.1b8 (accessed 24.05.18.).

Rochkind, S., Rousso, M., Nissan, M., Villarreal, M., Barr-Nea, L., Rees, D.G., 1989. Systemic effects of low-power laser irradiation on the peripheral and central nervous system, cutaneous wounds, and burns. Lasers Surg. Med. 9 (2), 174–182.

Rodrigo, S.M., Cunha, A., Pozza, D.H., Blaya, D.S., Moraes, J.F., Weber, J.B.B., et al., 2009. Analysis of the systemic effect of red and infrared laser therapy on wound repair. Photomed. Laser Surg. 27 (6), 929–935.

Rojas, J.C., Gonzalez-Lima, F., 2011. Low level light therapy of the eye and brain. Eye Brain 3, 49–67.

Salehpour, F., Mahmoudi, J., Kamari, F., Sadigh-Eteghad, S., Rasta, S.H., Hamblin, M.R., 2018. Brain photobiomodulation therapy: a narrative review. Mol. Neurobiol. 55 (8), 6601–6636.

Samoilova, K.A., Bogacheva, O.N., Obolenskaya, K.D., Blinova, M.I., Kalmykova, N.V., Kuzminikh, E.V., 2004. Enhancement of the blood growth promoting activity after exposure of volunteers to visible and infrared polarized light. Part I: Stimulation of human keratinocyte proliferation in vitro. Photochem. Photobiol. Sci. 3 (1), 96–101.

Steiner, R., 2011. Laser-tissue interactions. In: Raulin, C., Karsai, S. (Eds.), Laser and IPL Technology in Dermatology and Aesthetic Medicine. Springer-Verlag, Berlin, pp. 23–36.

Strangman, G.E., Li, Z., Zhang, Q., 2013. Depth sensitivity and source-detector separations for near infrared spectroscopy based on the Colin27 brain template. PLoS One 8 (8), e66319. Available from: https://doi.org/10.1371/journal.pone.0066319.

Thomas, D.D., Liu, X., Kantrow, S.P., Lancaster, J.R., 2001. The biological lifetime of nitric oxide: implications for the perivascular dynamics of NO and O_2. PNAS 98 (1), 355–360.

Wan, S., Parrish, J.A., Anderson, R.R., Madden, M.M., 1981. Transmittance of nonionizing radiation in human tissues. Photochem. Photobiol. 34 (6), 679–681.

Wong-Riley, M.T., Liang, H.L., Eells, J.T., Chance, B., Henry, M.M., Buchmann, E., et al., 2005. Photobiomodulation directly benefits primary neurons functionally inactivated by toxins: role of cytochrome c oxidase. J. Biol. Chem. 280 (6), 4761–4771.

Xuan, W., Agrawal, T., Huang, L., Gupta, G.K., Hamblin, M.R., 2015. Low-level laser therapy for traumatic brain injury in mice increases brain derived neurotrophic factor (BDNF) and synaptogenesis. J. Biophotonics 8 (6), 502–511.

Yip, K.K., Lo, S.C., Leung, M.C., So, S.K., Tang, C.Y., Poon, D.M., 2011. The effect of low-energy laser irradiation on apoptotic factors following experimentally induced transient cerebral ischemia. Neuroscience 190, 301–306.

Yu, W., Naim, J.O., McGowan, M., Ippolito, K., Lanzafame, R.J., 1997. Photomodulation of oxidative metabolism and electron chain enzymes in rat liver mitochondria. Photochem. Photobiol. 66 (6), 866–871.

Zhevago, N.A., Samoĭlova, K.A., Obolenskaia, K.D., Sokolov, D.I., 2005. Changes in cytokine content in the peripheral blood of volunteers after their exposure to polychromatic visible and infrared light. Tsitologiia 47 (5), 450–463. Russian – translated.

Zivin, J.A., Albers, G.W., Bornstein, N., Chippendale, T., Dahlof, B., Devlin, T., et al., 2009. Effectiveness and safety of transcranial laser therapy for acute ischemic stroke. Stroke 40 (4), 1359–1364.

Chapter 7

Light sources and dosimetry for the brain and whole body

James D. Carroll
Thor Photomedicine Ltd., Chesham, United Kingdom

7.1 Dose

Dosimetry in photobiomodulation usually describes the light parameters and application time to the surface of the body. The intention is that the light will reach the pathology/anatomical target, which is either subcutaneous or deep within the body or the brain. However, as light enters the tissue, the intensity is reduced because the light is scattered and absorbed by the cells it passes through, so the intensity becomes weaker the further the target is from the body surface. For this reason, there is a limit to how far light will penetrate tissue and deliver a sufficient intensity to have a biological effect. For transcranial PBM to be effective, the irradiation parameters (wavelength, power, beam area, irradiance, and pulse parameters) need to be within certain ranges and applied for a suitable amount of time. Clinical studies are typically 1–30 minutes for transcranial treatments (Cassano et al., 2018; Boonswang et al., 2012), and these may be applied multiple times [typically 1–18 treatment sessions (Disner et al., 2016; Naeser et al., 2016)], though there is one published case study on a patient who had been in a persistent vegetative state for 8 months. He received 146 treatments (twice a day for 73 days) and then started showing signs of movement (Nawashiro et al., 2012). Treatment intervals for most other successful studies ranged from 1 to 3 times a week (Naeser et al., 2016; Vargas et al., 2017).

If the wrong irradiation parameters are used or applied for the wrong irradiation time then treatment will not be effective. If the irradiance is too low and/or the time is too short then there is no significant effect, alternatively if the irradiance is too high and/or the treatment time is too long then the benefit is lost and sometimes inhibitory effects are seen (Huang et al., 2009). Unfortunately many researchers fail to accurately measure and report these parameters so the literature is unreliable on the topic of dose. This is due in part to a poor appreciation of the relevance of these parameters by authors and reviewers, but even with the best intentions, scientists frequently publish significant errors, and omit important parameters. Beam measurements require specialist instruments that need to be set up, operated by, and interpreted by optical engineers or physicists (Jenkins and Carroll, 2011; Hadis et al., 2016).

7.2 Irradiation parameters: wavelength (nm)

Light are packets of electromagnetic energy that also have a wave-like property. Wavelength is expressed in nanometers (nm); and light is visible from approximately 400 to 750 nm. From approximately 750 to 1500 nm the light is defined as near infrared (Hecht and Williams, 1922; Graham and Hartline, 1935). Wavelength not only defines color or place on the spectrum but also the energy of the photon. A red photon has an electric potential value of approximately 2 eV whereas a blue photon would be approximately 3 eV and so on as the wavelengths get shorter or longer (Fig. 7.1). This characteristic of light determines if an atom or molecule will absorb the photon because there is an energy value for the electron fields that make up the atom. When a photon has an electrical value that matches an electric value in an atom, then absorption is possible.

Wavelength determines which molecules will absorb or scatter the light, and therefore how deep the light penetrates. PBM devices are typically within the range 600 nm–1000 nm as there are many absorption peaks for cytochrome c oxidase in that range (Karu, 2010; Sommer et al., 2001), and they penetrate tissues well; many clinical trials have been successful in these wavelengths (although not around 720/730 nm) (Liang et al., 2008; Wu et al., 2012).

Photobiomodulation in the Brain. DOI: https://doi.org/10.1016/B978-0-12-815305-5.00007-5

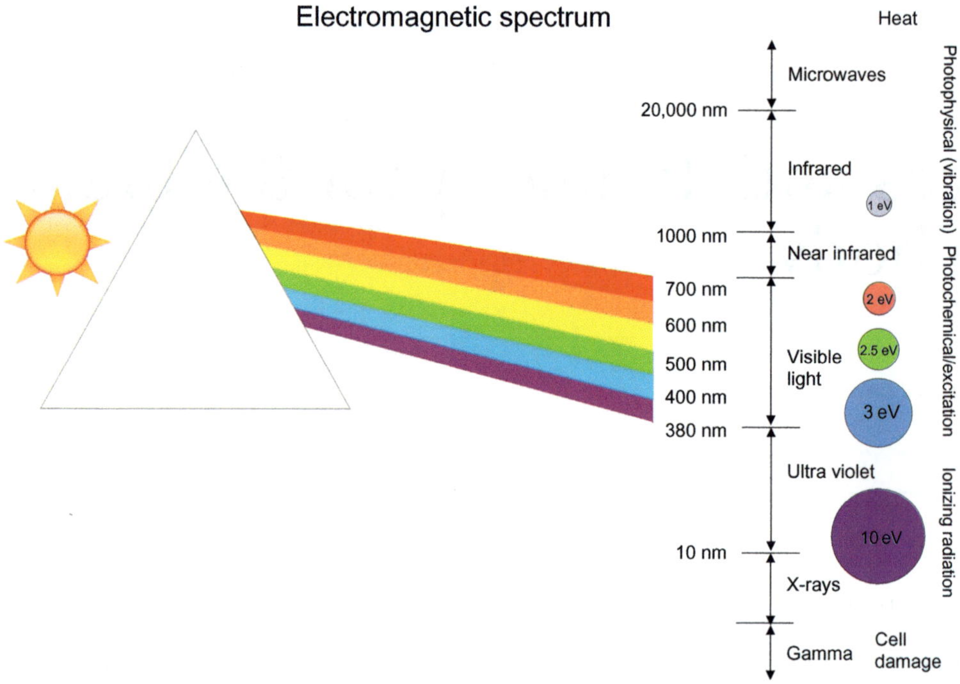

FIGURE 7.1 Electromagnetic spectrum showing wavelengths and corresponding energy levels.

However, there is some contention about wavelengths around ~920–1000 nm as they are more strongly absorbed by water so they penetrate less (because humans and animals are ~70% water). These wavelengths have not yet been shown to be absorbed by cytochrome c oxidase, but somehow they do increase ATP production (Benedicenti et al., 2008) and they also work clinically (Gur et al., 2004; Bjordal et al., 2008). Wang et al. (2017) demonstrated that 980 nm affects temperature-gated calcium ion channels, while 810 nm largely affects mitochondrial cytochrome c oxidase. Wavelengths around 980 nm seem to need a lower dose but do not penetrate well, and there is a risk of tissue heating, so it is recommended to use ~920–1000 nm with caution. There is another penetration window around 1050–1300 nm and there are several clinical successful studies published using 1064 nm (Disner et al., 2016; Vargas et al., 2017).

7.3 Penetration

In a study on the brains of human cadavers, Tedford et al. compared the penetration of 660, 808, and 940 nm laser light. It was confirmed that 808 nm was superior in brain tissue penetration and that 5 W continuous wave laser with a beam diameter of 30 mm (7 cm^2), (707 mW/cm^2) applied transcranially could penetrate the scalp, skull, meninges, and brain to a depth of 40 mm from the skull (Tedford et al., 2015). Australian researchers are overcoming light penetration limitations with optical fibers implanted in the brains of monkeys to deliver light to the substantia nigra the treatment of Parkinson's Disease (El Massri et al., 2017).

There is another good penetration window around 1050–1300 nm (Smith, 1991) that works well (Disner et al., 2016; Vargas et al., 2017), although the absorbing chromophores and therapeutic mechanism yet to be confirmed. However, there are many papers demonstrating clinical benefits in this wavelength range (Vargas et al., 2017; Hwang et al., 2016). Despite the best experimental penetration data coming from high power lasers, good clinical evidence has been found for low power light emitting diodes (LEDs) (Naeser et al., 2014; Cassano et al., 2018). Indirect treatments (distal not transcranial), to the carotid artery, to acupuncture points, up the nostrils, and even to the whole body have also shown some effect on the brain (El Massri et al., 2018). An in vivo study has shown laser treatments to the tibia release mesenchymal stem cells into the blood stream and they in turn ameliorate disease progression in a mouse model of Alzheimer's disease (Farfara et al 2015), so there may be more to treating the brain than just treating the brain.

7.4 Power Watts (W)

The amount of light energy (J) emitted per second (1 W = 1 J/s). When a light source is pulsed, both the peak and the average power should be reported. See section 7.7 Pulses.

7.5 Beam spot size (cm^2)

This is the beam area incident on the surface of the body. It is not the light source aperture area as is often assumed. The beam area is required for calculating irradiance (also known as power density) W/cm^2 and for calculating dose (fluence) J/cm^2, but the beam area is difficult to measure and is probably the most misunderstood and most frequently misreported parameter (Jenkins and Carroll, 2011). This is because diode laser beams are not necessarily round (often they are elliptical) and the beam intensity distribution is not eve; laser beams are usually brighter in the middle and gradually weaker toward the edge (Gaussian distribution). These factors seem to be unappreciated by most researchers and manuscript reviewers so errors are frequently made when reporting the beam area. For example, many assume that the aperture of a device defines the beam size; it probably never does. The correct way to measure the beam area is with a beam profiler and to report the $1/e^2$ area (Fred and Dickey, 2014). This is a job for an optical engineer/physicist, not a doctor/therapist or lab technician. When reading PBM research papers, be skeptical about the reported beam area, and therefore the irradiance and dose (fluence, J/cm^2) that are reported unless the workers describe how the beam area or irradiance was measured (Omura, 1983; Nikolaev, 1986; El Massri et al., 2017).

7.6 Irradiance (W/cm^2)

Often called "power density," irradiance is the intensity of the beam and is the product of power (W)÷beam area (cm^2) (see section 7.5 Beam spot size). It is sometimes described as the dose rate or fluence rate because it defines the speed at which the dose is being delivered. There is a speed limit to how fast an effective dose of PBM can be delivered. If it is delivered too quickly or too slowly the desired effect may not occur, even if the fluence (dose) is correct. Studies that have experimented with delivering a dose with a range of different irradiances consistently find that lower irradiances are more effective when the same fluence is applied (Oron et al., 2001; Lanzafame et al., 2007; Jang and Lee, 2012). These could end up being very long treatments (over an hour), so for practical reasons the minimal irradiance may not be a realistic option. There is minimum threshold irradiance and a minimum fluence (dose) at the anatomical target (pathology) below, which no effect may be seen. Unfortunately irradiance is often not reported, or it is misreported because of the difficulty with measuring beam area (Jenkins and Carroll, 2011; Fred and Dickey, 2014). Read more about measurement of the beam area above. Assuming you trust the parameters reported, then how much irradiance is required? Preclinical work and cadaver measurements to support an ischemic stroke clinical trial determined that an irradiance of 7–10 mW/cm^2 at the injured brain cortex was safe and effective, but that does not tell us what irradiance is at the scalp. Schiffer et al. performed a small study on 10 patients with major depression and anxiety. The treatment consisted a LED, 810 nm, with an irradiance of 250 mW/cm^2 when projected 4 mm from the skin, for 240 seconds (4 minutes); the fluence per site was 60 J/cm^2, and there were two sites on the forehead (F3 and F4 EEG placement sites). Based on their estimated penetration to the light to the dura (3.7%), they calculated the irradiance to be 9.5 mW/cm^2 and the fluence to be 2.1 J/cm^2 (Schiffer et al., 2009). It is not possible to determine the beam area from the published paper, however, from photographs it appears to be a couple of centimeters in diameter and using that assumption it can be calculated that the power was $\sim$1 W. Multiple transcranial studies (RCTs) at the University of Texas at Austin have been performed using a similar set of parameters using 1064 nm laser, 3.4 W, beam area of 13.6 cm^2, 0.25 W/cm^2 irradiance; the various studies ranged from 4 to 10 minutes treatments (240 to 600 seconds) which equals 60–150 J/cm^2. Using these parameters they achieved consistently good effects measuring a range of emotional and cognitive endpoints (Hwang et al., 2016; Barrett and Gonzalez-Lima, 2013; Tian et al., 2016).

7.7 Pulses

It is clear from animal studies that some pulse regimes have effects that are different from continuous beams. Oron et al. (2012) showed in an acute mouse traumatic brain injury (TBI) study that 100 Hz was superior to continuous beams, Ando et al. showed in an acute mouse TBI study that 10 Hz was superior to 100 Hz or continuous treatment (Ando et al., 2011). Several studies have gone on to use 10 Hz but there have been no further experiments comparing alternative pulse regimes to 10 Hz. It is conceivable that it is not only the pulse frequency that makes a difference but also the pulse width, so this also needs to be understood (Fig. 7.2).

When a beam is pulsed, the energy delivered will be reduced because the beam spends sometime in the off position. The amount of energy lost depends on the duty cycle and this can be calculated. A 50% duty cycle means energy delivered will be 50% less, a 90% duty cycle means energy delivered will be 10% less. If the beam is pulsed then the

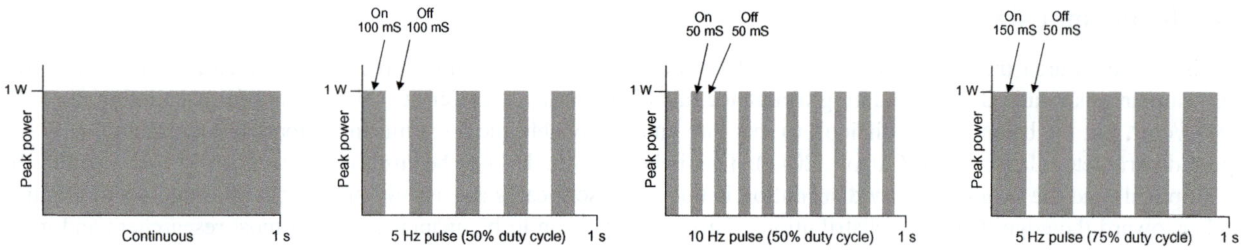

FIGURE 7.2 Different pulse structures with varying repetition rates and duty cycles.

FIGURE 7.3 Schematic of the waves in coherent and noncoherent light.

peak power, the duty cycle, and average power should be reported. If the pulse structure is expressed as a % (or ratio): the calculation is peak power (W) × percentage pulse time on (%) = average power (W).

Example: If the peak power is 1 W × percentage pulse time on 50%, then the average power 0.5 W.

7.8 Coherence

Laser light is coherent; LED light is not coherent (Fig. 7.3). It is often argued that laser light should achieve better results, however, there is no convincing clinical evidence of this hypothesis. On the other hand, there are hundreds of studies using noncoherent LEDs to good effect. It is true that that laser speckles persist deep in tissue, and this observation is an indication that coherence remains (Naeser et al., 2014); however, it is widely accepted that the coherent properties of laser light are biologically unimportant when treating superficial tissues, though there are claims that some additional (therapeutic) effects from coherent light should occur when treating deep anatomical targets (i.e., for low back pain and the brain). The theory is that because the dimensions of speckle patterns coincide with the dimensions of mitocondria, it is speculated that the intensity gradients produced by these speckles may help improve clinical effects in deep tissues where irradiance is low. No clinical trials have been published to date to confirm or refute this claim (Corazza et al., 2007; Zalevsky and Belkin, 2011). A few attempts have been made to compare coherent laser light versus noncoherent LED light; some showed coherent laser light to be superior, others showed noncoherent LED light to be superior to laser, but because not all parameters were identical the results are unreliable.

7.9 Time, energy, and fluence

Having established suitable irradiation parameters (wavelength, irradiance, pulsing), they must be applied for an adequate amount of time. If the wrong irradiation parameters are used or applied at the wrong irradiation time then the treatment will be ineffective. There are two common expressions of dose: energy (J) or energy density (fluence J/cm^2) both referred to as dose. These are not the same and neither of which are adequate by themselves.

$$\text{Energy (J)Calculated as: Power(W)} \times \text{time (seconds)}$$

Using joules as an expression of dose is potentially unreliable as it assumes an inverse relationship between power and time and it ignores irradiance. If a 1 W LED array is applied for 60 seconds, then 60 J has been delivered. What this does not tell you is the size of the area that is treated. A second problem is reciprocity. If the power is doubled and the time halved, the same energy may be applied but the results may be different (Lanzafame et al., 2007; Schindl et al., 2001). To be sure of replicating a successful treatment ideally the same power, beam area, and time should be used. Using more powerful devices as a way of reducing treatment time is not a reliable strategy.

7.10 Fluence (energy density) (J/cm^2)

$$\text{Calculated as: Power (W)} \times \text{time (seconds)} \div \text{beam area (cm}^2\text{)} = \text{Fluence (J/cm}^2\text{)}$$

Using fluence alone to describe the dose is also potentially unreliable as it assumes an inverse relationship between irradiance and time. Again, there is no reciprocity. If the power is doubled and the time halved the same fluence may be applied but the results may be different (Lanzafame et al., 2007; Schindl et al., 2001). If the beam area and power are halved, the irradiance remains the same, but the total energy applied will be halved and may not cover the whole pathology. To be sure of replicating a successful treatment, ideally the same power, beam area, and time should be used. Using more powerful light sources as a way of reducing treatment time is not a reliable strategy; using small laser beams can provide a high fluence but very little total energy delivered (Hadis et al., 2016).

7.11 Irradiation time (seconds)

Given the lack of reciprocity described previously, the safest way to record and prescribe PBM is to define all the irradiation parameters, then define the irradiation time and not rely on energy or fluence parameters alone. PBM treatment times for musculoskeletal pain are typically 30–150 seconds (Bjordal et al., 2008; Chow et al., 2009), but transcranial treatments are typically in the range of 4–30 minutes (Cassano et al., 2018; Hwang et al., 2016; Barrett and Gonzalez-Lima, 2013; Tian et al., 2016).

7.12 Number of treatments and treatment intervals (hours, days, or weeks)

The effects of different time intervals between individual treatments to the brain have not been explored. There is some evidence to suggest that this is an important parameter in other applications of photobiomodulation (Brondon et al., 2005). Clinical trials on chronic degenerative conditions (such as age-related macular degeneration, musculoskeletal pain, and non-healing wounds) have been successful being treated two or three times a week for many weeks (typically 3–6 weeks), whereas acute traumatic soft tissue injuries may need as little as one treatment immediately after injury or surgery.

Treatments to the brain for cognitive enhancement studies in healthy humans have been shown to work well with a single treatment through the duration of effect has not yet been explored (Tedford et al., 2015; Smith, 1991). There are no human studies on acute TBI, but in multiple rodent studies a single treatment works well for reducing memory/cognitive and motor deficits (Xuan et al., 2014; Lee et al., 2016). For humans, chronic or mild TBI responded well to a treatment plan of three times a week for 6 weeks (18 treatments total) (Naeser et al., 2014), and for major depressive disorder a treatment of twice a week for 8 weeks (16 treatments total) (Cassano et al., 2018); but as stated previously, the effects of different intervals (e.g., twice a day vs twice a week) between treatments to the brain has not been explored. Across the field of PBM a minimum of twice a week seems to be the minimum requirement for good outcomes.

Some negative effects were shown by Xuan et al. (2016). They showed that mice treated with PBM daily for 3 days after a TBI gave a significant improvement in neurological/cognitive function, but the same treatment delivered daily for 14 days gave significantly fewer benefits. They followed the mice 56 days post TBI and found the 14 day group showed worse neurological function and more activated glial cells than mice with no treatment at 2 weeks; though they improved over the next 6 weeks, and by 56 days were significantly better than the no treatment TBI mice, they were still not as good as the mice that received three treatments.

The opposite effect was found by El Massri et al. (2018). They explored the long-term effects of PBM and compared mice aged 3 months and 12 months. Five healthy aged mice (5 months old) were treated with PBM (whole body) 670 nm LED for 20 minutes daily for 7 months (210 consecutive treatments) from 5 months old to 12 months. Their brains were then compared with untreated controls—3- and 12-month-old mice. PBM was shown to be effective in reducing glial cell number in the basal ganglia. There were no deleterious effects on interneurons and their terminations, and they conclude that long-term treatment of PBM has beneficial effects on the aging brain.

The experiments were, of course, different not only in the number of treatments but also the disease model and dose. Xuan et al. (2016) treated TBI with 810 nm laser, CW, 18 J/cm^2, 25 mW/cm^2; El Massri et al. (2018) used old but healthy (uninjured) mice treated with 670 nm LED for 20 minutes. No other treatment parameters were reported so it is not possible to assess whether or not this may be an accumulated dose or accumulated dose rate phenomena. Note to Scientists, please report all parameters: wavelength, light source (laser, LED, or other), time, power, beam at the surface, irradiance, pulse regime, time, energy, fluence, number of treatments, and the interval between treatments.

7.13 Devices

There is a bewildering array of PBM devices ranging in price from $50 to $150,000, some emitting a single beam of 1 mW and others as much 95 W in a single beam. There are cluster arrays comprising just a few lasers or LEDs and others containing hundreds or thousands of emitters and delivering a combined output of 580 W of light in a capsule that treats the entire human body.

Ignorance, exaggeration, and misinformation are frequently disseminated by websites and company representatives. Claims of treating through clothing, magic therapeutic pulse frequencies for your aura, or the benefits of special coherence are frequent stories to lure the PBM novice. Shop around several suppliers, attend conferences, and work out who is trustworthy before making an expensive (but worthwhile) investment.

References

Ando, T., et al., 2011. Comparison of therapeutic effects between pulsed and continuous wave 810-nm wavelength laser irradiation for traumatic brain injury in mice. PLoS One 6 (10), e26212.

Barrett, D.W., Gonzalez-Lima, F., 2013. Transcranial infrared laser stimulation produces beneficial cognitive and emotional effects in humans. Neuroscience 230, 13–23.

Benedicenti, S., et al., 2008. Intracellular ATP level increases in lymphocytes irradiated with infrared laser light of wavelength 904 nm. Photomed. Laser Surg. 26 (5), 451–453.

Bjordal, J.M., et al., 2008. A systematic review with procedural assessments and meta-analysis of low level laser therapy in lateral elbow tendinopathy (tennis elbow). BMC Musculoskelet. Disord. 9, 75.

Boonswang, N.A., et al., 2012. A new treatment protocol using photobiomodulation and muscle/bone/joint recovery techniques having a dramatic effect on a stroke patient's recovery: a new weapon for clinicians. BMJ Case Rep. 2012.

Brondon, P., Stadler, I., Lanzafame, R.J., 2005. A study of the effects of phototherapy dose interval on photobiomodulation of cell cultures. Lasers Surg. Med. 36 (5), 409–413.

Cassano, P., et al., 2018. Transcranial photobiomodulation for the treatment of major depressive disorder. The ELATED-2 Pilot Trial. Photomed. Laser Surg.

Chow, R.T., et al., 2009. Efficacy of low-level laser therapy in the management of neck pain: a systematic review and meta-analysis of randomised placebo or active-treatment controlled trials. Lancet 374 (9705), 1897–1908.

Corazza, A.V., et al., 2007. Photobiomodulation on the angiogenesis of skin wounds in rats using different light sources. Photomed. Laser Surg. 25 (2), 102–106.

Disner, S.G., Beevers, C.G., Gonzalez-Lima, F., 2016. Transcranial laser stimulation as neuroenhancement for attention bias modification in adults with elevated depression symptoms. Brain Stimul. 9 (5), 780–787.

El Massri, N., et al., 2017. Photobiomodulation-induced changes in a monkey model of Parkinson's disease: changes in tyrosine hydroxylase cells and GDNF expression in the striatum. Exp. Brain Res. 235 (6), 1861–1874.

El Massri, N., et al., 2018. Photobiomodulation reduces gliosis in the basal ganglia of aged mice. Neurobiol. Aging 66, 131–137.

Farfara, D., et al., 2015. Low-level laser therapy ameliorates disease progression in a mouse model of Alzheimer's disease. J. Mol. Neurosci. 55 (2), 430–436.

Fred, M., Dickey, S.C.H., 2014. Laser Beam Shaping:Theory and Techniques, second ed. CRC Press.

Graham, C.H., Hartline, H.K., 1935. The response of single visual sense cells to lights of different wave lengths. J. Gen. Physiol. 18 (6), 917–931.

Gur, A., et al., 2004. Efficacy of 904 nm gallium arsenide low level laser therapy in the management of chronic myofascial pain in the neck: a double-blind and randomize-controlled trial. Lasers Surg. Med. 35 (3), 229–235.

Hadis, M.A., et al., 2016. The dark art of light measurement: accurate radiometry for low-level light therapy. Lasers Med. Sci. 31 (4), 789–809.

Hecht, S., Williams, R.E., 1922. The visibility of monochromatic radiation and the absorption spectrum of visual purple. J. Gen. Physiol. 5 (1), 1–33.

Huang, Y.Y., et al., 2009. Biphasic dose response in low level light therapy. Dose Response 7 (4), 358–383.

Hwang, J., Castelli, D.M., Gonzalez-Lima, F., 2016. Cognitive enhancement by transcranial laser stimulation and acute aerobic exercise. Lasers Med. Sci. 31 (6), 1151–1160.

Jang, H., Lee, H., 2012. Meta-analysis of pain relief effects by laser irradiation on joint areas. Photomed. Laser Surg. 30 (8), 405–417.

Jenkins, P.A., Carroll, J.D., 2011. How to report low-level laser therapy (LLLT)/photomedicine dose and beam parameters in clinical and laboratory studies. Photomed. Laser Surg. 29 (12), 785–787.

Karu, T.I., 2010. Multiple roles of cytochrome c oxidase in mammalian cells under action of red and IR-A radiation. IUBMB Life 62 (8), 607–610.

Lanzafame, R.J., et al., 2007. Reciprocity of exposure time and irradiance on energy density during photoradiation on wound healing in a murine pressure ulcer model. Lasers Surg. Med. 39 (6), 534–542.

Lee, H.I., et al., 2016. Pre-conditioning with transcranial low-level light therapy reduces neuroinflammation and protects blood-brain barrier after focal cerebral ischemia in mice. Restor. Neurol. Neurosci. 34 (2), 201–214.

Liang, H.L., et al., 2008. Near-infrared light via light-emitting diode treatment is therapeutic against rotenone- and 1-methyl-4-phenylpyridinium ion-induced neurotoxicity. Neuroscience 153 (4), 963–974.

Naeser, M.A., et al., 2014. Significant improvements in cognitive performance post-transcranial, red/near-infrared light-emitting diode treatments in chronic, mild traumatic brain injury: open-protocol study. J. Neurotrauma 31 (11), 1008–1017.

Naeser, M.A., et al., 2016. Transcranial, red/near-infrared light-emitting diode therapy to improve cognition in chronic traumatic brain injury. Photomed. Laser Surg. 34 (12), 610–626.

Nawashiro, H., et al., 2012. Focal increase in cerebral blood flow after treatment with near-infrared light to the forehead in a patient in a persistent vegetative state. Photomed. Laser Surg. 30 (4), 231–233.

Nikolaev, N.A., 1986. Therapeutic efficacy of laser and electropuncture reflexotherapy in correcting the initial manifestations of cerebral circulatory insufficiency. Zh. Nevropatol. Psikhiatr. Im. S S Korsakova 86 (1), 60–64.

Omura, Y., 1983. Non-invasive circulatory evaluation and electro-acupuncture & TES treatment of diseases difficult to treat in Western medicine. Acupunct. Electrother. Res. 8 (3-4), 177–256.

Oron, A., et al., 2012. Near infrared transcranial laser therapy applied at various modes to mice following traumatic brain injury significantly reduces long-term neurological deficits. J. Neurotrauma 29 (2), 401–407.

Oron, U., et al., 2001. Attenuation of infarct size in rats and dogs after myocardial infarction by low-energy laser irradiation. Lasers Surg. Med. 28 (3), 204–211.

Schiffer, F., et al., 2009. Psychological benefits 2 and 4 weeks after a single treatment with near infrared light to the forehead: a pilot study of 10 patients with major depression and anxiety. Behav. Brain Funct. 5, 46.

Schindl, A., Rosado-Schlosser, B., Trautinger, F., 2001. [Reciprocity regulation in photobiology. An overview]. Hautarzt 52 (9), 779–785.

Smith, K.C., 1991. The photobiological basis of low level laser radiation therapy. Laser Therapy 3 (1), 6.

Sommer, A.P., et al., 2001. Biostimulatory windows in low-intensity laser activation: lasers, scanners, and NASA's light-emitting diode array system. J. Clin. Laser Med. Surg. 19 (1), 29–33.

Tedford, C.E., et al., 2015. Quantitative analysis of transcranial and intraparenchymal light penetration in human cadaver brain tissue. Lasers Surg. Med. 47 (4), 312–322.

Tian, F., et al., 2016. Transcranial laser stimulation improves human cerebral oxygenation. Lasers Surg. Med. 48 (4), 343–349.

Vargas, E., et al., 2017. Beneficial neurocognitive effects of transcranial laser in older adults. Lasers Med. Sci. 32 (5), 1153–1162.

Wang, Y., et al., 2017. Photobiomodulation of human adipose-derived stem cells using 810nm and 980nm lasers operates via different mechanisms of action. Biochim. Biophys. Acta Gen. Subj. 1861 (2), 441–449.

Wu, Q., et al., 2012. Low-level laser therapy for closed-head traumatic brain injury in mice: effect of different wavelengths. Lasers Surg. Med. 44 (3), 218–226.

Xuan, W., et al., 2014. Transcranial low-level laser therapy enhances learning, memory, and neuroprogenitor cells after traumatic brain injury in mice. J. Biomed. Opt. 19 (10), 108003.

Xuan, W., Huang, L., Hamblin, M.R., 2016. Repeated transcranial low-level laser therapy for traumatic brain injury in mice: biphasic dose response and long-term treatment outcome. J. Biophotonics 9 (11–12), 1263–1272.

Zalevsky, Z., Belkin, M., 2011. Coherence and speckle in photomedicine and photobiology. Photomed. Laser Surg. 29 (10), 655–656.

Chapter 8

Mechanisms of photobiomodulation in the brain

Michael R. Hamblin[1,2]

[1]*Wellman Center for Photomedicine, Massachusetts General Hospital, Boston, MA, United States,* [2]*Department of Dermatology, Harvard Medical School, Boston, MA, United States*

8.1 Introduction

The underlying motivation for this book is the recent growth of interest in the use of photobiomodulation (PBM) as a therapy for diverse brain disorders. The application of PBM (or what used to be called low level laser therapy) to the brain started when it was first used for acute stroke (Streeter et al., 2004). Pioneering work by Uri Oron's group in Israel had shown that application of near-infrared laser (810 nm) to the head had pronounced beneficial effects in several different animal models of ischemic stroke (Hamblin, 2018; Lampl, 2007). These studies had investigated the ability of PBM to prevent long-term neurological damage following an induced acute stroke (Khuman et al., 2012) in various animal models. Typical parameters tested were 810 nm NIR light at 7.5 mW/cm^2 calculated as the power density at the level of the cortex. Mechanistic investigations that were carried out in the early studies concentrated on newly formed neuronal cells and migrating cells within the brain (Oron et al., 2006).

8.2 Molecular mechanisms of photobiomodulation

The first law of photobiology is that a photon must be absorbed by a specific molecular chromophore in order to have any biological effect. These chromophores that, broadly speaking, absorb different regions in the visible spectrum (blue, green, red, NIR) are shown in Fig. 8.1 and discussed below.

8.2.1 Mitochondria and cytochrome c oxidase

Cytochrome c oxidase (CCO) is the terminal enzyme (unit IV) in the electron transport chain situated in the outer mitochondrial membrane. The electron transport chain, through a series of redox reactions, facilitates the transfer of electrons across the inner membrane of the mitochondria. The net result of these electron transfer steps is to produce a proton gradient across the mitochondrial membrane that drives the activity of adenosine triphosphate (ATP) synthase (sometimes called unit V) that produces the high energy ATP from ADP. CCO mediates the transfer of electrons from cytochrome c to molecular oxygen. CCO is a complex protein, composed of 13 different polypeptide subunits, and also contains two heme centers and two copper centers. Each of these heme and copper centers can be either oxidized or reduced, giving 16 different oxidation states. Each of these oxidation states has a slightly different absorption spectrum, but CCO is almost unique amongst biological molecules in having a significant absorption in the near-infrared spectrum. In fact, Britton Chance estimated that over 50% of the absorption of NIR light by biological tissue could be attributed to this single enzyme as a chromophore (Cooper et al., 1999).

In many publications, CCO has been shown to be a biological photoacceptor and transducer of signals activated by light in the red and NIR regions of the spectrum (Karu, 2010; Wong-Riley et al., 2005). Specifically, absorption of the photons delivered in PBM seems to promote an increase in the availability of electrons for the reduction of molecular oxygen in the catalytic center of CCO, increasing mitochondrial membrane potential (MMP), and increasing levels of

Photobiomodulation in the Brain. DOI: https://doi.org/10.1016/B978-0-12-815305-5.00008-7

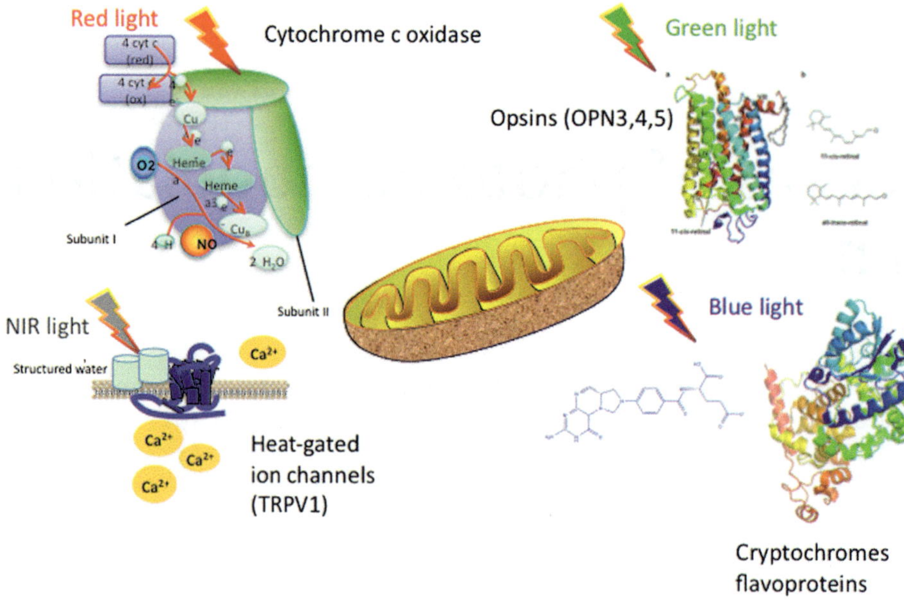

FIGURE 8.1 **Proposed chromophores for PBM that can absorb different colors of light.** It should be noted that there is considerable overlap between the chromophores, and that the NIR absorbed by structured water is likely to be a longer wavelength (>950 nm).

ATP, cyclic adenosine monophosphate, and reactive oxygen species (ROS); all of which indicate increased mitochondrial function, and can trigger initiation of cellular signaling pathways (de Freitas and Hamblin, 2016)

Although other units in the electron transfer chain, such as complexes I–IV and succinate dehydrogenase, also show increased activity as a result of PBM, CCO is still believed to be one of the primary photoacceptors. This notion is supported by the fact that low level light irradiation such as PBM causes increased oxygen consumption, and is bolstered by the fact that the majority of oxygen consumption occurs at complex IV, and moreover that the addition of sodium azide, a CCO inhibitor, abrogates the effects of PBM (Spitler et al., 2015; Nunez-Alvarez et al., 2017). Moreover, rho zero cells that lack functional mitochondria do not respond to PBM in the same way as their wild-type counterparts (Wu et al., 2014).

An alternate theory is that PBM works through the photodissociation of NO from CCO (Lane, 2006). Evidenced by the increased levels of NO following PBM, it has been proposed that NO photodissociation reverses the mitochondrial inhibition of cellular respiration that exists as a result of excessive NO binding (Moncada and Erusalimsky, 2002). According to this theory, NO could be photodissociated from its inhibitory binding sites at the heme iron and copper centers of CCO. Thus, NO could no longer compete with oxygen for binding to the catalytic metal centers allowing for an influx of oxygen, and therefore the enzymatic activity and respiration could return to baseline (see Fig. 8.2). NO can also be photoreleased from other intracellular sites, including nitrosylated myoglobin and hemoglobin, producing similarly positive effects (Lohr et al., 2009). Moreover nitrosylated thiols and a glutathione dinitrosyl iron complex can also be photodissociated (with a peak in the action spectrum at 670 nm) to release NO (Keszler et al., 2018). Poyton's laboratory reported a newly described function of CCO to catalyze nitrite-dependent NO synthesis, in other words to act as a nitrite reductase (Poyton and Ball, 2011). They found that both yeast and mouse brain mitochondrial CCO could produce NO over a wide range of oxygen concentrations, and that the rate of NO synthesis increased as the oxygen concentration decreased, becoming optimal under hypoxic conditions. PBM (590 ± 14 nm) increased NO production in an intensity-dependent fashion, but had no effect on oxygen consumption by CCO (Ball et al., 2011).

On a slightly broader scale, PBM is also believed to trigger retrograde mitochondrial signaling (Karu, 2008). This refers to signals and communications passing from the mitochondria to the nucleus of a cell, rather than vice versa. The aforementioned mitochondrial changes result in an altered mitochondrial ultrastructure, which drives the communications (Passarella and Karu, 2014). As a result, membrane permeability and ion flux at the cell membrane are altered, in turn, leading to the altered activity of activator protein-1 (AP1) and NFκB (Kaminski et al., 2012).

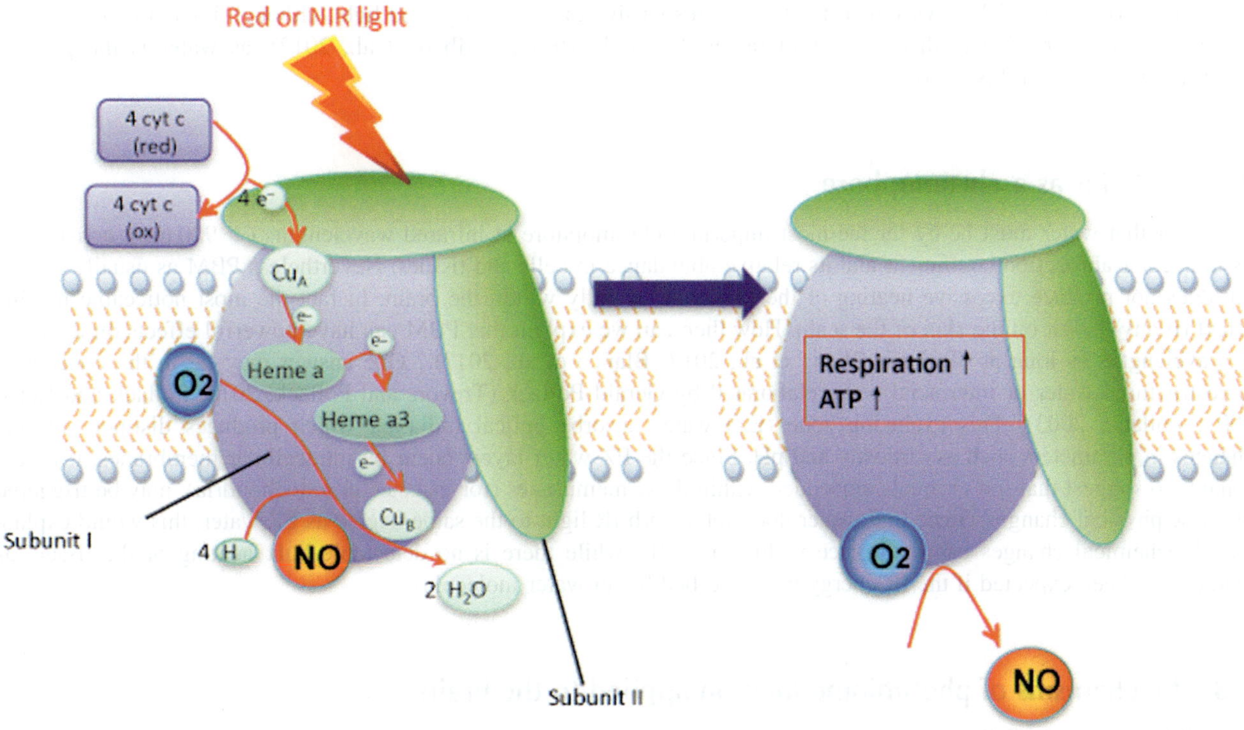

FIGURE 8.2 Proposed photodissociation of inhibitory nitric oxide from the Cu$_B$ center in CCO thus allowing respiration to resume and increasing ATP.

8.2.2 Opsins, flavins, and cryptochromes

While CCO is undoubtedly the most important chromophore in PBM, there is emerging evidence that other primary chromophores such as opsins, flavins, and cryptochromes, may mediate the biological absorption of light, particularly at shorter wavelengths (blue and green). Opsins contain a cis-retinaldehyde molecule as a chromophore that is photoisomerized to the all trans-isomer, thus producing a change in protein conformation and initiating a signaling cascade (Koyanagi and Terakita, 2014). Flavins and flavoproteins contain a chromophore such as riboflavin, flavin mononucleotide, or flavin adenine dinucleotide and can carry out redox reactions when excited by light (Losi and Gartner, 2011). Cryptochromes are a special subclass of flavoproteins that act as blue light receptors in plants, animals, and even humans (Chaves et al., 2011).

8.2.3 Light-gated ion channels

Although evidence proving that light-gated ion channels can be cited as mechanisms of action in PBM is sparse at the present time, it is gradually increasing. PBM is most likely to affect transient receptor potential (TRP) channels. First discovered in a Drosophila mutant as the mechanism responsible for the vision of insects, they are now known to be sensitive to light (Katz et al., 2017), in addition to a wide variety of other stimuli. TRP channels are calcium channels, and are modulated by phosphoinositides (Katz and Minke, 2018). Light-gated ion channels have attracted immense attention in the field of optogenetics (G et al., 2013). However, the majority of these studies employs ion channels similar to bacterially derived channel rhodopsin (G et al., 2013). The majority of research relating PBM to light-gated ion channels has been done by testing the TRPV vanilloid subfamily of TRP species. Evidence from studies done by various groups (Yang et al., 2007; Ryu et al., 2010; Albert et al., 2012; Gu et al., 2012) has led to the general consensus that TRP channels are most likely to be activated by green light. However, because green light lacks the same penetrating ability of infrared or near-infrared light, it lacks practical clinical application. However, Ryu et al. (2010) found that exposure to infrared (2780 nm) wavelength light attenuated TRPV1 activation, causing a decrease in generation of pain stimuli. A similar, but far less dramatic antinociceptive effect was also observed when TRPV4 was exposed to light of the same wavelength. TRPV4 was also shown to be responsive to 1875 nm pulsed light, although it cannot be ruled out that the results were due to thermal stimuli rather than light stimuli (Albert et al., 2012), as water is the primary absorber of infrared in this region.

8.2.4 Water as a chromophore

It is clear that water must be by far the most important chromophore at infrared wavelengths (>900 nm), considering its molecular absorption coefficient and its relative abundance in cells and tissues. Nevertheless PBM as usually carried out does not produce excessive heating of the tissues, especially within the brain. In fact the most noticeable heating effect (if any) is felt on the skin of the scalp. How then can we explain that PBM can have powerful effects on the brain at wavelengths as long as 1064 nm (Wang et al., 2017; Blanco et al., 2017).? One answer may lie in the concept of nanostructured water or interfacial water elaborated by Gerald Pollack (Trevors and Pollack, 2012; Pollack and Reitz, 2001; Pollack, 2003). This exclusion zone (EZ) water absorbs optical radiation which produces distinct physical changes in parameters such as viscosity and pH. Since the EZ water layers occur on intracellular membranes, it is reasonable to suggest that ion channels embedded within these membranes (for instance in mitochondria), may be triggered by these physical changes. Since bulk water does not absorb IR light to the same degree as EZ water, this would explain why biochemical changes can take place within the cells, while there is no detectable bulk heating of the tissue, as would have been expected if the IR energy was absorbed by all water molecules.

8.3 Mechanisms of photobiomodulation applied to the brain

As will be seen in the following paragraphs, a bewildering array of different mechanisms has been proposed to account for the benefits of tPBM on the brain. These are schematically shown in Fig. 8.3.

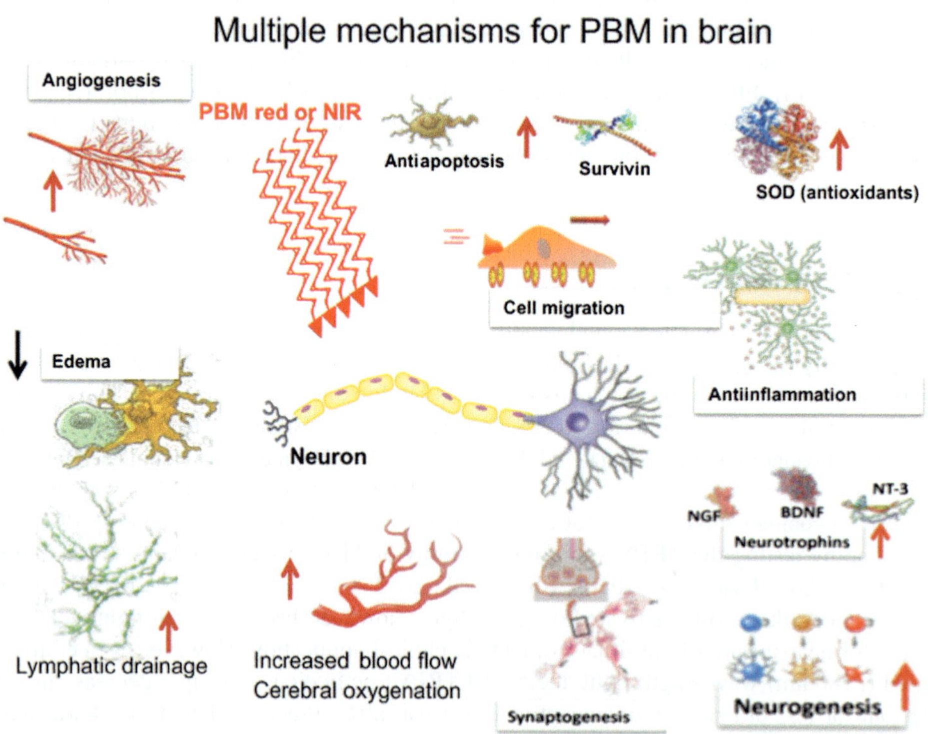

FIGURE 8.3 There are a large number of mechanisms for PBM in the brain that have been proposed, and many of which are discussed.

8.3.1 Metabolism

Improved metabolic functioning is one of the most easily recognizable effects of PBM, and increased intracellular ATP production is one the most strongly supported mechanisms of action. Using phosphorus magnetic resonance spectroscopy, significant increases in total nucleotide triphosphates, a marker of cellular energy availability, have been observed following PBM to healthy adult beagle dogs (Mintzopoulos et al., 2017). This strongly suggested that an increase in cellular ATP production has occurred as a result of PBM. Moreover, several preclinical studies have shown that the brain content of ATP was increased in experimental animals (mice or rats) subjected to PBM for various brain disorders (Ando et al., 2011; Salehpour et al., 2017). It is a general finding that mitochondrial dysfunction, inadequate supplies of ATP, and oxidative stress are contributory factors in almost all forms of brain disease (Kann, 2016). This has been reported for neurological conditions such as major depressive disorder (Cao et al., 2013), traumatic brain injury (Lyons et al., 2018), Parkinson's disease (Briston and Hicks, 2018), and Alzheimer's disease (Swerdlow et al., 2014).

8.3.2 Blood flow

One of the easiest changes to measure in animals, or especially in humans, that occurs after tPBM, is changes in cerebral blood flow and changes in oxygenation. This is because near-infrared spectroscopy is a noninvasive technique that has shown rapid growth in recent years. In fact, Wang et al. (2016) carried out this measurement on the forearms of human volunteers treated with a 1064 nm laser. They found that PBM induced significant increases of CCO concentration (Delta[CCO]) and oxygenated hemoglobin concentration (Delta[HbO]) on the treated site as the laser energy dose accumulated over time. A strong linear relationship between Delta[CCO] and Delta[HbO] was observed indicating a response in both oxygen supply and blood volume. Schiffer et al. (2009) tested tPBM using an 810 nm LED applied to the forehead for major depression and anxiety, and used an INVOS commercial system from Somanetics (Troy, MI) to measure cerebral hemoglobin (cHb) in left and right frontal and rCBF, in addition to the device's usual oxygen saturation output.

It has been suggested that the release of NO as a result of PBM is responsible for the increased cerebral blood flow (Lee et al., 2017). NO is a major neuronal signaling molecule which, among other functions, possesses the ability to trigger vasodilation. To do so, it first stimulates soluble guanylate cyclase to form cyclic-GMP (cGMP). The cGMP then activates protein kinase G, leading to the reuptake of Ca^{2+} and the opening of calcium-activated potassium channels. Due to the subsequent fall in concentration of Ca^{2+}, myosin light-chain kinase is prevented from phosphorylating the myosin molecule, causing the smooth muscle cells in the lining of blood vessels and lymphatic vessels to become relaxed (Charriaut-Marlangue et al., 2013). This vasodilation then promotes improved circulation, which in turn leads to improved cerebral oxygenation in a similar manner to that observed with pulsed electromagnetic fields (Bragin et al., 2015).

8.3.3 Neuroprotection

A wide variety of evidence suggests that PBM can be utilized for neuroprotection, essentially to protect cells from damage, to promote their survival and longevity and reverse apoptotic signaling processes. One way it achieves this result is by inhibiting the activity of glycogen synthase kinase 3β (GSK3β). To do so, it activates protein kinase B (AKT) which increases the phosphorylation level of its Ser9 residue, which then allows the N terminus of GSK3β to bind with its own binding site. One result of this is the accumulation and translocation to the nucleus of β-catenin, which ceases to be under-phosphorylated and therefore becomes more active when GSK3β activity is inhibited. Once allowed to accumulate in the nucleus, β-catenin relies on the increased TCF/LEF dependent transcriptional activity to promote cellular survival (Liang et al., 2012). This inhibition of GSK3β also helps to prevent apoptosis, the normal cell death that occurs as an organism grows. GSK3β is believed to act as a mediator between AKT and Bax, a protein which is translocated to the nucleus in the presence of proapoptotic stimuli to trigger the beginning of the process. However, when GSK3β is inhibited, AKT the communication pathway between AKT and Bax is cut off. As a result, Bax translocation cannot be signaled for and is thus inhibited (Zhang et al., 2010).

PBM also demonstrates neuroprotective qualities in the form of protection from senescence (Ling et al., 2014). It has been shown to activate the extracellular signal-related kinase (ERK)/forkhead box protein M1 (FOXM1) pathway. The FOXM1 protein regulates the progression from the G1 to the S phase of the cell cycle, and via the activation of the ERK/FOXM1 pathway, PBM leads to the greater translocation of ERK to the nucleus and the greater accumulation of FOXM1 in the nucleus. This, in turn, causes reduced expression of the p21 protein and mitotic arrest in the G1 phase, therefore slowing the overall progression of cellular senescence.

PBM has also been shown to be effective in protecting cells from the harmful effects of toxins (Eells et al., 2016). In a study done by Eells et al. (2003), irradiation with 670 nm light was successful in causing the recovery of retina function and the prevention of histological damage in rodent models exposed to methanol. This is likely due to the fact that methanol generates the toxic metabolite formic acid, an inhibitor of CCO, and PBM is a known stimulator of CCO. A study by Wong-Riley et al. (2005) on the effects of PBM posttetrodotoxin exposure produced similarly successful results, especially when models were irradiated with 670 and 830 nm light, the peaks of the CCO absorption spectrum. This further indicates that the antitoxin effect of PBM can be traced to its stimulation of CCO. PBM is also effective in prevention of the harmful effects associated with potassium cyanide. When pretreated with 670 nm light, Liang et al. (2006) found that neuronal expression of Bax induced by cyanide was decreased, preventing the subsequent apoptosis.

In addition, PBM has demonstrated the rather unique property of affecting cells in different states of health in different ways, essentially modifying the cell in whatever way might be necessary to promote its survival. For instance, in normal cells the absorption of light by CCO leads to an increase in MMP above baseline and a short surge in ROS production. However, in cells where MMP is low due to existing oxidative stress, excitotoxicity, or inhibition of electron transport, light absorption leads to an increase of MMP toward normal levels and a decrease of ROS production (Huang et al., 2013). Similarly, the typical response to PBM in healthy cells is an uptick in intracellular Ca^{2+} (Sharma et al., 2011) However, in cells that already contain excess Ca^{2+} (a phenomenon called excitotoxicity) PBM provokes the opposite reaction, in other words it lowers excessive levels of cellular calcium, thus promoting cell survival, lowering oxidative stress, and raising MMP back to normal (Huang et al., 2014)

8.3.4 Oxidative stress

Oxidative stress occurs when an imbalance exists between the production of ROS and the ability of the body to counteract their effects, which become harmful when they are in excess, via antioxidants. Many sources have linked oxidative stress to various neurological conditions, such as major depressive disorder (Roomruangwong et al., 2018) and traumatic brain injury (Rodriguez-Rodriguez et al., 2014), not to mention cardiovascular (Wu et al., 2014) and Alzheimer's diseases (Wang et al., 2014).

However, the situation is more complicated than at first appears, because large numbers of clinical trials of antioxidant therapy for all these diseases have failed (sometimes dismally) (Persson et al., 2014; Steinhubl, 2008). Apparently some level of oxidative stress is necessary for the optimum functioning of human beings, and removing all oxidative stress with supplementation with antioxidants can be counterproductive (Rahal et al., 2014). An important paper showed that the health giving benefits of exercise were removed when humans were administered antioxidants (Ristow et al., 2009).

Salehpour et al. (2018) showed that sleep deprivation in mice caused oxidative stress in the hippocampus and subsequent memory impairment. PBM with NIR (810 nm) was delivered (once a day for 3 days) transcranially to the head. Mice performed better on the Barnes maze and the what, -where, -which task; hippocampal levels of antioxidant enzymes were increased and oxidative stress biomarkers were decreased. In studies of the effect of PBM on traumatized muscle, PBM has been shown to be effective in regulating the amount of cytokine-inducible nitric oxide synthase (iNOS) produced by the cell. This is important because excessive amounts of iNOS can lead to the excessive production of NO, which would then signal increased production of the ROS/reactive nitrogen species (RNS) called peroxynitrite, leading to an increase in oxidative stress. Specifically, PBM could reduce peroxynitrite (Bartos et al., 2016), while still preserving the positive effects of other isoforms of NO synthase, such as endothelial nitric oxygen synthase (eNOS), which is the species primarily responsible for the vasodilating effects of PBM (Mungrue et al., 2002; Ahmed et al., 2011; Assis et al., 2013).

PBM has also been shown to stimulate increases in angiogenesis, leading to further improvements in blood flow. As demonstrated by Cury et al. (2013), PBM at 780 nm and 40 J/cm^2 triggered an increase in the expression of the protein HIF1α and of vascular endothelial growth factor, and a decrease in matrix metalloproteinase two activity, all of which were found to induce angiogenesis. Additionally, an in vitro study of the effects of red/NIR light on red blood cells showed that NIR light was quite effective in protecting red blood cells from oxidation (Walski et al., 2018), which is a common occurrence in brains compromised by conditions such as major depressive disorder (MDD) (Sarandol et al., 2007).

8.3.5 Antiinflammatory effects

Inflammation is one of the innate immune system's defenses against foreign bodies such as bacteria and viruses. On a cellular level, it occurs when the transcription factor NFκB is activated. While acute inflammation is positive, chronic inflammation can have very negative effects. Many diseases, including neurodegenerative diseases and mood disorders, can be traced at least in part to chronic inflammation.

One way PBM helps to quell inflammation is through the inhibition of the cyclo-oxygenase 2 (COX-2) enzyme. Lim et al. (2013) found that 635 nm light irradiation at low power was able to cause COX-2 inhibition by decreasing intracellular ROS. Inhibition of COX-2 via pharmaceutical means is widely supported at present, with COX-2 inhibitors making up a significant portion of the market for nonsteroidal antiinflammatory drugs (Yang and Gao, 2017). Using PBM, essentially the same result can be accomplished, just with a different stimulus.

PBM can also modulate cellular levels of free NFκB. NFκB is found in the cytosol bound to IκB, an inhibitor protein. Proinflammatory stimuli have the ability to activate IκB kinase, an upstream signaling regulator that causes the degradation of IκB. Once the IκB has been degraded, the NFκB is free to translocate to the nucleus, where it triggers the expression of proinflammatory genes. There is evidence that PBM can have opposite effects on NFκB depending on the type of cells and their activation state that is studied. Chen et al. (2011a) found that in normal fibroblasts PBM could activate NFκB via generation of low amounts of ROS from mitochondria that had been stimulated. The same group, however, found that in dendritic cells (another type of macrophage cell) which had been activated toward a M1 phenotype by Toll-like receptor agonists, that PBM could reduce proinflammatory cytokines (Chen et al., 2011b). Likewise Yamaura et al. (2009) found that the level of NFκB was reduced in activated rheumatoid arthritis derived synoviocytes that received PBM.

Additionally, PBM possesses the ability to modulate levels of cytokines, proteins that act as important signaling molecules for the immune system. PBM has been shown to modulate levels of both pro- and antiinflammatory cytokines, though for the reduction of inflammation, its ability to modulate levels of tumor necrosis factor and other proinflammatory cytokines is especially useful.

It should be noted that inflammation within the brain has distinct differences compared to inflammation in other parts of the body. In fact, the term neuroinflammation is commonly applied to the activation of microglia. Microglia are cells of the monocyte/macrophage lineage that act as the immune defense system in the central nervous system (Filiano et al., 2015). Microglia are constantly scavenging the CNS for plaques, damaged neurons and synapses, and infectious agents. Microglia are extremely sensitive to even small pathological changes in the CNS (Dissing-Olesen et al., 2007).

In common with other cells in the macrophage lineage, microglia can assume a diversity of phenotypes, and retain the capability to shift their function to maintain tissue homeostasis. Microglia can be activated by LPS or IFN-γ to an M1 phenotype that expresses proinflammatory cytokines and is able to kill microbial cells. On the other hand microglia can be activated by IL-4/IL-13 to an M2 phenotype for phagocytosis of debris, resolution of inflammation, and tissue repair. Increasing evidence suggests a role of metabolic reprogramming in the regulation of the innate inflammatory response (Orihuela et al., 2016). Studies have demonstrated that the M1 phenotype is often accompanied by a shift from oxidative phosphorylation to aerobic glycolysis for energy production (Haschemi et al., 2012). Under these conditions, energy demands would be associated with functional activities and cell survival and thus may serve to influence the contribution of microglia activation to various neurodegenerative conditions.

Since there is considerable evidence that PBM can activate the mitochondrial metabolism toward oxidative phosphorylation and away from aerobic glycolysis, this is a plausible reason why PBM may change the microglial phenotype from M1 toward M2 (Fernandes et al., 2015). The consequences of this shift would be that instead of M1 microglia that cannot dispose of substances such as beta-amyloid plaques, and pump out ROS and inflammatory cytokines, PBM-induced M2 microglia could clear the plaques, exert antiinflammatory and antioxidant effects and encourage tissue healing.

8.3.6 Neurogenesis

For many years, it was thought that the adult brain was incapable of growing new brain cells. Although it was realized that growing and developing brains in embryos, young animals and children must be capable of neurogenesis mediated by neural stem cells (NSCs) and neuroprogenitor cells; it was thought that this process had ceased in adulthood. The turning point in our perception was the discovery of adult neurogenesis and identification of cells that both in vitro and in vivo can function as NSCs, generating new neurons, glial cells, or both (Bergmann et al., 2015). The paradigm shift regarding the nature of NSCs and the potential of the postnatal brain to regenerate, opened the gates for new studies with a new outlook (Lepousez et al., 2015). Now the scientific community is engaged in not only an in-depth understanding of the adult brain neurogenesis and NSC functions, but also how they may be encouraged with novel treatment modalities (Kirschen et al., 2018). Experimental NSCs/NPs are detected by the incorporation of bromo-deoxyuridine

(BrdU) into the nuclei of dividing cells after infection into the animal at various times before sacrifice, that can be subsequently measured by an antibody (Zhang et al., 2015). However, it has been established that there are only a few well-defined areas of the brain (known as neurogenic niches) in which this neurogenesis is observed (Pozhilenkova et al., 2017). The most well-accepted neurogenic niches are the subgranular layer of the dentate gyrus of the hippocampus (Hu et al., 2015), and the subventricular zone (SVZ) of the lateral ventricles (Ribeiro Xavier et al., 2015). In order to be assured the BrdU positive cells are actually neurons, rather than glia or some other cell type, it is usual to stain them with a second antibody to NeuN (marker of mature neurons) or to Tuj-1 (beta tubulin class III) (Shen et al., 2016).

The first report of neurogenesis being stimulated by PBM delivered to the brain came from a study by Oron et al. (2006) who induced a stroke in rats, and treated them with PBM. The number of newly formed neuronal cells (BrdU-Tuj-1 double-positive) as well as migrating cells (doublecortin positive) was significantly elevated in the SVZ of the hemisphere ipsilateral to the induction of stroke when treated with PBM (Oron et al., 2006). A similar result was reported by Xuan et al. (2014) who treated mice that had suffered a TBI using PBM. They found that there was a significant increase in double-stained BrdU-NeuN (neuroprogenitor cells) in the dentate gyrus and in the SVZ at 1 week post-TBI but not at 4 weeks post-TBI. Increases in doublecortin (DCX) and TUJ-1 were also seen.

8.3.7 Synaptogenesis

One of the most notable and potentially significant effects of PBM on the brain discovered to date is its ability to promote synaptogenesis, also called neuroplasticity. This process is vitally important, as many brain conditions, including TBI, stroke, neurodegenerative diseases, and mood disorders can be traced, either partially or in full, to poor or aberrant neuronal connections in certain regions of the brain. If PBM possesses the ability to counter these effects by facilitating neural organization or reorganization, it could prove to be extremely promising as a novel method of treating such brain disorders.

One manner in which PBM promotes neuronal connectivity could be by up-regulation of brain derived neurotrophic factor (BDNF). It is a member of the neurotrophins class which also includes nerve growth factor, NT3, NT4, and GDNF (Bothwell, 2014). BDNF is a protein found in the nervous system, which helps to maintain existing neurons and encourage the growth of new neurons and new synapses. Specifically, it is believed to modulate dendritic structure to facilitate improved synaptic transmission (Yang et al., 2014). PBM has been shown to slow attenuation of BDNF via the ERK/CREB pathway, thus positively affecting dendritic morphogenesis and improved neuronal connectivity (Ando et al., 2011). BDNF is also a mediator of the protein synapsin-1, which improves synaptogenesis by accelerating the development of neuronal fibers and maintaining synaptic contacts (Barbieri et al., 2018). In a study done by Meng et al. (2013), denser branches and increased interconnectivity between fibers were observed in neural tissue of embryonic rats following irradiation with 780 nm light, indicating increased action of these proteins. BDNF has also been linked to improvements in neuroplastic adaptation, which is especially important in cases of traumatic brain injury and stroke (Wang et al., 2017).

If it can be conclusively shown that PBM stimulates neuroplasticity and synaptogenesis in humans as well as mice, then this opens the door to a wide range of clinical applications (Forrest et al., 2018). Impaired or aberrant neuroplasticity has been implicated in a wide range of brain disorders such as Alzheimer's (Nahum et al., 2013), psychiatric disorders (Kuhn et al., 2014), stroke (Felling and Song, 2015), TBI (Tomaszczyk et al., 2014), and addiction (Creed et al., 2015).

8.3.8 Stem cells

It should not be forgotten that when any kind of PBM light is shone onto a living animal, it is inevitable that some stem cells will be exposed to light. It is known that stem cells respond well to PBM in terms of proliferation and differentiation (Arany, 2016; Abrahamse and Hamblin, 2017). The stem cells may be located in the bone marrow underlying the tissue, and in the bones that are in the illuminated area. Farfara et al. (2015) showed that applying PBM to the bone marrow in the legs had a therapeutic effect in a mouse model of Alzheimer's disease. The same procedure had major therapeutic benefits for reducing the infarct area in heart attack models (Blatt et al., 2016; Tuby et al., 2011), and in ameliorating ischemic kidney injury (Oron et al., 2014).

8.3.9 Preconditioning

PBM preconditioning refers to the application of light to the body in advance of some insult, which subsequently damages the tissue (Agrawal et al., 2014). Preconditioning has been carried out with several other different physical interventions that can be regarded as causing mild levels of cell stress (Amin et al., 1995); the most commonly investigated has been ischemic preconditioning (Zhang et al., 2003). The term "remote preconditioning" refers to the systemic effects of restricting blood flow to a limb (Chen et al., 2018). Ischemic preconditioning has been shown to protect the brain in animal models of stroke (Hahn et al., 2011). Preconditioning with PBM has been well studied on athletic performance and muscle function (Ferraresi et al., 2016). PBM preconditioning was studied in a mouse model of ischemia/reperfusion brain injury induced by middle cerebral artery occlusion (Lee et al., 2017). The mice received PBM to the head twice a day for 2 days prior to cerebral ischemia. After reperfusion, the PBM group showed significantly smaller infarct, higher cerebral blood flow, and fewer behavioral deficits. There was significantly higher eNOS phosphorylation in PBM pretreated mice, possibly by stimulation of the phosphoinositide 3-kinase (PI3K)/Akt pathway.

8.3.10 Systemic effects

Another manner in which PBM affects the brain may be through broader systemic effects. That is, when PBM is used to target a specific area other than the brain, the brain can benefit remotely from the changes being caused. There is the possibility that PBM can trigger the secretion of a yet-unidentified extracellular signaling pathway in the circulating blood. One of the most convincing examples of the systemic effects of PBM was demonstrated by Johnstone et al. (2014). These workers studied a mouse model of Parkinson's disease caused by MPTP injection. They compared 670 nm light directed specifically at either the head or body. Indirect application (remote PBM or abscopal PBM) produced a significant rescue of tyrosine hydroxylase positive cells in the SNc at the low MPTP dose, but this protection was not as robust as that achieved by direct irradiation of the head. In another study (Kim et al., 2018) they showed that remote preconditioning with PBM (670 nm light to the mouse back) protected mice against a subsequent challenge with MPTP. Because remote PBM produced a similar degree of neuroprotection to ischemic preconditioning of the leg, they suggested the two approaches shared a common mechanism.

This possibility is evidenced by intranasal LED therapy, which is beginning to be used for brain applications. Although it has been proposed that intranasal LED application allows light to penetrate as far as the brain, an examination of the anatomy and optical parameters makes this somewhat unlikely. The Vielight transcranial LED device is used simultaneously with intranasal LEDs (Saltmarche et al., 2017). Nevertheless, it appears that intranasal LEDs used alone can have positive effects on brain function. Workers in China have used intranasal laser for treating insomnia (Wang, 2006; Xu et al., 2001), and found objective changes in EEG patterns (Xu et al., 2002a). The same group has reported the use of intranasal HeNe laser to treat Alzheimer's disease (Xu et al., 2002b), Parkinson's disease (Xu et al., 2003), and poststroke depression (Xu et al., 2002c).

8.3.11 Laser acupuncture

Laser acupuncture has been widely studied as an alternative to traditional Chinese needle acupuncture. Some of these beneficial reports include brain disorders. Laser acupuncture had beneficial effects (behavior and oxidative stress) in a mouse model of autism caused by valproic acid administration (Khongrum and Wattanathorn, 2015, 2017). Laser acupuncture improved the impaired memory in a rat model of Alzheimer's disease caused by administration of a cholinotoxin (Sutalangka et al., 2013). In a rat model of alcoholism, laser acupuncture decreased the acetylcholinesterase activity and the malondialdehyde level, but increased the activities of catalase, superoxide dismutase, and glutathione peroxidase in the hippocampus (Phunchago et al., 2015). In clinical studies, laser acupuncture in healthy volunteers produced resting state functional MRI changes in several different brain regions, suggesting it could improve cognitive function (Lv et al., 2016). Quah-Smith et al. (2013a) compared laser acupuncture with needle acupuncture in healthy volunteers using functional MRI, and found distinct differences in the activation of different brain regions. Laser acupuncture has been used to treat smoking cessation (Lim, 2018), and depression (Quah-Smith et al., 2013b) in patients.

8.4 Conclusion

One criticism that is often leveled at PBM for the brain is that no simple physical intervention can possibly exert so many beneficial changes in the brain. Nevertheless, as this chapter has shown, there is emerging evidence that all the mechanisms discussed may operate to one degree or another. Time will tell how important each individual mechanism is for all the different brain disorders that can be benefited by PBM, and which are covered in this book.

References

Abrahamse, H., Hamblin, M.R., 2017. Photomedicine and Stem Cells. IOP Science Ebooks (Morgan and Claypool Publishing), San Rafael, CA.

Agrawal, T., Gupta, G.K., Rai, V., Carroll, J.D., Hamblin, M.R., 2014. Pre-conditioning with low-level laser (light) therapy: light before the storm. Dose Response 12 (4), 619–649.

Ahmed, I., Bose, S.K., Pavese, N., Ramlackhansingh, A., Turkheimer, F., Hotton, G., et al., 2011. Glutamate NMDA receptor dysregulation in Parkinson's disease with dyskinesias. Brain 134 (Pt 4), 979–986.

Albert, E.S., Bec, J.M., Desmadryl, G., Chekroud, K., Travo, C., Gaboyard, S., et al., 2012. TRPV4 channels mediate the infrared laser-evoked response in sensory neurons. J. Neurophysiol. 107 (12), 3227–3234.

Amin, V., Cumming, D.V., Coffin, R.S., Latchman, D.S., 1995. The degree of protection provided to neuronal cells by a pre-conditioning stress correlates with the amount of heat shock protein 70 it induces and not with the similarity of the subsequent stress. Neurosci. Lett. 200 (2), 85–88.

Ando, T., Xuan, W., Xu, T., Dai, T., Sharma, S.K., Kharkwal, G.B., et al., 2011. Comparison of therapeutic effects between pulsed and continuous wave 810-nm wavelength laser irradiation for traumatic brain injury in mice. PLoS One 6 (10), e26212–e26220.

Arany, P.R., 2016. Photobiomodulation therapy: communicating with stem cells for regeneration? Photomed. Laser Surg. 34 (11), 497–499.

Assis, L., Moretti, A.I., Abrahao, T.B., de Souza, H.P., Hamblin, M.R., Parizotto, N.A., 2013. Low-level laser therapy (808 nm) contributes to muscle regeneration and prevents fibrosis in rat tibialis anterior muscle after cryolesion. Lasers Med. Sci. 28 (3), 947–955.

Ball, K.A., Castello, P.R., Poyton, R.O., 2011. Low intensity light stimulates nitrite-dependent nitric oxide synthesis but not oxygen consumption by cytochrome c oxidase: implications for phototherapy. J. Photochem. Photobiol. B 102 (3), 182–191.

Barbieri, R., Contestabile, A., Ciardo, M.G., Forte, N., Marte, A., Baldelli, P., et al., 2018. Synapsin I and Synapsin II regulate neurogenesis in the dentate gyrus of adult mice. Oncotarget 9 (27), 18760–18774.

Bartos, A., Grondin, Y., Bortoni, M.E., Ghelfi, E., Sepulveda, R., Carroll, J., et al., 2016. Pre-conditioning with near infrared photobiomodulation reduces inflammatory cytokines and markers of oxidative stress in cochlear hair cells. J. Biophotonics 9 (11–12), 1125–1135.

Bergmann, O., Spalding, K.L., Frisen, J., 2015. Adult neurogenesis in humans. Cold Spring Harb. Perspect. Biol. 7 (7), a018994.

Blanco, N.J., Maddox, W.T., Gonzalez-Lima, F., 2017. Improving executive function using transcranial infrared laser stimulation. J. Neuropsychol. 11 (1), 14–25.

Blatt, A., Elbaz-Greener, G.A., Tuby, H., Maltz, L., Siman-Tov, Y., Ben-Aharon, G., et al., 2016. Low-level laser therapy to the bone marrow reduces scarring and improves heart function post-acute myocardial infarction in the pig. Photomed. Laser Surg. 34 (11), 516–524.

Bothwell, M., 2014. NGF, BDNF, NT3, and NT4. Handb. Exp. Pharmacol. 220, 3–15.

Bragin, D.E., Statom, G.L., Hagberg, S., Nemoto, E.M., 2015. Increases in microvascular perfusion and tissue oxygenation via pulsed electromagnetic fields in the healthy rat brain. J. Neurosurg. 122 (5), 1239–1247.

Briston, T., Hicks, A.R., 2018. Mitochondrial dysfunction and neurodegenerative proteinopathies: mechanisms and prospects for therapeutic intervention. Biochem. Soc. Trans. 46, 829–842.

Cao, X., Li, L.P., Wang, Q., Wu, Q., Hu, H.H., Zhang, M., et al., 2013. Astrocyte-derived ATP modulates depressive-like behaviors. Nat. Med. 19 (6), 773–777.

Charriaut-Marlangue, C., Bonnin, P., Pham, H., Loron, G., Leger, P.L., Gressens, P., et al., 2013. Nitric oxide signaling in the brain: a new target for inhaled nitric oxide? Ann. Neurol. 73 (4), 442–448.

Chaves, I., Pokorny, R., Byrdin, M., Hoang, N., Ritz, T., Brettel, K., et al., 2011. The cryptochromes: blue light photoreceptors in plants and animals. Annu. Rev. Plant Biol. 62, 335–364.

Chen, A.C., Arany, P.R., Huang, Y.Y., Tomkinson, E.M., Sharma, S.K., Kharkwal, G.B., et al., 2011a. Low-level laser therapy activates NF-kB via generation of reactive oxygen species in mouse embryonic fibroblasts. PLoS One 6 (7), e22453.

Chen, A.C., Huang, Y.Y., Sharma, S.K., Hamblin, M.R., 2011b. Effects of 810-nm laser on murine bone-marrow-derived dendritic cells. Photomed. Laser Surg. 29 (6), 383–389.

Chen, G., Thakkar, M., Robinson, C., Dore, S., 2018. Limb remote ischemic conditioning: mechanisms, anesthetics, and the potential for expanding therapeutic options. Front. Neurol. 9, 40.

Cooper, C.E., Cope, M., Springett, R., Amess, P.N., Penrice, J., Tyszczuk, L., et al., 1999. Use of mitochondrial inhibitors to demonstrate that cytochrome oxidase near-infrared spectroscopy can measure mitochondrial dysfunction noninvasively in the brain. J. Cereb. Blood Flow Metab. 19 (1), 27–38.

Creed, M., Pascoli, V.J., Luscher, C., 2015. Addiction therapy. Refining deep brain stimulation to emulate optogenetic treatment of synaptic pathology. Science 347 (6222), 659–664.

Cury, V., Moretti, A.I., Assis, L., Bossini, P., Crusca Jde, S., Neto, C.B., et al., 2013. Low level laser therapy increases angiogenesis in a model of ischemic skin flap in rats mediated by VEGF, HIF-1alpha and MMP-2. J. Photochem. Photobiol. B 125, 164–170.

de Freitas, L.F., Hamblin, M.R., 2016. Proposed mechanisms of photobiomodulation or low-level light therapy. IEEE J. Sel. Top. Quantum Electron. 22 (3). Available from: http://dx.doi.org/10.1109/JSTQE.2016.2561201.

Dissing-Olesen, L., Ladeby, R., Nielsen, H.H., Toft-Hansen, H., Dalmau, I., Finsen, B., 2007. Axonal lesion-induced microglial proliferation and microglial cluster formation in the mouse. Neuroscience 149 (1), 112–122.

Eells, J.T., Henry, M.M., Summerfelt, P., Wong-Riley, M.T., Buchmann, E.V., Kane, M., et al., 2003. Therapeutic photobiomodulation for methanol-induced retinal toxicity. Proc. Natl. Acad. Sci. U.S.A. 100 (6), 3439–3444.

Eells, J.T., Gopalakrishnan, S., Valter, K., 2016. Near-infrared photobiomodulation in retinal injury and disease. Adv. Exp. Med. Biol. 854, 437–441.

Farfara, D., Tuby, H., Trudler, D., Doron-Mandel, E., Maltz, L., Vassar, R.J., et al., 2015. Low-level laser therapy ameliorates disease progression in a mouse model of Alzheimer's disease. J. Mol. Neurosci. 55 (2), 430–436.

Felling, R.J., Song, H., 2015. Epigenetic mechanisms of neuroplasticity and the implications for stroke recovery. Exp. Neurol. 268, 37–45.

Fernandes, K.P., Souza, N.H., Mesquita-Ferrari, R.A., Silva, D.F., Rocha, L.A., Alves, A.N., et al., 2015. Photobiomodulation with 660-nm and 780-nm laser on activated J774 macrophage-like cells: effect on M1 inflammatory markers. J. Photochem. Photobiol. B 153, 344–351.

Ferraresi, C., Huang, Y.Y., Hamblin, M.R., 2016. Photobiomodulation in human muscle tissue: an advantage in sports performance? J. Biophotonics 9 (11–12), 1273–1299.

Filiano, A.J., Gadani, S.P., Kipnis, J., 2015. Interactions of innate and adaptive immunity in brain development and function. Brain Res. 1617, 18–27.

Forrest, M.P., Parnell, E., Penzes, P., 2018. Dendritic structural plasticity and neuropsychiatric disease. Nat. Rev. Neurosci. 19 (4), 215–234.

G, N., Tan, A., Farhatnia, Y., Rajadas, J., Hamblin, M.R., Khaw, P.T., et al., 2013. Channelrhodopsins: visual regeneration & neural activation by a light switch. N. Biotechnol. 30, 461–474.

Gu, Q., Wang, L., Huang, F., Schwarz, W., 2012. Stimulation of TRPV1 by green laser light. Evid. Based Complement. Alternat. Med. 2012, 857123.

Hahn, C.D., Manlhiot, C., Schmidt, M.R., Nielsen, T.T., Redington, A.N., 2011. Remote ischemic per-conditioning: a novel therapy for acute stroke? Stroke 42 (10), 2960–2962.

Hamblin, M.R., 2018. Photobiomodulation for traumatic brain injury and stroke. J. Neurosci. Res. 96 (4), 731–743.

Haschemi, A., Kosma, P., Gille, L., Evans, C.R., Burant, C.F., Starkl, P., et al., 2012. The sedoheptulose kinase CARKL directs macrophage polarization through control of glucose metabolism. Cell Metab. 15 (6), 813–826.

Hu, M., Zhu, K., Chen, X.L., Zhang, Y.J., Zhang, J.S., Xiao, X.L., et al., 2015. Newly generated neurons at 2 months post-status epilepticus are functionally integrated into neuronal circuitry in mouse hippocampus. Exp. Neurol. 273, 273–287.

Huang, Y.Y., Nagata, K., Tedford, C.E., McCarthy, T., Hamblin, M.R., 2013. Low-level laser therapy (LLLT) reduces oxidative stress in primary cortical neurons in vitro. J. Biophotonics 6 (10), 829–838.

Huang, Y.Y., Nagata, K., Tedford, C.E., Hamblin, M.R., 2014. Low-level laser therapy (810 nm) protects primary cortical neurons against excitotoxicity in vitro. J. Biophotonics 7 (8), 656–664.

Johnstone, D.M., el Massri, N., Moro, C., Spana, S., Wang, X.S., Torres, N., et al., 2014. Indirect application of near infrared light induces neuroprotection in a mouse model of parkinsonism – an abscopal neuroprotective effect. Neuroscience 274, 93–101.

Kaminski, M.M., Roth, D., Sass, S., Sauer, S.W., Krammer, P.H., Gulow, K., 2012. Manganese superoxide dismutase: a regulator of T cell activation-induced oxidative signaling and cell death. Biochim. Biophys. Acta 1823 (5), 1041–1052.

Kann, O., 2016. The interneuron energy hypothesis: implications for brain disease. Neurobiol. Dis. 90, 75–85.

Karu, T.I., 2008. Mitochondrial signaling in mammalian cells activated by red and near-IR radiation. Photochem. Photobiol. 84 (5), 1091–1099.

Karu, T.I., 2010. Multiple roles of cytochrome c oxidase in mammalian cells under action of red and IR-A radiation. IUBMB Life 62 (8), 607–610.

Katz, B., Minke, B., 2018. The Drosophila light-activated TRP and TRPL channels – targets of the phosphoinositide signaling cascade. Prog. Retin. Eye Res. 66, 200–219.

Katz, B., Payne, R., Minke, B., 2017. TRP channels in vision. In: Emir, T.L.R. (Ed.), Neurobiology of TRP Channels. Frontiers in Neuroscience, Boca Raton, FL, pp. 27–63.

Keszler, A., Lindemer, B., Hogg, N., Weihrauch, D., Lohr, N.L., 2018. Wavelength-dependence of vasodilation and NO release from S-nitrosothiols and dinitrosyl iron complexes by far red/near infrared light. Arch. Biochem. Biophys. 649, 47–52.

Khongrum, J., Wattanathorn, J., 2015. Laser acupuncture improves behavioral disorders and brain oxidative stress status in the valproic acid rat model of autism. J. Acupunct. Meridian Stud. 8 (4), 183–191.

Khongrum, J., Wattanathorn, J., 2017. Laser acupuncture at HT7 improves the cerebellar disorders in valproic acid-rat model of autism. J. Acupunct. Meridian Stud. 10 (4), 231–239.

Khuman, J., Zhang, J., Park, J., Carroll, J.D., Donahue, C., Whalen, M.J., 2012. Low-level laser light therapy improves cognitive deficits and inhibits microglial activation after controlled cortical impact in mice. J. Neurotrauma 29 (2), 408–417.

Kim, B., Mitrofanis, J., Stone, J., Johnstone, D.M., 2018. Remote tissue conditioning is neuroprotective against MPTP insult in mice. IBRO Rep. 4, 14–17.

Kirschen, G.W., Kery, R., Ge, S., 2018. The hippocampal neuro-glio-vascular network: metabolic vulnerability and potential neurogenic regeneration in disease. Brain Plast. 3 (2), 129–144.

Koyanagi, M., Terakita, A., 2014. Diversity of animal opsin-based pigments and their optogenetic potential. Biochim. Biophys. Acta 1837 (5), 710–716.

Kuhn, M., Popovic, A., Pezawas, L., 2014. Neuroplasticity and memory formation in major depressive disorder: an imaging genetics perspective on serotonin and BDNF. Restor. Neurol. Neurosci. 32 (1), 25–49.

Lampl, Y., 2007. Laser treatment for stroke. Expert Rev. Neurother. 7 (8), 961–965.

Lane, N., 2006. Cell biology: power games. Nature 443 (7114), 901–903.

Lee, H.I., Lee, S.W., Kim, S.Y., Kim, N.G., Park, K.J., Choi, B.T., et al., 2017. Pretreatment with light-emitting diode therapy reduces ischemic brain injury in mice through endothelial nitric oxide synthase-dependent mechanisms. Biochem. Biophys. Res. Commun. 486 (4), 945–950.

Lepousez, G., Nissant, A., Lledo, P.M., 2015. Adult neurogenesis and the future of the rejuvenating brain circuits. Neuron 86 (2), 387–401.

Liang, H.L., Whelan, H.T., Eells, J.T., Meng, H., Buchmann, E., Lerch-Gaggl, A., et al., 2006. Photobiomodulation partially rescues visual cortical neurons from cyanide-induced apoptosis. Neuroscience 139 (2), 639–649.

Liang, J., Liu, L., Xing, D., 2012. Photobiomodulation by low-power laser irradiation attenuates Abeta-induced cell apoptosis through the Akt/GSK3beta/beta-catenin pathway. Free Radic. Biol. Med. 53 (7), 1459–1467.

Lim, R.C.H., 2018. Painless laser acupuncture for smoking cessation. Med. Acupunct. 30 (3), 159–162.

Lim, W., Kim, J., Kim, S., Karna, S., Won, J., Jeon, S.M., et al., 2013. Modulation of lipopolysaccharide-induced NF-kappaB signaling pathway by 635 nm irradiation via heat shock protein 27 in human gingival fibroblast cells. Photochem. Photobiol. 89 (1), 199–207.

Ling, Q., Meng, C., Chen, Q., Xing, D., 2014. Activated ERK/FOXM1 pathway by low-power laser irradiation inhibits UVB-induced senescence through down-regulating p21 expression. J. Cell Physiol. 229 (1), 108–116.

Lohr, N.L., Keszler, A., Pratt, P., Bienengraber, M., Warltier, D.C., Hogg, N., 2009. Enhancement of nitric oxide release from nitrosyl hemoglobin and nitrosyl myoglobin by red/near infrared radiation: potential role in cardioprotection. J. Mol. Cell Cardiol. 47 (2), 256–263.

Losi, A., Gartner, W., 2011. Old chromophores, new photoactivation paradigms, trendy applications: flavins in blue light-sensing photoreceptors. Photochem. Photobiol. 87 (3), 491–510.

Lv, J., Shi, C., Deng, Y., Lou, W., Hu, J., Shi, L., et al., 2016. The brain effects of laser acupuncture at thirteen ghost acupoints in healthy individuals: a resting-state functional MRI investigation. Comput. Med. Imaging Graph 54, 48–54.

Lyons, D.N., Vekaria, H., Macheda, T., Bakshi, V., Powell, D.K., Gold, B.T., et al., 2018. A mild traumatic brain injury in mice produces lasting deficits in brain metabolism. J. Neurotrauma .

Meng, C., He, Z., Xing, D., 2013. Low-level laser therapy rescues dendrite atrophy via upregulating BDNF expression: implications for Alzheimer's disease. J. Neurosci. 33 (33), 13505–13517.

Mintzopoulos, D., Gillis, T.E., Tedford, C.E., Kaufman, M.J., 2017. Effects of near-infrared light on cerebral bioenergetics measured with phosphorus magnetic resonance spectroscopy. Photomed. Laser Surg. 35 (8), 395–400.

Moncada, S., Erusalimsky, J.D., 2002. Does nitric oxide modulate mitochondrial energy generation and apoptosis? Nat. Rev. Mol. Cell Biol. 3 (3), 214–220.

Mungrue, I.N., Husain, M., Stewart, D.J., 2002. The role of NOS in heart failure: lessons from murine genetic models. Heart Fail. Rev. 7 (4), 407–422.

Nahum, M., Lee, H., Merzenich, M.M., 2013. Principles of neuroplasticity-based rehabilitation. Prog. Brain Res. 207, 141–171.

Nunez-Alvarez, C., Del Olmo-Aguado, S., Merayo-Lloves, J., Osborne, N.N., 2017. Near infra-red light attenuates corneal endothelial cell dysfunction in situ and in vitro. Exp. Eye Res. 161, 106–115.

Orihuela, R., McPherson, C.A., Harry, G.J., 2016. Microglial M1/M2 polarization and metabolic states. Br. J. Pharmacol. 173 (4), 649–665.

Oron, A., Oron, U., Chen, J., Eilam, A., Zhang, C., Sadeh, M., et al., 2006. Low-level laser therapy applied transcranially to rats after induction of stroke significantly reduces long-term neurological deficits. Stroke 37 (10), 2620–2624.

Oron, U., Tuby, H., Maltz, L., Sagi-Assif, O., Abu-Hamed, R., Yaakobi, T., et al., 2014. Autologous bone-marrow stem cells stimulation reverses post-ischemic-reperfusion kidney injury in rats. Am. J. Nephrol. 40 (5), 425–433.

Passarella, S., Karu, T., 2014. Absorption of monochromatic and narrow band radiation in the visible and near IR by both mitochondrial and non-mitochondrial photoacceptors results in photobiomodulation. J. Photochem. Photobiol. B 140, 344–358.

Persson, T., Popescu, B.O., Cedazo-Minguez, A., 2014. Oxidative stress in Alzheimer's disease: why did antioxidant therapy fail? Oxid. Med. Cell Longev. 2014, 427318.

Phunchago, N., Wattanathorn, J., Chaisiwamongkol, K., Muchimapura, S., Thukham-Mee, W., 2015. Acupuncture reduces memory impairment and oxidative stress and enhances cholinergic function in an animal model of alcoholism. J. Acupunct. Meridian Stud. 8 (1), 23–29.

Pollack, G.H., 2003. The role of aqueous interfaces in the cell. Adv. Colloid Interface Sci. 103 (2), 173–196.

Pollack, G.H., Reitz, F.B., 2001. Phase transitions and molecular motion in the cell. Cell Mol. Biol. (Noisy-le-grand). 47 (5), 885–900.

Poyton, R.O., Ball, K.A., 2011. Therapeutic photobiomodulation: nitric oxide and a novel function of mitochondrial cytochrome c oxidase. Discov. Med. 11 (57), 154–159.

Pozhilenkova, E.A., Lopatina, O.L., Komleva, Y.K., Salmin, V.V., Salmina, A.B., 2017. Blood-brain barrier-supported neurogenesis in healthy and diseased brain. Rev. Neurosci. 28 (4), 397–415.

Quah-Smith, I., Williams, M.A., Lundeberg, T., Suo, C., Sachdev, P., 2013a. Differential brain effects of laser and needle acupuncture at LR8 using functional MRI. Acupunct. Med. 31 (3), 282–289.

Quah-Smith, I., Smith, C., Crawford, J.D., Russell, J., 2013b. Laser acupuncture for depression: a randomised double blind controlled trial using low intensity laser intervention. J. Affect. Disord. 148 (2–3), 179–187.

Rahal, A., Kumar, A., Singh, V., Yadav, B., Tiwari, R., Chakraborty, S., et al., 2014. Oxidative stress, prooxidants, and antioxidants: the interplay. Biomed. Res. Int. 2014, 761264.

Ribeiro Xavier, A.L., Kress, B.T., Goldman, S.A., Lacerda de Menezes, J.R., Nedergaard, M., 2015. A distinct population of microglia supports adult neurogenesis in the subventricular zone. J. Neurosci. 35 (34), 11848–11861.

Ristow, M., Zarse, K., Oberbach, A., Kloting, N., Birringer, M., Kiehntopf, M., et al., 2009. Antioxidants prevent health-promoting effects of physical exercise in humans. Proc. Natl. Acad. Sci. U.S.A. 106 (21), 8665–8670.

Rodriguez-Rodriguez, A., Egea-Guerrero, J.J., Murillo-Cabezas, F., Carrillo-Vico, A., 2014. Oxidative stress in traumatic brain injury. Curr. Med. Chem. 21 (10), 1201–1211.

Roomruangwong, C., Anderson, G., Berk, M., Stoyanov, D., Carvalho, A.F., Maes, M., 2018. A neuro-immune, neuro-oxidative and neuro-nitrosative model of prenatal and postpartum depression. Prog. Neuropsychopharmacol. Biol. Psychiatry 81, 262–274.

Ryu, J.J., Yoo, S., Kim, K.Y., Park, J.S., Bang, S., Lee, S.H., et al., 2010. Laser modulation of heat and capsaicin receptor TRPV1 leads to thermal antinociception. J. Dent. Res. 89 (12), 1455–1460.

Salehpour, F., Ahmadian, N., Rasta, S.H., Farhoudi, M., Karimi, P., Sadigh-Eteghad, S., 2017. Transcranial low-level laser therapy improves brain mitochondrial function and cognitive impairment in D-galactose-induced aging mice. Neurobiol. Aging 58, 140–150.

Salehpour, F., Farajdokht, F., Erfani, M., Sadigh-Eteghad, S., Shotorbani, S.S., Hamblin, M.R., et al., 2018. Transcranial near-infrared photobiomodulation attenuates memory impairment and hippocampal oxidative stress in sleep-deprived mice. Brain Res. 1682, 36–43.

Saltmarche, A.E., Naeser, M.A., Ho, K.F., Hamblin, M.R., Lim, L., 2017. Significant improvement in cognition in mild to moderately severe dementia cases treated with transcranial plus intranasal photobiomodulation: case series report. Photomed. Laser Surg. 35 (8), 432–441.

Sarandol, A., Sarandol, E., Eker, S.S., Erdinc, S., Vatansever, E., Kirli, S., 2007. Major depressive disorder is accompanied with oxidative stress: short-term antidepressant treatment does not alter oxidative-antioxidative systems. Hum. Psychopharmacol. 22 (2), 67–73.

Schiffer, F., Johnston, A.L., Ravichandran, C., Polcari, A., Teicher, M.H., Webb, R.H., et al., 2009. Psychological benefits 2 and 4 weeks after a single treatment with near infrared light to the forehead: a pilot study of 10 patients with major depression and anxiety. Behav. Brain Funct. 5, 46.

Sharma, S.K., Kharkwal, G.B., Sajo, M., Huang, Y.Y., De Taboada, L., McCarthy, T., et al., 2011. Dose response effects of 810 nm laser light on mouse primary cortical neurons. Lasers Surg. Med. 43 (8), 851–859.

Shen, S.W., Duan, C.L., Chen, X.H., Wang, Y.Q., Sun, X., Zhang, Q.W., et al., 2016. Neurogenic effect of VEGF is related to increase of astrocytes transdifferentiation into new mature neurons in rat brains after stroke. Neuropharmacology 108, 451–461.

Spitler, R., Ho, H., Norpetlian, F., Kong, X., Jiang, J., Yokomori, K., et al., 2015. Combination of low level light therapy and nitrosyl-cobinamide accelerates wound healing. J. Biomed. Opt. 20 (5), 051022.

Steinhubl, S.R., 2008. Why have antioxidants failed in clinical trials? Am. J. Cardiol. 101 (10A), 14D–19DD.

Streeter, J., De Taboada, L., Oron, U., 2004. Mechanisms of action of light therapy for stroke and acute myocardial infarction. Mitochondrion 4 (5–6), 569–576.

Sutalangka, C., Wattanathorn, J., Muchimapura, S., Thukham-Mee, W., Wannanon, P., Tong-un, T., 2013. Laser acupuncture improves memory impairment in an animal model of Alzheimer's disease. J. Acupunct. Meridian Stud. 6 (5), 247–251.

Swerdlow, R.H., Burns, J.M., Khan, S.M., 2014. The Alzheimer's disease mitochondrial cascade hypothesis: progress and perspectives. Biochim. Biophys. Acta 1842 (8), 1219–1231.

Tomaszczyk, J.C., Green, N.L., Frasca, D., Colella, B., Turner, G.R., Christensen, B.K., et al., 2014. Negative neuroplasticity in chronic traumatic brain injury and implications for neurorehabilitation. Neuropsychol. Rev. 24 (4), 409–427.

Trevors, J.T., Pollack, G.H., 2012. Origin of microbial life hypothesis: a gel cytoplasm lacking a bilayer membrane, with infrared radiation producing exclusion zone (EZ) water, hydrogen as an energy source and thermosynthesis for bioenergetics. Biochimie 94 (1), 258–262.

Tuby, H., Maltz, L., Oron, U., 2011. Induction of autologous mesenchymal stem cells in the bone marrow by low-level laser therapy has profound beneficial effects on the infarcted rat heart. Lasers Surg. Med. 43 (5), 401–409.

Walski, T., Drohomirecka, A., Bujok, J., Czerski, A., Waz, G., Trochanowska-Pauk, N., et al., 2018. Low-level light therapy protects red blood cells against oxidative stress and hemolysis during extracorporeal circulation. Front. Physiol. 9, 647.

Wang, F., 2006. Therapeutic effect observation and nurse of intranasal low intensity laser therapy on insomnia. J. Commun. Med. 4, 58.

Wang, X., Wang, W., Li, L., Perry, G., Lee, H.G., Zhu, X., 2014. Oxidative stress and mitochondrial dysfunction in Alzheimer's disease. Biochim. Biophys. Acta 1842 (8), 1240–1247.

Wang, X., Tian, F., Soni, S.S., Gonzalez-Lima, F., Liu, H., 2016. Interplay between up-regulation of cytochrome-c-oxidase and hemoglobin oxygenation induced by near-infrared laser. Sci. Rep. 6, 30540.

Wang, Y., Huang, Y.Y., Wang, Y., Lyu, P., Hamblin, M.R., 2017. Photobiomodulation of human adipose-derived stem cells using 810nm and 980nm lasers operates via different mechanisms of action. Biochim. Biophys. Acta Gen. Subj. 1861 (2), 441–449. Available from: http://dx.doi.org/10.1016/j.bbagen.2016.10.008.

Wang, X., Tian, F., Reddy, D.D., Nalawade, S.S., Barrett, D.W., Gonzalez-Lima, F., et al., 2017. Up-regulation of cerebral cytochrome-c-oxidase and hemodynamics by transcranial infrared laser stimulation: a broadband near-infrared spectroscopy study. J. Cereb. Blood Flow Metab. 37 (12), 3789–3802.

Wong-Riley, M.T., Liang, H.L., Eells, J.T., Chance, B., Henry, M.M., Buchmann, E., et al., 2005. Photobiomodulation directly benefits primary neurons functionally inactivated by toxins: role of cytochrome c oxidase. J. Biol. Chem. 280 (6), 4761–4771.

Wu, J., Xia, S., Kalionis, B., Wan, W., Sun, T., 2014. The role of oxidative stress and inflammation in cardiovascular aging. Biomed. Res. Int. 2014, 615312.

Wu, S., Zhou, F., Wei, Y., Chen, W.R., Chen, Q., Xing, D., 2014. Cancer phototherapy via selective photoinactivation of respiratory chain oxidase to trigger a fatal superoxide anion burst. Antioxid. Redox Signal. 20 (5), 733–746.

Xu, C., Wang, L., Liu, J.H., Tan, Y., Li, Q., 2001. Endonasal low energy He-Ne laser treatment of insomnia. Qian Wei J. Med. Pharm. 18, 337–338.

Xu, C., Wu, Z., Wang, L., Shang, X., Li, Q., 2002a. The effects of endonasal low energy He-Ne laser treatment of insomniaon on sleep EEG. Prac. J. Med. Pharm. 19, 407–408.

Xu, C., Wang, L., Shang, X., Li, Q., 2002b. The treatment of Alzheimer's disease with hypoenergy He-Ne laser. Prac. J. Med. Pharm. 19, 647–648.

Xu, C., Wang, L., Lu, C., 2002c. Endonasal low energy He-Ne laser treatment of poststroke depression. Prac. J. Med. Pharm. 19, 893.

Xu, C., Lu, C., Wang, L., Li, Q., 2003. The effects of endonasal low energy He-Ne laser therapy on antioxydation of Parkinson's disease. Prac. J. Med. Pharm. 20, 816–817.

Xuan, W., Vatansever, F., Huang, L., Hamblin, M.R., 2014. Transcranial low-level laser therapy enhances learning, memory, and neuroprogenitor cells after traumatic brain injury in mice. J. Biomed. Opt. 19 (10), 108003.

Yamaura, M., Yao, M., Yaroslavsky, I., Cohen, R., Smotrich, M., Kochevar, I.E., 2009. Low level light effects on inflammatory cytokine production by rheumatoid arthritis synoviocytes. Lasers Surg. Med. 41 (4), 282–290.

Yang, Y., Gao, L., 2017. Celecoxib alleviates memory deficits by downregulation of COX-2 expression and upregulation of the BDNF-TrkB signaling pathway in a diabetic rat model. J. Mol. Neurosci. 62 (2), 188–198.

Yang, W.Z., Chen, J.Y., Yu, J.T., Zhou, L.W., 2007. Effects of low power laser irradiation on intracellular calcium and histamine release in RBL-2H3 mast cells. Photochem. Photobiol. 83 (4), 979–984.

Yang, J., Harte-Hargrove, L.C., Siao, C.J., Marinic, T., Clarke, R., Ma, Q., et al., 2014. proBDNF negatively regulates neuronal remodeling, synaptic transmission, and synaptic plasticity in hippocampus. Cell Rep. 7 (3), 796–806.

Zhang, H.X., Du, G.H., Zhang, J.T., 2003. Ischemic pre-conditioning preserves brain mitochondrial functions during the middle cerebral artery occlusion in rat. Neurol. Res. 25 (5), 471–476.

Zhang, L., Zhang, Y., Xing, D., 2010. LPLI inhibits apoptosis upstream of Bax translocation via a GSK-3beta-inactivation mechanism. J. Cell Physiol. 224 (1), 218–228.

Zhang, L., Li, H., Zeng, S., Chen, L., Fang, Z., Huang, Q., 2015. Long-term tracing of the BrdU label-retaining cells in adult rat brain. Neurosci. Lett. 591, 30–34.

Part II

Studies in animal models

Chapter 9

Transcranial photobiomodulation for stroke in animal models

Luis De Taboada[1] and Michael R. Hamblin[2,3]

[1]*Chief Technology Officer, LiteCure LLC, New Castle, DE, United States,* [2]*Wellman Center for Photomedicine, Massachusetts General Hospital, Boston, MA, United States,* [3]*Department of Dermatology, Harvard Medical School, Boston, MA, United States*

9.1 Introduction

Stroke is the third leading cause of death in the United States, after heart disease and cancer (Wolf, 1990). Each year, approximately 780,000 people experience a stroke. Approximately 600,000 of these are first attacks, and 180,000 are recurrent attacks (Rosamond et al., 2008). The 3-month mortality rate from ischemic stroke is approximately 12%. Internationally, millions of people have a first or recurrent stroke each year, and nearly a quarter of these people will die. Globally, stroke death rates vary widely; the highest rates are in Portugal, China, Korea, and most of Eastern Europe while the lowest rates are in Switzerland, Canada, and the United States.

Strokes can be classified into two major categories: ischemic and hemorrhagic. Ischemic strokes account for over 80% of all strokes. The most common cause of ischemic stroke is the blockage of an artery in the brain by a blood clot, thrombosis, embolism, or stenosis. Hemorrhagic strokes are the result of a cerebral artery rupture, which can lead to spasm of the artery and various degrees of bleeding into the brain. Until recently, the treatment of subjects with ischemic strokes was largely supportive, focusing on prevention and treatment of respiratory and cardiovascular complications. Common acute complications of stroke include pneumonia, urinary tract infection, and pulmonary embolism. Long-term morbidity in survivors of stroke is common, with difficulty walking in 20%, need for assistance in activities of daily living in 30%, and inability to return to work in 50%–70% of patients.

Ischemic stroke is a complex entity with multiple etiologies and variable clinical manifestations. Thromboses can form in the extracranial and intracranial arteries when the intimal surface is roughened and atherosclerotic plaque forms along the injured vessel. The endothelial injury permits platelets to adhere to the vessel wall and aggregate, then the coagulation cascade is activated and a thrombus develops at the site of the plaque. When the compensatory mechanism of collateral circulation fails, brain perfusion is compromised, leading to decreased oxygen and glucose delivery and neuronal cell death. During an embolic stroke, a clot travels to the brain via an artery. Cells in the core ischemic zone die within minutes. Ischemia impairs the neurologic function of the brain (see Fig. 9.1). Cells in the penumbra of the injury remain potentially salvageable for some hours (Durukan and Tatlisumak, 2007).

When an ischemic stroke occurs, the blood supply to the brain is interrupted, and brain cells are deprived of the glucose and oxygen they need to function. The human brain comprises only 2% of body weight but requires 20% of total oxygen consumption (Edvinsson and Krause, 2002). The brain requires this large amount of oxygen to generate sufficient ATP by oxidative phosphorylation to maintain and restore ionic gradients. One estimate suggests that the Na^+/K^+ATPase found on the plasma membrane of neurons consumes 70% of the energy supplied to the brain. This ion pump maintains the high intracellular K^+ concentration and the low intracellular Na^+ concentrations, necessary for the propagation of action potentials. After global ischemia, inhibition of mitochondrial ATP synthesis leads to most of the ATP being consumed within 2 minutes, causing neuronal plasma membrane depolarization, release of potassium into the extracellular space, and entry of sodium into the cells (Caplan, 2000). ATP depletion also prevents the plasma membrane Ca^{2+} ATPase from maintaining the very low concentrations of calcium that are normally present within each neuronal cell.

Photobiomodulation in the Brain. DOI: https://doi.org/10.1016/B978-0-12-815305-5.00009-9

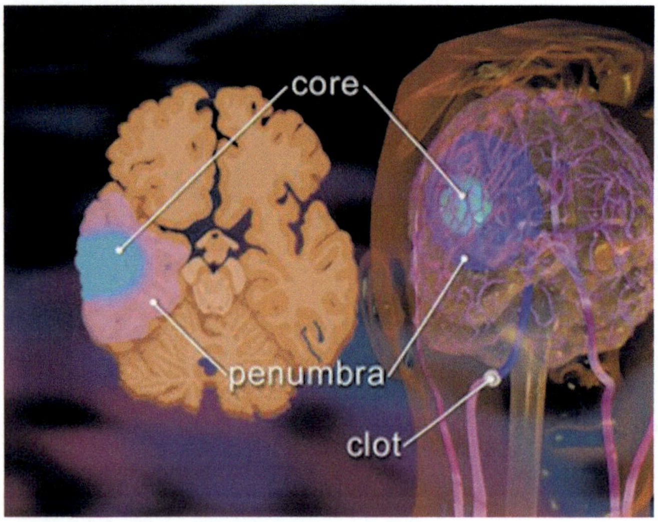

FIGURE 9.1 Acute ischemic stroke pathophysiology.

Strokes usually manifest themselves as focal injuries or infarcts, but diffuse axonal injury, similar to that seen in traumatic brain injury (TBI), can also occur (MacKenzie, 2015). Secondary injuries are attributable to further cellular damage that results from follow-on effects after the primary injury, and develop over a period of hours or days following the initial ischemic assault in a similar way to those seen in TBI (Beez et al., 2017). Secondary brain injury is mediated through excitotoxic cell death in which injured neurons depolarize and release glutamate (Palmer et al., 1993). Neighboring cells are in turn depolarized by the excessive glutamate concentrations and results in a vicious spiral of increasing glutamate concentration. Depolarized neurons suffer a huge influx of sodium and calcium but could survive if they had sufficient ATP to power the Na^+/K^+-ATPase pumps and handle this osmotic load. However, in the setting of metabolic failure they exhibit cytotoxic edema with eventual loss of viability (Pasantes-Morales and Tuz, 2006). An influx of calcium linked to delayed damage, a decrease in mitochondrial membrane potential and increased production of reactive oxygen species (ROS) are observed. Other biochemical processes exacerbating stroke include activation of astrocytes and microglia leading to increased cytokines contributing to inflammation and characterized by increased prostaglandin E2 (Ahmad et al., 2007). Excitotoxic cell death mediated by glutamate also affects glial cells as well as neurons, in particular oligodendrocytes (Matute et al., 2007). ROS have been implicated as key participants of other acute central nervous system injuries, such as TBI, spinal cord injury, and ischemia, as well as chronic neurodegenerative diseases (Ikeda and Long, 1990). Stroke causes physical disruption of neuronal tissue that sets into motion secondary damage resulting in death of additional tissue. Hypoperfusion and ischemia-induced increases in the formation of superoxide and nitric oxide (NO) have been reported after stroke, predicting a role for oxidant stress during damage (Cherian and Robertson, 2003). Cellular protection against these ROS involves an elaborate antioxidant defense system, including that associated with manganese superoxide dismutase.

Thrombolytic therapy is the only intervention of proven and substantial benefit for select patients with acute cerebral ischemia (Adams et al., 2005). The evidence-based for thrombolysis therapy includes 21 completed randomized controlled clinical trials enrolling 7152 patients, using various agents, doses, time windows, and intravenous or intra-arterial modes of administration (Sandercock et al., 2008).

Neuroprotection is defined as any strategy, or combination of strategies, that antagonizes, interrupts, or slows down the sequence of injurious biochemical and molecular events that, if left unchecked, would eventuate in irreversible ischemic injury to the brain. There have been an enormous variety of agents and strategies that have been tested clinically, each justified by a pathophysiological rationale. In all, approximately 165 ongoing or completed clinical trials have been published (Ginsberg, 2008) and as graphically summarized in Fig. 9.2. There has been an almost universal outcome of failure for all these trials, with the exceptions of demonstrated efficacy in thrombolysis and thrombectomy (Elgendy et al., 2016).

The beneficial effects of photobiomodulation (PBM) on cells and neurons in vitro, together with the demonstrated ability of transcranial NIR light to penetrate into the brain (Streeter et al., 2004), strongly suggested that transcranial PBM (tPBM) should be studied as a therapy for stroke. Based on these findings, it is thought that tPBM may have multiple mechanisms of action and could be beneficial in acute ischemic stroke (Lampl, 2007).

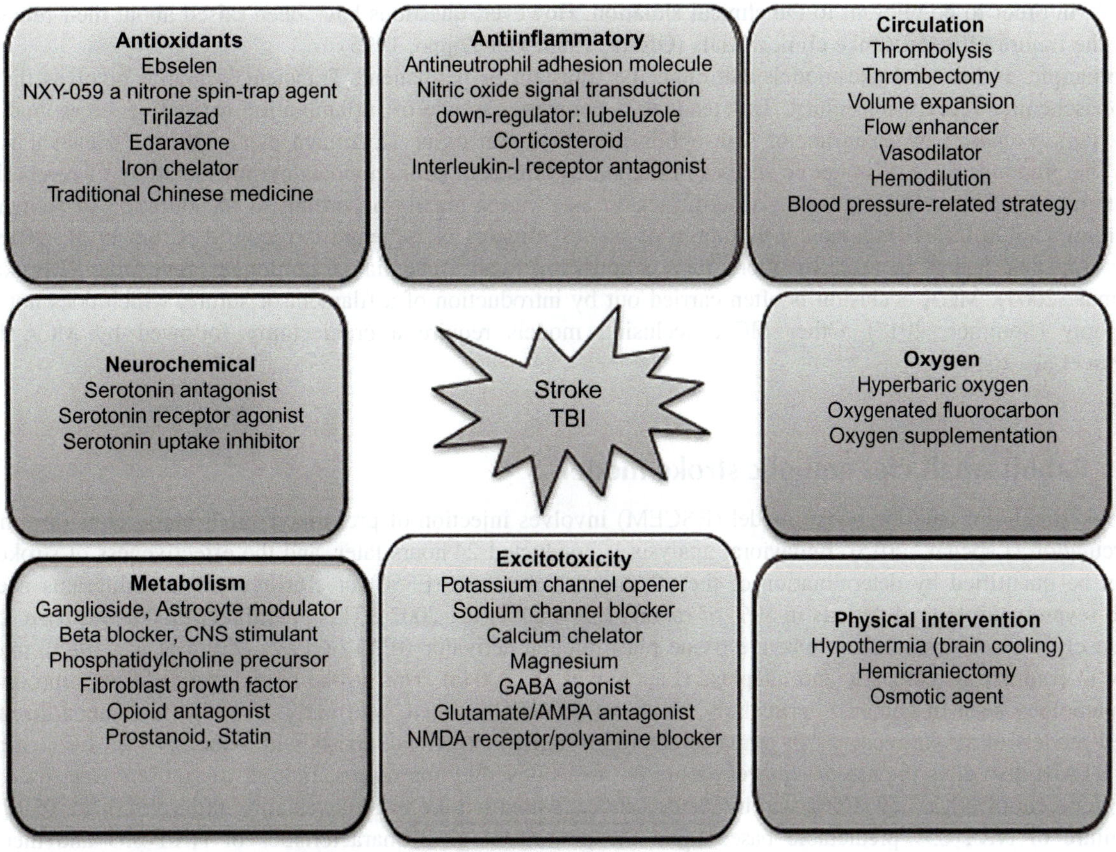

FIGURE 9.2 Clinical trials of pharmacological and physical therapies for stroke.

9.2 Animal models of stroke

Although there is no single animal model that identically mimics ischemic stroke in humans, the use of animal models is essential for development of therapeutic interventions for stroke. Outcome measures for animal models involve functional measures and infarct size evaluation. Recommendations from the Stroke Therapy Academic Industry Roundtable (STAIR) state that studies of new therapies in preclinical animal models should be conducted in rats followed by a second species, often primates (Lapchak et al., 2013; Stroke Therapy Academic Industry R, 1999).

9.2.1 Middle cerebral artery occlusion

Ischemia is typically induced by temporarily or permanently occluding the middle cerebral artery (MCA) in the brain. The MCA is the most often used blood vessel to simulate human stroke, as most human strokes are due to occlusion of this particular vessel or one of its branches (Philip et al., 2009). Animals that have been used in cerebral ischemia models for different purposes include rats, mice, gerbils, cats, rabbits, dogs, pigs, and nonhuman primates.

Although the initial scientific knowledge on stroke originally derived from higher species, now the majority of experiments are carried out in small rodents, such as rats and mice. Small rodents have lower costs and greater ethical acceptability compared to larger animals. The rat is the most commonly used animal in stroke studies (Mhairi Macrae, 1992; Chen et al., 1986). On the other hand, mice are most often employed for genetic modification and are mostly used in transgenic studies in the molecular pathophysiology of stroke (Cekanaviciute et al., 2014). Nonhuman primates have closer similarities with humans in terms of behavior and sensorimotor integration due to their gyrencephalic brains. It has been recommended that once a positive result is achieved from a drug study in small animals, the study should be later replicated in higher species before proceeding to clinical trials (Stroke Therapy Academic Industry R, 1999). Most animal stroke models were developed to induce cerebral ischemia within the parts of the brain supplied by the MCA, in order to be relevant to the clinical situation. However, questions have been raised about their relevance in light of the failure of many stroke clinical trials (Grotta, 1995; Del Zoppo, 1995).

In principle, ischemic stroke models can either be transient or permanent. Transient ischemia involves the occurrence of ischemia–reperfusion injury. This leads to a complex cascade of inflammation, oxidative stress, and excitotoxicity largely caused by an influx of neutrophils into the brain tissue (Enzmann et al., 2018). Transient ischemia reflects the preponderance of ischemic strokes in humans. Permanent stroke models permit the study of cerebral ischemia without the effect of reperfusion. Ischemic lesion size varies greatly according to the duration of ischemia. To obtain reproducible infarct volumes, a minimum of 60–90 minutes of ischemia is required (Crupi et al., 2018). It is well known that lesion induced by more than 3 hours of focal ischemia is no longer reversible (Durukan and Tatlisumak, 2007). MCA occlusion is often carried out by introduction of a filament or suture, which does not require craniectomy (Sommer, 2017). Other MCA occlusion models require a craniectomy followed by MCA ligation (Shmonin et al., 2014).

9.2.2 Rabbit small clot embolic stroke model

The rabbit small clot embolic stroke model (RSCEM) involves injection of preformed small blood clots into the cerebral circulation (Lapchak, 2015). Behavioral analysis is conducted 24 hours later, and the effectiveness of stroke therapies can be quantified by determination of the effective stroke dose (ES50) or clot amount in milligrams needed to produce severe neurological deficits in 50% of rabbits (Lapchak et al., 2002). This preclinical model was used to determine the efficacy of thrombolytic enzyme tissue plasminogen activator (tPA) in 1985 (Zivin et al., 1985), and more recently to compare tenecteplase and alteplase (Lapchak et al., 2004a). This stroke model most closely reproduces the stroke pathology seen in humans. Particularly, since, as STAIR indicated, "currently, there are no standardized, well-accepted models of stroke recovery in primates, although limited experience exists with baboons." As far as a primate model, STAIR also cites the use of squirrel monkeys, macaques, and marmosets. Indeed, marmosets were used during the development of NXY-059 (Cerovive) by Astra-Zeneca, which led the most devastating failure of NXY-059 [18,19]. Was failure of NXY-059 predictable based upon the pharmacological characteristics of NXY-059 and therapeutic window in preclinical stroke models? Yes, if one considers the chemical properties and pharmacological profile of NXY-059 in preclinical studies." Also, this year, 2010, is the silver (25th) anniversary of the rabbit small clot embolic stroke model (RSCEM) that was used to demonstrate that tPA could improve behavior or clinical rating scores [33] when administered following a stroke. There were also other reports using a rabbit embolic stroke model to study tPA pharmacology from Fisher and colleagues [34,35]. They showed that tPA induced partial or complete thrombus dissolution without the development of developed macroscopic cerebral hemorrhage. Moreover, Seibert and colleagues used a rabbit embolic stroke model to demonstrate that tPA can increase cerebral reperfusion and reduce infarct volume measured using the mitochondrial activity stain 2,3,5-triphenyltetrazolium chloride [36]. A similar result was also demonstrated by Gross et al. [37]. The RSCEM has been further characterized as a model where embolism using non-autologous clots leads to reduced cerebral blood flow, cell death, and decreased cortical energetics [38–40].

9.2.3 Photothrombotic stroke models

The photothrombotic stroke model involves injection of a photosensitizer (PS) which is usually an anionic xanthene dye such as rose bengal (RB) or erythrosine B (Liu et al., 2017). After injection, these water-soluble PS do not cross the blood–brain barrier and remain in the bloodstream, where they bind to the vascular endothelium, platelets, and other blood cells. In small rodents, a spot of laser light can be focused on the relatively thin skull bone, and the scattered light is absorbed by PS molecules after multiple scattering, causing shut down of the blood vessels in the illuminated area (Labat-gest and Tomasi, 2013). The wide absorption band of RB (480–580 nm) means that different light sources can be used such as filtered white light (560 nm), dye laser (562 nm), or solid-state laser (532 nm). Photothrombotic stroke shuts down several small cerebral vessels, unlike occlusion of only a single artery with middle cerebral artery occlusion (MCAO). Its advantages include the predictable and well-defined location of the ischemic lesion that is determined by the aiming of the laser beam at the predetermined brain region; the size of the ischemic lesion is determined by the light intensity and duration. Platelet aggregation and thrombosis-associated inflammation and apoptosis play a marked role in this model.

9.3 Photobiomodulation for ischemic stroke in MCAO models

The development of PBM for stroke starts with work from Uri Oron's laboratory in Israel. Oron had shown that PBM could have remarkable effects on heart attacks or myocardial infarction in various animal models. Oron et al., used two different animal models of heart attack (occlusion of coronary artery in both rats and dogs) (Oron et al., 2001b). They used three different power densities of an 810 nm laser applied to the infarcted area of the heart (open chest procedure) immediately after the heart attack. They found a biphasic dose response in rats. The best reduction in infarct area (60%) was seen with 6 mW/cm^2 with lesser improvements seen with 2.5 or 20 mW/cm^2. Dogs also showed a significant reduction in infarct area (4%). Oron's laboratory went on to publish three further papers on this remarkable discovery (Oron et al., 2001a; Yaakobi et al., 2001; Ad and Oron, 2001).

The company PhotoThera Inc. was formed by Jackson Streeter to commercialize this discovery; it was decided to concentrate on stroke because they realized that light could be delivered to the brain in a noninvasive transcranial manner, while light delivery to the heart would almost certainly require a surgical approach. The pathological similarities between the processes occurring after ischemic stroke and myocardial infarction leading to growth of a necrotic lesion in the brain or the heart were remarked upon (Streeter et al., 2004).

However, the first actual experiments on tPBM for ischemic stroke were carried out by Leung's laboratory in Hong Kong (Leung et al., 2002) These workers argued that ischemia–reperfusion injury is partly mediated by NO, while transforming growth factor beta 1 (TGF-b1) is neuroprotective in the stroke model. They used MCAO for 1 hour in Sprague-Dawley rats. tPBM was then applied to the head for different durations (1, 5, or 10 minutes). They used 660 nm (average power 8.8 mW, energy density for each minute of 2.64 J/cm^2, and pulse frequency of 10 kHz) applied directly to the affected area of the brain through a burr hole in the skull immediately after the end of the MCAO. The activity of NOS and the expression of TGF-b1 were evaluated in groups with different durations of PBM. The activity and expression of the three isoforms of NOS were significantly suppressed after tPBM, while expression of TGF-b1 was increased. These workers did not measure any neurological performance parameters.

De Taboada et al. published the first study using tPBM to treat MCAO-induced stroke in rats that involved neurological testing (De Taboada et al., 2006). Stroke was induced in 169 rats that were divided into four groups: control non-PBM, and three tPBM treated groups where an 808 nm laser was employed ipsilateral, contralateral, and on both sides of the head over the induced stroke. Rats were tested for neurological function using a modified neurological score (MNS) (Chen et al., 2001) 24 hours after MCAO. Rats that were scored above five on the MNS (signifying a marked neurological deficit) were included in the study and divided into four groups. The laser was employed transcranially on the shaved skin of the skull by placing the output surface of a custom designed fiber coupled optical assembly (4 mm diameter) onto the skin at two locations on the head (3 mm dorsal to the eye and 2 mm anterior to the ear) either ipsilateral, contralateral, or to both sides of the stroke. These locations had been determined from prior measurements to be sufficient to illuminate the entire ipsilateral brain hemisphere due to dispersion of the laser beam by the skin and the skull at that side. The laser irradiation power at the tip of the fiber optic was set to deliver a power density of 7.5 mW/cm^2 at the surface of the brain. A 2-minute illumination period delivered a calculated fluence of 0.9 J/cm^2 to the surface of the brain. In all three tPBM-treated groups, a marked and significant improvement in neurological deficits was evident at 14, 21, and 28 days poststroke relative to the nontreated group. The extent of improvement at 4 weeks poststroke in the tPBM treated rats was about twofold that of the nontreated rats (see Fig. 9.3).

The next study explored the timing of the tPBM (either 4 hours or 24 poststroke) (Oron et al., 2006). Strokes were induced by two different methods of accomplishing permanent MCAO. Experiment 1 used a craniotomy and MCA ligation, and experiment 2 used a noninvasive filament insertion via the carotid artery. Laser energy (808 nm) was delivered via a custom-designed fiber-coupled optical assembly (4 mm diameter beam) placed onto the shaved skin at two locations on the head (3 mm dorsal to the eye and 2 mm anterior to the ear) on the contralateral hemisphere to the stroke (Ilic et al., 2006), They had previously shown that illuminating the contralateral or both hemispheres made no difference to the functional outcome (De Taboada et al., 2006). The laser irradiation power at the tip of the fiber optic was set to deliver a power density of 7.5 mW/cm^2 at the surface of the brain. The duration of laser irradiation was 2 minutes (energy density of 0.9 J/cm^2) at each point on the skull. In experiment 1, the laser was CW, whereas in experiment 2, both CW and pulsed mode (70 Hz) were compared. Treatment at 4 hours did not show any significant improvement, but the improvement with tPBM at 24 hours was significant with CW laser being somewhat better than pulsed. The number of neuroprogenitor cells in the ipsilateral subventricular zone (determined by BrdU injection at day) was increased by CW tPBM. Markers of migrating neuroprogenitor cells such as TUJ1 (neuron-specific class III beta-tubulin) and DCX (doublecortin) were increased in the CW group.

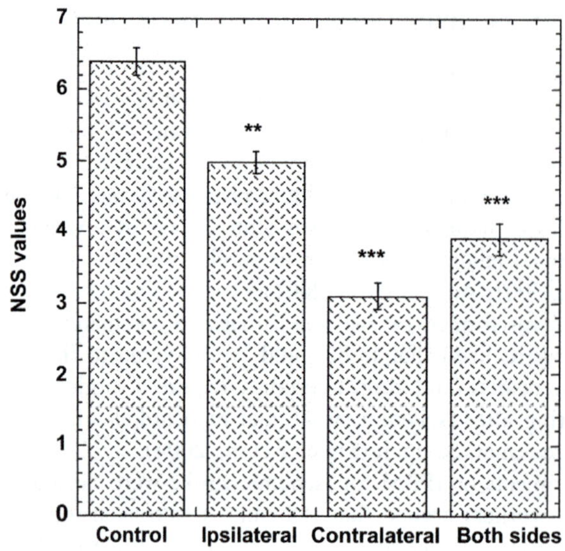

FIGURE 9.3 Neurological severity score of rats with MCAO stroke treated with PBM either ipsilateral or contralateral to the stroke location, or on both sides of the head (Oron et al., 2006). Note: $*p < 0.05$; $**p < 0.01$; $***p < 0.001$.

Lee et al. (2017c) carried out a study to test whether pretreatment with tPBM could protect against cerebral ischemia–reperfusion injury caused by MCAO in mice. They used a 610 nm LED probe (power density, 1.7 mW/cm^2; energy density, 2.0 J/cm^2; 4-mm diameter, 20 minutes) affixed to the head, twice a day for 2 days before MCAO. Twenty-four hours after MCAO, the mice underwent behavioral testing and then were sacrificed. The tPBM group showed significantly smaller infarct and edema volumes, significantly higher cerebral blood flow, and fewer behavioral deficits (wire grip test) compared to untreated injured mice. There were lower levels of endothelial nitric oxide synthase (eNOS) phosphorylation in the injured mouse brains, but significantly higher eNOS phosphorylation in tPBM-pretreated mice. The enhanced phospho-eNOS was inhibited by LY294002 [a phosphoinositide 3-kinase (PI3K) inhibitor], indicating that the effects of tPBM on the ischemic brain could be attributed to the upregulation of eNOS phosphorylation through the PI3K/Akt pathway. Moreover, no reductions in infarct or edema were observed in tPBM-pretreated eNOS-deficient mice.

9.4 Photobiomodulation for ischemic stroke using the RSCEM model

Lapchak et al., carried out the first study of tPBM in the RSCEM stroke model (Lapchak et al., 2004b). The main goal was to compare the effects of different doses of tPBM applied at different times following embolization. They compared a low dose different times between clot injection and tPBM. They also compared a low dose of 808 nm laser (7.5 mW/cm^2 for 2 minutes delivering 0.9 J/cm^2) with a high dose (25 mW/cm^2 for 10 minutes delivering 15 J/cm^2). They found a "therapeutic window" with the low dose ranging from 0 to 6 hours postembolization (with 6 hours being best), but a delay of 24 hours did not show a benefit (see Fig. 9.4). The high dose was also effective when administered at 1 or 6 hours poststroke. Laser treatment (25 mW/cm^2) did not affect the physiological parameters that were measured.

In a follow-up study, Lapchak et al. (2007) went on to compare continuous wave (CW) with pulsed wave tPBM at either 6 or 12 hours following embolization. They tested three different treatment regimens: (1) CW power density of 7.5 mW/cm^2; (2) PW1 using a pulse duration of 300 μs pulse at a frequency of 1 kHz; or (3) PW2 using a pulse duration of 2 ms at a frequency 100 Hz. Behavioral analysis was conducted 48 hours after embolization, allowing for the determination of the effective stroke dose (P50). Quantal dose–response analysis showed that PW1 and PW2 were significantly better than control and better than CW at 6-hour postembolization. At the 12-h postembolization treatment time, neither the CW nor the PW1 or PW2 regimens resulted in statistically significant improvement.

Lapchak and De Taboada went on to measure cortical adenosine 5′triphosphate (ATP) content following tPBM in the RSCEM model (Lapchak and De Taboada, 2010). Five minutes following embolization, rabbits were exposed to 2 minutes of tPBM using an 808 nm laser in either CW or PW mode. Three hours after embolization, the cerebral

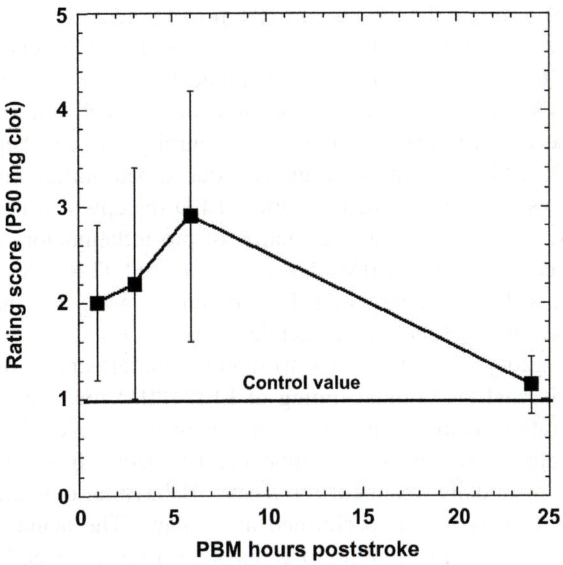

FIGURE 9.4 Clinical rating score in rabbits (RSCEM) in mice with photothrombotic stroke, calculated as mg of clot injected to produce serious impairment when PBM was delivered at different times later (Lapchak et al., 2004b).

cortex was excised and processed for the measurement of ATP content using a luciferin-luciferase assay. Embolization decreased cortical ATP content in the ischemic cortex by 45% compared to naive rabbits, a decrease that was attenuated by CW tPBM which resulted in a 41% increase in cortical ATP content compared to the sham embolized group ($P>.05$). The absolute increase in ATP content was 22.5% compared to naive rabbits. Following PW NILT, which delivered 5 (PW1) and 35 (PW2) times more energy than CW, they measured a 157% (PW1 $p = 0.0032$) and 221% (PW2 $p = 0.0001$) increase in cortical ATP content, respectively, compared to the sham embolized group.

Huisa et al. (2013) used the RSCEM model, to compare zero (sham), one, two, and three tPBM treatments carried out in the 5 hours after embolization. The single group received tPBM at 3 hours, the double treatment group received tPBM at 3 and 5 hours and the triple treatment group at 2, 3, and 4 hours postembolization. The triple treatment had 91% improvement when compared with the single treatment and 245% improvement when compared with the sham.

Meyer et al. (2016) used the RSCEM to study dose-escalating regimens of tPBM. They compared CW with PW at 10 or 100 Hz, and with power density values at the cortex ranging between 7.5 and 333 mW/cm^2, and a single treatment at 2 hours with a triple treatment at 2, 3, and 4 hours. A significant behavioral benefit was seen with triple tPBM pulsed at 100 Hz at 111 mW/cm^2 power density compared with the other tested regimens.

A study using the RLCEM asked whether the only FDA-approved treatment for stroke (tPA) could be safely combined with tPBM (Lapchak et al., 2008). tPA increases cerebral reperfusion, blood flow, and improves neurological performance. tPBM did not significantly alter hemorrhage incidence after embolization, but there was a trend for a modest reduction of hemorrhage volume (by 65%) in the tPBM-treated group compared with controls. Intravenous administration of tPA, using an optimized dosing regimen, significantly increased hemorrhage incidence by 160%. The tPA-induced increase in hemorrhage incidence was not significantly affected by tPBM, although there was a 30% decrease in hemorrhage incidence in combination-treated rabbits. There was no effect of tPBM on hemorrhage volume measured in tPA-treated rabbits and no effect of any treatment on 24-hour survival rate.

Lapchak and Boitano (2016a) asked whether there could be any benefit to the combination of tPA and tPBM. Using the RSCEM, they studied the effects of CW tPBM (7.5 mW/cm^2) alone or in combination with IV tPA (3.3 mg/kg) applied 1 hour postembolization. tPBM and tPA, used alone, significantly increased P50 values by 95% and 56%, respectively, over control. The tPBM–tPA combination increased P50 by 136% over control. Embolization reduced cortical ATP content by 39%; decreases that were attenuated by either tPBM or tPA treatment, while tPBM–tPA combination further enhanced cortical ATP levels 22% above that measured in naïve controls.

9.5 Photobiomodulation for ischemic stroke in photothrombotic model

Two groups have so far tested tPBM in photothrombotic stroke models.

A group in Korea led by HK Shin tested the effects of 610 nm LED therapy on ischemic brain damage in mice that had received a photothrombotic stroke (Lee et al., 2017a). They used a skin-adherent LED probe (power density, 1.7 mW/cm^2; energy density, 2.0 J/cm^2; 4-mm diameter, 20 minutes) twice a day for 3 days commencing at 4 hours postischemia. They affixed probes simultaneously at two locations on the head (the right midpoint of the parietal bone and the posterior midline of the seventh cervical vertebra). The control group was kept under isoflurane anesthesia for 20 minutes without LED application. LED therapy profoundly reduced neuroinflammatory responses including neutrophil infiltration and microglia activation in the ischemic cortex. LED therapy also decreased cell death and attenuated the NLRP3 inflammasome, in accordance with down-regulation of pro-inflammatory cytokines IL-1β and IL-18 in the ischemic brain. Moreover, mice receiving LED-tPBM showed suppressed TLR-2 levels, MAPK signaling, and NFκB activation. These findings suggested that by suppressing the inflammasomes, LED therapy can attenuate neuroinflammatory responses and reduce tissue damage following ischemic stroke.

In a follow-up study, the same group (Lee et al., 2017b) reported the effects on the long-term functional outcomes of the mice after cerebral ischemia, and the optimal timing of LED-tPBM for achieving the best functional recovery. Mice were assigned to a sham-operation control, ischemic (sham treatment), or one of three LED-tPBM groups depending on the time elapsing poststroke before treatment was commenced [immediately (acute), 4 days (subacute), or 10 days (delayed)]. In each case there was once daily treatment for 7 days. Behavioral outcomes were assessed 21 and 28 days postischemia, and histopathological analysis was performed at 28 days. The acute and subacute LED-tPBM groups showed a significant improvement in motor function up to 28 days postischemia (see Fig. 9.5), although no reduction in brain lesion size was observed. They observed proliferating cells (BrdU +) in the ischemic brain, and significant increases in BrdU + /GFAP + , BrdU + /DCX + , BrdU + /NeuN + , and CD31 + cells in the subacute LED-tPBM group. However, the BrdU + /Iba-1 + cell count was reduced in the subacute LED-tPBM group. Furthermore, brain-derived neurotrophic factor (BDNF) was significantly upregulated in the subacute LED-tPBM group. tPBM administered during the subacute stage had a positive impact on the long-term functional outcome, probably via neuronal and astrocyte proliferation, blood vessel reconstruction, and increased BDNF expression.

Zhang's group in Augusta, GA, used rats and investigated both behavioral deficits and neurogenesis (Yang et al., 2018). tPBM was delivered daily from day 1 to day 7, using a 2-minute CW exposure: 808 nm, 350 mW/cm^2, total 294 J at scalp level. The light was applied on the infarct injury area (1.8 mm anterior to the bregma and 2.5 mm lateral from the midline). Rats received intraperitoneal injections of 5-bromodeoxyuridine (BrdU) twice daily (50 mg/kg) from day 2 to 8 poststroke, and samples were collected at day 14. They showed that PBM significantly attenuated behavioral deficits and reduced the infarct volume. Further investigation showed that PBM remarkably enhanced neurogenesis and synaptogenesis, as evidenced by immunostaining of BrdU, Ki67, DCX, MAP2, spinophilin, and synaptophysin. Mechanistic studies suggested beneficial effects of PBM were accompanied by robust suppression of reactive gliosis

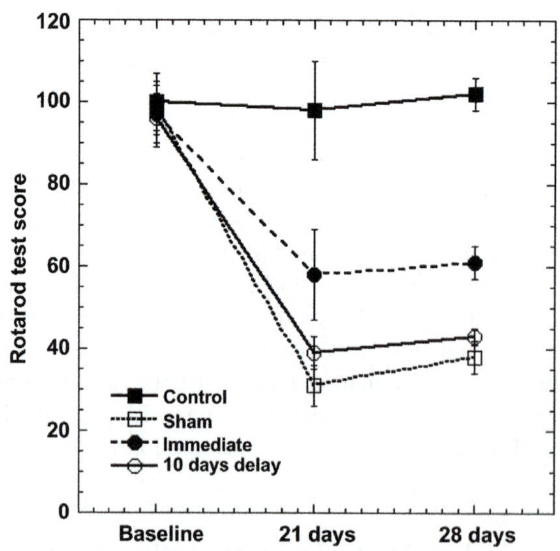

FIGURE 9.5 Rotarod test scores in mice with photothrombotic stroke treated with tPBM once a day for 7 consecutive days commencing either immediately or after 10 days delay (Lee et al., 2017b).

and proinflammatory cytokines. On the contrary, the release of antiinflammatory cytokines, cytochrome c oxidase activity, and ATP production in the peri-infarct regions was elevated following PBM treatment. Intriguingly, PBM could effectively switch the M1 microglial phenotype to an anti-inflammatory M2 phenotype.

9.6 Conclusion

This chapter has chosen to categorize the reports of tPBM in animal models of stroke, according to the method of inducing the strokes in the first place. This appears to be important in view of the disparate findings in some of the animal studies, and especially in light of the eventual failure of the NEST 3 clinical trial in patients (Lapchak and Boitano, 2016b; Zivin et al., 2014), which was based upon several of these preclinical reports. There are disagreements about whether CW or pulsed light is best, about what is the best time window to administer tPBM after the stroke, about what the repetition regimen should be if the decision is taken to administer multiple applications of tPBM, and about the wavelength of PBM (610 or 808 nm). In most reports initiating the treatment past a relatively early time point (a few hours after the stroke) is better than waiting until 24 hours; however one study showed the opposite (Oron et al., 2006). Most studies showed pulsed light to be better than CW, with one study showing the opposite (Oron et al., 2006). Most studies that looked at increasing doses, reported that higher doses were better than lower doses. The one thing that all the studies appear to agree upon, is that multiple applications are better than a single treatment, and it is worth noting that the NEST clinical trials were all based upon a single application of tPBM to the patients (Zivin et al., 2014).

The failure of the NEST trials has cast a shadow over the whole concept of treating acute stroke with tPBM. This failure is not unique to PBM, as mentioned above the literature is rife with reports of treatments that worked amazingly well in preclinical animal models of stroke, but eventually failed in clinical trials (Del Zoppo, 1995; del Zoppo, 1998). The failure of clinical trials to reflect preclinical studies may be partly because there is no broad agreement on which of the various animal models of stroke most closely resembles the human pathophysiology, and which could most closely predict the therapeutic outcomes in humans, and partly in the inherit limitations of generalizing tPBM's light dosimetry, which was remarkably effective in the small crania and brains of mice, rats, and rabbits, to the much larger crania and brains of humans. A limitation that led to the use of the most conservative, safe doses for first-in-man studies.

References

Ad, N., Oron, U., 2001. Impact of low level laser irradiation on infarct size in the rat following myocardial infarction. Int. J. Cardiol. 80 (2–3), 109–116.

Adams, H., Adams, R., Del Zoppo, G., Goldstein, L.B., 2005. Guidelines for the Early Management of Patients With Ischemic Stroke – 2005 guidelines update – A scientific statement from the Stroke Council of the American Heart Association/American Stroke Association. Stroke 36 (4), 916–923.

Ahmad, M., Ahmad, A.S., Zhuang, H., Maruyama, T., Narumiya, S., Dore, S., 2007. Stimulation of prostaglandin E2-EP3 receptors exacerbates stroke and excitotoxic injury. J. Neuroimmunol. 184 (1–2), 172–179.

Beez, T., Steiger, H.J., Etminan, N., 2017. Pharmacological targeting of secondary brain damage following ischemic or hemorrhagic stroke, traumatic brain injury, and bacterial meningitis – a systematic review and meta-analysis. BMC Neurol. 17 (1), 209.

Caplan, L.R. (Ed.), 2000. Caplan's Stroke: A Clinical Approach. third ed. Butterworth-Heinemann, Boston.

Cekanaviciute, E., Fathali, N., Doyle, K.P., Williams, A.M., Han, J., Buckwalter, M.S., 2014. Astrocytic transforming growth factor-beta signaling reduces subacute neuroinflammation after stroke in mice. Glia 62 (8), 1227–1240.

Chen, S.T., Hsu, C.Y., Hogan, E.L., Maricq, H., Balentine, J.D., 1986. A model of focal ischemic stroke in the rat: reproducible extensive cortical infarction. Stroke 17 (4), 738–743.

Chen, J., Sanberg, P.R., Li, Y., Wang, L., Lu, M., Willing, A.E., et al., 2001. Intravenous administration of human umbilical cord blood reduces behavioral deficits after stroke in rats. Stroke 32 (11), 2682–2688.

Cherian, L., Robertson, C.S., 2003. L-arginine and free radical scavengers increase cerebral blood flow and brain tissue nitric oxide concentrations after controlled cortical impact injury in rats. J. Neurotrauma 20 (1), 77–85.

Crupi, R., Di Paola, R., Esposito, E., Cuzzocrea, S., 2018. Middle cerebral artery occlusion by an intraluminal suture method. Methods Mol. Biol. 1727, 393–401.

Del Zoppo, G.J., 1995. Why do all drugs work in animals but none in stroke patients? 1. Drugs promoting cerebral blood flow. J. Intern. Med. 237 (1), 79–88.

del Zoppo, G.J., 1998. Clinical trials in acute stroke: why have they not been successful? Neurology 51 (3 Suppl. 3), S59–S61.

De Taboada, L., Ilic, S., Leichliter-Martha, S., Oron, U., Oron, A., Streeter, J., 2006. Transcranial application of low-energy laser irradiation improves neurological deficits in rats following acute stroke. Lasers Surg. Med. 38 (1), 70–73.

Durukan, A., Tatlisumak, T., 2007. Acute ischemic stroke: overview of major experimental rodent models, pathophysiology, and therapy of focal cerebral ischemia. Pharmacol. Biochem. Behav. 87 (1), 179–197.

Edvinsson, L., Krause, D.N., 2002. Cerebral Blood Flow and Metabolism, second ed. Lippincott, Williams and Wilkins, Philadelphia, PA, 521 pp.

Elgendy, I.Y., Mahmoud, A.N., Mansoor, H., Mojadidi, M.K., Bavry, A.A., 2016. Evolution of acute ischemic stroke therapy from lysis to thrombectomy: similar or different to acute myocardial infarction? Int. J. Cardiol. 222, 441–447.

Enzmann, G., Kargaran, S., Engelhardt, B., 2018. Ischemia-reperfusion injury in stroke: impact of the brain barriers and brain immune privilege on neutrophil function. Ther. Adv. Neurol. Disord. 11, 1756286418794184.

Ginsberg, M.D., 2008. Neuroprotection for ischemic stroke: past, present and future. Neuropharmacology 55 (3), 363–389.

Grotta, J., 1995. Why do all drugs work in animals but none in stroke patients? 2. Neuroprotective therapy. J. Intern. Med. 237 (1), 89–94.

Huisa, B.N., Chen, Y., Meyer, B.C., Tafreshi, G.M., Zivin, J.A., 2013. Incremental treatments with laser therapy augments good behavioral outcome in the rabbit small clot embolic stroke model. Lasers Med. Sci. 28 (4), 1085–1089.

Ikeda, Y., Long, D.M., 1990. The molecular basis of brain injury and brain edema: the role of oxygen free radicals. Neurosurgery 27 (1), 1–11.

Ilic, S., Leichliter, S., Streeter, J., Oron, A., DeTaboada, L., Oron, U., 2006. Effects of power densities, continuous and pulse frequencies, and number of sessions of low-level laser therapy on intact rat brain. Photomed. Laser Surg. 24 (4), 458–466.

Labat-gest, V., Tomasi, S., 2013. Photothrombotic ischemia: a minimally invasive and reproducible photochemical cortical lesion model for mouse stroke studies. J. Vis. Exp. (76), .

Lampl, Y., 2007. Laser treatment for stroke. Expert Rev. Neurother. 7 (8), 961–965.

Lapchak, P.A., 2015. A cost-effective rabbit embolic stroke bioassay: insight into the development of acute ischemic stroke therapy. Transl. Stroke Res. 6 (2), 99–103.

Lapchak, P.A., Boitano, P.D., 2016a. A novel method to promote behavioral improvement and enhance mitochondrial function following an embolic stroke. Brain Res. 1646, 125–131.

Lapchak, P.A., Boitano, P.D., 2016b. Transcranial near-infrared laser therapy for stroke: how to recover from futility in the NEST-3 Clinical Trial. Acta Neurochir. Suppl. 121, 7–12.

Lapchak, P.A., De Taboada, L., 2010. Transcranial near infrared laser treatment (NILT) increases cortical adenosine-5'-triphosphate (ATP) content following embolic strokes in rabbits. Brain Res. 1306, 100–105.

Lapchak, P.A., Araujo, D.M., Pakola, S., Song, D., Wei, J., Zivin, J.A., 2002. Microplasmin: a novel thrombolytic that improves behavioral outcome after embolic strokes in rabbits. Stroke 33 (9), 2279–2284.

Lapchak, P.A., Araujo, D.M., Zivin, J.A., 2004a. Comparison of Tenecteplase with Alteplase on clinical rating scores following small clot embolic strokes in rabbits. Exp. Neurol. 185 (1), 154–159.

Lapchak, P.A., Wei, J., Zivin, J.A., 2004b. Transcranial infrared laser therapy improves clinical rating scores after embolic strokes in rabbits. Stroke 35 (8), 1985–1988.

Lapchak, P.A., Salgado, K.F., Chao, C.H., Zivin, J.A., 2007. Transcranial near-infrared light therapy improves motor function following embolic strokes in rabbits: an extended therapeutic window study using continuous and pulse frequency delivery modes. Neuroscience 148 (4), 907–914.

Lapchak, P.A., Han, M.K., Salgado, K.F., Streeter, J., Zivin, J.A., 2008. Safety profile of transcranial near-infrared laser therapy administered in combination with thrombolytic therapy to embolized rabbits. Stroke 39 (11), 3073–3078.

Lapchak, P.A., Zhang, J.H., Noble-Haeusslein, L.J., 2013. RIGOR guidelines: escalating STAIR and STEPS for effective translational research. Transl. Stroke Res. 4 (3), 279–285.

Lee, H.I., Lee, S.W., Kim, N.G., Park, K.J., Choi, B.T., Shin, Y.I., et al., 2017a. Low-level light emitting diode (LED) therapy suppresses inflammasome-mediated brain damage in experimental ischemic stroke. J. Biophotonics 10 (11), 1502–1513.

Lee, H.I., Lee, S.W., Kim, N.G., Park, K.J., Choi, B.T., Shin, Y.I., et al., 2017b. Low-level light emitting diode therapy promotes long-term functional recovery after experimental stroke in mice. J. Biophotonics 10 (12), 1761–1771.

Lee, H.I., Lee, S.W., Kim, S.Y., Kim, N.G., Park, K.J., Choi, B.T., et al., 2017c. Pretreatment with light-emitting diode therapy reduces ischemic brain injury in mice through endothelial nitric oxide synthase-dependent mechanisms. Biochem. Biophys. Res. Commun. 486 (4), 945–950.

Leung, M.C., Lo, S.C., Siu, F.K., So, K.F., 2002. Treatment of experimentally induced transient cerebral ischemia with low energy laser inhibits nitric oxide synthase activity and up-regulates the expression of transforming growth factor-beta 1. Lasers Surg. Med. 31 (4), 283–288.

Liu, N.W., Ke, C.C., Zhao, Y., Chen, Y.A., Chan, K.C., Tan, D.T., et al., 2017. Evolutional characterization of photochemically induced stroke in rats: a multimodality imaging and molecular biological study. Transl. Stroke Res. 8 (3), 244–256.

MacKenzie, J.M., 2015. Axonal injury in stroke: a forensic neuropathology perspective. Am. J. Forensic Med. Pathol. 36 (3), 172–175.

Matute, C., Alberdi, E., Domercq, M., Sanchez-Gomez, M.V., Perez-Samartin, A., Rodriguez-Antiguedad, A., et al., 2007. Excitotoxic damage to white matter. J. Anat. 210 (6), 693–702.

Meyer, D.M., Chen, Y., Zivin, J.A., 2016. Dose-finding study of phototherapy on stroke outcome in a rabbit model of ischemic stroke. Neurosci. Lett. 630, 254–258.

Mhairi Macrae, I., 1992. New models of focal cerebral ischaemia. Br. J. Clin. Pharmacol. 34 (4), 302–308.

Oron, A., Oron, U., Chen, J., Eilam, A., Zhang, C., Sadeh, M., et al., 2006. Low-level laser therapy applied transcranially to rats after induction of stroke significantly reduces long-term neurological deficits. Stroke 37 (10), 2620–2624.

Oron, U., Yaakobi, T., Oron, A., Mordechovitz, D., Shofti, R., Hayam, G., et al., 2001a. Low-energy laser irradiation reduces formation of scar tissue after myocardial infarction in rats and dogs. Circulation 103 (2), 296–301.

Oron, U., Yaakobi, T., Oron, A., Hayam, G., Gepstein, L., Rubin, O., et al., 2001b. Attenuation of infarct size in rats and dogs after myocardial infarction by low-energy laser irradiation. Lasers Surg. Med. 28 (3), 204–211.

Palmer, A.M., Marion, D.W., Botscheller, M.L., Swedlow, P.E., Styren, S.D., DeKosky, S.T., 1993. Traumatic brain injury-induced excitotoxicity assessed in a controlled cortical impact model. J. Neurochem. 61 (6), 2015–2024.

Pasantes-Morales, H., Tuz, K., 2006. Volume changes in neurons: hyperexcitability and neuronal death. Contrib. Nephrol. 152, 221–240.

Philip, M., Benatar, M., Fisher, M., Savitz, S.I., 2009. Methodological quality of animal studies of neuroprotective agents currently in phase II/III acute ischemic stroke trials. Stroke 40 (2), 577–581.

Rosamond, W., Flegal, K., Furie, K., Go, A., Greenlund, K., Haase, N., et al., 2008. Heart disease and stroke statistics--2008 update: a report from the American Heart Association Statistics Committee and Stroke Statistics Subcommittee. Circulation 117 (4), e25–e146.

Sandercock, P., Lindley, R., Wardlaw, J., Dennis, M., Lewis, S., Venables, G., et al., 2008. The third international stroke trial (IST-3) of thrombolysis for acute ischaemic stroke. Trials 9 (1), 37.

Shmonin, A., Melnikova, E., Galagudza, M., Vlasov, T., 2014. Characteristics of cerebral ischemia in major rat stroke models of middle cerebral artery ligation through craniectomy. Int. J. Stroke 9 (6), 793–801.

Sommer, C.J., 2017. Ischemic stroke: experimental models and reality. Acta Neuropathol. 133 (2), 245–261.

Streeter, J., De Taboada, L., Oron, U., 2004. Mechanisms of action of light therapy for stroke and acute myocardial infarction. Mitochondrion 4 (5–6), 569–576.

Stroke Therapy Academic Industry R, 1999. Recommendations for standards regarding preclinical neuroprotective and restorative drug development. Stroke 30 (12), 2752–2758.

Wolf, P.A., 1990. An overview of the epidemiology of stroke. Stroke 21 (9 Suppl.), II4–II6.

Yaakobi, T., Shoshany, Y., Levkovitz, S., Rubin, O., Ben Haim, S.A., Oron, U., 2001. Long-term effect of low energy laser irradiation on infarction and reperfusion injury in the rat heart. J. Appl. Physiol. (1985) 90 (6), 2411–2419.

Yang, L., Tucker, D., Dong, Y., Wu, C., Lu, Y., Li, Y., et al., 2018. Photobiomodulation therapy promotes neurogenesis by improving post-stroke local microenvironment and stimulating neuroprogenitor cells. Exp. Neurol. 299 (Pt A), 86–96.

Zivin, J.A., Fisher, M., DeGirolami, U., Hemenway, C.C., Stashak, J.A., 1985. Tissue plasminogen activator reduces neurological damage after cerebral embolism. Science 230 (4731), 1289–1292.

Zivin, J.A., Sehra, R., Shoshoo, A., Albers, G.W., Bornstein, N.M., Dahlof, B., et al., 2014. NeuroThera(R) Efficacy and Safety Trial-3 (NEST-3): a double-blind, randomized, sham-controlled, parallel group, multicenter, pivotal study to assess the safety and efficacy of transcranial laser therapy with the NeuroThera(R) Laser System for the treatment of acute ischemic stroke within 24 h of stroke onset. Int. J. Stroke 9 (7), 950–955.

Chapter 10

Photobiomodulation in photothrombotic stroke

Lorelei Tucker, Luodan Yang, Yong Li and Quanguang Zhang

Department of Neuroscience and Regenerative Medicine, Medical College of Georgia, Augusta University, Augusta, GA, United States

Stroke is a leading cause of death, accounting for over 140,000 deaths each year in the United States alone (Yang et al., 2017). The most common form of stroke, ischemic stroke, causes the cessation or sharp reduction in cerebral blood flow, leading to hypoxic injury, disability, and death (Benjamin et al., 2017). Stroke is strongly age-related, with nearly 3/4 of strokes taking place after age 65. In an aging society, the impact of stroke is steadily increasing, and must be met with effective treatment strategies (Benjamin et al., 2017). Unfortunately, there is currently only one approved treatment for stroke, namely thrombolysis using tissue plasminogen activator (tPA). If administered quickly, tPA can be life-saving, but the time window is narrow and stroke must be accurately diagnosed, otherwise tPA treatment risks killing the patient (Kirmani et al., 2012). It is this limitation that demands the development of new therapeutic strategies that can be administered safely and over an extended duration after stroke. We believe that photobiomodulation (PBM) represents a potential agent to combat the damage caused by stroke. To test this hypothesis, we investigated the effects of PBM in a photothrombotic (PT) stroke model, a noninvasive rodent stroke model.

The PT stroke model was developed and introduced by Watson et al. (1985), as a reliable method of reproducible stroke induction in an animal model. The origins of this model, however, lie a few years prior in the form of various models performed in mice and cats that studied platelet aggregation after vascular insult (Rosenblum and El-Sabban, 1977, 1982). These models used the photoreactive properties of sodium fluorescein (NaFl), a fluorescent dye widely used in vascular studies, as a method to cause platelet aggregation and occlusion of the pial vasculature. NaFl was injected and allowed to circulate systemically, and subsequently a cranial window was established. Illumination was targeted at the pial vessels triggering a photochemical reaction, generating excited oxygen species that damaged the vascular epithelium. This damage initiated a thrombolytic response resulting in platelet aggregation and subsequent occlusion of the targeted vasculature (Rosenblum and El-Sabban, 1977, 1982). Watson et al., adapted this concept to a rat model, substituting the dye rose bengal (RB) for NaFl and forgoing the construction of the cranial window (Watson et al., 1985). RB was chosen for its significant advantages as a photosensitizer over NaFl, not the least of which is its exceptional singlet oxygen yield upon excitation. This pronounced potency allows for lower circulating concentrations of dye, minimizing adverse effects. Furthermore, the peak of absorption for RB is in the green range of the visible spectrum as opposed that of NaFl, which lies in the blue spectrum. The longer wavelength involved facilitates deeper light penetration into tissue, allowing for transcranial illumination in the PT model without the need for a partial craniectomy. Carbon black staining was used to determine the effect of RB stimulation on vascular integrity, showing rapid development of an infarct in the targeted region, primarily affecting cortical parts of the brain. To confirm the role of platelet aggregation in infarct development, ultrastructural analysis was performed via electron microscopy, revealing platelet activation in the damaged microvasculature. This was followed in the developing infarct with infiltration of peripheral macrophages observed in the days following the initial insult (Watson et al., 1985). Since its development, the PT model has been used in a plethora of studies, resulting in part from the relative ease of implementation.

The PT model is initiated with administration of RB, which can either be through intraperitoneal or intravenous injection. While the RB is circulating, the animal is anesthetized and securely mounted on a stereotaxic frame. After cleaning, the scalp is carefully incised and a light source, generally a laser or a cold white light, is positioned over a brain region of interest using stereotaxic coordinates to precisely deliver a repeatable infarct. It is imperative that temperature at the site of illumination be monitored, as there is a risk of causing heat damage to the brain which must be

Photobiomodulation in the Brain. DOI: https://doi.org/10.1016/B978-0-12-815305-5.00010-5

avoided, therefore the temperature at the skull is closely monitored throughout the illumination phase of the operation. Illumination is then initiated and is sustained for a specified duration; 20 minutes was specified in the original manuscript. After illumination is complete, the animal is returned to its cage and allowed to recover in darkness to reduce any potential peripheral skin damage via photosensitization due to residual circulating RB (Watson et al., 1985). Upon illumination, light interacts with circulating RB molecules, eliciting a Type II photochemical reaction wherein RB molecules absorb photons and transfer this newfound excess energy to ground state oxygen, converting it to its excited singlet state. Singlet state oxygen is incredibly reactive, causing oxidative damage to any nearby biological macromolecule. In the PT model, oxidative damage occurs to the vascular endothelium, triggering the aforementioned platelet activation and thrombolytic response. This process occurs rapidly, and is accompanied by cell death and local degradation of the blood–brain barrier (BBB). A core cortical infarct is generated quickly, with the insult extending to subcortical areas due to pressure generated by local edema. Around the core infarct lies a thin layer of at-risk tissue amenable to repair which corresponds to the penumbral region (Watson et al., 1985). It is this region with its susceptibility to potential recovery that is the main target of neuroprotective and neuroregenerative research.

Several variations of this model exist, each has been developed to investigate diverse mechanisms relevant to the heterogeneous nature of stroke, as observed in the clinic. One such modification is the PT ring model. This model was developed to address one of the most common issues leveled against the PT stroke model, the relatively small area of salvageable penumbra. Rather than illuminating via a solid laser beam or white light spot of the standard PT model, the ring model utilizes a ring-shaped laser output, generating a thin ring of core infarct surrounding a region of penumbra, effectively reversing the characteristic pattern of infarct and at-risk tissue of the original model (Wester et al., 1995). The extended field of penumbra allows greater depth investigations into the mechanisms of cell death and recovery in at-risk tissue. Work with this model has elucidated differences in antiapoptotic and apoptotic signaling between lightly damaged and highly damaged penumbral tissue, with high intensity irradiation, widely shifting the balance between cellular life and death toward apoptosis (Hu et al., 2004). Modifying this model further, it was found that a slightly thinner ring of illumination could allow for reperfusion in the penumbra region, addressing a limitation of previous applications of the PT model (Hu et al., 2001). Other variations of the PT model address limitations inherent not only to the PT model, but to animal stroke models at large. One example is an awake stroke model, developed to demonstrate the effects of anesthesia in stroke models. As with most surgical models, animal stroke models generally require deep anesthesia. Commonly used anesthetics, such as isoflurane, are steeped in controversy, as there is evidence that these agents can themselves protect against neurological insult in multiple brain injury models, albeit with much debate as to the relevance of these results (Jiang et al., 2017; Sun et al., 2015; Burchell et al., 2013; Kawaguchi et al., 2000; Warner, 2000). To determine whether these effects applied to the PT model and whether this protection may mask the effects of potential neuroprotective agents, a waking stroke model was developed, wherein a head-mounted fiber optic light source delivered illumination in free-roaming mice. Their results confirmed the suspicion that isoflurane may be masking the effects of experimental neuroprotective compounds, with the protective potential of two such compounds only emerging in the absence of isoflurane. Moreover, this variation has the advantage of more accurately modeling the situations in which stroke occurs, as most individuals do not suffer strokes while under anesthesia.

Waking, freely moving induction of stroke allowed study of functional outcomes and development of motor pathology in real time, as video monitoring revealed steadily deteriorating ambulatory behavior during PT induction (Seto et al., 2014). Another variation of this theme is a repetitive stroke model developed to study the impact of multiple successive strokes on cognitive decline (Schmidt et al., 2015). Recurring stroke is a discouragingly common occurrence after an initial stroke event, and is associated with greater levels of poststroke development of dementia. A recurrent stroke PT model wherein a second stroke is induced 10 days after the initial insult displayed progressive decline, accompanied with detrimental glial activation in the hippocampus (Schmidt et al., 2015). Further work with this variation of the PT model may be able to provide insight into strategies to protect against cognitive effects of multiple stroke events. These different iterations all offer different angles of attack at the pathology of stroke and expose different facets of neuroprotective mechanisms that could hold promise in future research and clinical practice.

Much like any experimental injury model, the PT stroke model has an array of benefits and drawbacks that must be taken into account when designing experiments, interpreting data, and extrapolating potential translational potential. One of the greatest advantages of the PT model is its repeatability and ease of application. Since the lesion is localized externally via light source placement, the infarct can be targeted with accuracy and precision with basic stereotaxic equipment. This fine control allows for specific regions to be targeted, allowing for finely specified functional outcomes to be generated and examined (Watson et al., 1985). These positive features are bolstered by a low mortality rate and a low skill barrier, as the model itself is minimally invasive and quick to perform, allowing researchers and technicians with less surgical experience to produce the model with confidence. Finally, the ring of penumbra, or core of penumbra in the ring model, is very amenable to neural repair mechanisms, a major target in stroke research (Sommer, 2017; Carmichael, 2005). Features of this penumbral at-risk region, however, rank among the main sources of criticism regarding the PT model. This penumbra is but a sliver in the standard model compared to a much larger region in both the middle cerebral artery occlusion model and human stroke. Furthermore, a lack of collateral blood flow separates this infarct from the reality of brain damage seen in human stroke. The ring model addresses some of these concerns by way of generating a larger penumbra area, although some criticisms still remain (Watson et al., 1985; Sommer, 2017). One further drawback is the nature of the infarct and its etiology. PT causes rapid cytotoxic and vascular edema uncharacteristic of the much slower developing vascular edema that typically presents in the occurrence of stroke in the clinic. Degradation of the BBB occurs in the PT model as in human stroke, but the development of this damage in PT stroke is much more rapid (Watson et al., 1985; Lee et al., 1996; van Bruggen et al., 1992). These features result from the nature of the infarct itself, as the photochemical lesion initiates with generation of excessive singlet oxygen production that generates the occlusion which is not mirrored in human stroke (Watson et al., 1985). Despite these drawbacks, the PT model serves well as a high throughput method of studying mechanisms of stroke recovery while minimizing animal usage.

The pathophysiology of PT stroke is multifactorial and complex, much like that of other animal models, and indeed the phenomena of human stroke they all are built to simulate. The hypoxic conditions induced by PT occlusion of cerebral vasculature causes cell death, occurring along a continuum of necrosis to apoptosis, within the core infarct and in the at-risk penumbral tissue in its immediate vicinity (Watson et al., 1985). Within the core infarct itself, apoptotic signaling dramatically increases, with mitochondrial apoptotic factors like cytochrome c, AIF, and Smac/DIABLO translocate to the cytosol, initiating the cell death cascade. In the at-risk penumbral region, however, apoptotic signaling is abated, to a degree, by upregulation of antiapoptotic cell signaling and activation of pro-repair signaling (Hu et al., 2004; Demyanenko et al., 2015). Therefore, the penumbra is an alluring target for experimental therapeutics, exploiting this potential for survival to potentially save and restore tissue. This potential, however, is stymied by a deleterious microenvironment characterized by oxidative stress, mitochondrial dysfunction, a compromised BBB, neuroinflammation, and reactive gliosis that all act in tandem to propagate pathology chronically (Hu et al., 2004; Demyanenko et al., 2015; Liguz-Lecznar et al., 2015; Vandeputte et al., 2011).

Astrocytes and microglia activate in response to this process, and begin to release into the microenvironment a mixture of inflammatory and antiinflammatory cytokines (Vandeputte et al., 2011; Ladwig et al., 2017) Alongside this neuroimmune response, peripheral macrophages infiltrate the infarct, facilitating breakdown of the BBB (Watson et al., 1985; Sommer, 2017). This furthers the neuroinflammatory response, perpetuating this destructive feedback cycle in a feed-forward manner. The pervasive influence of neuroinflammation contributes to and is exacerbated by the occurrence of widespread mitochondrial dysfunction that is also generated (Akbar et al., 2016). Alongside this destructive turn of events, neural repair is widely observed in the penumbral region in this model, and is associated with upregulation of proteins associated with mitophagy, neural navigation, and synaptic markers (Demyanenko et al., 2015). This restorative response happens in concert with the phenomenon of neurogenesis, much as in the case of human stroke (Gu et al., 2000, 2009). Just as in human stroke, however, the restorative potential of neurogenesis is limited. This appears, in part, to be an effect of the deleterious microenvironment imposed by the complex response to ischemic insult (Thored et al., 2006; Arvidsson et al., 2002). As the microenvironmental factors may, in fact, be the hurdle limiting the brain's endogenous repair mechanisms, it can be reasoned that targeting these factors may be the key to promoting neural repair (Thored et al., 2006; Arvidsson et al., 2002; Ahmed et al., 2016). Attempting to address such a wide breadth of pathologies at once, however, seems nearly overwhelming for any single treatment paradigm. Fortunately, the central role of mitochondrial dysfunction in stroke positions it as a promising target to confer multifactorial neuroprotection. As a mitochondrial targeted therapy, PBM is therefore suggested to be a treatment strategy with great protective and restorative potential.

PBM is a novel treatment paradigm that utilizes near-infrared (near-IR) light to modulate biological activity in a tissue or biological system of interest. Initially using laser technology shortly after its introduction, PBM was shown to accelerate hair growth in a rodent model of wound healing. Thereafter, PBM has been studied in several contexts. In recent years, extensive effort has been invested in the application of PBM in models of brain injury, as a potential noninvasive treatment strategy with minimal off-target effects, an alluring target in a field bereft of safe and effective therapies. PBM has shown potentially remarkable protective abilities against mitochondrial dysfunction and neuronal cell death in models of stroke, Parkinson's disease, Alzheimer's disease, and traumatic brain injury; although these benefits have yet to be successfully translated to the clinical setting (Berman et al., 2017; Lee et al., 2017; Oron and Oron, 2016; Xuan et al., 2016; Oueslati et al., 2015; Chung et al., 2012; Zivin et al., 2009). One of the key strengths that facilitate the utility of PBM in brain injury is the deep tissue penetrance conferred by the nature of near-IR light. The primary wavelengths used in PBM lie in a "sweet spot" in between the absorption spectra of water and that of natural chromophores, allowing light to pass through to be either scattered or absorbed by the primary proposed receptor in PBM, cytochrome c oxidase (CCO) (Chung et al., 2012). PBM administration results in immediate and persistent increases in ATP production and complex IV subunit activity, as well as significant resistance to oxidative stress. It has not been entirely elucidated how these effects are mediated, although some are attributed to NFKB signaling triggered by modest increases in ROS production, and some are attributed to the release of NO from CCO, stimulating respiration and vasomodulation resulting in increases in cerebral blood flow (Chung et al., 2012; Lee et al., 2017b; Salgado et al., 2015; Gonzalez-Lima et al., 2014).

Our previous work and the work of others investigating PBM in different models of ischemic brain injury, combined with the versatility of the PT model, informed our approach and decision to investigate PBM in the PT model. In our study, we performed the standard circular infarct PT model on adult male SD rats. We used a cold white light source administered to the skull via a fiber optic cable output, delivering light in a 6 mm diameter circle. Rats were injected intraperitoneally with RB and given 5 minutes for it to circulate systemically while anesthesia was induced. The infarct was situated on the skull 1.8 mm anterior to the bregma and 2.5 mm lateral from the midline, and illumination was sustained for 15 min, after which the animals were allowed to recover in the dark. PBM at 808 nm, delivering a power density of 25 mW/cm^2 at the cortical level was administered via diode laser for 2 minutes daily for 7 consecutive days, commencing the next day following PT stroke induction. For PBM treatment, a 1 cm^2 circular spot was focused on the rat's skull via fiber optic cable and expanding. Prior testing determined the penetrance of the laser through the rat cranium, allowing us to determine the proper laser power parameters necessary to deliver a consistent dosage. As shown in Fig. 10.1, animals were subjected to a battery of sensorimotor behavioral testing following stroke induction, with tests taking place at three different timepoints, beginning the day after PT stroke, including the cylinder test, balance beam test, and adhesive removal test. As we wished to investigate the effects of PBM on neurogenesis, BrdU, a synthetic nucleic acid analogue, was injected i.p. twice daily for 1 week beginning the second day after PT stroke. Following behavioral analysis, treatment, and BrdU injections, animals were sacrificed via transcardial perfusion, and brains were extracted for protein samples and histological analysis. For histological and immunostaining experiments, coronal sections were collected and analyzed for infarct volume, synaptic markers, neurogenesis, and glial activation. Brain homogenates were prepared for examination of mitochondrial activity, inflammatory cytokines, and microglial polarization using western blot and ELISA analysis. Data was analyzed via one-way and two-way ANOVA with posthoc Student–Newman–Keuls tests to determine differences between groups.

For patients and their families, the most apparent and valuable metric of stroke recovery is functional outcome. While the damage done to the brain is of critical importance, the average patient is likely more concerned with how their stroke will affect their day to day life. Functional recovery affects the quality of life, cost of auxiliary care, psychological well being of patients and that of their loved ones (King, 1996). As such, determining whether PBM may protect neurological function after stroke is imperative in determining its utility as a therapeutic strategy. In our work, we focused our behavioral evaluation on sensorimotor deficits that mirror the common disabilities faced after stroke. While subcortical structures are damaged in PT stroke, as in other models, the most robust and quantifiable deficits in the PT model involve sensorimotor deficits, often manifesting as and quantified by lateral asymmetry. These deficits can be measured via a variety of behavioral tasks that are specifically designed and finely tuned for this purpose, widely validated in the PT model and others. In our work, we used the adhesive removal test and the cylinder test, each investigating a distinct feature of neurological function (Schaar et al., 2010).

The adhesive removal test (Fig. 10.2A) is a rather straightforward task in which the animal must remove an adhesive strip affixed to its paw. It is widely used in stroke research due to its sensitivity to sensorimotor deficits and its ease of application (Schaar et al., 2010; Bouet et al., 2009). After placing the adhesive tape on the animal's forepaw, it is

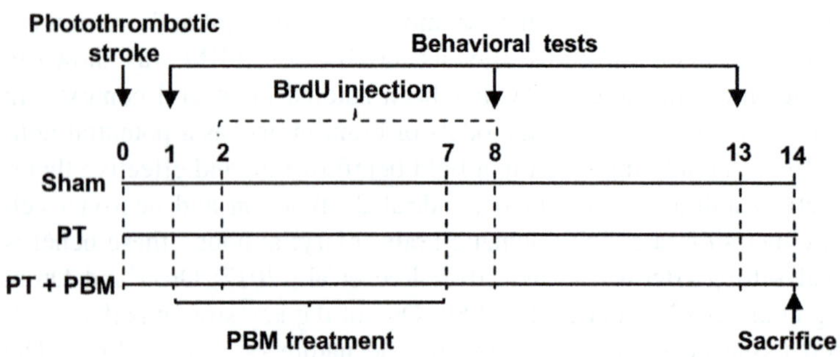

FIGURE 10.1 Schematic diagram of experimental protocol.

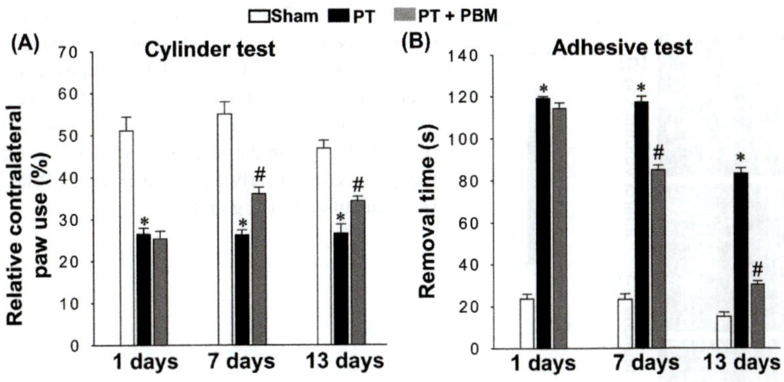

FIGURE 10.2 PBM reduces PT-induced sensorimotor deficits in the cylinder test (A) and adhesive test (B).

allowed to remove the tape under observation by a blinded researcher. The time taken to remove the strip from the paw is recorded, and is often compared between the paws contralateral and ipsilateral to the stroke insult to tease out somatosensory deficits (Bouet et al., 2009). In our study, we found, consistent with our previous work and that of others (Ahmed et al., 2016; Liguz-Lecznar et al., 2014), that PT insult in the somatosensory cortex led to significant increases in time taken to remove the adhesive strip, compared to uninjured sham animals. These deficits were observed for up to 13 days after the induction of the PT model. PBM treated animals, however, showed a marked progressive functional recovery. At day 1 after PT, the performance of PBM and PT control animals was indistinguishable, but as time progressed, a performance gap emerged between the two groups. At day 7 after PT, the removal time was lessened by a third in PBM animals, while at day 13 this gap had broadened to the point that PBM treated animals took less than half the time it took PT control animals to perform the same task. Although tape removal time in the PT control animals did recover over time, this finding was only noted at the 13 day mark. This apparent bolstering of functional recovery bodes well for the use of PBM as a potential stroke therapy, as do our findings on the cylinder test.

The cylinder test (Fig. 10.2B) is another simple and widely employed behavioral test, designed to reveal asymmetry in voluntary forepaw usage. While an individual uninjured rat may have a minor preference for one paw or the other, rats with unilateral sensorimotor cortex damage will prefer to use the unimpaired paw ipsilateral to the insult, reflecting motor impairments commonly observed in human stroke patients (King, 1996; Schaar et al., 2010). In the cylinder test, an animal is placed in a cylindrical transparent container and observed, undisturbed for a given duration. In this situation, the natural response of the rat is to rear and contact the walls of the cylinder with its forepaws. During testing, the number of times that the rat contacts each of its forepaws to the wall of the container are counted. As predicted, PT group rats displayed dramatically reduced rates of contralateral paw preference throughout the entirety of behavioral testing. PBM, in line with our adhesive removal test results, began ameliorating this deficit beginning on day 7 after PT stroke. This reflects the restorative potential of PBM on critically important functional recovery, in line with previous work of our group and others.

After stroke, the core infarct area is surrounded by at-risk tissue that, if left untreated, will expand. The increasing infarct will begin encroaching upon functional tissue, developing over time to eclipse a region far beyond the initial insult. As brain tissue is lost, the prognosis becomes poorer, so it is crucial to slow the advance of the infarct as much as possible (Geltman, 1984). The PT model is characterized by a small cortical infarct, localized by the placement of the source of illumination. In our study, we localized it over the right sensorimotor cortex, as seen in Fig. 10.3. The infarct (Fig. 10.3) developed by PT infarction was in line with the standard presentation of the model, as it did not damage the underlying hippocampus. We found that PBM treatment was able to reduce the size of the infarct by just over a third. When investigating neuronal survival in the peri-infarct region via the neuronal marker NeuN, PBM managed to rescue the stark loss of neurons in the region, with counts just shy of sham levels. This potent neuroprotection suggests further implications about the role PBM is playing in the context of PT stroke.

Neurons that survive the initial onslaught of ischemic assault are not yet in the clear. Though they may survive, the damage sustained may render them dysfunctional and may, in fact, set them up for long-term degeneration and programmed cell death. As such, it is important to consider the condition of the surviving neuronal population when investigating the protective properties of any putative therapy, PBM included. In our work we looked at two metrics of neuronal injury, synaptic markers, and dendritic integrity. MAP2 is a microtubule associated protein that is highly expressed in the dendrites and is often used to investigate neuronal damage (Kitagawa et al., 1989). Using immunohistochemistry, we observed a stark degradation of MAP2 in PT group animals (Fig. 10.4). There was a remarkable

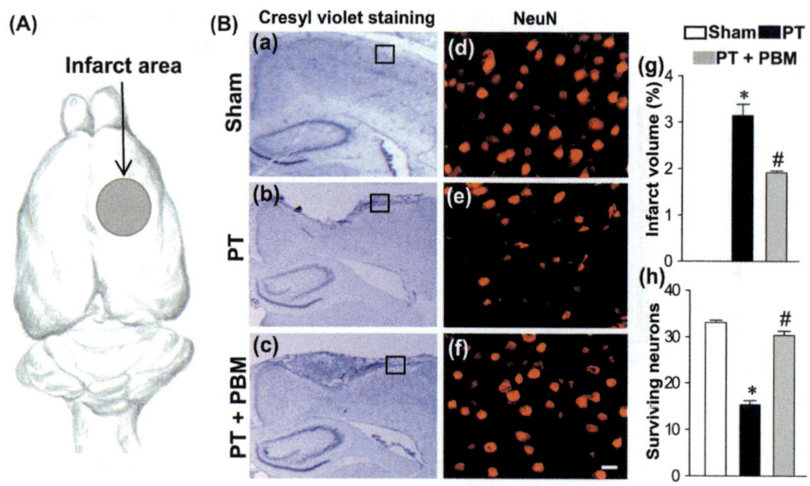

FIGURE 10.3 PBM reduces infarct volume and neuronal cell death after PT stroke. (A) depicts the location of PT infarct. PBM administration after PT reduces infarct volume (B, a–c). Neuronal cell count in the peri-infarct region (B, c–f), located in the black squares in (B, a–c), was preserved to nearly uninjured levels by PBM. Data quantified in (B, g and h).

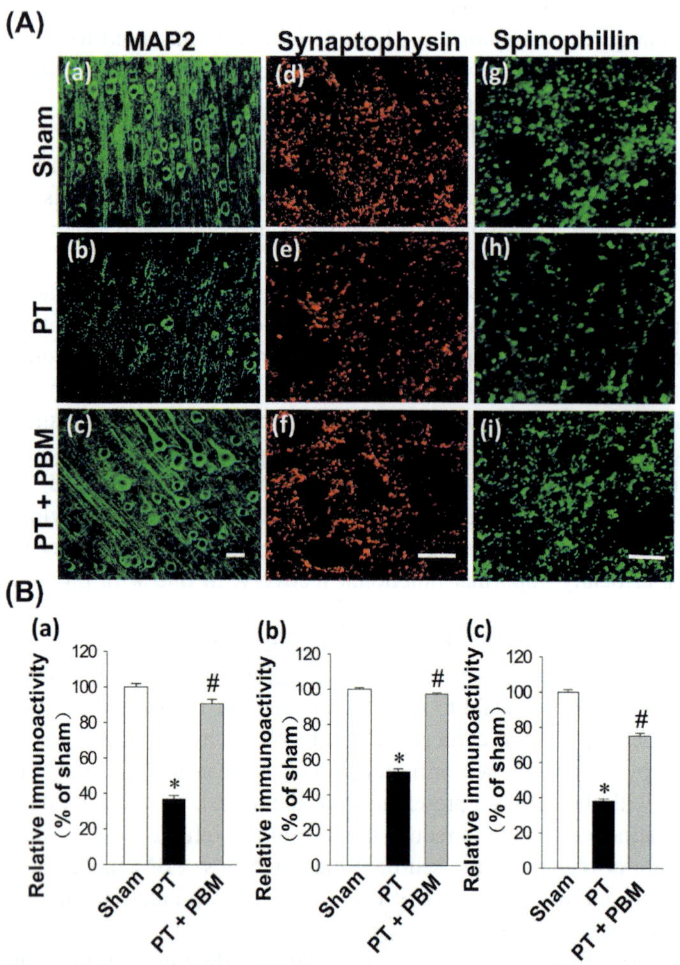

FIGURE 10.4 PBM treatment preserves dendritic and synaptic components in the peri-infarct area as evidenced by immunoreactivity of MAP2, synaptophysin, and spinophilin, respectively (A). Data is quantified in (B).

preservation of MAP2 immunofluorescence after PT conferred by PBM, with immunoreactivity 90% that of sham. Our results on synaptic markers complemented these results. Synaptic injury was investigated by the presence of the pre and postsynaptic markers synaptophysin and spinophilin, respectively. There was a sharp decrease observed in both synaptic markers after PT stroke. PBM completely ameliorated this decline in synaptophysin and brought spinophilin levels back to 80% of sham, double that of PT animals. These results indicate that PBM can not only prevent cell death in the penumbral region after PT stroke, but it can preserve the structure and integrity of these neurons, which is critical for their function and future viability.

Inflammation and reactive gliosis are common observations shared between human stroke and experimental animal stroke models, including the PT stroke model (Fluri et al., 2015; Burda and Sofroniew, 2014). After ischemic insult, the brain responds with a rapid neuroinflammatory response, with glial activation occurring followed by release of a complex mixture of both pro and antiinflammatory cytokines into the neuronal microenvironment. Pro-inflammatory cytokines accelerate cell death and promote further inflammation and infiltration of peripheral macrophages. Antiinflammatory cytokine signaling can counter the detrimental effects of inflammation and promote signaling pathways leading to neural repair (Burda and Sofroniew, 2014; Lambertsen et al., 2012). Neuroinflammation develops over time and, in PT stroke, generally reaches its maximum 2 days after induction (Vandeputte et al., 2011). Cytokines are potent modulators of inflammation, affecting a diverse set of cellular signaling responses including cell adhesion molecule expression, infiltration of peripheral macrophages, and cell death. Pro-inflammatory cytokine expression is commonly tied to infarct size, giving weight to the notion that treatments that target this process may be able to preserve intact penumbra (Lambertsen et al., 2012; Doll et al., 2014). In our work, we used sensitive ELISA assays to measure the levels of both pro-inflammatory cytokines IL-6, TNF-a, and IL-18, and antiinflammatory cytokines IL-1 and IL-10. As seen in Fig. 10.5, PT stroke promoted a vigorous pro-inflammatory response, significantly promoting expression of IL-6, TNF-a, and IL-18. PBM was able to substantially reduce this upregulation, with a near reversal in levels of both IL-6 and IL-18. Conversely, PT stroke caused a reduction in the levels of IL-1 and IL-10. Once again, this deleterious pattern was reversed to sham levels by PBM treatment. This shift in inflammatory signaling indicates that PBM may help fight against creation of a dangerous microenvironment for neurons in the penumbral region, as well as sensitive neuronal progenitor cell pools that may serve a role in neural repair.

Glial activation, or reactive gliosis, is a characteristic occurrence taking place after stroke and is observed in several brain injury, neurodegenerative, and psychiatric conditions. Astrocytes and microglia, once thought only to play a supporting role in the brain, are now well established to play an active role in brain function, in normal brain health, as well as a critical role in neural development. These cells change their role after brain insult, shifting to an "activated" phenotype that can promote inflammation, generate glial scars, phagocytose apoptotic bodies, and promote the recruitment of peripheral macrophages (Burda and Sofroniew, 2014). Though this process is a necessary immune response, prolonged neuroinflammation is detrimental to the brain and the recovery process. Glial activation is often identified via characteristic changes in cellular morphology. Reactive astrocytes display thicker, longer reaching processes, while

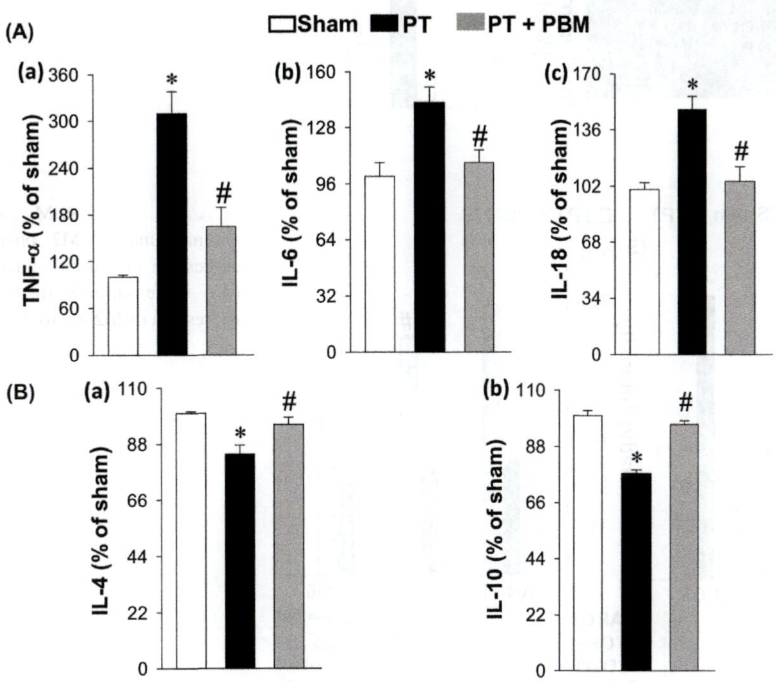

FIGURE 10.5 PBM reduces pro-inflammatory cytokine release (A) and increases antiinflammatory cytokine release (B) after PT stroke, as determined via ELISA.

activated microglia take on a rounded amoeboid morphology. Glial activation is also commonly detected by the astrocytic marker GFAP and the microglial marker Iba-1 (Kim et al., 2014). Represented in Fig. 10.6, we found using immunohistochemistry that PT stroke, much in line with our previous results and that of others, elevated levels of both Iba-1 and GFAP in the peri-infarct region, indicating significant induction of reactive gliosis. PBM was able to strikingly decrease this reaction, demonstrating a powerful role of PBM in ameliorating the activation of glial cells in the sensitive at-risk peri-infarct region.

Microglia, upon activation, differentiate into one of two activated phenotypes, M1 and M2. The M1 state is characterized by the release of pro-inflammatory cytokines, and can be identified in immunostaining via the markers CD32, CD86, and iNOS. M2 microglia, in contrast, release primarily antiinflammatory cytokines and are commonly identified by the markers ARG1, TGF-B, and CD206. After stroke, microglial polarization leans toward the M1 phenotype with only a small presence of M2 microglia (Ladwig et al., 2017; Lu et al., 2017; Hu et al., 2012). This is reflected in several of the animal models widely used in stroke research, including the PT model. Using Western blot analysis of brain homogenates, we found that PT stroke showed increased expression of M1 markers CD32, CD86, and iNOS relative to uninjured sham brain (Fig. 10.7). M2 markers ARG1, TGF-B, and CD206 were unaffected in PT animals. The overall

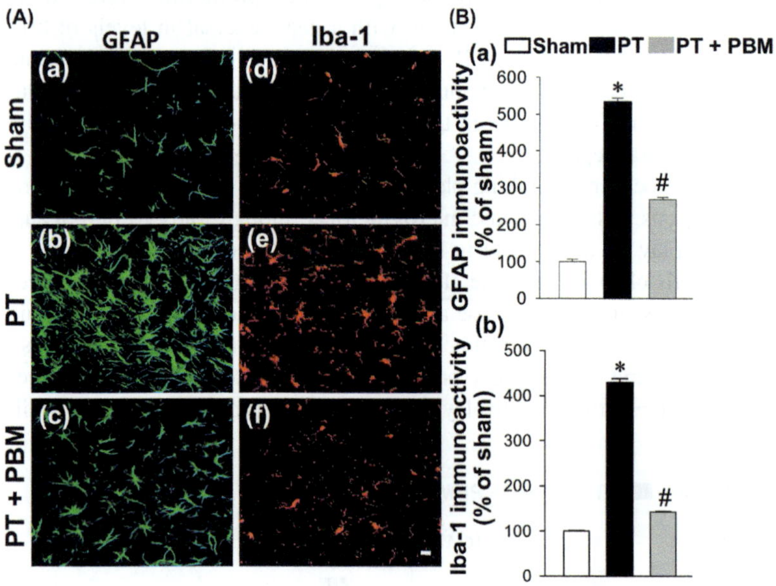

FIGURE 10.6 PBM decreases glial activation after PT stroke. After stroke microglia and astroctyes shift to an activated phenotype in response to infection or injury. As depicted in (A) both microglia and astrocytes, indicated by Iba-1 and GFAP, respectively, take on an activated phenotype in PT stroke animals. PBM reduces this activation. Data is quantified in (B).

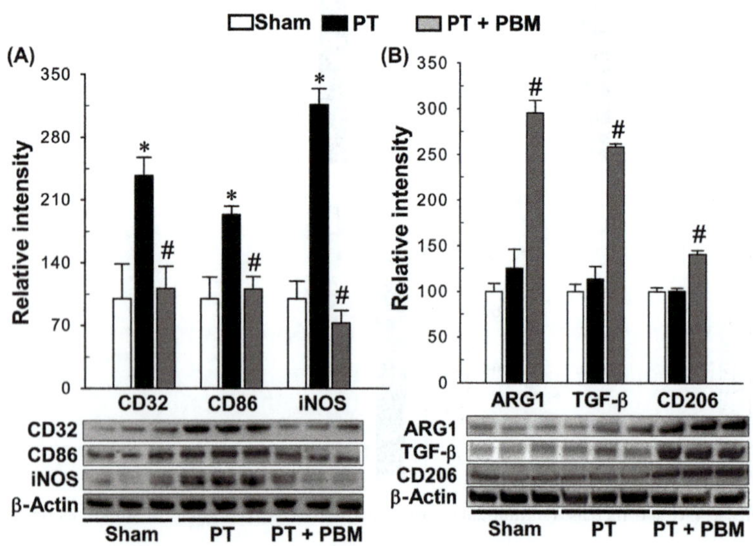

FIGURE 10.7 PBM shifts microglia to an antiinflammatory M2 phenotype. PT stroke induces expression of pro-inflammatory M1 microglial markers (A), while animals treated with PBM display high expression of M2 microglial markers (B).

pattern of expression was starkly reversed with administration of PBM, wherein M1 markers were unaffected and M2 markers were sharply upregulated. We believe this represents that PBM can lead to a shift from the pro-inflammatory M1 phenotype to the antiinflammatory M2 phenotype, falling in line with our results concerning inflammatory cytokines.

Mitochondrial dysfunction is a characteristic feature of human stroke and is seen as a uniting feature of neurodegenerative conditions. Ischemic brain injury triggers damage to mitochondrial components that, upon reperfusion, triggers electron leakage generating the production of excessive ROS. The state of excessive oxidative stress causes extensive cellular damage and perpetuates the pathological conditions brought on by ischemic injury (Akbar et al., 2016). Several well established models of animal stroke reflect this, and the PT stroke model is no exception (Akbar et al., 2016; Lu et al., 2016). In our work, we examined ATP production and CCO activity as metrics of mitochondrial function. We observed, shown in Fig. 10.8, that PT stroke animals displayed marked decreases in ATP content, with levels nearly half that of sham. PBM treatment was able to completely reverse this depletion. Mitochondrial CCO activity was likewise heavily impaired by PT stroke, and PBM was, once again, able to prevent this degradation, likely accounting for preservation ATP levels. Though reduction of ROS damage likely accounts for much of the benefits of PBM, the preservation of energy metabolism is critical for the neural repair and endogenous poststroke neurogenesis (Voloboueva and Giffard, 2011).

Neurogenesis in the adult brain has been observed in several animal stroke models, and is thought to promote neural repair after stroke (Voloboueva and Giffard, 2011). Unfortunately, most of the newly formed neurons rarely survive long-term and it is unknown the degree to which they functionally incorporate into existing neuronal circuitry (Voloboueva and Giffard, 2011). Our previous work and others suggest that promoting an amenable neuronal microenvironment can promote and sustain neurogenesis, so the previously detailed results in our study suggested that treatments like PBM may be able to do so (Ahmed et al., 2016; Wang et al., 2015). To investigate this possibility, we used the artificial thymine analogue BrdU and markers for immature neurons, Ki67 and DCX. BrdU incorporates into the DNA of proliferating cells and can be targeted by anti-BrdU antibodies, thereby highlighting cells that have divided during the period of BrdU administration (Wojtowicz and Kee, 2006). We observed in Fig. 10.9 a thin belt of neurogenesis surrounding the infarct with a breadth of about 1 mm in both PT and PBM group animals, with no cortical neurogenesis found in sham rats. The immunoreactivity of all three of these markers, however, was strikingly increased in PBM animals, indicating that PBM could bolster or support neurogenesis. To discern whether PBM was simply globally upregulating neurogenesis, we looked at the two established neurogenic niches in the adult brain the hippocampal dentate gyrus and the subventricular zone (Ming and Song, 2011). This analysis yielded no difference between the three groups, demonstrating that the observed effects on neurogenesis were specific to that stimulated by PT injury. We believe these results indicate that PBM is promoting the endogenous repair system in the brain, in part by promoting a favorable environment for the proliferation and differentiation of sensitive neural progenitor cells.

In summary, our work demonstrated that PBM treatment after PT could exert several interrelated beneficial effects on the neuronal microenvironment. We found that PBM reduced glial activation after PT stroke and shifted microglial polarization toward the antiinflammatory M2 state. This was reflected in the ability of PBM to substantially decrease the release of pro-inflammatory cytokines while bolstering that of antiinflammatory cytokines. Mitochondrial function was also preserved, with PBM mitigating the decline of CCO activity and ATP production. We believe that it is these

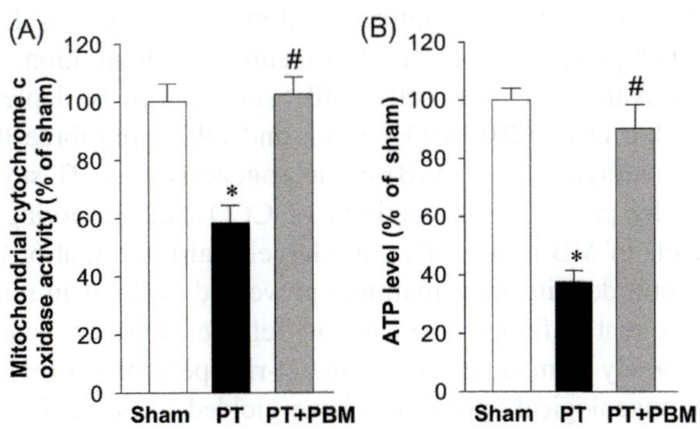

FIGURE 10.8 Mitochondrial function is preserved by PBM. Mitochondrial cytochrome c oxidase, complex IV in the electron transfer chain, is the rate-limiting step in oxidative phosphorylation and is a common target of mitochondrial damage. PBM promotes complex IV activity (A) and resultant ATP production (B) that is damaged after PT stroke.

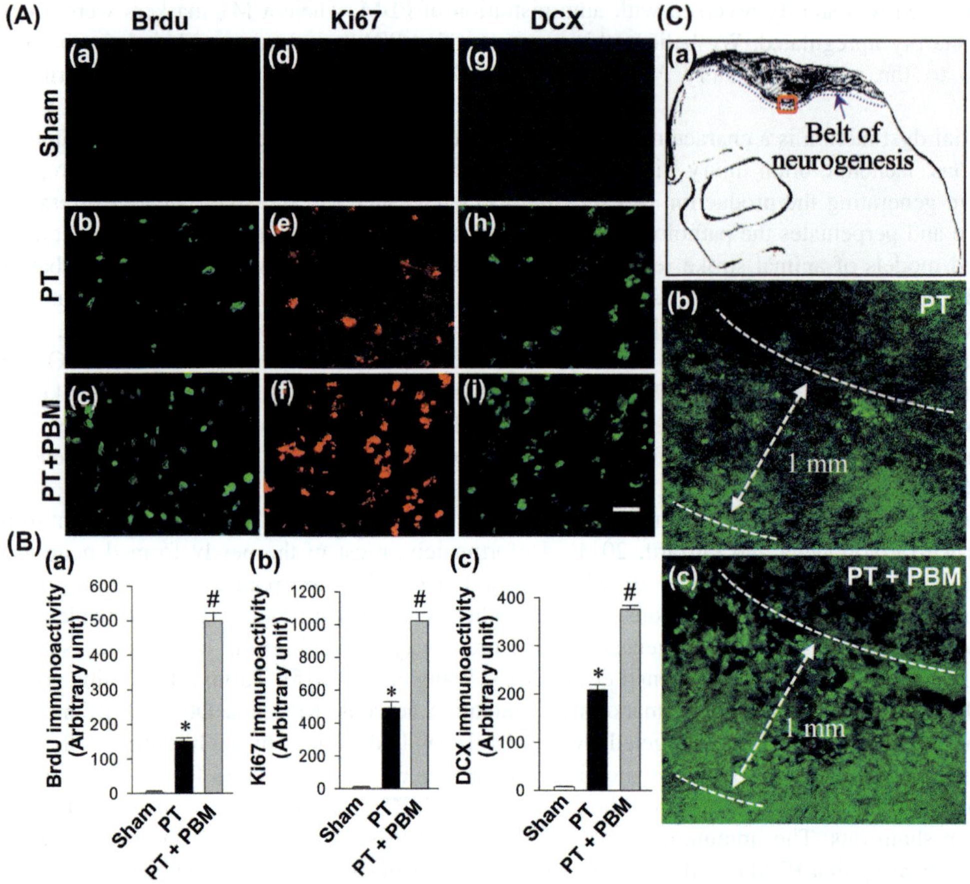

FIGURE 10.9 PBM promotes neurogenesis after PT stroke in the peri-infarct region. (C, a) illustrates the "belt of neurogenesis", a region where neurogenesis occurs after infarct induction, investigated in this study and the region imaged in (A) (red square). Brdu, an artificial analog of the nucleic acid thymidine, was administered in the days following stroke and is incorporated into the DNA of newly divided cells. BrdU and markers of immature neurons, Ki67 and DCX, were detected in PT-injured brains. This response was greatly amplified in PBM treated animals, suggesting.that PBM promotes repair mechanisms after PT stroke. Representative images (C, b & c) depict increased neuronal cell density in the aforementioned "belt of neurogenesis" in animals treated with PBM. Data is quantified in (B).

effects, in tandem then exerted potent neuroprotection in the sensitive at-risk penumbra region, protecting against PT-induced neuronal damage and cell death, decreasing infarct size. In this preserved region, we observed that PBM could promote the process of neurogenesis. Finally, these results were accompanied by improved behavioral and neurological outcome (Yang et al., 2018).

Persistent neurogenesis after stroke seems to be dependent on fostering the correct microenvironment in which the nascent neurons can thrive, an environment that is compromised after ischemic insult (Thored et al., 2006; Arvidsson et al., 2002). While inflammation is a potentiator of neural progenitor differentiation, the chronic inflammation associated with stroke can stifle the survival of the new cells while simultaneously depleting the extant progenitor pools (Gunton et al., 2017; Giannakopoulou et al., 2017). Our work and others reinforce this concept. Our previous study investigated the effects of the drug methylene blue (MB) on neurogenesis after PT stroke. MB shares similarities in its method of action as PBM, in part by promoting the activity of CCO and preserving mitochondrial function (Tucker et al., 2018). Accordingly, the effects of MB in the PT model largely mirrored that of PBM, conferring substantial neuroprotection and preventing behavioral deficits. MB treatment prevented declines in mitochondrial function and ameliorating reactive gliosis and subsequent inflammatory factor release. Further echoing PBM, these effects were accompanied by stark increases in newly formed neurons in the at-risk peri-infarct region (Yang et al., 2018).

Several other experimental pharmacological approaches have yielded further evidence that decreasing inflammation can be therapeutic in general, irrespective of the effects on neurogenesis. Caffeic acid phenethyl ester (CAPE), a compound derived from honeybee hives, was recently shown to exert remarkable antiinflammatory effects that resulted in decreases in infarct size. Treatment with CAPE administered after PT, decreased the expression of TNF-a, HIF-1a,

MCP1, and IL-1a, while upregulating heme oxygenase-1 (HO-1) and IL-10 (Hwang et al., 2018). PBM has been shown to increase expression of HO-1 in a model of diabetic retinopathy, suggesting that perhaps HO-1 may play a role in the effects of PBM in PT stroke, a topic warranting further investigation (Saliba et al., 2015). The flavonoid 2′-methoxy-6-methylflavone (2′MeO6MF) induced decreases in the levels of I IL-1β, INFy, and TNFa while improving functional recovery and reducing infarct size (Clarkson et al., 2018). In addition, the matrix metalloproteinase inhibitors, doxycycline and minocycline, have been shown to decrease infarct size alongside reduced levels of inflammatory mediators TNFa, MCP1, and IDO (Park et al., 2011). Various MMP isoforms have been shown to be downregulated by PBM, indicating yet another avenue at which PBM may be modulating the neuronal microenvironment to promote recovery (Ayuk et al., 2016). Even other nonpharmacological therapies that exhibit the same effects on inflammation, as does PBM, have demonstrated beneficial effects in the PT model. Application of low frequency-pulsed electromagnetic fields to PT stroke animals induced decreases of both IL-1β and MMP9, alongside decreases in caspase activation and promotion of apoptotic signaling. This produced reductions in both infarct size and behavioral deficits (Urnukhsaikhan et al., 2017). As well, MMP9 inhibition has been shown to promote the migration of newly formed progenitor cells (Kang et al., 2008). Melatonin effectively reduced the size of the PT infarct in a COX-1 dependent manner, with COX-1 KO animals showing no neuroprotection from the hormone (Zou et al., 2006). These and other investigations have demonstrated the importance of controlling inflammation in the PT model, reinforcing the importance of the antiinflammatory capacity of PBM (Liguz-Lecznar et al., 2015; Chang et al., 2017).

In relation to the effect of PBM on inflammation, glial activation and polarization is a target of PBM and several other experimental therapeutics tested in PT stroke. The liposoluble iron chelator 2,2′-dipyridyl (DP) resulted in reductions of astrocyte activation and HIF-1α, and increased the expression of HO-1. This was occurred in tandem with decreased infarct volume and BBB permeability (Demougeot et al., 2004). The sphingosine-1-phosphate analogue FTY720 decreased astrogliosis and resulted in improved behavioral outcome and larger postsynaptic densities near the peri-infarct region (Brunkhorst et al., 2013). That said, the role of astrocytic activation in PT is somewhat unclear, as astrocytic activation after PT can help reestablish the integrity of the BBB after PT stroke, facilitating the improvement of the neuronal microenvironment (Gliem et al., 2015). Mitigating microglial activation and shifting polarization toward the M2 state, however, seems positively correlated to improved outcomes after PT. In addition to previously mentioned results, alpha-linolenic acid demonstrates neuroprotective effects mediated, in part, by reduction in microglial activation (Liu et al., 2014). Moreover, shifts toward M2 polarization state via Hv1 proton channel KO reduced infarct size and multiple behavioral and neurological parameters associated with functional recovery (Tian et al., 2016). Macrophage-derived osteopontin was likewise observed to cause a M2-oriented shift and reductions in infarct size in a detailed investigation of microdomains of microglial polarization states (Ladwig et al., 2017). The effects of PBM on the PT model appear to be multifactorial, touching on a series of interrelated mechanisms that, taken together and individually, appear to promote neuroprotection.

Although our primary focus was on the ability of PBM to promote a healthy neuronal environment, PBM may promote neuroprotection and neural repair through other mechanisms. Preconditioning with PBM twice daily for 2 days prior to PT stroke was able to mitigate glial activation and downstream inflammatory signaling. As well, PBM preserved BBB integrity and prevented leukocyte infiltration into the peri-infarct cortex (Lee et al., 2016). Recent work also applied PBM administered via LED array on a mouse PT model. PBM, consistent with our findings, promoted neurogenesis and formation of new blood vessels in the cerebral cortex after PT stroke. Furthermore, LED PBM upregulated expression of BDNF, associated in other studies with neurogenesis in PT stroke (Lee et al., 2017a; Deng et al., 2017). Another strategy attempted in our lab was poststroke remote PBM treatment to the femur. Informed by Oron and Oron (2016), wherein 2 months of weekly PBM to the bone marrow reduced a beta load and cognitive decline in a transgenic AD rat; we treated the rats with PBM applied to the femur after PT stroke every other day for 34 days, excluding weekends. In unpublished data, this approach did not result in any appreciable effects to behavioral outcomes in the adhesive removal test, cylinder test, or grip strength test. This does, however, corroborate previous work showing that injection of human bone marrow mononuclear cells did not improve functional outcome after PT, although the authors of that study point out that the delayed nature of their treatment time course may have played a role (Minnerup et al., 2010). It is apparent, therefore, that the effects of PBM are multifaceted and in need of further mechanistic investigation.

The search for effective treatments for stroke has been challenging and fraught with difficulty. Several agents that have shown great promise in animal models have failed when implemented in clinical trials, and this does include PBM (Hacke et al., 2014; Xu and Pan, 2013). These failures could be a result of interspecies differences, pathophysiological variance between models and human stroke, or the heterogeneous nature of stroke itself. In the case of PBM, the trial relied on a single-dose paradigm administered up to 24 hours after stroke (Hacke et al., 2014). Our design was informed by this effort, focusing on the effects of successive treatment sessions in the days following stroke. We believe that a multiple dose treatment strategy may hold the key to successful neuroprotection in the clinic, although that question is left for future clinical trials to determine.

Stroke presents a daunting challenge to science, medicine, and society. The lives taken and affected by this condition will likely grow in the near future, expanding the breadth of its influence as a cruel effect of longer lifespan facilitated by medical advances. It seems imperative, then, to invest significant time and resources into remedies that can protect the brain against the hypoxic onslaught of stroke, sparing lives and improving the quality of life of patients and their loved ones. While this task seems vexingly difficult, it is our hope that novel therapeutics may soon fill that role. We believe that PBM may represent one such option, and it is our hope that our work and the work of others may bring this goal to fruition.

References

Ahmed, M.E., et al., 2016. Methylene Blue promotes cortical neurogenesis and ameliorates behavioral deficit after photothrombotic stroke in rats. Neuroscience 336, 39–48.

Akbar, M., et al., 2016. Mitochondrial dysfunction and cell death in neurodegenerative diseases through nitroxidative stress. Brain Res. 1637, 34–55.

Arvidsson, A., et al., 2002. Neuronal replacement from endogenous precursors in the adult brain after stroke. Nat. Med. 8 (9), 963–970.

Ayuk, S.M., Abrahamse, H., Houreld, N.N., 2016. The role of matrix metalloproteinases in diabetic wound healing in relation to photobiomodulation. J. Diabetes Res. 2016, 2897656.

Benjamin, E.J., et al., 2017. Heart disease and stroke statistics-2017 update: A report from the American Heart Association. Circulation 135 (10), e146–e603.

Berman, M.H., et al., 2017. Photobiomodulation with near infrared light helmet in a pilot, placebo controlled clinical trial in dementia patients testing memory and cognition. J. Neurol. Neurosci. 8 (1).

Bouet, V., et al., 2009. The adhesive removal test: a sensitive method to assess sensorimotor deficits in mice. Nat. Protoc. 4 (10), 1560–1564.

Brunkhorst, R., et al., 2013. FTY720 treatment in the convalescence period improves functional recovery and reduces reactive astrogliosis in photothrombotic stroke. PLoS One 8 (7), e70124.

Burchell, S.R., et al., 2013. Isoflurane provides neuroprotection in neonatal hypoxic ischemic brain injury. J. Investig. Med. 61 (7), 1078–1083.

Burda, J.E., Sofroniew, M.V., 2014. Reactive gliosis and the multicellular response to CNS damage and disease. Neuron 81 (2), 229–248.

Carmichael, S.T., 2005. Rodent models of focal stroke: size, mechanism, and purpose. NeuroRx 2 (3), 396–409.

Chang, L., et al., 2017. (+)-Borneol is neuroprotective against permanent cerebral ischemia in rats by suppressing production of proinflammatory cytokines. J. Biomed. Res. 31 (4), 306–314.

Chung, H., et al., 2012. The nuts and bolts of low-level laser (light) therapy. Ann. Biomed. Eng. 40 (2), 516–533.

Clarkson, A.N., et al., 2018. The flavonoid, 2'-methoxy-6-methylflavone, affords neuroprotection following focal cerebral ischaemia. J. Cereb. Blood Flow Metab. 271678X18755628.

Demougeot, C., et al., 2004. Cytoprotective efficacy and mechanisms of the liposoluble iron chelator 2,2'-dipyridyl in the rat photothrombotic ischemic stroke model. J. Pharmacol. Exp. Ther. 311 (3), 1080–1087.

Demyanenko, S.V., Panchenko, S.N., Uzdensky, A.B., 2015. Expression of neuronal and signaling proteins in penumbra around a photothrombotic infarction core in rat cerebral cortex. Biochemistry (Mosc.) 80 (6), 790–799.

Deng, G., et al., 2017. Delayed administration of guanosine improves longterm functional recovery and enhances neurogenesis and angiogenesis in a mouse model of photothrombotic stroke. Mol. Med. Rep. 15 (6), 3999–4004.

Doll, D.N., Barr, T.L., Simpkins, J.W., 2014. Cytokines: their role in stroke and potential use as biomarkers and therapeutic targets. Aging Dis. 5 (5), 294–306.

Fluri, F., Schuhmann, M.K., Kleinschnitz, C., 2015. Animal models of ischemic stroke and their application in clinical research. Drug Des. Dev. Ther. 9, 3445–3454.

Geltman, E.M., 1984. Infarct size as a determinant of acute and long-term prognosis. Cardiol. Clin. 2 (1), 95–103.

Giannakopoulou, A., Lyras, G.A., Grigoriadis, N., 2017. Long-term effects of autoimmune CNS inflammation on adult hippocampal neurogenesis. J. Neurosci. Res. 95 (7), 1446–1458.

Gliem, M., et al., 2015. Macrophage-derived osteopontin induces reactive astrocyte polarization and promotes re-establishment of the blood brain barrier after ischemic stroke. Glia 63 (12), 2198–2207.

Gonzalez-Lima, F., Barksdale, B.R., Rojas, J.C., 2014. Mitochondrial respiration as a target for neuroprotection and cognitive enhancement. Biochem. Pharmacol. 88 (4), 584–593.

Gu, W., Brannstrom, T., Wester, P., 2000. Cortical neurogenesis in adult rats after reversible photothrombotic stroke. J. Cereb. Blood Flow Metab. 20 (8), 1166–1173.

Gu, W., et al., 2009. Cell division in the cerebral cortex of adult rats after photothrombotic ring stroke. Stem Cell Res. 2 (1), 68–77.

Gunton, A.N., et al., 2017. Upregulation of inflammatory mediators in the ventricular zone after cortical stroke. Proteomics Clin. Appl. 11, 9–10.

Hacke, W., et al., 2014. Transcranial laser therapy in acute stroke treatment: results of neurothera effectiveness and safety trial 3, a phase III clinical end point device trial. Stroke 45 (11), 3187–3193.

Hu, X., et al., 2001. Progressive and reproducible focal cortical ischemia with or without late spontaneous reperfusion generated by a ring-shaped, laser-driven photothrombotic lesion in rats. Brain Res. Brain Res. Protoc. 7 (1), 76–85.

Hu, X.L., et al., 2004. Dynamic changes of the anti- and pro-apoptotic proteins Bcl-w, Bcl-2, and Bax with Smac/Diablo mitochondrial release after photothrombotic ring stroke in rats. Eur. J. Neurosci. 20 (5), 1177–1188.

Hu, X., et al., 2012. Microglia/macrophage polarization dynamics reveal novel mechanism of injury expansion after focal cerebral ischemia. Stroke 43 (11), 3063–3070.

Hwang, S.A., Kim, C.D., Lee, W.S., 2018. Caffeic acid phenethyl ester protects against photothrombotic cortical ischemic injury in mice. Korean J. Physiol. Pharmacol. 22 (1), 101–110.

Jiang, M., et al., 2017. Neuroprotection provided by isoflurane pre-conditioning and post-conditioning. Med. Gas Res. 7 (1), 48–55.

Kang, S.S., et al., 2008. Inhibition of matrix metalloproteinase-9 attenuated neural progenitor cell migration after photothrombotic ischemia. Brain Res. 1228, 20–26.

Kawaguchi, M., et al., 2000. Isoflurane delays but does not prevent cerebral infarction in rats subjected to focal ischemia. Anesthesiology 92 (5), 1335–1342.

Kim, J.Y., Kawabori, M., Yenari, M.A., 2014. Innate inflammatory responses in stroke: mechanisms and potential therapeutic targets. Curr. Med. Chem. 21 (18), 2076–2097.

King, R.B., 1996. Quality of life after stroke. Stroke 27 (9), 1467–1472.

Kirmani, J.F., et al., 2012. Advances in thrombolytics for treatment of acute ischemic stroke. Neurology 79 (13 Suppl. 1), S119–S125.

Kitagawa, K., et al., 1989. Microtubule-associated protein 2 as a sensitive marker for cerebral ischemic damage--immunohistochemical investigation of dendritic damage. Neuroscience 31 (2), 401–411.

Ladwig, A., et al., 2017. Osteopontin Augments M2 Microglia Response and Separates M1- and M2-Polarized Microglial Activation in Permanent Focal Cerebral Ischemia. Mediators Inflamm. 2017, 7189421.

Lambertsen, K.L., Biber, K., Finsen, B., 2012. Inflammatory cytokines in experimental and human stroke. J. Cereb. Blood Flow Metab. 32 (9), 1677–1698.

Lee, V.M., et al., 1996. Evolution of photochemically induced focal cerebral ischemia in the rat. Magnetic resonance imaging and histology. Stroke 27 (11), 2110–2118. discussion 2118–2119.

Lee, H.I., et al., 2016. Pre-conditioning with transcranial low-level light therapy reduces neuroinflammation and protects blood-brain barrier after focal cerebral ischemia in mice. Restor Neurol Neurosci 34 (2), 201–214.

Lee, H.I., et al., 2017a. Low-level light emitting diode therapy promotes long-term functional recovery after experimental stroke in mice. J. Biophotonics 10 (12), 1761–1771.

Lee, H.I., et al., 2017b. Pretreatment with light-emitting diode therapy reduces ischemic brain injury in mice through endothelial nitric oxide synthase-dependent mechanisms. Biochem. Biophys. Res. Commun. 486 (4), 945–950.

Liguz-Lecznar, M., et al., 2014. Functional assessment of sensory functions after photothrombotic stroke in the barrel field of mice. Behav. Brain Res. 261, 202–209.

Liguz-Lecznar, M., Zakrzewska, R., Kossut, M., 2015. Inhibition of Tnf-alpha R1 signaling can rescue functional cortical plasticity impaired in early post-stroke period. Neurobiol. Aging 36 (10), 2877–2884.

Liu, Y., et al., 2014. Linolenic acid provides multi-cellular protective effects after photothrombotic cerebral ischemia in rats. Neurochem. Res. 39 (9), 1797–1808.

Lu, Q., et al., 2016. Neuroprotective and functional improvement effects of methylene blue in global cerebral ischemia. Mol. Neurobiol. 53 (8), 5344–5355.

Lu, Y., et al., 2017. Treadmill exercise exerts neuroprotection and regulates microglial polarization and oxidative stress in a streptozotocin-induced rat model of sporadic Alzheimer's Disease. J. Alzheimers Dis. 56 (4), 1469–1484.

Ming, G.L., Song, H., 2011. Adult neurogenesis in the mammalian brain: significant answers and significant questions. Neuron 70 (4), 687–702.

Minnerup, J., et al., 2010. Intracarotid administration of human bone marrow mononuclear cells in rat photothrombotic ischemia. Exp. Transl. Stroke Med. 2 (1), 3.

Oron, A., Oron, U., 2016. Low-level laser therapy to the bone marrow ameliorates neurodegenerative disease progression in a mouse model of Alzheimer's Disease: a minireview. Photomed. Laser Surg. 34 (12), 627–630.

Oueslati, A., et al., 2015. Photobiomodulation suppresses alpha-synuclein-induced toxicity in an AAV-based rat genetic model of Parkinson's Disease. PLoS One 10 (10), e0140880.

Park, C.H., et al., 2011. Matrix metalloproteinase inhibitors attenuate neuroinflammation following focal cerebral ischemia in mice. Korean J. Physiol. Pharmacol. 15 (2), 115–122.

Rosenblum, W.I., El-Sabban, F., 1977. Platelet aggregation in the cerebral microcirculation: effect of aspirin and other agents. Circ. Res. 40 (3), 320–328.

Rosenblum, W.I., El-Sabban, F., 1982. Dimethyl sulfoxide (DMSO) and glycerol, hydroxyl radical scavengers, impair platelet aggregation within and eliminate the accompanying vasodilation of, injured mouse pial arterioles. Stroke 13 (1), 35–39.

Salgado, A.S., et al., 2015. The effects of transcranial LED therapy (TCLT) on cerebral blood flow in the elderly women. Lasers Med. Sci. 30 (1), 339–346.

Saliba, A., et al., 2015. Photobiomodulation mitigates diabetes-induced retinopathy by direct and indirect mechanisms: evidence from intervention studies in pigmented mice. PLoS One 10 (10), e0139003.

Schaar, K.L., Brenneman, M.M., Savitz, S.I., 2010. Functional assessments in the rodent stroke model. Exp. Transl. Stroke Med. 2 (1), 13.

Schmidt, A., et al., 2015. Progressive cognitive deficits in a mouse model of recurrent photothrombotic stroke. Stroke 46 (4), 1127–1131.

Seto, A., et al., 2014. Induction of ischemic stroke in awake freely moving mice reveals that isoflurane anesthesia can mask the benefits of a neuroprotection therapy. Front. Neuroenergetics 6, 1.

Sommer, C.J., 2017. Ischemic stroke: experimental models and reality. Acta Neuropathol. 133 (2), 245–261.

Sun, M., et al., 2015. Isoflurane preconditioning provides neuroprotection against stroke by regulating the expression of the TLR4 signalling pathway to alleviate microglial activation. Sci. Rep. 5, 11445.

Thored, P., et al., 2006. Persistent production of neurons from adult brain stem cells during recovery after stroke. Stem Cells 24 (3), 739–747.

Tian, D.S., et al., 2016. Deficiency in the voltage-gated proton channel Hv1 increases M2 polarization of microglia and attenuates brain damage from photothrombotic ischemic stroke. J. Neurochem. 139 (1), 96–105.

Tucker, D., Lu, Y., Zhang, Q., 2018. From mitochondrial function to neuroprotection-an emerging role for methylene blue. Mol. Neurobiol. 55, 5137–5153.

Urnukhsaikhan, E., et al., 2017. Neuroprotective effect of low frequency-pulsed electromagnetic fields in ischemic stroke. Appl. Biochem. Biotechnol. 181 (4), 1360–1371.

van Bruggen, N., et al., 1992. T2- and diffusion-weighted magnetic resonance imaging of a focal ischemic lesion in rat brain. Stroke 23 (4), 576–582.

Vandeputte, C., et al., 2011. Characterization of the inflammatory response in a photothrombotic stroke model by MRI: implications for stem cell transplantation. Mol. Imaging Biol. 13 (4), 663–671.

Voloboueva, L.A., Giffard, R.G., 2011. Inflammation, mitochondria, and the inhibition of adult neurogenesis. J. Neurosci. Res. 89 (12), 1989–1996.

Wang, Q., Yang, L., Wang, Y., 2015. Enhanced differentiation of neural stem cells to neurons and promotion of neurite outgrowth by oxygen-glucose deprivation. Int. J. Dev. Neurosci. 43, 50–57.

Warner, D.S., 2000. Isoflurane neuroprotection: a passing fantasy, again? Anesthesiology 92 (5), 1226–1228.

Watson, B.D., et al., 1985. Induction of reproducible brain infarction by photochemically initiated thrombosis. Ann. Neurol. 17 (5), 497–504.

Wester, P., et al., 1995. A photothrombotic 'ring' model of rat stroke-in-evolution displaying putative penumbral inversion. Stroke 26 (3), 444–450.

Wojtowicz, J.M., Kee, N., 2006. BrdU assay for neurogenesis in rodents. Nat. Protoc. 1 (3), 1399–1405.

Xu, S.Y., Pan, S.Y., 2013. The failure of animal models of neuroprotection in acute ischemic stroke to translate to clinical efficacy. Med. Sci. Monit. Basic Res. 19, 37–45.

Xuan, W., Huang, L., Hamblin, M.R., 2016. Repeated transcranial low-level laser therapy for traumatic brain injury in mice: biphasic dose response and long-term treatment outcome. J. Biophotonics 9 (11–12), 1263–1272.

Yang, Q., et al., 2017. Vital signs: recent trends in stroke death rates – United States, 2000–2015. MMWR Morb. Mortal. Wkly. Rep. 66 (35), 933–939.

Yang, L., et al., 2018. Photobiomodulation therapy promotes neurogenesis by improving post-stroke local microenvironment and stimulating neuroprogenitor cells. Exp. Neurol. 299 (Pt A), 86–96.

Zivin, J.A., et al., 2009. Effectiveness and safety of transcranial laser therapy for acute ischemic stroke. Stroke 40 (4), 1359–1364.

Zou, L.Y., et al., 2006. Melatonin reduces infarction volume in a photothrombotic stroke model in the wild-type but not cyclooxygenase-1-gene knockout mice. J. Pineal Res. 41 (2), 150–156.

Chapter 11

Remote photobiomodulation as a neuroprotective intervention—harnessing the indirect effects of photobiomodulation

Luke Gordon[1,2], Boaz Kim[1,2], Claudia Petrucco[1,2], Ji Yeon Kim[1,2,3], Patrick Benson[1,2], Jonathan Stone[1,2] and Daniel M. Johnstone[1,2]

[1]*Bosch Institute, University of Sydney, Sydney, NSW, Australia,* [2]*Discipline of Physiology, University of Sydney, Sydney, NSW, Australia,* [3]*School of Medicine, University of Queensland Centre for Clinical Research, Brisbane, QLD, Australia*

Over the past few decades there has been a dramatic increase in scientific research into photobiomodulation (PBM) therapy, both to treat injury or illness and to enhance normal function or performance. More recently, substantial focus has been placed on PBM of the brain and nervous system, with a range of preclinical studies and some clinical trials yielding promising results. Naturally, almost all studies have reported on the effects of PBM when light is targeted directly at the tissue under investigation. However, a small number of studies over the years, increasing in frequency in recent times, have provided evidence that the beneficial effects of PBM are not confined to the irradiated tissue. Instead, it appears that PBM can elicit systemic effects that promote protection of remote tissues. While the mechanisms remain to be understood, this phenomenon of a body-wide response to localized PBM treatment has far-reaching implications, both for our understanding of basic biology as well as the therapeutic application of PBM, particularly for difficult-to-irradiate organs such as the brain.

11.1 Transcranial photobiomodulation

In developing PBM therapy as a strategy to mitigate brain disease or enhance brain function, the simplest and most obvious approach for delivering light to the brain is to apply it externally (i.e., transcranially), at an intensity sufficient for an appropriate dose of light to penetrate the scalp and skull to reach the target brain tissue. As highlighted by many of the chapters in this book, this approach has yielded great success across a variety of small animal models of brain disease and in some small-cohort clinical trials. As just a snapshot, transcranial PBM has been shown to mitigate neuropathology or improve neurological function in animal models of stroke (Detaboada et al., 2006; Lapchak et al., 2004; Oron et al., 2006), traumatic brain injury (Oron et al., 2007; Wu et al., 2012; Xuan et al., 2013, 2016), Alzheimer's disease and frontotemporal dementia (De Taboada et al., 2011; Grillo et al., 2013; Lu et al., 2017; Purushothuman et al., 2013, 2014, 2015) and Parkinson's disease (El Massri et al., 2016a; Moro et al., 2013;Oueslati et al., 2015; Peoples et al., 2012; Reinhart et al., 2015, 2016a, 2017; Shaw et al., 2010, 2012). In human trials for acute ischemic stroke, transcranial PBM showed evidence of efficacy as a treatment in phase I and II clinical trials (Huisa et al., 2013; Lampl et al., 2007; Zivin et al., 2009) but failed an interim futility analysis in a phase III trial (Hacke et al., 2014), possibly due to a lack of dose optimization (Lapchak and Boitano, 2016). Small-cohort patient studies suggest that transcranial PBM improves neurological outcomes in patients with traumatic brain injury (Morries et al., 2015; Naeser et al., 2011), reduces depression and anxiety in patients with psychological disorders (Cassano et al., 2015; Schiffer et al., 2009), and improves cognition and emotional state in healthy individuals (Barrett and Gonzalez-Lima, 2013; Blanco et al., 2017).

Photobiomodulation in the Brain. DOI: https://doi.org/10.1016/B978-0-12-815305-5.00011-7

11.2 Limitations of transcranial photobiomodulation

Despite these successes in the laboratory and the clinic, a major barrier to clinical translation of transcranial PBM remains the delivery of sufficient light energy across the scalp and skull to the target brain tissue. Measurements in human postmortem tissue suggest that penetration across the skin and skull of far red to near-infrared wavelengths ranges from $\sim 1\%$ to 3% (depending on the thickness of the skull), and less than 1% of that light energy will penetrate 12 mm of brain tissue (Hart and Fitzgerald, 2016). The inverse relationship between light penetration and skull thickness has been confirmed in a different study comparing 800 nm laser light transmission across postmortem skulls from different species (Lapchak et al., 2015). Our own measurements in living mice of 670 nm LED light transmission across brain tissue indicated a 65% decline in light intensity per millimeter of brain tissue traversed (Moro et al., 2014). Together, these data suggest that transcranial PBM, while potentially a viable modality for treating the most superficial layers of the brain, is unlikely to deliver sufficient light energy to deeper brain structures without compromising the health of the overlying tissue (Hart and Fitzgerald, 2016; Johnstone et al., 2016). Nonetheless, it remains possible that irradiation of superficial brain structures with transcranial PBM may yield some symptomatic benefits even in conditions where degeneration primarily affects deep brain structures; or, given our growing understanding of the remote effects of PBM (discussed in detail below), that irradiation of the skull and superficial layers of the brain generates protective responses in deeper structures. This chapter explores what is known of the response of brain tissue to direct and remote PBM.

11.3 Alternative photobiomodulation treatment modalities

In recent times, several novel approaches intended to overcome this limitation of transcranial PBM and allow therapeutic doses of light to be delivered deeper into the brain have been explored.

11.3.1 Intracranial photobiomodulation

In collaboration with the team of Alim-Louis Benabid in Grenoble, France, several authors of this chapter have been involved in testing the safety and efficacy of an implantable device for emitting light within the brain (i.e., intracranial PBM). The inspiration for designing such a device was to enable effective delivery of PBM to vulnerable midbrain neurons involved in Parkinson's disease.

The initial design utilized a LED attached to an optical fiber, with a step index to limit the loss of irradiance along the length of the fiber, concentrating it at the fiber tip. The first safety study of this implant in mice and rats, in which the fiber tip was positioned in the lateral ventricle, revealed no behavioral deficits or tissue necrosis arising from intracranial PBM, other than minor mechanical damage resulting from implantation of the optical fiber (Moro et al., 2014). Importantly, when mice were exposed to the parkinsonian neurotoxin MPTP (1-methyl-4-phenyl-1,2,3,6-tetrahydropyridine), intracranial PBM (670 nm) protected against loss of dopaminergic cells in the midbrain. Stronger protective effects were observed when intracranial PBM was delivered intermittently (4×90 seconds over 2 days) rather than continuously (Moro et al., 2014). The neuroprotective efficacy of intracranial 670 nm PBM was confirmed in a rat model of Parkinson's disease, in which animals received a unilateral intrastriatal injection of the neurotoxin 6-hydroxydopamine (6-OHDA). In this model, intracranial PBM (2×90 seconds per day), delivered by optical fiber to the midline of the midbrain, significantly mitigated the loss of midbrain dopaminergic cells and the associated functional deficits, as assessed by apomorphine-induced rotational behavior (Reinhart et al., 2016b).

With safety and efficacy confirmed in rodent models, intracranial PBM was trialed in a primate model of Parkinson's disease. An optical fiber, coupled to a laser diode (670 nm), was implanted into the midline of the midbrain of adult macaques. Following a 14 day recovery period, macaques received intramuscular injections of MPTP to induce parkinsonism, with a subset receiving intermittent intracranial PBM (5 seconds ON/60 seconds OFF) throughout the 5 or 7 day injection period (Darlot et al., 2016). Daily evaluation of clinical signs revealed significantly less parkinsonism and motor dysfunction in MPTP macaques receiving intracranial PBM than in sham-treated MPTP macaques. Following a 3-week survival period, immunohistochemical analysis of the brain revealed significant mitigation of MPTP-induced midbrain neurodegeneration in PBM-treated animals, with cell counts similar to those of saline-injected controls, and more modest (though still significant) protection of dopaminergic terminations in the striatum (Darlot et al., 2016). Intracranial PBM also significantly reduced the level of MPTP-induced astrogliosis in both the midbrain and the striatum, and mitigated changes in microglial morphology (El Massri et al., 2016b).

Collectively, these results from various MPTP models of Parkinson's disease suggest that intracranial PBM is an effective modality for delivering therapeutic doses of light to deep brain structures. We eagerly await the results of early-phase clinical trials that will determine whether this treatment modality is safe and efficacious in human patients.

11.3.2 Intranasal photobiomodulation

An alternative, less-invasive treatment modality currently being developed by VieLight, Inc. is the intranasal delivery of light. A recent pilot study of five patients with mild to moderate cognitive impairment investigated the efficacy of daily PBM (810 nm LED light pulsed at 10 Hz), delivered by a combination of transcranial and intranasal modalities, over a 12 week period. Strikingly, participants showed a significant improvement in measures of cognitive function by the end of the 12 week treatment period, and a decline in cognitive performance following the cessation of treatment. These quantitative outcomes were mirrored by positive qualitative feedback from patients or their carers (Saltmarche et al., 2017). It will be interesting to see whether these exciting findings are confirmed in a large randomized controlled trial and whether, in addition to dementia, intranasal PBM holds promise for neurological conditions affecting deep brain structures.

11.4 Introducing "remote photobiomodulation"

Whether transcranial, intracranial, or intranasal, the approaches just discussed are intended to irradiate the target brain tissue directly. We now introduce the reader to an alternative treatment modality—"remote PBM"—in which light is targeted at a distal tissue, still with the purpose of providing protection of the brain or any other organ (Fig. 11.1). Understanding the rationale for attempting to treat the brain by irradiating a remotely-located tissue, requires an understanding of the history of the discovery of the indirect effects of PBM.

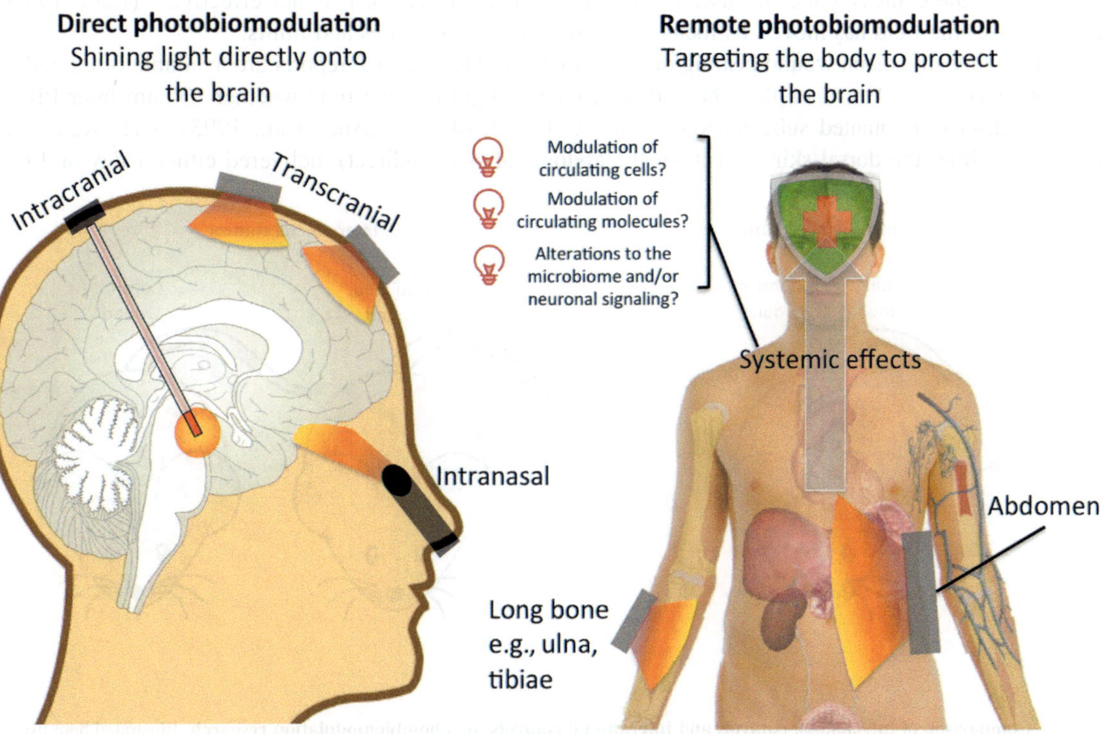

FIGURE 11.1 **Comparison of direct photobiomodulation and remote photobiomodulation.** The image on the left depicts different approaches for direct photobiomodulation of brain tissue, including transcranial, intracranial, and intranasal delivery. In contrast, remote photobiomodulation (depicted on the right) involves irradiating a peripheral target (e.g., bone marrow, abdomen) in order to stimulate a systemic protective response that confers protection to the brain. Some putative mechanisms are listed.

11.5 Discovering the indirect effects of photobiomodulation

While some early studies suggested that PBM can have systemic effects, it was the landmark work of Rochkind and colleagues that systematically demonstrated that unilateral PBM can produce bilateral healing (Rochkind et al., 1989). In their seminal 1989 paper, they utilized rat models to investigate the indirect effects of helium-neon laser PBM (632.8 nm) on cutaneous wounds, burn injuries, and nerve injuries. In rats with bilateral skin incisions along the back, daily PBM (7.6 J/cm^2) to the borders of one wound accelerated healing of both irradiated and nonirradiated wounds, relative to control rats receiving no PBM treatment. In rats with small burn injuries to both hind feet, daily PBM (10 J/cm^2) of one foot accelerated healing of both feet, relative to nonirradiated rats. In rats with a crush injury to both sciatic nerves, daily PBM (10 J/cm^2) to one nerve accelerated the recovery of function of both nerves. The bilateral effect of unilateral irradiation extended to the spinal cord, protecting motor neurons in both the left and right anterior horns against degeneration induced by sciatic nerve injury (Rochkind et al., 1989).

At the same time, Braverman and colleagues made similar observations in a rabbit model of skin wounds (Braverman et al., 1989). Two full skin thickness incisions were made on the back of each rabbit; one of the two wounds was irradiated with either 632.8 nm laser PBM (1.65 J/cm^2), 904 nm laser PBM (8.25 J/cm^2), or both PBM protocols. Importantly, like Rochkind and colleagues, Braverman's study included control animals that had been wounded but not subjected to PBM. And again, unilateral PBM increased tensile strength in both irradiated and nonirradiated wounds, relative to control animals that received no PBM treatment. The investigators speculated that PBM might induce the release of "circulating factors that affect the contralateral wound" (Braverman et al., 1989).

In addition to paving the way for further research into the indirect, even systemic, effects of PBM, these two studies highlighted the importance of including interanimal controls (i.e., animals receiving no PBM treatment) as opposed to simply intraanimal controls (i.e., an untreated site of an animal receiving localized PBM treatment) (Fig. 11.2). Taking wound healing as just one example, the majority of PBM experiments that included only intraanimal controls failed to show a difference in measures of healing between irradiated and nonirradiated wounds (Allendorf et al., 1997; Basford et al., 1986; Hunter et al., 1984; Kana et al., 1981; Sardari and Ahrari, 2016; Surinchak et al., 1983). Understanding that the reparative effects of PBM spread beyond the site of irradiation has helped reconcile these apparently negative findings with the more positive results seen when using interanimal controls, resolving a long-standing debate termed by Tiina Karu as "the endless game of 'laser biostimulation is effective vs. it is not effective'" (Karu, 1999b). In the subsequent decades the field has moved its focus to mechanisms, possibilities and limits.

Beyond the many studies on wound healing outlined in Table 11.1, several reports are available of the indirect effect of PBM in other contexts. For example, Abe and colleagues sought to determine whether 830 nm laser PBM affected the growth of gliomas implanted subcutaneously on the dorsum of mice (Abe et al., 1993). Mice were treated with PBM targeted at either the dorsal skin (direct) or the abdominal skin (indirect), delivered either 1 day or 14 days after

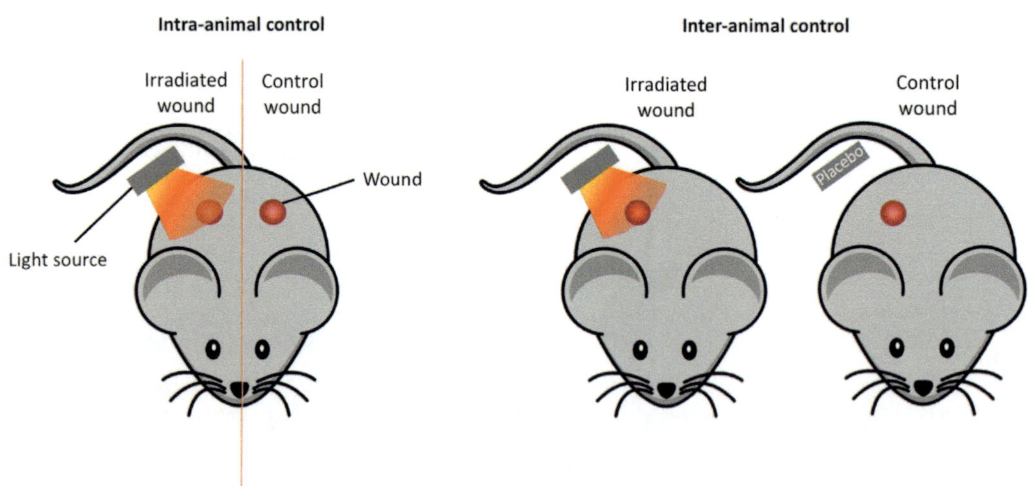

FIGURE 11.2 **Comparison of intraanimal controls and interanimal controls for photobiomodulation research.** Illustrated here using the example of PBM treatment of wounds, experiments employing intraanimal controls tend to create bilateral wounds but apply treatment unilaterally, so that the untreated wound serves as a control for the treated wound. This form of control ignores the possibility that localized PBM has indirect protective effects. In contrast, experiments employing interanimal controls, in which the control animal is wounded but receives no PBM treatment, account for the possibility that PBM has indirect protective effects.

TABLE 11.1 Studies of PBM for the treatment of cutaneous wounds—the effect of using interanimal versus intraanimal controls.

Reference	Model	Wavelength (nm)	Dose (J/cm^2)	Control (interanimal/ intraanimal)	Effect on wound healing relative to controls
Interanimal controls					
Lyons et al. (1987)	Hairless mice	632.8	1.22	Interanimal	+ ($P<.001$)
Yu et al. (1997)	Diabetic mouse	630	5	Interanimal	+ ($P<.01$)
Medrado et al. (2003)	Diabetic rats	670	4; 8	Interanimal	+ ($P<.001$, at 72 h)
Do Nascimento et al. (2004)	Healthy Wistar rat	670; 685	10	Interanimal	+
Maiya et al. (2005)	Diabetic rats	632.8	4.8	Interanimal	+ ($P<.0001$)
Carvalho et al. (2006)	Healthy and diabetic Wistar rat	632.8	4	Interanimal	+ ($P<.05$ for both diabetic and nondiabetic rats)
Rabelo et al. (2006)	Healthy and diabetic Wistar rats	632.8	10	Interanimal	+ ($P<.05$ for both diabetic and nondiabetic rats)
Rezende et al. (2007)	Healthy Wistar rat	830	1.3; 3	Interanimal	+ ($P<.05$)
Fekrazad et al. (2015)	Oral wound in Wistar rat	425; 532; 630	2	Interanimal	+ ($P<.001$)
Intraanimal controls					
Hunter et al. (1984)	Swine	632.8; 694.3	0.96	Intraanimal	– (ns)
Basford et al. (1986)	Swine	632.8; 254	1-4	Intraanimal	– (ns)
Van Cakenbergh (1986)	Male Swiss mice	835	Not specified	Intraanimal	– (ns)
Petersen et al. (1999)	Horse	830	2	Intraanimal	– (ns)
Sardari and Ahrari (2016)	Oral wound in Wistar rat	632	1	Intraanimal	– (ns)
Intra and intercontrols					
Kana et al. (1981)	Sprague Dawley rats	632.8; 514.5	4; 10; 20	Both	+ *Interanimal control* – *Intraanimal control* (some effect shown with 632.8 nm)
Surinchak et al. (1983)	Wounds in rabbit and rat	632.8	1.1; 2.2; 4.5	Both	+ *Interanimal control* – *Intraanimal control*

(Continued)

TABLE 11.1 (Continued)

Reference	Model	Wavelength (nm)	Dose (J/cm^2)	Control (interanimal/ intraanimal)	Effect on wound healing relative to controls
Rochkind et al. (1989)	Sprague-Dawley rats with cutaneous or burn wound	632.8	7.6	Both	+ *Interanimal control* − *Intraanimal control*
Braverman et al. (1989)	New Zealand white rabbits	632.8 (HeNe); 904 (IR)	1.65 (HeNe); 8.25 (IR)	Both	+ *Interanimal control* − *Intraanimal control*
Allendorf et al. (1997)	Sprague Dawley rats	632.8	1; 2; 4	Both	− *Interanimal control* − *Intraanimal control*
Byrnes et al. (2004)	Diabetic rats	632.8	4; 5; 7.2	Both	+ *Interanimal control* − *Intraanimal control*
Hopkins et al. (2004)	Human forearm bilateral wound	820	8	Both	+ *Interhuman control* − *Intrahuman control*
Gungormus and Akyol (2009)	Diabetic rats	808	10	Both	+ *Interanimal control* + *Intraanimal control*

" + " denotes positive effect, " − " denotes no effect, "ns" denotes not statistically significant.

tumor implantation. Both direct and indirect PBM substantially inhibited tumor growth when delivered 1 day after implantation; in contrast, indirect PBM administered 14 days after implantation appeared to enhance tumor growth (Abe et al., 1993). The mechanism underlying this seemingly discrepant finding is unclear, as is the robustness of the finding itself—we are not aware of any subsequent study that has reported accelerated tumor growth as a consequence of PBM.

Another interesting observation on the indirect effects of light energy comes from nanosecond laser therapy for the treatment of eye conditions. For example, 3 ns pulses of 532 nm laser light (0.15–0.45 mJ) has been trialed as a treatment for early age-related macular degeneration. Interestingly, unilateral treatment produced bilateral improvements, reducing pathology not only in the treated eye but also, to a lesser extent, in the untreated eye (Guymer et al., 2014). Similar observations have also been made in the context of treating glaucoma, with unilateral nanosecond laser therapy producing bilateral reductions in intraocular pressure (Rhodes et al., 2009). Exploration of potential mechanisms using a mouse model of retinopathy uncovered an increase in matrix metalloproteinase (MMP)-2 and -3 expression in the retinal pigmented epithelium of both treated and untreated eyes in mice receiving nanosecond laser therapy (Jobling et al., 2015).

11.6 The effects of photobiomodulation on stem cells

A separate line of research has contributed to the discovery that PBM has indirect effects, and provided some important mechanistic insights.

In pioneering studies, Tuby, Maltz, and Oron in Tel Aviv, sought to determine whether PBM could enhance proliferation of stem cells in culture. Bone marrow-derived cells that showed adherence to a culture dish (a property of stem cells) were subjected to a single 804 nm laser PBM treatment (50 mW/cm^2), at a total dose of 1 or 3 J/cm^2. Both PBM doses enhanced stem cell proliferation by at least twofold relative to sham-treated cells at all postirradiation time points assessed (Tuby et al., 2007). Next, they aimed to determine whether pretreating stem cells in culture with PBM produced better outcomes when those stem cells were subsequently transplanted in a disease model. Bone marrow-derived stem cells were treated with laser light as described above, at a total dose of 1 J/cm^2, and 24 hours later were implanted into the heart of a rat model of myocardial infarction (Tuby et al., 2009). While implantation of non-PBM-treated stem cells improved a number of cardiac measures relative to saline-injected mice, this effect was significantly augmented with PBM pretreatment. Specifically, implantation of PBM-treated stem cells produced a significant reduction in infarct size and an increase in angiogenesis and vascular endothelial growth factor relative to implantation with nontreated cells (Tuby et al., 2009), suggesting that outcomes from clinical stem cell therapies might be improved by treating cells with PBM prior to implantation. The next logical step was to determine whether PBM of the bone marrow in vivo could produce similar effects. Using a rat model of myocardial infarction, 804 nm laser PBM (total dose of 1 J/cm^2) was applied to the exposed tibia either 20 minutes or 4 hours after the induction of infarction (Tuby et al., 2011). Compared to sham-treated animals, rats treated with PBM of the tibia showed a significant reduction in infarct size and ventricular dilatation, in association with an increased number of c-kit$^+$ stem cells in the heart. Remarkably, PBM treatment of the tibia produced a stronger protective effect than direct PBM treatment of the heart (Tuby et al., 2011).

Subsequent studies have applied a similar approach to demonstrate that near-infrared laser PBM of the tibia mitigates ischemia-reperfusion kidney injury in rats (Oron et al., 2014), and cardiac damage following acute myocardial infarction in pigs (Blatt et al., 2016). Collectively, the results of these studies highlighted the remarkable potential of PBM in generating indirect, possibly body-wide, tissue protection, and point to bone marrow-derived stem cells as a key candidate for mediating this phenomenon.

11.7 Remote photobiomodulation as a neuroprotective intervention

Despite the growing evidence suggesting that localized PBM can have indirect, perhaps systemic, protective effects, it was only recently that studies investigated whether this protection extended to the brain.

11.7.1 Parkinson's disease

At the time when our team decided to test whether PBM targeted at the body might provide protection to the brain, we had been working for several years on transcranial PBM in rodent models of Parkinson's disease and Alzheimer's disease. While excited by the results, we were acutely aware of the inherent limitations of transcranial PBM. The possibility that PBM targeted at a peripheral tissue—more accessible than the brain—might elicit systemic mechanisms that provide protection of remote tissues, was of great interest to us. Yet we were wary of the inherent challenges that the blood–brain barrier imposes on transducing a peripheral signal to the brain; one assumption we had made (perhaps prematurely) is that remote PBM-induced protection is likely mediated by some circulating factor(s) that might need to cross the blood–brain barrier in order to elicit neuroprotection.

To test whether remote PBM induced neuroprotective actions, we performed the "helmet" experiment, in which the head of the animal was shielded by aluminum foil while the dorsum was irradiated with light. Utilizing a mouse model of acute Parkinson's disease, induced by intraperitoneal injection of the neurotoxin MPTP, we applied 670 nm LED light (50 mW/cm^2) to either the head or the dorsum of the animal in the form of a per-conditioning intervention, where PBM was delivered twice per day (4 J/cm^2 emitted light per treatment) on each MPTP injection day. Mice were allowed to survive for 7 days following MPTP injections.

Our initial finding from the helmet experiment was presented at the 2012 World Association for Laser Therapy (WALT) conference (Stone et al., 2013). In this initial experiment, mice were injected with a total of 50 mg/kg MPTP, administered on 2 consecutive days, and the primary outcome measure was the number of functional dopaminergic cells in the substantia nigra pars compacta (SNc), as determined by tyrosine hydroxylase (TH) immunolabelling and stereological analysis. As expected, MPTP induced loss of 35%–40% of the dopaminergic cells in the SNc. Consistent with our earlier observations, transcranial PBM provided significant neuroprotection of these cells, with mice receiving PBM of the head showing cell counts not significantly different from saline-injected controls. But the most striking result was for mice receiving PBM strictly targeting the body; despite no direct irradiation of the head, these mice showed significantly higher SNc dopaminergic cell counts than sham-treated MPTP animals, indicating that PBM of the body provides protection of the brain (Stone et al., 2013).

An expanded and more comprehensive version of the helmet study, incorporating more animals and additional MPTP dosages and glial cell counts, was published in 2014 (Johnstone et al., 2014). This study confirmed the previous findings that both transcranial PBM and remote PBM provide significant neuroprotection against parkinsonian insult with 50 mg/kg MPTP, preserving functional dopaminergic neurons in the SNc as assessed by TH immunohistochemistry. Although there was no significant difference in cell counts between the two groups, transcranial PBM appeared to provide more robust neuroprotection than remote PBM in this model. At higher MPTP doses (75 and 100 mg/kg), neither transcranial PBM nor remote PBM was capable of mitigating MPTP-induced damage to the SNc. Further, in mice receiving 75 mg/kg MPTP, we assessed astrocyte and microglia numbers in the SNc using immunohistochemical labeling of GFAP and IBA1, respectively. While MPTP intoxication appeared to have no significant effect on GFAP labeling, it did produce a significant increase in the number of $IBA1^+$ cells that was not mitigated by either transcranial PBM or remote PBM (Johnstone et al., 2014).

One limitation of the aforementioned studies was that PBM (whether remote or transcranial) was applied concurrent with MPTP intoxication, raising the possibility that the observed effects may not be due to PBM-induced neuroprotection per se, but may instead be the result of PBM interfering with the pharmacokinetics of MPTP. To address this limitation, we undertook a study in which remote PBM was delivered as a preconditioning intervention. Mice were treated with 670 nm LED light (50 mW/cm^2) to the body, once per day (4 J/cm^2 emitted light per treatment) for 2, 5, or 10 days. For all groups, MPTP injections (50 mg/kg total) did not commence until 24 hours after remote PBM treatments had been terminated. Similar to the per-conditioning regimen, preconditioning with remote PBM for 10 days mitigated MPTP-induced loss of dopaminergic cells in the SNc and associated abnormal neuronal activity in the caudate-putamen complex, as assessed by FOS immunohistochemistry (Ganeshan et al., 2019).

These initial findings laid the groundwork for the concept that PBM of peripheral tissues can protect the brain (Johnstone et al., 2015).

11.7.2 Alzheimer's disease

In parallel with our research into remote PBM using animal models of Parkinson's disease, Uri Oron's team in Tel Aviv were developing the concept of bone marrow-targeted PBM (described above) as a way of protecting multiple remote tissues, including the brain. Using the 5xFAD transgenic mouse model of Alzheimer's disease, the team applied PBM (808 nm, 1 J/cm^2) to the tibia six times over a two month period and assessed the effect of this treatment on mouse behavior and β-amyloid (Aβ) burden in the hippocampus (Farfara et al., 2015).

At 6 months of age, PBM-treated 5xFAD mice showed significantly improved performance on an object recognition task and fear conditioning task relative to sham-treated 5xFAD mice, in association with a significant reduction in the hippocampal burden of Aβ deposits. In addition, the authors conducted a study on bone marrow-derived MSCs that were exposed in vitro to either 1 J/cm^2 laser light or sham treatment. Laser-treated MSCs showed enhanced phagocytosis of $A\beta_{1\text{-}42}$ peptide compared to sham-treated MSCs (Farfara et al., 2015), which they interpreted as the possible involvement of bone marrow-derived MSCs in clearing Aβ from the brain and, through this process, partly mediating the neuroprotective effects of remote PBM in the context of AD.

11.7.3 Retinopathy

A recent study has demonstrated that the neuroprotective effects of remote PBM extend beyond the brain to another CNS structure: the retina. Mention has been made of the retina above, in the context of the bilateral effects of unilateral irradiation. In addition, Saliba and colleagues investigated the effects of PBM on diabetic retinopathy using a mouse model of streptozotocin-induced diabetes (Saliba et al., 2015). Commencing 4 weeks after the onset of diabetes, mice received PBM (670 nm, $\sim$5 J/cm^2 per day) daily for 10 weeks, using a large LED panel. Two treatment subsets were included: one in which the whole body and head were exposed to light, and a second in which the head was shielded from light using a lead covering.

In both diabetic groups receiving PBM, there was significantly reduced retinal superoxide production and leukostasis and improved visual function relative to sham-treated diabetic mice, suggesting irradiation of only the body with a therapeutic dose of PBM is sufficient to protect the retina against diabetes-induced changes. The authors also demonstrated that leukocytes isolated from PBM-treated diabetic mice were less cytotoxic to endothelial cells in culture than leukocytes isolated from sham-treated diabetic mice (Saliba et al., 2015), highlighting the possibility that modulation of leukocyte function might partly underlie the protective effects of remote PBM in this model.

11.8 The precedent: remote ischemic conditioning

While the concept of applying a treatment to a peripheral tissue in order to elicit protection of the brain might seem outlandish, it is not without precedent. Indeed, there is one other example of remote tissue conditioning that is more widely recognized and far better researched than remote PBM: remote ischemic conditioning (RIC).

RIC refers to an intervention in which brief cycles of ischemia-reperfusion are applied to one tissue (usually a limb) in order to induce protection of distal tissues (usually a critical organ). RIC has been extensively characterized in animal models and trialed clinically in various types of ischemia-related disease, yielding some promising results. Like remote PBM, RIC appears to provide body-wide, rather than tissue-specific protection; it also appears to confer benefit whether applied before, during, or after a physiological insult. A range of mechanisms, including humoral mediators and neurogenic signaling pathways, have been proposed to mediate this systemic effect, as reviewed in detail elsewhere (Kim et al., 2017). Since both RIC and PBM exhibit dose-response relationships that are not linear but instead biphasic, reflecting the phenomenon of hormesis (Mattson, 2008), it is likely that both interventions act as mild stressors, stimulating an adaptive stress response that enhances cellular and tissue resilience both locally and distally.

While most studies to date have investigated the capacity of RIC to protect against ischemia-related diseases, there is growing evidence that RIC might be more broadly applicable. For example, RIC has been shown in animal models to protect against systemic inflammation (Kim et al., 2014) and retinal damage induced by optic nerve transection or bright light exposure (Brandli et al., 2016; Liu et al., 2013). With respect to the brain, we have recently shown in an MPTP mouse model of Parkinson's disease that RIC, applied immediately before MPTP insult, mitigates loss of dopaminergic cells in the SNc (Kim et al., 2018).

At the time of writing, it is unclear whether RIC and remote PBM rely on similar or distinct mechanisms for mediating protection of the brain. One might expect that insights could be gained by combining these treatments; if both interventions activate similar pathways, no additive protective effect would be expected, whereas if each intervention activates a distinct set of pathways, one might expect to observe additive or even synergistic neuroprotective effects.

We sought to investigate this by combining RIC and remote PBM treatment of the MPTP mouse model of Parkinson's disease. While both RIC and remote PBM significantly mitigated loss of dopaminergic cells in the SNc when administered independently, there was no additive effect when combining these two interventions (Fig. 11.3) (Kim et al., 2018). This result suggests either that RIC and remote PBM share some common mechanistic pathways, or that a proportion of cells in the SNc are sufficiently damaged by MPTP to be refractory to neuroprotective intervention. One qualitative observation from this study, based on several unpublished outcome measures, was that not only was there no additive effect, but there was actually a hint of an antagonistic effect when combining RIC and remote PBM. Since both RIC and remote PBM exhibit a biphasic dose-response relationship, and both appear to act by stimulating endogenous stress response systems, we postulated that combining these two interventions pushed the total stress dose

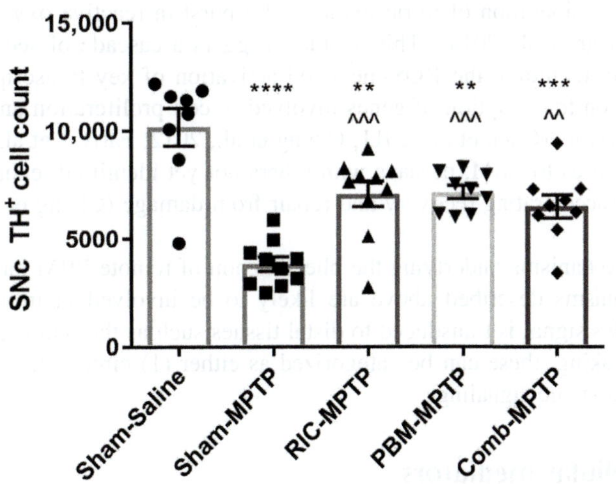

FIGURE 11.3 **Remote PBM and remote ischemic conditioning (RIC) both protect against MPTP-induced damage to the midbrain, but there is no additive effect when treatments are combined (Comb).** Data are stereological counts (mean ± SEM) of tyrosine hydroxylase-positive (TH^+) cells in the substantia nigra pars compacta (SNc). $^{**}P < .01$, $^{***}P < .001$, $^{****}P < .0001$ versus Sham-Saline; ^^$P < .01$, ^^^$P < .001$ versus Sham-MPTP. *Figure reproduced from Kim, B., Mitrofanis, J., Stone, J., Johnstone, D.M., 2018. Remote tissue conditioning is neuroprotective against MPTP insult in mice. IBRO Rep. 4, 14–17.*

to a suboptimal level, beyond the "point of diminishing returns" (Kim et al., 2018). Further experiments will be required to add substance to this speculation.

11.9 Peripheral tissue targets for remote photobiomodulation-induced neuroprotection

With the exception of the study by Farfara and colleagues on an Alzheimer's disease model, in which PBM was directed to the bone marrow in the tibia (Farfara et al., 2015), all published observations of remote PBM-induced neuroprotection to date have arisen from small animal studies in which the whole body of the animal has been exposed to light. This raises an important question for the clinical translation of remote PBM: is there an optimal peripheral tissue target?

As discussed earlier, the team of Uri Oron have generated substantial evidence that 804–808 nm laser PBM of the bone marrow, achieved through irradiation of the exposed tibia, produces improved outcomes in a number of disease models (Blatt et al., 2016; Oron et al., 2014; Tuby et al., 2011), including a model of Alzheimer's disease (Farfara et al., 2015). Identifying an optimal peripheral tissue target has also been an area of active investigation for our team over the past couple of years. Our unpublished findings in mouse and monkey models of MPTP-induced Parkinson's disease suggest that targeting the abdomen with PBM is particularly effective for inducing neuroprotection of the midbrain and mitigating clinical signs associated with MPTP intoxication. It is not yet clear whether there is a single particular organ in the abdominal cavity responsible for mediating this effect, or whether (as in studies of bone marrow) abdomen-targeted PBM stimulates MSCs in adipose tissue, or whether PBM might instead be influencing the brain through effects on the gut microbiome. It is also unclear whether abdomen-targeted PBM provides neuroprotection against other brain diseases, or even in non-MPTP models of Parkinson's disease. Projects are currently underway to address many of these questions. Whatever the answer, the successful identification of a peripheral target(s) associated with remote PBM-induced neuroprotection will be valuable for advancing this potential therapy to the clinic.

11.10 Mechanisms underlying remote photobiomodulation-induced protection

With the biological phenomenon of remote PBM now demonstrated in a number of experimental settings, the mechanisms of action of PBM can be considered from two complementary perspectives: (1) intracellular mechanisms and (2) systemic mechanisms.

Given the past focus on the direct effects of PBM on a cell or tissue target, considerable research has been done into the intracellular systems triggered by PBM, as discussed in other chapters of this book. Evidence from various experimental approaches indicates that cytochrome c oxidase, a key enzyme in the mitochondrial respiratory chain, is the primary photoacceptor of light at therapeutic PBM wavelengths (Karu, 1999a; Wong-Riley et al., 2005). The absorption of light by cytochrome c oxidase causes a redox change in the enzyme, leading to increased ATP production and associated effects on cAMP and Ca^{2+}, liberation of nitric oxide, and a burst in reactive oxygen species (ROS) (Chen et al., 2011; Chung et al., 2012; Farivar et al., 2014). This, in turn, triggers a cascade of secondary effects that persist long after light exposure has ceased, including the ROS-mediated activation of key transcription factors such as AP-1 and NF-kB and consequent effects on transcription of genes involved in cell proliferation and migration and in the production of cytokines and growth factors (Chen et al., 2011; Chung et al., 2012; Farivar et al., 2014). This complex network of intracellular molecular responses to PBM, perhaps with others not yet identified, enhances cell and tissue resilience, protecting against damage and accelerating recovery and repair from damage (Chung et al., 2012; Rojas and Gonzalez-Lima, 2011).

In contrast, the systemic mechanisms underlying the phenomenon of remote PBM have not yet been widely studied. While the intracellular mechanisms described above are likely to be involved in initiating the response to PBM, it remains unclear exactly how this signal is transduced to distal tissues such as the brain. There are many possible mediators of this effect; broadly speaking, these can be categorized as either (1) circulating cells, (2) circulating molecules, (3) microorganisms, or (4) neurogenic signaling.

11.10.1 Circulating cellular mediators

As described earlier in this chapter, one prime candidate for mediating remote PBM-induced protection is bone marrow-derived stem cells, in particular mesenchymal stem cells (MSCs). While MSCs were originally identified in the bone marrow, they have since been isolated from a range of tissues, including adipose tissue, endometrium, and dental tissues (Ullah et al., 2015). Experimental data consistently show that PBM enhances the proliferation of MSCs, as reviewed elsewhere (Fekrazad et al., 2016; Ginani et al., 2015). The team of Uri Oron has taken this one step further, showing that PBM of the bone marrow stimulates the proliferation and mobilization of MSCs in vivo, in association with an increased localization of MSCs at sites of injury (Tuby et al., 2011; Oron et al., 2014).

Of the possible cellular mediators of remote PBM-induced neuroprotection, MSCs are strong candidates; they can transmigrate across the blood–brain barrier (Matsushita et al., 2011; Simard and Rivest, 2004), home specifically to areas of tissue damage (Belema-Bedada et al., 2008; Karp and Leng Teo, 2009), and release myriad trophic factors that facilitate cell protection and repair (Glavaski-Joksimovic and Bohn, 2013). For example, systemic transplantation of MSCs prevents dopaminergic cell loss in rodent models of Parkinson's disease (Capitelli et al., 2014; Chao et al., 2009) and mitigates cognitive impairment and plaque pathology in a transgenic mouse model of Alzheimer's disease (Kim et al., 2012).

In addition to stem cells, there is evidence that PBM can also modulate circulating immune cells. For example, PBM has been shown to suppress immune cell activation in a rat model of spinal cord injury, in association with enhanced axonal regeneration and functional recovery (Byrnes et al., 2005). In a rat wound model, PBM has been shown to increase mast cell number and degranulation (El Sayed and Dyson, 1996), which will in turn enhance leukocyte recruitment.

11.10.2 Circulating molecular mediators

In addition to modulating circulating cells, it seems likely that PBM will also affect (either directly or indirectly) circulating molecules that can transduce protective effects to distant tissues, including the brain. The most obvious candidates would be cytokines and chemokines, and there is some evidence that PBM can influence these molecules in the setting of neural tissue injury. For example, whole body PBM treatment of an experimental autoimmune encephalomyelitis mouse model of multiple sclerosis resulted in down-regulation of pro-inflammatory cytokines (interferon-γ and tumor necrosis factor-α) and up-regulation of antiinflammatory cytokines (interleukin-4 and interleukin-10), in association with amelioration of disease (Muili et al., 2012). In a rat model of spinal cord injury, PBM modulated the expression of a number of genes encoding cytokines and chemokines at the site of injury (Byrnes et al., 2005).

Given the well-established effect of PBM on the mitochondria of cells (Chung et al., 2012), one candidate that warrants further investigation is the emerging class of circulating molecules referred to as mitokines. Mitokines are released into the circulation as a response to localized mitochondrial stress; this stress signal can then activate a mitochondrial stress response pathway in distal tissues (Durieux et al., 2011). Few mitokines have yet been identified, but one example is fibroblast growth factor 21, whose expression is increased in response to mitochondrial dysfunction and promotes protection against diet-induced obesity and insulin resistance (Kim et al., 2013). As PBM is widely considered to impose a mild stress on mitochondrial function (Kim et al., 2017), it is feasible that mitokines are key mediators in transducing the intracellular effects of localized PBM to distant tissues, such as the brain.

11.10.3 Modulation of the microbiome

The idea that the microbiome may mediate the beneficial systemic effects of PBM arises from our observations that abdomen-targeted PBM provides protection against MPTP-induced Parkinson's disease, from evidence that gut microbiota strongly influence brain pathology and motor deficits in Parkinson's disease (Heintz-Buschart et al., 2018; Sampson et al., 2016; Scheperjans, 2016; Scheperjans et al., 2015; Houser and Tansey, 2017), and from demonstrations that low-intensity light affects the proliferation of common strains of bacteria found in the gut, as reviewed elsewhere (Lubart et al., 2011). However, considerable work remains to determine whether modulation of the gut (or other) microbiome contributes to the neuroprotective effects of remote PBM—this is an area of active investigation for our team.

11.10.4 Neurogenic signaling

While we are not aware of any studies that have explicitly tested whether neurotransmission is involved in mediating the protective effects of PBM, the similarities between remote PBM and RIC warrant its inclusion as a candidate mechanism. As reviewed elsewhere (Kim et al., 2017), research into the mechanisms of RIC indicates that both humoral and neural pathways must be intact for complete RIC-induced protection to be achieved. One hypothesis for this interaction is that RIC activates peripheral nerves, which in turn induces the release of protective humoral factors. For example, abolishing innervation by the vagus nerve impacts the effectiveness of RIC-induced cardioprotection (Mastitskaya et al., 2016). Future studies should investigate whether the neuroprotective efficacy of remote PBM is similarly impaired when critical nerves are compromised.

11.11 Conclusion

Understandably, "direct PBM"—targeting light at the tissue one seeks to modulate or protect—has been the approach of choice in the overwhelming majority of PBM studies to date. Nonetheless, the emergence of "remote PBM"—harnessing the indirect effects of PBM by targeting light at one tissue to protect others—provides a feasible avenue to overcoming the practical barriers associated with light penetration to deep body tissues, and of particular interest to us, the deep regions of the brain (thalamus, basal ganglia, midbrain, and brainstem).

Research into neuroprotection with remote PBM is still in its infancy, and many questions remain to be answered. For example, is there an optimal peripheral tissue target for generalized neuroprotection, or does this vary for different brain disorders? What is the optimal remote PBM protocol for eliciting neuroprotection, with respect to wavelength, irradiance, frequency of treatment? Can treatment protocols be personalized to each patient, to account for different body compositions? Does our capacity to respond to remote PBM diminish with age? What are the mechanisms underlying the phenomenon of remote PBM-induced neuroprotection? And can knowledge of these mechanisms inform the development of biomarkers of treatment efficacy?

Despite (or, for the scientist, because of) these many unanswered questions, there is cause for excitement for remote PBM, since it ticks many boxes: it is safe, simple, pain-free, noninvasive, inexpensive, easy to administer, and well tolerated by the patient. Beyond its potential clinical utility, remote PBM highlights the complexity of animal physiology and the importance of understanding previously unsuspected interactions between bodily tissues in the search for effective neuroprotective interventions.

References

Abe, M., Fujisawa, K., Suzuki, H., Sugimoto, T., Kanno, T., 1993. Role of 830 nm low reactive level laser on the growth of an implanted glioma in mice. Keio J. Med. 42, 177–179.

Allendorf, J.D., Bessler, M., Huang, J., Kayton, M.L., Laird, D., Nowygrod, R., et al., 1997. Helium-neon laser irradiation at fluences of 1, 2, and 4 J/cm^2 failed to accelerate wound healing as assessed by both wound contracture rate and tensile strength. Lasers Surg. Med. 20, 340–345.

Barrett, D.W., Gonzalez-Lima, F., 2013. Transcranial infrared laser stimulation produces beneficial cognitive and emotional effects in humans. Neuroscience 230, 13–23.

Basford, J.R., Hallman, H.O., Sheffield, C.G., Mackey, G.L., 1986. Comparison of cold-quartz ultraviolet, low-energy laser, and occlusion in wound healing in a swine model. Arch. Phys. Med. Rehabil. 67, 151–154.

Belema-Bedada, F., Uchida, S., Martire, A., Kostin, S., Braun, T., 2008. Efficient homing of multipotent adult mesenchymal stem cells depends on FROUNT-mediated clustering of CCR2. Cell Stem Cell 2, 566–575.

Blanco, N.J., Saucedo, C.L., Gonzalez-Lima, F., 2017. Transcranial infrared laser stimulation improves rule-based, but not information-integration, category learning in humans. Neurobiol. Learn. Mem. 139, 69–75.

Blatt, A., Elbaz-Greener, G.A., Tuby, H., Maltz, L., Siman-Tov, Y., Ben-Aharon, G., et al., 2016. Low-level laser therapy to the bone marrow reduces scarring and improves heart function post-acute myocardial infarction in the pig. Photomed. Laser Surg. 34, 516–524.

Brandli, A., Johnstone, D.M., Stone, J., 2016. Remote ischemic preconditioning protects retinal photoreceptors: evidence from a rat model of light-induced photoreceptor degeneration. Invest. Ophthalmol. Vis. Sci. 57, 5302–5313.

Braverman, B., Mccarthy, R.J., Ivankovich, A.D., Forde, D.E., Overfield, M., Bapna, M.S., 1989. Effect of helium-neon and infrared laser irradiation on wound healing in rabbits. Lasers Surg. Med. 9, 50–58.

Byrnes, K.R., Barna, L., Chenault, V.M., Waynant, R.W., Ilev, I.K., Longo, L., et al., 2004. Photobiomodulation improves cutaneous wound healing in an animal model of type II diabetes. Photomed. Laser Surg. 22, 281–290.

Byrnes, K.R., Waynant, R.W., Ilev, I.K., Wu, X., Barna, L., Smith, K., et al., 2005. Light promotes regeneration and functional recovery and alters the immune response after spinal cord injury. Lasers Surg. Med. 36, 171–185.

Capitelli, C.S., Lopes, C.S., Alves, A.C., Barbiero, J., Oliveira, L.F., Da Silva, V.J., et al., 2014. Opposite effects of bone marrow-derived cells transplantation in MPTP-rat model of Parkinson's disease: a comparison study of mononuclear and mesenchymal stem cells. Int. J. Med. Sci. 11, 1049–1064.

Carvalho, P.T., Mazzer, N., Dos Reis, F.A., Belchior, A.C., Silva, I.S., 2006. Analysis of the influence of low-power HeNe laser on the healing of skin wounds in diabetic and non-diabetic rats. Acta Cir. Bras. 21, 177–183.

Cassano, P., Cusin, C., Mischoulon, D., Hamblin, M.R., De Taboada, L., Pisoni, A., et al., 2015. Near-infrared transcranial radiation for major depressive disorder: proof of concept study. Psychiatry J. 2015, 352979.

Chao, Y.X., He, B.P., Tay, S.S., 2009. Mesenchymal stem cell transplantation attenuates blood brain barrier damage and neuroinflammation and protects dopaminergic neurons against MPTP toxicity in the substantia nigra in a model of Parkinson's disease. J. Neuroimmunol. 216, 39–50.

Chen, A.C., Arany, P.R., Huang, Y.Y., Tomkinson, E.M., Sharma, S.K., Kharkwal, G.B., et al., 2011. Low-level laser therapy activates NF-kB via generation of reactive oxygen species in mouse embryonic fibroblasts. PLoS One 6, e22453.

Chung, H., Dai, T., Sharma, S.K., Huang, Y.Y., Carroll, J.D., Hamblin, M.R., 2012. The nuts and bolts of low-level laser (light) therapy. Ann. Biomed. Eng. 40, 516–533.

Darlot, F., Moro, C., El Massri, N., Chabrol, C., Johnstone, D.M., Reinhart, F., et al., 2016. Near-infrared light is neuroprotective in a monkey model of Parkinson disease. Ann. Neurol. 79, 59–75.

De Taboada, L., Yu, J., El-Amouri, S., Gattoni-Celli, S., Richieri, S., Mccarthy, T., et al., 2011. Transcranial laser therapy attenuates amyloid-beta peptide neuropathology in amyloid-beta protein precursor transgenic mice. J. Alzheimers Dis. 23, 521–535.

Detaboada, L., Ilic, S., Leichliter-Martha, S., Oron, U., Oron, A., Streeter, J., 2006. Transcranial application of low-energy laser irradiation improves neurological deficits in rats following acute stroke. Lasers Surg. Med. 38, 70–73.

Do Nascimento, P.M., Pinheiro, A.L., Salgado, M.A., Ramalho, L.M., 2004. A preliminary report on the effect of laser therapy on the healing of cutaneous surgical wounds as a consequence of an inversely proportional relationship between wavelength and intensity: histological study in rats. Photomed. Laser Surg. 22, 513–518.

Durieux, J., Wolff, S., Dillin, A., 2011. The cell-non-autonomous nature of electron transport chain-mediated longevity. Cell 144, 79–91.

El Massri, N., Johnstone, D.M., Peoples, C.L., Moro, C., Reinhart, F., Torres, N., et al., 2016a. The effect of different doses of near infrared light on dopaminergic cell survival and gliosis in MPTP-treated mice. Int. J. Neurosci. 126, 76–87.

El Massri, N., Moro, C., Torres, N., Darlot, F., Agay, D., Chabrol, C., et al., 2016b. Near-infrared light treatment reduces astrogliosis in MPTP-treated monkeys. Exp. Brain Res. 234, 3225–3232.

El Sayed, S.O., Dyson, M., 1996. Effect of laser pulse repetition rate and pulse duration on mast cell number and degranulation. Lasers Surg. Med. 19, 433–437.

Farfara, D., Tuby, H., Trudler, D., Doron-Mandel, E., Maltz, L., Vassar, R.J., et al., 2015. Low-level laser therapy ameliorates disease progression in a mouse model of Alzheimer's disease. J. Mol. Neurosci. 55, 430–436.

Farivar, S., Malekshahabi, T., Shiari, R., 2014. Biological effects of low level laser therapy. J. Lasers Med. Sci. 5, 58–62.

Fekrazad, R., Mirmoezzi, A., Kalhori, K.A., Arany, P., 2015. The effect of red, green and blue lasers on healing of oral wounds in diabetic rats. J. Photochem. Photobiol. B 148, 242–245.

Fekrazad, R., Asefi, S., Allahdadi, M., Kalhori, K.A., 2016. Effect of photobiomodulation on mesenchymal stem cells. Photomed. Laser Surg. 34, 533–542.

Ganeshan, V., Skladnev, N.V., Kim, J.Y., Mitrofanis, J, Stone, J., Johnstone, D.M., 2019. Pre-conditioning with remote photobiomodulation modulates the brain transcriptome and protects against MPTP insult in mice. Neuroscience 400, 8–97.

Ginani, F., Soares, D.M., Barreto, M.P., Barboza, C.A., 2015. Effect of low-level laser therapy on mesenchymal stem cell proliferation: a systematic review. Lasers Med. Sci. 30, 2189–2194.

Glavaski-Joksimovic, A., Bohn, M.C., 2013. Mesenchymal stem cells and neuroregeneration in Parkinson's disease. Exp. Neurol. 247, 25–38.

Grillo, S.L., Duggett, N.A., Ennaceur, A., Chazot, P.L., 2013. Non-invasive infra-red therapy (1072 nm) reduces beta-amyloid protein levels in the brain of an Alzheimer's disease mouse model, TASTPM. J. Photochem. Photobiol. B 123, 13–22.

Gungormus, M., Akyol, U.K., 2009. Effect of biostimulation on wound healing in diabetic rats. Photomed. Laser Surg. 27, 607–610.

Guymer, R.H., Brassington, K.H., Dimitrov, P., Makeyeva, G., Plunkett, M., Xia, W., et al., 2014. Nanosecond-laser application in intermediate AMD: 12-month results of fundus appearance and macular function. Clin. Exp. Ophthalmol. 42, 466–479.

Hacke, W., Schellinger, P.D., Albers, G.W., Bornstein, N.M., Dahlof, B.L., Fulton, R., et al., 2014. Transcranial laser therapy in acute stroke treatment: results of neurothera effectiveness and safety trial 3, a phase III clinical end point device trial. Stroke 45, 3187–3193.

Hart, N.S., Fitzgerald, M., 2016. A new perspective on delivery of red-near-infrared light therapy for disorders of the brain. Discov. Med. 22, 147–156.

Heintz-Buschart, A., Pandey, U., Wicke, T., Sixel-Doring, F., Janzen, A., Sittig-Wiegand, E., et al., 2018. The nasal and gut microbiome in Parkinson's disease and idiopathic rapid eye movement sleep behavior disorder. Mov. Disord. 33, 88–98.

Hopkins, J.T., Mcloda, T.A., Seegmiller, J.G., David Baxter, G., 2004. Low-level laser therapy facilitates superficial wound healing in humans: a triple-blind, sham-controlled study. J. Athl. Train 39, 223–229.

Houser, M.C., Tansey, M.G., 2017. The gut-brain axis: is intestinal inflammation a silent driver of Parkinson's disease pathogenesis? NPJ Parkinsons Dis. 3, 3.

Huisa, B.N., Stemer, A.B., Walker, M.G., Rapp, K., Meyer, B.C., Zivin, J.A., et al., 2013. Transcranial laser therapy for acute ischemic stroke: a pooled analysis of NEST-1 and NEST-2. Int. J. Stroke 8, 315–320.

Hunter, J., Leonard, L., Wilson, R., Snider, G., Dixon, J., 1984. Effects of low energy laser on wound healing in a porcine model. Lasers Surg. Med. 3, 285–290.

Jobling, A.I., Guymer, R.H., Vessey, K.A., Greferath, U., Mills, S.A., Brassington, K.H., et al., 2015. Nanosecond laser therapy reverses pathologic and molecular changes in age-related macular degeneration without retinal damage. FASEB J. 29, 696–710.

Johnstone, D.M., El Massri, N., Moro, C., Spana, S., Wang, X.S., Torres, N., et al., 2014. Indirect application of near infrared light induces neuroprotection in a mouse model of parkinsonism – an abscopal neuroprotective effect. Neuroscience 274, 93–101.

Johnstone, D.M., Mitrofanis, J., Stone, J., 2015. Targeting the body to protect the brain: inducing neuroprotection with remotely-applied near infrared light. Neural Regen. Res. 10, 349–351.

Johnstone, D.M., Moro, C., Stone, J., Benabid, A.L., Mitrofanis, J., 2016. Turning on lights to stop neurodegeneration: the potential of near infrared light therapy in Alzheimer's and Parkinson's Disease. Front. Neurosci. 9, 500.

Kana, J.S., Hutschenreiter, G., Haina, D., Waidelich, W., 1981. Effect of low-power density laser radiation on healing of open skin wounds in rats. Arch. Surg. 116, 293–296.

Karp, J.M., Leng Teo, G.S., 2009. Mesenchymal stem cell homing: the devil is in the details. Cell Stem Cell 4, 206–216.

Karu, T., 1999a. Primary and secondary mechanisms of action of visible to near-IR radiation on cells. J. Photochem. Photobiol. B 49, 1–17.

Karu, T.I., 1999b. A suitable model for wound healing: how many times are we to stumble over the same block? Lasers Surg. Med. 25, 283–284.

Kim, S., Chang, K.A., Kim, J., Park, H.G., Ra, J.C., Kim, H.S., et al., 2012. The preventive and therapeutic effects of intravenous human adipose-derived stem cells in Alzheimer's disease mice. PLoS One 7, e45757.

Kim, K.H., Jeong, Y.T., Oh, H., Kim, S.H., Cho, J.M., Kim, Y.N., et al., 2013. Autophagy deficiency leads to protection from obesity and insulin resistance by inducing Fgf21 as a mitokine. Nat. Med. 19, 83–92.

Kim, Y.H., Yoon, D.W., Kim, J.H., Lee, J.H., Lim, C.H., 2014. Effect of remote ischemic post-conditioning on systemic inflammatory response and survival rate in lipopolysaccharide-induced systemic inflammation model. J. Inflamm. (Lond.) 11, 16.

Kim, B., Brandli, A., Mitrofanis, J., Stone, J., Purushothuman, S., Johnstone, D.M., 2017. Remote tissue conditioning – an emerging approach for inducing body-wide protection against diseases of ageing. Ageing Res. Rev. 37, 69–78.

Kim, B., Mitrofanis, J., Stone, J., Johnstone, D.M., 2018. Remote tissue conditioning is neuroprotective against MPTP insult in mice. IBRO Rep. 4, 14–17.

Lampl, Y., Zivin, J.A., Fisher, M., Lew, R., Welin, L., Dahlof, B., et al., 2007. Infrared laser therapy for ischemic stroke: a new treatment strategy: results of the NeuroThera Effectiveness and Safety Trial-1 (NEST-1). Stroke 38, 1843–1849.

Lapchak, P.A., Boitano, P.D., 2016. Transcranial near-infrared laser therapy for stroke: how to recover from futility in the NEST-3 Clinical Trial. Acta Neurochir. Suppl. 121, 7–12.

Lapchak, P.A., Wei, J., Zivin, J.A., 2004. Transcranial infrared laser therapy improves clinical rating scores after embolic strokes in rabbits. Stroke 35, 1985–1988.

Lapchak, P.A., Boitano, P.D., Butte, P.V., Fisher, D.J., Holscher, T., Ley, E.J., et al., 2015. Transcranial near-infrared laser transmission (NILT) profiles (800 nm): systematic comparison in four common research species. PLoS One 10, e0127580.

Liu, X., Sha, O., Cho, E.Y., 2013. Remote ischemic postconditioning promotes the survival of retinal ganglion cells after optic nerve injury. J. Mol. Neurosci. 51, 639–646.

Lu, Y., Wang, R., Dong, Y., Tucker, D., Zhao, N., Ahmed, M.E., et al., 2017. Low-level laser therapy for beta amyloid toxicity in rat hippocampus. Neurobiol. Aging 49, 165–182.

Lubart, R., Lipovski, A., Nitzan, Y., Friedmann, H., 2011. A possible mechanism for the bactericidal effect of visible light. Laser Ther. 20, 17–22.

Lyons, R.F., Abergel, R.P., White, R.A., Dwyer, R.M., Castel, J.C., Uitto, J., 1987. Biostimulation of wound healing in vivo by a helium-neon laser. Ann. Plast. Surg. 18, 47–50.

Maiya, G.A., Kumar, P., Rao, L., 2005. Effect of low intensity helium-neon (He-Ne) laser irradiation on diabetic wound healing dynamics. Photomed. Laser Surg. 23, 187–190.

Mastitskaya, S., Basalay, M., Hosford, P.S., Ramage, A.G., Gourine, A., Gourine, A.V., 2016. Identifying the source of a humoral factor of remote (pre)conditioning cardioprotection. PLoS One 11, e0150108.

Matsushita, T., Kibayashi, T., Katayama, T., Yamashita, Y., Suzuki, S., Kawamata, J., et al., 2011. Mesenchymal stem cells transmigrate across brain microvascular endothelial cell monolayers through transiently formed inter-endothelial gaps. Neurosci. Lett. 502, 41–45.

Mattson, M.P., 2008. Hormesis defined. Ageing Res. Rev. 7, 1–7.

Medrado, A.R., Pugliese, L.S., Reis, S.R., Andrade, Z.A., 2003. Influence of low level laser therapy on wound healing and its biological action upon myofibroblasts. Lasers Surg. Med. 32, 239–244.

Moro, C., Torres, N., El Massri, N., Ratel, D., Johnstone, D.M., Stone, J., et al., 2013. Photobiomodulation preserves behaviour and midbrain dopaminergic cells from MPTP toxicity: evidence from two mouse strains. BMC Neurosci. 14, 40.

Moro, C., Massri, N.E., Torres, N., Ratel, D., De Jaeger, X., Chabrol, C., et al., 2014. Photobiomodulation inside the brain: a novel method of applying near-infrared light intracranially and its impact on dopaminergic cell survival in MPTP-treated mice. J. Neurosurg. 120, 670–683.

Morries, L.D., Cassano, P., Henderson, T.A., 2015. Treatments for traumatic brain injury with emphasis on transcranial near-infrared laser phototherapy. Neuropsychiatr. Dis. Treat. 11, 2159–2175.

Muili, K.A., Gopalakrishnan, S., Meyer, S.L., Eells, J.T., Lyons, J.A., 2012. Amelioration of experimental autoimmune encephalomyelitis in C57BL/6 mice by photobiomodulation induced by 670 nm light. PLoS One 7, e30655.

Naeser, M.A., Saltmarche, A., Krengel, M.H., Hamblin, M.R., Knight, J.A., 2011. Improved cognitive function after transcranial, light-emitting diode treatments in chronic, traumatic brain injury: two case reports. Photomed. Laser Surg. 29, 351–358.

Oron, A., Oron, U., Chen, J., Eilam, A., Zhang, C., Sadeh, M., et al., 2006. Low-level laser therapy applied transcranially to rats after induction of stroke significantly reduces long-term neurological deficits. Stroke 37, 2620–2624.

Oron, A., Oron, U., Streeter, J., De Taboada, L., Alexandrovich, A., Trembovler, V., et al., 2007. Low-level laser therapy applied transcranially to mice following traumatic brain injury significantly reduces long-term neurological deficits. J. Neurotrauma 24, 651–656.

Oron, U., Tuby, H., Maltz, L., Sagi-Assif, O., Abu-Hamed, R., Yaakobi, T., et al., 2014. Autologous bone-marrow stem cells stimulation reverses post-ischemic-reperfusion kidney injury in rats. Am. J. Nephrol. 40, 425–433.

Oueslati, A., Lovisa, B., Perrin, J., Wagnieres, G., Van Den Bergh, H., Tardy, Y., et al., 2015. Photobiomodulation suppresses alpha-synuclein-induced toxicity in an AAV-based rat genetic model of Parkinson's Disease. PLoS One 10, e0140880.

Peoples, C., Spana, S., Ashkan, K., Benabid, A.L., Stone, J., Baker, G.E., et al., 2012. Photobiomodulation enhances nigral dopaminergic cell survival in a chronic MPTP mouse model of Parkinson's disease. Parkinsonism Relat. Disord. 18, 469–476.

Petersen, S.L., Botes, C., Olivier, A., Guthrie, A.J., 1999. The effect of low level laser therapy (LLLT) on wound healing in horses. Equine Vet. J. 31, 228–231.

Purushothuman, S., Nandasena, C., Johnstone, D.M., Stone, J., Mitrofanis, J., 2013. The impact of near-infrared light on dopaminergic cell survival in a transgenic mouse model of parkinsonism. Brain Res. 1535, 61–70.

Purushothuman, S., Johnstone, D.M., Nandasena, C., Mitrofanis, J., Stone, J., 2014. Photobiomodulation with near infrared light mitigates Alzheimer's disease-related pathology in cerebral cortex – evidence from two transgenic mouse models. Alzheimers Res. Ther. 6, 2.

Purushothuman, S., Johnstone, D.M., Nandasena, C., Eersel, J., Ittner, L.M., Mitrofanis, J., et al., 2015. Near infrared light mitigates cerebellar pathology in transgenic mouse models of dementia. Neurosci. Lett. 591, 155–159.

Rabelo, S.B., Villaverde, A.B., Nicolau, R., Salgado, M.C., Melo Mda, S., Pacheco, M.T., 2006. Comparison between wound healing in induced diabetic and nondiabetic rats after low-level laser therapy. Photomed. Laser Surg. 24, 474–479.

Reinhart, F., Massri, N.E., Darlot, F., Torres, N., Johnstone, D.M., Chabrol, C., et al., 2015. 810nm near-infrared light offers neuroprotection and improves locomotor activity in MPTP-treated mice. Neurosci. Res. 92, 86–90.

Reinhart, F., El Massri, N., Johnstone, D.M., Stone, J., Mitrofanis, J., Benabid, A.L., et al., 2016a. Near-infrared light (670 nm) reduces MPTP-induced parkinsonism within a broad therapeutic time window. Exp. Brain Res. 234, 1787–1794.

Reinhart, F., Massri, N.E., Chabrol, C., Cretallaz, C., Johnstone, D.M., Torres, N., et al., 2016b. Intracranial application of near-infrared light in a hemi-parkinsonian rat model: the impact on behavior and cell survival. J. Neurosurg. 124, 1829–1841.

Reinhart, F., Massri, N.E., Torres, N., Chabrol, C., Molet, J., Johnstone, D.M., et al., 2017. The behavioural and neuroprotective outcomes when 670nm and 810nm near infrared light are applied together in MPTP-treated mice. Neurosci. Res. 117, 42–47.

Rezende, S.B., Ribeiro, M.S., Nunez, S.C., Garcia, V.G., Maldonado, E.P., 2007. Effects of a single near-infrared laser treatment on cutaneous wound healing: biometrical and histological study in rats. J. Photochem. Photobiol. B 87, 145–153.

Rhodes, K.M., Weinstein, R., Saltzmann, R.M., Aggarwal, N., Kooner, K.S., Petroll, W.M., et al., 2009. Intraocular pressure reduction in the untreated fellow eye after selective laser trabeculoplasty. Curr. Med. Res. Opin. 25, 787–796.

Rochkind, S., Rousso, M., Nissan, M., Villarreal, M., Barr-Nea, L., Rees, D.G., 1989. Systemic effects of low-power laser irradiation on the peripheral and central nervous system, cutaneous wounds, and burns. Lasers Surg. Med. 9, 174–182.

Rojas, J.C., Gonzalez-Lima, F., 2011. Low-level light therapy of the eye and brain. Eye Brain 3, 49–67.

Saliba, A., Du, Y., Liu, H., Patel, S., Roberts, R., Berkowitz, B.A., et al., 2015. Photobiomodulation mitigates diabetes-induced retinopathy by direct and indirect mechanisms: evidence from intervention studies in pigmented mice. PLoS One 10, e0139003.

Saltmarche, A.E., Naeser, M.A., Ho, K.F., Hamblin, M.R., Lim, L., 2017. Significant improvement in cognition in mild to moderately severe dementia cases treated with transcranial plus intranasal photobiomodulation: case series report. Photomed. Laser Surg. 35, 432–441.

Sampson, T.R., Debelius, J.W., Thron, T., Janssen, S., Shastri, G.G., Ilhan, Z.E., et al., 2016. Gut microbiota regulate motor deficits and neuroinflammation in a model of Parkinson's Disease. Cell 167, 1469–1480 e12.

Sardari, F., Ahrari, F., 2016. The effect of low-level helium-neon laser on oral wound healing. Dent. Res. J. (Isfahan) 13, 24–29.

Scheperjans, F., 2016. Gut microbiota, 1013 new pieces in the Parkinson's disease puzzle. Curr. Opin. Neurol. 29, 773–780.

Scheperjans, F., Aho, V., Pereira, P.A., Koskinen, K., Paulin, L., Pekkonen, E., et al., 2015. Gut microbiota are related to Parkinson's disease and clinical phenotype. Mov. Disord. 30, 350–358.

Schiffer, F., Johnston, A.L., Ravichandran, C., Polcari, A., Teicher, M.H., Webb, R.H., et al., 2009. Psychological benefits 2 and 4 weeks after a single treatment with near infrared light to the forehead: a pilot study of 10 patients with major depression and anxiety. Behav. Brain Funct. 5, 46.

Shaw, V.E., Spana, S., Ashkan, K., Benabid, A.L., Stone, J., Baker, G.E., et al., 2010. Neuroprotection of midbrain dopaminergic cells in MPTP-treated mice after near-infrared light treatment. J. Comp. Neurol. 518, 25–40.

Shaw, V.E., Peoples, C., Spana, S., Ashkan, K., Benabid, A.L., Stone, J., et al., 2012. Patterns of cell activity in the subthalamic region associated with the neuroprotective action of near-infrared light treatment in MPTP-treated mice. Parkinsons Dis. 2012, 296875.

Simard, A.R., Rivest, S., 2004. Bone marrow stem cells have the ability to populate the entire central nervous system into fully differentiated parenchymal microglia. FASEB J. 18, 998–1000.

Stone, J., Johnstone, D., Mitrofanis, J. 2013. The helmet experiment in Parkinson's disease: an observation of the mechanism of neuroprotection by near infra-red light. In: Proceedings of the 9th World Association for Laser Therapy Congress.

Surinchak, J.S., Alago, M.L., Bellamy, R.F., Stuck, B.E., Belkin, M., 1983. Effects of low-level energy lasers on the healing of full-thickness skin defects. Lasers Surg. Med. 2, 267–274.

Tuby, H., Maltz, L., Oron, U., 2007. Low-level laser irradiation (LLLI) promotes proliferation of mesenchymal and cardiac stem cells in culture. Lasers Surg. Med. 39, 373–378.

Tuby, H., Maltz, L., Oron, U., 2009. Implantation of low-level laser irradiated mesenchymal stem cells into the infarcted rat heart is associated with reduction in infarct size and enhanced angiogenesis. Photomed. Laser Surg. 27, 227–233.

Tuby, H., Maltz, L., Oron, U., 2011. Induction of autologous mesenchymal stem cells in the bone marrow by low-level laser therapy has profound beneficial effects on the infarcted rat heart. Lasers Surg. Med. 43, 401–409.

Ullah, I., Subbarao, R.B., Rho, G.J., 2015. Human mesenchymal stem cells - current trends and future prospective. Biosci. Rep. 35.

Van Cakenbergh, J., 1986. [Effect of the I.R. diode laser on wound healing]. Acta Belg. Med. Phys. 9, 89–91.

Wong-Riley, M.T., Liang, H.L., Eells, J.T., Chance, B., Henry, M.M., Buchmann, E., et al., 2005. Photobiomodulation directly benefits primary neurons functionally inactivated by toxins: role of cytochrome c oxidase. J. Biol. Chem. 280, 4761–4771.

Wu, Q., Xuan, W., Ando, T., Xu, T., Huang, L., Huang, Y.Y., et al., 2012. Low-level laser therapy for closed-head traumatic brain injury in mice: effect of different wavelengths. Lasers Surg. Med. 44, 218–226.

Xuan, W., Vatansever, F., Huang, L., Wu, Q., Xuan, Y., Dai, T., et al., 2013. Transcranial low-level laser therapy improves neurological performance in traumatic brain injury in mice: effect of treatment repetition regimen. PLoS One 8, e53454.

Xuan, W., Huang, L., Hamblin, M.R., 2016. Repeated transcranial low-level laser therapy for traumatic brain injury in mice: biphasic dose response and long-term treatment outcome. J. Biophotonics 9, 1263–1272.

Yu, W., Naim, J.O., Lanzafame, R.J., 1997. Effects of photostimulation on wound healing in diabetic mice. Lasers Surg. Med. 20, 56–63.

Zivin, J.A., Albers, G.W., Bornstein, N., Chippendale, T., Dahlof, B., Devlin, T., et al., 2009. Effectiveness and safety of transcranial laser therapy for acute ischemic stroke. Stroke 40, 1359–1364.

Chapter 12

Photobiomodulation for traumatic brain injury in mouse models

Michael R. Hamblin[1,2,3]

[1]*Wellman Center for Photomedicine, Massachusetts General Hospital, Boston, MA, United States,* [2]*Department of Dermatology, Harvard Medical School, Boston, MA, United States,* [3]*Harvard-MIT Division of Health Sciences and Technology, Cambridge, MA, United States*

12.1 Introduction

Traumatic brain injury (TBI) is caused by different types of trauma to the head, often resulting from road traffic accidents, assaults, falls, sports injuries, or blast injuries that can be inflicted in military conflicts. TBI is classified as mild (loss of consciousness 0–30 minutes, with an altered mental state <24 hours, and posttrauma amnesia <1 day); moderate (loss of consciousness 30 minutes to 24 hours, with an altered mental state >24 hours, and posttrauma amnesia >1–7 days); or severe (loss of consciousness >24 hours, with an altered mental state >24 hours, and posttrauma amnesia >7 days) (Blennow et al., 2016). There are three occurrences of TBI sustained each minute in the United States (Faul et al., 2010). Repeated mild episodes of TBI (also known as concussions) even without loss of consciousness, may have devastating cumulative effects (Kamins and Giza, 2016). Chronic traumatic encephalopathy is a recently recognized condition resulting from repeated head trauma, found in boxers, football players, and military personnel (McKee et al., 2016; Safinia et al., 2016). While many patients recover well from acute TBIs, many patients do not recover, and are left with life-changing deficits that can last for decades and can even be lifelong.

There is presently no accepted treatment for TBI, although some investigational approaches are being tested in both the acute (neuroprotection) and chronic (neurorehabilitation) settings (Loane and Faden, 2010). One of these novel approaches is photobiomodulation (PBM) (Hamblin, 2016a,b; Huang et al., 2012; Thunshelle and Hamblin, 2016).

12.2 Studies from other laboratories

Oron's group was the first (Oron et al., 2007) to demonstrate that a single exposure of the head of a mouse (a few hours after creation of a TBI lesion) to a NIR laser spot (808 nm) could improve neurological performance and reduce the size of the brain lesion. A weight-drop device was used to induce a closed-head TBI in the mice. An 808 nm diode laser with two energy densities calculated at the surface of the brain (1.2–2.4 J/cm^2 delivered by 2 minutes of irradiation with 200 mW laser power to the scalp) was delivered to the head 4 hours after TBI was induced. Neurobehavioral function was assessed by the neurological severity score (NSS). There was no significant difference in NSS between the power densities (10 vs 20 mW/cm^2) or significant differentiation between the control and laser treated group at early time points (24 and 48 hours) post-TBI. However, there was a significant improvement (27% lower NSS score) in the PBM group at times between 5 days and 4 weeks. The laser treated group also showed a smaller loss of cortical tissue than the sham group (Oron et al., 2007).

In another study (Oron et al., 2012) they varied the pulse parameters [continuous wave (CW), 100 or 600 Hz] and tested whether the transcranial photobiomodulation (tPBM) was equally effective when delivered at 4, 6, or 8 hours post-TBI. They first established that a calculated dose to the cortical surface of 1.2 J/cm^2 of 808 nm laser at 200 mW applied to the head, was more effective when delivered at 6 hours post-TBI than at 8 hours. They then selected an even shorter time post-TBI (4 hours) and compared CW with 100 and 600 Hz. At 56 days, more mice in the 100 Hz group (compared to the CW and 600 Hz groups) had fully recovered. The 600 Hz group had lower NSS scores than the CW

Photobiomodulation in the Brain. DOI: https://doi.org/10.1016/B978-0-12-815305-5.00012-9

and 100 Hz groups up to 20 days. Magnetic resonance imaging analysis demonstrated significantly smaller lesion volumes in PBM-treated mice compared to controls.

Khuman et al. (2012) delivered PBM (800 nm) either directly to the injured brain tissue (through the craniotomy) or transcranially in mice beginning 60–80 minutes after CCI-TBI. At a dose of 60 J/cm^2 (500 mW/cm^2) the mice showed increased performance in the Morris water maze (latency to the hidden platform, $P < .05$, and probe trial, $P < .01$) compared to nontreated controls. When PBM was delivered via open craniotomy there was reduced microgliosis at 48 hours (IbA-1 + cells, $P < .05$). Little or no effect of tPBM on postinjury cognitive function was observed using lower or higher doses, a 4-hour administration time point or 60 J/cm^2 at 7-day post-TBI.

Quirk et al. (2012) studied Sprague-Dawley rats who had received a severe controlled cortical impact (CCI) TBI and were divided into three groups: real TBI, sham surgery, and anesthetization only. Each group received either real or sham PBM consisting of 670 nm LED treatments of 15 J/cm^2, 50 mW/cm^2, 5 minutes, given two times per day for 3 days (chemical analysis) or 10 days (behavioral analysis using a TruScan nose-poke device). Significant differences in task entries, repeat entries, and task errors were seen in the TBI rats treated with PBM versus untreated TBI mice, and in sham surgery mice treated with PBM versus untreated sham mice. A statistically significant decrease was found in the proapoptotic marker Bax, and increases in the antiapoptotic marker Bcl-2 and reduced glutathione (GSH) levels in tPBM TBI mice.

Moreira et al., used a different model of TBI (Moreira et al., 2009). Wistar rats received a craniotomy, and a copper probe cooled in liquid nitrogen was applied to the surface of the brain to create a standardized cryogenic injury. They treated the rats with either a 780 or 660 nm laser at one of two different doses (3 or 5 J/cm^2) twice (once immediately after the injury and again 3 hours later). Rats were sacrificed 6 and 24 hours after the injury. The 780 nm laser was better at reducing levels of pro-inflammatory cytokines (TNFα, IL1β, IL6) particularly at early time points (Moreira et al., 2009). In a follow-up study using 3 J/cm^2 (Moreira et al., 2011) these workers reported on the healing of the injuries in these rats at time points 6 hours, 1, 7, and 14 days after the last irradiation. Cryogenic injury created focal lesions in the cortex characterized by necrosis, edema, hemorrhage, and inflammatory infiltrate. The most striking finding was: PBM-treated lesions showed less tissue loss than control lesions at 6 hours. During the first 24 hours the amount of viable neurons was significantly higher in the PBM groups. PBM reduced the amount of GFAP (glial fibrillary acidic protein, a marker of astrogliosis) and the numbers of leukocytes and lymphocytes, thus demonstrating its antiinflammatory effect.

12.3 Studies from the Hamblin laboratory

12.3.1 Closed-head traumatic brain injury study

Wu et al. (2012) first explored the effect of varying the laser wavelength on the efficiency of PBM on closed-head TBI in mice. Closed-head injury was induced via a weight-drop apparatus. Mice were randomly assigned to a PBM treatment group with a particular wavelength, or to a sham treatment TBI group as a control. To analyze the severity of the TBI, the NSS was measured and recorded. The injured mice were then treated with one of four different wavelengths of laser light (665, 730, 810, or 980 nm) at an energy density of 36 J/cm^2 directed onto the scalp. A single exposure of the mice was delivered at 4 hours post-TBI. The 665 and 810 nm laser groups showed significant improvement in NSS when compared to the control group between days 5 and 28. By contrast, the 730 and 980 nm laser groups did not show any significant improvement in NSS (Wu et al., 2012) (Fig. 12.1). The tissue chromophore, cytochrome c oxidase (CCO) is proposed to be responsible for the underlying photon absorption process that underlies many PBM effects. CCO has absorption bands (peaks) around 665 and 810 nm while it has a low absorption region (trough) at the wavelength of 730 nm (Karu et al., 2005). It should be noted that this particular study (Wu et al., 2012) found that the 980 nm did not produce the same positive effects as was observed with the 665 and 810 nm wavelengths. Nevertheless, previous studies did find that the 980 nm wavelength was an active wavelength for PBM (Anders et al., 2014), although another study in wound healing found that 980 nm was ineffective (Gupta et al., 2014). Wu et al., suggested that these dissimilar results may be explained by differences in the energy density, irradiance, etc. between the other studies and the Wu study (Wu et al., 2012). In particular, a much lower dose of 980 nm might have been effective in the closed-head TBI experiment had it been tested (Wang et al., 2017).

12.3.2 Pulsed versus continuous wave photobiomodulation for traumatic brain injury

Ando et al. (2011) next used the 810 nm wavelength produced by a Ga-Al-As diode laser delivered at similar parameters to those used in the Wu study (Wu et al., 2012), and varied the pulse modes of the laser. These modes consisted of either pulsed wave at 10 Hz or pulsed wave at 100 Hz (50% duty cycle) or CW laser. They used a different mouse

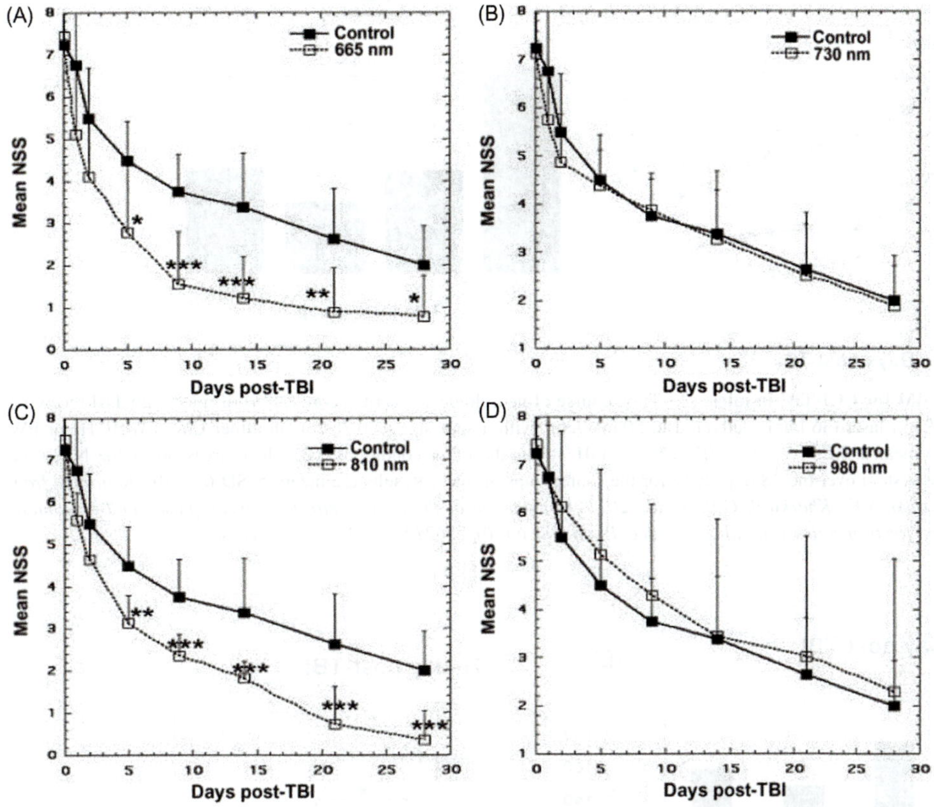

FIGURE 12.1 Effect of different laser wavelengths of tPBM in closed-head TBI in mice. (A) Sham-treated control versus 665 nm laser. (B) Sham-treated control versus 730 nm laser. (C) Sham-treated control versus 810 nm laser. (D) Sham-treated control versus 980 nm laser. Points are means of 8–12 mice and bars are SD. $*P<.05$; $**P<.01$; $***P<.001$ (one-way ANOVA). *Reprinted with permission from Wu, Q., Xuan, W., Ando, T., Xu, T., Huang, L., Huang, Y.Y., et al., 2012. Low-level laser therapy for closed-head traumatic brain injury in mice: effect of different wavelengths. Lasers Surg. Med. 44, 218–226.*

model of TBI that was induced with a CCI device. This device directly inflicts a controlled lesion onto the cortical surface via an open craniotomy. The mice were allowed to recover after the surgery, before being tested for NSS at 1 hour post-CCI. Mice that scored 7–8 on the NSS scale were included in the experiment. A single 810 nm PBM treatment with an average power density of 50 mW/m^2 and an energy density of 36 J/cm^2 (duration of 12 minutes) was given via tPBM to the closed-head in mice at 4 hours post-CCI. At 48 hours to 28 days post-TBI, all laser treated groups had significant decreases in the measured NSS when compared to the untreated TBI controls. Although all laser treated groups had similar NSS improvement rates up to day 7, the PW 10 Hz group began to show even greater improvement beyond this point as seen in Fig. 12.2. At day 28, the forced swim test for depression and anxiety was carried out, and showed a significant decrease in the immobility time for the PW 10 Hz group. In the tail suspension test, which also measures depression and anxiety, there was also a significant decrease in the immobility time at day 28, and also at day 1, in the PW 10 Hz group (Fig. 12.3).

12.3.3 Treatment repetition study

The next series of studies employed the same mouse CCI model as described above, and used the same 810 nm wavelength. However, in these studies (Xuan et al., 2013, 2014b, 2015, 2016), we tested the hypothesis that multiple daily applications of PBM would be more effective than a single application delivered 4 hours after CCI. However, because a single application of 36 J/cm^2 delivered at 50 mW/cm^2 was so effective (Ando et al., 2011), we decided to reduce the total energy density to test the effect of multiple applications. Therefore, we used 25 mW/cm^2 for the same 12 minutes to deliver 18 J/cm^2 given either in a single application (1 × PBM, 4 hours post-TBI), or given three times, once a day for 3 days commencing 4 hours post-TBI (3 × PBM), or given 14 times once a day for 14 days commencing 4 hours post-TBI (14 × PBM) (Xuan et al., 2013). We found that the 3 × PBM performed significantly better on NSS scores, than a single application (1 × PBM), over the period of 4 weeks postinjury (Fig. 12.4). However, we were surprised to observe that the same benefits were not observed in the 14 × PBM group mice. tPBM could reduce the number of neurons undergoing degeneration (shown via Fluoro Jade staining) (Xuan et al., 2013). Learning and memory as measured by the Morris water maze was also improved by tPBM (Xuan et al., 2014b)

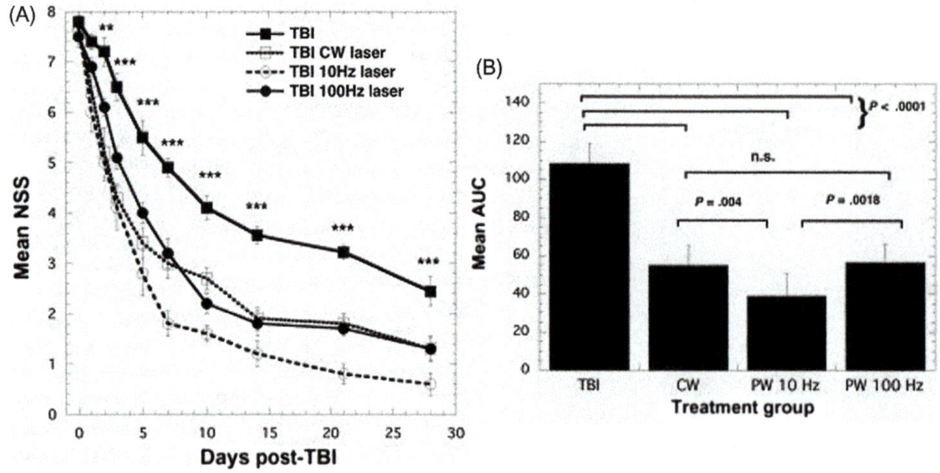

FIGURE 12.2 **Effects of pulsing in tPBM for CCI-TBI in mice.** (A) Time course of neurological severity score (NSS) of mice with TBI receiving either control (no lasertreatment), or 810 nm laser (36 J/cm^2 delivered at 50 mW/cm^2 with a spot size of 0.78 cm^2 in either CW, PW 10 Hz or PW 100 Hz modes. Results are expressed as mean ± S.E.M. $**P<0.01$, $***P<.001$ versus the other conditions. (B) Mean areas under the NSS-time curves in the two-dimensional coordinate system over the 28-day study for the four groups of mice. Results are means ± SD ($n = 10$). *Reprinted from Ando, T., Xuan, W., Xu, T., Dai, T., Sharma, S.K., Kharkwal, G.B., et al., 2011. Comparison of therapeutic effects between pulsed and continuous wave 810-nm wavelength laser irradiation for traumatic brain injury in mice. PLoS One 6 (10), e26212–e26220 (open access).*

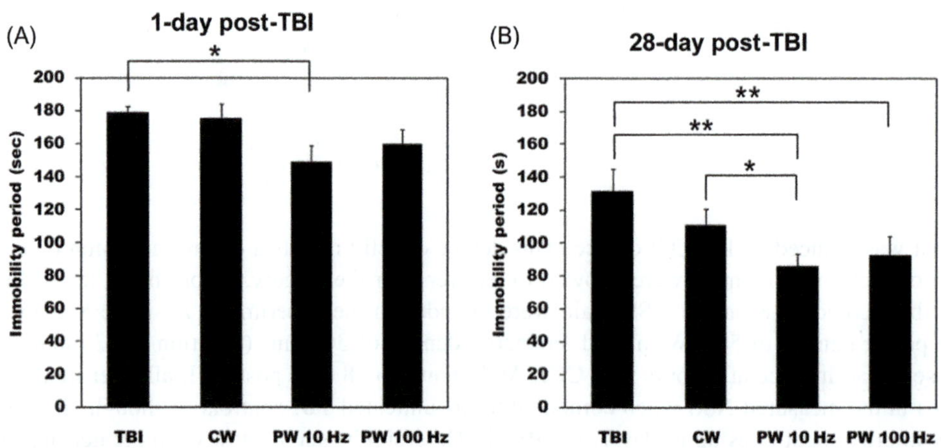

FIGURE 12.3 **Tail suspension test for depression and anxiety.** Tail suspension test (TST) showing immobility periods out of 360 s total test duration carried out at (A) 1 day; and (B) 28 days after TBI and tPBM. Values are in mean ± S.E.M. ($n = 10$). $*P<.05$; $**P<.01$.

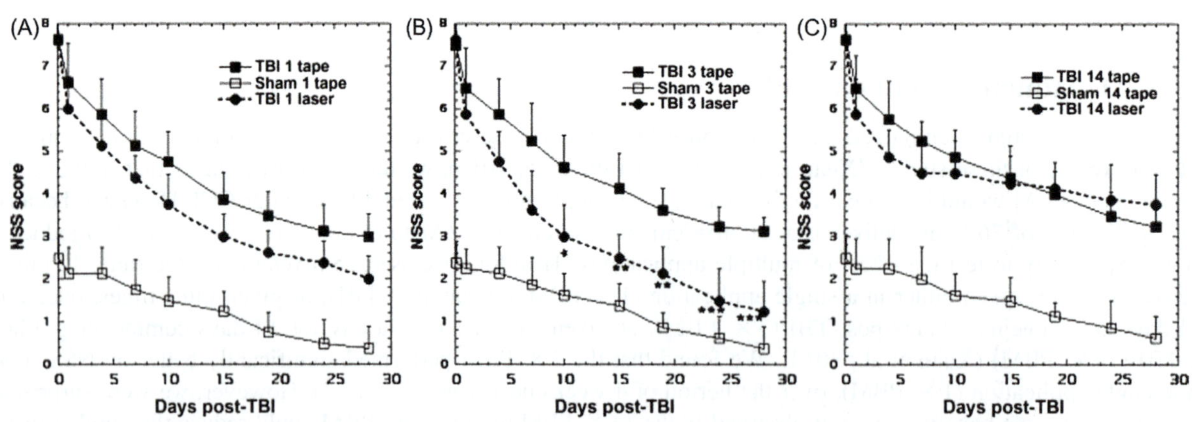

FIGURE 12.4 **NSS scores of the mice.** Mean ($n = 8$–14) NSS scores measured over 4 weeks of mice in nine groups consisting of sham TBI mice, sham control mice, tPBM TBI mice (18 J/cm^2 delivered at 25 mW/cm^2). (A) Mice and controls given 1 × -tPBM or 1 sham Tx at 4 h post-TBI; (B) mice and controls given 3 × -tPBM or 3 sham Tx; (C) mice and controls given 14 × - tPBM or 14 sham Tx. $*P<.05$; $**P<.01$; $***P<.001$. One-way ANOVA.

12.3.4 Photobiomodulation increases neurogenesis and neuroprogenitor cells in traumatic brain injury mice

We next moved on to show that tPBM could also increase bromodeoxyuridine (BrdU) positive cells in the vicinity of the brain lesion, suggesting that the neurological improvement may be partly explained by the process of neurogenesis (Bonfanti and Peretto, 2011). In recent years, adult neurogenesis has become a major topic of investigation in multiple classes of brain disease, including trauma (Zhang et al., 2011; Richardson et al., 2007), degenerative diseases (Mu and Gage, 2011), and psychiatric disorders (Hanson et al., 2011). Many studies have suggested (Beukelaers et al., 2012) that the areas of the mammalian brain that are particularly specialized to produce proliferating neuroprogenitor cells, are the subgranular layer of the dentate gyrus (DG) within the hippocampus (Masiulis et al., 2011), and the subventricular zone (SVZ) of the lateral ventricle (Bovetti et al., 2011). The process of neurogenesis may be important in this acute TBI model, where it can readily be seen after 4 weeks of follow-up that the neurological performance of the mice is improving, at the same time as the size of the cortical lesion is increasing (Xuan et al., 2013).

In one report, we described improvements in learning and memory in mice with severe CCI-TBI mice treated either one or three times with daily tPBM (Xuan et al., 2014b) (Fig. 12.5). Caspase-3 expression in the lesion was decreased at 4 days while BrdU-NeuN double-stained cells together with DCX and Tuj-1 staining were increased at 7 days. The evidence for hippocampal and SVZ-neurogenesis suggests that newly formed neurons may play a role in repairing the brain lesion and regaining adequate brain function.

The possible stimulation of neurogenesis by PBM is vitally important, as many brain conditions, not only TBI, but also neurodegenerative diseases and mood disorders can be traced, either partially or in full, to atrophy, cell death, and poor neuronal connections in certain regions of the brain. If PBM possesses the ability to counter these effects by facilitating neural regeneration, it could prove to be extremely promising as a novel method of treating such conditions.

Caspase-3 expression in the injured region at day 4 of the TBI induction is shown in Fig. 12.6. The TBI sustaining cohort is expressing significant amounts of the protein ($P < .001$ vs sham) at the lesion site, whereas laser treatments

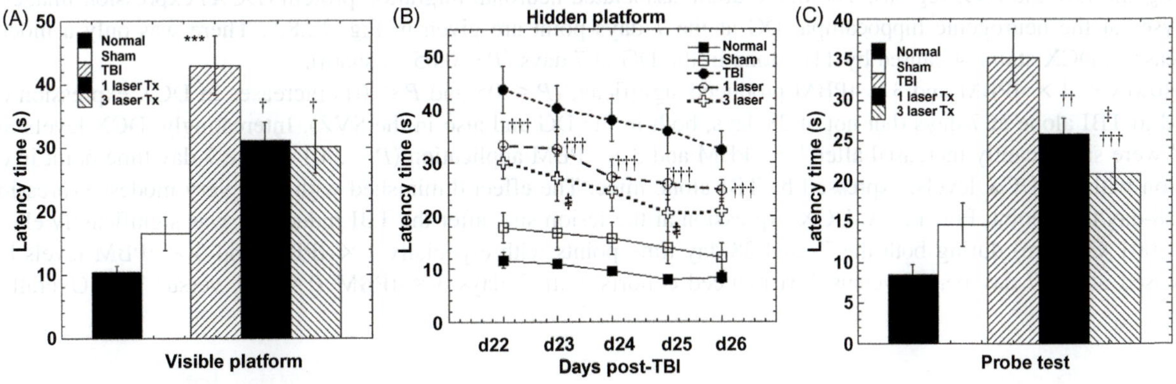

FIGURE 12.5 **Effect of tPBM on cognitive performance, learning and memory of TBI mice using Morris water maze test.** (A) visible platform test, (B) hidden platform test, (C) probe test ***$P < .001$ versus sham; †, ††, †††$P < .05$, .01, .001 versus TBI; ‡$P < .05$ versus TBI 1 laser.

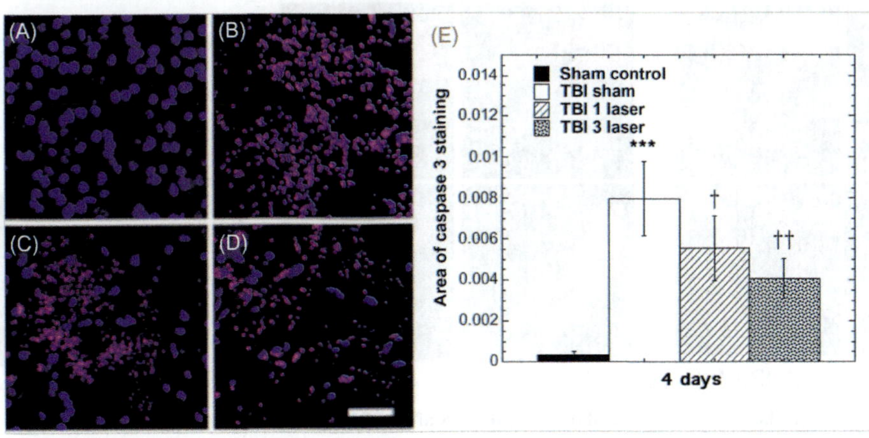

FIGURE 12.6 **Caspase-3 expression in perilesional cortex at 4 days post-TBI.** (A) sham control; (B) TBI sham; (C) 1 × -tPBM; (D) 3 × -tPBM; (E) quantification of caspase-3 staining. ***$P < 0.001$ versus sham; †$P < 0.05$; ††$P < 0.01$ versus TBI sham.

are showing a significant declining trend of its expression; 1 × -tPBM per day is $P < .05$ versus TBI and 3 × -tPBM per day is $P < 0.01$ versus TBI. These findings imply that tPBM can have a rapid cytoprotective effect by reducing apoptosis in the damaged lesion area. Moreover, the results are indicating that at the execution-stage of the cell apoptosis, the caspase activity has been stalled due to the tPBM effect.

To detect neurogenesis, we used the double labeling technique with costaining for BrdU and NeuN where BrdU labels newly generated cells whose DNA has replicated; and NeuN is a specific label for postmitotic mature neurons just before synaptic integration (Sharp et al., 2002). We concentrated on two areas in the mouse brain: the subgranular layer of the hippocampal DG and the SVZ of the lateral ventricle (Brus et al., 2013; Kim et al., 2011). Both these areas of the rodent brain are well known to give rise to neuroprogenitor cells both after TBI (Zhang et al., 2013) and after other interventions that have been tested for possible neuroprotective and neuro-regenerative effects (Acosta et al., 2010; Xiao et al., 2010). The two time-points of analysis were 7 and 28 days after the TBI induction. It should be noted that virtually all BrdU positive cells observed were yellow, indicating double staining (BrdU + NeuN), not green (BrdU alone) which would be expected from newly formed glial cells. The normalization of the BrdU-NeuN double-stained cells was done by counting the ratio of BrdU expression to DAPI staining (labeling cell nuclei). This was to correct for variation in the numbers of cells visible in different sections.

Our results indicated that some neurogenesis was caused by TBI alone in the DG both at 7 days and at 28 days (Fig. 12.7). However, significant ($P < .05$) further increases were generated by 1 × -tPBM ($P < .01$ vs TBI), 3 × -tPBM ($P < .001$ vs TBI), and ($P < .001$ vs sham) at 7 days. Note in Fig 12.7D (3 × tPBM) the double-stained cells are clearly visible lining the subgranular layer. At 28 days post-TBI, the PBM induction of neurogenesis in the DG appears to be more modest than it was at 7 days.

Double-cortin (DCX) is a microtubule-associated protein expressed by neuronal precursor cells and immature neurons in embryonic and adult cortical structures. Neuronal precursor cells begin to express DCX while actively dividing, and their neuronal daughter cells continue to express DCX for 2–3 weeks as the cells mature into neurons (von Bohlen Und Halbach, 2007). Fig. 12.8A shows our results for DCX expression at the two time points (days 7 and 28) at the neurogenic DG and SVZ regions. The microtubule-associated neuronal migration protein (DCX) expression images and analyses at the neurogenic hippocampal DG at the 7 days point are given in Fig. 12.8A. There was only a moderate increase in DCX staining caused by TBI alone in the DG at 7 days ($P < 0.05$ vs sham).

However, 1 × -tPBM and 3 × -tPBM produced significant ($P < .05$ and $P < .01$) increases in DCX expression compared to TBI alone at 7 days (but not at 28 days, both in the DG and also in the SVZ). Interestingly, DCX levels in the SVZ were significantly increased after 1 × -tPBM and 3 × -tPBM application ($P < .001$) at the 7-day time point in comparison with the DCX levels, expressed by TBI alone mice. The effect diminished with only very modest expression at the 28-day time point, Fig. 12.8A. DCX expression at the lesion site, after the TBI induction, was significantly elevated ($P < .001$ vs sham) during both the 7- and 28-day time points with especially 1 × -tPBM and 3 × -tPBM levels being increased at the 7-day period versus TBI induced cohorts, with 7 days 3 × -tPBM $P < .001$ versus TBI. Overall, this

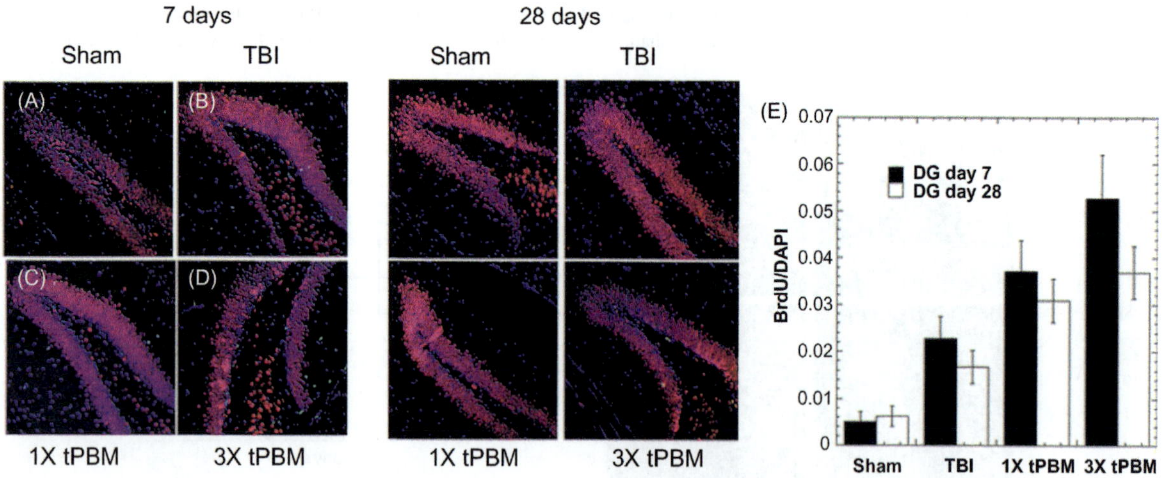

FIGURE 12.7 Neurogenesis in the dentate gyrus at day 7 and day 28 after PBM treatment. (A) sham control; (B) TBI sham; (C) 1 × -tPBM; (D) 3 × -tPBM; (E) quantification of neuroprogenitor cells.

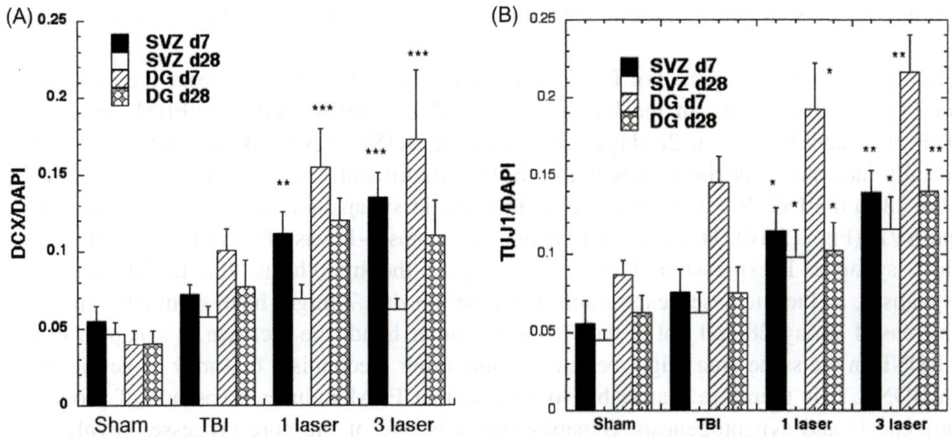

FIGURE 12.8 Double-cortin (DCX) and TUJ-1 expression in different brain regions (subventricular zone and dentate gyrus) analyzed at 7 and 28 days. (A) DCX/DAPI ratio; (B) TUJ-1/DAPI ratio. $*P<0.05$; $**P<0.01$; $***P<0.001$ versus sham.

remarkable increase in the level of DCX expressing cells may be indicating the elevated numbers of migrating neuroprogenitors and the PBM protective effects on the neuroprogenitors in migration.

TuJ-1 recognizes a neuron-specific class III beta-tubulin whose expression has been found to be an early molecular event in neuronal differentiation and is exhibited by a wide range of neuronal precursors (Memberg and Hall, 1995). Our results for Tuj-1 staining (used to distinguish the immature postmitotic neurons, just before synaptic integration) are given in the Fig. 12.8B. One can see that TBI alone only produced an increase in the DG at 7 days. However, 1 × -tPBM and 3 × -tPBM produced significant ($P<.05$ and $P<.01$, respectively) increases in Tuj-1 compared to TBI alone at both time points (days 7 and 28) in both neurogenic areas of the brain, SVZ and DG. One can see that the postmitotic TuJ-1-positive cells, generated throughout the period of cortical neurogenesis, were abundant in the 1 × and 3 × laser treated cohorts. 3 × -tPBM appeared to prolong the lifetime of the adult neuroprogenitors as indicated by the significant levels of TuJ-1 expression in the neurogenic hippocampal DG, even at day 28 after TBI induction ($P<.01$ vs TBI). In the other neurogenic hot spot, the SVZ, 3 × -PBM was also effective with $P<.01$ versus TBI at day 7 point and $P<.05$ versus TBI at 28 day point, where at day 28 only 3 × PBM exposed cohorts show elevated levels of TuJ-1 expressing neuroprogenitors.

12.3.5 Photobiomodulation increases BDNF and synaptogenesis in traumatic brain injury mice

One manner in which PBM stimulates neurogenesis and promotes improved neuronal connectivity is up-regulation of BDNF (brain derived neurotrophic factor). BDNF is a protein found in the nervous system that helps to maintain existing neurons and encourage the growth of newly formed neurons and synapses. Specifically, it is believed to modulate dendritic structures to facilitate improved synaptic transmission. BDNF plays a key role in modulating neurogenesis in the DG and SVZ, and is also a vital mediator in the process of synaptogenesis (Ambrogini et al., 2013). The synapsin-1 protein is involved in vesicle clustering, neurotransmitter release, axonal elongation, and maintenance of synaptic contacts. It is usually used as a marker of synaptic density (Ferreira et al., 2011). Studies have shown that BDNF is involved in the synthesis (Vaynman et al., 2006) and phosphorylation (Jovanovic et al., 1996, 2000) of synapsin-1. Some treatment approaches that enhance BDNF-related signaling via various mechanisms have been shown to restore neural connectivity that could facilitate neuroplastic changes leading to adaptive neural repair and consequently, to enhance the repair of cognitive deficits in both TBI and posttraumatic stress disorder (PTSD) (Kaplan et al., 2010). A recent study by Meng et al. suggested that PBM up-regulates BDNF via activation of ERK/CREB pathway and decreases neuronal loss and dendritic atrophy in mice with Alzheimer's disease (AD; Meng et al., 2013). Another study used male Wistar rats that were subjected to right sciatic nerve crush injury and were irradiated on a daily basis with helium-neon laser (collimated HeNe laser, continuous emission, wavelength: 632.8 nm, power density: 0.5 mW/cm^2, irradiation time: 20 seconds, energy density: 10 J/cm^2) during 7, 14, and 21 consecutive days, respectively. HeNe laser increased the mRNA expression of neurotrophic factors BDNF and nerve growth factor after 14 days of PBM, with peak expression at the 21st day. Additionally, HeNe laser reduced the inflammatory marker (inducible nitric oxide synthase iNOS) expression. The study opened a possibility of nerve regeneration using PBM (Gomes et al., 2012).

BDNF has also been linked to improvements in neuroplastic adaptation, which is especially important in cases of TBI and stroke (Wang et al., 2017).

In our study, we observed that the appropriate regimen of PBM (1 or 3 daily laser Tx) could increase the expression of BDNF in the SVZ at 7 and 28 days (Fig. 12.9). In the DG and the SVZ 1 × -tPBM and 3 × -tPBM, seemed to increase the expression of BDNF at 7 days but not at 28 days. No change in BDNF expression was observed in the lesion area. Furthermore, our results indicated that the expression of BDNF significantly decreased in SVZ at day 28 and completely diminished at day 28 in the DG. We also observed an increase in synapsin-1 with 3 × -tPBM treatment at day 28 in the lesion area and SVZ (Fig. 12.10). However, no change in synapsin-1 was observed in the DG. 1 × -tPBM treatment failed to induce synapsin-1 expression. Our results support the hypothesis that BDNF might be involved in the production of synapsin, as the increased expression of BDNF on day 7 might have stimulated the synthesis of synapsin which was increased at day 28 and not at day 7. On the other hand, the increase in synapsin in the lesion area with no increase in BDNF in the same area might be due to some other mechanism that may be acting independently or in synergism with BDNF. It is therefore reasonable to assume that PBM induces a cascade of processes that result in induction of neurogenesis and synaptogenesis. Synaptogenesis is one of the core processes involved in neuroplasticity, whereby the brain can remodel itself to take over functions that were previously performed by the damaged area of the brain. Moreover, many other brain disorders are characterized by abberant brain pathways that could be repaired by induction of neuroplasticity (He et al., 2018).

PBM has also proven to be effective as a tool for manipulating stem cells from other anatomical locations. For instance, Uri Oron showed that PBM delivered to the legs of mice in order to stimulate the bone marrow (BM), led to proliferation of mesenchymal stem cells (MSCs) and their homing to the ischemic heart suggesting its role in regenerative medicine (Blatt et al., 2016; Tuby et al., 2011). PBM could also stimulate MSCs of autologous BM in order to affect neurological behavior and β-amyloid burden in the progressive stages of an AD mouse model (Farfara et al., 2015). MSCs from wild-type mice stimulated with PBM to the legs showed increased ability to maturate toward a monocyte lineage and to increase phagocytosis activity toward soluble amyloid beta (Aβ). Furthermore, weekly PBM to the BM of AD mice for 2 months, starting at 4 months of age (progressive stage of AD), improved cognitive capacity

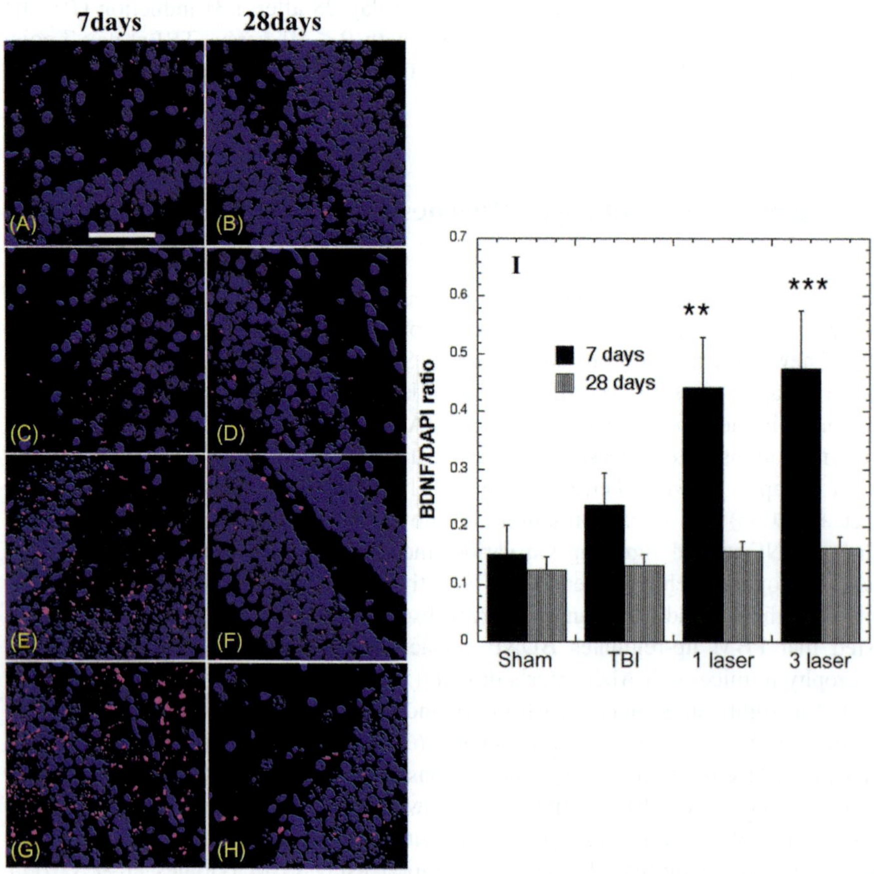

FIGURE 12.9 BDNF expression in the dentate gyrus at 7 and 28 days. (A) and (B) sham control; (C) and (D) TBI control; (E) and (F) 1 × -tPBM; (G) and (H) 3 × -tPBM; (I) quantification. $*P<0.05$; $**P<0.01$ versus sham.

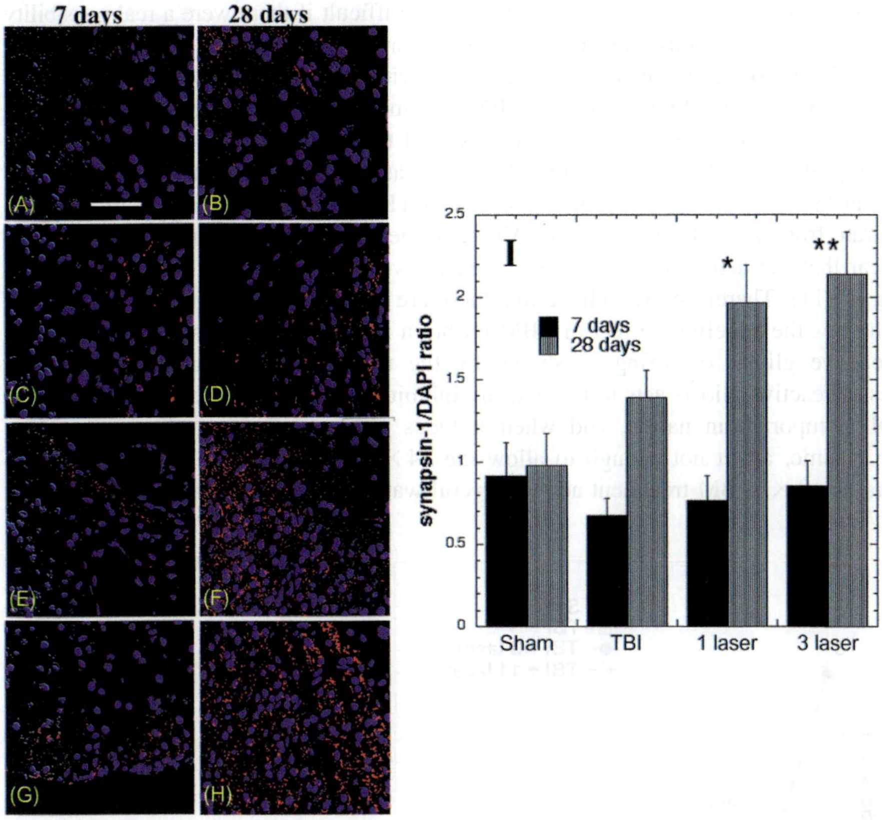

FIGURE 12.10 Synapsin-1 expression in the perilesional cortex at 7 and 28 days. (A) and (B) sham control; (C) and (D) TBI control; (E) and (F) 1 × -tPBM; (G) and (H) 3 × -tPBM; (I) quantification. $*P < 0.05$; $**P < 0.01$ versus sham.

and spatial learning, as compared to sham-treated AD mice. Histology revealed a significant reduction in Aβ brain burden (Oron and Oron, 2016).

Specifically, PBM has been shown to increase cell migration, differentiation, proliferation, and viability, all of which are important for the success of any sort of stem cell therapy (Abrahamse and Hamblin, 2017). In the brain, PBM has the potential to activate neural stem cells, which typically lay dormant in the brains of complex organisms once the organism has reached maturity (Hennessy and Hamblin, 2017). Once activated, the neural stem cells could facilitate the regeneration of damaged tissue. PBM additionally possesses the ability to provoke increased production of neuroprogenitor cells, which are similar in their functionality to neural stem cells and would have positive effects on neurogenesis.

12.3.6 The solution to the problem of 14 daily photobiomodulation treatments

Our previous studies showed that transcranial application of NIR laser to the head of mice was remarkably effective in mitigating the deleterious effects of acute serious/moderate TBI (Huang et al., 2012). Not only could tPBM improve neurological function (NSS) (Wu et al., 2012; Ando et al., 2011) and learning and memory (MWM) (Xuan et al., 2014b), but it could also beneficially affect several histological markers of brain repair (Xuan et al., 2014a,b). We compared the effects of three different treatment repetition regimens; a single PBM application delivered at 4 hours post-TBI; 3 PBM applications delivered on days 1–3 post-TBI; 14 daily PBM applications delivered on days 1–14 (Xuan et al., 2013). We reasoned that a human patient with acute TBI would be unlikely to receive just a single application of PBM, and therefore wished to see whether several repeated applications would be better than a single application. In order to do this we reduced the dose of PBM per treatment from 36 J/cm^2 of 810 nm laser delivered at 50 mW/cm^2 (which we had previously shown to be highly effective as a single treatment at 4 hours post-TBI (Ando et al., 2011) to 18 J/cm^2 of 810 nm laser delivered at 25 mW/cm^2 which took the same amount of time to deliver (12 minutes) (Xuan et al., 2013). In a different application of PBM (for arthritis of the knee in rats) we had previously shown (Castano et al., 2007) that the illumination time is an important parameter when optimizing a PBM regimen. In the study (Xuan et al., 2013) comparing 1, 3, and 14 daily PBM applications, we were somewhat surprised to find that 3 PBM applications were not only significantly better than a single PBM application, but also better than 14 consecutive daily applications. This could be interpreted as a worrying finding, as the clinical application of PBM for acute TBI would be more difficult if there were a real possibility of causing long-term damage to the brain by an injudicious selection of treatment parameters.

In the final study (Xuan et al., 2016) we followed the mice that were treated with the three different regimens (1 × -, 3 × -, and 14 × -tPBM) for as long as 8 weeks. We found that 3 × -PBM treatment improved neuromuscular performance and cognitive function (compared to untreated controls) at 4 weeks, and this improvement continued up to 8 weeks. However, when the PBM was repeated for 14 days (14 × -tPBM), there was a decline in cognitive function at 2 weeks and it was not until week 4 that the NSS caught up with the untreated TBI group. However, the 14 × -tPBM group improved at a relatively faster rate from week 4 up to week 8. Although the 14 × -tPBM group did not catch up to the 3 × -tPBM group, it does appear that the detrimental effect of the excessive 14 × -tPBM applications was only temporary rather than permanent (Fig. 12.11). Therefore, we believe that there are two processes occurring in the brain with the 14 × -tPBM applications. There is the beneficial effect of PBM on brain repair, and there is also, at the same time, a counteracting process of reactive gliosis occurring as shown by the rise in GFAP staining at 4 weeks (Fig. 12.12). Presumably, this process of reactive gliosis acts to temporarily inhibit the ongoing process of brain repair. However, the reactive gliosis is only temporary in nature, and when it fades at weeks 5–8, it allows the light-stimulated process of brain repair to resume, albeit not enough to allow the 14 × -tPBM group to catch up with the 3 × -tPBM group. We have shown that the 3 × -tPBM treatment acts in several ways to encourage brain repair in acute

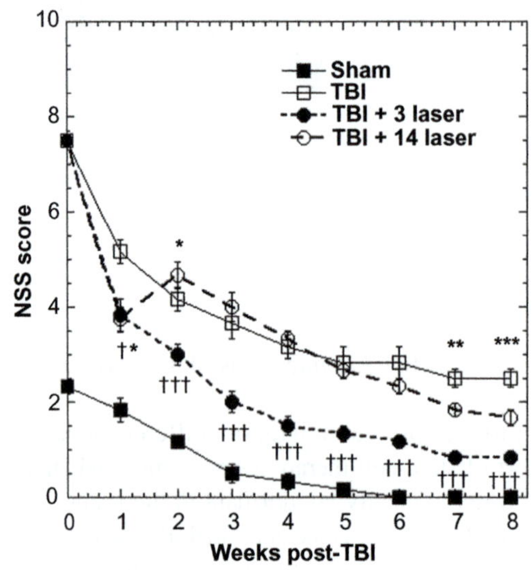

FIGURE 12.11 NSS scores in TBI mice treated with sham, 3 × -tPBM, or 14 × -tPBM and followed for 8 weeks. $*P<0.05$; $**P<0.01$; $***P<0.001$ versus TBI; $^{\dagger}P<0.05$; $^{\dagger\dagger\dagger}P<0.001$ versus TBI + 14 laser.

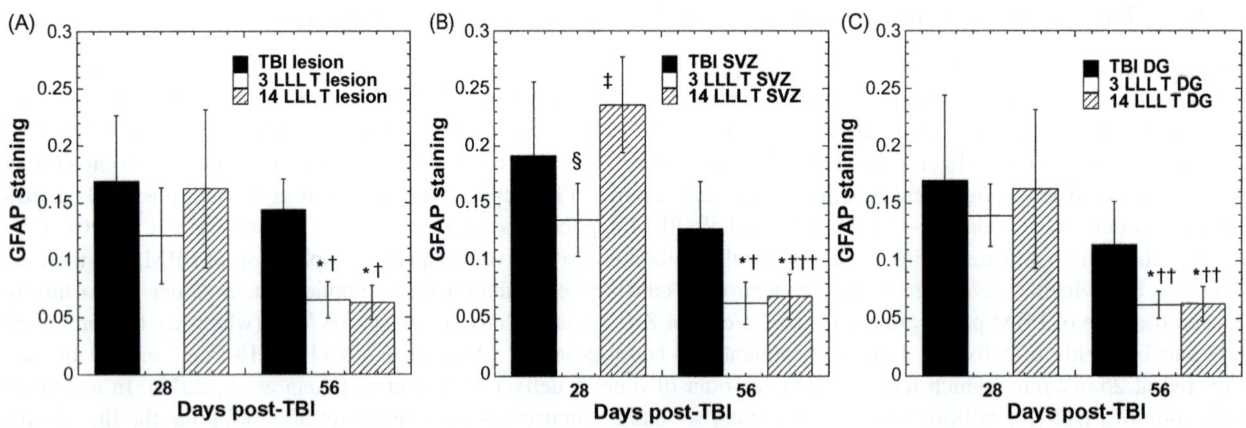

FIGURE 12.12 Glial fibrillary acidic protein (GFAP) expression in different brain regions of TBI mice treated with sham, 3 × -tPBM or 14 × -tPBM and analyzed at 4 or 8 weeks. (A) perilesional cortex; (B) subventricular zone; (C) dentate gyrus. $*P<0.05$ versus TBI; $^{\dagger}P<0.05$; $^{\dagger\dagger}P<0.01$; $^{\dagger\dagger\dagger}P<0.001$ versus same treatment at 28 days; $^{\S}P<0.05$ versus untreated TBI at 28 days; $^{\ddagger}P<0.05$ versus 3LLLT at 28 days.

TBI mice. It induces neurogenesis in the SVZ and the DG, together with increases in BDNF expression (both at 1 week) and increases synaptogesis in the perilesional cortex at week 4.

Reactive astrogliosis, is a physiological response of astrocytes seen in many brain disorders, including TBI, ischemic stroke, and diseases characterized by neuroinflammation and neurodegeneration (Pekny and Pekna, 2014). Glial cells actively interact with neurons and influence synaptic development and neuroplasticity through an array of secreted and contact-dependent signals (Jones, 2015). Compared with normal nonreactive astrocytes, reactive astrocytes show altered expression of many genes and exhibit distinct functions (Castejon, 2015). Increased expression of GFAP has been the most commonly employed molecular marker of reactive astrocytes (Eng et al., 2000). It is thought that reactive gliosis can act as somewhat of a "double-edged sword" in brain trauma (Pekny and Pekna, 2014). The beneficial function includes protecting neurons from dying (Li et al., 2008), detoxifying reactive oxygen species (Dringen et al., 2015), synthesizing vascular endothelial growth factor to encourage formation of new blood vessels (Beck and Plate, 2009), and limiting leukocyte infiltration (Christopherson et al., 2005). On the other hand, the deleterious effects of reactive gliosis include limiting synaptic regeneration (Wilhelmsson et al., 2004), limiting regeneration of axons (Overman et al., 2012), and decreasing formation of neuroprogenitor cells (Goldshmit et al., 2014). Since the literature shows that reactive gliosis can inhibit the processes we previously identified as being stimulated by tPBM in TBI mice (neurogenesis and synaptogenesis), our hypothesis appears at least plausible.

We did not measure markers of microglial activation in the brain slices, although in retrospect, it would have been a very good idea. Whalen's group (Khuman et al., 2012) showed that application of tPBM (800 nm) to mice with CCI-TBI, produced improvements in MWM performance. Importantly, they also showed that the expression of ionized calcium binding adaptor molecule 1 (iba-1, considered to be the best marker for activated microglia) was significantly reduced by PBM in brain sections at 2 days post-TBI.

In recent years the existence of the biphasic dose response (sometimes known as the "Arndt-Schulz law" or hormesis (Calabrese, 2013)) in PBM and photobiomodulation has become increasingly appreciated as a real and significant factor in designing PBM studies (Huang et al., 2009; Huang et al., 2011). This biphasic dose response has been demonstrated with several of the important parameters that are relevant to PBM. Demidova-Rice et al. (2007) showed (in a mouse model of excisional wound healing) that different fluences delivered at a constant power density (irradiance) showed a peak effectiveness at 2 J/cm^2, with both lower (1 J/cm^2) and higher (10 J/cm^2) fluences showing a lower efficacy, and a very high dose (50 J/cm^2) actually showing an inhibitory effect. Oron et al., found (Oron et al., 2001) that the same fluence (0.3 J/cm^2) delivered directly to the rat heart at 5 mW/cm^2, had a better effect at reducing infarct size after heart attack, than the same fluence delivered at a lower irradiance (2.5 mW/cm^2) or at a higher irradiance (20 mW/cm^2). As mentioned above, Xuan et al., found (Xuan et al., 2013) that repeating the tPBM application (for mouse TBI) 3 times daily was more effective than a single repetition, or than 14 daily repetitions. However, in the present case it appears that the deleterious effect of exceeding the optimum value in the dose response curve (number of treatments) is not exactly caused by too much of the same mechanism (this could be considered as over-stimulation) involved in the beneficial effect of a lesser number of applications, but rather may be caused by a second different mechanism (reactive gliosis) superimposed on top of the first mechanism (photostimulation of brain repair).

12.4 Conclusion

There is an accumulating body of evidence that tPBM is remarkably effective in a diverse range of animal models of TBI. Most of these animal models have been acute TBI modelsbecause chronic TBI in laboratory animal models has been much less studied. Therefore, at present it is uncertain how effective tPBM would be in animal models of chronic TB. However, in humans the situation is completely different, since most of the clinical studies have been in patients who are suffering the long-term consequences of a severe head injury sustained years if not decades ago (Naeser and Hamblin, 2015; Naeser et al., 2011, 2016).

The highly beneficial repair processes we have observed to occur in the mouse brain when treated with tPBM, do provide cause for optimism concerning the more widespread application of PBM in human brain disorders. If suppression of caspase-3, BDNF up-regulation, up-regulation of neurogenesis, stimulation of neuroprogenitor cell migration, stimulation of synaptogenesis, and neuroplasticity can be shown to be general attributes of tPBM, then the application of tPBM in humans may be impressively broad. Psychiatric disorders (major depressive disorder, suicidal ideation, major anxiety, PTSD, addiction, insomnia), neurodegenerative diseases (AD, Parkinson's disease, amyotrophic lateral sclerosis, frontotemporal dementia, vascular dementia, Lewy body dementia, primary progressive aphasia, chronic traumatic encephalopathy, Creutzfeldt – Jakob disease, Huntington's disease) and neurodevelopmental disorders (autism spectrum disorder and attention deficit hyperactivity disorder) are all possibilities.

References

Abrahamse, H., Hamblin, M.R., 2017. Photomedicine and Stem Cells. IOP Science Ebooks. (Morgan and Claypool Publishing), San Rafael, CA.

Acosta, S., Jernberg, J., Sanberg, C.D., Sanberg, P.R., Small, B.J., Gemma, C., et al., 2010. NT-020, a natural therapeutic approach to optimize spatial memory performance and increase neural progenitor cell proliferation and decrease inflammation in the aged rat. Rejuvenation Res. 13 (5), 581–588.

Ambrogini, P., Lattanzi, D., Ciuffoli, S., Betti, M., Fanelli, M., Cuppini, R., 2013. Physical exercise and environment exploration affect synaptogenesis in adult-generated neurons in the rat dentate gyrus: possible role of BDNF. Brain Res. 1534, 1–12.

Anders, J.J., Moges, H., Wu, X., Erbele, I.D., Alberico, S.L., Saidu, E.K., et al., 2014. In vitro and in vivo optimization of infrared laser treatment for injured peripheral nerves. Lasers Surg. Med. 46 (1), 34–45.

Ando, T., Xuan, W., Xu, T., Dai, T., Sharma, S.K., Kharkwal, G.B., et al., 2011. Comparison of therapeutic effects between pulsed and continuous wave 810-nm wavelength laser irradiation for traumatic brain injury in mice. PLoS One 6 (10), e26212–e26220.

Beck, H., Plate, K.H., 2009. Angiogenesis after cerebral ischemia. Acta Neuropathol. 117, 481–496.

Beukelaers, P., Vandenbosch, R., Caron, N., Nguyen, L., Moonen, G., Malgrange, B., 2012. Cycling or not cycling: cell cycle regulatory molecules and adult neurogenesis. Cell Mol. Life Sci. 69, 1493–1503.

Blatt, A., Elbaz-Greener, G.A., Tuby, H., Maltz, L., Siman-Tov, Y., Ben-Aharon, G., et al., 2016. Low-level laser therapy to the bone marrow reduces scarring and improves heart function post-acute myocardial infarction in the pig. Photomed. Laser Surg. 34 (11), 516–524.

Blennow, K., Brody, D.L., Kochanek, P.M., Levin, H., McKee, A., Ribbers, G.M., et al., 2016. Traumatic brain injuries. Nat. Rev. Dis. Primers 2, 16084.

Bonfanti, L., Peretto, P., 2011. Adult neurogenesis in mammals – a theme with many variations. Eur. J. Neurosci. 34 (6), 930–950.

Bovetti, S., Gribaudo, S., Puche, A.C., De Marchis, S., Fasolo, A., 2011. From progenitors to integrated neurons: role of neurotransmitters in adult olfactory neurogenesis. J. Chem. Neuroanat. 42 (4), 304–316.

Brus, M., Keller, M., Levy, F., 2013. Temporal features of adult neurogenesis: differences and similarities across mammalian species. Front. Neurosci. 7, 135.

Calabrese, E.J., 2013. Biphasic dose responses in biology, toxicology and medicine: accounting for their generalizability and quantitative features. Environ. Pollut. 182, 452–460.

Castano, A.P., Dai, T., Yaroslavsky, I., Cohen, R., Apruzzese, W.A., Smotrich, M.H., et al., 2007. Low-level laser therapy for zymosan-induced arthritis in rats: importance of illumination time. Lasers Surg. Med. 39 (6), 543–550.

Castejon, O.J., 2015. Biopathology of astrocytes in human traumatic and complicated brain injuries. Review and hypothesis. Folia Neuropathol. 53 (3), 173–192.

Christopherson, K.S., Ullian, E.M., Stokes, C.C., Mullowney, C.E., Hell, J.W., Agah, A., et al., 2005. Thrombospondins are astrocyte-secreted proteins that promote CNS synaptogenesis. Cell 120 (3), 421–433.

Demidova-Rice, T.N., Salomatina, E.V., Yaroslavsky, A.N., Herman, I.M., Hamblin, M.R., 2007. Low-level light stimulates excisional wound healing in mice. Lasers Surg. Med. 39 (9), 706–715.

Dringen, R., Brandmann, M., Hohnholt, M.C., Blumrich, E.M., 2015. Glutathione-dependent detoxification processes in astrocytes. Neurochem. Res. 40, 2570–2582.

Eng, L.F., Ghirnikar, R.S., Lee, Y.L., 2000. Glial fibrillary acidic protein: GFAP-thirty-one years (1969-2000). Neurochem. Res. 25 (9–10), 1439–1451.

Farfara, D., Tuby, H., Trudler, D., Doron-Mandel, E., Maltz, L., Vassar, R.J., et al., 2015. Low-level laser therapy ameliorates disease progression in a mouse model of Alzheimer's disease. J. Mol. Neurosci. 55 (2), 430–436.

Faul, M., Xu, L., Wald, M.M., Coronado, V., 2010. Traumatic Brain Injury in the United States: Emergency Department Visits, Hospitalizations and Deaths, 2002–2006. Centers for Disease Control and Prevention, National Center for Injury Prevention and Control, Atlanta, GA.

Ferreira, A.F., Real, C.C., Rodrigues, A.C., Alves, A.S., Britto, L.R., 2011. Short-term, moderate exercise is capable of inducing structural, BDNF-independent hippocampal plasticity. Brain Res. 1425, 111–122.

Goldshmit, Y., Frisca, F., Pinto, A.R., Pebay, A., Tang, J.K., Siegel, A.L., et al., 2014. Fgf2 improves functional recovery-decreasing gliosis and increasing radial glia and neural progenitor cells after spinal cord injury. Brain Behav. 4 (2), 187–200.

Gomes, L.E., Dalmarco, E.M., Andre, E.S., 2012. The brain-derived neurotrophic factor, nerve growth factor, neurotrophin-3, and induced nitric oxide synthase expressions after low-level laser therapy in an axonotmesis experimental model. Photomed. Laser Surg. 30 (11), 642–647.

Gupta, A., Dai, T., Hamblin, M.R., 2014. Effect of red and near-infrared wavelengths on low-level laser (light) therapy-induced healing of partial-thickness dermal abrasion in mice. Lasers Med. Sci. 29, 257–265.

Hamblin, M.R., 2016a. Shining light on the head: photobiomodulation for brain disorders. BBA Clin. 6, 113–124.

Hamblin, M.R., 2016b. Photobiomodulation or low-level laser therapy. J. Biophotonics 9 (11–12), 1122–1124.

Hanson, N.D., Owens, M.J., Nemeroff, C.B., 2011. Depression, antidepressants, and neurogenesis: a critical reappraisal. Neuropsychopharmacology 36 (13), 2589–2602.

He, W., Fong, P.Y., Leung, T.W.H., Huang, Y.Z., 2018. Protocols of non-invasive brain stimulation for neuroplasticity induction. Neurosci. Lett. Available from: https://doi.org/10.1016/j.neulet.2018.02.045.

Hennessy, M., Hamblin, M.R., 2017. Photobiomodulation and the brain: a new paradigm. J. Opt. 19 (1), 013003.

Huang, Y.Y., Chen, A.C., Carroll, J.D., Hamblin, M.R., 2009. Biphasic dose response in low level light therapy. Dose Response 7 (4), 358–383.

Huang, Y.Y., Sharma, S.K., Carroll, J.D., Hamblin, M.R., 2011. Biphasic dose response in low level light therapy – an update. Dose Response 9 (4), 602–618.

Huang, Y.Y., Gupta, A., Vecchio, D., de Arce, V.J., Huang, S.F., Xuan, W., et al., 2012. Transcranial low level laser (light) therapy for traumatic brain injury. J. Biophotonics 5 (11–12), 827–837. Available from: https://doi.org/10.1002/jbio.201200077.

Jones, O.D., 2015. Astrocyte-mediated metaplasticity in the hippocampus: help or hindrance? Neuroscience 309, 113–124.

Jovanovic, J.N., Benfenati, F., Siow, Y.L., Sihra, T.S., Sanghera, J.S., Pelech, S.L., et al., 1996. Neurotrophins stimulate phosphorylation of synapsin I by MAP kinase and regulate synapsin I-actin interactions. Proc. Natl. Acad. Sci. U.S.A. 93 (8), 3679–3683.

Jovanovic, J.N., Czernik, A.J., Fienberg, A.A., Greengard, P., Sihra, T.S., 2000. Synapsins as mediators of BDNF-enhanced neurotransmitter release. Nat. Neurosci. 3 (4), 323–329.

Kamins, J., Giza, C.C., 2016. Concussion-mild traumatic brain injury: recoverable injury with potential for serious sequelae. Neurosurg. Clin. N. Am. 27 (4), 441–452.

Kaplan, G.B., Vasterling, J.J., Vedak, P.C., 2010. Brain-derived neurotrophic factor in traumatic brain injury, post-traumatic stress disorder, and their comorbid conditions: role in pathogenesis and treatment. Behav. Pharmacol. 21 (5–6), 427–437.

Karu, T.I., Pyatibrat, L.V., Afanasyeva, N.I., 2005. Cellular effects of low power laser therapy can be mediated by nitric oxide. Lasers Surg. Med. 36 (4), 307–314.

Khuman, J., Zhang, J., Park, J., Carroll, J.D., Donahue, C., Whalen, M.J., 2012. Low-level laser light therapy improves cognitive deficits and inhibits microglial activation after controlled cortical impact in mice. J. Neurotrauma 29 (2), 408–417.

Kim, E.J., Ables, J.L., Dickel, L.K., Eisch, A.J., Johnson, J.E., 2011. Ascl1 (Mash1) defines cells with long-term neurogenic potential in subgranular and subventricular zones in adult mouse brain. PLoS One 6 (3), e18472.

Li, L., Lundkvist, A., Andersson, D., Wilhelmsson, U., Nagai, N., Pardo, A.C., et al., 2008. Protective role of reactive astrocytes in brain ischemia. J. Cereb. Blood Flow Metab. 28 (3), 468–481.

Loane, D.J., Faden, A.I., 2010. Neuroprotection for traumatic brain injury: translational challenges and emerging therapeutic strategies. Trends Pharmacol. Sci. 31 (12), 596–604.

Masiulis, I., Yun, S., Eisch, A.J., 2011. The interesting interplay between interneurons and adult hippocampal neurogenesis. Mol. Neurobiol. 44 (3), 287–302.

McKee, A.C., Alosco, M.L., Huber, B.R., 2016. Repetitive head impacts and chronic traumatic encephalopathy. Neurosurg. Clin. N. Am. 27 (4), 529–535.

Memberg, S.P., Hall, A.K., 1995. Dividing neuron precursors express neuron-specific tubulin. J. Neurobiol. 27 (1), 26–43.

Meng, C., He, Z., Xing, D., 2013. Low-level laser therapy rescues dendrite atrophy via upregulating BDNF expression: implications for Alzheimer's disease. J. Neurosci. 33 (33), 13505–13517.

Moreira, M.S., Velasco, I.T., Ferreira, L.S., Ariga, S.K., Barbeiro, D.F., Meneguzzo, D.T., et al., 2009. Effect of phototherapy with low intensity laser on local and systemic immunomodulation following focal brain damage in rat. J. Photochem. Photobiol. B 97 (3), 145–151.

Moreira, M.S., Velasco, I.T., Ferreira, L.S., Ariga, S.K., Abatepaulo, F., Grinberg, L.T., et al., 2011. Effect of laser phototherapy on wound healing following cerebral ischemia by cryogenic injury. J. Photochem. Photobiol. B 105 (3), 207–215.

Mu, Y., Gage, F.H., 2011. Adult hippocampal neurogenesis and its role in Alzheimer's disease. Mol. Neurodegener. 6, 85.

Naeser, M.A., Hamblin, M.R., 2015. Traumatic brain injury: a major medical problem that could be treated using transcranial, red/near-infrared LED photobiomodulation. Photomed. Laser Surg. 33, 443–446.

Naeser, M.A., Saltmarche, A., Krengel, M.H., Hamblin, M.R., Knight, J.A., 2011. Improved cognitive function after transcranial, light-emitting diode treatments in chronic, traumatic brain injury: two case reports. Photomed. Laser Surg. 29 (5), 351–358.

Naeser, M.A., Martin, P.I., Ho, M.D., Krengel, M.H., Bogdanova, Y., Knight, J.A., et al., 2016. Transcranial, red/near-infrared light-emitting diode therapy to improve cognition in chronic traumatic brain injury. Photomed. Laser Surg. 34 (12), 610–626.

Oron, A., Oron, U., 2016. Low-level laser therapy to the bone marrow ameliorates neurodegenerative disease progression in a mouse model of Alzheimer's Disease: a minireview. Photomed. Laser Surg. 34 (12), 627–630.

Oron, U., Yaakobi, T., Oron, A., Hayam, G., Gepstein, L., Rubin, O., et al., 2001. Attenuation of infarct size in rats and dogs after myocardial infarction by low-energy laser irradiation. Lasers Surg. Med. 28 (3), 204–211.

Oron, A., Oron, U., Streeter, J., de Taboada, L., Alexandrovich, A., Trembovler, V., et al., 2007. low-level laser therapy applied transcranially to mice following traumatic brain injury significantly reduces long-term neurological deficits. J. Neurotrauma 24 (4), 651–656.

Oron, A., Oron, U., Streeter, J., De Taboada, L., Alexandrovich, A., Trembovler, V., et al., 2012. Near infrared transcranial laser therapy applied at various modes to mice following traumatic brain injury significantly reduces long-term neurological deficits. J. Neurotrauma 29 (2), 401–407.

Overman, J.J., Clarkson, A.N., Wanner, I.B., Overman, W.T., Eckstein, I., Maguire, J.L., et al., 2012. A role for ephrin-A5 in axonal sprouting, recovery, and activity-dependent plasticity after stroke. Proc. Natl. Acad. Sci. U.S.A. 109 (33), E2230–E2239.

Pekny, M., Pekna, M., 2014. Astrocyte reactivity and reactive astrogliosis: costs and benefits. Physiol. Rev. 94 (4), 1077–1098.

Quirk, B.J., Torbey, M., Buchmann, E., Verma, S., Whelan, H.T., 2012. Near-infrared photobiomodulation in an animal model of traumatic brain injury: improvements at the behavioral and biochemical levels. Photomed. Laser Surg. 30 (9), 523–529.

Richardson, R.M., Sun, D., Bullock, M.R., 2007. Neurogenesis after traumatic brain injury. Neurosurg. Clin. N. Am. 18 (1), 169–181. xi.

Safinia, C., Bershad, E.M., Clark, H.B., SantaCruz, K., Alakbarova, N., Suarez, J.I., et al., 2016. Chronic traumatic encephalopathy in athletes involved with high-impact sports. J. Vasc. Interv. Neurol. 9 (2), 34–48.

Sharp, F.R., Liu, J., Bernabeu, R., 2002. Neurogenesis following brain ischemia. Brain Res. Dev. Brain Res. 134 (1–2), 23–30.

Thunshelle, C., Hamblin, M.R., 2016. Transcranial low-level laser (light) therapy for brain injury. Photomed. Laser Surg. 34 (12), 587–598.

Tuby, H., Maltz, L., Oron, U., 2011. Induction of autologous mesenchymal stem cells in the bone marrow by low-level laser therapy has profound beneficial effects on the infarcted rat heart. Lasers Surg. Med. 43 (5), 401–409.

Vaynman, S.S., Ying, Z., Yin, D., Gomez-Pinilla, F., 2006. Exercise differentially regulates synaptic proteins associated to the function of BDNF. Brain Res. 1070 (1), 124–130.

von Bohlen Und Halbach, O., 2007. Immunohistological markers for staging neurogenesis in adult hippocampus. Cell Tissue Res. 329 (3), 409–420.

Wang, Y., Huang, Y.Y., Wang, Y., Lyu, P., Hamblin, M.R., 2017. Photobiomodulation of human adipose-derived stem cells using 810nm and 980nm lasers operates via different mechanisms of action. Biochim. Biophys. Acta 1861, 441–449.

Wilhelmsson, U., Li, L., Pekna, M., Berthold, C.H., Blom, S., Eliasson, C., et al., 2004. Absence of glial fibrillary acidic protein and vimentin prevents hypertrophy of astrocytic processes and improves post-traumatic regeneration. J. Neurosci. 24 (21), 5016–5021.

Wu, Q., Xuan, W., Ando, T., Xu, T., Huang, L., Huang, Y.Y., et al., 2012. Low-level laser therapy for closed-head traumatic brain injury in mice: effect of different wavelengths. Lasers Surg. Med. 44, 218–226.

Xiao, X., Liu, Y., Qi, C., Qiu, F., Chen, X., Zhang, J., et al., 2010. Neuroprotection and enhanced neurogenesis by tetramethylpyrazine in adult rat brain after focal ischemia. Neurol. Res. 32 (5), 547–555.

Xuan, W., Vatansever, F., Huang, L., Wu, Q., Xuan, Y., Dai, T., et al., 2013. Transcranial low-level laser therapy improves neurological performance in traumatic brain injury in mice: effect of treatment repetition regimen. PLoS One 8 (1), e53454.

Xuan, W., Agrawal, T., Huang, L., Gupta, G.K., Hamblin, M.R., 2014a. Low-level laser therapy for traumatic brain injury in mice increases brain derived neurotrophic factor (BDNF) and synaptogenesis. J. Biophotonics 9999(9999).

Xuan, W., Vatansever, F., Huang, L., Hamblin, M.R., 2014b. Transcranial low-level laser therapy enhances learning, memory, and neuroprogenitor cells after traumatic brain injury in mice. J. Biomed. Optics 19 (10), 108003.

Xuan, W., Agrawal, T., Huang, L., Gupta, G.K., Hamblin, M.R., 2015. Low-level laser therapy for traumatic brain injury in mice increases brain derived neurotrophic factor (BDNF) and synaptogenesis. J. Biophotonics 8 (6), 502–511.

Xuan, W., Huang, L., Hamblin, M.R., 2016. Repeated transcranial low-level laser therapy for traumatic brain injury in mice: biphasic dose response and long-term treatment outcome. J. Biophotonics 9 (11–12), 1263–1272.

Zhang, C., Wu, H., Zhu, X., Wang, Y., Guo, J., 2011. Role of transcription factors in neurogenesis after cerebral ischemia. Rev. Neurosci. 22 (4), 457–465.

Zhang, L., Yan, R., Zhang, Q., Wang, H., Kang, X., Li, J., et al., 2013. Survivin, a key component of the Wnt/beta-catenin signaling pathway, contributes to traumatic brain injury-induced adult neurogenesis in the mouse dentate gyrus. Int. J. Mol. Med. 32 (4), 867–875.

Chapter 13

Photobiomodulation and mitochondria for traumatic brain injury in mouse models

Mei X. Wu[1,2,3] and Michael R. Hamblin[1,2,3]

[1]*Wellman Center for Photomedicine, Massachusetts General Hospital, Boston, MA, United States,* [2]*Department of Dermatology, Harvard Medical School, Boston, MA, United States,* [3]*Harvard-MIT Division of Health Sciences and Technology, Cambridge, MA, United States*

13.1 Introduction

Traumatic brain injury (TBI) results in cerebral structural damage and functional deficits due to both primary and secondary injuries. While the primary injury is caused instantly by the impact or blast inflicted during the incident, secondary brain injury evolves over time, providing a therapeutic window for prevention and treatment. It is believed that secondary brain injury results from a cascade of metabolic, cellular, and molecular events caused by inflammation, oxidative stress, perturbation of cellular calcium homeostasis, increased vascular permeability, mitochondrial dysfunction, glutamate excitotoxicity, and apoptosis (Bramlett and Dietrich, 2007). Hypoxia and its adverse effects on mitochondrial function may play a role in the initiation of this cascade, and could be targets for therapy or intervention (Longhi et al., 2007).

Despite representing only 2% of the total body mass, the brain consumes about 20% of the oxygen and 25% of the glucose that the human body produces or ingests and is therefore one of the most energy-hungry organs. The cellular source of energy (adenosine triphosphate, ATP), is mostly generated through oxidative phosphorylation in the mitochondria. The adequate function of mitochondria is pivotal for the survival and activity of brain cells, and even brief periods of hypoxia or glucose deprivation to the brain can adversely affect cerebral function within a few minutes, irreparably damaging neurons within several minutes (Budd, 1998). Abundant data from retrospective studies and prospective clinical trials have demonstrated brain hypoxia to be an early predictor of adverse outcomes after TBI (Chang et al., 2009; Oddo et al., 2011; Yan et al., 2014), because efficient ATP production by the mitochondrial respiratory chain relies on continuous oxygen supply.

13.2 IEX-1 in traumatic brain injury

The immediate early response gene (IEX-1) is an immediate-early gene that is induced by X-irradiation, ultraviolet radiation, growth factors such as EGF, tumor-promoting phorbol esters, by peptide growth factors such as pituitary adenylate cyclase activating peptide, by steroid hormones such as 1α,25-dihydroxyvitamin D3, and also during cellular differentiation (Schilling et al., 2001). The gene for IEX-1 encodes a protein with a predicted molecular weight of 17,000 daltons. The protein is posttranslationally modified by glycosylation (Kondratyev et al., 1996).

It is generally believed that the pathogenesis of secondary brain injury involves a complex cellular and molecular cascade in association with inadequate mitochondrial function, blood–brain barrier disruption, neuroinflammation, oxidative stress, cell death, and so on. Owing to the complicated pathogenesis of TBI, no single animal model has been able to recapitulate all the aspects of the pathogenesis observed in humans. IEX-1 plays a pivotal role in regulation of mitochondrial F_1F_0-ATPase activity (Campanella et al., 2008), in protection against apoptosis, and in resolution of

Photobiomodulation in the Brain. DOI: https://doi.org/10.1016/B978-0-12-815305-5.00013-0

inflammation. Null mutations in IEX-1 enhanced apoptosis (Shen et al., 2006), reduced ATP synthase activity (Shen et al., 2009), and prolonged inflammatory responses in a tissue-dependent manner in mouse models (Zhi et al., 2012). Furthermore, IEX-1 is up-regulated by NF-κB, and in turn, inhibits NF-κB activation as a negative feedback mechanism contributing to resolution of inflammation (Zhang et al., 2002; Wu, 2003).

In order to gain insights into a possible role for inadequate mitochondrial function in the initiation and progression of secondary brain injury, the effects of IEX-1 deficiency on the pathogenesis of secondary brain injury were investigated. We found that IEX-1 knockout (KO) mice were more susceptible to secondary brain injury than wild-type (WT) littermates following mild traumatic brain injury (mTBI), correlating with extensive neuronal cell death, prolonged neuroinflammation, and severe loss of brain tissue at and around the impact site. Strikingly, a single dose of PBM given at 4 hour post-TBI effectively prevented the secondary brain injury induced by IEX-1 deficiency. The finding underscores mitochondria to be a key player in both the pathogenesis of secondary brain injury and the effectiveness of PBM.

13.3 IEX-1 KO mice fail to fully recover from mild traumatic brain injury

To mimic mTBI in humans, we set up a closed-head impact model in which the hairless mouse head was directly hit by a pneumatic impact device set at a 2.0 mm punch depth. The severity of initial injury was assessed by the neurological severity score (NSS) at 1 hour post-TBI, and mice with scores of 4–6 were enrolled in the study. The impact force on the closed head brought about mild erythema on the impacted part of the scalp (Fig. 13.1A), but did not produce any structural damage to the skull, brain, or blood–brain barrier visibly when compared to sham controls (Fig. 13.1A). There was no evidence of acute hemorrhage either on the cortical surface or in the cortical matter underneath (Fig. 13.1A), resembling mild TBI described in humans. Assessment of neurological behavior of these mice revealed that the NSS diminished significantly from 4.3 ± 0.9 at 1 hour post-mTBI, to 1.0 ± 0.8 on day 14, and continuously to

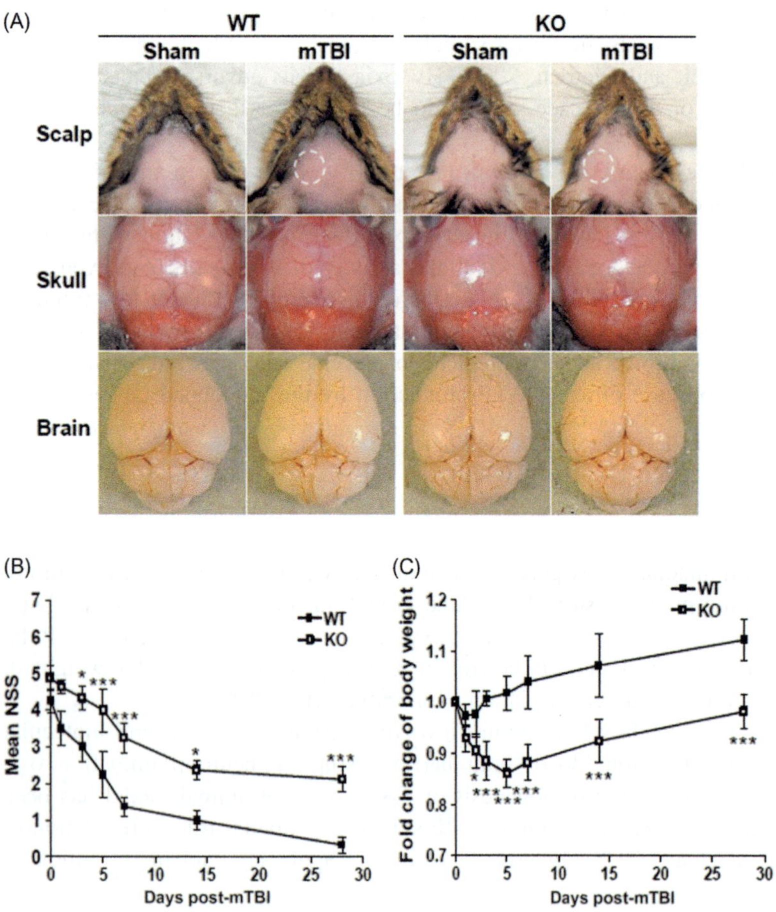

FIGURE 13.1 Failure of IEX-1 KO mice to fully recover from mTBI. (A) Representative photos of sham and mild injured brains of WT and KO mice. The photos were taken at 6 h after the mild injury showing erythema, marked in a white dash line circle, on the impact scalp (upper), with normal skull (middle), and brain (lower) underneath. (B) A time course of NSS after mTBI. NSS was assessed in WT and KO mice at 1 h and indicated days after mTBI and expressed as means ± SEM. (C) Body weight changes over time in injured mice. A health status of the animals was monitored by weighing the animals at the indicated days after mTBI and expressed as means ± SEM of body weight changes relative to a preinjury level that sets arbitrarily as 1. * $P < .05$ and *** $P < .001$ in the presence or absence of IEX-1 ($n = 9$ mice per group)

0.3 ± 0.5 on day 28 in WT mice (Fig. 13.1B), demonstrating the complete recovery of neurological function in most of the mice. In contrast, although IEX-1 KO mice had similar NSS values at 1 hour post-mTBI (4.9 ± 1.0), their recovery was largely delayed, and their NSS remained significantly higher at the end of the experimental period of 28 days (2.1 ± 1.0 vs 0.3 ± 0.5, $P < .001$) (Fig. 13.1B).

Moreover, the body weight of the WT mice decreased only slightly and briefly in the first day or two postinjury (Fig. 13.1C). Following the brief weight loss, the animals started regaining weight at day 3 and reached preinjury levels by day 4, followed by a steady gain body in weight above the preinjury levels thereafter. In contrast, IEX-1 KO mice displayed a precipitous loss of body weight during the first 5 consecutive days after the injury (Fig. 13.1C). After the KO mice reached the minimum body weight, they only began to regain body weight on day 7, this was 4 days later than WT mice, and gradually increased their body weight from then on, but never reached their preinjury weights up to the end of the experiment (Fig. 13.1C).

13.4 Histological alteration in IEX-1 KO mice after mild traumatic brain injury

Concurrent with the poor recovery in body weight after the injury, the gross morphology of the brain started to display an overt lesion at the effected impact site at day 7, which then exacerbated over time in KO mice (Fig. 13.2A). The lesion expanded and deepened on day 14, forming a large scar by day 28 (Fig. 13.2A). The discernible aggravation of the lesion in KO mice was in sharp contrast to the normal gross appearance of the WT brains during the entire experimental period (Fig. 13.2A). There was no evidence of any lesion at the impact site in WT mice over the entire course of the study at low magnification (Fig. 13.2B, upper panel). By contrast, KO mice demonstrated a steady increase in the lesion size, which was hardly visible at day 1, but was clearly seen at day 7, greatly expanded by day 14, and had aggressively spread from cortex into the hippocampus by day 28, resulting in a profound loss of brain tissue (Fig. 13.2B, lower panel). Quantitatively, average lesion areas (relative to the whole brain taken as 0% on day 1) were increased by 1.8% ± 1.2% at day 7, by 7.7% ± 1.3% at day 14, and by 12.8% ± 4.0% at day 28 (Fig. 13.2E).

At high magnification, healthy cell nuclei were relatively large consisting of several discernible nucleoli in the nucleoplasm in the sham-treated cerebral neocortex, regardless of whether or not IEX-1 was expressed (Fig. 13.2C, first panel). After mTBI, abundant necrotic cells (*unfilled arrows*) with pyknotic and condensed nuclei were uniformly present beneath the contusion site at 1 day postinjury and were indistinguishable in the presence or absence of IEX-1 (Fig. 13.2C, second panel). The discernible nucleoli could no longer be seen in these necrotic cells, probably owing to nuclear condensation. The injured cortical cells in the WT mice appeared to have mostly recovered from the damage at day 7, evidenced by a drastic decrease in the number of necrotic cells, concomitant with a robust increase in the number of the cells morphologically resembling those in the sham brain (Fig. 13.2C, third upper panel). The percentage of morphologically normal cells increased from 3.1% ± 1.5% at day 1, to 61.2% ± 13.0% at day 7, to 82.3% ± 11.5% at day 14, and reached 99.4% ± 0.4% at day 28 postinjury at the impact site in WT mice (Fig. 13.2F); clearly indicating robust regeneration of brain cells following the injury. In marked contrast to the robust recovery in WT mice, there was little evidence of recovery in the cell morphology in injured brain from IEX-1 KO mice (Fig. 13.2C, third to fifth lower panels). On the contrary, the percentage of normal cells at the impact area decreased steadily in the KO mice over the course of the experiment (Fig. 13.2F). Neuroinflammation, demonstrated by a large infiltration of leukocytes (*filled arrows*), was seen at the impact site, and which increased over the 7-day period. On days 14 and 28, the cortical tissue in the direct area of the impact was lost either partially or completely (Fig. 13.2, fourth to fifth panels), and peri-lesional tissues showed leukocyte infiltrates. The data suggest aggressive spreading of the lesion into the surrounding tissue.

Beneath the impact site, the necrotic cell death started to appear in the hippocampus at 7 days posttrauma, and increased over time, leading to severe hippocampal lesions by day 28 in KO mice (Fig. 13.2D, fifth lower panel). As shown in Fig. 13.2G, necrotic hippocampal cells increased from 0.3% ± 0.1% at day 1, to 10.2% ± 4.8% at day 7, to 17.4% ± 5.7% by day 14, and reached as high as 50.1% ± 6.3% at day 28 post-mTBI. Similar to initiation of neuroinflammation in the cortical area, we also observed necrotic cell death in the hippocampus prior to leukocyte infiltration, hinting that the inflammation may have caused the cell death in the surrounding tissue. There was no cell death or leukocyte infiltrate in the hippocampal region in WT mice, confirming that injury-induced cellular responses were confined within the cortical area (Fig. 13.2D, upper panel). The results strongly suggest the importance of IEX-1 in protection against secondary brain injury, either through its ability to protect against cell death, to resolve inflammation, or both.

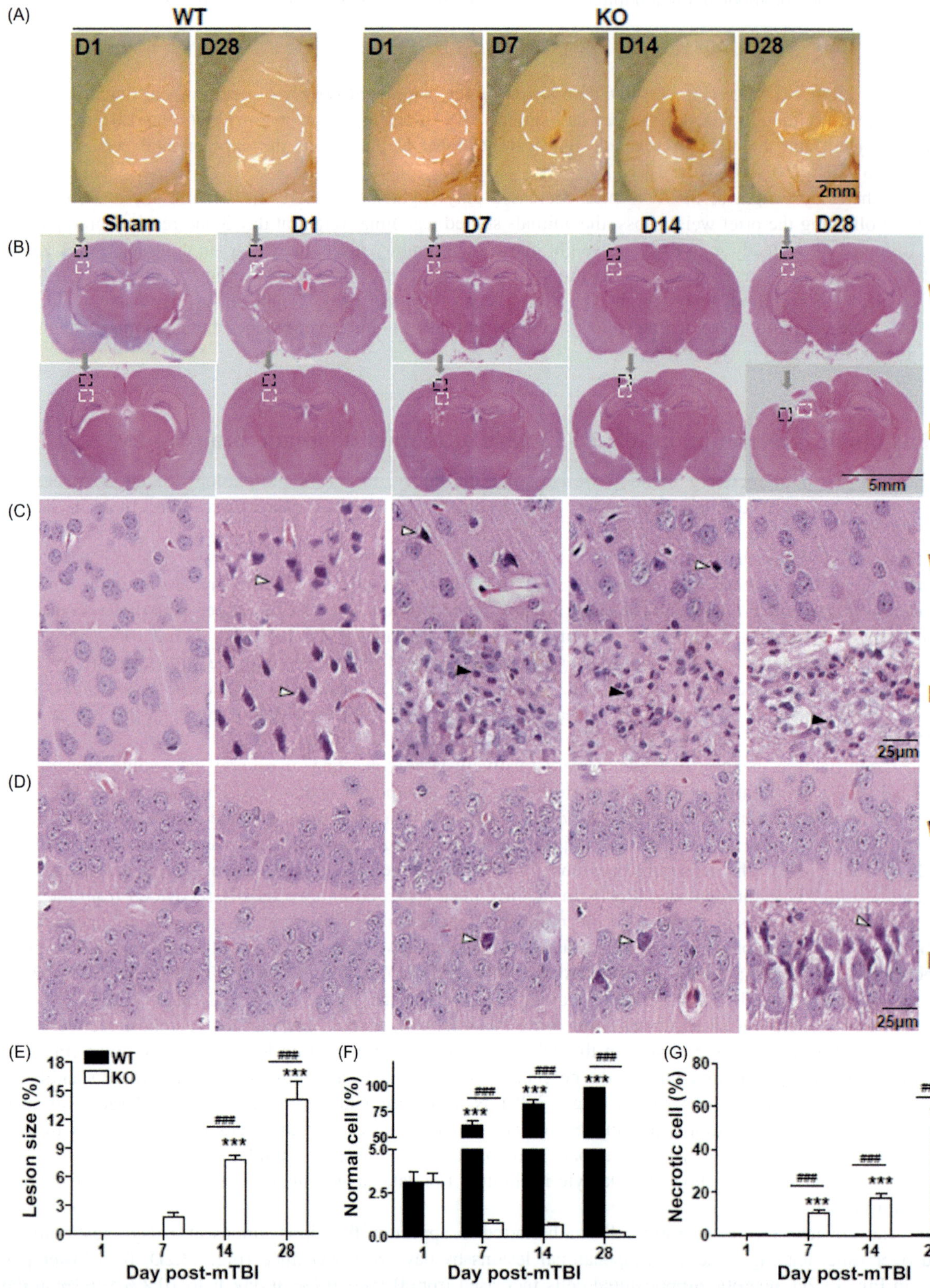

FIGURE 13.2 Secondary brain injury occurs in IEX KO mice but not in WT mice after mild TBI insult. (A) Gross morphology of the impact, left hemisphere of the brain. Note, a normal appearance on day 1 (D1) and day 28 (D28) of WT brain after trauma, but a lesion deteriorating over time in KO brain, marked by white line dash circles. (B–D) Histological examination of coronal sections stained with H&E from WT and KO brains at indicated days after mTBI. The impact site in section (B) is pointed by an arrow; the injured neocortex is highlighted in a dash black line square and enlarged in (C); and the hippocampus underneath is outlined by a dashed white-line square and magnified in (D). Unfilled arrows indicate one of the necrotic cells in each field and filled arrows indicate one of the infiltrated leukocytes. All data in the figures were representative of six mice per group. Percentages of a lesion size over the whole brain section in (B) were determined by image J and expressed as means ± SEM in (E) ($n = 6$ mice per group). Percentages of normal cells in (C) or necrotic cells in (D) relative to a total number of cells counted in the same fields were determined as detailed in materials and methods and shown in (F) and (G), respectively. $^{***}P < .001$ compared with Day 1 and $^{\#\#\#}P < .001$ in the presence or absence of IEX-1.

13.5 Inflammatory responses after mild traumatic brain injury

To confirm the high level of inflammatory responses in the absence of IEX-1, inflammatory mediators were assayed by qRT-PCR from tissue taken from the impact site at varying times after mTBI. mTBI resulted in up-regulation of IL-1β, IL-6, CCL2, CXCL10, and TNF-α in both WT and KO cortex at 6 hour post-mTBI (Fig. 13.3A), similar to previous investigations showing that neuroinflammation could be acutely provoked by TBI (Chiu et al., 2016). In comparison with WT mice, KO mice expressed significantly higher levels of IL-6, CCL2, CXCL10, and TNF-α, correlating with the more vigorous inflammatory responses in KO brain described in Fig. 13.2C. All the cytokines and chemokines dwindled down to baseline levels in WT mice by 28 days postinjury. However, four out of the five inflammatory mediators tested remained significantly elevated in injured KO cortex relatively to injured WT cortex or to sham KO cortex at 28 days postinjury (Fig. 13.3A). It was noticed that IL-1β, and to a lesser degree, IL-6 were higher in IEX-1 KO mice than WT mice at 6 hours after sham-treatment (Fig. 13.3A). The sham-treatment, included anesthesia, hair removal, and sham impact by setting the punch depth of device tip at 0 mm causes some stress to the mice, which might be better tolerated by the WT mice compared to the IEX-1 KO mice. IEX-1 is a stress-inducible gene that is up-regulated in response to a variety of stresses and its loss can compromise the stress-management capacity of the mice (Wu, 2003). In support of this hypothesis, no differences in the transcription levels of these cytokines were found between naïve WT and KO mice (Fig. 13.3A, naïve columns). In spite of increasing transcription of inflammatory cytokines, there was no evidence of leukocyte infiltration in sham brains (Fig. 13.2C), suggesting that sham-induced inflammation (if any) is transient in these mice.

The finding that IL-6, CCL2, CXCL10, and TNF-α were not restored to the WT level by 28 days after mTBI in KO mice suggested impaired resolution of inflammatory responses in these mice. To confirm that the aggravated neurological pathogenesis may be ascribed (at least in part) to elevated inflammatory responses induced by the absence of IEX-1, we carried out an additional experiment. LPS at a dose of 4 mg/kg body weight [which is capable of inducing a low degree of inflammatory response (Qin et al., 2007)] was administered to WT mice 1 hour after mTBI. LPS further worsened the neurobehavioral performance, and the deterioration was apparent as early as 1 hour after LPS injection, increasing the NSS from 4.8 ± 0.4 to 6.4 ± 0.5 (Fig. 13.3B, $P < .01$). The worsening effect on NSS lasted for more than 3 days in injured WT mice (5.0 ± 1.9 vs 1.8 ± 0.5 on day 3, $P < .001$, Fig. 13.3B), and subsided by day 7, but could be re-provoked by a second LPS injection (Fig. 13.3B). LPS alone did not interfere with the neurological function significantly in control uninjured mice under similar conditions (Fig. 13.3B). These data indicate that neuroinflammation is aggravated by null mutation of IEX-1, resulting in exacerbated leukocyte infiltration that can contribute significantly to initiation and progression of secondary brain injury.

13.6 Transcranial photobiomodulation for traumatic brain injury in IEX-1 Knockout Mice

Several studies have shown the antiinflammatory therapeutic effects of PBM on TBI and neurodegenerative diseases in some animal models (Hamblin, 2016a,b, 2018; Hennessy and Hamblin, 2017). We thus addressed whether PBM could prevent the neuroinflammation triggered by IEX-1 deficiency and protect the mice from secondary brain injury. A single application of PBM (give parameters) was noninvasively delivered onto the hairless scalp over the injured area at 4 hour post-mTBI. Levels of inflammatory cytokines and chemokines were measured by qRT-PCR in the injured tissues at 2 hours after the PBM. IL-1β, IL-6, CCL2, and CXCL10, but not TNF-α, all diminished drastically after PBM in these injured mice regardless of IEX-1 expression (Fig. 13.4, left panel). Although the level of CCL2 and CXCL10 expression appeared to be higher in KO mice than in WT after PBM when analyzed by student *t* test (data not shown), it was without statistical significance if the difference was tested by two-way ANOVA (Fig. 13.4, left panel). In sharp contrast to the decrease in these inflammatory mediators after PBM, TNF-α was up-regulated by PBM, with more pronounced effects in WT mice (4.5-fold) than those seen in KO mice (2.0-fold) at 6 hour post-mTBI (Fig. 13.4, left panel). The disparate effects of PBM on TNF-α gene expression gave rise to comparable levels of TNF-α in both WT and KO mice after light exposure. The high level of TNF-α expression was transient and returned to a sham level by 28 days in both WT and KO mice (Fig. 13.4, right panel). Besides TNF-α, IL-1β, IL-6, and CCL2, but not CXCL10, all diminished to the sham level by 28 days after PBM in both WT and KO mice. CXCL10 remained elevated in comparison with the sham controls at day 28, with an approximate 8-fold increase in WT mice and a 21-fold increase in KO mice (Fig. 13.4, right panel, $P < .01$)

PBM-mediated suppression of neuroinflammation was clearly associated with improved neurological performance as reflected by greatly lower NSS of the mice (Fig. 13.5A). The initial NSS assessed at 1 hour post-mTBI was comparable in the presence or absence of IEX-1 (Fig. 13.5A). The NSS of KO mice was significantly lower in the PBM-treated

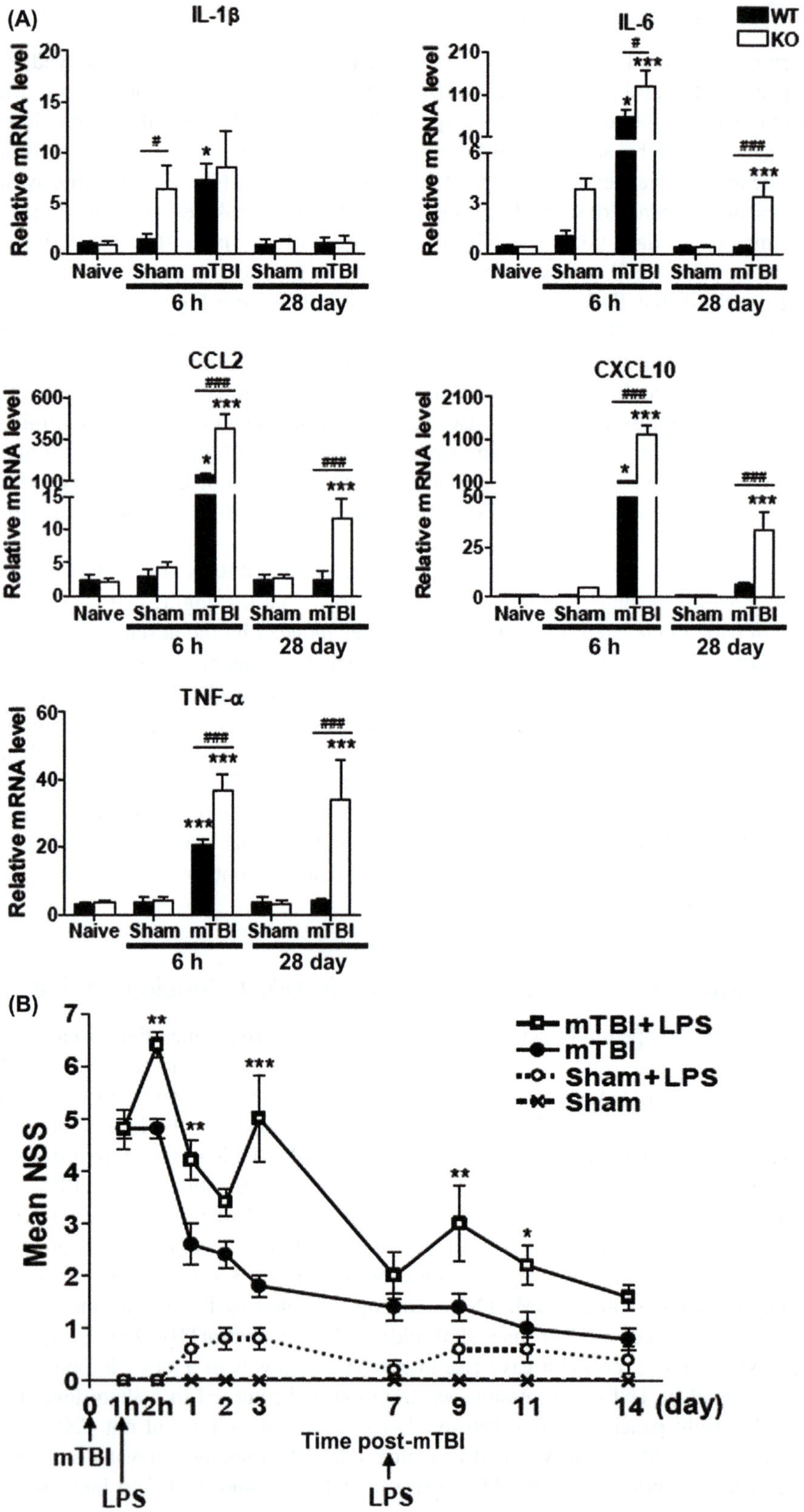

FIGURE 13.3 Exaggerated inflammatory responses in injured IEX-1 KO brain. (A) qRT-PCR analysis of pro-inflammatory mediators at indicated times after mTBI. Total RNA was isolated from the impact cortex in 6 h and/or peri-lesional area in 28 days after injury. IL-1β, IL-6, CCL2, CXCL10, and TNF-α mRNA levels were analyzed by qRT-PCR and normalized to β-actin. Results are shown as mean ± SEM ($n = 5$ mice per group). $^{*}P < P.05$ and $^{***}P < .001$ between injured brains and sham controls; and $^{\#}P < .05$ and $^{\#\#\#}P < .001$ in the presence or absence of IEX-1. (B) LPS-mediated detrimental effects on mTBI-induced neurobehavior. LPS was IP administered into WT mice at 1 h and 7 days after mTBI. NSS was assessed at indicated times as Fig. 13.1B. Results are presented as means ± SEM ($n = 5$ mice per group). $^{*}P < .05$, $^{**}P < .01$, and $^{***}P < .001$ in the presence or absence of LPS.

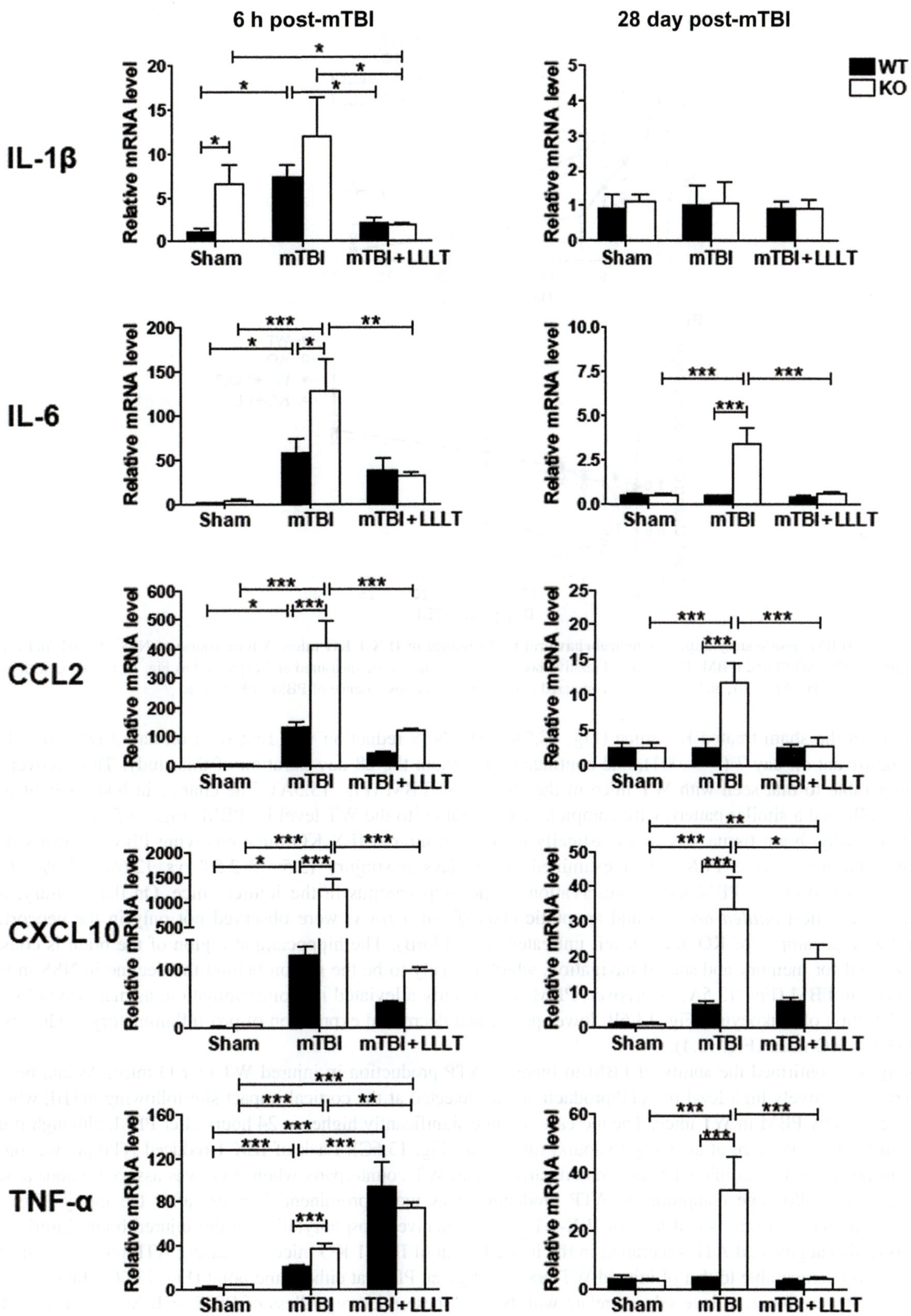

FIGURE 13.4 **Disparate influences of tPBM on the expression of various inflammatory mediators.** Pro-inflammatory cytokines and chemokines were analyzed by qRT-PCR in 2 h after PBM or 6 h after mTBI as well as 28 days postinjury as Fig. 13.3A. The data are expressed as means ± SEM ($n = 5$ mice per group). *$P < .05$, **$P < .01$, and ***$P < .001$ when mTBI or mTBI + PBM groups are compared with sham groups on the same animal background and at the same time point.

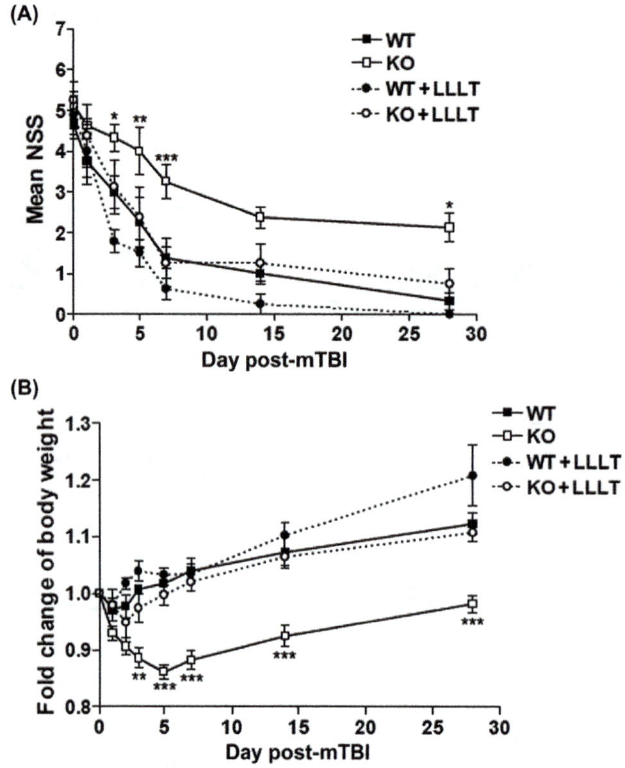

FIGURE 13.5 **tPBM significantly improves neurobehavioral performance in IEX-1 KO mice.** A time course of NSS (A) and fold changes of body weight (B) after mTBI and tPBM. The NSS and relative body weight changes were measured and expressed as Fig. 13.1B and C with nine mice in each group. $^*P<.05$, $^{**}P<.01$, and $^{***}P<.001$ compared in the presence versus absence of PBM in KO mice.

group than in the sham-treated KO mice (Fig. 13.5A). The NSS reduction was first seen on day 3 ($P<.05$), became highly significant on day 7 ($P<.001$), and continued throughout the 28-day duration of this study. This recovery trend was comparable to that seen with WT mice in the absence of PBM (Fig. 13.5A). The change in body weight in these KO mice followed a similar pattern with complete normalization to the WT level by PBM (Fig. 13.5B).

Histologically, brain tissue loss was drastically reduced in injured IEX-KO mice receiving PBM as opposed to the mice that did not receive PBM when examined at 28 days postinjury (5.5% ± 2.5% vs 12.6% ± 1.9%, $P<.01$, Fig. 13.6A). Importantly, PBM led to preservation of the hippocampus in the injured mice. On the contrary, a great number of necrotic (*unfilled arrows*) and apoptotic cells (*filled arrows*) were observed not only in the neocortex but also in the hippocampus of KO mice, if left untreated (Fig. 13.6B). The hippocampal region of the brain is considered to be essential for memory and spatial navigation, which is likely to be the reason behind the decline in NSS in the KO mice receiving PBM (Fig. 13.5A). Moreover, PBM significantly alleviated neuroinflammation, as manifested by diminished infiltration of leukocytes (Fig. 13.6B, lower panel) and decreased expression of pro-inflammatory mediators at the impact site in KO mice (Fig. 13.4).

Finally, we confirmed the ability of PBM to increase ATP production in injured WT or KO mice. As can be seen in Fig. 13.6C, a relatively high level of ATP production was detected at the cortical impact site following mTBI, which was further elevated by PBM in WT mice. The increase became significantly higher at 24 hours after PBM, although it did not reach statistical significance at the early (6 hours) time point (Fig. 13.6C). Lack of IEX-1 reduced ATP production in the sham controls as well as in injured brains in comparison with WT counterparts when ATP was assayed 6 hour postinjury (Fig. 13.6C, $P<.05$). The compromised ATP production was more prominent 24 hours after the injury (Fig. 13.6C, $P<.001$). The result confirms that lack of IEX-1 impairs oxidative phosphorylation in the injured brain. Similar to WT mice, PBM also augmented ATP generation in the injured brain of IEX-1 KO mice, increasing ATP production in the KO mice to a level comparable to that of injured WT brains not given PBM at either time point (Fig. 13.6C). Interestingly, the levels of ATP were found to inversely correlate with NSS in these mice regardless of PBM or IEX-1 expression. Highly statistical correlations were seen between ATP levels measured at 6 hour postinjury and improved neurobehavioral performance over the course of the study, with a $R^2=0.9283$ ($P<.0001$) on day 3, $R^2=0.9512$ ($P<.0001$) on day 7, $R^2=0.8533$ ($P<.0001$) on day 14, or $R^2=0.7732$ ($P<.001$) on day 28 postinjury (Fig. 13.6D). Similarly, the ATP level measured at 24 hour postinjury was also highly correlated with the reduced NSS in these animals, with $R^2=0.9271$

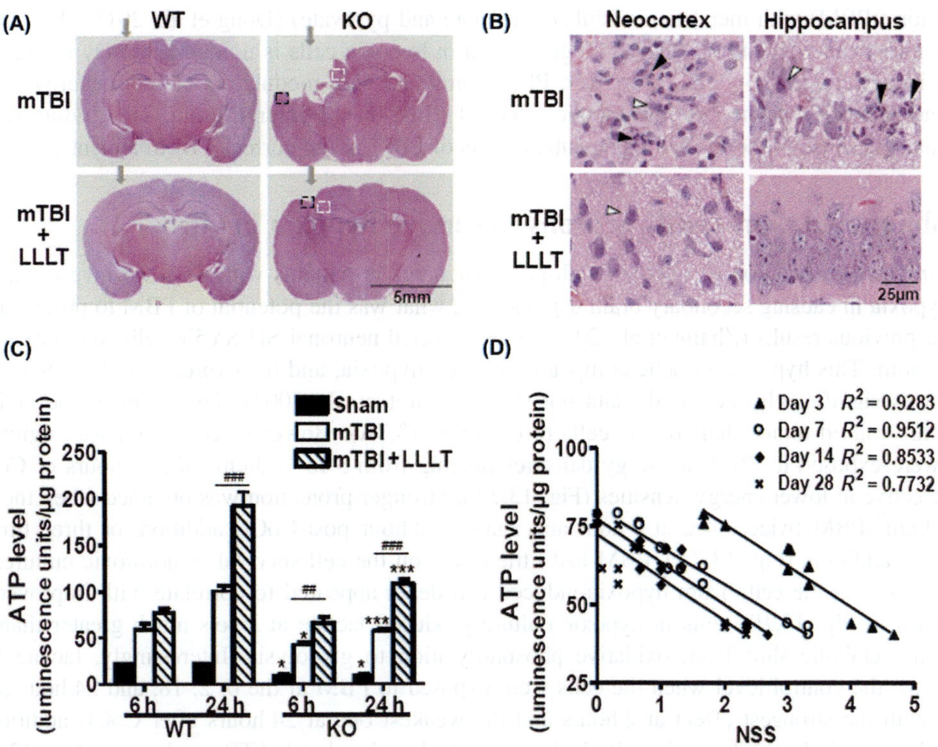

FIGURE 13.6 **Prevention of IEX-1 deficiency-induced cortical tissue loss by a single application of tPBM.** (A) Representative histological coronal sections from WT and KO mice with mTBI (upper) or the injured mice receiving PBM (lower). The sections were prepared in 28 days postinjured and arrows indicate the impact sites. Highlighted areas in the neocortex (*a black dash line square*) and hippocampus (*a white dash line square*) in KO mice are shown on a high magnification (B). Representative necrotic cells with pyknotic and condensed nuclei are marked with unfilled arrows and apoptotic cells with pyknotic and fragmented nuclei are labeled with filled arrows, respectively, in (B). The data are representative of six mice per group. (C and D) Cortical ATP production was low in the absence compared to the presence of IEX-1, but higher after PBM, regardless of IEX-1 expression (C). WT and KO mice were subjected to mTBI, followed 4 h later by PBM. At 6 and 24 h after mTBI, the cortical lysates were collected for ATP measurement, and relative ATP levels were expressed as means ± SEM ($n = 6$ mice per group). $^{*}P < .05$ and $^{***}P < .001$ in the presence or absence of IEX-1; and $^{\#\#}P < .01$ and $^{\#\#\#}P < .001$ in the presence or absence of PBM. Correlations between 6 h ATP levels and NSS measured at indicated days regardless of IEX-1 expression are analyzed by coefficient of determination (R^2) in (D). Each symbol represents the data from individual groups with 12 mice in each group. Among the 12 animals, 6 animals were sacrificed at 6 h after mTBI for ATP measurement, and the remaining 6 mice were monitored for neurological performance at indicated days.

($P < .0001$) at day 3, $R^2 = 0.7165$ ($P < .001$) at day 7, $R^2 = 0.8086$ ($P < .0001$) at day 14, and $R^2 = 0.6004$ ($P < .01$) at day 28 postinjury, respectively (data not shown). These observations demonstrate a causal relationship between sufficient ATP production at early stages after mTBI and the long-term neurological recovery.

It has been long appreciated that neuroinflammation is responsible for both beneficial and detrimental effects on secondary brain injury. It was found that neither antiinflammatory nor enhanced inflammatory treatment could significantly ameliorate the injury experimentally or even clinically (Finnie, 2013). PBM-mediated suppression of inflammation may be one of the key factors contributing to the protection against secondary brain damage in mice (Khuman et al., 2012). But, it is not known whether the protection is due to selective up-regulation of TNF-α expression by PBM at early stages of TBI, while still inhibiting the expression of other antiinflammatory mediators. TNF-α-deficient mice exhibited an early functional improvement but failed to fully recover over time post-TBI (Bermpohl et al., 2007), raising the intriguing possibility the effect of TNF-α has on brain injury also depends on the presence of other pro-inflammatory cytokines. For instance, TNF-α was found to synergistically enhance the neurotoxic effects of IL-1β, as both cytokines share many of the same physiological effects (Stahel et al., 2000). Therefore, inhibition of IL-1β, but up-regulation of TNF-α, may tilt the balance toward TNF-α being more effective. The selective modulation of inflammatory cytokine production is difficult to achieve by current antiinflammatory based therapies, emphasizing the unique ability of PBM to prevent secondary brain damage.

13.7 Combination of photobiomodulation and metabolic modulation

The inverse correlation of ATP synthesis with brain tissue loss that we demonstrated in the IEX-1 KO brain (Zhang et al., 2014) hinted at a pivotal role for hypoxia in secondary brain damage. We therefore tested the combination of transcranial photobiomodulation (tPBM) with metabolic modulators (lactate and pyruvate) (Dong et al., 2015). The ability of PBM to improve mitochondrial function and enhance ATP generation in hypoxic cells is unique, and may be one of the mechanisms underlying its effectiveness in TBI. Moreover, PBM combined with modulators of brain energy metabolism could additively or synergistically enhance the therapeutic effect of PBM on the injured brain. The finding would be of high clinical significance, because of the limited therapeutic effects of PBM in the current TBI treatments.

13.8 Photobiomodulation assists neurons to survive hypoxia in vitro

Because of the frequent association of hypoxia with poor outcomes in patients with TBI, we investigated whether there was a role for hypoxia in causing secondary brain damage, and what was the potential of PBM to protect against the damage based on our previous results (Zhang et al., 2014). We subjected neuronal SH-SY5Y cells to hypoxia by inclusion of $CoCl_2$ in the medium. This hypoxic mimetic compound induced hypoxia, and the expression of hypoxia-inducible factor 1-alpha (HIF1) was significantly increased (data not shown) (Piret et al., 2002). Treatment of the cells with 150 μM $CoCl_2$ for 48 hours caused almost half of the cells to die (Fig. 13.7A). However, cell death was significantly reduced when the cells were exposed to PBM at energy densities ranging from 3 to 10 J/cm^2 after 2 hours of $CoCl_2$ incubation. PBM was not effective at lower energy densities (Fig. 13.7A). Stronger protection was obtained when the cells were illuminated with 3 J/cm^2 PBM twice, once at 0 hour and again at 2 hour post-$CoCl_2$ addition, or three times at 0, 2, and 18 hour post-$CoCl_2$ addition. (Fig. 13.7A). PBM had little effect on the cell survival in normoxic cultures (control). The ability of PBM to protect the cells from hypoxia-induced cell death appeared to correlate with improved mitochondrial function. As shown in Fig. 13.7B, cells in hypoxic culture produced lactate at levels much greater than in controls, in agreement with a metabolic shift from oxidative phosphorylation to glycolysis. Interestingly, lactate production was reduced to or below the control level when the cells were exposed to PBM at the 0, 2, 18, and 24 hour time points after $CoCl_2$ addition, with the strongest effect at 2 hours and the weakest one at 24 hours after $CoCl_2$ addition (Fig. 13.7B). Due to the metabolic switch, the hypoxic cells had extremely low levels of ATP production (Fig. 13.7C) and robust increases in reactive oxygen species (ROS) production (Fig. 13.7D). The compromised mitochondrial function caused by a decrease in oxygen was partially alleviated by PBM treatment,as evidenced by the increase in ATP levels in a light dose-dependent manner (Fig. 13.7C) and the suppression of ROS generation (Fig. 13.7D) in hypoxic cells. Strikingly, although PBM sustained mitochondrial activity and mitochondrial membrane potential ($\Delta\Psi m$) (Fig. 13.7E), it did not alter the level of HIF1α expression (data not shown), confirming mitochondria to be the direct target of PBM. It is noteworthy that PBM only improved mitochondrial functions under hypoxia, but it was without effect in normoxic cultures, in agreement with the generally accepted protective role of PBM against cellular stress.

13.9 Photobiomodulation suppresses apoptosis induced by hypoxia

Cells undergoing apoptosis are characterized by activated caspase-3, and activated caspase-3 was much higher in the presence of $CoCl_2$. Activated caspase-3 was significantly reduced following PBM exposure (Fig. 13.8A). We postulated that PBM increased the mitochondrial membrane potential thus preventing cytochrome c leakage, and reducing caspase-3 activation and apoptosis. To confirm this, cytochrome c (*green*) and mitochondria (Mitotracker *red*) were separately costained, and the orange overlap was diminished after $CoCl_2$ incubation, giving largely nonoverlapping green and red colors, suggesting cytochrome c was released from mitochondria (Fig. 13.8B). As expected, PBM blunted cytochrome c leakage and helped to retain the cytochrome c inside the mitochondria in hypoxic cells (Fig. 13.8B). To further confirm the ability of PBM to prevent cytochrome c release from mitochondria, neurons were treated with two different apoptosis-inducing drugs: ABT-737 and PAC-1 that act upstream and downstream of Lampl (2007) cytochrome c release, respectively. Annexin V staining revealed that ABT-737 induced about 35% death in the cells after 24 hours treatment, but the death was reduced substantially after PBM treatment (Fig. 13.8C and D). By contrast, PBM did not show any effect on apoptosis induced by PAC-1. The results were consistent with the ability of PBM to sustain mitochondrial membrane potential and reduce apoptosis through prevention of cytochrome c leakage.

13.10 Hypoxia accelerates, but photobiomodulation protects against secondary brain injury

Damage to blood vessels including capillaries in the brain causes an immediate and dramatic decrease of cerebral blood flow (CBF) that can last for days (Lampl, 2007). We found partial or complete damage to many small blood vessels in the hippocampal region 3 hours after the injury (Fig. 13.9A). On average, around a third of the small blood vessels displayed varying degrees of damage, whereas no such damaged vessels were found in the control hippocampus (Fig. 13.9A). Conceivably, a decrease of CBF could disrupt the oxygen supply to the brain and result in cerebral

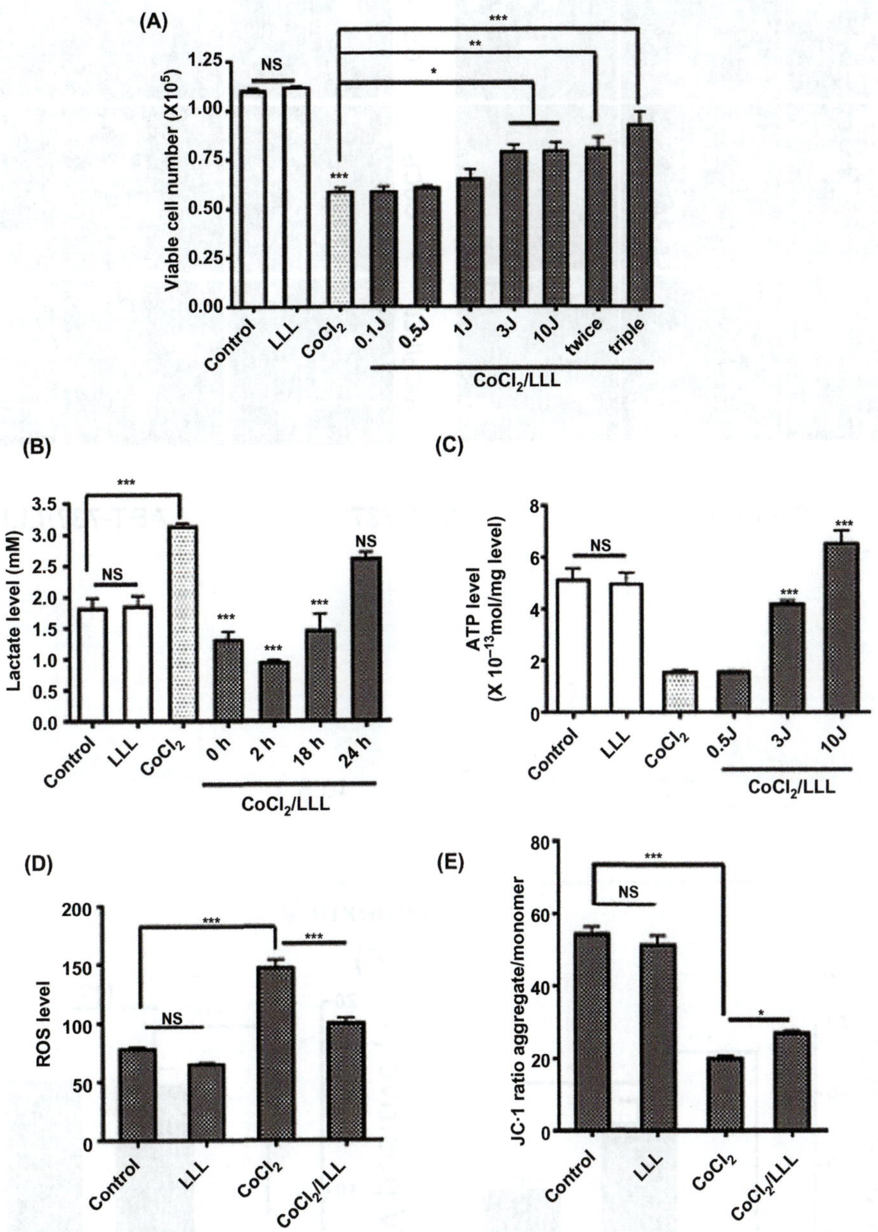

FIGURE 13.7 PBM maintains mitochondrial functions under hypoxia. (A) Prevention of hypoxia-induced cell death by PBM. SH-SY5Y cells were treated with $CoCl_2$ for 2 h followed by PBM illumination at indicated doses: twice, treatment with PBM at 3 J/cm^2 in 0 and 2 h post-$CoCl_2$ incubation and triple, PBM treatment at 0, 2, and 18 h post-$CoCl_2$ incubation. (B) Lactate production. The cells were treated with or without $CoCl_2$ and then illuminated with 3 J/cm^2 PBM at indicated times after $CoCl_2$ administration, and lactate accumulation was measured 48 h after $CoCl_2$ administration. (C) Effects of $CoCl_2$ and PBM on ATP production. The cells were subject to $CoCl_2$ treatment, followed by an indicated dose of PBM treatment in 2 h. ATP levels were measured 1 h after PBM illumination. ROS production (D) and mitochondrial membrane potential (E) were measured 4 h after $CoCl_2$ addition or 2 h after PBM treatment. The ratio of JC-1 aggregate (red) to monomer (green) fluorescence was calculated as measurement of mitochondrial membrane potential. All results were expressed as means ± SEM. $n = 8$; $*P < .05$, $**P < .01$, $***P < .001$, and *NS*, nonsignificance, respectively.

hypoxia. Although this pathological scenario is widely accepted, experimental evidence is somewhat lacking. We thus introduced hypoxia to the injured brain and compared resultant histological changes. To select an inducer for hypoxia, SH-SY5Y cells were treated with $CoCl_2$ or oxyrase, and hypoxia was measured by a hypoxia probe containing a nitro (NO_2) moiety that can be reduced in hypoxic cells leading to a release of the fluorescence probe. While oxyrase consumed oxygen rapidly and effectively inside and outside the cells, $CoCl_2$ must enter cells to be effective, which proved to be difficult and insufficient in tissues, in marked contrast to cells, and thus oxyrase was selected for subsequent studies (Gottschald et al., 2010). As shown in Fig. 13.9B, hypoxia was uniformly and specifically detected in hypoxic cultures as early as 20 minutes of oxyrase incubation. For in vivo study, oxyrase was applied topically to the cortex at the

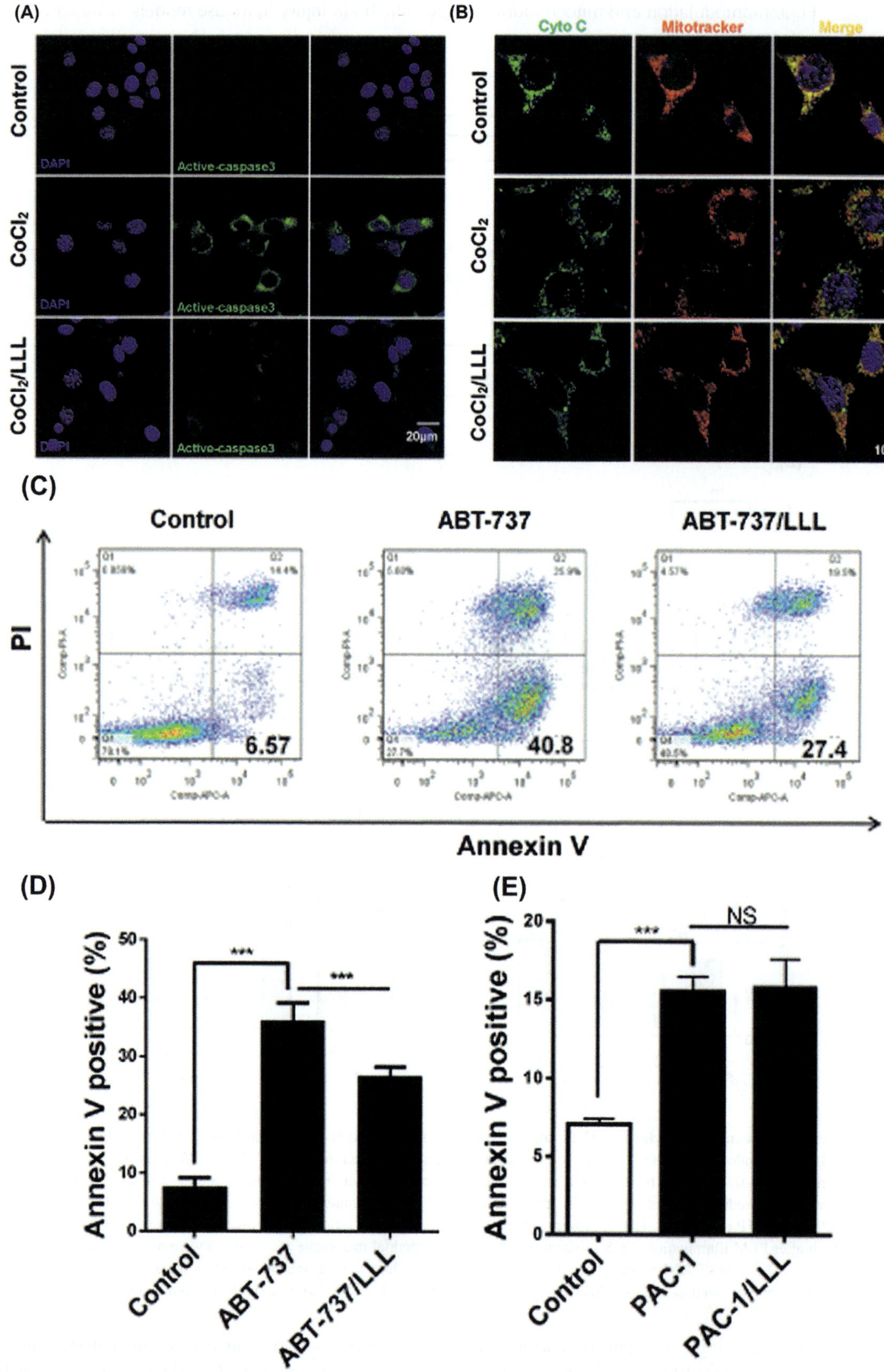

FIGURE 13.8 **PBM prevents cytochrome c release in hypoxic cells.** Representative images of active caspase-3 (A) and colocalization of cytochrome c and mitochondria (B). SH-SY5Y cells were cultured in complete medium (control) or in the medium containing $CoCl_2$ for 2 h, after which the cells were illuminated with PBM at 3 J/cm^2 as above. Caspase-3 activation and cytochrome c were identified by specific antibodies (*green*), cell nuclei were marked by DAPI (*blue*), and mitochondria were labeled by MitoTracker (*red*). Mitochondrial colocalization of cytochrome c is indicated by a merged orange color, arising from green and red in (B). PBM suppresses apoptosis induced by ABT-737 (C and D), but not by PAC-1 (E). SH-SY5Y cells were treated either with ABT-737 alone or along with PBM (3 J/cm^2) illumination (C and D). The cells were also treated similarly with PAC-1 alone or along with PBM illumination (E). The treated cells were stained with Annexin-V and PI in 24 h and analyzed by flow cytometry. Representative flow cytometry profiles are given in (C). Annexin V-positive and PI-negative cells are early apoptotic cells. Mean percentages ± SEM of early apoptotic cells are summarized in (D) and (E). $n = 8$; $***P < .001$; and *NS*, nonsignificance.

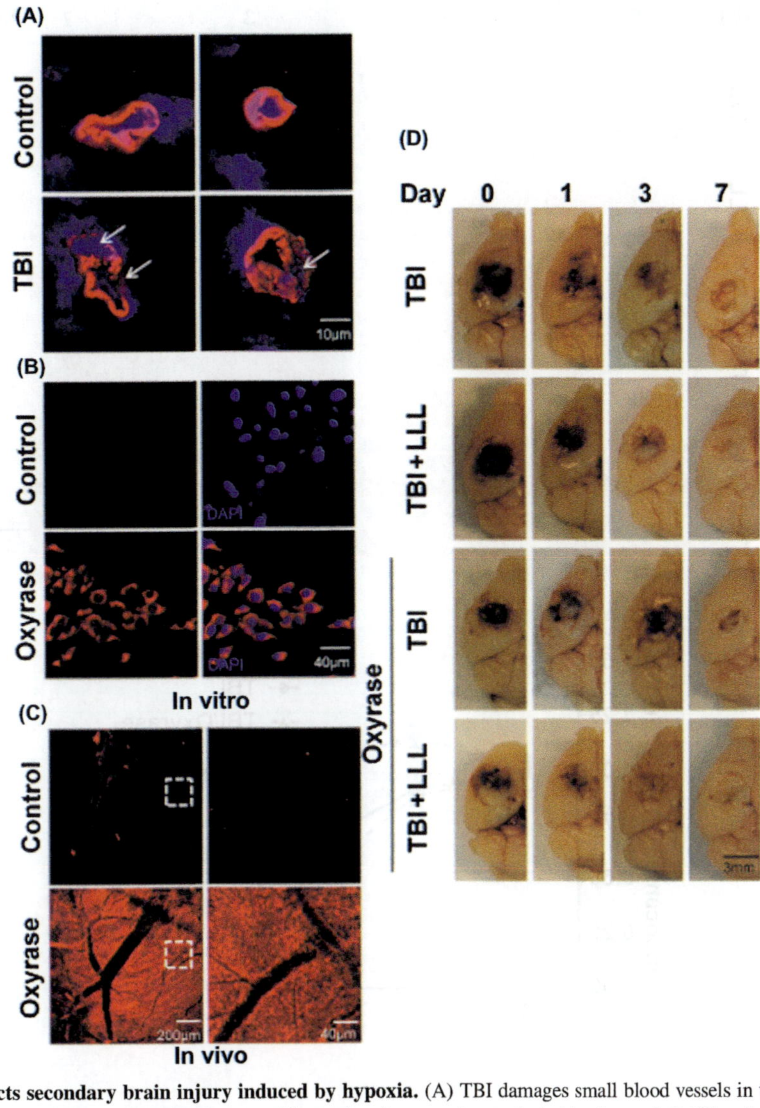

FIGURE 13.9 **PBM protects secondary brain injury induced by hypoxia.** (A) TBI damages small blood vessels in the hippocampus. Blood vessels in the hippocampus were identified by α-actin antibody, a specific marker for smooth muscle in blood vessel. DAPI stained cell nuclei in the vessel wall and arrows indicated the damage in the vessel. Shown in (A) are two representative small blood vessels from each group. Induction of hypoxia in SH-SY5Y cells (B) and injured brain (C). SH-SY5Y cells were incubated with oxyrase for 20 min and stained with a hypoxia-sensitive probe. The resultant red fluorescence in cytoplasma confirmed hypoxia in almost all the cells in the presence of oxyrase but not in the control (B). Likewise, strong hypoxia probe was identified in coronal sections of oxyrase-treated brain, but not in untreated brain (C). The area highlighted by a dashed white square in the left panel was enlarged in the right panel. (D) Gross morphology of the impact brain over 7 days of the experiment. The detrimental effects of hypoxia and beneficial effects of PBM on the injured brain were analyzed on the day of the injury (day 0), and days 1, 3, and 7 postinjury. (E) Histological examination of injured brain at indicated days after TBI. The hippocampus was outlined by the dashed black line: DG, dentate gyrus and CA, cornu amonis. The hippocampus ratios of the lesion side to the opposite side in the same mice in E were determined by image J and expressed as means ± SEM in (F).

injured site of the brain by dropping artificial cerebrospinal fluid (De Taboada et al., 2011) containing oxyrase. After 20 minutes of oxyrase dropping, a hypoxia probe was added to the injured site. The cranial window was closed by a glass cover and monitored by two-photon confocol microscopy. Severe hypoxia was observed at the site of oxyrase application, not at the control site, as shown by bright red fluorescence over the tissue (Fig. 13.9C). The exogenous hypoxia resulted in expansion and deepening of the lesion from day 1 and throughout the entire experimental period (Fig. 13.9D).

Histologically, the lesion treated with the hypoxic inducer was aggressively spreading from the cortex into the hippocampus and eventually led to an almost complete loss of the hippocampal dentate gyrus within 7 days (Fig. 13.9E). In the controls free of oxyrase, however, the lesion was contained mainly within the cortex and only a small surface portion of the hippocampus was affected, leaving the majority of the hippocampal dentate gyrus intact (Fig. 13.9E, upper). PBM treatment restrained the spreading of the lesion and completely protected not only the dentate gyrus, but also the entire hippocampus from injury, whether or not oxyrase was applied, with more pronounced effects in the presence of oxyrase. After PBM illumination, the lesions progressed much slower than non-PBM-treated controls both in the presence and absence of

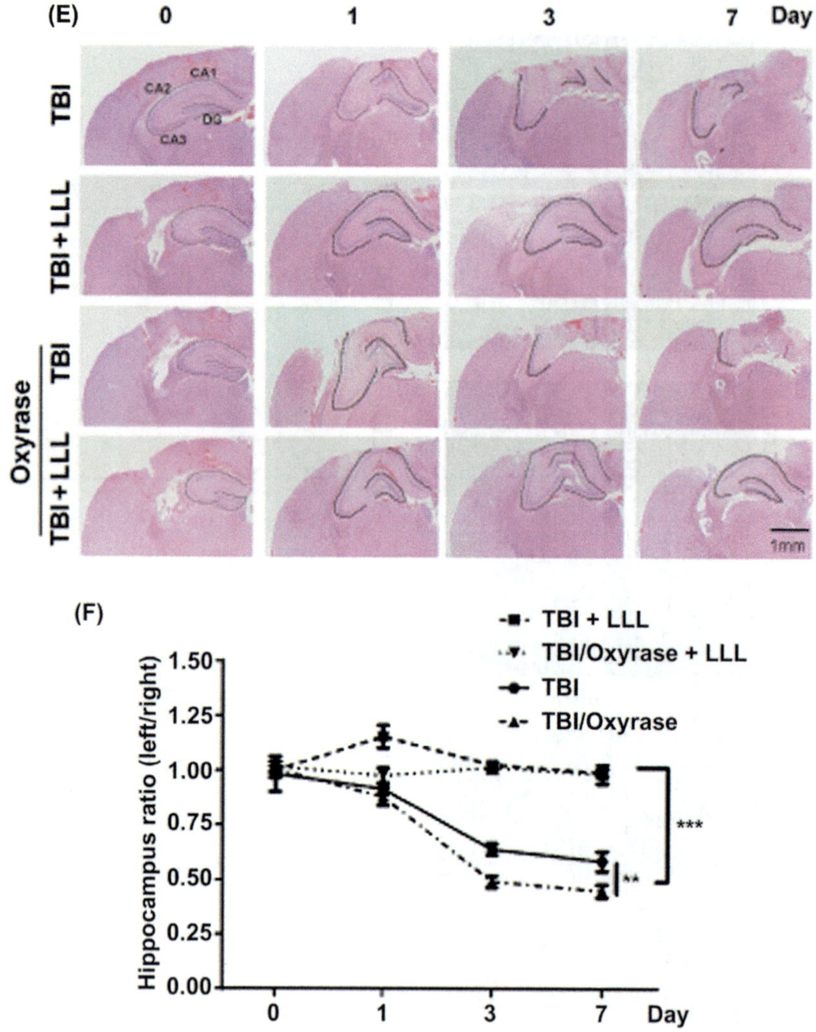

FIGURE 13.9 *(Continued)*

oxyrase (Fig. 13.9D and E). To quantify the severity of hippocampal lesions, we compared the size of hippocampus in the traumatic side with that in the opposite side. The volume of the hippocampus diminished considerably at day 3 and tissue loss reached a level as high as 50% on day 7 compared to the uninjured site, with a more profound loss in the presence of oxyrase (Fig. 13.9F). Strikingly, PBM treatment completely prevented the loss of hippocampal tissues both in the presence and absence of oxyrase. Notably, there was a slight increase in the size of hippocampus on day 1 in the injured brain following PBM treatment, due to edema. Edema also occurred in the other three groups and was embedded in the measurement owing to a loss of the hippocampal tissue in the groups. In other words, the loss of the hippocampal tissues was under-estimated in these groups. The results verify the detrimental effects of hypoxia on the pathogenesis of secondary brain injury and the ability of PBM to protect against secondary brain injury induced by hypoxia and other causes.

13.11 Mitochondrial functions are additively improved by the combination of photobiomodulation with lactate or pyruvate

To amplify therapeutic efficacy of PBM, we combined PBM with other mitochondria-improving metabolic agents like glucose, lactate, and pyruvate which are all substrates of the mitochondrial tricarboxylic acid (TCA) cycle. These substrates may be able to further augment oxidative energy metabolism under hypoxic conditions when combined with PBM. As shown in Fig. 13.10A, none of the three substrates raised ATP production in normoxic cultures, but they exhibited variable influences in hypoxic cultures, with a more predominant effect of lactate on ATP production in the culture. However, when combined with PBM, glucose or pyruvate increased oxidative phosphorylation in hypoxic

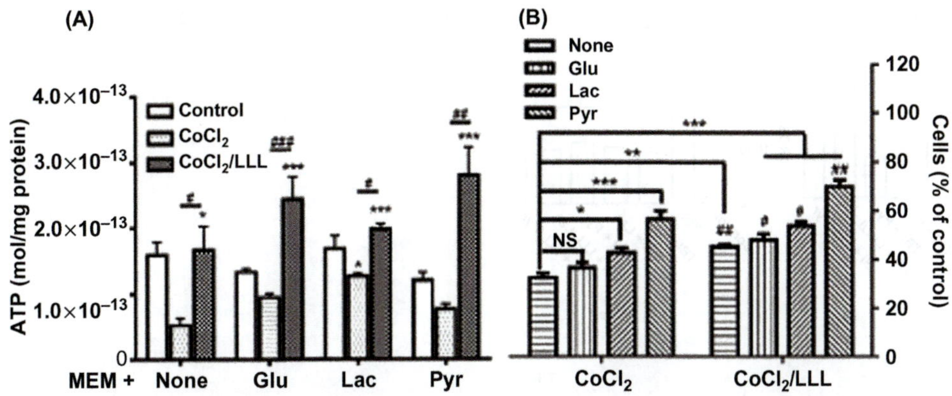

FIGURE 13.10 **Combination of PBM and lactate, pyruvate, or glucose improves mitochondrial functions of hypoxic cells.** SH-SY5Y cells were cultured in medium alone (none) or supplemented with glucose (Glu), pyruvate (Pyr), or lactate (Lac). To some of these cultures $CoCl_2$ was added, along with or without PBM illumination after 2 h of $CoCl_2$ incubation. ATP was measured 1 h following PBM treatment and expressed as obituary luminescence units normalized by protein concentrations (A). Viable cells were determined 48 h later and expressed as % of control cells cultured in medium alone (B). Note: $CoCl_2$ suppressed cell growth by more than 50% in medium alone control. *, **, and ***$P < .05$, .01, or .001 compared with $CoCl_2$ without treatment group, #, ##, and ###$P < .05$, .01, or .001 respectively, in the presence or absence of PBM, $n = 8$.

cultures more than lactate (Fig. 13.10A). Moreover, a combination of PBM with glucose, pyruvate, or lactate, protected cells from hypoxia-induced death significantly better than any single modality. Notably, the effect of PBM on cell survival was stronger in the presence of pyruvate than in the presence of glucose or lactate (Fig. 13.10B). These data suggest that PBM and energy metabolic modulators together could maximize ATP production and cell survival under a condition of hypoxia (Dong et al., 2015).

13.12 Photobiomodulation and lactate or pyruvate together fully protect the hippocampal tissue and its function

We showed that lactate or pyruvate was more effective than glucose in augmentation of PBM-mediated ATP production in vitro (Fig. 13.10). These two substrates were further evaluated in vivo in injured brain. Along with severe tissue loss, the injured site produced 75% less ATP than healthy brain and the impairment was only modestly restored by PBM, lactate or pyruvate alone (Fig. 13.11A). However, a combination of PBM and lactate or pyruvate synergistically or additively increased ATP formation, bringing it to a healthy level in the injured brain. Concurrent with reduced ATP production, the injured brain displayed a robust increase in ROS production that was however effectively suppressed by PBM illumination (Fig. 13.11B). Unlike the effects on ATP production, a combination of PBM with either pyruvate or lactate (Fig. 13.11B) did not further suppress ROS production. We next compared the morphological changes after single or combinational treatments. Histological examination revealed severe lesion concomitant with substantial brain tissue loss after TBI (Fig. 13.11C, *arrows*). The lesion was repaired modestly by treatment with PBM, lactate, or pyruvate alone as suggested by a less severe brain tissue loss measured by gross morphology (Fig. 13.11C) or lesion size (Fig. 13.11D). On the contrary, treatments of PBM and lactate or pyruvate resulted in a faster recovery and less brain tissue loss than any noncombinational treatment (Fig. 13.11C and D). TBI mice treated with PBM plus lactate or pyruvate recovered fully in 7 days post-TBI, whereas severe cortical lesion remained apparent in other groups of mice. Moreover, TBI–induced cortical lesion and inflammation aggressively spread from the cortex to hippocampus over 3 days post-TBI, causing severe neuron death (*white arrow*) and inflammation (*black arrow*) in the hippocampus in untreated control mice (Fig. 13.11E). Under similar conditions, the lesion progression into the hippocampus was completely blocked by treatment with PBM, along with either lactate or pyruvate (Fig. 13.11E), in marked contrast to only modest alleviation mediated by PBM, lactate, or pyruvate alone (Fig. 13.11E).

The hippocampal region of brain is considered to be essential for memory and spatial navigation. These activities were assessed by Morris water maze tests after two weeks of the initial brain injury. During the 6-day tests, it took a long escape time for untreated mice to find the hidden platform, and there was no significant improvement over the days of training in the mice, suggesting poor memory and learning (Fig. 13.11F), in agreement with a severe loss of the hippocampal tissue in the mice (Fig. 13.11E). PBM alone did not significantly shorten the escape latency in most of TBI mice. In contrast, the combinational treatment fully restored cognitive ability of TBI mice to a normal level (Fig. 13.11F). The ability of PBM and a metabolic substrate in restoration of the normal cognitive behaviors of the mice confirms full protection of hippocampal neurons from brain injury.

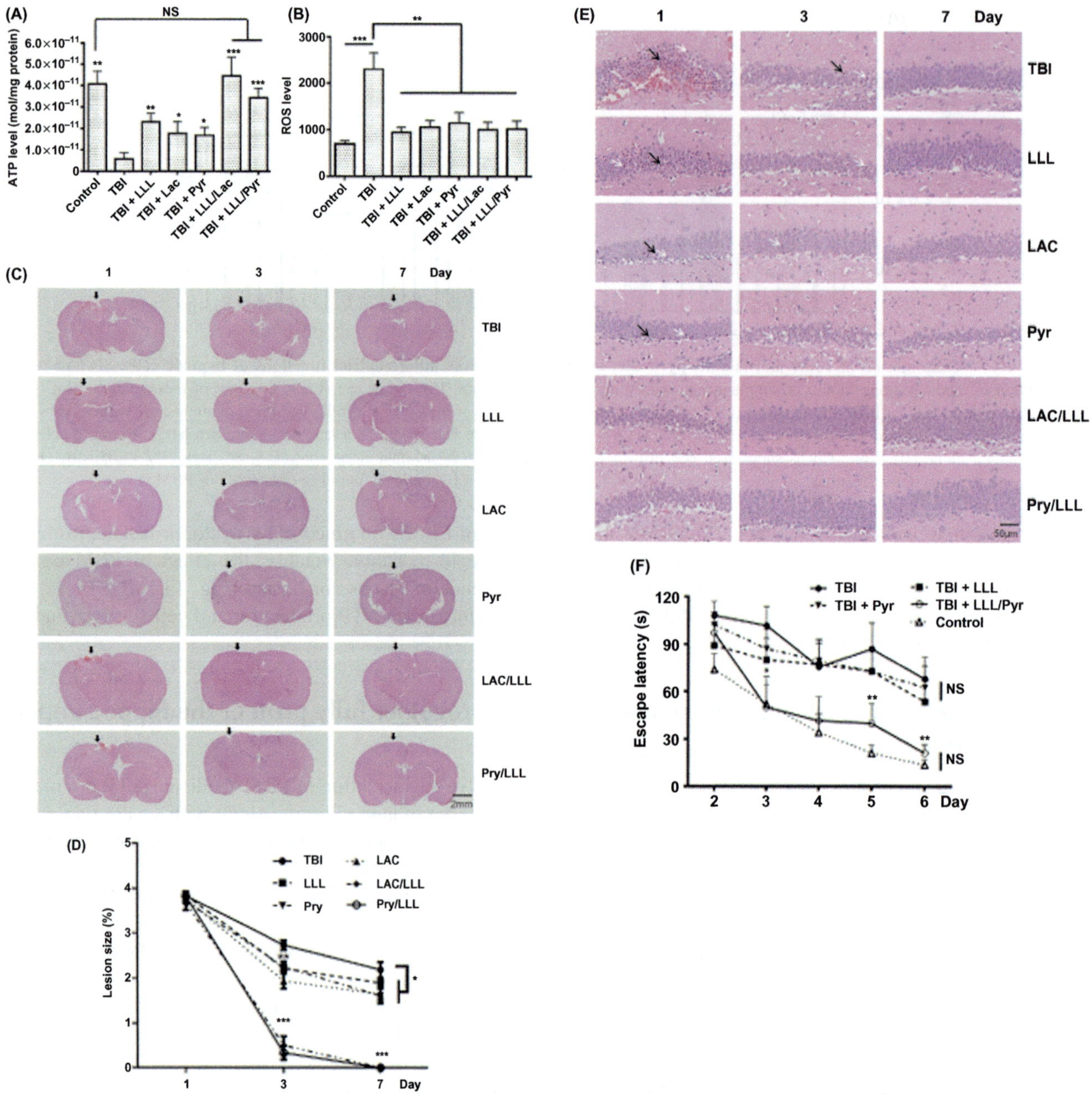

FIGURE 13.11 **PBM in combination with lactate or pyruvate fully protects the hippocampus from secondary damage in TBI.** ATP (A) and ROS (B) production in injured cortex. Cortical ATP and ROS were measured 5 h post-TBI with indicated treatments. Combinational treatments, as opposed to single treatment, sufficiently elevated the level of cortical ATP production in the injured brain to a normal level (A). ROS production was suppressed robustly by PBM alone, which was not furthered by combination with any substrate (B). (C) Histological examination of injured brains. The mice were intraperitoneally administered lactate or pyruvate 1 h post-TBI, or exposed to PBM 4 h after injury, or treated with both protocols. Coronal views showed a loss of brain tissues around injured site (*arrow*) over time following TBI, but the loss was effectively prevented by combinational therapy (Lac/PBM and Pyr/PBM), not by single therapy (PBM, Lac, or Pyr) when compared to untreated controls (TBI) (C). Quantitative analysis of the lesion sizes was carried out using image J and expressed as mean percentages ± SEM relative to the whole brain section (D). The region of the hippocampal dentate gyrus (DG) was analyzed on a high magnification (E). The black arrows indicate one of the infiltrated leukocytes, and the white arrows indicate one of the necrotic cells in each field. (F) Combinational therapy, but not single therapy protects the memory and learning functions of injured mice. Mice were subject to TBI, followed with indicated treatments as (C). After 2 weeks of the initial brain injury, these TBI-experienced mice were evaluated for their learning and memory by Morris water maze test. Latency to reach the platform over 6 test days was reduced in the mice after combinational treatment but not single treatment or no treatment. Data are expressed mean ± SEM. $n = 8$ for (A and B) $*P < .05$, $**P < .01$, and $***P < .001$ compared with TBI group, or $n = 9$ for (C–E); $*P < .05$, $**P < .01$, and $***P < .001$; and *NS*, nonsignificant.

13.13 Conclusion

Despite promising results from preclinical studies, potential TBI treatments have not yet translated into successful outcomes in clinical trials. Further studies are urgently needed to elucidate the mechanisms underlying secondary brain damage and the specificity of individual treatments. Our previous investigations showed that mice bearing inadequate activity of mitochondria as a sequel of IEX-1 null mutation were prone to secondary brain injury (Zhang et al., 2014). The observation underlines the role of mitochondrial activity in protection against secondary brain damage, because neurons are one of the most oxygen-sensitive cells types. Damage to blood vessels, especially capillaries, occurs frequently at the site of the injured brain, which drastically disrupts oxygen supply and reduces CBF. Severe and/or prolonged reductions in CBF lead to deprivations of oxygen and glucose, causing cerebral hypoxia and inadequate mitochondrial function. Data from retrospective studies and prospective clinical trials have shown that a high level of brain hypoxia is always linked to adverse outcomes for TBI. However, the heterogeneous pathophysiological alterations following TBI make it difficult to determine the precise effect of hypoxia on secondary brain damage. The subjection of TBI mice to systemic hypoxia aggravated neuronal death and enlarged the lesion size (Ishige et al., 1987), but it was not known whether local cerebral hypoxia that simulated brain injury in the clinics was also detrimental to TBI. This new TBI model provided direct experimental evidence with respect to adverse effects of hypoxia on secondary brain injury.

Confirmation of the adverse effect of hypoxia on secondary brain damage reinforces the importance of mitochondrial functions in protection of injured brain from secondary damage. The ability of PBM to sustain mitochondrial functions in the injured brain may account for part of the benefit of PBM in TBI treatment which has been demonstrated in this volume. Hypoxia increases cytochrome c release from mitochondria. Upon release from mitochondria, cytochrome c activates Apaf-1, caspase-9, and caspase-3, executing apoptosis via the intrinsic mitochondria-dependent pathway. By induction of apoptosis with two different agents, we showed that PBM could protect cells from apoptosis induced by ABT-737 that abrogated the antiapoptotic function of the Bcl-2 family upstream of cytochrome c release, but not PAC-1 that activates caspase-3 directly. The finding is consistent with the ability of PBM to sustain mitochondrial membrane potential. These investigations uncover the mechanism of PBM in protecting hypoxic neurons from apoptosis and in prevention of brain tissue loss.

The primary energy source for brain cells is glucose and it is transported from the blood into the brain and metabolized to lactate or pyruvate. Lactate and pyruvate can enter mitochondria and serve as substrates for the TCA cycle and oxidative phosphorylation (Belanger et al., 2011). For many years, lactate has traditionally been thought as a useless dead end product of anaerobic metabolism which could sometimes be harmful. Lactate elevation in the brain is associated with cerebral ischemic damage (Rehncrona et al., 1981). However, lactate and pyruvate can readily cross the blood–brain barrier and enter the TCA cycle (Schurr et al., 1988), being preferential oxidative energy substrates over glucose for neurons (Matsumoto et al., 1994). The beneficial effect of exogenous glucose, lactate, and pyruvate has been demonstrated in some TBI models (Moro et al., 2016; Shijo et al., 2015; Bouzat et al., 2014), but these metabolic substrates only modestly reduce the lesion size, and did not have any apparent functional protection in our model. But, when lactate or pyruvate was combined with PBM illumination, mitochondrial functions were improved either additively or synergistically, giving rise to a faster recovery and less brain tissue loss compared to PBM, pyruvate, or lactate alone. Most importantly, the hippocampal region was fully protected from the injury by combinational treatments, whereas in the controls the lesion in the cerebral cortex spreads into the hippocampus underneath. It is well known that the hippocampus takes a central role in the consolidation of information relating to short-term and long-term memory, as well as spatial navigation (Eichenbaum and Cohen, 1988). Although there are some neural stem cells in the adult mammalian hippocampus with the capacity to differentiate into neurons, astrocytes, and oligodendrocytes, differentiation of the cells hardly leads to fully functional recovery of the hippocampus, as shown by several investigations (Rolando and Taylor, 2014). Upon damage to the hippocampus, a severe loss of memory and difficulty in establishing new memories takes place. Therefore, the ability to protect the hippocampus from secondary damage is important. TBI is a complicated disease and the current investigation suggests that combinational treatment may yield a better outcome than a single treatment. The finding is clinically significant because of limited success in the current clinical treatment of TBI. Moreover, it is well known that mitochondrial dysfunction is involved in many other brain diseases, such as Alzheimer's, Parkinson's, amyotrophic lateral sclerosis, schizophrenia, etc. (Correia et al., 2010).

References

Belanger, M., Allaman, I., Magistretti, P.J., 2011. Brain energy metabolism: focus on astrocyte-neuron metabolic cooperation. Cell Metab. 14 (6), 724–738.

Bermpohl, D., You, Z., Lo, E.H., Kim, H.H., Whalen, M.J., 2007. TNF alpha and Fas mediate tissue damage and functional outcome after traumatic brain injury in mice. J. Cereb Blood Flow Metab. 27 (11), 1806–1818.

Bouzat, P., Sala, N., Suys, T., Zerlauth, J.B., Marques-Vidal, P., Feihl, F., et al., 2014. Cerebral metabolic effects of exogenous lactate supplementation on the injured human brain. Intensive Care Med. 40 (3), 412–421.

Bramlett, H.M., Dietrich, W.D., 2007. Progressive damage after brain and spinal cord injury: pathomechanisms and treatment strategies. Prog. Brain Res. 161, 125–141.

Budd, S.L., 1998. Mechanisms of neuronal damage in brain hypoxia/ischemia: focus on the role of mitochondrial calcium accumulation. Pharmacol. Ther. 80 (2), 203–229.

Campanella, M., Casswell, E., Chong, S., Farah, Z., Wieckowski, M.R., Abramov, A.Y., et al., 2008. Regulation of mitochondrial structure and function by the F1Fo-ATPase inhibitor protein, IF1. Cell. Metab. 8 (1), 13–25.

Chang, J.J., Youn, T.S., Benson, D., Mattick, H., Andrade, N., Harper, C.R., et al., 2009. Physiologic and functional outcome correlates of brain tissue hypoxia in traumatic brain injury. Crit. Care Med. 37 (1), 283–290.

Chiu, C.C., Liao, Y.E., Yang, L.Y., Wang, J.Y., Tweedie, D., Karnati, H.K., et al., 2016. Neuroinflammation in animal models of traumatic brain injury. J. Neurosci. Methods 272, 38–49.

Correia, S.C., Carvalho, C., Cardoso, S., Santos, R.X., Santos, M.S., Oliveira, C.R., et al., 2010. Mitochondrial preconditioning: a potential neuroprotective strategy. Front. Aging Neurosci. 2.

De Taboada, L., Yu, J., El-Amouri, S., Gattoni-Celli, S., Richieri, S., McCarthy, T., et al., 2011. Transcranial laser therapy attenuates amyloid-beta peptide neuropathology in amyloid-beta protein precursor transgenic mice. J. Alzheimers Dis. 23 (3), 521–535.

Dong, T., Zhang, Q., Hamblin, M.R., Wu, M.X., 2015. Low-level light in combination with metabolic modulators for effective therapy of injured brain. J. Cereb. Blood Flow Metab.

Eichenbaum, H., Cohen, N.J., 1988. Representation in the hippocampus: what do hippocampal neurons code? Trends Neurosci. 11 (6), 244–248.

Finnie, J.W., 2013. Neuroinflammation: beneficial and detrimental effects after traumatic brain injury. Inflammopharmacology 21 (4), 309–320.

Gottschald, O.R., Malec, V., Krasteva, G., Hasan, D., Kamlah, F., Herold, S., et al., 2010. TIAR and TIA-1 mRNA-binding proteins co-aggregate under conditions of rapid oxygen decline and extreme hypoxia and suppress the HIF-1alpha pathway. J. Mol. Cell Biol. 2 (6), 345–356.

Hamblin, M.R., 2016a. Photobiomodulation and the brain – has the light dawned? The Biochemist 38 (6), 24–33.

Hamblin, M.R., 2016b. Shining light on the head: Photobiomodulation for brain disorders. BBA Clin. 6, 113–124.

Hamblin, M.R., 2018. Photobiomodulation for traumatic brain injury and stroke. J. Neurosci. Res. 96, 731–743.

Hennessy, M., Hamblin, M.R., 2017. Photobiomodulation and the brain: a new paradigm. J. Opt. 19 (1), 013003.

Ishige, N., Pitts, L.H., Berry, I., Carlson, S.G., Nishimura, M.C., Moseley, M.E., et al., 1987. The effect of hypoxia on traumatic head injury in rats: alterations in neurologic function, brain edema, and cerebral blood flow. J. Cereb. Blood Flow Metab. 7 (6), 759–767.

Khuman, J., Zhang, J., Park, J., Carroll, J.D., Donahue, C., Whalen, M.J., 2012. Low-level laser light therapy improves cognitive deficits and inhibits microglial activation after controlled cortical impact in mice. J. Neurotrauma 29 (2), 408–417.

Kondratyev, A.D., Chung, K.N., Jung, M.O., 1996. Identification and characterization of a radiation-inducible glycosylated human early-response gene. Cancer Res. 56 (7), 1498–1502.

Lampl, Y., 2007. Laser treatment for stroke. Expert Rev. Neurother. 7 (8), 961–965.

Longhi, L., Pagan, F., Valeriani, V., Magnoni, S., Zanier, E.R., Conte, V., et al., 2007. Monitoring brain tissue oxygen tension in brain-injured patients reveals hypoxic episodes in normal-appearing and in peri-focal tissue. Intensive Care Med. 33 (12), 2136–2142.

Matsumoto, K., Yamada, K., Kohmura, E., Kinoshita, A., Hayakawa, T., 1994. Role of pyruvate in ischaemia-like conditions on cultured neurons. Neurol. Res. 16 (6), 460–464.

Moro, N., Ghavim, S.S., Harris, N.G., Hovda, D.A., Sutton, R.L., 2016. Pyruvate treatment attenuates cerebral metabolic depression and neuronal loss after experimental traumatic brain injury. Brain Res. 1642, 270–277.

Oddo, M., Levine, J.M., Mackenzie, L., Frangos, S., Feihl, F., Kasner, S.E., et al., 2011. Brain hypoxia is associated with short-term outcome after severe traumatic brain injury independently of intracranial hypertension and low cerebral perfusion pressure. Neurosurgery 69 (5), 1037–1045. discussion 45.

Piret, J.P., Mottet, D., Raes, M., Michiels, C., 2002. CoCl2, a chemical inducer of hypoxia-inducible factor-1, and hypoxia reduce apoptotic cell death in hepatoma cell line HepG2. Ann. N.Y. Acad. Sci. 973, 443–447.

Qin, L., Wu, X., Block, M.L., Liu, Y., Breese, G.R., Hong, J.S., et al., 2007. Systemic LPS causes chronic neuroinflammation and progressive neurodegeneration. Glia 55 (5), 453–462.

Rehncrona, S., Rosen, I., Siesjo, B.K., 1981. Brain lactic acidosis and ischemic cell damage: 1. Biochemistry and neurophysiology. J. Cereb. Blood Flow Metab. 1 (3), 297–311.

Rolando, C., Taylor, V., 2014. Neural stem cell of the hippocampus: development, physiology regulation, and dysfunction in disease. Curr. Top. Dev. Biol. 107, 183–206.

Schilling, D., Pittelkow, M.R., Kumar, R., 2001. IEX-1, an immediate early gene, increases the rate of apoptosis in keratinocytes. Oncogene 20 (55), 7992–7997.

Schurr, A., West, C.A., Rigor, B.M., 1988. Lactate-supported synaptic function in the rat hippocampal slice preparation. Science 240 (4857), 1326–1328.

Shen, L., Guo, J., Santos-Berrios, C., Wu, M.X., 2006. Distinct domains for anti- and pro-apoptotic activities of IEX-1. J. Biol. Chem. 281 (22), 15304–15311.

Shen, L., Zhi, L., Hu, W., Wu, M.X., 2009. IEX-1 targets mitochondrial F1Fo-ATPase inhibitor for degradation. Cell Death Differ. 16 (4), 603–612.

Shijo, K., Ghavim, S., Harris, N.G., Hovda, D.A., Sutton, R.L., 2015. Glucose administration after traumatic brain injury exerts some benefits and no adverse effects on behavioral and histological outcomes. Brain Res. 1614, 94–104.

Stahel, P.F., Shohami, E., Younis, F.M., Kariya, K., Otto, V.I., Lenzlinger, P.M., et al., 2000. Experimental closed head injury: analysis of neurological outcome, blood-brain barrier dysfunction, intracranial neutrophil infiltration, and neuronal cell death in mice deficient in genes for pro-inflammatory cytokines. J. Cereb. Blood Flow Metab. 20 (2), 369–380.

Wu, M.X., 2003. Roles of the stress-induced gene IEX-1 in regulation of cell death and oncogenesis. Apoptosis 8 (1), 11–18.

Yan, E.B., Satgunaseelan, L., Paul, E., Bye, N., Nguyen, P., Agyapomaa, D., et al., 2014. Post-traumatic hypoxia is associated with prolonged cerebral cytokine production, higher serum biomarker levels, and poor outcome in patients with severe traumatic brain injury. J. Neurotrauma 31 (7), 618–629.

Zhang, Q., Zhou, C., Hamblin, M.R., Wu, M.X., 2014. Low-level laser therapy effectively prevents secondary brain injury induced by immediate early responsive gene X-1 deficiency. J. Cereb. Blood Flow Metab. 34, 1391–1401.

Zhang, Y., Schlossman, S.F., Edwards, R.A., Ou, C.N., Gu, J., Wu, M.X., 2002. Impaired apoptosis, extended duration of immune responses, and a lupus-like autoimmune disease in IEX-1-transgenic mice. Proc. Natl. Acad. Sci. U.S.A. 99 (2), 878–883.

Zhi, L., Ustyugova, I.V., Chen, X., Zhang, Q., Wu, M.X., 2012. Enhanced Th17 differentiation and aggravated arthritis in IEX-1-deficient mice by mitochondrial reactive oxygen species-mediated signaling. J. Immunol. 189 (4), 1639–1647.

Chapter 14

Photobiomodulation for depression in animal models

Farzad Salehpour[1,2], Javad Mahmoudi[1], Saeed Sadigh-Eteghad[1] and Paolo Cassano[3,4,5]

[1]Neurosciences Research Center, Tabriz University of Medical Sciences, Tabriz, Iran, [2]ProNeuroLIGHT LLC, Phoenix, AZ, United States, [3]Department of Psychiatry, Harvard Medical School, Boston, MA, United States, [4]Depression Clinical and Research Program, Department of Psychiatry, Massachusetts General Hospital, Boston, MA, United States, [5]Center for Anxiety and Traumatic Stress Disorders, Department of Psychiatry, Massachusetts General Hospital, Boston, MA, United States

14.1 Introduction

Recently, photobiomodulation (PBM) has been proposed as an innovative, potential treatment for a wide range of neurological and psychiatric disorders (Salehpour et al., 2018b). Brain PBM, using the transcranial irradiation method, is still experimental and yet an increasingly common approach in centers specializing in PBM (Rojas and Gonzalez-Lima, 2013). In transcranial PBM, owing to partial transmission of photons through the scalp and skull, only a small amount of light energy reaches the cortical surface, yet enough to produce cellular, neurophysiological, and therapeutic effects. Cytochrome c oxidase (COX), the main PBM photoacceptor, is the terminal enzyme in the mitochondrial electron-transport chain; its peak absorption encompasses both red and near-infrared (NIR) light, approximately 600–880 nm wavelength (Karu and Kolyakov, 2005). However, due to greater penetration through the skull of the NIR over the visible spectrum, wavelengths above 800 nm are most frequently used in transcranial PBM studies (Lapchak, 2012). Transcranial PBM increases regional cerebral blood flow (CBF) in depressed patients (Schiffer et al., 2009) and augments cerebral energy metabolism in healthy subjects (Wang et al., 2017). Increase in cerebral ATP levels and focal increase in regional CBF potentially contribute to recovery in mood disorders (Kato et al., 1992). In this chapter, we summarize the animal studies testing the antiantidepressant effects of transcranial PBM in various models of depression.

14.2 Major depressive disorder

14.2.1 The extent of the problem

The term depression refers both to a transient mood state experienced by almost all people at some time during life and to clinically diagnosed syndromes such as major depressive disorder (MDD). MDD is a debilitating disorder that includes abnormalities of mood, sleep, appetite, energy, and cognitive and psychomotor functions (Fava and Kendler, 2000). Moreover, feelings of guilt, anxious states, and repeated thoughts of death and suicide are common among patients who suffer from MDD (Nestler et al., 2002). MDD has a lifetime prevalence of 16.2% in the world's adult population and is twice as common in women as in men (Kessler et al., 2007). In addition, estimates of the point prevalence of MDD are 2% in children and 4%–8% in adolescents (Costello et al., 2003). MDD is known to pose an elevated risk of suicide (Manji et al., 2001).

14.2.2 Pathophysiology of major depressive disorder

In terms of neurobiology and pathophysiology of MDD, hereafter we reviewed some of the most accredited theories.

Photobiomodulation in the Brain. DOI: https://doi.org/10.1016/B978-0-12-815305-5.00014-2

14.2.2.1 Neurotransmitter systems

The monoaminergic neurotransmitter systems are thought to play a central role in the MDD pathology. The alterations of monoaminergic systems [5-hydroxytryptamine (serotonin), norepinephrine (NE), and dopamine (DA)] throughout the network of limbic, striatal, and prefrontal cortical circuits are related to emotional and cognitive manifestations of MDD (Manji et al., 2001). A strong body of evidence has shown reduced levels of serotonin and its metabolites in blood, cerebrospinal fluid (CSF), and postmortem brain tissue in MDD patients. Deficiency of serotonin in the neuronal synapse is associated with a variety of neurobehavioral dysfunctions such as changes in mood, sleep, and eating; aggressive behaviors and suicide attempts (Jacobsen et al., 2012). Studies on neuronal pathways have also shown a key role of NE in executive functions (cognition), social interactions, and enhancement of reward and motivation during pleasurable activities, such as eating and having sex (Manji et al., 2001). In depressed patients, low CSF levels of a DA metabolite, the homovanillic acid, were associated with an increased risk for suicidal behavior (Roy et al., 1989). Diminished interest or pleasure (also called anhedonia), a cardinal manifestation of MDD, has been related to dysfunctions in the reward system, and in particular to dysfunction in the DA system (Der-Avakian and Markou, 2012). Animal models of depression have revealed altered DA release in the nucleus accumbens (Di Chiara and Tanda, 1997) and altered DA receptor expression within limbic regions (Dziedzicka-Wasylewska et al., 1997). In addition, decreased striatal dopaminergic activity was observed in depressed patients (Pruessner et al., 2004).

Studies have suggested that an imbalance of amino acid neurotransmitters, glutamate and aminobutyric acid (GABA), signal results in neural over- or underactivation, and subsequent atrophy of neurons and loss of glia in the cortical and limbic structures (Charney et al., 2013). Studies have also shown alterations of glutamate levels in blood, CSF, or brain tissues of MDD patients (Yüksel and Öngür, 2010). Studies using proton magnetic resonance spectroscopy (MRS) have also demonstrated decreased levels of Glx (a composite of glutamate and glutamine) in different brain regions of depressed patients, including the dorsolateral, dorsomedial, and ventromedial prefrontal cortex (PFC), anterior cingulate, amygdala, and hippocampus (Yüksel and Öngür, 2010). With respect to GABA, depressed patients showed lower plasma, CSF, and brain GABA levels than nondepressed comparison individuals (Sanacora and Saricicek, 2007). Studies using proton MRS have also reported decreased levels of GABA in occipital and frontal lobes of brain in depressed patients (Sanacora and Saricicek, 2007). Moreover, extensive loss of GABA-ergic interneurons in the dorsolateral PFC (Maciag et al., 2010) and lower levels of glutamic acid decarboxylase (a key enzyme in the synthesis of GABA) (Karolewicz et al., 2010) have been found in MDD patients.

14.2.2.2 Cerebral blood flow

The coupling of CBF and neuronal cells' metabolic needs is considered to be necessary for normal function in the brain (Dirnagl et al., 1993). Studies in patients with MDD have revealed a low CBF in the PFC, paralimbic, bilateral temporal, and anterior parietal regions (Galynker et al., 1998). These abnormalities in regional blood flow are likely related to the low brain energy metabolism found in MDD patients (Cassano et al., 2016; Moore et al., 1997; Moylan et al., 2013; Volz et al., 1998). Nitric oxide (NO), an inorganic and gaseous neurotransmitter, mediates CBF regulation (Dirnagl et al., 1993; Selley, 2004; Toda et al., 2009), and as Dirnagl et al. (1993) have shown, blockade of NO production through inhibition of NO synthase (NOS) reduces regional CBF by approximately 50% in rats. Given its ability for inducing continual and potent basal vasodilator tone and given its short half-life, NO is a suitable mediator for the coupling of regional CBF and neuronal activity, with both high temporal and spatial resolution (Dirnagl et al., 1993). Postmortem studies conducted by Bernstein et al., demonstrated that NOS immunopositive neurons are significantly reduced in hypothalamus of subjects with depression (Bernstein et al., 1998). NO deficiency, which not only likely alters brain perfusion, but also is expected to influence the release of regulatory peptides including vasopressin, oxytocin, and CRH, which in turn might contribute to disorders such as depression (Bernstein et al., 1998; Purba et al., 1996; Raadsheer et al., 1995). Given these preclinical and clinical evidence, we can speculate on the enhancement of cerebral NO content as a mechanism for amelioration of depressive symptoms.

14.2.2.3 Cerebral bioenergetics

According to neuroimaging findings, glucose consumption levels are reduced in the brain of MDD patients and antidepressant regimens are able to revert these abnormalities (Cassano et al., 2016). Mitochondria have roles in cellular major activities including energy generation, broadly metabolism, apoptotic cell death, and intracellular signaling (Tobe, 2013). The putative role of mitochondria in the generation of energy is of great neuropsychiatric importance because it is well documented that mitochondrial dysfunction is associated with pathological and behavioral changes observed in MDD. Increased levels of proinflammatory cytokine in neuroinflammation may disrupt mitochondrial function through suppression of mitochondrial respiratory chain enzymes (Samavati et al., 2008; Stadler et al., 1992;

Zell et al., 1997). In addition, excessive generation of reactive oxygen species (ROS) impairs ATP production through oxidative damage to mitochondrial structural and enzymatic components (Halliwell, 2006; Wagner et al., 1990). Gardner et al. (2003) for the first time detected biochemical abnormalities occurring in mitochondria of depressed subjects; so far mitochondrial ATP production (MAPRs) rate and mitochondrial respiratory chain enzyme ratios (NAD-cytochrome-c-reductase/COX and succinate-cytochrome reductase/COX) were significantly reduced (Gardner et al., 2003). In the same study, the authors found a significant correlation between low MAPRs and endorsement of somatic anxiety, "psychasthenia" (phobias, obsessions, compulsions, or excessive anxiety), and suspicion at the Karolinska Scales of Personality (KSP), indicating patient's vulnerability to psychopathology. Moreover, these psychosomatic manifestations have been reported in subjects with mitochondrial diseases (Chinnery and Turnbull, 1997). Besides, impaired mitochondrial function negatively affects neurogenesis and cell survival (Voloboueva and Giffard, 2011), ultimately leading to aberrant hippocampal remodeling in MDD patients (Kempermann, 2002).

14.2.2.4 Oxidative stress

Neurons are amongst the most susceptible cells that are affected by either free radicals overproduction or deficiencies in antioxidant scavenging systems. This condition leads to oxidative stress burden which significantly impacts the function of neuronal components including lipids, proteins, and nucleic acids (Moretti et al., 2012; Wang and Michaelis, 2010). Evidence suggests that oxidative stress may have a crucial role in the pathology of psychiatric conditions such as schizophrenia, bipolar disorder, anxiety, and MDD (Pandya et al., 2013). Studies have shown that exposure to chronic and unpredictable stress induces depression-like behaviors in rodents (Moretti et al., 2012), while also enhancing cerebral oxidative stress (Fontella et al., 2005; Lucca et al., 2009). It seems that such effects in part are mediated through an imbalance between prooxidant (lipid peroxidation) and antioxidant agents [i.e., total antioxidant capacity and levels of antioxidant enzymes, including superoxide dismutase (SOD), catalase (CAT) and glutathione peroxidase (GPx)] (Fontella et al., 2005; Moretti et al., 2012). In this respect, Moretti et al. (2012) showed that administration of antioxidant regimens effectively improved motivational behaviors of mice under repeated unpredictable stress, while also reducing oxidative burden in their hippocampus and cortex. Clinical studies evaluating oxidative stress status in MDD patients showed deficits in blood levels of SOD, CAT, and GPx and increased levels of lipid peroxidation compared to healthy subjects (Bilici et al., 2001; Khanzode et al., 2003; Ozcan et al., 2004). Taken together, it seems that alterations in oxidative stress parameters might contribute to the pathogenesis of stress-related disorders, such as MDD.

14.2.2.5 Neuroinflammation

There is abundant evidence that shows inflammatory responses are involved in the pathophysiology of MDD. The correlation between depression and neuronal inflammatory responses—macrophage theory of depression—was first hypothesized by Smith (1991). According to his hypothesis, the increase in cytokines levels [in particular interleukin (IL) IL-1] released by macrophages could be accompanied by depression. In the brain, microglial cells respond to inflammatory signals through production of proinflammatory cytokines including IL-1β, IL-2, IL-6, interferon-γ (IFN-γ), and tumor necrosis factor-α (TNF-α) (Dowlati et al., 2010; Kim et al., 2016). These cytokines not only facilitate inflammatory responses (Kim and Maes, 2003) but also contribute to memory formation, synaptic plasticity, as well as neurotransmitter metabolism and mood control (Du et al., 2008; Villanueva, 2013). In general, cytokines have low homeostatic concentrations in the brain (Pitossi et al., 1997); imbalances of cytokines are associated with immunological and psychological changes (Anisman et al., 2002). For instance, four consecutive weeks of chronic stress in rats increased cerebral levels of proinflammatory cytokines (IL-1β, TNF-α, and IL-6) and simultaneously reduced antiinflammatory cytokines (TGF-β and IL-10), while also inducing depressive-like behaviors and slowing neurogenesis (You et al., 2011). Moreover, systemic administration of the lipopolysaccharide and/or cytokines, such as IL-1 and TNF-α, induces long-lasting neuroinflammation characterized by elevation of proinflammatory cytokines (Dunn et al., 2005; Kent et al., 1992; Qin et al., 2007). Neuroinflammation leads to sickness behavior in rodents such as anhedonia, appetite loss, reduced exploratory and motor function as well as impaired sexual and cognitive function (Dunn et al., 2005; Krishnan and Nestler, 2008; Maes et al., 2009). Notably, occurrence of imbalance between pro- and antiinflammatory cytokines in favor of cerebral inflammation leads to activation of the hypothalamic-pituitary-adrenal axis, to an increase in central monoamine catabolism and to mood dysregulation (Licinio et al., 2007; Loftis et al., 2010).

14.2.2.6 Neurotrophic factors and neurogenesis

Neurotrophic factors are a group of small proteins that mediate survival, maturation, and differentiation of neurons by binding to specific kinase receptors. They regulate neurogenesis, synaptic plasticity, and neuronal repair, as well as maintenance of neuronal connectivity (Nasrolahi et al., 2018). Among these factors, glial cell-derived neurotrophic factor (GDNF), nerve growth factor (NGF), and brain-derived neurotrophic factor (BDNF) have attracted significant attention for their regulatory effects on neurogenesis during treatment of MDD (Berry et al., 2012; Karege et al., 2002; Lin and Tseng, 2015).

Neurogenesis is a dynamic and continual process occurring upon environmental stimuli, and is characterized by the differentiation and survival of neural progenitor cells into newly born neurons (Haughey et al., 2002; Kim et al., 2008). These newborn neurons form synapses with preexisting neurons and extend the functional connectivity of the neuronal networks (Taupin, 2006). Stress is a precipitant factor of MDD down-regulates hippocampal neurogenesis (Sen et al., 2008) and results in subsequent hippocampal mal-adaptive changes and atrophy (Campbell et al., 2004; Malberg et al., 2000; Watanabe et al., 1992). Interestingly, a reduction in hippocampal size has been reported both in stressed animals (Watanabe et al., 1992) as well as in the brain of depressed patients (Campbell et al., 2004; MacMaster and Kusumakar, 2004). In contrast, antidepressant regimens avert this effect possibly through inducing neurotrophic expression and subsequent neurogenesis (Karege et al., 2002; Malberg et al., 2000; Sen et al., 2008; Taliaz et al., 2010). Clinical findings have shown lower levels of GDNF (Diniz et al., 2013), NGF (Wiener et al., 2015), as well as BDNF (Karege et al., 2002) in MDD patients than healthy subjects. Beside this, a reduced amount of neurotrophic factors may be linked to symptoms such as locomotor slowing, agitation, and anxiety observed in patients with MDD (Karege et al., 2002). Notably, preclinical studies with rodents have shown that infusion of BDNF into the cerebral structures including midbrain, hippocampus, or lateral ventricles induce antidepressant effects (Duman and Monteggia, 2006; Sen et al., 2008). Naumenko et al. (2014) have demonstrated that intracerebral delivery of NGF improved spatial memory impairment in antidepressant sensitive cataleptics mice; however, it did not produce an antidepressant effect. These findings suggest a role for NGF and BDNF in the regulation of brain plastic changes to overcome environmental stimuli (Berry et al., 2012).

14.2.3 Animal models of depression and photobiomodulation studies

Psychiatric syndrome of MDD cannot be entirely recapitulated in animal models, however, some aspects of this syndrome have been mimicked in rodents, and in several instances improved with antidepressant treatment (Krishnan and Nestler, 2011). Animal models used in neuroscience and psychiatric studies should be valid in terms of sensitivity to change elicited by known antidepressants and in terms of selectivity for the specific animal species in use, and they should be simple and easy to replicate (Porsolt et al., 1991).

14.2.3.1 Pharmacological models

There are a variety of pharmacological agents such as reserpine, apomorphine, yohimbine, and clonidine that have been used for depression modeling in animal studies (Porsolt et al., 1991). However, the reserpine model is the most frequent pharmacological approach for this purpose. Reserpine is an irreversible inhibitor of the vesicular monoamine transporter 2, which diminishes brain monoamines (Antkiewicz-Michaluk et al., 2014). It has been shown that reserpine in acute (Huang et al., 2004) and chronic (Antkiewicz-Michaluk et al., 2014) administrations could recapitulate behavioral aspects of depression in animals. High doses of drug (6 or 8 mg/kg) could mimic the disease phenomena in periods of 24–48 hours (Huang et al., 2004; Minor and Hanff, 2015), while lower doses (0.1 or 0.2 mg/kg) show the same phenomena after 14 days of intraperitoneal (i.p.) administration in rodents (Ikram and Haleem, 2017; Mohammed, 2016).

A study conducted by Mohammed showed beneficial effects of PBM on chronic reserpine (0.2 mg/kg i.p. for 14 days) model of depression. PBM effectively improved depression induced by reserpine as indicated by behavioral and electrophysiological parameters (Mohammed, 2016).

14.2.3.2 Restraint stress

Psychological models of depression are valid models since they entirely rely on innate social behavior (Krishnan and Nestler, 2011). This group of depression models are suitable candidates for chronic but not acute treatments (Porsolt et al., 1991). One of the most important paradigms of psychological models is restraint stress. In this model, animals are restricted in narrow tubes, restrainers, or cages, for a period of 15 minutes to 6 hours a day (Stepanichev et al., 2014). This model depends on daily restraint time and may take 2 weeks to 2 months for induction (Lee et al., 2013; Nagata et al., 2009). Recently, Xu et al. (2017) demonstrated that brain PBM therapy effectively improves depression-like behaviors and impaired molecular signaling in the restraint stress (2 hours every day for 2 weeks) model of mice.

14.2.3.3 Chronic mild stress

Stressful conditions are considered as a major determinant for the development of mood disorders in humankind (Kendler et al., 1999). Likewise, exposure to chronic stress or even to a single stress episode may cause specific depression-like behavioral changes in rodents (Stepanichev et al., 2014). The chronic mild stress (CMS) paradigm includes a variety of mild unpredictable stressors in a random order over several weeks (Willner, 2017). The stressors include isolation and paired housing, food deprivation, exposure to an empty water bottle, cold stress, overnight stroboscopic illumination, cage tilting at 45 degrees, soil bedding, white noise, restraint and forced swimming (Overstreet, 2012). The study conducted by Salehpour et al. (2016) revealed protective effects of transcranial PBM therapy on CMS-induced (different stress situations for 4 weeks) depressive-like behaviors. Similarly, another study provided by Wu et al. (2012) showed that NIR laser could ameliorate depression symptoms in CMS (mild, unpredictable stressors for 8 weeks) model.

14.2.3.4 Transgenic models

Among all models of depression, genetic models are amenable to the study of its underlying mechanisms. It has been shown that the mutation of Abelson helper integrationsite-1 (AHI1) gene is associated with neurodevelopmental and psychiatric disorders (Ren et al., 2014). The evidence demonstrated that conditional Ahi1 gene knockout could cause depressive-like behaviors in mice. This might result from the dysregulation of the TrkB signaling pathway (Xu et al., 2010). In this respect, Xu et al. (2017) showed that transcranial PBM could improve behavioral outcomes in the aforementioned genetic model of depression.

14.2.3.5 Traumatic brain injury-induced depression

There is evidence that posttraumatic stress disorder is common following traumatic brain injury (TBI) (Perez-Garcia et al., 2018). Also, depression is the most frequent psychiatric manifestation after TBI (Fleminger et al., 2003). Among TBI models, fluid percussion, controlled cortical-impact, weight-drop impact acceleration, and blast models are widely used in rodents (Xiong et al., 2013). Although TBI-induced depression has not been fully studied in animal models, Ando et al. (2011) study showed that NIR PBM could ameliorate depression-like behavior in a TBI model induced by a controlled cortical-impact device.

14.2.3.6 Other models

Some inconsistency in data has been reported by different depression studies, which could be a consequence of differences of stress paradigms or of their realization in different laboratories. Additional rodent models of depression such as social stress, early life stress, learned helplessness, or fear conditioning in PBM studies could further our understanding of the pathogenesis of depression and of the underlying protective mechanisms of PBM.

14.2.4 Behavioral tests used in depression and photobiomodulation studies

14.2.4.1 Forced swimming test

The forced swimming test (FST) is a preclinical behavioral test in rodents for assessment of antidepressant drugs, compounds, and approaches that are aimed at preventing depressive-like states (Can et al., 2012a). In this test, a rat or mouse is forced to swim in a narrow cylinder filled with water from which there is no escape. The cylinder diameter and height differ depending on animal species. Different studies have reported one or two steps training for this test, however, currently, the one-step test is more utilized, which lasts 6–10 minutes. After an initial period of forceful movement, the animal assumes a characteristic immobile posture. In this condition, the animal moves only when necessary to keep its head above the water, which is considered as "behavioral despair." Immobility time in FST is the depression index in animals (Castagné et al., 2010). This test has been frequently used in depression and PBM studies and antidepressant effects of different doses and types of low-level NIR light have been documented with this task (Mohammed, 2016; Salehpour et al., 2016; Wu et al., 2012).

14.2.4.2 Tail suspension test

The tail suspension test (TST) is a behavioral task for the screening of potential antidepressant agents in mice (Mahmoudi et al., 2015). This test is based on the fact that short-term, inescapable stress of suspension of animals by their tails leads to an immobile posture, and antidepressant agents typically reduce the frequency and duration of immobility (Cryan et al., 2005). Although sophisticated modes of delivery improve the quality of data in TST, basically, the procedure only requires a suspension bar and tape (Can et al., 2012b). The task is a one-step test and duration is usually 6 minutes. The TST as a behavioral task has been used in the PBM and depression mice models conducted by Xu et al. (2017) and Ando et al. (2011).

14.3 Photobiomodulation therapy

14.3.1 Introduction to photobiomodulation therapy

PBM therapy, also known as, low-level laser/light therapy (LLLT), has been introduced as an innovative modality for the stimulation of biological processes (Hamblin and Demidova, 2006). PBM involves exposing cells or tissues to coherent laser or noncoherent light emitting diode (LED) photons in the wavelength range of red to NIR (600–1200 nm) (Chung et al., 2012). PBM therapy applies light at lower power and energy densities (total irradiance of $<700\ mW/cm^2$ and fluencies of $0.04-120\ J/cm^2$ at scalp surface) as compared to other forms of light application in medicine that are used for ablation, cutting, and thermally coagulating tissue (Chung et al., 2012). In order to improve cerebral function, many light delivering approaches have been used in the literature, including transcranial (Rojas and Gonzalez-Lima, 2013), intracranial (Moro et al., 2014), and intranasal irradiation methods (Saltmarche et al., 2017). Brain PBM therapy using transcranial irradiation method is a new approach in specialized settings (Salehpour et al., 2018b). In this method, owing to partial transmission of photons through the scalp and skull, an amount of light energy—sufficient to produce therapeutic effects—could reach the cortical surface and in some instances penetrate into deeper structures of the brain (Morries et al., 2015). The potential neurostimulatory benefits of PBM therapy have been reported in a variety of preclinical and clinical studies including animal models of TBI (Xuan et al., 2015), ischemic stroke (Lee et al., 2017b), Alzheimer diseases (AD) (Blivet et al., 2018), Parkinson disease (PD) (Darlot et al., 2016), and aging (Salehpour et al., 2017), as well as patients with TBI (Naeser et al., 2014), acute stroke (Zivin et al., 2009), dementia (Saltmarche et al., 2017), and healthy individuals (Blanco et al., 2017). Furthermore, the beneficial effects of the NIR PBM therapy in psychiatric disorders, and more specifically on symptoms of depression and anxiety, have been recently shown (Cassano et al., 2015, 2016).

14.3.2 Mechanisms of photobiomodulation therapy

14.3.2.1 Molecular and cellular action mechanisms

At a molecular level, the absorption of red/NIR light by the unit IV of the mitochondrial respiratory chain (COX), and the subsequent increase of its activity are considered as the main mechanism of PBM (Hamblin, 2017b). In pathological conditions, unhealthy or hypoxic cells might have more amounts of NO, which partially inhibits mitochondrial respiration by displacement of O_2 from COX. It is suggested that PBM leads to photodissociation of NO from the COX, which in turn results in an increase of mitochondrial membrane potential (MMP), intracellular Ca^{2+} ions, proton gradient, O_2 consumption, and subsequent increase in ATP production (Hamblin and Demidova, 2006; Karu et al., 2005b). During the ATP synthesis, O_2 molecule acts as the last electron acceptor in the electron-transport chain and is converted to water molecule. Part of the O_2 that is metabolized generates ROS as natural by-products. This brief burst of low-intensity mitochondrial ROS generation following PBM acts as a redox signal and initiates a retrograde signaling pathway from the mitochondria to the nucleus (de Freitas and Hamblin, 2016). Brief and modest levels of ROS activates several transcription factors including NF-κB that collectively upregulate many protective genes and resulting in long-lasting effects on cells, including proliferation, survival, and migration (Song et al., 2003). Moreover, NIR PBM can differentiate host stem cells to promote tissue regeneration via ROS-mediated activation of transforming growth factor–β1 (TGF-β1) signaling pathway (Arany et al., 2014).

14.3.2.2 Neuroprotective mechanisms

14.3.2.2.1 Cerebral blood flow

As mentioned above, NO is a well-known powerful vasodilator, which could be released by photodissociation process from its binding sites in COX during PBM therapy (Karu et al., 2005a). It is believed that changes in the cerebral perfusion following PBM therapy are associated with a release of NO and a consequential increase of CBF. Preclinical and clinical studies have shown that red/NIR light irradiation to specific areas of the brain potentially increase the neuronal NO content and improve regional CBF (Lee et al., 2017a; Salgado et al., 2015; Uozumi et al., 2010). It has been shown that red LED light irradiation significantly increases NO synthesis by modulating COX/NO activity (Ball et al., 2011). According to the animal studies, transcranial NIR irradiation at 808 nm increases CBF in the illuminated/nonilluminated hemispheres as well as cortical NO content in the rat (Uozumi et al., 2010). Also, it has been shown that pretreatment with 610 nm LED light results in an acute increase in CBF in mouse cerebral ischemic model (Lee et al., 2017a). With respect to clinical studies, transcranial LED therapy (at 850 nm), bilaterally to the forehead, increased CBF in the left anterior frontal lobe of a patient in a vegetative state (Nawashiro et al., 2012). Likewise, a study in MDD patients showed that transcranial LED irradiation (at 810 nm) to the forehead could partially increase the prefrontal CBF immediately after treatment (Schiffer et al., 2009). Moreover, increased CBF in the left middle cerebral artery and the basilar artery has been reported following transcranial LED therapy (at 627 nm) in healthy individuals (Salgado et al., 2015).

14.3.2.2.2 Cerebral bioenergetics

Neural tissue is very rich in mitochondria and has a great dependence on mitochondrial ATP (Schwarz, 2013). As noted earlier, the primary mechanism of action of red/NIR PBM therapy is through the increase of ATP production by boosting mitochondrial enzyme activity and bioenergetic function. LED light at wavelengths of 670 and 830 nm remarkably upregulated mitochondrial COX activity in cultured visual cortical neurons (Wong-Riley et al., 2005). Also, an 633 nm LED light irradiation of rat head was able to significantly increase the COX activity in the whole brain (Rojas et al., 2008) as well as—in case of selective irradiation—in specific brain regions such as PFC (Rojas et al., 2012) and superior colliculus (Rojas et al., 2008). Likewise, the restoration of mitochondrial COX expression patterns in the neocortex and hippocampus have been shown in mouse AD model, following long-term LED therapy (at 670 nm) (Purushothuman et al., 2014). In addition, LED therapy (at 808 nm) significantly increased the COX activity in the PFC (Xu et al., 2017) and hippocampus (Lu et al., 2017) in vivo.

In terms of ATP content, it has been reported that an 808 nm transcranial laser therapy increases cerebral ATP levels in various mouse models including transgenic (De Taboada et al., 2011) and amyloid-β (Aβ)-induced (Lu et al., 2017) AD models as well as in depression model (Xu et al., 2017). An 830 nm laser irradiation directly to the parietal cortical neurons of naive rat resulted in an increase of ATP/ADP ratio (Mochizuki-Oda et al., 2002). In addition, it has been shown that an 808 nm transcranial laser therapy, whether continuous wave (CW) (Lapchak and Boitano, 2016) or 100-Hz pulsed wave (PW) (Lapchak and De Taboada, 2010) mode, increases cortical ATP levels in a rabbit embolic stroke model. Also, an 808 nm laser light significantly improved cerebral bioenergetics in the naive dog, as evaluated by MRS (Mintzopoulos et al., 2017).

Since red/NIR light stimulates COX enzyme activity and accelerates electron transfer rate, it can be expected that following PBM therapy, cerebral mitochondrial oxidative phosphorylation and O_2 consumption will increase. Indeed, a study in naive rat brain has shown that transcranial LED therapy (at 660 nm) can increase O_2 consumption in the PFC (Rojas et al., 2012). Furthermore, transcranial laser therapy (at 808 nm) resulted in an increase in O_2 consumption in the whole brain of transgenic AD mice (De Taboada et al., 2011). To date, only limited clinical studies in healthy subjects have been designed to evaluate the effect of transcranial NIR PBM on cerebral metabolism and blood O_2 supply. Laser irradiation (at 1064 nm) transcranially to the right forehead resulted in a rapid increase of oxidized COX and subsequent increase in CBF, blood volume, and blood oxygenation (Wang et al., 2017). Likewise, it has been shown that the application of aforementioned laser to the center and right side of the forehead improves cerebral oxygenation in both right and left PFC (Tian et al., 2016).

14.3.2.2.3 Neuronal antioxidant defence

It is well established that mitochondria are the main source of free radicals (ROS), and excessive ROS production affects neuronal tissue in part by damaging their mitochondrial function (Bhat et al., 2015). On the other hand, studies have suggested that beneficial effects of PBM are in part linked to modest amounts of mitochondrial ROS generation (Chen et al., 2009). It is believed that absorption of red/NIR light at the optimum fluence (energy density at target tissue) of $<10\ J/cm^2$ produces a brief burst of low-level mitochondrial ROS accompanied by a rise in MMP (Sharma et al., 2011). Indeed, the physiological levels of ROS have a crucial role in mediating cell signaling pathways involved in cell survival and proliferation (Sena and Chandel, 2012). Mitochondrial low-intensity ROS production allows mutual communication between mitochondria and the cytosol and/or the nucleus and influences gene expression through the induction of transcription factors (Zhang et al., 2001). It has been postulated that mitochondrial ROS induced by PBM at low doses could stimulate the expression of genes related to antioxidant defense systems, through the ROS-mediated signaling pathways (Song et al., 2003). In vitro studies have shown the neuroprotective effects of PBM against oxidative stress in several neurotoxic models induced by Aβ (Yang et al., 2010), H_2O_2 (Huang et al., 2013), $CoCl_2$ (Dong et al., 2015; Huang et al., 2013), and rotenone (Huang et al., 2013). The antioxidant properties of NIR PBM have been shown in some animal studies. A study in a mouse model of sleep deprivation showed an improvement in hippocampal total antioxidant capacity and activities of antioxidant enzymes such as SOD and GPx, as well as decrease in ROS and malondialdehyde contents following acute treatment with 810 nm laser (Salehpour et al., 2018a). Moreover, suppression of NOS isoforms activity (endothelial, neuronal, and inducible) following 660 nm laser irradiation has been put forward as a possible mechanism responsible for PBM regulation of oxidative stress in a rat model of cerebral ischemia (Leung et al., 2002).

14.3.2.2.4 Neuroinflammation

In response to different types of neuronal insult, proinflammatory markers including chemokines, cytokines (TNF-α and several ILs), NO, and ROS are excessively released by microglia. One of the interesting benefits of red/NIR PBM that has recently emerged, is its pronounced antiinflammatory action. Several studies have shown an antineuroinflammatory action of PBM in various animal models of brain disorder. A study in a rat model of cryogenic brain injury indicated that 660 or 780 nm laser irradiation directly to the lesion site decreased levels of IL-1β at 24 hours postinjury. However, this study found no changes in brain TNF-α levels after PBM with mentioned lasers at both time points (Moreira et al., 2009). Transcranial NIR laser at 810 nm down-regulated the expression of some proinflammatory chemokine such as CC-chemokine ligand 2 (CCL2) and CXC-chemokine ligand 10 (CXCL10) in mice brain at both time points of 6 hours and 28 days postclosed head injury. However, this work showed disparate short- and long-term inflammatory responses in terms of cytokine expression. NIR laser reduced cerebral levels of IL-1β at 6 hours and of TNF-α at 28 days, but increased TNF-α levels at 6 hours (Zhang et al., 2014). Transcranial LED therapy using 610 nm light modulated ischemia-induced inflammatory response in mice via reduction of IL-1β and IL-18 levels (at 72 hours postischemia) (Lee et al., 2017b). Pretreatment with 610 nm LED light also suppressed neuroinflammation in mice through inhibition of cortical TNF-α and IL-1β production at 24 hours poststroke (Lee et al., 2016). Furthermore, daily administration of transcranial 808 nm laser for 7 days, suppressed cortical proinflammatory cytokines, TNF-α, IL-6, and IL-18 and elevated antiinflammatory cytokines, IL-4 and IL-10 in rat at 14 days poststroke (Yang et al., 2018).

It has been proposed that PBM-induced modulation of NO, ROS, cyclic AMP (cAMP), and Ca^{2+} are involved in the antiinflammatory effects of red/NIR light (Hamblin, 2017a). In addition, NIR laser light decreases proinflammatory cytokines and subsequent inflammatory responses, possibly mediated by cAMP and inhibited NF-κB signaling pathway (Chen et al., 2011). Taken together, it could be stated that the antineuroinflammatory actions of PBM may at least partly be due to its ability to modulate microglial activity and a subsequent reduction in inflammatory mediators (Lee et al., 2017b; Yang et al., 2018).

14.3.2.2.5 Neurotrophic factors and neurogenesis

It is believed that improved expression of neurotrophins such as BDNF and NGF may account for observations of stimulation of hippocampal neurogenesis and synaptogenesis (Telerman et al., 2011). In addition, improved cerebral BDNF expression could contribute to reduced atrophy and cell loss in the hippocampus and PFC in MDD (Martinowich et al., 2007).

To date, among the several members of the family of neurotrophic factors (neurotrophins), recent literature has been focused on the stimulatory effect of PBM on BDNF, NGF, and GDNF. PBM (at 632.8 nm) has been shown to induce intracellular inositol trisphosphate (IP3) receptor activation resulting in intracellular Ca^{2+} releases and consequent activation of ERK/CREB pathway, which eventually improved BDNF expression in primary cultured neurons (Yan et al., 2017). Laser irradiation using the aforementioned wavelength, also rescued dendritic atrophy and neuronal survival through activation of the ERK/CREB/BDNF pathway in Aβ-treated neurons (Meng et al., 2013). A study in a rat neurotoxicity model has shown a remarkable improvement in BDNF expression in the occipital cortex following 670 nm transcranial laser therapy (Ghanbari et al., 2017). Intracranial LED therapy with 670 nm light also increased GDNF expression in the striatum region of the brain in monkey model of PD (El Massri et al., 2017).

Moreover, some studies indicated that PBM could promote neurogenesis and synaptogenesis, and exert a positive effect on the neural progenitor cells differentiation, in turn, reduce long-term neurological deficits following brain injury. In this respect, in a notable series of studies in mice TBI model, Xuan et al. have shown that transcranial PBM (at 810 nm) markedly promoted neurogenesis and upregulated migrating neural progenitor cells, as well as BDNF in the dentate gyrus (DG) and subventricular zone (SVZ) regions, and stimulated synaptogenesis and neuroplasticity in the cortex (Xuan et al., 2013, 2014, 2015, 2016). Also, studies in the rat stroke models have shown that transcranial PBM (at 808 nm) can induce neurogenesis and migration of neural progenitor cells in the SVZ region (Oron et al., 2006) and inhibit stroke-induced dendritic and synaptic injury as well as enhance cortical neurogenesis (Yang et al., 2018). Hippocampal atrophy and neurogenesis deficits in the DG have been shown in MDD (Campbell and MacQueen, 2004; Mueller et al., 2010). Based on the previously mentioned evidence, it could be postulated that PBM might benefit depressive brain structure and function in terms of neurogenesis and synaptogenesis.

14.3.2.2.6 Cerebral neurotransmitters

As mentioned earlier, an imbalance between the three main monoamine neurotransmitters namely, serotonin, DA, and NE in the CNS has been implicated in the pathophysiology of MDD (Bressan and Crippa, 2005; Elhwuegi, 2004;

Moret and Briley, 2011). Although direct evidence has not yet been provided for the effects of PBM therapy on neurotransmitters in depression models, regulation of some brain neurotransmitters following red (632.8 nm) laser irradiation has been observed in two different studies. Shu-Zhi and Li-Hua showed that low-level laser to the rat brain decreased striatal DA, serotonin, and glutamic acid, and increased striatal GABA levels in the caudate nucleus. Furthermore, they also showed that irradiation to the frontal region decreased serotonin and its metabolites and increased aspartic acid and GABA levels (Shu-Zhi and Li-Hua, 1982). Increased serotonin and GABA, as well as decreased glutamate in the rat striatum and hippocampus has been also shown following red laser irradiation (Lombard et al., 1990). Moreover, red and NIR light (808 and 830 nm) modulated levels of different amino acid neurotransmitters in the brain cortical and hippocampus regions (Ahmed et al., 2008; Radwan et al., 2009). It is possible that brain PBM therapy stimulates the neuronal mitochondria which in turn lead to enhanced ATP production with subsequent improvement in synaptic transmission and neurotransmitters. Further studies are required to test the feasibility of this modality in regulation of the neurotransmitters, which are involved in mood disorders.

14.3.3 Translational photobiomodulation studies in depression animal models

Several in vivo studies have shown that transcranial NIR PBM therapy has antidepressant effects on various animal models of depression. Of the six depression studies presented here describing PBM therapy in mice (Ando et al., 2011; Xu et al., 2017) and rats (Mohammed, 2016; Salehpour et al., 2016; Tanaka et al., 2011; Wu et al., 2012), four reported beneficial effects in animal depression models, one in depression following brain trauma (Ando et al., 2011), and one in naïve animals (Tanaka et al., 2011). Table 14.1 summarizes the published animal studies in the depression models treated with transcranial PBM therapy.

In 2011, Tanaka et al. (2011) applied the multiwavelength infrared light (600–1600 nm) to the naïve rat head and evaluated the possible antidepressant and anxiolytic effects. After 10 consecutive days of irradiation, depressive-like behavior significantly decreased as indicated by increased mobility time in FST. Both acute (single session) and chronic (10 sessions) administration of infrared light increased number of BrdU-positive cells, as a marker for neurogenesis, in cornus ammonis (CA1) of the hippocampus. In addition, anxiety-like behaviors markedly decreased in the chronic group as assessed by elevated plus maze (EPM) and light/dark tests.

In 2011, Ando et al. (2011) for the first time attempted to evaluate possible antidepressant benefits of NIR laser (810 nm, CW and PW at 10- or 100-Hz modes) on depressive symptoms after TBI in a mouse model. After one session treatment with a cortical fluence of $\sim$2.2 J/cm^2 (36 J/cm^2 on the scalp), their results from neurobehavioral performances at 4 weeks posttraumatic event showed a superiority of 10-Hz mode over 100-Hz and CW modes in the reduction of immobility time in FST and TST.

In 2012, Wu et al. (2012) studied the beneficial effects of NIR laser (810 nm, PW at 100-Hz mode) in a rat CMS depression model and compared results of PBM therapy with pharmacotherapy. They used much higher fluence (120 J/cm^2 on the scalp) and treatment sessions (nine sessions) as compared to Ando et al.'s work (2011). According to their results, antidepressant effect of NIR laser was as effective as Fluoxetine as represented by significant improvements in depressive-like behaviors in FST at 3 weeks posttreatment, including decreased immobility and increased swimming time.

In 2016, Salehpour et al. (2016) compared antidepressant effects of transcranial PBM therapy using two different wavelengths (630 and 810 nm) at 10-Hz PW mode with pharmacotherapy in a rat model of depression based on unpredictable CMS. Their results showed that after 12 sessions PBM therapy with a cortical fluence of $\sim$1.2 J/cm^2 per each session, only 810 nm laser had antidepressant-like effects similar to Citalopram as indicated by increased swimming time and decreased immobility time in FST. In addition, 630 nm laser decreased serum cortisol levels and 810 nm laser increased body weight gain at the end of the experiment.

In 2016, Mohammed (2016) used an 804 nm laser at CW mode in a rat pharmacological depression model and showed a significant reduction in immobility time in FST and modulation of electrocorticogram spectra after 14 days irradiation using six points on the scalp (38.4 J/cm^2 per point). In addition, the laser power of 80 mW was reported as the best therapeutic power compared to 200 and 400 mW within the same irradiation time, for the amelioration of depressive-like symptoms. Given this, the biphasic response of low, intermediate, and high levels of laser irradiation on the animal's swimming activity in FST showed that the optimum (38.4 J/cm^2) and highest (190.8 J/cm^2) dosage of laser irradiation had stimulatory and inverse effects, respectively.

Finally, in 2017, Xu et al. (2017) showed antidepressant-like effects of 808 nm laser at CW mode in restraint stress and genetic mutation mouse models of depression. Although mitochondrial complex I–III levels did not exhibit any significant change in brain PFC, hippocampus, and hypothalamus regions, significant improvements in ATP synthesis

TABLE 14.1 Reports of transcranial photobiomodulation therapy used for major depressive disorder in animal models.

Date	Model	Species	Light source	Wavelengths	Treatment parameters	Treatment approach/ irradiation site	Findings	References
2011	No induction of depression	Rat	Infrared emitter	600–1600 nm	Single dose for 3 min (acute) Multiple dose for 3 min daily for 10 sessions (chronic)	Transcranially; Whole head	Decreased depressive-like behavior in chronic group as indicated by: increased mobility time in FST Decreased anxiety-like behaviors in chronic group as indicated by: increase of entry into and time spent in open arm in EPM; increase of entry into and time spent in light chamber in light/dark test Increased number of BrdU-positive cells, as a marker for neurogenesis, in CA1 of the hippocampus in acute and chronic groups	Tanaka et al. (2011)
2011	Depression following TBI	Mouse	Laser, DioDent Micro 810, HOYA ConBio (Fremont, CA, USA)	810 nm	3.5 W, 50 mW/cm^2, 36 J/cm^2, CW or 10 or 100-Hz PW (duty cycle; 50%), single session for 12 min	Transcutaneously	Improved body weight gain, decreased immobility time in FST and TST at week 4	Ando et al. (2011)
2012	CMS model	Rat	Laser, PhotoThera, Inc. (Carlsbad, CA, USA)	810 nm	350 mW, 120 J/cm^2, 100-Hz PW (duty cycle; 20%), 2 min per session, for 9 sessions	Transcranially; At the midline of the dorsal surface in region between eyes and ears (shaved scalp)	Improved body weight gain, decreased immobility and increased swimming time in FST at week 3	Wu et al. (2012)
2016	CMS model	Rat	Laser, Mustang 2000 + (Moscow, Russia)	630 or 810 nm	89 mW/cm^2 and 562 mW/cm^2, 6 J/cm^2 and 2.8 J/cm^2 with corresponding wavelengths of 630 and 810 nm, respectively. 10-Hz PW (duty cycle; 50%), 12 sessions	Transcranially; At the midline of the dorsal surface in prefrontal region (shaved scalp)	Improved body weight gain, decreased immobility and increased swimming time in FST, decreased blood cortisol and glucose levels	Salehpour et al. (2016)
2016	Pharmacological model (reserpine, 0.2 mg/kg i.p.)	Rat	Laser, Lasotronic Inc. (Zug, Switzerland)	804 nm	80 mW, 640 mW/cm^2, 230 J/cm^2 (total points of 6 on the scalp), CW, 6 min per session, for 14 sessions	Transcranially; At six points arranged symmetrically, three on each side of the skull (shaved scalp)	Decreased immobility and increased swimming time in FST, regulated the EEG in all wave frequency bands except for theta	Mohammed (2016)
2017	Space restriction model, Ahi1 KO model	Mouse	Laser, quartz-silica, fiber, Shenzhen Fuzhe Technology Co., Ltd. (Shenzhen, China)	808 nm	30 mW, 23 mW/cm^2, 41.4 J/cm^2, CW, 30 min per session, for 28 sessions	Transcranially; Fiber optics on the top of the head (shaved scalp)	Decreased immobility time in FST and TST in both models at day 28, increased ATP synthesis in PFC, increased mitochondrial complex IV level in PFC, increased mitochondrial complex IV activity in PFC	Xu et al. (2017)

Ahi1 KO, Abelson helper integration site-1 knockout; *ATP*, adenosine triphosphate; *BrdU*, Bromodeoxyuridine; *CA1*, cornus ammonis; *CMS*, chronic mild stress; *CW*, continues wave; *EPM*, elevated plus maze; *EEG*, electroencephalogram; *FST*, forced swimming test; *NSS*, neurological severity score; *PW*, pulsed wave; *PFC*, prefrontal cortex; *TBI*, traumatic brain injury; *TST*, tail suspension test.

and enhanced mitochondrial complex IV level and the activity within the PFC region were observed following 28 days irradiation using a daily fluence of 41.4 J/cm^2 (on the scalp). Besides molecular insights, their neurobehavioral findings demonstrated reductions in depressive-like performance in FST and TST after 28 days treatment for restraint stress model and after 14 days treatment for genetic mutation model.

Different values of light fluence at scalp surface from 2.8 to 190.8 J/cm^2 per session were used in transcranial PBM therapy for animal models of depression. It seems that the fluencies in the range about 1–2 J/cm^2 on the cortical surface per each session are enough for effective antidepressant-like benefits in transcranial PBM therapy in rodent depression model. Moreover, in terms of pulse frequency, beneficial effects of PW—compared to CW—laser light at frequencies ranging from 10- to 100-Hz in the improvement of spatial memory in AD mice model (De Taboada et al., 2011) and neurobehavioral performance of TBI mice model (Ando et al., 2011) have been reported. Likewise, both studies of Salehpour et al. (2016) and Ando et al. (2011) in animal depression models, also revealed significant antidepressant effects of pulsed light therapy, particularly 10-Hz PW mode.

In the majority of experimental transcranial PBM studies, irradiation of animals' head is performed in the alert, unanesthetized animal. In order to ensure animal stability, manual head holding and use of restraint devices are two common options. In the manual method, due to the animal unpredictable movements and possible moving away of the head from the irradiation area, a portion of exposure light might be wasted and insufficient dose might be delivered to the target region. In addition, both methods of stabilization of the animal induce an extra stress to the animal and could be a potential confounding factor in the psychiatric animal studies. Therefore, an application of high-power lasers is suggested in order to decrease the irradiation period and in order to minimize the animal extra stress during transcranial PBM therapy.

14.4 Conclusions and future outlook

Over the past decade, applications of transcranial PBM therapy as a noninvasive and safe treatment approach for depression have been studied in different animal models, and to date, only very few adverse effects have been reported for this natural light-based therapy in the literature. Brain PBM therapy improves cerebral metabolic capacity and blood flow, stimulates neurogenesis and synaptogenesis, regulates neurotransmitters, and provides neuroprotection via antiinflammatory and antioxidant signaling in neuropathological conditions.

It should be noted that rodent models of depression are less expensive, and require considerably less time and effort to achieve results than human clinical studies. It is recommended that future in vivo investigation of antidepressant effects of PBM also use other depression models such as social stress, early life stress, learned helplessness, and fear conditioning; since these models could add further insights on the action mechanism and therapeutic effects of PBM.

In order to ameliorate cognitive deficits, functional outcomes, and mood disturbances in MDD, potential strategies include the combination therapy of PBM with antidepressant drugs, metabolic modulators, and other conventional treatment modalities [e.g., transcranial magnetic stimulation (TMS), transcranial direct current stimulation (tDCS), deep brain stimulation (DBS), and electroconvulsive therapy (ECT)] as well as with cognitive improvement methods (e.g., psychotherapy, aerobic exercise, and environmental enrichment).

Owing to the beneficial effects of brain PBM therapy in MDD, new explorations for other depressive disorders or subtypes such as treatment-resistant depression, subsyndromal depression, chronic MDD, premenstrual dysphoric disorder, bipolar depression, disruptive mood dysregulation disorder, and postpartum (or perinatal) MDD might well emerge in the future.

References

Ahmed, N.A.E.H., Radwan, N.M., Ibrahim, K.M., Khedr, M.E., El Aziz, M.A., Khadrawy, Y.A., 2008. Effect of three different intensities of infrared laser energy on the levels of amino acid neurotransmitters in the cortex and hippocampus of rat brain. Photomed. Laser. Surg. 26, 479–488.

Ando, T., Xuan, W., Xu, T., Dai, T., Sharma, S.K., Kharkwal, G.B., et al., 2011. Comparison of therapeutic effects between pulsed and continuous wave 810-nm wavelength laser irradiation for traumatic brain injury in mice. PLoS One 6, e26212.

Anisman, H., Hayley, S., Turrin, N., Merali, Z., 2002. Cytokines as a stressor: implications for depressive illness. Int. J. Neuropsychopharmacol. 5, 357–373.

Antkiewicz-Michaluk, L., Wąsik, A., Możdżeń, E., Romańska, I., Michaluk, J., 2014. Antidepressant-like effect of tetrahydroisoquinoline amines in the animal model of depressive disorder induced by repeated administration of a low dose of reserpine: behavioral and neurochemical studies in the rat. Neurotox. Res. 26, 85–98.

Arany, P.R., Cho, A., Hunt, T.D., Sidhu, G., Shin, K., Hahm, E., et al., 2014. Photoactivation of endogenous latent transforming growth factor–β1 directs dental stem cell differentiation for regeneration. Sci. Transl. Med. 6, 238ra269.

Ball, K.A., Castello, P.R., Poyton, R.O., 2011. Low intensity light stimulates nitrite-dependent nitric oxide synthesis but not oxygen consumption by cytochrome c oxidase: implications for phototherapy. J. Photochem. Photobiol. B. 102, 182–191.

Bernstein, H.G., Stanarius, A., Baumann, B., Henning, H., Krell, D., Danos, P., et al., 1998. Nitric oxide synthase-containing neurons in the human hypothalamus: reduced number of immunoreactive cells in the paraventricular nucleus of depressive patients and schizophrenics. Neuroscience 83, 867–875.

Berry, A., Bindocci, E., Alleva, E., 2012. NGF, brain and behavioral plasticity. Neural. Plast. 2012, 784040.

Bhat, A.H., Dar, K.B., Anees, S., Zargar, M.A., Masood, A., Sofi, M.A., et al., 2015. Oxidative stress, mitochondrial dysfunction and neurodegenerative diseases; a mechanistic insight. Biomed. Pharmacother. 74, 101–110.

Bilici, M., Efe, H., Koroglu, M.A., Uydu, H.A., Bekaroglu, M., Deger, O., 2001. Antioxidative enzyme activities and lipid peroxidation in major depression: alterations by antidepressant treatments. J. Affect Disord. 64, 43–51.

Blanco, N.J., Maddox, W.T., Gonzalez-Lima, F., 2017. Improving executive function using transcranial infrared laser stimulation. J. Neuropsychol. 11, 14–25.

Blivet, G., Meunier, J., Roman, F.J., Touchon, J., 2018. Neuroprotective effect of a new photobiomodulation technique against Aβ 25–35 peptide–induced toxicity in mice: novel hypothesis for therapeutic approach of Alzheimer's disease suggested. Alzheimers. Dement. 4, 54–63.

Bressan, R.A., Crippa, J.A., 2005. The role of dopamine in reward and pleasure behaviour–review of data from preclinical research. Acta Psychiatr. Scand. 111, 14–21.

Campbell, S., MacQueen, G., 2004. The role of the hippocampus in the pathophysiology of major depression. J. Psych. Neurosci. 29, 417–426.

Campbell, S., Marriott, M., Nahmias, C., MacQueen, G.M., 2004. Lower hippocampal volume in patients suffering from depression: a meta-analysis. Am. J. Psych. 161, 598–607.

Can, A., Dao, D.T., Arad, M., Terrillion, C.E., Piantadosi, S.C., Gould, T.D., 2012a. The mouse forced swim test. J. Vis. Exp. Available from: https://doi.org/10.3791/3638.

Can, A., Dao, D.T., Terrillion, C.E., Piantadosi, S.C., Bhat, S., Gould, T.D., 2012b. The tail suspension test. J. Vis. Exp. 59, e3769.

Cassano, P., Cusin, C., Mischoulon, D., Hamblin, M.R., De Taboada, L., Pisoni, A., et al., 2015. Near-infrared transcranial radiation for major depressive disorder: proof of concept study. Psychiatr. J. 2015, 352979.

Cassano, P., Petrie, S.R., Hamblin, M.R., Henderson, T.A., Iosifescu, D.V., 2016. Review of transcranial photobiomodulation for major depressive disorder: targeting brain metabolism, inflammation, oxidative stress, and neurogenesis. Neurophotonics 3, 031404.

Castagné, V., Moser, P., Roux, S., Porsolt, R.D., 2010. Rodent models of depression: forced swim and tail suspension behavioral despair tests in rats and mice. Curr. Protoc. Pharmacol. 5.8.1–5.8.14.

Charney, D.S., Buxbaum, J.D., Sklar, P., Nestler, E.J., 2013. Neurobiology of Mental Illness. Oxford University Press.

Chen, A.C.-H., Huang, Y.-Y., Arany, P.R., Hamblin, M.R., 2009. Role of reactive oxygen species in low level light therapy. Mechanisms for Low-Light Therapy IV. Int. Soc. Opt. Phot. 716502.

Chen, A.C.-H., Huang, Y.-Y., Sharma, S.K., Hamblin, M.R., 2011. Effects of 810-nm laser on murine bone-marrow-derived dendritic cells. Photomed. Laser. Surg. 29, 383–389.

Chinnery, P.F., Turnbull, D.M., 1997. Mitochondrial medicine. QJM: Monthly J. Assoc. Physic. 90, 657–667.

Chung, H., Dai, T., Sharma, S.K., Huang, Y.-Y., Carroll, J.D., Hamblin, M.R., 2012. The nuts and bolts of low-level laser (light) therapy. Ann. Biomed. Eng. 40, 516–533.

Costello, E.J., Mustillo, S., Erkanli, A., Keeler, G., Angold, A., 2003. Prevalence and development of psychiatric disorders in childhood and adolescence. Arch. Gen. Psych. 60, 837–844.

Cryan, J.F., Mombereau, C., Vassout, A., 2005. The tail suspension test as a model for assessing antidepressant activity: review of pharmacological and genetic studies in mice. Neurosci. Biobehav. Rev. 29, 571–625.

Darlot, F., Moro, C., Massri, N., Chabrol, C., Johnstone, D.M., Reinhart, F., et al., 2016. Near-infrared light is neuroprotective in a monkey model of Parkinson disease. Ann. Neurol. 79, 59–75.

de Freitas, L.F., Hamblin, M.R., 2016. Proposed mechanisms of photobiomodulation or low-level light therapy. IEEE J. Sel. Top. Quantum Electron. 22, 348–364.

De Taboada, L., Yu, J., El-Amouri, S., Gattoni-Celli, S., Richieri, S., McCarthy, T., et al., 2011. Transcranial laser therapy attenuates amyloid-β peptide neuropathology in amyloid-β protein precursor transgenic mice. J. Alzheimers. Dis. 23, 521–535.

Der-Avakian, A., Markou, A., 2012. The neurobiology of anhedonia and other reward-related deficits. Trends Neurosci. 35, 68–77.

Di Chiara, G., Tanda, G., 1997. Blunting of reactivity of dopamine transmission to palatable food: a biochemical marker of anhedonia in the CMS model? Psychopharmacology (Berl.) 134, 351–353.

Diniz, B.S., Teixeira, A.L., Machado-Vieira, R., Talib, L.L., Gattaz, W.F., Forlenza, O.V., 2013. Reduced serum nerve growth factor in patients with late-life depression. Am. J. Geriat. Psych. 21, 493–496.

Dirnagl, U., Lindauer, U., Villringer, A., 1993. Role of nitric oxide in the coupling of cerebral blood flow to neuronal activation in rats. Neurosci. Lett. 149, 43–46.

Dong, T., Zhang, Q., Hamblin, M.R., Wu, M.X., 2015. Low-level light in combination with metabolic modulators for effective therapy of injured brain. J. Cereb. Blood. Flow. Metab. 35, 1435–1444.

Dowlati, Y., Herrmann, N., Swardfager, W., Liu, H., Sham, L., Reim, E.K., et al., 2010. A meta-analysis of cytokines in major depression. Biol. Psych. 67, 446–457.

Du, J., Creson, T.K., Wu, L.-J., Ren, M., Gray, N.A., Falke, C., et al., 2008. The role of hippocampal GluR1 and GluR2 receptors in manic-like behavior. J. Neurosci. 28, 68–79.

Duman, R.S., Monteggia, L.M., 2006. A neurotrophic model for stress-related mood disorders. Biol. Psych. 59, 1116–1127.

Dunn, A.J., Swiergiel, A.H., de Beaurepaire, R., 2005. Cytokines as mediators of depression: what can we learn from animal studies? Neurosci. Biobehav. Rev. 29, 891–909.

Dziedzicka-Wasylewska, M., Willner, P., Papp, M., 1997. Changes in dopamine receptor mRNA expression following chronic mild stress and chronic antidepressant treatment. Behav. Pharmacol. 8, 607–618.

El Massri, N., Lemgruber, A.P., Rowe, I.J., Moro, C., Torres, N., Reinhart, F., et al., 2017. Photobiomodulation-induced changes in a monkey model of Parkinson's disease: changes in tyrosine hydroxylase cells and GDNF expression in the striatum. Exp. Brain Res. 235, 1861–1874.

Elhwuegi, A.S., 2004. Central monoamines and their role in major depression. Prog. Neuropsychopharmacol. Biol. Psychiatry 28, 435–451.

Fava, M., Kendler, K.S., 2000. Major depressive disorder. Neuron 28, 335–341.

Fleminger, S., Oliver, D.L., Williams, W.H., Evans, J., 2003. The neuropsychiatry of depression after brain injury. Neuropsychol. Rehabil. 13, 65–87.

Fontella, F.U., Siqueira, I.R., Vasconcellos, A.P., Tabajara, A.S., Netto, C.A., Dalmaz, C., 2005. Repeated restraint stress induces oxidative damage in rat hippocampus. Neurochem. Res. 30, 105–111.

Galynker, I.I., Cai, J., Ongseng, F., Finestone, H., Dutta, E., Serseni, D., 1998. Hypofrontality and negative symptoms in major depressive disorder. J. Nucl. Med. 39, 608–612.

Gardner, A., Johansson, A., Wibom, R., Nennesmo, I., von Dobeln, U., Hagenfeldt, L., et al., 2003. Alterations of mitochondrial function and correlations with personality traits in selected major depressive disorder patients. J. Affect Disord. 76, 55–68.

Ghanbari, A., Ghareghani, M., Zibara, K., Delaviz, H., Ebadi, E., Jahantab, M.H., 2017. Light-Emitting Diode (LED) therapy improves occipital cortex damage by decreasing apoptosis and increasing BDNF-expressing cells in methanol-induced toxicity in rats. Biomed. Pharmacother. 89, 1320–1330.

Halliwell, B., 2006. Oxidative stress and neurodegeneration: where are we now? J. Neurochem. 97, 1634–1658.

Hamblin, M.R., 2017a. Mechanisms and applications of the anti-inflammatory effects of photobiomodulation. AIMS Biophys. 4, 337.

Hamblin, M.R., 2017b. Mechanisms and mitochondrial redox signaling in photobiomodulation. Photochem. Photobiol. 94, 199–212.

Hamblin, M.R., Demidova, T.N., 2006. Mechanisms of low level light therapy. Mechanisms for low-light therapy. Int. Soc. Opt. Photon. 614001.

Haughey, N.J., Nath, A., Chan, S.L., Borchard, A.C., Rao, M.S., Mattson, M.P., 2002. Disruption of neurogenesis by amyloid beta-peptide, and perturbed neural progenitor cell homeostasis, in models of Alzheimer's disease. J. Neurochem. 83, 1509–1524.

Huang, Q.-j, Jiang, H., Hao, X.-l, Minor, T.R., 2004. Brain IL-1beta was involved in reserpine-induced behavioral depression in rats. Acta Pharmacol. Sin. 25, 293–296.

Huang, Y.Y., Nagata, K., Tedford, C.E., McCarthy, T., Hamblin, M.R., 2013. Low-level laser therapy (LLLT) reduces oxidative stress in primary cortical neurons in vitro. J. Biophoton. 6, 829–838.

Ikram, H., Haleem, D.J., 2017. Repeated treatment with reserpine as a progressive animal model of depression. Pak. J. Pharm. Sci. 30, 897–902.

Jacobsen, J.P., Medvedev, I.O., Caron, M.G., 2012. The 5-HT deficiency theory of depression: perspectives from a naturalistic 5-HT deficiency model, the tryptophan hydroxylase 2Arg439His knockin mouse. Phil. Trans. R Soc. B 367, 2444–2459.

Karege, F., Perret, G., Bondolfi, G., Schwald, M., Bertschy, G., Aubry, J.M., 2002. Decreased serum brain-derived neurotrophic factor levels in major depressed patients. Psychiatry Res. 109, 143–148.

Karolewicz, B., Maciag, D., O'Dwyer, G., Stockmeier, C.A., Feyissa, A.M., Rajkowska, G., 2010. Reduced level of glutamic acid decarboxylase-67 kDa in the prefrontal cortex in major depression. Int. J. Neuropsychopharmacol. 13, 411–420.

Karu, T., Kolyakov, S., 2005. Exact action spectra for cellular responses relevant to phototherapy. Photomed. Laser Ther. 23, 355–361.

Karu, T.I., Pyatibrat, L.V., Afanasyeva, N.I., 2005a. Cellular effects of low power laser therapy can be mediated by nitric oxide. Lasers Surg. Med. 36, 307–314.

Karu, T.I., Pyatibrat, L.V., Kolyakov, S.F., Afanasyeva, N.I., 2005b. Absorption measurements of a cell monolayer relevant to phototherapy: reduction of cytochrome c oxidase under near IR radiation. J. Photochem. Photobiol. B. 81, 98–106.

Kato, T., Takahashi, S., Shioiri, T., Inubushi, T., 1992. Brain phosphorous metabolism in depressive disorders detected by phosphorus-31 magnetic resonance spectroscopy. J. Affect Disord. 26, 223–230.

Kempermann, G., 2002. Regulation of adult hippocampal neurogenesis – implications for novel theories of major depression. Bipolar. Disord. 4, 17–33.

Kendler, K.S., Karkowski, L.M., Prescott, C.A., 1999. Causal relationship between stressful life events and the onset of major depression. Am. J. Psychiatry 156, 837–841.

Kent, S., Bluthe, R.M., Kelley, K.W., Dantzer, R., 1992. Sickness behavior as a new target for drug development. Trends Pharmacol. Sci. 13, 24–28.

Kessler, R.C., Angermeyer, M., Anthony, J.C., De Graaf, R., Demyttenaere, K., Gasquet, I., et al., 2007. Lifetime prevalence and age-of-onset distributions of mental disorders in the World Health Organization's World Mental Health Survey Initiative. World Psychiatry. 6, 168.

Khanzode, S.D., Dakhale, G.N., Khanzode, S.S., Saoji, A., Palasodkar, R., 2003. Oxidative damage and major depression: the potential antioxidant action of selective serotonin re-uptake inhibitors. Redox Rep. Communicat. Free Rad. Res. 8, 365–370.

Kim, S.J., Son, T.G., Park, H.R., Park, M., Kim, M.S., Kim, H.S., et al., 2008. Curcumin stimulates proliferation of embryonic neural progenitor cells and neurogenesis in the adult hippocampus. J. Biol. Chem. 283, 14497–14505.

Kim, Y.K., Maes, M., 2003. The role of the cytokine network in psychological stress. Acta Neuropsychiatr. 15, 148–155.

Kim, Y.K., Na, K.S., Myint, A.M., Leonard, B.E., 2016. The role of pro-inflammatory cytokines in neuroinflammation, neurogenesis and the neuroendocrine system in major depression. Prog. Neuropsychopharmacol. Biol. Psychiatry 64, 277–284.

Krishnan, V., Nestler, E.J., 2008. The molecular neurobiology of depression. Nature 455, 894.

Krishnan, V., Nestler, E.J., 2011. Animal models of depression: molecular perspectives. Molecular and Functional Models in Neuropsychiatry. Springer, pp. 121–147.

Lapchak, P.A., 2012. Transcranial near-infrared laser therapy applied to promote clinical recovery in acute and chronic neurodegenerative diseases. Expert. Rev. Med. Devices. 9, 71–83.

Lapchak, P.A., Boitano, P.D., 2016. A novel method to promote behavioral improvement and enhance mitochondrial function following an embolic stroke. Brain Res. 1646, 125–131.

Lapchak, P.A., De Taboada, L., 2010. Transcranial near infrared laser treatment (NILT) increases cortical adenosine-5′-triphosphate (ATP) content following embolic strokes in rabbits. Brain Res. 1306, 100–105.

Lee, B., Sur, B., Park, J., Kim, S.-H., Kwon, S., Yeom, M., et al., 2013. Chronic administration of baicalein decreases depression-like behavior induced by repeated restraint stress in rats. Korean J. Physiol. Pharmacol. 17, 393–403.

Lee, H.I., Lee, S.-W., Kim, S.Y., Kim, N.G., Park, K.-J., Choi, B.T., et al., 2017a. Pretreatment with light-emitting diode therapy reduces ischemic brain injury in mice through endothelial nitric oxide synthase-dependent mechanisms. Biochem. Biophys. Res. Commun. 486, 945–950.

Lee, H.I., Lee, S.W., Kim, N.G., Park, K.J., Choi, B.T., Shin, Y.I., et al., 2017b. Low-level light emitting diode (LED) therapy suppresses inflammasome-mediated brain damage in experimental ischemic stroke. J. Biophoton. 10, 1502–1513.

Lee, H.I., Park, J.H., Park, M.Y., Kim, N.G., Park, K.-J., Choi, B.T., et al., 2016. Pre-conditioning with transcranial low-level light therapy reduces neuroinflammation and protects blood-brain barrier after focal cerebral ischemia in mice. Restor. Neurol. Neurosci. 34, 201–214.

Leung, M.C., Lo, S.C., Siu, F.K., So, K.F., 2002. Treatment of experimentally induced transient cerebral ischemia with low energy laser inhibits nitric oxide synthase activity and up-regulates the expression of transforming growth factor-beta 1. Lasers Surg. Med. 31, 283–288.

Licinio, J., Mastronardi, C., Wong, M.L., 2007. Pharmacogenomics of neuroimmune interactions in human psychiatric disorders. Exp. Physiol. 92, 807–811.

Lin, P.Y., Tseng, P.T., 2015. Decreased glial cell line-derived neurotrophic factor levels in patients with depression: a meta-analytic study. J. Psychiatr. Res. 63, 20–27.

Loftis, J.M., Huckans, M., Morasco, B.J., 2010. Neuroimmune mechanisms of cytokine-induced depression: current theories and novel treatment strategies. Neurobiol. Dis. 37, 519–533.

Lombard, A., Rossetti, V., Cassone, M., Urciuoli, R., Rolfo, P., 1990. Neurotransmitter content and enzyme activity variations in rat brain following in vivo He-Ne laser irradiation. Proc. Round Table Basic Appl. Res. Photobiol. Photochem. 10–11.

Lu, Y., Wang, R., Dong, Y., Tucker, D., Zhao, N., Ahmed, M.E., et al., 2017. Low-level laser therapy for beta amyloid toxicity in rat hippocampus. Neurobiol. Aging. 49, 165–182.

Lucca, G., Comim, C.M., Valvassori, S.S., Reus, G.Z., Vuolo, F., Petronilho, F., et al., 2009. Effects of chronic mild stress on the oxidative parameters in the rat brain. Neurochem. Int. 54, 358–362.

Maciag, D., Hughes, J., O'Dwyer, G., Pride, Y., Stockmeier, C.A., Sanacora, G., et al., 2010. Reduced density of calbindin immunoreactive GABAergic neurons in the occipital cortex in major depression: relevance to neuroimaging studies. Biol. Psychiatry 67, 465–470.

MacMaster, F.P., Kusumakar, V., 2004. Hippocampal volume in early onset depression. BMC. Med. 2, 2.

Maes, M., Yirmyia, R., Noraberg, J., Brene, S., Hibbeln, J., Perini, G., et al., 2009. The inflammatory & neurodegenerative (I&ND) hypothesis of depression: leads for future research and new drug developments in depression. Metab. Brain. Dis. 24, 27–53.

Mahmoudi, J., Farhoudi, M., Talebi, M., Sabermarouf, B., Sadigh-Eteghad, S., 2015. Antidepressant-like effect of modafinil in mice: evidence for the involvement of the dopaminergic neurotransmission. Pharmacol. Rep. 67, 478–484.

Malberg, J.E., Eisch, A.J., Nestler, E.J., Duman, R.S., 2000. Chronic antidepressant treatment increases neurogenesis in adult rat hippocampus. J. Neurosci. 20, 9104–9110.

Manji, H.K., Drevets, W.C., Charney, D.S., 2001. The cellular neurobiology of depression. Nat. Med. 7, 541.

Martinowich, K., Manji, H., Lu, B., 2007. New insights into BDNF function in depression and anxiety. Nat. Neurosci. 10, 1089.

Meng, C., He, Z., Xing, D., 2013. Low-level laser therapy rescues dendrite atrophy via upregulating BDNF expression: implications for Alzheimer's disease. J. Neurosci. 33, 13505–13517.

Minor, T.R., Hanff, T.C., 2015. Adenosine signaling in reserpine-induced depression in rats. Behav. Brain. Res. 286, 184–191.

Mintzopoulos, D., Gillis, T.E., Tedford, C.E., Kaufman, M.J., 2017. Effects of near-infrared light on cerebral bioenergetics measured with phosphorus magnetic resonance spectroscopy. Photomed. Laser. Surg. 35, 395–400.

Mochizuki-Oda, N., Kataoka, Y., Cui, Y., Yamada, H., Heya, M., Awazu, K., 2002. Effects of near-infra-red laser irradiation on adenosine triphosphate and adenosine diphosphate contents of rat brain tissue. Neurosci. Lett. 323, 207–210.

Mohammed, H.S., 2016. Transcranial low-level infrared laser irradiation ameliorates depression induced by reserpine in rats. Lasers. Med. Sci. 31, 1651–1656.

Moore, C.M., Christensen, J.D., Lafer, B., Fava, M., Renshaw, P.F., 1997. Lower levels of nucleoside triphosphate in the basal ganglia of depressed subjects: a phosphorous-31 magnetic resonance spectroscopy study. Am. J. Psychiatry 154, 116–118.

Moreira, M.S., Velasco, I.T., Ferreira, L.S., Ariga, S.K.K., Barbeiro, D.F., Meneguzzo, D.T., et al., 2009. Effect of phototherapy with low intensity laser on local and systemic immunomodulation following focal brain damage in rat. J. Photochem. Photobiol. B. 97, 145–151.

Moret, C., Briley, M., 2011. The importance of norepinephrine in depression. Neuropsychiatr. Dis. Treat. 7, 9–13.

Moretti, M., Colla, A., de Oliveira Balen, G., dos Santos, D.B., Budni, J., de Freitas, A.E., et al., 2012. Ascorbic acid treatment, similarly to fluoxetine, reverses depressive-like behavior and brain oxidative damage induced by chronic unpredictable stress. J. Psychiatr. Res. 46, 331–340.

Moro, C., Massri, N.E., Torres, N., Ratel, D., De Jaeger, X., Chabrol, C., et al., 2014. Photobiomodulation inside the brain: a novel method of applying near-infrared light intracranially and its impact on dopaminergic cell survival in MPTP-treated mice. J. Neurosurg. 120, 670–683.

Morries, L.D., Cassano, P., Henderson, T.A., 2015. Treatments for traumatic brain injury with emphasis on transcranial near-infrared laser phototherapy. Neuropsychiatr. Dis. Treat. 11, 2159.

Moylan, S., Maes, M., Wray, N.R., Berk, M., 2013. The neuroprogressive nature of major depressive disorder: pathways to disease evolution and resistance, and therapeutic implications. Mol. Psychiatry 18, 595–606.

Mueller, S.G., Schuff, N., Yaffe, K., Madison, C., Miller, B., Weiner, M.W., 2010. Hippocampal atrophy patterns in mild cognitive impairment and Alzheimer's disease. Hum. Brain. Mapp. 31, 1339–1347.

Naeser, M.A., Zafonte, R., Krengel, M.H., Martin, P.I., Frazier, J., Hamblin, M.R., et al., 2014. Significant improvements in cognitive performance post-transcranial, red/near-infrared light-emitting diode treatments in chronic, mild traumatic brain injury: open-protocol study. J. Neurotrauma 31, 1008–1017.

Nagata, K., Nakashima-Kamimura, N., Mikami, T., Ohsawa, I., Ohta, S., 2009. Consumption of molecular hydrogen prevents the stress-induced impairments in hippocampus-dependent learning tasks during chronic physical restraint in mice. Neuropsychopharmacology 34, 501.

Nasrolahi, A., Mahmoudi, J., Akbarzadeh, A., Karimipour, M., Sadigh-Eteghad, S., Salehi, R., et al., 2018. Neurotrophic factors hold promise for the future of Parkinson's disease treatment: is there a light at the end of the tunnel? Rev. Neurosci. 29, 475–489.

Naumenko, V.S., Kondaurova, E.M., Bazovkina, D.V., Tsybko, A.S., Ilchibaeva, T.V., Khotskin, N.V., et al., 2014. Effect of GDNF on depressive-like behavior, spatial learning and key genes of the brain dopamine system in genetically predisposed to behavioral disorders mouse strains. Behav. Brain. Res. 274, 1–9.

Nawashiro, H., Wada, K., Nakai, K., Sato, S., 2012. Focal increase in cerebral blood flow after treatment with near-infrared light to the forehead in a patient in a persistent vegetative state. Photomed. Laser. Surg. 30, 231–233.

Nestler, E.J., Barrot, M., DiLeone, R.J., Eisch, A.J., Gold, S.J., Monteggia, L.M., 2002. Neurobiology of depression. Neuron 34, 13–25.

Oron, A., Oron, U., Chen, J., Eilam, A., Zhang, C., Sadeh, M., et al., 2006. Low-level laser therapy applied transcranially to rats after induction of stroke significantly reduces long-term neurological deficits. Stroke 37, 2620–2624.

Overstreet, D.H., 2012. Modeling depression in animal models. Psychiatr Disord. Springer, pp. 125–144.

Ozcan, M.E., Gulec, M., Ozerol, E., Polat, R., Akyol, O., 2004. Antioxidant enzyme activities and oxidative stress in affective disorders. Int. Clin. Psychopharmacol. 19, 89–95.

Pandya, C.D., Howell, K.R., Pillai, A., 2013. Antioxidants as potential therapeutics for neuropsychiatric disorders. Prog. Neuropsychopharmacol. Biol. Psychiatry 46, 214–223.

Perez-Garcia, G., De Gasperi, R., Sosa, M.A.G., Perez, G.M., Otero-Pagan, A., Tschiffely, A., et al., 2018. PTSD-related behavioral traits in a rat model of blast-induced mTBI are reversed by the mGluR2/3 receptor antagonist BCI-838. eNeuro 5, ENEURO. 0357-0317.2018.

Pitossi, F., del Rey, A., Kabiersch, A., Besedovsky, H., 1997. Induction of cytokine transcripts in the central nervous system and pituitary following peripheral administration of endotoxin to mice. J. Neurosci. Res. 48, 287–298.

Porsolt, R., Lenegre, A., McArthur, R., 1991. Pharmacological models of depression. Animal Models in Psychopharmacology. Springer, pp. 137–159.

Pruessner, J.C., Champagne, F., Meaney, M.J., Dagher, A., 2004. Dopamine release in response to a psychological stress in humans and its relationship to early life maternal care: a positron emission tomography study using [11C] raclopride. J. Neurosci. Nurs. 24, 2825–2831.

Purba, J.S., Hoogendijk, W.J., Hofman, M.A., Swaab, D.F., 1996. Increased number of vasopressin- and oxytocin-expressing neurons in the paraventricular nucleus of the hypothalamus in depression. Arch. Gen. Psychiatry 53, 137–143.

Purushothuman, S., Johnstone, D.M., Nandasena, C., Mitrofanis, J., Stone, J., 2014. Photobiomodulation with near infrared light mitigates Alzheimer's disease-related pathology in cerebral cortex–evidence from two transgenic mouse models. Alzheimers Res. Ther. 6, 2.

Qin, L., Wu, X., Block, M.L., Liu, Y., Breese, G.R., Hong, J.S., et al., 2007. Systemic LPS causes chronic neuroinflammation and progressive neurodegeneration. Glia 55, 453–462.

Raadsheer, F.C., van Heerikhuize, J.J., Lucassen, P.J., Hoogendijk, W.J., Tilders, F.J., Swaab, D.F., 1995. Corticotropin-releasing hormone mRNA levels in the paraventricular nucleus of patients with Alzheimer's disease and depression. Am. J. Psychiatry 152, 1372–1376.

Radwan, N.M., Ahmed, N.A.E.H., Ibrahim, K.M., Khedr, M.E., Aziz, M.A., Khadrawy, Y.A., 2009. Effect of infrared laser irradiation on amino acid neurotransmitters in an epileptic animal model induced by pilocarpine. Photomed. Laser. Surg. 27, 401–409.

Ren, L., Qian, X., Zhai, L., Sun, M., Miao, Z., Li, J., et al., 2014. Loss of Ahi1 impairs neurotransmitter release and causes depressive behaviors in mice. PLoS One 9, e93640.

Rojas, J.C., Bruchey, A.K., Gonzalez-Lima, F., 2012. Low-level light therapy improves cortical metabolic capacity and memory retention. J. Alzheimers. Dis. 32, 741–752.

Rojas, J.C., Gonzalez-Lima, F., 2013. Neurological and psychological applications of transcranial lasers and LEDs. Biochem. Pharmacol. 86, 447–457.

Rojas, J.C., Lee, J., John, J.M., Gonzalez-Lima, F., 2008. Neuroprotective effects of near-infrared light in an in vivo model of mitochondrial optic neuropathy. J. Neurosci. 28, 13511–13521.

Roy, A., De Jong, J., Linnoila, M., 1989. Cerebrospinal fluid monoamine metabolites and suicidal behavior in depressed patients: a 5-year follow-up study. Arch. Gen. Psychiatry 46, 609–612.

Salehpour, F., Ahmadian, N., Rasta, S.H., Farhoudi, M., Karimi, P., Sadigh-Eteghad, S., 2017. Transcranial low-level laser therapy improves brain mitochondrial function and cognitive impairment in D-galactose–induced aging mice. Neurobiol. Aging. 58, 140–150.

Salehpour, F., Farajdokht, F., Erfani, M., Sadigh-Eteghad, S., Shotorbani, S.S., Hamblin, M.R., et al., 2018a. Transcranial near-infrared photobiomodulation attenuates memory impairment and hippocampal oxidative stress in sleep-deprived mice. Brain Res. 1682, 36–43.

Salehpour, F., Mahmoudi, J., Kamari, F., Sadigh-Eteghad, S., Rasta, S.H., Hamblin, M.R., 2018b. Brain photobiomodulation therapy: a narrative review. Mol. Neurobiol. 55, 6601–6636.

Salehpour, F., Rasta, S.H., Mohaddes, G., Sadigh-Eteghad, S., Salarirad, S., 2016. Therapeutic effects of 10-Hz pulsed wave lasers in rat depression model: a comparison between near-infrared and red wavelengths. Lasers Surg. Med. 48, 695–705.

Salgado, A.S., Zângaro, R.A., Parreira, R.B., Kerppers, I.I., 2015. The effects of transcranial LED therapy (TCLT) on cerebral blood flow in the elderly women. Lasers. Med. Sci. 30, 339–346.

Saltmarche, A.E., Naeser, M.A., Ho, K.F., Hamblin, M.R., Lim, L., 2017. Significant improvement in cognition in mild to moderately severe dementia cases treated with transcranial plus intranasal photobiomodulation: case series report. Photomed. Laser. Surg. 35, 432–441.

Samavati, L., Lee, I., Mathes, I., Lottspeich, F., Huttemann, M., 2008. Tumor necrosis factor alpha inhibits oxidative phosphorylation through tyrosine phosphorylation at subunit I of cytochrome c oxidase. J. Biol. Chem. 283, 21134–21144.

Sanacora, G., Saricicek, A., 2007. GABAergic contributions to the pathophysiology of depression and the mechanism of antidepressant action. CNS. Neurol. Disord. Drug. Targets. 6, 127–140.

Schiffer, F., Johnston, A.L., Ravichandran, C., Polcari, A., Teicher, M.H., Webb, R.H., et al., 2009. Psychological benefits 2 and 4 weeks after a single treatment with near infrared light to the forehead: a pilot study of 10 patients with major depression and anxiety. Behav. Brain Funct. 5, 46.

Schwarz, T.L., 2013. Mitochondrial trafficking in neurons. Cold Spring Harb. Perspect. Biol. 5, a011304.

Selley, M.L., 2004. Increased (E)-4-hydroxy-2-nonenal and asymmetric dimethylarginine concentrations and decreased nitric oxide concentrations in the plasma of patients with major depression. J. Affect Disord. 80, 249–256.

Sen, S., Duman, R., Sanacora, G., 2008. Serum BDNF, depression and anti-depressant medications: meta-analyses and implications. Biol. Psychiatry 64, 527–532.

Sena, L.A., Chandel, N.S., 2012. Physiological roles of mitochondrial reactive oxygen species. Mol. Cell 48, 158–167.

Sharma, S.K., Kharkwal, G.B., Sajo, M., Huang, Y.Y., De Taboada, L., McCarthy, T., et al., 2011. Dose response effects of 810 nm laser light on mouse primary cortical neurons. Lasers Surg. Med. 43, 851–859.

Shu-Zhi, L., Li-Hua, W., 1982. Effects of a low power laser beam guided by optic fiber on rat brain striatal monoamines and amino acids. Neurosci. Lett. 32, 203–208.

Smith, R.S., 1991. The macrophage theory of depression. Med. Hypotheses 35, 298–306.

Song, S., Zhang, Y., Fong, C.-C., Tsang, C.-H., Yang, Z., Yang, M., 2003. cDNA microarray analysis of gene expression profiles in human fibroblast cells irradiated with red light. J. Invest. Dermatol. 120, 849–857.

Stadler, J., Bentz, B.G., Harbrecht, B.G., Di Silvio, M., Curran, R.D., Billiar, T.R., et al., 1992. Tumor necrosis factor alpha inhibits hepatocyte mitochondrial respiration. Ann. Surg. 216, 539–546.

Stepanichev, M., Dygalo, N.N., Grigoryan, G., Shishkina, G.T., Gulyaeva, N., 2014. Rodent models of depression: neurotrophic and neuroinflammatory biomarkers. Biomed. Res. Int. 2014, 932757.

Taliaz, D., Stall, N., Dar, D.E., Zangen, A., 2010. Knockdown of brain-derived neurotrophic factor in specific brain sites precipitates behaviors associated with depression and reduces neurogenesis. Mol. Psychiatry 15, 80–92.

Tanaka, Y., Akiyoshi, J., Kawahara, Y., Ishitobi, Y., Hatano, K., Hoaki, N., et al., 2011. Infrared radiation has potential antidepressant and anxiolytic effects in animal model of depression and anxiety. Brain Stimul. 4, 71–76.

Taupin, P., 2006. Neurogenesis and the effect of antidepressants. Drug Target Insights 1, 13–17.

Telerman, A., Lapter, S., Sharabi, A., Zinger, H., Mozes, E., 2011. Induction of hippocampal neurogenesis by a tolerogenic peptide that ameliorates lupus manifestations. J. Neuroimmunol. 232, 151–157.

Tian, F., Hase, S.N., Gonzalez-Lima, F., Liu, H., 2016. Transcranial laser stimulation improves human cerebral oxygenation. Lasers Surg. Med. 48, 343–349.

Tobe, E.H., 2013. Mitochondrial dysfunction, oxidative stress, and major depressive disorder. Neuropsychiatr. Dis. Treat. 9, 567–573.

Toda, N., Ayajiki, K., Okamura, T., 2009. Cerebral blood flow regulation by nitric oxide in neurological disorders. Can. J. Physiol. Pharmacol. 87, 581–594.

Uozumi, Y., Nawashiro, H., Sato, S., Kawauchi, S., Shima, K., Kikuchi, M., 2010. Targeted increase in cerebral blood flow by transcranial near-infrared laser irradiation. Lasers Surg. Med. 42, 566–576.

Villanueva, R., 2013. Neurobiology of major depressive disorder. Neural. Plast. 2013, 873278.

Voloboueva, L.A., Giffard, R.G., 2011. Inflammation, mitochondria, and the inhibition of adult neurogenesis. J. Neurosci. Res. 89, 1989–1996.

Volz, H.P., Rzanny, R., Riehemann, S., May, S., Hegewald, H., Preussler, B., et al., 1998. 31P magnetic resonance spectroscopy in the frontal lobe of major depressed patients. Eur. Arch. Psychiatry Clin. Neurosci. 248, 289–295.

Wagner, K.R., Kleinholz, M., Myers, R.E., 1990. Delayed decreases in specific brain mitochondrial electron transfer complex activities and cytochrome concentrations following anoxia/ischemia. J. Neurol. Sci. 100, 142–151.

Wang, X., Michaelis, E.K., 2010. Selective neuronal vulnerability to oxidative stress in the brain. Front. Aging Neurosci. 2, 12.

Wang, X., Tian, F., Reddy, D.D., Nalawade, S.S., Barrett, D.W., Gonzalez-Lima, F., et al., 2017. Up-regulation of cerebral cytochrome-c-oxidase and hemodynamics by transcranial infrared laser stimulation: a broadband near-infrared spectroscopy study. J. Cereb. Blood. Flow. Metab. 37, 3789–3802.

Watanabe, Y., Gould, E., Cameron, H.A., Daniels, D.C., McEwen, B.S., 1992. Phenytoin prevents stress- and corticosterone-induced atrophy of CA3 pyramidal neurons. Hippocampus 2, 431–435.

Wiener, C.D., de Mello Ferreira, S., Pedrotti Moreira, F., Bittencourt, G., de Oliveira, J.F., Lopez Molina, M., et al., 2015. Serum levels of nerve growth factor (NGF) in patients with major depression disorder and suicide risk. J. Affect Disord. 184, 245–248.

Willner, P., 2017. The chronic mild stress (CMS) model of depression: history, evaluation and usage. Neurobiol. Stress 6, 78–93.

Wong-Riley, M.T., Liang, H.L., Eells, J.T., Chance, B., Henry, M.M., Buchmann, E., et al., 2005. Photobiomodulation directly benefits primary neurons functionally inactivated by toxins role of cytochrome c oxidase. J. Biol. Chem. 280, 4761–4771.

Wu, X., Alberico, S.L., Moges, H., De Taboada, L., Tedford, C.E., Anders, J.J., 2012. Pulsed light irradiation improves behavioral outcome in a rat model of chronic mild stress. Lasers Surg. Med. 44, 227–232.

Xiong, Y., Mahmood, A., Chopp, M., 2013. Animal models of traumatic brain injury. Nat. Rev. Neurosci. 14, 128.

Xu, X., Yang, H., Lin, Y.-F., Li, X., Cape, A., Ressler, K.J., et al., 2010. Neuronal Abelson helper integration site-1 (Ahi1) deficiency in mice alters TrkB signaling with a depressive phenotype. Proc. Natl. Acad. Sci. 107, 19126–19131.

Xu, Z., Guo, X., Yang, Y., Tucker, D., Lu, Y., Xin, N., et al., 2017. Low-level laser irradiation improves depression-like behaviors in mice. Mol. Neurobiol. 54, 4551–4559.

Xuan, W., Agrawal, T., Huang, L., Gupta, G.K., Hamblin, M.R., 2015. Low-level laser therapy for traumatic brain injury in mice increases brain derived neurotrophic factor (BDNF) and synaptogenesis. J. Biophoton. 8, 502–511.

Xuan, W., Huang, L., Hamblin, M.R., 2016. Repeated transcranial low-level laser therapy for traumatic brain injury in mice: biphasic dose response and long-term treatment outcome. J. Biophoton. 9, 1263–1272.

Xuan, W., Vatansever, F., Huang, L., Hamblin, M.R., 2014. Transcranial low-level laser therapy enhances learning, memory, and neuroprogenitor cells after traumatic brain injury in mice. J. Biomed. Opt. 19, 108003.

Xuan, W., Vatansever, F., Huang, L., Wu, Q., Xuan, Y., Dai, T., et al., 2013. Transcranial low-level laser therapy improves neurological performance in traumatic brain injury in mice: effect of treatment repetition regimen. PLoS One 8, e53454.

Yan, X., Liu, J., Zhang, Z., Li, W., Sun, S., Zhao, J., et al., 2017. Low-level laser irradiation modulates brain-derived neurotrophic factor mRNA transcription through calcium-dependent activation of the ERK/CREB pathway. Lasers. Med. Sci. 32, 169–180.

Yang, L., Tucker, D., Dong, Y., Wu, C., Lu, Y., Li, Y., et al., 2018. Photobiomodulation therapy promotes neurogenesis by improving post-stroke local microenvironment and stimulating neuroprogenitor cells. Exp. Neurol. 299, 86–96.

Yang, X., Askarova, S., Sheng, W., Chen, J., Sun, A.Y., Sun, G.Y., et al., 2010. Low energy laser light (632.8 nm) suppresses amyloid-β peptide-induced oxidative and inflammatory responses in astrocytes. Neuroscience 171, 859–868.

You, Z., Luo, C., Zhang, W., Chen, Y., He, J., Zhao, Q., et al., 2011. Pro- and anti-inflammatory cytokines expression in rat's brain and spleen exposed to chronic mild stress: involvement in depression. Behav. Brain. Res. 225, 135–141.

Yüksel, C., Öngür, D., 2010. Magnetic resonance spectroscopy studies of glutamate-related abnormalities in mood disorders. Biol. Psychiatry 68, 785–794.

Zell, R., Geck, P., Werdan, K., Boekstegers, P., 1997. TNF-alpha and IL-1 alpha inhibit both pyruvate dehydrogenase activity and mitochondrial function in cardiomyocytes: evidence for primary impairment of mitochondrial function. Mol. Cell. Biochem. 177, 61–67.

Zhang, Q., Zhou, C., Hamblin, M.R., Wu, M.X., 2014. Low-level laser therapy effectively prevents secondary brain injury induced by immediate early responsive gene X-1 deficiency. J. Cereb. Blood. Flow. Metab. 34, 1391–1401.

Zhang, Z., Oliver, P., Lancaster Jr, J.R., Schwarzenberger, P.O., Joshi, M.S., et al., 2001. Reactive oxygen species mediate tumor necrosis factor alpha-converting, enzyme-dependent ectodomain shedding induced by phorbol myristate acetate. FASEB J. 15, 303–305.

Zivin, J.A., Albers, G.W., Bornstein, N., Chippendale, T., Dahlof, B., Devlin, T., et al., 2009. Effectiveness and safety of transcranial laser therapy for acute ischemic stroke. Stroke 40, 1359–1364.

Chapter 15

Transcranial photobiomodulation treats Alzheimer's disease in amyloid-β protein precursor transgenic mice

Luis De Taboada[1] and Michael R. Hamblin[2,3]

[1]*Chief Technology Officer, LiteCure LLC, New Castle, DE, United States,* [2]*Wellman Center for Photomedicine, Massachusetts General Hospital, Boston, MA, United States,* [3]*Department of Dermatology, Harvard Medical School, Boston, MA, United States*

15.1 Introduction

There is evidence suggesting that the primary mitochondrial chromophore or photoacceptor molecule for photobiomodulation (PBM) is cytochrome-c-oxidase (CCO) (Wong-Riley et al., 2005; Karu, 2010). The CCO enzyme complex contains two copper centers, CuA and CuB, with the CuA center having a broad wavelength absorption peak between 800–830 nm in its oxidized form. CCO is a terminal enzyme in the cellular respiratory chain that is located in the inner mitochondrial membrane. It plays a central role in the bioenergetics of eukaryotic cells by moving protons across the inner membrane and driving the formation of ATP by oxidative phosphorylation (Tuner and Hode, 2002). A possible mechanism of action resulting in neuroprotection following transcranial photobiomodulation (tPBM), is an increase in ATP formation (Lapchak and De Taboada, 2010), which then leads to the preservation of compromised tissue in the ischemic penumbra. PBM can reduce apoptosis by altering mitochondrial signaling molecules, and upregulating antiapoptotic proteins (Bcl2 and survivin) thus enhancing neuroprotection (Blivet et al., 2018) and neurorecovery (Janzadeh et al., 2016). Since AD has been linked to mitochondrial dysfunction (Lynn et al., 2010), tPBM may be a viable treatment for AD (Hamblin, 2016). Therefore, we hypothesized that the efficacy of tPBM may be mediated by enhanced mitochondrial function (Trimmer et al., 2009).

Amyloid-β (Aβ)-containing senile plaques represent one of the neuropathological hallmarks of AD with considerable effort, having been expended in understanding the relationship of Aβ and Aβ-containing senile plaques to AD pathophysiology (Harrington, 2012). Much of this work has focused on the biosynthesis of Aβ and factors that influence its production and deposition. Aβ peptides are primarily generated via internal proteolysis of its precursor, the amyloid-β protein precursor (AβPP), giving rise to two peptides of either 40 or 42 amino acids (De Strooper and Annaert, 2000). In addition to Aβ-containing senile plaques, a variety of neuronal cytoskeletal alterations are prominent features of AD neuropathology. These features include hyper-phosphorylated-tau-containing neurofibrillary tangles, dystrophic neurites present in senile plaques, and loss and deterioration of synapses (Selkoe, 1999). Whether these abnormal features are the result of or cause of neuronal loss is still controversial (Jack et al., 2016). Regardless of the precise mechanism, this neuronal and synaptic loss leads to cognitive decline (De-Paula et al., 2012).

Early onset autosomal dominant AD is directly linked to mutations in one of several genes: AβPP, presenilin 1 (PS1), or presenilin 2 (PS2) (Levy-Lahad et al., 1995; Lin et al., 2000). In addition, several risk factor genes, most notably the apolipoprotein E 4 allele, alter the risk for later onset AD (Strittmatter et al., 1993). It is therefore clear that mutations or polymorphisms in several genes can lead to similar AD phenotypes (Sancesario and Bernardini, 2018). Secretases act on APP to cleave the protein into three fragments. Sequential cleavage by β-secretase (BACE) and γ-secretase produces the AβP peptide fragment that aggregates into clumps called plaques. If α-secretase acts on APP first instead of BACE, no amyloid-β is formed because α-secretase recognizes a target protein sequence closer to the cell surface than BACE. The nonpathogenic middle fragment formed by an α/γ cleavage sequence is called P3 (John, 2006). In addition, the γ-secretase enzyme (a complex of PS1 and PS2) cleaves the transmembrane domain to release

Photobiomodulation in the Brain. DOI: https://doi.org/10.1016/B978-0-12-815305-5.00015-4

the Aβ peptide and the carboxyl terminus fragment (Weihofen et al., 2002). Altered functions of any of these enzymes can lead to enhanced production of Aβ peptide, which may contribute to AD pathogenesis. A number of studies have shown that mutations in the APP gene or in presenilins result in increased β-secretase cleavage and production of both Aβ1-40 and Aβ1-42 (Selkoe, 2001).

Transgenic (Tg) mice that overexpress mutant familial AD APP genes have contributed to an understanding of AD pathology, and support the amyloid cascade hypothesis (Kokjohn and Roher, 2009). Although many sophisticated mice APP models exist, none perfectly recapitulate AD cellular and behavioral pathology. The morphological resemblance to AD amyloidosis is impressive, but fundamental biophysical and biochemical properties of the APP/Aβ produced in Tg mice differ substantially from those of humans. The greater resilience of Tg mice in the presence of substantial Aβ burdens suggests that levels and protein isoforms that are deleterious to human neurons, are not as noxious to these mice. We assessed the effects of tPBM in a hAPPwt mouse model (Hook et al., 2009) by determining the effects of tPBM on the reduction in Aβ peptide levels, and inflammatory markers, on the halting and reversal of amyloid deposition. Moreover, the key goal of AD therapy, that is, behavioral improvement was tested (De Taboada et al., 2011).

15.2 Study design

The experimental design was described in detail in our previous publication (De Taboada et al., 2011). One hundred male AβPP transgenic mice at 3 months of age, were randomly divided into five groups. Mice received tPBM or sham treatment three times a week for 6 months. The optical parameters were: spot size, 3 mm diameter; laser wavelength, 810 nm; peak radiant powers of 40, 200 or 400 mW; laser modulation, either CW or pulsed with 2 ms pulse duration at 100 Hz. Group 1 was sham, group 2 was CW (40 mW), group 3 was pulsed (P1) 40 mW, group 4 was pulsed (P2) 200 mW, and group 5 was pulsed (P3) 400 mW. Mice were tested once in the MWM between days 176 and 180. Mice were sacrificed on day 180 and the brains were removed for analysis. The amyloid load in the brain was measured by immunohistochemistry using thin sections, levels of Aβ peptides and cytokines in cerebrospinal fluid (CSF), plasma and brain were measured by ELISA, ATP was measured by a luciferin/luciferase assay, and oxygen consumption by mitochondrial fractions with a Clarke electrode.

15.3 Transcranial photobiomodulation improves cognitive performance as measured by Morris Water Maze

It is important when studying new treatments in mouse models of AD to include a cognitive performance test in the outcome measures, as previous studies relying solely on amyloid peptide or plaque measurements have not translated into effective clinical treatments (Mehta et al., 2017).

In the present study, we used the Morris Water Maze (MWM) which is probably the most well-validated behavioral test for spatial memory and learning that can be applied to rodent models (Edwards et al., 2014). Fig. 15.1 shows the improvements in cognitive performance as measured by the MWM. The same pattern was evident in both MWM outcome measures in which the pulsed light at 200 mW was the best, with the pulsed at 40 and 400 mW less effective and the CW light least effective (but still significantly better than the sham). The improvement in time spent in target quadrant (memory) with the P2 regimen was impressive (threefold increase).

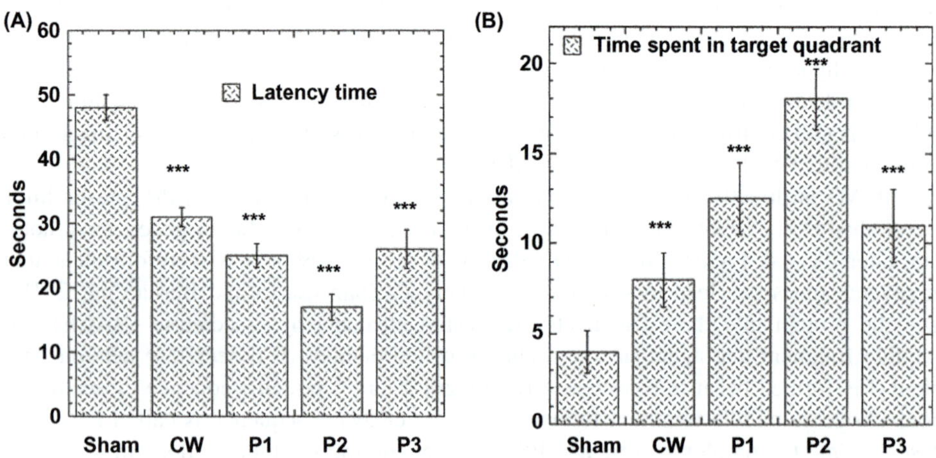

FIGURE 15.1 Effect of tPBM on Morris Water Maze behavioral studies. (A) Effect of tPBM on Morris water maze latency time (s). (B) Effect of tPBM on Morris water maze time spent in the target quadrant (s). Mean6SEM (n520) ***P,.001, **P,.01, *P,.05.

15.4 Transcranial photobiomodulation lowers the amyloid load in brain and reduces levels of Aβ peptides in brain, cerebrospinal fluid, and plasma

We next proceeded to measure the levels of amyloid load and the levels of Aβ peptides in brain, CSF, and plasma as shown in Fig. 15.2. The pattern was similar for all four measures, that is, P2 < P1 ≅ P3 < CW < sham. By and large the reductions in the brain were larger than the reductions in the CSF and in the plasma.

15.5 Transcranial photobiomodulation reduces inflammation in the brain

One important feature of AD that should be affected by any potential treatment, is the neuroinflammation that is so characteristic of the disease progression. This inflammation is hypothesized to arise from activated microglia that are failing to remove the accumulating amyloid plaque (Cai et al., 2014). We measured the levels of three prototypical M1 microglial inflammatory cytokines (interleukin-1beta IL-1β; (B) tumor necrosis factor alpha, TNFα; (C) transforming growth factor beta, TGFβ) in brain samples by ELISA. The results are shown in Fig. 15.3. It can be seen that the pattern was remarkably similar for all three cytokines. There was a large increase in the transgenic AD mice compared to their age matched wild-type counterparts, the CW tPBM produced a modest decrease, the P1 and P3 regimens produced a further decrease, while in every case the P2 regimen produced the largest decrease.

A three-way vicious cycle of inflammation can be formed between Aβ accumulation, activated microglia, and microglial inflammatory mediators, which enhances further Aβ deposition and increases neuroinflammation. Since PBM can reverse the microglial activation phenotype from M1 (cytokines) to M2 (phagocytosis), this may not only reduce M1 inflammatory cytokines, but also allow the M2 microglia to dispose of the accumulated plaque.

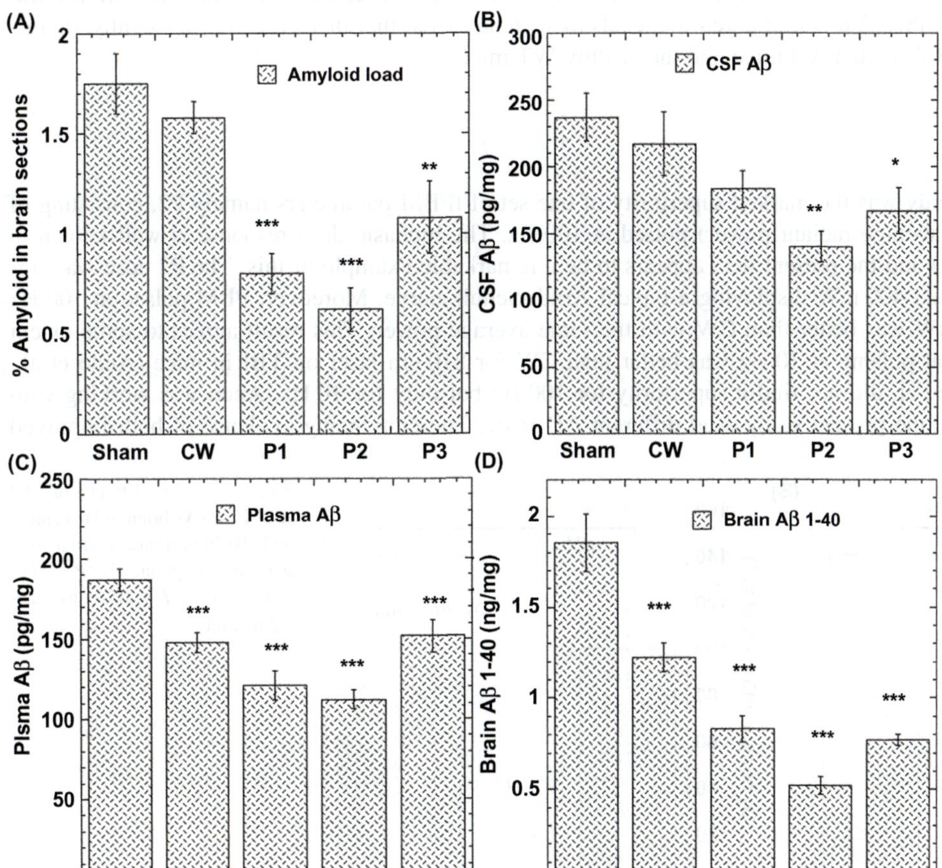

FIGURE 15.2 Effect of tPBM on (A) amyloid load in brain, (B) level of Aβ peptides in CSF, (C) level of Aβ peptides in plasma, (D) level of Aβ1-40 peptide in brain. $^{***}P < .001$, $^{**}P < .01$, $^{*}P < .05$.

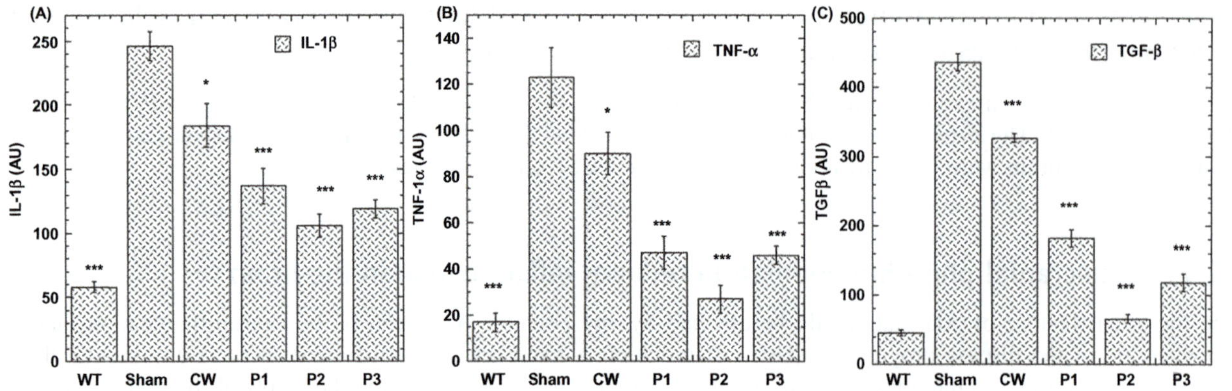

FIGURE 15.3 Effect of tPBM on (A) interleukin-1beta, (B) tumor necrosis factor alpha, (C) transforming growth factor beta, in brain samples measured by ELISA. $***P<.001$, $**P<.01$, $*P<.05$.

15.6 Transcranial photobiomodulation improves mitochondrial function in the brain

One of the earliest established and most robust mechanisms of PBM was the increase in mitochondrial ATP synthesis (Pastore et al., 1996) and a consequent increase in oxygen consumption due to respiration (Pastore et al., 1994). Since it is accepted that an important feature of AD brains is mitochondrial dysfunction (Picone et al., 2014), it made sense to measure ATP and mitochondrial oxygen consumption in the AD mouse brain samples after the tPBM protocols. We only used the P2 tPBM protocol since by now, we had convinced ourselves that this was clearly the best set of parameters. The results are shown in Fig. 15.4.

The transgenic mice had a marked decrease in both brain ATP levels (only one-third of the WT type mice) and in mitochondrial oxygen consumption (less than half the WT mice). However, in both cases (brain ATP and brain mitochondrial oxygen consumption), the P2 tPBM restored the levels to such an extent that there was no statistically significant difference between the tPBM treated AD mice and the healthy WT mice.

15.7 Discussion

One interesting result of this study was the marked superiority of one set of tPBM parameters namely P2, consisting of 200 mW peak power (40 mW average radiant power) pulsed at 100 Hz. The biphasic dose response is well known in PBM (Huang et al., 2009, 2011) and the present data appears to be a remarkable example of this. The P2 dose was significantly superior to both 40 and 400 mW displaying a typical Arndt-Schulz curve. Moreover, PBM pulsed at 100 Hz with an average power of 40 mW was better than CW with the same average power. This result agrees to some extent with Ando et al. who found that pulsing at 10 Hz was better than CW for 810 nm laser on TBI in mice (Ando et al., 2011). However, while, Ando et al. did not find a superiority for 100 Hz but only for 10 Hz. Ando was working with mice suffering from TBI, while the present study used different AD mice. The convincing evidence of both improved

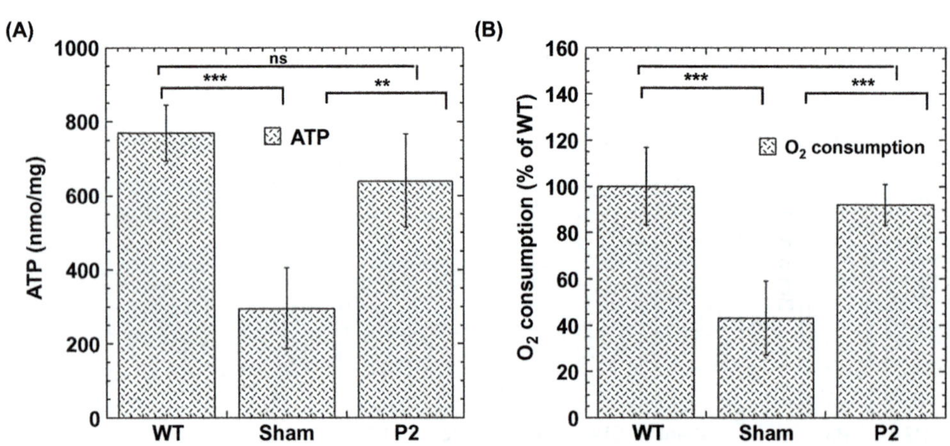

FIGURE 15.4 Effect of P2 tPBM on (A) brain ATP content and (B) brain mitochondrial oxygen consumption. $***P<.001$, $**P<.01$, $*P<.05$. ns = not significant.

cognitive performance together combined with biochemical measurement of Aβ and amyloid plaque load do suggest that AD will be treatable with PBM.

15.8 Conclusion

Since this study was first published in 2011 (De Taboada et al., 2011), there have been a few preliminary reports published describing case studies of human patients with AD or dementia being benefitted by PBM. Saltmarche et al. (2017) reported a case series of five patients with mild to moderately severe dementia or possible Alzheimer's disease (AD) with Mini-Mental State Exam (MMSE) baseline scores of 10–24. Patients were treated with 810 nm, 10 Hz pulsed LED devices combining transcranial plus intranasal PBM to treat the cortical nodes of the fault mode network (DMN, bilateral mesial prefrontal cortex, precuneus/posterior cingulate cortex, angular gyrus, and hippocampus) for 12 weeks of active treatment as well as a follow-up no-treatment, 4-week period. There was significant improvement after 12 weeks of PBM (MMSE, $P < .003$; ADAS-cog, $P < .023$). Increased function, better sleep, fewer angry outbursts, less anxiety, and wandering were reported post-PBM.

Berman et al. (2017) carried out a small pilot double blind, placebo-controlled trial ($n = 11$) 6 active, 3 controls, and 2 dropouts assessing the effect of 28 consecutive daily, six minute sessions of transcranial NIR PBM using 1060–1080 nm LEDs. Results showed improvement in executive functioning; clock drawing, immediate recall, praxis memory, visual attention, and task switching (Trails A&B) as well as a trend for improved EEG amplitude and connectivity measures.

Maksimovich (2011) used a different approach. He reported a series of 46 patients aged 34–79 (average 65) with a history of AD, who received endovascular surgery leading to transcatheter revascularization and recovery of collateral and microvascular circulation of the brain by means of low-energy transluminal laser irradiation. Patients had a positive outcome evidenced by a prolonged decline in dementia symptoms, and a decrease in cognitive impairment.

Taken together, the present mouse study added to the three preliminary clinical studies discussed above, suggests that PBM may indeed be a highly encouraging treatment approach ahead of the expected "epidemic of Alzheimer's disease" (Trempe and Lewis, 2018).

References

Ando, T., Xuan, W., Xu, T., Dai, T., Sharma, S.K., Kharkwal, G.B., et al., 2011. Comparison of therapeutic effects between pulsed and continuous wave 810-nm wavelength laser irradiation for traumatic brain injury in mice. PLoS One 6 (10), e26212.

Berman, M.H., Halper, J.P., Nichols, T.W., Jarrett, H., Lundy, A., Huang, J.H., 2017. Photobiomodulation with near infrared light helmet in a pilot, placebo controlled clinical trial in dementia patients testing memory and cognition. J. Neurol. Neurosci. 8, 1.

Blivet, G., Meunier, J., Roman, F.J., Touchon, J., 2018. Neuroprotective effect of a new photobiomodulation technique against Abeta25-35 peptide-induced toxicity in mice: Novel hypothesis for therapeutic approach of Alzheimer's disease suggested. Alzheimers Dement. (N.Y.) 4, 54–63.

Cai, Z., Hussain, M.D., Yan, L.J., 2014. Microglia, neuroinflammation, and beta-amyloid protein in Alzheimer's disease. Int. J. Neurosci. 124 (5), 307–321.

De-Paula, V.J., Radanovic, M., Diniz, B.S., Forlenza, O.V., 2012. Alzheimer's disease. Subcell Biochem. 65, 329–352.

De Strooper, B., Annaert, W., 2000. Proteolytic processing and cell biological functions of the amyloid precursor protein. J. Cell Sci. 113 (Pt 11), 1857–1870.

De Taboada, L., Yu, J., El-Amouri, S., Gattoni-Celli, S., Richieri, S., McCarthy, T., et al., 2011. Transcranial laser therapy attenuates amyloid-beta peptide neuropathology in amyloid-beta protein precursor transgenic mice. J. Alzheimers Dis. 23 (3), 521–535.

Edwards, S.R., Hamlin, A.S., Marks, N., Coulson, E.J., Smith, M.T., 2014. Comparative studies using the Morris water maze to assess spatial memory deficits in two transgenic mouse models of Alzheimer's disease. Clin. Exp. Pharmacol. Physiol. 41 (10), 798–806.

Hamblin, M.R., 2016. Shining light on the head: photobiomodulation for brain disorders. BBA Clin. 6, 113–124.

Harrington, C.R., 2012. The molecular pathology of Alzheimer's disease. Neuroimaging Clin. N. Am. 22 (1), 11–22.

Hook, V.Y., Kindy, M., Reinheckel, T., Peters, C., Hook, G., 2009. Genetic cathepsin B deficiency reduces beta-amyloid in transgenic mice expressing human wild-type amyloid precursor protein. Biochem. Biophys. Res. Commun. 386 (2), 284–288.

Huang, Y.Y., Chen, A.C., Carroll, J.D., Hamblin, M.R., 2009. Biphasic dose response in low level light therapy. Dose Response 7 (4), 358–383.

Huang, Y.Y., Sharma, S.K., Carroll, J.D., Hamblin, M.R., 2011. Biphasic dose response in low level light therapy – an update. Dose Response 9 (4), 602–618.

Jack Jr, C.R., Knopman, D.S., Chetelat, G., Dickson, D., Fagan, A.M., et al., 2016. Suspected non-Alzheimer disease pathophysiology--concept and controversy. Nat. Rev. Neurol. 12 (2), 117–124.

Janzadeh, A., Nasirinezhad, F., Masoumipoor, M., Jameie, S.B., Hayat, P., 2016. Photobiomodulation therapy reduces apoptotic factors and increases glutathione levels in a neuropathic pain model. Lasers Med. Sci. 31 (9), 1863–1869.

John, V., 2006. Human beta-secretase (BACE) and BACE inhibitors: progress report. Curr. Top. Med. Chem. 6 (6), 569–578.

Karu, T.I., 2010. Multiple roles of cytochrome c oxidase in mammalian cells under action of red and IR-A radiation. IUBMB Life 62 (8), 607–610.

Kokjohn, T.A., Roher, A.E., 2009. Amyloid precursor protein transgenic mouse models and Alzheimer's disease: understanding the paradigms, limitations, and contributions. Alzheimers Dement. 5 (4), 340–347.

Lapchak, P.A., De Taboada, L., 2010. Transcranial near infrared laser treatment (NILT) increases cortical adenosine-5′-triphosphate (ATP) content following embolic strokes in rabbits. Brain Res. 1306, 100–105.

Levy-Lahad, E., Wasco, W., Poorkaj, P., Romano, D.M., Oshima, J., Pettingell, W.H., et al., 1995. Candidate gene for the chromosome 1 familial Alzheimer's disease locus. Science 269 (5226), 973–977.

Lin, X., Koelsch, G., Wu, S., Downs, D., Dashti, A., Tang, J., 2000. Human aspartic protease memapsin 2 cleaves the beta-secretase site of beta-amyloid precursor protein. Proc. Natl. Acad. Sci. U.S.A. 97 (4), 1456–1460.

Lynn, B.C., Wang, J., Markesbery, W.R., Lovell, M.A., 2010. Quantitative changes in the mitochondrial proteome from subjects with mild cognitive impairment, early stage, and late stage Alzheimer's disease. J. Alzheimers Dis. 19 (1), 325–339.

Maksimovich, I., 2011. Endovascular low-energy laser radiation effect on dyscirculatory angiopathy of Alzheimer's type in the treatment of Alzheimer's disease. Alzheimer's Dement. J. Alzheimer''s Assoc. 7, S791–S793.

Mehta, D., Jackson, R., Paul, G., Shi, J., Sabbagh, M., 2017. Why do trials for Alzheimer's disease drugs keep failing? A discontinued drug perspective for 2010-2015. Expert Opin. Investig. Drugs 26 (6), 735–739.

Pastore, D., Greco, M., Petragallo, V.A., Passarella, S., 1994. Increase in <--H + /e- ratio of the cytochrome c oxidase reaction in mitochondria irradiated with helium-neon laser. Biochem. Mol. Biol. Int. 34 (4), 817–826.

Pastore, D., Di Martino, C., Bosco, G., Passarella, S., 1996. Stimulation of ATP synthesis via oxidative phosphorylation in wheat mitochondria irradiated with helium-neon laser. Biochem. Mol. Biol. Int. 39 (1), 149–157.

Picone, P., Nuzzo, D., Caruana, L., Scafidi, V., Di Carlo, M., 2014. Mitochondrial dysfunction: different routes to Alzheimer's disease therapy. Oxid. Med. Cell. Longev. 2014, 780179.

Saltmarche, A.E., Naeser, M.A., Ho, K.F., Hamblin, M.R., Lim, L., 2017. Significant improvement in cognition in mild to moderately severe dementia cases treated with transcranial plus intranasal photobiomodulation: case series report. Photomed. Laser Surg. 35 (8), 432–441.

Sancesario, G.M., Bernardini, S., 2018. Alzheimer's disease in the omics era. Clin. Biochem. 59, 9–16.

Selkoe, D.J., 1999. Translating cell biology into therapeutic advances in Alzheimer's disease. Nature 399 (6738 Suppl.), A23–A31.

Selkoe, D.J., 2001. Presenilin, Notch, and the genesis and treatment of Alzheimer's disease. Proc. Natl. Acad. Sci. U.S.A. 98 (20), 11039–11041.

Strittmatter, W.J., Saunders, A.M., Schmechel, D., Pericak-Vance, M., Enghild, J., Salvesen, G.S., et al., 1993. Apolipoprotein E: high-avidity binding to beta-amyloid and increased frequency of type 4 allele in late-onset familial Alzheimer disease. Proc. Natl. Acad. Sci. U.S.A. 90 (5), 1977–1981.

Trempe, C.L., Lewis, T.J., 2018. It's never too early or too late-end the epidemic of Alzheimer's by preventing or reversing causation from pre-birth to death. Front. Aging Neurosci. 10, 205.

Trimmer, P.A., Schwartz, K.M., Borland, M.K., De Taboada, L., Streeter, J., Oron, U., 2009. Reduced axonal transport in Parkinson's disease cybrid neurites is restored by light therapy. Mol. Neurodegener. 4, 26.

Tuner, J., Hode, L., 2002. Laser Therapy – Clinical Practice and Scientific Background. Prima Books, Grängesberg.

Weihofen, A., Binns, K., Lemberg, M.K., Ashman, K., Martoglio, B., 2002. Identification of signal peptide peptidase, a presenilin-type aspartic protease. Science 296 (5576), 2215–2218.

Wong-Riley, M.T., Liang, H.L., Eells, J.T., Chance, B., Henry, M.M., Buchmann, E., et al., 2005. Photobiomodulation directly benefits primary neurons functionally inactivated by toxins: role of cytochrome c oxidase. J. Biol. Chem. 280 (6), 4761–4771.

Chapter 16

Low-level laser therapy to the bone marrow: a new therapeutic approach to neurodegenerative diseases

Amir Oron[1] and Uri Oron[2]

[1]*Department of Orthopedic Surgery, Kaplan Medical Center, Rehovot, Israel,* [2]*Department of Zoology, George S. Wise Faculty of Life Sciences and Sagol School of Neuroscience, Tel Aviv University, Tel Aviv, Israel*

Regenerative capacity following injury or ischemic event is confined to non mammalian vertebrates. In particular, fish and primitive amphibians can regenerate organs like the heart, brain, and limbs. However, mammals have limited capacity to restore organs flowing injury, in organs such as the liver and skeletal muscles, but practically no ability to regenerate organs like the heart or brain postischemic event or injury. The mammalian heart including the human heart, for example, has a negligible capacity to regenerate following damage or an acute ischemic event like myocardial infarction (MI). This is due to the very low level of cardiomyocyte proliferation and the limited number of cells expressing stem-cell marker proteins. Stem-cells-based therapy was suggested as a potential solution to the above situation. In recent years, cell-based therapy for organ repair has undergone a rapid transition from basic science research to clinical reality (Mummery et al., 2010). There are several central issues pertaining to the use of cell implantation in stem-cell therapy: the number of implanted stem cells has to be high since there is massive cell death following implantation or injection of cells into the heart or the blood circulation. Another central issue in stem-cell implantation for organ repair is the creation of a receptive cell environment in the ischemic organ. Several factors (e.g., inhibition of inflammation and apoptosis, secretion of cell growth factors etc.) are necessary for optimal cell implantation. The injected cells may have to migrate from the circulating blood to the ischemic niche. They can then remain active and secrete growth factor, exerting a paracrine effect on the ischemic tissue (Gnecchi et al., 2005). We were trying to overcome some of the issues of stem-cell therapy mentioned above by using a new approach of delivering low-level laser therapy (LLLT) to autologous bone marrow (BM), which is enriched with stem cells and various progenitor cells, in order to induce those cells for the benefit of ischemic organs or organ going through a process of degeneration. The approach of LLLT to the BM could be beneficial for neurodegenerative diseases was also examined by us recently. Bone-marrow stem cells have the ability to populate the entire central nervous system into fully differentiated parenchymal microglia. It has also been suggested that blood-derived microglia, and not their resident counterparts, have the ability to eliminate amyloid deposits by a cell-specific phagocytic mechanism (Simard et al., 2006). Defining efficient ways by which to stimulate BM stem cells to migrate to the brain might thus offer a crucial step in stem-cell therapy in Alzheimer's disease (AD). We therefore postulate that LLLT to the tibia bone of the hind-leg in an AD animal model showing profound neuronal death in the brain, will have two main beneficial effects by activating BM derived microglia toward clearance of the neurotoxic amyloid-beta (Aβ) oligomers and fibrils in addition to induction of neurogenesis towards neuroprotection. In this communication the ability of LLLT to stimulate mesenchymal stem cells (MSCs) and other cells in autologous BM and enhance their capacity to infiltrate the brain, clear β-amyloid, and improve cognition is reviewed.

AD affects more than 18 million people worldwide and is characterized by progressive memory deficits, cognitive impairment, and personality changes. The main cause of AD is generally attributed to the increased production and accumulation of Aβ, in association with neurofibrillary tangle formation that leads to neuronal death. This process mainly affects the hippocampus which is critical for learning and memory (Selkoe, 2004). The presence of stem cells in

Photobiomodulation in the Brain. DOI: https://doi.org/10.1016/B978-0-12-815305-5.00016-6

this structure has led to an increased interest in the phenomenon of adult neurogenesis and its role in hippocampal functioning (Morgan, 2007). Many known factors impacting neurogenesis in the hippocampus are implicated in the pathogenesis of AD (Rodriguez and Verkhratsky, 2011). Since neurogenesis is modifiable, stimulation of this process, or the potential use of stem cells, recruited endogenously or implanted by transplantation, has been speculated as a possible treatment for neurodegenerative disorders such as AD. It was previously demonstrated that the source of brain cells might be either local from the subventricular zone of the forebrain or the subgranular zone of the hippocampus. The possibility that stem cells can be derived from a peripheral zone of the brain such as from the BM, was also explored (Uccelli et al., 2008). In the same study it was shown that after exposure of these cells to neurogenic differentiation conditions their neurogenic phenotype was expressed indicating that these cells might be another source of neural glial cells for cell-based therapies. The BM features several different types of pluripotent cells: hematopoietic stem cells, MSCs, endothelial progenitor cells, side population cells, and multipotent adult progenitor cells. Like other stem cells, MSCs are capable of multilineage differentiation from a single-cell and in vivo functional reconstitution of injured tissues. One of the properties of stem cells is their capacity to migrate to one or more appropriate micro-environment (Devine et al., 2001). Certain stem cells are able to exit their production site and circulate in the blood before reseeding in their target tissues. For MSCs, the nature of homing sites and their kinetics in circulation peripheral blood is still under debate. However, following infusion MSCs have been found in multiple tissues, leading to the hypothesis that they have the ability to home and adjust their differentiation pathways to diverse tissue micro-environments (Liechty et al., 2000). It was recently shown that intracerebral transplantation of bone-marrow-derived MSCs into the brain of an induced AD model reduced the Aβ protein levels and accelerated the activation of microglia cells when compared to sham-transplanted animals. Furthermore, it was suggested that blood-derived microglia like cells have the ability to eliminate amyloid deposits by means of a cell-specific phagocytic mechanism (Simard et al., 2006).

LLLT has been found to have photobiomodulation (PBM) effect on various biological processes (Oron, 2006; Karu, 2007). The PBM of processes in the brain by LLLT has been addressed by several studies. Transcranially applied LLLT has been shown to have beneficial effects on rats, rabbits, and humans poststroke (Lapchak et al., 2004; Oron et al., 2006 ; Lampl et al., 2007). Furthermore, LLLT applied multiple times transcranially to AD mouse had improved neurological function in those mice over nonlaser treated ones (De Taboada et al., 2011).

PBM of stem cells or progenitor cells by LLLT has not been extensively studied. Low-level laser application to adipose derived stem cells attenuated their activity (Mevula et al., 2010). Laser application to normal human neural progenitor cells significantly increases ATP production in these cells (Oron et al., 2007). LLLT applied to autologous BM in rats after MI led to a 79% significant reduction of the extent of scarring in the heart post-MI as well as inducing cardiogenesis along the border line of the infarcted area (Tuby et al., 2011, 2013a,b). This phenomenon could be partially attributed to a higher extent of the laser-induced MSC from the BM being mobilized to the infarcted area in the heart. In a recent study using the porcine model of MI, it was found that applying LLLT to their BM caused significant (68%) reduction of scarring relative to nonlaser treated pigs (Blatt et al., 2016). Furthermore, in the same study, the left ventricular ejection fraction in the laser treated pigs was found to be significantly higher than in the nonlaser treated ones.

The rationale behind the attempt to use LLLT to induce the entire cells in the BM was that one cannot significantly affect the complex processes post-MI or ischemic injury in other organs with a single type of stem cells. Macrophages have been shown recently to have a crucial role in mitigating the scarring process post-MI. It may be hypothesized that LLLT may induce concomitantly various types of cells in the BM that will increase in number in the blood circulation following their enhanced proliferation or mobilization from the BM. These cells will probably, eventually, to a certain extent and under certain circumstances, home in on the ischemic zone in the ischemic or degenerated organs. In a recent study (Blatt et al., 2016) we tried to determine whether LLLT to the BM can activate a beneficial immune response or induce stem cells to home on in the brain at progressive stages of an AD mouse model.

We first intended to evaluate the ability of laser-treated MSCs to phagocytose Aβ proteins. MSCs were isolated and then seeded in 24-well culture plates at a concentration of 1.3×10^6 cell/cm^2 for 1 week. The cultured MSCs were exposed to the Ga-Al-As laser for 20 seconds at a power density of 50 mW/cm^2 to yield 1.0 J/cm^2 energy density. Another set of six well plates containing MSCs were sham-exposed (control) to the laser (cells treated as above, but the laser was not turned on). The laser-treated and the control MSCs were left in the incubator for 3 days postlaser treatment and then incubated until 70% confluence. For phagocytosis of Aβ, MSCs cells were labeled with anti-CD11b antibody, and the percentage of Aβ phagocytosis was analyzed by fluorescence-activated cell sorting (FACS).

A significant ($P = .041$) 35% increase in phagocytosis of Aβ (1−42) was found in the cells that were laser treated as compared to the nonlaser-treated cells. Furthermore, a significant ($P < .0001$) 10% increase was detected in the CD11b activation marker of monocyte-derived cells. These results suggest that laser application to monocytes or other cell types that demonstrate phagocytotic activity, among the MSC population in the BM, can cause significant activation of these cells and, hence, enhance their capacity to uptake specifically Aβ proteins accumulated in the brain in AD mice. Indeed the ability of macrophages to act as phagocytes was found to be modulated under the application of LLLT (Gavish et al., 2008). Thus in our recent study (Farfara et al., 2015) we demonstrated a significant elevation in the activation of immune cells, detected by CD11b in MSC, following LLLT. Furthermore, we showed an increase in MSC reactivity to phagocytose soluble neurotoxic Aβ, which tends to lead to toxic oligomers in the brain. It can be hypothesized that these cells can be activated in the BM, migrate to the circulating blood, and then infiltrate to the brain and reduce the amyloid burden there. It was previously suggested that the migration of peripherally-derived mononuclear cells leads to clearance of amyloid load in an AD mouse model and improves cognition (Simard et al., 2006). Moreover, several studies have suggested that activation of peripheral monocyte-derived macrophages, BM derived cells, and microglia can play a role in clearance of brain amyloids in an AD mouse model (Butovsky et al., 2007; Frenkel et al., 2008).

Following the findings regarding the enhanced capacity of the laser-induced CD11b positive cells in the BM to phagocytose neurotoxic soluble Aβ in vitro, we conducted an in vivo study in an AD mice model. For this mouse model, 5X FAD transgenic male mice (Tg6799) generations were used. 5X FAD mice demonstrate amyloid burden starting from 2 months of age. By the age of 4 months, this mouse model features a high amyloid load, commencing in the cortex and expanding to the hippocampus. At 6 months of age mice demonstrate a robust amyloid burden in both the cortex and the hippocampus. In this study we sought to investigate the effect of LLLT at the progressive stage of the disease. Four-month-old AD mice were therefore treated every 10 days for 2 months with LLLT (totaling six treatments) to their BM. A tunable laser (Ga-Al-As laser, wavelength 808 nM) with power output of maximum of 400 mW was used. LLLT to the BM was performed as described previously (Tuby et al., 2011). The power density of the laser beam was 10 mW/cm^2 on the BM and duration of irradiation 100 seconds to yield 1.0 J/cm^2 energy density. Control mice underwent the same procedure as the laser-irradiated group but the laser was not turned on. Mice were divided into three groups: AD mice treated with LLLT every 10 days; sham-operated AD mice; and intact WT mice of the same strain as the AD mice. At the age of 6 months, mice were subjected to behavioral and cognitive tests and then sacrificed and the brain fixed and processed for β-amyloid burden in the brain. Two neurobehavioural tests were used: object recognition test (ORT) and contextual fear-conditioning test (FCT). ORT is distinguished by more time spent interacting with the novel object. Memory was operationally defined by the discrimination index for the novel object as the proportion of time the mice spent investigating the novel object in comparison with the familiar one. In the FCT test male mice were subjected to an unconditioned electric stimulus in a presession training. Twenty-four hours later, FCT was measured by scoring freezing behavior (the absence of all but respiratory movement) for 180 seconds using a freeze-frame automated scoring system. Following the neurobehavioural tests the brains (left hemisphere) were cut into sagittal sections with a cryostat at −20°C, and used for histological examination. The slices were stained with congo-red staining and anti-Aβ, and visualized by fluorescence microscopy for quantification of amyloid depositions. The hippocampal Aβ burden was presented as the percentage of insoluble total Aβ and congo-red positive region of the entire hippocampus region.

The results from the ORT revealed that application of LLLT to the BM of AD mice significantly elevated the percentage of time spent near the new object almost to the level of that of the WT mice. WT 6-month-old mice demonstrated an average of 73% ± 4.11% of their time spent around a new object. This value was significantly ($P < .01$) reduced to 47.3% ± 5.6% in the group of 6-month-old nonlaser-treated AD mice, indicating a significant memory loss in the latter. However, the average percentage of time spent near a new object in the laser-treated mice was 68.7% ± 3.1%, suggesting that the multiple LLLT application to the BM of AD mice had led to recovery of their memory loss. There was no statistical difference in the time spent near a new object between the WT mice and the AD mice treated by LLLT. The FCT of nonlaser-treated AD mice showed poorer cognitive abilities (11.6 ± 4.6 seconds) than the WT mice (71.1 ± 4.6 seconds). Laser-treated AD mice showed a significantly increased freezing time of 40.4 ± 5.28 seconds, compared to nonlaser-treated mice. These results indicate a significantly enhanced cognitive ability and memory-gain in the laser-treated BM mice, compared to nonlaser-treated mice. Amyloid burden in the brains of the AD mice after 2 months of treatment with the LLLT was also found to correlate with the in vitro and behavior tests. The percentage of Aβ burden as revealed from the histology in the hippocampus region of the nonlaser-treated mice was 180 ± 15, while in the laser-treated mice there was a significant reduction of 68% ($P < .05$) in Aβ burden relative to the control mice.

Thus, in this study (Farfara et al., 2015) it has been demonstrated that LLLT applied to autologous BM induces MSC cell activation toward phagocytosis of toxic Aβ, leading to a cognitive function improvement in an AD mouse model. It was found that LLLT treatment led to a significant reduction in brain amyloid load following a short period of treatment, starting at the late progressive disease stage. Furthermore, LLLT treatment improved cognitive behavior in the laser-treated AD mice as compared to nontreated mice at an advanced and progressive stage of AD disease. These results also correlated with the general beneficial effect of applying LLLT transcranially to AD mice (De Taboada et al., 2011). However, in the current study, laser was applied for a shorter period of time and at a less frequent application of LLLT than in the study by De Taboada et al. (2011). In addition, LLLT was applied in this study to autologous BM cells as a target organ that is remote from the impaired brain.

The amyloid burden (at 6 months) in the AD mouse model was found to be significantly reduced in the LLLT-treated mice as compared to control mice The results of the behavioral tests in this study are in concert with a reduction of amyloid burden in the brain. They indicate a significantly improved cognitive ability and memory in the laser-treated mice over the nonlaser-treated ones. It should be noted that regarding the ORT, the mice that received multiple applications of LLLT to the BM between 4 and 6 months of age demonstrated a significant improvement that reached the level of the cognitive ability of the WT mice.

The present study also has clinical relevance. The safety of LLLT application (at a similar power density as in the current study) in experimental animals and in double-blind studies in humans postacute stroke has been reported (Lampl et al., 2007). Moreover, we recently demonstrated that LLLT application even at higher power densities to the BM of mice did not cause any histological changes in various organs over a period of almost their entire life-span (Tuby et al., 2013a,b). Thus it may be assumed that LLLT to the BM in humans will be safe. Our ability to demonstrate that LLLT application to the BM improves cognitive brain function and reduces plaque concentration in the brain of AD mice, even when treatment is commenced at a progressive stage, is of significance. It suggests that LLLT could be applied to humans with AD, which is usually only diagnosed when already at a progressive stage.

The novel approach presented here, of the use of stem cells for cell therapy to the infarcted heart, avoids the need to isolate stem cells, to grow them in vitro and to inject them back into the patients. It also avoids the massive loss of cells involved in cell implantation/injection due to insufficient seeding of cells or cell death shortly after implantation. The approach in the present study also overcomes the problem of growing autologeous stem cell cultures, determination of the optimal amount of implanted cells, and the optimal timing for their delivery.

In conclusion, our results indicate a novel approach of applying LLLT to autologous BM of AD mice, which induces the production of stem cells and immune cells which are then recruited to the brain, demonstrating the possibility of the patient's own abilities to initiate a regenerative response in an organ, leading to a marked beneficial effect. LLLT thus offers a potential therapeutic strategy in treating symptoms of AD and perhaps also other neurodegenerative diseases.

Acknowledgment

The authors wish to thank Ms. Naomi Paz for editorial assistance.

References

Blatt, A., Tuby, H., Maltz, L., Siman Tov, Y., Ben-Aharon, G., Kopel, L., et al., 2016. Low-level laser therapy to the bone marrow reduces scarring and improves heart function post myocardial infarction in the porcine model. Photomed. Laser Surg. 34, 516–524.

Butovsky, O., Kunis, G., Koronyo-Hamaoui, M., Schwartz, M., 2007. Selective ablation of bone marrow-derived dendritic cells increases amyloid plaques in a mouse Alzheimer's disease model. Eur. J. Neurosci. 26 (2), 413–416.

De Taboada, L., Yu, J., El-Amouri, S., et al., 2011. Transcranial laser therapy attenuates amyloid-beta peptide neuropathology in amyloid-beta protein precursor transgenic mice. J. Alzheimers Dis. 23 (3), 521–535.

Devine, S.M., Bartholomew, A.M., Mahmudm, N., et al., 2001. Mesenchymal stem cells are capable of homing to the bone marrow of non-human primates following systemic infusion. Exp. Hematol. 29 (2), 244–255.

Farfara, D., Tuby, H., Trudler, D., Doron, E., Maltz, L., Frenkel, D., et al., 2015. Low-level laser therapy ameliorates disease progression in a mouse model of Alzheimer's disease. J. Mol. Neurosci. 55, 430–436.

Frenkel, D., Puckett, L., Petrovi, S.S., et al., 2008. A nasal proteosome adjuvant activates microglia and prevents amyloid deposition. Ann. Neurol. 53, 591–601.

Gavish, L., Perez, L.S., Reissma, P., Gertz, S.D., 2008. Irradiation with 780 nm diode laser attenuates inflammatory cytokines but upregulates nitric oxide in lipopolysaccharide-stimulated macrophages: implications for the prevention of aneurysm progression. Lasers Surg. Med. 40 (5), 371–378.

Gnecchi, M., He, H., Liang, O.D., Melo, L.G., Morello, F., Mu, H., et al., 2005. Paracrine action accounts for marked protection of ischemic heart by Akt-modified mesenchymal stem cells. Nat. Med. 367–368.

Karu, T., 2007. Ten Lectures on Basic Science of Laser Phototherapy. Prima Books, Gragesberg.

Lampl, Y., Zivin, J.A., Fisher, M., et al., 2007. Infrared laser therapy for ischemic stroke: a new treatment strategy: results of the NeuroThera Effectiveness and Safety Trial-1 (NEST-1). Stroke 38 (6), 1843–1849.

Lapchak, P.A., Wei, J., Zivin, J.A., 2004. Transcranial infrared laser therapy improves clinical rating scores after embolic strokes in rabbits. Stroke 35 (8), 1985–1988.

Liechty, K., MacKenzie, W.T.C., Shaaban, A.F., et al., 2000. Human mesenchymal stem cells engraft and demonstrate site-specific differentiation after in utero transplantation in sheep. Nat. Med. 6 (11), 1282–1286.

Morgan, D., 2007. Amyloid, memory and neurogenesis. Exp. Neurol. 205 (2), 330–335.

Mummery, C.L., Davis, R.P., Krieger, J.E., 2010. Challenges in using stem cells for cardiac repair. Sci. Trans. Med. 2 (27), 1–5.

Mvula, B., Moore, T.J., Abrahamse, H., 2010. Effect of low-level laser irradiation and epidermal growth factor on adult human adipose-derived stem cells. Lasers Med. Sci. 25 (1), 33–39.

Oron, A., Oron, U., Chen, J., et al., 2006. Low-level laser therapy applied transcranially to rats after induction of stroke significantly reduces long-term neurological deficits. Stroke 37 (10), 2620–2624.

Oron, U., 2006. Photoengineering of tissue repair in skeletal and cardiac muscles. Photomed. Laser Surg. 24, 111–120.

Oron, U., Ilic, S., De Taboada, L., Streeter, J., 2007. Ga-As (808 nm) laser irradiation enhance ATP production in human neuronal cells in culture. Photomed. Laser Surg. 25 (3), 180–182.

Rodriguez, J.J., Verkhratsky, A., 2011. Neurogenesis in Alzheimer's disease. J. Anat. 219 (1), 78–89.

Selkoe, D.J., 2004. Cell biology of protein misfolding: the examples of Alzheimer's and Parkinson's diseases. Nat. Cell Biol. 6 (11), 1054–1061.

Simard, A.R., Soulet, D., Gowing, G., Julien, J.P., Rivest, S., 2006. Bone marrow-derived microglia play a critical role in restricting senile plaque formation in Alzheimer's disease. Neuron 49 (4), 489–502.

Tuby, H., Maltz, L., Oron, U., 2011. Induction of autologous mesenchymal stem cells in the bone marrow by low-level laser therapy has profound beneficial effects on the infarcted rat heart. Lasers Surg. Med. 43, 401–409.

Tuby, H., Yaakobi, T., Maltz, L., Delarea, Y., Sagi-Assif, O., Oron, U., 2013a. Effect of autologous mesenchymal stem cells induced by low level laser therapy on cardiogenesis in the infarcted area following myocardial infarction in rats. J. Biomed. Sci. Eng. 6, 24–31.

Tuby, H., Hertzberg, E., Maltz, L., Oron, U., 2013b. Long-term safety of low-level laser therapy at different power densities and single or multiple applications to the bone marrow in mice. Photomed. Laser Surg. 31 (6), 269–273.

Uccelli, A., Moretta, L., Pistoia, V., 2008. Mesenchymal stem cells in health and disease. Nat. Rev. Immunol. 8 (9), 726–736.

Chapter 17

The experimental evidence for photobiomodulation-induced cellular and behavioral changes in animal models of Parkinson's disease: a template for translation to patients

Nabil El Massri and John Mitrofanis
Department of Anatomy, University of Sydney, Sydney, NSW, Australia

17.1 Introduction

The current clinical treatment paradigms for Parkinson's disease patients, although effective in treating the motor signs of the disease, offer little in terms of neuroprotection, the slowing or stopping of the disease pathology. In this chapter, we consider the experimental evidence in a range of different animal models of Parkinson's disease for neuroprotection, reductions in gliosis, expression of trophic growth factors, changes in functional brain activity and behavioral improvements offered by low-level light therapy or photobiomodulation ($\lambda = 600-1070$ nm). Finally, we will consider all this experimental evidence in relation to a translation to patients. First, we provide a brief overview of Parkinson's disease, the current treatment options available, and the various animal models developed of this disease.

17.2 Parkinson's disease and animal models

Parkinson's disease is a well-known motor disorder with distinct cardinal signs of resting tremor, akinesia, bradykinesia, and lead-pipe rigidity (Bergman and Deuschl, 2002; Jankovic and Poewe, 2012). These signs develop after disruption to the basal ganglia, a group of nuclei that lie deep in the forebrain and brainstem and includes the striatum, globus pallidus, subthalamic nucleus, substantia nigra (pars compacta and reticulata), zona incerta, and the pedunculopontine tegmental nucleus (Rinne, 1993; Blandini et al., 2000; Bergman and Deuschl, 2002; Fig. 17.1). The basal ganglia collectively do not necessarily make a movement, but together with areas of frontal cortex, play an important role in planning and/or programming a movement for primary motor cortex and subsequently the spinal cord has more direct access to the somatic muscles (Monchi et al., 2006; Fig. 17.1). One may view the basal ganglia and overlying frontal cortex as the "brains" behind a movement and the primary motor cortex and spinal cord as the "brawn" driving the movement.

What essentially happens in Parkinson's disease is that there is a loss of neurones mainly from the substantia nigra pars compacta (SNc) of the midbrain and their terminations in the striatum (Rinne, 1993; Blandini et al., 2000; Bergman and Deuschl, 2002). This loss of dopamine from the basal ganglia circuitry then triggers a cascade of abnormal activity in some of the nuclei, in particular the subthalamic nucleus, manifesting in the distinct clinical signs of the disease (Blandini et al., 2000; Bergman and Deuschl, 2002; Jankovic and Poewe, 2012).

The factors that generate the loss of the nigral dopaminergic neurones are not clear, but a central one relates to mitochondrial dysfunction (Fig. 17.2; Fukae et al., 2007; Exner et al., 2012). This dysfunction has been linked to either an exposure to toxin (e.g., pesticide) or, in small number of cases, to a defective gene (e.g., PINK1, α-synuclein, parkin; Corti and Brice, 2013). More recently, some authors have proposed that Parkinson's disease results after abnormal or

Photobiomodulation in the Brain. DOI: https://doi.org/10.1016/B978-0-12-815305-5.00017-8

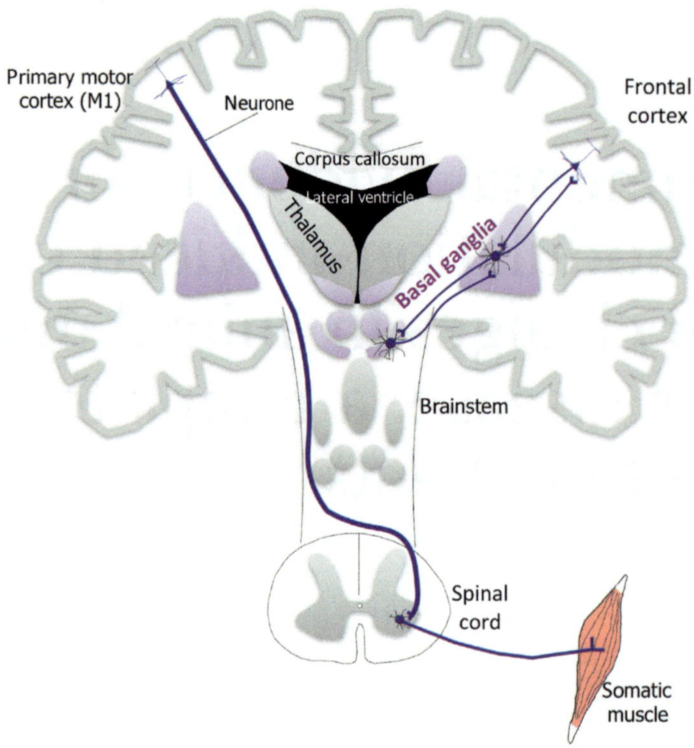

FIGURE 17.1 Schematic section through brain and spinal cord, indicating the major neural centers. The structures in blue and individual cells. The purple structures are nucleus of the basal ganglia. Parkinson's disease results after lesion to the basal ganglia. The basal ganglia do not make a movement, but together with frontal cortex, play a role to help plan and program move for primary motor cortex and spinal cord, that act more directly on the somatic muscles.

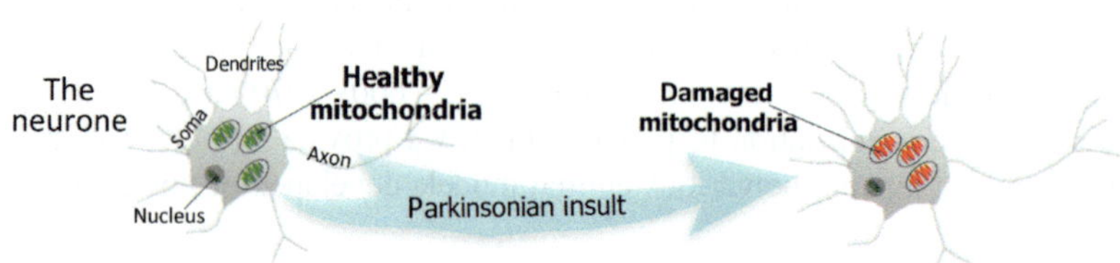

FIGURE 17.2 Schematic diagram of individual neurones showing their major constituent parts. The factor that generates loss of cells in Parkinson's disease is not clear, but it appears to be due to mitochondrial dysfunction, as a result of either exposure to toxin (e.g., pesticide) or in small number of cases, a defective gene. More recently, some authors suggest the loss may result of abnormal/mis-folding protein conformations, and becoming self-propagating across brain, like prions (e.g., α-synuclein).

mis-folding protein conformations (e.g., α-synuclein) that become self-propagating across brain, like prions (Brettschneider et al., 2015; Goedert, 2015).

Parkinson's disease is characterized by not only a progressive loss of dopaminergic neurones in the SNc and their terminations in the striatum, but also a massive gliosis, particularly of astrocytes and microglia, the two major forms of glia (McGeer and McGeer, 1998, 2008; Barcia et al., 2003). Gliosis is the hypertrophy or proliferation of glial cells in response to forms of neural damage, including neurodegeneration. Historically, gliosis has been associated with toxic, detrimental effects on neurones, by either inhibiting axonal regeneration by forming glial scars and/or secreting pro-inflammatory cytokines and other neurotoxic products. More recently, gliosis has also been associated with more beneficial effects after damage, for example, with the release of neuroprotective agents such as glial derived neurotrophic factor (GDNF). The relationship between toxic and beneficial function is complex, being dependent on an array of different factors and molecular signaling mechanisms, and appears to change with time period after the insult (McGeer and McGeer, 1998, 2008; Barcia et al., 2003; Hamby and Sofroniew, 2010; Halliday and Stevens, 2011; Pekny et al., 2014; Pekny and Pekna, 2014; Verkhratsky et al., 2014; Burda et al., 2016).

The current treatment for Parkinson's disease patients involves dopamine replacement drug therapy, which aims to replace the lost dopamine from the system. When the efficacy of drug therapy reduces after a few years, patients are selected for deep brain stimulation, which aims to correct the abnormal activity of some of the basal ganglia nuclei (e.g., subthalamic nucleus, zona incerta, or globus pallidus) (Benabid et al., 2009). Both of these treatments are very effective at treating the signs of the disease, at least initially in the case of the drug therapy, but they both suffer in that they do not slow or stop the progression of the disease. They are not disease-modifying or neuroprotective (Olanow et al., 2008; Jankovic and Poewe, 2012; Schapira et al., 2014). Hence, there is a real need for the development of new neuroprotective treatments that stop disease progression; and many of the more recent ones target the ailing mitochondria in the disease state, to help the mitochondria resist or offset the parkinsonian insult (Chaturvedi and Beal, 2008).

There are a number of different types of animal models for Parkinson's disease that have been developed over the years, perhaps more than any other major disease, including for example, Alzheimer's disease and schizophrenia. The bulk of these models have been developed to cause a degeneration of nigral dopaminergic neurones and/or to generate motor behavioral and functional deficit models in a range of different animals, from mouse to monkey. There are also several in vitro models using dopaminergic cell cultures (Martínez-Morales and Liste, 2012; Xicoy et al., 2017). Broadly, all these models fall either into toxin-induced [e.g., rotenone, 6OHDA (6 hydroxydopamine), MPTP (methyl-4-phenyl-1,2,3,6-tetrahydropyridine)], or transgenic (e.g., α-synuclein) models (Schober, 2004; Blandini and Armentero, 2012; Blesa et al., 2012; Bové and Perier, 2012). Individually, each of these models do not replicate "perfectly" the disease state in humans, although when considered together, they can provide a more than useful picture of the human condition, and in particular, as a test of any potential treatment (Torres et al., 2017).

17.3 Photobiomodulation

The beneficial outcomes of low-level light therapy or photobiomodulation ($\lambda = 600-1070$ nm), hinge on the concept that photons can stimulate chemical changes within neurones, that light energy can be converted to metabolic energy with subsequent influence on intrinsic neuronal function and survival (Eells et al., 2004; Karu, 2010; Rojas and Gonzalez-Lima, 2011; Khan and Arany, 2015; Hamblin, 2016). Such a concept has traditionally been met by many clinicians and scientists with much skepticism and "raised eyebrows." However, when one considers that all plants use light energy to fuel many of their cellular functions, is it so difficult to accept that animals use light energy also? We live by light from the sun and it drives much of our vision and good health (e.g., vitamin D production by skin), so why not use the light energy for cell function and as a means of cell survival (Rojas and Gonzalez-Lima, 2011; Mitrofanis, 2017).

Over the last two decades or so, a rich vein of research has provided much to reduce the skepticism and lower the raised eyebrows of many colleagues. Although there is still much to discover, and many more still to convince, this research has provided us with a clearer view of how photobiomodulation works at a cellular and systemic level (Eells et al., 2004; Karu, 2010; Rojas and Gonzalez-Lima, 2011; Khan and Arany, 2015; Hamblin, 2016). Two general modes of action pf photobiomodulation have been described. First, is direct stimulation (Fig. 17.3), which relies on the light hitting the cell "flush." Light is absorbed by a chromophore, that lies within a photoreceptor molecule and responsible for light absorption. A main photoacceptor that has been identified in cytochrome c oxidase, whose activation generates an increase in mitochondrial function, including an increase in electron transfer in respiratory chain, mitochondrial membrane potential, reactive oxygen species, and adenosine triphosphate (ATP) energy. This then leads to activation of a number of transcription factors in the nucleus that prompts the expression of various stimulatory and protective genes that relate to many beneficial cellular features including neurogenesis, synaptogenesis, increase in growth factors [e.g., brain derived growth factor (BDNF) and GDNF] and neuroprotection (Eells et al., 2004; Karu, 2010; Rojas and Gonzalez-Lima, 2011; Khan and Arany, 2015; Hamblin, 2016). In essence, direct stimulation by light enhances intrinsic mechanisms that help cell survival, stimulating intrinsic or compensatory systems within neurones to improve their overall survival (Johnstone et al., 2016; Mitrofanis, 2017). It should be noted that cytochrome c oxidase is the main chromophore at wavelengths 600–700 and 760–940 nm, but there are indications that there may be other chromophores that absorb light at other wavelengths. For example, 980 nm has been reported to activate temperature-gated ion channels on cell membrane to increase intracellular calcium in fat tissue (Hamblin, 2016).

A second mode of action by photobiomodulation is indirect stimulation (Fig. 17.4). This type of stimulation relies on a "middle-man," such as the immune and/or stem cell system (Tuby et al., 2011; Liebert et al., 2014; Johnstone et al., 2014, 2016; Oron and Oron, 2016; Mitrofanis, 2017). Several previous studies have shown that if photobiomodulation is applied to one region of the body (e.g., thigh; A, Fig. 17.4), then there is recovery and beneficial effects evident in another part of the body (e.g., brain; B, Fig. 17.4). It has been suggested that photobiomodulation stimulates immune and/or stem cells (or indeed other molecules) in the circulation that then swarm to the region of damage or distress and provide benefit, by decreasing pro-inflammatory and increasing antiinflammatory cytokines. It is not known if the indirect stimulation, using a middle-man, also activates the same intrinsic mechanisms as the direct stimulation.

A key question is which of these two modes of action is the most likely. In truth, they probably work together, because there is good evidence for both. From a distance, one would think that because the direct stimulation goes

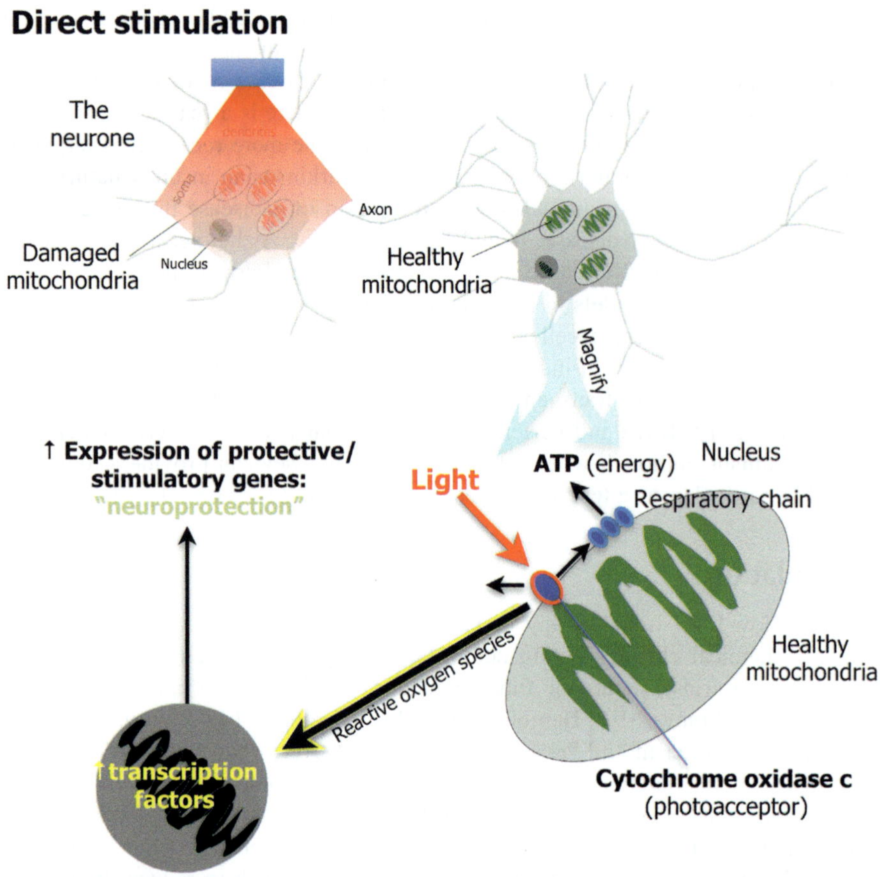

FIGURE 17.3 Schematic diagram summarizing major features of mechanism of direct stimulation of photobiomodulation. When applied directly onto neurones, light is absorbed by chromophore, which lies within a photoreceptor molecule and responsible for light absorption. A main photoacceptor is cytochrome c oxidase, that then increases electron transfer in respiratory chain, mitochondrial membrane potential, and adenosine triphosphate (ATP) energy. There is also an increase in reactive oxygen species that then generates activation of transcription factors in nucleus that leads to increases in expression of various stimulatory and protective genes that relate to neuroprotection.

straight to the source, it probably is more efficacious and consistent than the indirect stimulation, which relies on other system(s)—a middle-man—to get to the source. This reliance on another system may result in a weaker overall beneficial effect. Indeed, there is evidence indicating that although direct and indirect stimulation offer beneficial effects, direct stimulation has a stronger impact than indirect (e.g., Johnstone et al., 2014).

A key feature of photobiomodulation is that it has the ability to affect neurons in different states of health in different ways, essentially modifying the cell in whatever way might be necessary to promote its survival (Hamblin, 2016). For instance, in healthy neurones, the light absorption by cytochrome c oxidase leads to an increase in mitochondrial membrane potential and reactive oxygen species. However, in distressed neurones where there is a decrease in mitochondrial membrane potential and a massive increase in reactive oxygen species produced from dysfunctional mitochondria, light absorption leads to an increase in mitochondrial membrane potential and a decrease in reactive oxygen species. Similarly, a typical response to light in healthy neurones is an increase in intracellular calcium, while in distressed neurones that already contain excessive calcium, for example, with excitotoxicity, light provokes the opposite reaction.

In summary, photobiomodulation has been shown to have a clear impact on cell function and survival, acting mainly through a chromophore on the mitochondrial membrane that results in an increase in ATP energy and activation of various stimulatory and protective genes. In the context of Parkinson's disease, in which mitochondrial dysfunction is so central to the neurodegenerative process (and that there is no current treatment paradigm that addresses this dysfunction) the use of photobiomodulation would appear an ideal therapy for trial in this disease.

In the sections that follow, we will consider the experimental evidence for photobiomodulation generating beneficial outcomes in various animal models of Parkinson disease, from flies to rats and from mice to monkeys. In particular, we will explore whether photobiomodulation results in neuroprotection, a reduction in gliosis, an expression of growth factors, a restoration of the normal functional activity in the basal ganglia and/or an improvement in motor behavior and reduction in clinical signs.

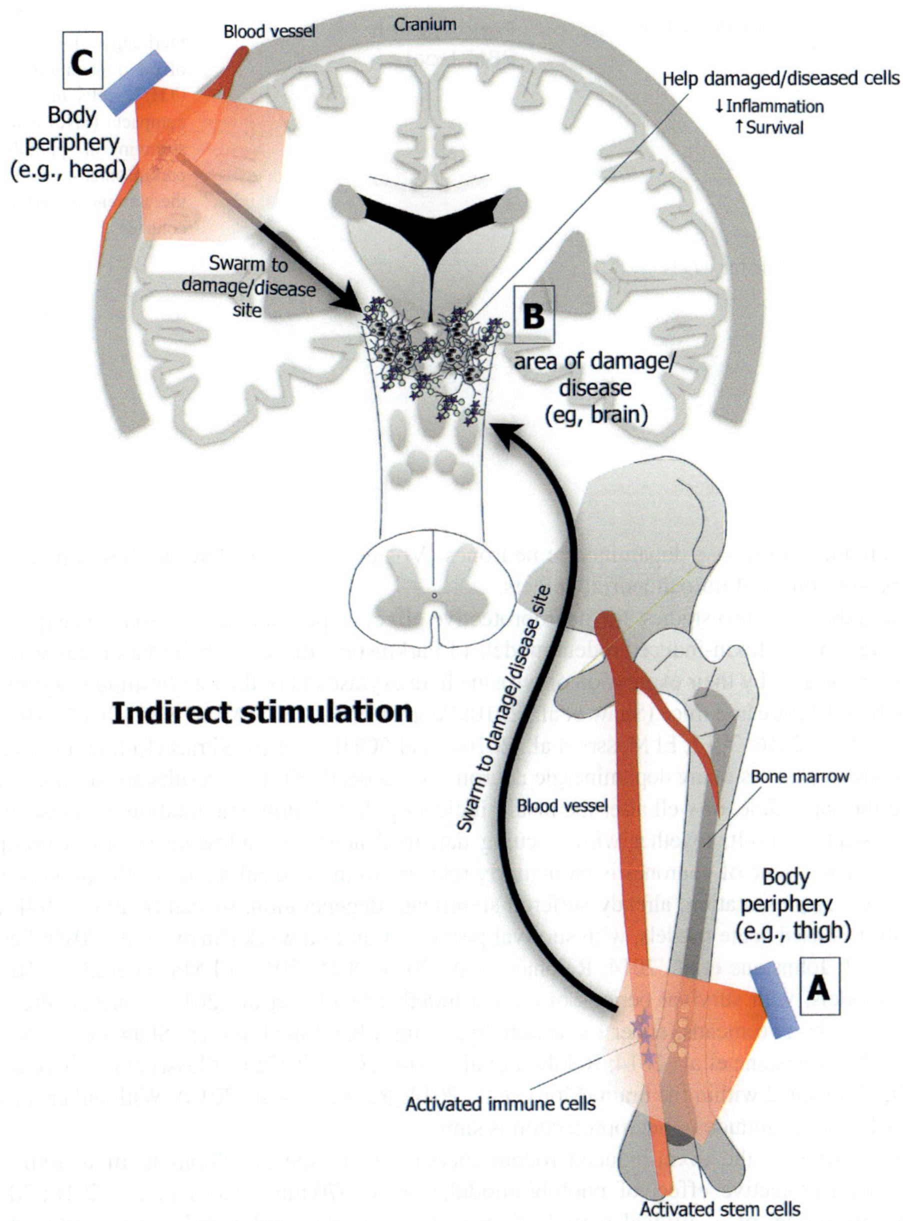

FIGURE 17.4 Schematic diagram summarizing major features of mechanism of indirect stimulation of photobiomodulation. When light is applied to one region of body (e.g., thigh (A) or head (C)), it helps distressed cells in another (e.g., brain (B)), perhaps using a "middle-man" in the circulation, such as immunes/stem cells. Light may stimulate these circulatory cells, and they swarm to the area of damage and help neurones survive. This method is the likely means of stimulation after transcranial photobiomodulation in humans, activating circulatory cells from vessels in the head.

17.4 Neuroprotection

The pioneer studies reporting a neuroprotective effect of photobiomodulation using a Parkinson's disease model were undertaken in vitro (Liang et al., 2008; Ying et al., 2008). These studies showed that photobiomodulation (670 nm) reduces cell death, increases ATP content, and decreases levels of oxidative stress in rat striatal and cortical neurones exposed to the parkinsonian toxins rotenone and MPP^+ (1-methyl-4-phenylpyridium) in vitro. A few years later, in cultures of human neuroblastoma neurones engineered to overexpress α-synuclein, photobiomodulation (810 nm) was reported to increase mitochondrial function and reduce oxidative stress after MPP^+ (1-methyl-4-phenylpyridinium) exposure (Trimmer et al., 2009). In addition, in hybrid neurones bearing mitochondrial DNA from Parkinson's disease patients, mitochondrial movement along axons improves substantially after photobiomodulation (Trimmer et al., 2009). There has also been a more recent in vitro investigation examining the effect of photobiomodulation (808 nm) on

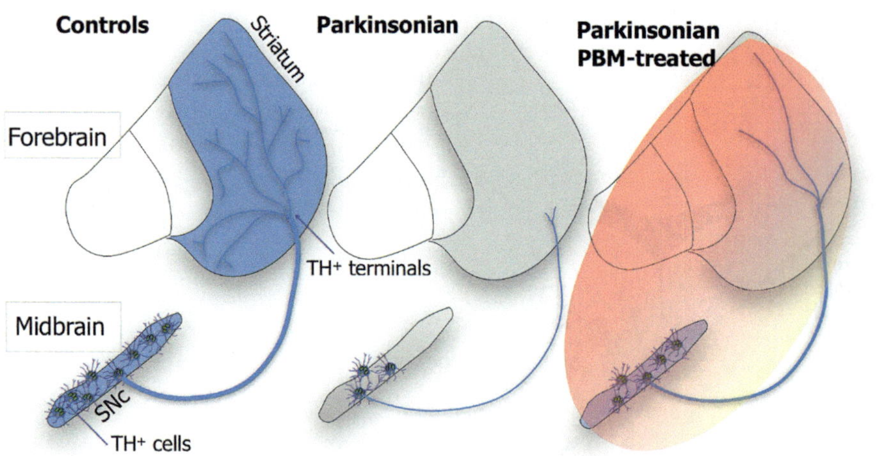

FIGURE 17.5 Schematic diagrams summarizing major features of neuroprotection of the dopaminergic [tyrosine hydroxylase $(TH)^+$] cells in the substantia nigra pars compacta (SNc) and their terminals in the striatum in photobiomodulation (PBM) parkinsonian-treated animals compared to the untreated parkinsonian ones and the controls.

drosophila pink-1 mutants and mouse dopaminergic neurones (Vos et al., 2013). Here, in these mutants, photobiomodulation rescues major systemic and mitochondrial defects.

Following on from these in vitro studies, the neuroprotective effect of photobiomodulation was explored in vivo. These first in vivo studies were on the toxin-induced rodent models of Parkinson's disease. Their main focus was the dopaminergic neurones of the SNc, identified by their expression of tyrosine hydroxylase (TH), the rate limiting enzyme in dopamine production (Fig. 17.5). In MPTP-treated mice (Shaw et al., 2010; Peoples et al., 2012; Moro et al., 2013, 2014; Johnstone et al., 2014; Reinhart et al., 2014, 2016, 2017; El Massri et al., 2016a) and 6OHDA-lesioned rats (Reinhart et al., 2015), photobiomodulation (670 or 810 nm) saves many dopaminergic neurones from death. Further, results are similar whether the therapy is applied before, at the same time, or well after the insult, indicating that photobiomodulation both conditions healthy neurones to resist a subsequent insult, together with rescuing damaged neurones following an insult (Peoples et al., 2012; Reinhart et al., 2016). The rescue of neurones is particularly relevant to the clinical reality of the parkinsonian condition, in which individuals have, at presentation, already suffered significant degeneration, so that treatment follows neuronal loss. The bulk of these studies used acute models, with survival periods of up to a week (Shaw et al., 2010; Peoples et al., 2012; Moro et al., 2013, 2014; Johnstone et al., 2014; Reinhart et al., 2014, 2016, 2017; El Massri et al., 2016a), although some used more chronic models, with survival periods of over a month (Peoples et al., 2012). Further, the application of the photobiomodulation was by two means, either extracranially, using a hand-held device (Shaw et al., 2010; Peoples et al., 2012; Moro et al., 2013; Johnstone et al., 2014; Reinhart et al., 2014, 2016, 2017; El Massri et al., 2016a), or intracranially, using an optical fiber implanted within the brain (Moro et al., 2014; Reinhart et al., 2015). With either means of application in these rodent models, the magnitude of neuroprotection is similar.

These promising results in the toxin-induced rodent models led to the development of a toxin-induced, primate model, testing the neuroprotective effect of photobiomodulation at 670 nm (Darlot et al., 2016; Moro et al., 2016, 2017). In this subacute model of a survival period of up to three weeks, parkinsonism was induced after a series of MPTP injections and some animals, at the same time as the injections, were treated intracranially with photobiomodulation. In this series of experiments, the number of neurones in the SNc was assessed by both their expression of TH, so-called functional neuroprotection, and by their Nissl staining, so-called true neuroprotection. All of the photobiomodulation-treated MPTP monkeys had a greater number of surviving nigral neurones, both TH^+ and Nissl-stained, compared to those that were left untreated (Darlot et al., 2016). There was both true and functional neuroprotection. In addition to this analysis in the SNc, the density of dopaminergic terminations was examined in the major output zone of the SNc, the striatum. Here, in the striatum, there were more terminals in the photobiomodulation-treated animals compared to those that were left untreated (Darlot et al., 2016). These studies also tested the efficacy of different doses of photobiomodulation. A dose-response effect was clear, with animals treated with a lower dose (25–35 J: Darlot et al., 2016) showing a stronger neuroprotective effect than those treated with a higher one (125 J: Moro et al., 2016). It should be noted that there was no evidence of toxicity by photobiomodulation from the intracranial device, notwithstanding the fact the light source was implanted within the brain, and hence being applied directly on neural tissue (Darlot et al., 2016; Moro et al., 2016, 2017).

There have also been explorations of the neuroprotective effect in a transgenic rodent model. In the K369I transgenic mouse model of frontotemporal dementia, which also shows parkinsonian signs and a chronic and progressive degeneration of dopaminergic neurones (TH^+) in the SNc over a period of 5–6 months, photobiomodulation (670 nm) decreased oxidative stress and hyperphosphorylated tau and increased dopaminergic cell survival in the SNc (Purushothuman et al., 2013).

There is also evidence for photobiomodulation-induced neuroprotection in an α-synuclein genetic rat model of Parkinson's disease (Oueslati et al., 2015). In this model, rats were injected with an adeno-associated virus inducing overexpression of α-synuclein and were treated daily with photobiomodulation (808 nm) for about a month. The treated animals had more dopaminergic neurones (TH^+) in the SNc and terminations in the striatum compared to the untreated ones.

In summary, there is a wealth of experimental evidence for neuroprotection in a wide range of animal models of Parkinson's disease, from toxin-induced mouse, rat, and monkey models to transgenic fly, mouse, and rat models. In all these models of the disease, of which include both acute and chronic varieties, photobiomodulation (670 and 808–810 nm) has been reported to offer neuroprotection in both the SNc (both true and functional) and the striatum of the basal ganglia.

17.5 Gliosis

There is evidence that photobiomodulation has an impact, not only on the ailing neurones affected by the disease, but also on the hypertrophied glial cells (Fig. 17.6A). Previous studies have shown that photobiomodulation (670 nm) affects the MPTP-induced gliosis in both the SNc and striatum of mice (Johnstone et al., 2014; El Massri et al., 2016a) and monkeys (El Massri et al., 2016b), the two areas where clear neuroprotection is evident (Shaw et al., 2010; Darlot et al., 2016). In general, photobiomodulation reduces the number and size of glial cells, both measures of their reactive state, particularly in the striatum, the major synaptic zone of the dopaminergic neurones (see Section 17.2). In both mice and monkeys, these photobiomodulation-induced outcomes are stronger among astrocytes than microglia, but this may be due to the acute nature of the MPTP model that has been used; at some point, beyond the timeline of the experimental survival period of the model, microglia may show a stronger response to photobiomodulation (Johnstone et al., 2014; El Massri et al., 2016a,b).

It is not known at this stage if the overall photobiomodulation-induced reduction in gliosis is due to the light acting on the glial cells directly or secondary to the neuroprotection. If acting directly on the glial cells, the photobiomodulation could stimulate a neuroprotective role for these cells, perhaps by triggering various intrinsic cellular mechanisms, resulting in an increase of their secretion of antiinflammatory agents and a decrease of their pro-inflammatory ones (McGeer and McGeer, 2008). This in turn, would result in a greater survival of dopaminergic neurones in the SNc and their terminations in the striatum (El Massri et al., 2016b).

In summary, photobiomodulation influences the glial response to parkinsonian insult. Notwithstanding whether this influence was direct or secondary to the neuroprotection, it raises the possibility of glial cells as a future therapeutic target of photobiomodulation.

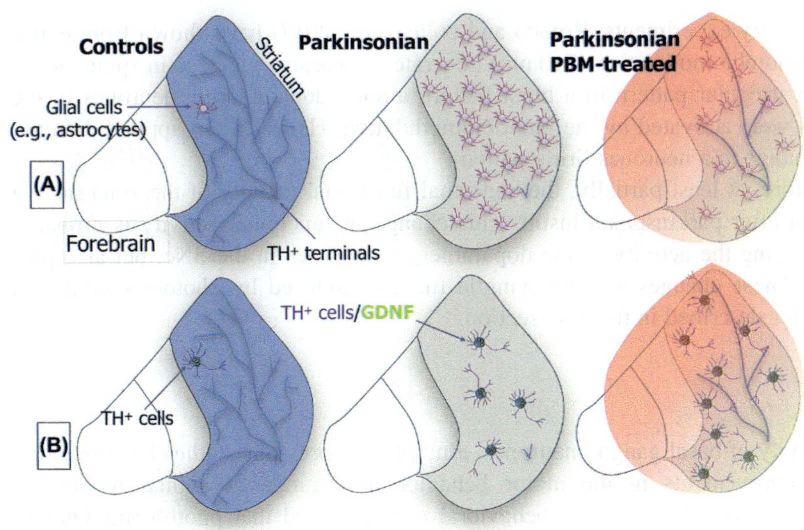

FIGURE 17.6 Schematic diagrams summarizing the reduction in glial cells (A), in particular astrocytes, and the increases in TH (tyrosine hydroxylase)$^+$ cell number and glial derived neurotrophic factor (GDNF) expression (B) in the striatum in photobiomodulation (PBM) parkinsonian-treated animals compared to the untreated parkinsonian ones and the controls. The patterns of TH^+ terminals in the striatum are shown also.

17.6 Growth factors

Photobiomodulation has been associated with the expression of various growth factors in nervous tissue. These factors are important in the regeneration and regrowth of damaged neurones. For example, and with regard to basal ganglia and Parkinson's disease, previous studies have shown that both GDNF (Gash et al., 2005; Orme et al., 2013) and BDNF (Du et al., 1995), increase the number of dopaminergic neurones in cell culture and in the SNc. Further, in vivo application of GDNF generates in a substantial increase in TH^+ cell number in the striatum after MPTP insult (Palfi et al., 2002; Sebastián et al., 2007).

Recent studies in a MPTP-treated monkey model have shown that photobiomodulation (670 nm) prompts the expression of GDNF across the striatum (Fig. 17.6B). This expression has been suggested to have a two-fold action. First, to help damaged dopaminergic afferents regrow and establish new synaptic contacts and second, to switch-on TH expression in many striatal cells, which helps to restore dopamine levels in the striatum after MPTP insult (El Massri et al., 2017). It is not known at this stage whether the photobiomodulation-induced GDNF expression in the striatum derives from the neurones and/or the surrounding glia, in particular astrocytes (Sandhu et al., 2009; d'Anglement de Tassigny et al., 2015).

Photobiomodulation has also been associated with an increase in BDNF expression after traumatic brain injury (810 nm; Xuan et al., 2013), together with increasing dendritic morphogenesis and neural connectivity in embryonic rats (633 nm; Meng et al., 2013). In fact, BDNF, together with GDNF, are expressed in the striatum of 6OHDA-lesioned rats after injection of viral vectors and both trophic factors are effective in improving behavior and offering neuroprotection (Sun et al., 2005).

In summary, there is evidence that the beneficial effects of photobiomodulation, at least in part, may be mediated through the expression of the growth factors, GDNF and BDNF. These factors could prompt the expression of the dopaminergic phenotype and have a trophic action, triggering the regrowth and synaptogenesis of damaged axons. These features raise the idea of a new therapeutic direction, one linking photobiomodulation with growth factors and the development of a potential therapeutic approach.

17.7 Functional activity

In an effort to understand whether any of the neuroprotection evident in experimental models is "useful," Shaw and colleagues examined the patterns of Fos expression in some of the basal ganglia nuclei in an MPTP-treated mouse model (Shaw et al., 2012). Fos protein is released by neurones after they have been stimulated (and it is a well-established measure of neuronal activity); the more Fos expression there is in a nucleus, as revealed by immunohistochemistry, then the more activated its neurones are. Two nuclei of the basal ganglia were examined, the zona incerta and the subthalamic nucleus, both of which have been shown to be overactive in parkinsonian cases and thence major targets of deep brain stimulation (Plaha et al., 2006; Benabid et al., 2009). In mice, the MPTP-induced increase in Fos expression, was reduced substantially after photobiomodulation (670 nm). This reduction did not quite reach control levels, indicating that the restoration was partial, and was attributed to the neuroprotection of dopaminergic neurones in the SNc (Shaw et al., 2012).

In addition to these Fos immunohistochemistry experiments, Romeo and colleagues (2017) have shown from extracellular electrophysiological recordings that photobiomodulation (710 nm) generated a massive increase in spontaneous firing rate and a shift in firing from a slight irregular pattern to a burst pattern among dopaminergic neurones in the SNc. Further, that previously silent neurones were activated by the photobiomodulation. Hence, when applied directly, photobiomodulation can induce a dramatic change to a neurones firing pattern.

In summary, photobiomodulation can restore, at least partially, the abnormal functional activity of the zona incerta and subthalamic nucleus of the basal ganglia after parkinsonian insult. This change in functional activity is probably due to the photobiomodulation, not only changing the activity of the dopaminergic neurones in the SNc, but also protecting many of them from the toxic insult. These changes to basal ganglia function induced by photobiomodulation may well underlie the improved motor behavior described in the next section.

17.8 Behavior

Following on from changes in functional activity of basal ganglia neurones, a number of previous studies have reported that photobiomodulation results in clear improvements in the motor behavior in a range of animal models of Parkinson's disease. In an MPTP-treated mouse model, open-field behavioral tests revealed that photobiomodulation (670 and 810 nm) improved various parameters of locomotion, for example, mobility and velocity (Whelan et al., 2008;

Moro et al., 2013; Reinhart et al., 2014, 2016, 2017). Further, in a transgenic α-synuclein mouse model, photobiomodulation (670 nm) has been reported to delay disease progression and reduced the severity of the disease phenotype (Quirk et al., 2012). In a 6OHDA-lesioned hemiparkinsonian rat model of the disease, there was markedly reduced apomorphine-induced rotational behavior after photobiomodulation (670 nm; Reinhart et al., 2015). There is also evidence that photobiomodulation rescues flight defects in pink-1 mutant flies (Vos et al., 2013).

Perhaps the most compelling evidence for an improved motor behavioral outcome after photobiomodulation has been shown in the subacute MPTP-treated monkey model of the disease (Darlot et al., 2016; see above). In this model, a modified Schneider scale of clinical assessment was used, one that measures a range of parameters, from tremor to akinesia, and from general activity to rigidity (Ashkan et al., 2007; Wallace et al., 2007). Using this clinical assessment, all of the photobiomodulation-treated (670 nm) MPTP monkeys had reduced clinical signs compared to untreated monkeys. Further, all of the treated monkeys had considerably more locomotive movement, as assessed in an open-field behavioral test. These improvements in clinical signs and movement were still evident up to three weeks after the short, 5 day period of photobiomodulation (see above), indicating that the therapeutic effects are long-lasting and not confined to the periods of treatment application (Darlot et al., 2016). A similar effect has been reported in the MPTP-treated mouse model, in which the beneficial effects of photobiomodulation were evident days after the last treatment (Reinhart et al., 2016).

In summary, there are clear indications that photobiomodulation improves motor behavior in a number of animal models of Parkinson's disease, from flies to rodents to monkeys; in the monkey model, there is also a clear reduction in the clinical signs of the disease, many of which are apparent in patients. These improvements are likely to reflect changes in the functional activity of the basal ganglia, generated by a neuroprotection of dopaminergic neurones in the SNc by photobiomodulation (see above).

17.9 Translation to patients

We have outlined a wealth of beneficial cellular and behavioral outcomes by photobiomodulation from many experimental studies on a range of different animal models of Parkinson's disease. Notwithstanding this large body of experimental evidence, there is still much resistance and skepticism among some colleagues as to the potential effectiveness of photobiomodulation in patients. Many neurologists for instance, would consider 670 nm red light devices more useful as decorations on their Christmas trees, rather than explore their effect on patients (quote from Catherine Hamilton; https://redlightsonthebrain.blog/). For some, there are major issues in translating all of this experimental evidence to the clinic. As we see it, these "issues" involve the following.

First, as mentioned above, the concept of light inducing a chemical and metabolic change in neurones appears difficult to accept or to grasp by some colleagues. On the other hand, these colleagues, mainly because of their training, have no trouble accepting the concept that a substance (e.g., a drug) can induce a change through a series of receptors. Following along these lines, one should consider that photobiomodulation works on a similar principal, that it has receptors (i.e., chromophores) that, when activated, prompt a series of intrinsic cellular activities beneficial to the cell's survival (Karu, 2013). The "hard" scientific evidence that photobiomodulation does indeed change neuronal activity and influence neuronal survival against insult has become irrefutable.

Second, it is often the case that any result on an animal model of Parkinson's disease is met with much suspicion, with the argument that no model mimics faithfully the human condition. In particular, no model reflects exactly the neuropathology (e.g., Lewy bodies), the chronic progressive nature (e.g., slow and relentless degeneration of neurones), the topography of degeneration (e.g., across different transmitter cell groups in brainstem), and the clinical signs and symptoms (e.g., no model reproduces effectively the resting tremor) of the disease in humans. Further, most cases in humans are idiopathic (no known cause), while the animal models are all "induced" whether by either toxin or genetic mutation (Schober, 2004; Olanow et al., 2008; Blandini and Armentero, 2012; Blesa et al., 2012; Bové and Perier, 2012; Jankovic and Poewe, 2012; Schapira et al., 2014; Torres et al., 2017). Notwithstanding these suspicions, many of the animal models do, in fact, get it "right" in many respects. For example, they can generate a loss of neurones in the main zone of pathology (i.e., SNc and striatum), create comparable patterns of abnormal activity in basal ganglia nuclei (e.g., subthalamic nucleus), reproduce many of the major clinical signs, particularly in the nonhuman primates (e.g., rigidity, akinesia), and finally, replicate closely the chronic progressive nature of degeneration. If one trials agents and gathers data from several models (i.e., toxin-induced and transgenic), then one can create a more than worthwhile overall picture of the human condition (Torres et al., 2017). Such experimental data is, at the very least, worthy of consideration as a template for translation to humans. Photobiomodulation has been tested in a wide range of animal models of Parkinson's disease, from toxin-induced mouse, rat, and monkey to transgenic fly, mouse, and rat models. In all models, of which include both acute and chronic varieties, photobiomodulation offers neuroprotection, restores functional activity, improves motor behaviors and reduces clinical signs, reduces gliosis and induces expression of trophic growth factors.

Third, if one considers the transcranial approach, then the issue of penetration is raised, namely, how can external light reach the main zone of pathology in the very deep lying brainstem, covered by hair (in most cases), skin, bone, thick meninges and a mass of brain tissue. This is a valid issue, because most studies have noted that light can only penetrate 20–30 mm through body tissues, and the brainstem lies some 80–100 mm below the cranial surface (Johnstone et al., 2016; Hamblin, 2016; Mitrofanis, 2017). The simple response to this issue is that it does not have to penetrate all that distance. Photobiomodulation may influence the survival of cells in the brainstem through an indirect stimulation, using a middle-man in the circulation, for example, the immune or stem cell systems (see above; C, Fig. 17.4). Photobiomodulation would only have to reach the blood vessels of the cranium, of which there are many, and these are certainly within reach by an external device (C, Fig. 17.4). It should be noted that if one considers application by the intracranial method, then there would be no penetration issues, because the photobiomodulation is delivered directly onto the neurones from an optical fiber device implanted within the brainstem (Darlot et al., 2016).

In summary, there are solid arguments to offset the bulk of the suspicion and skepticism of some colleagues with regard to the use of photobiomodulation therapy in Parkinson's disease patients. The translation from experimental to clinical is well-founded and certainly worthy of consideration.

17.10 Conclusion

Photobiomodulation has emerged as an effective neuroprotective agent in a wide range of animal models of Parkinson's disease, including toxin-induced and/or transgenic fly, rodent, and monkey models. There are beneficial outcomes by photobiomodulation in each of these animal models, from neuroprotection to functional and behavioral improvements and from reductions in gliosis to the expression of trophic growth factors. This body of experimental evidence in animal models is overwhelming and it should serve as a template for translation to the clinic.

References

Ashkan, K., Wallace, B.A., Mitrofanis, J., Pollo, C., Brard, P.-Y., Fagret, D., et al., 2007. SPECT imaging, immunohistochemical and behavioural correlations in the primate models of Parkinson's disease. Parkinsonism Relat. Disord. 13, 266–275. Available from: https://doi.org/10.1016/j.parkreldis.2006.10.009.

Barcia, C., Barreiro, A.F., Poza, M., Herrero, M.-T., 2003. Parkinson's disease and inflammatory changes. Neurotox. Res. 5, 411–417. Available from: https://doi.org/10.1007/BF03033170.

Benabid, A.L., Chabardes, S., Mitrofanis, J., Pollak, P., 2009. Deep brain stimulation of the subthalamic nucleus for the treatment of Parkinson's disease. Lancet Neurol. 8, 67–81. Available from: https://doi.org/10.1016/S1474-4422(08)70291-6.

Bergman, H., Deuschl, G., 2002. Pathophysiology of Parkinson's disease: from clinical neurology to basic neuroscience and back. Mov. Disord. 17 (Suppl. 3), S28–S40.

Blandini, F., Armentero, M.-T., 2012. Animal models of Parkinson's disease. FEBS J. 279, 1156–1166. Available from: https://doi.org/10.1111/j.1742-4658.2012.08491.x.

Blandini, F., Nappi, G., Tassorelli, C., Martignoni, E., 2000. Functional changes of the basal ganglia circuitry in Parkinson's disease. Prog. Neurobiol. 62, 63–88.

Blesa, J., Phani, S., Jackson-Lewis, V., Przedborski, S., 2012. Classic and new animal models of Parkinson's disease. J. Biomed. Biotechnol. 2012, 845618. Available from: https://doi.org/10.1155/2012/845618.

Bové, J., Perier, C., 2012. Neurotoxin-based models of Parkinson's disease. Neuroscience 211, 51–76. Available from: https://doi.org/10.1016/j.neuroscience.2011.10.057.

Brettschneider, J., Tredici, K.D., Lee, V.M.-Y., Trojanowski, J.Q., 2015. Spreading of pathology in neurodegenerative diseases: a focus on human studies. Nat. Rev. Neurosci. 16, 109–120. Available from: https://doi.org/10.1038/nrn3887.

Burda, J.E., Bernstein, A.M., Sofroniew, M.V., 2016. Astrocyte roles in traumatic brain injury. Exp. Neurol. 275 (Pt 3), 305–315. Available from: https://doi.org/10.1016/j.expneurol.2015.03.020.

Chaturvedi, R.K., Beal, M.F., 2008. Mitochondrial approaches for neuroprotection. Ann. N.Y. Acad. Sci. 1147, 395–412. Available from: https://doi.org/10.1196/annals.1427.027.

Corti, O., Brice, A., 2013. Mitochondrial quality control turns out to be the principal suspect in parkin and PINK1-related autosomal recessive Parkinson's disease. Curr. Opin. Neurobiol. 23, 100–108. Available from: https://doi.org/10.1016/j.conb.2012.11.002.

d'Anglemont de Tassigny, X., Pascual, A., López-Barneo, J., 2015. GDNF-based therapies, GDNF-producing interneurons, and trophic support of the dopaminergic nigrostriatal pathway. Implications for Parkinson's disease. Front. Neuroanat 9, 10. Available from: https://doi.org/10.3389/fnana.2015.00010.

Darlot, F., Moro, C., El Massri, N., Chabrol, C., Johnstone, D.M., Reinhart, F., et al., 2016. Near-infrared light is neuroprotective in a monkey model of Parkinson disease. Ann. Neurol. 79, 59–75. Available from: https://doi.org/10.1002/ana.24542.

Du, X., Stull, N.D., Iacovitti, L., 1995. Brain-derived neurotrophic factor works coordinately with partner molecules to initiate tyrosine hydroxylase expression in striatal neurons. Brain Res. 680, 229–233.

Eells, J.T., Wong-Riley, M.T.T., VerHoeve, J., Henry, M., Buchman, E.V., Kane, M.P., et al., 2004. Mitochondrial signal transduction in accelerated wound and retinal healing by near-infrared light therapy. Mitochondrion 4, 559–567. Available from: https://doi.org/10.1016/j.mito.2004.07.033.

El Massri, N., Johnstone, D.M., Peoples, C.L., Moro, C., Reinhart, F., Torres, N., et al., 2016a. The effect of different doses of near infrared light on dopaminergic cell survival and gliosis in MPTP-treated mice. Int. J. Neurosci. 126, 76–87. Available from: https://doi.org/10.3109/00207454.2014.994063.

El Massri, N., Moro, C., Torres, N., Darlot, F., Agay, D., Chabrol, C., et al., 2016b. Near-infrared light treatment reduces astrogliosis in MPTP-treated monkeys. Exp. Brain Res. 1–8. Available from: https://doi.org/10.1007/s00221-016-4720-7.

El Massri, N., Lemgruber, A.P., Rowe, I.J., Moro, C., Torres, N., Reinhart, F., et al., 2017. Photobiomodulation-induced changes in a monkey model of Parkinson's disease: changes in tyrosine hydroxylase cells and GDNF expression in the striatum. Exp. Brain Res. 1–14. Available from: https://doi.org/10.1007/s00221-017-4937-0.

Exner, N., Lutz, A.K., Haass, C., Winklhofer, K.F., 2012. Mitochondrial dysfunction in Parkinson's disease: molecular mechanisms and pathophysiological consequences. EMBO J. 31, 3038–3062. Available from: https://doi.org/10.1038/emboj.2012.170.

Fukae, J., Mizuno, Y., Hattori, N., 2007. Mitochondrial dysfunction in Parkinson's disease. Mitochondrion 7, 58–62. Available from: https://doi.org/10.1016/j.mito.2006.12.002.

Gash, D.M., Zhang, Z., Ai, Y., Grondin, R., Coffey, R., Gerhardt, G.A., 2005. Trophic factor distribution predicts functional recovery in parkinsonian monkeys. Ann. Neurol. 58, 224–233. Available from: https://doi.org/10.1002/ana.20549.

Goedert, M., 2015. Alzheimer's and Parkinson's diseases: the prion concept in relation to assembled Aβ, tau, and α-synuclein. Science 349, 1255555. Available from: https://doi.org/10.1126/science.1255555.

Halliday, G.M., Stevens, C.H., 2011. Glia: initiators and progressors of pathology in Parkinson's disease. Mov. Disord. 26, 6–17. Available from: https://doi.org/10.1002/mds.23455.

Hamblin, M.R., 2016. Shining light on the head: photobiomodulation for brain disorders. BBA Clin. 6, 113–124. Available from: https://doi.org/10.1016/j.bbacli.2016.09.002.

Hamby, M.E., Sofroniew, M.V., 2010. Reactive astrocytes as therapeutic targets for CNS disorders. Neurotherapeutics 7, 494–506. Available from: https://doi.org/10.1016/j.nurt.2010.07.003.

Jankovic, J., Poewe, W., 2012. Therapies in Parkinson's disease. Curr. Opin. Neurol. 25, 433–447. Available from: https://doi.org/10.1097/WCO.0b013e3283542fc2.

Johnstone, D.M., El Massri, N., Moro, C., Spana, S., Wang, X.S., Torres, N., et al., 2014. Indirect application of near infrared light induces neuroprotection in a mouse model of parkinsonism – an abscopal neuroprotective effect. Neuroscience 274, 93–101. Available from: https://doi.org/10.1016/j.neuroscience.2014.05.023.

Johnstone, D.M., Moro, C., Stone, J., Benabid, A.-L., Mitrofanis, J., 2016. Turning on lights to stop neurodegeneration: the potential of near infrared light therapy in Alzheimer's and Parkinson's disease. Front. Neurosci 9, 500. Available from: https://doi.org/10.3389/fnins.2015.00500.

Karu, T., 2010. Mitochondrial mechanisms of photobiomodulation in context of new data about multiple roles of ATP. Photomed. Laser Surg. 28, 159–160. Available from: https://doi.org/10.1089/pho.2010.2789.

Karu, T., 2013. Is it time to consider photobiomodulation as a drug equivalent? Photomed. Laser Surg. 31, 189–191. Available from: https://doi.org/10.1089/pho.2013.3510.

Khan, I., Arany, P., 2015. Biophysical approaches for oral wound healing: emphasis on photobiomodulation. Adv. Wound Care (New Rochelle) 4, 724–737. Available from: https://doi.org/10.1089/wound.2014.0623.

Liang, H.L., Whelan, H.T., Eells, J.T., Wong-Riley, M.T.T., 2008. Near-infrared light via light-emitting diode treatment is therapeutic against rotenone- and 1-methyl-4-phenylpyridinium ion-induced neurotoxicity. Neuroscience 153, 963–974. Available from: https://doi.org/10.1016/j.neuroscience.2008.03.042.

Liebert, A., Bicknell, B., Adams, R., 2014. Protein conformational modulation by photons: a mechanism for laser treatment effects. Med. Hypotheses YMEHY 7445. Available from: https://doi.org/10.1016/j.mehy.2013.12.009.

Martínez-Morales, P.L., Liste, I., 2012. Stem cells as in vitro model of Parkinson's disease. Stem Cells Int. 2012, 980941. Available from: https://doi.org/10.1155/2012/980941.

McGeer, P.L., McGeer, E.G., 1998. Glial cell reactions in neurodegenerative diseases: pathophysiology and therapeutic interventions. Alzheimer Dis. Assoc. Disord. 12 (Suppl. 2), S1–S6.

McGeer, P.L., McGeer, E.G., 2008. Glial reactions in Parkinson's disease. Mov. Disord. 23, 474–483. Available from: https://doi.org/10.1002/mds.21751.

Meng, C., He, Z., Xing, D., 2013. Low-level laser therapy rescues dendrite atrophy via upregulating BDNF expression: implications for Alzheimer's disease. J. Neurosci. 33, 13505–13517. Available from: https://doi.org/10.1523/JNEUROSCI.0918-13.2013.

Mitrofanis, J., 2017. Why and how does light therapy offer neuroprotection in Parkinson's disease? Neural. Regen. Res. 12, 574–575.

Monchi, O., Petrides, M., Strafella, A.P., Worsley, K.J., Doyon, J., 2006. Functional role of the basal ganglia in the planning and execution of actions. Ann. Neurol. 59, 257–264.

Moro, C., Torres, N., El Massri, N., Ratel, D., Johnstone, D.M., Stone, J., et al., 2013. Photobiomodulation preserves behaviour and midbrain dopaminergic cells from MPTP toxicity: evidence from two mouse strains. BMC Neurosci. 14, 40. Available from: https://doi.org/10.1186/1471-2202-14-40.

Moro, C., El Massri, N., Torres, N., Ratel, D., De Jaeger, X., Chabrol, C., et al., 2014. Photobiomodulation inside the brain: a novel method of applying near-infrared light intracranially and its impact on dopaminergic cell survival in MPTP-treated mice. J. Neurosurg. 120, 670–683. Available from: https://doi.org/10.3171/2013.9.JNS13423.

Moro, C., El Massri, N., Darlot, F., Torres, N., Chabrol, C., Agay, D., et al., 2016. Effects of a higher dose of near-infrared light on clinical signs and neuroprotection in a monkey model of Parkinson's disease. Brain Res. 1648 (Part A), 19–26. Available from: https://doi.org/10.1016/j.brainres.2016.07.005.

Moro, C., Torres, N., Arvanitakis, K., Cullen, K., Chabrol, C., Agay, D., et al., 2017. No evidence for toxicity after long-term photobiomodulation in normal non-human primates. Exp. Brain Res. 235, 3081–3092. Available from: https://doi.org/10.1007/s00221-017-5048-7.

Olanow, C.W., Kieburtz, K., Schapira, A.H.V., 2008. Why have we failed to achieve neuroprotection in Parkinson's disease? Ann. Neurol. 64 (Suppl. 2), S101–110. Available from: https://doi.org/10.1002/ana.21461.

Orme, R.P., Bhangal, M.S., Fricker, R.A., 2013. Calcitriol imparts neuroprotection in vitro to midbrain dopaminergic neurons by upregulating GDNF expression. PLoS One 8. Available from: https://doi.org/10.1371/journal.pone.0062040.

Oron, A., Oron, U., 2016. Low-level laser therapy to the bone marrow ameliorates neurodegenerative disease progression in a mouse model of Alzheimer's Disease: a minireview. Photomed. Laser Surg. 34, 627–630. Available from: https://doi.org/10.1089/pho.2015.4072.

Oueslati, A., Lovisa, B., Perrin, J., Wagnières, G., van den Bergh, H., Tardy, Y., et al., 2015. Photobiomodulation suppresses alpha-synuclein-induced toxicity in an AAV-based rat genetic model of Parkinson's disease. PLoS One 10, e0140880. Available from: https://doi.org/10.1371/journal.pone.0140880.

Palfi, S., Leventhal, L., Chu, Y., Ma, S.Y., Emborg, M., Bakay, R., et al., 2002. Lentivirally delivered glial cell line-derived neurotrophic factor increases the number of striatal dopaminergic neurons in primate models of nigrostriatal degeneration. J. Neurosci. 22, 4942–4954.

Pekny, M., Pekna, M., 2014. Astrocyte reactivity and reactive astrogliosis: costs and benefits. Physiol. Rev. 94, 1077–1098. Available from: https://doi.org/10.1152/physrev.00041.2013.

Pekny, M., Wilhelmsson, U., Pekna, M., 2014. The dual role of astrocyte activation and reactive gliosis. Neurosci. Lett. 565, 30–38. Available from: https://doi.org/10.1016/j.neulet.2013.12.071.

Peoples, C., Shaw, V.E., Stone, J., Jeffery, G., Baker, G.E., Mitrofanis, J., 2012. Survival of dopaminergic amacrine cells after near-infrared light treatment in MPTP-treated mice. ISRN Neurol. 2012, 850150. Available from: https://doi.org/10.5402/2012/850150.

Plaha, P., Ben-Shlomo, Y., Patel, N.K., Gill, S.S., 2006. Stimulation of the caudal zona incerta is superior to stimulation of the subthalamic nucleus in improving contralateral parkinsonism. Brain 129, 1732–1747. Available from: https://doi.org/10.1093/brain/awl127.

Purushothuman, S., Nandasena, C., Johnstone, D.M., Stone, J., Mitrofanis, J., 2013. The impact of near-infrared light on dopaminergic cell survival in a transgenic mouse model of parkinsonism. Brain Res. 1535, 61–70. Available from: https://doi.org/10.1016/j.brainres.2013.08.047.

Quirk, B.J., Desmet, K.D., Henry, M., Buchmann, E., Wong-Riley, M., Eells, J.T., et al., 2012. Therapeutic effect of near infrared (NIR) light on Parkinson's disease models. Front. Biosci. (Elite Ed.) 4, 818–823. Available from: https://doi.org/22201916.

Reinhart, F., Massri, N.E., Darlot, F., Torres, N., Johnstone, D.M., Chabrol, C., et al., 2014. 810nm near-infrared light offers neuroprotection and improves locomotor activity in MPTP-treated mice. Neurosci. Res. Available from: https://doi.org/10.1016/j.neures.2014.11.005.

Reinhart, F., El Massri, N., Darlot, F., Moro, C., Costecalde, T., Peoples, C.L., et al., 2015. Evidence for improved behaviour and neuroprotection after intracranial application of near infrared light in a hemi-parkinsonian rat model. J Neurosurg 27, 1–13. PMID: 26613166.

Reinhart, F., El Massri, N., Johnstone, D.M., Stone, J., Mitrofanis, J., Benabid, A.-L., et al., 2016. Near-infrared light (670 nm) reduces MPTP-induced parkinsonism within a broad therapeutic time window. Exp. Brain Res. Available from: https://doi.org/10.1007/s00221-016-4578-8.

Reinhart, F., Massri, N.E., Torres, N., Chabrol, C., Molet, J., Johnstone, D.M., et al., 2017. The behavioural and neuroprotective outcomes when 670 nm and 810 nm near infrared light are applied together in MPTP-treated mice. Neurosci. Res. Available from: https://doi.org/10.1016/j.neures.2016.11.006. PMID: 27871905.

Rinne, J.O., 1993. Nigral degeneration in Parkinson's disease. Mov. Disord 8 (Suppl. 1), S31–S35.

Rojas, J., Gonzalez-Lima, F., 2011. Low-level light therapy of the eye and brain. Eye Brain 3, 49–67.

Romeo, S., Vitale, F., Viaggi, C., di Marco, S., Aloisi, G., Fasciani, I., et al., 2017. Fluorescent light induces neurodegeneration in the rodent nigrostriatal system but near infrared LED light does not. Brain Res. 1662, 87–101. Available from: https://doi.org/10.1016/j.brainres.2017.02.026.

Sandhu, J.K., Gardaneh, M., Iwasiow, R., Lanthier, P., Gangaraju, S., Ribecco-Lutkiewicz, M., et al., 2009. Astrocyte-secreted GDNF and glutathione antioxidant system protect neurons against 6OHDA cytotoxicity. Neurobiol. Dis. 33, 405–414. Available from: https://doi.org/10.1016/j.nbd.2008.11.016.

Schapira, A.H.V., Olanow, C.W., Greenamyre, J.T., Bezard, E., 2014. Slowing of neurodegeneration in Parkinson's disease and Huntington's disease: future therapeutic perspectives. Lancet 384, 545–555. Available from: https://doi.org/10.1016/S0140-6736(14)61010-2.

Schober, A., 2004. Classic toxin-induced animal models of Parkinson's disease: 6-OHDA and MPTP. Cell Tissue Res. 318, 215–224. Available from: https://doi.org/10.1007/s00441-004-0938-y.

Sebastián, W.S., Guillén, J., Manrique, M., Belzunegui, S., Ciordia, E., Izal-Azcárate, A., et al., 2007. Modification of the number and phenotype of striatal dopaminergic cells by carotid body graft. Brain 130, 1306–1316. Available from: https://doi.org/10.1093/brain/awm061.

Shaw, V.E., Spana, S., Ashkan, K., Benabid, A.-L., Stone, J., Baker, G.E., et al., 2010. Neuroprotection of midbrain dopaminergic cells in MPTP-treated mice after near-infrared light treatment. J. Comp. Neurol. 518, 25–40. Available from: https://doi.org/10.1002/cne.22207.

Shaw, V.E., Peoples, C., Spana, S., Ashkan, K., Benabid, A.-L., Stone, J., et al., 2012. Patterns of cell activity in the subthalamic region associated with the neuroprotective action of near-infrared light treatment in MPTP-treated mice. Parkinsons Dis. 2012, 296875. Available from: https://doi.org/10.1155/2012/296875.

Sun, M., Kong, L., Wang, X., Lu, X., Gao, Q., Geller, A.I., 2005. Comparison of the capability of GDNF, BDNF, or both, to protect nigrostriatal neurons in a rat model of Parkinson's disease. Brain Res. 1052, 119–129. Available from: https://doi.org/10.1016/j.brainres.2005.05.072.

Torres, N., Molet, J., Moro, C., Mitrofanis, J., Benabid, A.L., 2017. Neuroprotective surgical strategies in Parkinson's Disease: role of preclinical data. Int. J. Mol. Sci. 18. Available from: https://doi.org/10.3390/ijms18102190.

Trimmer, P.A., Schwartz, K.M., Borland, M.K., De Taboada, L., Streeter, J., Oron, U., 2009. Reduced axonal transport in Parkinson's disease cybrid neurites is restored by light therapy. Mol. Neurodegener. 4, 26. Available from: https://doi.org/10.1186/1750-1326-4-26.

Tuby, H., Maltz, L., Oron, U., 2011. Induction of autologous mesenchymal stem cells in the bone marrow by low-level laser therapy has profound beneficial effects on the infarcted rat heart. Lasers Surg. Med. 43, 401–409. Available from: https://doi.org/10.1002/lsm.21063.

Verkhratsky, A., Parpura, V., Pekna, M., Pekny, M., Sofroniew, M., 2014. Glia in the pathogenesis of neurodegenerative diseases. Biochem. Soc. Trans. 42, 1291–1301. Available from: https://doi.org/10.1042/BST20140107.

Vos, M., Lovisa, B., Geens, A., Morais, V.A., Wagnières, G., van den Bergh, H., et al., 2013. Near-infrared 808 nm light boosts complex IV-dependent respiration and rescues a Parkinson-related pink1 model. PLoS One 8 (e78562). Available from: https://doi.org/10.1371/journal.pone.0078562.

Wallace, B.A., Ashkan, K., Heise, C.E., Foote, K.D., Torres, N., Mitrofanis, J., et al., 2007. Survival of midbrain dopaminergic cells after lesion or deep brain stimulation of the subthalamic nucleus in MPTP-treated monkeys. Brain 130, 2129–2145. Available from: https://doi.org/10.1093/brain/awm137.

Whelan, H.T., DeSmet, K.D., Buchmann, E.V., Henry, M.M., Wong-Riley, M., Eells, J.T., et al., 2008. Harnessing the cell's own ability to repair and prevent neurodegenerative disease. SPIE Newsroom 24, 1–3.

Xicoy, H., Wieringa, B., Martens, G.J.M., 2017. The SH-SY5Y cell line in Parkinson's disease research: a systematic review. Mol. Neurodegener. 12, 10. Available from: https://doi.org/10.1186/s13024-017-0149-0.

Xuan, W., Vatansever, F., Huang, L., Wu, Q., Xuan, Y., Dai, T., et al., 2013. Transcranial low-level laser therapy improves neurological performance in traumatic brain injury in mice: effect of treatment repetition regimen. PLoS One 8, e53454. Available from: https://doi.org/10.1371/journal.pone.0053454.

Ying, R., Liang, H.L., Whelan, H.T., Eells, J.T., Wong-Riley, M.T.T., 2008. Pretreatment with near-infrared light via light-emitting diode provides added benefit against rotenone- and MPP + -induced neurotoxicity. Brain Res. 1243, 167–173. Available from: https://doi.org/10.1016/j.brainres.2008.09.057.

Further reading

Sonnier, L., Pen, G.L., Hartmann, A., Bizot, J.-C., Trovero, F., Krebs, M.-O., et al., 2007. Progressive loss of dopaminergic neurons in the ventral midbrain of adult mice heterozygote for Engrailed1. J. Neurosci. 27, 1063–1071. Available from: https://doi.org/10.1523/JNEUROSCI.4583-06.2007.

Chapter 18

Effects of near-infrared low-level laser stimulation on neuronal excitability

Ljubica M. Konstantinović[1,2] and Saša R. Filipović[3]

[1]*Department of Rehabilitation, Faculty of Medicine, University of Belgrade, Belgrade, Serbia,* [2]*Clinic for Rehabilitation "Dr Miroslav Zotović", Belgrade, Serbia,* [3]*Institute for Medical Research, University of Belgrade, Belgrade, Serbia*

18.1 Introductory remarks

Low level laser (LLL) irradiation with red, near-infrared (NIR), and infrared (IR) lights[1] has been shown to produce changes at molecular, cellular, and tissue levels, with these physiological effects referred to as photobiomodulation (PBM). There is an "optical window" roughly covering the red and NIR parts of the optical spectrum (650–1200 nm) where tissue penetration of the light is maximal, while the likelihood for heating effect is not particularly high. Due to these features, NIR LLL has attracted particular attention for use in brain PBM (Salehpour et al., 2018).

It is well-established that photons in the red-NIR wavelengths (i.e., 600–1200 nm) are absorbed by chromophores in cytochrome C oxidase (CCO) and stimulate mitochondrial metabolism, which is then manifested in raised mitochondrial membrane potential, higher oxygen consumption, production of reactive oxygen species (ROS), and increase of adenosine triphosphate (ATP) (Karu et al., 1995; Mochizuki-Oda et al., 2002; Wong-Riley et al., 2005; de Freitas and Hamblin, 2016). The possible molecular mechanism for this stimulation of CCO activity could be the photodissociation of inhibitory nitric oxide (NO) that binds to the heme or copper centers in the CCO (Karu et al., 2005; Lane, 2006). In addition, it is proposed that for longer wavelength NIR-LLL (e.g., 940 nm and above), the alternative chromophore could be water. A small increase in vibrational energy by a water cluster formed in or on a sensitive protein, such as a heat-gated ion channel,[2] could be sufficient to perturb the tertiary protein structure, thus opening the channel and allowing modulation of intracellular calcium levels (Damodaran, 2015). Moreover, changes in intracellular signaling molecules such as calcium ions, ROS, and redox sensitive transcription factors like (e.g., NF-kB) are also thought to mediate the effects of light. They cause transcription changes which leads to upregulation of cytoprotective gene products such as antioxidant enzymes, heat shock proteins, and antiapoptotic proteins with important cellular roles. One of the most upregulated genes after NF-kB activation is the antioxidant mitochondrial superoxide dismutase (Chen et al., 2011).

Overall, it has been found that LLL NIR boosts cellular energy metabolism and other vital cellular protective functions. This has led to its extensive use in promotion of wound healing and treatment of inflammation; its use in reduction of ischemic neural damage has attracted a fair amount of attention too (Hashmi et al., 2010). However, the aforementioned metabolic and other changes may also affect the excitability of neural tissue, which could open the way for use of the LLL NIR in neuromodulation as well.

1. The classification of the International Commission on Illumination (CIE): Infra-red A (IR-A): 780–1400 nm; IR-B: 1400–3000 nm; and IR-C: 3000 nm–1 mm. In PBM literature, stimulation with IR-A wavelengths is typically referred as NIR stimulation, while stimulation with IR-B and IR-C wavelengths is typically referred as infrared stimulation.

2. Superfamily of transient receptor potential (TRP) channels. The member of this superfamily is TRPV (vanilloid subfamily) and TRPV1 (capsaicin receptor).

Photobiomodulation in the Brain. DOI: https://doi.org/10.1016/B978-0-12-815305-5.00018-X

18.2 Neuronal excitability—experimental results

Neuronal excitability can be defined, depending on the level of detail, as the readiness of a nerve cell or a neural circuit to respond to a stimulus. The response is typically in the form of an action potential (AP), a transient change of electrical charge (polarization) of the neuronal membrane. The AP can be measured either individually, at the level of an individual nerve cell, or as sum of APs in form of a compound action potential (CAP) or an evoked potential (EP), at the level of group of neurons or neural circuits. Also, responses of central nervous system neural circuits can be measured at the level of their peripheral outputs, such as muscle contraction (using electromyography).

18.2.1 Effects on peripheral nerves

The effects of the LLL NIR on neuronal electrophysiological features have been investigated extensively at the level of the peripheral nerves predominantly related to the pain control (reviewed in Chow et al., 2011). The studies have been carried out mostly in animals (e.g., rat, mouse, cat, dog, rabbit, guinea pig) and to a lesser degree in humans. Given the focus on pain, the main outcome variable has been conduction velocity (CV), while some studies assessed also CAPs, and somato-sensory/pain EPs. The excitability per se has rarely been a subject of peripheral nerve studies.

In animal studies, NIR LLL has been applied in several ways: transcutaneously, to exposed nerves in situ, to isolated nerves, or to nerve cell cultures (e.g., Tsuchiya et al., 1993, 1994; Wakabayashi et al., 1993; Mezawa et al., 1988). Generally, regardless whether LLL was applied in pulsed or continuous wave mode, slowing of CV, reduction of CAP amplitude, and inhibition of neuronal discharges in ascending neuronal structures (e.g., ipsilateral dorsal root, trigeminal nucleus) have been consistent findings. The effects seem to have been dose dependent since they were stronger with longer duration of LLL irradiation.

In human studies, the LLL has been applied almost exclusively transcutaneously. The results have been less consistent than in animal studies. Generally, continuous wave LLL, red (780 nm), and NIR (typically 820–830 nm), applied over different peripheral nerves (i.e., sural, median, and radial) have been shown so slow CV at the corresponding nerve regardless of the wavelength and LLL parameters (i.e., intensity, duration) (e.g., Baxter et al., 1993, 1994; Kramer and Sandrin, 1993; Lowe et al., 1994; Hadian and Moghagdam, 2003). In contrast, pulsed LLL has shown variable effects which seems to be highly dependent on the irradiation parameters (wavelength, pulse frequency, duration, intensity). For example, pulsed 830 nm LLL at various frequencies (7, 9, 12, 73 Hz, 5 kHz) and different energy doses did not slow CV of the median or sural nerves (Lowe et al., 1995; Walsh et al., 2000). On the other hand, pulsed (73 Hz) LLL with longer wavelength (904 nm) applied over superficial radial nerve slowed CV when applied for 120 seconds, but not when applied for only 20 seconds (Greathouse et al., 1985). Moreover, even 830 nm LLL (140 mW, 5.1 J/cm^2) when applied with higher pulse frequency (1500 Hz) slowed CV in the sural nerve; however, when applied at different energy doses (30 mW, 2.55 J/cm^2; and 400 mW, 7.65 J/cm^2) it did not slow down (Cambier et al., 2000).

In summary, the NIR LLL irradiation of the peripheral nerves seems to tend to cause inhibitory effects mainly affecting the nerve CV. However, the results are not always consistent and may depend on LLL stimulation parameters. Unfortunately, the excitability of the peripheral neurons has not been the primary aim of investigations, and still little is known of the subject.

18.2.2 Effects on brain

The results of the peripheral nerve studies are hard to apply to the study of the possible LLL effects on the central nervous system. This is not only since excitability per se has been subject of almost no peripheral nerve studies, but also because of the much higher order of complexity of the central neural structures. Central nervous system consists of vast variety of neural cells with different morphology, metabolism, and functional features. Moreover, these cells do not work only in isolation, but as parts of complex functional networks with numerous mutual interactions. On top of this, the cerebral cortex, which is effectively the only accessible structure for brain PBM, is organized in a morphologically complex interplay of gyri and sulci making precise understanding of the effects of the laser light beams even more difficult.

There are only a few studies that investigated electrophysiological effects of the transcranial LLL stimulation (TLS) with NIR light, and they were all conducted in humans.

In our study (Konstantinović et al., 2013) we evaluated the effects on motor cortex (M1) excitability of 5 minutes TLS with pulsedmode (3 kHz) NIR-LLL (905 nm, power density 50 mW/cm^2, total dose 15 J/cm^2) applied over M1. The M1 excitability change was measured by changes in the amplitudes of the motor evoked potentials (MEPs) in a hand muscle elicited by transcranial magnetic stimulation (TMS) over the same M1 spot (Fig. 18.1). The TLS caused suppression of M1 excitability (i.e., reduction of MEP amplitudes in comparison to baseline) lasting for up to 30 minutes. The maximum suppression was within the interval 10–20 minutes following the TLS. In more than 70% of participants, MEP amplitudes were below the baseline level for most of the 30 minutes.

The TLS induced reduction of excitability in the first 10 minutes post-TLS period was negatively correlated with baseline threshold TMS intensity for eliciting MEP at rest (rest motor threshold—RMT). Given that RMT is directly proportional to the distance from the coil to the cortical surface (McConnell et al., 2001; Stokes et al., 2007), smaller TLS induced MEP suppression in participants with higher RMT may be explained in the same way—the larger distance from the laser diode to the cortical surface in participants with higher RMT may have caused smaller TLS effects.

In the extension of the study we tried to see whether the TLS suppression effect may be dose-dependent (unpublished data). For that purpose we enrolled four subjects with absent MEP suppression following TLS. We also attempted to enroll several subjects with weak suppression but very high RMT; unfortunately only one was available. By increasing the NIR-LLL pulse frequency from 3 to 5 kHz we delivered more energy within the same stimulation time. In all subjects tested the MEP amplitudes following stronger TLS were smaller than in the first part of the study (Fig. 18.2). The difference was significant. Even more, all four initial nonresponders had clear suppression for the most of the 30 minutes post-TLS period. The results suggest a dose-dependency of the TLS effect.

Using 810 nm continuous wave diverging laser beam (power density of 500 mW/cm^2 on the surface of the skin, resulting in less than 5 mW/cm^2 cortical fluence ($\sim$1 J total energy)) for 10 minutes, Chaieb et al. (2015) essentially confirmed our findings regarding reduced M1 excitability following TLS, lasting for up 30 minutes. This is of note given potentially different modes of action between pulsed-mode and continuous lasers on neural tissue. In addition, using paired-pulse TMS, they tested intrinsic M1 inhibition and facilitation in more details. Paired-pulse stimulation involves a conditioning stimulus (CS) followed by a test stimulus (TS) given through the same coil, and the MEP amplitudes following conditioned TS are compared to those produced by the TS alone as a reference (baseline) (Kujirai et al., 1993). The CS is typically subthreshold for eliciting MEP, while TS is of intensity which when applied alone produces stable MEP response. At short interstimulus intervals (ISI) of 1–6 ms the MEP to TS is reduced—short-latency

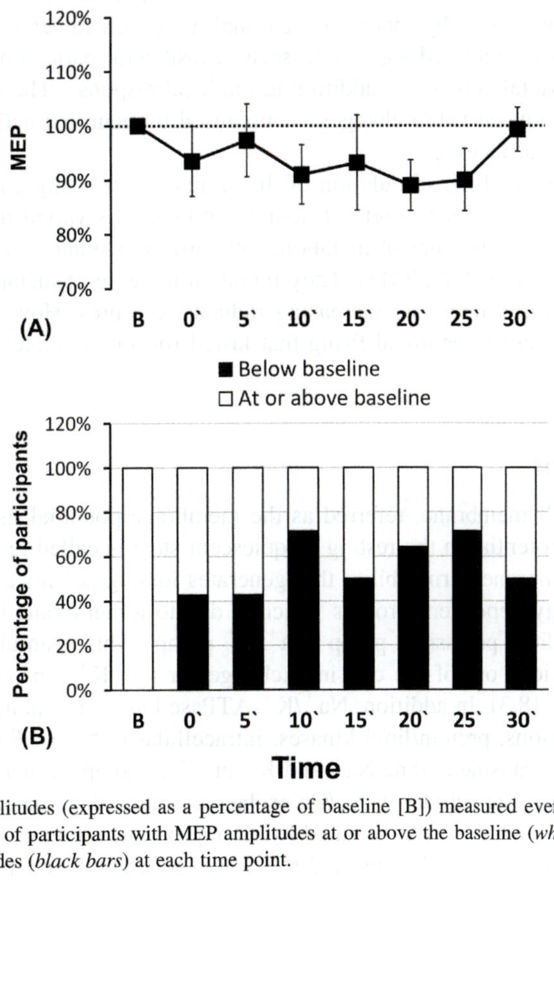

FIGURE 18.1 (A) Average MEP amplitudes (expressed as a percentage of baseline [B]) measured every 5 min following LLL NIR; vertical lines are standard errors (SE). (B) Percentage of participants with MEP amplitudes at or above the baseline (*white bars*) contrasted with number of participants with below baseline MEP amplitudes (*black bars*) at each time point.

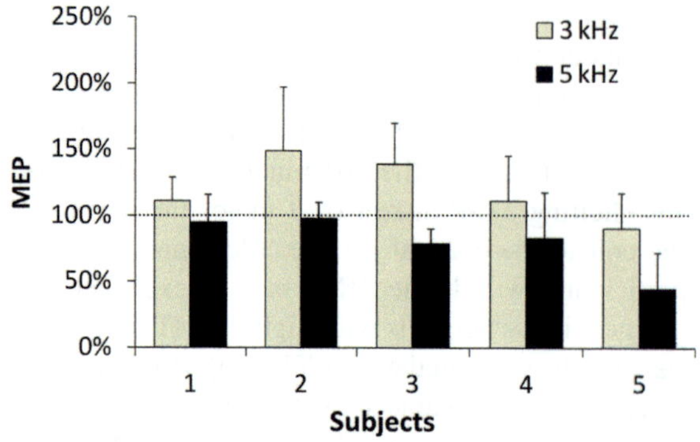

FIGURE 18.2 Comparison of average post-LLL relative MEP sizes (relative to baseline) between the two LLL intensities. First four subjects were nonresponders in the main experiment, but became responders when higher intensity LLL was applied. Subject five, who had relatively high rMT, was responder in the main experiment, but in the add-on experiment his MEP size became considerably more suppressed.

intracortical inhibition (SICI), while at ISI of 6–30 ms MEP to TS gets larger in comparison to the baseline—intracortical facilitation (ICF). Chaieb et al. (2015) found increase in SICI (i.e., even smaller conditioned TS MEP) and decrease in ICF (i.e., smaller increase in conditioned TS MEP) immediately after the TLS, but not 30 minutes later. In contrast, long-latency intracortical inhibition (LICI), which was assessed in a similar manner but with longer ISI (50–200 ms) (Valls-Sole et al., 1992) was not affected by TLS. SICI and ICF are localized phenomena, linked to the stimulated area. The observed changes can be explained by increased resistance to depolarization by TLS targeted neuronal cells. In contrast, LICI depends on a wider spread network, part of which may not have been affected by TLS.

Using continuous wave 810 nm diode laser (power density 204–236 mW/cm^2, light fluence 55–63 J/cm^2) applied over dorsolateral prefrontal cortex for 4.5 minutes, Nawashiro et al. (2017) showed in six subjects consistent increase in functional magnetic resonance imaging (fMRI) blood-oxygen-level-dependent response directly under the laser probe. Given the duration of the stimulation and duration of the fMRI acquisition, authors were of the opinion that the observed change in blood flow was caused by increased neuronal activation rather than laser induced NO release. In addition, for most of the subjects, the increased signal was seen in ipsilateral parietal area too. This could suggest TLS induced activation of the fronto-parietal network in addition to the local response. The network activation would be further proof that the observed TLS induced fMRI changes were caused by neuronal activation rather than laser induced local release of NO and other humoral factors.

Moreover, one of the rare studies of LLL stimulation of the animal brain, using pulsed (50–200 Hz) IR (1875 nm) light delivered directly to the rat somatosensory cortex (radiant exposure was varied from 0.01 to 0.55 J/cm^2), investigated local changes in blood flow (as a measure of metabolic activation) together with single unit electrophysiological recordings from the targeted area (Cayce et al., 2011). They found an increase of metabolic activation of the stimulated cortical area with faster laser repetition rates and increasing radiant exposures. However, single unit recordings indicated a statistically significant decrease in neuronal firing that lasted for approximately 1.5–2.0 seconds following the end of illumination.

18.3 Proposed mechanisms

The voltage difference across a cell membrane, referred as the membrane potential, is the core determinant of a neuron's excitability. The membrane potential in the resting or quiescent state is called resting membrane potential (RMP) and is maintained by selective membrane permeability that generates ionic gradients across the membrane. Generation of ionic gradient is an active energy dependent process which is due to a membrane bound enzyme system, the Na^+/K^+-ATPase, also known as the sodium potassium pump (Na^+/K^+ pump). This is an ubiquitously expressed membrane protein that pumps out three Na^+ ions out of the cell in exchange for two K^+ ions entering the cell, using one ATP molecule as an energy source (Fig. 18.3). In addition, Na^+/K^+-ATPase has an ion-pumping independent receptor function that influences protein interactions, protein/lipid kinases, intracellular Ca^{2+} oscillations, and ROS generation (Cui and Xie, 2017). Maintenance of the transmembrane Na^+ and K^+ gradients supports neuronal excitability, but also regulates intracellular Ca^{2+} by controlling the activity of Na/Ca exchanger and voltage-gated calcium channels.

When an excitable cell, such as a neuron gets activated by an appropriate stimulus, a momentary reversal in electrical potential across membrane occurs—an AP (Bean, 2007). The generation and propagation of these changes in

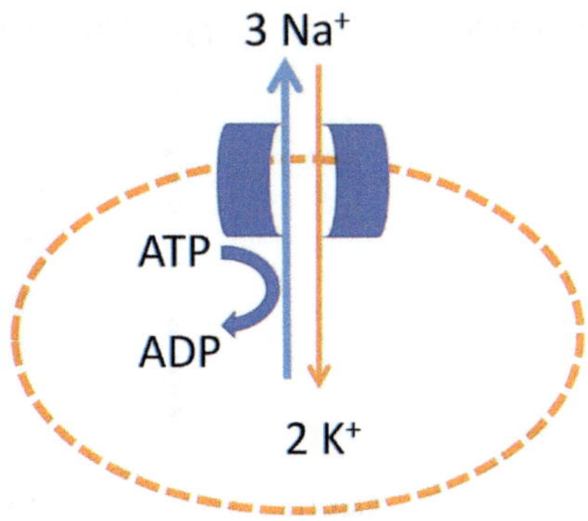

FIGURE 18.3 Schematic presentation of sodium-potassium pump.

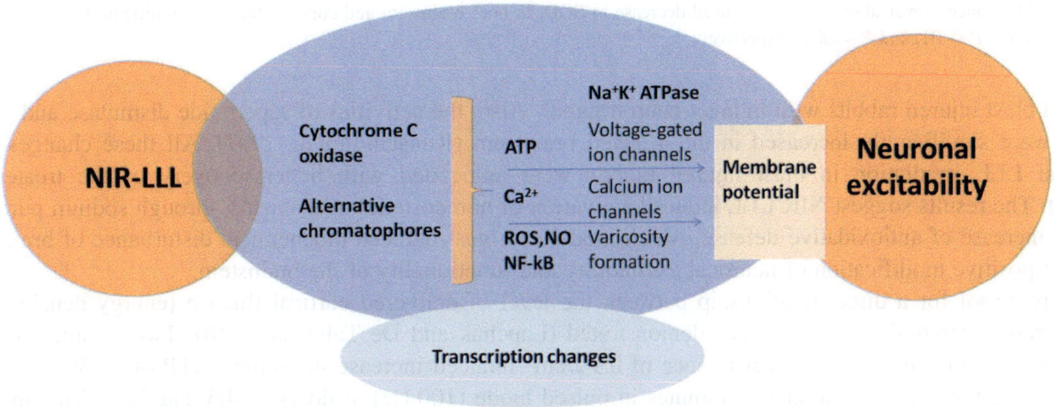

FIGURE 18.4 Possible mechanisms of interaction between NIR-LLL and neuronal excitability.

membrane potentials is the essence of encoding and transmitting information in nervous system. The mechanism behind the process is opening and closing of the voltage-gated sodium and potassium channels (Hodgkin and Huxley, 1952). Voltage-gated channels are sensitive to small changes in membrane potential. Biophysical characteristics of the channels, including activation and inactivation voltage ranges and properties, may be modulated by phosphorylation via various messenger systems. Sodium channels open in response to depolarization, admitting Na^+, and driving the membrane voltage positive during the initial fast rising phase of the AP. However, they get inactivated quickly allowing for the K^+ channels to restore the membrane potential to rest levels in the second, down-going phase of the AP (Armstrong and Hille, 1998). To maintain and restore ionic gradients, a sufficient amount of oxygen in the cells is needed to generate adequate quantity of ATP by oxidative phosphorylation. The neural membranes in the brain especially require a large amount of oxygen with estimation that Na^+/K^+-ATPase found on the plasma membrane of neurons consumes more than 50% of the energy supplied to the brain. To meet this need, cortical and subcortical tissues are rich in mitochondria which supply the ATP necessary to support axonal transport and to maintain membrane stability and excitability.

Several mechanisms of energy and cellular metabolism involved in maintenance of membrane potential, as the main factor in neuronal excitability, are also known to be targets of NIR LLL (Fig. 18.4). There are experimental data from peripheral nerve and transcranial brain LLL stimulation studies supporting this assumption.

Transcutaneous continuous LLL irradiation at 830 nm (60 mW, 6–15 seconds, 0.9 J/cm^2) applied over rat saphenous nerve increased activity of Na^+/K^--ATPase (Kudoh, 1989). Moreover, a few studies observed an increase in ATP levels with subsequent hyperpolarization of the nerve membrane which in turn led to increased amplitude of the AP following LLL irradiation (Nissan et al., 1986; Rochkind et al., 1986).

Our previous studies have shown that transcranial NIR pulsed LLL irradiation at 904 nm (5 kHz, 3 J/cm^2) applied over the upper cervical region, in the anatomical projection of the brainstem, caused the significant increase of the sodium pump substrate uptake rate with increased affinity to ATP in the cortical and brainstem tissue of both, intact

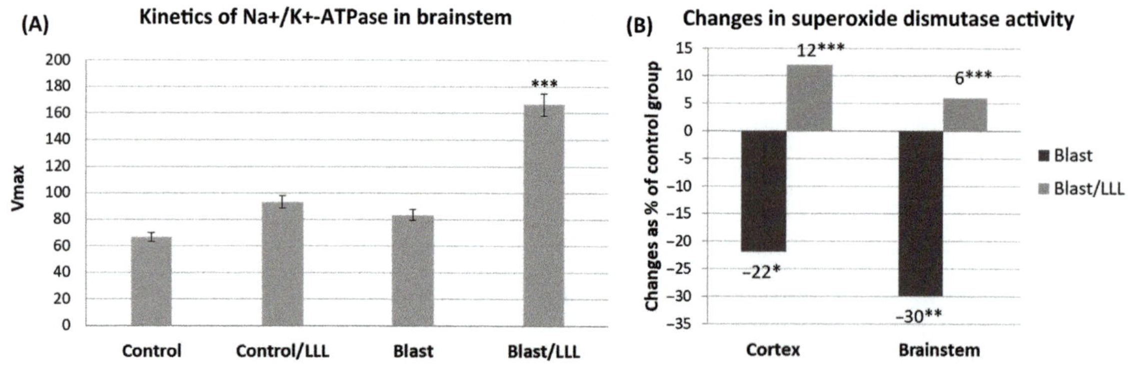

FIGURE 18.5 Results of our experiments on transcranial NIR LLL effects on blast injured rabbits (Konstantinović, 1997). (A) Changes in kinetics of NaK-ATPase in brainstem, measured by Malachite-green test and additional enzyme kinetic analysis by Linewear Burk method, show an increase of substrate reuptake by sodium pump following LLL, particularly pronounced in blast injured animals. (B) Activity of superoxide dismutase (SOD), measured by spectrophotometry is shown for two groups of blast injured rabbits as percentage change in comparison to corresponding controls—results show LLL induced reversal of the blast induced decrease in SOD, in both brainstem and cortex. Statistical significance is marked with *, **, and *** for $P < .05$, $P < .01$, and $P < .001$, respectively.

rabbits and blast injured rabbits with indirect neurotrauma.[3] Also, the activities of superoxide dismutase and glutathione reductase were significantly increased in the affected brainstem (Konstantinović, 1997). All these changes following transcranial LLL irradiation in blast injured rabbits were associated with better recovery of the treated animals (Fig. 18.5). The results suggest NIR LLL induced activation of homeostatic mechanisms, through sodium pump upregulation and increase of antioxidative defense, which modulated blast-induced biochemical disturbance of brainstem with consequent positive modification of neuronal excitability and functionality of the brainstem.

Recently, proof for a direct relationship between the level of delivered cortical fluence (energy density) and ATP content in rabbit cerebral cortex has been demonstrated (Lapchak and De Taboada, 2010). Two minutes of NIR LLL (808 nm) in continues mode at cortical fluence of 0.9 J/cm^2 induced increase in cortical ATP of 22.5% in comparison to control naive rabbits, while after two minutes in pulsed mode (100 Hz) at doses of 4.5 and 31.5 J/cm^2 increase was of 41% and 77%, respectively. However, the relationship between the dose of LLL illuminations and metabolic changes seems not to be linear, but appears to follows biphasic response pattern—at lower levels LLL was found to induce a significant increase in ATP and mitochondrial membrane potential, but with the dose increase the effect was found to be reversed causing suppression in both metabolic variables (Huang et al., 2011).

Although, it may seem paradoxical at first sight, the finding that NIR LLL increases neuronal energy metabolism and yet induces inhibition of excitability when applied transcranially, may be well understood if the physiology of membrane potential maintenance is considered. Namely, as discussed previously, maintenance of the RMP is an energy demanding process. It can be assumed that increased availability of intracellular energy sources boosts the activity of Na^+/K^+-ATPase and other membrane potential maintenance mechanisms, which in consequence makes membrane potential more resistant to externally induced changes, such as TMS pulses used in assessing M1 excitability.

In addition, cortical nerve cells do not work in isolation, but in rich interactions with neurons in their neighborhood as well as with distant but functionally related neurons. Therefore, the excitability of a single cortical neuron is a result of not only its intrinsic characteristics and state, but also of the net effects of the impulses coming from all the other connected neurons. Moreover, transcranial brain LLL illumination can never be focused enough to target an isolated neuron. The illumination/stimulation always encompasses a cluster of neurons with various functional interactions. In the case of M1 excitability assessment by TMS the output variable, the muscle twitch, is generated straight on by AP coming from large cortical pyramidal cells. However, the TMS generally does not directly affect pyramidal cells in M1, but exercises its impact through a network of small cortical interneurons with modulatory effects on pyramidal cells' RMP and excitability. In that context, inhibitory interneurons (Markram et al., 2004) are of particular interest since it may be that M1 inhibition, which has been observed in TMS studies, has resulted from LLL induced increase in

3. Pulmonary blast wave with nominal peak pressure of 304 kPa was generated in laboratory conditions, using air driven shock tube to the lung of rabbits. Pulmonary blast injuries can induce local, systemic and cerebral response. Cerebral response is manifests with brain edema, disturbances of activity of enzymes of antioxidative defense (superoxide dismutase—SOD, glutathione reductase—GR) and functional activity of sodium pump.

activation of inhibitory interneurons. The inhibitory interneurons occupy more superficial cortical layers and thus are exposed to the main bulk of LLL illumination that manages to pass the scalp structure and reaches cortical surface.

18.4 Future directions

Number of issues related to the NIR LLL effects on neuronal, and particularly CNS, excitably are still open. The results so far demonstrate an obvious effect of LLL illumination on neuronal excitability which is dependent on the illumination parameters. The mechanism of the effect seems to be related to the LLL effects on the cellular metabolism. Further investigations are needed of the interplay between cellular metabolism and neuronal excitability, as well as of the differential effects of the various LLL parameters. Also, more understanding is required of the effects of the LLL irradiation on the cortical micro- and macrocircuits.

Acknowledgment

The work on this article was supported in part by project grant (#175012) from the Ministry for Education, Science, and Technological Development of Republic of Serbia.

References

Armstrong, C.M., Hille, B., 1998. Voltage-gated ion channels and electrical excitability. Neuron 20, 371–380.

Baxter, G.D., Allen, A.J., Bell, J., Ravey, J., Diamantopoulos, C., 1993. Effects of 830 nm continuous wave laser diode irradiation on median nerve function in normal subjects. Laser Surg. Med. 13, 597–604.

Baxter, G.C., Walsh, D.M., Allen, J.M., Lowe, A.S., Bell, A.J., 1994. Effects of low intensity infrared laser irradiation upon conduction in the human median nerve in vivo. Exp. Physiol. 79, 227–234.

Bean, B.P., 2007. The action potential in mammalian central neurons. Nat. Rev. Neurosci. 8, 451–465.

Cambier, D., Blom, K., Witvrouw, E., Ollevier, G., De Muynck, M., Vanderstraeten, G., 2000. The influence of low intensity infrared laser irradiation on conduction characteristics of peripheral nerve: a randomised, controlled, double blind study on the sural nerve. Lasers Med. Sci. 15, 195–200.

Cayce, J.M., Friedman, R.M., Jansen, E.D., Mahavaden-Jansen, A., Roe, A.W., 2011. Pulsed infrared light alters neural activity in rat somatosensory cortex in vivo. NeuroImage 57 (1), 155–166. Available from: https://doi.org/10.1016/j.neuroimage.2011.03.084.

Chaieb, L., Antal, A., Masurat, F., Paulus, W., 2015. Neuroplastic effects of transcranial near-infrared stimulation (tNIRS) on the motor cortex. Front. Behav. Neurosci. 9, 147. Available from: https://doi.org/10.3389/fnbeh.2015.00147.

Chen, A.C., Arany, P.R., Huang, Y.Y., Tomkinson, E.M., Sharma, S.K., Kharkwal, G.B., et al., 2011. Low-level laser therapy activates NF-kB via generation of reactive oxygen species in mouse embryonic fibroblasts. PLoS One 6, e22453.

Chow, R., Armati, P., Laakso, E.L., Bjordal, J.M., Baxter, G.D., 2011. Inhibitory effects of laser irradiation on peripheral mammalian nerves and relevance to analgesic effects: a systematic review. Photomed. Laser Surg. 29, 365–381.

Cui, X., Xie, Z., 2017. Protein interaction and Na/K-ATPase-mediated signal transduction. Molecules 22, 990.

Damodaran, S., 2015. Water at biological phase boundaries: its role in interfacial activation of enzymes and metabolic pathways. Subcell. Biochem. 71, 233–261.

De Freitas, L.F., Hamblin, M.R., 2016. Proposed mechanisms of photobiomodulation or low-level light therapy. IEEE J. Sel. Top. Quantum Electron. 22, 7000417.

Greathouse, D.G., Currier, D.P., Gilmore, R.L., 1985. Effects of clinical infrared laser on superficial radial nerve conduction. Phys. Ther. 65, 1184–1187.

Hadian, M., Moghagdam, B., 2003. The effects of low power laser on electrophysiological parameters of sural nerve in normal subjects: a comparison between 670 and 780 nm wavelengths. Acta Med. Iran 41, 138–142.

Hashmi, J.T., Huang, Y.Y., Osmani, B.Z., Sharma, S.K., Naeser, M.A., Hamblin, M.R., 2010. Role of low-level laser therapy in neurorehabilitation. PM R 2 (12Suppl. 2), S292–S305. Available from: https://doi.org/10.1016/j.pmrj.2010.10.013.

Hodgkin, A.L., Huxley, A.F., 1952. A quantitative description of membrane current and its application to conduction and excitation in nerve. J. Physiol. 117, 500–544.

Huang, Y.Y., Sharma, S.K., Carroll, J., Hamblin, M.R., 2011. Biphasic dose response in low level light therapy – an update. Dose Response 9, 602–618.

Karu, T., Pyatibrat, L., Kalendo, G., 1995. Irradiation with He-Ne laser increases ATP level in cells cultivated in vitro. J. Photochem. Photobiol. B 27, 219–223.

Karu, T.I., Pyatibrat, L.V., Afanasyeva, N.I., 2005. Cellular effects of low power laser therapy can be mediated by nitric oxide. Lasers Surg. Med. 36, 307–314.

Kramer, J.F., Sandrin, M., 1993. Effect of low-power laser and white light on sensory conduction rate of the superficial radial nerve. Physiother. Canada 45, 165–170.

Konstantinović, L.M., 1997. The Mechanism of Effects of Low-Power Laser of Various Wavelengths on Indirect Neurotrauma (Ph.D. thesis), Military Medical Academy, Belgrade.

Konstantinović, L.M., Jelić, M.B., Jeremić, A., Stevanović, V.B., Milanović, S.D., Filipović, S.R., 2013. Transcranial application of near-infrared low-level laser can modulate cortical excitability. Lasers Surg. Med. 45 (10), 648–653. Available from: https://doi.org/10.1002/lsm.22190.

Kudoh, C., Inomata, K., Okajima, K., Motegi, M., Ohshiro, T., 1989. Effects of 830nm Gallium Aluminium Garsenide diode laser radiation on rat saphenous nerve sodium-potassium-adenosine riphosphatase activity: a possible pain attenuation mechanism examined. Laser. Ther. 1, 63–67.

Kujirai, T., Caramia, M.D., Rothwell, J.C., Day, B., Thompson, P.D., Ferbert, A., et al., 1993. Cortico cortical inhibition in human motor cortex. J. Physiol. 471, 501–519. Available from: https://doi.org/10.1113/jphysiol.1993.sp019912.

Lane, N., 2006. Cell biology: power games. Nature 443, 901–903.

Lapchak, P.A., De Taboada, L., 2010. Transcranial near infrared laser treatment (NILT) increases cortical adenosine-5′-triphosphate (ATP) content following embolic strokes in rabbits. Brain Res. 1306, 100–105.

Lowe, A., Baxter, G., Walsh, D., Allen, J., 1994. Effect of low intensity laser (830 nm) irradiation on skin temperature and antidromic conduction latencies in the human median nerve: relevance of radiant exposure. Lasers Surg. Med. 14, 40–46.

Lowe, A., Baxter, G., Walsh, D., Allen, J., 1995. The relevance of pulse repetition rate and radiant exposure to the neurophysiological effects of low-intensity laser (820 nm/pulsed wave) irradiation upon skin temperature and antidromic conduction latencies in the human median nerve. Lasers Med. Sci. 10, 253–259.

Markram, H., Toledo-Rodriguez, M., Wang, Y., Gupta, A., Silberberg, G., Wu, C., 2004. Interneurons of the neocortical inhibitory system. Nat. Rev. Neurosci. 10, 793–807.

McConnell, K.A., Nahas, Z., Shastri, A., Lorberbaum, J.P., Kozel, F.A., Bohning, D.E., et al., 2001. The transcranial magnetic stimulation motor threshold depends on the distance from coil to underlying cortex: a replication in healthy adults comparing two methods of assessing the distance to cortex. Biol. Psychiatry 49, 454–459.

Mezawa, S., Iwata, K., Naito, K., Kamogawa, H., 1988. The possible analgesic effect of soft-laser irradiation on heat nociceptors in the cat tongue. Arch. Oral Biol. 33, 693–694.

Mochizuki-Oda, N., Kataoka, Y., Cui, Y., Yamada, H., Heya, M., Awazu, K., 2002. Effects of near-infra-red laser irradiation on adenosine triphosphate and adenosine diphosphate contents of rat brain tissue. Neurosci. Lett. 323, 207–210.

Nawashiro, H., Sato, S., Kawauchi, S., Takeuchi, S., Nagatani, K., Yoshihara, N., et al., 2017. Blood-oxygen-level-dependent (BOLD) functional magnetic resonance imaging (fMRI) during transcranial near-infrared laser irradiation. Brain Stimul. 6, 1136–1138. Available from: https://doi.org/10.1016/j.brs.2017.08.010.

Nissan, M., Rochkind, S., Razon, N., Bartal, A., 1986. HeNe laser irradiation delivered transcutaneously: its effect on the sciatic nerve of rats. Lasers Surg. Med. 6, 435–438.

Rochkind, S., Nissan, M., Razon, N., Schwartz, M., Bartal, A., 1986. Electrophysiological effect of HeNe laser on normal and injured sciatic nerve in the rat. Acta Neurochir. (Wien) 83, 125–130.

Salehpour, F., Mahmoudi, J., Kamari, F., Sadigh-Eteghad, S., Rasta, S.H., Hamblin, M.R., 2018. Brain photobiomodulation therapy: a narrative review. Mol. Neurobiol. [Epub ahead of print] . Available from: https://doi.org/10.1007/s12035-017-0852-4.

Stokes, M.G., Chambers, C.D., Gould, I.C., English, T., McNaught, E., McDonald, O., et al., 2007. Distance-adjusted motor threshold for transcranial magnetic stimulation. Clin. Neurophysiol. 118, 1617–1625.

Tsuchiya, K., Kawatani, M., Takeshige, C., Sato, T., Matsumoto, I., 1993. Diode laser irradiation selectively diminishes slow component of axonal volleys to dorsal roots from the saphenous nerve in the rat. Neurosci. Lett. 161, 65–68.

Tsuchiya, K., Kawatani, M., Takeshige, C., Matsumoto, I., 1994. Laser irradiation abates neuronal responses to nociceptive stimulation of rat-paw skin. Brain Res. Bull. 34, 369–374.

Valls-Sole, J., Pascual-Leone, A., Wassermann, E.M., Hallett, M., 1992. Human motor evoked responses to paired transcranial magnetic stimuli. Electroencephalogr. Clin. Neurophysiol. 85, 355–364.

Wakabayashi, H., Hamba, M., Matsumoto, K., Tachibana, H., 1993. Effect of irradiation by semiconductor laser on responses evoked in trigeminal caudal neurons by tooth pulp stimulation. Lasers Surg. Med. 13, 605–610.

Walsh, D., Baxter, G., Allen, J., 2000. Lack of effect of pulsed low-intensity infrared (820 nm) laser irradiation on nerve conduction in the human superficial radial nerve. Lasers Surg. Med. 26, 485–490.

Wong-Riley, M.T., Liang, H.L., Eells, J.T., Chance, B., Henry, M.M., Buchmann, E., et al., 2005. Photobiomodulation directly benefits primary neurons functionally inactivated by toxins: role of cytochrome c oxidase. J. Biol. Chem. 280, 4761–4771.

Chapter 19

Photobiomodulation for multiple sclerosis in animal models

M.A. Tolentino and J.A. Lyons
College of Health Sciences, University of Wisconsin-Milwaukee, Milwaukee, WI, United States

19.1 Introduction

Photobiomodulation therapy (PBMT) mediated by visible and near infrared (VIS/NIR) light between 600 and 1100 nm is showing great therapeutic potential for the treatment of chronic inflammation (Whelan et al., 2003; Dutta et al., 2006; Dutta and Trapp, 2007), retinal diseases (Albarracin et al., 2011; Eells et al., 2003), and neurodegenerative diseases, including Parkinson disease (Willis and Turner, 2007), Alzheimer Disease (Grillo et al., 2013), and stroke (Lampl et al., 2007). As an autoimmune, demyelinating disease with accompanying neurodegeneration, the therapeutic application of PBMT to multiple sclerosis (MS) could prove to be an effective treatment strategy against this chronic, debilitating disease. Experimental autoimmune Encephalomyelitis (EAE) is the primary animal model for MS. Like MS, EAE is an autoimmune demyelinating disease with associated neurodegeneration. Thus the EAE model provides a unique model to study the therapeutic potential of PBMT for MS as well as a model system to understand the mechanisms of action responsible for the observed clinical effects of PBMT.

19.2 Experimental autoimmune encephalomyelitis and multiple sclerosis

Multiple sclerosis. MS is a chronic autoimmune, neurodegenerative disease that affects approximately 400,000 people in the United States and 2.5 million worldwide (MS Overview, 2015; Browne et al., 2014). The disease is restricted to the myelinated tissues of the central nervous system (CNS), including the brain, optic nerves, and spinal cord; and characterized by destruction of the myelin sheath, axons, and neurons leading the disruption of the transmission of the nerve impulses from the CNS to the periphery (MS Overview, 2015). Signs and symptoms of MS are variable and unpredictable over the course of disease and may include anxiety, bowel problems, depression, pain, numbness, sexual dysfunction, nystagmus, stiffness, tremor, loss of balance, ataxia, muscle weakness, and paralysis (MS Overview, 2015). While not usually fatal, MS reduces life expectancy by 7–14 years (Scalfari et al., 2013).

Persons with MS usually present relapsing-remitting (RRMS) disease upon diagnosis. RRMS is characterized by the sudden appearance of symptoms (termed attacks, relapses, or exacerbations) which remain for a few days or months and then disappear for long periods of time (months or years, remission periods). Patients typically progress to limited mobility, termed secondary progressive MS, within 5–10 years of diagnosis. A rarer form of MS affecting approximately 10% of persons with MS is primary progressive MS, characterized by a steady decline in neurological function, without relapse or remission. Current therapeutic strategies are approved and affective for RRMS. There are currently no effective therapeutic strategies for preventing disease progression.

The pathogenesis of MS is not fully understood and is multifactorial. Factors implicated in disease onset and pathogenesis include place of birth, race, environmental, infectious agents, and genetic and epigenetic factors (Marrie, 2004). The current paradigm describes MS as an autoimmune disease mediated by pro-inflammatory, myelin-reactive CD4 + T cells, particularly during RRMS disease (Steinman, 2014). A role for mitochondrial dysfunction and reactive oxygen species (ROS) and reactive nitrogen species (RNS) throughout the disease process is also recognized (Steinman and Zamvil, 2016). Together, the autoreactive immune response and mitochondrial dysfunction contribute to the long-term disability associated with MS. Current therapeutic strategies effectively target the immune response but do not offer the

Photobiomodulation in the Brain. DOI: https://doi.org/10.1016/B978-0-12-815305-5.00019-1

TABLE 19.1 Experimental autoimmune encephalomyelitis in various mouse strains.

Mouse strain	Antigen for EAE induction	Disease course
SJL	MBP PLP MOG	Relapsing-remitting 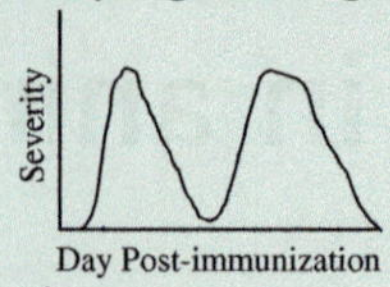
C57BL/6	MOG	Chronic 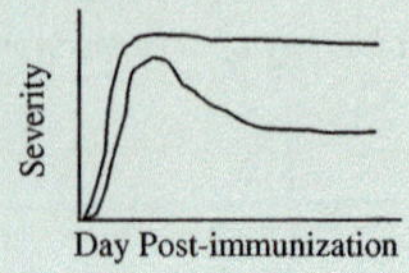

necessary neuroprotection to prevent disease progression. Thus it is important to focus on the development of treatment to treat neurodegeneration, improve mitochondrial dysfunction, reduce axon damage, and promote remyelination.

Experimental autoimmune encephalomyelitis. EAE is the primary animal model for MS. Upon immunization of susceptible species and strains with myelin proteins (i.e., antigens), animals present clinical signs reminiscent of MS, including ascending muscle weakness and paralysis. Current studies typically utilize inbred mice immunized with myelin basic protein (MBP), myelin oligodendrocyte glycoprotein (MOG), or proteolipid protein (PLP). The two most common mouse strains used in EAE are the SJL mouse, presenting a relapsing-remitting disease course, and C57BL/6 mice, presenting a chronic disease course. SJL mice are considered a model of early MS, while C57/BL/6 mice are considered a model for later stage disease (Table 19.1).

EAE also resembles MS histopathologically, characterized by mononuclear infiltration of activated T cells, B cells, and macrophages from the periphery into the CNS, forming the typical demyelinated lesions, or plaques, associated with pathology. Resident astrocytes and microglia also contribute to lesion formation and disease pathogenesis. The clinical, histopathological, and mechanistic similarities between MS and EAE make the EAE model a valuable tool for investigating disease pathogenesis and novel therapeutic strategies for the treatment of MS (Croxford et al., 2011).

Pathogenesis of EAE/MS. EAE and MS are described as autoimmune diseases, mediated by CD4 + T helper (Th) cells specific for myelin proteins (e.g., MBP, PLP, MOG) (Steinman, 2014; Steinman and Zamvil, 2016; Croxford et al., 2011). This paradigm is based on passive transfer studies in the EAE model, demonstrating disease is induced in naïve mice following passive transfer of activated, myelin-reactive CD4 + T cells isolated from mice immunized with appropriate myelin proteins (Stromnes and Goverman, 2006; Fritz and Zhao, 1996; Mannara et al., 2012).

EAE/MS is initiated by the activation of naïve myelin-reactive CD4 + T cells in the periphery (Miljkovic and Spasojevic, 2013). In the EAE model, cell priming is initiated by immunization with myelin protein in adjuvant. The mechanism of immune priming in persons with MS is unclear, but a number of mechanisms have been proposed, including: molecular mimicry (Wucherpfennig and Strominger, 1995; Wucherpfennig, 2001; Berer and Krishnamoorthy, 2014), bystander activation (Berer and Krishnamoorthy, 2014), and neoepitope generation within the CNS with subsequent release into the periphery (Casaccia-Bonnefil et al., 2008). T cells are activated in the peripheral lymph nodes, and once activated, migrate to the CNS, and cross the blood–brain barrier to enter the CNS. In the CNS, autoreactive T cells are reactivated by resident microglia and infiltrating macrophages, amplifying the inflammatory infiltrate within the CNS and setting the stage for the demyelination leading to oligodendrocyte loss, axonal degeneration, and the progression of clinical disease (Miljkovic and Spasojevic, 2013; Goverman, 2009).

A fundamental aspect of EAE/MS pathogenesis that PBMT would potentially target is the pro-inflammatory immune response that is responsible for the onset and propagation of disease pathogenesis. CD4 + T helper (Th) cells phenotypically differentiate into a variety of pro-inflammatory, antiinflammatory, and regulatory populations dependent on the conditions under which they are activated (Zhu et al., 2010; Luckheeram et al., 2012) (Fig. 19.1).

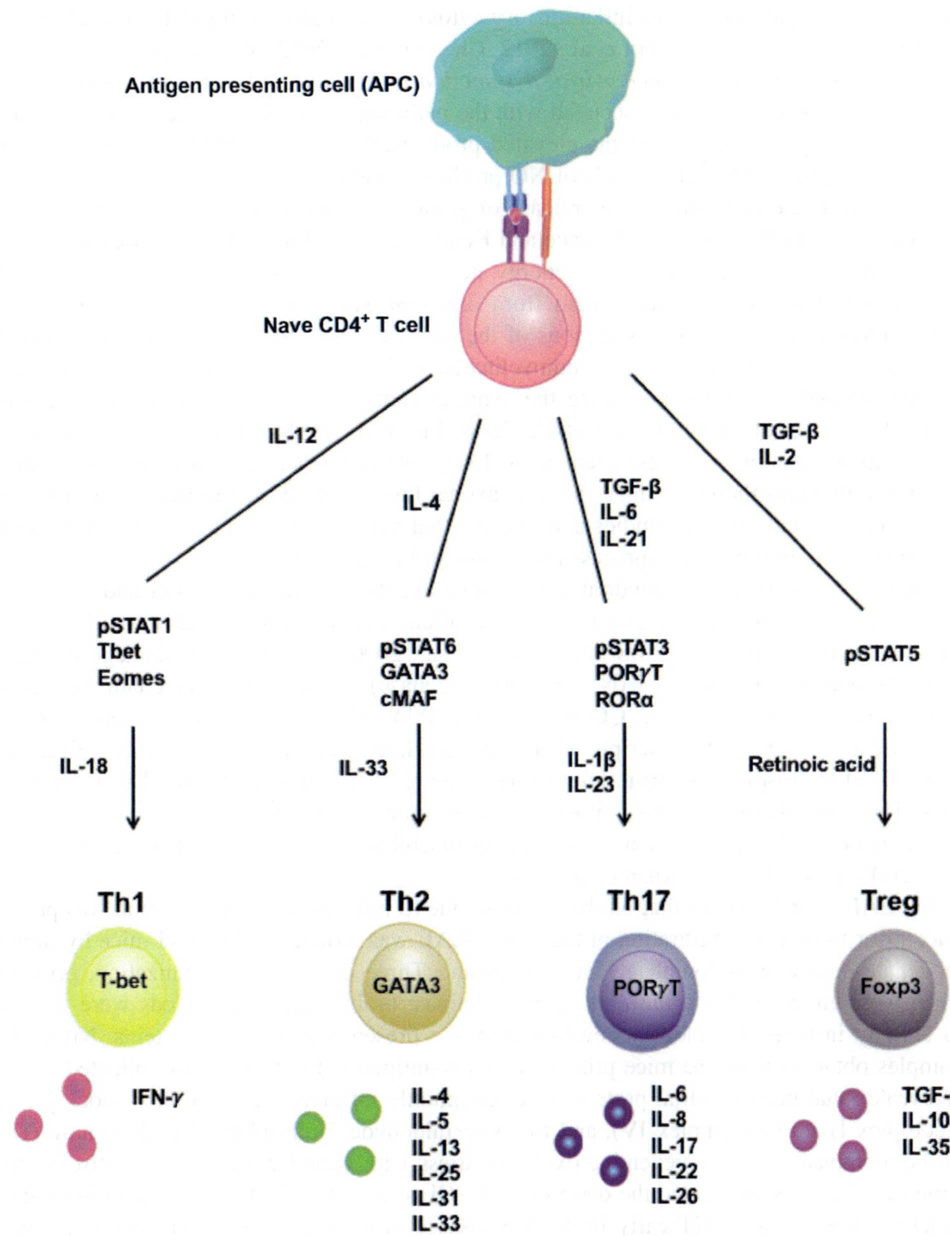

FIGURE 19.1 Differentiation of CD4 + Th subpopulations. CD4 + Th cells control the immune response by secreting cytokines that control the differentiation and activation of immune cells, leading to pro-inflammatory or antiinflammatory immune responses. Prior to activation, CD4 + Th cells exist in a pluripotent precursor state. The differentiation of particular Th cell subpopulations is controlled by the presence of secreted factors and the subsequent activation of transcription factor pathways. In part, the therapeutic efficacy of photobiomodulation is mediated through modulation of these transcription factor pathways.

Pro-inflammatory Th1 cells, secreting IFN-γ and interleukin (IL-2) and Th17 cells, secreting IL-17 and IL-23, are recognized as pathogenic for EAE/MS. Without the infiltration of Th1 and Th17 cells into the CNS, EAE/MS cannot occur (Yednock et al., 1992; Johnson, 2007). Conversely, antiinflammatory Th2 cells, secreting IL-4, IL-5, and IL-13, and regulatory T cells (Treg) secreting IL-10 and transforming growth factor-β are protective against disease. Previous work demonstrated that MS patients have decreased Treg cells with impaired function, and this is considered a contributing factor to the onset of disease (Viglietta et al., 2004; Kumar et al., 2006). The induction of a Th2 response, and/or the presence of functional Treg cells will inhibit the pro-inflammatory Th1 and Th17 cells, protecting against active disease (Mondal et al., 2012; Kohm et al., 2003; McGeachy et al., 2005). Indeed, resolution of chronic inflammation by PBMT is associated with the down-regulation of pro-inflammatory cytokines, and also up-regulate antiinflammatory cytokines by VIS/NIR light (Lim et al., 2008, 2009; Choi et al., 2012; Cheong et al., 2012; Muili et al., 2012, 2013).

Nitrosative and oxidative stress and mitochondrial dysfunction in MS and EAE. Multiple sources of ROS and RNS contribute to the oxidative/nitrosative stress associated with the pathology of EAE/MS. Microglial activation is common in CNS pathologies, including MS, leading to the elevated production of NO via iNOS. In vitro experiments showed that the activation of microglia and the high levels of NO produced lead to the inhibition of cellular respiration (depletion of ATP) in cocultured neurons, leading to the release of glutamate from the neurons and contributing to cell death by excitotoxicity (Bal-Price and Brown, 2001; Trabace and Kendrick, 2000; McNaught and Brown, 1998).

Axons are rich in mitochondria to support the energy needs of nerve signal transduction (Su et al., 2009). The healthy myelinated structure of axons protects them from oxidative stress as consequence of nerve transduction by sequestering mitochondria into the myelinated sections of the axon. On the other hand, ion channels important for the signal transduction are concentrated at the small, unmyelinated nodes of Ranvier. However, with demyelination, the mitochondria and ion channels are distributed along the axon, disrupting nerve transduction and exposing the axon to toxic ROS/RNS (De Stefano et al., 1998; Gonen et al., 2000; Fu et al., 1998; Khan et al., 2005). A consequence of demyelination is axonal transection and subsequent axonal degeneration associated with long-term disability (Trapp et al., 1999). Contrary to previous thoughts suggesting that axonal loss was restricted to later stages of disease, magnetic resonance spectrophotometry and imaging studies demonstrate that axonal degeneration is present in absence of active inflammation, beginning early in the disease process (Steinman and Zamvil, 2016).

Studies in MS demonstrated that mitochondrial dysfunction and the generation of ROS and RNS contribute to the pathology and progression of disease. Generation ROS by the inflammatory response leads to oxidative damage in mitochondrial DNA (mtDNA) in active chronic MS plaques (Lu et al., 2000). The oxidative damage in mtDNA correlates with a reduction in the activity of complex I (NADH dehydrogenase) of the respiratory chain and an increase in the activity of complex IV (cytochrome c oxidase, CCO) (Lu et al., 2000). Microarray analysis on postmortem motor cortex samples from MS patients showed that transcripts of 26 mitochondrial genes were reduced, as well as the activities of the complexes I and III of the respiratory chain. Interestingly, the reduction in mitochondrial gene expression was present just in neurons. These results indicate that mitochondrial dysfunction in demyelinated axons leads to reduced ATP production and participates in the progression to permanent disability in MS (Dutta et al., 2006; De Stefano et al., 1998; Gonen et al., 2000; Fu et al., 1998; Khan et al., 2005).

Studies using the EAE model showed that oxidative stress due to mitochondrial dysfunction was present in the CNS of immunized mice prior to immune infiltration of the tissue. EAE was induced in DBA/1J mice by immunization with spinal cord homogenate emulsion in complete Freund adjuvant. Three days after immunization, prior to the onset of clinical signs, mice were euthanized and the optic nerves, retinas, brains, and spinal cords were collected (Qi et al., 2006). The ROS activity in tissue samples was evaluated by the presence of nitrated proteins. Nitrated proteins were detected in all samples obtained from the mice primed with autoantigen but not in tissues collected from control mice. Further analysis revealed that mitochondrial proteins were commonly nitrated, including heat shock protein 70, NADH dehydrogenase (complex I), CCO (complex IV), and the glyceraldehyde 3-phosphate dehydrogenase (Qi et al., 2006). These authors further demonstrated that preventing oxidative stress within the CNS early in the disease process, prior to immune infiltration into the CNS, prevented the onset of EAE (Qi et al., 2006, 2007). Thus targeting mitochondrial dysfunction and oxidative stress with PBMT early in the MS disease process could have profound effects on preventing disease progression.

19.3 Photobiomodulation therapy for the treatment of experimental autoimmune encephalomyelitis/multiple sclerosis

Photobiomodulation therapy. PBMT is emerging as a viable treatment strategy for chronic inflammatory (Eells et al., 2004; Arany et al., 2007; Whelan et al., 2003) and neurodegenerative conditions, including Parkinson disease (Johnstone et al., 2014; Shaw et al., 2010), Alzheimer Disease (Grillo et al., 2013), stroke (Lampl et al., 2007), and retinal diseases (Albarracin et al., 2011; Eells et al., 2006, 2008). The current hypothesis states that photobiomodulation is induced following absorption VIS/NIR light by the photoacceptor, Cytochrome c Oxidase (CCO), present in the mitochondria to improve mitochondrial function and initiate signal transduction to mediate gene transcription; ultimately leading to down-regulation of pro-inflammatory mediators, up-regulation of antiinflammatory factors, and clinical improvement in the aforementioned conditions (Eells et al., 2004; Chen et al., 2011; Chung et al., 2011). As discussed above, the pathology of MS and the EAE model is due to a combination of pro-inflammatory factors and mitochondrial dysfunction. Thus PBMT would be hypothesized to be effective for the treatment of MS and the EAE model an ideal system to test this hypothesis.

Efficacy of PBMT in the EAE model. Our laboratory was the first to demonstrate efficacy of PBMT in the EAE model (Muili et al., 2012, 2013). Using the MOG, peptide 35−55 (MOG35−55)-induced model in C57BL/6 mice, a model of chronic disease, we demonstrated clinical improvement induced by PBMT with 670 nm LED arrays (5 J/cm^2, 0.028 W) administered by suppression (Fig. 19.2A) and double treatment (Fig. 19.2B) protocols, compared to sham-treated (restraint stress) animals. Our conclusions were subsequently supported by others (Goncalves et al., 2015). Using lasers emitting at 660 nm (0.030 W, 10 J/cm^2, continuous wave, 0.06 cm^2 beam area) or 904 nm (0.07 W, 3 J/cm^2, pulsed beam) (60 ns pulse, 0.10 cm^2 beam area), these authors demonstrated clinical improvement using a suppression protocol in the C57BL/6 model of EAE.

As noted above, the balance of pro-inflammatory and antiinflammatory cytokines is a potent regulator of the EAE disease process. With the clinical amelioration of disease in treated mice, we would expect to note a decrease in pro-inflammatory cytokines and an up-regulation of antiinflammatory cytokines. Indeed, this shift in the inflammatory response was noted in lymph node cells and within the spinal cord of mice receiving PBMT by either the suppression protocol (Fig. 19.3) or double treatment protocol (Fig. 19.4). Mice receiving PBMT demonstrated decreased production pro-inflammatory mediators, including interferon-γ, tumor necrosis factor-α (Figs. 19.3A and B and 19.4A and B) and increased production of antiinflammatory cytokines, IL-4 and IL-10 (Figs. 19.3C and D and 19.4C and D) (Muili et al., 2012).

The shift in nitrosative-oxidative stress and antioxidant mechanisms noted with PBMT would be expected to contribute to clinical improvement in the EAE model. As with the immune response, decreased production of nitric oxide by lymph node cells (Fig. 19.5A) and decreased expression of inducible nitric oxide synthase (iNOS) are noted in the spinal cord (Fig. 19.5B) of treated mice (Muili et al., 2013). iNOS is the NOS isoform responsible for producing large amounts of nitric oxide (NO) during a pro-inflammatory immune response, and it contributes to the pathology of EAE/MS. An assessment of antioxidant production demonstrated increased expression of the mitochondrial isoforms of superoxide dismutase (SOD2; Fig. 19.6A and B) and peroxiredoxin (PRDX6; Fig. 19.6C and D) in the spinal cords and lymph nodes of mice receiving PBMT (Fig. 19.6). Predicted by this shift in pro-oxidant versus antioxidant mediators, and important to the therapeutic efficacy of PBMT for chronic MS, we also noted protection from apoptosis in the CNS of treated mice (Fig. 19.7) (Muili et al., 2013).

Considerations for the translation of PBMT to MS clinical practice. The clinical course and pathogenic mechanisms of MS are heterogeneous (Steinman, 2014; Steinman and Zamvil, 2016; Sospedra and Martin, 2005). For example, the role of the immune response in contributing the relapsing-remitting disease course of MS is recognized (Steinman, 2014). As demonstrated by us and others, PBMT acts in part by exerting an effect on the immune response, regulating the expression of cytokines known to contribute to MS pathogenesis. Likewise, NO plays both beneficial and

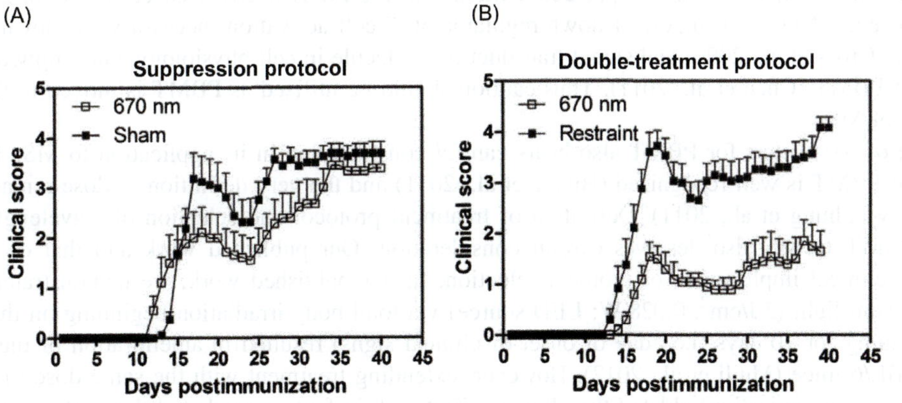

FIGURE 19.2 Photobiomodulation at 670 nm reduces clinical score in EAE mice. C57BL/6 mice were immunized with MOG_{35-55} to induce EAE, then treated with 670 nm LED arrays (5 J/cm^2, 0.028 W) or no light treatment sham-treated (restraint) group. (A) Suppression protocol: mice received 670 nm light on the dorsal surface beginning the day after immunization and continuing for 10 days. There was a reduction in the clinical score during the acute episode compared with sham-treated mice. (B) Double treatment protocol: Light treatment was applied for 7 days starting the day of onset of disease, followed for a 7-day rest period and then a 7-day treatment period. This treatment protocol produced an improved clinical effect which was sustained throughout the experiment (Muili et al., 2012).

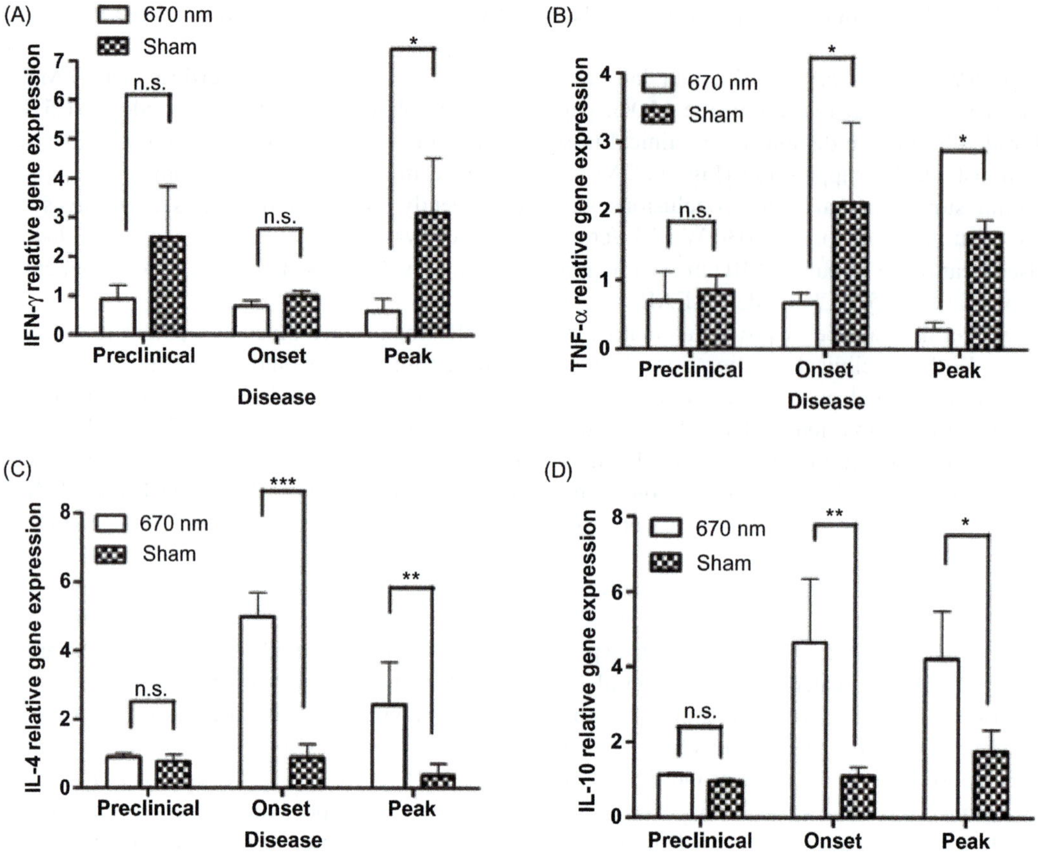

FIGURE 19.3 Suppression treatment protocol modulates cytokine production in the CNS over the course of EAE. Spinal cord tissue was isolated over the course of the experiments from EAE mice treated with the suppression protocol. QPCR was performed to determine changes in cytokine gene expression. Pro-inflammatory cytokines IFN-γ (A) and TNF-α (B) were significantly down-regulated in comparison with sham-treated mice. Antiinflammatory cytokines IL-4 (C) and IL-10 (D) were significantly up-regulated in comparison with sham-treated mice. **P*, .05, ***P*, .01, ****P*, .0001, n.s., not significant (Muili et al., 2012).

pathogenic roles in the pathology of MS and other conditions. The role of iNOS and NO in EAE/MS pathology is recognized (Dutta et al., 2006; Dutta and Trapp, 2007; Steinman and Zamvil, 2016; Qi et al., 2006, 2007; Cross et al., 1996, 2006). However, NO is an important down-regulator of T cell activation, necessary for down-regulation of the immune response (Cross et al., 2006) and signal transduction molecule in cell physiology, and implicated in the mechanism of action of PBMT (Chen et al., 2011). Thus caution should be utilized as PBMT is moved to the clinical setting for the treatment of MS.

The selection of parameters for PBMT also bears careful consideration in its application to MS. The biphasic dose response curve of PBMT is well-recognized (Huang et al., 2011) and the very definition of dose remains debated in the PBMT community (Chung et al., 2011). Definition of treatment protocol and selection of wavelength for the clinical application of PBMT to MS also deserves careful consideration. Our published work and that of Goncalves et al., demonstrates the clinical implications of protocol selection. In our published work, we demonstrated that application once daily of 670 nm light (5 J/cm^2, 0.028 W; LED source) via total body irradiation beginning on the day of immunization and continuing for 10 days (i.e., day of onset of clinical signs) resulted in amelioration of the acute episode of EAE in the C57BL/6 mice (Muili et al., 2012). However, extending treatment with the same dose through the peak of disease (15 days postimmunization) ablated the clinical effect and, in fact, exacerbated the acute episode in these mice (Fig. 19.8). Our preliminary data suggest that longer wavelengths of light may exacerbate the acute episode in some cases. Conversely, Goncalves et al. demonstrated clinical efficacy with 660 nm light delivered via laser to multiple areas (0.030 W, 10 J/cm^2, continuous wave, 0.06 cm^2 beam area) beginning with immunization and continuing for 30 days. Understanding the differences in protocol utilized in the response to PBMT applied, that resulted in these dramatically different results is important to avoiding exacerbation of disease as PBMT is applied to persons with MS.

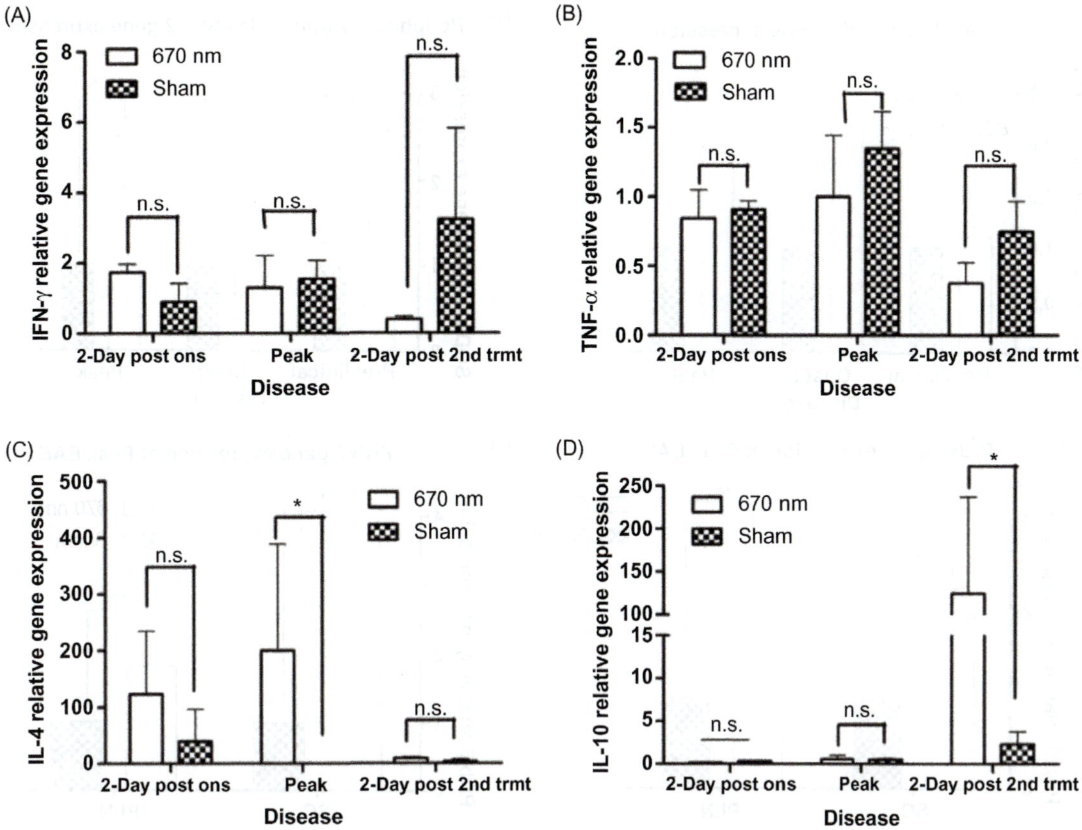

FIGURE 19.4 Double treatment protocol modulates cytokine production in the CNS over the course of EAE. Spinal cord tissue was isolated over the course of EAE from mice treated with the double treatment protocol. QPCR was performed to determine changes in cytokine gene expression. Pro-inflammatory cytokines (A) IFN-γ and (B) TNF-α did not show significant changes in comparison with sham-treated mice at any point over the course of the experiment. Antiinflammatory cytokines (C) IL-4 and (D) IL-10 were significantly up-regulated during the peak and the chronic stage of disease respectively in comparison to sham-treated mice. *P, .05, **P, .01, ***P, .0001, n.s., not significant (Muili et al., 2012).

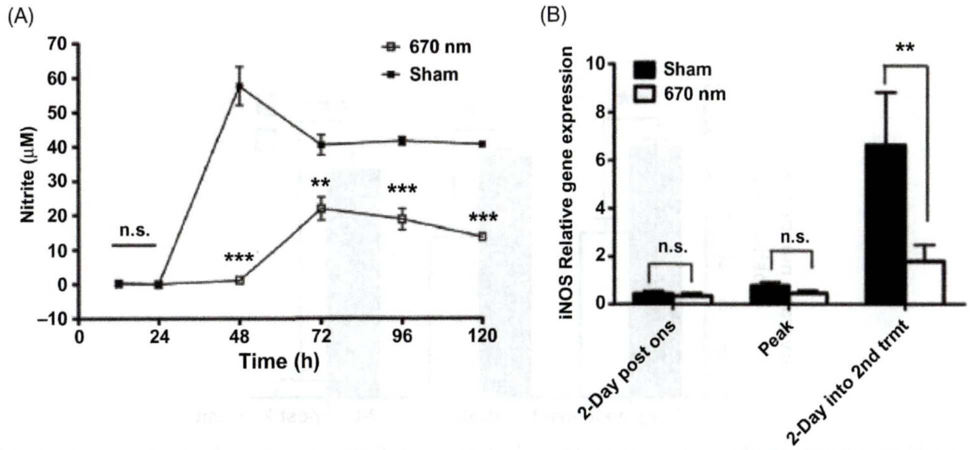

FIGURE 19.5 Photobiomodulation at 670 nm modulates nitrosative stress in vitro and in vivo in the EAE model. (A) Peripheral lymph nodes were isolated from MOG_{35-55} immunized C57BL/6 mice 10 days postimmunization (dpi), then used in cell suspension incubated with MOG_{35-55}. Cells were treated with 670 nm every 24 h for a total of 120 h. Cells with no light treatment were used as control (sham group). Nitrite levels were measured on supernatants every 24 h throughout the experiment by the Griess reaction assay. 670 nm treatments produced a significant reduction in nitrate levels in comparison with control. **P, .01, ***P, .0001 by two-way ANOVA. (B) MOG_{35-55} immunized C57BL/6 mice were subjected to double treatment protocol or strain treatment (sham group). Spinal cords were collected during the course of the experiment to measure change in iNOS gene expression. Double treatment protocol produced a significant reduction in iNOS expression during the chronic stage of diseases. **P, .05 by two-way ANOVA (Muili et al., 2013).

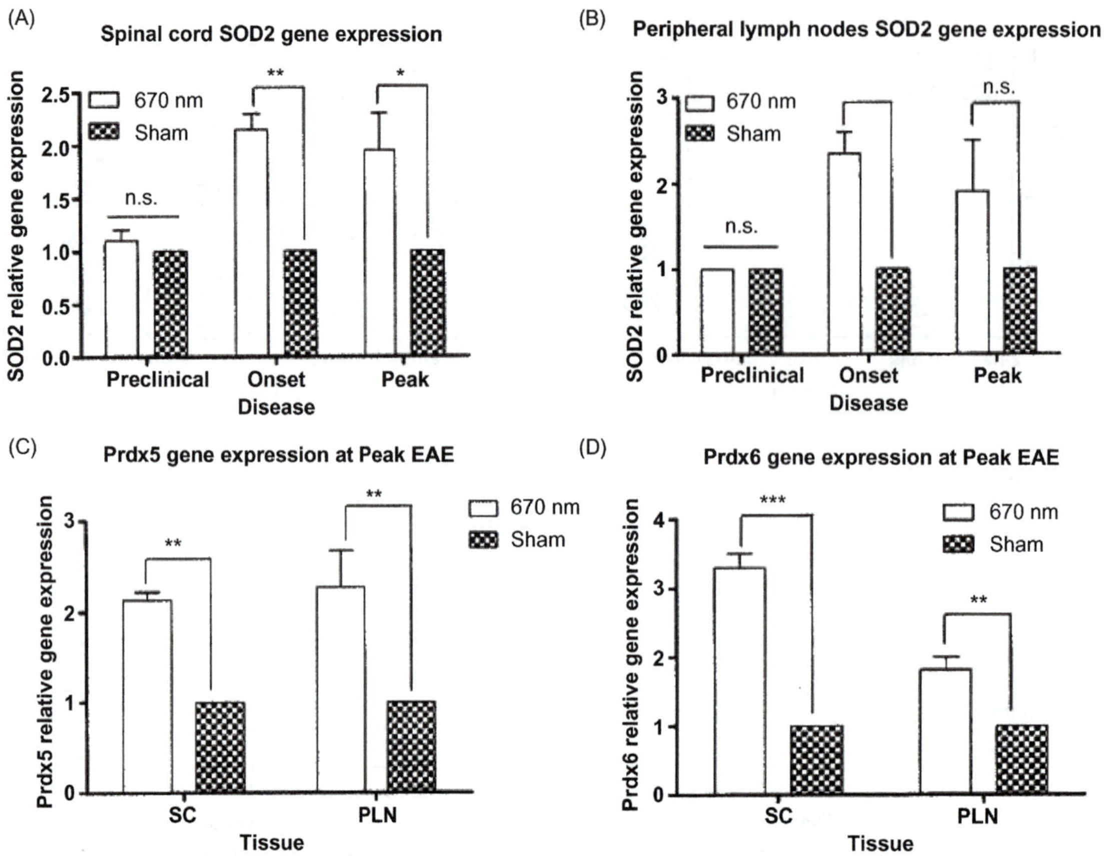

FIGURE 19.6 **Photobiomodulation at 670 nm up-regulates antioxidants SOD2, Prdx5, and Prdx6 in EAE mice.** Spinal cords and peripheral lymph nodes were isolated from immunized MOG_{35-55} C 57BL/6 mice receiving the suppression protocol over the course of the experiment. SOD2 was significantly up-regulated in (A) spinal cord and (B) peripheral lymph nodes during the onset and peak of disease. (C) Prdx5 and (D) Prdx6 were significantly up-regulated in spinal cords and peripheral lymph nodes at the peak of disease. $*P < .05$, $**P < .01$, $***P < .001$ by two-way ANOVA.

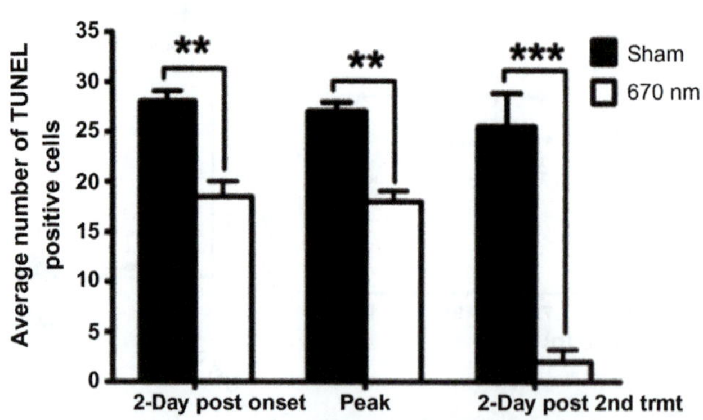

FIGURE 19.7 **Photobiomodulation at 670 nm reduces apoptosis in EAE mice.** EAE was induced by immunizing mice C57BL/6 mice with MOG_{35-55} Immunized mice were subjected to the double treatment protocol or sham treatment. Spinal cord tissue was isolated over the course of the experiment for assessment of apoptosis. PBMT reduced the number of apoptotic cells present in microsections of spinal cords stained by TUNEL assay. The number of apoptotic cells was reduced throughout the experiment in comparison with restrained group. *P, .05, **P, .01, ***P, .001, n.s., not significant by two-way ANOVA (Muili et al., 2013).

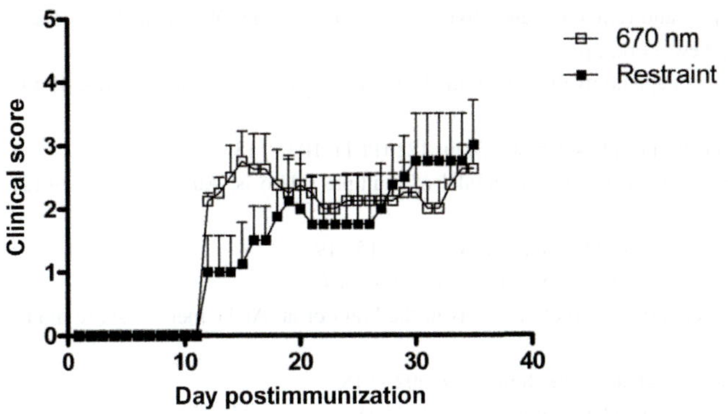

FIGURE 19.8 PBMT administered during the induction of clinical disease and continued through peak disease of the acute episode exacerbates clinical disease. C57BL/6 mice were immunized with MOG_{35-55} to induce EAE, then treated with 670 nm LED arrays (5 J/cm^2, 0.028 W) or no light treatment (sham treatment; restraint group). Treatment began the day after immunization and continuing for 15 days, through the peak of the acute episode.

19.4 Conclusion and future directions

PBMT is a promising therapeutic strategy for the treatment of MS. Data in the EAE model demonstrate the therapeutic potential of PBMT and provide evidence of protective mechanisms. Given that a number of therapeutic agents are available for the treatment of RRMS and are considered standard of care, but do not provide lasting benefit and come with significant side effects, consideration of PBMT as an adjunct therapy is warranted. The current therapeutic agents are largely immunomodulatory in action. While PBMT offers the same immune modulation, the additional amelioration of oxidative stress and neuroprotection could improve the therapeutic outcome of persons with MS. PBMT could also prove beneficial for symptom management in MS, including optic neuritis and muscle fatigue (Leal-Junior, 2015; Larkin-Kaiser et al., 2015, 2016). Careful consideration of treatment parameters, including dose, wavelength, and application will contribute to the adoption of PBMT as standard-of-care therapy for the treatment of MS.

References

Albarracin, R., Eells, J., Valter, K., 2011. Photobiomodulation protects the retina from light-induced photoreceptor degeneration. Invest. Ophthalmol. Vis. Sci. 52 (6), 3582–3592.

Arany, P.R., et al., 2007. Activation of latent TGF-beta1 levels in laser-enhanced oral wound healing. Wound Repair Regen. 15, 866–874.

Bal-Price, A., Brown, G.C., 2001. Inflammatory neurodegeneration mediated by nitric oxide from activated glia-inhibiting neuronal respiration, causing glutamate release and excitotoxicity. J. Neurosci. 21 (17), 6480–6491.

Berer, K., Krishnamoorthy, G., 2014. Microbial view of central nervous system autoimmunity. FEBS Lett. 588 (22), 4207–4213.

Browne, P., et al., 2014. Atlas of Multiple Sclerosis 2013: a growing global problem with widespread inequity. Neurology 83 (11), 1022–1024.

Casaccia-Bonnefil, P., Pandozy, G., Mastronardi, F., 2008. Evaluating epigenetic landmarks in the brain of multiple sclerosis patients: a contribution to the current debate on disease pathogenesis. Prog. Neurobiol. 86 (4), 368–378.

Chen, A.C., et al., 2011. Low-level laser therapy activates NF-kB via generation of reactive oxygen species in mouse embryonic fibroblasts. PLoS One 6 (7), e22453.

Cheong, K.A., et al., 2012. Irradiation of light emitting diode at 850nm inhibits T cell-induced cytokine expression. J. Dermatol. Sci. 65 (1), 27–37.

Choi, H., et al., 2012. Inflammatory cytokines are suppressed by light-emitting diode irradiation of P. gingivalis LPS-treated human gingival fibroblasts: inflammatory cytokine changes by LED irradiation. Lasers Med. Sci. 27 (2), 459–467.

Chung, H., et al., 2011. The nuts and bolts of low-level laser (light) therapy. Ann. Biomed. Eng. 40 (2), 516–533.

Cross, A.H., et al., 1996. Inducible nitric oxide synthase gene expression and enzyme activity correlate with disease activity in adoptively-transferred murine EAE. J. Neuroimmunol. 71, 145–153.

Cross, A.H., Ramsbottom, M.J., Lyons, J.-A., 2006. NOS2 regulates cytokine production and VLA-4 expression in experimental autoimmune encephalomyelitis. J. Neuroimmunol. 173 (1–2), 79–86.

Croxford, A.L., Kurschus, F.C., Waisman, A., 2011. Mouse models for multiple sclerosis: historical facts and future implications. Biochim. Biophys. Acta 1812 (2), 177–183.

De Stefano, N., et al., 1998. Axonal damage correlates with disability in patients with relapsing-remitting multiple sclerosis. Results of a longitudinal magnetic resonance spectroscopy study. Brain 121 (Pt 8), 1469–1477.

Dutta, R., Trapp, B.D., 2007. Pathogenesis of axonal and neuronal damage in multiple sclerosis. Neurology 68 (22 Suppl. 3), S22–S31. discussion S43–S54.

Dutta, R., et al., 2006. Mitochondrial dysfunction as a cause of axonal degeneration in multiple sclerosis patients. Ann. Neurol. 59 (3), 478–489.

Eells, J.T., et al., 2003. Therapeutic photobiomodulation for methanol-induced retinal toxicity. Proc. Natl. Acad. Sci. U.S.A. 100 (6), 3439–3444.

Eells, J.T., et al., 2004. Mitochondrial signal transduction in accelerated wound and retinal healing by near-infrared light therapy. Mitochondrion 4, 559–567.

Eells, J.T., et al., 2006. Near–infrared light therapy for retinitis pigmentosa. Invest. Ophthalmol. Vision Sci. 47, p. E-Abstract 1022.

Eells, J.T., et al., 2008. Photobiomodulation in the treateatment of retinal injury and retinal degenerative diseases. In: Tata, D., Waynant, R.W. (Eds.), Light Activated Tissue Regeneration and Therapy. Springer Publishing, New York, pp. 39–51.

Fritz, R.B., Zhao, M.L., 1996. Active and passive experimental autoimmune encephalomyelitis in strain 129/J (H-2b) mice. J. Neurosci. Res. 45 (4), 471–474.

Fu, L., et al., 1998. Imaging axonal damage of normal-appearing white matter in multiple sclerosis. Brain 121 (Pt 1), 103–113.

Goncalves, E., et al., 2015. Low-level laser therapy ameliorates disease progression in a mouse model of multiple sclerosis. Autoimmunity 49 (2), 132–142.

Gonen, O., et al., 2000. Total brain N-acetylaspartate: a new measure of disease load in MS. Neurology 54 (1), 15–19.

Goverman, J., 2009. Autoimmune T cell responses in the central nervous system. Nat. Rev. Immunol. 9 (6), 393–407.

Grillo, S.L., et al., 2013. Non-invasive infra-red therapy (1072 nm) reduces beta-amyloid protein levels in the brain of an Alzheimer's disease mouse model, TASTPM. J. Photochem. Photobiol. B 123, 13–22.

Huang, Y.-Y., et al., 2011. Biphasic dose response in low level light therapy--an update. Dose Reponse 9, 602–618.

Johnson, K.P., 2007. Natalizumab (Tysabri) treatment for relapsing multiple sclerosis. Neurologist 13 (4), 182–187.

Johnstone, D., et al., 2014. Indirect application of near infrared light induces neuroprotection in a mouse model of Parkinsonism – an abscopal neuroprotective effect. Neuroscience 74, 93–101.

Khan, O., et al., 2005. Axonal metabolic recovery and potential neuroprotective effect of glatiramer acetate in relapsing-remitting multiple sclerosis. Mult. Scler. 11 (6), 646–651.

Kohm, A.P., Carpentier, P.A., Miller, S.D., 2003. Regulation of experimental autoimmune encephalomyelitis (EAE) by CD4 + CD25 + regulatory T cells. Novartis Found. Symp. 252, 45–52. discussion 52–54, 106–114.

Kumar, M., et al., 2006. CD4 + CD25 + FoxP3 + T lymphocytes fail to suppress myelin basic protein-induced proliferation in patients with multiple sclerosis. J. Neuroimmunol. 180 (1–2), 178–184.

Lampl, Y., et al., 2007. Infrared laser therapy for ischemic stroke: a new treatment strategy results of the NeuroThera effectiveness and Safety Trial-1 (NEST-1). Stroke 38 (6), 1843–1849.

Larkin-Kaiser, K.A., et al., 2015. Near-infrared light therapy to attenuate strength loss after strenuous resistance exercise. J. Athl. Train. 50 (1), 45–50.

Larkin-Kaiser, K.A., et al., 2016. Photobiomodulation delays the onset of skeletal muscle fatigue in a dose-dependent manner. Lasers Med. Sci. 31 (7), 1325–1332.

Leal-Junior, E.C.P., 2015. Photobiomodulation therapy in skeletal muscle: from exercise performance to muscular dystrophies. Photomed. Laser Surg. 33 (2), 53–54.

Lim, J.H., et al., 2008. The effects of daily irradiation with polychromatic visible polarized light on human lymphocyte populations. Photomed. Laser Surg. 26 (4), 361–366.

Lim, J.H., et al., 2009. The effects of light-emitting diode irradiation at 610 nm and 710 nm on murine T-cell subset populations. Photomed. Laser Surg. 27 (5), 813–818.

Lu, F., et al., 2000. Oxidative damage to mitochondrial DNA and activity of mitochondrial enzymes in chronic active lesions of multiple sclerosis. J. Neurol. Sci. 177 (2), 95–103.

Luckheeram, R., et al., 2012. CD4 + T Cells: differentiation and functions. Clin. Dev. Immunol. 2012.

MS Overview. Multiple Sclerosis Association of America, Update June 15, 2015. http://www.mymsaa.org/about-ms/overview/#TypesofMS.

Mannara, F., et al., 2012. Passive experimental autoimmune encephalomyelitis in C57BL/6 with MOG: evidence of involvement of B cells. PLoS One 7 (12), e52361.

Marrie, R.A., 2004. Environmental risk factors in multiple sclerosis aetiology. Lancet Neurol. 3 (12), 709–718.

McGeachy, M.J., Stephens, L.A., Anderton, S.M., 2005. Natural recovery and protection from autoimmune encephalomyelitis: contribution of CD4 + CD25 + regulatory cells within the central nervous system. J. Immunol. 175 (5), 3025–3032.

McNaught, K.S., Brown, G.C., 1998. Nitric oxide causes glutamate release from brain synaptosomes. J. Neurochem. 70 (4), 1541–1546.

Miljkovic, D., Spasojevic, I., 2013. Multiple sclerosis: molecular mechanisms and therapeutic opportunities. Antioxid. Redox Signal. 19 (18), 2286–2334.

Mondal, S., et al., 2012. Protection of Tregs, suppression of Th1 and Th17 cells, and amelioration of experimental allergic encephalomyelitis by a physically-modified saline. PLoS One 7 (12), e51869.

Muili, K.A., et al., 2012. Amelioration of experimental autoimmune encephalomyelitis in C57BL/6 mice by photobiomodulation induced by 670 nm light. PLoS One 7 (1), e30655.

Muili, K.A., et al., 2013. Photobiomodulation induced by 670 nm light ameliorates MOG35-55 induced EAE in female C57BL/6 mice: a role for remediation of nitrosative stress. PLoS One 8 (6), e67358.

Qi, X., et al., 2006. Mitochondrial protein nitration primes neurodegeneration in experimental autoimmune encephalomyelitis. J. Biol. Chem. 281 (42), 31950–31962.

Qi, X., et al., 2007. Suppression of mitochondrial oxidative stress provides long-term neuroprotection in Experimental Autoimmune Encephalomyelitis. Invest. Ophthalmol. Vision Sci. 48, 681–691.

Scalfari, A., et al., 2013. Mortality in patients with multiple sclerosis. Neurology 81 (2), 184–192.

Shaw, V.E., et al., 2010. Neuroprotection of midbrain dopaminergic cells in MPTP-treated mice after near-infrared light treatment. J. Comp. Neurol. 518, 25–40.

Sospedra, M., Martin, R., 2005. Immunology of multiple sclerosis. Annu. Rev. Immunol. 23, 683–747.

Steinman, L., 2014. Immunology of relapse and remission in multiple sclerosis. Annu. Rev. Immunol. 32, 257–281.

Steinman, L., Zamvil, S.S., 2016. Beginning of the end of two-stage theory purporting that inflammation then degeneration explains pathogenesis of progressive multiple sclerosis. Curr. Opin. Neurol. 29 (3), 340–344.

Stromnes, I.M., Goverman, J.M., 2006. Passive induction of experimental allergic encephalomyelitis. Nat. Protocols 1, 1952–1960.

Su, K.G., et al., 2009. Axonal degeneration in multiple sclerosis: the mitochondrial hypothesis. Curr. Neurol. Neurosci. Rep. 9 (5), 411–417.

Trabace, L., Kendrick, K.M., 2000. Nitric oxide can differentially modulate striatal neurotransmitter concentrations via soluble guanylate cyclase and peroxynitrite formation. J. Neurochem. 75 (4), 1664–1674.

Trapp, B.D., Ransohoff, R., Rudick, R., 1999. Axonal pathology in multiple sclerosis: relationship to neurologic disability. Curr. Opin. Neurol. 12 (3), 295–302.

Viglietta, V., et al., 2004. Loss of functional suppression by CD4 + CD25 + regulatory T cells in patients with multiple sclerosis. J. Exp. Med. 199 (7), 971–979.

Whelan, H.T., et al., 2003. Effect of NASA light-emitting diode irradiation on molecular changes for wound healing in diabetic mice. J. Clin. Laser Med. Surg. 21 (2), 67–74.

Willis, G.L., Turner, E.J.D., 2007. Primary and secondary features of Parkinson's disease improve with strategic exposure to bright light: a case series study. Chronobiol. Int. 24 (3), 521–537.

Wucherpfennig, K.W., 2001. Mechanisms for the induction of autoimmunity by infectious agents. J. Clin. Invest. 108 (8), 1097–1104.

Wucherpfennig, K.W., Strominger, J.L., 1995. Molecular mimicry in T cell-mediated autoimmunity: viral peptides activate human T cell clones specific for myelin basic protein. Cell 80 (5), 695–705.

Yednock, T.A., et al., 1992. Prevention of experimental autoimmune encephalomyelitis by antibodies against a4b1integrin. Nature 356, 63–66.

Zhu, J., Yamane, H., Paul, W., 2010. Differentiation of effector CD4 T cell populations. Annu. Rev. Immunol. 28, 445–489.

Chapter 20

Hepatic encephalopathy and photobiomodulation: experimental models and clinical features

Natalia Arias[1,2], Juan Díaz González[3,4], Alberto Martín Pernía[3,4] and Jorge L. Arias[2,5]

[1]*Department of Basic and Clinical Neuroscience, Institute of Psychiatry, Psychology and Neuroscience, King's College London, London, United Kingdom,* [2]*INEUROPA (Instituto de Neurociencias del Principado de Asturias), Oviedo, Spain,* [3]*Área de Tecnología Electrónica de la Universidad de Oviedo, Gijon, Spain,* [4]*Grupo de Electrónica para la Innovación Industrial, Gijon, Spain,* [5]*Laboratory of Neuroscience, Department of Psychology, University of Oviedo, Asturias, Spain*

20.1 Introduction

We are celebrating the 50th anniversary of photobiomodulation (PBM), also known as low-level laser therapy, discovered around the mid-1960s by Endre Mester (1967), and based on red light irradiation by laser to stimulate wound healing and hair regrowth. This was the beginning of a new research field seeking future treatments based on laser biostimulation. Laser biostimulation offered the possibility of reaching deep areas of the body without having to injure or cause harsh lesions to the patient. This would open a pathway for the application of a less expensive technique, in which lasers could be integrated into simpler apparatus.

However, some problems arose when researchers wanted to determine the possible mechanisms that might support the use of this PBM technique. In this regard, Wong-Rilley et al. (2001) applied red light with wavelengths between 670 and 830 nm, which were considered to be the most effective, in cultured rat visual cortical neurons. Their results revealed that the greatest effectiveness was achieved in the activation of cytochrome c oxidase (CCO). These in vitro culture studies allowed measuring the responses to the different wavelengths and verifying the powers needed to obtain the maximum performance in CCO activity (Wong-Rilley et al., 2005), supposedly the enzyme on which the applied light would act.

The reason to consider CCO as the target molecule of PBM, is that it functions as the main chromophore, absorbing the incident light within the mitochondria (Eells et al., 2003). CCO is also called Unit IV of the respiratory chain, which develops in the inner membrane of the mitochondria. This is a complex molecule consisting of 13 separate protein subunits containing two different copper (Cu) centers and two heme centers, which can be reduced or oxidized and thus, transfer four protons to the oxygen molecule to form two molecules of water, generating ATP molecules (Karu and Kolyakov,2005; Wong-Rilley et al., 2005). Thanks to Pastore et al. (2000), we know that, in vitro, this molecule is stimulated with 633 nm laser light, although, more accurately, on the basis of the reduction of Cu, this occurs between 620 and 760 nm, whereas oxidation occurs between 680 and 825 nm (Henderson and Morries, 2015); but certainly, clinical applications that would need to reach deep areas, such as the brain, with PBM would require deep penetrating light. For this purpose, we must not only penetrate hair, skin, and bones, but also the meninges and vessels, all surrounded by a large amount of fluids that include blood and, not least, water. All of these structures absorb or disperse light, so their penetration will vary depending on the structures. Therefore, the true cellular mechanism of action is still the object of speculation, and it is thought that this mechanism is not only due to the activation of CCO, but also to the photodissociation of inhibitory nitric oxide, which can bond with the copper and heme centers of the CCO enzyme and thereby

Photobiomodulation in the Brain. DOI: https://doi.org/10.1016/B978-0-12-815305-5.00020-8

increase the mitochondrial membrane potential (Yu et al., 2015). However, other mechanisms may be also involved (Caterina and Pang, 2016; Hamblin, 2018; Poyton and Ball, 2011). It has also been reported that the primary action of PBM is to provide energy to the electrons for the respiratory chain, but as side effects, there would be a chain of possible reactions due to the activation of enzymatic pathways that would affect the metabolic capacity, gene expression for mitogenic and repair signaling, cytoskeleton processing, and protein expression and translocation (González-Lima et al., 2014; Salehpour et al., 2018).

To address this question, we have developed a test bench to measure the power density after the laser light has passed through different structures. The optical power meter used is a PM160 by ThorLabs. By using this test bench it is possible to determine the percentage of optical power that manages to penetrate the different structures tested (Fig. 20.1). With this apparatus it was possible to measure how much optical power passes through different structures of the rat head. Several lasers with different wavelengths were studied. We observed that, with lower wavelengths, a higher penetration was not achieved; the greatest achieved was approximately at 700 nm, with a peak at the level of 1000 nm, which was also found by Giacci et al. (2014). But the most remarkable finding was that the absorption caused by the skull, skin, and hair (Fig. 20.2) follows the same pattern of penetration as when we placed the lasers on the brain itself (Fig. 20.3).

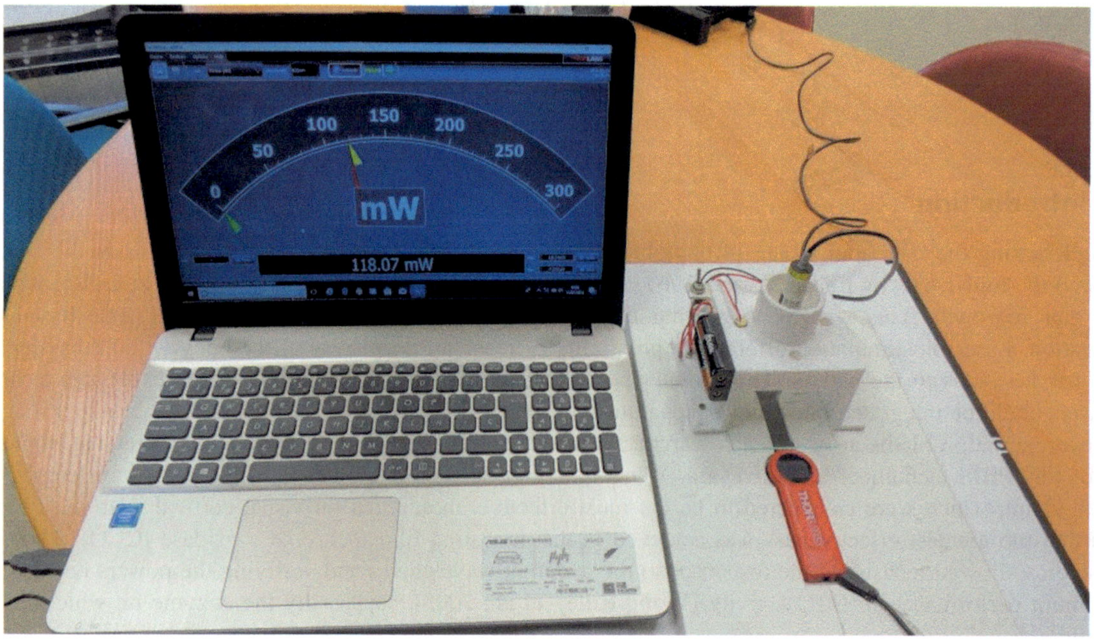

FIGURE 20.1 Test bench developed to measure optical power density.

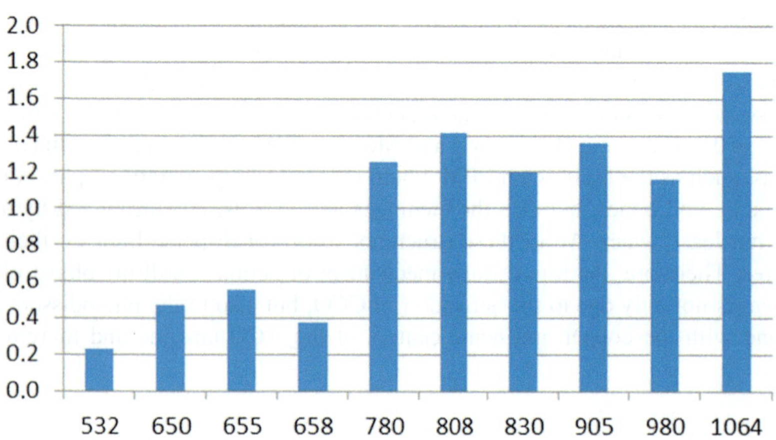

FIGURE 20.2 Percentage of optical power as a function of the wavelength of the laser used, measured after penetrating hair, skin, and bone.

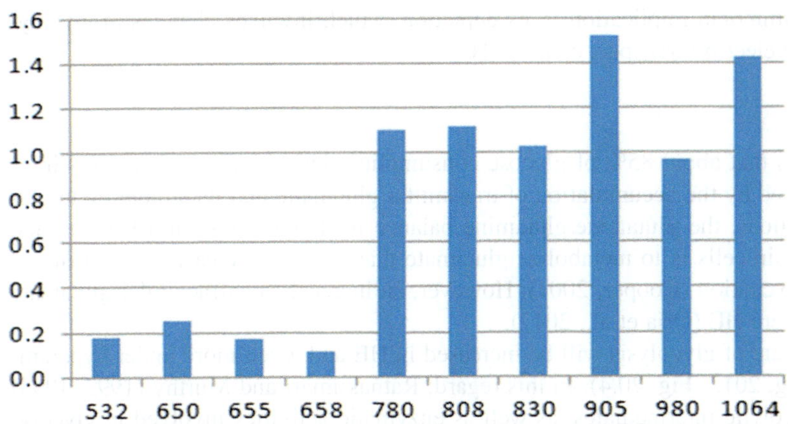

FIGURE 20.3 Percentage of optical power as a function of the wavelength of the laser used, after penetrating the brain.

We wondered what the real penetration of these doses of light would be needed to obtain the stimulation or inhibition in deep areas of the brain, in our case, the nucleus accumbens (ventral area of the striatum). Moreover, another limitation as Hamblin (2018) stated, is how to reach the accurate intensity of 3 J/cm which is able to trigger the mitochondrial membrane potential. Regarding this, we proposed not only to apply a certain wavelength of laser, but even to combine a set of lasers that could stimulate a large area of the rat head. Perhaps, using this approach we would be able to increase the activity in such deep areas of the brain as the nucleus accumbens.

20.2 What is hepatic encephalopathy?

By definition, hepatic encephalopathy (HE) is a neuropsychiatric disorder, which is caused by a brain dysfunction driven by liver insufficiency and/or portosystemic shunting. HE manifests as a wide spectrum of neurological or psychiatric abnormalities ranging from subclinical alterations to coma (Vilstrup et al., 2014). Therefore, the variability in HE presentation makes it difficult to diagnose, treat, or manage, especially at the early stages. Despite this, the impairment of brain function caused by this disorder is potentially reversible, making HE an attractive area of study for clinicians, pharmacologists, and researchers.

The pathogenesis of HE has not yet been clearly defined, although several factors are considered to be responsible for the development and progression of the disease. For many years, the focus has been the increased levels of ammonia (Felipo and Butterworth, 2002), while more recently, other factors such as inflammation (Jalan et al., 2004) and oxidative stress (Bosoi and Rose, 2013) have gained more attention.

20.2.1 The contribution of ammonia

Ammonia is an ubiquitous byproduct of the metabolism of nitrogen-containing compounds, and its accumulation has been seen in a number of metabolic disorders (Burton, 2000). The gastrointestinal tract is the primary source of ammonia generation in two ways: (1) direct, ammonia is liberated from the breakdown products of dietary protein and metabolism of circulating glutamine by the action of glutaminase; (2) indirect, by the action of the gut microbiome upon urea and ingested food.

However, the physiological balance of ammonia production and clearance is disrupted at multiple levels in patients with cirrhosis, resulting in hyperammonemia. Patients with cirrhosis have more urease-active bacteria in the gut, which leads to increased nitrogenous products (Hansen and Vilstrup, 1985). Moreover, glutamine undergoes degradation by the kidney (secreting 70% of the ammonia into the urine). When the liver fails to clear the ammonia generated, other organs such as the brain and muscles are able to detoxify the system through the glutamine synthesis (Rose, 2012). In line with this, an associated problem is the sarcopenia (muscle volume depletion) in these patients, which contributes to the development of hyperammonemia and encephalopathy.

Ammonia is a neurotoxin at elevated concentrations. Most ammonia exists in an ionic form (NH_4^+) and crosses biological membranes through potassium channels. Due to the permeability of ammonia across the blood–brain barrier, toxic levels in the brain can easily be reached. Ammonia interferes with cerebral functions via various mechanisms, some being the induction of oxidative and nitrosative stress, which then leads to mitochondrial dysfunction and cerebral energy failure. This cerebral energy metabolic alteration has been observed in acute and chronic HE and other hyperammonemic disorders, highlighting the ammonia implication in its depletion, which involves changes in glycolysis, the tricarboxylic acid cycle (TCA), and the electron transport chain (ETC).

20.2.1.1 The glycolytic rate

It has been suggested (Dienel and Hertz, 2001) that about 85% of glucose consumption during brain activation is initiated by aerobic glycolysis in astrocytes, triggered by the accumulation of transmitter glutamate and its amidation to glutamine. It is well-known that under HE conditions, the glutamate/glutamine balance is altered between astrocytes and neurons. The role of this transport between brain cells is to metabolize glutamate that has been released from neurons through astrocytes, preventing neuronal over-excitation (Cooper, 2001). However, an increase in extracellular glutamate has been documented to be associated with severe HE (Oria et al., 2012).

Regarding this, it can be expected that the rate of glycolysis will be increased in HE and, even more under hyperammonemic conditions (Rama Rao and Norenberg, 2012; Fig. 20.4). In this regard, Ratnakumari and Murthy (1992, 1993) have shown an increase in brain levels of glycolytic intermediates, as well as enzymatic activities involved in glycolysis, including key enzymes such as phosphofructokinase, aldolase, glyceraldehyde-3-phosphate dehydrogenase, and pyruvate kinase in an animal model of hyperammonemia. Furthermore, an increased rate of brain glycolysis was also demonstrated in a rodent model of congenital chronic hyperammonemia (Ratnakumari et al., 1992).

Consequently, the increase in pyruvate should be accompanied by an enhancement of the operational rate of the TCA cycle. However, this did not occur in HE and hyperammonemia; instead of being channeled into the TCA cycle, the excess of glycolytic pyruvate seems to be converted to lactate (Rama Rao and Norenberg, 2012).

20.2.1.2 Lactate overproduction

The lactate uptake represents up to 25% of the rate of glucose oxidation. However, under HE conditions, there is an increase in the local release of lactate in comparison to its uptake (Dienel and Hertz, 2001). This increase in local lactate cannot be cleared in order to maintain the intracellular redox state, so it is reflected in increased blood and brain lactate levels in HE patients (Walsh et al., 1999).

It is important to highlight that the accumulation of lactate instead of pyruvate transformation into acetyl CoA, which enters the TCA, seems to be due to the inhibitory effect of ammonia on pyruvate dehydrogenase (PDH), which is the enzyme mediating this reaction (Zwingmann et al., 2003).

The direct impact of increased lactate concentration is the decrease in pyruvate levels, which is accompanied by an alteration in operational rate of the TCA cycle and in the ETC due to a diminished availability of reducing equivalents (NAD and NADH) (Fig. 20.4). Altogether, it will negatively impact the subsequent production of ATP.

20.2.1.3 Crisis in the tricarboxylic acid cycle

Apart from the inhibition of PDH, it has been shown that ammonia inhibits other enzymes such as α-ketoglutarate dehydrogenase (α-KGDH) (Lai and Cooper, 1991) or isocitrate dehydrogenase (Zwingmann et al., 2003), being both TCA cycle enzymes. The inhibition of α-KGDH produces the depletion of α-KG metabolite, which is in line with the decreased α-KG levels found in rat models (both acute and chronic) of HE (Zwingmann et al., 2003).

Such inhibition of α-KGDH and the subsequent depletion of α-KG levels could profoundly affect the operational rate of the TCA cycle and, subsequently, the ETC. Taken together, it appears that a reduction in α-KGDH activity, the rate limiting enzyme in the TCA cycle, may adversely affect cerebral bioenergetics in acute and chronic HE (Fig. 20.4). However, other studies have shown that levels of α-KG do not seem to be altered in portocaval-shunted rats (another model of chronic HE) (Hawkins and Mans, 1990), and unchanged activity of α-KGDH in patient's brain with HE (Lavoie et al., 1987) just adds uncertainty to the explanation of the failed TCA cycle under HE conditions.

20.2.1.4 Failure in oxidative phosphorylation

Oxidative phosphorylation is the process whereby precursors from the TCA cycle lead to the phosphorylation of ADP through the ETC, producing ATP. In order to produce energy (ATP), reducing equivalents such as NAD and NADH need to be transferred from the cytosol into mitochondria through the malate-aspartate shuttle (Ratnakumari and Murthy, 1989).

Hindfelt et al. (1977) suggested that the decreased levels of glutamate observed in animals with a portocaval-shunt (model of chronic HE) could explain the dysfunction of this shuttle and, consequently the alteration of the ETC. Furthermore, it has been demonstrated that ammonia inhibits state III respiration in brain mitochondria in normal and

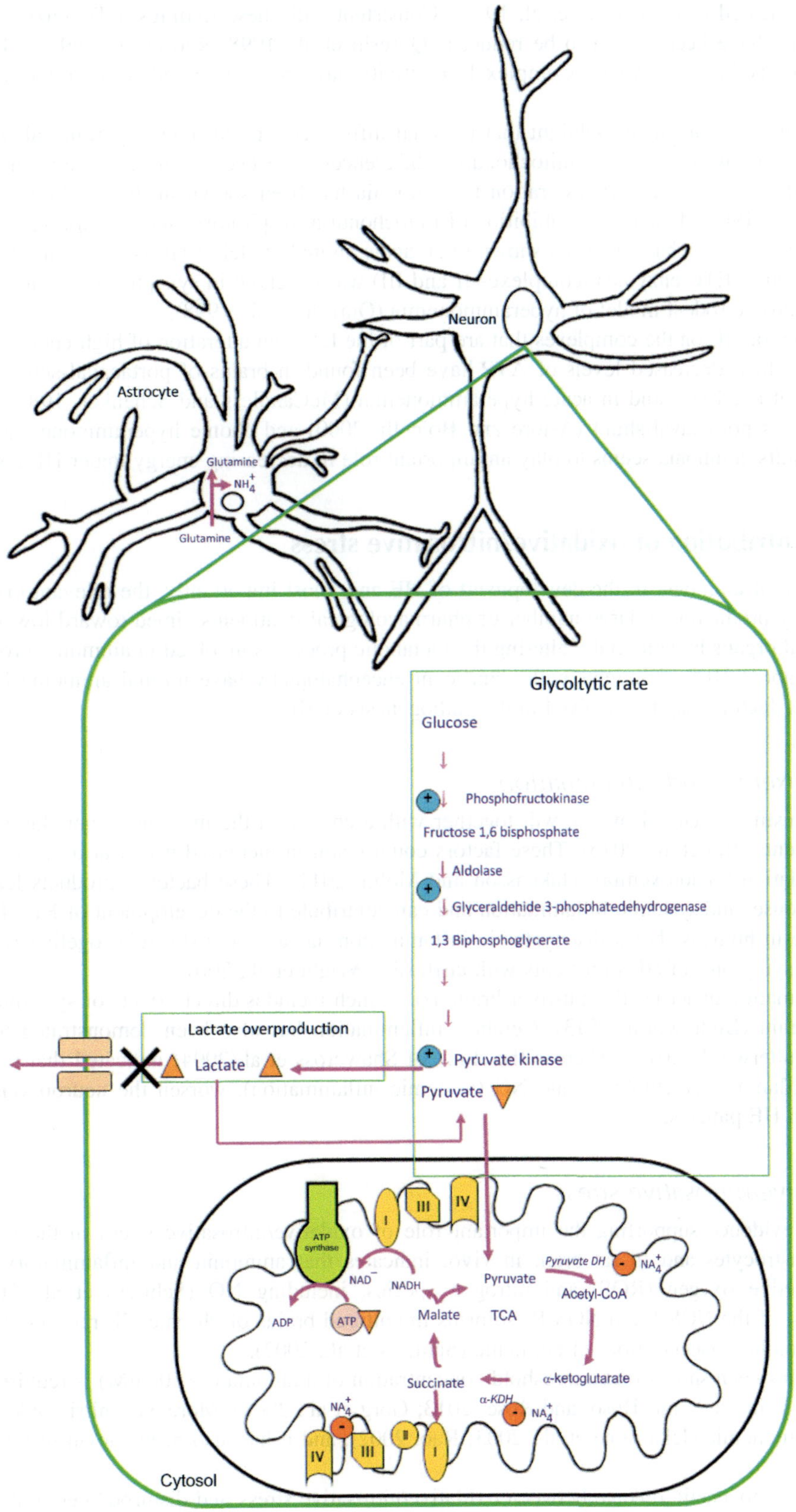

FIGURE 20.4 Schematic diagram representing the main mitochondrial dysfunction in HE. Inhibitory effects of ammonia in glycolisis, TCA, and oxidative phosphorilation have been highlighted. After entering the cells, ammonia can not only inhibit key enzymes for the above-mentioned processes, but a depletion of energy (ATP) may even occur as a consequence.

acute ammonia-intoxicated rats (Kosenko et al, 1997). Consistent with these findings, ETC enzymes included in complexes II, III, and IV have been shown to be reduced (Qureshi et al., 1998; Rao et al., 1997). Moreover, changes in brain metabolic activity based on CCO (Complex IV) activity have been observed in minimal HE rats (Arias et al., 2015).

Regarding this, it is important to highlight that regional differences in the brain regarding altered ETC and differences between synaptic and nonsynaptic mitochondrial differences have been found in several models of HE. In line with this, higher inhibition of state III respiration by ammonia has been shown in the cerebral cortex of normal rats (McKhann and Tower, 1961). In addition, inhibition of mitochondrial respiratory chain complexes (I, III, and IV) in the cerebellum and cerebral cortex has also been shown in an acute animal model of HE (Boer et al., 2009). Additionally, a higher reduction of other ETC enzymes (complexes II and III) was observed in synaptosomes compared to nonsynaptic mitochondria in a chronic rodent model of hyperammonemia (Qureshi et al., 1998).

Due to the impact of HE on the complexes that are part of the ETC, an alteration of high energy metabolites such as ATP is expected. In fact, decreased levels of ATP have been found in brains of portacaval-shunted rats infused with ammonia (Hindfelt et al., 1977) and in acute hyperammonemia (McCandless and Schenker, 1981) and chronic animal models of HE such as portocaval-shunt (Astore and Boicelli, 2000) and chonic hyperammonemia (Rao et al., 1997). Regarding these results, ammonia seems to play an important role in decreasing energy under HE conditions.

20.2.2 The contribution of oxidative/nitrosative stress

Due to the relevance of ammonia in the development of HE and, most importantly, the relevant effect of ammonia on mitochondrial energy production, a large number of pharmacological treatments aimed toward lowering hyperammonemia in the peripheral organs by potentially altering the metabolic processes involved in ammonia production and detoxification. However, about 10% of patients with significant encephalopathy have normal ammonia levels (Stahl, 1963), suggesting that other factors may be involved in the pathogenesis of HE.

20.2.2.1 Endotoxemia and inflammation

The presence of intestinal bacterial overgrowth together with a change in the microbiota population has recently been reported in HE patients (Rai et al., 2015). These factors could result in increased bacterial translocation and release of endotoxins in circulation (endotoxemia) (Hakansson and Molin, 2011). These bacterial products lead to the activation of the immune response, and systemic inflammation and can contribute to the development of HE (Tranah et al., 2013). Moreover, evidence in humans shows that systemic inflammation has a synergistic role together with ammonia in the exacerbation of the symptoms of HE in patients with cirrhosis (Wright et al., 2007).

Systemic inflammation underlies the gut-liver-brain axis, which includes direct effects of systemic pro-inflammatory molecules in the brain (Butterworth, 2013). Cerebral inflammation has also been demonstrated to contribute to the induction of HE (Butterworth, 2011), whereas the study of Shawcross et al.(2004) indicated that inflammatory mediators, such as pro-inflammatory cytokines and NO (systemic inflammation), worsen the neuropsychological effects of hyperammonemia in HE patients.

20.2.2.2 Oxidative/nitrosative stress

There is plenty of evidence supporting the important role of oxidative/nitrosative stress in the pathogenesis of HE. Work in cultured astrocytes and in rat brain in vivo, indicates that ammonia and inflammatory cytokines (among others), induce reactive oxygen (ROS) and nitrogen species, including NO (Schliess et al., 2006). Additionally, increased expression of the NOS-I and NOS-II isoforms in cultured brains of chronic HE rat model has been observed in exposure to millimolar concentrations of ammonia (Schliess et al., 2002).

Furthermore, it has been shown that a threshold concentration of ammonia ($>500\ \mu M$) is required for the induction of the oxidative stress in the brain (Bosoi and Rose, 2013; Gorg et al., 2008). Moreover, nitric oxide (NO) is increased in HE and hyperammonemia (Jekabsone et al., 2003; Rao, 2002), and it has also been shown to inhibit CCO (Bolaños et al., 1994).

Another critical pathogenetic linkage between oxidative/nitrosative stress and disturbed cerebral energy metabolism appears to be the mitochondrial permeability transition (mPT). This mPT leads to an alteration of the mitochondrial membrane potential, which involves not only a lack of pumping out of protons by the ETC, even preventing the movement of metabolites, but also impairment of oxidative phosphorylation, increased levels of ROS, and ultimately, decreased levels of ATP (Bustamante et al., 2011). Furthermore, other studies demonstrated that oxidative/nitrosative stress can also be a consequence of mPT (Votyakova and Reynolds, 2005; Zorov et al., 2006).

It is well-known that the generation of mPT is Ca^{2+} dependent (Kobayashi et al., 2003). In this regard an important effect of increased ammonia levels is the rise in Ca^{2+} levels, as has been investigated in cultured cells (Norenberg et al., 2008). Indeed, the addition of calcium chelators effectively mitigated ammonia-induced mitochondrial injury (Norenberg et al., 2008).

In this regard, Albrecht and Norenberg (2006) proposed that, once ammonia enters the brain, it is removed by astrocytes and converted to glutamine. The accumulation of glutamine has been proven in HE patients (Bjerring et al., 2008). Glutamine is then taken up by astrocytic mitochondria and converted back to glutamate, with ammonia being responsible for the initiation of mPT (Witt et al., 2017).

The increased glutamate, together with ammonia, generation of mPT, and alterations of the ETC lead to decreased levels of ATP and, ultimately, to decreased brain metabolic activity, which will have an impact on the cognition of HE patients. In animal models induced by thioacetamide, it has been demonstrated that brain levels of ATP have been reduced by 32% compared to controls (Rama Rao and Norenberg, 2012). Moreover, flow cytometric analysis revealed a significant depolarization of mitochondrial membrane in bile duct ligated rats (chronic HE model) and an inability to maintain proton gradient across the inner mitochondrial membrane (Dhanda et al., 2018). Bustamante et al. (2011) also reported a collapse in mitochondrial membrane potential in the hippocampus of animals suffering from minimal HE.

Finally, another important finding is that increased mPT leads to the appearance of a nonspecific "mega-channel" that is permeable to solutes under 1.5 kDa (Zoratti and Szabo, 1995). Bustamante et al. (2011) have shown that, in an animal model of MHE, 15% of total cytochrome c-oxidase was released. This result will explain the decrease of brain metabolic activity in an animal model of MHE when measured by CCO histochemistry (Arias et al., 2016).

PBM has been shown to enhance the metabolic capacity of neurons in culture through the photostimulation of CCO activity (Rojas et al., 2012). Based on its effects on energy metabolism, it is proposed that PBM will also affect brain metabolism and could modulate higher order cognitive functions.

20.3 Photobiomodulation for hepatic encephalopathy

In this chapter we have shown that mitochondrial dysfunction plays a pivotal role in the etiology of HE, which involves reduced glycolysis, alterations of TCA cycle, glycolysis, and oxidative phosphorylation, which finally drives the cells to overproduction of ROS, the induction of mPT leading to loss of mitochondrial membrane potential, the release of CCO into the cytosol and, ultimately, reduction of ATP production (Salehpour et al., 2018). In order to explain the potential mechanisms of the action through which PBM has a beneficial effect on energy metabolism, the main hypothesis relies on the ability of PBM to excite CCO and dissociate NO from the enzyme. As we discussed above, (nitrosative stress inhibited ETC) the dissociation of NO will recover the mitochondrial membrane potential and proton gradient, thereby improving ETC functioning and increasing ATP production (Hamblin, 2008). The second hypothesis is that PBM (red light irradiation) could suppress NOS isoforms (Leung et al., 2002) and increase antioxidant function (Lu et al., 2017), avoiding the neurotoxic effect of prolonged and elevated NO cell exposure, being both hypotheses complementary.

At the same time, PBM affects the production of ROS and increases intracellular Ca^{2+} (Salehpour et al., 2018). Firstly, the overproduction of ROS has been described previously in HE, which involves increased levels of pro-inflammatory cytokines through the activation of NF-κB (Gloire et al., 2006). Under these circumstances, PBM inhibits NF-κB signaling pathways, reducing inflammatory processes (Chen et al., 2011; Hamblin, 2017). Several studies have demonstrated the antiinflammatory effects of PBM in animal models of brain injury (Moreira et al., 2009; Zhang et al., 2014) and postischemia (Lee et al., 2017). Secondly, Ca^{2+} is a versatile second messenger (Lavi et al., 2003; Lan et al., 2012) involved in transcriptional regulation via PKA, MAPK, and CaMKs (Fields et al., 2005). Moreover, an increase in intracellular Ca^{2+} can initiate the Ras/ERK cascade (Rosen et al., 1994), producing a long-lasting effect on cells, which suggests the potential expression of some pro-oncogenes.

Furthermore, several studies have shown decreased brain metabolism associated with HE in animal models (Arias et al., 2012, 2013, 2015). In this regard, PBM seems promising due to its potential for increasing CCO expression 24 hours after its delivery in animals under nonbehavioral conditions (Morries et al., 2015; Rojas et al., 2012). Moreover, this result highlights the potential of PBM to create a nonimmediate improvement in neuronal metabolic capacity.

Moreover, we want to highlight that further studies should be carried out on the differential effect between distributed and massive interventions. These results would help to understand the importance of dosing schedules in order to achieve favorable results in a therapeutic intervention for HE patients. In fact, this dose fractionation has been tested in vitro and in vivo, and it has been shown to be highly effective to prevent neuronal degeneration. In addition, it has also been shown that PBM fractionation protocols that include prophylactic doses given before neurotoxic metabolic lesions are also effective to prevent neurodegeneration (Isomura et al., 2004; Zhuo et al., 2014). It has also been proposed that photobiomodulation effects on the brain will enhance already established intrinsic networks (Kringelbach et al., 2011; Xiao et al., 2015). We suggest that PBM with weak light at a specific wavelength could promote CCO activity in cortical and subcortical regions involved in altered networks. Then, PBM could allow the re-establishment of normal functioning of these brain networks and might alleviate HE pathological conditions (Kanzaki et al., 2013; Lim et al., 2011; Lubart et al., 2006).

Finally, another important aspect is to attempt to understand PBM regional brain effects under learning conditions. It is known that PBM potentially reaches all brain regions, but will it selectively enhance those with higher energy demands due to task-dependent activation? This effect could be due to the previously demonstrated fact that light absorption by CCO is maximized in the presence of a mixed valence enzyme state that is more likely to be found with higher respiratory chain electron flow (Rojas et al., 2012). In turn, this condition is expected to occur in states of increased energy consumption, such as those of neuronal networks activated during a particular task. In this regard, Arias et al. (2016) showed that animals with MHE under PBM treatment improved their behavior when compared to control animals. So, PBM action will not only improve enzyme activity and reduce stress, but even cognitive improvements can be expected.

Acknowledgment

This research was supported by the Princedom of Asturias GRUPIN14-088, MICINN: PSI2015-73111-EXP and PSI 2017-83893-R.

References

Albrecht, J., Norenberg, M.D., 2006. Glutamine: a Trojan horse in ammonia neurotoxicity. Hepatology 44, 788–794.

Arias, N., Méndez, M., Arias, J., Arias, J.L., 2012. Brain metabolism and spatial memory are affected by portal hypertension. Metab. Brain Dis. 27, 183–191.

Arias, N., Méndez, M., Fidalgo, C., Aller, M.A., Arias, J., Arias, J.L., 2013. Mapping metabolic brain activity in three models of hepatic encephalopathy. Int. J. Hypertens. 390872, 1–7.

Arias, N., Méndez, M., Gómez-Lázaro, E., Azpiroz, A., Arias, J.L., 2015. Main target of minimal hepatic encephalopathy: morphophysiological, inflammatory and metabolic view. Physiol. Behav. 149, 247–254.

Arias, N., Méndez, M., Arias, J.L., 2016. Low-light-level therapy as a treatment for minimal hepatic encephalopathy: behavioural and brain assessment. Lasers Med. Sci. 31, 1717–1726.

Astore, D., Boicelli, C.A., 2000. Hyperammonemia and chronic hepatic encephalopathy: an in vivo PMRS study of the rat brain. MAGMA 10, 160–166.

Bjerring, P.N., Hauerberg, J., Frederiksen, H.J., Jorgensen, L., Hansen, B.A., Tofteng, F., et al., 2008. Cerebral glutamine concentration and lactate-pyruvate ratio in patients with acute liver failure. Neurocrit. Care 9, 3–7.

Boer, L.A., Panatto, J.P., Fagundes, D.A., Bassani, C., Jeremias, I.C., Daufenbach, J.F., et al., 2009. Inhibition of mitochondrial respiratory chain in the brain of rats after hepatic failure induced by carbon tetrachloride is reversed by antioxidants. Brain Res. Bull. 80, 75–78.

Bolaños, J.P., Peuchen, S., Heales, S.J., Land, J.M., Clark, J.B., 1994. Nitric oxide-mediated inhibition of the mitochondrial respiratory chain in cultured astrocytes. J. Neurochem. 63, 910–916.

Bosoi, C.R., Rose, C.F., 2013. Oxidative stress: a systemic factor implicated in the pathogenesis of hepatic encephalopathy. Metab. Brain Dis. 28, 175–178.

Burton, B.K., 2000. Urea cycle disorders. Clin. Liver Dis. 4, 815–830.

Bustamante, J., Lores-Arnaiz, S., Tallis, S., Roselló, D.M., Lago, N., Lemberg, A., et al., 2011. Mitochondrial dysfunction as a mediator of hippocampal apoptosis in a model of hepatic encephalopathy. Mol. Cell Biochem. 354, 231–240.

Butterworth, R.F., 2011. Neuroinflammation in acute liver failure: mechanisms and novel therapeutic targets. Neurochem. Int. 59, 830–836.

Butterworth, R.F., 2013. The liver-brain axis in liver failure: neuroinflammation and encephalopathy. Nat. Rev. Gastroenterol. Hepatol. 10, 522–528.

Caterina, M.J., Pang, Z., 2016. TRP chanels in skin biology and pathophysiology. Pharmaceuticals (Basel) 9, 77.

Chen, A.C., Arany, P.R., Huang, Y.Y., Tomkinson, E.M., Sharma, S.K., Kharkwal, G.B., 2011. Low-level laser therapy activates nf-kb via generation of reactive oxygen species in mouse embryonic fibroblasts. PLoS One 6, e22453.

Cooper, A.J., 2001. Role of glutamine in cerebral nitrogen metabolism and ammonia neurotoxicity. Ment. Retard. Dev. Disabil. Res. Rev. 7, 280–286.

Dhanda, S., Sunkaria, A., Halder, A., Sandhir, R., 2018. Mitochondrial dysfunctions contribute to energy deficits in rodent model of hepatic encephalopathy. Metab. Brain Dis. 33, 209–223.

Dienel, G.A., Hertz, L., 2001. Glucose and lactate metabolism during brain activation. J. Neurosci. Res. 66, 824–838.

Eells, J.T., Henry, M.M., Summerfelt, P., Wong-Riley, M.T., Buchmann, E.V., Kane, M., et al., 2003. Therapeutic photobiomodulation for metanol-induced retinal toxicity. Proc. Natl. Acad. Sci. U.S.A. 100, 3439–3444.

Felipo, V., Butterworth, R.F., 2002. Neurobiology of ammonia. Prog. Neurobiol. 67, 259–279.

Fields, R.D., Lee, P.R., Cohen, J.E., 2005. Temporal integration of intracellular Ca^{2+} signaling networks in regulating gene expression by action potentials. Cell Calcium 37, 433–442.

Giacci, M.K., Wheeler, L., Lovett, S., Dishington, E., Majda, B., Bartlett, C.A., et al., 2014. Differential effects of 670 and 830 nm red near infrared irradiation therapy: a comparative study of optic nerve injury, retinal degeneration, traumatic brain and spinal cord injury. PLoS One 9, e104565.

Gloire, G., Legrand-Poels, S., Piette, J., 2006. NF-κB activation by reactive oxygen species: fifteen years later. Biochem. Pharmacol. 72, 1493–1505.

González-Lima, F., Barksdale, B.R., Rojas, J.C., 2014. Mitochondrial respiration as a target for neuroprotection and cognitive enhancement. Biochem. Pharmacol. 88, 584–593.

Gorg, B., Qvartskhava, N., Keitel, V., Bidmon, H.J., Selbach, O., Schliess, F., et al., 2008. Ammonia induces RNA oxidation in cultured astrocytes and brain in vivo. Hepatology 48, 567–579.

Hakansson, A., Molin, G., 2011. Gut microbiota and inflammation. Nutrients 3, 637–682.

Hamblin, M.R., 2008. The role of nitric oxide in low level light therapy. Biomedical Optics (BiOS). Int. Soc. Optics Photonics 6846, 684602–684611.

Hamblin, M.R., 2017. Mechanisms and applications of the anti-inflammatory effects of photobiomodulation. AIMS Biophys. 4, 337–361.

Hamblin, M.R., 2018. Mechanisms and mitochondrial redox signaling in photobiomodulation. J. Photochem. Photobiol. 94, 199–212.

Hansen, B.A., Vilstrup, H., 1985. Increased intestinal hydrolysis of urea in patients with alcoholic cirrhosis. Scand. J. Gastroenterol. 20, 346–350.

Hawkins, R.A., Mans, A.M., 1990. Cerebral function in hepatic encephalopathy. Adv. Exp. Med. Biol. 272, 1–22.

Henderson, T.A., Morries, L.D., 2015. Near-infrared photonic energy penetration: can infrared phototherapy effectively reach the human brain? Neuropsychiatr. Dis. Treat. 11, 2191–2208.

Hindfelt, B., Plum, F., Duffy, T.E., 1977. Effect of acute ammonia intoxication on cerebral metabolism in rats with portacaval shunts. J. Clin. Invest. 59, 386–396.

Isomura, H., Fujie, K., Shibata, K., Inoue, N., Iizuka, T., Takebe, G., et al., 2004. Bone metabolism and oxidative stress in postmenopausal rats with iron overload. Toxicology 197, 93–100.

Jalan, R., Olde Damink, S.W., Hayes, P.C., Deutz, N.E., Lee, A., 2004. Pathogenesis of intracranial hypertension in acute liver failure: inflammation, ammonia and cerebral blood flow. J. Hepatol. 41, 613–620.

Jekabsone, A., Ivanoviene, L., Brown, G.C., Borutaite, V., 2003. Nitric oxide and calcium together inactivate mitochondrial complex I and induce cytochrome c release. J. Mol. Cell Cardiol. 35, 803–809.

Kanzaki, H., Shinohara, F., Kajiya, M., Kodama, T., 2013. The Keap1/Nrf2 protein axis plays a role in osteoclast differentiation by regulating intracellular reactive oxygen species signaling. J. Biol. Chem. 288, 23009–23020.

Karu, T.I., Kolyakov, S.F., 2005. Exact action spectra for cellular responses relevant to phototherapy. Photomed. Laser Surg. 23, 355–361.

Kobayashi, T., Kuroda, S., Tada, M., Houkin, K., Iwasaki, Y., Abe, H., 2003. Calcium-induced mitochondrial swelling and cytochrome c release in the brain: its biochemical characteristics and implication in ischemic neuronal injury. Brain Res. 960, 62–70.

Kosenko, E., Felipo, V., Montoliu, C., Grisolia, S., Kaminsky, Y., 1997. Effects of acute hyperammonemia in vivo on oxidative metabolism in nonsynaptic rat brain mitochondria. Metab. Brain Dis. 12, 69–82.

Kringelbach, M.L., Green, A.L., Aziz, T.Z., 2011. Balancing the brain: resting state networks and deep brain stimulation. Front. Integr. Neurosci. 5, 8.

Lai, J.C., Cooper, A.J., 1991. Neurotoxicity of ammonia and fatty acids: differential inhibition of mitochondrial dehydrogenases by ammonia and fatty acyl coenzyme A derivatives. Neurochem. Res. 16, 795–803.

Lan, C.C., Wu, S.B., Wu, C.S., Shen, Y.C., Chiang, T.Y., Wei, Y.H., et al., 2012. Induction of primitive pigment cell differentiation by visible light (helium-neon laser): a photoacceptor-specific response not replicable by UVB irradiation. J. Mol. Med. (Berl.) 90, 321–330.

Lavi, R., Shainberg, A., Friedmann, H., Shneyvays, V., Rickover, O., Eichler, M., et al., 2003. Low energy visible light induces reactive oxygen species generation and stimulates an increase of intracellular calcium concentration in cardiac cells. J. Biol. Chem. 278, 40917–40922.

Lavoie, J., Giguere, J.F., Layrargues, G.P., Butterworth, R.F., 1987. Activities of neuronal and astrocytic marker enzymes in autopsied brain tissue from patients with hepatic encephalopathy. Metab. Brain Dis. 2, 283–290.

Lee, H.I., Lee, S.W., Kim, N.G., Park, K.J., Choi, B.T., Shin, Y.I., et al., 2017. Low-level light emitting diode (LED) therapy suppresses inflammasome-mediated brain damage in experimental ischemic stroke. J. Biophotonics 10, 1502–1513.

Leung, M.C., Lo, S.C., Siu, F.K., So, K.F., 2002. Treatment of experimentally induced transient cerebral ischemia with low energy laser inhibits nitric oxide synthase activity and up-regulates the expression of transforming growth factor-beta 1. Lasers Surg. Med. 31, 283–288.

Lim, W.B., Kim, J.S., Ko, Y.J., Kwon, H., Kim, S.W., Min, H.K., et al., 2011. Effects of 635nm light-emitting diode irradiation on angiogenesis in CoCl(2)-exposed HUVECs. Lasers Surg. Med. 43, 344–352.

Lu, Y., Wang, R., Dong, Y., Tucker, D., Zhao, N., Ahmed, M.E., et al., 2017. Low-level laser therapy for beta amyloid toxicity in rat hippocampus. Neurobiol. Aging 49, 165–182.

Lubart, R., Lavi, R., Friedmann, H., Rochkind, S., 2006. Photochemistry and photobiology of light absorption by living cells. Photomed. Laser Surg. 24, 179–185.

McCandless, D.W., Schenker, S., 1981. Effect of acute ammonia intoxication on energy stores in the cerebral reticular activating system. Exp. Brain Res. 44, 325–330.

McKhann, G.M., Tower, D.B., 1961. Ammonia toxicity and cerebral oxidative metabolism. Am. J. Physiol. 200, 420–424.

Moreira, M.S., Velasco, I.T., Ferreira, L.S., Ariga, S.K.K., Barbeiro, D.F., Meneguzzo, D.T., et al., 2009. Effect of phototherapy with low intensity laser on local and systemic immunomodulation following focal brain damage in rat. J. Photochem. Photobiol. B Biol. 97, 145–151.

Morries, L.D., Cassano, P., Henderson, T.A., 2015. Treatments for traumatic brain injury with emphasis on transcranial near-infrared laser phototherapy. Neuropsychiatr. Dis. Treat. 11, 2159–2175.

Norenberg, M.D., Rao, K.V.R., Jayakumar, A.R., 2008. Signaling factors in the mechanism of ammonia neurotoxicity. Metab. Brain Dis. 24, 103–117.

Oria, M., Romero-Giménez, J., Arranz, J.A., Riudor, E., Raguer, N., Córdoba, J., 2012. Ornithine phenylacetate prevents disturbances of motor-evoked potentials induced by intestinal blood in rats with portacaval anastomosis. J. Hepatol. 56, 109–114.

Pastore, D., Greco, M., Passarella, S., 2000. Specific helium-neon laser sensitivity of the purified cytochrome c oxidase. Int. J. Radiat. Biol. 76, 863–870.

Poyton, R.O., Ball, K.A., 2011. Therapeutic photobiomodulation: nitric oxide and a novel function of mitochondrial cytochrome c oxidase. Discov. Med. 11, 154–159.

Qureshi, K., Rao, K.V., Qureshi, I.A., 1998. Differential inhibition by hyperammonemia of the electron transport chain enzymes in synaptosomes and non-synaptic mitochondria in ornithine transcarbamylase-deficient spf-mice: restoration by acetyl-L-carnitine. Neurochem. Res. 23, 855–861.

Rai, R., Saraswat, V.A., Dhiman, R.K., 2015. Gut microbiota: its role in hepatic encephalopathy. J. Clin. Exp. Hepatol. 5, S29–S36.

Rama Rao, K.V., Norenberg, M.D., 2012. Brain energy metabolism and mitochondrial dysfunction in acute and chronic hepatic encephalopathy. Neurochem. Int. 60, 697–706.

Rao, K.V., Mawal, Y.R., Qureshi, I.A., 1997. Progressive decrease of cerebral cytochrome C oxidase activity in sparse-fur mice: role of acetyl-L-carnitine in restoring the ammonia-induced cerebral energy depletion. Neurosci. Lett. 224, 83–86.

Rao, V.L., 2002. Nitric oxide in hepatic encephalopathy and hyperammonemia. Neurochem. Int. 41, 161–170.

Ratnakumari, L., Murthy, C.R., 1989. Activities of pyruvate dehydrogenase, enzymes of citric acid cycle, and aminotransferases in the subcellular fractions of cerebral cortex in normal and hyperammonemic rats. Neurochem. Res. 14, 221–228.

Ratnakumari, L., Murthy, C.R., 1992. In vitro and in vivo effects of ammonia on glucose metabolism in the astrocytes of rat cerebral cortex. Neurosci. Lett. 148 (1-2), 85–88.

Ratnakumari, L., Murthy, C.R., 1993. Response of rat cerebral glycolytic enzymes to hyperammonemic states. Neurosci. Lett. 161, 37–40.

Ratnakumari, L., Qureshi, I.A., Butterworth, R.F., 1992. Effects of congenital hyperammonemia on the cerebral and hepatic levels of the intermediates of energy metabolism in spf mice. Biochem. Biophys. Res. Commun. 184, 746–751.

Rojas, J.C., Bruchey, A.K., Gonzalez-Lima, F., 2012. Low-level light therapy improves cortical metabolic capacity and memory retention. J. Alzheimers Dis. 32, 741–752.

Rose, C.F., 2012. Ammonia-lowering strategies for the treatment of hepatic encephalopathy. Clin. Pharmacol. Ther. 92, 321–331.

Rosen, L.B., Ginty, D.D., Weber, M.J., Greenberg, M.E., 1994. Membrane depolarization and calcium influx stimulate MEK and MAP kinase via activation of Ras. Neuron 12, 1207–1221.

Salehpour, F., Mahmoudi, J., Kamari, F., Sadigh-Eteghad, S., Rasta, S.H., Hamblin, M.R., 2018. Brain Photobiomodulation therapy: a narrative review. Mol. Neurobiol. Available from: https://doi.org/10.1007/s12035-017-0852-4.

Schliess, F., Görg, B., Fischer, R., Desjardins, P., Bidmon, H.J., Herrmann, A., et al., 2002. Ammonia induces MK-801-sensitive nitration and phosphorylation of protein tyrosine residues in rat astrocytes. FASEB J. 16, 739–741.

Schliess, F., Gorg, B., Haussinger, D., 2006. Pathogenetic interplay between osmotic and oxidative stress: the hepatic encephalopathy paradigm. Biol. Chem. 387, 1363–1370.

Shawcross, D.L., Davies, N.A., Williams, R., Jalan, R., 2004. Systemic inflammatory response exacerbates the neuropsychological effects of induced hyperammonemia in cirrhosis. J. Hepatol. 40, 247–254.

Stahl, J., 1963. Studies of the blood ammonia in liver disease. Its diagnostic, prognostic, and therapeutic significance. Ann. Intern. Med. 58, 1–24.

Tranah, T.H., Vijay, G.K., Ryan, J.M., Shawcross, D.L., 2013. Systemic inflammation and ammonia in hepatic encephalopathy. Met. Brain Dis. 28, 1–5.

Vilstrup, H., Amodio, P., Bajaj, J., Cordoba, J., Ferenci, P., Mullen, K.D., et al., 2014. Hepatic encephalopathy in chronic liver disease: 2014 Practice Guideline by the American Association for the Study of Liver Diseases and the European Association for the Study of the Liver. Hepatology 60, 715–735.

Votyakova, T.V., Reynolds, I.J., 2005. Ca^{2+}-induced permeabilization promotes free radical release from rat brain mitochondria with partially inhibited complex I. J. Neurochem. 93, 526–537.

Walsh, T.S., McLellan, S., Mackenzie, S.J., Lee, A., 1999. Hyperlactatemia and pulmonary lactate production in patients with fulminant hepatic failure. Chest 116, 471–476.

Witt, A.M., Larsen, F.S., Bjerring, P.N., 2017. Accumulation of lactate in the rat brain during hyperammonaemia is not associated with impaired mitochondrial respiratory capacity. Metab. Brain Dis. 32, 461–470.

Wong-Rilley, M.T., Bai, X., Buchmann, E., Whelan, H.T., 2001. Light-emitting diode treatment reversers the effect of TTX on cytochrome oxidase in neurons. Neuroreport 12, 3033–3037.

Wong-Rilley, M.T., Liang, H.L., Eells, J.T., Chance, B., Henry, M.M., Buchmann, E., et al., 2005. Photobiomodulation direcctly benefits primary neurons functionally inactivated by toxins: role of cytochrome c oxidase. J. Biol. Chem. 280, 4761–4771.

Wright, G., Davies, N.A., Shawcross, D.L., Hodges, S.J., Zwingmann, C., Brooks, H.F., et al., 2007. Endotoxemia produces coma and brain swelling in bile duct ligated rats. Hepatology 45, 1517–1526.

Xiao, H., Yang, Y., Xi, J.H., Chen, Z.Q., 2015. Structural and functional connectivity in traumatic brain injury. Neural. Regen. Res. 10, 2062–2071.

Yu, Z., Liu, N., Zhao, J., Li, Y., McCarthy, T.J., Tedford, C.E., et al., 2015. Near infrared radiation rescues mitocondrial dysfunction in cortical neurons after oxygen-glucose deprivation. Metab. Brain Dis. 30, 491–496.

Zhang, Q., Zhou, C., Hamblin, M.R., Wu, M.X., 2014. Low-level laser therapy effectively prevents secondary brain injury induced by immediate early responsive gene X-1 deficiency. J. Cereb. Blood Flow Metab. 34, 1391–1401.

Zhuo, Y., Gauthier, J.Y., Black, W.C., Percival, M.D., Duong, L.T., 2014. Inhibition of bone resorption by the cathepsin K inhibitor odanacatib is fully reversible. Bone 67, 269–280.

Zoratti, M., Szabo, I., 1995. The mitochondrial permeability transition. Biochim. Biophys. Acta 1241, 139–176.

Zorov, D.B., Juhaszova, M., Sollott, S.J., 2006. Mitochondrial ROS-induced ROS release: an update and review. Biochim. Biophys. Acta 1757, 509–517.

Zwingmann, C., Chatauret, N., Leibfritz, D., Butterworth, R.F., 2003. Selective increase of brain lactate synthesis in experimental acute liver failure: results of a [H-C] nuclear magnetic resonance study. Hepatology 37, 420–428.

Further reading

Skowronska, M., Zielińska, M., Wójcik-Stanaszek, L., Ruszkiewicz, J., Milatovic, D., Aschner, M., et al., 2012. Ammonia increases paracellular permeability of rat brain endothelial cells by a mechanism encompassing oxidative/nitrosative stress and activation of matrix metalloproteinases. J. Neurochem. 121, 125–134.

Chapter 21

Photobiomodulation in animal models of retinal injury and disease

Janis T. Eells

Department of Biomedical Sciences, University of Wisconsin-Milwaukee, Milwaukee, WI, United States

21.1 Introduction

The retina is a neurosensory tissue lining the inner surface of the back of the eyeball. It is comprised of several cell layers which include the sensory neurons that respond to light (photoreceptors) and intricate neural circuits that perform the initial stages of image processing. (Kolb, 2003; Sung and Chuang, 2010) (Fig. 21.1). Retinal image processing occurs through circuits involving five classes of cells: photoreceptors, bipolar cells, amacrine cells, horizontal cells, and ganglion cells. These processes collectively amplify, extract, and compress signals to preserve relevant information before it gets transmitted to the midbrain and the thalamus through the axons of the ganglion cells that form the optic nerve (ON).

The retina is one of most bioenergetically active tissues in the human body. The inner layers of the retina including the retinal ganglion cells (RGCs) exhibit a high metabolic rate associated with all central nervous system (CNS) neurons. The oxygen consumption rate of the photoreceptor layer is several times higher. Rod and cone photoreceptor cells within the retina are rich in mitochondria concentrated in their inner segments. These mitochondria provide the ATP required by the ionic pumps that drive the dark current. Modulation of the dark current by light is the beginning of vision (Kolb, 2003; Sung and Chuang, 2010; Yu and Cringle, 2005). As a result, photoreceptors consume more oxygen per gram of tissue weight than any cell in the body, making the retina one of the highest oxygen-consuming tissues in the human body (Yu and Cringle, 2005). This intense oxidative phosphorylation in photoreceptor cell inner segments, coupled with high concentrations of polyunsaturated fatty acids in their outer segments renders the retina susceptible to oxidative stress and lipid peroxidation (Winkler, 1981). In addition, photosensitizers in the retina cause an increase in oxidative stress when exposed to visible light (Jarrett and Boulton, 2012). Normally, oxidative damage is minimized by endogenous antioxidants and repair systems. With aging and/or retinal disease there is an increase in mitochondrial dysfunction and oxidative damage. Mitochondrial dysfunction and oxidative damage to the neuronal and nonneuronal components of the retina have been implicated in many forms of retinal injury and degeneration (Eells et al., 2016; Gouras et al., 2016; Fisher and Ferrington, 2018) (Fig. 21.2).

Mitochondrial repair and attenuation of oxidative stress are critical to the long-term survival of the retina. As a consequence, therapeutic strategies directed toward improving mitochondrial integrity and function and reducing oxidative stress have considerable potential for the treatment of retinal disease. Treatment using low-intensity far-red to near-infrared (FR/NIR) light has been shown to act on mitochondria-mediated signaling pathways to preserve mitochondrial function, attenuate oxidative stress, stimulate the production of cytoprotective factors, and prevent neuronal death. (Eells et al., 2004; Chung et al., 2012) (Fig. 21.3). FR/NIR photons penetrate the brain, retina and ON and this treatment, commonly known as photobiomodulation (PBM) has documented efficacy in the treatment of retinal aging, injury, and retinal degenerative disease (Fitzgerald et al., 2013; Geneva, 2016). Research on photobiomodulation using FR/NIR light in animal models of retinal injury and disease is the focus of this chapter.

Photobiomodulation in the Brain. DOI: https://doi.org/10.1016/B978-0-12-815305-5.00021-X

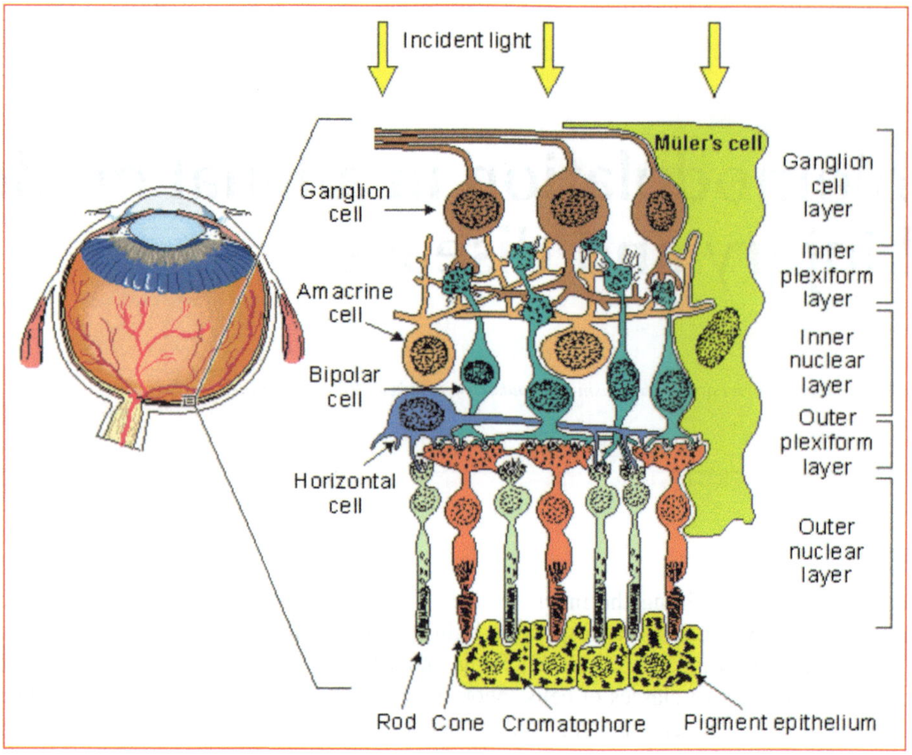

FIGURE 21.1 Anatomy of the human retina.

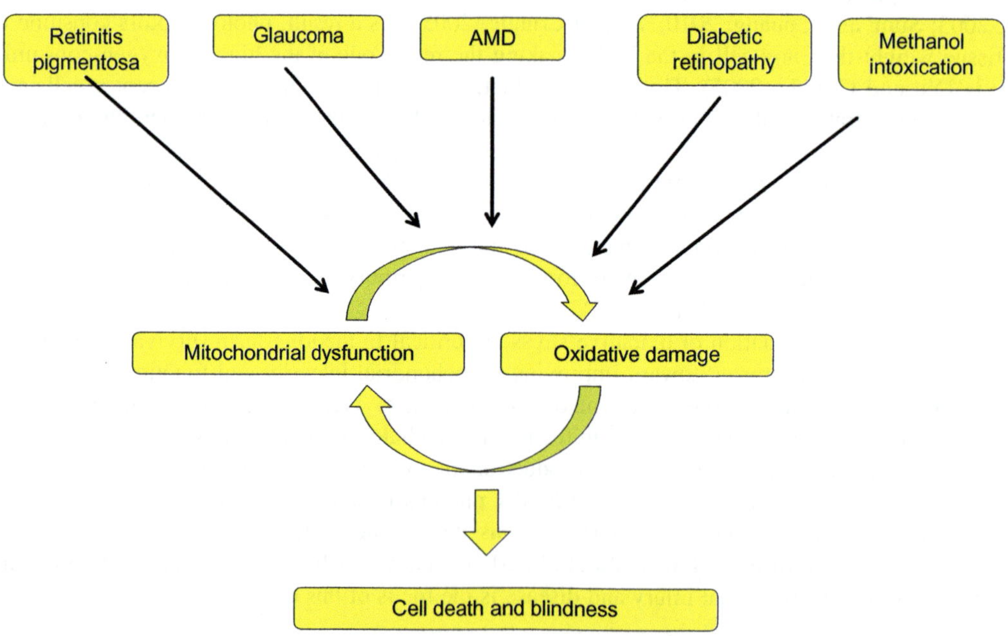

FIGURE 21.2 Mitochondrial dysfunction plays an important role in retinal injury and disease.

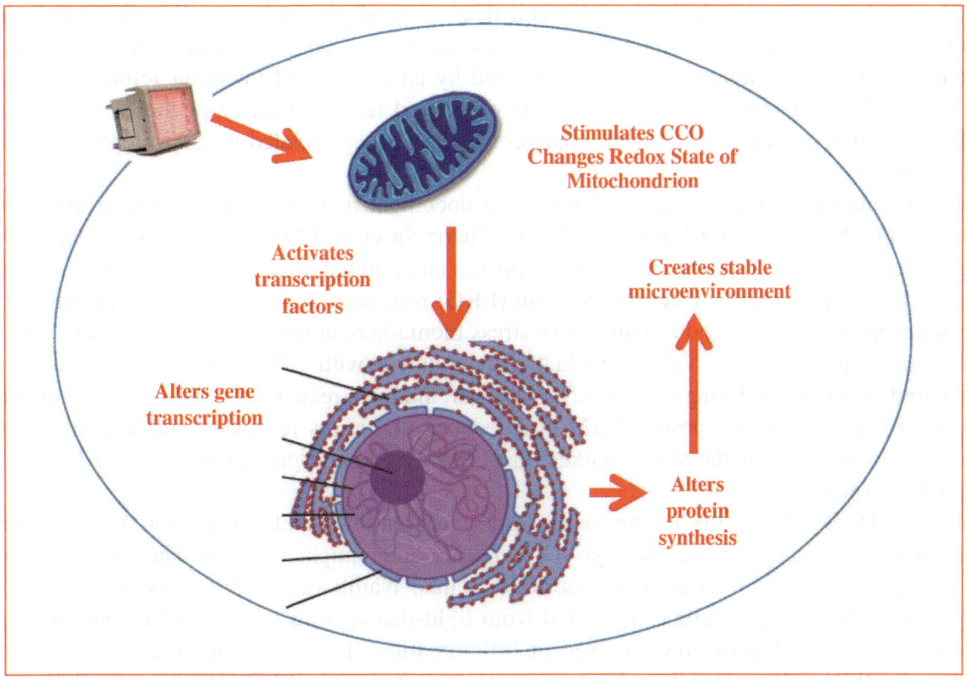

FIGURE 21.3 FR/NIR photons stimulate mitochondrial cytochrome c oxidase and activate protective intracellular pathways.

21.2 Methanol intoxication

Eells et al. (2003) reported the first direct link between the actions of FR light on mitochondrial bioenergetics in vitro and retinoprotection in vivo using an established model of retinal mitochondrial toxicity, methanol intoxication. Accumulation of formic acid generated from methanol oxidation after methanol poisoning produces toxic injury to the retina and ON, resulting in blindness (Eells et al., 2003). An acute toxic exposure to methanol results in an increase in tissue formic acid concentrations, metabolic acidosis, and visual toxicity within 72 hours of ingestion (Seme et al., 1999). Formic acid is the toxic metabolite in methanol intoxication. Formic acid inhibits the key mitochondrial enzyme, cytochrome c oxidase (CCO), by reversibly binding to the same site as cyanide and azide (Seme et al., 1999; Eells et al., 2003; Wong-Riley et al., 2005). Studies were undertaken to test the hypothesis that 670 nm PBM using light-emitting diode (LED) arrays would protect the retina against the toxic actions of methanol-derived formic acid in a rodent model of methanol toxicity and would improve the recovery of retinal function following methanol intoxication. Using the electroretinogram as a sensitive indicator of retinal function, these studies demonstrated that three brief 670 nm LED treatments (160 seconds at 25 mW/cm^2 producing a fluence of 4 J/cm^2 at the surface of the eye) (Spectralife, Quantum Devices Inc., Barneveld, WI) delivered at 5, 25, and 50 hours of methanol intoxication, attenuated the retinotoxic effects of methanol-derived formate. The authors documented a significant protection against formate-induced retinal dysfunction during the course of intoxication and recovery of retinal function following intoxication in 670 nm light-treated, methanol-intoxicated rats. 670 nm light also protected the retina from the histopathologic changes induced by methanol-derived formate. These findings provided the first link between the actions of FR to NIR light on mitochondrial oxidative metabolism in vitro and retinoprotection in vivo. Moreover, they were the first studies to suggest that FR/NIR PBM had therapeutic potential for the treatment of retinal injury and disease.

21.3 Bright light-induced retinal damage

Another method of retinal damage is exposure to excessive white light which causes photoreceptor damage and death (Grimm and Reme, 2013). Oxidative damage produced by photo-oxidation of the photoreceptor outer segments is widely accepted as the initiating event in light-induced retinal damage (LD) (Demontis et al., 2002). The lesions produced by LD are characterized by photoreceptor cell death, retinal pigment epithelial (RPE) cell damage, Muller cell gliosis, and disruption of the outer limiting membrane (Hao et al., 2002). In addition to these structural changes, there is the induction of an inflammatory state characterized by an invasion of the outer retina by activated microglia (Levine et al., 2014). This progressive degeneration has been used to model many of the factors contributing to the expansion of the degenerative area, similar to the changes observed in age-related macular degeneration (AMD) (Rutar et al., 2010, 2011, 2012).

Protection by photobiomodulation against LD has been documented by several research groups (Qu et al., 2010; Albarracin et al., 2011; Albarracin and Valter, 2012a,b). Albarracin et al. (2011), first showed that 670 nm light (9 J/ cm^2) administered before, during, or after bright-light exposure ameliorated the damaging effects of LD. PBM protected photoreceptor function as measured by electroretinogram (ERG) responses and morphology. This protection involved a reduction in photoreceptor cell death and inflammatory stress biomarkers in the retina and a reduction in microglial and macrophage invasion (Albarracin and Valter, 2012a,b). Pretreatment with 670 nm light proved to be most effective against LD compared to treatment during or after LD. However, animals treated with PBM post-LD also recovered photoreceptor function by 1-month postexposure. PBM also reduced cell stress responses and inflammation in the retina. PBM mitigated the upregulation of the stress marker, glial fibrillary acidic protein (GFAP), in Müller glial cells and reduced microglial activation.

At the molecular level, LD results in increased lipid peroxidation in photoreceptor outer segments, therefore resulting in morphological and functional damage (Organisciak and Vaughan, 2010). The oxidative stress caused by LD also increases gene expression of endogenous antioxidants. Natoli et al. (2010) explored the neuroprotective potential of FR/NIR pretreatment in retinas harvested from light-damaged rats using DNA microarray analysis. They showed that gene expression in pathways involved in inflammation and cell death were downregulated by 670 nm light in LD retinas. In this study, experimental groups were protected against LD (1000 lux for 24 hours) by pretreatment with 670 nm light (9 J/cm^2 at the eye, daily for 5 days). Quantitative real-time PCR analysis of 14 selected genes was used to validate the microarray results. LD caused the regulation of 175 entities (genes and noncoding RNAs, ncRNAs) beyond criterion levels ($P < .05$ in comparison with controls, fold-change >2). 670 nm light pretreatment reduced the expression of 126 of these 175 LD-regulated entities below criterion. In addition, 670 nm light altered 67 entities not regulated by LD. A high proportion of the entities regulated by LD (>90%) were known genes. By contrast, ncRNAs were prominent among the entities regulated by 670 nm light in its neuroprotective role (62%). One of the most highly modulated genes identified in this study was Ccl2, a potent chemokine involved in the recruitment of monocytes, T-cells, and dendritic cells to sites of tissue injury. This chemokine family has also been implicated in the pathogenesis of AMD.

Müller glial cells play an important role in the maintenance and normal function of the photoreceptors and other elements of the retina. Albarracin and Valter (2012a,b) examined the effects of pretreatment with 670 nm red light on Müller cells using the bright light-induced rat model of retinal degeneration. Müller cell-specific markers were used to assess structural and functional changes in this cell type seven days later. Changes in gene (Edn2, LIF, TNF-α) and protein (S100β, Vimentin, LIF, iNOS, GS, Cyclin-D1) levels and localization were evaluated using RT-qPCR and immunohistochemistry, respectively. 670 nm light pretreatment ameliorated the bright light-induced alterations in the expression of Müller cell-specific markers for structure, stress, metabolism, and inflammation. Thus, PBM preconditioning may promote neuroprotective effects in the retina from light-induced damage, possibly through pathways regulating the roles of Müller cells in maintaining retinal homeostasis.

PBM has been shown to downregulate TNFα, a key pro-inflammatory cytokine, in two independent studies in the eye. Albarracin and Valter reported a reduction in TNFα gene expression in the retinas of LD rats pretreated with 670 nm PBM using PCR (Albarracin and Valter, 2012a,b). In the second study, Kokkinopoulos et al. (2013), showed that 670 nm irradiation reduced TNFα immunoreactivity in the retinas of aged mice. They further showed a reduction in the recruitment of IBA-1 positive macrophages to the outer retina and a reduction in C3b and C3d immunoreactivity in Bruch membrane. All of these changes are consistent with a down-regulation of inflammatory responses by PBM.

Complement activation is associated with the pathogenesis of AMD (Zipfel et al., 2010). Complement activation also occurs following LD (Rutar et al., 2011, 2012). 670 nm light pretreatment (9 J/cm^2) reduced the expression of complement components and receptors in the retina following LD (Rutar et al., 2010). Moreover, there was a reduction in the recruitment of C-3 expressing microglia/macrophages in the retina following 670 nm light treatment and a concomitant reduction in the biomarker of oxidative damage 4-hydroxynonenal (4-HNE). These findings indicate the 670 nm light pretreatment attenuates oxidative damage to photoreceptors and reduces inflammation, which may reduce the stimulation of the complement cascade and, thus, protect photoreceptors.

21.4 Diabetic retinopathy

Diabetic retinopathy is a common long-term complication of diabetes. This retinopathy is characterized by damage to the vasculature and neurons, and in severe cases, loss of vision itself (Rao and Dlouhy, 2012; Saliba et al., 2015). Although, the pathogenesis of diabetic retinopathy is incompletely understood, a reduction in hyperglycemia has been shown to exert positive effects on the development and progression of diabetic retinopathy. However, in many patients, maintenance of glycemic control is difficult, therefore effective therapies are needed to inhibit the retinopathy (Stewart, 2016).

The laboratory of Tim Kern has extensively investigated the therapeutic potential of PBM in diabetic retinopathy (Tang et al., 2013). The effects of 670 nm light treatment (240 seconds, 25 mW/cm^2, 6 J/cm^2) on pathologic changes relevant to the development of diabetic retinopathy in vivo and in vitro were studied. In streptozotocin (STZ)-diabetic rats, 670 nm PBM attenuated diabetes-induced abnormalities of retinal function and reduced RGC death. PBM also decreased retinal superoxide generation and inhibited diabetes-induced abnormalities in the ERG, RGC viability, superoxide generation, leukostasis, and expression of MnSOD and ICAM-1 in vivo. In cultured retinal cells incubated in diabetes-like high concentrations of glucose (30 mM), PBM attenuated oxidative stress, expression of inflammatory biomarkers, and improved cell survival in RGC5 (immortalized retinal ganglion cells) and 661 W (immortalized photoreceptor-like cells) cells. The authors concluded that 670 nm PBM is a simple adjunct therapy to attenuate the development of diabetic retinopathy.

In another series of studies, the Kern lab (Saliba et al., 2015) tested whether 670 nm PBM would exert beneficial effects in another species (mice), in the presence of heavy pigmentation (C57Bl/6 J), as an intervention therapy. Their model system included diabetic mice treated with an inhibitor of the antioxidant enzyme, heme oxygenase 1 (HO-1). These mice were then exposed to 670 nm light while their eyes were blocked from direct exposure of the light. Mice treated with 670 nm light showed both neuronal and vascular beneficial effects, and that this effect is mediated at least in part systemically.

21.5 Retinitis pigmentosa

Retinitis pigmentosa (RP) is a heterogeneous group of blinding retinal diseases with the common feature of photoreceptor degeneration that is commonly associated with single-gene mutations (Wang et al., 2005). Numerous investigations have provided considerable information regarding the genetic bases of these degenerations, and several previous studies have provided evidence that, in many forms of retinal dystrophy, oxidative damage to mitochondria is a step in the death of photoreceptors (Campochiaro et al., 2015).

The therapeutic efficacy and mechanism of action of 670 nm PBM were investigated in a rodent model of RP, the P23H rat (Kirk et al., 2013). In this rodent model of human disease, the transgene is a rhodopsin gene engineered to mimic a mutation that causes an autosomal dominant form of human RP common in North America. P23H rat pups were treated once per day during the critical period of photoreceptor development (between p10 and p25) with a 670 nm LED array (180 seconds treatments at 50 mW/cm^2; fluence 9 J/cm^2) (Quantum Devices Inc., Barneveld, WI). Sham-treated rats were restrained, but not exposed to 670 nm light. In the first series of studies, rats were treated from postnatal day (p) 16–20. The status of the retina was determined at p22 by assessment of mitochondrial function, oxidative stress, and cell death. In a second series of studies, rat pups were treated from p10 to p25. Retinal status was assessed at p30 by measuring photoreceptor function by ERG and retinal morphology by spectral domain Optical coherence tomography. 670 nm light treatment increased retinal mitochondrial cytochrome oxidase activity. 670 nm light treatment also attenuated photoreceptor cell loss and improved photoreceptor function. These data suggest that PBM protects photoreceptors in the developing P23H retina by augmenting mitochondrial bioenergetics.

21.6 Aging and age-related macular degeneration

AMD is the most common cause of untreatable blindness among older adults worldwide (Ehrlich et al., 2008). Visual loss in AMD is due, in large part, to age-dependent impairment of RPE function. Although, the pathogenesis of AMD has not been completely elucidated, considerable evidence supports a role for mitochondrial dysfunction, oxidative stress, and inflammation in its onset and development (Beckman and Ames, 1998; Gouras et al., 2016). In addition to mitochondrial dysfunction and oxidative stress, inflammation is a common feature in both retinal aging and in AMD. The laboratory of Glen Jeffery in the Institute of Ophthalmology at University College in London has examined the effects of 670 nm light on retinal aging and on AMD in mouse models. They have shown 670 nm light treatment, in doses ranging from 4 to 7 J/cm^2 in the aged mouse retina increases mitochondrial membrane potential and reduces retinal inflammation (Kokkinopoulos et al., 2013).

Additional studies investigated the effects of PBM in a mouse model of AMD, the complement factor H knockout mouse (*Cfh*$^{-/-}$) (Begum et al., 2013). In this model, retinal inflammation and AB deposition occur, resulting in loss of retinal function. For these studies 670 nm PBM was by LED arrays mounted on the sides of the animal cages (360 seconds at a dose of 7.2 J/cm^2 twice per day for 14 days) rather than directly focused on the retina. PBM-treated animals exhibited significant increases in CCO, a critical mitochondrial enzyme regulating oxidative phosphorylation. Complement component C3, an inflammatory marker in the outer retina was downregulated as were key biomarkers of retinal stress, vimentin and GFAP. The authors concluded that 670 nm PBM is effective in reducing retinal inflammation likely due to CCO activation in mice with a genotype similar to that in 50% of AMD patients even when brief exposures are delivered via environmental LED arrays. The efficacy revealed here supports current early stage clinical trials of 670 nm in AMD patients.

Studies by Calaza et al. (2015), tested the link between aging, retinal ATP, and complement system polymorphisms by examining ATP concentrations in normal aging eyes and those of complement factor H knockout mice (*Cfh*$^{-/-}$). The *Cfh*$^{-/-}$ mouse is a widely used murine model of AMD (Coffey et al., 2007). The authors observed a premature decline in retinal ATP in the aging *Cfh*$^{-/-}$ mouse and also changes in the expression of the mitochondrial heat shock protein, Hsp60. 670 nm PBM (40 mW/cm^2 for 90 seconds daily for 5 days; 3.6 J/cm^2 per treatment) corrected the ATP decline in *Cfh*$^{-/-}$ mice and shifted Hsp60 labeling pattern. The decline in retinal ATP concentrations in *Cfh*$^{-/-}$ mice occurs prior to the development of the ocular phenotype involving inflammation and photoreceptor dysfunction. These findings suggest that inflammation, Aβ deposition, and reduced retinal function identified at 12 months in this mouse model, may be related to a reduction in mitochondrial bioenergetics that can be corrected by PBM.

21.7 Retinopathy of prematurity

Retinopathy of prematurity (ROP) is a disorder of the developing retina and a serious sight-threatening complication of supplemental oxygen therapy in premature neonates (Flynn et al., 1991). ROP is a two-stage disease resulting from abnormal development of the retinal vasculature caused by a disruption of the normal oxygen environment in the developing retina. In the first stage there are higher systemic oxygen concentrations than encountered in utero, suppressing the development of the retinal vasculature. This is followed in the second stage by retinal hypoxia and the release of hypoxia-stimulated factors that stimulate neovascularization when the neonate begins breathing normal air (Flynn et al., 1991).

Current therapeutic interventions for ROP include laser photocoagulation or cryotherapy to target the angiogenic aspect of the disease. These interventions are invasive, expensive, and only partially effective. Natoli et al. (2013), investigated the therapeutic efficacy of 670 nm light in two well-established rodent models of oxygen-induced retinopathy (OIR), hyperoxia exposed mice, and hyperoxia exposed rats. These OIR models take advantage of the fact that normal retinal vascularization occurs ex utero in rodents, and thus replicates the incomplete vascular development of the retinal vasculature in premature infants.

Animals were treated with 670 nm light delivered by LED arrays at a dose of 9 J/cm^2 (50 mW/cm^2 × 180 seconds) once daily during exposure to hyperoxia (p7–p17, mouse; p0–p18, rat). PBM protected the retina from hyperoxia in both models. PBM reduced vaso-obliteration, neovascularization, and retinal hemorrhage. PBM also preserved retinal vascular branching architecture and reduced neuronal cell death. In addition, the authors observed in the rat model that PBM reduced the incidence of lung pathology indicative of a systemic benefit. Although the mechanism of 670 nm PBM protection of the retinal vasculature in OIR is unclear, the authors hypothesize that 670 nm light activates mitochondrial metabolism thus promoting the consumption of excess oxygen in the hyperoxic phase of the disease. They further suggest that PBM protects retinal glial cells and reduces oxidative stress.

21.8 Optic nerve injury

ON injury can be induced by intravitreal injection of rotenone to inhibit mitochondrial complex I (Zhang et al., 2002). Using this rodent model of toxic optic neuropathy, Rojas et al. (2008) reported on the neuroprotective actions of 633 nm light. Pigmented rats received single bilateral intravitreal doses of rotenone or rotenone plus different dosage regimens of 633 nm light (3.6 J/cm^2 for 3 or 6 days). Treatment effects were evaluated using behavioral testing, histology, and neurochemistry. Rotenone induced a decrease in visual function compared with vehicle-treated controls. Behavioral impairment correlated with a decrease in retinal and visual pathway metabolic activity, retinal nerve fiber layer thickness, and ganglion cell layer cell density. These changes were prevented by light treatments given after rotenone in a dose-dependent manner (the most effective total dose was six 3.6 J/cm^2 doses of 633 nm delivered once per day for 6 days following rotenone treatment). Whole-brain cytochrome oxidase and superoxide dismutase activities were also increased in light-treated subjects in a dose-dependent manner, suggesting an in vivo transcranial effect of PBM. In whole-brain membrane isolates, PBM prevented the rotenone-induced decrease in cell respiration. The results show that PBM can effectively prevent the neurotoxic effects of rotenone suggesting therapeutic benefits for neurodegenerative disorders associated with mitochondrial dysfunction.

Another model of ON injury involves examination of secondary degeneration through partial transection of the ON (Fitzgerald et al., 2010). Traumatic injury to the CNS is often accompanied by the spreading damage of secondary degeneration, resulting in further loss of neurons and function. In this model of secondary degeneration, axons of RGCs in the ventral ON are spared from initial dorsal injury, however they are vulnerable to secondary degeneration mediated by oxidative stress (Fitzgerald et al., 2010, 2013; Cummins et al., 2013). Using this partial injury model, Fitzgerald et al. (2010), have demonstrated that PBM (WARP10 LED array, 670 nm) reduced oxidative stress in areas of ON vulnerable to secondary degeneration. Visual function was also restored to normal by 670 nm light treatment as assessed using optokinetic nystagmus and the Y-maze pattern discrimination task, thus providing evidence that 670-nm light attenuates oxidative stress and improves function in the CNS after traumatic injury in vivo.

21.9 Glaucoma

Glaucoma is a leading cause of irreversible blindness worldwide (Weinreb et al., 2014). It is associated with RGC degeneration leading to damage to the ON and visual field loss (Weinreb et al., 2014). Intraocular pressure (IOP) is currently the only modifiable risk factor for glaucoma, although RGC and vision loss can continue in patients despite well-controlled IOP (Coleman and Kodjebacheva, 2009). Studies in a rodent model of glaucoma produced by elevated IOP have provided evidence that 670 nm light focused through the pupil can attenuate the negative effects induced by ischemia (Del Olmo-Aguado et al., 2016). Elevation of IOP induces a state of retinal ischemia and results in increased expression of stress proteins including GFAP, HO-1, and mTORC1 culminating in RGC cell death. These negative effects of ischemia were significantly reduced by 670 nm light treatment (165 mW/cm^2). The authors concluded that low fluences of 670 nm light focused on the retina for a short period of time are sufficient to attenuate an insult of raised IOP to the rat retina.

21.10 Conclusion and future directions

Taken as a whole, these studies in experimental models of retinal injury and disease show that FR/NIR PBM improves mitochondrial function, reduces oxidative stress, and modulates inflammatory mediators leading to decreased apoptosis and retinoprotection. Moreover, an increasing number of clinical investigations also support the therapeutic efficacy of FR/NIR PBM in the treatment of AMD and Diabetic retinopathy (Tang et al., 2014; Merry et al., 2017). Further studies are necessary to characterize the effect of PBM on the human retina and to define safe protocols for the application of this novel therapy to mechanistically complex diseases.

Acknowledgment

The author would like to acknowledge support from the National Institutes of Health (R43-EY025892, P30-EY01931), the Defense Advanced Research Projects Agency (N66001-01-1-8969, N66001-03-1-8906), the Foundation Fighting Blindness (T-PC-0604-0256, TA-NE-0606-0348-UWI, TA-NP-0709-0465 UWI), the International Retinal Research Foundation and Fight for Sight.

References

Albarracin, R., Valter, K., 2012a. 670 nm red light preconditioning supports Muller cell function: evidence from the white light-induced damage model in the rat retina. Photochem. Photobiol. 88 (6), 1418–1427.

Albarracin, R.S., Valter, K., 2012b. Treatment with 670-nm light protects the cone photoreceptors from white light-induced degeneration. Adv. Exp. Med. Biol. 723, 121–128.

Albarracin, R., Eells, J., Valter, K., 2011. Photobiomodulation protects the retina from light-induced photoreceptor degeneration. Invest. Ophthalmol. Vis. Sci. 52, 3582–3592.

Beckman, K.B., Ames, B.N., 1998. The free radical theory of aging matures. Physiol. Rev. 78 (2), 547–581.

Begum, R., Pwoner, M.P., Hudson, N., Hogg, C., Jeffery, G., 2013. Treatment with 670 nm light up regulates cytochrome c oxidase expression and reduces inflammation in an age-related macular degeneration model. PLoS One 8, e57828.

Calaza, K.C., Kam, J.H., Hogg, C., Jeffery, G., 2015. Mitochondrial decline precedes phenotype development in the complement factor H mouse model of retinal degeneration but can be corrected by near infrared light. Neurobiol. Aging 36 (10), 2869–2876.

Campochiaro, P.A., Strauss, R.W., Lu, L., Hafiz, G., Wolfson, Y., Shah, S.M., et al., 2015. Is there excess oxidative stress and damage in eyes of patients with retinitis pigmentosa? Antioxid. Redox Signal. 23 (7), 643–648.

Chung, H., Dai, T., Sharma, S.K., Huang, Y.Y., Carroll, J.D., Hamblin, M.R., 2012. The nuts and bolts of low-level laser (light) therapy. Ann. Biomed. Eng. 40 (2), 516–533.

Coffey, P.J., Gias, C., McDermott, C.J., Lundh, P., Pickering, M.C., Sethi, C., et al., 2007. Complement factor H deficiency in aged mice causes retinal abnormalities and visual dysfunction. Proc. Natl. Acad. Sci. U.S.A. 104 (42), 16651–16656.

Coleman, A.L., Kodjebacheva, G., 2009. Risk factors for glaucoma needing more attention. Ophthalmol. J. 3, 38–42.

Cummins, N., Bartlett, C.A., Archer, M., Bartlett, E., Hemmi, J.M., Harvey, A.R., et al., 2013. Changes to mitochondrial ultrastructure in optic nerve vulnerable to secondary degeneration in vivo are limited by irradiation at 670 nm. BMC Neurosci. 14, 98.

Del Olmo-Aguado, S., Nunez-Alvarez, C., Osborne, N., 2016. Red light of the visual spectrum attenuates cell death in culture and retinal ganglion cell death *in situ*. Acta Ophthalmol. e481.

Demontis, G.C., Longoni, B., Marchiafava, P.L., 2002. Molecular steps involved in light-induced oxidative damage to retinal rods. Invest. Ophthalmol. Vis. Sci. 43 (7), 2421–2427.

Eells, J.T., Henry, M.M., Summerfelt, P., Wong-Riley, M.T., Buchmann, E.V., Kane, M., et al., 2003. Therapeutic photobiomodulation for methanol-induced retinal toxicity. Proc. Natl. Acad. Sci. U.S.A. 100 (6), 3439–3444.

Eells, J.T., Wong-Riley, M.T., VerHoeve, J., Henry, M., Buchman, E.V., Kane, M.P., et al., 2004. Mitochondrial signal transduction in accelerated wound and retinal healing by near-infrared light therapy. Mitochondrion 4 (5–6), 559–567.

Eells, J.T., Gopalakrishnan, S., Valter, K., 2016. Near-infrared photobiomodulation in retinal injury and disease. Adv. Exp. Med. Biol. 854, 437–441.

Ehrlich, R., Harris, A., Kheradiya, N.S., Winston, D.M., Ciulla, T.A., Wirostko, B., 2008. Age-related macular degeneration and the aging eye. Clin. Interv. Aging 3 (3), 473–482.

Fisher, C.R., Ferrington, D.A., 2018. Perspective in AMD pathobiology: a bioenergetic crisis in the RPE. Invest. Ophthalmol. Vis. Sci. 59, 41–47.

Fitzgerald, M., Bartlett, C.A., Payne, S.C., Hart, N.S., Rodger, J., Harvey, A.R., et al., 2010. Near infrared light reduces oxidative stress and preserves function in CNS tissue vulnerable to secondary degeneration following partial transection of the optic nerve. J. Neurotrauma 27 (11), 2107–2119.

Fitzgerald, M., Hodgetts, S., VanDenHeuvel, C., Natoli, R., Hart, N.S., Valter, K., et al., 2013. Red/near-infrared irradiation therapy for treatment of central nervous system injuries and disorders. Rev. Neurosci. 24 (2), 205–226.

Flynn, J.T., Bancalari, E., Snyder, E.S., Goldberg, R.N., Feuer, W., Cassady, J., et al., 1991. A cohort study of transcutaneous oxygen tension and the incidence and severity of retinopathy of prematurity. Trans. Am. Ophthalmol. Soc. 89, 77–92. discussion 92–95.

Geneva, I.I., 2016. Photobiomodulation for the treatment of retinal diseases: a review. Int. J. Ophthalmol. 9 (1), 145–152.

Gouras, P., Ivert, L., Neuringer, M., Nagasaki, T., 2016. Mitochondrial elongation in the macular RPE of aging monkeys, evidence of metabolic stress. Graefes Arch. Clin. Exp. Ophthalmol. 254 (6), 1221–1227.

Grimm, C., Reme, C.E., 2013. Light damage as a model of retinal degeneration. Methods Mol. Biol. 935, 87–97.

Hao, W., Wenzel, A., Obin, M.S., Chen, C.K., Brill, E., Krasnoperova, N.V., et al., 2002. Evidence for two apoptotic pathways in light-induced retinal degeneration. Nat. Genet. 32, 254–260.

Jarrett, S.G., Boulton, M.E., 2012. Consequences of oxidative stress in age-related macular degeneration. Mol. Aspects Med. 33 (4), 399–417.

Kirk, D.K., Gopalakrishnan, S., Schmitt, H., Abroe, B., Stoehr, M., Dubis, A., et al., 2013. Photobiomodulation reduces photoreceptor death and regulates cytoprotection in early states of P23H retinal dystrophy. In: Proc. SPIE BIOS, San Francisco, USA, SPIE.

Kokkinopoulos, I., Colman, A., Hogg, C., Heckenlively, J., Jeffery, G., 2013. Age-related retinal inflammation is reduced by 670 nm light via increased mitochondrial membrane potential. Neurobiol. Aging 34 (2), 602–609.

Kolb, H., 2003. How the retina works. Am. Sci. 91, 28–35.

Levine, E.S., Zam, A., Pechko, A., Wang, X., Fitzgerald, P., Pugh, E.N., et al., 2014. Rapid light-induced activation of retinal microglia in mice lacking Arrestin-1. Vision Res 102, 71–79.

Merry, G.F., Munk, M.R., Dotson, R.S., Walker, M.G., Devenyi, R.G., 2017. Photobiomodulation reduces drusen volume and improves visual acuity and contrast sensitivity in dry age-related macular degeneration. Acta Ophthalmol. 95, 270–277.

Natoli, R., Zhu, Y., Valter, K., Bisti, S., Eells, J., Stone, J., 2010. Gene and noncoding RNA regulation underlying photoreceptor protection: microarray study of dietary antioxidant saffron and photobiomodulation in rat retina. Mol. Vis. 16, 1801–1822.

Natoli, R., Valter, K., Barbosa, M., Dahlstrom, J., Rutar, M., Kent, A., et al., 2013. 670nm photobiomodulation as a novel protection against retinopathy of prematurity: evidence from oxygen induced retinopathy models. PLoS One 8 (8), e72135.

Organisciak, D.T., Vaughan, D.K., 2010. Retinal light damage: mechanisms and protection. Prog. Retin. Eye Res. 29, 113–134.

Qu, C., Cao, W., Fan, Y., Lin, Y., 2010. Near-infrared light protect the photoreceptor from light-induced damage in rats. Adv. Exp. Med. Biol. 664, 365–374.

Rao, R.C., Dlouhy, B.J., 2012. Diabetic retinopathy. N. Engl. J. Med. 367 (2), 184.

Rojas, J.C., Lee, J., John, J.M., Gonzalez-Lima, F., 2008. Neuroprotective effects of near-infrared light in an in vivo model of mitochondrial optic neuropathy. J. Neurosci. 28 (50), 13511–13521.

Rutar, M., Provis, J.M., Valter, K., 2010. Brief exposure to damaging light causes focal recruitment of macrophages, and long-term destabilization of photoreceptors in the albino rat retina. Curr. Eye Res. 35 (7), 631–643.

Rutar, M., Natoli, R., Valter, K., Provis, J.M., 2011. Early focal expression of the chemokine Ccl2 by Muller cells during exposure to damage-inducing bright continuous light. Invest. Ophthalmol. Vis. Sci. 52 (5), 2379–2388.

Rutar, M., Natoli, R., Albarracin, R., Valter, K., Provis, J., 2012. 670-nm light treatment reduces complement propagation following retinal degeneration. J. Neuroinflammation 9, 257.

Saliba, A., Du, Y., Liu, H., Patel, S., Roberts, R., Berkowitz, B.A., et al., 2015. Photobiomodulation mitigates diabetes-induced retinopathy by direct and indirect mechanisms: evidence from intervention studies in pigmented mice. PLoS One 10 (10), e0139003.

Seme, M.T., Summerfelt, P., Henry, M.M., Neitz, J., Eells, J.T., 1999. Formate-induced inhibition of photoreceptor function in methanol intoxication. J. Pharmacol. Exp. Ther. 289 (1), 361–370.

Sung, C.-H., Chuang, J.-Z., 2010. The cell biology of vision. J. Cell Biol. 190, 953–963.

Stewart, M.W., 2016. Treatment of diabetic retinopathy: recent advances and unresolved challenges. World J. Diabetes 7 (16), 333–341.

Tang, J., Du, Y., Lee, C.A., Talahalli, R., Eells, J.T., Kern, T.S., 2013. Low-intensity far-red light inhibits early lesions that contribute to diabetic retinopathy: in vivo and in vitro. Invest. Ophthalmol. Vis. Sci. 54 (5), 3681–3690.

Tang, J., Herda, A.A., Kern, T.S., 2014. Photobiomodulation in the treatment of patients with non-center-involving diabetic macular oedema. Br. J. Ophthalmol. 98 (8), 1013–1015.

Wang, D.Y., Chan, W.M., Tam, P.O., Baum, L., Lam, D.S., Chong, K.K., et al., 2005. Gene mutations in retinitis pigmentosa and their clinical implications. Clin. Chim. Acta 351 (1–2), 5–16.

Weinreb, R.N., Aung, T., Medeiros, F.A., 2014. The pathophysiology and treatment of glaucoma: a review. JAMA 311 (18), 1901–1911.

Winkler, B.S., 1981. Glycolytic and oxidative metabolism in relation to retinal function. J. Gen. Physiol. 77 (6), 667–692.

Wong-Riley, M.T., Liang, H.L., Eells, J.T., Chance, B., Henry, M.M., Buchmann, E., et al., 2005. Photobiomodulation directly benefits primary neurons functionally inactivated by toxins: role of cytochrome c oxidase. J. Biol. Chem. 280 (6), 4761–4771.

Yu, D.Y., Cringle, S.J., 2005. Retinal degeneration and local oxygen metabolism. Exp. Eye Res. 80 (6), 745–751.

Zhang, X., Jones, D., Gonzalez-Lima, F., 2002. Mouse model of optic neuropathy caused by mitochondrial complex I dysfunction. Neurosci. Lett. 326 (2), 97–100.

Zipfel, P.F., Lauer, N., Skerka, C., 2010. The role of complement in AMD. Adv. Exp. Med. Biol. 703, 9–24.

Further reading

Barja, G., 2002. Rate of generation of oxidative stress-related damage and animal longevity. Free Radic. Biol. Med. 33 (9), 1167–1172.

Hartnett, M.E., Lane, R.H., 2013. Effects of oxygen on the development and severity of retinopathy of prematurity. J. AAPOS 17 (3), 229–234.

Huang, Y.Y., Sharma, S.K., Carroll, J., Hamblin, M.R., 2011. Biphasic dose response in low level light therapy – an update. Dose Response 9 (4), 602–618.

Jackson, G.R., Owsley, C., 2003. Visual dysfunction, neurodegenerative diseases, and aging. Neurol. Clin. 21 (3), 709–728.

Johnstone, D., Massri, N., Moro, C., et al., 2014. Indirect application of near infrared light induces neuroprotection in a mouse model of parkinsonism—an abscopal neuro-protective effect. Neuroscience 274, 93–101.

Karu, T., 2010. Mitochondrial mechanisms of photobiomodulation in context of new data about multiple roles of ATP. Photomed. Laser Surg. 28 (2), 159–160.

Wang, X., Tian, F., Soni, S.S., Gonzalez-Lima, F., Liu, H., 2016. Interplay between up-regulation of cytochrome-c-oxidase and hemoglobin oxygenation induced by near-infrared laser. Sci. Rep. 6, 30540.

Zhang, R., Mio, Y., Pratt, P.F., Lohr, N., Warltier, D.C., Whelan, H.T., et al., 2009. Near infrared light protects cardiomyocytes from hypoxia and reoxygenation injury by a nitric oxide dependent mechanism. J. Mol. Cell Cardiol. 46 (1), 4–14.

Chapter 22

Transcranial photobiomodulation therapy for pain: animal models, dosimetry, mechanisms, perspectives

Marcelo Victor Pires de Sousa[1], Nathali Cordeiro Pinto[2] and Elisabeth Mateus Yoshimura[3]

[1]*Bright Photomedicine Ltd., São Paulo, Brazil,* [2]*Physiotherapy, Bright Photomedicine Ltd., São Paulo, Brazil,* [3]*Institute of Physics, Laboratory of Radiation Dosimetry and Medical Physics, University of São Paulo, São Paulo, Brazil*

22.1 Introduction

This chapter summarizes some preclinical research that supports the importance of transcranial photobiomodulation therapy (TPT) for pain. It describes the evidence for the analgesic effect, translational studies regarding to neuroscience and neurophysiology of pain, the dosimetry of transcranial light irradiation, the mechanisms and effects of TPT, and discusses the impact for human health and the future perspectives in this area.

In principle, photobiomodulation therapy (PBMT) can reduce pain by two distinct mechanisms: (1) light can interact directly with neurons, or (2) light can have an antiinflammatory action which leads to an analgesic effect. In the first mechanism, PBMT, using relatively high energy densities, is often used to elicit a direct analgesic effect due to neuromodulation. In the second method, light with appropriate wavelength and energy density activates signaling pathways, leading to a cascade of metabolic effects and a reduction in inflammatory markers such as prostaglandins and interleukins. These markers of inflammation stimulate C-fibers; therefore, their inhibition leads to pain attenuation. Usually PBMT is applied directly over the painful area causing temporary inhibition of axonal transport in small diameter nerve fibers (Aδ and C); notwithstanding, we shall demonstrate that transcranial irradiation may be more efficient and adequate in several situations.

The efficacy of photobiomodulation (PBM) therapy in clinical relief of pain has been confirmed in several literature reviews and systematic studies: chronic neck pain (Karu, 2014), tendonitis (Chow et al., 2009), chronic injuries to ligaments (Bjordal et al., 2008), lower back pain (Bjordal et al., 2003; Yousefi-Nooraie et al., 2008), and myofascial pain (Kingsley et al., 2014). For clinical pain relief the usual wavelengths are in the red range ($\lambda = 632.8$ and 670 nm) and in the near infrared (NIR, $\lambda = 780$; 810–830; 904 nm) (Karu, 2014). When the target cells of PBM are neurons, the process can be termed photoneuromodulation. Photoneuromodulation, with the central nervous system (CNS) as a target, alters the release of mediators of nociception resulting in decrease of pain.

Photoneuromodulation generates three types of reactions: neurochemical, neurobiological, and modulation of brain networks; all of which can have consequent analgesic effects. Preclinical models of TPT are most useful to study these reactions. Neurochemical TPT effects are enzymatic reactions elicited by photon-absorption during light exposure; usually, it can be demonstrated in vitro as well as in vivo. Neurobiological TPT effects consist of a longer-lasting sequence of biological events occurring in vivo after light stimulation. It can take minutes, hours, or days for some of these events to happen, and the environment of the whole organism may include activation of all the possible neurobiological events. Brain network effects are higher-order neurophysiological consequences of neurochemistry and neurobiology interacting with the functional neuroanatomical networks, and include cognitive processes which have a high demand for metabolic energy.

Noninvasive modulation of brain functions using TPT is a novel and appealing neurotherapeutic concept with many potential applications (Pires de Sousa et al., 2018). The dosimetry in this type of PBM is a challenge since the target tissue (the brain) is located beneath the skin, skull, and other tissues. Therefore, the dosimetry and the choice of irradiation

Photobiomodulation in the Brain. DOI: https://doi.org/10.1016/B978-0-12-815305-5.00022-1

parameters are even more critical than in PBMT applied to other parts of the body. Specifically, to make the translation from an animal to another different animal, and then to humans; the size, shape, optical characteristics of tissue in the photon trajectory must be considered to select the appropriate irradiation parameters. To preform TPT, due to the optical window in tissue, the use of monochromatic light in the red-to-near-infrared wavelengths is far more important than in usual PBM. So, the correct choice of power density (mW/cm^2), total energy (J), area (cm^2), and exposure time (s) is even more critical to modulate brain function in a safe and nonthermal manner.

TPT is a promising, noninvasive neuroprotective method to treat pain (despite the difficulties described above) to undergo translation to clinical trials. Furthemore, the remarkable results found for TPT as a noninvasive treatment for brain injuries or diseases brings confidence in its use as an analgesic method. TPT improved motor recovery after stroke in rats (Masoumipoor et al., 2014) and humans (Alayat et al., 2014); it significantly reduced recovery time in traumatic brain injuries (TBIs) (Navratil and Dylevsky, 1997), with little evidence of side effects (Walker, 1983). Encouraging results have been obtained for some degenerative diseases of the CNS such as familial amyotrophic lateral sclerosis (Jimbo et al., 1998), Parkinson's disease (Cambier et al., 2000), and Alzheimer's disease (Lampl, 2007).

The evidence for pain reduction connected with modulation of neural activity and release of neurotransmitters, in addition to the recent findings of positive outcomes of TPT in many brain diseases led us to hypothesize that TPT could modulate brain activity related to the perception of painful stimuli. In this report we quantified the pain attenuation and photoneuromodulation of biochemical markers after TPT in several models of pain in mice.

Future directions of TPT for pain—theoretical, biochemical, preclinical, clinical trials, and their implications for future global health—are also discussed in this chapter.

22.2 Pain—a major problem for human health

Chronic pain is one of the most common conditions reported to healthcare professionals, particularly by older patients (Reid et al., 2015). Prevalence of chronic pain increases through adult life, reaching a peak around the seventh decade (Macfarlane, 2016) (Fig. 22.1). In a nationwide survey of older adults in the United States, 52.8% reported having experienced bothersome pain in the preceding month (Reid et al., 2015). Similar findings have been reported in studies

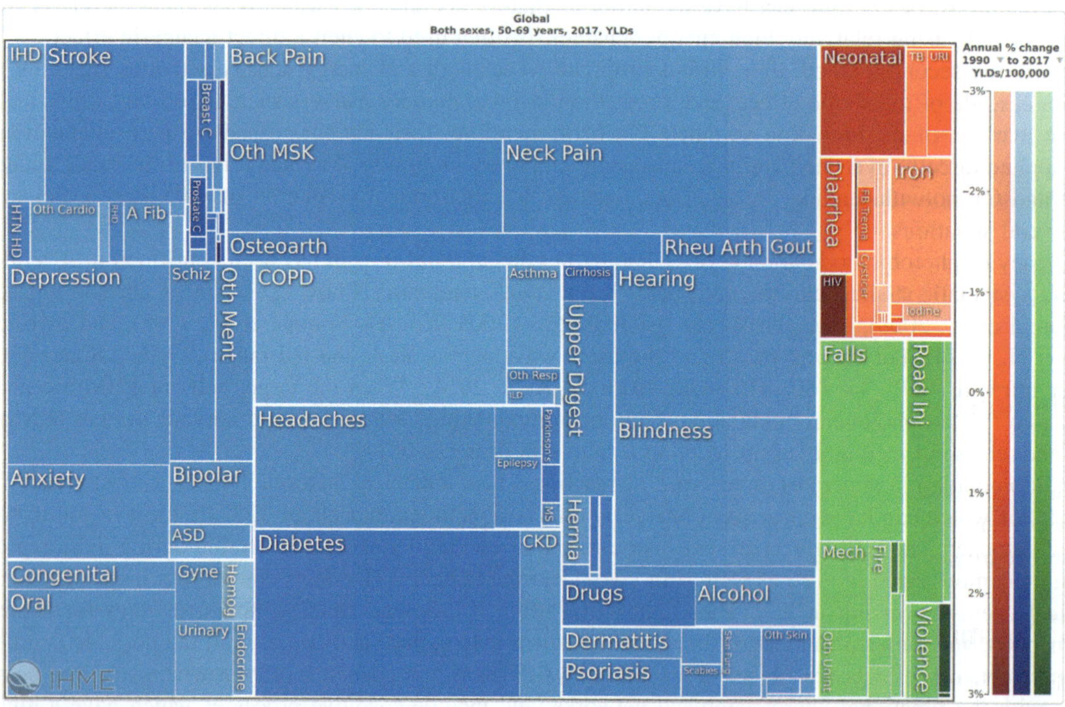

FIGURE 22.1 Data visualization for the Global Burden of Diseases. This chart represents the years lived with disability (YLD), in 2017, globally, both sexes, 50–69 years old. The area is proportional to YLD of each disease. Blue, red,and green represents, respectively, noncommunicable diseases, communicable diseases, injuries. As the reader may be interested in checking the Institute for Health Metrics and Evaluation's website (https://vizhub.healthdata.org/gbd-compare/); pain is a direct cause of more than 25% of YLD, and many other causes, like fall, depression, and diabetes are also related to pain.

conducted in Europe, Asia, and Australia and developing countries (Reid et al., 2015). In 2011, approximately 100 million US adults suffered from pain at a cost of approximately $600 billion a year (Worley, 2016).

Over the past decade, there has been a substantial rise in the use of prescription medications for the management of patients with chronic pain (Martel et al., 2015). In addition to opioids, the concurrent prescription of several other analgesic and nonanalgesic medications is very common, including antidepressants, anxiolytics/sedatives, anticonvulsants, muscle relaxants, and nonsteroidal antiinflammatory drugs (NSAIDs) (Martel et al., 2015).

Despite the potential benefits of each of these medications for the management of patients with pain, it is well-known that the combination of a wide range of medications may lead to several adverse side effects, including nausea, dizziness, headaches, constipation, and weakness (Martel et al., 2015). Unintentional overdose of paracetamol is an important cause of hepatotoxicity and NSAIDs have established gastrointestinal, cardiovascular, and renal risks, which increase with age (Reid et al., 2015). Opioids are the most potent analgesics available and their use to ameliorate chronic pain is still controversial because of their side effects (Furlan et al., 2006). The physical tolerance they produce (with the related withdrawal reactions and possibility of addiction) and the anxiety engendered by the disapproval of regulatory bodies is increasingly making patients unwilling to take opioids (Furlan et al., 2006). Studies indicate that opioids were discontinued in 48% of the patients, mostly because of poorly tolerated side effects, including constipation, changes of mental status, and nausea (Reid et al., 2015).

These medication side effects are frequently observed in clinical settings and represent a complex pain management issue (Martel et al., 2015). While patients and clinicians await the development of more effective drugs to treat pain, many patients continue to receive ineffective treatments or are unable to gain access to appropriate treatments (Worley, 2016). The suffering experienced by these patients continues to result in greater medical costs, in an enormous loss of productivity, and in a significant reduction in quality of life (Worley, 2016; Martel et al., 2015).

22.3 Transcranial photobiomodulation therapy—a multidisciplinar solution for pain

Careful surveillance to monitor for toxicity and efficacy of drugs is critical, given that advancing age increases the risk for adverse effects. A multimodal approach is strongly recommended—emphasizing a combination of both pharmacologic and nonpharmacologic treatments (Makris et al., 2014).

Therefore, nonpharmacological treatments are an essential component of any comprehensive pain management program but are rarely available to many patients. NIR light is used to treat a variety of conditions such as muscle pain, wounds, neuropathic pain, and headache without side effects (Cassano et al., 2016). Recently, remarkable results have been found using TPT as a noninvasive treatment for brain injuries or diseases, such as treatment of depression, anxiety, cognitive impairment, and improved motor recovery after stroke in rats and humans. The use of TPT is safe and well-tolerated by patients. Moreover, Pires de Sousa et al. (2016) showed the effectiveness of photoneuromodulation using transcranial NIR to suppress nociception in mice (Pires de Sousa et al., 2016).

Due to the subjective evaluation of pain by patients, and to the effects of many interactions of a social and personal nature on an individual's perception of pain, the use of animal models is imperative for evaluating the efficacy of a treatment for chronic pain. Therefore, animal models furnish a solid basis for the use of PBM as an additional tool to combat pain.

22.4 Photoneuromodulation: dosimetry, mechanisms, and therapeutics in translational research

PBMT, formerly known as low-level laser therapy (LLLT), is a noninvasive treatment without side effects that has been shown to have multiple beneficial therapeutic effects including pain relief. When PBM acts on neurons, it may be termed photoneuromodulation. When PBM is performed through the skull it is considered transcranial PBM.

22.4.1 Dosimetry

Positive clinical outcomes from TPT are highly dependent upon the precise and accurate quantification of photons that can cross the skull and be absorbed at specific tissue targets (Lampl et al., 2007; Desmet et al., 2006; Byrnes et al., 2005; Oron et al., 2001; Detaboada et al., 2006; Ilic et al., 2006; Lapchak and Araujo, 2007). Several theoretical, computational, preclinical, and clinical considerations remain necessary to fully understand light distribution and its effects in the brain after a transcranial irradiation. The first part of this section will deal with the pathway of photons to the brain and the second part deals with the pathway inside the brain.

22.4.1.1 Transcranial photon penetration in animals and humans

Red and NIR light penetrates the scalp, skull, and some brain tissues in humans and animals (Detaboada et al., 2006; Ilic et al., 2006; Lapchak et al., 2004; Zhang et al., 2000; Mochizuki-Oda et al., 2002). Lasers (800 nm, with power densities ranging between 200 – 700 mW/cm^2, coupled into an optical fiber) were used to determine photon penetration profiles through the skulls as a function of power density (Fig. 22.2). These measurements were performed with skulls from three animal species and human calvaria and the percent of photons that penetrated through the skulls was evaluated. The data represent penetration profiles through mouse, rat, rabbit, and human desiccated dry calvaria.

22.4.1.2 Optical properties of brain

The brain is a particularly complex PBM target, given its location, anatomy, physiology, and heterogeneity of tissues, metabolism, and functions. The neural component is the most usual target region for TPT since it is more superficial and has the highest metabolism, consequently it is expected to have higher susceptibility to neurostimulation by PBM. Moreover, photon penetration through tissues is a critical aspect of TPT.

A cluster of light-emitting diodes (LED) in contact with the scalp is a common light source in TPT, and it is thought that this modality would require higher transmittance in comparison with a laser (Hamblin and Marcelo de Sousa, 2016). LED helmet montages and arrays can be built with ergonomic considerations to fit comfortable upon the whole-head. In contrast, some applications will need localized absorption of photons for a specific individual structure within the brain. This may be advantageous for boosting cell functions in specific nodes within dysfunctional neural networks in which connectivity could be otherwise impaired with broad irradiation.

Light intensity decreases due to absorption and scattering of photons while penetrating into biological tissue. This attenuation can be roughly modeled by Beer-Lambert's law, therefore, the optical properties of scalp, bone, dura, brain

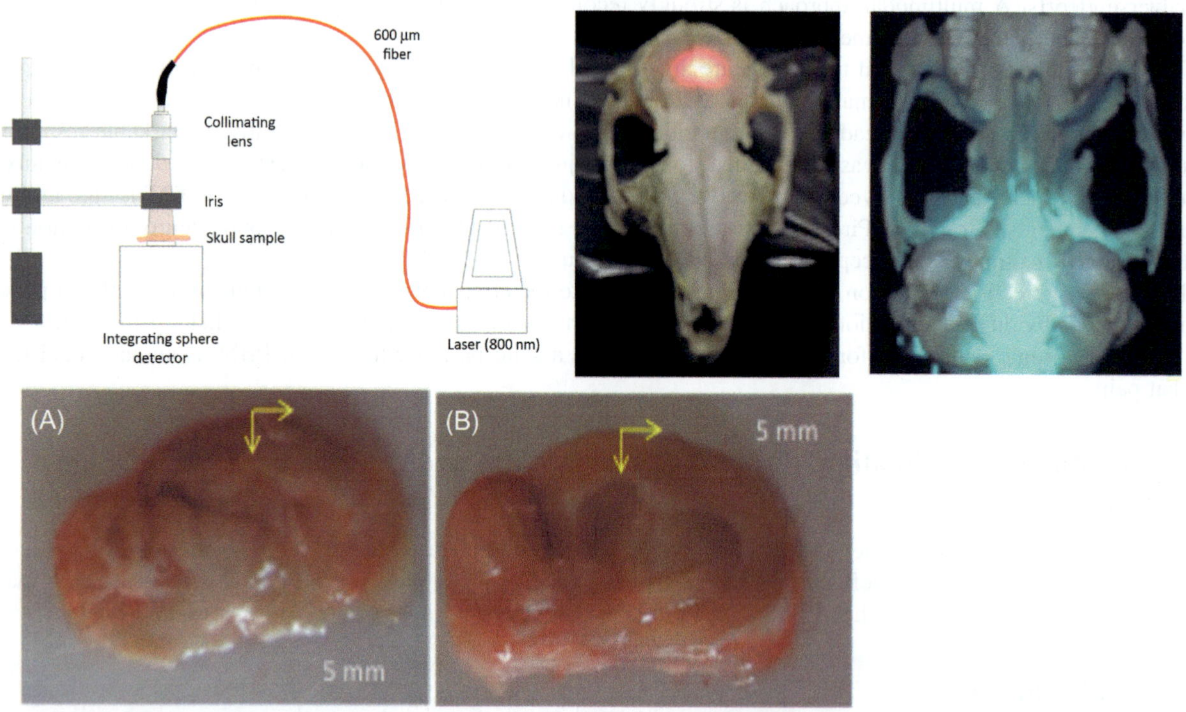

FIGURE 22.2 Top left: A diagrammatic view of the experimental set-up to measure light penetration through skulls and calvaria of various species. Top right: The laser tracking diameter on the surface of a skull from a New Zealand white rabbit. For most studies, we measured penetration profiles through the skull at bregma. Bottom (A and B): The brains were sectioned by sagittal cut producing two slices with 2.0 mm thickness. *Top right: Adapted from Lapchak, P.A., Butte, P., Rajput, P.S., 2016. Difficult path to treating acute ischemic stroke patients with transcranial near-infrared laser therapy. In: Handbook of Low-Level Laser Therapy. Pan Stanford, pp. 777–796 (Lapchak et al., 2016). Bottom (A and B): Adapted from Sousa, M.V., Prates, R., Kato, I.T., Sabino, C.P., Suzuki, L.C., Ribeiro, M.S., et al., 2012b. Laser scattering by transcranial rat brain illumination. In: Biophotonics: Photonic Solutions for Better Health Care III, vol. 8427. International Society for Optics and Photonics, p. 842728.*

cortex, and deeper brain regions must be taken in account to calculate the desired photon penetration inside the brain (Hamblin and Marcelo de Sousa, 2016).

Therefore, it is necessary to choose the parameters of the light that better suit the given optical properties of the biological tissues. The wavelength is the most important parameter since we must use red to NIR wavelengths to achieve higher penetration. Photons at wavelengths between 630 and 800 nm can travel up to 28 mm even in tissue layers with relatively low transparencies such as skin, connective tissue, muscle, bone, and spinal cord. The cranium shows absorption in the red to NIR spectrum. Postmortem human specimens have helped to estimate the transmittance through the skull (Jacques, 2013). Approximately 2% of NIR light (laser, 1064 nm, 250 mW/cm^2, 60 J/cm^2) passes through the adult human supraorbital frontal bone. Furthermore, tissue thickness and vascularization decrease light penetration, but in humans and other small animals it remains sufficient to carry out PBM inside the brain when red and NIR light is used (Zhang et al., 2000).

The unique optical properties of the brain are also relevant for deciding targets and irradiation sites for transcranial stimulation. High transmittance has been documented across the entire rat brain (from dorsal to ventral surface) at 8.5% (Mochizuki-Oda et al., 2002). Nevertheless, white matter scatters most of the incident photons resulting in a short penetration depth (Lapchak and Araujo, 2007). Transmittance of the gray matter is approximately twice that of the white matter. Fluence and power density are also important parameters influencing dose-response. The high number of potential combinations of such parameters may imply that finding the right combination needed to elicit analgesic effect is a computational task. Consequently, achieving a specified fluence at a desired depth and target tissue should be solved by algorithms.

Sousa et al. (2013) used two diode lasers (808 and 660 nm), and a high-resolution camera (14.7 megapixels, 56 pixels/mm) to measure the relative transmittance through adult rat (Rattus novergicus) brain slices. Moreover, the 90-degree scattered light from different regions (hippocampus, cerebellum, frontal cortex, and coronal view of hippocampus) of rat brain allowed the quantification of the total attenuation (Sousa et al., 2012b). Comparing images produced with NIR and red light one could notice that NIR penetrates deeper than red light. This suggests that NIR is more useful for PBM in deeper parts of the brain. The tissue boundaries and the differences of attenuation inside the brain are qualitatively clear in the pictures.

The analysis of intensity profiles of light clearly demonstrates that there are great differences in optical characteristics between different brain tissues. They indicate a relative high scattering from skin, and a low scattering from bone. It is also observed that the transmission decreases exponentially with cranial thickness at irradiation point. In brain surface there is a high scattering and reflectance demonstrating great differences in optical properties between this tissue and bone. Far from brain surface, light attenuation is approximately exponential:

$$I(z) = I_0 e^{-\mu_t z}$$

So, we can obtain the total attenuation coefficient (μ_t) by:

$$\mu_t = -\frac{\ln\left(\frac{I(z)}{I_0}\right)}{z}$$

where z, is the depth of scattered photon and $I(z)$ is the intensity scattered at that point. Each region of the brain has specific optical properties, so it has different attenuation coefficient. Comparing profiles, we found differences in exponential coefficient for each point (hippocampus, cerebellum, and frontal cortex), in sagittal cuts; and differences for each direction (sagittal and coronal) in hippocampus; and to red and NIR light, we can see clearly that red light is more attenuated to all these brain substructures (Fig. 22.3).

22.4.2 Mechanisms

The mechanism of pain reduction by PBM in the peripheral nervous system (PNS) is supported by reports such as: (1) Increased activity of acetylcholinesterase in the synapses (Pires de Sousa, 2016) (acetylcholinesterase degrades the nociceptive neurotransmitter acetylcholine); (2) Increase in serotonin synthesis (Safavi et al., 2008) and β-endorphin synthesis (Ribeiro et al., 2012); (these are neurotransmitters related to pain relief); (3) temporary suppression of action potentials evoked by bradykinin (Rochkind, 2009) (bradykinin is a substance present in inflammatory processes); (4) Inhibition of Na^+/K^+-ATPase (Barbosa et al., 2008), (Na^+/K^+-ATPase is responsible for maintaining the resting membrane potential of the neuron). The speed of transmission is reduced and the latency period is increased in nerves continuously irradiated by laser ($\lambda = 830$ nm) (Sousa et al., 2012a).

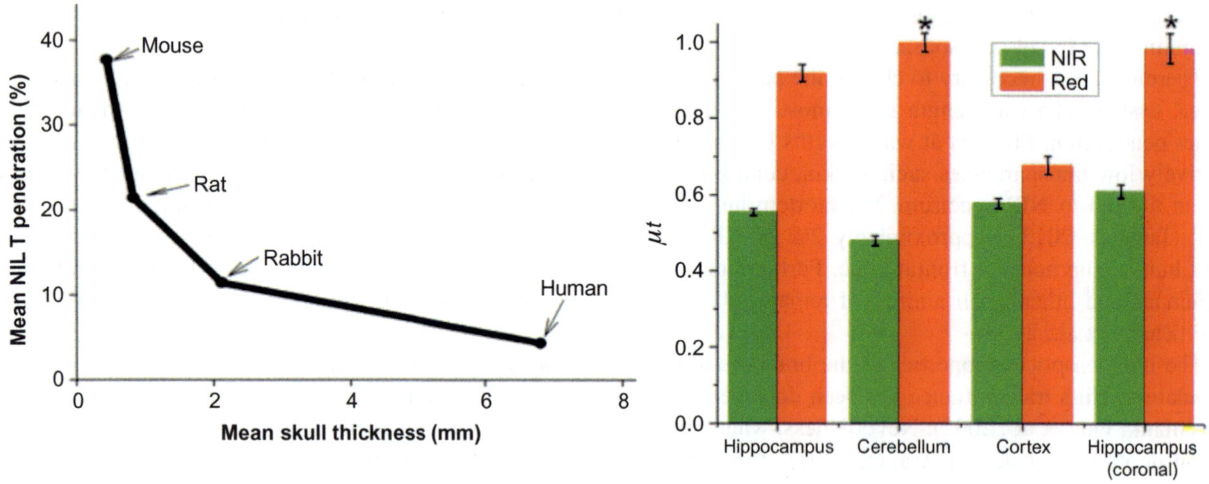

FIGURE 22.3 Left: Light penetration as a direct function of average skull thickness. From left to right are mouse, rat, rabbit, and human skulls measured at bregma in all species. The graph shows a nonlinear correlation between penetration and thickness. Right: Light attenuation for red and NIR wavelengths in different brain spots. Attenuation is statistically the equal (*) for red light in cerebellum and hippocampus (coronal). *Left: Adapted from Lapchak, P.A., Butte, P., Rajput, P.S., 2016. Difficult path to treating acute ischemic stroke patients with transcranial near-infrared laser therapy. In: Handbook of Low-Level Laser Therapy. Pan Stanford, pp. 777–796 (Lapchak et al., 2016). Right: Adapted from Sousa, M.V., Prates, R., Kato, I.T., Sabino, C.P., Suzuki, L.C., Ribeiro, M.S., et al., 2012b. Laser scattering by transcranial rat brain illumination. In: Biophotonics: Photonic Solutions for Better Health Care III, vol. 8427. International Society for Optics and Photonics, p. 842728.*

The primary photoreceptors for red and NIR light are mitochondria. As cortical neurons are exceptionally rich in mitochondria, it is very likely that brain cells are ideally set up to respond to light therapy. The absorption of photons by mitochondrial cytochrome c oxidase (CCO) leads to an increase in electron transport, and thus an increase in adenosine triphosphate (ATP), cyclic adenosine monophosphate (AMP), NO, ROS, and Ca^{2+}, enhancing cellular energy supply, and stimulating signal transduction. These biochemical changes lead to macroscopic effects with medical benefits such as promotion of cellular proliferation, wound healing, decrease in inflammation, and pain inhibition. When the target cells are neurons, the process can be termed photoneuromodulation.

ATP is a very important molecule since it is the direct energy source in cells; it is synthesized by the mitochondria during the process of oxidative phosphorylation. Several studies on cultured cells and in isolated mitochondria demonstrated the influence of red and NIR low-level light in mitochondrial physiology and its ability to increase ATP synthesis. Increasing the amount of ATP could be indirectly responsible for mediating analgesic process since its leads to production and release of endogenous analgesics like prostatic acid phosphatase (PAP).

A possible mechanism of action on the nervous tissue has been provided: PBM induces massive neurite sprouting and outgrowth in cultured neuronal, as well as Schwann cell proliferation. Also, PBM may enhance recovery of neurons from injury by altering mitochondrial oxidative metabolism and guiding neuronal growth cones, perhaps due to the interaction with cytoplasmic proteins and, particularly, due to the enhancement of actin polymerization at the leading axon edge.

Chow et al. (2007) described changes in dorsal horn neuronal activity and increased cellular stress in small and medium caliber neurons after PBM in the dorsal root ganglion (DRG). Another mechanism that has been suggested for suppression of nociception is the formation of varicosities in the tubulin cytoskeleton. Chow et al. proposed that temporary structural changes were generated in neurons that absorbed the photons. Such changes in the architecture of the neurons were called varicosities, involving clustering of mitochondria and disruption of microtubules. They were associated with reduced mitochondrial membrane potential and the blockade of axonal transport in small diameter fibers.

TPT immediately induced a dose-dependent enhancement of oxygen consumption in the cerebral cortex of naive rats. It was demonstrated by in situ quenching oximetry appointing decrease in local concentration of oxygen of 5% ± 1% and 15.8% ± 2% after energy densities respectively of 1 and 5 J/cm^2 (Anders et al., 2014). In addition, photoneuromodulation of cytochrome oxidase activates not only increase of oxygen consumption, but also release of nitric oxide (NO) (Takhtfooladi and Sharifi, 2015). These primary PBM neurochemical effects can be observed in vitro, whereas a myriad of secondary neurobiological effects occurs only in vivo. An important neurobiological effect of TPT is to increase brain ATP production, which results from increasing the mitochondrial process of oxidative phosphorylation via stimulation of CCO (Rochkind et al., 2009). Neurobiological effects, such as increasing in prefrontal cortex CCO activity, were observed even 24 hours after light exposure. Subjects treated with 10.9 J/cm^2 had CCO activity 13% higher compared to nontreated control subjects. Interestingly, higher LLLT doses of 21.6 and 32.9 J/cm^2 were less effective, showing a typical biphasic dose-response curve for PBM (Anders et al., 2014). Dose-response of neurobiological effect of TPT was observed due to whole-brain levels of CCO activity and cytosolic and mitochondrial superoxide dismutase activity (Serafim et al., 2012).

22.4.3 Therapeutic effects

The therapeutic benefits of PBM therapy in nerve repair and restoration of neural functions has been demonstrated in a wide range of injuries that damage nerves (Anders et al., 2014). There are well documented effects of PBM therapy for improvement of mitochondrial metabolism (ATP synthesis) in neurons; Wallerian degeneration postinjury (decreased retrograde), reduction of inflammation and edema; increase in axonal length and remyelination rates as well as increase in the number of axons and Schwann cells. For example, the effects of PBM therapy lead to decreased expression of pro-inflammatory factors such as IL-1β, IL-6, IL-10, COX-2, IFN-γ as well as increased expression of antiinflammatory cytokines such as IL-2, IL-4, IL-13. These collective cellular effects result in significant functional recovery of somatosensory and motor postinjury responses. Furthermore, it has been noted that PBM therapy improves microcirculation and perfusion in injured tissues via vasodilatation and angiogenesis. In addition, it is known that PBM therapy reduces the levels of pain-causing (nociceptive) neuropeptides such as substance P and reduces the conduction velocity and/ or the amplitude of compound action potentials of neurons. In summary, PBM directly modulates neuropathic pain responses.

Nerve growth factor (NGF) can play a role in the regeneration and growth of axonal processes to promote the survival of sensory neurons and reverse myelin degeneration. Vascular endothelial growth factor (VEGF), a potent growth factor for angiogenesis, also plays an important role in the proliferation of Schwann cells and in the improvement of nerve repair and motor performance. PBM can suppress inflammation-induced TNF-a, IL-1b, and HIF-1a accumulation to control neuropathic pain in diabetic mice and can elevate VEGF and NGF levels in injured nerves. Thus, functional recovery and nerve regeneration are improved. PBM therapy in animals with injured peripheral nerves prevents an increase in microglia expressing markers of the pro-inflammatory phenotype and causes a shift of the protective, antiinflammatory M2 phenotype (Anders et al., 2014).

Janzadeh et al. (2016) studied the effects of PBMT (660 nm) on neuropathic pain, in a chronic constriction injury (CCI) model for 2 weeks. CCI decreased the pain threshold, 2-week PBMT significantly increased mechanical and thermal threshold, decreased P2X3 expression, and increased bcl2 expression; however, PBM did not have any effect on Bax expression. Although PBMT increased ROS generation, it increased antioxidants such as glutathione. Increase in Bcl2 is another mitochondrial protection mechanism for cell survival and the pain relief and the decrease in P2X3 expression confirm it.

TPT has not been associated with any neurohistological or behavioral adverse effects. Preclinical data using TPT to stimulate the brain in vivo support that it is safe. Adverse behavioral effects can be induced only by massive energy densities higher (100 times) than those used to induce beneficial effects, but even the adverse effects of high doses can be attenuated if the total energy is delivered using intermittent energy pulses (Ilic et al., 2006).

22.4.4 Irradiation of nervous system: peripheral versus central

Light with appropriate parameters, including wavelength, energy density, intensity, time, and other dose components, activates signaling pathways inflammatory markers such as prostaglandins and interleukins. These markers of inflammation act as a stimulus for C-fibers; therefore, their inhibition leads to pain attenuation. Treatment with PBM, using relatively high-energy densities, is often used to elicit an analgesic effect. Light is usually applied to the painful area and is thought to cause temporary inhibition of axonal transport in small diameter nerve fibers (Aδ and C).

Recent research efforts have demonstrated that optical radiation can be used to stimulate neurons. With the development of compact light sources to evoke neural responses, it has been proved that stimulation with optical radiation is spatially selective. Pain reduction by photoneuromodulation may occur through more than one mechanism, and there may be fundamental differences between photoneuromodulation of the CNS and the PNS. Moreover, the amount of light energy necessary for effective photoneuromodulation in vivo and in vitro is likely to be different.

The mechanism of pain reduction through PBM in the PNS is supported by reports such as: (1) increased activity of acetylcholinesterase in the synapses (acetylcholinesterase degrades the nociceptive neurotransmitter acetylcholine); (2) increased serotonin synthesis and β-endorphin synthesis; (these are neurotransmitters related to pain relief); (3) temporary suppression of action potentials evoked by bradykinin (bradykinin is a substance present in inflammatory processes); (4) inhibition of the expression of Na^+/K^+-ATPase (Na^+/K^+-ATPase is responsible for maintaining the resting membrane potential of the neurons). The speed of transmission is reduced and the latency period is increased in nerves continuously irradiated by a NIR laser ($\lambda = 830$ nm).

Recently, remarkable results have been found using transcranial laser therapy (TLT) as a noninvasive treatment for brain injuries or diseases. TLT improved motor recovery after stroke in rats and humans and significantly reduced recovery time in TBIs, with little evidence of side effects. Sousa et al. (2018) showed that NIR (810 nm) laser irradiation to the lower back of mice reversibly increased up to three times the pain threshold in the hind paw, with a peak at 2–3 hours post-PBM (Pires de Sousa et al., 2018). 24hours after PBM, the pain thresholds returned to basal values. A dose response was observed, with 6 and 30 J/cm^2 being effective and 1.2 J/cm^2 being ineffective. Irradiation of the head, neck, and ipsilateral paw was also effective, unlike the irradiation of the abdomen, tail, or contralateral paw. There was no tolerance developed to the analgesic effects of PBM, as the treatment was equally effective when repeated daily for one week. The mGluR1 levels were reduced and PAP and tubulin-positive varicosities were increased as shown by immunofluorescence of DRG at 3 hours post-PBM. The data suggest that PBM to the DRG could be tested in human patients with chronic peripheral pain.

Various markers have been used in the CNS to monitor analgesic effects. Glutamate is an excitatory neurotransmitter present in the CNS and is correlated with synaptic plasticity and cell death and chronic pain. Glutamate binds to the ionotropic glutamate receptors (NMDA, AMPA), and to mGluR (G protein-coupled) lowering the excitation threshold of synaptic transmission. The concentration of free glutamate in the extracellular environment and the number of mGluR on the synapses determine the degree of excitatory stimulus. Glutamate may become toxic if the concentration rises excessively high. When the concentration of glutamate increases in the spinal cord dorsal horn, AMPA receptors are activated and rapidly depolarize the membranes of neurons in this region. Therefore, there is a decrease in the threshold to trigger the transmission of pain signal to higher-order neurons. When NMDA receptors are activated by glutamate the intensity of the stimuli response is potentiated and long-term potentiation is elicited, leaving the organism in a hyperalgesia state. The clearance of glutamate from the synaptic region is extremely important to maintain the capacity of neurons to respond to different stimuli.

PAP is an enzyme whose transmembrane isoform is a marker of nociceptive stimuli. PAP is present mainly in the nonpeptide afferent fibers of type C nociceptors and can dephosphorylate extracellular AMP producing adenosine and downregulate painful sensations. The analgesic effect of PAP only occurs in the presence of adenosine receptor type A1. Type A1 receptors are highly expressed in nociceptive fibers of PNS and in the CNS, promoting inhibition of calcium cellular intake and reducing glutamate release. Consequently, PAP modulates nociceptive responses related to thermal and mechanical stimuli. When PAP is found in higher concentrations, it indicates that there was a decrease in nociceptive signaling. Thus, PAP is an important marker for analgesia in studies with assessment of nociception and chronic pain models.

The experiments described by Pires de Sousa et al. (2016) showed the effectiveness of photoneuromodulation using transcranial NIR ($\lambda = 808$ nm) to suppress nociception in mice (Pires de Sousa et al., 2016). The assessment of pain threshold in mice demonstrated that photoneuroinhibition of pain was a temporary and reversible process, with a peak between 2 and 3 hours and the return of basal values about 24 hours after transcranial laser irradiation. Other nociception tests showed that the attenuation of pain due to photoneuromodulation in CNS occurred in various parts of the animal body (front paws, hind paws, and tail) and for different kinds of stimuli such as mechanical stress, cold, heat, and inflammation. This paper (Pires de Sousa et al., 2016) evidenced the efficacy of TLT to reduce pain sensation in the whole body evoked by various types of stimuli, by the laser irradiation to the brain cortex. The outcomes were supported by evidence of photoneuromodulation of neuro-markers related to nociception.

The effectiveness of photoneuromodulation depends on the dose used, and the effects of transcranial irradiation take some time to be noticed, as evidenced by Sousa's researches. PBM with different energy densities can significantly reduce painful dysfunctions, such as those produced by neuropathy, acting in the stimulation of the release of beta endorphin, this being a neurotransmitter responsible for the analgesia. De Andrade et al. (2017), evidenced that IR light (808 nm) in mice sciatic nerve acts positively on the reduction and control of neuropathic pain. They also noted that higher energy density, as 20 and 40 J/cm^2, is more effective, besides stimulating greater production of β-endorphin (de Andrade et al., 2017).

22.5 Photoneuromodulation of glutamate receptors, prostatic acid phophatase and adenosine triphosphate

Sousa et al. performed an extensive study about TPT in mouse models for pain. The TPT was performed using a diode laser (810 nm, 1 cm^2, and 300 mW/cm^2, continuous wave) equipped with an optical fiber in contact with the mouse scalp. Parameters were set up taking into account the attenuation of light passing through the layers of skin and skull bone as described by Pires de Sousa (Pires de Sousa et al., 2018). The experimental groups were classified and named according to the energy density and consequently time exposure. Thus, groups GL24, GL120, and GC received respectively 7.2 J/cm^2 in 24 seconds, 36 J/cm^2 in 120 seconds, sham application in 120 seconds. Each group consisted of 15 animals: 5 animals were used to quantify the pain threshold; 5 for nociceptive tests (cold, heat, formalin); and 5 for the quantification of ATP, immunofluorescence staining and H&E.

22.5.1 Behavioral evaluation of pain

The mice that received real TPT showed an increase in the pain threshold, while the control group remained with the same basal pain threshold. The therapeutic effect of laser irradiation was not seen immediately, but increased with time until a maximum effect (pain threshold up to threefold higher than baseline), which was reached at 3 hours post-TPT (Fig. 22.4).

The latency period until the mouse reacted in response to low temperature exposure (cold plate pain model) was increased (from 30 to 161 seconds) by TPT indicating a much higher tolerance to cold-induced pain. A similar result was measured in response to high temperature (tail-flick pain model using radiant heat), in which the latency period increased from 4.1 to 30.1 seconds. TPT decreased the degree of inflammatory pain (formalin injection model) 3- and 30-fold during acute and tonic phases respectively (Fig. 22.5).

22.5.2 Neurochemical and neurobiological evidences of analgesic effect

The immunofluorescence images captured by confocal microscopy were used to quantify neuromarkers. TPT increased ATP concentration in 90%, allowing the organism to product more PAP (endogenous analgesic), but had no morphological effect on tubulins (cytoskeleton). TPT also decreased metabotropic glutamate receptors leading to reduced conduction of nociceptive stimuli because of the reduced frequency of action potentials (Fig. 22.6).

This study presented evidence of the efficacy of TPT to reduce pain sensation in the whole body evoked by various types of stimuli. The laser irradiation to the brain cortex, specifically to the somatosensory cortex, was supported by evidence of photoneuromodulation of the expression of neuromakers related to nociception.

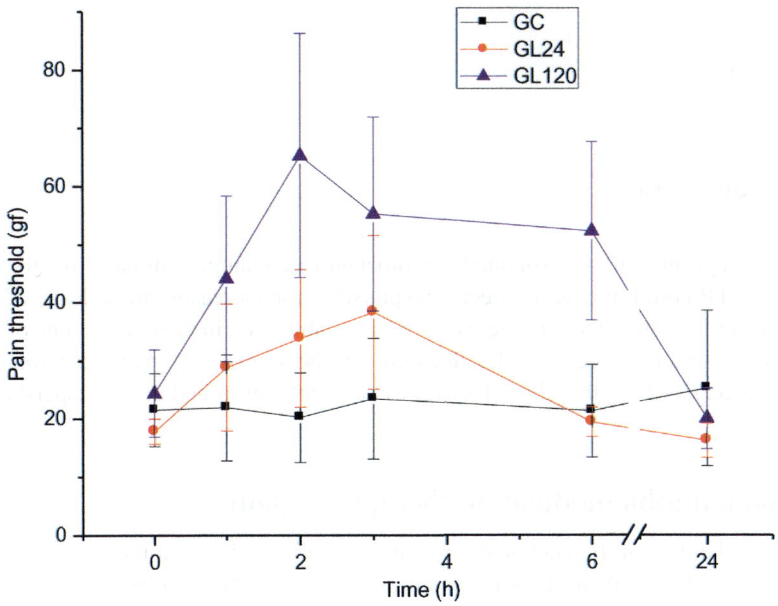

FIGURE 22.4 Average pain threshold and standard deviations (SD) before irradiation and 1, 2, 3, 6, 24 h after the irradiation for the control group (sham irradiation) and those receiving transcranial irradiation of 7.2 or 36 J. Maximum average threshold occurred between 2 and 3 h after irradiation. After 24 h the pain threshold returned to baseline in all groups.

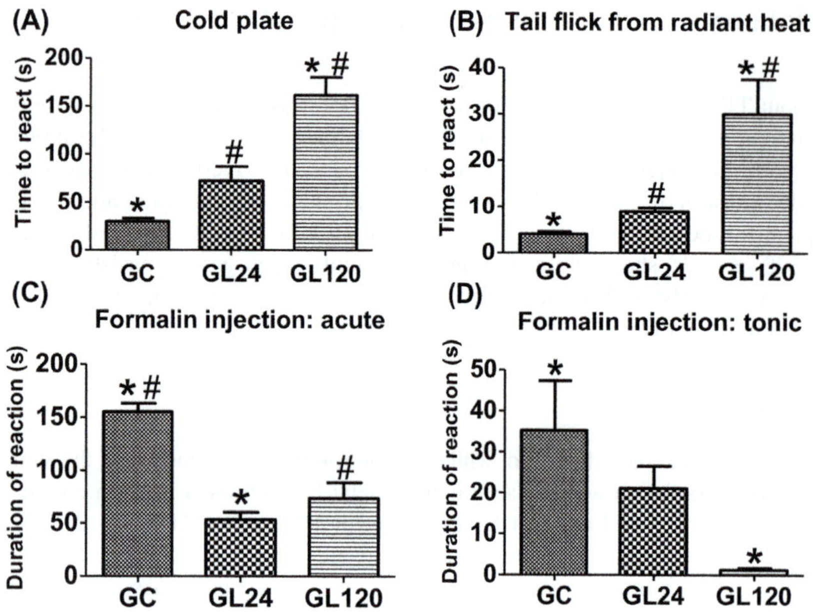

FIGURE 22.5 Mean and standard deviation (SD) of the groups ($n = 5$) for nociception tests: (A) time to reaction of the mice in response to low temperature stimulus to the foot; (B) time to reaction of the mice in response to high temperature stimulus to the tail; (C) time in which the mice remain with the inflamed paw raised during the acute phase after formalin injection; (D) time in which the mice remain with the inflamed paw raised during the tonic phase after formalin injection. The symbols * and # indicate pairs of statistically different groups, $P < .05$.

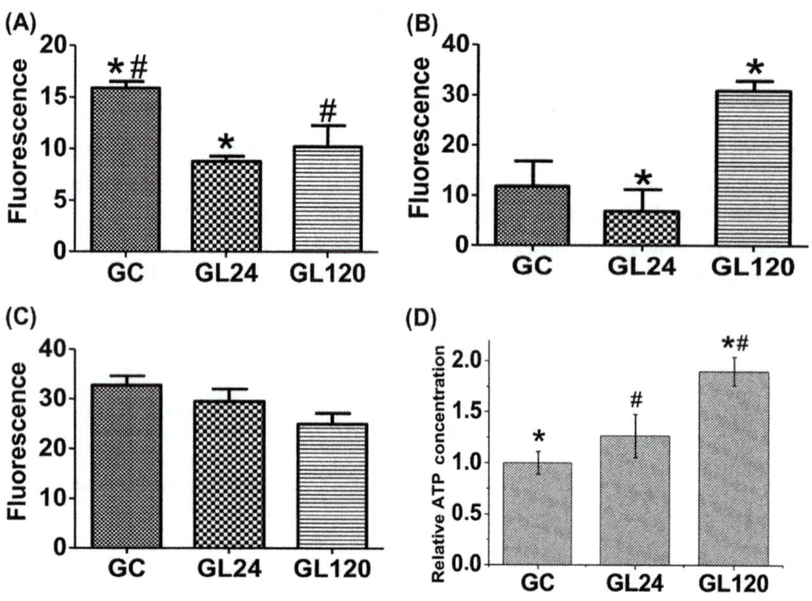

FIGURE 22.6 Mean and standard deviation (SD) of the groups ($n = 5$) for neuromarkers fluorescence and relative ATP concentration. (A) Glutamate, (B) PAP, (C) Tubulin, (D) ATP. The symbols * and # indicate pairs of statistically different groups, $P < .05$.

The mechanisms for the suppression of nociception can be explained by biochemical changes initiated by the absorption of photons. Increasing the amount of ATP could also be indirectly responsible for increasing the release of the endogenous analgesic PAP since PAP synthesis requires enough energy to be available. An increased amount of endogenous analgesics PAP could reduce pain signaling; moreover, the decrease in the number of pain receptors (mGluR) reinforces this antinociception route. It became clear that the effectiveness of photoneuromodulation depends on the dose and takes some time to occur.

22.6 Future directions of transcranial photobiomodulation therapy for pain

In this section we present some future directions—theoretical, biochemical, preclinical, clinical trials and their implications for future global health—that may occur for TPT for pain. This fascinating new intervention method for pain, (TPT) has a broad range of applicability since it improves mitochondrial respiration and also stimulates neural networks in a noninvasive way. Improvement of brain mitochondrial respiration due to TPT has evidence as discussed above and constitutes a promising neurotherapeutic principle, but still some challenges must be overcome.

The depth of transcranial penetration of photonic energy of different wavelengths should be precisely computed by software. The neurochemical reactions should be precisely understood to predict the neurobiological effects that are elicited, and consequently produce only the desired outcomes without side effects. The effects on brain networks are important and can be a source of beneficial results, but on the other hand it is important to avoid any possibility of undesired effects.

Most preclinical research, using animals to test translational approaches relies on mice or rats, however, as has been discussed above, attention must be paid to the conversion of small animal doses into human doses. Clinical trials for TPT must develop special protocols to take into account the complexity of photobiomodulation. It is necessary to pay attention to radiation (or light) interactions with tissue, similar to radiology; and overall dosimetry similar to pharmacology.

22.7 Conclusion

Photoneuromodulation has been applied in humans for stroke, TBI, neurodegenerative diseases, depression, and even for augmentation of cognitive brain functions in healthy adults. It has been observed that low optical power at the brain surface is sufficient to obtain beneficial results without any brain heating or any damaging effects. A major difficulty faced by transcranial PBM (same as for transcranial photodiagnosis) is the impossibility of delivering light to deep regions of the CNS while maintaining a low intensity at the surface. Nevertheless, we believe that transcranial photoneuromodulation will be further studied and developed as a complementary modality for treating various types of pain in humans and animals. We believe that this chapter covering studies in animal models could serve as a theoretical and experimental basis for many scientific, translational, and clinical studies about transcranial photoneuromodulation to treat pain to overcome today's pressing medical issues.

References

Alayat, M.S.M., et al., 2014. Long-term effect of high-intensity laser therapy in the treatment of patients with chronic low back pain: a randomized blinded placebo-controlled trial. Lasers Med. Sci. 29, 1065–1073.

Anders, J.J., Moges, H., Wu, X., Erbele, I.D., Alberico, S.L., Saidu, E.K., et al., 2014. In vitro and in vivo optimization of infrared laser treatment for injured peripheral nerves. Lasers Surg. Med. 46 (1), 34–45.

Barbosa, A.M., Villaverde, A.B., Guimaraes-Souza, L., Ribeiro, W., Cogo, J.C., Zamuner, S.R., 2008. Effect of low-level laser therapy in the inflammatory response induced by Bothrops jararacussu snake venom. Toxicon 51 (7), 1236–1244.

Bjordal, J.M., et al., 2003. A systematic review of low level laser therapy with location-specific doses for pain from chronic joint disorders. Aust. J. Physiother. 49, 107–116.

Bjordal, J.M., et al., 2008. A systematic review with procedural assessments and meta-analysis of low level laser therapy in lateral elbow tendinopathy (tennis elbow). BMC Musculoskeletal Disord. 9, 75.

Byrnes, K.R., Waynant, R.W., Ilev, I.K., Wu, X., Barna, L., Smith, K., et al., 2005. Light promotes regeneration and functional recovery and alters the immune response after spinal cord injury. Lasers Surg. Med. 36 (3), 171–185.

Cambier, D., et al., 2000. The influence of low intensity infrared laser irradiation on conduction characteristics of peripheral nerve: a randomised, controlled, double blind study on the sural nerve. Lasers Med. Sci. 15, 195–200.

Cassano, P., Petrie, S.R., Hamblin, M.R., Henderson, T.A., Iosifescuh, D.V., 2016. Review of transcranial photobiomodulation for major depressive disorder: targeting brain metabolism, inflammation, oxidative stress, and neurogenesis. Neurophotonics 3 (3), 031404.

Chow, R.T., et al., 2007. 830 nm laser irradiation induces varicosity formation, reduces mitochondrial membrane potential and blocks fast axonal flow in small and medium diameter rat dorsal root ganglion neurons: implications for the analgesic effects of 830 nm laser. J. Peripher. Nerv. Syst. 12, 28–39.

Chow, R.T., et al., 2009. Efficacy of low-level laser therapy in the management of neck pain: a systematic review and meta-analysis of randomised placebo or active-treatment controlled trials. Lancet 374, 1897–1908.

de Andrade, A.L.M., Bossini, P.S., do Canto De Souza, A.L.M., Sanchez, A.D., Parizotto, N.A., 2017. Effect of photobiomodulation therapy (808 nm) in the control of neuropathic pain in mice. Lasers Med. Sci. 32 (4), 865–872.

Desmet, K.D., Paz, D.A., Corry, J.J., Eells, J.T., Wong-Riley, M.T., Henry, M.M., et al., 2006. Clinical and experimental applications of NIR-LED photobiomodulation. Photomed. Laser Surg. 24 (2), 121–128.

Detaboada, L., Ilic, S., Leichliter-Martha, S., Oron, U., Oron, A., Streeter, J., 2006. Transcranial application of low-energy laser irradiation improves neurological deficits in rats following acute stroke. Lasers Surg. Med. 38 (1), 70–73.

Furlan, A.D., Sandoval, J.A., Mailis-Gagnon, A., Tunks, E., 2006. Opioids for chronic noncancer pain: a meta-analysis of effectiveness and side effects. CMAJ 174 (11), 1589–1594.

Hamblin, M.R., Pires de Sousa, M.V., Agrawal, T. (Eds.), 2016. Handbook of Low-Level Laser Therapy. CRC Press.

Ilic, S., Leichliter, S., Streeter, J., Oron, A., DeTaboada, L., Oron, U., 2006. Effects of power densities, continuous and pulse frequencies, and number of sessions of low-level laser therapy on intact rat brain. Photomed. Laser Surg. 24 (4), 458–466.

Jacques, S.L., 2013. Optical properties of biological tissues: a review. Phys. Med. Biol. 58 (11), R37.

Janzadeh, A., Nasirinezhad, F., Masoumipoor, M., Jameie, S.B., Hayat, P., 2016. Photobiomodulation therapy reduces apoptotic factors and increases glutathione levels in a neuropathic pain model. Lasers Med. Sci. 31 (9), 1863–1869.

Jimbo, K., et al., 1998. Suppressive effects of low-power laser irradiation on bradykinin evoked action potentials in cultured murine dorsal root ganglion cells. Neurosci. Lett. 240, 93–96.

Karu, T.I., 2014. Cellular and molecular mechanisms of photobiomodulation (low-power laser therapy). IEEE J. Sel. Top. Quantum Electron. 20, 143–148.

Kingsley, J.D., et al., 2014. Low-level laser therapy as a treatment for chronic pain. Front. Physiol. 5.

Lampl, Y., 2007. Laser treatment for stroke. Expert Rev. Neurother. 7, 961–965.

Lampl, Y., Zivin, J.A., Fisher, M., Lew, R., Welin, L., Dahlof, B., et al., 2007. Infrared laser therapy for ischemic stroke: a new treatment strategy: results of the NeuroThera Effectiveness and Safety Trial-1 (NEST-1). Stroke 38 (6), 1843–1849.

Lapchak, P.A., Araujo, D.M., 2007. Advances in ischemic stroke treatment: neuroprotective and combination therapies. Expert Opin. Emerg. Drugs 12 (1), 97–112.

Lapchak, P.A., Wei, J., Zivin, J.A., 2004. Transcranial infrared laser therapy improves clinical rating scores after embolic strokes in rabbits. Stroke 35 (8), 1985–1988.

Lapchak, P.A., Butte, P., Rajput, P.S., 2016. Difficult path to treating acute ischemic stroke patients with transcranial near-infrared laser therapy. Handbook of Low-Level Laser Therapy. Pan Stanford, pp. 777–796.

Macfarlane, G.J., 2016. The epidemiology of chronic pain. Pain 157, 2158–2159.

Makris, U.E., Abrams, R.C., Gurland, B., Reid, M.C., 2014. Management of persistent pain in the older patient a clinical review. JAMA 312 (8), 825–836.

Martel, M.O., Finan, P.H., Dolman, A.J., Subramanian, S., Edwards, R.R., Wasan, A.D., et al., 2015. Self-reports of medication side effects and pain-related activity interference in patients with chronic pain: a longitudinal cohort study. Pain 156 (6), 1092–1100.

Masoumipoor, M., et al., 2014. Effects of 660 nm low level laser therapy on neuropathic pain relief following chronic constriction injury in rat sciatic nerve. Arch. Neurosci. 1, 76–81.

Mochizuki-Oda, N., Kataoka, Y., Cui, Y., Yamada, H., Heya, M., Awazu, K., 2002. Effects of near-infra-red laser irradiation on adenosine triphosphate and adenosine diphosphate contents of rat brain tissue. Neurosci. Lett. 323 (3), 207–210.

Navratil, L., Dylevsky, I., 1997. Mechanisms of the analgesic effect of therapeutic lasers in vivo. Laser Ther. 9, 33–39.

Oron, U., Yaakobi, T., Oron, A., Mordechovitz, D., Shofti, R., Hayam, G., et al., 2001. Low-energy laser irradiation reduces formation of scar tissue after myocardial infarction in rats and dogs. Circulation 103 (2), 296–301.

Pires de Sousa, M.V., 2016. Chapter 1 What is low-level laser (light) therapy? Handbook of Low-Level Laser Therapy. Pan Stanford Publishing Pte. Ltd, pp. 1–16.

Pires de Sousa, M.V., Ferraresi, C., Kawakubo, M., Kaippert, B., Yoshimura, E.M., Hamblin, M.R., 2016. Transcranial low-level laser therapy (810 nm) temporarily inhibits peripheral nociception: photoneuromodulation of glutamate receptors, prostatic acid phophatase, and adenosine triphosphate. Neurophotonics 3 (1), 015003.

Pires de Sousa, M.V., Kawakubo, M., Ferraresi, C., Kaippert, B., Yoshimura, E.M., Hamblin, M.R., 2018. Pain management using photobiomodulation: mechanisms, location, and repeatability quantified by pain threshold and neural biomarkers in mice. J. Biophotonics. 11, e201700370.

Reid, M.C., Eccleston, C., Pillemer, K., 2015. Management of chronic pain in older adults. BMJ 350, h532.

Ribeiro, D.A., Paiotti, A.P., Medalha, C.C., 2012. Dual role of cyclooxygenase-2 during tissue repair induced by low level laser therapy: an intriguing issue. J. Cosmet. Laser Ther. 14 (4), 184–188.

Rochkind, S., 2009. Phototherapy in peripheral nerve regeneration: from basic science to clinical study. Neurosurg. Focus 26 (2), E8.

Rochkind, S., Geuna, S., Shainberg, A., 2009. Phototherapy in peripheral nerve injury: effects on muscle preservation and nerve regeneration. Int. Rev. Neurobiol. 87, 445–464.

Safavi, S.M., Kazemi, B., Esmaeili, M., Fallah, A., Modarresi, A., Mir, M., 2008. Effects of low-level He-Ne laser irradiation on the gene expression of IL-1beta, TNF-alpha, IFN-gamma, TGF-beta, bFGF, and PDGF in rat's gingiva. Lasers Med. Sci. 23 (3), 331–335.

Serafim, K.G., Ramos Sde, P., de Lima, F.M., Carandina, M., Ferrari, O., Dias, I.F., et al., 2012. Effects of 940 nm light-emitting diode (led) on sciatic nerve regeneration in rats. Lasers Med. Sci. 27 (1), 113–119.

Sousa, M.V., Prates, R., Kato, I.T., Sabino, C.P., Suzuki, L.C., Ribeiro, M.S., et al., 2012a. Laser scattering by transcranial rat brain illumination. SPIE Photonics Europe. International Society for Optics and Photonics.

Sousa, M.V., Prates, R., Kato, I.T., Sabino, C.P., Suzuki, L.C., Ribeiro, M.S., et al., 2012b. Laser scattering by transcranial rat brain illumination, Biophotonics: Photonic Solutions for Better Health Care III, vol. 8427. International Society for Optics and Photonics, p. 842728.

Sousa, M.V., Prates, R., Kato, I.T., Sabino, C.P., Yoshimura, T.M., Suzuki, L.C., et al., March 2013. Inhomogeneity in optical properties of rat brain: a study for LLLT dosimetry, Mechanisms for Low-Light Therapy VIII, vol. 8569. International Society for Optics and Photonics, p. 856905.

Takhtfooladi, M.A., Sharifi, D., 2015. A comparative study of red and blue light-emitting diodes and low-level laser in regeneration of the transected sciatic nerve after an end to end neurorrhaphy in rabbits. Lasers Med. Sci. 30 (9), 2319–2324.

Walker, J., 1983. Relief from chronic pain by low power laser irradiation. Neurosci. Lett. 43, 339–344.

Worley, S.L., 2016. New directions in the treatment of chronic pain. National pain strategy will guide prevention, management, and research. P T 41 (2), 107–114.

Yousefi-Nooraie, R., et al., 2008. Low Level Laser Therapy for Nonspecific Low–Back Pain. John Wiley & Sons, Inc., Hoboken, NJ.

Zhang, Q., Ma, H., Nioka, S., Chance, B., 2000. Study of near infrared technology for intracranial hematoma detection. J. Biomed. Opt. 5 (2), 206–213.

Part III

Cinical studies

Chapter 23

The challenge of effectively translating transcranial near-infrared laser therapy to treat acute ischemic stroke

Paul A. Lapchak
Neurocore LLC, Pomona, CA, United States

23.1 Introduction

Photobiomodulation (PBM) is a potentially useful method to promote cytoprotection and repair the stroke-damaged brain (Lapchak, 2010a; Naeser and Hamblin, 2011). The wavelength specific process (Anders et al., 1993; Castro-e-Silva et al., 2003; Mochizuki-Oda et al., 2002), may be optimal when the brain is irradiated with near infrared (NIR) irradiation in the 800−830 nm range (Desmet et al., 2006; Nissan et al., 1986; Byrnes et al., 2005; Oron et al., 2001; Ad and Oron, 2001; De Taboada et al., 2006; Ilic et al., 2006; Lampl et al., 2007; Lapchak and Araujo, 2007), but recent studies have discovered that a wide range of wavelengths and treatment modes can produce beneficial effects for a variety of disorders (Chung et al., 2012; Huang et al., 2009, 2012; Naeser et al., 2011; Xuan et al., 2014, 2016; Thunshelle and Hamblin, 2016; Hamblin, 2016; Tatmatsu-Rocha et al., 2016; Fernandes et al., 2015). PBM photon energy primarily in the NIR range of the electromagnetic spectrum, is nonionizing, and does not have the hazards typically associated with ultraviolet (UV) light. Moreover, NIR PBM produces little thermal activity during short periods of photon transmission. It has been suggested that irradiation with specific NIR wavelengths (i.e., 800−830 nm) can penetrate the scalp, skull, and brain tissues (Zhang et al., 2000; De Taboada et al., 2006; Ilic et al., 2006; Lapchak et al., 2004), but penetration can be limited (Lapchak et al., 2015).

PBM directly affects mitochondrial metabolism and function (Eells, 2003; Agrawal et al., 2014; Chung et al., 2012; Ferraresi et al., 2015; Hamblin, 2010, 2016; Naeser and Hamblin, 2011; Thunshelle and Hamblin, 2016; Wu et al., 2010), via the primary mitochondrial chromophore for transcranial laser therapy (TLT), the Cytochrome c oxidase (COX) (Eells, 2003; Desmet et al., 2006; Karu, 2010; Drochioiu, 2010) complex (Agrawal et al., 2014; Chung et al., 2012; Ferraresi et al., 2015; Hamblin, 2010, 2016; Naeser and Hamblin, 2011; Thunshelle and Hamblin, 2016; Wu et al., 2010), which is located in the inner mitochondrial membrane. COX is an enzyme complex that contains two copper centers, Cu_A and Cu_B. The Cu_A center has a broad absorption peak with a maximum absorption at 830 nm. TLT at 808 nm stimulates COX activity, and regulates cellular bioenergetics by driving the formation of ATP by oxidative phosphorylation (Lapchak and De Taboada, 2010; Huang et al., 2009; Lapchak and Boitano, 2016a). An important observation by Uozumi et al. (2010) showed that there was a positive correlation between TLT-induced nitric oxide synthase activity and nitric oxide (NO) production, which may be reflected by enhanced cerebral blood flow (CBF). Recently, Wang et al. demonstrated a relationship between laser-induced stimulation and brain hemodynamics (Wang et al., 2017). Taken together, TLT can be cytoprotective by enhancing mitochondrial function, and increasing blood flow to the stroke-damaged brain. These should be beneficial properties of TLT to treat stroke in which there is mitochondrial dysfunction and reduced CBF.

23.2 NeuroThera effectiveness and safety trial (NEST): from transcranial laser therapy efficacy to NEST futility

Fig. 23.1 provides a direct comparison of the modified Rankin scale (mRS) results from the NeuroThera Effectiveness and Safety Trial (NEST) clinical trials.

Photobiomodulation in the Brain. DOI: https://doi.org/10.1016/B978-0-12-815305-5.00023-3

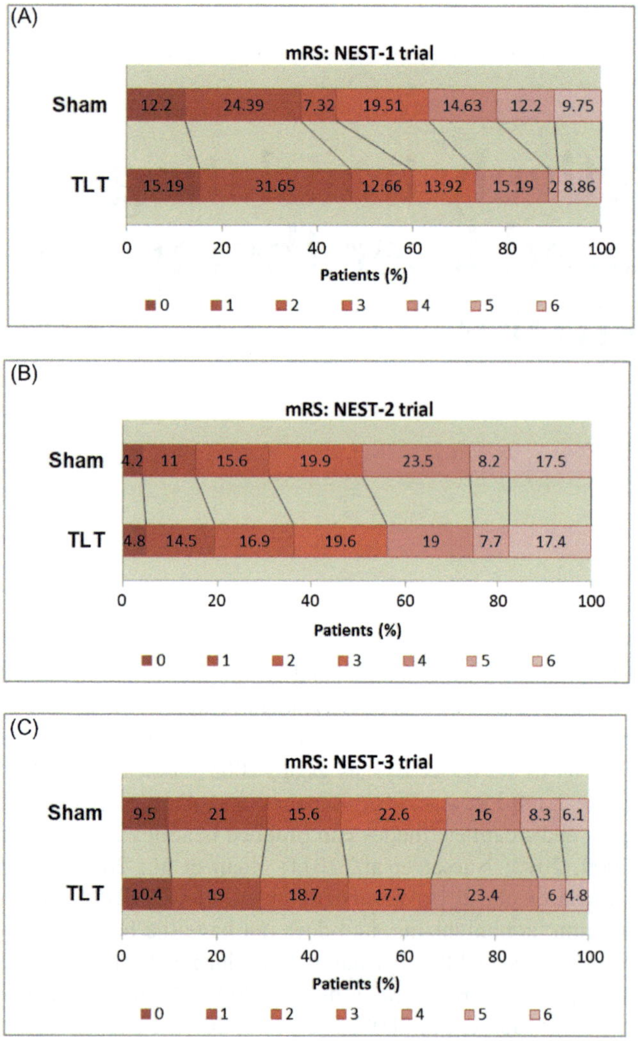

FIGURE 23.1 Direct comparison of mRS plots for patients treated with TLT in (A) NEST-1, (B) NEST-2, and (C) NEST-3. Data is shown as mRS, 0–6 (6 defined as dead) and 0, 2 defined as efficacy for 100% of the patient population in each trial. Sham is TLT not turned on, and TLT is the active group. mRS in NEST-1 and NEST-3 were extracted from publication by Lampl et al. (2007) and Hacke et al. (2014) in NEST-2 and was reconstructed from available data published in Zivin et al. (2009).

23.2.1 NeuroThera effectiveness and safety trial-1

The NEST-1 prospective, double blinded and biased randomized (2:1) clinical trial studied TLT in 120 patients 40–85 years of age (mean 68.5–70.2). Patients were randomized and heads shaved to reveal scalp before TLT because the wavelength of TLT used in the studies did not effectively penetrate hair. Patients were treated within 24 hours of ischemic stroke onset with baseline National Institutes of Health Stroke Scale (NIHSS) scores between 7 and −22 (Lampl et al., 2007) (mean time ~16 hours). The primary standardized endpoints were the NIHSS and the mRS commonly used in stroke clinical trials (Turcato et al., 2017; Chang et al., 2017; Naess et al., 2016; Dubuc et al., 2015; Kwah and Diong, 2014; Sartor et al., 2013; Leira et al., 2013; Berthier et al., 2013; Olavarria et al., 2011; Nilanont et al., 2010). The specific endpoint criteria used in NEST-1 was the following:

1. *NIHSS*: Three categories of values of the NIHSS were used. The three NIHSS strata were 7–10, 11–15, and 16–22. Since the NIHSS scale is not an interval scale, the categories were used to reduce potential heterogeneity. NIHSS outcome was collapsed into a binary NIH outcome, in which successful treatment could be measured in two ways: as a 90-day NIHSS score 0 to 1 or as a decrease in score (change) of 9 or more points from baseline to 90 days.
2. *mRS*: The mRS 90-day outcome was also measured in two ways: The first used the standard seven-category ordinal form, analyzed across the whole distribution of scores on the 0–6 mRS scale (full mRS), whereas the alternate used a binary mRS with scores of 0, 2 considered as positive or successful outcome and scores of 3, 6 as negative or treatment failure.

TABLE 23.1 NEST-1 trial enrollment and outcome.

Criteria	Treatment groups-double-blind and randomized	
	NEST-1 (Lampl et al., 2007)	
	Placebo (41)	TLT (79)
Age range	40–85 (mean 68.5)	40–85 (mean 70.2)
Sex	26 males	43 males
	15 female	36 females
Time to treatment (TT) (hours:minutes)	4:05–23:22 (mean 16:20)	2:00–23:56 (mean 16:56)
Stroke severity • NIHSS	 (mean 10)	 (mean 11)
Stroke outcome • NIHSS • stratified (by severity) • stratified (by TT) • mRS • binary outcome • stratified (by TT) • full • stratified (by TT)	 51%	 70% $P = .048$ $P = .035$ $P = .043$ $P = .034$ $P = .026$ $P = .020$
Mortality rate	9.8%	8.9%
Serious adverse events	36.6%	25.3%

Unlike the rabbit embolic stroke studies in which a single application was sufficient to encompass the complete brain and for sustained behavioral improvement (Lapchak and Boitano, 2016b; Lapchak, 2010a, 2012; Lapchak and De Taboada, 2010; Lapchak et al., 2004, 2008), for the human experiment, because of the size differences of the brain and skull, the probe was applied to 20 predetermined equally-spaced points on the shaved head for a duration of 2 minutes at each spot, with a power density of 10 mW/cm^2 and estimated energy density of 1.2 J/cm^2. This treatment design reportedly allowed for the treatment to encompass the complete cortex and allow for limited photon penetration of many brain regions.

The outcome of the NEST-1 study was quite promising as shown in Table 23.1 and Fig. 23.1. There were more patients in the active TLT-treatment group with successful outcomes than controls, as measured prospectively on the NIHSS and mRS. In this neuroprotection clinical trial, Lampl and colleagues (2007) reported that TLT-treated patients showed greater improvement from baseline NIHSS to 90 day NIHSS and mRS ($P = .021$) than the fully randomized sham-treated group did in which the laser device was not "turned on," but the probe was placed at all 20 positions on the laser skull template (Table 23.1).

Logistic regression analysis of limited data from this study, which controlled for age, sex, and time to treatment and baseline NIHSS enrollment score showed that more than 1/3 of patients, 38% in fact, of TLT patients improved by nine points or greater for a final NIHSS score of 0–1 and. For the binary outcome on the mRS, 0, 2 versus 2, 6, 60% of TLT-treated patients had a positive outcome ($P = 0.034–0.043$), and the TLT treatment was safe.

Mortality rates and serious adverse events did not differ significantly between the two treatment groups. The observation of efficacy was important from this trail, even though the clinical trial was originally designed to be a TLT safety study.

23.2.2 NeuroThera effectiveness and safety trial-2

NEST-1 incentivized the rapid enrollment of the NEST-2 trial to study TLT efficacy in a larger stroke patient population; the phase III trial was conducted in 660 stroke victims (Zivin et al., 2009). The protocol of the trial was similar to

TABLE 23.2 NEST-2 trial enrollment and outcome.

Criteria	Treatment groups-double-blind and randomized	
	NEST-2 (Zivin et al., 2009)	
	Placebo (327)	TLT (331)
Age range	42–92 (mean 71.5)	19–95 (mean 61.2)
Sex	189 male	183 male
	138 female	148 female
Time to treatment (TT) (hours:minutes)	2:30–23:54 (mean 14:43)	2:42–23:54 (mean 14:38)
Stroke severity • NIHSS	(mean 13.2)	(mean 13.1)
Stroke outcome • NIHSS • stratified (severity) • stratified (TT)	30.9%	36.3% $P = .094$ (NIHSS 7–15, $P = .044$)
Mortality rate	17.5%	17.4%
Serious adverse events	41.8%	37.8%

the NEST-1 trial except the primary endpoint was the mRS. Because of the inconclusive results of overall analysis, NIHSS was also used for additional analysis to stratify and eventually tease out significance (i.e., stratification of results according to enrollment baseline). The mRS endpoint criteria used in the NEST-2 trial measured primary efficacy outcome on a dichotomous mRS scoring scale evaluated 90 days after TLT administration with efficacy success measured as mRS 0, 2 and failure or lack of efficacy as mRS 3, 6.

Surprisingly, NEST-2 efficacy did not reach statistical significance for all patients enrolled in the trial ($P = .094$) (Table 23.2 and Fig. 23.1), and the trial was not considered to be positive. This finding was quite unexpected in light of NEST-1 trial results. However, post-hoc analysis indicated that it only moderately affected AIS patients with NIHSS 7–15 upon enrollment achieved better performance at 90 days. This did reach statistical significance for the particular patient group ($P = .044$). Thus, using prespecified subgroup analyses, the study found that a subgroup with NIHSS 7–15 benefited from the TLT, but patients with high baseline NIHSS scores were not improved; possibly due to the severity of brain damage and lack of penumbral substrate to save.

23.2.3 NeuroThera effectiveness and safety trial-3

The NEST-3 clinical trial design reported in the literature is similar to the NEST-2 clinical trial (see Table 23.2) (Hacke et al., 2014). The trial proposed to enroll 1000 patients within 24 hours of a stroke with a NIHSS baseline score between 7 and 17, the range in which there was a trend for beneficial effects of TLT observed in NEST-1 and subgroup of NEST-2 (Lampl et al., 2007; Zivin et al., 2009), but more specifically in the NEST-2 trial where TLT did not offer benefit in stroke patients with an NIHSS baseline score >16 points. However, the trial was fraught with problems (see Hacke et al., 2014), and only 630 patients were randomized to two groups (Table 23.3 and Fig. 23.1).

Oddly, but probably a sign of the times, the NEST-3 trial design did not include patients treated with the standard-of-care therapy, tPA (within 3–4.5 hours of a stroke), because the TT chosen by the steering committee investigators was 4.5–24 hours, outside of the range of efficacious tPA administration. Moreover, since this trial was conducted between 2010 and 2012, this precluded study with now approved state-of-the art embolectomy or thrombectomy, endovascular approaches effectively remove the embolus in large vessel-occluded patients. This has now proven to be beneficial in a select patient population as demonstrated in a series of recent trials (Berkhemer et al., 2015; Goyal et al., 2015; Campbell et al., 2015; Jovin et al., 2015; Saver et al., 2015; Rebello et al., 2016; Lapchak, 2015b).

The result of NEST-3 was disappointing to patients, and the stroke community at large, but it may have been expected. There is an ominous statement made by Hacke et al. (2014) in the stroke paper reporting the results of the

TABLE 23.3 NEST-3 trial enrollment and outcome.

Criteria	Treatment groups-double-blind and randomized	
	NEST-3 (Hacke et al., 2014)	
	630 Patients	
	Placebo (316)	TLT (314)
Age range	65 + 11 (mean 65)	66 + 10 (mean 66)
Sex	199 male	198 male
	117 female	116 female
Time to treatment (TT) (hours:minutes)	Prespecified groups <8 h, >8–16 h, >16–24	
Stroke severity • NIHSS	(mean 10)	(mean 10)
Stroke outcome • mRS 0, 2 • mRS 3, 6	46.1% 46.9%	48.1% 47.1%
Mortality rate	6.1%	4.8%
Serious adverse events	28.0%	20.9%

study. The authors state, "that only a few of the 20 skull sites used for light application are likely to be adjacent to penumbral tissue...requires a certain degree of credulity." Thus, after the third trial, the developers and investigators even questioned their own rationale for the 20 spot treatment of the stroke patient "skull" and "brain." Moreover, there are other significant problems with attempting to irradiate the human brain with PBM transcranially, and they will be discussed in Section 23.4.

23.3 Translational stroke research in the embolic stroke rabbit model

TLT has been studied and tested extensively in New Zealand white rabbits using the rabbit small clot embolic stroke model (RSCEM) (Lapchak et al., 2004) per current research guidelines for the conduct and development of therapies to treat stroke (Lapchak, 2013, 2017; Lapchak et al., 2013; Landis et al., 2012). During the initial development of TLT, the RSCEM was used because it has multiple advantages over ischemia models used in traditional translational research. These include:

- The use of nonautologous blood clots for embolization. This is important because there has yet to be a human case of stroke caused by the placement of a suture in the middle cerebral artery reported in the peer-reviewed literature.
- Embolization or clot-push into the brain occurs in the absence of anesthetics, which avoids the possibility of combination therapy effects.
- Inclusion of a heterogeneous population of stroke subjects and a well-defined clinically relevant behavioral endpoint (Lapchak, 2010b, 2011, 2015a; Turner et al., 2011; Lapchak and Boitano, 2017; Lyden and Lapchak, 2012) (see also Zivin and Grotta, 1990; Zivin et al., 1985).

23.3.1 Preclinical efficacy

TLT has been shown to be efficacious as a method to promote neuronal function following a stroke based upon translational studies measuring clinical function. This novel method of behavioral improvement was first demonstrated in the RSCEM (Lapchak et al., 2004) using TLT in a continuous wave pattern (power density of 7.5 mW/cm^2 and a fixed cortical fluence of 0.9 J/cm^2), similar to that used in NEST-1 to NEST-3 (Lapchak, 2010a; Lapchak and De Taboada, 2010; Lapchak et al., 2007; De Taboada et al., 2011; Lapchak and Boitano, 2016a). Initial studies to provide proof of concept PBM efficacy showed that TLT was effective when applied transcranially following embolization

(Lapchak et al., 2004), but this data has an interesting caveat related to skull thickness that was not discussed in the original article (see Section 23.4). The TT widow in embolized rabbits was wide (up to 6 hours), but shorter than that used in the NEST clinical trials (up to 24 hours). When TLT-induced behavioral improvement, the effect was sustained for up to 21 days after a single TLT treatment (Lapchak et al., 2004, 2007).

23.4 What went wrong in NeuroThera effectiveness and safety trials?

While statistically significant efficacy was noted in NEST-1 and in a subset of NEST-2 patients using low power density CW TLT. There is no evidence that TLT was ever optimized for use in humans. Photon transmission characteristics through the skull have been a topic of some debate for many years, and still is an important area of research.

Initial transmission study experience included TLT penetration of the rabbit skull and brain. The studies as reported by Lapchak et al. (2004) indicated that the "thin" skull allowed penetration up to a depth of 25–30 mm, and the TLT penetration would encompass the majority of the brain if placed on the skin surface posterior to bregma at midline. However, using rats, with thin skulls, another study (De Taboada et al., 2006) suggested that there would be an estimated power density drop from 10 mW/cm^2 (1.2 J/cm^2) at the cortex of the rat brain to approximately 7.5 μW/cm^2 (0.9 mJ/cm^2) at 18 mm depth from the cortical surface.

Even though the studies described above in published reports demonstrated that 800 nm CW TLT can penetrate the rat and rabbit skull and brain (Zhang et al., 2000; De Taboada et al., 2006; Ilic et al., 2006; Lapchak et al., 2004), a recent systematic multispecies study demonstrated emphasized that insufficient levels of laser light penetrates the thick human skull (Lapchak et al., 2015; Tedford et al., 2015) at the power density used for clinical stroke studies. The decrease in laser light penetration has also been emphasized by a number of researchers in the field (Yaroslavsky et al., 2002; Pitzschke et al., 2015a,b).

In the NEST-2 trial publication, the investigators also indicate that TLT will only penetrate the brain to approximately 20 mm depth (Zivin et al., 2009) using the CW design (10 mW/cm^2; 1.2 J/cm^2), and with the manual placement of the probe at 20 prespecified positions, there may be insufficient coverage of all structures underlying the cortex with CW TLT. Readers should note that this is even documented in the NEST-3 clinical trial paper (Hacke et al., 2014). This is also discussed by Wan et al. (1981) and Naeser et al. (2011). The investigators mathematically calculated that a power density of 8 J/cm^2 applied to the head will only allow for penetration of 2%–3% of photons 1 cm from the scalp surface, and only 0.2%–0.3% of photons will reach 2 cm depth. Thus, the overall penetration of CW administered photons to significant depths in the human brain is almost negligible. The multispecies skull penetration conducted by Lapchak and colleagues (Lapchak et al., 2015) documented the extensive decrease in transmission in one studies *mouse* > *rat* > *rabbit* > *human* skulls. Using skulls from three animal species, using a wavelength of 800 nm and a surface power density of 700 mW/cm^2, TLT decreased from 40.10% (mouse) to 21.24% (rat) to 11.36% (rabbit) as skull thickness measured at bregma, increased from 0.44 mm in mouse to 0.83 mm in rat and then 2.11 mm in rabbit. Furthermore, with human calvaria, where the mean thickness ranged from 5.91 to 7.19 mm, only 4.18%–4.24% of applied photons were transmitted through the skull.

Thus, perhaps it is not surprising that the NEST trials were met with lack of significant efficacy when the three studies are viewed in retrospect, and the investigators had the photon transmission knowledge at their disposal, but did not optimize the TLT delivery treatment regimen to ensure that photons could be delivered in a reproducible way to all all cortical layers and subcortical structures that would be damaged by ischemia.

23.5 Conclusions and commentary: should transcranial laser therapy be further considered as an approach to treat stroke?

Extensive preclinical data from a community of investigators suggests that TLT has bona fide effects on brain that promote behavioral recovery in neurodegeneration models (Thunshelle and Hamblin, 2016; Hamblin, 2016; Ando et al., 2013; Huang et al., 2012; Lapchak et al., 2017; Lapchak, 2010a, 2012). Since it is clear that 808 nm CW TLT does not effectively penetrate the human skull, novel methods to apply photons to repair damaged brain tissue are required.

Most preclinical stroke studies used a single CW TLT treatment between 5 minutes and 6 hours following a stroke, and data showed that this treatment regimen could promote long-lasting behavioral recovery for up to 21 days (Lapchak et al., 2004). However, the NEST clinical trials, which also used a single CW treatment within 24 hours of a stroke, also used standardized stroke scales to measure efficacy at a terminal 90-day point. The reasoning and scientific rationale for these choices has not been clearly documented, since there was no preclinical basis for the use of a single dose with an extended measure behavioral or clinical endpoint.

Rationale design for the next TLT study needs to address:

- *TLT dosing: CW or pulse mode (improved penetration profile?)*
- *TLT dosing: single dose or repeated dosing (frequency)*
- *TLT in combination with standard-of-care therapy: rt-PA and/or embolectomy (benefit of neuroprotection and enhanced NO and CBF with reperfusion therapies)*
- *Baseline enrollment: clinical scale (NIHSS and/or mRS) and salvageable penumbra (Alberta stroke program early CT score (ASPECTS) (Liebeskind et al., 2014; Pexman et al., 2001)*
- *Study endpoint: clinical scale (NIHSS and/or mRS) and/or penumbra/infact size*

Thus, while TLT should be further pursued for stroke, a rationally-designed trial and method of photon application needs to be developed based upon preclinical and translational data.

References

Ad, N., Oron, U., 2001. Impact of low level laser irradiation on infarct size in the rat following myocardial infarction. Int. J. Cardiol. 80, 109–116.

Agrawal, T., Gupta, G.K., Rai, V., Carroll, J.D., Hamblin, M.R., 2014. Pre-conditioning with low-level laser (light) therapy: light before the storm. Dose Response 12, 619–649.

Anders, J.J., Borke, R.C., Woolery, S.K., VAN De merwe, W.P., 1993. Low power laser irradiation alters the rate of regeneration of the rat facial nerve. Lasers Surg. Med. 13, 72–82.

Ando, T., Sato, S., Kobayashi, H., Nawashiro, H., Ashida, H., Hamblin, M.R., et al., 2013. Low-level laser therapy for spinal cord injury in rats: effects of polarization. J. Biomed. Opt. 18, 098002.

Berkhemer, O.A., Fransen, P.S.S., Beumer, D., et al., 2015. A randomized trial of intraarterial treatment for acute ischemic stroke. N. Engl. J. Med. 372, 11–20.

Berthier, E., Decavel, P., Vuillier, F., Verlut, C., Moulin, T., DE Bustos medeiros, E., 2013. Reliability of NIHSS by telemedicine in non-neurologists. Int. J. Stroke 8, E11.

Byrnes, K.R., Waynant, R.W., Ilev, I.K., Wu, X., Barna, L., Smith, K., et al., 2005. Light promotes regeneration and functional recovery and alters the immune response after spinal cord injury. Lasers Surg. Med. 36, 171–185.

Castro-e-Silva JR., O., Zucoloto, S., Marcassa, L.G., Marcassa, J., Kurachi, C., Melo, C.A., et al., 2003. Spectral response for laser enhancement in hepatic regeneration for hepatectomized rats. Lasers Surg. Med. 32, 50–53.

Campbell, B.C., Mitchell, P.J., Kleinig, T.J., Dewey, H.M., Churilov, L., Yassi, N., et al., 2015. Endovascular therapy for ischemic stroke with perfusion-imaging selection. N. Engl. J. Med. 372, 1009–1018.

Chang, W.H., Sohn, M.K., Lee, J., Kim, D.Y., Lee, S.G., Shin, Y.I., et al., 2017. Long-term functional outcomes of patients with very mild stroke: does a NIHSS score of 0 mean no disability? An interim analysis of the KOSCO study. Disabil. Rehabil. 39, 904–910.

Chung, H., Dai, T., Sharma, S.K., Huang, Y.Y., Carroll, J.D., Hamblin, M.R., 2012. The nuts and bolts of low-level laser (light) therapy. Ann. Biomed. Eng. 40, 516–533.

De Taboada, L., Ilic, S., Leichliter-Martha, S., Oron, U., Oron, A., Streeter, J., 2006. Transcranial application of low-energy laser irradiation improves neurological deficits in rats following acute stroke. Lasers Surg. Med. 38, 70–73.

De Taboada, L., Yu, J., El-Amouri, S., Gattoni-Celli, S., Richier, I.S., Mccarth, Y.T., et al., 2011. Transcranial laser therapy attenuates amyloid-β peptide neuropathology in amyloid-β protein precursor transgenic mice. J. Alzheimer's Dis. 23, 521–535.

Desmet, K.D., Paz, D.A., Corry, J.J., Eells, J.T., Wong-Riley, M.T., Henry, M.M., et al., 2006. Clinical and experimental applications of NIR-LED photobiomodulation. Photomed. Laser Surg. 24, 121–128.

Drochioiu, G., 2010. Laser-induced ATP formation: mechanism and consequences. Photomed. Laser Surg. 28, 573–574.

Dubuc, V., Choi, P., Hill, M.D., Coutts, S.B., 2015. Can 90-day NIHSS be used for outcome assessment in TIA and minor stroke studies? Cerebrovasc. Dis. 40, 97–98.

Eells, J.T., 2003. Therapeutic photobiomodulation for methanol-induced retinal toxicity. Proc. Natl. Acad. Sci. U.S.A. 100, 3439.

Fernandes, K.P., Souza, N.H., Mesquita-Ferrari, R.A., Silva dde, F., Rocha, L.A., Alves, A.N., et al., 2015. Photobiomodulation with 660-nm and 780-nm laser on activated J774 macrophage-like cells: effect on M1 inflammatory markers. J. Photochem. Photobiol. B 153, 344–351.

Ferraresi, C., Kaippert, B., Avci, P., Huang, Y.Y., De sousa, M.V., Bagnato, V.S., et al., 2015. Low-level laser (light) therapy increases mitochondrial membrane potential and ATP synthesis in C2C12 myotubes with a peak response at 3-6 h. Photochem. Photobiol. 91, 411–416.

Goyal, M., Demchuk, A.M., Menon, B.K., Eesa, M., Rempel, J.L., Thornton, J., et al., 2015. Randomized assessment of rapid endovascular treatment of ischemic stroke. N. Engl. J. Med. 372, 1019–1030.

Hacke, W., Schellinger, P.D., Albers, G.W., Bornstein, N.M., Dahlof, B.L., Fulton, R., et al., 2014. Transcranial laser therapy in acute stroke treatment: results of Neurothera Effectiveness and Safety Trial 3, a Phase III Clinical End Point Device Trial. Stroke 45, 3187–3193.

Hamblin, M.R., 2010. Introduction to experimental and clinical studies using low-level laser (light) therapy (LLLT). Lasers Surg. Med. 42, 447–449.

Hamblin, M.R., 2016. Photobiomodulation or low-level laser therapy. J. Biophotonics 9, 1122–1124.

Huang, Y.Y., Chen, A.C., Carroll, J.D., Hamblin, M.R., 2009. Biphasic dose response in low level light therapy. Dose Response 7, 358–383.

Huang, Y.Y., Gupta, A., Vecchio, D., De arce, V.J., Huang, S.F., Xuan, W., et al., 2012. Transcranial low level laser (light) therapy for traumatic brain injury. J. Biophotonics 5, 827–837.

Ilic, S., Leichliter, S., Streeter, J., Oron, A., De Taboada, L., Oron, U., 2006. Effects of power densities, continuous and pulse frequencies, and number of sessions of low-level laser therapy on intact rat brain. Photomed. Laser Surg. 24, 458–466.

Jovin, T.G., Chamorro, A., Cobo, E., De miquel, M.A., Molina, C.A., Rovira, A., et al., 2015. Thrombectomy within 8 hours after symptom onset in ischemic stroke. N. Engl. J. Med. 372, 2296–2306.

Karu, T., 2010. Mitochondrial mechanisms of photobiomodulation in context of new data about multiple roles of ATP. Photomed. Laser Surg. 28, 159–160.

Kwah, L.K., Diong, J., 2014. National Institutes of Health Stroke Scale (NIHSS). J. Physiother. 60, 61.

Lampl, Y., Zivin, J.A., Fisher, M., Lew, R., Welin, L., Dahlof, B., et al., 2007. Infrared laser therapy for ischemic stroke: a new treatment strategy: results of the NeuroThera effectiveness and safety trial-1 (NEST-1). Stroke 38, 1843–1849.

Landis, S.C., Amara, S.G., Asadullah, K., Austin, C.P., Blumenstein, R., Bradley, E.W., et al., 2012. A call for transparent reporting to optimize the predictive value of preclinical research. Nature 490, 187–191.

Lapchak, P.A., 2010a. Taking a light approach to treating acute ischemic stroke patients: transcranial near-infrared laser therapy translational science. Ann. Med. 42, 576–586.

Lapchak, P.A., 2010b. Translational stroke research using a rabbit embolic stroke model: a correlative analysis hypothesis for novel therapy development. Transl. Stroke Res. 1, 96–107.

Lapchak, P.A., 2011. A clinically relevant rabbit embolic stroke model for acute ischemic stroke therapy development: mechanisms & targets. In: Lapchak, P.A., Zhang, J.H. (Eds.), Translational Stroke Research: From Target Selection to Clinical Trials. Springer, USA.

Lapchak, P.A., 2012. Transcranial near-infrared laser therapy applied to promote clinical recovery in acute and chronic neurodegenerative diseases. Expert Rev. Med. Devices 9, 71–83.

Lapchak, P.A., 2013. Recommendations and practices to optimize stroke therapy: developing effective translational research programs. Stroke 44, 841–843.

Lapchak, P.A., 2015a. A cost-effective rabbit embolic stroke bioassay: insight into the development of acute ischemic stroke therapy. Transl. Stroke Res. 6, 99–103.

Lapchak, P.A., 2015b. Critical early thrombolytic and endovascular reperfusion therapy for acute ischemic stroke victims: a call for adjunct neuroprotection. Transl. Stroke Res. 6, 345–354.

Lapchak, P.A., 2017. Translational stroke research opportunities and a strategy to develop effective cytoprotection. Transl. Stroke. Res. 8, 318–321.

Lapchak, P.A., Araujo, D.M., 2007. Advances in ischemic stroke treatment: neuroprotective and combination therapies. Expert Opin. Emerg. Drugs 12, 97–112.

Lapchak, P.A., Boitano, P.D., 2016a. A novel method to promote behavioral improvement and enhance mitochondrial function following an embolic stroke. Brain Res 1646, 125–131.

Lapchak, P.A., Boitano, P.D., 2016b. Transcranial near-infrared laser therapy for stroke: how to recover from futility in the NEST-3 clinical trial. Acta Neurochir. Suppl. 121, 7–12.

Lapchak, P.A., Boitano, P.D., 2017. Reflections on neuroprotection research and the path toward clinical success. In: Lapchak, P.A., Zhang, J.H. (Eds.), Neuroprotective Therapy for Stroke and Ischemic Disease. Springer International Publishing, Cham.

Lapchak, P.A., De Taboada, L., 2010. Transcranial near infrared laser treatment (NILT) increases cortical adenosine-5′-triphosphate (ATP) content following embolic strokes in rabbits. Brain Res. 1306, 100–105.

Lapchak, P.A., Wei, J., Zivin, J.A., 2004. Transcranial infrared laser therapy improves clinical rating scores after embolic strokes in rabbits. Stroke 35, 1985–1988.

Lapchak, P.A., Salgado, K.F., Chao, C.H., Zivin, J.A., 2007. Transcranial near-infrared light therapy improves motor function following embolic strokes in rabbits: an extended therapeutic window study using continuous and pulse frequency delivery modes. Neuroscience 148, 907–914.

Lapchak, P.A., Han, M.K., Salgado, K.F., Streeter, J., Zivin, J.A., 2008. Safety profile of transcranial near-infrared laser therapy administered in combination with thrombolytic therapy to embolized rabbits. Stroke 39, 3073–3078.

Lapchak, P.A., Zhang, J.H., Noble-Haeusslein, L.J., 2013. RIGOR guidelines: escalating STAIR and STEPS for effective translational research. Transl. Stroke Res. 4, 279–285.

Lapchak, P.A., Boitano, P.D., Butte, P.V., Fisher, D.J., Holscher, T., Ley, E.J., et al., 2015. Transcranial Near-Infrared Laser Transmission (NILT) Profiles (800 nm): systematic comparison in four common research species. PLoS One 10, e0127580.

Lapchak, P.A., Butte, P., Rajput, P.S., 2017. Difficult path to treating acute ischemic stroke patients with transcranial near-infrared laser therapy. In: Hamblin, M.R., De sousa, M.V.P., Agrawal, T. (Eds.), Handbook of Low-Level Laser Therapy. Pan Stanford Publishing Pte, Great Britain.

Leira, E.C., Coffey, C.S., Jorge, R.E., Morton, S.M., Froehler, M.T., Davis, P.H., et al., 2013. The NIHSS supplementary motor scale: a valid tool for multidisciplinary recovery trials. Cerebrovasc. Dis. 36, 69–73.

Liebeskind, D.S., Jahan, R., Nogueira, R.G., Jovin, T.G., Lutsep, H.L., Saver, J.L., et al., 2014. Serial Alberta Stroke Program early CT score from baseline to 24 hours in Solitaire Flow Restoration with the Intention for Thrombectomy study: a novel surrogate end point for revascularization in acute stroke. Stroke 45, 723–727.

Lyden, P., Lapchak, P., 2012. Sisyphus and translational stroke research. Sci. Transl. Med. 4, 156ps20.

Mochizuki-Oda, N., Kataoka, Y., Cui, Y., Yamada, H., Heya, M., Awazu, K., 2002. Effects of near-infra-red laser irradiation on adenosine triphosphate and adenosine diphosphate contents of rat brain tissue. Neurosci. Lett. 323, 207–210.

Naeser, M.A., Hamblin, M.R., 2011. Potential for transcranial laser or LED therapy to treat stroke, traumatic brain injury, and neurodegenerative disease. Photomed. Laser Surg. 29, 443–446.

Naeser, M.A., Saltmarche, A., Krengel, M.H., Hamblin, M.R., Knight, J.A., 2011. Improved cognitive function after transcranial, light-emitting diode treatments in chronic, traumatic brain injury: two case reports. Photomed. Laser Surg. 29, 351–358.

Naess, H., Kurtz, M., Thomassen, L., Waje-Andreassen, U., 2016. Serial NIHSS scores in patients with acute cerebral infarction. Acta Neurol. Scand. 133, 415–420.

Nilanont, Y., Komoltri, C., Saposnik, G., Cote, R., Di legge, S., Jin, Y., et al., 2010. The Canadian Neurological Scale and the NIHSS: development and validation of a simple conversion model. Cerebrovasc. Dis. 30, 120–126.

Nissan, M., Rochkind, S., Razon, N., Bartal, A., 1986. HeNe laser irradiation delivered transcutaneously: its effect on the sciatic nerve of rats. Lasers Surg. Med. 6, 435–438.

Olavarria, V.V., Delgado, I., Hoppe, A., Brunser, A., Carcamo, D., Diaz-Tapia, V., et al., 2011. Validity of the NIHSS in predicting arterial occlusion in cerebral infarction is time-dependent. Neurology 76, 62–68.

Oron, U., Yaakobi, T., Oron, A., Mordechovitz, D., Shofti, R., Hayam, G., et al., 2001. Low-energy laser irradiation reduces formation of scar tissue after myocardial infarction in rats and dogs. Circulation 103, 296–301.

Pexman, J.H., Barber, P.A., Hill, M.D., Sevick, R.J., Demchuk, A.M., Hudon, M.E., et al., 2001. Use of the Alberta Stroke Program Early CT Score (ASPECTS) for assessing CT scans in patients with acute stroke. AJNR Am. J. Neuroradiol. 22, 1534–1542.

Pitzschke, A., Lovisa, B., Seydoux, O., Haenggi, M., Oertel, M.F., Zellweger, M., et al., 2015a. Optical properties of rabbit brain in the red and near-infrared: changes observed under in vivo, postmortem, frozen, and formalin-fixated conditions. J. Biomed. Opt. 20, 25006.

Pitzschke, A., Lovisa, B., Seydoux, O., Zellweger, M., Pfleiderer, M., Tardy, Y., et al., 2015b. Red and NIR light dosimetry in the human deep brain. Phys. Med. Biol. 60, 2921–2937.

Rebello, L.C., Haussen, D.C., Grossberg, J.A., Belagaje, S., Lima, A., Anderson, A., et al., 2016. Early endovascular treatment in intravenous tissue plasminogen activator-ineligible patients. Stroke 47, 1131–1134.

Sartor, E.A., Albright, K., Boehme, A.K., Morales, M.M., Shaban, A., Grotta, J.C., et al., 2013. The NIHSS score and its components can predict cortical stroke. J. Neurol. Disord. Stroke 2, 1026.

Saver, J.L., Goyal, M., Bonafe, A., Diener, H.C., Levy, E.I., Pereira, V.M., et al., 2015. Stent-retriever thrombectomy after intravenous t-PA vs. t-PA alone in stroke. N. Engl. J. Med. 372, 2285–2295.

Tatmatsu-Rocha, J.C., Ferraresi, C., Hamblin, M.R., Damasceno maia, F., Do nascimento, N.R., Driusso, P., et al., 2016. Low-level laser therapy (904nm) can increase collagen and reduce oxidative and nitrosative stress in diabetic wounded mouse skin. J. Photochem. Photobiol. B 164, 96–102.

Tedford, C.E., Delapp, S., Jacques, S., Anders, J., 2015. Quantitative analysis of transcranial and intraparenchymal light penetration in human cadaver brain tissue. Lasers Surg. Med. 47, 312–322.

Thunshelle, C., Hamblin, M.R., 2016. Transcranial low-level laser (light) therapy for brain injury. Photomed. Laser Surg. 34, 587–598.

Turcato, G., Cervellin, G., Cappellari, M., Bonora, A., Zannoni, M., Bovi, P., et al., 2017. Early function decline after ischemic stroke can be predicted by a nomogram based on age, use of thrombolysis, RDW and NIHSS score at admission. J. Thromb. Thrombolysis 43, 394–400.

Turner, R., Jickling, G., Sharp, F., 2011. Are underlying assumptions of current animal models of human stroke correct: from STAIRS to high hurdles? Transl. Stroke Res. 2, 138–143.

Uozumi, Y., Nawashiro, H., Sato, S., Kawauchi, S., Shima, K., Kikuchi, M., 2010. Targeted increase in cerebral blood flow by transcranial near-infrared laser irradiation. Lasers Surg. Med. 42, 566–576.

Wan, S., Parrish, J.A., Anderson, R.R., Madden, M., 1981. Transmittance of nonionizing radiation in human tissues. Photochem. Photobiol. 34, 679–681.

Wang, X., Tian, F., Reddy, D.D., Nalawade, S.S., Barrett, D.W., Gonzalez-Lima, F., et al., 2017. Up-regulation of cerebral cytochrome-c-oxidase and hemodynamics by transcranial infrared laser stimulation: a broadband near-infrared spectroscopy study. J. Cereb. Blood Flow Metab. 37, 3789–3802.

Wu, Q., Huang, Y.-Y., Dhital, S., Sharma, S.K., Chen, A.C.-H., Whalen, M.J., et al., 2010. Low level laser therapy for traumatic brain injury. SPIE Proc. 7552, 755206-1-8.

Xuan, W., Vatansever, F., Huang, L., Hamblin, M.R., 2014. Transcranial low-level laser therapy enhances learning, memory, and neuroprogenitor cells after traumatic brain injury in mice. J. Biomed. Opt. 19, 108003.

Xuan, W., Huang, L., Hamblin, M.R., 2016. Repeated transcranial low-level laser therapy for traumatic brain injury in mice: biphasic dose response and long-term treatment outcome. J. Biophotonics 9, 1263–1272.

Yaroslavsky, A.N., Schulze, P.C., Yaroslavsky, I.V., Schober, R., Ulrich, F., Schwarzmaier, H.J., 2002. Optical properties of selected native and coagulated human brain tissues in vitro in the visible and near infrared spectral range. Phys. Med. Biol. 47, 2059–2073.

Zhang, Q., Ma, H., Nioka, S., Chance, B., 2000. Study of near infrared technology for intracranial hematoma detection. J. Biomed. Opt. 5, 206–213.

Zivin, J.A., Grotta, J.C., 1990. Animal stroke models. They are relevant to human disease. Stroke 21, 981–983.

Zivin, J.A., Fisher, M., Degirolami, U., Hemenway, C.C., Stashak, J.A., 1985. Tissue plasminogen activator reduces neurological damage after cerebral embolism. Science 230, 1289.

Zivin, J.A., Albers, G.W., Bornstein, N., Chippendale, T., Dahlof, B., Devlin, T., et al., 2009. Effectiveness and safety of transcranial laser therapy for acute ischemic stroke. Stroke 40, 1359–1364.

Chapter 24

Effects of photobiomodulation on traumatic brain injury: proposed clinical assessment

Sherry Fox[1] and Victoria Campbell[2]

[1]*BioCare Systems, Inc., Parker, CO, United States, Colorado BioScience Association, Denver, CO, United States, National Association of Laser Therapy, Baltimore, MD, United States, LumiWave NIR Therapy Device, Operation Stand Tall Against TBI – A Non-Profit Organization, Calhan, CO, United States,* [2]*RaVive Health, Inc., Operation Stand Tall Against TBI – A Non-Profit Organization, Calhan, CO, United States, National Association of Social Workers, Washington, D.C., United States, American Psychological Association, Washington, D.C., United States, Campbell Method for Treating TBI*

24.1 Introduction

Concussion, or mild traumatic brain injury (mTBI), has become a headline issue in the world of sports and the military today. More than 250,000 service members suffered a traumatic brain injury (TBI) between 2000 and 2012 according to the Veterans Affairs Department (Gerberding and Binder, 2015). mTBI is caused by sudden concussive impact events on the brain. Resulting damage includes severed neuron synapse connections, neuronal apoptosis, microbleeding, and ischemia. Clinical manifestations of damage include cognitive impairment, mood disorders, headaches, disrupted sleep patterns, and inappropriate behaviors (Tanielian and Jaycox, 2008). Concussion and mTBI have been shown to reduce blood flow to the brain (Bonne, 2003; Maugans et al., 2012). A correlation has been shown in studies between blood flow and brain function; impeded blood flow reduces the brain's ability to operate (Villringer and Dirnagle, 1995). Estimates are that from 10% to 30% of concussion patients experience prolonged symptoms, described as post-concussion syndrome (PCS). Some PCS patients have persistent hypoperfusion in concussed areas of the brain as revealed by single-photon emission computed tomography (Agrawal et al., 2005). Such hypoperfusion may impede neuronal function by starving cells of glucose and oxygen, thus impacting mitochondrial function and cell respiration.

As detailed in an independent study, "Acute Effects of Near-Infrared Light Therapy on Brain State in Healthy Subjects," the diagnosis of concussion is one of the main challenges faced by neurologists.

> *The study indicates that quantitative electroencephalography (qEEG) has been shown to be a viable tool for the diagnosis of concussion/mTBI based on several measurable factors. These measures include P300 brain speed, reaction time, and amplitude. Amplitude is a measure of the voltage levels of neural cells. The typical amplitude for healthy functioning cells as measured from the scalp can range from 10 to 100 microvolts (lV). P300 is an event-related potential (ERP) component of EEG elicited in the process of decision making. The signal is often used as a metric of cognitive function in decision-making processes.*
>
> Grover et al. (2017)

Reaction time is an overall measure of the speed of the brain response to stimuli and reflects the response efficiency of the brain at a given time. Improvement in these measures over time or treatment may indicate brain state changes, which favor repair of concussive damage (Duff, 2004).

Photobiomodulation in the Brain. DOI: https://doi.org/10.1016/B978-0-12-815305-5.00024-5

24.2 Definition and statistics—traumatic brain injury

TBI can be defined as an event that occurs when an external mechanical force causes the brain to become dysfunctional (Savitsky et al., 2016). The diagnostic and statistical manual (DSM), version five (5), classifies TBI and its neuropsychiatric relatives within the framework of neurocognitive disorders (NCDs) (American Psychiatric Association (APA), 2013). Per the APA, the NCD's are "renamed and reframed" diagnostic criteria for all conditions within the framework except for delirium (APA, 2013). The NCDs are conditions in which the individual experiences impaired cognitive functioning, which is not the result of congenital or early developmental origin (APA, 2013). Within the DSM-5, TBI and its associated neuropsychiatric co-morbid conditions, such as mood distortion, anxiety, thought and impulse issues, substance use/abuse disorders, and a range of personality issues (Zuckerman et al., 2017), are all explored and taken into consideration in expanded detail (Wortzel and Arciniegas, 2014). The DSM-5 also provides criteria related to diagnosing an injury event as a TBI (APA, 2013).

Below are the specific DSM-5 diagnostic criteria for mild NCD:

1. Evidence of modest cognitive decline from a previous level of performance in one or more cognitive domains (complex attention, executive function, learning, and memory, language, perceptual-motor, or social cognition) based on:
   a. The concern of the individual, a knowledgeable informant, or the clinician that there has been a significant decline in cognitive function; and
   b. A substantial impairment in cognitive performance, preferably documented by standardized neuropsychological testing or, in its absence, another quantified clinical assessment.
2. The cognitive deficits do not interfere with independence in everyday activities (but the greater effort, compensatory strategies, or accommodation may be required).
3. The cognitive deficits do not occur exclusively in the context of a delirium.
4. Another mental disorder does not better explain the cognitive deficits (e.g., major depressive disorder, schizophrenia) (APA, 2013).

Below are the specific diagnostic criteria per the DSM-5 for major NCD:

1. Evidence of significant cognitive decline from a previous level of performance in one or more cognitive domains (complex attention, executive function, learning and memory, language, perceptual-motor, or social cognition) based on:
   a. The concern of the individual, a knowledgeable informant, or the clinician that there has been a significant decline in cognitive function; and
   b. A substantial impairment in cognitive performance, preferably documented by standardized neuropsychological testing or, in its absence, another quantified clinical assessment.
2. The cognitive deficits interfere with independence in everyday activities (i.e., at a minimum, requiring assistance with complex instrumental activities of daily living).
3. The cognitive deficits do not occur exclusively in the context of a delirium.
4. Another mental disorder does not better explain the cognitive deficits (e.g., major depressive disorder, schizophrenia) (APA, 2013).

Finally, below is the specific diagnostic criteria for major or mild NCD due to a TBI:

1. The criteria are met for major or mild NCD.
2. There is evidence of a TBI—that is, an impact to the head or other mechanisms of rapid movement or displacement of the brain within the skull, with one or more of the following:
   a. Loss of consciousness.
   b. Posttraumatic amnesia.
   c. Disorientation and confusion.
   d. Neurological signs (e.g., neuroimaging demonstrating injury; new onset of seizures; a marked worsening of a preexisting seizure disorder; visual field cuts; anosmia; hemiparesis).
3. The NCD presents immediately after the occurrence of the TBI or immediately after recovery of consciousness and persists past the acute postinjury period (APA, 2013).

Within the official diagnostic criteria for TBI is a rating scale, which depending on the variables will classify a TBI as mild, moderate, severe, or penetrating (Carlozzi et al., 2015).

The signs and symptoms of TBI vary from person to person, and between each incident of injury. The most common signs and symptoms are:

Headaches or neck pain that will not go away;
Difficulty remembering, concentrating, or making decisions;
Slowness in thinking, speaking, acting, or reading;
Getting lost or easily confused;
Feeling tired all the time, having no energy or motivation;
Mood changes (feeling angry for no reason);
Changes in sleep patterns (sleeping too much or too little);
Lightheadedness, dizziness, or loss of balance;
Urge to vomit (nausea);
Increased sensitivity to lights, sounds, or distractions;
Blurred vision or eyes that tire easily;
Loss of sense of smell or taste; and
Ringing in the ears (Kristman et al., 2014).

The signs and symptoms of a TBI in children are similar but not exclusive to the list mentioned above. Signs and symptoms often observed in children who have experienced a TBI event include:

Tiredness or listlessness;
Irritability or crankiness (unable to stop crying or be consoled);
Changes in eating;
Changes in the way that the child plays;
Differences in performance at school;
Lack of interest in favorite toys or activities;
Loss of new skills, such as potty training;
Loss of balance or unsteady walking; or
Vomiting (Babcock et al., 2013).

24.3 Developmental aspects

When exploring the developmental aspects and ramifications that could be manifested after a TBI incident, it is essential to take into consideration at what age the injury occurred and what was its severity. One study examined the prognosis of children within four age groups to explore the ramifications of a TBI from a perspective of whether children in one group compared to another were able to show greater recovery based on their age when the injury occurred (Garcia et al., 2015). Initially, the hypothesis was that children in the younger age groups would demonstrate a poorer outcome for recovery; in actuality, what was found, was that children in the middle age group established the poorest outcomes. The researchers hypothesized that this corresponded with a critically important period for brain development and cognitive development at the time when the brain sustained its injury (Garcia et al., 2015).

From the information that is available through research, and from a developmental aspect, much of the brain's ability to recover naturally, post incident, is dependent on the age of the individual and the severity of the injury; not to mention the importance of providing interventions that would assist with ongoing developmental challenges resulting from a childhood TBI.

24.4 Physiological components

From a physiological point of view, a TBI is a nondegenerative, noncongenital insult to the brain resulting from a peripheral mechanical force that may well potentially lead to permanent or transient impairment, of a host of cerebral functions, and may or may not include a diminished or altered state of consciousness (Narotam et al., 2014). In relation to physiological components that could be altered as the result of a TBI, and specific to vision capabilities, the patient can potentially suffer partial or total loss of vision, weakness of the eye muscles and/or double vision, blurred vision, difficulty judging distance, nystagmus, or photophobia (Petraglia et al., 2014).

From an auditory perspective, when someone experiences a traumatic event that could result in a TBI, they may also experience physiological deficits such as a decrease or loss of hearing, tinnitus, or increased sensitivity to sound (Astafiev et al., 2016). The patient's sense of smell and taste could also be diminished (Astafiev et al., 2016).

Finally, from a physiological perspective, the patient may experience seizure activity, physical paralysis or spasticity, chronic pain, problems with bowel and bladder control, sleep issues, loss of stamina, challenges in regulating their body temperature, and potential menstrual difficulties (Astafiev et al., 2016). This is by no means an exhaustive list of physiological aspects related to the incurrence of a TBI, keeping in mind that all traumatic brain injuries are different and how the injury impacts the patient is also individualized.

24.5 Psychological manifestations

The psychological manifestations related to a TBI impact not only the patient but also the patient's extended support systems. There are both behavioral and emotional consequences of TBI that can include personality changes, anger and frustration issues, emotional distress, difficulties in social relations, and executive function difficulties (Bolzenius et al., 2015).

One of the challenges, when patients incur a TBI, is that this is considered to be an invisible injury, and yet the ramifications are still far-reaching. Families often experience confusion over the personality changes that occur with their loved one; overnight a patient could go from being a person who was outgoing and active, to a person who is withdrawn and could even be experiencing suicidal ideation. For the patient, the inability to manage their frustration and anger combined with a general reduction in patience and tolerance can exacerbate the physiological aspects of the injury.

Emotional distress can be observed as symptoms of depression or anxiety that can commonly follow a TBI (Andruszkow et al., 2014). The emotional changes could result directly from the area of the brain that incurred the injury and the subsequent changes that occurred as a result of the injury (Rosenberg et al., 2015). To take into consideration, it must also be conceptualized that the emotional distress may result in issues such as problems adjusting to cognitive and behavioral challenges that result post-TBI (Lengenfelder et al., 2015). From an emotional aspect, anxiety may be discernible as general nervousness or restlessness such as in a panic attack (Rosenberg et al., 2015).

24.6 Sociological implications

Social communication and the impact on social relationships in the aftermath of a TBI is an area that requires a sense of conceptualization. This area of challenge can create a barrier to recovery for the patient, further stressing the relationships that exist within the patient's support channels. Social communication requires several skills that often function in a collaborative nature. These skills include both verbal and nonverbal abilities, social communication, being able to start and end a conversation, being able to select and change conversation topics, turn-taking, asking for clarification, maintaining eye contact, speaking at an appropriate rate, and using gestures (Togher et al., 2016).

Patients with a TBI could experience problems with skills coupled with effective interpersonal communication along with difficulties understanding nuances that occur within social relationships (Togher et al., 2016). For example, patients who have experienced a TBI may demonstrate a decrease in sensitivity to social norms and a decrease in sensitivity to the emotions of others (Neumann et al., 2014).

The cognitive challenges that are often related to a TBI, and that could be a factor to hinder social communication, include: attention and concentration issues, memory problems, executive functioning problems, and impaired social cognition (Gordon and Duff, 2016).

24.7 Causation

There are several ways in which a TBI can occur. Most frequently this injury is a result of a blow or jolt to the head or body (Couch and Stewart, 2015); however, an object penetrating the skull can also result in a TBI (Couch and Stewart, 2015).

Accurately assessing the prevalence of TBI is a difficult task. The Centers for Disease Control and Prevention report that there are approximately 1.5 million people a year within the United States who suffer from a TBI; roughly 50,000 individuals die from TBI each year, and approximately 85,000 people survive with long-term disabilities (Ma et al., 2014). Overall, the causes of TBI vary and are diverse in nature, with the most common contributing incidents resulting from car accidents, falls, and firearms (Roozenbeek et al., 2013). The cause of a TBI can be associated with the environment that the person is exposed to. For example, military members are more likely to incur a TBI from a blast exposure or fall, whereas civilian personnel are less likely to incur a TBI from an explosion and are more likely to incur a TBI from a motor vehicle accident.

When exploring the mechanisms of injury related to TBI, one must take into consideration the different types of brain injury: open head injury, closed head injury, deceleration injuries, chemical or toxic injury, hypoxia, tumors, infections, and stroke (Bramlett and Dietrich, 2015).

24.8 Treatment approaches

When exploring the treatment approaches that are currently available for addressing traumatic brain injuries, what is sadly evident is that presently there is no existing treatment that will heal the condition. The universal practice, which is the current standard within the professional field, is for practitioners to treat the behavioral and emotional issues that are the result of the condition (Bergersen et al., 2017). Behavioral management strategies are a conventional approach to treating TBI in combination with techniques to address stress and anger management difficulties (Bergersen et al., 2017). Emotional liability is another consideration when working to include the support systems for the individual who has incurred the injury. Emotional liability is the cost that can be measured and the impact that caring for a loved one with a TBI can have on the patient's support systems; techniques that address personal issues of self-control, impulsivity, and potential poor judgment, such as self-awareness from a patient perspective with regards to the changes that occur in the advent of a brain injury (McMillan and Wood, 2017). In addition to treating the behaviors and symptoms, the following are being explored as treatment options:

24.9 Most common treatments recommended

Those suffering from TBI are limited to the number of therapy options that are currently provided. Found below is a list of some of the more popular alternative therapies for treating posttraumatic stress disorder (PTSD) and TBI. Several are supported by substantial clinical evidence, while others currently lack rigorous scientific backing. It is critical to address not only the physical aspect of injury associated with TBI but also the psychological ramifications. Think of the brain as the hardware of a computer, when this mechanism is injured or broken it is unable to effectively incorporate the new software (skill based or psychological interventions) that it is provided.

Hyperbaric medicine, also known as hyperbaric oxygen therapy, is the medical use of oxygen at a level higher than atmospheric pressure. The equipment required consists of a pressure chamber, which may be of rigid or flexible construction, and a means of delivering 100% oxygen.

Transcranial magnetic stimulation (TMS) is a noninvasive method used to stimulate small regions of the brain. During a TMS procedure, a magnetic field generator, or coil, is placed near the head of the person receiving the treatment. The coil produces small electric currents in the region of the brain just under the coil via electromagnetic induction.

High-performance neurofeedback is a pulsed, extremely low-strength electric current that is effective in reducing the symptoms of anxiety, depression, TBI, PTSD, addiction, and attention deficit disorder.

Cranial electrotherapy stimulation is a form of noninvasive brain stimulation that applies a small, pulsed electric current across a person's head with the intention of treating a variety of conditions such as anxiety, depression, and insomnia.

Biofeedback is the process of gaining greater awareness of many physiological functions, primarily, using instruments that provide information on the activity of those same systems, with a goal of being able to manipulate them at will. Some of the processes that can be controlled include brainwaves, muscle tone, skin conductance, heart rate, and pain perception.

Pharmacological intervention is commonly used when managing the immediate and long-term consequences of such injuries; clinicians have many pharmacological options, including psychostimulants, antidepressants, antiparkinsonian agents, and anticonvulsants.

Photobiomodulation (PBM) is any of several therapeutic techniques that employ low-level laser or light-emitting diode light to relieve pain or to heal wounds.

Near infrared (NIR) light as a form of PBM has specific mechanisms by which tissue repair and regeneration is accelerated. The chief effect is a release of nitric oxide (NO) from the endothelial of vascular tissue and oxyhemoglobin. NO is a potent vasodilator, increasing blood flow and lymph flow to tissues (Wecksler et al., 2004).

This speeds the healing of damaged tissues and improves the nutrient environment of recovering tissues. Also, ATP production is triggered by NIR light. This is the direct energy source for running cellular functions, including repair and repair signaling.

There is also evidence for NIR triggering stem cell signaling mechanisms. Clinical observations of NIR light in wound healing show acceleration of the cell proliferation and modeling phases of healing. Minimal scarring is observed compared to untreated tissue repair, suggesting that the process is more efficient and orderly than seen in untreated tissue.

Some impressive work out of UC Berkeley demonstrates the role of intercellular signaling in stem cell engraftment. Investigators found that mice of advanced age tend to heal with more scarring than younger mice. Investigators conjoined genetically identical mice by the skin—one old, the other young. The animals thus shared a common blood circulation by which chemical signals could be exchanged. When both mice were given muscle nicks, the older mouse healed with the same minimal scarring as the younger mouse. We believe NIR light may have a role in triggering these stem cell signaling messengers (Grover et al., 2017).

In our opinion, the most promising emerging treatment option for TBI is PBM using NIR light therapy, which has been shown to stimulate blood flow in damaged peripheral tissue (Radegran and Saltin, 1999). Applying this treatment to improve neural tissue blood flow is a natural inference from this observed mechanism of action. Recent studies on TBI patients show that NIR light therapy may improve postconcussion symptoms. Results of in vitro studies, combined with human clinical studies, show improved blood flow and possible neurogenesis resulting from treatment (Grover et al., 2017). Several studies demonstrate that NIR light wavelengths penetrate to significant depths through cranial tissues. One study showed the half-time of exponential decay of 800 nm NIR light corresponds to photon migration over a distance of 4 cm (Chance et al., 1988).

In another study using optical path length measurement on adult and newborn infant heads using phase-resolved optical spectroscopy, the investigators measured mean path lengths of NIR light of 26.48 and 19.96 cm, respectively. Recent human case studies suggest that NIR light can help in the healing of acute concussions and chronic traumatic brain injuries. Some subjects report antianxiety effects after treatment, lasting up to three days suggesting that there may be a change induced by NIR light (Campbell, 2014).

24.10 Results

NIR light shows promise as a potential treatment modality for mTBI. Elucidating the underlying mechanisms of action of NIR light on neural tissue is an important step in optimizing treatment regimens. A possible related mechanism might be NIR light stimulation of neuronal energy metabolism, at least for tissues functioning suboptimally. Amplitude improvements in subjects displaying low-voltage readings initially suggest that cell respiratory activation may be stimulated by NIR light, resulting in an electrical brain state change favoring reaction time improvement. Evidence points to NIR light having no effect on already healthy cells and tissues. This would align with a treatment for symptomatic mTBI. In brief, evidence suggests that NIR works in the brain by increasing blood flow and assisting fuel and oxygen delivery, increasing drainage thus lowering swelling, clearing out the proteins and toxins, and finally, by stimulating brain cell regeneration and restoring brain function. Beyond uncovering the mechanisms of NIR light effect vis-à-vis reaction time on healthy tissue in an acute setting, the next step suggested is to measure longer-term NIR light treatment of diagnosed mTBI patients to determine whether the model of stimulating unhealthy tissue to return to normal functionality holds true in a clinical setting.

24.11 Discussion

A concussion initiates a complex cascade of metabolic events, leading to perturbation of delicate neuronal homeostatic balances. Starting from neurotoxicity, energetic metabolism disturbance caused by the initial mitochondrial dysfunction seems to be the main biochemical explanation for most postconcussive signs and symptoms (Signoretti et al., 2011). An immediate consequence of brain trauma is abrupt, indiscriminate release of neurotransmitters and unchecked ionic fluxes, leading to increased potassium and calcium pump activity to restore membrane potential. This results in significant increase in glucose demand to manage the pump activity. At the same time, cerebral blood flow is significantly decreased in the damaged regions for up to 10 days in the acute concussion setting. This exacerbates the brain's ability to mobilize glucose and clear ions, leading to a period of depressed metabolism that limits mitochondrial function and causes a buildup of calcium, which, in turn, hastens cell death. PCS describes a variety of chronic symptoms resulting from an initial mTBI event. Cerebral blood flow is linked to neural activity (Ances, 2004). NIR light has been shown to stimulate the release of nitric oxide, a potent endogenous vasodilator resulting in increased blood flow. This has been shown to occur in the absence of thermal effects, using lasers to deliver the NIR light (Lohr et al., 2009). Additional mechanisms appear to be at work. Karu (1989) showed that NIR light is absorbed by components of the cell respiratory chain, causing activation leading to changes in the redox status of both the mitochondria and the cytoplasm. Radiation in the NIR range was shown to initiate a response in the cell membrane through effects on the calcium channels. This in turn influences membrane permeability and ion transport (Karu, 1989). This suggests a possibl e effect of NIR light on the concussed neural tissue. Indeed, Ca^{++} waves are prominent in synaptic and extracellular processes that influence coherent neuronal firings measured by EEG (Ingber et al., 2014). Further, Lapchak and De Taboada (2010) demonstrated that transcranial NIR light treatment increases ATP content through cytochrome C oxidase photoacceptors in mitochondria in a rabbit model. This, in turn, can energize brain cell electrical activity (Lapchak and De Taboada, 2010). Investigators have noted that light therapy has the most pronounced effect on poorly functioning cells and little, if any, effect on normally functioning cells. For example, light therapy appears to stimulate healing of trophic skin ulcers, whereas the effect on normally healing wounds is insignificant. The implication is that normally functioning cells are not accelerated in their activity by NIR light but rather are left in their normal functioning state. However, if a cell is operating below optimal levels, then NIR light stimulates it to revert to an optimal, functional state. This is thought to be accomplished through the already cited stimulation of cell respiratory chains, leading ultimately to altering Ca^{++} flux that, in turn, affects the levels of cyclic nucleotides, which modulates DNA and RNA syntheses and resulting cell proliferation. In this study, subjects displaying normal-range amplitudes showed little effect from a single NIR light treatment. However, subjects displaying subnormal amplitudes showed a fairly large change in amplitude after treatment. The authors believe that acute activation of neuronal energetic metabolism in suboptimally functioning cells through biostimulation of the cell respiratory chain, coupled with increased blood flow bringing additional glucose and oxygen, may explain this finding.

24.12 Future clinical trials for the treatment of traumatic brain injury

In a scientific publication by FDA staff we found the following:

> *In support of the Traumatic Brain Injury Endpoint Development (Ted) Initiative, the FDA Commissioner's Fellow research project will focus on the development of a cross-Center and cross-Agency team of subject matter experts that would help facilitate an approach for adoption of brain disease-specific open data standards in order to improve the quality, efficiency and cost-effectiveness of TBI clinical trials. Traumatic Brain Injury (TBI) is a major medical problem. Each year in the United States, at least 1.7 million people suffer TBI; it is a contributing factor in a third of all injury-related US deaths. In the military, TBI is one of the most common causes of injury and disability in active duty service members. Researchers have been actively working on better ways to diagnose and treat TBI, but at present, we have no "cures" for TBI, or even very good ways to diagnose if it has happened, or its severity. As such, the FDA has not approved any therapeutic drug, medical device, or diagnostic tool for patients suffering from mild and moderate TBIs. At the root of this, is a lack of scientific and consensus-driven endpoints that can be used in the clinical trials needed to support approval of these products. The TED Initiative, funded through the Congressionally Directed Medical Research Program (CDMRP), will receive 18 Million Dollars over the next five years, with their sole goal of developing these endpoints, and for these endpoints to be in agreement with FDA.*

Endpoints for TBI studies could realize in many forms, including clinical outcome assessments, patient-centered outcomes, imaging, and blood-based biomarkers. FDA has many tools available, including the drug development tool and medical device development tool to help facilitate qualification of these potential endpoints. The FDA also is working with the research and clinical community to develop better-designed clinical studies so new medical products can be developed (Manley et al., 2017).

24.13 Conclusion

In conclusion, TBI is a pervasive injury that occurs in a variety of settings and across the populace. To date, numerous studies have explored the way to diagnose this condition best. However, there have been only a handful of very small studies investigating the efficacy of PBM for treating TBI or similar chronic brain disease states. Thus far, there is insufficient data for conclusive evidence to support one empirically based treatment that will heal the mind and the body. In our opinion, currently there is high awareness of the problem and advancements in diagnostics have improved, but little progress has been forthcoming on a definitive solution. As noted above, the FDA has not yet cleared or approved any standalone medical products that are intended to diagnose or treat TBI specifically. More work is yet to be done to address this global concern.

References

Agrawal, D., Gowda, N.K., Bal, C.S., Pant, M., Mahapatra, A.K., 2005. Is medial temporal injury responsible for pediatric postconcussion syndrome? A prospective controlled study with single-photon emission computerized tomography. J. Neurosurg. 102 (2 Suppl.), 167–171.

American Psychiatric Association, 2013. Diagnostic and Statistical Manual of Mental Disorders, fifth ed. American Psychiatric Publishing, Arlington, VA.

Andruszkow, H., Deniz, E., Urner, J., Probst, C., Grün, O., Lohse, R., et al., 2014. Physical and psychological long-term outcome after traumatic brain injury in children and adult patients. Health Qual. Life Outcomes 12 (1), 26.

Ances, B.M., 2004. Coupling of changes in cerebral blood flow with neural activity: what must initially dip must come back up. J. Cereb. Blood Flow Metab. 24, 1–6.

Astafiev, S.V., Zinn, K.L., Shulman, G.L., Corbetta, M., 2016. Exploring the physiological correlates of chronic mild traumatic brain injury symptoms. NeuroImage 11, 10–19.

Babcock, L., Byczkowski, T., Wade, S.L., Ho, M., Mookerjee, S., Bazarian, J.J., 2013. Predicting postconcussion syndrome after mild traumatic brain injury in children and adolescents who present to the emergency department. JAMA Pediatr. 167 (2), 156–161.

Bergersen, K., Halvorsen, J.Ø., Tryti, E.A., Taylor, S.I., Olsen, A., 2017. A systematic literature review of psychotherapeutic treatment of prolonged symptoms after mild traumatic brain injury. Brain Inj. 31, 279–289.

Bolzenius, J.D., Roskos, P.T., Salminen, L.E., Paul, R.H., Bucholz, R.D., 2015. Cognitive and self-reported psychological outcomes of blast-induced mild traumatic brain injury in veterans: a preliminary study. Appl. Neuropsychol. 22 (2), 79–87.

Bonne, O., 2003. Cerebral blood flow in chronic symptomatic mild traumatic brain injury. Psychiatry Res. 124, 141–152.

Bramlett, H.M., Dietrich, W.D., 2015. Long-term consequences of traumatic brain injury: current status of potential mechanisms of injury and neurological outcomes. J. Neurotrauma 32 (23), 1834–1848.

Campbell, V. (2014, October 15). [personal interview [S.Fox].

Carlozzi, N.E., Kirsch, N.L., Kisala, P.A., Tulsky, D.S., 2015. An examination of the Wechsler Adult Intelligence Scales, (WAIS-IV) in individuals with complicated mild, moderate and Severe traumatic brain injury (TBI). Clin. Neuropsychol. 29 (1), 21–37.

Chance, B., Leigh, J.S., Miyake, H., Smith, D.S., Nioka, S., Greenfeld, R., et al., 1988. Comparison of time-resolved and unresolved measurements of deoxyhemoglobin in brain. Proc. Natl. Acad. Sci. U.S.A. 85, 4971–4975.

Couch, J., Stewart, K., 2015. Definition of and comparison of the "Headache Problem" in veterans of the Iraq (OIF) and Afghanistan (OEF) wars with and without traumatic brain injury (TBI) using parameters of headache frequency and intensity. Neurology 84 (14 Suppl.), P1.314.

Duff, J., 2004. The usefulness of quantitative EEG (QEEG) and Neurotherapy in the assessment and treatment of postconcussion syndrome. Clin. EEG Neurosci. 35, 198–209.

Garcia, D., Hungerford, G.M., Bagner, D.M., 2015. Topical review: negative behavioral and cognitive outcomes following traumatic brain injury in early childhood. J. Pediatr. Psychol. 40 (4), 391–397.

Gerberding, J.L., Binder, S., 2015. Report to Congress on Mild Traumatic Brain Injury in the United States: Steps to Prevent a Serious Public Health Problem. National Center for Injury Protection & Control; Division of Unintentional Injury Prevention, Atlanta, GA.

Gordon, R.G., Duff, M.C., 2016. Incorporating principles of the collaborative contextualized intervention approach with the empirical study of learning and communication in traumatic brain injury. Aphasiology 30 (12), 1461–1482.

Grover, F., Weston, J., Weston, M., 2017. Acute effects of near-infrared light therapy on brain state in healthy subjects as quantified by qEEG measures. Photomed. Laser Surg. 35 (3), 136–141.

Ingber, L., Pappalepore, M., Stesiak, R.R., 2014. Electroencephalographic field influence on calcium momentum waves. J. Theor. Biol. 343, 138–153.

Karu, T., 1989. Laser Biostimulation: a photobiological phenomenon. J. Photochem. Photobiol. 3, 638–640.

Kristman, V.L., Borg, J., Godbolt, A.K., Salmi, L.R., Cancelliere, C., Carroll, L.J., et al., 2014. Methodological issues and research recommendations for prognosis after mild traumatic brain injury: results of the International Collaboration on Mild Traumatic Brain Injury Prognosis. Arch. Phys. Med. Rehabil. 95 (3), S265–S277.

Lapchak, P.A., De Taboada, L., 2010. Transcranial near infrared laser treatment (NILT) increases cortical adenosine-5c-triphosphate (ATP) content following embolic strokes in rabbits. Brain Res. 1306, 100–105.

Lengenfelder, J., Arjunan, A., Chiaravalloti, N., Smith, A., DeLuca, J., 2015. Assessing frontal behavioral syndromes and cognitive functions in traumatic brain injury. Appl. Neuropsychol. 22 (1), 7–15.

Lohr, N.L., Keszler, A., Pratt, P., Bienengraber, M., Warltier, D.C., Hogg, N., 2009. Enhancement of nitric oxide release from nitrosyl hemoglobin and nitrosyl myoglobin by red/near infrared radiation: potential role in cardioprotection. J. Mol. Cell Cardiol. 47, 256–263.

Ma, V.Y., Chan, L., Carruthers, K.J., 2014. Incidence, prevalence, costs, and impact on disability of common conditions requiring rehabilitation in the United States: stroke, spinal cord injury, traumatic brain injury, multiple sclerosis, osteoarthritis, rheumatoid arthritis, limb loss, and back pain. Arch. Phys. Med. Rehabil. 95 (5), 986–995.

Manley, G.T., MacDonald, C.L., Markowitz, A., Stephenson, D., Robbins, A., Gardner, R.C., et al., 2017. The traumatic brain injury endpoints development (TED) initiative: progress on a public-private regulatory collaboration to accelerate diagnosis and treatment traumatic brain injury. Neurotrauma 34 (19), 2721–2730.

Maugans, T.A., Farley, C., Altaye, M., Leach, J., Cecil, K.M., 2012. Pediatric sports-related concussion produces cerebral blood flow alterations. Pediatrics 129, 2606–2612.

McMillan, T.M., Wood, R.L. (Eds.), 2017. Neurobehavioural Disability and Social Handicap Following Traumatic Brain Injury. Psychology Press.

Narotam, P.K., Morrison, J.F., Schmidt, M.D., Nathoo, N., 2014. Physiological complexity of acute traumatic brain injury in patients treated with a brain oxygen protocol: utility of symbolic regression in predictive modeling of a dynamical system. J. Neurotrauma 31 (7), 630–641.

Neumann, D., Zupan, B., Malec, J.F., Hammond, F., 2014. Relationships between alexithymia, affect recognition, and empathy after traumatic brain injury. J. Head Trauma Rehabil. 29 (1), E18–E27.

Petraglia, A.L., Dashnaw, M.L., Turner, R.C., Bailes, J.E., 2014. Models of mild traumatic brain injury: translation of physiological and anatomic injury. Neurosurgery 75, S34–S49.

Radegran, G., Saltin, B., 1999. Nitric oxide in the regulation of vasomotor tone in human skeletal muscle. Am. J. Physiol. 276, H1951–H1960.

Rosenberg, H., Dethier, M., Kessels, R.P., Westbrook, R.F., McDonald, S., 2015. Emotion perception after moderate-severe traumatic brain injury: the valence effect and the role of working memory, processing speed, and nonverbal reasoning. Neuropsychology 29 (4), 509.

Roozenbeek, B., Maas, A.I., Menon, D.K., 2013. Changing patterns in the epidemiology of traumatic brain injury. Nat. Rev. Neurol. 9 (4), 231–236.

Savitsky, B., Givon, A., Rozenfeld, M., Radomislensky, I., Peleg, K., 2016. Traumatic brain injury: it is all about definition. Brain Inj. 30 (10), 1194–1200.

Signoretti, S., Lazzarino, G., Tavazzi, B., Vagnozzi, R., 2011. The pathophysiology of concussion. Phys. Med. Rehabil 3 (10 Suppl 2), S359–S368.

Tanielian, T., Jaycox, L.H., 2008. Invisible Wounds of War: Psychological and Cognitive Injuries, Their Consequences, and Services to Assist Recovery. Rand Corp., Santa Monica, CA, pp. 13–14, ISBN 978-0-8330-4454-9.

Togher, L., McDonald, S., Tate, R., Rietdijk, R., Power, E., 2016. The effectiveness of social communication partner training for adults with severe chronic TBI and their families using a measure of perceived communication ability. NeuroRehabilitation 38 (3), 243–255.

Villringer, A., Dirnagle, U., 1995. Coupling of brain activity and cerebral blood flow: basis of functional neuroimaging. Cerebrovasc. Brain Metab. Rev. 7, 240–276.

Wecksler, S., Mikhailovsky, A., Ford, P.C., 2004. Photochemical production of nitric oxide via two-photon excitation with NIR light. J. Am. Chem. Soc. 126, 13566–13567.

Wortzel, H.S., Arciniegas, D.B., 2014. The DSM-5 approach to the evaluation of traumatic brain injury and its neuropsychiatric sequelae. NeuroRehabilitation 34 (4), 613–623.

Zuckerman, A., Ram, O., Ifergane, G., Matar, M.A., Sagi, R., Ostfeld, I., et al., 2017. Controlled low-pressure blast-wave exposure causes distinct behavioral and morphological responses modeling mild traumatic brain injury, post-traumatic stress disorder, and comorbid mild traumatic brain injury–post-traumatic stress disorder. J. Neurotrauma 34 (1), 145–164.

Chapter 25

Transcranial, red/near-infrared light-emitting diode therapy for chronic traumatic brain injury and poststroke aphasia: clinical studies

Margaret A. Naeser[1,2], Paula I. Martin[1,2], Michael D. Ho[1], Maxine H. Krengel[1,2], Yelena Bogdanova[1,3], Jeffrey A. Knight[1,3,4], Andrea Fedoruk[1], Michael R. Hamblin[5,6,7] and Bang-Bon Koo[8]

[1]*VA Boston Healthcare System, Boston, MA, United States,* [2]*Department of Neurology, Boston University School of Medicine, Boston, MA, United States,* [3]*Department of Psychiatry, Boston University School of Medicine, Boston, MA, United States,* [4]*National Center for PTSD—Behavioral Sciences Division, VA Boston Healthcare System, Boston, MA, United States,* [5]*Wellman Center for Photomedicine, Massachusetts General Hospital, Boston, MA, United States,* [6]*Department of Dermatology, Harvard Medical School, Boston, MA, United States,* [7]*Harvard-MIT Division of Health Sciences and Technology, Cambridge, MA, United States,* [8]*Bio-imaging Informatics Lab, Department of Anatomy and Neurobiology, Boston University School of Medicine, Boston, MA, United States*

25.1 Traumatic brain injury

25.1.1 Introduction to traumatic brain injury

Traumatic brain injury (TBI) is a major medical problem worldwide, and in the United States three TBIs occur every minute (Faul et al., 2013). Approximately 1.7 million patients are evaluated annually; over 5 million Americans are living with TBI-related disabilities. The annual cost is between $60 and $76.5 billion (Faul et al., 2013). Closed-head, mild TBI (mTBI) is the most common (75%), and persistent cognitive dysfunction occurs in 5%–22% of these cases. mTBI is associated with loss of consciousness (LOC) 30 minutes or less (including no LOC), and with a period of altered mental status that could include posttraumatic amnesia—memory gaps or confusion lasting up to 24 hours.

25.1.2 Sports-related traumatic brain injury

Cognitive dysfunction associated with sports-related mTBI is of increasing concern, both for males and females, including children (McCrea et al., 2003). Within the past 10 years, a diagnosis of concussion in high school sports has increased annually, by 16.5% (Lincoln et al., 2011). With each successive concussion, there is a cumulative effect (Cantu, 2006; McAllister et al., 2012) including a prolonged period of recovery, and a progressively increased risk for reinjury (Guskiewicz et al., 2003; Wall et al., 2006). Postseason verbal learning scores were observed to be lower than expected in 24% of college athletes who had participated in contact sports (vs only 3.6%, in noncontact sports); and the greater the number of head impacts sustained, the slower the reaction time on IMPACT testing (McAllister et al., 2012). Among collegiate football players, those playing in the offensive linemen positions are at the greatest risk (Baugh et al., 2015).

25.1.3 Traumatic brain injury in soldiers and veterans

Closed-head, blast injury is the signature injury of soldiers returning from Iraq and Afghanistan as part of Operation Enduring Freedom/Operation Iraqi Freedom (OEF/OIF) (Hoge et al., 2008; Warden, 2006). The cognitive sequelae,

Photobiomodulation in the Brain. DOI: https://doi.org/10.1016/B978-0-12-815305-5.00025-7

recovery, and rehabilitation are of increasing concern (Bogdanova and Verfaellie, 2012). Estimates are as high as 320,000, regarding those who have returned with TBI (Lew et al., 2008; Tanielian and Jaycox, 2008). Posttraumatic stress disorder (PTSD) is also a major concern with OEF/OIF soldiers who have experienced mTBI (Vasterling et al., 2009). One estimate is that 28% of those diagnosed with mTBI also report clinical levels of PTSD symptoms. A dose-response gradient for exposure to blast/combination mTBI on clinical levels of PTSD symptoms has been observed (Kontos et al., 2013). The incidence of mTBI with comorbidities of PTSD and depression is higher in the military, than in civilians (Schneiderman et al., 2008).

25.1.4 Diffuse axonal injury and white matter abnormalities on magnetic resonance imaging scans

In most cases with closed-head mTBI, focal lesions are not present on CT scan, nor on structural magnetic resonance imaging (MRI) (Mittl et al., 1994; Provenzale, 2010; Van Boven et al., 2009). However, 30% of cases without abnormality on structural scans have abnormalities in white matter using diffusion tensor imaging (DTI) MRI scans (Bazarian et al., 2007).

Diffuse axonal injury (DAI) or traumatic axonal injury, has been recognized as one of the main consequences of closed-head mTBI (Medana and Esiri, 2003). DAI results when shearing, stretching, and/or angular forces pull on axons and small vessels (Johnson et al., 2013a; Mendez et al., 2005; Smith et al., 2003). The frontal lobes, including medial and lateral prefrontal cortex areas, are especially vulnerable to damage following mTBI where both linear and angular, acceleration/deceleration effects occur—for example, whiplash in a motor vehicle accident (MVA), concussive blast force, or even direct impacts to the head (Goldstein et al., 2012; Laplaca and Prado, 2010; McDonald et al., 2002). This leads to impaired axonal transport, focal axonal swelling and (after several hours) may result in axonal disconnection (Hurley et al., 2004). All severities of TBI can result in a degree of axonal damage (Kraus et al., 2007). Functionally, the reduction or loss of interconnectivity produces the cognitive, emotional and behavioral problems observed following TBI (Niogi and Mukherjee, 2010).

25.1.5 Development of neurodegenerative disease posttraumatic brain injury

White matter degeneration and inflammation and may be present years after only a single TBI (Johnson et al., 2013a, b). Increased microglial activity may persist long-term post-TBI (Gentleman et al., 2004; Ramlackhansingh et al., 2011); persistent neuroinflammation may be a mechanistic link between TBI and the development of neurodegenerative disease, including Alzheimer disease (AD). Damaged axons may serve as a source of the AD-associated protein amyloid-beta (Johnson et al., 2010; Tran et al., 2011).

Chronic traumatic encephalopathy (CTE), a progressive tau protein-linked neurodegenerative disease, is believed to develop in part, from repeated head impacts (McKee et al., 2009, 2010). Symptoms include cognitive dysfunction, progressive irritability, suicidal ideation and dementia. It may develop years after the head trauma occurred. CTE has been documented in US military veterans exposed to blast injury in Iraq and Afghanistan (Goldstein et al., 2012).

25.1.6 Functional brain imaging in traumatic brain injury

Since 1999, mTBI cases have been observed in functional neuroimaging studies to have alterations in neural activation during performance of cognitive tasks (McAllister et al., 1999). While accuracy during a working memory (WM) task did not differ for the mTBI cases at 1 month post-mTBI, versus healthy adults, there was more widespread bifrontal and biparietal activation for the mTBI cases, even in the initial level of WM load. This pattern of abnormal hyperactivation was present even at 1 year post-mTBI (McAllister et al., 2002, 2006). While the mTBI cases showed no significant postconcussive symptoms (PCS) at the 1-year follow-up, they continued to show mildly slower reaction speed relative to healthy adults.

25.1.7 Resting-state, functional-connectivity magenetic resonance imaging in traumatic brain injury

Using a different type of fMRI scan—for example, resting-state functional-connectivity MRI (rs-fcMRI) scans with healthy adults, have demonstrated specific neural networks that function in a wide-spread, but temporally-coordinated manner (with a very low frequency oscillations—e.g., 0.01−0.08 Hz) while the subject is resting quietly in the MRI scanner without external, task-related stimuli (Raichle et al., 2001). One of these intrinsic networks, the default mode network (DMN) consists of the mesial prefrontal cortex (mPFC); the precuneus and posterior cingulate cortex (precu/pCC); infero-lateral parietal cortices (parts of angular gyrus areas), and medial temporal lobe/hippocampus (Greicius et al., 2009). The DMN, including the precu/PCC in particular, shows rapid and highly reactive deactivation in normals, during attentionally-demanding tasks where the lateral fronto-parietal, dorsal attention network shows concomitant, increased activation. Abnormality in the DMN (with failure to deactivate) has been observed in mTBI cases (Bonnelle et al., 2011; Johnson et al., 2012; Mayer et al., 2011).

Abnormalities have also been observed with mTBI cases in two additional intrinsic networks important for normal brain function during attention-related and cognitive tasks. The first is the salience network (SN) (Bonnelle et al., 2012). The SN controls the DMN, and the SN consists of the anterior insulae (AI), the presupplementary motor areas (preSMAs), and the dorsal anterior cingulate cortex (dACC) areas. The SN is critical for normal executive function and inhibition (Beckmann et al., 2005; Seeley et al., 2007; Sridharan et al., 2008). Brain cortex in the anterior parts of the SN (preSMAs and dACC) promote inhibitory control (deactivation) of the posterior parts of the DMN (precun/PCC), particularly during tasks that require inhibition and rapid switching for success. The SN is important for signaling the need to change behavior (Menon and Uddin, 2010). The second, additional rs-fcMRI network is the central executive network (CEN). This network comprises dorsolateral prefrontal cortex (DLPFC) areas and the infero-parietal sulci (Laird et al., 2011; Menon, 2011). The CEN is particularly important, as the name implies, for executive function, including cognitive manipulation of temporal information, WM, processing speed, reasoning, problem solving, planning and execution, task flexibility, and multitasking.

25.1.8 Cognitive dysfunction in traumatic brain injury

The most common complaints of cases with mTBI are in the domains of attention/concentration and WM—that is, the ability to hold information in mind, and to manipulate it in light of incoming material (Levin et al., 2013; Stuss et al., 1985). At 6 months postinjury, indices of executive function have been found to predict persistence of postconcussive syndrome, in mild and moderate TBI patients (Hartikainen et al., 2010). One of the most debilitating sequelae of mTBI is the failed attempt to reestablish family and work relationships (Chew and Zafonte, 2009). Due to the diffuse nature of damage, however, no single behavioral outcome measure captures the multidimensional nature of TBI outcome (Zafonte et al., 2009).

25.1.9 Sleep disturbances in traumatic brain injury

In addition to the cognitive and psychosocial issues post-TBI, there are problems with sleep (Bloomfield et al., 2010; Bryan, 2013; Ruff et al., 2009). An estimated 53% of individuals with TBI report sleep disturbances (Mathias and Alvaro, 2012). Sleep problems may exacerbate symptoms of TBI (Ouellet et al., 2015), increase neuropsychiatric symptoms (depression, anxiety) post-TBI (Rao et al., 2014), and interfere with participation in rehabilitation treatment (Gilbert et al., 2015). Although the etiology of TBI-related sleep problems is currently under investigation, research has indicated neurobiological factors, specifically, impairment in the function of neural circuits involved in sleep/wake regulation (Saper et al., 2001; Faraguna et al., 2008). Among many factors, poor sleep would disrupt the normal, and necessary clearing of metabolites including beta-amyloid and other potentially neurotoxic waste products that accumulate during the awake central nervous system (Xie et al., 2013).

25.1.10 Pharmacologic treatments for traumatic brain injury

Trials of pharmacologic treatments for TBI have been mostly unsuccessful (Narayan et al., 2002; Zafonte et al., 2009). There are some pharmacologic interventions available for systemic and intracranial changes associated with moderate and severe TBI, with few controlled systematized studies for pharmacologic treatment of cognitive impairment (Lee et al., 2005). There was a large clinical trial (COBRIT) for moderate, severe, and complicated mTBI, which utilized the pharmacologic agent, Citicoline (Zafonte et al., 2009). Citicoline was evaluated as a neuroprotective agent and in TBI has previously demonstrated some efficacy in secondary measures for stroke, and smaller TBI studies. In 2012, the results of the COBRIT study showed the use of citicoline compared with a placebo for 90 days did not result in improvement in functional and cognitive status (Zafonte et al., 2012).

There are currently no pharmacologic treatments for mTBI secondary injury or for prevention of cognitive and behavioral problems associated with mTBI (Kan et al., 2012; Loane and Faden, 2010). Further investigation is warranted to examine the effects of cholinesterase inhibitors, with preliminary evidence suggesting improvement in attentional difficulties and mixed results for memory treatment (Chew and Zafonte, 2009). Regarding executive dysfunction, no conclusions can be drawn for improvement by pharmacologic intervention post-TBI (Lee et al., 2005).

McAllister et al. (2011b) examined the effect of Bromocriptine, a dopamine D2 receptor agonist on WM performance in healthy controls versus mTBI patients. Bromocriptine was associated with improved WM performance only in healthy controls, not mTBI patients. Imaging showed that mTBI patients were not able to recruit WM task-specific regions of interest. These results suggest that mTBI patients may have altered response to dopamine. In another study, McAllister et al. (2011a) found the opposite effect, using Guanfacine, an alpha-2 adrenergic receptor agonist. Guanfacine was found to selectively improve WM performance in mTBI but not in healthy controls. In the mTBI group, increased activation was observed within a WM task-specific region of interest. This pharmacologic agent may be a promising pharmacologic agent to test hypotheses about the neural mechanisms of cognitive dysfunction after mTBI.

25.1.11 Cognitive rehabilitation therapies for traumatic brain injury

Review of the effectiveness of current cognitive and behavioral treatments to improve executive function after TBI has shown limited evidence for the efficacy of cognitive rehabilitation (Boelen et al., 2011; Cicerone et al., 2011). Executive dysfunction continues to present a challenge to the rehabilitation process (Cicerone et al., 2006). Behavioral treatment approaches for TBI patients attempt to maximize the patient's behavioral functioning by working with residual brain-based capacities, but injured, nonfunctioning brain cells may limit their potential for success. Treatments are needed that directly target injured brain cells to improve the functioning of underlying brain systems (including functional-connectivity in networks such as the DMN, SN, and CEN) that regulate attention, executive function, memory, emotions, and behavior. Effective treatments to improve cognition in individuals with TBI are currently lacking, and these are urgently needed for veterans, as well as nonveterans. Transcranial red/NIR PBMT is a promising clinical research method to fill these needs.

25.2 Photobiomodulation for chronic traumatic brain injury

25.2.1 Transcranial light-emitting diode treatment performed at home, to improve cognition in chronic, mild traumatic brain injury—case reports

Naeser et al. (2011) described two case reports with chronic, mTBI where cognition improved following tPBM treatment with red/NIR transcranial LEDs (tLED). The LEDs were applied to forehead and scalp areas including midline and bilateral frontal, temporal, parietal, and occipital areas. Each red/NIR LED cluster head (MedX Health, Toronto) had a 5.35 cm diameter, and was 500 mW containing 61 diodes (9, at 633 nm, and 52, at 870 nm). The power density was 22.2 mW/cm^2, and 13 J/cm^2 (CW) was applied to the scalp from each LED cluster head for 9 minutes 45 seconds (estimated 0.4 J/cm^2 to surface brain cortex).

Patient 1 (a 59 year old female, professor of web design) began tLED treatments 7 years after a closed-head mTBI from a MVA. Pre-tLED, her ability to sustain attention (computer work) lasted only 20 minutes. After 8 weekly tLED treatments, her sustained attention on the computer increased to 3 hours. She reported that if she stopped treating for more than 2 weeks, she regressed.

Patient 2 (52 year old female, high-ranking, retired military officer) had a history of closed-head mTBIs (sports/military, and recent fall onto concrete from a swing). Her structural MRI brain scan showed moderate fronto-parietal atrophy for her age. Pre-tLED, she was on medical disability for 5 months. After 4 months of nightly tLED treatments at home, her medical disability was discontinued; she returned to work full-time as an executive consultant with an international technology-consulting firm. Neuropsychological (NP) testing after 9 months of tLED indicated significant improvement (+2 SD) in memory, and (+1 SD) in executive function (inhibition, inhibition accuracy), and as well as reduction in PTSD. Patient 2 reported that if she stopped treating for more than 1 week, she regressed. Both patients have continued home treatments with tLED for at least 5 years; there have been no adverse events or negative side effects. Patient 1 is lost to follow-up, and Patient 2 continues with home treatments.

25.2.2 Transcranial light-emitting diode treatment to improve cognition in chronic, mild traumatic brain injury—open protocol, group study

Naeser et al. (2014) conducted a pilot, open-protocol study with a larger number of chronic, mTBI patients to examine whether tLED with the same red/NIR LED cluster heads (MedX Health, Toronto) described above, could improve cognition when a systematic treatment protocol was used. Eleven chronic, mTBI participants (26–62 years of age, 6 M) with nonpenetrating brain injury and persistent cognitive dysfunction were treated for 18 outpatient sessions (Monday, Wednesday, Friday, for 6 weeks), starting at 10 months to 8 years post-mTBI (MVA or sports-related; and one participant, improvised explosive device blast injury). Four had a history of multiple concussions. Each LED cluster head (5.35 cm diameter, 500 mW, 22.2 mW/cm^2) was applied for 9 minutes 45 seconds to each of 11 scalp placements (13 J/cm^2, CW). LEDs were placed on the midline from front-to-back hairline; and bilaterally on frontal, parietal, and temporal areas. Six LED cluster heads were applied simultaneously, held in place with a soft nylon cap. Each participant was treated in a recliner chair. See Naeser et al. (2014) for specific LED placements. These placements covered cortical nodes located on the DMN, SN, and CEN. It was hypothesized that these placements would increase ATP and improve focal, regional cerebral blood flow (rCBF) in these cortical areas. NP testing was performed pre-LED, and at 1 week, and 1 and 2 months after the 18th tLED treatment.

25.2.3 Results

A significant linear trend was observed for the effect of tLED treatment over time for the Stroop (color word interference) test for executive function, trial 3 inhibition ($P = .004$); Stroop, trial 4 inhibition switching ($P = .003$); California Verbal Learning Test (CVLT) II, alternating versions, total trials 1–5 ($P = .003$); and CVLT-II, long delay (20 minutes) free recall ($P = .006$) (see Fig. 25.1). Participants reported improved sleep, and fewer PTSD symptoms, if present (Fig. 25.2). Participants and family reported better ability to perform social, interpersonal, and occupational functions.

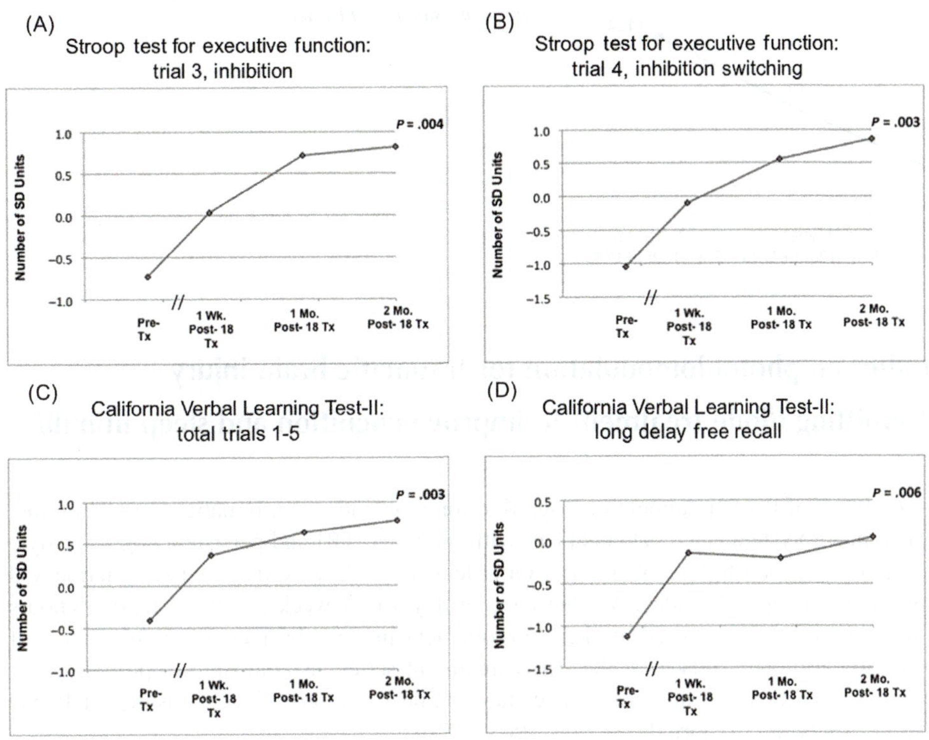

FIGURE 25.1 Graphs showing a significant linear trend over time, for the effect of transcranial LED treatments in mTBI on: (A) Stroop (color-word interference test) for executive function: Trial 3, inhibition ($P = .004$); (B) Stroop, trial 4 inhibition switching ($P = .003$); (C) California Verbal Learning Test (CVLT-II), total trials 1–5 ($P = .003$); and (D) CVLT-II, long felay (20 min) free recall ($P = .006$). *Reprinted with authors' permission, Naeser, M.A., Zafonte, R., Krengel, M.H., Martin, P.I., Frazier, J., Hamblin, M.R., et al., 2014. Significant improvements in cognitive performance post-transcranial, red/near-infrared light-emitting diode treatments in chronic, mild traumatic brain injury: open-protocol study. J. Neurotrauma 31 (11), 1008–1017.*

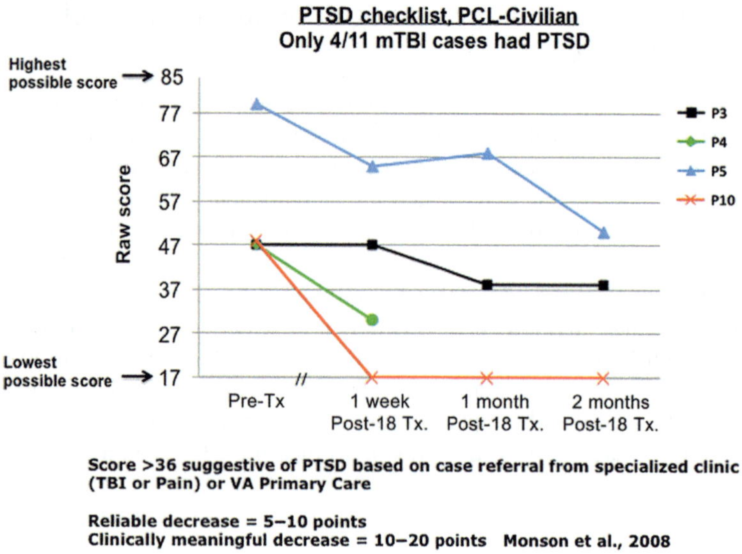

FIGURE 25.2 Four of the eleven mTBI cases treated in the Naeser et al. (2014) study, also had posttraumatic stress disorder (PTSD). All four cases showed a clinically meaningful or reliable decrease in symptoms of PTSD, after the transcranial, red/near-infrared LED treatment series. *Reprinted with authors' permission from Naeser M.A., Martin P.I., Ho M.D., Krengel M.H., Bogdanova Y., Knight J.A., et al., Transcranial, Red/near-infrared light-emitting diode (LED) therapy for chronic, traumatic brain injury, Photomed. Laser Surg. 34 (12), 2016, 610–626.*

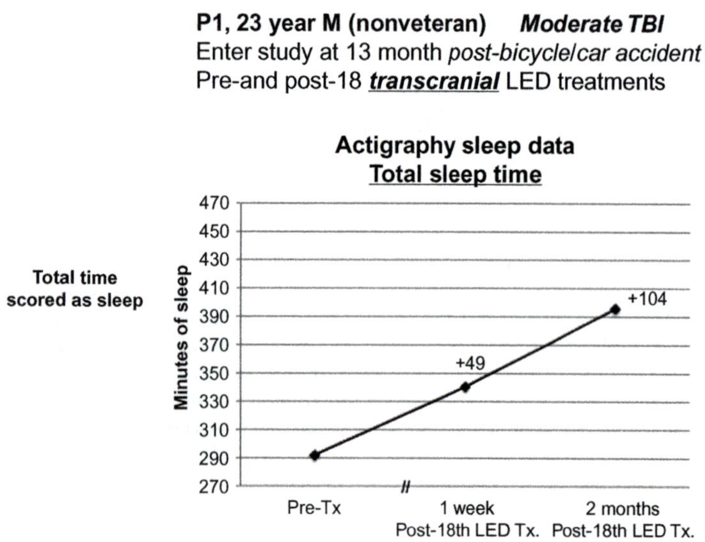

FIGURE 25.3 Time spent asleep increased by about 1 h after 18 transcranial red/NIR LED treatments. The Actigraphy watch was worn 24/7 for 1 week, at each time point. This participant with moderate TBI also improved on executive function and verbal memory. *Reprinted from Bogdanova, Y., Martin, P.I., Ho, M.D., Krengel, M.H., Ho, V.T., Yee, M.K., Knight, J.A., Hamblin, M.R., Naeser, M.A. (2014). LED therapy improves sleep and cognition in chronic moderate TBI: pilot case studies. Abstract. Archives of Physical Medicine and Rehabilitation, 95 (10), e77. doi:10.1016/j.apmr.2014.07.247, with permission from Elsevier.*

25.3 Ongoing current studies on photobiomodulation for traumatic brain injury

25.3.1 Transcranial light-emitting diode treatment to improve cognition and sleep in mild traumatic brain injury

Bogdonova et al. (2014) studied the effects of tLED treatment on cognitive function and sleep, in patients with chronic, moderate TBI at the VA Boston Healthcare System. Two patients (1 F) with moderate TBI and persistent cognitive dysfunction (at least 2 SD below average on one, or 1 SD below average on at least two NP tests of executive function and memory) received 18 sessions of tLED therapy (Monday, Wednesday, Friday for 6 weeks, with at least 48 hours between treatments). Both cases treated with tLED showed marked improvement in sleep by increasing an average of 1 hour per night (measured with Actigraphy), at 1 week post the tLED treatment series, as compared to pre-tLED. P1 also improved in executive function, verbal memory, and sleep efficiency; while P2 improved on measures of PTSD (PCL-M), and depression. No adverse events were reported (see Figs. 25.3–25.6).

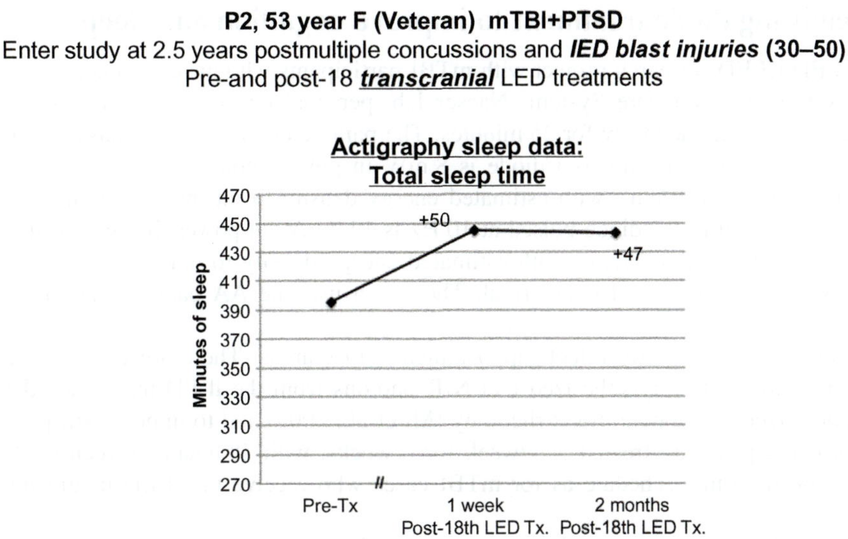

FIGURE 25.4 Time spent asleep, increased by about 1 h after 18 transcranial red/NIR LED treatments. The Actigraphy watch was worn 24/7 for 1 week, at each time point. This participant also had reduced symptoms of PTSD and depression post-tLED, see Figs. 25.5 and 25.6. *Reprinted from Bogdanova, Y., Martin, P.I., Ho, M.D., Krengel, M.H., Ho, V.T., Yee, M.K., Knight, J. A., Hamblin, M.R., Naeser, M.A. (2014). LED therapy improves sleep and cognition in chronic moderate TBI: pilot case studies. Abstract. Archives of Physical Medicine and Rehabilitation, 95 (10), e77. doi:10.1016/j.apmr.2014.07.247, with permission from Elsevier.*

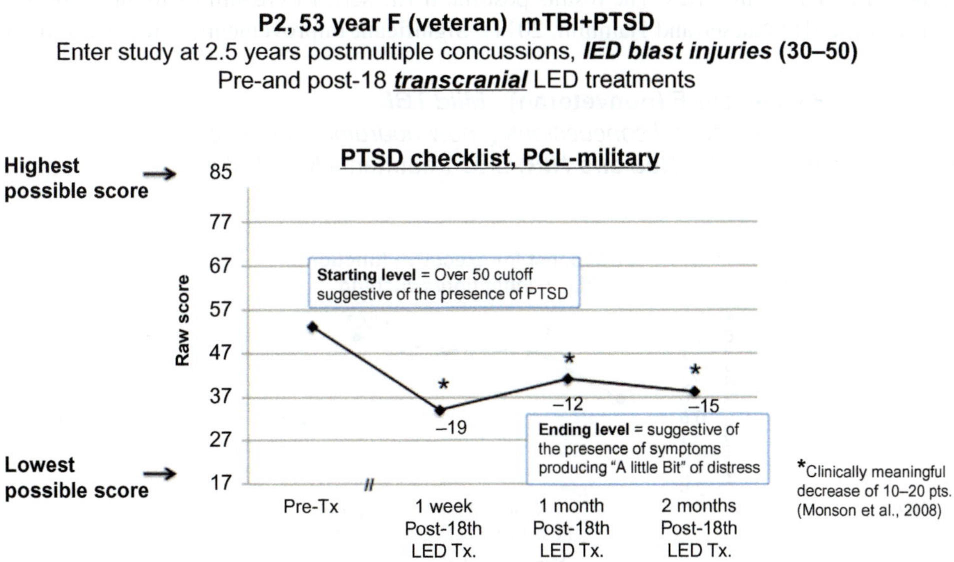

FIGURE 25.5 Reduced PTSD symptoms after 18 transcranial red/NIR LED treatments. The changes showed a clinically meaningful decrease in PTSD symptoms at 1 week and at 2 months post-tLED, compared to pre-tLED. *Reprinted from Bogdanova, Y., Martin, P.I., Ho, M.D., Krengel, M.H., Ho, V.T., Yee, M.K., Knight, J.A., Hamblin, M.R., Naeser, M.A. (2014). LED therapy improves sleep and cognition in chronic moderate TBI: pilot case studies. Abstract. Archives of Physical Medicine and Rehabilitation, 95 (10), e77. doi:10.1016/j.apmr.2014.07.247, with permission from Elsevier.*

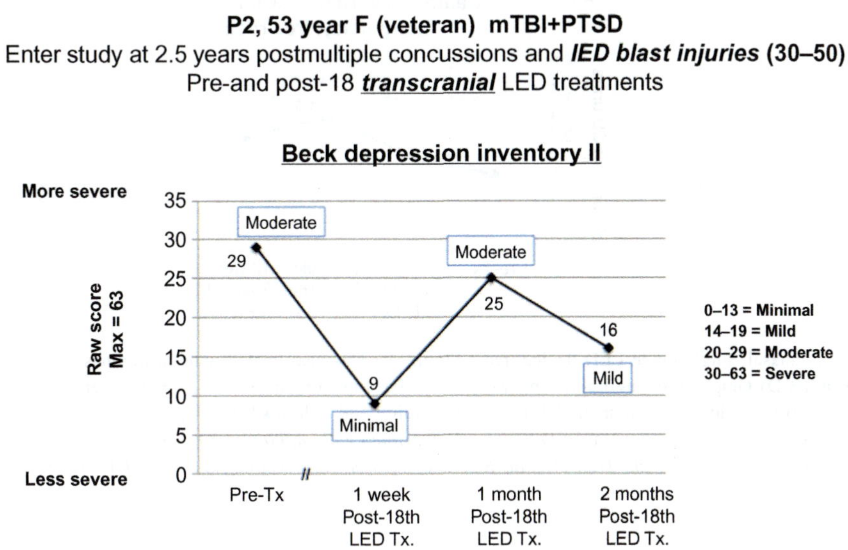

FIGURE 25.6 Reduced levels of depression after 18 transcranial red/NIR LED treatments in this female veteran with mTBI + PTSD. Pre-tLED, the depression level was rated as moderate. The depression level was rated as minimal at 1 week post the 18th treatment. Although the depression returned to moderate at 1 month post the 18th tLED treatment, it was rated as only mild at 2 months after the last tLED treatment. Results might have been more consistent, if she had access to LED devices for home treatment. *Reprinted from Bogdanova, Y., Martin, P.I., Ho, M.D., Krengel, M.H., Ho, V.T., Yee, M.K., Knight, J. A., Hamblin, M.R., Naeser, M.A. (2014). LED therapy improves sleep and cognition in chronic moderate TBI: pilot case studies. Abstract. Archives of Physical Medicine and Rehabilitation, 95 (10), e77. doi:10.1016/j.apmr.2014.07.247, with permission from Elsevier.*

25.3.2 Intranasal (only) light-emitting diode treatment to improve cognition and sleep

A pilot, open-protocol, intranasal (only) LED (iLED) research project with mTBI participants who have chronic, cognitive dysfunction was initiated at the VA Boston Healthcare System (Naeser lab, personal observation). Two, small diodes (one clipped into each nostril) were used simultaneously for 25 minutes. The parameters of each intranasal diode (Vielight, Toronto) are as follows: (1) The red, 633 nm intranasal diode is 8 mW in power (continuous wave, CW); beam spot size on contact, 1 cm^2; power density, 8 mW/cm^2; with estimated energy density to mucosa in 25 minutes, 12 J/cm^2 (see Fig. 25.7A). (2) The NIR 810 nm, intranasal diode pulsed at 10 Hz is 14.2 mW in power; beam spot size on contact, 1 cm^2; power density pulsed at 10 Hz, 14.2 mW/cm^2; with estimated energy density to mucosa in 25 minutes, 10.65 J/cm^2. Both iLEDs are noninvasive, painless, and nonthermal. They each use one AA battery (1.4 V) for power.

It is hypothesized that the NIR photons might indirectly affect hippocampal cortex areas. There are connections from the olfactory bulbs to the hippocampal areas; it is hypothesized that NIR photons from the iLED may reach the olfactory bulbs. The red photons are hypothesized to improve blood rheology (Mi et al., 2004) and to improve sleep by increasing melatonin (Zhao et al., 2012). Participants are treated 3 × /week for 6 weeks; with 48 hours between treatments. The pre- and post-iLED testing is on the same schedule as for mTBI cases who receive the tLED treatments (Naeser et al., 2016).

One mTBI participant (a 24 year old female) with a history of four sports-related concussions (two snow boarding and two field hockey) received the iLED treatment series. The results post the iLED series were similar to those for our tLED studies (Naeser et al., 2011, 2014, 2016; Naeser and Hamblin, 2011). Significant improvements were observed on

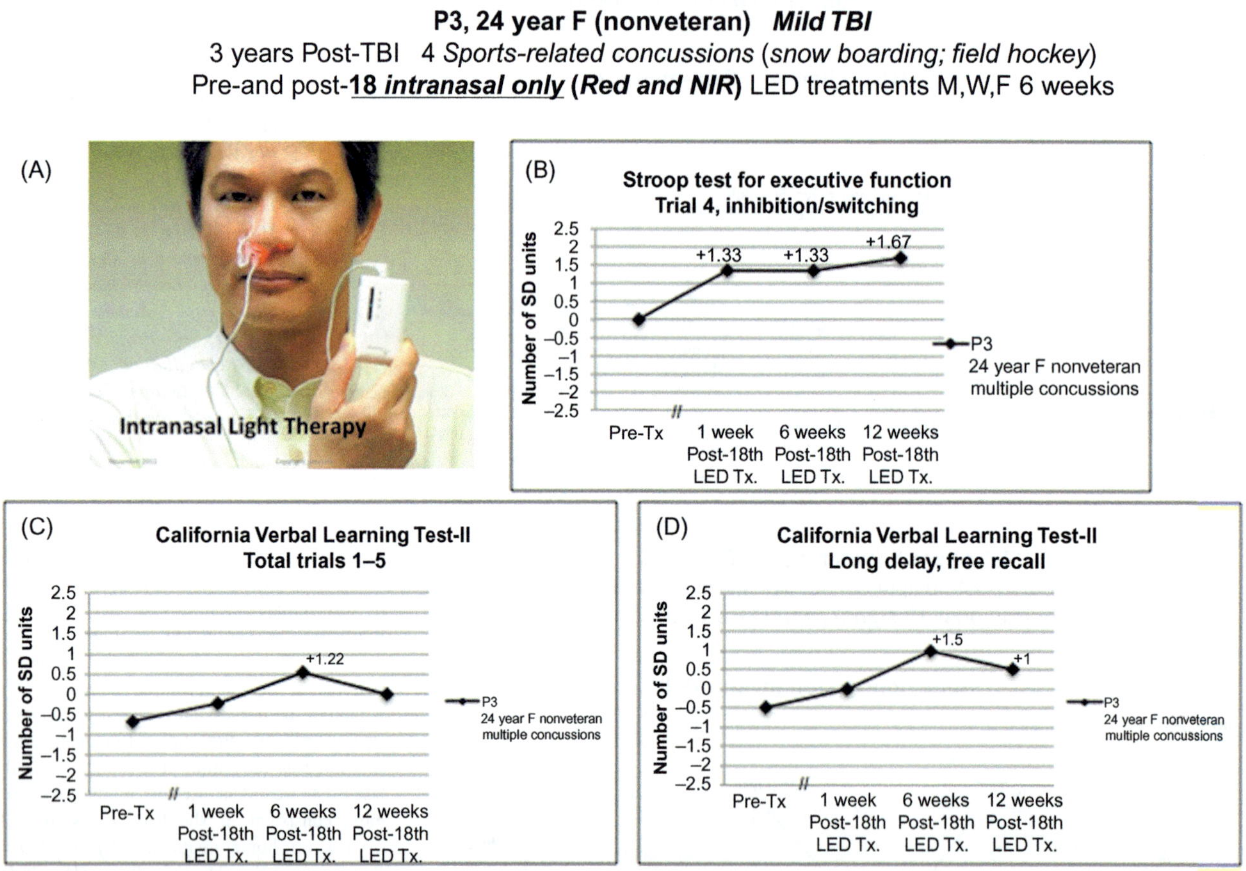

FIGURE 25.7 (A) Figure showing Vielight, Inc., Intranasal, single diode LED device, red 633nm. She was also treated with an intranasal, single diode LED, NIR 810nm, pulsed at 10Hz, not shown. (B-D) Graphs showing improvements in executive function and verbal memory after the 18th intranasal LED treatment in this 24 year old female who had a history of four mild TBIs (snowboarding and field hockey). Improvements were +1 SD or more, on these tests at 1 week, 6 and/or 12 weeks post the 18th intranasal LED treatment. *Reprinted with authors' permission from Naeser M. A., Martin P.I., Ho M.D., Krengel M.H., Bogdanova Y., Knight J.A., et al., Transcranial, Red/near-infrared light-emitting diode (LED) therapy for chronic, traumatic brain injury, Photomed. Laser Surg. 34 (12), 2016, 610–626.*

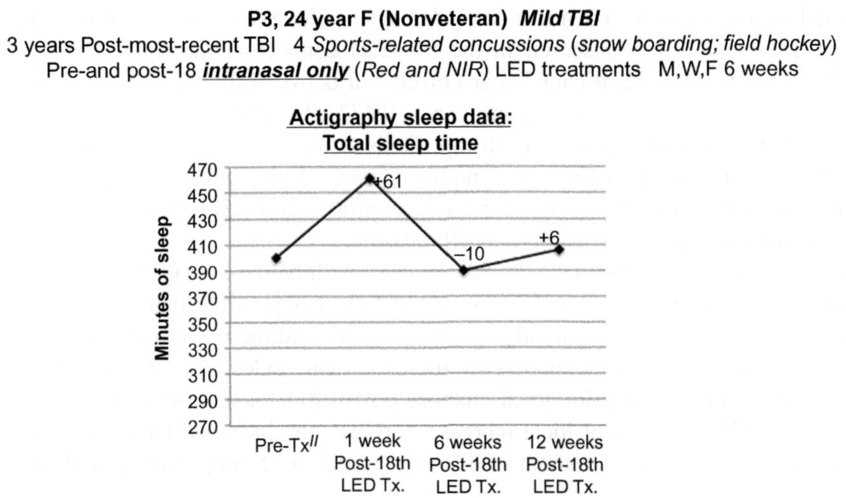

FIGURE 25.8 Time spent asleep increased by about 1 h after 18 intranasal red/NIR LED treatments. The actigraphy watch was worn 24/7 for 1 week, at each time point. Gains were made at 1 week post the intranasal treatments (but not sustained at 6 and 12 weeks post the last intranasal treatment). However, she was able to stop all sleep medications at 12 weeks post the last intranasal treatment. Additional intranasal LED home treatments should be considered for this participant for continued improvement. *Reprinted with authors' permission from Naeser M.A., Martin P.I., Ho M.D., Krengel M.H., Bogdanova Y., Knight J.A., et al., Transcranial, Red/near-infrared light-emitting diode (LED) therapy for chronic, traumatic brain injury, Photomed. Laser Surg. 34 (12), 2016, 610–626.*

tasks of executive function (Fig. 25.7B), verbal memory (Fig. 25.7C and D), attention and verbal fluency, at 1, 6, and 12 weeks post the 18th iLED treatment. At one-week post the 18th iLED treatment, the participant's average total sleep time had increased by 61 minutes per night (Fig. 25.8), and her sleep efficiency (total sleep time/total time in bed) had increased by 11%. Her sleep efficiency at 12 weeks post-iLED was 5% above pre-iLED levels, and she reported no longer using any sleep medication that she had previously been using regularly. It is possible that her sleep parameters would have continued to improve even more, if she had the opportunity for self-administered home treatment with the iLEDs. There were no negative side effects or complications.

25.4 Discussion, photobiomodulation for traumatic brain injury

Major findings from the above-mentioned, open-protocol tLED studies to improve cognition in chronic mTBI cases were significant improvements in executive function, and in verbal learning and memory post-tLED. In the Naeser et al. (2014) study, the cases had experienced persistent cognitive dysfunction, ranging from 10 months to 8 years, prior to the tLED treatments. As is common with mTBI cases, heterogeneity was present among the 11 cases (Naeser et al., 2014), including 4 cases with a history of multiple concussions. These findings are discussed separately below, and possible mechanisms associated with beneficial effects post-tLED are offered.

25.4.1 Executive function, and relationship to resting-state, functional-connectivity magenetic resonance imaging networks (default mode network and salience network)

In the area of executive function (Stroop, color-word interference test, trial 4, inhibition switching, the most difficult of the Stroop tasks), there was variability in the entry levels across the mTBI cases. For example, in 5/9 cases (56%), the pre-tLED levels were at least -1.0 SD below average; whereas 4/9 entered with average scores (age- and education-adjusted norms). All five cases who entered with below-average scores on the Stroop trial 4, improved by $+1-+4.5$ SD at 2 months post-tLED.

In the Bonnelle et al. (2012) study, variability in performance on Stroop, inhibition s,witching was also observed among a large number of TBI cases who were studied with rs-fcMRI, fMRI and DTI (Bonnelle et al., 2012). In that study, 20/46, 43% also performed poorly on the stop-signal reaction task (SSRT) with slower response inhibition (higher SSRT response times). These cases with slower reaction times in the NoGo condition were observed to have failure in deactivating the DMN, particularly the precu/PCC portion. Other parts within the DMN including mPFC and left hippocampus were also observed to have less deactivation. Failure to properly deactivate the DMN during cognitive tasks that require rapid shifting of attention and inhibition has also been observed in other studies with TBI cases (Bonnelle et al., 2011; Mayer et al., 2011).

The Bonnelle et al. (2012) study, however, also observed abnormalities in the SN in their TBI cases; particularly among those cases who failed to deactivate the precu/PCC during the SSRT. The SN consists of the AI, preSMA, and dACC; and the SN regulates activity in the DMN. For example, these authors observed that failure to deactivate precun/PCC during the SSRT was predicted by the amount of white matter damage in the SN—that is, significantly lower fractional anisotropy (FA) values on DTI scans were observed in the SN tract connecting the right AI to the preSMA and dACC in these TBI cases. They observed a significant linear correlation between the right AI-preSMA/dACC tract FA, with the amount of precu/PCC activation during stopping ($r = -0.472$, $P < .0005$). Furthermore, and most relevant to results in the present study, they observed that on the Stroop (Delis et al., 2004) that "the FA within the rAI–preSMA/dACC tract (corrected for age and whole-brain white matter damage) was significantly correlated with the inhibition/switching versus combined color naming and word reading contrast score (Spearman one tailed $r = -0.265$, $P = .029$, $n = 52$)." The SN was not correlated with any other NP measure.

Thus, at least for the five mTBI cases in the Naeser et al. (2014) study who entered with below-average Stroop, inhibition switching scores, but who also improved by at least $+1 - +4.5$ SD, post-tLED, it is possible that the red/NIR photons from the LED cluster heads that had been placed extracranially (particularly on midline placements) affected some nodes within the SN, and/or the DMN, thus improving function of these cortical nodes, and/or connections between these nodes. Possible underlying mechanisms and physiological changes post-tLED, such as increase of ATP in hypoxic/compromised cells that are part of the DMN or SN; or local increase in rCBF, are discussed later. These are cellular changes that could have supported the behavioral improvement in executive function (Stroop trial 4, inhibition switching) post-tLED.

Transcranial LED and fMRI research in our lab has observed focal, increased activation in targeted cortical areas on fMRI scans, post-18 transcranial red/NIR LED treatments with chronic, left-hemisphere stroke patients (Naeser et al., 2012). Thus, our results with stroke patients suggest that the LED placement loci may have a focal effect, subjacent to the LED placement locations. Although no rs-fcMRI, or task-related fMRI studies were part of our pilot studies with chronic mTBI cases, specific LED placements may have had a beneficial, focal effect on specific cortical nodes within the SN and the DMN.

25.4.2 Specific transcranial light-emitting diode placements may affect specific parts of the salience network and default mode network in traumatic brain injury cases

Specific tLED placements that may have treated specific nodes within the SN include the following: (1) The tLED placement on the midline of face, centered over the forehead and front hairline, likely targeted the left and right dACC nodes within the SN. (2) The tLED placement on the midline, vertex of the head likely also targeted the left and right preSMA nodes within the SN. (3) The tLED placements on the left and right temple areas may have reached the AI within the SN. This is unknown, however, due to the greater depth of the anterior insula.

Specific LED placements that may have treated specific nodes within the DMN include the following: (1) The tLED placement on the midline of face, centered over the forehead and front hairline likely targeted the left and right mPFC nodes within the DMN. (2) The tLED placement on the scalp midline, superior to the external occipital protuberance (and half-way toward the vertex), likely targeted the left and right precuneus areas, part of the precu/PCC node of the DMN. (3) The tLED placements, posterior and superior to each ear, likely targeted the left and right infero-parietal sulcus (angular gyrus areas), also nodes within the DMN.

Appropriate increased activation in the CEN explained below, would also be important for improved behavior on executive function, Stroop, trial 4, inhibition switching. The increased activation in the CEN can only occur, however, when there is appropriate decreased activation in the precun/PCC.

25.4.3 Verbal learning and memory, and relationship to resting-state, functional-connectivity magenetic resonance imaging (central executive network)

The CVLT is a verbal WM task where increased activation on task-related fMRI is associated with DLPFC, and/or fronto-parietal areas (Smith and Jonides, 1998; Curtis and D'Esposito, 2003). WM is associated with the CEN on rs-fcMRI (Beckmann et al., 2005). The CEN is a fronto-parietal system consisting primarily of the DLPFC and posterior, parietal cortex (PPC). In addition to WM, the CEN is important for high-level cognitive functions such as planning, decision-making, and control of attention. During WM tasks such as the CVLT, the CEN should be activated, however, the DMN should be deactivated, with a coordinated toggling back and forth between the two networks.

In the Naeser et al. (2014) study, all three cases who entered with scores at least -1 SD below average on the CVLT, total trials 1–5, improved by $+1$ to $+2$ SD at 2 months post-tLED. Also, a total of five out of seven cases who entered with scores at least −1 SD below average, on the CVLT, long delay free recall, improved by +1 to +3.5 SD at 2 months post-tLED.

25.4.4 Specific transcranial light-emitting diode placements may affect specific parts of the central executive network in traumatic brain injury cases

Although no rs-fcMRI, or task-specific fMRI studies were part of this pilot study, specific tLED placements may have had a beneficial, focal effect on specific nodes within the CEN. These include the following: (1) The tLED placements, located immediately posterior to the left and right front hairline, likely targeted the DLPFC areas; and (2) the tLED placements, located posterior and superior to each ear, likely targeted the inferior parietal cortex/PPC (angular gyrus) areas. Each of these fronto-parietal LED placement areas represent nodes within the CEN.

The two patients who had showed no change on the Stroop, trial 4, nhibition switching at 2 months post-tLED (P7, P9), both showed improvement, however, on the CVLT tests—for example, +1.5 SD on the total trials, 1−5; and +1 SD on the long delay dree recall. Thus, every patient who entered the Naeser et al. (2014) study (regardless of severity level at entry) improved by at least +1 SD on either the Stroop, trial 4, inhibition switching (most cases, seven out of nine); or on the CVLT tests (seven cases improved on the CVLT long delay free recall). P7 and P9 who had not improved on Stroop, trial 4, inhibition switching improved by at least 1 SD on the CVLT tests—for example, total trials 1−5, and on long delay free recall. P7 and P9 had suffered multiple concussions, and the disruption of white matter pathways and functional-connectivity patterns for these two mTBI cases was likely quite variable.

25.4.5 Depression

There was only a trend for significant change in depression at the 1-week post-tLED testing ($P = .045$), and not an overall linear effect at 2 months post-tLED. A total of only five cases had entered the study with moderate or severe depression. The pattern of initial reduction in depression at 1-week post-tLED in 4/5 of these cases (but not an overall lasting change at 1 or 2 months post-tLED), is similar to the results observed in the Schiffer et al. (2009) study with 10 severe depression cases, where depression was significantly reduced at 2 weeks post a single, NIR tLED treatment to the left and right forehead areas, but scores returned toward baseline at 4 weeks post-tLED. In both the Schiffer et al. (2009) and Naeser et al. (2014) studies, however, most of the post-tLED depression scores did not return to the pre-LED levels. Some studies have suggested that there is downregulation of neurogenesis in the hippocampus in major depression (Hanson et al., 2011). The potential role of tLED (or iLED) to upregulate neurogenesis via the hippocampus requires further study. Our data, as well as that from Schiffer et al. (2009) suggest, however, that continued, on-going tLED treatments, would seem to be necessary for sustained reduction in depression.

25.4.6 Posttraumatic stress disorder relationship to intrinsic networks, default mode network and salience network

Impaired response inhibition has been observed in veterans who had mTBI plus PTSD (or mTBI without PTSD) (Swick et al., 2012). These authors observed more errors on NoGo trials in both groups, compared to controls. In the Naeser et al. (2014) study, four of the mTBI cases had PCL-C scores pre-tLED, that were suggestive of PTSD and three of these four cases also had pre-tLED, Stroop trial 4, inhibition switching scores that were at least −2 SD at entry (P4, 5, 10). (The fourth case, P3, entered with an average score of 0 SD.) All four of these cases reported reduced PTSD symptoms by a "reliable decrease," or a "clinically meaningful decrease," post-tLED (Fig. 25.2) and all four of these cases improved by +1 SD to +2 SD on the Stroop trial 4, inhibition switching, post-tLED. Thus, improved inhibition could also have had a beneficial effect on level of PTSD, post-tLED.

Abnormalities in the DMN have been reported in studies with patients who have PTSD (Daniels et al., 2010; Sripada et al., 2012). Sripada and colleagues observed "reduced functional-connectivity within the DMN (between DMN seeds and other DMN regions) including rostral ACC/vmPFC; and increased connectivity within the SN (between insula seeds and other SN regions) including the amygdala." In addition, there was an increased cross-network connectivity where DMN seeds showed elevated connectivity with SN regions including the insula; and SN seeds exhibited increased connectivity with DMN regions including the hippocampus. Their results suggested a dominance of "threat-sensitive circuitry in PTSD, even in task-free conditions. Disequilibrium between large-scale networks subserving salience detection versus internally focused thought may be associated with PTSD pathophysiology." Thus, the four participants who experienced reduced PTSD symptoms post-tLED therapy that included treatment of nodes within the DMN and the SN, may have experienced less PTSD symptoms due to better modulation between these two important intrinsic networks. Studies with rs-fcMRI pre- and post-tLED, would be necessary to further examine the potential relationship between decrease in PTSD and LED placements.

Menon (2011) has proposed a "triple network model of aberrant saliency mapping and cognitive dysfunction in psychopathology" that is present in psychiatric as well as neurological disorders. The three large-scale brain networks are the DMN, the SN, and CEN. Results from the Naeser et al. (2014) tLED study with chronic, mTBI cases suggest that all three of these networks may have been abnormal pre-LED, but modulated in function, post a series of 18 tLED treatments that likely targeted nodes within each of these three networks. Stevens et al. (2012) studied 12 separate resting-state networks in 30 mTBI cases (13–136 days postinjury), and observed abnormal functional-connectivity in every brain network, including visual processing, motor, limbic, and numerous circuits believed to underlie executive cognition; some connections were decreased, and some increased. "PCS severity was linked to abnormal regional connectivity within nearly every brain network identified, particularly the anterior cingulate." Thus, one of the most important tLED placement areas may be the LED placement area located at the midline of the face, centered over the front hairline and the forehead, likely targeting the dACC, of the SN; and the mPFC, of the DMN. The LED placement area targeting the precuneus (DMN), and the LED placement areas targeting the DLPFC areas (CEN) are also among the most important placement areas.

25.4.7 Weak connections between cortical nodes within intrinsic neural networks

Several rs-fcMRI studies with TBI patients have suggested that the intrinsic neural networks continue to be present postinjury, but their connections are weak (Marquez de la Plata et al., 2011; Nakamura et al., 2009; Zhang et al., 2012). Some of the aberrant, or weak, functional connections may persist for months, or even years postinjury (Slobounov et al., 2012). Menon (2011) has commented that weak anatomical connectivity within and acrossnetwork nodes can compromise dynamic interaction of the core networks, all of which can result in abnormal psychiatric and neurological behavior.

25.4.8 Mechanisms and cellular effects, post-red/near-infrared transcranial light-emitting diode

While the specific mechanisms and cellular effects involved with improved cognitive function post-tLED are not entirely known, a few possible mechanisms are listed below.

1. There may be increased ATP, especially in cells comprising the large-scale, intrinsic networks (DMN, SN, CEN), where large demands for energy are constant. Increased ATP would improve cellular respiration, oxygenation, and function of these hypoxic/comprised cells.
2. There may be increased vasodilation/rCBF in surface cortical areas, subjacent to the scalp placements of the LED cluster heads. Previous animal studies using transcranial NIR LED to treat acute TBI (Khuman et al., 2012) have suggested that the mitochondria are the primary target, where absorption of photons by cytochrome c oxidase releases bound nitric oxide (diffusing it outside the cell, promoting local vasodilation). Two previous studies with humans have reported an increase of rCBF in cortical areas, subjacent to LED placements on the forehead (Schiffer et al., 2009; Nawashiro et al., 2012).
3. There is a potential increase in antioxidant enzymess. The red/NIR photons induce redox-sensitive transcription factors such as nuclear factor-kappa B that promote gene transcription (Chen et al., 2011). The strong antioxidant, mitochondrial superoxide dismutase, is one of the most upregulated genes after NF-kB activation (Sompol et al., 2006). Another highly upregulated gene after NF-kB activation and after LED/photobiomodulation (PBM) is heat-shock protein 70, a molecular chaperone for protein molecules that prevents misfolding and unwanted protein aggregation, especially at the telomeres of the DNA (Zhang et al., 1994).
4. There may be decreased inflammation, post-tLED. Photons in the red/NIR wavelengths reduce inflammation (Rojas and Gonzalez-Lima, 2011). The study by Khuman et al. (2012) showed that transcranial, NIR PBM therapy had an antiinflammatory effect, inhibiting microglial activation, when treating acute TBI in mice. Many reports demonstrate that red/NIR photons reduce COX-2 expression levels and reduce prostaglandins in multiple animal models, as well as in vitro (Khuman et al., 2012; Aimbire et al., 2005).

5. There may be increased sleep, which was observed in a quantitative manner using the Actiwatch system, in our pilot studies at the VA Boston Healthcare System. The improved sleep was observed in TBI cases who received the tLED treatment series, or the iLED series only (Naeser et al., 2016; Bogdanova et al., 2017).
6. One of the most important considerations may be the concept that tLED PBM may improve brain function by stimulating neurogenesis and synaptogenesis. These notions are based on small animal studies using tPBM to treat acute, severe TBI in mice (Xuan et al., 2014a,b). Two key sites for adult human neurogenesis include the subventricular zone of the lateral ventricles, and the subgranular layer of the dentate gyrus in the hippocampus (Eriksson et al., 1998). "Neurogenesis persists in the adult mammalian brain, where it can be stimulated by physiological factors, such as growth factors and environmental enrichment, and by pathological processes, including ischemia..." (Jin et al., 2006).

25.5 Photobiomodulation to improve language in chronic aphasia, due to left hemisphere stroke

25.5.1 Stroke-aphasia

Aphasia is a language disorder due to acquired brain damage primarily in the left hemisphere (LH; stroke, trauma, tumor, or neurodegenerative disease). LH damage produces aphasia for the vast majority of cases, including right- and left-handers (Naeser and Borod, 1986). In aphasia, the ability to communicate by oral or written language, or both, is affected (Goodglass, 1993). Aphasia is further defined as "...a multi-modality disorder represented by a variety of impairments in auditory comprehension, reading, oral-expressive language, and writing...but it cannot be explained by dementia, sensory loss, or motor dysfunction" (Rosenbek et al., 1989).

When aphasia is caused by stroke, it typically results from injury to an extended network of cortical and subcortical structures perfused by the left middle cerebral artery (Fama and Turkeltaub, 2014), but can also extend into brain structures perfused by the left anterior or posterior cerebral arteries. Lesion size is not always a factor in the prognosis for language recovery (Lazar et al., 2008), but rather lesion site is critically important. For example, deep, smaller white matter lesions located near borders of the left lateral ventricle are often associated with severe limitation in speech output, and the extent of these white matter lesions are predictive for potential of speech recovery (phrase length, etc.) (Naeser et al., 1989; Naeser and Palumbo, 1994). Lesion size has been observed, however, to be associated with general ability to name pictures in chronic Wernicke's aphasia cases (Naeser and Palumbo, 1994), and in anomia aphasia cases (Kertesz, 1979).

In the chronic stage poststroke onset (greater than 6 months), approximately 50%–60% of aphasia cases have communicative impairment (Pedersen et al., 2004). In patients with chronic aphasia, improved language following various therapies has been associated with new activation in perilesional and remaining LH areas (Leger et al., 2002; Martin et al., 2009; Meinzer et al., 2008; Small et al., 1998). In fact, Heiss and Thiel (2006) have suggested that for better long-term recovery, right hemisphere (RH) recruitment may be less efficient than restoring the LH network.

The RH may play a role in supporting some recovery, however, when there is a great deal of damage to LH language areas (Fernandez et al., 2004); new RH activation has been observed following speech-language therapy in some studies (Cherney and Small, 2006; Crosson et al., 2005; Marcotte and Ansaldo, 2010).

Presence of over-activation in RH inferior frontal and motor/sensory areas for the mouth, in the chronic stage poststroke has been associated with poor, hesitant, agrammatic (nonfluent) speech (Naeser et al., 2004; Postman-Caucheteux et al., 2010; Rosen et al., 2000). Our research with repetitive transcranial magnetic brain stimulation (rTMS) in chronic, nonfluent aphasia cases observed significant improvement in picture-naming and phrase length following suppression of a portion of the RH Boca's area (pars triangularis, in R inferior frontal gyrus) (Naeser et al., 2005). Functional MRI studies before and after the rTMS series showed less activation in the RH frontal areas, and increased activation in the LH perilesional areas post-rTMS (Martin et al., 2009).

The relative contribution of the LH perilesional areas (vs the RH homologous language areas) to recovery from aphasia in stroke has recently been reviewed (Fama and Turkeltaub, 2014; Shah et al., 2013). Results from recent studies have continued to support the role of the LH perilesional areas in better recovery. In the chronic stage poststroke (1 year or more), better language recovery was associated with fMRI activation patterns in the LH areas, including increased activation in the left frontal and parietal areas, bilateral cerebellum, and decreases in the RH superior temporal areas. Stroke patients with poor language recovery had increased activation in the RH (Saur et al., 2006; Szaflarski et al., 2013). While most studies have supported a beneficial role for increased LH activation in better aphasia recovery, some areas within the RH may also contribute. Less is known, however, regarding the role of the RH, and additional studies are required (Shah et al., 2013).

Whether recovery in aphasia is mediated primarily from LH perilesional areas, or from RH language homologues (or both), the above-mentioned studies all suggest there is potential for brain reorganization and improved language in chronic, poststroke aphasia (Crosson et al., 2009; Price and Crinion, 2005).

25.5.2 Importance of specific light-emitting diode placement areas on the scalp to treat aphasia, in chronic stroke

Naeser et al. (2012) carried out a pilot, research treatment study using tLED to treat nonfluent aphasia. Two R-handed, chronic, nonfluent aphasia patients with single, LH stroke were treated with tLED (P1, 67YrM, 18 years poststroke; P2, 59YrM, 12 years poststroke). They were each treated with two, separate LED treatment methods, identical for each case.

Prior to any LED treatments, each patient was tested three times (separate visits) for entry/baseline language abilities with the Boston Diagnostic Aphasia Exam (BDAE) (Goodglass and Kaplan, 1983), and the Philadelphia Naming Test (PNT) (Roach et al., 1996). Each patient was also tested after the LED series was completed, at 1-2 weeks, and again at 1-2 months after the final LED treatment. Significant change was defined as $\pm$ 2 SD from baseline.

Structural MRI scans showed LH lesion sites compatible with nonfluent aphasia, even though P2 had primarily subcortical lesion (Naeser et al., 1989; Naeser and Palumbo, 1994). Overt-naming task fMRI scans (3-Tesla Philips Achieva) were obtained for P1, pre- and post- each tLED treatment series (first series, bilateral; and second series, left hemisphere only, explained below). The hemodynamic delay method of Martin et al. (2005) was used.

The MedX Health (Toronto) LED cluster heads (FDA-cleared, nonsignificant risk device) were applied to the head. Each LED cluster head was 5.35 cm diameter (9 red, 633 nm diodes; 52 NIR, 870 nm diodes); 22.48 cm^2 in size; 500 mW; 22.2 mW/cm^2 power density; pulsed wave 146 Hz (duty cycle, 80% on). Each aphasia patient was first treated with bilateral, LED placements, using the same tLED placement loci used with chronic TBI cases explained above (Naeser et al., 2014). Two months later, each patient was treated with LH only, tLED placements.

25.5.3 Bilateral transcranial light-emitting diode treatment method

The following midline/mid-sagittal areas, plus L and R cortical areas were treated: Frontal poles (LEDs were placed on the forehead, centered above each eyebrow but below the front hairline); inferior frontal gyri (IFG); and anterior temporal lobes (on temple areas); middle frontal gyri (MFG) (on the pupil line, immediately posterior to front hairline); superior temporal gyri, STG (immediately superior to ear tip); inferior parietal areas (posterior–superior to ear tip); lower sensori-motor (S-M) cortex, mouth areas (immediately posterior to LED placement for MFG); and midline (front-hairline to back-hairline placements) including anterior cingulate cortex and ventral medial prefrontal cortex (at center front hairline); supplementary motor areas, SMAs (near vertex); and precuneus areas (superior to occipital protuberance, halfway towards the vertex), alternating with inferior to the occipital protuberance, at every other treatment.

Six, 5.35 cm diameter LED cluster heads were applied simultaneously (Set A, followed by Set B), 13 J/cm^2 (12 minutes 11 seconds per set, and 146 Hz, PW). It was estimated that 0.4 J/cm^2 reached brain cortex. The LED therapy is noninvasive, painless, and nonthermal. Patients were treated in a recliner chair. The LED cluster heads were held in place with a soft nylon elastic cap. Each patient received 18 tLED treatments (M,W,F) for 6 weeks, followed by posttesting.

25.5.4 Left hemisphere only, transcranial light-emitting diode treatment method

Only some of the same LED placement areas listed above in the bilateral tLED treatment method were used, and the LED cluster heads were placed only on LH areas. No midline areas were treated, including no placement of an LED cluster head on the L&R SMAs (near vertex). No LEDs were placed on the R side of the head.

The same LED device was used, however, with application of an energy density of 39 J/cm^2 (36 minutes, 33 seconds, PW 146 Hz) per LED cluster head applied to the LH only, placement loci (L side of head). It was estimated that 1.2 J/cm^2 reached brain cortex. Each patient received 18 tLED treatments (M, W, F) for 6 weeks, followed by posttesting.

25.5.5 Results

Unexpectedly, post the bilateral LED treatment series, there was significant decrease in ability to name pictures of objects for P1; and no significant change for P2. Conversely, post the unilateral LH only, LED treatment series there was a significant increase in ability to name pictures for each stroke patient. On the PNT, P1 improved from 25.33 (SD, 1.53) at baseline, to 29, at 1 month post the LH only, LED treatment series. P2 almost doubled his score on the PNT, improving from 5 (SD, 2) at baseline, to a score of 9, at 1 week post the LH only, LED treatments.

Changes on the overt-naming fMRI scans post each tLED treatment series paralleled changes in ability to name pictures for P1. Post the bilateral LED treatment series, there was increased bilateral activation including L peri-lesional S-M, mouth area; L & R SMAs and R fronto-temporo-parietal areas. High activation in R frontal and the R SMA has been associated with poor naming and poor language recovery in aphasia (Martin et al., 2009; Karbe et al., 1998; Weiduschat et al., 2011). P1 had shown a significant decrease in picture naming at 1 month post the bilateral LED treatment series.

Conversely, post the unilateral LH only, LED treatment series, the overt-naming, task-fMRI scans for P1, especially at 2 weeks post-tLED treatment, showed increased activation in the ipsi-lesional, LH side only (the side treated with LEDs), and in the L SMA (despite no direct treatment to the L SMA area itself). There are known connections between the LH language network and L SMA. Activation in the LH included L peri-lesional, S-M mouth areas, as well as the L SMA. There was little or no activation in R frontal, temporal, parietal contra-lesional areas, or the R SMA; these RH areas were not treated with the LED cluster heads in the unilateral LH only, LED treatment series. Activation in the L SMA (along with absence of activation in the R SMA) was likely associated with the significant increase in naming that was present, post the unilateral LH only, LED treatment series for P1 (Martin et al., 2009; Karbe et al., 1998; Weiduschat et al., 2011).

The tLED treatments appeared to target local, cortical neural networks subjacent to the extra-cranial LED placement loci. Trancranial LED therapy with appropriate wavelengths, energy density (J/cm^2) and power density (mW/cm^2), has potential to modulate brain plasticity and promote recovery in stroke, as well as other central nervous system disorders. The Naeser et al. (2012) study was the first report of fMRI documentation for a focal, hemispheric effect following scalp application of red/NIR LED therapy to treat stroke.

After the second tLED treatment series (left side of the head, to treat the LH only), there was significant increase in picture naming for both stroke cases; P1 showed LH perilesional activation. Thus, these data suggest that in left-hemisphere stroke patients with aphasia, the tLED should be applied to the left side of the head only, in order to have a beneficial effect. A general recommendation in treating acute or chronic stroke patients would be to only use tLED placements that are ipsilesional to the side of the brain where the stroke took place; not contralesional or bilateral.

Perhaps even better results would have been found for the NEST-1 and NEST-2 trials with tPBM to treat acute stroke patients (Stemer et al., 2010) if, instead of treating both the left and the right sides of the head (irrespective of hemispheric side of the stroke), the tPBM placements had been restricted to the ipsilesional side of the head. Our better language results following LED placements only on the LH side of the head (ipsilesional to the LH stroke), is supportive of fMRI research studies reviewed above, where chronic aphasia patients with increased LH activation (not RH activation) had better language recovery (Saur et al., 2006; Szaflarski et al., 2013). Although the PhotoThera studies used a single measure for overall stroke severity, the NIH Stroke Severity Scale pre- and post-tPBM, the same idea would apply to overall stroke recovery—that is, application of the tPBM may well produce better results overall, if the tPBM is applied only to the ipsilesional side. Our later studies with LH stroke patients with aphasia observed better naming ability if in addition to the LH, LED placements, two midline, cortical nodes of the DMN were added—for example, mPFC, located at midline, high forehead at the hairline; and precuneus, located midline, halfway between vertex and the occipital protuberance; and if 26 J/cm^2 was used, instead of 13 J/cm^2 (Ho et al., 2016). More research studies that examine the short-term and long-term effects of tPBM placements are needed.

25.5.6 Photobiomodulation to treat primary progressive aphasia, a neurodegenerative disease

Naeser et al. (2012) also reported on the effect of tLED to treat one aphasia patient with the neurodegenerative disease, PPA. Patients with PPA present with symptoms of aphasia for at least 2 years before diagnosis (however, no stroke has taken place); cognitive/memory disturbances are present only in later stages. Impaired naming is the common deficit among the three variants of PPA: nonfluent (nfv), logopenic (lv), and semantic (Kirshner, 2010).

The etiology of PPA is often in the clinical spectrum related to frontotemporal lobar degeneration (FTLD) (Grossman, 2010). Estimates of the prevalence of FTLD are 2.7–15.0 per 100,000. FTLD is the second most common dementia behind AD (below the age of 65) and the fourth most common dementia in industrialized nations (Reilly et al., 2010). Twenty to forty percent of FTLD cases have PPA (Grossman, 2010). The average age of onset is in the late 50s, although a wide range is reported (20–82). Survival is approximately 7 years (range 2 to 8 + years). The prognosis is highly variable, there is no gender bias and there are no known environmental risk factors (Rosso et al., 2003). There is no known treatment for PPA (Rosso et al., 2003; Gorno-Tempini et al., 2011; Mesulam, 2007).

The three variants of PPA (nfv, lv, semantic) are associated with cortical thinning in specific language-related regions of the brain, primarily in the LH—for example, inferior frontal (nfv), posterior temporo-parietal (lv), or anterior temporal (semantic) (Gorno-Tempini et al., 2011; Sapolsky et al., 2010). Data are inconclusive regarding the histopathological basis for the variants of PPA (Grossman, 2010). PPA is typically associated with one of three classes of pathology: tau-positive immunoreactivity, more common in nfv PPA; amyloid plaques and neurofibrillary tangles, more common in lv PPA; or frontotemporal lobar dementia–ubiquitin immunoreactivity (FTLD-U), more common in semantic PPA (Forman et al., 2006; Lipton et al., 2004; Mann et al., 1993).

One PPA patient diagnosed with lv PPA (similar to Wernicke's and conduction aphasia) was treated with tLED, using the same LED equipment described above. She received only the bilateral LED placements (LH and RH sides, and the entire midline/mid-sagittal placements from front hairline to back hairline, including the vertex, bilateral SMAs). She was treated three times a week (M,W,F) for 6 weeks.

After the bilateral LED series, there was significant increase in sentence-level auditory comprehension (BDAE) at 1 week post the tLED treatment series (but not at 1 and 2 months post-tLED). There was no impairment in picture naming, at any time. Her overt-naming fMRI scans showed increased right and left perisylvian activation (but not SMA), post the tLED treatment series. It was noted that there was increased L temporo-parietal lobe activation on the fMRI scan at 1 week post-tLED. These are areas that had showed significant cortical thinning on the structural MRI scan in this case; compatible with her lv (Wernicke's/conduction fluent aphasia) (see Fig. 25.9). The new L temporo-parietal activation may have been compatible with her significant improvement in sentence-level, auditory comprehension at 1 week post-tLED. This improvement was not presentat the 1 month and 2 month follow-up language testing. Also, the increased L temporo-parietal activation that was present at 1 week post-tLED on the overt naming fMRI scan, was no longer present on the 1 month and 2 month follow-up fMRI scans. This rapid deterioration of the improved auditory comprehension and increased temporo-parietal activation that was present on fMRI at 1 week post-tLED, without continued benefit at 1 and 2 months post-tLED, suggests that this process was likely related to the progressive nature of the neurodegenerative disease. A case such as this would require continued tLED treatments, preferably at home, especially to the LH only, and to the midline mPFC and to the precuneus, midline nodes of the DMN.

In summary, there was significant improvement in sentence-level, auditory comprehension at 1 week post-tLED in this lv PPA case, along with increased LH activation in LH temporo-parietal areas where she had significant cortical thinning. These results suggest that additional tLED studies should be undertaken with PPA cases, however, the tLED placements should be on the LH only (plus two midline DMN placements, mesial prefrontal and precuneus), and the tLED treatments may need to be continued, indefinitely. The goal would be to improve language and quality of life for as long as possible, in the PPA cases. These tLED treatments can be performed in the home (Naeser et al., 2011).

25.6 Photobiomodulation for possible chronic traumatic encephalopathy

A case study (Martin et al., 2018) was carried out involving a retired, professional football player (65 year old, middle linebacker; 700 tackles in college; thousands of subconcussive hits), PhD in Sports Physiology, with a 4- to 10-year history of cognitive decline and emotional outbursts. The following treatment regimens were carried out: (1) In-office tLED treatment series: 6 weeks (three times a week), with 500 mW (MedX Health), red/NIR LED cluster heads (633 nm/870 nm), 26 J/cm^2 per LED placement, same placements as in previous TBI studies (Naeser et al., 2014; Naeser et al., 2016). (2) 12 weeks without tLED. (3) At-home LED treatment 12 weeks, with NIR tLEDs and iLED (Vielight Neuro Gamma, pulsed at 40 Hz), where only the cortical node areas of the DMN were treated—for example, bilateral, mesial prefrontal, precuneus, angular gyrus, and hippocampal areas. The NIR diodes were 25–100 mW, fluence 15–60 J/cm^2, plus an extra, 8 mW red, 633 nm, intranasal (CW) single LED.

Posttesting at 1 week after the in-office LED series showed +2 SD on 5 NP tests (+1 SD, 9); 1 month later, +2 SD, 3 tests (+1 SD, 12); but at 2 months, +2 SD, only on 1 test (+1 SD, 5). The same pattern was observed regarding emotional outbursts—for example, PTSD (PCL-C) scores, pre-58, improved to 25 and 32 at 1 week and 1 month; but 47 at 2 months; and with depression, Beck depression scores, pre-24, improved to 1 and 0 at 1 week and 1

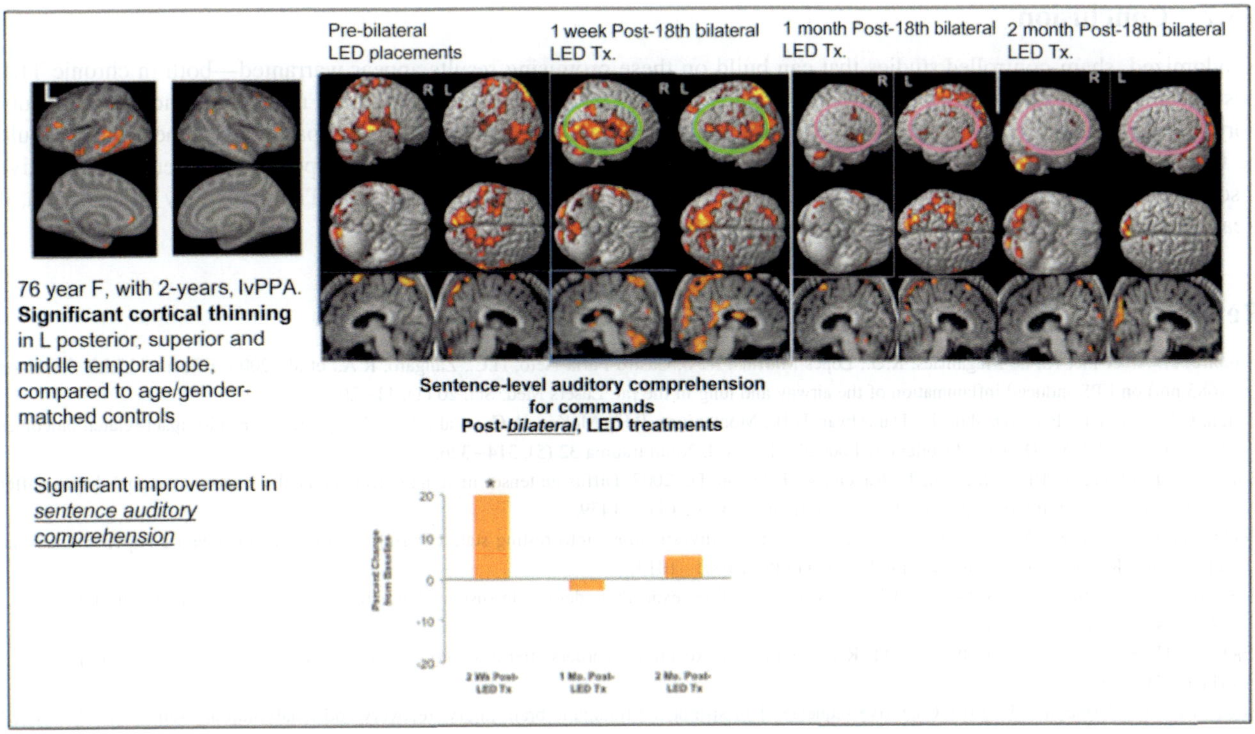

FIGURE 25.9 Language data and fMRI scans for a primary progressive aphasia (PPA) patient who received 18 tLED treatments, with bilateral LED placements, 13 J/cm^2, pulsed wave, 146 Hz, 10 min per placement, 11 placements. These are the same LED placement areas used with our original TBI patients (Naeser et al., 2014). This PPA case showed significant improvements in sentence-level auditory comprehension, at the 1 week, posttesting time point. However, at 1 month and at 2 months later, the beneficial effect had worn off. On overt-naming task fMRI, she initially showed increased activation in the left and right temporal lobes at 1 week post the 18th LED treatment (*green circles*). However, at 1 month and at 2 months, there was no increased activation in the left and right temporal lobes, and the effect of increased activation on the fMRI scans had worn off (*pink circles*). Her language problems were associated with a progressive neurodegenerative disease—for example, PPA. She likely had beta-amyloid, abnormal protein deposits in the left temporal lobe, where major cortical thinning was present (see *red areas* of cortical atrophy, in upper left corner of figure). If improvements are initially observed, patients who have a progressive neurodegenerative disease will likely need to continue with LED home treatments, perhaps life-long. See section below, on using transcranial and intranasal LED home treatments in retired, professional football player number 1. Significant improvements were again observed after 3 months of at-home LED therapy in his case. He continues with LED home treatments 10 months later and anecdotally reports doing well.
This PPA patient may do even better, if the LED placements are only over the left hemisphere (and two midline cortical nodes of the DMN—mesial prefrontal cortex, and high-parietal/precuneus). *Naeser, M.A., Ho, M., Martin, P.I., Treglia, E., Krengel, M.H., Hamblin, M.R., et al., 2012. Improved language after scalp application of red/near-infrared light-emitting diodes: pilot study supporting a new, noninvasive treatment for chronic aphasia. Procedia Soc. Behav. Sci. 61, 138–139.*

month, but 9, at 2 months. After a 3-month no treatment period, he started treating at home with the NIR LED device described above (neuro gamma), and an extra, red intranasal diode.

After 3 months of at-home LED treatments, there were again significant improvements versus baseline, +2 SD on six tests (+1 SD, 10); reduced PTSD, PCL-C, 23; and no depression. The football player's rs-fcMRI scans showed parallel changes—that is, increased functional connectivity after the in-office LED treatment series, and after the at-home LED series, but decreased, after the 3-month no treatment period. Decline after the no treatment period in this case, was similar to observations with patients who have a progressive neurodegenerative disease—for example, our PPA patient (Naeser et al., 2012), explained above; and five dementia cases treated with tLED plus iLED (Saltmarche et al., 2017). This pattern was not seen with our original mTBI cases whose improvements post-tLED remained stable, or even improved further at 2 months post-tLED (Naeser et al., 2014, 2016) (see Fig. 25.1). Those with a progressive neurodegenerative disease may require continued at-home LED treatments, to maintain previous gains.

A second, retired, professional football player (57 year old, cornerback) was also treated with the in-office tLED treatment series. He was treated using the Thor helmet, on the same treatment schedule. His improvements at 1 week and at 1 month post-tLED are similar to football player number 1, with improved cognition, reduced emotional outbursts and reduced depression. In addition, he was able to discontinue two narcotic pain medications, at 1 month post-tLED. There was also reduction of tinnitus and improved vestibular function. He is currently doing tLED home treatments, similar to football player number 1.

25.7 Conclusion

Randomized, sham-controlled studies that can build on these promising results appear warranted—both in chronic TBI, poststroke aphasia, and athletes possibly developing CTE. If these behavioral results are replicated, and there are supportive MRI studies, then perhaps additional patient populations (e.g., mild cognitive impairment, dementia/AD) could be treated with red/NIR tLED therapy and/or iLEDs (especially in the earlier stages of a progressive neurodegenerative disorder). In addition, the possibility of prevention of long-term cognitive dysfunction in TBI cases may be an option, if treatments can be provided earlier postinjury.

References

Aimbire, F., Albertine, R., de Magalhaes, R.G., Lopes-Martins, R.A., Castro-Faria-Neto, H.C., Zangaro, R.A., et al., 2005. Effect of LLLT Ga-Al-As (685 nm) on LPS-induced inflammation of the airway and lung in the rat. Lasers Med. Sci. 20 (1), 11–20.

Baugh, C.M., Kiernan, P.T., Kroshus, E., Daneshvar, D.H., Montenigro, P.H., McKee, A.C., et al., 2015. Frequency of head-impact-related outcomes by position in NCAA Division I Collegiate Football Players. J. Neurotrauma 32 (5), 314–326.

Bazarian, J.J., Zhong, J., Blyth, B., Zhu, T., Kavcic, V., Peterson, D., 2007. Diffusion tensor imaging detects clinically important axonal damage after mild traumatic brain injury: a pilot study. J. Neurotrauma 24 (9), 1447–1459.

Beckmann, C.F., DeLuca, M., Devlin, J.T., Smith, S.M., 2005. Investigations into resting-state connectivity using independent component analysis. Phil. Trans. R. Soc. Lond. Series B, Biol. Sci. 360 (1457), 1001–1013.

Bloomfield, I.L., Espie, C.A., Evans, J.J., 2010. Do sleep difficulties exacerbate deficits in sustained attention following traumatic brain injury? J. Int. Neuropsychol. Soc. 16 (1), 17–25.

Boelen, D.H., Spikman, J.M., Fasotti, L., 2011. Rehabilitation of executive disorders after brain injury: are interventions effective? J. Neuropsychol. 5 (Pt 1), 73–113.

Bogdanova, Y., Verfaellie, M., 2012. Cognitive sequelae of blast-induced traumatic brain injury: recovery and rehabilitation. Neuropsychol. Rev. 22 (1), 4–20.

Bogdanova, Y., Ho, V.T., Martin, P.I., Ho, M.D., Yee, M.K., Hamblin, M.R., et al., 2017. Treatment for cognitive dysfunction and sleep in chronic TBI: randomized controlled pilot trial. Abstract. Arch. Phys. Med. Rehabil. 98 (10), e122–e123.

Bogdanova, Y., Martin, P.I., Ho, M.D., Krengel, M.H., Ho, V.T., Yee, M.K., et al., 2014. LED therapy improves sleep and cognition in chronic moderate TBI: pilot case studies. Abstract. Arch. Phys. Med. Rehabil. 95 (10), e77. Available from: https://doi.org/10.1016/j.apmr.2014.07.247.

Bonnelle, V., Leech, R., Kinnunen, K.M., Ham, T.E., Beckmann, C.F., De Boissezon, X., et al., 2011. Default mode network connectivity predicts sustained attention deficits after traumatic brain injury. J. Neurosci. 31 (38), 13442–13451.

Bonnelle, V., Ham, T.E., Leech, R., Kinnunen, K.M., Mehta, M.A., Greenwood, R.J., et al., 2012. Salience network integrity predicts default mode network function after traumatic brain injury. Proc. Natl. Acad. Sci. U.S.A. 109 (12), 4690–4695.

Bryan, C.J., 2013. Repetitive traumatic brain injury (or concussion) increases severity of sleep disturbance among deployed military personnel. Sleep 36 (6), 941–946.

Cantu, R.C., 2006. Concussion classification: ongoing controversy. In: Slobounov, S., Sebastianelli, W. (Eds.), Foundations of Sports-Related Brain Injuries. Springer Science and Business Media, Inc., New York, NY.

Chen, A.C., Arany, P.R., Huang, Y.Y., Tomkinson, E.M., Sharma, S.K., Kharkwal, G.B., et al., 2011. Low-level laser therapy activates NF-kB via generation of reactive oxygen species in mouse embryonic fibroblasts. PLoS One 6 (7), e22453.

Cherney, L.R., Small, S.L., 2006. Task-dependent changes in brain activation following therapy for nonfluent aphasia: discussion of two individual cases. J. Int. Neuropsychol. Soc. 12 (6), 828–842.

Chew, E., Zafonte, R.D., 2009. Pharmacological management of neurobehavioral disorders following traumatic brain injury—a state-of-the-art review. J. Rehabil. Res. Dev. 46 (6), 851–879.

Cicerone, K., Levin, H., Malec, J., Stuss, D., Whyte, J., 2006. Cognitive rehabilitation interventions for executive function: moving from bench to bedside in patients with traumatic brain injury. J. Cogn. Neurosci. 18 (7), 1212–1222.

Cicerone, K.D., Langenbahn, D.M., Braden, C., Malec, J.F., Kalmar, K., Fraas, M., et al., 2011. Evidence-based cognitive rehabilitation: updated review of the literature from 2003 through 2008. Arch. Phys. Med. Rehabil. 92 (4), 519–530.

Crosson, B., Moore, A.B., Gopinath, K., White, K.D., Wierenga, C.E., Gaiefsky, M.E., et al., 2005. Role of the right and left hemispheres in recovery of function during treatment of intention in aphasia. J. Cogn. Neurosci. 17 (3), 392–406.

Crosson, B., Moore, A.B., McGregor, K.M., Chang, Y.L., Benjamin, M., Gopinath, K., et al., 2009. Regional changes in word-production laterality after a naming treatment designed to produce a rightward shift in frontal activity. Brain Lang. 111 (2), 73–85.

Curtis, C.E., D'Esposito, M., 2003. Persistent activity in the prefrontal cortex during working memory. Trends Cogn. Sci. 7 (9), 415–423.

Daniels, J.K., McFarlane, A.C., Bluhm, R.L., Moores, K.A., Clark, C.R., Shaw, M.E., et al., 2010. Switching between executive and default mode networks in posttraumatic stress disorder: alterations in functional connectivity. J. Psychiatry Neurosci. 35 (4), 258–266.

Delis, D.C., Kramer, J.H., Kaplan, E., Holdnack, J., 2004. Reliability and validity of the Delis-Kaplan Executive Function System: an update. J. Int. Neuropsychol. Soc. 10 (2), 301–303.

Eriksson, P.S., Perfilieva, E., Bjork-Eriksson, T., Alborn, A.M., Nordborg, C., Peterson, D.A., et al., 1998. Neurogenesis in the adult human hippocampus. Nat. Med. 4 (11), 1313–1317.

Fama, M.E., Turkeltaub, P.E., 2014. Treatment of poststroke aphasia: current practice and new directions. Semin. Neurol. 34 (5), 504–513.

Faraguna, U., Vyazovskiy, V.V., Nelson, A.B., Tononi, G., Cirelli, C., 2008. A causal role for brain-derived neurotrophic factor in the homeostatic regulation of sleep. J. Neurosci. 28 (15), 4088–4095.

Faul, M., Xu, L., Wald, M.M., Coronado, V.G., 2013. Traumatic Brain Injury in the United States: Emergency Department Visits, Hospitalizations, and Deaths. CDC, Atlanta, GA [updated March 27, 2013]. Available from: http://www.cdc.gov/traumaticbraininjury/statistics.html.

Fernandez, B., Cardebat, D., Demonet, J.F., Joseph, P.A., Mazaux, J.M., Barat, M., et al., 2004. Functional MRI follow-up study of language processes in healthy subjects and during recovery in a case of aphasia. Stroke 35 (9), 2171–2176.

Forman, M.S., Farmer, J., Johnson, J.K., Clark, C.M., Arnold, S.E., Coslett, H.B., et al., 2006. Frontotemporal dementia: clinicopathological correlations. Ann. Neurol. 59 (6), 952–962.

Gentleman, S.M., Leclercq, P.D., Moyes, L., Graham, D.I., Smith, C., Griffin, W.S., et al., 2004. Long-term intracerebral inflammatory response after traumatic brain injury. Forensic Sci. Int. 146 (2–3), 97–104.

Gilbert, K.S., Kark, S.M., Gehrman, P., Bogdanova, Y., 2015. Sleep disturbances, TBI and PTSD: implications for treatment and recovery. Clin. Psychol. Rev. 40, 195–212.

Goldstein, L.E., Fisher, A.M., Tagge, C.A., Zhang, X.L., Velisek, L., Sullivan, J.A., et al., 2012. Chronic traumatic encephalopathy in blast-exposed military veterans and a blast neurotrauma mouse model. Sci. Transl. Med. 4 (134), 134ra60.

Goodglass, H., 1993. Understanding Aphasia. Academic Press, San Diego.

Goodglass, H., Kaplan, E., 1983. Assessment of Aphasia and Related Disorders. Lea and Febiger, Philadelphia.

Gorno-Tempini, M.L., Hillis, A.E., Weintraub, S., Kertesz, A., Mendez, M., Cappa, S.F., et al., 2011. Classification of primary progressive aphasia and its variants. Neurology 76 (11), 1006–1014.

Greicius, M.D., Supekar, K., Menon, V., Dougherty, R.F., 2009. Resting-state functional connectivity reflects structural connectivity in the default mode network. Cereb. Cortex 19 (1), 72–78.

Grossman, M., 2010. Primary progressive aphasia: clinicopathological correlations. Nat. Rev. Neurol. 6 (2), 88–97.

Guskiewicz, K.M., McCrea, M., Marshall, S.W., Cantu, R.C., Randolph, C., Barr, W., et al., 2003. Cumulative effects associated with recurrent concussion in collegiate football players: the NCAA Concussion Study. JAMA 290 (19), 2549–2555.

Hanson, N.D., Owens, M.J., Nemeroff, C.B., 2011. Depression, antidepressants, and neurogenesis: a critical reappraisal. Neuropsychopharmacology 36 (13), 2589–2602.

Hartikainen, K.M., Waljas, M., Isoviita, T., Dastidar, P., Liimatainen, S., Solbakk, A.K., et al., 2010. Persistent symptoms in mild to moderate traumatic brain injury associated with executive dysfunction. J. Clin. Exp. Neuropsychol. 32 (7), 767–774.

Heiss, W.D., Thiel, A., 2006. A proposed regional hierarchy in recovery of post-stroke aphasia. Brain Lang. 98 (1), 118–123.

Ho, M., Martin, P., Yee, M., Koo, B., Baker, E., Hamblin, M., 2016. Increased functional connectivity in default mode network associated with application of transcranial, light-emitting diodes to treat chronic aphasia: case series. J. Int. Neuropsychol. Soc. 22, 229.

Hoge, C.W., McGurk, D., Thomas, J.L., Cox, A.L., Engel, C.C., Castro, C.A., 2008. Mild traumatic brain injury in U.S. Soldiers returning from Iraq. N. Engl. J. Med. 358 (5), 453–463.

Hurley, R.A., McGowan, J.C., Arfanakis, K., Taber, K.H., 2004. Traumatic axonal injury: novel insights into evolution and identification. J. Neuropsychiatry Clin. Neurosci. 16 (1), 1–7.

Jin, K., Wang, X., Xie, L., Mao, X.O., Zhu, W., Wang, Y., et al., 2006. Evidence for stroke-induced neurogenesis in the human brain. Proc. Natl. Acad. Sci. U.S.A. 103 (35), 13198–13202.

Johnson, B., Zhang, K., Gay, M., Horovitz, S., Hallett, M., Sebastianelli, W., et al., 2012. Alteration of brain default network in subacute phase of injury in concussed individuals: resting-state fMRI study. Neuroimage 59 (1), 511–518.

Johnson, V.E., Stewart, W., Smith, D.H., 2010. Traumatic brain injury and amyloid-beta pathology: a link to Alzheimer's disease? Nat. Rev. Neurosci. 11 (5), 361–370.

Johnson V.E., Stewart W., Smith D.H. Axonal pathology in traumatic brain injury. Exp. Neurol. 2013a 246 35-43.

Johnson, V.E., Stewart, J.E., Begbie, F.D., Trojanowski, J.Q., Smith, D.H., Stewart, W., 2013b. Inflammation and white matter degeneration persist for years after a single traumatic brain injury. Brain 136 (Pt 1), 28–42.

Kan, E.M., Ling, E.A., Lu, J., 2012. Microenvironment changes in mild traumatic brain injury. Brain Res. Bull. 87 (4–5), 359–372.

Karbe, H., Thiel, A., Weber-Luxenburger, G., Herholz, K., Kessler, J., Heiss, W.D., 1998. Brain plasticity in poststroke aphasia: what is the contribution of the right hemisphere? Brain Lang. 64 (2), 215–230.

Kertesz, A., 1979. Aphasia and Associated Disorders: Taxonomy, Localization and Recovery. Grune and Stratton, New York.

Khuman, J., Zhang, J., Park, J., Carroll, J.D., Donahue, C., Whalen, M.J., 2012. Low-level laser light therapy improves cognitive deficits and inhibits microglial activation after controlled cortical impact in mice. J. Neurotrauma 29 (2), 408–417.

Kirshner, H.S., 2010. Frontotemporal dementia and primary progressive aphasia: an update. Curr. Neurol. Neurosci. Rep. 10 (6), 504–511.

Kontos, A.P., Kotwal, R.S., Elbin, R.J., Lutz, R.H., Forsten, R.D., Benson, P.J., et al., 2013. Residual effects of combat-related mild traumatic brain injury. J. Neurotrauma 30 (8), 680–686.

Kraus, M.F., Susmaras, T., Caughlin, B.P., Walker, C.J., Sweeney, J.A., Little, D.M., 2007. White matter integrity and cognition in chronic traumatic brain injury: a diffusion tensor imaging study. Brain 130 (Pt 10), 2508–2519.

Laird, A.R., Fox, P.M., Eickhoff, S.B., Turner, J.A., Ray, K.L., McKay, D.R., et al., 2011. Behavioral interpretations of intrinsic connectivity networks. J. Cogn. Neurosci. 23 (12), 4022–4037.

Laplaca, M.C., Prado, G.R., 2010. Neural mechanobiology and neuronal vulnerability to traumatic loading. J. Biomech. 43 (1), 71–78.

Lazar, R.M., Speizer, A.E., Festa, J.R., Krakauer, J.W., Marshall, R.S., 2008. Variability in language recovery after first-time stroke. J. Neurol. Neurosurg. Psychiatry 79 (5), 530–534.

Lee, H., Kim, S.W., Kim, J.M., Shin, I.S., Yang, S.J., Yoon, J.S., 2005. Comparing effects of methylphenidate, sertraline and placebo on neuropsychiatric sequelae in patients with traumatic brain injury. Human Psychopharmacol. 20 (2), 97–104.

Leger, A., Demonet, J.F., Ruff, S., Aithamon, B., Touyeras, B., Puel, M., et al., 2002. Neural substrates of spoken language rehabilitation in an aphasic patient: an fMRI study. Neuroimage 17 (1), 174–183.

Levin, H.S., Li, X., McCauley, S.R., Hanten, G., Wilde, E.A., Swank, P., 2013. Neuropsychological outcome of mTBI: a principal component analysis approach. J. Neurotrauma 30 (8), 625–632.

Lew, H.L., Vanderploeg, R.D., Moore, D.F., Schwab, K., Friedman, L., Yesavage, J., et al., 2008. Overlap of mild TBI and mental health conditions in returning OIF/OEF service members and veterans. J. Rehabil. Res. Dev. 45 (3), xi–xvi.

Lincoln, A.E., Caswell, S.V., Almquist, J.L., Dunn, R.E., Norris, J.B., Hinton, R.Y., 2011. Trends in concussion incidence in high school sports: a prospective 11-year study. Am. J. Sports Med. 39 (5), 958–963.

Lipton, A.M., White 3rd, C.L., Bigio, E.H., 2004. Frontotemporal lobar degeneration with motor neuron disease-type inclusions predominates in 76 cases of frontotemporal degeneration. Acta Neuropathol. 108 (5), 379–385.

Loane, D.J., Faden, A.I., 2010. Neuroprotection for traumatic brain injury: translational challenges and emerging therapeutic strategies. Trends Pharmacol. Sci. 31 (12), 596–604.

Mann, D.M., South, P.W., Snowden, J.S., Neary, D., 1993. Dementia of frontal lobe type: neuropathology and immunohistochemistry. J. Neurol. Neurosurg. Psychiatry 56 (6), 605–614.

Marcotte, K., Ansaldo, A.I., 2010. The neural correlates of semantic feature analysis in chronic aphasia: discordant patterns according to the etiology. Semin. Speech Lang. 31 (1), 52–63.

Marquez de la Plata, C.D., Garces, J., Shokri Kojori, E., Grinnan, J., Krishnan, K., Pidikiti, R., et al., 2011. Deficits in functional connectivity of hippocampal and frontal lobe circuits after traumatic axonal injury. Arch. Neurol. 68 (1), 74–84.

Martin, P., Ho, M., Bogdanova, Y., Krengel, M., Knight, J., Hamblin, M.R., et al., 2018. LED therapy improves functional connectivity and cognition in professional football player with TBI: Case study (Abstract). Arch. Phys. Med. Rehabil. 99 (10), e104–e105.

Martin, P.I., Naeser, M.A., Doron, K.W., Bogdan, A., Baker, E.H., Kurland, J., et al., 2005. Overt naming in aphasia studied with a functional MRI hemodynamic delay design. Neuroimage 28 (1), 194–204.

Martin, P.I., Naeser, M.A., Ho, M., Doron, K.W., Kurland, J., Kaplan, J., et al., 2009. Overt naming fMRI pre- and post-TMS: two nonfluent aphasia patients, with and without improved naming post-TMS. Brain Lang. 111 (1), 20–35.

Mathias, J.L., Alvaro, P.K., 2012. Prevalence of sleep disturbances, disorders, and problems following traumatic brain injury: a meta-analysis. Sleep Med. 13 (7), 898–905.

Mayer, A.R., Mannell, M.V., Ling, J., Gasparovic, C., Yeo, R.A., 2011. Functional connectivity in mild traumatic brain injury. Human Brain Map. 32 (11), 1825–1835.

McAllister, T.W., Saykin, A.J., Flashman, L.A., Sparling, M.B., Johnson, S.C., Guerin, S.J., et al., 1999. Brain activation during working memory 1 month after mild traumatic brain injury: a functional MRI study. Neurology 53 (6), 1300–1308.

McAllister, T.W., Sparling, M.B., Flashman, L.A., McDonald, B.C., Wishart, H., Saykin, A.J., 2002. Working memory activatio patterns one month and one year after mild traumatic brain injury: a longitudinal fMRI study [abstract P51]. J. Neuropsychiatry Clin. Neurosci. 14 (93), 116.

McAllister, T.W., Flashman, L.A., McDonald, B.C., Saykin, A.J., 2006. Mechanisms of working memory dysfunction after mild and moderate TBI: evidence from functional MRI and neurogenetics. J. Neurotrauma 23 (10), 1450–1467.

McAllister, T.W., McDonald, B.C., Flashman, L.A., Ferrell, R.B., Tosteson, T.D., Yanofsky, N.N., et al., 2011a. Alpha-2 adrenergic challenge with guanfacine one month after mild traumatic brain injury: altered working memory and BOLD response. Int. J. Psychophysiol. 82 (1), 107–114.

McAllister, T.W., Flashman, L.A., McDonald, B.C., Ferrell, R.B., Tosteson, T.D., Yanofsky, N.N., et al., 2011b. Dopaminergic challenge with bromocriptine one month after mild traumatic brain injury: altered working memory and BOLD response. J. Neuropsychiatry Clin. Neurosci. 23 (3), 277–286.

McAllister, T.W., Flashman, L.A., Maerlender, A., Greenwald, R.M., Beckwith, J.G., Tosteson, T.D., et al., 2012. Cognitive effects of one season of head impacts in a cohort of collegiate contact sport athletes. Neurology 78 (22), 1777–1784.

McCrea, M., Guskiewicz, K.M., Marshall, S.W., Barr, W., Randolph, C., Cantu, R.C., et al., 2003. Acute effects and recovery time following concussion in collegiate football players: the NCAA Concussion Study. JAMA 290 (19), 2556–2563.

McDonald, B.C., Flashman, L.A., Saykin, A.J., 2002. Executive dysfunction following traumatic brain injury: neural substrates and treatment strategies. NeuroRehabilitation 17 (4), 333–344.

McKee, A.C., Cantu, R.C., Nowinski, C.J., Hedley-Whyte, E.T., Gavett, B.E., Budson, A.E., et al., 2009. Chronic traumatic encephalopathy in athletes: progressive tauopathy after repetitive head injury. J. Neuropathol. Exp. Neurol. 68 (7), 709–735.

McKee, A.C., Gavett, B.E., Stern, R.A., Nowinski, C.J., Cantu, R.C., Kowall, N.W., et al., 2010. TDP-43 proteinopathy and motor neuron disease in chronic traumatic encephalopathy. J. Neuropathol. Exp. Neurol. 69 (9), 918–929.

Medana, I.M., Esiri, M.M., 2003. Axonal damage: a key predictor of outcome in human CNS diseases. Brain 126 (Pt 3), 515–530.

Meinzer, M., Flaisch, T., Breitenstein, C., Wienbruch, C., Elbert, T., Rockstroh, B., 2008. Functional re-recruitment of dysfunctional brain areas predicts language recovery in chronic aphasia. Neuroimage 39 (4), 2038–2046.

Mendez, C.V., Hurley, R.A., Lassonde, M., Zhang, L., Taber, K.H., 2005. Mild traumatic brain injury: neuroimaging of sports-related concussion. J. Neuropsychiatry Clin. Neurosci. 17 (3), 297–303.

Menon, V., 2011. Large-scale brain networks and psychopathology: a unifying triple network model. Trends Cogn. Sci. 15 (10), 483–506.

Menon, V., Uddin, L.Q., 2010. Saliency, switching, attention and control: a network model of insula function. Brain Struct. Funct. 214 (5–6), 655–667.

Mesulam, M.M., 2007. Primary progressive aphasia: a 25-year retrospective. Alzheimer Dis. Assoc. Disord. 21 (4), S8–S11.

Mi, X.Q., Chen, J.Y., Liang, Z.J., Zhou, L.W., 2004. In vitro effects of helium-neon laser irradiation on human blood: blood viscosity and deformability of erythrocytes. Photomed. Laser Surg. 22 (6), 477–482.

Mittl, R.L., Grossman, R.I., Hiehle, J.F., Hurst, R.W., Kauder, D.R., Gennarelli, T.A., et al., 1994. Prevalence of MR evidence of diffuse axonal injury in patients with mild head injury and normal head CT findings. AJNR Am. J. Neuroradiol. 15 (8), 1583–1589.

Monson, C.M., Gradus, J.L., Young-Xu, Y., et al., 2008. Change in posttraumatic stress disorder symptoms: do clinicians and patients agree? Psychol. Assess. 20 (2), 131–138. Available from: https://doi.org/10.1037/1040-3590.20.2.131. PMID: 18557690.

Naeser, M.A., Borod, J.C., 1986. Aphasia in left-handers: lesion site, lesion side, and hemispheric asymmetries on CT. Neurology 36 (4), 471–488.

Naeser, M.A., Hamblin, M.R., 2011. Potential for transcranial laser or LED therapy to treat stroke, traumatic brain injury, and neurodegenerative disease. Photomed. Laser Surg. 29 (7), 443–446.

Naeser, M.A., Palumbo, C.L., 1994. Neuroimaging and language recovery in stroke. J. Clin. Neurophysiol. 11 (2), 150–174.

Naeser, M.A., Palumbo, C.L., Helm-Estabrooks, N., Stiassny-Eder, D., Albert, M.L., 1989. Severe nonfluency in aphasia. Role of the medial subcallosal fasciculus and other white matter pathways in recovery of spontaneous speech. Brain 112 (Pt 1), 1–38.

Naeser, M.A., Martin, P.I., Baker, E.H., Hodge, S.M., Sczerzenie, S.E., Nicholas, M., et al., 2004. Overt propositional speech in chronic nonfluent aphasia studied with the dynamic susceptibility contrast fMRI method. Neuroimage 22 (1), 29–41.

Naeser, M.A., Martin, P.I., Nicholas, M., Baker, E.H., Seekins, H., Kobayashi, M., et al., 2005. Improved picture naming in chronic aphasia after TMS to part of right Broca's area: an open-protocol study. Brain Lang. 93 (1), 95–105.

Naeser, M.A., Saltmarche, A., Krengel, M.H., Hamblin, M.R., Knight, J.A., 2011. Improved cognitive function after transcranial, light-emitting diode treatments in chronic, traumatic brain injury: two case reports. Photomed. Laser Surg. 29 (5), 351–358.

Naeser, M.A., Ho, M., Martin, P.I., Treglia, E., Krengel, M.H., Hamblin, M.R., et al., 2012. Improved language after scalp application of red/near-infrared light-emitting diodes: pilot study supporting a new, noninvasive treatment for chronic aphasia. Procedia Soc. Behav. Sci. 61, 138–139.

Naeser, M.A., Zafonte, R., Krengel, M.H., Martin, P.I., Frazier, J., Hamblin, M.R., et al., 2014. Significant improvements in cognitive performance post-transcranial, red/near-infrared light-emitting diode treatments in chronic, mild traumatic brain injury: open-protocol study. J. Neurotrauma 31 (11), 1008–1017.

Naeser, M.A., Martin, P.I., Ho, M.D., Krengel, M.H., Bogdanova, Y., Knight, J.A., et al., 2016. Transcranial, Red/near-infrared light-emitting diode (LED) therapy for chronic, traumatic brain injury. Photomed. Laser Surg. 34 (12), 610–626.

Nakamura, T., Hillary, F.G., Biswal, B.B., 2009. Resting network plasticity following brain injury. PLoS One 4 (12), e8220.

Narayan, R.K., Michel, M.E., Ansell, B., Baethmann, A., Biegon, A., Bracken, M.B., et al., 2002. Clinical trials in head injury. J. Neurotrauma 19 (5), 503–557.

Nawashiro, H., Wada, K., Nakai, K., Sato, S., 2012. Focal increase in cerebral blood flow after treatment with near-infrared light to the forehead in a patient in a persistent vegetative state. Photomed. Laser Surg. 30 (4), 231–233.

Niogi, S.N., Mukherjee, P., 2010. Diffusion tensor imaging of mild traumatic brain injury. J. Head Trauma Rehabil. 25 (4), 241–255.

Ouellet, M.C., Beaulieu-Bonneau, S., Morin, C.M., 2015. Sleep-wake disturbances after traumatic brain injury. Lancet Neurology 14 (7), 746–757.

Pedersen, P.M., Vinter, K., Olsen, T.S., 2004. Aphasia after stroke: type, severity and prognosis. The Copenhagen aphasia study. Cerebrovasc. Dis. 17 (1), 35–43.

Postman-Caucheteux, W.A., Birn, R.M., Pursley, R.H., Butman, J.A., Solomon, J.M., Picchioni, D., et al., 2010. Single-trial fMRI shows contralesional activity linked to overt naming errors in chronic aphasic patients. J. Cogn. Neurosci. 22 (6), 1299–1318.

Price, C.J., Crinion, J., 2005. The latest on functional imaging studies of aphasic stroke. Curr. Opin. Neurol. 18 (4), 429–434.

Provenzale, J.M., 2010. Imaging of traumatic brain injury: a review of the recent medical literature. AJR Am. J. Roentgenol. 194 (1), 16–19.

Raichle, M.E., MacLeod, A.M., Snyder, A.Z., Powers, W.J., Gusnard, D.A., Shulman, G.L., 2001. A default mode of brain function. Proc. Natl. Acad. Sci. U.S.A. 98 (2), 676–682.

Ramlackhansingh, A.F., Brooks, D.J., Greenwood, R.J., Bose, S.K., Turkheimer, F.E., Kinnunen, K.M., et al., 2011. Inflammation after trauma: microglial activation and traumatic brain injury. Ann. Neurol. 70 (3), 374–383.

Rao, V., McCann, U., Han, D., Bergey, A., Smith, M.T., 2014. Does acute TBI-related sleep disturbance predict subsequent neuropsychiatric disturbances? Brain Inj. 28 (1), 20–26.

Reilly, J., Rodriguez, A.D., Lamy, M., Neils-Strunjas, J., 2010. Cognition, language, and clinical pathological features of non-Alzheimer's dementias: an overview. J. Commun. Disord. 43 (5), 438–452.

Roach, A., Schwartz, M.F., Martin, N., Grewal, R.S., Brecher, A., 1996. The Philadelphia Naming Test: scoring and rationale. Clin. Aphasiol. 24, 121–133.

Rojas, J.C., Gonzalez-Lima, F., 2011. Low-level light therapy of the eye and brain. Eye Brain 3, 49–67.

Rosen, H.J., Petersen, S.E., Linenweber, M.R., Snyder, A.Z., White, D.A., Chapman, L., et al., 2000. Neural correlates of recovery from aphasia after damage to left inferior frontal cortex. Neurology 55 (12), 1883–1894.

Rosenbek, J.C., LaPointe, L.L., Wertz, R.T., 1989. Aphasia: A Clinical Approach. PRO-ED, Austin, TX.

Rosso, S.M., Donker Kaat, L., Baks, T., Joosse, M., de Koning, I., Pijnenburg, Y., et al., 2003. Frontotemporal dementia in The Netherlands: patient characteristics and prevalence estimates from a population-based study. Brain 126 (Pt 9), 2016–2022.

Ruff, R.L., Ruff, S.S., Wang, X.F., 2009. Improving sleep: initial headache treatment in OIF/OEF veterans with blast-induced mild traumatic brain injury. J. Rehabil. Res. Dev. 46 (9), 1071–1084.

Saltmarche, A.E., Naeser, M.A., Ho, K.F., Hamblin, M.R., Lim, L., 2017. Significant improvement in cognition in mild to moderately severe dementia cases treated with transcranial plus intranasal photobiomodulation: case series report. Photomed. Laser Surg. 35 (8), 432–441.

Saper, C.B., Chou, T.C., Scammell, T.E., 2001. The sleep switch: hypothalamic control of sleep and wakefulness. Trends Neurosci. 24 (12), 726–731.

Sapolsky, D., Bakkour, A., Negreira, A., Nalipinski, P., Weintraub, S., Mesulam, M.M., et al., 2010. Cortical neuroanatomic correlates of symptom severity in primary progressive aphasia. Neurology 75 (4), 358–366.

Saur, D., Lange, R., Baumgaertner, A., Schraknepper, V., Willmes, K., Rijntjes, M., et al., 2006. Dynamics of language reorganization after stroke. Brain 129 (Pt 6), 1371–1384.

Schiffer, F., Johnston, A.L., Ravichandran, C., Polcari, A., Teicher, M.H., Webb, R.H., et al., 2009. Psychological benefits 2 and 4 weeks after a single treatment with near infrared light to the forehead: a pilot study of 10 patients with major depression and anxiety. Behav. Brain Funct. 5, 46.

Schneiderman, A.I., Braver, E.R., Kang, H.K., 2008. Understanding sequelae of injury mechanisms and mild traumatic brain injury incurred during the conflicts in Iraq and Afghanistan: persistent postconcussive symptoms and posttraumatic stress disorder. Am. J. Epidemiol. 167 (12), 1446–1452.

Seeley, W.W., Menon, V., Schatzberg, A.F., Keller, J., Glover, G.H., Kenna, H., et al., 2007. Dissociable intrinsic connectivity networks for salience processing and executive control. J. Neurosci. 27 (9), 2349–2356.

Shah, P.P., Szaflarski, J.P., Allendorfer, J., Hamilton, R.H., 2013. Induction of neuroplasticity and recovery in post-stroke aphasia by non-invasive brain stimulation. Front. Human Neurosci. 7, 888.

Slobounov, S., Gay, M., Johnson, B., Zhang, K., 2012. Concussion in athletics: ongoing clinical and brain imaging research controversies. Brain Imaging Behav. 6 (2), 224–243.

Small, S.L., Flores, D.K., Noll, D.C., 1998. Different neural circuits subserve reading before and after therapy for acquired dyslexia. Brain Lang. 62 (2), 298–308.

Smith, D.H., Meaney, D.F., Shull, W.H., 2003. Diffuse axonal injury in head trauma. J. Head Trauma Rehabil. 18 (4), 307–316.

Smith, E.E., Jonides, J., 1998. Neuroimaging analyses of human working memory. Proc. Natl. Acad. Sci. U.S.A. 95 (20), 12061–12068.

Sompol, P., Xu, Y., Ittarat, W., Daosukho, C., Clair, D.S., 2006. NF-kappaB-Associated MnSOD induction protects against beta-amyloid-induced neuronal apoptosis. J. Mol. Neurosci. 29 (3), 279–288.

Sridharan, D., Levitin, D.J., Menon, V., 2008. A critical role for the right fronto-insular cortex in switching between central-executive and default-mode networks. Proc. Natl. Acad. Sci. U.S.A. 105 (34), 12569–12574.

Sripada, R.K., King, A.P., Welsh, R.C., Garfinkel, S.N., Wang, X., Sripada, C.S., et al., 2012. Neural dysregulation in posttraumatic stress disorder: evidence for disrupted equilibrium between salience and default mode brain networks. Psychosom. Med. 74 (9), 904–911.

Stemer, A.B., Huisa, B.N., Zivin, J.A., 2010. The evolution of transcranial laser therapy for acute ischemic stroke, including a pooled analysis of NEST-1 and NEST-2. Curr. Cardiol. Rep. 12 (1), 29–33.

Stevens, M.C., Lovejoy, D., Kim, J., Oakes, H., Kureshi, I., Witt, S.T., 2012. Multiple resting state network functional connectivity abnormalities in mild traumatic brain injury. Brain Imaging Behav. 6 (2), 293–318.

Stuss, D.T., Ely, P., Hugenholtz, H., Richard, M.T., LaRochelle, S., Poirier, C.A., et al., 1985. Subtle neuropsychological deficits in patients with good recovery after closed head injury. Neurosurgery 17 (1), 41–47.

Swick, D., Honzel, N., Larsen, J., Ashley, V., Justus, T., 2012. Impaired response inhibition in veterans with post-traumatic stress disorder and mild traumatic brain injury. J. Int. Neuropsychol. Soc. 18 (5), 917–926.

Szaflarski, J.P., Allendorfer, J.B., Banks, C., Vannest, J., Holland, S.K., 2013. Recovered vs. not-recovered from post-stroke aphasia: the contributions from the dominant and non-dominant hemispheres. Restor. Neurol. Neurosci. 31 (4), 347–360.

Tanielian, T., Jaycox, L.H. (Eds.), 2008. Invisible Wounds of War: Psychological and Cognitive Injuries, Their Consequences, and Services to Assist Recovery. RAND Corporation, Santa Monica, CA.

Tran, H.T., LaFerla, F.M., Holtzman, D.M., Brody, D.L., 2011. Controlled cortical impact traumatic brain injury in 3xTg-AD mice causes acute intra-axonal amyloid-beta accumulation and independently accelerates the development of tau abnormalities. J. Neurosci. 31 (26), 9513–9525.

Van Boven, R.W., Harrington, G.S., Hackney, D.B., Ebel, A., Gauger, G., Bremner, J.D., et al., 2009. Advances in neuroimaging of traumatic brain injury and posttraumatic stress disorder. J. Rehabil. Res. Dev. 46 (6), 717–757.

Vasterling, J.J., Verfaellie, M., Sullivan, K.D., 2009. Mild traumatic brain injury and posttraumatic stress disorder in returning veterans: perspectives from cognitive neuroscience. Clin. Psychol. Rev. 29 (8), 674–684.

Wall, S.E., Williams, W.H., Cartwright-Hatton, S., Kelly, T.P., Murray, J., Murray, M., et al., 2006. Neuropsychological dysfunction following repeat concussions in jockeys. J. Neurol. Neurosurg. Psychiatry 77 (4), 518–520.

Warden, D., 2006. Military TBI during the Iraq and Afghanistan wars. J. Head Trauma Rehabil. 21 (5), 398–402.

Weiduschat, N., Thiel, A., Rubi-Fessen, I., Hartmann, A., Kessler, J., Merl, P., et al., 2011. Effects of repetitive transcranial magnetic stimulation in aphasic stroke: a randomized controlled pilot study. Stroke 42 (2), 409–415.

Xie, L., Kang, H., Xu, Q., Chen, M.J., Liao, Y., Thiyagarajan, M., et al., 2013. Sleep drives metabolite clearance from the adult brain. Science 342 (6156), 373–377.

Xuan, W., Agrawal, T., Huang, L., Gupta, G.K., Hamblin, M.R., 2014a. Low-level laser therapy for traumatic brain injury in mice increases brain derived neurotrophic factor (BDNF) and synaptogenesis. J. Biophotonics 9999, 9999.

Xuan, W., Vatansever, F., Huang, L., Hamblin, M.R., 2014b. Transcranial low-level laser therapy enhances learning, memory, and neuroprogenitor cells after traumatic brain injury in mice. J. Biomed. Opt. 19 (10), 108003.

Zafonte, R., Friedewald, W.T., Lee, S.M., Levin, B., Diaz-Arrastia, R., Ansel, B., et al., 2009. The citicoline brain injury treatment (COBRIT) trial: design and methods. J. Neurotrauma 26 (12), 2207–2216.

Zafonte, R., Hammond, F., Dennison, A., Chew, E., 2009. Pharmacotherapy to enhance arousal: what is known and what is not. Prog. Brain Res. 177, 293–316.

Zafonte, R.D., Bagiella, E., Ansel, B.M., Novack, T.A., Friedewald, W.T., Hesdorffer, D.C., et al., 2012. Effect of citicoline on functional and cognitive status among patients with traumatic brain injury: Citicoline Brain Injury Treatment Trial (COBRIT). JAMA 308 (19), 1993–2000.

Zhang, K., Johnson, B., Gay, M., Horovitz, S.G., Hallett, M., Sebastianelli, W., et al., 2012. Default mode network in concussed individuals in response to the YMCA physical stress test. J. Neurotrauma 29 (5), 756–765.

Zhang, Y.H., Takahashi, K., Jiang, G.Z., Zhang, X.M., Kawai, M., Fukada, M., et al., 1994. In vivo production of heat shock protein in mouse peritoneal macrophages by administration of lipopolysaccharide. Infect Immun. 62 (10), 4140–4144.

Zhao, J., Tian, Y., Nie, J., Xu, J., Liu, D., 2012. Red light and the sleep quality and endurance performance of Chinese female basketball players. J. Athlet. Training 47 (6), 673–678.

Chapter 26

Photobiomodulation as a potential therapeutic strategy for improving cognitive and functional outcomes in traumatic brain injury

Thomas J. Covey[1], David W. Shucard[1], Melissa Meynadasy[1], Thomas Mang[2] and Praveen R. Arany[3]

[1]*Division of Cognitive and Behavioral Neurosciences, Department of Neurology, Jacobs School of Medicine and Biomedical Sciences, University at Buffalo, Buffalo, NY, United States,* [2]*Department of Oral and Maxillofacial Surgery, School of Dental Medicine, University at Buffalo, Buffalo, NY, United States,* [3]*Department of Oral Biology and Biomedical Engineering, School of Dental Medicine, University at Buffalo, Buffalo, NY, United States*

26.1 Introduction

Traumatic brain injury (TBI) can result from blunt force trauma to the head, penetrating head injuries, or acceleration/deceleration forces (i.e., whiplash). It is a common occurrence in automobile accidents, unintentional falls, sports related injuries, physical violence, and war-zone injuries in military personnel. Faul et al. (2010) estimated that there are 1.7 million cases of TBI each year in the United States, and the incidence of TBI is estimated to be even higher internationally (Roozenbeek et al., 2013). In the acute/primary phase of TBI, which is in the immediate aftermath of the direct physical damage due to insult to the head/brain, symptoms can include loss or alteration of consciousness or arousal, disorientation or confusion, and impaired memory (including loss of memory for the event itself; Marr and Coronado, 2004; see also a review by Arciniegas, 2011). The development of secondary symptoms in TBI is dependent on the initial severity of brain insult, whether brain injury was focal or diffuse, and emergent pathological processes. Secondary symptoms include affective, social, occupational, and cognitive problems (see Mechtler et al., 2014). Due to the heterogeneity of the neuropathology associated with TBI, there is substantial variability in the severity of impairment and in the neurological domains that are affected across individual patients.

Determining TBI severity often requires a multidimensional approach that includes clinical evaluation, cognitive testing, and neuroimaging. The Glasgow Coma Scale (GCS) is an example of one clinical measure that is used to evaluate neurological severity in patients with head injury (Langfitt, 1978; Teasdale and Jennet, 1974). This scale grades patients on eye opening, and verbal and motor responsiveness. The GCS serves as a clinical demarcation tool to categorize patients into mild, moderate, or severe TBI ranges (GCS scores of 13 or greater, 9–12, or 8 or less, respectively). In moderate and severe forms of TBI, secondary neurological/cognitive symptoms and disability can persist for long periods of time (i.e., greater than six months), and impaired cognition in multiple domains is common, including in higher-level cognitive domains such as intellectual functioning and reasoning ability (Dikmen et al., 2009). In contrast, major cognitive impairment and disability are typically resolved in mild-to-moderate TBI within several months after the initial injury (for a review, see Rabinowitz and Levin, 2014). However, there may be subtle cognitive disturbances that persist even in mild TBI, particularly in patients with relatively low cognitive reserve, which is (in part) an individual's aptitude for compensatory brain function in the face of brain injury (Oldenburg et al., 2015). Premorbid cognitive reserve may be an important factor contributing to variance in cognitive outcomes in TBI patients.

Despite the heterogeneity of neuropathology found in TBI, there are some generalities in the cognitive domains and brain regions that are commonly impacted in head injury patients. Deficits in executive functions, attention, working memory, processing speed, and new learning and long-term memory are commonly observed (Kinnunen et al., 2011;

Photobiomodulation in the Brain. DOI: https://doi.org/10.1016/B978-0-12-815305-5.00026-9

McDonald et al., 2002; Rabinowitz and Levin, 2014). Cognitive impairment in mild and moderate TBI has been associated with depressive symptoms, which are among the commonly observed neuropsychiatric sequelae of TBI (Rapaport et al., 2005). Cognitive impairment is also predictive of occupational outcomes for TBI patients (Benedictus et al., 2010). Thus, the cognitive deficits associated with TBI can significantly affect an individual's life. Identification of potential therapies that can improve cognitive outcomes in TBI is therefore critically important. Over the course of the past decade, evidence has accumulated that photobiomodulation (PBM) methods can be effectively applied to patients with TBI. The present review will summarize and discuss this evidence, with a focus on the potential for PBM interventions to improve cognitive outcomes in TBI patients.

A variety of treatment approaches have been developed that may potentially improve the cognitive deficits associated with TBI. The adoption of evidence-based guidelines by clinicians and improved treatment protocols has resulted in improved outcomes for TBI patients overall (Carpenter et al., 2015), but mortality rates have not decreased, and the incidence of TBI, in general, appears to have increased in conjunction with longer lifespans and the increased prevalence of motor vehicles worldwide (Roozenbeek et al., 2013). With respect to cognitive dysfunction following TBI, pharmacological interventions have yielded mixed findings overall (Gruenbaum et al., 2016), with some evidence suggesting that dopaminergic medications (Writer and Schillerstrom, 2009) and donepezil (cholinesterase inhibitor, Rees et al., 2007) may show some promise for the treatment of cognitive problems. Previous reviews have noted that on the whole, pharmacological interventions may have limited application for the secondary injury associated with TBI (Naeser and Hamblin, 2015; Zhang et al., 2014). Cognitive rehabilitation training may benefit some patients with TBI, although these results have also been mixed (Carney et al., 1999; Cicerone et al., 2011; Morries et al., 2015). Further, cognitive and behavioral rehabilitation strategies may rely primarily on recruiting remaining cognitive resources to compensate for disrupted brain function. However, there may be an upper limit to the effectiveness of this treatment in that it does not focus on the restoration of functionality to damaged brain regions (as noted in Naeser and Hamblin, 2015). It has also been suggested that exercise may have positive benefits on cognitive performance in TBI patients, but again the results in this literature have been mixed, and more methodologically rigorous studies are needed to understand the extent of the impact of exercise on cognitive outcomes in TBI (Morris et al., 2016).

Noninvasive brain stimulation methods are emerging as potentially viable treatment options in patients with TBI. These methods include transcranial magnetic stimulation (TMS), transcranial direct current stimulation (tDCS), and PBM (Demirtas-Tatlided et al., 2012; Villamar et al., 2012). TMS and tDCS induce changes in the activity of the underlying cortex via the application of a magnetic pulse or a low amplitude direct current through the scalp, respectively. The use of low dose biophotonics treatments has been termed PBM therapy, previously termed low-level light/laser therapy (Anders et al., 2015). PBM therapy involves the transcranial (and sometimes intranasal) application of light energy to the brain via low-level lasers or light-emitting diodes (LEDs). PBM interventions in TBI have most commonly, if not exclusively, used wavelengths in the red through near-infrared spectrum (Salehpour et al., 2018b; Tsai and Hamblin, 2017). These brain stimulation methods can potentially induce neuroplasticity in networks underlying cognitive performance, which could thereby result in the resolution of cognitive issues associated with TBI. Brain stimulation methods differ from other intervention strategies such as cognitive rehabilitation and exercise in that they are theoretically better suited to rescue functionally disrupted/damaged network circuitry, as opposed to strengthening only neural circuitry that has been spared from damage. From a methodological and applications standpoint, brain stimulation interventions have better regional specificity in comparison to other treatment interventions (including pharmacologic treatments), since brain stimulation approaches can directly target specific cortical regions that are (1) found to be most affected by TBI, as revealed by clinical neuroimaging; and (2) associated with disrupted cognitive domains on a patient-by-patient basis. However, note that these different noninvasive brain stimulation methods have important differences in their purported mechanisms of action, their possible side effects, and the extent to which they have been examined in different clinical populations.

With respect to TMS and tDCS, there are some potential safety concerns. For example, the most common side effect of TMS is that it can potentially cause headaches over repeated sessions (Taylor et al., 2018). TMS can also potentially induce seizures (Wassermann, 2000; but note it is relatively uncommon, Taylor et al., 2018), and patients with TBI may be more susceptible to this possibility due to their injury. Further examination of the potential safety risks associated with these methods has been recommended (Demirtas-Tatlided et al., 2012; Villamar et al., 2012). The risks associated with the use of PBM methods in healthy and clinical populations, on the other hand, are thought to be minimal (Ilic et al., 2006; Lapchak et al., 2008; Moro et al., 2017; Naeser and Hamblin, 2015). Furthermore, the putative mechanisms that are targeted by PBM therapy overlap with key aspects of the neuropathology underlying cognitive/behavioral impairment associated with TBI.

In the following sections of this review, the rationale for PBM as an intervention that can target the relevant neuropathology and improve cognitive and functional outcomes in TBI patients will be discussed. Several lines of evidence support this preliminary view. Studies that have examined the efficacy of PBM treatment in animal models of TBI will be reviewed initially. These studies provide a basis for the specific neurobiological mechanisms that are targeted by PBM in preclinical models of TBI. Literature that has examined the effects of PBM on cognitive performance in healthy individuals will also be summarized, as this work provides a basis for the potential efficacy of PBM for improving cognitive outcomes. Studies that have examined PBM in clinical TBI patients will then be reviewed. To date, the work in human TBI patients has been largely case-based. The initial findings are promising, but still preliminary. Throughout this review, particular emphasis will be placed on discussing the effects of PBM on cognitive outcomes. The final section of the review will provide a general summary and discussion, with recommendations for future directions.

26.2 Neuropathology of traumatic brain injury

An important theoretical consideration for the appropriateness of PBM therapy in treating TBI is whether the neuropathological processes underlying TBI can be targeted by PBM in the first place. In this section, aspects of the neuropathology underlying TBI will be described briefly, starting with large-scale neural network disruption in TBI, and then moving to an overview of the cellular and molecular pathology. With respect to gross functional alterations, frontal cortical regions and white matter networks are very commonly disrupted in TBI. Due to the nature of the types of accidents that often result in head injuries, focal cortical contusions and axonal shearing (traumatic axonal injury produced by acceleration/deceleration forces) are often present (Cicerone et al., 2006). Frontal/prefrontal networks are the neural substrates of executive functions. Executive functions can be described as a set of higher-level cognitive processes that include planning, problem solving, working memory, response inhibition, and top-down attentional control. The high incidence of frontal disturbances in TBI can account for the prevalence of executive dysfunction in this clinical population (McDonald et al., 2002).

In addition to direct cortical injury, damage to white matter pathways can disrupt functional interactions among cortical networks. Multiple studies have found that a network of brain regions known to be activated when individuals are not actively engaged in a cognitive task or purposeful behavior, (i.e., the "resting state") is disrupted in TBI. This resting state brain network can be measured via functional magnetic resonance imaging (fMRI) and is referred to as the default mode network. Dysregulation of default mode network activity can interfere with cognitive performance, as this network is thought to be switched off during cognitive engagement. TBI patients with traumatic axonal injury, compared to healthy controls, exhibited lower interhemispheric functional connectivity in hippocampus and anterior cingulate cortex during rest, as well as disrupted recruitment in dorsolateral prefrontal cortex (DLPFC; Marquez de la Plata et al., 2011). Johnson et al. (2012) examined differences in resting state brain activity between athletes that were diagnosed with mild TBI due to a sports related concussion and athletes that did not have a previous concussion. Notably, the athletes with mild TBI in this study had been recently cleared to return to play and were considered asymptomatic at the time of their fMRI scan. Overall resting state functional connectivity was found to be decreased in mild TBI athletes compared to non-TBI individuals, and DLPFC had reduced connectivity with posterior cingulate cortex and bilateral lateral parietal cortex. Reduced overall connectivity to other structures was also found for lateral parietal cortex (but note that a seemingly discrepant finding of increased connectivity to other regions for the medial prefrontal cortex was found in mild TBI compared to normal individuals). Furthermore, this study also found that lower connectivity between left DLPFC and left lateral parietal cortex was associated with an increased number of concussions. These findings suggest that even in individuals with mild TBI that are considered clinically asymptomatic, disrupted brain activity can still be present in regions that comprise the default mode network.

The cellular and molecular pathology that underlies neuronal dysfunction can potentially be addressed via PBM therapy. In the immediate aftermath of injury, during the acute phase of TBI, damage to neuronal tissue leads to excess glutamate release and alterations in ionic flux. Ionic pumps that restore the voltage gradient to baseline levels require a large amount of adenosine triphosphate (ATP) energy to drive this process, depleting ATP stores. Biomechanical forces also result in axonal injury, as noted above, and impaired neural signaling. Axonal dysfunction and disrupted neurotransmission in particular have been linked to impaired cognitive performance (Giza and Hovda, 2014). Excitotoxicity can lead to necrosis and/or apoptosis, depending on the severity and other characteristics of the injury (Werner and Engelhard, 2007). During the secondary phase of TBI, neuroinflammatory cascades involving astrocytes and microglia are also engaged, which can result in multiphasic neuroprotective and neurotoxic responses (Karve et al., 2016). Astrocytes and microglia support homeostatic functions, mediate immune responses, and are involved in functions of the blood–brain barrier. They can also modify neural signaling and synaptic transmission. Of particular relevance to the discussion of PBM is the presence of mitochondrial dysfunction in TBI, which results in increased free radicals and oxidative stress, apoptosis, and disrupted ATP production (Heibert et al., 2015). In the following section, the appropriateness of PBM therapy to address these pathological processes in TBI will be discussed.

26.3 Putative targets of photobiomodulation therapy in traumatic brain injury

A number of previous reviews have comprehensively discussed the biological mechanisms of PBM therapy in the brain (e.g., Gonzalez-Lima and Barrett, 2014; Hamblin, 2016; Hennessy and Hamblin, 2017; Rojas and Lima, 2013; Salehpour et al., 2018a,b; Tsai and Hamblin, 2017); as well as the use of PBM in TBI specifically (Hamblin, 2017; Li et al., 2015; Naeser et al., 2016). PBM therapy has been described by Hamblin as "the use of red or near-infrared light to stimulate, heal, regenerate, and protect tissue that has either been injured, is degenerating, or else is at risk of dying" (Hamblin, 2016). The red-to-near-infrared range of light encompasses wavelengths from approximately 600–1100 nm that can be absorbed by a photoacceptor in cytochrome c oxidase, which is known to be one of the primary biological targets of PBM. This enzyme has a key role in the mitochondrial respiratory cascade; and increased activation of cytochrome c oxidase via PBM is thought to modulate nitric oxide and reactive oxygen species, stabilize ATP production, and increase cerebral blood flow (see De Freitas and Hamblin, 2016; Hamblin, 2016; Pastore et al., 2000). These processes in turn can improve functionality of ATP-dependent ion pumps that regulate the voltage gradient across the cell membrane (Konstantinovic et al., 2013), and lead to a cascade of downstream activation of transcription factors that can produce sustained effects, such as promoting synaptic modulation (De Freitas and Hamblin, 2016). Transcranial delivery of light energy through the skull to the cortex has also been shown to be feasible. Laser light in the infrared range (i.e., 810–980 nm) administered at an overall power of 10–15 W can penetrate human skull and tissue to a depth of 3 cm in the cortex, with a fluence range of 0.9–15.0 J/cm^2 suggested as a therapeutic dosage for patients with TBI (Henderson and Morries, 2015). Tedford et al. (2015) also found that in human cadaver brain tissue, 808 nm laser light penetrates to a depth in the brain of 4–5 cm, even when passing through scalp, skull, and meninges. The 808 nm wavelength was also found to be superior in terms of tissue penetration in comparison to 660 and 940 nm wavelength light.

The neurobiological targets of PBM treatment overlap with many of the neuropathologic mechanisms related to mitochondrial dysfunction that are thought to exist in TBI, such as disrupted cerebral blood flow and ATP production, oxidative stress, and apoptosis. Recent literature suggests a considerable relationship between mitochondrial functions and neurocognitive processes in general (Lomeli et al., 2017; Picard and McEwen, 2014). Hara et al. (2014) found that poorer working memory performance was associated with greater mitochondrial stress in the prefrontal cortex of monkeys, and mitochondrial morphology that was consistent with oxidative stress was related to synaptic dysfunction. Mitochondrial dysfunction has also been associated with age-related declines in neurocognitive performance (Currais, 2015). There is growing evidence that modulation of mitochondrial function via transcranial PBM therapy may prevent or minimize neurocognitive decline due to aging (de la Torre, 2017). This convergent evidence provides a basis for the notion that targeting mitochondrial pathology could also improve cognitive outcomes in patients with TBI. A number of studies in healthy individuals (highlighted below) have reported evidence that PBM treatments may indeed enhance neurocognitive performance. The mechanism underlying this effect is suggested to involve modulation of mitochondrial function, via photosensitive cytochrome c oxidase and its downstream targets. The two subsequent sections will review studies that have examined the effects of PBM on these biological targets in animal models of TBI.

26.4 Treatment parameters and biological targets of photobiomodulation in animal models of traumatic brain injury

Several recent review papers have already comprehensively covered the biological mechanisms of PBM therapy in animal models of TBI (Hamblin, 2017; Hennessy and Hamblin, 2017; Li et al., 2015). Here, although there will be some discussion of the underlying biological mechanisms, the focus will be on the impact of PBM therapy on neurological and cognitive outcomes in rodent models of TBI. In this section, studies that have provided insight about the treatment parameters and biological mechanisms will be discussed, with an emphasis on their relationship to behavioral/motoric neurological outcomes in TBI animal models. In the next section, the relationship between treatment parameters and biological targets in animal models will be discussed with respect to cognitive outcomes more specifically. These studies are summarized in Table 26.1, which provides an overview of the animal model methods, treatment parameters, study groups, and general findings. In terms of a fundamental discussion of PBM dosimetry, there are specific device parameters that are typically used to describe treatment protocols. These include the irradiance or power

TABLE 26.1 Effects of photobiomodulation in animal models of traumatic brain injury.

Study	Animal model	Photobiomodulation protocol and groups	Cognitive and neurological findings
Ando et al. (2011)	Male BALB/c mice. CCI with craniotomy (left parietal/frontoparietal cortex location).	810 nm laser, 50 mW/cm^2, 36 J/cm^2, 12 min duration, 4 h post-TBI, transcranial administration to injured side of head. Study groups: CW ($n=10$); 10 Hz pulsed wave ($n=10$); 100 Hz pulsed wave ($n=10$); TBI, no treatment ($n=10$); no treatment sham-operated control ($n=3$).	NSS was lower in treatment groups compared to sham control group.The 10 Hz pulse wave group showed the lowest NSS overall, had best outcome on the forced swim test and tail suspension test, and had lower lesion volume than the control group.
Esenaliev et al. (2018)	Male Sprague-Dawley rats. CHI over right hemisphere of the head.	808 nm laser, pulsed (20 Hz), 300 J/cm^2, 5 min duration, 1 h postinjury; generates low-level optoacoustic waves, referred to as nanopulsed laser therapy (NPLT), administered transcranially at site of injury. Study groups: TBI with treatment, $n=16$; TBI without treatment, $n=15$; sham group, $n=18$ (anesthetized, no CHI).	Treatment group had similar performance on the Morris water maze compared to sham group, and better performance than untreated TBI group. Treatment was associated with reduced apoptosis and increased neurogenesis.
Giacci et al. (2014)	Adult male Sprague-Dawley rats. Lateral fluid percussion brain injury model. Other clinical models also assessed in the study: partial optic nerve injury, female PVG hooded rats, $n=36$; retinal degeneration, albino Sprague-Dawley rats, $n=36$.	LED administered 30 min a day for 7 days postinjury, transcranial administration above injury site Study groups: sham uninjured, $n=5$; injured untreated controls, $n=6$; 670 nm treated, 28.4 J/cm^2, $n=5$; 830 nm treated, 22.6 J/cm^2, $n=5$.	TBI rats that received 670 and 830 nm treatment showed similar performance on motor and sensory outcomes, and were better on these outcomes compared to sham controls. Lesion volumes were not different between groups at 7 days postinjury.
Khuman et al. (2012)	Male C57BL/6 mice (3 months). CCI with craniotomy over left temporo-parietal cortex.	800 nm laser, multiple treatment groups: Open craniotomy, treatment administered at 60–80 min postinjury: 2 min duration, 30 J/cm^2, 250 mW/cm^2, $n=7$; 2 min duration, 60 J/cm^2, 500 mW/cm^2, $n=22$; 2 min duration, 120 J/cm^2, 1000 mW/cm^2, $n=10$; 7 min duration, 105 J/cm^2, 250 mW/cm^2, $n=7$; 7 min duration, 210 J/cm^2, 500 mW/cm^2, $n=10$; sham injured with craniotomy, no treatment ($n=43$). Transcranial treatment groups, all received 2 min duration, 60 J/cm^2, 500 mW/cm^2: single treatment at 60–80 min postinjury ($n=12$ per group); single treatment at 4 h postinjury ($n=9$ per group); one treatment a day over 7 days, starting at 60–80 min postinjury ($n=10$ per group).	60 J/cm^2 dose at 60–80 min postinjury (transcranial or open craniotomy delivery) elicited better Morris water maze performance compared to sham-treatment controls. 120 J/cm^2 at 60–80 min (open craniotomy) and transcranial 60 J/cm^2 for 7 days also produced improved performance on MWM compared to controls. No significant effects of treatment on motor function or lesion volume.
Moreira et al. (2009)	Adult male Wistar rats. Cryogenic brain injury model with craniotomy over frontal-parietal cortex.	CW used for all treatments. Laser treatment was applied to two points—at the injured site following lesion, and transcranially after closing craniotomy, 3 h apart. Study Groups ($n=10$ per group): 660 nm laser, 3 J/cm^2 per point (3 s); 660 nm laser, 5 J/cm^2 per point (5 s); 780 nm laser, 3 J/cm^2 per point (3 s); 780 nm laser, 5 J/cm^2 per point (5 s); no treatment.	No cognitive/behavioral analyses. Treatment conditions had differential effects on inflammatory cytokines.

(Continued)

TABLE 26.1 (Continued)

Study	Animal model	Photobiomodulation protocol and groups	Cognitive and neurological findings
Moreira et al. (2011)	Adult male Wistar rats. Cryogenic brain injury model with craniotomy over frontal-parietal cortex.	780 nm laser, CW, power output = 40 mW, energy density of 3 J/cm^2, 3 s duration, treatment applied at two points—at the injured site following lesion, and transcranially after closing craniotomy, 3 h apart. Study groups ($n = 20$ per group): no treatment sham control; irradiated condition.	No cognitive/behavioral analyses. Treatment group had smaller lesions and lower concentration of inflammatory markers compared to sham controls.
Oron et al. (2007)	Male Sabra mice. CHI above parietal cortex.	808 nm laser, two different doses: 10 mW/cm^2, 1.2 J/cm^2; and 20 mW/cm^2, 2.4 J/cm^2). For both doses, 2 min duration, CW, transcranial administration at midline, 4 mm caudal to coronal suture of skull. Study groups: sham control, $n = 8$; treated mice, $n = 16$.	Treated groups exhibited better NSS outcomes than sham group from 5 to 28 days postinjury. Lesion volume was lower at follow-up in the treatment groups compared to sham group.
Oron et al. (2012)	Male Sabra mice. CHI above parietal cortex.	808 nm laser, 10 mW/cm^2, 1.2 J/cm^2, transcranial administration at midline, 4 mm caudal to coronal suture of skull. Study groups: Experiment 1: CW at 6 h postinjury, $n = 6$; CW at 8 h postinjury, $n = 7$; sham controls, $n = 7$. Experiment 2: pulsed (100 Hz) at 4 h postinjury, $n = 7$; pulsed (600 Hz) at 4 h postinjury, $n = 6$; CW at 4 h postinjury, $n = 6$; sham control group, $n = 6$.	Treated groups had better NSS outcomes from 5 to 28 days postinjury than sham group. Higher percentage of mice in pulsed wave compared to continuous wave group had full recovery to NSS = 0. Treated groups had lower lesion volume at 56 days post-TBI compared to sham group.
Quirk et al. (2012)	Sprague-Dawley rats. CCI over left parietal cortex.	670 nm LED, 50 mW/cm^2, 15 J/cm^2, twice per day for 10 days, 5 min duration per treatment, transcranial administration at top of head. Study groups (104 animals used, mortality of 10): CCI with treatment; CCI without treatment; sham surgery with treatment; sham surgery without treatment; anesthetization only with treatment; anesthetization only without treatment.	Treatment improved outcomes for goal-seeking behavior (e.g., nose pokes to baited areas). Changes were also observed for biomarkers of apoptotic function in the treatment groups compared to controls.
Wu et al. (2010)	Adult male BALB/c mice. CHI, 1 mm lateral to midline in the mid-coronal plane.	670 or 810 or 980 nm laser, 150 mW/cm^2, 36 J/cm^2, 4 min duration, transcranial administration to irradiate whole brain. Study groups: 670 nm at 4 h postinjury, $n = 8$; 810 nm at 4 h postinjury, $n = 8$; 980 nm at 4 h postinjury; sham operated control, $n = 8$.	670 and 810 nm treated groups had improved NSS outcomes compared to controls. 980 nm treated group did not show a significant difference from controls on NSS outcomes. 670 and 810 nm range treatment groups had lower lesion size compared to controls.
Wu et al. (2012)	Adult male BALB/c mice. CHI, 1 mm lateral to midline in the mid-coronal plane.	665, 730, 810, or 980 nm laser, 150 mW/cm^2, 36 J/cm^2, 4 min duration, administered at 4 h postinjury, transcranial administration at spot on top of head over sutures. Study groups: 665 nm ($n = 8$), 730 nm ($n = 8–12$), 810 nm ($n = 11$), 980 nm ($n = 8–12$); sham-treated TBI control, $n = 9$.	670 and 810 nm range treatment improved NSS outcomes compared to controls, whereas 730 and 980 nm treatment did not produce significantly better outcomes than sham controls. 670 and 810 nm range treatment groups had fewer brain defects compared to controls.

Xuan et al. (2013)	Adult male BALB/c mice. CCI over right parietotemporal cortex with craniotomy.	810 nm, CW, 25 mW/cm^2, 18 J/cm^2, 12 min duration, treatments beginning at 4 h postinjury, transcranial administration centrally on top of the head. Study groups ($n = 16$ per group): TBI single treatment; TBI 3 consecutive days of treatment; TBI 14 consecutive days of treatment; sham-TBI (no CCI) with single treatment; sham-TBI 3 consecutive days of treatment; sham-TBI 14 consecutive days of treatment; TBI, single sham treatment; TBI 3 sham treatments; TBI 14 sham treatments; sham-TBI sham-treatment; real TBI sham-treatment.	Single and 3 day treatment groups had lower NSS and better motor performance on wire-grip and motion test compared to controls. This effect was not observed in 14 day treatment group. Single and 3 day treatment groups had lower lesion volume, fewer markers of degenerating neurons, increased markers of neurogenesis, compared to controls.
Xuan et al. (2014)	Young adult male BALB/c mice (8 weeks). CCI with craniotomy over temporo-parietal cortex.	810 nm laser, 18 J/cm^2, 25 mW/cm^2, 12 min duration, transcranial administration at top of head. Study groups ($n = 16$ per group): sham TBI, sham laser; TBI, sham laser; TBI, single laser treatment; TBI, 3 laser treatments.	Treatment groups had significantly better performance than controls on MWM. Three treatments produced more pronounced effects than a single treatment. Treatment groups had reduced expression of apoptotic markers, and increased expression of neuroprogenitor markers.
Xuan et al. (2015)	Male BALB/c mice (6-8 weeks). CCI with craniotomy, over right frontal-parietal cortex.	810 nm laser, CW, 50 mW/cm^2, 36 J/cm^2, treatment started at 4 h post-TBI, 12 min duration, transcranial administration to injured side of the head. Study groups ($n = 10$ per group): sham TBI; TBI, sham treatment; TBI, single treatment; TBI, 3 treatments on consecutive days.	Treatment groups had improved NSS outcomes compared to sham treated group, and 3 treatment group had fastest improvement in NSS. Brain derived neurotrophic factor was upregulated in the hippocampus and subventricular zone following treatment.
Xuan et al. (2016)	Male BALB/c mice (6–8 weeks). CCI with craniotomy over temporo-parietal cortex.	810 nm laser, CW, 12 min duration, 18 J/cm^2, 25 mW/cm^2, transcranial administration at top of head. Study groups ($n = 24$ per group): sham surgery; TBI without treatment; TBI with 3 consecutive days of treatment; TBI with 14 consecutive days of treatment.	Three and 14 day TBI treatment groups had better NSS performance than sham treatment TBI group at 1 week postinjury. Three day treatment TBI group had more improvement that was sustained for 2–8 weeks postinjury, compared to other groups. Transient changes in biomarkers of reactive gliosis were associated with treatment.
Zhang et al. (2014)	Adult WT and IEX-1 KO mice (129 Sv/C57BL/6). CHI to left lateral scalp.	810 nm laser, pulsed (10 Hz), 150 mW/cm^2, 36 J/cm^2, 4 min duration, one treatment applied 4 h postinjury, transcranial administration on contusion site on head. Study Groups: WT sham injury; WT mTBI; WT mTBI with treatment; KO sham injury; KO mTBI; KO mTBI with treatment.	Treatment resulted in improved outcomes in WT, and rescued neurologic function (NSS) in IES-1 KO. Pronounced secondary injury processes observed in IEX-1 KO mice were rescued by laser treatment.

Notes: *CW*, Continuous wave; *CCI*, controlled cortical impact; *CHI*, closed head injury; *KO*, knockout; *MWM*, Morris Water Maze; *NSS*, neurological severity score; *TBI*, traumatic brain injury.

density (W/cm^2), time of treatment and the fluence (J/cm^2), often called the dose. The irradiance is a complex parameter to accurately describe, as the actual treatment probe output and power of the device can be variably delivered to the target tissue surface based on the distance and, often, a sweeping or scanning motion of the probe head. These should be carefully documented for accurate dose assessments. Further, the power can be delivered as a continuous wave or as a pulsed beam which has significant implications for the nonthermal nature of this treatment. The time of treatment is relatively easy to document precisely and the cumulative product of the two determines the total energy density of fluence. The reader is referred to several detailed PBM reviews on the accurate description and reporting of these parameters (Arany, 2016; Jenkins and Carroll, 2011; Tuner and Jenkins, 2016). In the following sections, we have made a concerted effort to review the literature paying specific attention to the PBM treatment parameters reported in individual studies.

As described above, PBM therapy can reduce apoptosis, modulate mitochondrial function, and improve blood flow, in addition to modulating a number of other processes. Additionally, multiple lines of evidence suggest that PBM can stimulate synaptogenesis and neurogenesis, as reflected by up-regulation of brain-derived neurotrophic factor (BDNF) following PBM (Hamblin, 2017; Xuan et al., 2015). Greater serum BDNF levels have been associated with higher memory performance scores in TBI patients at 6 and 12 months postinjury (Faila et al., 2016), which implies that the capacity for neural/synaptic plasticity is related to better outcomes. Recently it has been suggested that a therapeutic strategy that mimics BDNF may offer opportunity for neurorehabilitation in TBI patients (Wurzelmann et al., 2017). This line of work provides an example of an important goal of PBM research in animal models of TBI—that is, identifying important biological targets of PBM (e.g., BDNF), and then determining whether they are of clinical relevance with respect to the pathology and symptoms (cognitive and otherwise) associated with TBI.

In general, the rodent models of TBI can be classified in two ways. One approach for modeling TBI has been to damage exposed cortex directly. This involves creating an incision in the scalp and then performing a craniotomy (under anesthetization) over a specified brain region, such as, for example, left parietal cortex (e.g., Ando et al., 2011). In the area where the brain is exposed, a controlled cortical impact (CCI) is then initiated, usually using a pneumatic piston device (but see below for another approach, cryogenic brain injury). This method allows for precision in the localization of brain damage. Following the injury, PBM treatment can be applied directly to the brain, or the treatment is applied transcranially after the skull opening has been replaced/filled and the scalp is sutured. Another approach has been to administer a closed head injury, which is often accomplished with a weight drop device (e.g., Oron et al., 2012). This is typically done by securing a small cone in place at the specified injury site. A cylindrical weight falls onto the cone, initiating an impact that results in brain injury. With this approach, histological follow-up can help determine the extent of cortical injury, as the specific brain region affected may not be as precisely controlled as with a CCI after craniotomy. However, an advantage of the closed head approach is that it may be a more clinically relevant model, as it may more closely reflect the types of closed-head injuries often sustained by TBI patients, in comparison to direct impact to exposed cortex (Flierl et al., 2009).

Multiple outcome measures for neurological and cognitive performance have been utilized to examine the effects of PBM treatment in rodent models of TBI. The neurological severity score (NSS) has been the most commonly used behavioral symptom outcome measure, beginning with the work done by Oron et al. (2007), and then thereafter in a number of other studies (Ando et al., 2011; Oron et al., 2012; Wu et al., 2010, 2012; Xuan et al., 2013, 2014, 2015; Zhang et al., 2014). The NSS is a 10 item scale that allows for a standardized, replicable assessment of neurologic function in rodent models. Items on the scale include the presence of mono- or hemiparesis; inability to walk on beams of varying widths (3, 2, and 1 cm wide); inability to balance on beams of varying widths (1 cm wide, 0.5 cm round stick); inability to walk straight; loss of startle behavior; and loss of seeking behavior. While many of these items assess some aspect of motor function, this scale takes into account motivation-related behaviors as well (e.g., seeking behavior). A point is awarded for each item that the animal fails, with a minimum score of 0 and a maximum of 10. Scores along this range are used to categorize the severity of the brain injury: near-fatal (9–10), severe (approximately 7–8), moderate (approximately 5–6), or mild (approximately 4 or less). This categorization scheme may provide roughly analogous demarcations, perhaps, to the GCS that is used in clinically diagnosed TBI patients. The use of this metric by research groups such as Hamblin and colleagues allows for direct comparison across studies that may use different methods for TBI models or PBM treatment parameters.

A number of studies have found that NSS outcomes are improved following PBM therapy in rodent models of TBI. In many of these studies, the effects of PBM treatment are examined across different treatment parameters. Key features of the treatment parameters that have been manipulated include the dosage (irradiation and/or fluence), the frequency mode (pulsed vs continuous), the target wavelength, and the timing of the treatment protocol (duration and quantity of treatment sessions). In Oron et al. (2007), mice were assessed following closed head injury via weight drop at a scalp site above parietal cortex. An 808 nm laser (continuous wave) treatment took place 4 hours post-TBI, and it was applied by placing the tip of the fiber optic on the surface of the scalp such that the entire brain could be illuminated for a duration of 2 minutes. There were three groups of mice in the study: one laser treatment group received a dosage of 10 mW/cm^2 (2-minute duration, 1.2 J/cm^2), another laser treatment group received a dosage of 20 mW/cm^2 (2-minute duration, 2.4 J/cm^2), and a sham group underwent the same procedures but did not receive any laser irradiation. Although there were no differences among groups on NSS for up to 48 hours post-TBI, from 5 days post-TBI on (up to 28 days) individuals in the laser treatment groups exhibited significantly lower NSS than the sham group mice. Furthermore, lesion volume was significantly greater in the sham group at the damaged site than in the laser treatment groups.

This work was followed up by a study from the same group (Oron et al., 2012) that examined the effects of either pulsed or continuous frequency modes (808 nm laser) on NSS outcomes and lesion burden. The same closed head injury TBI model was used. This study found that NSS was lower in the PBM treated mice at 5–28 days post-TBI compared to the sham group, replicating the prior results. In addition, full recovery to an NSS of 0 was observed in a higher percentage of mice that received the pulsed wave compared to the continuous wave mode of PBM treatment (although note that these differences were descriptive due to sample sizes ranging from 6 to 7 mice per group). At 56 days post-TBI, the pulsed and continuous wave treated groups both had significantly lower lesion volume compared to the sham group, and they did not significantly differ from each other. These results suggest that both pulsed and continuous wave modes of PBM, with 808 nm wavelength laser, can potentially improve neurological outcomes (i.e., NSS) and reduce lesion burden post-TBI. Along similar lines, Ando et al. (2011) examined the effects of 810 nm laser delivered across several different continuous/pulsed mode conditions. They used a CCI with craniotomy over left parietal/frontal-parietal cortex in mice, and treatment was administered at 4 hours post-TBI for a 12 minutes duration (50 mW/cm^2, 36 J/cm^2). There were four different treatment conditions: continuous wave mode, 10 Hz pulsed wave mode, 100 Hz pulsed wave mode, and no treatment. NSS was lower in the laser treatment groups compared to the nontreatment group following TBI, with the 10 Hz pulse mode group showing the lowest scores overall. The 10 Hz pulse mode group also had the most pronounced positive outcome on the forced swim test (depression analog) and the tail suspension test (depression/anxiety analog), and was the only PBM treatment group that had significantly lower lesion volume than the control group.

Hamblin and colleagues have included the NSS as an outcome measure in several studies that have investigated the effects of different PBM wavelength parameters. In Wu et al. (2010), moderate-to-severe TBI was simulated by adjusting the height of weight drop impact in a closed head model such that NSS ranged from 6 to 8. Laser treatment was applied at 4 hours post-TBI, with wavelengths at 670, 810, or 980 nm (150 mW/cm^2, 36 J/cm^2, 4-minute duration); these treatment groups were compared to a sham control group. Mice that received 670 or 810 nm PBM treatment had significant improvements in NSS compared to controls over the course of follow-up testing, whereas the 980 nm treatment group did not. The 670 and 810 nm treatment groups also had significantly lower lesion size compared to the control group, as well as no difference in lesion size in comparison to each other. In a subsequent study, Wu et al. (2012) examined the effect of 665, 730, 810, and 980 nm wavelength laser treatment in the same TBI model (closed head, NSS moderate-to-severe). The other treatment parameters (fluence, administration timing, and duration) were the same as in their previous study (Wu et al., 2010). Again, treatments that used 665 nm (670 nm range) and 810 nm laser resulted in significantly better NSS follow-up scores compared to controls. The 730 and 980 nm treatment groups did not exhibit significant differences from controls. The authors suggested that these findings could be due to the sensitivity of photoacceptors in cytochrome c oxidase being tuned within frequency ranges that include 665 and 810 nm.

Multiple subsequent studies have focused on the 810 nm frequency range while manipulating the duration/quantity of treatments across groups. Xuan et al. (2013) examined different treatment regimens at this wavelength (810 nm, continuous wave, 25 mW/cm^2, 18 J/cm^2, 12-minute duration) in mice that underwent CCI with craniotomy over temporal-parietal cortex. Laser treatment began at 4 hours post-TBI. Treatment groups were as follows: single treatment (1 day); 3 consecutive days of treatment; or 14 consecutive days of treatment. Compared to control groups, mice in the 1 and 3 day treatment groups had lower NSS, lower lesion volume, fewer degenerating neurons (less Fluoro-Jade staining), and possibly increased neurogenesis (more BrdU-positive cells) at up to 4 weeks post-TBI. In contrast, the group that received 14 consecutive days of PBM treatment did not exhibit significant differences from controls. In a follow-up study that sought to replicate and extend this work, groups that received either the 3-day treatment or 14-consecutive-day treatment protocols were assessed at up to 8 weeks postinjury, instead of only up to 4 weeks (Xuan et al., 2016). Here the authors sought to determine whether there were latent or multiphasic effects in the 14-day treatment protocol that were not observed in the first study due to a limitation in follow-up assessment time. In this latter study, the authors found that both treated groups (3- and 14-day treatments) had better NSS performance than the sham treatment TBI group at 1 week postinjury (Xuan et al., 2016). The group that received 14 consecutive treatments did not have better

NSS than the sham-TBI group from 2 to 6 weeks postinjury, but they did show better performance than the sham-TBI group from 7 to 8 weeks postinjury. However, in contrast, the TBI group that received three treatments showed better NSS performance than both TBI sham and 14-day treatment TBI groups and this improvement was sustained from 2 to 8 weeks post-TBI. These findings indicate possible multiphasic, dynamic changes in the time period postinjury with the 14-day treatment regimen. The authors suggested that an extended PBM treatment regimen may have led to transient changes in reactive gliosis, as evidenced by transient changes in a biomarker of this process (glial fibrillary acidic protein) over the 8-week follow-up in the 14-day treatment TBI group.

In Xuan et al. (2015), CCI with craniotomy was used to induce severe TBI in mice, and the 1- and 3-day laser treatment conditions with 810 nm were again used, but the dosage was doubled to 36 J/cm^2 (continuous wave, 50 mW/cm^2) in comparison to Xuan et al. (2013). Similar to the previous work, NSS was found to be reduced in both PBM treatment groups compared to controls, but the group that received three treatments showed an earlier improvement in NSS, and tended to have lower scores overall. In addition, the laser treated groups exhibited differences from controls on several biomarkers of neurogenesis (such as BDNF) and synaptogenesis. Overall, the studies that have evaluated varying treatment regimens indicate that 3 days of consecutive treatments, starting at 4 hours post-TBI, may be the most consistently effective treatment regimen in moderate-to-severe TBI rodent models. However, all treatment regimens still showed some level of effectiveness compared to control groups, and there is other work (described below) that indicates a higher frequency of treatment may also be effective in TBI.

While many of these studies have evaluated PBM in TBI models that simulate moderate-to-severe TBI, as reflected by NSS in the 7–8 range, Zhang et al. (2014) sought to examine the effects of PBM treatment in a model that simulated mild TBI, with a focus on secondary injury processes. TBI was induced in mice by a pneumatic impact to hairless scalp. Mice with NSS in the range of 4–6 at 1 hour post-TBI were included in the study, thus approximating mild TBI. Wild type mice and mice with an immediate early gene X-1 (IEX-1) knockout were examined. IEX-1 is critically involved in mitochondrial regulation, and it was therefore hypothesized to be important for regulating secondary pathology and symptoms post-TBI. An 810 nm laser treatment (150 mW/cm^2, 36 J/cm^2, 10 Hz pulsed wave, 4-minute duration) was delivered 4 hours post-TBI. Sham treated IEX-1 knockouts had pronounced secondary injury that was not observed in wild type mice, which included cell death, brain atrophy, and inflammation. PBM treatment rescued ATP functionality in IEX-1 knockouts, and pronounced secondary injury processes were not observed. Furthermore, PBM effectively rescued neurologic symptoms, as measured by NSS, in IES-1 knockouts that received treatment. The findings of Zhang et al. (2014) are consistent with the notion that mitochondrial functions can help minimize secondary pathology in TBI. These results suggest that in individuals with TBI that have disrupted mitochondrial function, the extent of secondary pathology can potentially be minimized with PBM treatment, which in turn could minimize the extent of behavioral and cognitive dysfunction.

PBM treatment methods using a variety of other wavelengths, in addition to those noted above, have been examined in several other studies. Giacci et al. (2014) compared 670 and 830 nm LED treatments (28.4 and 22.6 J/cm^2, respectively) in Sprague-Dawley rat models of different central nervous system injuries, one of which was a lateral percussion TBI model. In that group, craniotomy was performed over parietal cortex and a pressure pulse impacted the exposed brain. PBM treatment was then administered transcranially for 30 minutes, once per day for 7 days after injury. Following TBI, rats treated with either 670 or 830 nm wavelength light did not differ significantly from sham-treated controls on motor ability as assessed by rotarod balance, or sensory ability as assessed by the bilateral asymmetry test (e.g., latency of removal of sticky label applied to forepaws). There was also no significant difference between treatment and sham groups on lesion volume. Quirk et al. (2012) also used a 670 nm LED treatment, but their study yielded positive findings. In their study, TBI was induced with a CCI and craniotomy over left parietal cortex. In the treatment groups, 670 nm LED (5 minutes, 50 mW/cm^2, 15 J/cm^2) was administered twice per day for up to 10 days (some animals were sacrificed after 72 hours). This treatment resulted in changes to biomarkers of apoptotic function (indicating decreased apoptosis) compared to controls. The treatment groups, compared to controls, also had improved outcomes for goal-seeking behavior such as nose pokes to baited areas. Thus, while Giacci et al. (2014) reported null findings for behavioral tests and lesion volume following 670 nm treatment, Quirk et al. (2012) reported positive findings for behavioral measures and apoptosis outcomes following 670 nm treatment, in agreement with other studies (e.g., Wu et al., 2012). Differences in the treatment parameters, the TBI model methods, and/or the outcome measures used across these studies could possibly account for these discrepancies.

In Moreira et al. (2009), brain injury was induced cryogenically. Following craniotomy of the skull over frontal-parietal cortex, a lesion was induced in the exposed brain of Wistar rats via a copper probe immersed in liquid nitrogen. This study manipulated both the laser frequency (red vs infrared) and the dosage. An initial treatment was delivered at two points immediately following the lesion, before the craniotomy was sealed, and then a second treatment was delivered three hours later. Rats received one of the following treatment conditions (continuous wave for all treatments): 660 nm at 3 J/cm^2 (3-second irradiation), 660 nm at 5 J/cm^2 (5-second irradiation), 780 nm at 3 J/cm^2 (3 seconds), 780 nm at 5 J/cm^2 (5 seconds), or sham control. The results of this study were not unequivocal, with different PBM treatment parameters differentially affecting the concentration of different inflammatory cytokines. For example, greater stability of Interleukin-1 beta levels were observed in 660 nm, 3 J/cm^2, and 780 nm, 5 J/cm^2 groups, compared to sham and other groups. In a subsequent study (Moreira et al., 2011), this same group examined the effects of 780 nm laser (continuous wave, 3 J/cm^2, 3-second duration, two irradiations at 3-hour intervals) again in a cryogenic brain injury rat model. Compared to sham controls, the group that underwent PBM treatment had smaller lesions, and these lesions had a lower concentration of inflammatory markers. While the findings are preliminary, they are consistent with the conception that the beneficial effects of PBM treatment in TBI may occur through the modulation of postinjury inflammatory processes.

26.5 Effects of photobiomodulation on cognitive performance in animal models of traumatic brain injury

The studies in animal models of TBI that are discussed above have used a variety of behavioral tasks to measure neurological outcomes that are based primarily on motor-related functions, some motivated behaviors (e.g., food seeking), and depression/anxiety analogs (e.g., forced swim test). The NSS has been particularly useful for measuring general neurologic severity in preclinical TBI models, and its widespread use in PBM treatment research enables comparison across studies. However, relatively few studies of PBM treatment in TBI models have used tests that are designed to assess specific cognitive modalities. Four studies thus far have used the Morris water maze (MWM, see Table 26.1; Esenaliev et al., 2018; Khuman et al., 2012; Xuan et al., 2014; 2016), which is a test of spatial learning and memory commonly used in rodents (Bromley-Brits et al., 2011), with validity as a test of learning and memory dysfunction in preclinical models of TBI (Tucker et al., 2018). While the MWM has multiple variations, generally the animal is placed into one of four quadrants in a circular pool, and must find the location of a hidden or visible platform. The first trial usually entails the animal swimming to a visible, well-marked platform, and then in follow-up trials that platform may be visible but just beneath the surface of the water, or the water can be made opaque so that the platform is hidden. When the platform is hidden, it is thought that animals must rely on spatial memory for the platform location learned on previous trials/days. The latency to the platform is typically the measure of interest, with an upper time limit before the trial is terminated (e.g., 90 seconds). In addition, there is often a probe trial in which there is no platform, and the amount of time spent in the quadrant where the platform was located in previous trials can be measured.

Khuman et al. (2012) examined the effects of PBM treatment on MWM performance (and other measures) in mice that had a CCI with craniotomy over the left temporal-parietal cortex. In their study, some mice received 800 nm laser treatment directly to the brain, during open craniotomy, whereas other mice received the laser treatment transcranially (bone replaced and skin sutured). Mice were split into a variety of different treatments and TBI/sham condition combinations ($n = 239$ total mice). In the open craniotomy groups, mice received a single treatment within 60–80 minutes postinjury at one of five possible doses, ranging from 30 J/cm^2 (250 mW/cm^2 for 2 minutes) to 210 J/cm^2 (500 mW/cm^2 for 7 minutes). There was also a sham treatment open craniotomy group. For the transcranial treatment groups, 60 J/cm^2 (500 mW/cm^2) was applied for 2 minutes, but the timing of the treatments after TBI was varied. The mice received either a single treatment at 60–80 minutes post-TBI, a single treatment at 4 hours post-TBI, or one treatment per day over 7 days (starting 60–80 minutes post-TBI). MWM trials were completed on days 7–10 postinjury, and the primary outcome measures were the latency to finding the platform on hidden platform trials and the time spent in the target quadrant for probe trials. The authors found that the 60 J/cm^2 dose that was delivered within 60–80 minutes postinjury, delivered either directly via open craniotomy or transcranially, was effective in eliciting significantly better MWM performance compared to sham-treatment controls on the hidden platform and probe trials of the task. Mice that received 120 J/cm^2 at 60–80 minutes via open craniotomy, and mice that received the transcranial 60 J/cm^2 treatment for 7 days, also had improved probe trial performance compared to controls. Transcranial treatment at 4 hours post-TBI, and other dosages in the open craniotomy treatment groups, did not result in significant performance differences in comparison to controls. Furthermore, sham-injured mice (no-TBI) that received PBM treatments did not exhibit improved performance over nontreated controls. This study also did not find any changes in motor function (wire grip test) and lesion volume following PBM treatment, in contrast to previous work (e.g., Oron et al., 2007), a point noted by the authors. The authors attribute the differences in their findings in comparison to other studies to several possibilities. Oron et al. (2007) and others used the NSS, which may be a more comprehensive and sensitive measure of motoric impairment than the wire grip test that they used to assess motor function. Also, the treatment dosage used in Khuman et al. (2012) was higher than in other previous studies. Despite some differences in findings in comparison to other studies, the work from Khuman et al. highlights the importance of appropriate dosage and timing of PBM intervention when attempting to maximize cognitive outcomes. Their work indicated that an intervention within approximately 60–80 minutes postinjury resulted in the best outcomes on spatial learning/memory following cortical injury in mice.

Xuan et al. (2014) used the MWM as a primary outcome measure, and assessed changes in neurogenesis in the putative brain regions underlying spatial learning and memory. They used a CCI with craniotomy over temporal-parietal cortex to induce TBI symptoms in adult mice, as done previously by this same group (see Xuan et al., 2013). An 810 nm laser treatment (18 J/cm^2, 25 mW/cm^2) was transcranially administered. One group of mice received a single treatment at 4 hours postinjury, and another group received 3 days of treatment (once a day) starting at 4 hours postinjury. Performance on MWM was evaluated from 21 to 27 days postinjury. Both treatment groups had significantly better performance than controls on visible platform, hidden platform, and probe trials of the MWM. Further, three treatments appeared to result in more pronounced effects at several time points of the hidden platform trials, and on probe trials, compared to a single treatment in TBI mice. Both treatment groups also exhibited improved motor function on the weight grip and motion test. Note that the mice in this study had NSS in the 7–8 range, which is analogous to severe TBI. The mice were sacrificed at 7 or 28 days postinjury to assess immunofluorescence markers of apoptosis and neurogenesis. The PBM treatment groups, compared to nontreated TBI mice, had reduced expression of caspase-3, which is a protein involved in apoptosis, at the lesion site. Neuroprogenitor cell markers (BrdU, doublecortin, TUJ-1) were also increased in the dentate gyrus of the hippocampus and in the subventricular zone (at both 7 and 28 days postinjury at both sites) in treated groups compared to sham and nontreated TBI mice. The hippocampus and subventricular zone are key sites for neurogenesis, particularly in rodent models, and the hippocampus is a primary brain region associated with spatial learning and memory in mice (e.g., Clelland et al., 2009). The work by Xuan et al. (2014) provides evidence that PBM treatment in a rodent model of severe TBI can induce neurogenesis in the subventricular zone and hippocampus (in addition to reducing apoptotic events in the lesion site).

This study also contrasts with Khuman et al. (2012), who did not observe improved performance on the MWM in mice that received PBM treatment at 800 nm at 4 hours post-TBI. Only mice that received treatment at 60–80 minutes postinjury showed improved MWM performance in that study. Other than differences in the fluence and irradiation parameters, another notable difference between these studies was the duration of time of each individual treatment. In Khuman et al. the duration of contact and laser treatment was 2–7 minutes, and in Xuan et al. (2014) the duration was 12 minutes. Another factor that may account for the discrepant findings is that the inclusion criteria of these studies were different. Recall that in Xuan et al. (2014), only mice with severe TBI (NSS from 7 to 8) were included, whereas it is not certain what the specific range of neurologic impairment may have been in Khuman et al. (NSS was not used). This raises an important point about the conclusions that can be drawn when specific parameters of a PBM treatment are or are not found to be effective. The observed outcomes following the administration of different PBM treatment parameters may depend upon the experimental context (e.g., TBI model method, timing of treatment, etc.).

A recent study has examined the paired effects of transcranial 808 nm laser treatment and ultrasound waves on MWM performance (in addition to other measures) in a rat model of blast-induced TBI (Esenaliev et al., 2018). The pairing of these two potential treatment modalities was termed nano-pulsed laser therapy (NPLT) by the authors. In this study, TBI was induced in a group of Sprague-Dawley rats with a barrel impact into the right hemisphere of the head. The animals were anesthetized initially, and then this was discontinued once the animal was positioned under the impact device. The traumatic impact was initiated as soon as a withdrawal response was detected following repeated paw pinches. Rats in this group were then randomly assigned to either receive the treatment or receive no treatment. A sham-injury control group was also included, in which the animals were anesthetized and positioned under the device, but the blast injury was not carried out. For the TBI rats, the NPLT technique was applied to the site in which the blast injury was induced for a 5-minute duration at 1 hour postinjury. The technique involved 10 nanosec pulses of 808 nm light (20 Hz, 300 J/cm^2 over the 5-minute duration), which also generates low-level optoacoustic waves. From 6 to 10 days postinjury, animals were tested on the MWM. Animals underwent one trial of learning on the MWM, and a second trial was initiated after the learning trial, wherein the conditions remained the same. The latency to the platform on the second trial was the measure of interest. Second trial latency was significantly longer in the TBI group that did not receive treatment at 7 days postinjury in comparison to the sham-injured controls and the TBI group that received NPLT. The TBI group that received NPLT and the sham-injured controls did not differ on their performance on the MWM at any time-point. This pattern of findings, in which NPLT appeared to rescue performance following TBI, was also observed at 1 day postinjury for a beam balance test, and at 1 and 2 days postinjury for a beam walking test.

In addition to the behavioral findings, Esenaliev et al. (2018) found that, compared to nontreated groups, animals that were treated with NPLT also exhibited evidence of reduced apoptosis (e.g., downregulation of caspase-3) and increased neurogenesis (e.g., upregulation of BDNF). Compared to sham treated TBI animals, NPLT treated animals also had increased BDNF in the hippocampus and exhibited additional evidence of neural proliferation in the dentate gyrus. These results are in line with the findings of Xuan et al. (2014) noted above. Xuan et al. (2014) and Esenaliev et al. (2018), taken together, indicate that forms of PBM that include 808–810 nm range laser treatment can reduce apoptosis and induce neural proliferation in hippocampus. These effects provide a compelling mechanistic basis for the observed benefits of PBM treatment on spatial learning and memory, as evidenced by treatment-related benefits to performance outcomes on the MWM.

26.6 Enhancement of cognitive performance in healthy individuals with photobiomodulation treatment

A number of previous studies have examined whether PBM treatments can improve cognitive performance and modify brain function in normal, healthy human subjects. This work is summarized in Table 26.2 (see initial portion of table, studies that are not bolded). A series of studies from Gonzalez-Lima and colleagues have focused on the effects of PBM on cognitive processes that are associated with frontal cortical function, such as attentional control, category learning, and working memory. Their research group has reported improved performance following PBM treatment on the psychomotor vigilance task (PVT) and on the delayed match-to-sample (DMS) task (Barrett and Gonzalez-Lima, 2013; Hwang et al., 2016). The PVT is a sustained attention task in which participants must initiate a button press as quickly as possible whenever a target stimulus is visually presented on the screen. The target is presented randomly throughout the sequence of trials, and thus the participant must maintain engagement of attention across each trial of the task in order to detect the target stimuli. During the DMS task, participants are first presented with a complex visual stimulus (encoding phase), and then several seconds later they are presented with two stimulus options, one of which matches and one of which does not match the previously presented item (memory retrieval phase). This task is thought to engage short-term/working memory functions. Generally, short-term memory is the ability to maintain information in mind over a short time period (second to minutes); working memory is a related cognitive construct, but also entails the manipulation or executive control of information held in short-term memory (Baddeley, 2010). Sustained attention and working memory processes are both often categorized under the umbrella of executive functions, and they both can be affected in TBI patients.

Barrett and Gonzalez-Lima (2013) examined the effects of PBM treatment on the PVT and DMS task in healthy young adults. These cognitive tasks were administered immediately before (pretest) and immediately after the PBM procedure (posttest). PBM treatment was administered transcranially to the scalp above the right frontal pole (forehead sites medial and lateral to FP2, using the 10–20 EEG electrode placement system). A continuous wave laser application was used with 1064 nm wavelength, irradiance of 250 mW/cm^2, and fluence of 60 J/cm^2. The total duration of the treatment intervention was 8 minutes. For the control group, a similar procedure was used, but the approximate cumulative energy dosage was one-twelfth of what was applied to the treatment group. Participants in the treatment group, compared to controls, exhibited significantly faster reaction time on the PVT task at posttest, as well as faster response latency during the DMS task. The treatment group also had significantly better posttest accuracy on the DMS task, compared to controls. However, the effect for DMS accuracy may have been driven to some extent by a drop in performance from pre-to-posttest in the control group, as opposed to an improvement in performance from pre-to-posttest in the treatment group. This particular set of effects may have been due in part to testing fatigue, since the pretest, treatment, and posttest occurred all within the same laboratory session. The authors suggested that the effect of PBM in this instance could be that it reduced testing fatigue in treated study participants. In addition to the cognitive outcomes, the treatment group had more positive subjective effect scores at 2 weeks following the treatment compared to controls, although again this effect appeared to be driven primarily by a general decrease in mood from pretest to the 2-week follow-up in the control group, with more stability in pretest-to-follow-up mood in the treatment group. The authors speculated that decreased effect scores in the control group at the 2-week follow-up could be attributed to the fact that the study participants completed testing over the course of an academic semester, and may have had lower scores at follow-up due to end-of-semester stressors. So, similar to the rationale for the cognitive performance findings, the authors suggested that the effect of PBM treatment may have been to stabilize what otherwise would have been decreased mood at the follow-up.

TABLE 26.2 Effects of photobiomodulation on cognitive and functional performance in healthy human subjects and patients with traumatic brain injury.

Study	Human sample	Photobiomodulation protocol and groups	Cognitive and neurological findings
Barrett and Gonzalez-Lima (2013)	Young adults aged 18-35.	Treatment group, $n = 20$: 1064 nm laser, CW, 250 mW/cm^2, 60 J/cm^2, single treatment of 8 min duration, transcranial administration at right forehead sites (above right frontal pole). Sham-treatment control group: $n = 20$: Underwent the same protocol, but with 1/12th of the cumulative energy dosage.	Compared to controls, the treatment group had reduced response speed on the psychomotor vigilance task and delayed match to sample task, and better accuracy on delayed match to sample task following treatment.
Blanco et al. (2017a)	Young adults, mean age = 20.4 years (SD = 1.64).	Treatment group, $n = 15$: 1064 nm laser, CW, 250 mW/cm^2, 60 J/cm^2, single treatment of 8 min duration, transcranial administration at right forehead sites (above right frontal pole). Placebo group, $n = 15$: Underwent the same protocol, but with 1/12th of the cumulative energy dosage.	Compared to placebo controls, treatment group had better accuracy and learning on the Wisconsin Card Sorting Task. No differences were observed between groups on speed of rule learning, or overall response speeds.
Blanco et al. (2017b)	Young adults, mean age = 19 years (SD = 1.91).	Treatment: 1064 nm laser, CW, 250 mW/cm^2, 60 J/cm^2, single treatment of 8 min duration (4 mins per two locations), transcranial administration at right forehead sites (above frontal pole). Placebo control: Underwent the same protocol, but with 1/12th of the cumulative energy dosage Study groups: Active treatment rule-based task condition, $n = 32$; placebo rule-based condition, $n = 28$; active treatment information-integration condition, $n = 29$; placebo information-integration condition, $n = 29$.	Participants in the active-treatment, rule-based group had better performance compared to the placebo group. No differences were observed between active treatment information-integration group and placebo.
Chaieb et al. (2015)	Young adults, age range 18–35 years.	Treatment, $n = 15$: 810 nm laser, CW, 500 mW/cm^2 (estimated cortical fluence of about 1 J total energy), single treatment of 10 min duration, transcranial administration over primary motor cortex. Sham treatment, $n = 13$: 30 s of stimulation with same parameters.	Reduced excitability of motor evoked potentials in treatment group compared to sham group. No significant differences were observed between groups on a serial reaction time task.
Grover et al. (2017)	Individuals in an age range of 16–65, (seven of 31 subjects were 18 years or younger).	903 nm LED, 16.67 mW/cm^2, 20 J/cm^2, single treatment of 20 min duration, transcranial administration over occipital, left and right temporal, frontal, and parietal locations (cap configuration). Condition 1: Pretreatment measures obtained, treatment administered, posttreatment measures obtained. Condition 2: At 2–4 months, 18 participants returned, pretreatment measures obtained, rest period, posttreatment measures obtained.	Treatment condition produced greater improvement in reaction time to target trials of the task. No significant differences on event-related potential P300 component amplitude at whole-group level. Subgroup analyses found that there was a larger increase in P300 amplitude from pre- to posttreatment for individuals with low baseline P300 amplitude.

Hwang et al. (2016)	Young adults, age range 18–30 years.	1064 nm laser, CW, 250 mW/cm^2, 60 J/cm^2, single treatment of 8 min duration (alternated between two locations), transcranial administration at two locations on right forehead (medial and lateral to right frontal pole). Study Groups ($n = 15$ per group): Laser therapy sham exercise; acute high-intensity aerobic exercise and sham laser; combined laser therapy and acute exercise; sham exercise and sham laser.	Groups that received laser treatment, exercise treatment, or both were better than controls on change in psychomotor vigilance task reaction time and delayed match to sample task accuracy (pre- to posttreatment).
Konstantinovic et al. (2013)	Adults, mean age 35.0 years, (SD = 11.2), all female.	Treatment (study sample $n = 14$): 905 nm laser, pulsed (3 kHz), 50 mW/cm^2, 3 J/cm^2, single treatment of 60 min duration, transcranial administration at scalp over primary motor cortex.	Reduced excitability of primary motor cortex following the laser treatment, which lasted up to 30 min posttreatment.
Moghadam et al. (2017)	Young adults, age range 18–24 years.	Treatment group, $n = 17$: 850 nm LED, CW, 285 mW/cm^2, 60 J/cm^2, single treatment of 2.5 min duration, transcranial administration over right frontal pole. Sham group, $n = 17$: 5 s of stimulation with same parameters, amounting to 1/12th cumulative dosage.	Treatment group had better accuracy, faster response speed, and better task efficiency scores on go/no-go task at posttreatment compared to sham group.
Tian et al. (2016)	Experiment 1: Adults with mean age of 24.7 years (SD = 5.3), $n = 9$. Experiment 2: Adults of mean age 24.6 years (SD = 5.6), $n = 9$; six individuals were participants in both studies.	Treatment: 1064 nm laser, CW, 0.25 W/cm^2, 13.75 J/cm^2 per cycle, stimulation was 10 1 min cycles (55 s of laser stimulation, 5 s without laser stimulation), transcranial administration. Experiment 1: Laser administered on center of forehead; Experiment 2: Laser administered on right side of forehead.	Laser treatment in both experiments (right and center of forehead) resulted in an increase in oxygenated hemoglobin, and decrease deoxygenated hemoglobin in brain.
Wang et al. (2017)	Individuals with a mean age of 31 years (SD = 13.7).	Treatment condition: 1064 nm laser, CW, 0.25 W/cm^2, 13.75 J/cm^2 per cycle, stimulation was 8 1 min cycles (55 s of laser stimulation in each cycle, 5 s without stimulation), transcranial administration at right forehead. Placebo condition: Laser device applied to forehead at 0.1 W, laser aperture covered so no light was released, no stimulation; $n = 11$ completed placebo and treatment conditions.	Compared to placebo, treatment resulted in significant increase to oxygenated hemoglobin and up-regulation of cytochrome c oxidase in the brain.
Hesse et al. (2015)	Case series: Five TBI patients with disorders of consciousness (four unresponsive wakefulness/minimal consciousness, one subacute with akinetic mutism), age range 21–71.	Treatment: 785 nm laser, CW, 10 mW/cm^2, 6 J per diode, 10 min duration, 5 days a week for 6 weeks, transcranial administration, 30 total sessions.	All patients exhibited some improvements in level of interaction following intervention onset, including improvements in nonverbal interaction, visual pursuit, and coma recovery scale scores.
Hipskind et al., (2018)	Twelve chronic TBI patients, all military veterans (note: TBIs were not all due to military events), age range 21–55.	Two neoprene therapy pads wrapped around the upper portion of the head, with 220 near-infrared (850 nm) and 180 red (629 nm) LEDs. Pulsed power output of 3.3W, power density of 6.4 mW/cm2, 7.7 J/cm2; peak power density of 18.3 mw/cm2. 18 total sessions, applied for 20 minutes, three times a week for 6 weeks.	Following treatment, patients had significant improvement on several neuropsychological tests, including multiple measures of the California Verbal Learning Test (2nd edition), and multiple measures of processing speed from the Wechsler Adult Intelligence Scale (fourth edition; symbol search, coding). Eight of the twelve patients also had increased regional cerebral blood flow following training, as measured by single photon emission computed tomography (SPECT).

(*Continued*)

TABLE 26.2 (Continued)

Study	Human sample	Photobiomodulation protocol and groups	Cognitive and neurological findings
Morries et al. (2015)	Case series: Ten TBI patients, chronic mild-to-moderate TBI, sequential new referrals.	Treatment parameters used: 810/980 nm laser, dual wave, or 810 nm single laser device; power range of 10–15 W; fluence ranged from 55 to 81 J/cm^2; 10 to 20 treatment sessions, 8 to 10 min per treatment; transcranial administration to different areas across different participants, including bilateral frontal, left temporal, bilateral temporal areas.	Patients reported a variety of subjective improvements following treatment, including for physical symptoms, cognition, mood, and anxiety.
Naeser et al. (2011)	Case study, two patients (female) with chronic TBI, 59 and 52 years of age.	Patient 1 treatment: 870 nm LED, CW, 25.8 mW/cm^2, 8–20 J/cm^2, weekly transcranial treatment, 5 min 10 s to 12 min 54 s, transcranial administration on left and right forehead; after 7 months, continued self-administration of treatment for years with at-home device. Patient 2 treatment: 633 nm LED, CW, 22.2 mW/cm^2, 9.3–13.3 J/cm^2, 7–10 min per area, daily applications over a 4-week period, transcranial administration over bilateral forehead, frontal, parietal, temporoparietal areas; continued treatment for up to 4 months.	Following treatment, patient 1 reported subjective improvements in sustained attention and quality of life. Following treatment, patient 2 had improved performance on the Stroop test and logical memory subtest of the Wechsler Memory Scale –revised, in addition to improvements in sleep and self-regulation.
Naeser et al. (2014)	Case series: Eleven patients with chronic, mild TBI, mean age of 44.3 years (SD = 13.7).	633 and 870 nm LED, CW, 22.2 mW/cm^2, 13 J/cm^2 per head placement, 18 treatments completed over a 6-week period, 20 min duration per session (10 min for two different head placements), transcranial administration to multiple scalp sites on head.	Following treatment intervention, patients had improved performance on the Stroop and California Verbal Learning Test (2nd edition). Significant changes were not observed for other neuropsychological tests.
Naeser et al. (2016)	Reported preliminary findings from ongoing work: Ongoing study 1: Chronic moderate TBI, two patients (male and female). Ongoing study 2: Single mild TBI patient (24 year old female).	Ongoing study 1: 18 sessions of transcranial LED treatment, over 6 weeks. Ongoing Study 2: Device 1: 633, nm LED, CW, 8 mW/cm^2, 12 J/cm^2; Device 2: 810 nm LED, pulsed (10 Hz), 14.2 mW/cm^2, 10.65 J/cm^2; 18 sessions of intranasal LED treatment, over 6 weeks.	Ongoing study 1: Improvements in sleep patterns following treatment, one patient showed improvements in cognitive tests. Ongoing study 2: Patient exhibited improvement in cognitive performance and sleep patterns following treatment.
Poiani et al. (2018)	Planned randomized double-blind trial, moderate and severe TBI patients, planned $n = 36$ (18–60 years for inclusion), with $n = 12$ as of time of report.	Treatment condition: 632 nm LED, 830 mW over skull surface, 3.74 J/cm^2, 18 treatment sessions over 6 weeks, 30 min duration per treatment. Sham condition: Same as treatment procedure, but power is less than 1 mW.	Results pending. Methodology includes comprehensive neuropsychological outcome measures, with Stroop performance as the primary outcome.

Notes: *CW*, continuous wave; *fluence*, J/cm^2; *LED*, light emitting diode; *power density*, mW/cm^2; *TBI*, traumatic brain injury.

A similar experimental protocol was used by Hwang et al. (2016) to examine young adult participants, with the inclusion of additional study groups that underwent an acute aerobic exercise treatment. The scalp locations targeted, wavelength (1064 nm), irradiance, and fluence were all the same as in Barrett and Gonzalez-Lima (2013). Participants in groups that received PBM treatment, exercise treatment, or both treatments were significantly different from controls (sham exercise and sham PBM) on the change in PVT reaction time and DMS accuracy from pretest-to-posttest. However, the control participants exhibited an increase in RT on the PVT from pretest-to-posttest, and a decrease in accuracy on the DMS task from pretest-to-posttest (see Table 3 of Hwang et al., 2016). Again, the pretest, treatment, and posttest all occurred at the same session, and thus poorer performance on posttest compared to pretest measures in the control group can potentially be explained by testing fatigue. Similar to the previous study, the authors suggested that the treatment interventions may have protected against this fatigue. One piece of evidence that lends credence to this interpretation is that cognitive performance was similar between the study groups at baseline, as determined by scores on the Kaufman brief intelligence test. Further, DMS task accuracy and PVT RT were also similar between groups at pretest (although, they were not similar for DMS response latency at pretest). The overall findings from this study and Barrett and Gonzalez-Lima (2013) suggest the possibility that targeting right prefrontal cortex with PBM treatment can enhance or maintain aspects of sustained attention and short-term/working memory performance, and may potentially help compensate for the effects of mental fatigue.

This same research group also examined the effectiveness of PBM therapy on improving performance on the Wisconsin Card Sorting Task (WCST; Blanco et al., 2017a) and a category learning task (Blanco et al., 2017b) in young adults. This is a logical extension of the previous work from this group, as the WCST and other tasks that place an emphasis on categorization are strongly associated with frontal cortical networks. These studies used the same treatment and placebo parameters as the studies described above, with the exception of a slight adjustment to the placement of the laser diode on the scalp (lower and upper forehead locations, instead of medial and lateral). During the WCST, a set of four reference cards are presented, each with items of varying shape, color, and quantity (1–4 items). While the reference cards are continuously shown, individual target cards are presented which also have items with varying shape, color, and quantity. Accuracy on each trial of the task is determined by whether or not the participant can identify the reference card that matches the correct rule (i.e., matches the shape, color, or quantity of the target card), which is arbitrarily selected and changed after 10 consecutive correct trials. However, the subject is not explicitly told the rule, but they must instead learn the rule based on feedback about the accuracy of their responses. For example, if the first rule was based on color, feedback about trial accuracy would be based on whether they selected the reference card that matched the color of the target card. Ten consecutive correct responses are used as the threshold to indicate that the rule was discerned and properly applied by the subject. After 10 consecutive correct responses the rule is changed without explicitly informing the participant, and the new rule must then be learned. This task reflects executive function and cognitive flexibility, processes that are directly linked to frontal cortical function (Mountain and Snow, 1993). In Blanco et al. (2017a) participants either received PBM treatment or a sham treatment prior to being tested on the WCST. Participants that received the PBM treatment had significantly better overall accuracy than the sham group (85.5% compared to 79.9% accuracy, respectively). Furthermore, the PBM treatment group learned the second rule in the set of trials significantly faster than the sham group. The two groups did not significantly differ on how quickly they learned the other rules, although the treatment group showed a tendency for faster rule learning on the third and fourth rules of the task. The treatment and placebo groups did not differ on overall response speeds. Thus, while this study found that there were apparent benefits of PBM treatment on performance on the WCST, these benefits did not extend to every metric of task performance.

Blanco et al. (2017b) examined a relatively large sample size of young adults ($n = 118$, split into four groups, two treatment groups and two sham groups) using a protocol that was very similar to those of their prior studies, with a few notable changes. First, the PBM treatment was applied to two regions that overlapped with FP2 and right frontal (F4, F8) scalp sites. Second, the outcome measure was a category learning task that differed from the WCST. For the task in this study, Gabor patterns (visual grating stimuli) were displayed on the screen, and participants had to determine the category of these patterns based on the thickness and orientation of the grating patterns. Some participants were placed in the "information-integration" based category group. Participants in this group had to categorize the stimuli based on rules that are difficult to verbalize and understand explicitly, and therefore involve implicit rule-learning systems. Other participants were assigned to the rule-based structure group, in which the pattern categories can be explicitly described and verbalized. Rule-based learning is thought to be particularly related to frontal executive functions, in contrast to the more implicitly driven information-integration rule structures. Thus the hypothesis here was that treatment of right prefrontal cortex with PBM would selectively enhance performance in the rule-based group, but not the information-integration group. The findings yielded confirmatory evidence for this hypothesis. Participants in the information-integration PBM treatment group did not have significantly different performance from the placebo/sham group, whereas participants in the rule-based PBM treatment group exhibited significantly better performance compared to the sham group. Again, these findings reinforce the notion that enhancement of executive-mediated cognitive performance can be achieved via PBM which specifically targets prefrontal cortical regions.

Other researchers have also examined the effects of PBM on the right frontal pole in young adults. Moghadam et al. (2017) used an LED treatment with 850 nm wavelength (continuous wave, 285 mW/cm^2, 60 J/cm^2) administered for a total of 2.5 minutes. A sham treatment group received a fraction of the same dosage, but with the same protocol (one-twelfth cumulative dose, in line with the studies above from Ganzalez-Lima and colleagues). Before and after treatment, participants completed a go/no-go task, in which they had to perform a button press when certain letters were displayed on a computer monitor (go trials), but inhibit their response when other letter types were presented (no-go trials). The go/no-go task provides measurement of selective/sustained attention and inhibitory control, which, as noted already above, have major neural underpinnings in frontal cortical networks. Participants in the PBM treatment group had significantly better accuracy, faster response speed, and better task efficiency scores (rapid and accurate responding together) at post compared to pretest on the task. These improvements were not found in the sham group.

The results of this body of work suggests that PBM treatment applied transcranially to right prefrontal cortex can improve performance on executive function tests of inhibitory control, selective and sustained attention, category discrimination, and working memory. Since frontally-associated cognitive impairment and pathology are commonly observed in TBI, these findings are relevant with respect to understanding the potential for PBM as a therapeutic intervention in TBI. Other studies have examined the effects of PBM on brain function more directly in healthy individuals, and this work has revealed additional information about the possible neural mechanisms that mediate these effects. Multiple studies have examined the effects of PBM on motor evoked potentials, for example. Konstantinovic et al. (2013) applied 905 nm laser transcranially to a scalp location above primary motor cortex for 60 seconds (pulsed wave, 50 mW/cm^2, cumulative fluence = 15 J/cm^2) in healthy adults. Motor evoked potentials were elicited with TMS and measured via electromyography (EMG) at the first dorsal interosseous muscle of the right hand, before PBM application and at multiple time points afterwards. The authors found reduced excitability of primary motor cortex following the PBM application, which lasted up to 30 minutes posttreatment. While the precise implications of this effect were not determined, the authors noted that one potential interpretation of this finding was that the PBM intervention may have increased resistance in pyramidal neurons in motor cortex, thus reducing excitability of these neurons in response to TMS.

Chaieb et al. (2015) applied 810 nm laser (continuous wave, 500 mW/cm^2, estimated cortical fluence of about 1 J total energy) transcranially to primary motor cortex for 10 minutes in young adult subjects, and motor evoked potentials were induced with TMS and measured at baseline and after PBM treatment. Participants received both an active treatment condition (full 10 minutes of PBM) and a sham condition treatment (only 30 seconds of stimulation), randomly counterbalanced across participants. Performance on a serial reaction time task (SRTT) was also assessed. The SRTT was completed while the intervention procedure was taking place, during the ten minutes in which subjects underwent the active treatment or the sham procedure. This task requires participants to respond with button presses based on the sequence of four dots presented on a computer screen. The sequences follow patterns that the participant is not informed of, and thus these patterns and their response mappings can be learned implicitly over the course of the task. Learning the stimulus-response mappings corresponds with faster response speed over the course of the task block. Therefore, the authors were able to assess the impact of PBM treatment to M1 on implicit motor learning. Similar to the findings of Konstantinovic et al. (2013), the results of Chaieb et al. indicated reduced excitability of motor evoked potentials after PBM treatment that was sustained for up to 30 minutes thereafter. Despite changes in the excitability of motor cortex, no significant differences were observed between the treatment and sham conditions on SRTT performance. Thus, improvements in implicit learning were not observed concurrently with PBM-mediated changes in motor cortex excitability. It may be necessary to stimulate additional cortical/subcortical regions to improve implicit learning, such as precuneus, cingulate gyrus, and caudate nucleus (Yang and Li, 2012; see also above discussion with respect to Blanco et al., 2017b). Still another possibility is that a different wavelength may have been more effective, such as the 1064 nm wavelength, which was found to elicit positive cognitive performance effects by others (e.g., Hwang et al., 2016). Furthermore, the experimental design may have obscured potential effects of the PBM treatment, as the administration of the treatment while the task was ongoing may not have provided sufficient time for the biological mechanisms to confer their full effects on performance.

Grover et al. (2017) examined the effects of PBM on brain activity indexed by quantitative electroencephalography (EEG). Specifically, they examined reaction time performance and EEG activity obtained for an oddball target-detection paradigm. During the task, participants were presented with a series of consecutive auditory tone stimuli.

These stimuli consisted of two possible tone categories. One tone category was a nontarget stimulus (low auditory tone) that occurred relatively frequently during the task, and the other tone category was considered a target stimulus (high auditory tone) that occurred infrequently during the task (rare). Participants had to click a mouse each time they detected the rarely occurring target tone. During the oddball, target tones elicit a "P300" event-related potential (ERP) component, which is derived by signal-averaging the stimulus-locked EEG data. This component is a positive amplitude peak elicited at approximately 300–500 ms after the onset of the target stimulus (Polich, 2007). Participants in Grover et al. (31 healthy individuals between the ages of 16–65) completed the oddball task, underwent a PBM treatment session (20-minute duration), and then completed a follow-up oddball task immediately after PBM treatment, all within the same testing session. Furthermore, 18 participants returned for a second visit 2–4 months later where they completed the oddball twice with 20 minutes of rest between the two testing sessions (no PBM treatment). This served as the study control. The PBM treatment was delivered with a cap that covered occipital, bilateral temporal, frontal, and parietal regions, using a wavelength of 903 nm (16.67 mW/cm^2, 20 J/cm^2). The treatment condition produced a greater improvement in reaction time to target trials during the task. There were no statistically significant differences between the treatment and control arm of the study with respect to P300 amplitude and latency at the whole-group level. The authors also performed additional subgroup analyses in which they examined individuals in the treatment and control arm with initially low amplitude. This subanalysis indicated that in individuals with low baseline amplitude, there was a significantly larger increase in amplitude from pre-to-posttest for the treatment compared to control arms. This is relevant with respect to TBI, as other researchers have found reduced P3 amplitude in mild-to-moderate TBI patients compared to healthy controls during the oddball (Elting et al., 2005) and other similar tasks (Duncan et al., 2005). Thus, it can be speculated that patients with TBI may fall within the subgroup of individuals that would exhibit changes to the P300 following PBM. Furthermore, EEG/ERP measures such as the P300 index brain activity associated with activity of pyramidal cortical neurons, similar to the neural sources of the MEP measures noted above (Chaieb et al., 2015; Konstantinovic et al., 2013). These studies, taken as a whole, indicate that activity of pyramidal cortical neurons can be reliably measured to index the effects of PBM. Moreover, PBM interventions that are applied to the scalp are likely to impact these neurons in particular, since pyramidal neurons traverse outer cortical layers.

Recent work points toward several biological mechanisms of PBM in the brain that can help account for the link between PBM treatment and improved cognitive performance. Low-level laser stimulation using a wavelength of 1064 nm at forehead regions has been associated with increased oxygenated hemoglobin and decreased deoxygenated hemoglobin concentration (Tian et al., 2016; Wang et al., 2017). Wang et al. (2017) found that application of 1064 nm laser at the right forehead (continuous wave, 0.25 W/cm^2, 13.75 J/cm^2 per cycle, 8 one minute cycles with 55 seconds of stimulation per cycle) resulted in up-regulation of cytochrome c oxidase as determined by near-infrared spectroscopy measurements. The authors suggested that the biochemical processes that are relevant to this finding include cytochrome c oxidase mediated proton pump activation, nitric oxide release, and ATP synthase, among other mechanisms of action. These near-infrared spectroscopy studies offer a neurobiological basis for the other studies summarized above that have found enhanced cognitive performance following PBM treatment. Furthermore, the 1064 nm wavelength used by Tian et al. (2016) and Wang et al. (2017) is the same wavelength used by the majority of the studies that have examined cognitive changes following PBM treatment in healthy subjects. Thus it is likely that the findings of these studies reflect common mechanisms that are photosensitive at this wavelength. The enhancement of mitochondrial function via cytochrome c oxidase, which could enhance neuronal performance, is a reasonable mechanism for the improved performance on executive function tasks following PBM treatment to frontal cortical regions. These findings are informative for understanding the mechanisms and potential efficacy of PBM therapy in clinical populations with cognitive impairment, such as in patients with TBI.

26.7 Effects of photobiomodulation therapy on cognitive outcomes in traumatic brain injury patients

The preclinical studies of PBM therapy in rodent models of TBI have been instrumental in determining the mechanisms of action that may underlie improved outcomes following PBM. They have also provided preliminary comparisons of the efficacy of treatment parameters, and information about the potential of PBM for improving motoric/neurologic outcomes (e.g., as with NSS) and learning/memory functions (i.e., MWM assessment). The work outlined in the previous section also indicates that PBM can improve cognitive performance in human subjects as well. Work in humans with TBI has been underway to determine whether the promising results in animal models of TBI can extend to clinical applications. The latter portion of Table 26.2 highlights the relevant studies in this area (see the bolded references).

Much of this work has been reviewed previously by Naeser et al. (2016). But research on PBM therapy in humans with TBI is still in its early stages. The initial studies have been promising, but findings are still preliminary and have primarily consisted of case studies or case-series studies. A study by Naeser et al. (2011) was the first to report on the effects of PBM treatment in patients with TBI. They reported findings for two patients. One patient had sustained a closed head injury during a motor vehicle accident. This patient had a high premorbid IQ and functionality (e.g., she was a member of Mensa), but after sustaining a TBI she reported experiencing cognitive problems. She underwent PBM therapy at 7 years post-TBI. An 870 nm LED device (continuous wave, 25.8 mW/cm^2, 8–20 J/cm^2) was applied to areas on the left and right forehead for a total time of five minutes (10 seconds per area on the forehead). Following this treatment, she reported improved ability for sustained attention working at a computer. She continued treatments thereafter, with adjustments made to the treatment parameters as the intervention continued (i.e., additional scalp sites, increased dosage). After 8 weeks of treatment, she reported substantial improvement in sustained attention working at a computer, in comparison to before beginning treatment (a reported change from 20 minutes to 3 hours). The patient continued to self-administer treatments thereafter for upwards of 6 years with an at-home device, and reported continued ability to sustain attention and an improved quality of life compared to before beginning the PBM therapy.

The case report for the second patient in Naeser et al. (2011) provided additional quantitative measurement of pre- and post-PBM treatment outcomes. She had sustained her TBI with an injury from impacting a concrete surface on the back of the head, and reported subjective cognitive complaints afterwards. She also was on medical disability. Over a year after her TBI, she completed neuropsychological testing, and then began PBM treatment with a 633 nm LED device. Treatment was applied to multiple areas on the scalp, including over forehead, frontal, parietal, and parietal-temporal regions, with a range of timing and dosage levels (continuous wave, 22.2 mW/cm^2, 7–10 minutes per area). Nine months after beginning treatment, she was tested a second time on neuropsychological tests. She exhibited improved performance on multiple measures of the Stroop task, a test of response inhibition; and on the logical memory performance subtests of the Wechsler Memory Scale—revised. Scores on these tests ranged from +1 to +2 standard deviation improvements from pre- to posttreatment.

Subsequent reports examined the efficacy of PBM therapy using a case series approach. Morries et al. (2015) reported on a series of ten patients that were referred to their clinic for mild-to-moderate TBI. Patients were treated with an 810/980 nm laser device or with an 810 nm laser (fluence ranged from 55 to 81 J/cm^2). The treatments varied from 10 to 20 sessions total, with a time per treatment ranging from 8 to 10 minutes. The locations of treatments varied across patients, and included bilateral temporal regions, and frontal/forehead applications. TBI had been acquired in a variety of manners, including concussive blast injuries, motor vehicle accidents, abuse, hypoxia, etc. Patients also varied in the length of time that they had TBI before undergoing PBM therapy, from less than a month to over 20 years. A number of patients also had comorbid posttraumatic stress disorder, major depressive disorder, and/or generalized anxiety disorder. Prior to treatment, five patients reported the presence of suicidal thoughts, and all patients had sleep disturbances of some kind. After treatment, no patients reported suicidal thoughts, and all patients reported that their sleep disturbances had resolved. Further, baseline to follow-up scores on tests of subjective depressive symptoms were improved by a statistically significant amount, with scores at follow-up in the nondepressed range on average. Patients also reported subjective improvements in cognition and mood, and vocational outcomes were also positive at follow-up. Note, however, that cognitive improvements were not formally assessed with any neuropsychological measures, but were based on subjective patient report and observable life changes (such as vocational outcomes).

Hesse et al. (2015) examined the effects of PBM therapy in five TBI patients with severe disorders of consciousness. Patients with disorders of consciousness are in a minimally responsive or conscious state. The revised Coma Recovery Scale (r-CRS), which provides information about the responsiveness to sensory stimuli, was administered at various time points at baseline and postintervention. Additional outcome measures included the Barthel Index and modified Rankin Scale, which assess activities of daily living and caregiver ratings. The PBM intervention was a 785 nm laser (continuous wave, 10 mW/cm^2) that was applied to forehead locations for 10 minutes, across 30 total sessions. Patients were examine from 21 days prior to the intervention, and then up to 70 days after beginning the PBM therapy. All patients showed some improvements in their level of interaction from baseline to follow-up. For example, all patients had improved nonverbal interaction and visual pursuit over the course of the intervention. All patients also exhibited improved scores on the r-CRS. In comparison, changes on the modified Rankin Scale and Barthel Index were minimal.

To date, there have been only three studies that have directly assessed neurocognitive function formally with neuropsychological/psychometric tests that reflect specific cognitive domains. One was the case study that reported effects of PBM therapy in two patients, described above, from Naeser et al. (2011). Another was a case-series study from Naeser et al. (2014) that utilized a battery of cognitive tests that are commonly used to assess neuropsychological performance in cognitively impaired clinical populations, such as TBI. Eleven patients with mild TBI were entered into the study, with treatments beginning from 10 months to 8 years post-TBI. The PBM treatment devices included 633 nm (9 diodes) and 870 nm (52 diodes) wavelength LED light, continuous wave with a power density of 22.2 mW/cm^2. Treatment was completed for 18 sessions over a 6-week period, with two 10-minute treatments per session. Treatment was administered to a variety of scalp locations during each of the two 10-minute treatments, which included areas that would target multiple relevant brain networks (e.g., the default mode network), and frontal cortical regions. Patients were tested on a battery of neuropsychological tests at baseline and at posttreatment, and then subsequently at 1- and 2-month follow-ups.

Since this is one of only a few studies to date that has reported in depth, multimodal cognitive outcomes following PBM therapy in TBI, the individual tests that this study used and their specificity for assessing distinct cognitive domains will be outlined. Neuropsychological testing in Naeser et al. (2014) included the Stroop test, California Verbal Learning Test-II (CVLT-II), the trail makingt test from the Delis-Kaplan Executive Function System (DKEFS), the Controlled Oral Word Association Test (COWAT), and the digit span (forwards and backwards) from the Wechsler Adult Intelligence Scale (WAIS-IV). These tests have been identified by previous studies as being sensitive to neurocognitive impairment in TBI (Fork et al., 2005; Millis et al., 2001; Vanderploeg et al., 2005). The Stroop test is a measure of response inhibition (Scarpina and Tagini, 2017). Inhibitory control is measured on this test during trials that require the individual to say the colors of a word (or another visual item) out loud. However, on some items of the test, there is a mismatch between the color and the actual text of the word (e.g., the word "green" written in red colored ink, which would require the response "red" from the participant). This requires inhibitory control to override the prepotency to respond with the text of the word, instead of the color. The CVLT-II is a verbal learning and memory test comprised of several parts (Delis et al., 2000). Short- and long-term memory are assessed with both free recall and cued recognitions tests. A list of words is read to the participant a total of five times, and the participant must immediately recall the words after each list. Thus, it is possible to determine how well the participant learns the list over repeated presentations (five learning trials). The individual is tested on recall and recognition of the list of words again after a short-delay, during which a second list of words is presented for one learning trial, thus measuring short-delay recall after interference. Individuals are then tested again after a 20-minute delay in order to determine long-term free recall and cued recognition of the initially learned material. The trail making test (both the DKEFS version and other forms of the test) is a measure of processing speed and task-shifting (Yochim et al., 2007). In one form of the test, the individual must trace a line through a sequence of numbers (i.e., 1-2-3-4, etc.) that are arranged pseudo-randomly on the page. In another form of the test, participants must alternate between numbers and letters (i.e., 1-A-2-B, etc.), which requires the executive function of task-shifting. The COWAT is thought to be an executive task that measures verbal fluency, and requires individuals to generate as many words as possible within a 1-minute time period that begin with a specific letter (Loonstra et al., 2001). Finally, the digit span task has both forward and backward conditions (Heinly et al., 2005). In the digit span forward, the examiner reads a string of numbers out loud, and the examinee must then immediately recall them out loud, in order. This task allows for measurement of the maximum number of items that the participant can correctly recall, and thus it is an index of short-term memory capacity. In the digit span backward, the examiner again reads a string of numbers out loud, but this time the examinee must say the letters back in the reverse order. This requires the mental manipulation of information in memory, and is therefore often construed to be reflective of working memory, which entails both executive control and memory span.

In Naeser et al. (2014), changes in performance trends from baseline to follow-up sessions were assessed with one-way repeated measures analysis of variance. They found that TBI patients had improved cognitive performance following PBM therapy (parameters described above) on Stroop test measures and on CVLT-II learning and long-delay free recall. However, changes from baseline to follow-up were not statistically significant for the trail making test, the COWAT, or the digit span. Patients were also assessed on depression and pain outcomes. There was significant reduction of depressive symptoms at 1 week after treatment, only; and there were no changes in pain outcomes as evidenced by the visual analog scale for pain. Overall, this work provides a template for the use of multimodal neuropsychological outcome measures in studies of PBM therapy in TBI. While this study provides an important proof of concept and intriguing initial evidence for the effectiveness of PBM interventions for improving neurocognitive outcomes in TBI, the authors note that interpretation of the findings requires caution. The same can be said of all of the work that has been reported thus far that has examined PBM therapy in clinical TBI. No studies of this type have included a control comparison group, for example. Comparison of TBI patients to a group of healthy individuals without TBI diagnosis or history would enable better characterization of baseline cognitive impairment, and clinically specific PBM effects. Comparison of TBI patients that receive a specific PBM intervention to a group of patients that receive a placebo treatment (possibly using a crossover design) is essential for understanding the true effects that PBM may have on cognitive outcomes. Studies with a randomized placebo-control group scheme, with large sample sizes and multimodal cognitive assessments are yet to be completed.

The initial reports from several ongoing studies, which include both case series and an open-protocol pilot study, continue to hint at promising outcomes. Naeser et al. (2016) reported improved outcomes in two TBI patient cases in some of their ongoing work, including improved cognitive outcomes in one of the patients. This research group also found that another patient that received intranasal application of LED treatment, as part of an ongoing pilot study, showed improved performance on the Stroop and CVLT-II from baseline to follow-up. Poiani et al. (2018) have published a report of an ongoing randomized, double-blind trial with moderate and severe TBI patients. They plan to test a total of 36 patients, and 12 had completed the study protocol as of the time of their report. The PBM intervention for this study is an LED helmet that emits 632 nm light, with an estimated dosage of 3.74 J/cm^2 per treatment session. Patients in their study complete 18 treatment sessions at 30 minutes per session. The sham group receives the same procedure, but the power is less than 1 mW, as opposed to a total of 830 mW in the active treatment group. This would be the first study examining PBM in TBI patients that will have included a sham-treatment control group. This trial also utilizes a full battery of neuropsychological tests. Depression and anxiety symptoms are being assessed with the Beck Depression Inventory (BDI-II) and Beck Anxiety Inventory (BAI), respectively. Cognitive tests in the study include the trail making test, Stroop test, COWAT, and digit span, which were all included in the previous Naeser et al. (2014) pilot study. Additional tests in Poiani et al. include: the five-point test, an executive function measure of visuospatial fluency; the symbol digit modalities test (SDMT), a measure of information processing speed; the Rey Auditory Verbal Learning Test (RAVLT), which is comparable to the CVLT-II and serves as a measure of verbal learning and memory; the sequence numbers and letters task, a measure of working memory; and the Rey-Osterrieth Complex Figure Test (ROCF), a measure of visuospatial cognitive/executive functions. This study will provide important information about the effects of PBM therapy on cognition in TBI. Replicating the findings of Naeser et al. (2014) would entail that the PBM treatment is effective in improving outcomes on the Stroop and the RAVLT (similar to the CVLT-II), using a different wavelength (although still close in range). It will also be interesting to see if outcomes are not improved for other tests, such as the trail making test, COWAT, and digit span, since significantly improved group-wise outcomes were not observed for these tests in Naeser et al. (2014). The speculation is that consistent findings of improved outcomes in particular types of tests—for example, Stroop tests of inhibition, and tests of verbal learning and memory—would point toward specificity of PBM interventions for particular cognitive domains in TBI. Recently, Hipskind et al. (in press) have reported improvements in neuropsychological test performance in a sample of twelve TBI patients that underwent eighteen sessions of PBM therapy (629 nm and 850 nm LED light, pulsed, 6.4 mW/cm^2, 7.7 J/cm^2, 20 minutes per session). They found significant improvements from baseline to follow-up on multiple measures of the CVLT-II, as well as on tests of the WAIS-IV that measure the speed of information processing. The CVLT-II results from this study provide evidence for replication of previous findings along these lines in Naeser et al. (2014). Note also that Naeser et al. and Hipskind et al. both used red and near-infrared light at similar wavelengths for the PBM treatment. Hipskind et al. also found that eight of the twelve TBI patients exhibited enhanced cerebral blood flow (as measured by SPECT) following PBM treatment. Additional work is required to determine whether these preliminary findings can be consistently replicated, and to determine if PBM-induced enhancement of cognitive performance and brain function can transfer to other aspects of everyday functioning in TBI patients.

26.8 Summary and future directions

The literature examining the effects of PBM interventions on outcomes in TBI includes positive findings in a number of areas. The preclinical studies have laid the foundation for the rationale behind the use of PBM as a therapeutic intervention for TBI. These studies have highlighted several key targets and mechanisms that underlie the effects of PBM treatment in TBI. These mechanisms center on the modulation of mitochondrial activity via photosensitive activation of cytochrome c oxidase (De Freitas and Hamblin, 2016; Hamblin, 2016). The downstream effects of this are thought to involve improved mitochondrial function and the promotion of neural plasticity, including neurogenesis and synaptogenesis (Hamblin, 2017; Xuan et al., 2013, 2015). Neural plasticity is thought to be critically important for the rehabilitation and recovery of functional and cognitive outcomes following brain injury (Berlucchi, 2011; Kleim, 2011). Indeed, PBM interventions have been effective in improving neurological performance outcomes in rodent models of TBI, as evidenced by NSS findings (e.g., Oron et al., 2007, 2012). The NSS places emphasis on motoric and motivational behaviors, and not necessarily cognitive performance per se. Other studies have also found improved spatial learning and memory, as measured by the MWM, following PBM treatment in rodent models of TBI (Esenaliev et al., 2018; Khuman et al., 2012; Xuan et al., 2014), and these improvements correspond with neuroplasticity in the hippocampus and subventricular zone (Esenaliev et al., 2018; Xuan et al., 2014).

The research examining PBM therapy in clinical studies of TBI is still in its early stages. Case studies have yielded positive findings, including reports of improved cognitive performance following PBM treatment (Hesse et al., 2015; Morries et al., 2015; Naeser et al., 2011). Further, there is preliminary evidence that PBM therapy can improve outcomes on tests of processing speed, inhibitory control and verbal learning and memory in patients with TBI (Hipskind et al, in press; Naeser et al., 2014). However, no studies to date have included a control comparison group that has sufficiently addressed the possibility of placebo and motivational effects. The work thus far has been predominantly case-based with low sample sizes, and blinding procedures have not been used. There are no randomized clinical trials that provide level I or level II evidence that PBM is an effective treatment of cognitive impairment in TBI at this point in time. Additional studies are currently underway that are examining multimodal neurocognitive outcomes in larger samples of TBI patients and include control/placebo groups (Naeser et al., 2016; Poiani et al., 2018). Well-designed, placebo-controlled studies of this type are needed for the development of a consensus understanding of the effects of PBM on cognitive outcomes in TBI.

Studies of the effects of PBM on cognitive performance in healthy individuals can provide clues about the potential for PBM to remediate cognitive symptoms in TBI. There is evidence that PBM treatment can potentially reduce cognitive fatigue, and enhance sustained attention, working memory performance (Barrett and Gonzalez-Lima, 2013; Hwang et al., 2016), complex stimulus categorization (Blanco et al., 2017a,b), and inhibitory control (Moghadam et al., 2017). Coincidentally, these cognitive domains are among the variety of executive control abilities that can be impaired in TBI (Barker-Collo et al., 2015; Rabinowitz and Levin, 2014). Changes in brain function have also been observed following PBM treatment in healthy individuals. This includes changes in motor evoked potentials (Chaieb et al., 2015; Konstantinovic et al., 2013) and the concentration of oxygenated and deoxygenated hemoglobin in the brain (Tian et al., 2016; Wang et al., 2017). Near-infrared spectroscopy measurements indicate up-regulation of cytochrome c oxidase in the brain—which is thought to be a primary mechanism of PBM—following treatment in healthy individuals. Furthermore, positive results have been observed with the application of PBM treatments in clinical populations other than TBI. For example, patients with dementia have shown improved performance following PBM treatment on tests of executive functioning, attention, and task-switching (Berman et al., 2017; Saltmarche et al., 2017). Of course, TBI is associated with its own unique set of pathological processes and profile of cognitive impairments. It is therefore possible (perhaps likely) that PBM treatments may yield differential findings for TBI in comparison to other study groups, such as healthy individuals or other clinical populations. Still, when considering the findings of PBM-related cognitive improvement in healthy individuals, other clinical populations, preclinical models, and individuals with TBI, the totality of this work points toward PBM as a potentially effective intervention strategy for improving cognitive outcomes in TBI. But there are many outstanding questions and concerns that warrant extensive additional research.

The optimal parameters of PBM treatment in TBI have not been fully determined, but the growing body of work has provided some results that are suggestive of parameter specific efficacy. In animal models of TBI, both continuous and pulsed modes have been effective, although some studies suggest that certain pulsed modes may have an advantage (Ando et al., 2011; Oron et al., 2012). Studies that have directly compared different wavelengths suggest that 670 and 810 nm wavelengths may have an advantage over other frequencies (Wu et al., 2010, 2012), but other wavelengths, such as 780 nm, have also yielded positive results in other studies (Moreira et al., 2011). There is also evidence of multiphasic responses that are dependent on the number of treatments, as Xuan et al. (2013) found that while 1 or 3 days of PBM treatment may be more effective than longer treatment protocols, there may be improvements at extended follow-up (e.g., 56 days post-TBI) even with longer treatment that included 14 days (Xuan et al., 2016). Human TBI studies have applied treatments at 785 nm (Hesse et al., 2015), 810 nm (Morries et al., 2015), or used multi-unit devices with both 633 and 870 nm wavelengths (Naeser et al., 2011, 2014). Thus far, no studies in TBI patients have used the 1064 nm frequency band that has been shown to elicit effects in healthy individuals in the studies by Gonzalez-Lima and colleagues. PBM interventions in TBI patients have also generally been completed across multiple treatment sessions (e.g., 10–20 treatments, or even up to several years in reported case studies).

Future work in clinically diagnosed TBI patients should eventually directly compare different PBM treatment protocols. Experimental manipulations can be placed on the wavelength used, the energy dosage (irradiation and fluence), and the frequency and duration of treatments, taking cues from the preclinical work and studies in other healthy/clinical human subject populations that have been done before. As noted above, placebo-control groups and large sample sizes are also necessary in future work to determine the true efficacy of PBM in TBI. With larger samples in subsequent studies, it will be possible to also determine whether specific subtypes of TBI patients are particularly responsive to PBM therapy. There may be different dosages and treatment schedules that are optimal for mild, moderate, and severe categories of TBI. The phase of injury at which to administer PBM treatment is also an important consideration. One potential hypothesis, for example, is that PBM application at the acute phase of injury would be more effective at reducing long-term, sustained impairments than interventions in the secondary stage of TBI, particularly when the initial trauma is severe. Heterogeneity of pathology in TBI and variance in baseline cognitive reserve may also contribute to individual differences in the responsiveness to PBM treatment. Improved experimental design will further enable the assessment of synergistic effects when other treatment modalities (e.g., pharmacologic intervention, cognitive rehabilitation, exercise) are combined with PBM. With an appropriate experimental design, statistical approaches such as multiple linear regression modeling can be utilized to determine whether specific factors are strongly or weakly predictive of treatment outcomes.

Another important step moving forward will be the use of advanced neuroimaging outcome measures. Preclinical studies have used histopathological measures to assess changes in biological mechanisms in animal models of TBI following PBM, but only one study to date has examined neuroimaging outcomes following PBM in clinically diagnosed TBI patients (i.e., Hipskind et al., in press). Conventional neuroimaging techniques that are used to identify clinically relevant pathology in TBI include computed tomography and magnetic resonance imaging methods. Advanced neuroimaging measures in TBI include susceptibility-weighted imaging, which is sensitive to micro-hemorrhages; diffusion tensor imaging (DTI) methods that are sensitive to the integrity of white matter microstructure; magnetic resonance spectroscopy, which can assess metabolic functions; and fMRI (and also positron emission tomography), which measures changes in blood oxygen that is reflective of regional activity (for a review, see Mechtler et al., 2014). Electrophysiological measures such as quantitative EEG can also provide neural indices of large-scale cognitive network activity in TBI with high temporal resolution (on a millisecond timescale). ERP components can be derived by signal averaging ongoing EEG activity that is obtained during cognitive tasks (Duncan et al., 2011). These different neuroimaging methods can serve to analyze distinct outcomes related to TBI. EEG/ERP and fMRI measures can provide information about the neural underpinnings of cognitive dysfunction when they are studied in the context of cognitive performance. For example, a number of studies have found that TBI patients, compared to controls, have alterations in P300 ERP component amplitude/latency, which is reflective of dysfunction in stimulus categorization and evaluation processes (Duncan et al., 2011); and fMRI measures have been used to identify disruption in the default mode network in TBI (Marquez de la Plata et al., 2011). Methods such as magnetic resonance spectroscopy (and near-infrared spectroscopy, see discussion regarding Wang et al., 2017 above), may be particularly useful since this can measure metabolic changes associated with mitochondrial regulation, a key target of PBM therapy. Single photon emission computed tomography (SPECT) has been a useful tool for determining changes in cerebral blood flow following PBM in TBI patients (Hipskind et al., in press). DTI measures have also been useful in measuring disconnectivity of white matter in TBI (Fagerholm et al., 2015; Xiao et al., 2015), and microstructural white matter disturbances have been correlated with poorer cognitive performance (Niogi et al., 2008). Integrating these neuroimaging techniques into future work as outcome measures is essential for determining the neural underpinnings of PBM interventions in TBI.

To properly evaluate cognitive outcomes following PBM treatment in TBI, future work should utilize neuropsychological testing protocols that assess multiple cognitive domains, and that includes tests that are clinically relevant. Initial work has identified tasks of complex processing speed, the Stroop task, and the CVLT-II as potentially sensitive to the effects of PBM in patients with TBI (Hipskind et al., in press; Naeser et al., 2014). While there is not a consensus neuropsychological testing battery in TBI that can be used as a complete set of "gold standard" assessments, there are testing batteries that are commonly used, and there are specific cognitive domains that are consistently impacted in TBI. Rabinowitz and Levin (2014) list the following set of cognitive executive functions that are commonly affected in TBI: "memory acquisition and retrieval; top-down control of attention; planning; judgment; cognitive aspects of decision-making." Computerized cognitive testing methods are also commonly used to assess cognitive performance in TBI, and these include the automated neuropsychological assessment metric (ANAM4), the central nervous system—vital signs (CNS-VS), the CogState battery, and the immediate postconcussion assessment and cognitive testing (ImPACT) battery. However, the construct validity and symptomatic sensitivity of these computerized batteries may be more limited in comparison to standard neuropsychological assessment, as administered by a clinician (Arrieux et al., 2017). A brief exam is often used clinically, such as the mini-mental state exam (Pangman et al., 2000) or the Montreal Cognitive Assessment (Nasreddine et al., 2005), but tests of this nature are meant to provide a gross overview of impairment, and not a fine-grained assessment (Kosaka, 2006). Striking the right balance in terms of including study measures that are common to clinical practice, and measures that may also be useful from a more experimental perspective, is an important study design consideration.

26.9 Conclusion

There is emerging evidence that PBM may be an effective intervention strategy for treating cognitive impairment in patients with TBI. A cascade of biological mechanisms that relate to mitochondrial function are associated with PBM interventions in rodent animal models of TBI. There is also evidence that cognitive performance is enhanced in healthy individuals that undergo PBM treatment. The preliminary work that has examined PBM therapy in clinically diagnosed TBI patients is promising, and indicates the potential for PBM to improve cognitive outcomes and other metrics of everyday functioning. This initial research has laid the groundwork for the use of PBM as a potential treatment in TBI. Additional research is now warranted and necessary to continue to carry this line of work forward. In addition to continued development of case-series and pilot research, future studies should also use experimental designs with sufficient sample sizes to measure the effects of interest, placebo-controlled groups, and robust measurement of neurocognitive outcomes. As work of this kind is completed, the efficacy of PBM for treating cognitive impairment in TBI can be compared across different TBI patient subgroups, and also linked with other treatment modalities.

References

Anders, J.J., Lanzafame, R.J., Arany, P.R., 2015. Low-level light/laser therapy versus photobiomodulation therapy. Photomed. Laser Surg. 33 (4), 183–184.

Ando, T., Xuan, W., Xu, T., Dai, T., Sharma, S.K., Kharkwal, G.B., et al., 2011. Comparison of therapeutic effects between pulsed and continuous wave 810-nm wavelength laser irradiation for traumatic brain injury in mice. PLoS One 6 (10), e26212.

Arany, P.R., 2016. Craniofacial wound healing with photobiomodulation therapy: new insights and current challenges. J. Dent. Res. 95 (9), 977–984.

Arciniegas, D.B., 2011. Clinical electrophysiologic assessments and mild traumatic brain injury: state-of-the-science and implications for clinical practice. Int. J. Psychophysiol. 82, 41–52.

Arrieux, J., Cole, W.R., Ahrens, A.P., 2017. A review of the validity of computerized neurocognitive assessment tools in mild traumatic brain injury assessment. Concussion 2 (1). Available from: https://doi.org/10.2217/cnc-2016-0021.

Baddeley, A., 2010. Working memory. Curr. Biol. 20 (4), R136–R140.

Barker-Collo, S., Jones, K., Theadom, A., Starkey, N., Dowell, A., McPherson, K., et al., 2015. Neuropsychological outcomes and its correlates in the first year after adult mild traumatic brain injury: a population-based New Zealand study. Brain Inj. 29 (13–14), 1604–1616.

Barrett, D.W., Gonzalez-Lima, F., 2013. Transcranial infrared laser stimulation produces beneficial cognitive and emotional effects in humans. Neuroscience 230, 13–23.

Benedictus, M.R., Spikman, J.M., van der Naalt, J., 2010. Cognitive and behavioral impairment in traumatic brain injury related to outcome and return to work. Arch. Phys. Med. Rehabil. 91, 1436–1441.

Berlucchi, G., 2011. Brain plasticity and cognitive neurorehabilitation. Neuropsychol. Rehabil. 21 (5), 560–578.

Berman, M., Halper, J.P., Nichols, T.W., Jarrett, H., Lundy, A., Huang, J.H., 2017. Photobiomodulation with near infrared light helmet in a pilot, placebo controlled clinical trial in dementia patients testing memory and cognition. J. Neurol. Neurosci. 8 (1). Available from: https://doi.org/10.21767/2171-6625.1000176.

Blanco, N.J., Maddox, W.T., Gonzalez-Lima, F., 2017a. Improving executive function using transcranial infrared laser stimulation. J. Neuropsychol. 11, 14–25.

Blanco, N.J., Saucedo, C.L., Gonzalez-Lima, F., 2017b. Transcranial infrared laser stimulation improves rule-based, but not information-integration, category learning in humans. Neurobiol. Learn. Mem. 139, 69–75.

Bromley-Brits, K., Deng, Y., Song, W., 2011. Morris water maze test for learning and memory deficits in Alzheimer's disease model mice. J. Vis. Exp. 53. Available from: https://doi.org/10.3791/2920.

Carney, N., Chesnut, R.M., Maynard, H., Mann, N.C., Patterson, P., Helfand, M., 1999. Effect of cognitive rehabilitation on outcomes for persons with traumatic brain injury: a systematic review. J. Head Trauma Rehabil. 14 (3), 277–307.

Carpenter, K.L.H., Czosnyka, M., Jalloh, I., Newcombe, V.F.J., Helmy, A., Shannon, R.J., et al., 2015. Systemic, local, and imaging biomarkers of brain injury: more needed, and better use of those already established. Front. Neurol. 6 (1). Available from: https://doi.org/10.3389/fneur.2015.00026.

Chaieb, L., Antal, A., Masurat, F., Paulus, W., 2015. Neuroplastic effects of transcranial near-infrared stimulation (tNIRS) on the motor cortex. Front. Behav. Neurosci. 9 (147). Available from: https://doi.org/10.3389/fnbeh.2015.00147.

Cicerone, K., Levin, H., Malec, J., Stuss, D., Whyte, J., 2006. Cognitive rehabilitation interventions for executive function: moving from bench to bedside in patients with traumatic brain injury. J. Cogn. Neurosci. 18 (7), 1212–1222.

Cicerone, K.D., Langenbahn, D.M., Braden, C., Malec, J.F., Kalmar, K., Fraas, M., et al., 2011. Evidence-based cognitive rehabilitation: updated review of the literature from 2003 through 2008. Arch. Phys. Med. Rehabil. 92, 519–530.

Clelland, C.D., Choi, M., Romberg, C., Clemenson Jr., G.D., Fragniere, A., Tyers, P., et al., 2009. A functional role for adult hippocampal neurogenesis in spatial pattern separation. Science 325 (5937), 210–213.

Currais, A., 2015. Ageing and inflammation – a central role for mitochondria in brain health and disease. Ageing Res. Rev. 21, 30–42.

de la Torre, J.C., 2017. Treating cognitive impairment with transcranial low level laser therapy. J Photochem. Photobiol. B: Biology 168, 149–155.

De Freitas, L.F., Hamblin, M.R., 2016. Proposed mechanisms of photobiomodulation or low-level light therapy. IEEE J. Sel. Top. Quant. Elect. 22 (3). Available from: https://doi.org/10.1109/JSTQE.2016.2561201.

Delis, D.C., Kramer, J.H., Kaplan, E., Ober, B.A., 2000. Manual for the California Verbal Learning Text, (CVLT-II). The Psychological Corporation, San Antonio, TX.

Demirtas-Tatlided, A., Vahabzadeh-Hagh, A.M., Bernabeu, M., Tormos, J.M., Pascual-Leone, A., 2012. Noninvasive brain stimulation in traumatic brain injury. J. Head Trauma Rehabil. 27 (4), 274–292.

Dikmen, S.S., Corrigan, J.D., Levin, H.S., Machamer, J., Stiers, W., Weisskopf, M.G., 2009. Cognitive outcome following traumatic brain injury. J. Head Trauma Rehabil. 24 (6), 430–438.

Duncan, C.C., Kosmidis, M.H., Mirsky, A.F., 2005. Closed head injury-related information processing deficits: an event-related potential analysis. Int. J. Psychophysiol. 58, 133–157.

Duncan, C.C., Summers, A.C., Perla, E.J., Coburn, K.L., Mirsky, A.F., 2011. Evaluation of traumatic brain injury: brain potential is diagnosis, function, and prognosis. Int. J. Psychophysiol. 82, 24–40.

Elting, J.W., van der Naalt, J., van Weerden, T.W., De Keyser, J., Maurits, N.M., 2005. P300 after head injury: psuedodelay caused by reduced P3A amplitude. Clin. Neurophysiol. 116, 2606–2612.

Esenaliev, R.O., Petrov, I.Y., Petrov, Y., Guptarak, J., Boone, D.R., Mocciaro, E., et al., 2018. Nano pulsed laser therapy is neuroprotective in a rat model of blast-induced neurotrauma. J. Neurotrauma 35, 1510–1522.

Fagerholm, E.D., Hellyer, P.J., Scott, G., Leech, R., Sharp, D.J., 2015. Disconnection of network hubs and cognitive impairment after traumatic brain injury. Brain 138, 1696–1709.

Faila, M.D., Juengst, S.B., Arenth, P.M., Wagner, A.K., 2016. Preliminary associations between brain-derived neurotrophic factor, memory impairment, functional cognition, and depressive symptoms following severe TBI. Neurorehabil. Neural. Repair 30 (5), 419–430.

Faul, M., Xu, L., Wald, M.M., Coronado, V.G., 2010. Traumatic Brain Injury in the United States: Emergency Department Visits, Hospitalizations and Deaths 2002-2006. U.S. Department of Health and Human Services, Centers for Disease Control and Prevention.

Flierl, M.A., Stahel, P.F., Beauchamp, K.M., Morgan, S.J., Smith, W.R., Shohami, E., 2009. Mouse closed head injury model induced by a weight-drop device. Nat. Protoc. 4, 1328–1337.

Fork, M., Bartels, C., Ebert, A.D., Grubich, C., Synowitz, H., Wallesch, C.W., 2005. Neuropsychological sequelae of diffuse traumatic brain injury. Brain Inj. 19 (2), 101–108.

Giacci, M.K., Wheeler, L., Lovett, S., Dishington, E., Majda, B., Bartlett, C.A., et al., 2014. Differential effects of 670 and 830 nm red near infrared irradiation therapy: a comparative study of optic nerve injury, retinal degeneration, traumatic brain and spinal cord injury. PLoS One 9 (8), e104565.

Giza, C.C., Hovda, D.A., 2014. The new neurometabolic cascade of concussion. Neurosurgery 75, S24–S33.

Gonzalez-Lima, F., Barrett, D.W., 2014. Augmentation of cognitive brain functions with transcranial lasers. Front. Syst. Neurosci. 8 (36). Available from: https://doi.org/10.3389/fnsys.2014.00036.

Grover Jr., F., Weston, J., Weston, M., 2017. Acute effects of near infrared light therapy on brain state in healthy subjects as quantified by qEEG measures. Photomed. Laser Surg. 35 (3), 136–141.

Gruenbaum, S.E., Zlotnik, A., Gruenbaum, B.F., Hersey, D., Bilotta, F., 2016. Pharmacologic neuroprotection for functional outcomes after traumatic brain injury: a systematic review of the literature. CNS Drugs 30 (9), 791–806.

Hamblin, M.R., 2016. Shining light on the head: photobiomodulation for brain disorders. BBA Clin. 6, 113–124.

Hamblin, M.R., 2017. Photobiomodulation for traumatic brain injury and stroke. J. Neurosci. Res. 96 (4), 731–743.

Hara, Y., Yuk, F., Puri, R., Janssen, W.G.M., Raapp, P.R., Morrison, J.H., 2014. Presynaptic mitochondrial morphology in monkey prefrontal cortex correlates with working memory and is improved with estrogen treatment. Proc. Natl. Acad. Sci. 111, 486–491.

Heibert, J.B., Shen, Q., Thimmesch, A.R., Pierce, J.D., 2015. Traumatic brain injury and mitochondrial dysfunction. Am. J. Med. Sci. 350 (2), 132–138.

Heinly, M.T., Greve, K.W., Bianchini, K.J., Love, J.M., Brennan, A., 2005. WAIS digit span-based indicators of malingered neurocognitive dysfunction: classification accuracy in traumatic brain injury. Assessment 12 (4), 429–444.

Henderson, T.A., Morries, L.D., 2015. Near-infrared photonic energy penetration: can infrared phototherapy effectively reach the human brain? Neuropsychiatr. Dis. Treat. 11, 2191–2208.

Hennessy, M., Hamblin, M.R., 2017. Photobiomodulation and the brain: a new paradigm. J. Opt. 19 (1). Available from: https://doi.org/10.1088/2040-8986/19/1/013003.

Hesse, S., Werner, C., Byhahn, M., 2015. Transcranial low-level laser therapy may improve alertness and awareness in traumatic brain injured subjects with severe disorders of consciousness: a case series. J. Neurol. Neurosci. 6 (2). Available from: https://doi.org/10.21767/2171-6625.100010.

Hipskind S.G., Grover F.L., Fort T.R., Helffenstein D., Burke T.J., Quint S.A., et al., (in press). Pulsed transcranial red/near-infrared light therapy using light-emitting diodes improves cerebral blood flow and cognitive function in veterans with chronic traumatic brain injury: a case series, Photomed. Laser Surg., epub ahead of print.

Hwang, J., Castelli, D.M., Gonzalez-Lima, F., 2016. Cognitive enhancement by transcranial laser stimulation and acute aerobic exercise. Lasers Med. Sci. 31 (6), 1151–1160.

Ilic, S., Leichliter, S., Streeter, J., Oron, A., DeTaboada, L., Oron, U., 2006. Effects of power densities, continuous and pulse frequencies, and number of sessions of low-level laser therapy on intact rat brain. Photomed. Laser Surg. 24 (4), 458–466.

Jenkins, P.A., Carroll, J.D., 2011. How to report low-level laser therapy (LLLT)/photomedicine dose and beam parameters in clinical and laboratory studies. Photomed. Laser Surg. 29 (12), 785–787.

Johnson, B., Zhang, K., Gay, M., Horovitz, S., Hallett, M., Sebastianelli, W., et al., 2012. Alteration of brain default network in subacute phase of injury in concussed individuals: resting-state fMRI study. NeuroImage 59, 511–518.

Karve, I.P., Taylor, J.M., Crack, P.J., 2016. The contribution of astrocytes and microglia to traumatic brain injury. Br. J. Pharmacol. 173, 692–702.

Khuman, J., Zhang, J., Park, J., Carroll, J.D., Donahue, C., Whalen, M.J., 2012. Low-level laser light therapy improves cognitive deficits and inhibits microglial activation after controlled cortical impact in mice. J. Neurotrauma 29, 408–417.

Kinnunen, K.M., Greenwood, R., Powell, J.H., Leech, R., Hawkins, P.C., Bonnelle, V., et al., 2011. White matter damage and cognitive impairment after traumatic brain injury. Brain 134, 449–463.

Kleim, J.A., 2011. Neural plasticity and neurorehabilitation: teaching the new brain old tricks. J. Commun. Disord. 44, 521–528.

Konstantinovic, L.M., Jelic, M.B., Jeremic, A., Stevanovic, V.B., Milanovic, S.D., Filipovic, S.R., 2013. Transcranial application of near-infrared low-level laser can modulate cortical excitability. Lasers Surg. Med. 45, 648–653.

Kosaka, B., 2006. Neuropsychological assessment in mild traumatic brain injury: a clinical overview. BC Med. J. 48 (9), 447–452.

Langfitt, T.W., 1978. Measuring the outcome from head injuries. J. Neurosurg. 48 (5), 673–678.

Lapchak, P.A., Han, M.K., Salgado, K.F., Streeter, J., 2008. Safety profile of transcranial near-infrared laser therapy administered in combination with thrombolytic therapy in embolized rabbits. Stroke 39, 3073–3078.

Li, S., Zaninotto, A.L., Neville, I.S., Paiva, W.S., Nunn, D., Fregni, F., 2015. Clinical utility of brain stimulation modalities following traumatic brain injury: current evidence. Neuropsychiatr. Dis. Treat. 11, 1573–1586.

Lomeli, N., Di, K., Czerniawski, J., Guzowski, J.F., Bota, D.A., 2017. Cisplatin-induced mitochondrial dysfunction is associated with impaired cognitive function in rats. Free Radic. Biol. Med. 102, 274–286.

Loonstra, A.S., Rarlow, A.R., Sellers, A.H., 2001. COWAT metanorms across age, education, and gender. Appl. Neuropsychol. 8 (3), 161–166.

Marquez de la Plata, C.D., Carces, J., Kojori, E.S., Grinnan, J., Krishnan, K., Pidikiti, R., et al., 2011. Deficits in functional connectivity of hippocampal and frontal lobe circuits after traumatic axonal injury. Arch. Neurol. 68 (1), 74–84.

Marr, A., Coronado, V.G., 2004. Central Nervous System Injury Surveillance Data Submission Standards—2002. National Center for Injury Control and Prevention, Centers for Disease Control and Prevention, Atlanta.

McDonald, B.C., Flashman, L.A., Saykin, A.J., 2002. Executive dysfunction following traumatic brain injury: neural substrates and treatment strategies. NeuroRehabilitation. 17, 333–344.

Mechtler, L.L., Shastri, K.K., Crutchfield, K.E., 2014. Advanced neuroimaging of mild traumatic brain injury. Neurol. Clin. 32, 31–58.

Millis, S.R., Rosenthal, M., Novack, T.A., Shere, M., Nick, T.G., Kreutzer, J.S., et al., 2001. Long-term neuropsychological outcome after traumatic brain injury. J. Head Trauma Rehabil. 16 (4), 343–355.

Moghadam, H.S., Nazari, M.A., Jahan, A., Mahmoudi, J., Salimi, M.M., 2017. Beneficial effects of transcranial light emitting diode (LED) therapy on attentional performance: an experimental design. Iran. Red Crescent Med. J. 19 (5), e44513.

Moreira, M.S., Velasco, I.T., Ferreira, L.S., Ariga, S.K.K., Barbeiro, D.F., Meneguzzo, D.T., et al., 2009. Effect of phototherapy with low intensity laser on local and systemic immunomodulation following focal brain damage in rat. J. Photochem. Photobiol. B 97, 145–151.

Moreira, M.S., Velasco, I.T., Ferreira, L.S., Ariga, S.K.K., Abatepaulo, F., Grinberg, L.T., et al., 2011. Effect of laser phototherapy on wound healing following cerebral ischemia by cryogenic injury. J. Photochem. Photobiol. B 105, 207–215.

Moro, C., Torres, N., Arvanitakis, K., Cullen, K., Chabrol, C., Agay, D., et al., 2017. No evidence for toxicity after long-term photobiomodulation in normal non-human primates. Exp. Brain Res. 235, 3081–3092.

Morries, L.D., Cassano, P., Henderson, T.A., 2015. Treatments for traumatic brain injury with emphasis on transcranial near-infrared laser phototherapy. Neuropsychiatr. Dis. Treat. 11, 2159–2175.

Morris, T., Gomes Osman, J., Tormos Munoz, J.M., Costa Miserachs, D., Pascual Leone, A., 2016. The role of physical exercise in cognitive recovery after traumatic brain injury: a systemic review. Restor. Neurol. Neurosci. 34 (6), 977–988.

Mountain, M.A., Snow, W.G., 1993. Wisconsin card sorting test as a measure of frontal pathology: a review. Clin. Neuropsychol. 7 (1), 108–118.

Naeser, M.A., Hamblin, M.R., 2015. Traumatic brain injury: a major medical problem that could be treated using transcranial, red/near-infrared LED photobiomodulation. Photomed. Laser Surg. 33 (9), 443–446.

Naeser, M.A., Saltmarche, A., Krengel, M.H., Hamblin, M.R., Knight, J.A., 2011. Improved cognitive function after transcranial light-emitting diode treatments in chronic, traumatic brain injury: two case reports. Photomed. Laser Surg. 29 (5), 351–358.

Naeser, M.A., Zafonte, R., Krengel, M.H., Martin, P.I., Frazier, J., Hamblin, M.R., et al., 2014. Significant improvements in cognitive performance post-transcranial, red-near-infrared light-emitting diode treatments in chronic, mild traumatic brain injury: open-protocol study. J. Neurotrauma 31, 1008–1017.

Naeser, M.A., Marting, P.I., Ho, M.D., Kregnel, M.H., Bogdanova, Y., Knight, J.A., et al., 2016. Transcranial, red/near-infrared light-emitting diode therapy to improve cognition in chronic traumatic brain injury. Photomed. Laser Surg. 34 (12), 610–626.

Nasreddine, Z.S., Phillips, N.A., Bedirian, V., Charbonneau, S., Whitehead, V., Collin, I., et al., 2005. The Montreal Cognitive Assessment, MoCA: a brief screening tool for mild cognitive impairment. J. Am. Geriatr. Soc. 53 (4), 695–699.

Niogi, S.N., Mukherjee, P., Ghajar, J., Johnson, C., Kolster, R.A., Sarkar, R., et al., 2008. Extent of microstructural white matter injury in postconcussive syndrome correlates with impaired cognitive reaction time: a 3T diffusion tensor imaging study of mild traumatic brain injury. AJNR Am. J. Neuroradiol. 29 (5), 967–973.

Oldenburg, C., Lundin, A., Edman, G., Nygren-de-Boussard, C., Bartfai, A., 2015. Cognitive reserve and persistent post-concussion symptoms – a prospective mild traumatic brain injury (mTBI) cohort study. Brain Inj. 30 (2), 146–155.

Oron, A., Oron, U., Streeter, J., Taboada, L.D., Alexandrovich, A., Trembovler, V., et al., 2007. Low-level laser therapy applied transcranially to mice following tramatic brain injury significantly reduces long-term neurological deficits. J. Neurotrauma 24 (4), 651–656.

Oron, A., Oron, U., Streeter, J., Taboada, L.D., Alexandrovich, A., Trembovler, V., et al., 2012. Near infrared transcranial laser therapy applied at various modes to mice following traumatic brain injury significantly reduces long-term neurological deficits. J. Neurotrauma 29, 401–407.

Pangman, V.C., Sloan, J., Guse, L., 2000. An examination of psychometric properties of the Mini-Mental State Examination and the Standardized Mini-Mental State Examination: implications for clinical practice. Appl. Nurs. Res. 13 (4), 209–213.

Pastore, D., Greco, M., Passarella, S., 2000. Specific helium-neon laser sensitivity of the purified cytochrome c oxidase. Int. J. Radiat. Biol. 76 (6), 863–870.

Picard, M., McEwen, B.S., 2014. Mitochondria impact brain function and cognition. Proc. Natl. Acad. Sci. 111 (1), 7–8.

Poiani, G.C.R., Zaninotto, A.L., Carneiro, A.M.C., Zangaro, R.A., Salgado, A.S.I., Parreira, R.B., et al., 2018. Photobiomodulation using low-level laser therapy (LLLT) for patients with chronic traumatic brain injury: a randomized controlled trial study protocol. Trials 19 (17). Available from: https://doi.org/10.1186/s13063-017-2414-5.

Polich, J., 2007. Updating P300: an integrative theory of P3a and P3b. Clin. Neurophysiol. 118, 2128–2148.

Quirk, B.J., Torbey, M., Buchmann, E., Verma, S., Whelan, H.T., 2012. Near-infrared photobiomodulation in an animal model of traumatic brain injury: improvements at the behavioral and biochemical levels. Photomed. Laser Surg. 30 (9), 523–529.

Rabinowitz, A.R., Levin, H.S., 2014. Cognitive sequelae of traumatic brain injury. Psychiatr. Clin. North Am. 37 (1), 1–11.

Rapaport, M.J., McCullagh, S., Shammi, P., Feinstein, A., 2005. Cognitive impairment associated with major depression following mild and moderate traumatic brain injury. J. Neuropsychiatry Clin. Neurosci. 17 (1), 61–65.

Rees, L., Marshall, S., Hartridge, C., Mackie, D., Weiser, M., 2007. Cognitive interventions post acquired brain injury. Brain Inj. 21 (2), 161–200.

Rojas, J.C., Gonzalez-Lima, F., 2013. Neurological and psychological applications of transcranial lasers and LEDs. Biochem. Pharmacol. 86, 447–457.

Roozenbeek, B., Maas, A.I.R., Menon, D.K., 2013. Changing patterns in the epidemiology of traumatic brain injury. Nat. Rev. Neurol. 9, 231–236.

Salehpour, F., Farajdokht, F., Erfani, M., Sadigh-Eteghad, S., Shotorbani, S.S., Hamblin, M.R., et al., 2018a. Transcranial near-infrared photobiomodulation attenuates memory impairment and hippocampal oxidative stress in sleep-deprived mice. Brain Res. 1682, 36–43.

Salehpour, F., Mahmoudi, J., Kamari, F., Sadigh-Eteghad, S., Rasta, S.H., Hamblin, M.R., 2018b. Brain photobiomodulation therapy: a narrative review. Mol. Neurobiol. 55 (8), 6601–6636.

Saltmarche, A.E., Naeser, M.A., Ho, K.F., Hamblin, M.R., Lim, L., 2017. Significant improvement in cognition in mild to moderately severe dementia cases treated with transcranial plus intranasal photobiomodulation: case series report. Photomed. Laser Surg. 35 (8), 432–441.

Scarpina, F., Tagini, S., 2017. The Stroop Color and Word Test. Front. Psychol. 8, 557. Available from: https://doi.org/10.3389/fpsyg.2017.00557.

Taylor, R., Galvez, V., Loo, C., 2018. Transcranial magnetic stimulation (TMS) safety: a practical guide for psychiatrists. Australas. Psychiatry 26 (2), 189–192.

Teasdale, G., Jennet, B., 1974. Assessment of coma and impaired consciousness: a practical scale. Lancet 304 (7872), 81–84.

Tedford, C.E., DeLapp, S., Jacques, S., Anders, J., 2015. Quantitative analysis of transcranial and intraparenchymal light penetration in human cadaver brain tissue. Lasers Surg. Med. 47, 312–322.

Tian, F., Hase, S.N., Gonzalez-Lima, F., Liu, H., 2016. Transcranial laser stimulation improves human cerebral oxygenation. Lasers Surg. Med. 48, 343–349.

Tsai, S.R., Hamblin, M.R., 2017. Biological effects and medical applications of infrared radiation. J. Photochem. Photobiol. B 170, 197–207.

Tucker, L.B., Velosky, A.G., McCabe, J.T., 2018. Applications of the Morris water maze in translational traumatic brain injury research. Neurosci. Biobehav. Rev. 88, 187–200.

Tuner, J., Jenkins, P.A., 2016. Parameter reproducibility in photobiomodulation. Photomed. Laser Surg. 34 (3), 91–92.

Vanderploeg, R.D., Curtiss, G., Belanger, H.G., 2005. Long-term neuropsychological outcomes following mild traumatic brain injury. J. Int. Neuropsychol. Soc. 11 (3), 228–236.

Villamar, M.F., Portilla, A.S., Fregni, F., Zafonte, R., 2012. Noninvasive brain stimulation to modulate neuroplasticity in traumatic brain injury. Neuromodulation 15, 326–338.

Wang, X., Tian, F., Reddy, D.D., Nalawade, S.S., Barrett, D.W., Gonzalez-Lima, F., et al., 2017. Up-regulation of cerebral cytochrome-c-oxidase and hemodynamics by transcranial infrared laser stimulation: a broadband near-infrared spectroscopy study. J. Cereb. Blood Flow Metab. 37 (12), 3789–3802.

Wassermann, E.M., 2000. Side effects of repetitive transcranial magnetic stimulation. Depress. Anxiety 12, 124–129.

Werner, C., Engelhard, K., 2007. Pathophysiology of traumatic brain injury. Br. J. Anaesth. 99 (1), 4–9.

Writer, B.W., Schillerstrom, J.E., 2009. Psychopharmacological treatment for cognitive impairment in survivors of traumatic brain injury: a critical review. J. Neuropsychiatry Clin. Neurosci. 21 (4), 362–370.

Wu, Q., Huang, Y.Y., Dhital, S., Sharma, S.K., chen, A.C.H., Whalen, M.J., et al., 2010. Low level laser therapy for traumatic brain injuryIn: Hamblin, M.R., Waynant, R.W., Anders, J. (Eds.), SPIE, San Fransisco, CA, pp. 755206–755208.

Wu, Q., Xuan, W., Ando, T., Xu, T., Huang, L., Huang, Y.Y., et al., 2012. Low-level laser therapy for closed-head traumatic brain injury in mice: effect of different wavelengths. Lasers Surg. Med. 44 (3), 218–226.

Wurzelmann, M., Romeika, J., Sun, D., 2017. Therapeutic potential of brain-derived nerotrophic factor (BDNF) and a small molecular mimics of BDNF for traumatic brain injury. Neural Regen. Res. 12 (1), 7–12.

Xiao, H., Yang, Y., Xi, J.H., Chen, Z.Q., 2015. Structural and functional connectivity in traumatic brain injury. Neural Regen. Res. 10 (12), 2062–2071.

Xuan, W., Vatansever, F., Huang, L., Wu, Q., Xuan, Y., Dai, T., et al., 2013. Transranial low-level laser therapy improves neurological performance in traumatic brain injury in mice: effect of treatment repetition regimen. PLoS One 8 (1), e53454.

Xuan, W., Vatansever, F., Huang, L., Hamblin, M.R., 2014. Transcranial low-level laser therapy enhances learning, memory, and neuroprogenitor cells after traumatic brain injury in mice. J. Biomed. Opt. 19 (10), 108003.

Xuan, W., Agrawal, T., Huang, L., Gupta, G.K., Hamblin, M.R., 2015. Low-level laser therapy for traumatic brain injury in mice increases brain derived neurotrophic factor (BDNF) and synaptogenesis. J. Biophotonics 8 (6), 502−511.

Xuan, W., Huang, L., Hamblin, M.R., 2016. Repeated transcranial low-level laser therapy for traumatic brain injury in mice: biphasic dose response and long-term treatment outcome. J. Biophotonics 9 (11−12), 1263−1272.

Yang, J., Li, P., 2012. Brain networks of explicit and implicit learning. PLoS One 7 (8), e42993.

Yochim, B., Baldo, J., Nelson, A., Delis, D.C., 2007. D-KEFS Trail Making Test performance in patients with lateral prefrontal cortex lesions. J. Int. Neuropsychol. Soc. 13 (4), 704−709.

Zhang, Q., Zhou, C., Hamblin, M.R., Wu, M.X., 2014. Low-level laser therapy effectively prevents secondary brain injury induced by immediate early responsive gene X-1 deficiency. J. Cereb. Blood Flow Metab. 34, 1291−1401.

Chapter 27

Advanced neuroimaging methods for assessment of low-level light therapy

Suk-tak Chan[1,2], Maria Gabriela Longo[3], Eva-Maria Ratai[1,2,3] and Rajiv Gupta[1,3]

[1]*Harvard Medical School, Massachusetts General Hospital, Boston, MA, United States,* [2]*Department of Radiology, Athinoula A. Martinos Center for Biomedical Imaging, Massachusetts General Hospital, Boston, MA, United States,* [3]*Department of Radiology, Neuroradiology Division, Massachusetts General Hospital, Boston, MA, United States*

27.1 Introduction

Neuropathological conditions being targeted by light therapy generally constitute a set of heterogeneous diseases. As a result, clinical outcome measures that monitor the effect of light therapy are highly variable. A high degree of intrinsic variability—much of it stemming from methodologic reasons such as an individual's perception of pain or other symptoms—unfortunately necessitate a large cohort size in order to demonstrate a statistically significant benefit of any therapeutic intervention (Tolias and Bullock, 2004). Because only a limited number of patients are available at a single site for study of any therapy, multisite studies become necessary. One of the conclusion from expert panels in this space is that primary outcome measures based on validated biomarkers are critically important in study design (Saatman et al., 2008). Use of specific biomarkers allows one to establish, using a much smaller study population that the intervention (in our case low-level light therapy or LLLT) is engaging one or more of the mechanistic targets of the therapy. These studies can and so confirm safety and provide pilot clinical data (as a secondary outcome), both of which are needed to design a large phase III study using primary clinical outcomes.

In this chapter, we describe neuroimaging biomarkers that can be employed to assess the efficacy of light therapy. We show that the known mechanisms of light therapy can be interrogated directly using noninvasive magnetic resonance imaging (MRI). Before we embark on the description of imaging methods and associated biomarkers, it is important to understand the mechanistic basis of light therapy, and the evidence that exists for its efficacy. This discussion will motivate the design of noninvasive imaging-based biomarkers that can engage the appropriate physiologic pathways via which light therapy works.

27.2 Known mechanisms of light therapy

Neither the molecular nor the downstream functional mechanisms that drive the therapeutic benefit of LLLT in the brain are fully understood. However, research over the past decade has identified some of the key molecular and functional mechanisms driving near infrared (NIR) biostimulation. These effects are believed to originate from light absorption by mitochondrial organelles. Cytochrome c oxidase (CCO), located on the inner membrane, is a likely target. CCO is a large trans-membrane protein complex and is a unit IV of the respiratory electron transport chain (Capaldi, 2012). Absorption of NIR light by CCO induces upregulation of messenger molecules including reactive oxygen species, nitric oxide (NO), and adenosine triphosphate. These signaling molecules activate transcription factors including NF-κβ and AP-1 that enter the nucleus and cause transcription of a range of new gene products (Fig. 27.1).

Some of the downstream effects of LLLT in the brain have been characterized in preclinical research. Vascular and neuroprotective functions have both been strongly associated with LLLT in the brain (Garavello et al., 2004; Corazza et al., 2007; Bossini et al., 2009). Vascular changes may be associated with light-induced increase in tissue NO. Increased levels of NO can improve blood flow, raise tissue oxygenation, recruit inflammatory cells, and induce

Photobiomodulation in the Brain. DOI: https://doi.org/10.1016/B978-0-12-815305-5.00027-0

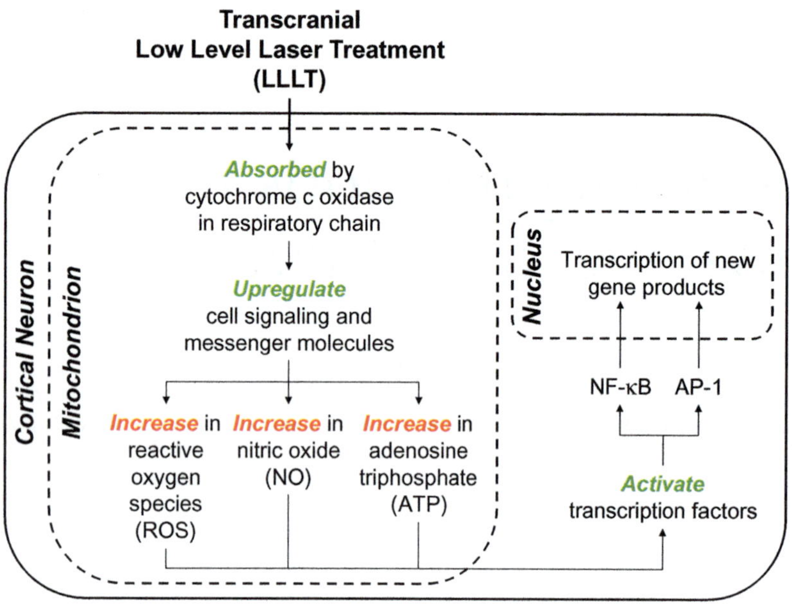

FIGURE 27.1 Molecular mechanisms of LLLT. *Adapted from Huang, Y.-Y., Gupta, A., Vecchio, D., de Arce, V.J., Huang, S.F., Xuan, W., et al., 2012. Transcranial low level laser (light) therapy for traumatic brain injury. J. Biophotonics 5 (11–12), 827–837.*

angiogenesis (Fukumura and Jain, 1998; Ziche and Morbidelli, 2000; Antunes et al., 2004; Lohr et al., 2009; Zhang et al., 2009). The mediators of light-induced neuroprotection may include the inducible proteins survivin (Hemvani et al., 2005), Bcl2, heat shock proteins (Coombe et al., 2001), and superoxide dismutase (Malinovskaya et al., 2008). The neuroprotective effect of LLLT has been demonstrated in studies of cortical neurons challenged by cyanide (Liang et al., 2006), tetrodotoxin (Wong-Riley et al., 2005), and methanol (Eells et al., 2003). Neuroimaging methods can be used to interrogate these vascular and neuroprotective mechanisms by noninvasive means.

27.3 Preclinical evidence for light therapy

Over the past decade multiple studies have demonstrated that acute LLLT improves functional recovery from traumatic brain injury (TBI) in animal models. In one study conducted at our institution, the investigators measured the neurological severity score (NSS) of mice exposed to a closed head diffuse axonal injury (DAI) model of TBI and treated acutely (4 hours after TBI) with LLLT at four optical wavelengths (Wu et al., 2012). The NSS includes measures of neurological motor and cognitive function. Animals given LLLT at 810 and 665 nm functionally improved relative to sham treated animals (no LLLT) or LLLT at 730 and 980 nm (Fig. 27.1). The beneficial effect of the 810 and 665 nm therapy results from their overlap with presumed cellular targets of LLLT (Capaldi, 2012). Histological examinations of lesion sizes also demonstrate improved recovery in 810 nm treated animals relative to untreated controls. Similar studies have been conducted by other investigators. Each reports improvement associated with LLLT when delivered at 665 or 810 nm.

Further evidence of LLLT efficacy in treating acute brain trauma comes from stroke studies. Ischemic stroke carries many pathologic similarities to moderate and severe TBI. Using both middle cerebral artery occlusion and small clot embolic stroke models in rabbits, LLLT in the 810 nm band resulted in statistically significant improvements in behavioral performance, neurological function, and histological evidence of suppressed NOS activity, upregulation of TGF-b1, and increased neurogenesis.

27.4 Clinical evidence of light therapy efficacy

Positive preclinical results motivated a first clinical study of LLLT for stroke. The results of that study, NEST-1, were published in 2007 (Lampl et al., 2007). This double-blinded placebo controlled study tested the efficacy and safety of LLLT across 120 patients. Light was delivered through a handheld device placed against the shaved scalp. The device was moved to 20 predetermined locations, and held at each location for a duration of 2 minutes. The treated group received LLLT at an average of 16 hours after stroke. Patients were evaluated based on the NIH stroke severity (NIHSS) scale at baseline, +5, +30, +60, and +90 days after stroke. The treatment group in this study showed improved NIHSS over this time-frame. In a follow-up study of 660 patients (NEST-2) (Zivin et al., 2009), there was a trend toward therapeutic effect of LLLT but statistical significance was not obtained. The inability of the study to achieve significance has been attributed to an overly broad study population that included older patients (> 80 years), severe stroke patients, and patients experiencing a second stroke (despite an outcome based on return to normal rather than a return to their condition prior to the second stroke). When controlling for these variables, LLLT for stroke was shown to induce statistically significant benefits (Huisa et al., 2013). A third study (NEST-3) is currently underway.

In addition to providing clinical evidence that transcranial LLLT is effective in treating brain trauma, the NEST-1 and -2 studies demonstrated a very strong safety profile for the intervention. Detailed safety analyses were performed in both studies and both concluded that LLLT was not associated with any risks within the stroke population of the studies.

27.5 Evidence for transcranial delivery of light

Like the NEST-1 and -2 clinical studies, our proposed study will rely on transcranial delivery of NIR light into the brain. To validate this approach, the amount of light that can be delivered into the brain should be compared against that need required to achieve a therapeutic effect. The former is relatively straightforward to approximate from the extensive literature of light propagation in tissue, while the latter must be estimated from available preclinical and clinical data.

Among the multiple measures of light "strength," fluence (defined as energy per unit area) is most relevant in LLLT. With an assumed light fluence on the scalp, we can use the optical properties of tissue to predict the fluence at varying depths within the cortex. For example, for an incident fluence of 42 J/cm^2 (equivalent to that provided by our proposed device), approximately 3% of this or 1.3 J/cm^2 reaches the cortical surface. This is calculated from measurements of scalp/skull transmission in the NIR using cadavers (Wan et al., 1981). Within the cortex, NIR light transmission has reported to be 10% (Matcher et al., 1997; Haeussinger et al., 2011). We can then estimate the light fluence at 1 cm depth in the cortex (2 cm deep overall) as 0.13 J/cm^2 (assuming 42 J/cm^2 on the scalp).

The light fluence needed to induce a therapeutic benefit in humans is not known, and we must extrapolate from preclinical data. An in vitro study demonstrated neuronal changes with a fluence as low as 0.035 J/cm^2 and peaking at 3 J/cm^2 (Sharma et al., 2011). Multiple animal studies have demonstrated therapeutic benefit with fluences at the cortical surface ranging from 1 to 3 J/cm^2 (Huang et al., 2012). Finally, the positive NEST-1 and -2 clinical studies used a similar fluence as planned in this study. These data suggest fluence levels predicted in the outer centimeter of the cortex (0.13–1.3 J/cm^2) may be sufficient. This therapeutic volume is illustrated in Fig. 27.2.

While these calculations suggest sufficient NIR light can be delivered, many questions on dosimetry cannot be answered with existing data. In rodent models used in preclinical efficacy studies, NIR light is able to penetrate through the entire brain, while the human brain will remain confined to the first 1–2 cm. It is not known how this partial exposure will affect LLLT efficacy. It is also not known if signaling across long distances within the brain can extend a therapeutic effect beyond the volume of exposed tissues. The proposed pilot study will help to answer some of these questions.

27.6 Neuroimaging methods

27.6.1 Computed tomography

Computed tomography (CT) is good for structural imaging. It is the first-line of imaging for patients with moderate or severe TBI in acute setting (Radiology, 2015). Additionally to the necessity of surgical intervention assessment, the early head CT could also be relevant for the prognosis evaluation (Yuh et al., 2013). However, it is a low sensitivity exam, the complementation with magnetic resonance (MR) being sometimes necessary, especially for the evaluation of the LLLT neuroimaging targets.

27.6.2 Magnetic resonance imaging

This imaging should be performed as soon as is safe from a clinical point of view. The ultimate goal is to acquire images in the acute stage (<7 days after de trauma); subacute stage (21 days after the trauma); and chronic stage (3 months after the trauma). Subjects should be asked to lie still in a supine position for the duration of the study. Head motion should be minimized during image acquisition by the use of a dedicated head coil with holder. No contrast agent or other pharmacological substance other than ones clinically required by the patient need to be given. Fig. 27.3 shows the most common sequences used and their proposes. In the next pages, we will describe the fundamentals of the sequences and how they are analyzed.

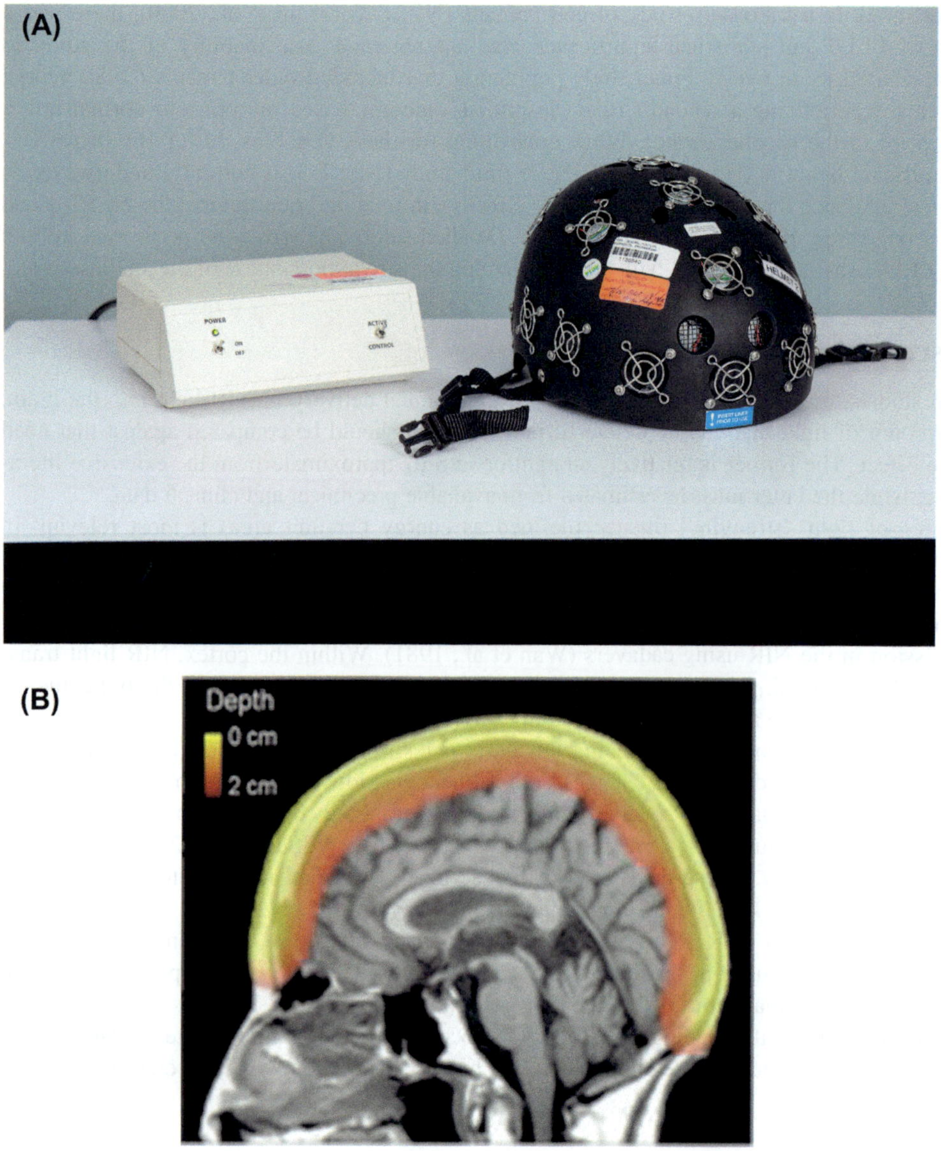

FIGURE 27.2 (A) LLLT helmet containing 360 LEDs on the inner surface to transcranially deliver NIR to the brain. Over the 20 min procedure, the helmet delivers approximately 43 J/cm^2 of energy, of which approximately 3% or 1.3 J/cm^2 reaches the brain cortex. (B) The volume receiving therapeutic levels of NIR light is illustrated over an MR anatomical image. Therapeutic levels of NIR light can reach depths of 2 cm beyond the skull and 1 cm beyond the cortical surface.

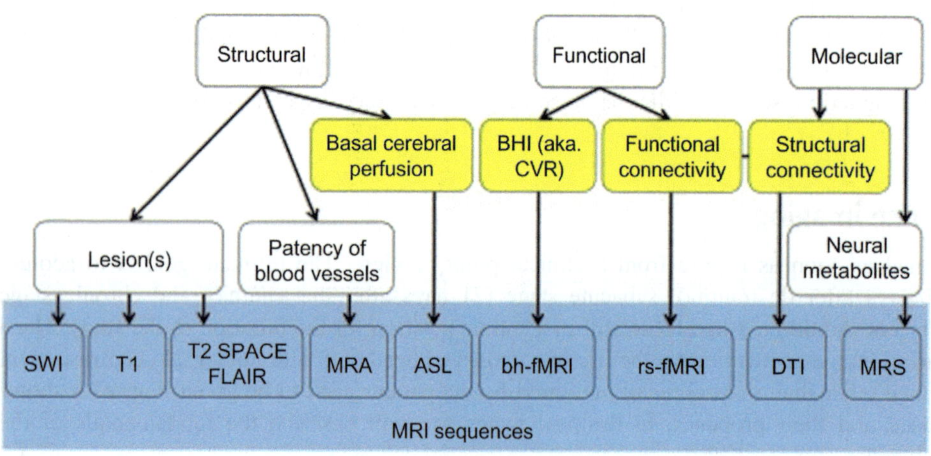

FIGURE 27.3 Quantitative MRI sequences used to study the structural, functional, and molecular changes of the brain. *ASL*, arterial spin labeling; *bh-fMRI*, breathhold functional MRI; *DTI*, diffusion tensor imaging; *MRA*, magnetic resonance angiography; *MRS*, magnetic resonance spectroscopy; *rs-fMRI*, resting state functional MRI; *SWI*, susceptibility weighted imaging.

27.7 Structural imaging

The structural imaging is used primarily to detect larger structural lesions and to study anatomic aspects of the subjects (Fig. 27.4). For this propose, the most indicated sequences are: (1) MP-RAGE; (2) T2 SPACE fluid-attenuated inversion-recovery (FLAIR); (3) susceptibility weighted imaging (SWI); and (4) magnetic resonance angiography (MRA). The first three sequences are important for the lesion detection, while the MRA is used to determine the vessels' patency.

The MP-RAGE is a volumetric T1 sequence used to evaluate especially the anatomy and the presence of acute bleed in the context of the TBI (Brant-Zawadzki and Gillan, 1992). The degree of anatomical detail that could be extracted from this sequence allows the evaluation of changes in the brain structural size and thickness over time (DeCarli et al., 2005; Dickie et al., 2013)—a potential target for the assessment of the LLLT. The FLAIR sequence (2) uses a long inversion time leading to suppression of the cerebrospinal fluid (CSF), in a T2-weighted sequence. This property allows a better evaluation of the lesions in the gray-white matter interface or closer to the CSF—typically the DAI (Ashikaga et al., 1997). The SWI (3) uses tissue magnetic susceptibility differences to infer the iron content and generates different contrasts. The rupture of the microvasculature during the trauma with deposition of blood components leads to inhomogeneity in the tissue. Therefore the SWI is especially indicated to the identification of microbleeds (Haacke et al., 2009; Wang et al., 2014). The MRA (4) sequence used is the time-of-flight, which consists of an MR technique that identifies

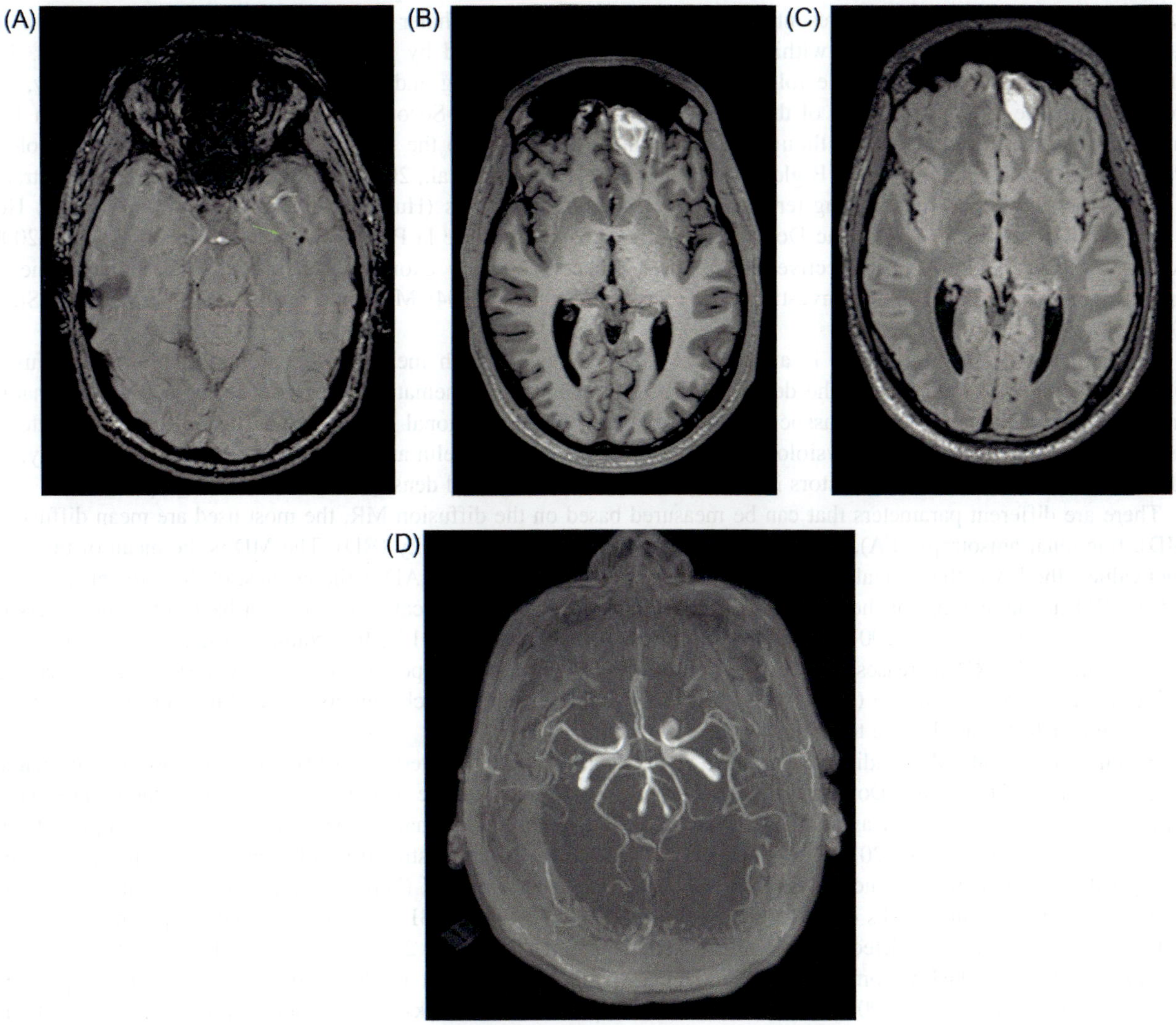

FIGURE 27.4 Imaging from different patients with moderate TBI. Few microbleeds in left temporal lobe (*arrow*) were found on SWI (A), the other sequences of this patient showed no abnormalities. Axial MP-RAGE (B) and axial FLAIR (C) from the same patient showing intraparenchymal hemorrhage in the left frontal lobe in a patient with moderate TBI in the subacute stage. Example of TOF (D) used to determine the vessels' patency.

the flow inside of the vessels, without the use of contrast. The spins in the blood flowing into an imaging section are unsaturated leading to higher signal compared with the adjacent tissues (Stepansky et al., 2008).

There are a large number of findings that could be detected secondary to the TBI. Due to a necessity to standardize these evaluations, the National Institute of Neurologic Disorders and Stroke in association with the Department of Defense developed the neuroimaging common data elements panel—a standard form to describe the most prevalent findings due to TBI (Duhaime et al., 2010). In a recent study that reviewed 834 structural MRIs from a cohort of military service members with a history of TBI, the most prevalent findings described were: brain atrophy, contusion, DAI, encephalomalacia, intracranial hemorrhage, microhemorrhage, subarachnoid hemorrhage, subdural hematoma-subcortical, skull fracture, T2 hyperintensity (Riedy et al., 2016).

In a clinical setting, the structural MR is not indicated as an initial study, being reserved especially for short-term follow-up imaging after the TBI or in cases of neurologic deterioration, delayed recovery or persistent unexplained deficits (Radiology, 2015). However, the last studies have been demonstrating the importance of this study in the orientations of the severity of the TBI also to predict the prognosis (Yuh et al., 2013; Wang et al., 2014; Riedy et al., 2016). The presence of any positive finding related to the trauma classifies the patient as at least moderate TBI (Riedy et al., 2016).

27.8 Diffusion imaging

DAI is a biomarker of acute TBI, present in 40%–50% of patients (Meythaler et al., 2001; Wang, 2008). Primary cytoskeletal damage in DAI is visible within hours of injury, characterized by focal neurofilament misalignments. The effects of the trauma continue in the following hours leading to swelling and expansion of the axon. Eventually, this process can cause the disconnection of the axon (Arfanakis et al., 2002). Secondary injury processes occur within 1–3 days of the primary injury and are thought to contribute significantly to the semiacute and chronic axonal pathology (Pettus, 1994; Maxwell et al., 1997; Bigler, 2001; Gaetz, 2004; Wilson et al., 2004). Multiple studies have demonstrated a correlation between DAI and long-term functional outcome measures (Huisman et al., 2004; Benson, 2007; Hou et al., 2007; Ichord et al., 2007; Mac Donald et al., 2007a,b; Marquez de la Plata et al., 2007; Yanagawa et al., 2009; Skandsen et al., 2011). Neuroprotective strategies to reduce secondary axonal degeneration, or even to regenerate injured axons, are being actively investigated (Tolias and Bullock, 2004; Marklund et al., 2006; Sayeed and Stein, 2009; Girard et al., 2012).

Diffusion tensor imaging (DTI) is a structural MR technique, which measures the properties of water diffusion (Fig. 27.5) (Yendiki et al., 2011). The degree of diffusion, which is mathematically represented by a tensor, is characterized by three eigenvectors and associated diffusivity in three orthogonal directions. Diffusion in white matter is restricted and anisotropic due to physiological obstacles posed by the myelin and axon membranes. More generally, the diffusion properties reflect many factors including the myelination, axonal density, and cellular integrity.

There are different parameters that can be measured based on the diffusion MR, the most used are mean diffusivity (MD), fractional anisotropy (FA), axial diffusivity (AD), and radial diffusivity (RD). The MD is the mean of the three eigenvalues; the FA is the normalized variance of the three eigenvalues; the AD is the greatest of the three eigenvalues; and the RD in the average of the two lesser eigenvalues. Different biological processes can have different effects on these values (Jellison et al., 2004). A model proposed by Aung et al. (2013), for example, suggests that during the myelin damage the RD increases (as the water diffusivity increases in the perpendicular direction); the AD decreases during the acute axonal damage (due to neurofilament misalignments); and chronic axonal and myelin impairment will increase either RD and AD (due to changes in the extracellular water).

Recent experimental data indicate that white matter abnormalities detected using DTI correlate closely with pathological evidence of DAI (Mac Donald et al., 2007a,b). DTI evidence of white matter damage is more marked in subjects with moderate and severe TBI as compared with mild TBI, with lasting changes detectable years after injury (Kraus et al., 2007; Newcombe et al., 2011). In general, a decrease in FA is demonstrated in different white matter tracts especially in the corpus callosum and its associated fibers in patients with TBI (Wilde et al., 2006; Akpinar et al., 2007; Yuan et al., 2007; Ewing-Cobbs et al., 2008). A recent meta-analysis in TBI patients suggested the potential utility of DTI of corpus callosum for detection of white matter damage (Aoki et al., 2012). Changes in DTI parameters can be used as a predictor of TBI outcome (Parvizi and Damasio, 2003; Huisman et al., 2004; Sidaros et al., 2008). A prospective study, by Sidaros et al. (2008) evaluated severe TBI patients at 8 weeks and 12 months after injury. They found that decreased regional FA at 8 weeks was predictive of unfavorable outcome at 12 months.

However, the pattern of the FA fluctuation over the stages of the trauma is still a research target (Lange et al., 2012; Ling et al., 2012; Kim et al., 2013). Some reasons are the postprocessing strategy (whole-brain voxel-wise × ROI

(A)

(B)

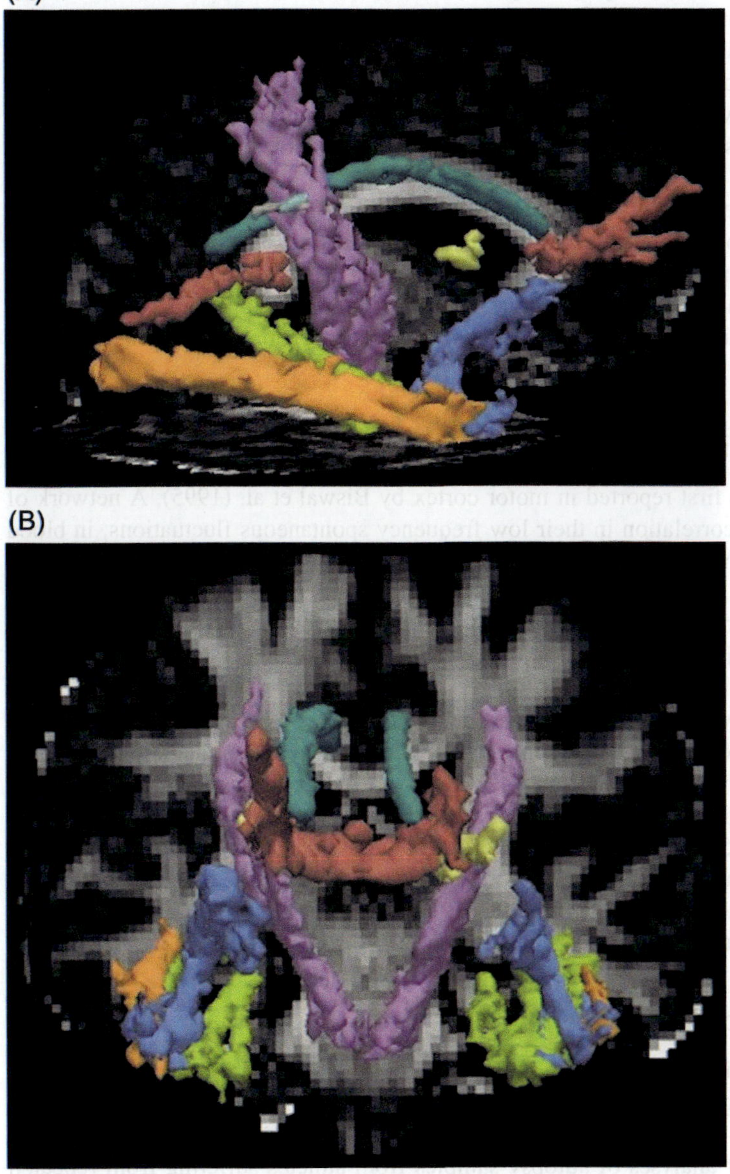

FIGURE 27.5 Sagittal and coronal images (A and B) showing a 3D reconstruction of the white matter tracts, process from a diffusion MR sequence using TRACULA (TRActs Constrained by UnderLying Anatomy) tool (Yendiki et al., 2011).

analysis); acquisition parameters; differences between scanners; just to name a few. These different results between studies limit the individual use of the DTI/FA analysis. One approach that could decrease the bias in the longitudinal evaluation of the same subject over time. This kind of analysis can decrease the error to around 5% (Ling et al., 2012; Yendiki et al., 2016).

27.9 Perfusion imaging

TBI-induced impairment in cerebral blood flow (CBF) likely plays a significant role in the acute response of the brain. Changes in blood flow can result directly from trauma, but also may result from secondary processes such as inflammation, pericyte dysfunction (Yemisci et al., 2009), increased intracranial pressure (Servadei, 2011), disrupted autoregulation, or angiogenesis. An arterial spin labeling (ASL) MR sequence has been used as a noninvasive tool to assess resting CBF/basal cerebral perfusion. Several perfusion studies on humans suggested the early contribution of perfusion deficits in the pathogenesis and regional selectivities in the disorders including TBI. Previous resting CBF studies using ASL perfusion MRI technique showed global CBF increase in patients with acute TBI (Doshi et al., 2015), suggesting a compensatory process of cerebral blood supply to vascular injury. In patients with chronic moderate and severe TBI, global CBF reduction was reported with more prominent regional hypoperfusion found in posterior cingulate cortices, thalami, and multiple locations in the frontal cortices where the greatest volume losses occurred (Kim et al., 2010). The authors suggested that structural lesions contributed to chronic CBF changes. Since CBF plays a critical role in the maintenance of neuronal integrity, reduced CBF was also found to be associated with poorer white matter integrity. In patients with chronic TBI, decreased CBF was found significantly associated with reduced cingulum FA (Clark et al., 2017).

Basal cerebral perfusion was found to be altered in early versus chronic phases of TBI injury and in mild versus severe degrees of injury. However, it is difficult to distinguish a deficit due to a primary vascular injury from a functional deficit resulting from cellular injury that might not be sufficient to cope with the cerebral metabolic demand under challenge. To assess any potential functional deficit in the resting condition and under challenge, resting state functional connectivity imaging and cerebrovascular reactivity (CVR) assessment using functional imaging under hypercapnic challenge have been recently used.

27.10 Resting state functional connectivity imaging

Functional connectivity of resting human brain was first reported in motor cortex by Biswal et al. (1995). A network of brain regions that show a high degree of temporal correlation in their low frequency spontaneous fluctuations, in blood oxygenation or flow, are considered to be functionally connected. To date, several brain networks including default mode, executive control, salience, dorsal attention, auditory, sensorimotor, and visual networks have been identified in resting state functional connectivity data (Buckner et al., 2011; Yeo et al., 2011; Choi et al., 2012; Raichle, 2015). fMRI-BOLD has been commonly used because of its sensitivity in blood oxygenation and the high temporal resolution of data acquisition for the whole brain.

Changes of resting state functional connectivity at acute, subacute, and chronic stages after TBI have previously been reported (Rosenthal et al., 2018); except the consistent hyperconnectivity was found in acute stage of TBI, both increased and decreased resting state connectivity were found in the later stages postinjury. The hyperconnectivity at acute stage may be due to compensatory brain region recruitment or hyperactivity associated with abnormal response after injury (Mayer et al., 2011; Bharath et al., 2015; Czerniak et al., 2015; Iraji et al., 2016; Xiong et al., 2016), while the conflicting connectivity changes in subacute and chronic stages may be explained by increased cognitive effort in order to compensate for continued deficits in function (Broglio et al., 2012; Bharath et al., 2015; Westfall et al., 2015), or lag of activation or incomplete activation resulting in deficits (Vakhtin et al., 2013).

27.11 Functional imaging using hypercapnic challenges

MRI-based measures of cerebrovascular dysfunction serve as a biomarker of acute LLLT. Damage to small and medium-sized cerebral blood vessels is a well-recognized consequence of TBI (Tomlinson, 1970; Graham et al., 2002). It has been confirmed through histological analysis of animal models and human autopsy samples (Graham et al., 2002; McKee et al., 2009; Goldstein et al., 2012). Recent analyses of autopsy samples from athletes suffering from repeated mild TBI have demonstrated accumulation of phosphorylated tau in the perivascular regions (McKee et al., 2009). Neuroimaging has revealed a cerebrovascular dysfunction after TBI that persists for months and years (Furuya et al., 2003; Menon, 2006). Deficits in CVR has been demonstrated in multiple rodent models of TBI (Baranova et al., 2008; Wei et al., 2009; Gao et al., 2010; Oda et al., 2011) and shown to correlate with behavioral deficit (Wei et al., 2009). Mechanistic studies have implicated dysfunction in NO regulated pathways as a trigger in cerebrovascular dysfunction. Using rat model of TBI, it was shown that nitric oxide synthase is attenuated, while endothelial reactivity to nitric oxide was retained (Wei et al., 2009). This suggests that interventions such as LLLT, which upregulate NO signaling, may encounter some of the observed cerebrovascular dysfunction associated with TBI.

As neuronal activity increases in an area of the brain, hemodynamic responses cause an overcompensation of blood flow to the region, resulting in increased signal from a local change in deoxy:oxyhemoglobin ratio. Patients with TBI may have dysregulation in the coupling of hemodynamic response with neuronal activity. In the resting condition without any challenges, connectivity analysis of spontaneous physiological fluctuations manifested as signal changes with low signal-to-noise ratio may not have adequate sensitivity to study cerebrovascular dysfunction and it often requires averaging the results from subject groups (Mutch et al., 2015). Recent advancements in MR imaging enable noninvasive measures of CVR in humans under hypercapnic challenges using low dose carbon dioxide administration and breathhold task. Increased carbon dioxide is commonly used as vasoactive stimulus in the assessment of CVR

TABLE 27.1 Magnetic resonance spectroscopy (MRS)—metabolites and their function after TBI (Ashwal et al., 2006; Marino et al., 2011; Xu et al., 2011; Lama et al., 2014; Xiong et al., 2014; Croall et al., 2015; Brown et al., 2018).

Metabolites	Function
N-Acetylaspartate (NAA)	Neuronal marker found exclusively in the brain (predominantly in neurons in the adult brain). It is used to determine the *integrity of either white and gray matter (axonal and neuronal)*. Decreases after the brain injury, especially in the subacute stage of TBI. The persistence of the NAA/Cr depletion in the chronic stage may represent a clinical neuronal dysfunction and long-term disability.
Choline (Cho)	Biomarker for cell membrane turnover, as it represents the degradation products of cellular membrane and myelin. The increase of Cho after the TBI is related to the *presence of membrane damage*.
Creatine (Cr)	Cell energy metabolism. Cr + Phosphocreatine, commonly referred to as Cr, is often used as the *internal standard*, as is stable in most diseases and with age.
Lactate (Lac)	Biomarker of anaerobic glycolysis, for example, due to mitochondrial impairment following TBI. As a general rule, excessive lactate accumulation correlates with poor prognosis. Some authors consider the presence of lactate as *evidence of postinjury ischemia*.
Myo-inositol (mI)	Most studies report *an elevation of the concentration of the m following TBI*. The persistence of the peak of the mI may indicate the presence of gliosis.
Glutamine and Glutamate (Glx)	Glutamate is the most important excitatory neurotransmitter. Using standard MRS sequences at 1.5T or 3T, it is not possible to measure Glutamate separately from the Glutamine. However, changes in the concentration of the Glx indicates *osmotic stress and changes in the neuronal metabolism*. An excess of Glutamate may be neurotoxic. Most studies demonstrate an increase in the Glx after trauma, with normalization overtime.

(Fierstra et al., 2013; Pillai and Mikulis, 2015). Reduced CVR in response to CO_2 challenges has been found in patients with chronic mild, moderate, and severe TBI (Chan et al., 2014; Mutch et al., 2014, 2016a,b; Kenney et al., 2016; Amyot et al., 2018). Global, gray matter and white matter CVR are reported to be a more reliable and potentially useful biomarker relative to basal perfusion in TBI patients (Amyot et al., 2018).

27.12 Magnetic resonance spectroscopy

Magnetic resonance spectroscopy (MRS) is a noninvasive method that is able to provide the chemical information of a tissue base on the principle that protons in different molecules have different resonance frequencies. The results of the MRS are provided through a graphic where the horizontal line represents the resonance frequency of each metabolite and the area under the curve represents the concentration of the metabolites (Ashwal et al., 2006; Marino et al., 2011; Xiong et al., 2014). In Table 27.1, we describe the main metabolites studied and their importance in the evaluation of the TBI patients.

Funding

Gupta's research fund was partially supported by the following grants: Air Force Contract Number FA8650-17-C-9113; Army USAMRAA Joint Warfighter Medical Research Program, Contract No. W81XWH-15-C-0052; Congressionally Directed Medical Research Program W81XWH-13-2-0067.

References

Akpinar, E., Koroglu, M., Ptak, T., 2007. Diffusion tensor MR imaging in pediatric head trauma. J. Comput. Assist. Tomogr. 31 (5), 657–661.

American College of Radiology, 2015. ACR Appropriateness Criteria, pp. 1–18.

Amyot, F., Kenney, K., Moore, C., Haber, M., Turtzo, L.C., Shenouda, C., et al., 2018. Imaging of cerebrovascular function in chronic traumatic brain injury. J. Neurotrauma 35 (10), 1116–1123.

Antunes, F., Boveris, A., Cadenas, E., 2004. On the mechanism and biology of cytochrome oxidase inhibition by nitric oxide. Proc. Natl. Acad. Sci. U.S. A. 101 (48), 16774–16779.

Aoki, Y., Inokuchi, R., Gunshin, M., Yahagi, N., Suwa, H., 2012. Diffusion tensor imaging studies of mild traumatic brain injury: a meta-analysis. J. Neurol. Neurosurg. Psychiatry 83 (9), 870–876.

Arfanakis, K., Haughton, V.M., Carew, J.D., Rogers, B.P., Dempsey, R.J., Meyerand, M.E., 2002. Diffusion tensor MR imaging in diffuse axonal injury. AJNR 23 (5), 794–802.

Ashikaga, R., Araki, Y., Ishida, O., 1997. MRI of head injury using FLAIR. Neuroradiology 39 (4), 239–242.

Ashwal, S., Babikian, T., Gardner-Nichols, J., Freier, M.C., Tong, K.A., Holshouser, B.A., 2006. Susceptibility-weighted imaging and proton magnetic resonance spectroscopy in assessment of outcome after pediatric traumatic brain injury. Arch. Phys. Med. Rehabil. 87 (12 Suppl. 2), S50–S58.

Aung, W.Y., Mar, S., Benzinger, T.L., 2013. Diffusion tensor MRI as a biomarker in axonal and myelin damage. Imaging Med. 5 (5), 427–440. Available at: http://www.futuremedicine.com/doi/abs/10.2217/iim.13.49.

Baranova, A.I., Wei, E.P., Ueda, Y., Sholley, M.M., Kontos, H.A., Povlishock, J.T., 2008. Cerebral vascular responsiveness after experimental traumatic brain injury: the beneficial effects of delayed hypothermia combined with superoxide dismutase administration. J. Neurosurg. 109 (3), 502–509.

Benson, R., 2007. Global white matter analysis of diffusion tensor images is predictive of injury severity in traumatic brain injury. J. Neurotrauma 24 (3), 446–459.

Bharath, R., Munivenkatappa, A., Gohel, S., Panda, R., Saini, J., Rajeswaran, J., et al., 2015. Recovery of resting brain connectivity ensuing mild traumatic brain injury. Front. Hum. Neurosci. 9, 513.

Bigler, E., 2001. Distinguished Neuropsychologist Award Lecture 1999. The lesion(s) in traumatic brain injury: implications for clinical neuropsychology. Arch. Clin. Neuropsychol. 16 (2), 95–131.

Biswal, B., Yetkin, F.Z., Haughton, V.M., Hyde, J.S., 1995. Functional connectivity in the motor cortex of resting human brain using echo-planar MRI. Magn. Reson. Med. 34 (4), 537–541.

Bossini, P.S., Fangel, R., Habenschus, R.M., Renno, A.C., Benze, B., Zuanon, J.A., et al., 2009. Low-level laser therapy (670 nm) on viability of random skin flap in rats. Lasers Med. Sci. 24 (2), 209–213.

Brant-Zawadzki, M., Gillan, G.D.,N.W., 1992. MP RAGE: a three-dimensional, T1-weighted, gradient-echo sequence – initial experience in the brain. Radiology 182 (3), 769–775.

Broglio, S., Eckner, J.T., Paulson, H.L., Kutcher, J.S., 2012. Cognitive decline and aging: the role of concussive and subconcussive impacts. Exerc. Sport Sci. Rev. 40 (3), 138–144.

Brown, M., Baradaran, H., Christos, P.J., Wright, D., Gupta, A., Tsiouris, A.J., 2018. Magnetic resonance spectroscopy abnormalities in traumatic brain injury: a meta-analysis. J. Neuroradiol. 45 (2), 123–129. Elsevier Masson SAS.

Buckner, R., Krienen, F.M., Castellanos, A., Diaz, J.C., Yeo, B.T., 2011. The organization of the human cerebellum estimated by intrinsic functional connectivity. J. Neurophysiol. 106 (5), 2322–2345.

Capaldi, R., 2012. Structure and function of cytochrome c oxidase. Annu. Rev. Biochem. 59 (1), 569–596.

Chan, S., Evans, K.C., Rosen, B.R., Song, T.Y., Kwong, K.K., 2014. A case study of magnetic resonance imaging of cerebrovascular reactivity: a powerful imaging marker for mild traumatic brain injury. Brain Inj. 1–5.

Choi, E., Yeo, B., Buckner, R., 2012. The organization of the human striatum estimated by intrinsic functional connectivity. J. Neurophysiol. 108 (8), 2242–2263.

Clark, A., Bangen, K.J., Sorg, S.F., Schiehser, D.M., Evangelista, N.D., McKenna, B., et al., 2017. Dynamic association between perfusion and white matter integrity across time since injury in Veterans with history of TBI. Neuroimage Clin. 14, 308–315.

Coombe, A., Ho, C.T., Darendeliler, M.A., Hunter, N., Philips, J.R., Chapple, C.C., et al., 2001. The effects of low level laser irradiation on osteoblastic cells. Clin. Orthodont. Res. 4 (1), 3–14.

Corazza, A.V., Jorge, J., Kurachi, C., Bagnato, V.S., 2007. Photobiomodulation on the angiogenesis of skin wounds in rats using different light sources. Photomed. Laser Surg. 25 (2), 102–106.

Croall, I., Smith, F.E., Blamire, A.M., 2015. Magnetic resonance spectroscopy for traumatic brain injury. Topics Magn. Reson. Imaging 24 (5), 267–274.

Czerniak, S., Sikoglu, E.M., Liso Navarro, A.A., McCafferty, J., Eisenstock, J., Stevenson, J.H., et al., 2015. A resting state functional magnetic resonance imaging study of concussion in collegiate athletes. Brain Imaging Behav. 9 (2), 323–332.

DeCarli, C., Massaro, J., Harvey, D., Hald, J., Tullberg, M., Au, R., et al., 2005. Measures of brain morphology and infarction in the framingham heart study: establishing what is normal. Neurobiol. Aging 26 (4), 491–510.

Dickie, D.A., Job, D.E., Gonzalez, D.R., Shenkin, S.D., Ahearn, T.S., Murray, A.D., et al., 2013. Variance in brain volume with advancing age: implications for defining the limits of normality. PLoS One 8 (12), 1–12.

Doshi, H., Wiseman, N., Liu, J., Wang, W., Welch, R.D., O'Neil, B.J., et al., 2015. Cerebral hemodynamic changes of mild traumatic brain injury at the acute stage. PLoS One 10 (2), e0118061.

Duhaime, A.C., Gean, A.D., Haacke, E.M., Hicks, R., Wintermark, M., Mukherjee, P., et al., 2010. Common data elements in radiologic imaging of traumatic brain injury. Arch. Phys. Med. Rehabil. 91 (11), 1661–1666.

Eells, J.T., Henry, M.M., Summerfelt, P., Wong-Riley, M.T., Buchmann, E.V., Kane, M., et al., 2003. Therapeutic photobiomodulation for methanol-induced retinal toxicity. Proc. Natl. Acad. Sci. 100 (6), 3439–3444.

Ewing-Cobbs, L., Prasad, M.R., Swank, P., Kramer, L., Cox Jr, C.S., Fletcher, J.M., et al., 2008. Arrested development and disrupted callosal microstructure following pediatric traumatic brain injury: relation to neurobehavioral outcomes. Neuroimage 42 (4), 1305–1315.

Fierstra, J., Sobczyk, O., Battisti-Charbonney, A., Mandell, D.M., Poublanc, J., Crawley, A.P., et al., 2013. Measuring cerebrovascular reactivity: what stimulus to use? J. Physiol. 591 (23), 5809–5821.

Fukumura, D., Jain, R.K., 1998. Role of nitric oxide in angiogenesis and microcirculation in tumors. Cancer Metastasis Rev. 17 (1), 77–89.

Furuya, Y., Hlatky, R., Valadka, A.B., Diaz, P., Robertson, C.S., 2003. Comparison of cerebral blood flow in computed tomographic hypodense areas of the brain in head-injured patients. Neurosurgery 52 (2), 340–345.

Gaetz, M., 2004. The neurophysiology of brain injury. Clin. Neurophysiol. 115 (1), 4–18.

Gao, G., Oda, Y., Wei, E.P., Povlishock, J.T., 2010. The adverse pial arteriolar and axonal consequences of traumatic brain injury complicated by hypoxia and their therapeutic modulation with hypothermia in rat. J. Cerebral Blood Flow Metab. 30 (3), 628–637.

Garavello, I., Baranauskas, V., da Cruz-Höfling, M.A., 2004. The effects of low laser irradiation on angiogenesis in injured rat tibiae. Histol. Histopathol. 19 (1), 43–48.

Girard, C., et al., 2012. Axonal regeneration and neuroinflammation: roles for the translocator protein 18 kDa. J. Neuroendocrinol. 24 (1), 71–81.

Goldstein, L., Liu, S., Adams, D., Lacroix, C., Sinéus, M., Boucher, C., et al., 2012. Chronic traumatic encephalopathy in blast-exposed military veterans and a blast neurotrauma mouse model. Sci. Transl. Med. 4 (134), 134ra60.

Graham, D.I., Gennarelli, T.A., McIntosh, T.K., 2002. In: Graham, E.D.I., London, M.J.E. (Eds.), Greenfield's Neuropathology, CRC Press.

Haacke, E.M., Mittal, S., Wu, Z., Neelavalli, J., Cheng, Y.C., 2009. Susceptibility-weighted imaging: technical aspects and clinical applications, part 1. AJNR 30 (1), 19–30.

Haeussinger, F.B.F., Heinzel, S., Hahn, T., Schecklmann, M., Ehlis, A.C., Fallgatter, A.J., 2011. Simulation of near-infrared light absorption considering individual head and prefrontal cortex anatomy: implications for optical neuroimaging. Audio, Transactions of the IRE Professional Group on 6 (10), e26377.

Hemvani, N., Sadashiv, C., Bhagwanani, N.S., 2005. Helium-neon and nitrogen laser irradiation accelerates the phagocytic activity of human monocytes. Photomed. Laser Surg. 23 (6), 571–574.

Hou, D., Tong, K.A., Ashwal, S., Oyoyo, U., Joo, E., Shutter, L., et al., 2007. Diffusion-weighted magnetic resonance imaging improves outcome prediction in adult traumatic brain injury. J. Neurotrauma 24 (10), 1558–1569.

Huang, Y.-Y., Gupta, A., Vecchio, D., de Arce, V.J., Huang, S.F., Xuan, W., et al., 2012. Transcranial low level laser (light) therapy for traumatic brain injury. J. Biophotonics 5 (11–12), 827–837.

Huisa, B.N., et al., 2013. Transcranial laser therapy for acute ischemic stroke: a pooled analysis of NEST-1 and NEST-2. Int. J. Stroke 8, 315–320.

Huisman, T., Schwamm, L.H., Schaefer, P.W., Koroshetz, W.J., Shetty-Alva, N., Ozsunar, Y., et al., 2004. Diffusion tensor imaging as potential biomarker of white matter injury in diffuse axonal injury. AJNR 25 (3), 370–376.

Ichord, R., Naim, M., Pollock, A.N., Nance, M.L., Margulies, S.S., Christian, C.W., 2007. Hypoxic-ischemic injury complicates inflicted and accidental traumatic brain injury in young children: the role of diffusion-weighted imaging. J. Neurotrauma 24 (1), 106–118.

Iraji, A., Chen, H., Wiseman, N., Welch, R.D., O'Neil, B.J., Haacke, E.M., et al., 2016. Compensation through functional hyperconnectivity: a longitudinal connectome assessment of mild traumatic brain injury. Neural. Plast. 4072402.

Jellison, B.J., Field, A.S., Medow, J., Lazar, M., Salamat, M.S., Alexander, A.L., 2004. Diffusion tensor imaging of cerebral white matter: a pictorial review of physics, fiber tract anatomy, and tumor imaging patterns. AJNR 25 (3), 356–369.

Kenney, K., Amyot, F., Haber, M., Pronger, A., Bogoslovsky, T., Moore, C., et al., 2016. Cerebral vascular injury in traumatic brain injury. Exp. Neurol. 275 (3), 353–366.

Kim, J., Whyte, J., Patel, S., Avants, B., Europa, E., Wang, J., et al., 2010. Resting cerebral blood flow alterations in chronic traumatic brain injury: an arterial spin labeling perfusion FMRI study. J. Neurotrauma 27 (8), 1399–1411.

Kim, N., Branch, C.A., Kim, M., Lipton, M.L., 2013. Whole brain approaches for identification of microstructural abnormalities in individual patients: comparison of techniques applied to mild traumatic brain injury. PLoS One 8, e59382.

Kraus, M.F., Susmaras, T., Caughlin, B.P., Walker, C.J., Sweeney, J.A., Little, D.M., 2007. White matter integrity and cognition in chronic traumatic brain injury: a diffusion tensor imaging study. Brain 130 (10), 2508–2519.

Lama, S., Auer, R.N., Tyson, R., Gallagher, C.N., Tomanek, B., Sutherland, G.R., 2014. Lactate storm marks cerebral metabolism following brain trauma. J. Biol. Chem. 289 (29), 20200–20208.

Lampl, Y., Zivin, J.A., Fisher, M., Lew, R., Welin, L., Dahlof, B., et al., 2007. Infrared laser therapy for ischemic stroke: a new treatment strategy. Stroke 38 (6), 1843–1849.

Lange, R.T., Iverson, G.L., Brubacher, J.R., Madler, B., Heran, M.K., 2012. Diffusion tensor imaging findings are not strongly associated with post-concussional disorder 2 months following mild traumatic brain injury. J. Head Trauma Rehabil. 27, 188–198.

Liang, H.L., Whelan, H.T., Eells, J.T., Meng, H., Buchmann, E., Lerch-Gaggl, A., et al., 2006. Photobiomodulation partially rescues visual cortical neurons from cyanide-induced apoptosis. Neuroscience 139 (2), 639–649.

Ling, J.M., Peña, A., Yeo, R.A., Merideth, F.L., Klimaj, S., Gasparovic, C., et al., 2012. Biomarkers of increased diffusion anisotropy in semi-acute mild traumatic brain injury: a longitudinal perspective. Brain 135 (4), 1281–1292.

Lohr, N.L., Keszler, A., Pratt, P., Bienengraber, M., Warltier, D.C., Hogg, N., 2009. Enhancement of nitric oxide release from nitrosyl hemoglobin and nitrosyl myoglobin by red/near infrared radiation: potential role in cardioprotection. J. Mol. Cell. Cardiol. 47 (2), 256–263.

Mac Donald, C.L., Dikranian, K., Song, S.K., Bayly, P.V., Holtzman, D.M., Brody, D.L., 2007a. Detection of traumatic axonal injury with diffusion tensor imaging in a mouse model of traumatic brain injury. Exp. Neurol. 205 (1), 116–131.

Mac Donald, C.L., Dikranian, K., Bayly, P., Holtzman, D., Brody, D., 2007b. Diffusion tensor imaging reliably detects experimental traumatic axonal injury and indicates approximate time of injury. J. Neurosci. 27 (44), 11869–11876.

Malinovskaya, S.L., Monich, V.A., Artifeksova, A., 2008. Effect of low-intensity laser irradiation and wideband red light on experimentally ischemized myocardium. Bull. Exp. Biol. Med. 145 (5), 573–575.

Marino, S., Ciurleo, R., Bramanti, P., Federico, A., De Stefano, N., 2011. 1H-MR spectroscopy in traumatic brain injury. Neurocrit. Care 14 (1), 127–133.

Marklund, N., Bakshi, A., Castelbuono, D.J., Conte, V., McIntosh, T.K., 2006. Evaluation of pharmacological treatment strategies in traumatic brain injury. Curr. Pharm. Design 13, 1645–1680.

Marquez de la Plata, C., Ardelean, A., Koovakkattu, D., Srinivasan, P., Miller, A., Phuong, V., et al., 2007. Magnetic resonance imaging of diffuse axonal injury: quantitative assessment of white matter lesion volume. J. Neurotrauma 24 (4), 591–598.

Matcher, S.J.S., Cope, M.M., Delpy, D.T., 1997. In vivo measurements of the wavelength dependence of tissue-scattering coefficients between 760 and 900 nm measured with time-resolved spectroscopy. Appl. Opt. 36 (1), 386–396.

Maxwell, W.L., Povlishock, J.T., Graham, D.L., 1997. A mechanistic analysis of nondisruptive axonal injury: a review. J. Neurotrauma 14 (7), 419–440.

Mayer, A., Mannell, M.V., Ling, J., Gasparovic, C., Yeo, R.A., 2011. Functional connectivity in mild traumatic brain injury. Human Brain Map. 32 (11), 1825–1835.

McKee, A.C., Cantu, R.C., Nowinski, C.J., Hedley-Whyte, E.T., Gavett, B.E., Budson, A.E., et al., 2009. Chronic traumatic encephalopathy in athletes: progressive tauopathy after repetitive head injury. J. Neuropathol. Exp. Neurol. 68 (7), 709–735.

Menon, D., 2006. Brain ischaemia after traumatic brain injury: lessons from $15O_2$ positron emission tomography. Curr. Opin. Crit. Care 12 (2), 85–89.

Meythaler, J.M., Peduzzi, J.D., Eleftheriou, E., Novack, T.A., 2001. Current concepts: diffuse axonal injury-associated traumatic brain injury. Arch. Phys. Med. Rehabil. 82 (10), 1461–1471.

Mutch, W., Ellis, M.J., Graham, M.R., Wourms, V., Raban, R., Fisher, J.A., et al., 2014. Brain MRI CO_2 stress testing: a pilot study in patients with concussion. PLoS One, 9(7) e102181.

Mutch, W., Ellis, M.J., Graham, M.R., Wourms, V., Raban, R., Fisher, J.A., et al., 2015. Brain magnetic resonance imaging CO stress testing in adolescent postconcussion syndrome. J. Neurosurg. 1–13.

Mutch, W., Ellis, M., Ryner, L., Ruth Graham, M., Dufault, B., Gregson, B., et al., 2016a. Brain magnetic resonance imaging CO_2 stress testing in adolescent postconcussion syndrome. J. Neurosurg. 125 (3), 648–660.

Mutch, W., Ellis, M., Ryner, L., Morissette, M., Pries, P.J., Dufault, B., et al., 2016b. Longitudinal brain magnetic resonance imaging CO_2 stress testing in individual adolescent sports-related concussion patients: a pilot study. Front. Neurol. 7, 107.

Newcombe, V., Chatfield, D., Outtrim, J., Vowler, S., Manktelow, A., Cross, J., et al., 2011. Mapping traumatic axonal injury using diffusion tensor imaging: correlations with functional outcome. PLoS One, 6(5) e19214.

Oda, Y., Gao, G., Wei, E.P., Povlishock, J.T., 2011. Combinational therapy using hypothermia and the immunophilin ligand FK506 to target altered pial arteriolar reactivity, axonal damage, and blood-brain barrier dysfunction after traumatic brain injury in rat. J. Cereb. Blood Flow Metab. 31 (4), 1143–1154.

Parvizi, J., Damasio, A., 2003. Neuroanatomical correlates of brainstem coma. Brain 126 (7), 1524–1536.

Pettus, E., 1994. Traumatically induced altered membrane permeability: its relationship to traumatically induced reactive axonal change. J. Neurotrauma 11 (5), 507–522.

Pillai, J., Mikulis, D., 2015. Cerebrovascular reactivity mapping: an evolving standard for clinical functional imaging. AJNR 36 (1), 7–13.

Raichle, M., 2015. The brain's default mode network. Annu. Rev. Neurosci. 38, 433–447.

Riedy, G., Senseney, J.S., Liu, W., Ollinger, J., Sham, E., Krapiva, P., et al., 2016. Findings from structural MR imaging in military traumatic brain injury. Radiology 279 (1), 207–215.

Rosenthal, S., Gray, M., Fatima, H., Sair, H.I., Whitlow, C.T., 2018. Functional MR imaging: blood oxygen level-dependent and resting state techniques in mild traumatic brain injury. Neuroimaging Clin. N. Am. 28 (1), 107–115.

Saatman, K., Duhaime, A.C., Bullock, R., Maas, A.I., Valadka, A., Manley, G.T., et al., 2008. Classification of traumatic brain injury for targeted therapies. J. Neurotrauma 719–738.

Sayeed, I., Stein, D.G., 2009. Progesterone as a neuroprotective factor in traumatic and ischemic brain injury. Prog. Brain Res. 175, 219–237.

Servadei, F., 2011. Clinical value of decompressive craniectomy. N. Engl. J. Med. 364, 1558–1559.

Sharma, S.K., Kharkwal, G.B., Sajo, M., Huang, Y.Y., De Taboada, L., McCarthy, T., et al., 2011. Dose response effects of 810 nm laser light on mouse primary cortical neurons. Lasers Surg. Med. 43 (8), 851–859.

Sidaros, A., Engberg, A.W., Sidaros, K., Liptrot, M.G., Herning, M., Petersen, P., et al., 2008. Diffusion tensor imaging during recovery from severe traumatic brain injury and relation to clinical outcome: a longitudinal study. Brain 131 (2), 559–572.

Skandsen, T., Kvistad, K.A., Solheim, O., Lydersen, S., Strand, I.H., Vik, A., 2011. Prognostic value of magnetic resonance imaging in moderate and severe head injury: a prospective study of early MRI findings and one-year outcome. J. Neurotrauma 28 (5), 691–699.

Stepansky, F., Hecht, E.M., Rivera, R., Hirsh, L.E., Taouli, B., Kaur, M., et al., 2008. Dynamic MR angiography of upper extremity vascular disease: pictorial review. Radiographics 28 (1), e28.

Tolias, C.M., Bullock, R., 2004. Critical appraisal of neuroprotection trials in head injury: what have we learned? NeuroRX 1 (1), 71–79.

Tomlinson, B., 1970. Brain-stem lesions after head injury. J. Clin. Pathol. Suppl. (R. Coll. Pathol.) 4, 154–165.

Vakhtin, A.A., Calhoun, V.D., Jung, R.E., Prestopnik, J.L., Taylor, P.A., Ford, C.C., 2013. Changes in intrinsic functional brain networks following blast-induced mild traumatic brain injury. Brain Inj. 27 (11), 1304–1310.

Wan, S., Parrish, J.A., Anderson, R.R., Madden, M., 1981. Transmittance of nonionizing radiation in human tissues. Photochem. Photobiol. 34 (6), 679–681.

Wang, J., 2008. Diffusion tensor tractography of traumatic diffuse axonal injury. Arch. Neurol. 65 (5), 619–626.

Wang, X., Wei, X.E., Li, M.H., Li, W.B., Zhou, Y.J., Zhang, B., et al., 2014. Microbleeds on susceptibility-weighted MRI in depressive and non-depressive patients after mild traumatic brain injury. Neurol. Sci. 35 (10), 1533–1539.

Wei, E.P., Hamm, R.J., Baranova, A.I., Povlishock, J.T., 2009. The long-term microvascular and behavioral consequences of experimental traumatic brain injury after hypothermic intervention. J. Neurotrauma 26 (4), 527–537.

Westfall, D., West, J.D., Bailey, J.N., Arnold, T.W., Kersey, P.A., Saykin, A.J., et al., 2015. Increased brain activation during working memory processing after pediatric mild traumatic brain injury (mTBI). J. Pediatr. Rehabil. Med. 8 (4), 297–308.

Wilde, E., Chu, Z., Bigler, E.D., Hunter, J.V., Fearing, M.A., Hanten, G., et al., 2006. Diffusion tensor imaging in the corpus callosum in children after moderate to severe traumatic brain injury. J. Neurotrauma 23 (10), 1412–1426.

Wilson, S., Raghupathi, R., Saatman, K.E., MacKinnon, M.A., McIntosh, T.K., Graham, D.I., 2004. Continued in situ DNA fragmentation of microglia/macrophages in white matter weeks and months after traumatic brain injury. J. Neurotrauma 21 (3), 239–250.

Wong-Riley, M.T.T., Liang, H.L., Eells, J.T., Chance, B., Henry, M.M., Buchmann, E., et al., 2005. Photobiomodulation directly benefits primary neurons functionally inactivated by toxins. J. Biol. Chem. 280 (6), 4761–4771.

Wu, Q., Xuan, W., Ando, T., Xu, T., Huang, L., Huang, Y.Y., et al., 2012. Low-level laser therapy for closed-head traumatic brain injury in mice: effect of different wavelengths. Lasers Surg. Med. 44 (3), 218–226.

Xiong, K., Zhu, Y., Zhang, W., 2014. Diffusion tensor imaging and magnetic resonance spectroscopy in traumatic brain injury: a review of recent literature. Brain Imaging Behav. 8 (4), 487–496.

Xiong, K., Zhang, J.N., Zhang, Y.L., Zhang, Y., Chen, H., Qiu, M.G., 2016. Brain functional connectivity and cognition in mild traumatic brain injury. Neuroradiology 58 (7), 733–739.

Xu, S., Zhuo, J., Racz, J., Shi, D., Roys, S., Fiskum, G., et al., 2011. Early microstructural and metabolic changes following controlled cortical impact injury in rat: a magnetic resonance imaging and spectroscopy study. J. Neurotrauma 28 (10), 2091–2102.

Yanagawa, Y., Sakamoto, T., Takasu, A., Okada, Y., 2009. Relationship between maximum intracranial pressure and traumatic lesions detected by T2*-weighted imaging in diffuse axonal injury. J. Trauma 66 (1), 162–165.

Yemisci, M., Gursoy-Ozdemir, Y., Vural, A., Can, A., Topalkara, K., Dalkara, T., 2009. Pericyte contraction induced by oxidative-nitrative stress impairs capillary reflow despite successful opening of an occluded cerebral artery. Nat. Med. 15 (9), 1031–1037.

Yendiki, A., Panneck, P., Srinivasan, P., Stevens, A., Zöllei, L., Augustinack, J., et al., 2011. Automated probabilistic reconstruction of white-matter pathways in health and disease using an atlas of the underlying anatomy. Front. Neuroinform. 5 (23).

Yendiki, A., Reuter, M., Wilkens, P., Rosas, H.D., Fischl, B., 2016. Joint reconstruction of white-matter pathways from longitudinal diffusion MRI data with anatomical priors. Neuroimage 15 (127), 277–286.

Yeo, B., Krienen, F.M., Sepulcre, J., Sabuncu, M.R., Lashkari, D., Hollinshead, M., et al., 2011. The organization of the human cerebral cortex estimated by intrinsic functional connectivity. J. Neurophysiol. 106 (3), 1125–1165.

Yuan, W., Holland, S.K., Schmithorst, V.J., Walz, N.C., Cecil, K.M., Jones, B.V., et al., 2007. Diffusion tensor MR imaging reveals persistent white matter alteration after traumatic brain injury experienced during early childhood. AJNR 28 (10), 1919–1925.

Yuh, E., Mukherjee, P., Lingsma, H.F., Yue, J.K., Ferguson, A.R., Gordon, W.A., et al., 2013. MRI improves 3-month outcome prediction in mild traumatic brain injury. Ann. Neurol. 73 (2), 224–235.

Zhang, R., Mio, Y., Pratt, P.F., Lohr, N., Warltier, D.C., Whelan, H.T., et al., 2009. Near infrared light protects cardiomyocytes from hypoxia and reoxygenation injury by a nitric oxide dependent mechanism. J. Mol. Cell. Cardiol. 46 (1), 4–14.

Ziche, M., Morbidelli, L., 2000. Nitric oxide and angiogenesis. J. Neuro-Oncol. 50 (1), 139–148.

Zivin, J.A., Albers, G.W., Bornstein, N., Chippendale, T., Dahlof, B., Devlin, T., et al., 2009. Effectiveness and safety of transcranial laser therapy for acute ischemic stroke. Stroke 40 (4), 1359–1364.

Chapter 28

Treatment of traumatic brain injury with near-infrared light

Larry D. Morries[1] and Theodore A. Henderson[1,2]

[1]Neuro-Laser Foundation, Centennial, CO, United States, [2]The Synaptic Space, Centennial, CO, United States

28.1 Background

Since 2004, traumatic brain injury (TBI) has exploded in the press and in scientific literature. The number of research articles on TBI was less than 400 annually for many years, but increased to over 1000 in 2005, reached over 2000 annually in 2012, and reached 3,940 in 2018. Issues of diagnosis, incidence, pathophysiological consequences, and long-term consequences have dominated the literature. Very little has been published on effective treatments (Morries et al., 2015; Henderson and Morries, 2015a). This situation is quite disturbing in the face of the growing incidences and the realization that many do not fully recover from even mild TBI/concussion. Research has demonstrated that a single concussion can have lasting effects and multiple subconcussive impacts can result in an accumulation of pathological changes (Bigler and Maxwell, 2012; Peskind et al., 2013). A recent study showed 21% of children who sustain a concussion have persistent symptoms (Grubenhoff et al., 2014).

Herein, we will review key information about TBI and its impact on human life. We will highlight certain vulnerable populations. The use of near-infrared (NIR) light energy as a tool for the treatment of TBI will be discussed in detail with review of the current literature and an update on the ongoing work of the Neuro-Laser Foundation.

28.1.1 Definition

TBI is a traumatically induced physiological disruption of brain function. TBI is graded in unequal divisions as mild, moderate, or severe. These grades are determined by symptoms within the first few hours after injury and scoring using the Glasgow Coma Score (GCS). The GCS is a readily available and widely used acute assessment tool for determining likely brain injury. It correlates well with moderate–to-severe TBI, but often underestimates mild TBI. The instrument requires assessing whether the eyes are open or not, whether the patient makes verbal responses, and whether the patient moves and/or follows commands to move.

Mild TBI is manifested by at least one of the following documented within 24–72 hours of a head injury:

- any loss of consciousness less than 30 minutes;
- any loss of memory for events immediately before or after the injury;
- any alteration of mental status at the time of the injury (e.g., feeling dazed, disoriented, or confused);
- focal neurological deficit(s) that may or may not be transient;
- GCS ≥ 13;
- posttraumatic amnesia (PTA) not greater than 24 hours (American Congress of Rehabilitation Medicine Head Injury Interdisciplinary Special Interest Group, 1993) (Kay et al. 1993 and State of Colorado, 2012).

Moderate TBI is manifested by at least one of the following:

- altered state or loss of consciousness for greater that 30 minutes,
- an initial GCS of 9–12,
- evidence of trauma on anatomical neuroimaging,

Photobiomodulation in the Brain. DOI: https://doi.org/10.1016/B978-0-12-815305-5.00028-2

- persistent focal neurological deficit(s),
- PTA greater than 24 hours.

Severe TBI is manifested by symptoms associated with moderate TBI and a GCS < 9.

The World Health Organization defines postconcussion syndrome as "persistence of a constellation of physical, cognitive, emotional, and sleep symptoms beyond the usual recovery period after a concussion" (WHO, 2010), including three or more of the following after head injury: headache, dizziness, fatigue, irritability, insomnia, reduced tolerance of stress, concentration difficulty, or memory difficulty.

28.1.2 Incidence

The incidence of documented TBI is steadily increasing despite attention to the potential hazards of even mild TBI/concussion. The Centers for Disease Control and Prevention estimated that 1.5 million Americans sustained documented TBI annually in 2000 (Bazarian et al., 2005). The incidence rose to 1.7 million brain injuries annually in 2006 (Vaishnavi et al., 2009; Faul et al., 2010). Despite the increased awareness over the past two decades, the rate of TBI has increased from 521.0 per 100,000 in 2001 to 823.7 per 100,000 in 2010 (Pervez et al., 2018). Current estimates of the prevalence of TBI among veterans range from 9.6% to 20% (Logan et al., 2013) with an estimated total of more than 379,000 cases of TBI among military personnel since 2000 (DOD Worldwide Numbers for TBI, 2018). A large number of mild TBI/concussions go unreported. The current estimates of the combined number of sports-related concussions and brain injuries in the United States are 1.6–3.8 million annually (Gilchrist et al., 2011; Noble and Hesdorffer, 2013; Selassie et al., 2013). While many recover from mild TBI/concussion with no sequalae, roughly 15%–20% of reported cases will have persistent symptoms (Pervez et al., 2018). This conservatively represents a staggering 760,000 cases of persistent mild TBI per year. The annual costs of concussion and TBI were estimated in 2010 at $11.5 billion in direct medical costs and $64.8 billion in lost wages/productivity (Humphreys et al., 2013).

28.1.3 Vulnerable populations

The increased vulnerability of certain populations—women, the elderly, and the young—is emerging. Common risk factors for persistent symptoms after mild TBI/concussion include: of the female gender, younger age, history of prior concussion, history of migraine headaches, history of psychiatric disorders, and history of learning disabilities (Colvin et al., 2009; Pervez et al., 2018).

28.1.3.1 Women

Women appear to have a more serious neurological response to concussive events. For example, in a sample of 2340 high school and collegiate athletes, 155 athletes sustained a concussion in a single season (Broshek et al., 2005). The female athletes had greater declines in reaction time (RT) relative to baseline measures that were obtained preseason. The female athletes reported more subjective adverse events from the concussion and objective measures showed greater cognitive impairment (Broshek et al., 2005). Among lacrosse players with concussions, female players had significantly greater neurocognitive impairment on computerized testing and were more likely to demonstrate protracted recovery (Sandel et al., 2017). The incidence of concussions among women playing a given sport is 1.4–3.7 times greater than for men playing the same sport, including the less violent sports (Gessel et al., 2007; Lincoln et al., 2011; Covassin et al., 2013).

28.1.3.2 Elderly

The elderly are much more likely to be admitted to the hospital after a TBI of any severity (Peschman et al., 2011). Falls represent the largest cause of TBI in the elderly, followed by motor vehicle accidents (MVAs) (Fu et al., 2017). Despite having less severe injuries in general, the elderly have higher mortality and worse functional outcomes (Susman et al., 2002; Styrke et al., 2007; Richmond et al., 2011; Ramanathan et al., 2012). One contributing factor is that elderly have a considerably higher incidence of subdural hematoma following TBI of any severity. Risks factors for subdural hematoma in the elderly include: atrophy, concomitant strain on the dural vessels, extensive use of aspirin and anticoagulant therapy, balance problems, and concomitant medications with effects on balance (Shapey et al., 2016).

28.1.3.3 Children

Children make up fully one-third of all TBI cases. Children involved in sports have a higher rate of head injury compared to collegiate players, most likely reflecting inexperience or uncoordination. The immature brain is considerably more vulnerable to injury due to the incomplete myelination of large portions of the cerebral cortex, particularly the frontal lobes (Prins and Giza, 2012). Furthermore, research in animal models indicate that TBI leads to more profound changes in growth, neuroplasticity, and metabolism in the young brain (Babikian et al., 2010). In the United States, 65% of cases of concussion occur in sports and recreational activities (Davis Moore et al., 2018). It should be noted that 70% of all US football players are under age 14 (Cobb et al., 2013). A helmet sensor study of 50 child football players, ages 9–12, revealed over 11,900 helmet impacts in one season (Cobb et al., 2013). Urban and colleagues (2013) reported there were 16,500 impacts among 40 high school football players in one season. Evidence indicates that football players who started tackle football before age 12 had greater cognitive impairment as they age (Alosco et al., 2017).

28.1.4 Symptoms

TBI results in a wide spectrum of neurological, psychiatric, cognitive, and emotional consequences. In part, the variation is related to the severity grade of the injury (mild, moderate, severe TBI). Furthermore, the diversity of sequalae can be related to the areas of the brain that are injured, the severity of the injury (highly variable within the classification of mild and "moderate"), and the evolution of the injury over time due to neuroinflammatory processes (Kumar and Loane, 2012; Ziebell and Morganti-Kossmann, 2010). Additional mechanisms thought to underlie the damage of TBI include decreased mitochondrial function, calcium and magnesium dysregulation, excitotoxicity, disruption of neural networks, free radical-induced damage, excessive nitric oxide, ischemia, and damage to the blood brain barrier (Bigler and Maxwell, 2012). Together, these can contribute to a progression of the damage over time.

Patients with mild TBI can experience: headaches, visual disturbances, dizziness, cognitive impairment, loss of executive skills, memory impairment, fatigue, impulsivity, impaired judgment, emotional outbursts, anxiety, and depression (Lew, 2005; Kennedy et al., 2007; Kashluba et al., 2008; Vaishnavi, et al., 2009). The situation can be further complicated by secondary and/or comorbid posttraumatic stress disorder (PTSD), depression, and anxiety, (Fann et al., 2004; Jorge et al., 2004; Vasterling et al. 2006; Kennedy et al., 2007; Lew et al., 2008) which can have symptoms that overlap with those described above. PTSD, depression, and/or anxiety appear to be increasingly likely in the context of repetitive concussive or subconcussive brain injury (Bryan and Clemans, 2013; Prins et al., 2013).

In moderate-to-severe TBI, all of these symptoms can be present, along with neurological impairment of motor function, speech function, vision, hearing, and visuospatial skills depending on the portion of the brain that is injured. Furthermore, executive function problems can be expected due to the high probability of injury to the frontal lobes.

28.2 Diagnostic workup

We have found that the diagnostic workup of the patient is a vital part of the treatment process. Not only does the diagnostic evaluation provide key symptoms and signs to follow during the treatment course, but it can also uncover other clinical conditions that warrant treatment. Our experience has revealed that many patients with TBI who have been in treatment at other centers come to our facilities with missed diagnoses. It is not unusual for a patient with TBI to have a missed and untreated fracture. The reverse has also been true. Patients who have been involved in a severe MVA have often reported no diagnostic workup for brain injury, although they received extensive clinical interventions for multiple fractures. Our belief is that patients deserve a comprehensive examination and we routinely refer patients with unaddressed orthopedic problems to our orthopedic colleagues.

28.2.1 Neurological and physical evaluation

The initial history should be focused on the index traumatic event. However, an expanded history of the patient provides valuable insight and can often uncover previously unidentified disabilities. All complaints should be identified and a comprehensive history of the index traumatic event and any prior traumas should be obtained. General medical history and a psychiatric history should be gathered. The psychosocial functioning of the patient often changes dramatically following a TBI. These aspects of the history are vital to addressing all of the patient's needs.

A careful history of headaches should be obtained in the context of TBI. Headaches are a frequent (90% in our clinic) complaint for patients with TBI. History should include:

- onset of severe headaches with vomiting and neck stiffness;
- headaches accompanied by loss of consciousness, slow pulse, or increased drowsiness;
- headaches accompanied by progressive failure of vision;
- associated memory loss, disturbance, or a definite change in personality;
- associated convulsions or seizures;
- sudden change in a chronic headache pattern;
- continually worsening headaches over several days;
- early morning headaches sufficient to awaken the individual from sleep;
- headaches of subacute onset that are significantly aggravated by the Valsalva maneuver or coughing or sneezing;
- highly localized chronic pain.

Since polytrauma is the rule rather than the exception in cases of TBI, the initial physical examination should be a thorough trauma examination, which can be split into two appointments for the sake of time and the physical endurance of the patient.

The neurological examination should be equally thorough. The clinician should have specialized training in neurology or neurological sciences. The examination should include: mental status, cranial nerve function, motor status with each limb independently assessed, sensory status with each limb independently assessed, balance function, coordination, gait and station.

Mental status examination includes both formal and informal observations. Formal evaluation of mental status is detailed in the section 28.2.5 on Questionnaires and Cognitive Testing below. Additional attention should be provided to alertness, ability to provide a coherent history, social and/or behavioral decorum, hygiene, attention, concentration, memory, affect, mood, thought process, judgment, insight, and ability to follow directions. The patient's recounting of their history should be monitored for evidence of abnormal thought content, such as hallucinations, delusions, paranoia, etc. The patient should also be specifically queried about the presence of hallucinations or delusions. Examiners should be aware that abnormal cognitive function, memory, or arithmetic skill may occur in the setting of chronic pain, fatigue, medication use, or sleep deprivation—all common experiences for patients with TBI. In addition, attention to preexisting or acute psychiatric disorders, including PTSD, can also influence the findings on examination or response to near-infrared light therapy (NILT). In our clinic, we have treated patients with preexisting bipolar disorder, who experienced hypomania following multiwatt NIR laser therapy. In cases where a comorbid diagnosis is suspected, it is often beneficial to refer the patient to a collaborating psychiatrist for evaluation and treatment.

Cranial nerve examination must include careful evaluation of olfactory function. Multiple studies show a rate of anosmia in cases of TBI between 16% and 30% (Swann et al., 2006; Atighechi et al., 2013; Schofield et al., 2014; Proskynitopoulos et al., 2016; Bratt et al., 2018). The proximity of the olfactory bulb to the frontal lobes (immediately subjacent) and the kinetics of rotational forces during an impact render the olfactory bulb and tract vulnerable to injury. These same factors created a correlation between frontal lobe injury and injury of central olfactory structures. Olfactory impairment (anosmia, dysosmia) appears to correlate with injury severity, although the literature has yielded mixed results. Most studies concur that the presence of olfactory dysfunction is correlated with increased likelihood of cognitive or neuropsychological impairment. Moreover, in cases of concussion/mild TBI, the presence of olfactory dysfunction may signal a more severe injury of the frontal lobes (Schofield et al., 2014). Unfortunately, recovery of olfactory function may be limited. In patients with severe olfactory dysfunction in the acute setting, little recovery is found (Bakker et al., 2016). Among those with less severe olfactory dysfunction, significant improvement can be observed; however, it is usually incomplete. Curiously, many patients are unaware of the sensory impairment, and olfaction is not routinely tested in patients with TBI and rarely considered in those patients with concussion/mild TBI. Hence, inclusion of olfaction questions in the history and a formal olfactory examination are encouraged. Clues can be gathered from history, particularly in terms of the taste of food. The University of Pennsylvania's smell identification test (UPSIT) is a very useful tool for assessing olfaction, although it can be time-consuming. The UPSIT consists of four test booklets containing 10 odorants each, yielding a total of 40 distinct measures. The examiner scratches the surface of the capsule with a pencil, which releases the odor. The examiner then provides four odor name options from which the patient may chose. The UPSIT has high reliability ($r = 0.94$) (Proskynitopoulos et al., 2016). The "sniffin' sticks" test kit (Hummel et al., 1997) is another commercially available method of testing olfaction. The test involves 16 common odors (orange, leather, cinnamon, menthol, banana, lemon, licorice, garlic, coffee, apple, pineapple, rose, fish, anise, clove, and

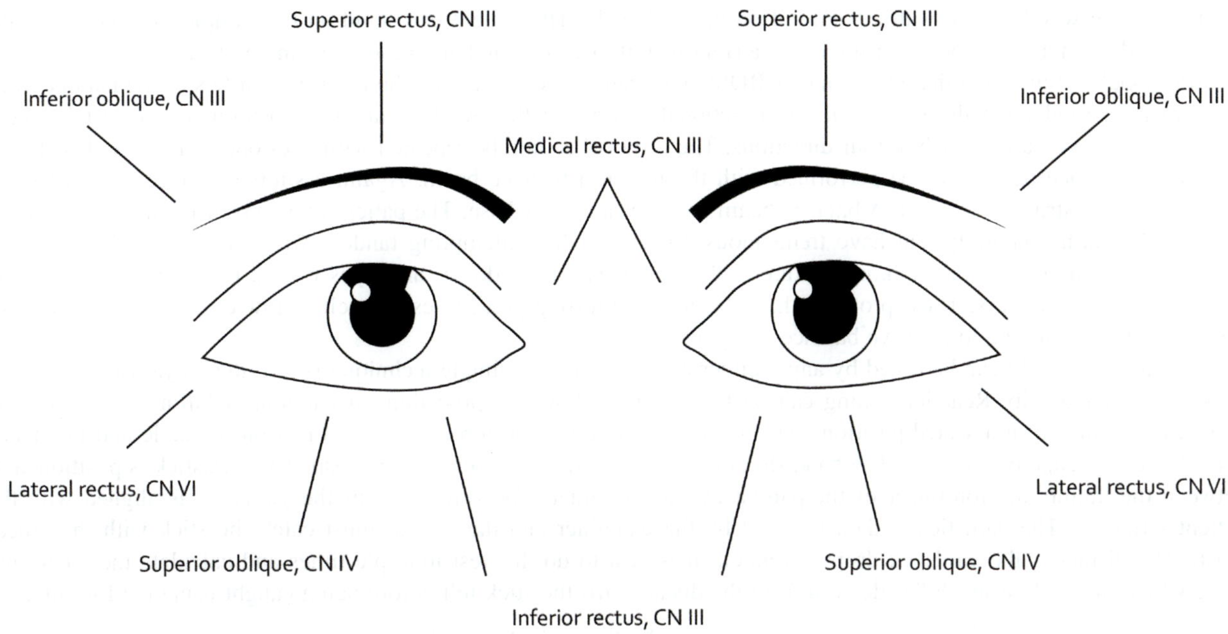

FIGURE 28.1 Diagram illustrated the extraocular muscles and corresponding cranial nerves.

turpentine) using a multiple forced choice design. Metrics include odor threshold, odor discrimination, and odor identification (Hummel et al., 2007). There are several other olfactory tests in the literature.

Testing of the cranial nerves is part of a standard neurological examination. Careful attention must be given to each cranial nerve in patients who have sustained a TBI. Visual acuity with a Snellen chart, pupillary reflex, convergence, oculomotor function (see Fig. 28.1), presence of nystagmus, presence of diplopia, facial nerve function, ptosis, uvula and tongue position, and audition provide information on the integrity of each cranial nerve. The trigeminal nerve can be tested with a sensory examination of the face. Each of the three divisions of the trigeminal nerve should be tested (mandibular, maxillary, and ophthalmic).

The motor and sensory examination may reveal limb-specific dysfunction related to peripheral nerve disease—another often-missed clinical condition in trauma patients. Range of motion of the cervical spine, upper and lower extremities, and lumbar spine should be tested. Sensory examination can be conducted with nothing more than light touch and a source of vibration, such as a tuning fork. Both upper and lower extremities should be included in the sensory examination. Unilateral motor or sensory signs may provide clues to the laterality of the brain injury. Severity of these findings may indicate other physical/orthopedic/extremity injuries. Attention should be given to gait, foot pronation or supination, and evidence of injuries which have not healed. In our clinic, we have noted extremity injuries that have resulted in restricted range of motion, gait disturbance, and/or sensory loss, yet were not listed in prior exams. The most extreme example was a patient with an unfused hip fracture, years after the injury. The patient's gait was dramatically impaired; yet, she had never undergone radiological examination of her hips.

28.2.2 Balance testing

Balance function is an important area to assess in TBI. There are numerous computerized balance board testing machines available; however, these are not required to perform a thorough balance examination. In the absence of these machines/computerized units, there are simple clinical maneuvers that can be performed. A minitrampoline or a BOSU ball can provide an unstable base, allowing a reproducible testing protocol. Balance should be assessed with eyes open initially and a spotter to avoid falls. A two-leg stand should be tested first. Then one-leg stands should be attempted. Trampolines have different amounts of instability depending on where on the trampoline the patient is standing. Closer to the edge of the trampoline, the base will be more stable. Maximal instability is achieved at the center of the trampoline. The balance assessment should include testing with the eyes closed as well. This is the truest test of the vestibular/proprioception system when the visual cues are removed. A simple wooden tilt board can be constructed with a

25 cm × 40 cm wooden base with a 5 cm rail on the underside. The patient can stand on this simple tilt board with the rail oriented in an anterior-posterior direction, a right-to-left direction, and at a 45-degree angle also.

Additional testing using the trampoline or BOSU ball can include assessing the righting abilities of the patient when pressure is exerted upon the torso. Balance functional reserve in the two-leg standing position on an unstable base should include pressure in each of four directions. The protocol should be repeated with eyes open and eyes closed.

Further balance testing can be performed with the use of a balance beam. Again, a simple balance beam of some 12 ft. can be constructed of wood. A balance beam provides a narrow base. The patient may appear to have a reasonable gait when walking naturally, but have tremendous difficulty when attempting tandem walk on a beam. The balance beam testing emphasizes the walking posture with eyes ahead, eyes down, and eyes closed (advanced), to ascertain body sway, and weakness. These protocols have been useful testing procedures, as well as low-cost treatment exercises for patient home training to improve balance and gait.

Reaction time (RT) can be tested by any number of procedures. A simple technique is to utilize a yardstick or meter stick with a hand grip. Reaction timing can be tested with a low-cost procedure of catching a falling yardstick. The patient is positioned in a seated position near the edge of a table. The forearms should rest on the table and the hands extend over the edge of the table. The hand should be positioned in an open pincer grasp. The yardstick is positioned in between the thumb and forefinger of the patient by the examiner. The zero mark of the yardstick is aligned with the patient's fingers. The yardstick is then released by the examiner and the patient must catch the stick with the pincer grasp. The distance "the yardstick drops" is noted. It is best to do this test multiple times and calculate the mean distance which the stick drops. RT is derived from the distance (*d*) the stick fell before being caught using the formula:

(i). $RT = \text{sqrt}(2d/9.8\ \text{m/s}^2)$, where sqrt is square root and $9.8\ \text{m/s}^2$ is the acceleration of gravity.

28.2.3 Dysautonomia

Concussion and/or TBI are known to affect cognitive function, but can also alter other physiological systems, including the cardiovascular and the autonomic nervous systems. Estimates are that 10%–35% of patients with TBI will experience some degree of dysautonomia (Baguley et al., 1999). Concussed athletes have exaggerated sympathetic nervous activity and increased heart rates when compared with controls. Cerebral autoregulation and cerebral blood flow are disturbed after concussion, which may explain why symptoms reappear or worsen with physical exertion or other stressors that increase blood pressure. Noting blood pressure and pulse in both sitting and standing positions is recommended. In more severe TBI, there is increased risk for sympathetic storms, thyroid storms, and serotonin syndrome.

28.2.4 Cervicogenic headaches

Studies of concussive head injuries indicate a high rate of involvement of cervical spine in the injury process. Stretching and injury to the ligaments of the upper cervical spine and craniocervical junction can lead to pain and discomfort in the neck, as well as headaches. These cervicogenic headaches are a common complication in patients with TBI. Examination for restricted cervical mobility—in flexion, extension, and rotation, mobility with or without pain when slight downward pressure is applied to the vertex, point tenderness along the cervical spine, particularly at C4 transverse process, C2/3 joint, and path of greater occipital nerve; and pain or restricted movement of upper trapezius, scalenes, levator scapulae, and pectoralis minor may support the diagnosis of cervicogenic headache. Factor analysis has shown that upper cervical joint dysfunction (particularly C1/2), and shortened pectoralis minor muscle length were able to discriminate a group of subjects with cervicogenic headache from a group with migraine headache and controls with 80% sensitivity (Zito et al., 2006). Physical therapy and exercise can be more helpful than nonsteroidal antiinflammatories or opiate pain medications in cervicogenic headaches. Soft tissue massage also has benefits. We have found that NILT to be highly effective at relieving the pain and restricted mobility associated with cervicogenic headaches. Applied multiwatt NIR laser therapy to the posterior cervical spine and lateral cervical spinal areas, as well as along the course of the upper trapezius, levator scapulae, and posterior cervical musculature appears to reduce inflammation and associated pain. We often combine the NILT with a program of antigravity exercises at home.

28.2.5 Questionnaires and cognitive testing

We use a number of questionnaires and paper/pencil tests in our assessment of patients with TBI.

The Rivermead Post-Concussion Symptoms Questionnaire (RPQ) (King et al., 1995) is particularly useful for quantifying symptoms commonly found in patients with TBI. Patients rank symptoms on a scale of 0–4 wherein 0 = symptom not present to 4 = a severe problem. The symptoms assessed include: headache, dizziness, nausea, sleep disturbance, fatigue, irritability, depression, frustration, memory, impaired concentration, trouble thinking, blurred vision, light/noise sensitivity, diplopia and restlessness. Two scoring methods are utilized. The first score, RPQ-3, is the sum of the first three items. A score greater than 6 (maximal score = 12) is considered positive. The second score, RPQ-13, is the sum of the remaining 13 items. A score greater than 25 (maximal score = 52) is considered positive. However, the RPQ is not a diagnostic tool. It is an assessment tool. The RPQ's reliability has been questioned as the symptoms included do not correlate to a single factor in factor analysis (Eyres et al., 2005). Yet, the RPQ has proven useful and demonstrates significant improvement in scores during the treatment, which correlates with the patient's subjective improvement.

Another commonly used test for concussion/mild TBI symptoms is the ImPACT test. This test is widely used in sports and is subject to limitations based on efforts in the baseline testing. Nevertheless, in the clinic we are typically not seeing patients prior to the TBI. The ImPACT has been useful for tracking symptom changes with treatment and for assessing the effects of additional head injuries (Note: patients who have had a TBI in the past have a higher incidence of new concussive events). This computerized neurocognitive assessment allows an alternative method of assessing RT.

The Montreal Cognitive Assessment (MoCA) is a widely used assessment instrument for detecting cognitive impairment (Nasreddine et al., 2005) with a maximal score of 30 points. While typically used for assessing for dementia or mild cognitive impairment, it contains many components which assess functions which are often disrupted in TBI. The MoCA assesses short-term recall, which is often disrupted in TBI. It examines visuospatial abilities with a clock-drawing task and a three-dimensional figure drawing. A trail-making B task is included, which assesses executive functioning. A phonemic fluency task is included, which we find is often impaired following a TBI. A serial subtraction task challenges attention, concentration, and working memory. The MoCA also includes an animal naming task, a sentence repetition task, abstract thinking task, and assessment of orientation to place and time. The MoCA is a quick assessment that combines several of the best cognitive tests into a 7–10-minute examination. The MoCA has a number of versions with different three-dimensional objects to draw, different words to recall, and other differences. The use of these various versions lessens the chance of test-retest effects. The MoCA is available in multiple languages, which has been useful.

In our clinic, we have noted that memory, verbal fluency, and performance on the trail-making B task are most often impaired. Patients can rarely recall five out of five items at 5 minutes. Patients also rarely can think of 11 or more words beginning with a particular letter. Animal naming and abstract thinking ("how is an apple and an orange alike?") are not often affected in our patients. Most patients score below 26, which is the cutoff for normal cognitive function.

The Quick Inventory of Depression Symptomology-Self Report 16 (QIDS) is a very useful depression questionnaire which is experienced positively by the patients (Rush et al., 2003; Trivedi et al., 2004). Other depression questionnaires appear daunting or repetitive to many patients. The QIDS includes questions about primary, middle, and terminal phase insomnia, duration of sleep, mood, appetite, weight changes, concentration, self-perception, suicidal ideation, anhedonia, energy, restless, and motor retardation. A total score of 27 is possible. A score of 6–10 indicates mild depression, while 11–15 indicates moderate depression. A score of 16 or over indicates severe depression.

The Zung Self-Rating Depression Scale (ZungD) includes 20 items, half of which are positively worded and half are negatively worded (Zung, 1965). The patient scores each sentence from 1 ("a little of the time") to 4 ("most of the time"). The ZungD assesses mood, sleep, crying spells, appetite, sexual interest, constipation, tachycardia, fatigue, clarity of thought, sense of usefulness, and suicidality. Many patients dislike the forced choice nature of the ZungD, as they are not able to select "never" as an option. The ZungD score greater than 50 is considered consistent with depression. The ZungD is a well-established and well-normed assessment for depression.

The Zung Self-Rating Anxiety Scale (ZungA) also includes 20 items with a similar positive and negative wording (Zung, 1971). The patient scores each sentence from 1 ("a little of the time") to 4 ("most of the time"). The ZungA assesses anxiousness, fear, panic symptoms, headache, trembling, weakness, tachycardia, dizziness, numbness, urinary frequency, sleep, and nightmares. We find the ZungA useful because there is significant overlap between symptoms in chronic TBI and/or postconcussive syndrome and the items assessed in the ZungA

28.2.5.1 Patient Diary

The Patient Diary instrument was developed based on questions found in standardized questionnaires already in clinical practice in the United States and elsewhere. Instruments for psychological symptoms, pain, energy, and quality of life which were unrestricted in their availability were reviewed. Key questions were selected and rewritten to promote easier comprehension for the lay person and to require a personal written or verbal response. Key features of the Patient Diary are:

- a simplified format easily understood by the layperson, asking for a response on a Likert scale (graded plus or minus, 5 points each direction from no change to significant change).
- inquiring about general health status, diet, nutrition, sleep, smoking, vaping, drug use, etc. Internal vs external reliability challenged.
- Questions requiring a written response on exercise – cardiovascular, stressing, kinetic vs static.
- Simple, layperson language questions about depression, anxiety, suicidal ideation, frustration, relationships, and communications.
- Questions covering body aches, pains and other somatic complaints, energy, etc.
- Questions about Headache types, often directly related to trauma, as well as changes in hearing or vision. The Patient Diary, was completed prior to treatment and then on a weekly basis, throughout the course of treatment. An additional diary (Spousal or significant other) was also used on a weekly base, to provide for an external historian's observations on the patient's response to care. Questions were similar, but placed in the second person syntax.

28.2.6 Neuroimaging

Neuroimaging has long been utilized to evaluate TBI. Computed tomography (CT) scans are routinely used as an initial evaluation tool for TBI. Unfortunately, the sensitivity of CT is very low for mild TBI. CT scans can be useful for screening for intracranial hemorrhage or skull fracture, but offer little information on the brain parenchyma. The majority of CT scans in mild TBI are normal (Raji and Henderson, 2018; Chamard and Lichtenstein, 2018). For example, in a combined sample of over 4000 mild TBI cases, roughly 5%–10% had abnormal CT scans (Haydel et al., 2000). These positive cases were associated with headache, vomiting, increased age, alcohol or drug intoxication, anterograde amnesia, head or neck lacerations, or seizures (Borczuk, 1995; Miller et al., 1997; Haydel et al., 2000). There is extensive, high quality evidence to support the conclusion that CT is not predictive of functional recovery (Amyot et al., 2015). Similarly, anatomical magnetic resonance imaging (MRI) is most often negative in concussion/mild TBI (Abu-Judeh et al., 2000; Stamatakis et al., 2002; Shin et al., 2006). This reflects the fact that in concussion/mild TBI, the anatomical changes are subtle or undetectable.

Some valuable information may be obtained from a type of MRI that purportedly reveals the integrity of axonal pathways. Based on the movement of water molecules within the linearly constrained confines of the neuronal axon, diffusion tensor imaging (DTI) indirectly identifies areas of damaged white matter. In mild TBI, DTI can reveal widespread changes in water diffusion out of proportion to the injury (Asken et al., 2018). Difficulties in getting consistent protocols for this technically challenging technique have limited its applicability in the clinic (Douglas et al., 2015). DTI depends upon statistical analysis, as opposed to visual reading. Different statistical techniques yield different results, rendering DTI unsuitable for interpretation in the individual patient. Although research studies indeed find group differences, there is inadequate normative data to establish diagnostic or prognostic stratification in the individual patient (Amyot et al., 2015). In contrast, functional neuroimaging is highly consistent and is designed to specifically detect areas of abnormal brain function.

Functional neuroimaging refers to visualizing functional processes in the brain. These processes derive from the electrochemical activity of the neurons, synaptic transmission, and the metabolic processes that support these processes. Information flow along and between neurons is based on the movement of ions across the cell membrane and, to generate the energy required for pumping ions, neurons depend largely on glucose and oxygen supplied by the blood on a moment-by-moment basis. Neurons can control the localized flow of blood by molecular signaling mediated by the neurotransmitter glutamate via the surrounding astrocytes. Glutamate raises the calcium ion concentration in astrocytes, thereby generating arachidonic acid from membrane lipids which can either dilate or constrict blood arterioles to increase or decrease blood flow (Sestini, 2007). The details are not important here. What is important is that when blood enters the brain it doesn't course indiscriminately through the organ's vessels; instead it is selectively channeled to specific regions in a need-based fashion. As a result, perfusion to cortical regions and even areas as small as a single cortical column adjusts moment-to-moment to meet the constantly changing metabolic demands of the neurons (Cox et al., 1993). Certain areas of the brain become more active during a given task, requiring more energy, which is provided by an almost instantaneous change in local perfusion to provide the necessary glucose.

Functional neuroimaging techniques take advantage of this intimate association between neuronal activity, metabolic activity, and local perfusion. For example, fluoro-deoxyglucose positron emission tomography visualizes metabolism using an analogue of glucose which accumulates over a 20-minute period and provides an integration of activity over that time period. Perfusion single photon emission computed tomography (SPECT) provides a visual representation of local cerebral perfusion that is integrated over a short time frame of 2 minutes. Functional MRI visualizes local changes in venous blood flow due to the slight magnetic properties of deoxyhemoglobin.

Extensive literature now exists to support the widespread use of perfusion SPECT brain imaging as a tool for evaluating TBI, whether mild, moderate, or severe. As described below, SPECT is superior to CT and MRI in identifying and assessing the extent of TBI. Moreover, SPECT also has both positive and negative prognostic value. In contrast, functional MRI suffers from similar limitations to DTI. The analysis depends upon statistical analysis. The methods of acquisition and analysis vary widely creating disparate results. Currently, functional MRI can reveal group differences, but it is not useful in the clinical delineation of TBI or prognostic stratification in the individual patient (Amyot et al., 2015; Raji and Henderson, 2018).

Henderson and colleagues produced a systematic review of all published studies on SPECT in the evaluation of TBI (Raji et al., 2014). The review provided level IIA evidence (at least one randomized controlled trial) for the utility of brain SPECT in TBI. The review covered 52 cross-sectional studies and 19 longitudinal studies representing a total of 2634 individuals over 30 years of literature. The literature strongly supported the superiority of SPECT to CT scans for the assessment of TBI in all forms. For example, one study of 228 patients found that CT missed or underestimated brain lesions found on SPECT (Abu-Judeh et al., 2000). Similarly, SPECT is more sensitive for TBI in all forms than anatomical MRI. The same study above included MRI exams that were negative, although 24% of cases had a frontal lobe injury (Abu-Judeh et al., 2000). In a comparison study by Stamatakis and colleagues (2002), TBI patients who received both MRI and SPECT scans within 2 weeks of injury were studied. Based on statistical parametric analysis, SPECT scans detected more lesions and greater lesion volume than the anatomical MRI (Stamatakis et al., 2002).

In a key prospective study evaluating the sensitivity and specificity of SPECT for mild TBI, Jacobs and colleagues (1994, 1996) evaluated 167 patients with head trauma caused by MVAs using serial SPECT scans over 12 months. Jacobs and colleagues carefully excluded any patients with past head trauma, past positive neuroimaging findings, epilepsy or other neurological conditions, psychiatric conditions, drug abuse or alcohol abuse. All of the patients had negative CT scans or MRI scans at the time of the injury. Initial SPECT scans were performed within 1 week of injury. If a patient had a positive initial SPECT scan, then they had repeat SPECT scans at 6 and 12 months. All patients underwent a complete neurological examination, memory testing, concentration testing, and thorough questioning concerning post-concussive symptoms based on a standardized protocol. This same clinical assessment was repeated at 3, 6, and 12 months. The patients who had negative SPECT scans had repeat neuropsychological testing at 12 months. While they did not perform statistical parametric analysis, Jacobs and colleagues had three separate physicians read the SPECT scans and score them according to the presence, absence, and severity of TBI. A positive initial SPECT scan had a sensitivity of 100% and a specificity of 85% for predicting persistent neuropsychological deficits consistent with mild TBI at 12 months (Jacobs et al., 1994, 1996). Patients who had a normal SPECT scan within 1 week of a head injury had a 92% chance of being symptom-free at 3 months and a 100% chance of complete neuropsychological recovery at 12 months (100% negative predictive value). These findings led the European Association of Nuclear Medicine to render their opinion that SPECT has prognostic value in TBI (Tatsch et al., 2002).

One important implication of this work is that a negative SPECT scan shortly after initial injury predicts there will be no long-term functional deficits. This cannot be said of other imaging modalities such as conventional CT or MRI. Similarly, Laatsch and colleagues ()(1997) found an abnormal baseline SPECT correlated strongly with abnormal neuropsychological testing (Laatsch et al., 1997). The systematic review by Henderson and colleagues included 18 cross-sectional studies (81%) showing correlation between abnormal SPECT findings and neuropsychological deficits (Raji et al., 2014).

A more recent pair of studies by Henderson and his colleagues highlighted the usefulness of SPECT in TBI when areas of the default mode network are examined. One study of over 20,000 subjects showed that SPECT can distinguish TBI from PTSD with 80%–100% sensitivity and an average of 70% specificity (Amen et al., 2015). The study examined two groups—one group was a small cohort of patients with TBI or PTSD and closely matched controls, and the other was a large ($N = 7505$) comorbid cohort with TBI versus a large ($N = 1077$) comorbid cohort with PTSD with considerable diversity compared to a large ($N = 11{,}147$) cohort of patients with other diagnoses (Amen et al., 2015). This study was replicated in a smaller cohort of 196 veterans with TBI, PTSD, or both (Raji et al., 2015). The accuracy of SPECT in diagnosis of TBI and its differentiation from PTSD or normal controls was between 83% and 94%.

In other words, even in cases with significant comorbidity, assessment of the areas of the default mode network by SPECT allowed accurate identification of TBI.

SPECT also has been used to assess treatment outcomes in TBI. Laatsch and colleagues (1999) showed a correlation between positive behavioral changes with cognitive behavioral therapy and increases in cerebral blood flow on serial SPECT. Two studies employed serial SPECT scans to demonstrate increased cerebral blood flow in response to hyperbaric oxygen therapy. The first study (Harch et al., 2012) examined 13 patients who were exposed to blast injury 1–5 years before treatment, while the second study (Boussi-Gross et al., 2013) examined 56 patients at 1–5 years after mild TBI. Serial SPECT scans were used to show that a multifactorial lifestyle program in a cohort of retired NFL players led to increased cerebral blood flow in the frontal lobes (Amen et al., 2011). We have also shown that perfusion changes seen by SPECT correlated with clinical response to transcranial multiwatt NIR phototherapy (Henderson and Morries, 2015b).

In summary, perfusion SPECT brain imaging provides a useful and widely available tool for assessing TBI—both in terms of severity and location of areas of involvement. The latter aspect becomes extremely important in the context of transcranial NIR treatment, given that a laser is a focused and directional mode of treatment. In addition, SPECT provides a valuable method of assessing changes in neurophysiology in response to treatment.

28.3 Treatment of traumatic brain injury with near-infrared light therapy

28.3.1 Overview

Extensive literature now indicates that deploying NIR light of wavelengths between 600 and 1200 nm is critical to the effectiveness of infrared phototherapy (Karu and Kolyakov, 2005; Lapchak, 2010). These wavelengths are absorbed by cytochrome-c-oxidase in the mitochondrial respiratory chain, which leads to an increase in adenosine triphosphate (ATP) production (Karu and Kolyakov, 2005; Henderson and Morries, 2015a). Simply increasing ATP in wounded or underperfused cells may be sufficient to stimulate cells in areas of injury or metabolic derangement (Wu et al., 2013); however, tissue culture and animal studies implicate secondary molecular and cellular events (Chung et al., 2012; Henderson and Morries, 2015a; Salehpour et al., 2018). NIR light at these wavelengths also appears to alter nitric oxide levels (Chung et al., 2012; Huang et al., 2009; Morries et al., 2015; Henderson and Morries, 2015a), which may have beneficial downstream effects on vasodilation and calcium ion homeostasis. NIR light likewise appears to modulate reactive oxygen and reactive nitrogen species and activate early response genes (Chung et al., 2012; Huang et al., 2009; Morries et al., 2015; Henderson and Morries, 2015a). NIR light also has been shown to activate nuclear factor kappa B, which is a redox-sensitive transcription factor (Chen et al., 2011). This pro-survival transcription factor modulates the expression of numerous genes, including ones involved in inflammation, early response, and cell survival.

NIR phototherapy also has been shown to activate numerous genes in both the mitochondria and the cell nucleus, with an increased transcription of over 100 genes (Chung et al., 2012; Huang et al., 2009; Morries et al., 2015; Henderson and Morries, 2015a). In particular, cell-survival genes and neural differentiation factors are transcribed (Kushibiki et al., 2013). Increased synaptogenesis has been observed in animal models (Chung et al., 2012; Henderson and Morries, 2015a; Salehpour et al., 2018). NIR light also increases the production of numerous growth factors (Henderson, 2016) and upregulates several inflammatory mediators in animal models (Huang et al., 2009; Chung et al., 2012). Examples of key growth factors stimulated by NIR light include nerve growth factor, brain-derived neurotrophic factor (BDNF), transforming growth factor-beta, and vascular endothelial growth factor, which may contribute to late brain remodeling after TBI (Schwartz et al., 2002; Mirsky et al., 2002; Lubart et al., 2005; Frank et al., 2004; Szymanska et al., 2013; von Leden et al., 2013). For example, a fivefold increase in nerve growth factor mRNA transcription occurred after irradiation of skeletal muscle cell culture with 633 nm NIR light (Schwartz et al., 2002).

Data suggest that transcranial NILT can increase the process of neurogenesis in adult mice with stroke or TBI. For example, increased numbers of neuroprogenitor cells have been demonstrated in both the dentate gyrus of the hippocampus and in the subventricular zone of the lateral ventricle of mouse models after treatment with 810 nm transcranial NIR light at 18 J/cm^2 (Oron et al., 2006, 2007; Xuan et al., 2014a,b; Henderson, 2016). Other studies provide evidence that transcranial NILT may increase the process of synaptogenesis (Xuan et al., 2014a,b). Together, these processes may aid in the neuroplasticity responsible for neural repair and improved function in cases of chronic TBI (Chung et al., 2012; Henderson and Morries, 2015a; Salehpour et al., 2018).

Additional clues to the molecular mechanisms underlying the beneficial effects of NILT have been ascertained. Using a null mutation mouse model for the immediate early response gene IEX-1, Zhang and colleagues (2014) investigated the role of this pivotal protein. IEX-1 upregulates nuclear factor kappa-B (NF-kappaB), exerts antiapoptotic effects, exerts antiinflammatory effects, and regulates mitochondrial ATPase activity. Knock-out mice lacking IEX-1 demonstrate an exaggerated response to even mild brain injury with prolonged inflammation, extensive apoptosis, and larger resulting brain lesions. However, a single treatment with 810 nm NIR light from a diode laser at 150 mW/cm^2 for 4 minutes (36 J/cm^2) applied to the scalp resulted in amelioration of all of the excessive pathogenesis in these IEX-1 knock-out mice. Mice treated with NIR light had lesions that were indistinguishable for wild-type mice given the same brain injury. The authors believe this reflects the impact of NIR light on inflammatory processes, rather than some of the downstream effects of NF-kappaB. They reach this conclusion in part because wild-type animals given the same brain injury and injected with lipopolysaccharide (to invoke an inflammatory response) also had an exaggerated degree of pathological changes (Zhang et al., 2014). This elegant study uncovers a key player in the actions of NIR phototherapy; however, the influence of NIR upon other pathways downstream from IEX-1 cannot be ruled out.

Notably, these molecular and cellular changes appear to persist for considerably longer than the duration of NIR light application itself, when delivered at appropriate wavelengths and amplitudes (Yip et al., 2011; Lapchak, 2012; Wu et al., 2013). For example, low-level light therapy (LLLT) in the red and NIR range with a power density of 0.9–36 J/cm^2 applied in a single treatment at 24 hours poststroke in animal models yielded a reduction in neurological deficits, as well as histochemical evidence of neuron proliferation and migration which were evident weeks later (Yip et al., 2011; Wu et al., 2013). A single application of LLLT with light-emitting diodes (LEDs) in rodent models of TBI had similar prolonged benefits (Ando et al., 2011; Wu et al., 2012; Xuan et al., 2013). In many studies, a delay of 1–4 weeks was noted, consistent with a progressive neurological regeneration cascade set in motion by the NIR light exposure (Chung et al., 2012; Henderson and Morries, 2015a; Morries et al., 2015).

28.3.2 Review of the literature

28.3.2.1 Preclinical studies

Oron and colleagues (2007) conducted the first mouse studies of NILT for TBI. They employed a single application of 808 nm NIR energy from a 200 mW Ga-As laser emitter 4 hours postinjury. They observed little change in neurological impairment of the treated mice compared to controls on postinjury days 1–4; however, on day 5, the treated animals demonstrated progressively less neurological impairment. While the neurological function of treated animals at day 5 was not compared to day 28 statistically, there appeared to be a 17% decrease in neurological impairment over this time period. At day 28, the mice were sacrificed and brain lesion volume was determined. They found that a single treatment with 808 nm NIR energy resulted in a marked reduction in lesion size (12% volume to 1.4%) by 28 days (Oron et al., 2007).

Wu and colleagues (2010) examined a similar mouse model of closed-head injury by weight drop. As a group, these animals had somewhat more severe neurological impairment following injury [neurological severity scale (NSS) = 6–8 vs 4–6]. Mice received either a single dose of 670 nm NIR, a single dose of 810 nm NIR, a single dose of 980 nm NIR, or sham. The authors used fiber optic laser emitters. They report delivering an energy density of 36 J/cm^2 using a power density of 150 mW/cm^2 for a duration of 4 minutes. In animals treated with 670 nm laser, NSS scores were significantly less after 9 days. By 28 days, the treated mice showed an 89% improvement in NSS scores while sham-treated mice showed a 72% improvement. Mice treated with 810 nm NIR light showed a more robust behavioral response. By day 1, the NSS scores of 810-nm treated mice were statistically significantly different from sham-treated mice. At 9 days, a 68% drop in NSS scores was observed in 810-nm treated mice (vs 48% in sham cases). By 28 days, the 810-nm treated mice had NSS scores that were 95% reduced from acute postinjury scores. Sham controls showed a 72% improvement. Mice treated with 980 nm NIR, showed no significant difference from sham-treated animals throughout the 28 days postinjury (Wu et al., 2010). Despite the more profound initial NSS scores of the animals in this study, the size of the histological lesion at baseline was considerably smaller compared to that reported by Oron and colleagues (2007). Nonetheless, a 50% reduction in lesion volume was reported (Wu et al., 2010).

It is at this point that confusion entered the field. McCarthy and colleagues (2010) used an 808 nm gallium aluminum arsenide (GaAlA) diode laser and delivered NIR at 70 mW yielding a power density of 268 J/cm^2 to rat brain as an assessment of the safety of the device for potential human use. The authors purported from unpublished data that they used this device to deliver 10 mW/cm^2 or 1.2 J/cm^2 to the human cortical surface. This represents less than half of 1% of the light energy delivered to the scalp reaching the subjacent cortical surface. Curiously, a later study by Oron and colleagues (2012), using the same device from the same company, precisely described measurement of transmission of NIR light energy through the parietal bone of fresh mouse skull. The GaAlA 808 mn diode laser was used to deliver 21 mW to the skull surface for 2 minutes. and yielded 10 mW and 1.2 J/cm^2 on the subjacent internal surface of the mouse skull. Yet, the human clinical trials utilized the same settings and 2 minutes treatment interval and purported to deliver the same 1.2 J/cm^2 to the cortical surface of human brain (Lapchak, 2012) despite the much greater thickness of human scalp and skull compared to the thickness of mouse skull alone (Oron et al., 2012). These clinical trials, known as the Neuro-Thera effectiveness and safety trials (NEST1-3), eventually were terminated due to lack of efficacy.

Wu and colleagues (2012) examined additional wavelengths for clinical benefit in a replication of their earlier work. Using a closed-head injury mouse model, they created a moderate TBI (NSS = 6–8). They tested single treatments of 150 mW/cm^2 with 665, 730, 810, and 980 nm laser diodes. Changes in NSS were followed over 28 days. Mice treated with 730 and 980 nm NIR light showed no significant behavioral differences compared to sham-treated animals. In contrast, mice treated with 665 nm NIR energy showed a rapid reduction in neurological symptoms. By day 5, differences were apparent. Similarly, mice treated with 810 nm NIR light showed rapid and robust changes in neurological impairment. Notably, there was no discernable difference in neurological status of animals treated with 665 and 810 nm NIR light (Wu et al., 2012).

Ando and colleagues (2011) assessed the effect of pulsing with 810 nm NIR light on a mouse model of TBI. They found that a single application of continuous wave 810 nm NIR light applied within 4 hours of injury resulted in significant functional recovery based on NSS scores. This replicated the results of Wu and colleagues (2010) and Oron and colleagues (2007). In addition, they found that 810 nm light pulsed at 100 Hz had a similar benefit as continuous NIR light. They also found that light pulsed at 10 Hz yielded additional functional recovery (Ando et al., 2011). Interestingly, they examined the brains of sham controls and treated mice histopathologically and found different results from the work of Oron and colleagues (2007). They also examined earlier timepoints (2, 15, and 28 days). While Oron and colleagues had found the lesion volume was 1.2% of total cortical volume at 28 days, Ando and colleagues found lesion volume to be 16% at 28 days after NIR treatment. In other words, one group found a 90% reduction in lesion volume while the other group found only a 15% reduction in lesion volume. Pulsing appeared to be beneficial as the group treated with 10 Hz pulsed NIR light had a lesion only 13% of the cortical volume which represents a 43% reduction in lesion volume.

Oron and colleagues (2012) examined the effect of different pulsing rates using 808 nm NIR light on a mouse model of TBI. Using a GaAlA laser diode, NIR energy was delivered at 10 mW/cm^2 in continuous mode and at 100 or 600 Hz. These experimental groups were compared to a sham-treated group. The NSS scores after head injury were 4–6. Regardless of the pulse frequency (0, 100, 600 Hz), laser-treated animals showed behavioral improvement. However, those treated at 600 Hz showed delayed neurobehavioral response. While 26% of the mice treated with continuous laser showed complete recovery, 67% of those treated with 100 Hz laser showed complete recovery. Those treated with 600 Hz were similar to those treated with continuous wave (28% response rate) (Oron et al., 2012).

Repeated NIR treatments appear to have some benefit, but the frequency and number of treatments are critical factors. While a single NIR application had benefit, daily applications for 3 days yielded much greater neurological benefit (Xuan et al., 2013) with smaller lesion size, fewer degenerating neurons, more proliferating cells, increased synaptogenesis (Xuan et al., 2014a,b) and greater levels of BDNF (Xuan et al., 2014a,b) compared to a single treatment in a mouse model. Similarly, one or three daily treatments yielded increased neurogenesis in the hippocampus and subventricular zone, as well as improved performance on learning tasks (Xuan et al, 2014a,b). In contrast, daily treatment for 7 days (Quirk et al., 2012) or 14 days (Xuan et al., 2013, 2016) showed no difference from controls. NIR energy densities in the range of 0.9–36 J/cm^2 resulted in significant biochemical and behavioral changes (Oron et al., 2007, 2012; Wu et al., 2010; Ando et al., 2011; Xuan et al, 2013, 2014a,b; Khuman et al., 2012).

Khuman and colleagues (2012) assessed different NIR energy doses in a transcranial and novel open-craniotomy approach in a mouse direct cortical impact model. Groups of animals received NIR energy via direct illumination of the exposed cortex – a 2 minutes treatment with either 250, 500, 1000 mW/cm^2 or a 7 minutes treatment with either 250 or 500 mW/cm^2 – utilizing a 800 nm diode laser. Other groups received NIR energy transcranially at 60 J/cm^2 (500 mW/cm^2 × 2 minutes), but received multiple daily treatments for 7 days, or one treatment 1 hour or 4 hours after injury. From this complicated regimen, a small number of observations can be noted. Animals that received 60 J/cm^2 NIR energy, whether transcranially in a single treatment or by direct illumination of the cortex (500 mW/cm^2 × 2 minutes) showed the most robust improvements in behavioral and learning measures. Animals treated with daily NIR energy for 7 days, had considerably less benefit compared to single treatment mice.

The outcome of the multiple-treatment group in this study warrants further consideration, particularly in light of the collected works of Xuan and colleagues on the matter. These data suggest that too much NIR energy may lead to a loss of benefit. The question is how much of *what* is too much? If we accept that NIR acts on the mitochondria, increasing ATP production and reducing reactive oxygen species, then it can be argued that too much NIR energy leads to oxidative stress. There is evidence that exists proving multiple treatments in a short period of time can magnify the production of free radicals. Indeed, in a cell culture model, free radical production was almost 300% greater following a second irradiation with 808 nm NIR light (Kujawa et al., 2014). However, ample evidence suggests that NIR also activates early response genes leading to increased production of BDNF and antiapoptotic genes. In this pathway, the timing of NIR stimulation may be important. The pathway of BDNF stimulation (Henderson, 2016) is multistepped and can be potentially saturated or inhibited at any one of these steps by multiple treatments. Fluidity of cell membranes is affected by NIR irradiation at 808 and 905 nm. Repeated exposures lead to decrease fluidity and impaired membrane ion pumps (Kujawa et al., 2004). An alternate hypothesis suggested by Xuan and colleagues (2016) is that excessive number of treatments induces reactive gliosis, which temporarily inhibits brain repair processes. Thus, it may be that the frequency of treatments over a given time period may be a limiting factor to benefit from NIR irradiation, which means that the *frequency* of dosing is the element that is too much versus the *actual dose* given in a single treatment. Further work is needed to clarify this question in the models of TBI, but our clinical experience supports the benefit of spacing treatments at appropriate intervals.

Esenaliev and colleagues (2018) recently reported on a treatment that utilizes nano-pulsed 808 nm NIR light to generate optoacoustic (ultrasound) waves in a rat closed-head injury model. The proprietary device is incompletely described but appears to utilize 10 nanosecond pulse duration (100 MHz), at 1 J/cm^2 and was utilized to administer nano-pulsed NIR light for 5 minutes in a single treatment at 1 hour after injury. Animals treated in this acute model showed significantly better motor performance on a balance beam within 1 day. These differences became less distinct by day 10. Numbers of caspase-3 positive (apoptotic) cells were significantly reduced in nano-pulsed NIR-treated rats. Levels of BDNF were elevated in treated animals, as was evidence of neuroproliferation (Esenaliev et al., 2018).

Quirk and colleagues (2012) examined a different closed-head injury model in rats and also assessed the impact of multiple treatments. Using 670 nm LEDs at 50 mW/cm^2, the authors administered two treatments lasting 5 minutes per day for 3 days. Animals were studied over the subsequent 10 days. Notably, the authors observed fairly minimal behavioral differences between sham-treated rats and NIR-treated rats.

28.3.2.2 Clinical studies

In a double-blind study of healthy volunteers, NILT improved on memory and attention (Barrett and Gonzalez-Lima, 2013). With only one application of NIR to the right forehead, targeting the right frontal pole of the cerebral cortex (Brodmann's area 9 and 10), the authors reported an improvement on psychomotor vigilance and visual memory when tested 2 weeks after the treatment. The device was a class IV laser CG-5000 (Cell Gen Therapeutics, Dallas TX, USA) and the parameters were as follows: wavelength 1064 nm, irradiance 250 mW/cm^2, fluence 60 J/cm^2, with 4 minutes exposure per site (2 sites).

In a case report on two patients with TBI, Naeser and colleagues (2011) reported NILT improved sustained attention, memory, and executive functions. Both patients were treated with an instrument with three separate LED cluster heads. The parameters used for the treatment were the following: NIR wavelength 870 and 633 nm (red light), irradiance 22.2–25.8 mW/cm^2, fluence 13.3 J/cm^2, with 10 minutes exposure per site. LED cluster heads were positioned sequentially over the bilateral forehead, temples, parietal region, high frontal region, and the vertex. An LED cluster was also placed on the foot. The first patient was 7 years post-TBI and had significant postconcussive symptoms. The patient received weekly treatments of 8–20 J/cm^2 over 7 months and then switched to daily treatments at home. The patient continued home treatments for over 6 years. Notably, the patient only had transient benefit from this protocol. If the patient stopped treatment, then symptoms returned within 2 weeks (Naeser et al., 2011). The second patient received daily treatments with a similar device delivering 9.3–13.3 J/cm^2. Within 4 months, most of the patient's symptoms had improved (decreased memory, poor sleep, emotional dysregulation, irritability), and the patient returned to work. This patient also noted that symptoms returned if treatments were stopped for more than 1 week.

The authors (Naeser et al., 2011) recognized that the LED device only delivered a tiny fraction of NIR energy to the cortical surface—they estimated 0.24 J/cm^2 (3%)—and a much smaller amount to the depth of 2 cm—they estimated 0.024 J/cm^2 (0.3%). Also, the authors had no neuroimaging to localize the lesion to which they were trying to direct NIR energy. It cannot be assumed that the traumatic lesion to the brain, which would be seen only with functional imaging (the MRI of patient 1 was negative; generalized slight cortical atrophy was reported in patient 2), was located subjacent to the LED cluster head. Nor can it be assumed that any NIR directly reached the site of traumatic lesion. Moreover, the application of NIR energy to the foot has no direct effect on the brain. The authors suggest a corollary to acupuncture points and further hypothesize that the NIR energy is having an effect on blood flow in the frontal lobe. Given that NIR light does not penetrate significantly through 2 mm of hair (Henderson and Morries, 2019), it is unclear if any energy was delivered to the vertex or the parietal regions in these patients who had not apparently shaved their heads.

The same group reported on a cohort of eleven subjects with TBI who had persistent cognitive dysfunction and were treated with a similar NILT protocol (Naeser et al., 2014). The eleven subjects received NILT with a three LED cluster heads device (MedX Health Model 1100, Toronto, ON, Canada). The parameters used for the treatment were the following: NIR wavelength 870 and 633 nm (red light), irradiance 22.2 mW/cm^2, fluence 13 J/cm^2, approximate time 10 minutes per site. The NILT light was applied three times per week for 6 weeks (18 sessions), on 11 sites for 10 minutes per site (the total duration of each session was 20 minutes). The sites on the skull were chosen on the midline, and bilaterally on frontal, parietal, and temporal areas (Naeser et al., 2014). At the follow-up neuropsychological testing, NILT had a significant effect on attention, inhibition, and inhibition switching in the Stroop task; similarly improved verbal learning and memory, as well as enhanced long-delay free recall on the California Verbal Learning Test (Naeser et al., 2014). The same concerns outlined in the previous paragraph stand for this study. It is unclear if LED emitters can deliver sufficient energy at a depth of 1–2 cm to induce any neurological changes. Moreover, the LEDs were positioned over the subjects' hair, which would have drastically reduced NIR light penetration.

Eight subjects, from the same cohort (Naeser et al., 2014), were identified as having mild, moderate, or severe depression based on the Beck depression inventory–II total score (range: 15–34). The three cases, who entered the study with only mild depression, remained the same after NILT treatment. Results for the five cases with moderate-severe depression were as follows: two moderate cases improved to mild/minimal depression 8 weeks after the end of the NILT series, and one severe case improved to moderate depression. Two moderate or severe depression cases remained the same after 8 weeks of follow-up from the last NILT session (Naeser et al., 2014).

Cassano and colleagues (2015) described an open-label trial of four patients treated for depression with an LED-based device. The device was a GaAlA laser at 808 nm at 5 W applied for 2 minutes per site to multiple sites on the forehead. Hamilton Depression Scale scores decreased from 19.8 + 4.4 (severe depression) to 13 + 5.35 (mild to moderate depression). Herein, the device is a laser emitter applied to scalp at the forehead without intervening hair. Cassano and colleagues (2018) recently published a double-blind, sham-controlled extension of these findings. In a sample of 13 completers, Hamilton depression scale scores decreased ($p = 0.031$).

28.3.2.3 *Our Clinical Experience*

A device with considerably greater power was used in an open-label study we published in 2015 (Morries and colleagues, 2015). Prompted by the transient nature of the symptom relief described in studies utilizing LEDs to treat TBI, the observations by Lapchak (2010), and laboratory studies exploring the penetration of NIR light through tissue (Henderson and Morries, 2015a), we utilized a laser with 10 W, and later 15 W, to treat patients with TBI. We currently utilize a laser diode device with a maximal output of 10 W in 810 nm and 10 W in 980 nm. The device can deliver dual wavelength treatment utilizing a fiber optic cable and a handheld device. One or both wavelengths can be delivered in a pulsed fashion. The aperture of the device is 3.5 cm. At a typical setting of 10 W of 810 nm light and 3 W of 980 nm light, the device delivers 3.71 W/cm^2. The parameters for treatment were as follow: wavelength 810 and 980 nm, irradiance 3.71 W/cm^2, fluence at skin 55–125 J/cm^2 with 5 minutes exposure per area measuring 87.5–125 cm^2. Given our previous work in NIR energy penetration into tissue (Henderson and Morries, 2015a), we estimate that the fluence delivered to tissue at 3 cm depth is 1.65–3.7 J/cm^2. Note that animal models of both stroke and TBI indicate NIR energy densities in the range of 0.9–36 J/cm^2 yielded significant biochemical and behavioral changes (Morries et al., 2015).

Patient treatment protocols are established and are based on history, location of lesions derived from perfusion SPECT scans, patient size, weight, head circumference, bone structure, presence of hair, and, most importantly, condition or conditions to be treated. Patients are started at a low dose using 810 nm wavelength only, for the first two visits with gradually increasing dose to the target dose. Thereafter, the 810 nm settings remain steady, while the 980 nm wavelength is gradually increased and monitored by the doctor and therapist. The maximum dosage level is that which is therapeutic and comfortable for the patient. Treatments generally are 20 minutes in duration. Treatments are initially administered three times per week, but this decrease to two times per week after the first six treatments.

The therapy report is an integral part of the treatment process to accurately record and catalog exact dosage levels for each visit, location and size of areas treated (cm^2), time, total joules, wavelength settings, and continuous or pulsed wave settings (Hz or ms on/off). Review of medications and contraindications occurs prior to each treatment. Patient response to previous treatment is reviewed prior to each treatment. Patient diaries are reviewed weekly.

The total number of treatments in a course is generally based on patient symptoms, but is approximately 20 treatments. At that time, we consider whether a rest period is appropriate, which is typically for a period of 2–3 months before any additional therapy is implemented. During the rest periods, exercise and balance training programs that supplement multiwatt NILT therapy are continued.

Our initial open-label trial involved 10 subjects with mild TBI, several of whom had experienced loss of consciousness at the time of the injury. After a course of 10 treatments of NILT (20 treatments in four patients), each patient experienced significant clinical improvement with reduction or resolution of many of their symptoms. Six of the patients had frequent headaches, which resolved in three patients and were less frequent and less severe in the remaining three patients. All 10 patients experienced sleep disturbance, which resolved in all cases. Nine of the patients had irritability, which resolved in all cases. Five patients had cognitive problems which resolved in all but one case. Short-term memory was impaired in four cases and this symptom resolved with multiwatt NILT (Morries et al., 2015).

Anxiety and depression were quite common in these patients with mild TBI. All 10 patients endorsed anxiety and two endorsed hyperarousal. These symptoms resolved in all, but one patient. Nine patients endorsed depression. Seven patients had completed Beck depression scales prior to treatment and had mean scores of 25.3 ± 12.1. After treatment, the Beck depression scale scores dropped to 12 ± 6.5 (nondepressed range). This represented a significant decrease. While 50% of the patients endorsed suicidal ideation prior to treatment, none of the patients continued to have suicidal thoughts after treatment. The patients reported improved cognitive ability and a desire to return to meaningful work. Five of the six unemployed patients returned to work. The two patients who were Iraq/Afghanistan veterans found new careers in highly skilled trades (Morries et al., 2015).

We recently reported an open-label clinical trial showing multiwatt NILT effective for depression symptoms (Henderson and Morries, 2017). Out of a group of 39 patients, 36 experienced a treatment response. Overall, 92% responded to the multiwatt NILT and 82% remitted from their depressive symptoms ($p = 6.45 \times 10^{-13}$). The time to response and time to remission were notable. Patients saw benefit often within four treatments (less than two weeks). It was demonstrated that resolution of depressive symptoms could be achieved with only eight treatments (less than three weeks). The time to response (with exceptions already noted) was often more rapid than that found for typical antidepressants, usually 6–8 weeks. An 82% treatment remission rate is superior to that described for oral antidepressants (Thase et al., 2001; Connolly and Thase, 2011) and to that described by Schiffer and colleagues (60%) in 10 patients treated with low-power NIR light (Schiffer et al., 2009); and Cassano and colleagues (50%) in 4 patients (Cassano et al., 2015) using somewhat higher power (5 W) laser diodes for short intervals. In addition, our posttreatment interviews for up to 55 months posttreatment show a large proportion of the patients remained free of depression symptoms. Therefore, these benefits do not appear to be transient, as was observed in studies of low-power NIR phototherapy.

Since the first paper describing our clinical protocol was published (Morries et al., 2015), we have treated over 150 patients. With few exceptions, the patients have seen substantial clinical improvement. Remarkably, the improvement of symptoms is both rapid and persistent, in contrast to results obtained with LED devices. Patients see improvement in sleep, headache, hyperarousal, irritability, cognitive difficulties, impaired concentration, impaired memory, anxiety, PTSD symptoms, depression, and suicidal ideation (Fig. 28.2). Improved sleep, with resolution of primary and middle

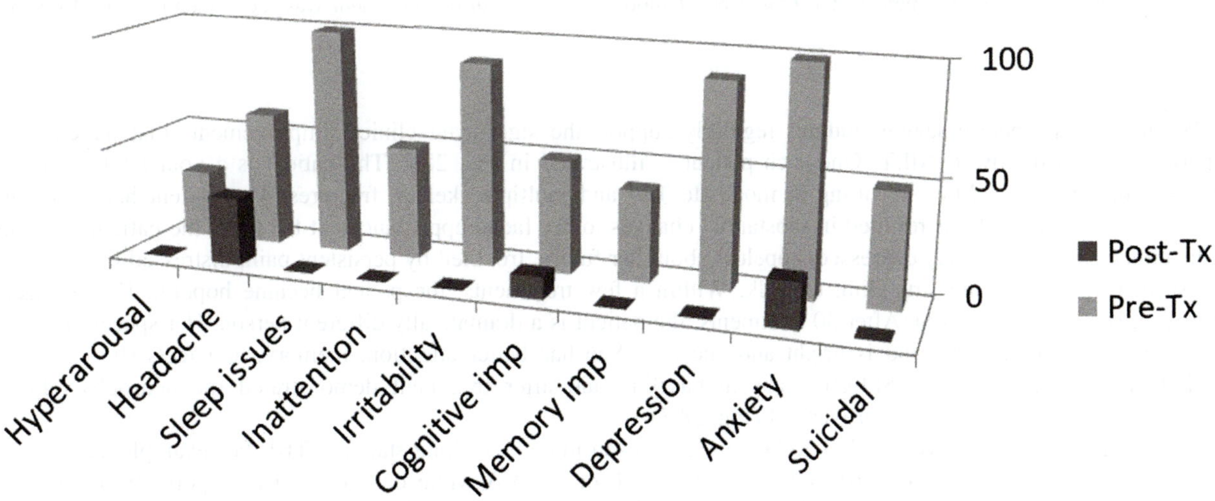

FIGURE 28.2 Clinical improvement of specific symptoms following a course of multiwatt NILT.

insomnia, is reported by a majority of patients within the first 10 treatments. Headaches show a similar rapid response to multiwatt NILT, often prior to the 10th treatment. As we found in our study of depression (Henderson and Morries, 2017), symptoms such as irritability, low mood, suicidal ideation, and hopelessness respond robustly and rapidly to multiwatt NILT.

Physiological signs such as impaired balance, visual disturbance, and RT delays also improve over the course of 20 multiwatt NILT treatments. Similarly, photophobia and hyperacusis show improvement with this treatment protocol. Patients note decreased sound and light sensitivity within even a few treatments.

Symptoms associated with PTSD, such as intrusive thoughts, exaggerated startle reflex, recurring memories, and nightmares in those patients who have been so affected have shown a vigorous response to multiwatt NILT. Some of our patients had recent traumatic injuries with PTSD symptoms, while others had been struggling with symptoms for years. For example, a veteran patient had been in therapy at the Veterans Administration facility for over 20 years related to injuries and TBI received in the line of duty over 50 years ago. Psychotherapy, group therapy, antidepressant medication, and eye movement desensitization and reintegration had not been successful. After the 10th multiwatt NILT treatment, this veteran reported he no longer had intrusive thoughts, exaggerated startle, or nightmares. He felt his PTSD was resolved. Notably, he has remained symptom-free for the two intervening years since his 10th treatment.

Clinical vignette

The patient was an active manager of a nonprofit organization involved in organizing many events and managing a large team of employees. The patient was involved in a MVA on April 8, 2017 and her head was thrown against the head restraint. She did not lose consciousness, but was immediately confused, dizzy, and complained of a headache. Within a few hours, she was in intense pain and vomiting. She was taken to the emergency room and underwent a CT scan. The scan was read as negative. By the following morning, the patient developed photosensitivity, tinnitus, increased headache, and neck pain. Her vision became "fuzzy" and she noted confusion and memory issues. She soon developed dizziness, numbness in right arm and hand, and confusion. Within days, her sleep deteriorated with primary and middle insomnia and nightmares. The patent became unable to perform her job, as she could not think clearly or concentrate. Her ability to compose writing was drastically impaired. She was incapacitated by headaches, fatigue, and photosensitivity.

On May 5, 2017, we examined the patient. A MoCA was performed and the patient scored 23 out of 30, indicating cognitive impairment. Her Rivermead score was 25. Her QIDS score was 9, indicating mild depression. A brain perfusion SPECT scan was ordered and it revealed extensive hypoperfusion in the orbitofrontal cortices and the bilateral temporal lobes. Shortly thereafter, the patient began treatments of the bilateral frontal and temporal regions with multiwatt NILT. Following the first treatment which encompassed only her frontal regions, the patient noted reduction of her intense headache pain. Following her fourth treatment, the patient noted improvement in her sleep with resolution of nightmares and middle insomnia. Primary insomnia remained a problem until after her 12th treatment. Headaches resolved completely after her seventh treatment. Irritability, low mood, anxiety, and hopefulness improved over the course of the first 10 treatments. By the 12th treatment, the patient reported feeling a much more positive life attitude. Symptoms of pain had resolved. Her sound and light sensitivity had resolved. She noted improved memory and concentration. After her 15th treatment, her MoCA score was 28, her Rivermead score was 11, and her QIDS score was 2. She reported feeling "joyful" and energetic. She was able to return to her work, which she greatly enjoyed. She completed her 20th treatment on August 18, 2017 and has been symptom-free for over 18 months. A repeat SPECT scan showed modest improvement in the frontal and temporal cortices. Further treatment was recommended based on the scan results.

We have found neuroimaging findings regularly support the significant clinical improvements that patients have experienced with multiwatt NILT. One such patient is illustrated in Fig. 28.2. The patient is a young female patient who was involved in a MVA resulting in moderate TBI and multiple skeletal fractures. The patient had substantial facial bone fractures and this resulted in substantial changes to her facial appearance. At the time, the patient first came to the clinic, she was severely depressed, hopeless about her future, troubled by persistent pain, distressed by her persistent speech dysarthria, and struggling to walk. Within a few treatments, the patient became hopeful. Her headaches improved. Her mood improved. After 30 treatments, the patient is a dramatically different person. Her speech is considerably less dysarthric. Her mood is bright and cheerful. She has better attention, memory, and feels she can "think much better." Perfusion brain SPECT scans at baseline and after treatment demonstrated the neurophysiological changes that reflect her clinical improvement (Fig. 28.3).

We also have found multiwatt NILT to be beneficial even in cases of long-standing TBI. For example, another veteran patient received a moderate TBI in 1983 as the result of a MVA while on active duty. Reportedly, he was in a coma for 2 months. Initially after the TBI, the patient had marked left hemiparesis and nonfluent speech. He made

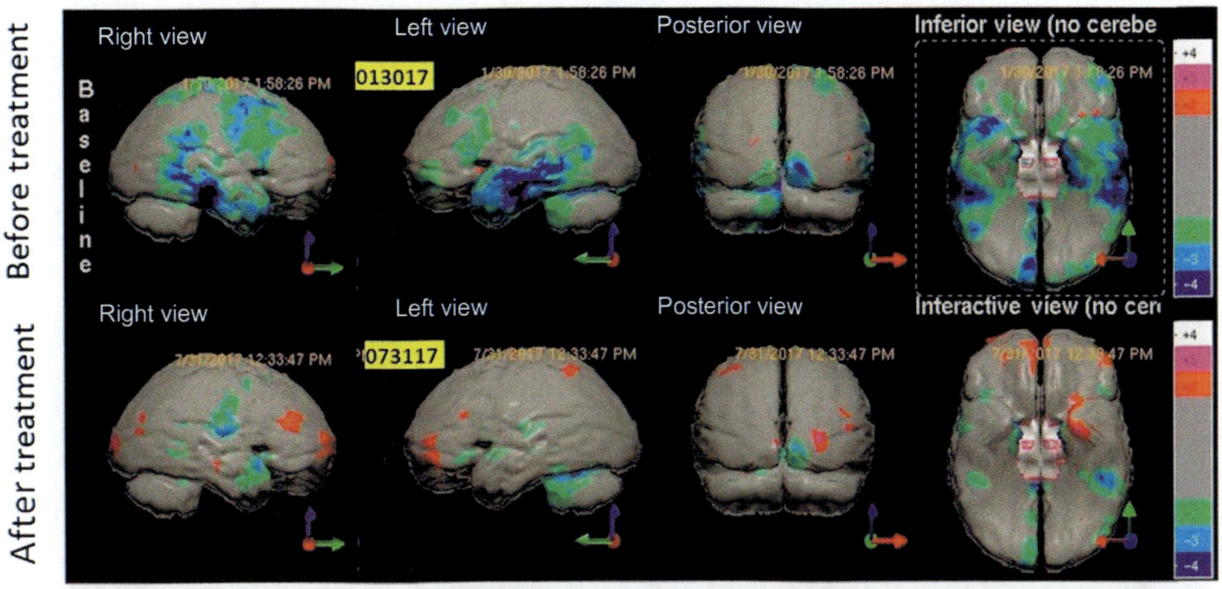

FIGURE 28.3 Pre- and posttreatment SPECTsScans in TBI patient after 20 NILT applications. SPECT data from a TBI patient compared with a normative database ($N = 64$). Top row: Data (01/30/17) before treatment. Bottom Row: Data (07/31/17) after treatment. Areas of decreased function in right posterior frontal and bilateral temporal lobes show significant improvement following NILT treatment. A map of statistically significant differences from the normative database was generated using Segami Inc. software. The color scale indicates (1) gray for areas that did not differ significantly from the normative database, (2) green for areas that were 2–3 standard deviations (SDs) below the mean of the normative database, (3) light blue for areas that were 3–4 SDs below the normative mean, and (4) dark blue for areas that were greater than 4 SDs below the mean of the normative database.

significant recovery over the intervening years, but still experienced difficulties with speech and gait. He complained of low mood, irritability, poor sleep, and frequent recurrent headaches. On initial examination, the patient had moderate dysarthria and a tendency to emphasize atypical portions of words. His gait was asymmetrical with an outward swing of the right leg. His flow of ideas was nonsequential and somewhat difficult to follow. He tended to clang words and insisted on reading some of his poetry. His writings were based on acronyms and had a clanging quality to them. Anatomical MRI and baseline perfusion SPECT scans show injury in the right temporal lobe and portions of his right frontal cortex (see Fig. 28.4).

After 20 treatments with multiwatt NILT, the patient showed significant improvement. His speech was much less garbled. In the 6 months after treatment, his speech improved further. His gait improved considerably with no appreciable outward swing of the right leg. He reported that his sleep, mood, and judgment improved. His recurrent headaches resolved. He was less erratic. The patient was able to write and speak in a much more coherent fashion. The character of his poetry became more unrhymed iambic pentameter or dramatic monologue. Fig. 28.3 illustrates the patient's neuroimaging. Scarring is evident on anatomical MRI. The repeat SPECT scan demonstrated very limited changes in perfusion within the foci of TBI (right temporal and right lateral frontal cortices); however, the penumbra regions adjacent to these foci showed notable improvement. In addition, areas of more distant cortex, which had previously shown hypoperfusion, were considerably better perfused. The patient has retained these improvements in the subsequent 4 years since his treatment.

At follow-up intervals every 6 months posttreatment, we have now followed patients treated with multiwatt NILT for up to 8 years after treatment. Patients report sustained improvements in symptoms. Repeat or follow-up perfusion SPECT imaging in numerous patients has shown significant increases in perfusion in injured areas of the brain and overall improved cortical function following similar courses of multiwatt NILT. In our most recent follow-up survey, 100% of patients reported sustained benefit. The majority noted continued good quality sleep (78%). The patients surveyed noted persistent improvement in concentration (3.4 out of 5 on a 1–5 scale), memory (3.4 out of 5), mood (3.5 out of 5), anxiety (3.0 out of 5), and balance (3.7 out of 5). Roughly 50% of patients noted increased frequency of headaches since the end of treatment. These results (albeit informal) underscore a vital difference between low-level infrared therapy and multiwatt NILT—patients benefit from sustained symptom relief in most cases. Also, it is notable that patients achieved symptom reduction or resolution in 10–20 treatments (duration 20-minute each) in the majority of cases. This is considerably more rapid than that reported for low-level infrared therapy, in which months of treatments, each lasting one hour or more, is required to see relief of the symptoms of TBI.

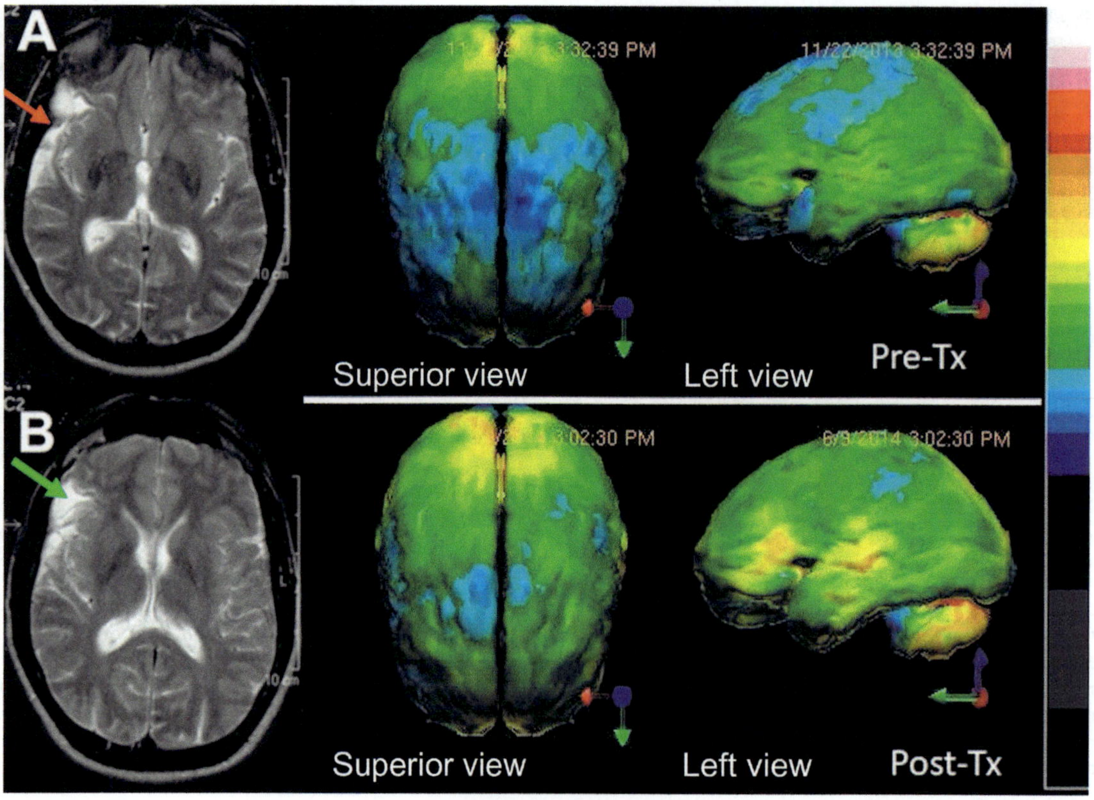

FIGURE 28.4 Pre- and posttreatment SPECT scans in long-term TBI patient after 20 NILT applications. (A and B) The two anatomical MRI images illustrate persistent damage of the right temporal lobe and right dorsolateral frontal lobe. Superior and left lateral views of the SPECT scan are provided. Both scans were normalized to cerebellar perfusion and were coprocessed using Segami Inc. software. The color scale was scaled relative to the patient's mean cerebral perfusion. Mean blood flow (72% of patient's maximal blood flow (MBF)) is in yellow. Color shifts occur at approximately every 0.5 SD (3%) relative to the patient's mean. The area of glial scarring in the right hemisphere remained unchanged (not shown). However, surrounding areas (superior views) and the contralateral cerebral cortices (left lateral view) demonstrated increased perfusion from 45% of MBF to 54% MBF (posterior frontal) and from 54% MBF to 66% MBF or more (left frontal and temporal).

28.4 Conclusion

As of yet, the mechanisms of action of NILT in treating TBI are not entirely delineated. Moreover, as we suggest elsewhere (Henderson and Morries, 2019), the mechanisms responsible for symptomatic response in patients may be different from those induced in laboratory animals. Issues of tissue-light interactions and the challenge of penetrating the much thicker human scalp and skull may have a key role in determining how transcranial NILT works in the clinic. The work of the Neuro-Laser Foundation using multiwatt NILT has demonstrated significant and persistent clinical improvement among patients with TBI and other neurological conditions. Patients have substantial improvement in the key symptoms of TBI, which endure for months to years in our clinical follow-up. Moreover, we have been able to demonstrate persistent neurophysiological changes utilizing functional neuroimaging. More work is needed in this area and a double-blind, placebo-controlled clinical trial of multiwatt NILT is conspicuously absent.

As we move forward in the exploration of transcranial NILT, several key concepts need to be kept in mind. The critical issue of assuring that adequate photonic energy reaches the injured areas of the brain cannot be overemphasized. Evaluation of patients with TBI needs to be comprehensive. All too often, the focus of prior clinicians has been on the TBI and not trauma to the rest of the body or vice versa. Attention to the character of headaches and the potential for a cervicogenic mechanism underlying persistent posttraumatic headaches is warranted. Psychometric testing is strongly encouraged to track patient response. These testing scores also can be helpful to the patient to track his/her response to treatment. Exercise is a powerful adjunct to multiwatt NILT and any treatment for TBI. We encourage exercise and balance training to be included in all TBI programs. Older patients may respond more slowly to treatment and an additional 20%–30% of treatments may be necessary. Additional rest periods may be helpful in the treatment regimen for geriatric patients. As William Osler once said, "Listen to your patient; he is telling you the diagnosis." It is vital to observe your patient. Listen to what they have to say. Treat them with respect.

Acknowledgment

The authors would like to acknowledge the technical assistance of Stefa Grabchenko and Ms. Jeannie Rylands.

References

Abu-Judeh, H.H., Parker, R., Aleksic, S., Singh, M.L., Naddaf, S., Atay, S., et al., 2000. SPECT brain perfusion findings in mild or moderate traumatic brain injury. Nucl. Med. Rev. Cent. East Eur. 3 (1), 5–11.

Alosco, M.L., Kasimis, A.B., Stamm, J.M., Chua, A.S., Baugh, C.M., Daneshvar, D.H., et al., 2017. Age of first exposure to American football and long-term neuropsychiatric and cognitive outcomes. Transl. Psychiatry 7 (9), e1236. 19.

Amen, D.G., Wu, C., Taylor, D., Willeumier, K., 2011. Reversing brain damage in former NFL players: implications for traumatic brain injury and substance abuse rehabilitation. J. Psychoactive Drugs 43 (1), 1–5.

Amen, D.G., Raji, C.A., Willeumier, K., Taylor, D., Tarzwell, R., Newberg, A., et al., 2015. Functional neuroimaging distinguishes posttraumatic stress disorder from traumatic brain injury in focused and large community datasets. PLoS One 10 (7), e0129659.

Amyot, F., Arciniegas, D.B., Brazaitis, M.P., Curley, K.C., Diaz-Arrastia, R., Gandjbakhche, A., et al., 2015. A review of the effectiveness of neuroimaging modalities for the detection of traumatic brain injury. J. Neurotrauma 32 (22), 1693–1721.

Ando, T., Xuan, W., Xu, T., Dai, T., Sharma, S.K., Kharkwal, G.B., et al., 2011. Comparison of therapeutic effects between pulsed and continuous wave 810-nm wavelength laser irradiation for traumatic brain injury in mice. PLoS One 6 (10), e26212.

Asken, B.M., DeKosky, S.T., Clugston, J.R., Jaffee, M.S., Bauer, R.M., 2018. Diffusion tensor imaging (DTI) findings in adult civilian, military, and sport-related mild traumatic brain injury (mTBI): a systematic critical review. Brain Imaging Behav. 12 (2), 585–612.

Atighechi, S., Zolfaghari, A., Baradaranfar, M., Dadgarnia, M., 2013. Estimation of sensitivity and specificity of brain magnetic resonance imaging and single photon emission computed tomography in the diagnosis of olfactory dysfunction after head traumas. Am. J. Rhinol. Allergy 27, 403–406.

Babikian, T., Prins, M.L., Cai, Y., Barkhoudarian, G., Hartonian, I., Hovda, D.A., et al., 2010. Molecular and physiological responses to juvenile traumatic brain injury: focus on growth and metabolism. Dev. Neurosci. 32 (5–6), 431–441.

Baguley, I., Nicholls, J., Felmingham, K., Crooks, J., Gurka, J., Wade, L., 1999. Dysautonomia after traumatic brain injury: a forgotten syndrome? J. Neurol. Neurosurg. Psychiatry 67 (1), 39–43.

Bakker, K., Catroppa, C., Anderson, V., 2016. Recovery of olfactory function following pediatric traumatic brain injury: a longitudinal follow-up. J. Neurotrauma 33 (8), 777–783.

Barrett, D.W., Gonzalez-Lima, F., 2013. Transcranial infrared laser stimulation produces beneficial cognitive and emotional effects in humans. Neuroscience 230, 13–23.

Bazarian, J.J., McClung, J., Shah, M.N., Cheng, Y.T., Flesher, W., Kraus, J., 2005. Mild traumatic brain injury in the United States 1998-2000. Brain Inj. 19 (2), 85–91.

Bigler, E.D., Maxwell, W.L., 2012. Neuropathology of mild traumatic brain injury: relationship to neuroimaging findings. Brain Imaging Behav. 6 (2), 108–136. Available from: https://doi.org/10.1007/s11682-011-9145-0.

Borczuk, P., 1995. Predictors of intracranial injury in patients with mild head trauma. Ann. Emerg. Med. 25 (6), 731–736.

Boussi-Gross, R., Golan, H., Fishlev, G., Bechor, Y., Volkov, O., Bergan, J., et al., 2013. Hyperbaric oxygen therapy can improve post concussion syndrome years after mild traumatic brain injury – randomized prospective trial. PLoS One 8 (11), e79995. 15.

Bratt, M., Skandsen, T., Hummel, T., Moen, K.G., Vik, A., Nordgård, S., et al., 2018. Frequency and prognostic factors of olfactory dysfunction after traumatic brain injury. Brain Inj. 9, 1–7.

Broshek, D.K., Kaushik, T., Freeman, J.R., Erlanger, D., Webbe, F., Barth, J.T., 2005. Sex differences in outcome following sports-related concussion. J. Neurosurg. 102 (5), 856–863.

Bryan, C.J., Clemans, T.A., 2013. Repetitive traumatic brain injury, psychological symptoms, and suicide risk in a clinical sample of deployed military personnel. JAMA Psychiatry 70 (7), 686–691.

Cassano, P., Cusin, C., Mischoulon, D., Hamblin, M.R., De Taboada, L., Pisoni, A., et al., 2015. Near-infrared transcranial radiation for major depressive disorder: proof of concept study. Psychiatry J. 2015, 352979. Available from: https://doi.org/10.1155/2015/352979.

Cassano, P., Petrie, S.R., Mischoulon, D., Cusin, C., Katnani, H., Yeung, A., et al., 2018. Transcranial photobiomodulation for the treatment of major depressive disorder. The ELATED-2 Pilot Trial. Photomed. Laser Surg. Available from: https://doi.org/10.1089/pho.2018.4490Oct 20.

Chamard, E., Lichtenstein, J.D., 2018. A systematic review of neuroimaging findings in children and adolescents with sports-related concussion. Brain Inj. 12, 1–16.

Chen, A.C., Arany, P.R., Huang, Y.Y., Tomkinson, E.M., Sharma, S.K., Kharkwal, G.B., et al., 2011. Low-level laser therapy activates NF-kB via generation of reactive oxygen species in mouse embryonic fibroblasts. PLoS One 6 (7), e22453.

Chung, H., Dai, T., Sharma, S.K., Huang, Y.Y., Carroll, J.D., Hamblin, M.R., 2012. The nuts and bolts of low-level laser (light) therapy. Ann. Biomed. Eng. 40 (2), 516–533.

Cobb, B.R., Urban, J.E., Davenport, E.M., Rowson, S., Duma, S.M., Maldjian, J.A., et al., 2013. Head impact exposure in youth football: elementary school ages 9-12 years and the effect of practice structure. Ann. Biomed. Eng. 41 (12), 2463–2473.

Colvin, A.C., Mullen, J., Lovell, M.R., West, R.V., Collins, M.W., Groh, M., 2009. The role of concussion history and gender in recovery from soccer-related concussion. Am. J. Sports Med. 37 (9), 1699–1704.

Connolly, K.R., Thase, M.E., 2011. If at first you don't succeed: a review of the evidence for antidepressant augmentation, combination and switching strategies. Drugs 71 (1), 43–64.

Covassin, T., Moran, R., Wilhelm, K., 2013. Concussion symptoms and neurocognitive performance of high school and college athletes who incur multiple concussions. Am. J. Sports Med. 41 (12), 2885–2889.

Cox, S.B., Woolsey, T.A., Rovainen, C.M., 1993. Localized dynamic changes in cortical blood flow with whisker stimulation corresponds to matched vascular and neuronal architecture of rat barrels. J. Cereb. Blood Flow Metab. 13, 899–913.

DOD Worldwide Numbers for TBI. Defense and Veterans Brain Injury Center, Silver Spring, MD. Available from: <http://dvbic.dcoe.mil/dod-worldwide-numbers-tbi> (accessed 25.01.18.).

Davis Moore, R., Kay, J.J., Ellemberg, D., 2018. The long-term outcomes of sport-related concussion in pediatric populations. Int. J. Psychophysiol. pii: S0167-8760(18)30173-9.

Douglas, D.B., Iv, M., Douglas, P.K., Anderson, A., Vos, S.B., Bammer, R., et al., 2015. Diffusion tensor imaging of TBI: potentials and challenges. Top. Magn. Reson. Imaging 24 (5), 241–251.

Esenaliev, R.O., Petrov, I.Y., Petrov, Y., Guptarak, J., Boone, D.R., Mocciaro, E., et al., 2018. Nano-pulsed laser therapy is neuroprotective in a rat model of blast-induced neurotrauma. J. Neurotrauma . Available from: https://doi.org/10.1089/neu.2017.5249.

Eyres, S., Carey, A., Gilworth, G., Neumann, V., Tennant, A., 2005. Construct validity and reliability of the Rivermead Post-Concussion Symptoms Questionnaire. Clin. Rehabil. 19 (8), 878–887.

Fann, J.R., Burington, B., Lenonetti, A., Jaffe, K., Katon, W.J., Thompson, R.S., 2004. Psychiatric illness following traumatic brain injury in an adult health maintenance organization population. Arch. Gen. Psychiatry 61 (1), 53–61.

Faul, M., Xu, L., Wald, M., Coronado, V.G., 2010. Traumatic brain injury in the United States. U.S. Department of Health and Human Services Report. Available from: <http://www.cdc.gov/TraumaticBrainInjury/> (accessed 02.03.14.).

Frank, S., Oliver, L., Lebreton-De, C.C., Moreau, C., Lecabellec, M.T., Michel, L., et al., 2004. Infrared radiation affects the mitochondrial pathway of apoptosis in human fibroblasts. J. Invest. Dermatol. 123 (5), 823–831.

Fu, W.W., Fu, T.S., Jing, R., McFaull, S.R., Cusimano, M.D., 2017. Predictors of falls and mortality among elderly adults with traumatic brain injury: a nationwide, population-based study. PLoS One 12 (4), e0175868.

Gessel, L.M., Fields, S.K., Collins, C.L., Dick, R.W., Comstock, R.D., 2007. Concussions among United States high school and collegiate athletes. J. Athl. Train. 42 (4), 495–503.

Gilchrist, J., Thomas, K.E., Xu, L., McGuire, L.C., Coronado, V.G., 2011. Nonfatal sports and recreation related traumatic brain injuries among children and adolescents treated in emergency departments in the United States, 2001-2009. Morb. Mort. Wkly. Rep. 60 (39), 1337. Available from: http://www.cdc.gov/mmwr/pdf/wk/mm6039.pdf (accessed 03.06.18.).

Grubenhoff, J.A., Deakyne, S.J., Brou, L., Bajaj, L., Comstock, R.D., Kirkwood, M.W., 2014. Acute concussion symptom severity and delayed symptom resolution. Pediatrics 134 (1), 54–62.

Harch, P.G., Andrews, S.R., Fogarty, E.F., Amen, D., Pezzullo, J.C., Lucarini, J., et al., 2012. A phase I study of low-pressure hyperbaric oxygen therapy for blast-induced post-concussion syndrome and post-traumatic stress disorder. J. Neurotrauma 29 (1), 168–185.

Haydel, M.J., Preston, C.A., Mills, T.J., Luber, S., Blaudeau, E., DeBlieux, P.M., 2000. Indications for computed tomography in patients with minor head injury. N. Engl. J. Med. 343 (2), 100–105.

Henderson, T.A., 2016. Multi-watt near infrared light therapy as a neuroregenerative treatment for TBI. Neural. Regen. Res. 11 (4), 563–565.

Henderson, T.A., Morries, L.D., 2015a. Near-infrared photonic energy penetration: can infrared phototherapy effectively reach the human brain? Neuropsychiatr. Dis. Treat. 11, 2191–2208.

Henderson, T.A., Morries, L.D., 2015b. SPECT perfusion imaging demonstrates improvement of traumatic brain injury with transcranial near-infrared laser phototherapy. Adv. Mind Body Med. 29 (4), 27–33.

Henderson, T.A., Morries, L.D., 2017. Multi-watt near-infrared phototherapy for the treatment of comorbid depression: an open-label single-arm study. Front. Psychiatry 29 (8), 187.

Henderson, T.A., Morries, L.D., 2019. Near-infrared photonic energy penetration—principles and practice. In: Hamblin, M.R., Huang, Y.Y. (Eds.), Photobiomodulation in the Brain. Elsevier, pp. 67–88.

Huang, Y.Y., Chen, A.C., Carroll, J.D., Hamblin, M.R., 2009. Biphasic dose response in low level light therapy. Dose Response 7 (4), 358–383.

Hummel, T., Sekinger, B., Wolf, S.R., Pauli, E., Kobal, G., 1997. 'Sniffin' Sticks': olfactory performance assessed by the combined testing of odor identification, odor discrimination and olfactory threshold. Chem. Senses 22 (1), 39–52.

Hummel, T., Kobal, G., Gudziol, H., Mackay-Sim, A., 2007. Normative data for the "Sniffin' Sticks" including tests of odor identification, odor discrimination, and olfactory thresholds: an upgrade based on a group of more than 3,000 subjects. Eur. Arch. Otorhinolaryngol. 264 (3), 237–243.

Humphreys, I., Wood, R.L., Phillips, C.J., Macey, S., 2013. The costs of traumatic brain injury: a literature review. Clinicoecon. Outcomes Res. 5, 281–287.

Jacobs, A., Put, E., Ingels, M., Bossuyt, A., 1994. Prospective evaluation of Technetium-99m-HMPAO SPECT in mild and moderate traumatic brain injury. J. Nucl. Med. 35 (6), 947–948.

Jacobs, A., Put, E., Ingels, M., Bossuyt, A., 1996. One-year follow-up of Technetium-99m-HMPAO SPECT in mild head injury. J. Nucl. Med. 37, 1605–1609.

Jorge, R.E., Robinson, R.G., Moser, D., Tateno, A., Crespo-Facorro, B., Arndt, S., 2004. Major depression following traumatic brain injury. Arch. Gen. Psychiatry 61 (1), 42–50.

Karu, T.I., Kolyakov, S.F., 2005. Exact action spectra for cellular responses relevant to phototherapy. Photomed. Laser Surg. 23 (4), 355–361.

Kashluba, S., Hanks, R.A., Casey, J.E., Millis, S.R., 2008. Neuropsychologic and functional outcome after complicated mild traumatic brain injury. Arch. Phys. Med. Rehabil. 89 (5), 904–911.

Kay, T., Harrington, D.E., Adams, R., Anderson, T., Berrol, S., Cicerone, K., et al., 1993. Definition of mild traumatic brain injury: American Congress of Rehabilitation Medicine. J. Head Trauma Rehabil. 8 (3), 86–87.

Kennedy, J.E., Jaffee, M.S., Leskin, G.A., Stokes, J.W., Leal, F.O., Fitzpatrick, P.J., 2007. Posttraumatic stress disorder and posttraumatic stress disorder-like symptoms and mild traumatic brain injury. J. Rehabil. Res. Dev. 44 (7), 895–920.

Khuman, J., Zhang, J., Park, J., Carroll, J.D., Donahue, C., Whalen, M.J., 2012. Low-level laser light therapy improves cognitive deficits and inhibits microglial activation after controlled cortical impact in mice. J. Neurotrauma 29, 408–417.

King, N.S., Crawford, S., Wenden, F.J., Moss, N.E., Wade, D.T., 1995. The Rivermead Post Concussion Symptoms Questionnaire: a measure of symptoms commonly experienced after head injury and its reliability. J. Neurol. 242 (9), 587–592.

Kujawa, J., Zavodnik, L., Zavodnik, I., Buko, V., Lapshyna, A., Bryszewska, M., 2004. Effect of low-intensity (3.75-25 J/cm^2) near-infrared (810 nm) laser radiation on red blood cell ATPase activities and membrane structure. J. Clin. Laser Med. Surg. 22 (2), 111–117.

Kujawa, J., Pasternak, K., Zavodnik, I., Irzmański, R., Wróbel, D., Bryszewska, M., 2014. The effect of near-infrared MLS laser radiation on cell membrane structure and radical generation. Lasers Med. Sci. 29 (5), 1663–1668.

Kumar, A., Loane, D.J., 2012. Neuroinflammation after traumatic brain injury: opportunities for therapeutic intervention. Brain Behav. Immun. 26 (8), 1191–1201.

Kushibiki, T., Hirasawa, T., Okawa, S., Ishihara, M., 2013. Regulation of miRNA expression by low-level laser therapy (LLLT) and photodynamic therapy (PDT). Int. J. Mol. Sci. 14 (7), 13542–13558.

Laatsch, L., Jobe, T., Sychra, J., Lin, Q., Blend, M., 1997. Impact of cognitive rehabilitation therapy on neuropsychological impairments as measured by brain perfusion SPECT: a longitudinal study. Brain Inj. 11 (12), 851–863.

Laatsch, L., Pavel, D., Jobe, T., Lin, C., Quintana, J.C., 1999. Incorporation of SPECT imaging in a longitudinal cognitive rehabilitation therapy programme. Brain Inj. 13 (8), 555–570.

Lapchak, P.A., 2010. Taking a light approach to treating acute ischemic stroke patients: transcranial near-infrared laser therapy translational science. Ann. Med. 42 (8), 576–586.

Lapchak, P.A., 2012. Transcranial near-infrared laser therapy applied to promote clinical recovery in acute and chronic neurodegenerative diseases. Expert Rev. Med. Devices 9 (1), 71–83.

Lew, H.L., 2005. Rehabilitation needs of an increasing population of patients: traumatic brain injury, polytrauma, and blast-related injuries. J. Rehabil. Res. Dev. 42 (4), xiii–xvi.

Lew, H.L., Vanderploeg, R.D., Moore, D.F., Schwab, K., Friedman, L., Yesavage, J., et al., 2008. Overlap of mild TBI and mental health conditions in returning OIF/OEF service members and veterans. J. Rehabil. Res. Dev. 45 (3), xi–xvi.

Lincoln, A.E., Caswell, S.V., Almquist, J.L., Dunn, R.E., Norris, J.B., Hinton, R.Y., 2011. Trends in concussion incidence in high school sports: a prospective 11-year study. Am. J. Sports Med. 39 (5), 958–963.

Logan, B.W., Goldman, S., Zola, M., Mackey, A., 2013. Concussive brain injury in the military: September 2001 to the present. Behav. Sci. Law 31 (6), 803–813.

Lubart, R., Eichler, M., Lavi, R., Friedman, H., Shainberg, A., 2005. Low-energy laser irradiation promotes cellular redox activity. Photomed. Laser Surg. 23 (1), 3–9.

McCarthy, T.J., De Taboada, L., Hildebrandt, P.K., Ziemer, E.L., Richieri, S.P., Streeter, J., 2010. Long-term safety of single and multiple infrared transcranial laser treatments in Sprague-Dawley rats. Photomed. Laser Surg. 28 (5), 663–667.

Miller, E.C., Holmes, J.F., Derlet, R.W., 1997. Utilizing clinical factors to reduce head CT scan ordering for minor head trauma patients. J. Emerg. Med. 15 (4), 453–457.

Mirsky, N., Krispel, Y., Shoshany, Y., Maltz, L., Oron, U., 2002. Promotion of angiogenesis by low energy laser irradiation. Antioxid. Redox Signal. 4 (5), 785–790.

Morries, L.D., Cassano, P., Henderson, T.A., 2015. Treatments for traumatic brain injury with emphasis on transcranial near infrared laser phototherapy. Neuropsychiatr. Dis. Treat. 11, 2159–2175.

Naeser, M.A., Saltmarche, A., Krengel, M.A., Hamblin, M.R., Knight, J.A., 2011. Improved cognitive function after transcranial, light-emitting diode treatments in chronic, traumatic brain injury: two case reports. Photomed. Laser Surg. 29 (5), 351–358.

Naeser, M.A., Zafonte, R., Krengel, M.H., Martin, P.I., Frazier, J., Hamblin, M.R., et al., 2014. Significant improvements in cognitive performance post-transcranial, red/near-infrared light-emitting diode treatments in chronic, mild traumatic brain injury: open-protocol study. J. Neurotrauma 31 (11), 1008–1017.

Nasreddine, Z.S., Phillips, N.A., Bédirian, V., Charbonneau, S., Whitehead, V., Collin, I., et al., 2005. The Montreal Cognitive Assessment, MoCA: a brief screening tool for mild cognitive impairment. J. Am. Geriatr. Soc. 53 (4), 695–699.

Noble, J.M., Hesdorffer, D.C., 2013. Sport-related concussions: a review of epidemiology, challenges in diagnosis, and potential risk factors. Neuropsychol. Rev. 23 (4), 273–284.

Oron, A., Oron, U., Chen, J., Eilam, A., Zhang, C., Sadeh, M., et al., 2006. Low level laser therapy applied transcranially to rats after induction of stroke significantly reduces long-term neurological deficits. Stroke 37, 2620–2624.

Oron, A., Oron, U., Streeter, J., de Taboada, L., Alexandrovich, A., Trembovler, V., et al., 2007. Low-level laser therapy applied transcranially to mice following traumatic brain injury significantly reduces long-term neurological deficits. J. Neurotrauma 24, 651–656.

Oron, A., Oron, U., Streeter, J., De Taboada, L., Alexandrovich, A., Trembovler, V., et al., 2012. Near infrared transcranial laser therapy applied at various modes to mice following traumatic brain injury significantly reduces long-term neurological deficits. J. Neurotrauma 29 (2), 401–407.

Pervez, M., Kitagawa, R.S., Chang, T.R., 2018. Definition of traumatic brain injury, neurosurgery, trauma orthopedics, neuroimaging, psychology, and psychiatry in mild traumatic brain injury. Neuroimaging Clin. N. Am. 28 (1), 1–13.

Peschman, J., Neideen, T., Brasel, K., 2011. The impact of discharging minimally injured trauma patient: does age play a role in trauma admission? J. Trauma 70 (6), 1331–1336.

Peskind, E.R., Brody, D., Cernak, I., McKee, A., Ruff, R.L., 2013. Military- and sports-related mild traumatic brain injury: clinical presentation, management, and long-term consequences. J. Clin. Psychiatry 74 (2), 180–188.

Prins, M.L., Giza, C.C., 2012. Repeat traumatic brain injury in the developing brain. Int. J. Dev. Neurosci. 30 (3), 185–190.

Prins, M.L., Alexander, D., Giza, C.C., Hovda, D.A., 2013. Repeated mild traumatic brain injury: mechanisms of cerebral vulnerability. J. Neurotrauma 30 (1), 30–38.

Proskynitopoulos, P.J., Stippler, M., Kasper, E.M., 2016. Post-traumatic anosmia in patients with mild traumatic brain injury (mTBI): a systematic and illustrated review. Surg. Neurol. Int. 7 (Suppl. 10), S263–S275.

Quirk, B.J., Torbey, M., Buchmann, E., Verma, S., Whelan, H.T., 2012. Near-infrared photobiomodulation in an animal model of traumatic brain injury: improvements at the behavioral and biochemical levels. Photomed. Laser Surg. 30 (9), 523–529.

Raji, C.A., Henderson, T.A., 2018. PET and single-photon emission computed tomography in brain concussion. Neuroimaging Clin. N. Am. 28 (1), 67–82.

Raji, C.A., Tarzwell, R., Pavel, D., Schneider, H., Uszler, M., Thornton, J., et al., 2014. Clinical utility of SPECT neuroimaging in the diagnosis and treatment of traumatic brain injury: a systematic review. PLoS One 9 (3), e91088. Available from: https://doi.org/10.1371/journal.pone.0091088.

Raji, C.A., Willeumier, K., Taylor, D., Tarzwell, R., Newberg, A., Henderson, T.A., et al., 2015. Functional neuroimaging with default mode network regions distinguishes PTSD from TBI in a military veteran population. Brain Imaging Behav. 9 (3), 527–534.

Ramanathan, D.M., McWilliams, N., Schatz, P., Hillary, F.G., 2012. Epidemiological shifts in elderly traumatic brain injury: 18-year trends in Pennsylvania. J. Neurotrauma 29 (7), 1371–1378.

Richmond, R., Aldaghlas, T.A., Burke, C., Rizzo, A.G., Griffen, M., Pullarkat, R., 2011. Age: is it all in the head? Factors influencing mortality in elderly patients with head injuries. J. Trauma 71 (1), E8–E11.

Rush, A.J., Trivedi, M.H., Ibrahim, H.M., Carmody, T.J., Arnow, B., Klein, D.N., et al., 2003. The 16-Item Quick Inventory of Depressive Symptomatology (QIDS), clinician rating (QIDS-C), and self-report (QIDS-SR): a psychometric evaluation in patients with chronic major depression. Biol. Psychiatry 54 (5), 573–583.

Salehpour, F., Mahmoudi, J., Kamari, F., Sadigh-Eteghad, S., Rasta, S.H., Hamblin, M.R., 2018. Brain photobiomodulation therapy: a narrative review. Mol. Neurobiol. Available from: https://doi.org/10.1007/s12035-017-0852-4.

Sandel, N.K., Schatz, P., Goldberg, K.B., Lazar, M., 2017. Sex-based differences in cognitive deficits and symptom reporting among acutely concussed adolescent lacrosse and soccer players. Am. J. Sports Med. 45 (4), 937–944.

Schiffer, F., Johnston, A.L., Ravichandran, C., Polcari, A., Teicher, M.H., Webb, R.H., et al., 2009. Psychological benefits 2 and 4 weeks after a single treatment with near infrared light to the forehead: a pilot study of 10 patients with major depression and anxiety. Behav. Brain Funct. 5, 46.

Schofield, P.W., Moore, T.M., Gardner, A., 2014. Traumatic brain injury and olfaction: a systematic review. Front. Neurol. 5, 5.

Schwartz, F., Brodie, C., Appel, E., Kazimirsky, G., Shainberg, A., 2002. Effect of helium/neon laser irradiation on nerve growth factor synthesis and secretion in skeletal muscle cultures. J. Photochem. Photobiol. B 66 (3), 195–200.

Selassie, A.W., Wilson, D.A., Pickelsimer, E.E., Voronca, D.C., Williams, N.R., Edwards, J.C., 2013. Incidence of sport-related traumatic brain injury and risk factors of severity: a population-based epidemiologic study. Ann. Epidemiol. 23 (12), 750–756.

Sestini, S., 2007. The neural basis of functional neuroimaging signal with positron and single-photon emission tomography. Cell Mol. Life Sci. 64 (14), 1778–1784.

Shapey, J., Glancz, L.J., Brennan, P.M., 2016. Chronic subdural haematoma in the elderly: is it time for a new paradigm in management? Curr. Geriatr. Rep. 5, 71–77.

Shin, Y.B., Kim, S.-J., Kim, I.-J., Kim, Y.K., Kim, D.S., Park, J.H., et al., 2006. Voxel-based statistical analysis of cerebral blood flow using Tc-99m ECD brain SPECT in patients with traumatic brain injury: group and individual analyses. Brain Inj. 20 (6), 661–667.

Stamatakis, E.A., Wilson, J.T., Hadley, D.M., Wyper, D.J., 2002. SPECT imaging in head injury interpreted with statistical parametric mapping. J. Nucl. Med. 43 (4), 476–483.

State of Colorado – Division of Labor and Employement, Division of Worker's Compensation. Traumatic Brain Injury Medical Treatment Guidelines. Revised November 26, 2012. <https://www.colorado.gov/pacific/sites/default/files/Rule_17_Exhibit_10_Traumatic_Brain_Injury.pdf> (accessed 24.05.18.).

Styrke, J., Stålnacke, B.M., Sojka, P., Björnstig, U., 2007. Traumatic brain injuries in a well-defined population: epidemiological aspects and severity. J. Neurotrauma 24 (9), 1425–1436.

Susman, M., DiRusso, S.M., Sullivan, T., Risucci, D., Nealon, P., Cuff, S., et al., 2002. Traumatic brain injury in the elderly: increased mortality and worse functional outcome at discharge despite lower injury severity. J. Trauma 53 (2), 219–223.

Swann, I.J., Bauza-Rodriguez, B., Currans, R., Riley, J., Shukla, V., 2006. The significance of post-traumatic amnesia as a risk factor in the development of olfactory dysfunction following head injury. Emerg. Med. J. 23, 618–621.

Szymanska, J., Goralczyk, K., Klawe, J.J., Lukowicz, M., Michalska, M., Goralczyk, B., et al., 2013. Phototherapy with low-level laser influences the proliferation of endothelial cells and vascular endothelial growth factor and transforming growth factor-beta secretion. J. Physiol. Pharmacol. 64 (3), 387–391.

Tatsch, K., Asenbaum, S., Bartenstein, P., Catafau, A., Halldin, C., Pilowsky, L.S., et al., 2002. European Association of Nuclear Medicine Procedure Guidelines for brain perfusion SPECT using 99mTc-labelled radiopharmaceuticals. Eur. J. Nucl. Med. 29, BP36–BP42.

Thase, M.E., Entsuah, A.R., Rudolph, R.L., 2001. Remission rates during treatment with venlafaxine or selective serotonin reuptake inhibitors. Br. J. Psychiatry 178, 234–241.

Trivedi, M.H., Rush, A.J., Ibrahim, H.M., Carmody, T.J., Biggs, M.M., Suppes, T., et al., 2004. The Inventory of Depressive Symptomatology, Clinician Rating (IDS-C) and Self-Report (IDS-SR), and the Quick Inventory of Depressive Symptomatology, Clinician Rating (QIDS-C) and Self-Report (QIDS-SR) in public sector patients with mood disorders: a psychometric evaluation. Psychol. Med. 34 (1), 73–82.

Urban, J.E., Davenport, E.M., Golman, A.J., Maldjian, J.A., Whitlow, C.T., Powers, A.K., et al., 2013. Head impact exposure in youth football: high school ages 14 to 18 years and cumulative impact analysis. Ann. Biomed. Eng. 41 (12), 2474–2487.

Vaishnavi, S., Rao, V., Fann, J.R., 2009. Neuropsychiatric problems after traumatic brain injury: unraveling the silent epidemic. Psychosomatics 50 (3), 198–205.

Vasterling, J.J., Proctor, S.P., Amoroso, P., Kane, R., Heeren, T., Franz, M., 2006. Neuropsychological outcomes of Army personnel following deployment to the Iraq War. JAMA 296 (5), 519.

von Leden, R.E., Cooney, S.J., Ferrara, T.M., Zhao, Y., Dalgard, C.L., Anders, J.J., et al., 2013. 808 nm wavelength light induces a dose-dependent alteration in microglial polarization and resultant microglial induced neurite growth. Lasers Surg. Med. 45 (4), 253–263.

WHO (World Health Organization). International Statistical Classification of Diseases and Related Health Problems, 2010. 10th Revision. <http://apps.who.int/classifications/icd10/browse/2010/en> (accessed 24.05.18.).

Wu, H.M., Huang, S.C., Vespa, P., Hovda, D.A., Bergsneider, M., 2013. Redefining the pericontusional penumbra following traumatic brain injury: evidence of deteriorating metabolic derangements based on positron emission tomography. J. Neurotrauma 30 (5), 352–360.

Wu, Q., Huang, Y.-Y., Dhital, S., Sharma, S.K., Chen, A.C.-H., Whalen, M.J., et al., 2010. Low level laser therapy for traumatic brain injury. In: Hamblin, M.R., Waynant, R.W., Anders, J. (Eds.), Mechanisms for Low-Light Therapy V. SPIE, San Francisco, CA, pp. 755206–755208.

Wu, Q., Xuan, W., Ando, T., Xu, T., Huang, L., Huang, Y.Y., et al., 2012. Low-level laser therapy for closed-head traumatic brain injury in mice: effect of different wavelengths. Lasers Surg. Med. 44, 218–226.

Xuan, W., Vatansever, F., Huang, L., Wu, Q., Xuan, Y., Dai, T., et al., 2013. Transcranial low-level laser therapy improves neurological performance in traumatic brain injury in mice: effect of treatment repetition regimen. PLoS One 8 (1), e53454.

Xuan, W., Agrawal, T., Huang, L., Gupta, G.K., Hamblin, M.R., 2014a. Low-level laser therapy for traumatic brain injury in mice increases brain derived neurotrophic factor (BDNF) and synaptogenesis. J. Biophotonics . Available from: https://doi.org/10.1002/jbio.201400069.

Xuan, W., Vatansever, F., Huang, L., Hamblin, M.R., 2014b. Transcranial low-level laser therapy enhances learning, memory, and neuroprogenitor cells after traumatic brain injury in mice. J. Biomed. Opt. 19 (10), 108003.

Xuan, W., Huang, L., Hamblin, M.R., 2016. Repeated transcranial low-level laser therapy for traumatic brain injury in mice: biphasic dose response and long-term treatment outcome. J. Biophotonics 9 (11–12), 1263–1272.

Yip, K.K., Lo, S.C., Leung, M.C., So, S.K., Tang, C.Y., Poon, D.M., 2011. The effect of low-energy laser irradiation on apoptotic factors following experimentally induced transient cerebral ischemia. Neuroscience 190, 301–306.

Zhang, Q., Zhou, C., Hamblin, M.R., Wu, M.X., 2014. Low-level laser therapy effectively prevents secondary brain injury induced by immediate early responsive gene X-1 deficiency. J. Cereb. Blood Flow Metab. 34 (8), 1391–1401.

Ziebell, J.M., Morganti-Kossmann, M.C., 2010. Involvement of pro- and anti-inflammatory cytokines and chemokines in the pathophysiology of traumatic brain injury. Neurotherapeutics 7 (1), 22–30.

Zito, G., Jull, G., Story, I., 2006. Clinical tests of musculoskeletal dysfunction in the diagnosis of cervicogenic headache. Man Ther. 11 (2), 118–129.

Zung, W.W., 1965. A self-rating depression scale. Arch. Gen. Psychiatry 12, 63–70.

Zung, W.W., 1971. A rating instrument for anxiety disorders. Psychosomatics 12 (6), 371–379.

Chapter 29

Photobiomodulation: a novel approach to treating Alzheimer's disease

Lew Lim[1], Genane Loheswaran[1], Reza Zomorrodi[2], Anita Saltmarche[3] and Linda Chao[4]

[1]*Vielight Inc., Toronto, ON, Canada,* [2]*Temerty Centre for Therapeutic Brain Intervention, Centre for Addiction and Mental Health, Toronto, ON, Canada,* [3]*Saltmarche Health & Associates Inc., Orangeville, ON, Canada,* [4]*Departments of Radiology & Biomedical Imaging and Psychiatry, University of California, San Francisco, CA, United States*

29.1 Introduction

Dementia is a major neurocognitive condition that interferes with both cognitive function and everyday activities. Alzheimer's disease (AD) is the most common type of dementia in elderly individuals worldwide (Wilson et al., 2012). The disease is characterized by neurodegeneration, histopathological changes in the brain, severe cognitive decline and in many cases, neuropsychiatric symptoms. The key histopathological features in AD are extracellular amyloid-containing plaques and intracellular tangles of hyperphosphorylated tau protein, which are often recognized to distinguish AD from other types of dementia. Despite decades of research on AD, the causative mechanisms of the disease are largely unclear, and an effective treatment remains elusive.

This chapter presents photobiomodulation (PBM) as a promising and novel approach to treating the disease covering the following areas:

- Analysis of the failure of pharmacotherapies in treating AD;
- Discussion on the importance of mitochondrial health in AD;
- Reasons why delivering light at certain frequencies has greater potential for an effective treatment;
- How these PBM studies fare in dementia/AD clinical investigations;
- Discussion on the effect of key PBM parameters to the outcomes—the wavelength power, treatment time, anatomical locations and pulse rate;
- Validation with cognitive tests such as the mini mental status exams (MMSE) and AD assessment scale for cognition (ADAS-cog) and clinical reports;
- Supporting data from imaging reports that include electroencephalography (EEG), arterial spin-labeled (ASL) perfusion, resting state functional magnetic resonance imaging (rs-fMRI);
- Discussion on the progress of investigations into PBM as a AD intervention;
- Potential uses of EEG measures to identify brainwave signatures of the disease, allowing personalization of treatments in the future, in brain-computer interface systems by adjusting selected PBM parameters.

29.2 Pharmacotherapies for Alzheimer's disease

Most of the resources and attention into AD research have been invested into pharmacotherapies. Five pharmacotherapies approved by the Food and Drug Administration (FDA) currently in use are Donepezil, Galantamine and Rivastigmine, Memantine and Namzaric (a combination of Donepezil and Memantine). However, none of these drugs are considered effective beyond the first few months of the therapy. Donepezil, Galantamine, and Rivastigmine all fall within the class of cholinesterase inhibitors. Cholinesterase inhibitors increase levels of acetylcholine, a neurotransmitter that plays an important role in cognition and memory (Hasselmo, 2006) by preventing the breakdown of the acetylcholine by acetylcholinesterase. The biochemical actions produce a temporary compensatory effect for the memory and other cognitive deficits associated with AD. Memantine is a NMDA receptor antagonist that is approved

Photobiomodulation in the Brain. DOI: https://doi.org/10.1016/B978-0-12-815305-5.00029-4

for clinical use for patients with moderate to severe AD (Hellweg et al., 2012). The use of Memantine is targeted at decreasing the damage to the glutamatergic system through excitotoxicity in the later stages of AD. Lastly, Namzaric, the combination of Donepezil and Memantine, is also used to treat symptoms in patients with moderate to severe AD. However, while these medications target some of the symptoms of AD, none are disease-modifying.

29.3 Pathophysiology of Alzheimer's disease

29.3.1 Amyloid cascade hypothesis

The amyloid cascade hypothesis has been the dominant hypothesis over the last decades. Patients with AD have a build-up of amyloid-β (Aβ) in their brains, which form the senile plaques, widely recognized as a pathological hallmark of AD. The transmembrane glycoprotein, amyloid precursor protein (APP), is metabolically processed to produce either amyloidogenic (i.e., Aβ forming) or nonamyloidogenic products (Fraser et al., 1997). Mutations in the APP gene (Goate et al., 1991) and the genes that form the products that cleave APP (i.e., secretase enzymes) play an important role in the amyloid cascade hypothesis (Sherrington et al., 1995). If the enzyme α-secretase cleaves APP proteolytically, a neuroprotective fragment called sAPPα is formed, preventing the formation of Aβ (Pearson and Peers, 2006). However, if APP is sequentially cleaved by β-secretase and then γ-secretase, the formation of Aβ occurs. The apolipoprotein E (ApoE) type 4 has been identified as the most significant known risk factor for AD (Corder et al., 1993).

The amyloid cascade hypothesis provided many drug targets for AD without success to date. When drugs were able to remove some of the amyloid plaque load, they did not translate into an improvement in symptoms of AD (Herrup, 2015; Iqbal et al., 2014; Karran and Hardy, 2014; Swerdlow et al., 2014). In recent years, the amyloid cascade hypothesis has been criticized as an overly simplistic view of a complex disease (Herrup, 2015).

29.3.2 Neurofibrillary tangles

In addition to accumulation of senile plaques formed by the accumulation of Aβ, neurofibrillary tangles (NFTs) are also characteristic of the pathology of AD. NFT are intracellular structures formed out of hyperphosphorylated microtubule-associated tau proteins. These proteins self-aggregate to become insoluble forms known as straight and paired helical filaments. The accumulation of tau in the neurons begins prior to the formation of NFT, suggesting that there is an early imbalance in the activity of protein kinases and phosphatases in AD (Brion, 1998). Several drug trials have focused on tau-based targets, including tau protein, tau phosphorylation, tau oligomerization, tau degradation and tau-based vaccination. To date, none of these trials have been successful.

29.3.3 Other protein targets

Aβ plaques and NFT consist of other proteins in addition to their primary constituents of Aβ and tau respectively. These have also been considered as pharmacotherapy targets. Aβ plaques contain other proteins that encompass and infiltrate the plaques. These proteins include proteins found in astrocytes, microglia, dystrophic neuritis, and amyloid binding proteins (i.e., Ubiquitin, apolipoprotein E, and clusterin). Similarly, NFT have been found to contain tau binding proteins (i.e., cytoskeletal proteins, kinases, and heat shock proteins (Mandelkow and Mandelkow, 2012).

29.4 The odds against a monotherapy

The failure rate for Alzheimer's pharmacotherapy-based clinical trials stands at 99%, with no new Alzheimer's therapies having gained US FDA approval since 2003 (Gardner, 2017). These pharmacotherapies have largely focused on specific targets in various proteins and enzymes underlying various pathological pathways in AD. However, it has clearly been established that the pathophysiology of AD is complex with multiple protein pathogeneses. This complex nature of the disease may explain why monotherapies have not been successful in treating AD. For this reason, drug trials are now shifting their focus to the early biochemical phase of AD, prior to the emergence of clinical symptoms, leaving diagnosed AD cases with no hope (Dubois et al., 2016).

29.5 Mitochondrial cascade hypothesis of Alzheimer's disease

More recently, the mitochondrial cascade hypothesis of AD has been receiving increasing attention as a key mechanism explaining the pathogenesis of AD. The mitochondria are vital to neuronal function as they supply cellular energy, in the form of ATP. Furthermore, mitochondria provide the energy for synaptic plasticity, the mechanism by which the brain adapts to experience or use and encodes information at the synaptic level. Mitochondrial dysfunction is thought to play a key role in the impairments in synaptic plasticity seen in AD (Castello and Soriano, 2014; Drachman, 2014; Morris and Berk, 2015; Spires-Jones and Knafo, 2012). They are found in abundance in synaptic terminals. Impairments in mitochondrial function are associated with synaptic dysfunction, decreased synaptic and neuronal outgrowth and apoptosis (Du et al., 2012; Friedland-Leuner et al., 2014; Swerdlow et al., 2014).

Mitochondrial dysfunctions are well-documented in the brains of patients with AD (Friedland-Leuner et al., 2014; Gibson and Shi, 2010). Their dysfunction is a major cause for deficits in cerebral glucose metabolism, which occurs in the brains of patients with AD, in brain regions associated with memory such as the hippocampus and entorhinal cortex (Kapogiannis and Mattson, 2011). The deficits occur well in advance of presentations of the clinical symptoms. Mitochondrial dysfunctions observed in AD are expressed as decreased mitochondrial enzyme activity, decreased activity of complexes of the respiratory chain, and excessive levels of reactive oxygen species (ROS), producing increased oxidative stress.

In summary, the reductions in energy metabolism, increased oxidative stress, and synaptic dysfunction embody a common final pathway of all risk factors (genetic and nongenetic) for the development of AD (Friedland-Leuner et al., 2014; Leuner et al., 2012; Muller et al., 2010), leading to the development of the mitochondrial cascade hypothesis. The mitochondrial cascade hypothesis states that: (1) A person's baseline mitochondrial function is determined by gene inheritance; (2) A combination of environmental and inherited factors govern the rate at which the age-related alterations in mitochondrial function and efficiency develop; (3) Both the mitochondrial function at baseline and the rate of change in mitochondrial function influence a person's chronology of AD (Swerdlow et al., 2014).

It has been demonstrated that mitochondrial dysfunction can push APP processing toward the formation of Aβ production (Gabuzda et al., 1994; Gasparini et al., 1997; Webster et al., 1998), suggesting that mitochondrial dysfunction is a factor driving the amyloid cascade. The presence of increased oxidative stress and slightly increased Aβ due to individual risk factors leads to mitochondrial dysfunction well before the formation of Aβ deposits. Increases in ROS over time result in further damage to mitochondria, resulting in a self-feeding feedback loop leading to increases in the production of Aβ and further mitochondrial damage. Furthermore, mitochondrial dysfunction has been found to produce inflammation (Lopez-Armada et al., 2013) and affect tau phosphorylation (Blass et al., 1990a,b; Szabados et al., 2004). The theory posits that mitochondrial dysfunction is the leading pathomechanism that causes neurodegeneration and AD-associated deficits (Swerdlow et al., 2014).

Given the involvement of mitochondrial impairment as the key driving force underlying the decline seen in aging to AD, therapies to counteract mitochondrial dysfunction in patients with AD hold promise as a treatment for the disease. However, the development of pharmacotherapies to target mitochondrial dysfunction is severely lagging. While some agents that may improve mitochondrial function have been explored in preclinical and in vitro studies, very few, if any clinical trials have been conducted. There is an urgent need to identify therapies that are effective at treating mitochondrial dysfunction. Photobiomodulation could be the intervention that meets this need.

29.6 Photobiomodulation and mitochondrial function

PBM, also known as low-level light therapy, is a biostimulation technique that has shown promise in treating a number of conditions, including dementia and AD. The most well investigated mechanism of action of PBM is its fundamental effect on mitochondrial function (for review, Hamblin, 2016). PBM has been demonstrated to increase the activity of complexes in the electron transport chain of mitochondria, comprising complex I, II, III, IV and succinate dehydrogenase. In complex IV, the enzyme cytochrome c oxidase (CCO), functions as photo acceptor as well as transducer. CCO specifically accepts and transduces light in the red (620–700 nm) and the near-infrared (780–1110 nm) spectrums, the wavelengths of lights processed in PBM (de Freitas and Hamblin, 2016). The process increases the amount of ATP produced, as well as cyclic adenosine monophosphate (cAMP) and ROS (Wu et al., 2014).

The increase in ATP increases the activity of ion channels, regulates cAMP and calcium, which results in the stimulation of diverse biological cascades (Farivar et al., 2014; Passarella and Karu, 2014) and activates up to 110 genes, which then themselves lead to the prolongation of the production of energy by the mitochondria (Lane, 2006).

PBM increases ROS formation that is transient and at low levels. This is thought to activate mitochondrial signaling pathways that have antioxidant, antiapoptotic, and cytoprotective effects on cells (Waypa et al., 2016). A number of cellular mechanisms are in involved in sensing excessive levels of ROS, and respond by activating transcription factors which produce increased antioxidant defenses, preserving homeostasis (Bindoli and Rigobello, 2013).

In addition to increasing levels of ATP and cAMP, it has been observed that PBM results in an increase in nitric oxide (NO) levels, dissociated from CCO when photons are absorbed by CCO (Hamblin, 2016; Hashmi et al., 2010; Poyton and Ball, 2011). The dissociation of NO from CCO leads to the enhancement of ATP production and acts as a vasodilator as well as a dilator of lymphatic flow, and can signal to activate a number of beneficial cellular pathways (Hamblin, 2016; Passarella and Karu, 2014).

29.7 Photobiomodulation in animal models of Alzheimer's disease

Media are eager to report AD advances in animal studies produced out of traditional research sources, forgetting that PBM had reported a major one several years ago. A recent study that made headlines, reported that slowly reducing the level of the enzyme β-site amyloid precursor protein cleaving enzyme 1 through genetic modification in mice as they aged, either prevented or reversed the formation of amyloid plaques in the brain (Hu et al., 2018). In comparison, a reduction in the amyloid-β peptide neuropathology in a mice model of AD in response to near-infrared PBM was reported by De Taboada et al., 6 years earlier. Administration of transcranial laser therapy of 808 nm wavelength three times per week at various doses to an amyloid β protein precursor on the transgenic mouse model beginning at three months of age, resulted in a significant reduction in amyloid load and improved behavior (De Taboada et al., 2011). Expression of inflammatory markers was reduced, producing a decrease in the activity of β-secretase, leading to reduced Aβ plaque count.

More recently, PBM delivery of 670 nm light to the skull of two different mouse models of AD resulted in a reduction in AD-related neuropathology in the cerebral cortex (Purushothuman et al., 2014). The two mouse models used in the study were the K3 tau transgenic model and the APPsew/PSEN1dE9 (APP/PSI) transgenic model. The K3 model is engineered to develop high levels of hyperphosphorylated tau and NFTs, and present cognitive deficits. The APP/PSI transgenic mouse model had high levels of Aβ and amyloid plaques and significant cognitive impairments. After the mice developed symptoms, they were treated with 670 nm light from light emitting diodes (LEDs) delivered to the skull in a 90-second cycle for 5 days per week for 4 weeks. The outcomes were compared against a control/wild-type breed. For the K3 mice, the PBM treatment produced a decrease in levels of hyperphosphorylated tau, NFTs, and markers of oxidative stress. In the neocortex and hippocampus, these levels were reduced to those seen in the control mice. Notably, the expression of the mitochondrial CCO was restored. Similarly, in the APP/PSI mice, the treatment with PBM produced a decrease in the number and size of the Aβ plaques.

Together, findings from these studies warranted human clinical studies for PBM to treat AD as the next stage of investigations.

29.8 Human clinical studies of photobiomodulation on dementia and Alzheimer's

Human studies involving PBM on dementia and AD subjects are relatively recent. At the time of this writing, the following studies have been published:

- Saltmarche et al. (2017), involving five participants assessed over 12 weeks in a case series report presenting significant improvement using transcranial and intranasal PBM devices.
- Berman et al. (2017), assessed 11 subjects with a transcranial PBM helmet over a short period of 28 days with some tests and electroencephalogram (EEG) readings that the authors interpret as improvement trends.
- Zomorrodi et al. (2017), reported on a moderately impaired AD case over 12 weeks who presented significant improvement in cognition within days, along with significant changes in EEG measures.

The following studies have been registered with the clinicaltrials.gov website and are ongoing at the time of writing as of March 31,2018:

- Chao LL, University of California, San Francisco, applying measures of cognitive, Alzheimer's disease assessment scale cognitive subscale (ADAS-cog), behavioral neuropsychiatric inventory (NPI), 3 Tesla resting state functional and arterial spin labeling (ASL) and perfusion magnetic resonance imaging (MRI) on eight participants with dementia (Chao, 2017).
- Lim et al., Vielight Inc., in a randomized double-blind pilot study involving 40 moderate to severe AD participants over 12 weeks, applying the severe impairment battery (SIB) scale as the primary endpoint, along with Alzheimer's

disease cooperative study—activities of daily living for severe Alzheimer's disease (ADCS-ADL-Sev) and neuropsychiatric inventory (NPI) as secondary endpoints (Lim, 2017).

The following is being mobilized to commence in spring 2018:

- Fischer et al., St. Michael's Hospital, Toronto, in a randomized double-blind pivotal study involving 228 patients with moderate to severe AD over eight sites with a duration of 6 months for each patient. The main primary and secondary endpoints are similar to the ongoing pilot study above.
- Chao and Rojas. University of California, San Francisco, in a sham-controlled pilot study of 14 patients with biomarker confirmed diagnoses of AD. The pilot study will have a duration of 4 months for patients randomized to active PBM and an optional 4 months of open-label PBM use for patients randomized to sham PBM. The primary endpoints will be measures of cognitive and behavioral function. The secondary endpoints will be measures of fluid (i.e., blood and cerebral spinal fluid) biomarkers of neuroinflammation, neurodegeneration, neurotrophic factors, and AD pathology (e.g., Aβ42, Aβ42/Aβ40, total- and hyperphosphorylated tau).

Further details of completed and partially completed studies where available at the time of this writing are presented in more detail below.

29.8.1 Saltmarche et al. (2017)

The study was aimed at evaluating the effect of near-infrared PBM on five patients with mild to moderately severe dementia or possible AD using the Vielight Neuro as the primary delivery device (Saltmarche et al., 2017) (see Fig. 29.1).

The near-infrared light (810 nm) was pulsed at 10 Hz frequency via LEDs, delivered to the hubs of the default mode network (DMN). The case series followed the participants over 12 weeks of active treatment and a 4-week follow-up period with no treatment. Cognitive impairment was assessed using the MMSE and the ADAS-cog scales. Patients were administered PBM using an intranasal device at home daily and were administered transcranial-intranasal PBM in the clinical site on a weekly basis. Following 12 weeks of treatment with PBM, significant improvement in cognition was observed, as assessed using the MMSE ($P < .003$) and ADAS-cog ($P < .023$). See Fig. 29.2 for the MMSE scores.

The 12-week treatment also resulted in increased function, better sleep, reduced anxiety, and fewer angry outbursts. No adverse effects were reported. Significant declines in cognitive functioning were observed during the 4-week no-treatment period, suggesting that maintenance treatment would be required. This study demonstrated for the first time that near-infrared PBM can produce significant improvements in cognition in patients with mild to moderately severe dementia cases with possible AD.

29.8.2 Zomorrodi et al. (2017)

This is a follow-up study to Saltmarche et al., above, introducing EEG to investigate the case of a randomly recruited AD patient. Certain modifications were made to the protocols of the earlier study, incorporated into the Vielight Neuro Gamma device: (1) The pulse rate was changed from 10 Hz (alpha) to 40 Hz (gamma). (2) Patients were given the headset to use at home for once a night, six nights a week. (3) Fewer LEDs were used, but more precisely targeted at the hubs of the DMN, and with more power (Zomorrodi et al., 2017).

The outcomes were even more significant than any of the participants of the previous study. Significant behavioral changes were already observed from the second day of treatment, such as improvement in eye contact from January 2010 to September 2, 2010. By the second day, the patient was emerging from silence and starting to hold meaningful conversations, and able to write. By the third week, he has regained most of his quality of life, with much improved ability to communicate. These are shown in Table 29.1.

Based on MMSE and ADAS-cog scales, the greatest rate of improvements were experienced in the first 3 weeks. Investigators in this study reported data over a total of 17 weeks. As presented in Fig. 29.3, MMSE increased from 21 at baseline to 26 at week 17 and ADAS-cog improved from 35 to 25. They covered 3 weeks for the ADCS-ADL scores, which showed improvements from 43 to 58.

The above global increase in EEG power was observed over a duration of 3 weeks. Measured during each of the few minutes of treatment, the investigators observed significant and consistent acute changes/entrainment during each 20-minute treatment.

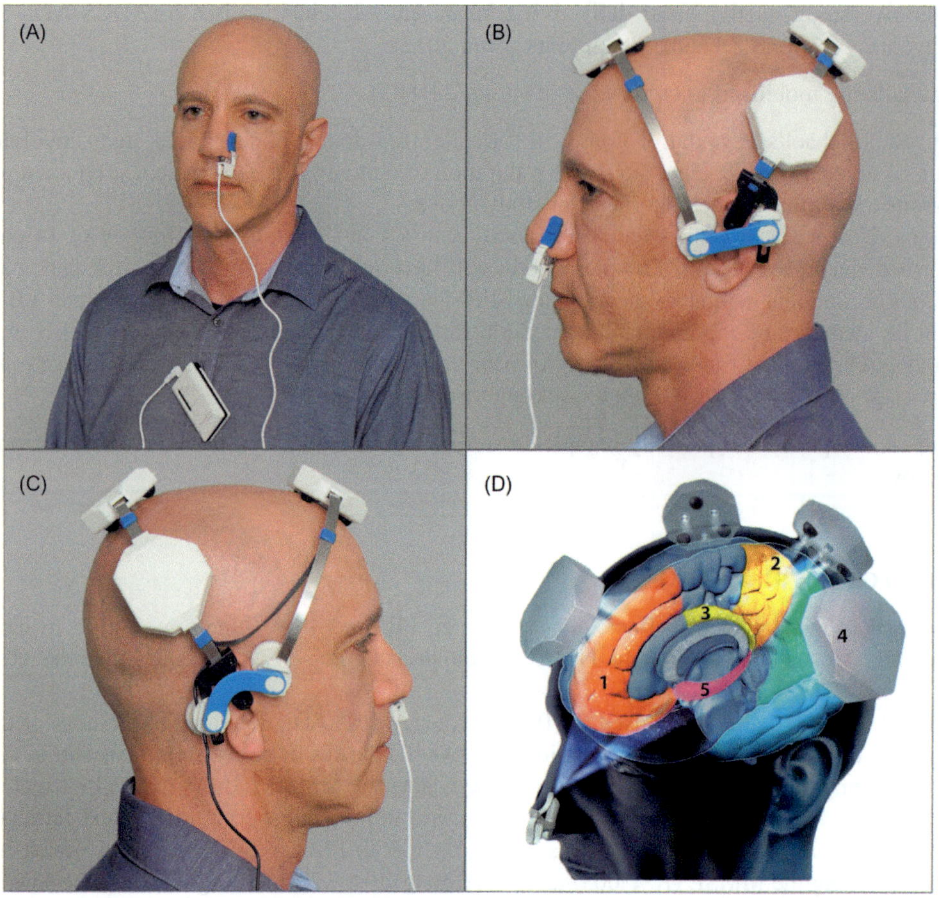

FIGURE 29.1 Photographs of Vielight 810 and Neuro illustrating correct device positions for treatment, and corresponding targeted network hubs. (A) Vielight 810; (B) Vielight Neuro, left view; (C) Vielight Neuro, right view; (D) Targeted default network nodes: (1) Mesial prefrontal cortex, (2) Precuneus, (3) Posterior cingulate cortex, (4) Inferior parietal lobe, (5) Hippocampus.

During each PBM treatment, significant elevation in the higher frequency oscillations in gamma, beta, and alpha were observed. On the other hand, the theta and delta oscillations were reduced. These are illustrated in Fig. 29.4.

In this case report, delivery of near infrared light at 810 nm to the hubs of the DMN, pulsed at 40 Hz produced significant improvements in:

- cognition (measured over 17 weeks);
- daily living and quality of life factors (measured over 3 weeks);
- electrophysiological baseline power over 3 weeks, across all oscillations;
- acute short-term entrainment after each treatment, elevating the power of gamma, beta, and alpha oscillations; and attenuating the power of theta and delta oscillations.

Outcomes were rapid and significant, noticeable within days, continuous and sustained over 3 weeks. The metrics continued to improve over the 17 weeks of the study. No negative side effects were observed.

The lessons learned here have been carried over to future AD studies, two are ongoing at this time, one of them, by Chao et al. at UCSF has interim data at this time.

29.8.3 Ongoing study—Chao (2018)

This study sought to replicate and expand on the findings of Saltmarche et al. (above), by using 3 Tesla ASL, perfusion MRI, and rs-fMRI to measure changes in brain perfusion and resting state functional connectivity. Like the Zomorrodi et al. study, patients randomized to immediate (IM) treatment were given the Vielight Neuro Gamma device to use at home. However, unlike the Zomorrodi et al. study, the frequency of PBM treatments was once every other day for

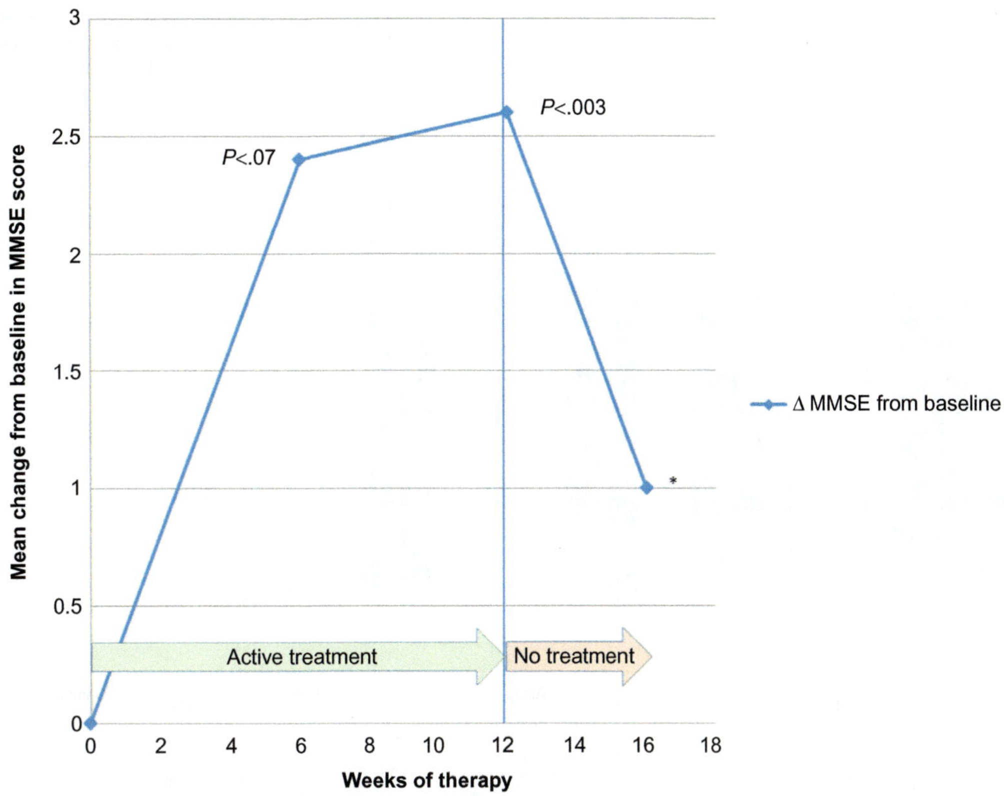

FIGURE 29.2 Mean change from baseline in MMSE scores.

TABLE 29.1 Changes in selected ordinal categorical normal functioning variables.

	Treatment					
Variable	**Day 0**	**1**	**2**	**7**	**14**	**21**
Eye contact	1	1	9.2	9.8	9.8	9.8
Demeanor	1	1	3	5	7	8
Motor skill—writing	0	0	4	7	7	
Motor skill—other	0	0	4	7	7	8
Reading	0	0	0	5	6	8
Email	0	0	0	4	7	8
Orientation	1	1	0	4	5	8
Long term memory	4	4	4	5	6	8
Short term memory	2	2	4	4	5	6
Clarity	4	4	5	6.5	6	6
Critical and abstract thinking	1	1	3	6	8	8
Conversation	1	1	4	6	6	7
Mood	1	1	3	4	7	8

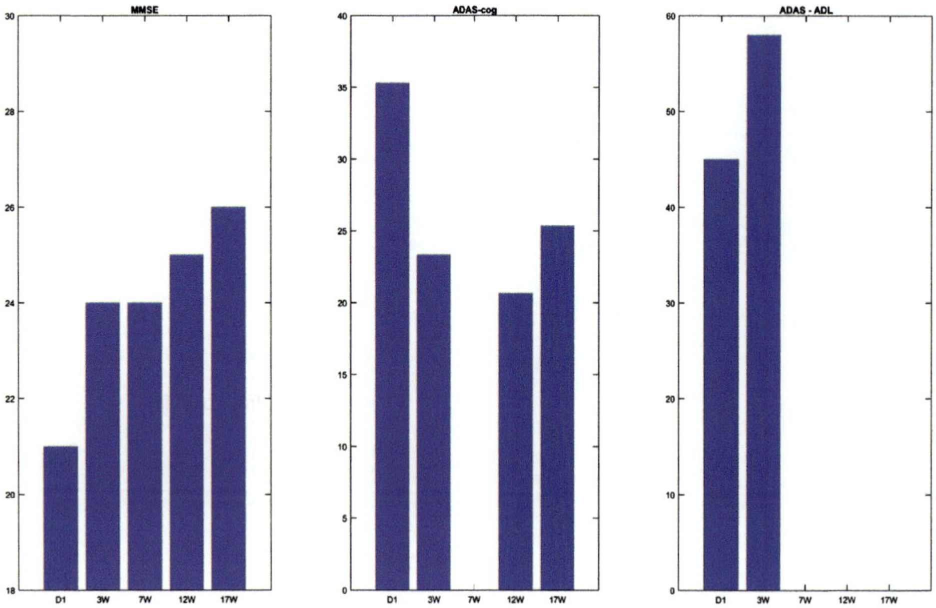

FIGURE 29.3 MMSE, ADAS-cog over 17 weeks and ADCS-ADL over 3 weeks.

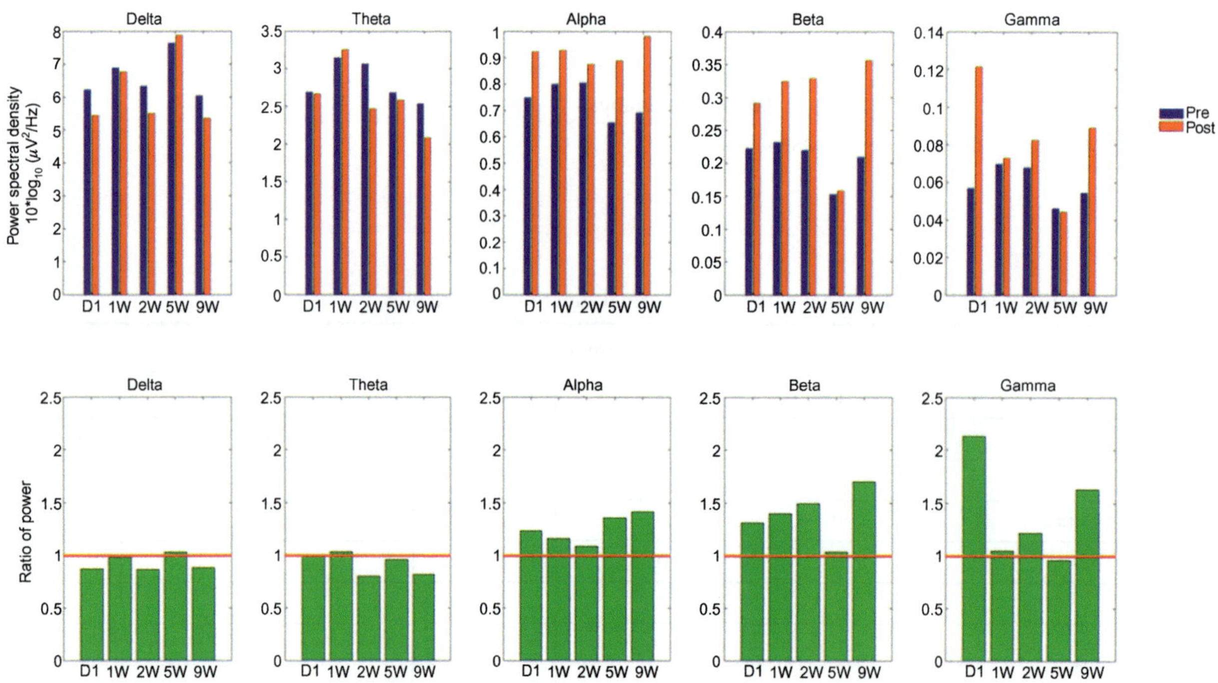

FIGURE 29.4 Acute short-term changes in brain oscillations.

Note: The study utilized a sham device in the week 5 measurement, and there was no change in the gamma and beta power spectral density, which supports the theory that the brain responds to an active (but not to sham) Neuro Gamma device. There were also no significant changes in the theta and delta oscillations. The change in alpha oscillations was expressed partly because the subject closed his eyes during the sessions.

12 weeks. Patients randomized to the wait-list (WL) group were followed for 12 weeks of usual care. At the time of this writing, eight patients with dementia have completed the pilot trial. Four patients were randomized to the IM group while four were randomized to WL. The two groups did not differ significantly on demographic variables or MMSE at baseline (see Table 29.2).

After 12 weeks, patients randomized to the IM group experienced significant improvements in cognitive function (using ADAS-cog) and behavioral symptoms (using NPI, see Fig. 29.5). In contrast, cognition and behavioral

TABLE 29.2 Demographic and baseline MMSE of pilot participants.

	IM group	WL group
N	4	4
Age, years (SD)	76.8 (12.0)	79.3 (5.7)
Education, years (SD)	19.0 (1.2)	18.0 (1.6)
Male: female ratio	2:2	2:2
Baseline MMSE (SD)	22.3 (7.3)	21.3 (5.7)

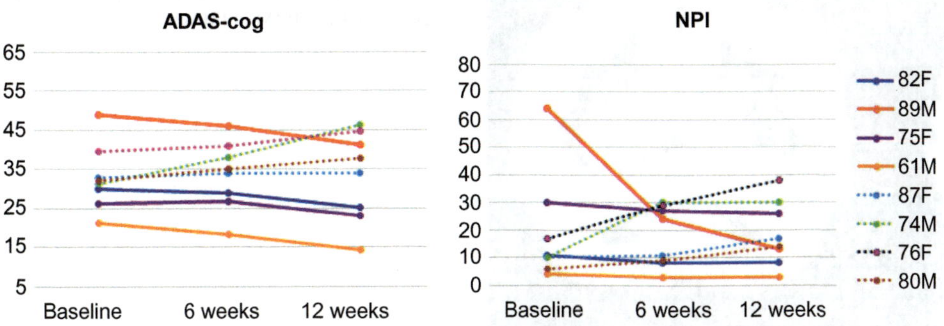

FIGURE 29.5 Plots of cognitive (ADAS-cog) and behavior (NPI) changes over 12 weeks in eight pilot PBM subjects. Lower scores on both measures indicate better function. IM subjects (*solid lines*) improved over 12 weeks while WL subjects (*dashed lines*) declined.

symptoms declined in the WL subjects. A repeated measures ANCOVA, with group (IM vs WL) as the between-subject measure, time (baseline, 6 and 12 weeks) as the with-in subject measure and age, education and baseline MMSE as covariates revealed a significant group by time interaction for ADAS-cog ($F_{1,3} = 33.35$, $P = .01$) and NPI ($F_{1,3} = 18.01$, $P = 0.02$).

WL subjects had the option of receiving 12 weeks of PBM treatments after completing 12 weeks of usual care. Cognitive and behavioral assessments were reassessed at 18 and 24 weeks, however these subjects did not undergo further MRIs. Fig. 29.6 shows that two WL subjects who declined during 12 weeks of usual care improved cognitively and behaviorally after they started to use the Vielight device (indicated by *yellow shading* in Fig. 29.6). Paired *t*-tests revealed significant differences in NPI scores at baseline and 12 weeks ($t = -15.0$, $df = 1$, $P = .04$), at 12 and 24 weeks ($t = 15.0$, $df = 1$, $P = .04$), but not at baseline and 24 weeks ($t = 0.0$, $df = 1$, $P = 1.0$) in these two WL subjects. Although paired *t*-test revealed no significant difference in ADAS-cog scores from baseline to 12 weeks, 12 to 24 weeks, or from baseline to 24 weeks, Fig. 29.6 suggests this is because the female WL subject's ADAS-cog scores changed minimally over the 24 weeks. Nevertheless, it is noteworthy that her ADAS-cog score after 12 weeks of using the Vielight device (30.67 at 24 weeks) was lower (i.e., better) than it was at baseline.

As noted previously, the LEDs of the Vielight Neuro Gamma device target the hubs of the DMN. Research suggests that AD patients not only have reduced functional connectivity within the DMN (Mevel et al., 2011), but they also have reduced resting glucose metabolism and exhibit eventual atrophy in the DMN nodes that progresses with disease severity (Minoshima et al., 1997; Scahill et al., 2002; Thompson et al., 2003). Amyloid beta plaques have also been observed to selectively accumulate within DMN modes, even prior to symptom onset (Klunk et al., 2004). After 12 weeks of PBM treatments, there was increased perfusion in the posterior cingulate and bilateral parietal cortex inferior temporal, and left frontal cortex (see Fig. 29.7). There was also increased functional connectivity within DMN in the four IM patients (see Fig. 29.8). Specifically, there was greater connectivity between the hippocampus, lateral parietal cortex, and posterior cingulate cortex (PCC) in the IM group at week 12 compared to baseline.

29.8.4 Discussion on the clinical studies

The results of the clinical have been promising to date. However, a PBM intervention is not ready to be considered an effective treatment for AD until it passes the same standard of scrutiny placed on the pharmacotherapy trials.

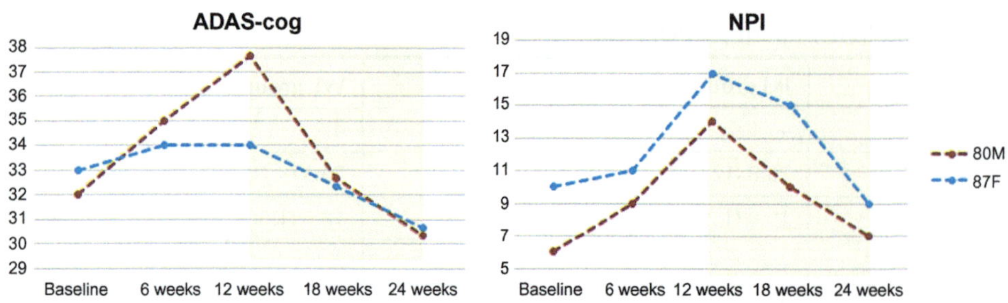

FIGURE 29.6 Change in ADAS-cog and NPI scores in two WL subjects during 12 weeks of usual care (i.e., no PBM) and after starting PBM (*shaded yellow*). Lower scores = better cognitive and behavioral function.

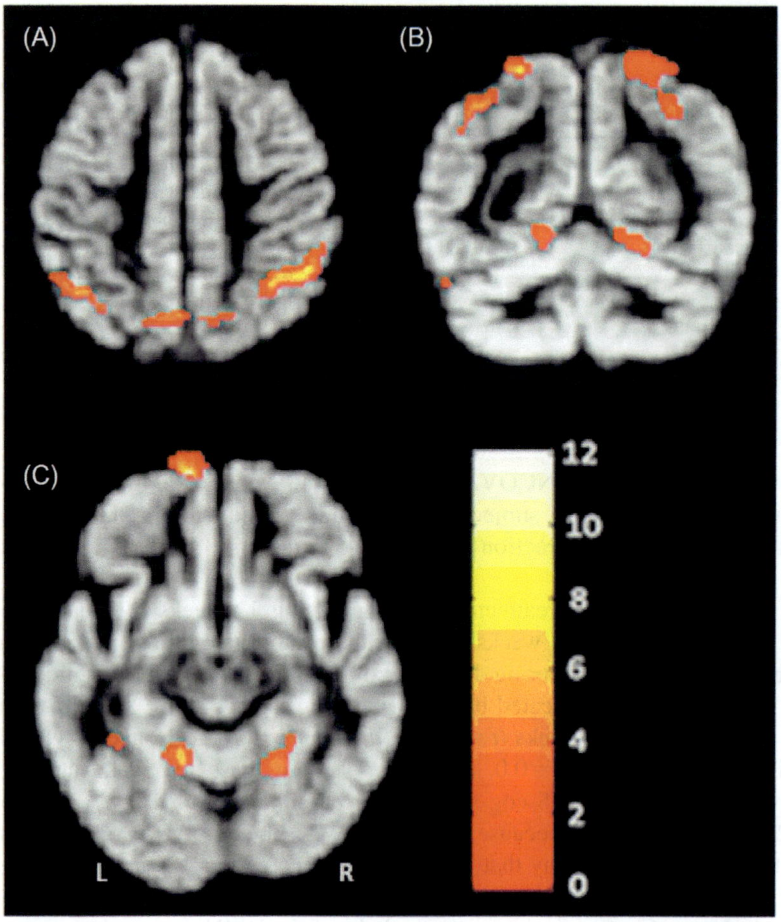

FIGURE 29.7 Increased perfusion in the posterior cingulate and bilateral parietal cortex (A and B), inferior temporal (B and C), and left frontal (C) cortex after 12 weeks of PBM in IM subjects.

Considering the expensive failures of pharmacotherapy trials, such a claim would be considered extraordinary, requiring to pass gold-standard tests. These tests would call for much larger number of participants, producing statistically significant results in double-blind studies. These are the factors partially incorporated into the ongoing pilot study sponsored by Vielight Inc. (Lim, 2017) and then into a pivotal trial with its primary clinical trial site at the St. Michael's Hospital in Toronto, also sponsored by Vielight Inc.

29.9 Key parameters

The combination of several parameters are likely to be the major contributors to the potential success of PBM to treat AD, some of these are novel in the PBM field, particularly the significance of the DMN and the pulse rate.

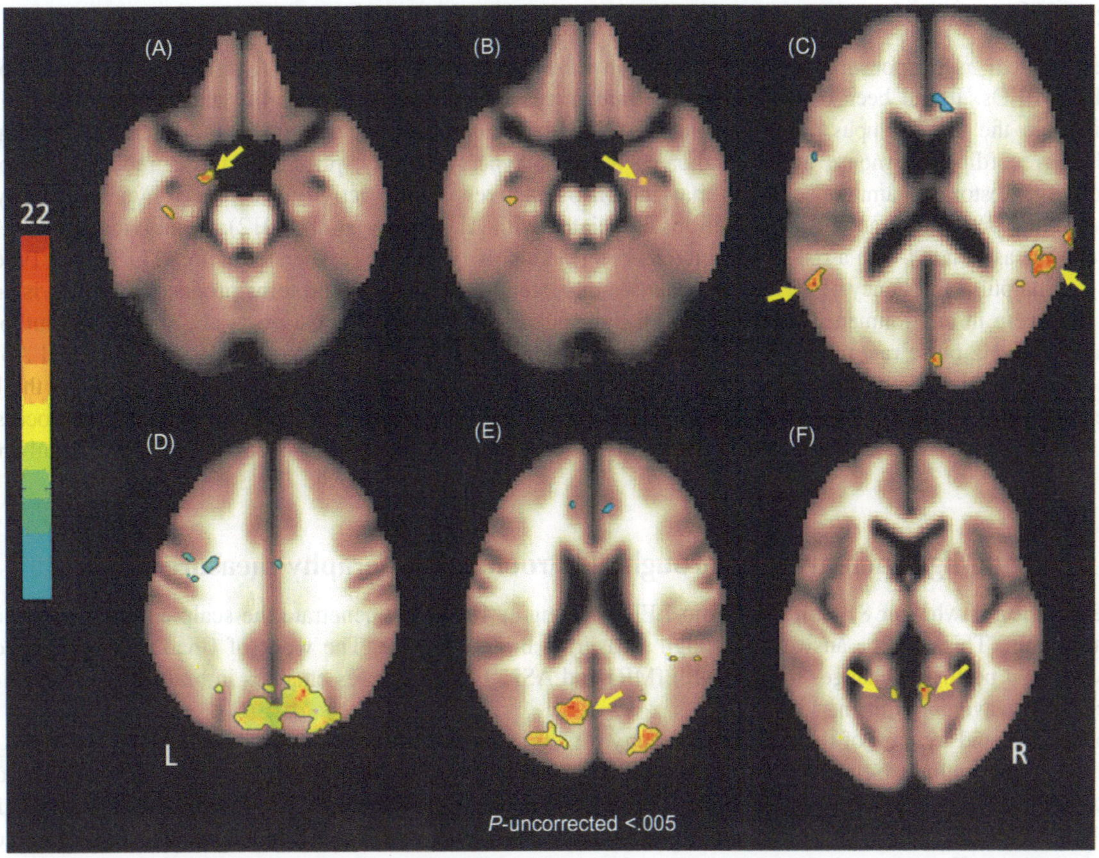

FIGURE 29.8 Increased functional connectivity within DMN network after 12 weeks of PBM in four IM subjects. There was greater connectivity between the left (A) and right (B) hippocampus and lateral parietal cortex (C) with a seed in the posterior cingulate cortex, and greater connectivity between the posterior cingulate cortex with seeds in the left (D and E) and right (F) lateral parietal cortex.

29.9.1 The default mode network

The DMN is a large-scale brain network involved that is particularly active when the brain is in a state of wakeful rest. This network includes the mesial prefrontal cortex, the PCC, the hippocampus, the precuneus, the inferior parietal lobe, and the temporal lobe, as presented in Fig. 29.1D. The DMN is involved with a number of cognitive functions, including autobiographical memory, memory consolidation, and self-referential thought (Andrews-Hanna et al., 2007). This network is of particular relevance for AD as the mesial prefrontal cortex, the medial temporal lobe, and particularly the hippocampus are involved in mediating episodic moment processing. In AD, an impairment in episodic memory is one of the first symptoms observed (Greicius et al., 2004).

Significant disruptions in the DMN have been reported in patient with AD (Beason-Held, 2011; Binnewijzend et al., 2012; Buckner et al., 2009; Damoiseaux et al., 2012; Greicius et al., 2004). Investigations using resting-state functional connectivity magnetic resonance imaging demonstrated decreased functional connectivity between the PCC/precuneus and the hippocampus (Greicius et al., 2004). Disruptions in the DMN have also been observed in patients with mild cognitive impairment (Binnewijzend et al., 2012; Petrella et al., 2011; Sorg et al., 2007), in patients with a family history of AD (Fleisher et al., 2009) and in patients with the APOE4 allele, which is considered a significant genetic risk factor for late onset AD (Filippini et al., 2009; Patel et al., 2013; Westlye et al., 2011). Age-related connectivity alterations in the DMN are more profound in patients with AD than healthy controls (Jones et al., 2011). Furthermore, a correlation has been reported between the anatomical distribution of amyloid plaques, neurotrophy, and alterations in glucose metabolism (Buckner et al., 2005). Given the significant role that the DMN plays in the pathophysiology of AD, this network represents an important neuroanatomical target for treatment with PBM.

29.9.2 Pulse rate of 40 Hz

The latest clinical trials sponsored by Vielight Inc. involve the delivery of 810 nm light at pulse frequency of 40 Hz from all the LEDs. It had been observed that aberrant increases in network excitability and compensatory inhibitory mechanisms in the hippocampus may contribute to Aβ-induced neurological deficits in mouse models (Palop et al., 2007). EEG recordings in mouse models indicated network hypersynchrony, primarily during reduced gamma oscillatory activity. Restoring gamma oscillation may inhibit overactive synaptic activity and reduce hypersynchrony, memory deficits, and premature mortality—conditions associated with AD (Verret et al., 2012). In individuals with AD, this phenomenon is associated with a risk for an increased formation of Aβ protein associated with AD (Selkoe, 1996).

The gamma pulse frequency of 40 Hz has been demonstrated to attenuate Aβ proteins production in the visual cortex of mice that were in environments illuminated with light pulsing at that rate (Iaccarino et al., 2016). The authors theorized that the 40 Hz pulse rate modifies microglia into the noninflammatory state that engulfs the unwanted Aβ protein deposits. When 40 Hz pulsing light were optogenetically induced in the hippocampus, Aβ peptide levels in the location also attenuated significantly. From the data we hypothesize that Aβ is attenuated in the brain regions that process pulsed light at 40 Hz, and if we can target 40 Hz to the right areas such as the DMN, it could be an impactful treatment for AD.

29.10 Proving light penetration through electroencephalography measures

Now the question is whether 810 nm light from LEDs pulsing at 40 Hz can penetrate the scalp and subdermal layers to entrain the brain. This may be answered through measuring brain response. The work of Chao et al. presented above has shown blood perfusion changes imaged with fMRI and ASL. EEG measures changes in brain wave oscillations and can teach brain wave entrainment with PBM.

To prove this and to learn of the electrophysiology effects, a double-blind EEG study was carried out on 20 healthy subjects using the Vielight Neuro Gamma (this paper is being prepared by Zomorrodi et al. for publication at the time of this writing). The power spectrum and connectivity analysis were performed to assess the difference in the induced change between active and sham intervention. The analysis revealed that PBM delivered by the device produced frequency dependent effects on endogenous brain activity. A reduction of low-frequency power (i.e., delta and theta) and an increase of high-frequency power (i.e., alpha, beta, and gamma) were observed following active stimulation, replicating the earlier case report for the moderately impaired AD patient (Zomorrodi et al., 2017). Comparison between active and sham groups demonstrated a significant difference in delta 1–4 Hz; $t = -2.53$ $P = .0203$), theta (4–7 Hz; $t = -3.18$ $P = .0049$), alpha (8–12 Hz; $t = 4.26$ $P = .0004$), beta (12–30 Hz; $t = 3.02$ $P = .0070$), and gamma (30–50 Hz; $t = 3.84$, $P = .0011$). The weighted phase lag index and graph theory measures revealed significant change ($P < .001$) between active and sham stimulation, limited to the alpha and gamma frequencies. Findings from this study provide early evidence that PBM modulates cortical oscillations that impact brain connectivity in a pulse frequency-dependent manner. The study confirms that near infrared light with the right parameters can penetrate the brain sufficiently enough to produce significant electrophysiological changes.

29.11 Electroencephalography as a tool for developing Alzheimer's disease therapies

EEG is a potentially valuable tool in understanding the impairments in brain activity in individuals with AD. It is noninvasive and used for recording brain waves from the surface of the scalp (Swartz and Goldensohn, 1998). EEG allows for the measurement of neuronal electrical activity noninvasively, both spontaneous and event-related activity, over the complete surface of the brain. One of the key areas of brain activities presented with EEG are brain oscillations of various frequencies. These brain oscillations are the result of thousands of neurons synchronizing their activity and play an important functional role. Oscillatory activities in neural systems may play a functional role, and abnormalities in neural synchronization mechanisms might be involved in the pathophysiology of several neuropsychiatric disorders (Uhlhaas and Singer, 2006). The brain oscillations of various frequencies are traditionally grouped into bands corresponding the physiological characteristics of the given band. The bands include delta (0.1–4 Hz), theta (4–8 Hz), alpha (8–12 Hz), beta (12–30 Hz) and gamma (30–70 Hz) (Basar et al., 2013). Delta waves can occur either in the cortex or the thalamus (Maquet et al., 1997). They are commonly observed during sleep (Rasch and Born, 2013). Delta waves also play an important role in motivational drive (Knyazev, 2007) and stimulate the release of hormones including prolactin and growth stimulating hormone (Brandenberger, 2003). Theta waves play an important role in various forms of learning and memory (Berry and Thompson, 1978; Liebe et al., 2012; Macrides et al., 1982) and are also important for synaptic plasticity (Greenstein et al., 1988; Hyman et al., 2003; Larson et al, 1986). Theta oscillations are observed during REM sleep, but not during the deeper stages of sleep (Vanderwolf, 1969; Winson, 1974). Alpha waves were originally thought to originate primarily from the occipital cortex during a state of restful relaxation with eyes closed (Basar et al, 1997; Pfurtscheller et al., 1996). More recent evidence suggests that alpha waves represent functional inhibition through event-related synchronization (Jensen and Mazaheri, 2010; Klimesch, 2012; Klimesch et al., 2007). Beta waves are associated with active concentration during a wakeful state (Baumeister et al., 2008; Neuper and Pfurtscheller, 2001). Over the motor cortex, beta waves occur during muscle contractions in isotonic movements (Baker, 2007). Beta waves become suppressed before and during changes in movements (Baker, 2007). Gamma waves are associated with a number of behaviors including attention (Jensen et al., 2007), working memory (Howard et al., 2003), and visual perception (Beauchamp et al., 2012). Recently, growing attention has been given to studying the gamma wavelengths due to its association with AD, particularly at 40 Hz (Iaccarino et al., 2016) as discussed above.

Abnormalities in brain oscillations are very common in neurological, neuropsychiatric, and neurodegenerative conditions. Numerous studies have reported increases in theta and delta activities and decreases in alpha and beta activities in patients with AD (Babiloni et al., 2004, 2013; Brenner et al., 1986; Coben et al., 1983, 1985; Giaquinto and Nolfe, 1986; Huang et al., 2000; Jeong, 2004). During rest, a decrease of posterior alpha power is observed in patients with AD (Babiloni et al., 2004, 2013; Huang et al., 2000; Jeong, 2004). This decrease in alpha power is correlated with severity of AD and cognitive impairments (Jeong, 2004). Furthermore, in examining coherence, a measure of the degree of association between two frequency bands, a decrease in coherence between alpha and beta bands are frequently found in patients with AD (Dunkin et al., 1994; Leuchter et al., 1987). Most importantly, these abnormalities are correlated with the severity of the disease (Kowalski et al., 2001).

29.12 Pulsed photobiomodulation as a potential treatment modality

We have observed that PBM at 40 Hz consistently elevates the power of alpha, beta, and delta, and reduces the power of theta and delta (Zomorrodi et al., 2018). It follows that this pulse rate has the potential to treat the AD by altering the associated brain EEG patterns. This would overturn characteristic brain oscillation patterns of the AD population observed by Assenza et al. (2017) – low power in alpha and high gamma, and high power in delta and theta. Throughout the years, Babiloni et al. (2004, 2013) has reported that AD downshifts the alpha frequency peak, corresponding to a worsening of cognitive performance.

Apart from 40 Hz, there is no report of clinical tests for other pulse frequencies. It may be possible that there is a broad range, including 40 Hz that may achieve similar results.

29.13 The future of photobiomodulation as a treatment for Alzheimer's disease

To date, the evidence we have observed from the use of PBM to treat AD has been significant and very promising as presented in this chapter. However, any claim that is considered as extraordinary as this, needs evidence beyond reproach. To achieve this call for pivotal randomized double-blind clinical trials. In this respect, the upcoming pivotal trial sponsored by Vielight Inc. will be carefully watched. Successful data from the trial provides the long-awaited breakthrough in the search for an effective treatment for AD. However, it also gives the field of PBM the level of credibility that has been long overdue.

EEG representation of brain response to the brain has been broadly predictable. As presented above, pulse rates can influence functions as represented by brain waves patterns. In the clinical studies, participants with AD symptoms have responded to PBM differently. Further research using different parameters such as pulse rate, LED power, and location may reveal how these variables can be adjusted to personalize treatment according to EEG presentations. This may further magnify outcomes with PBM that have been observed so far.

References

Andrews-Hanna, J.R., Snyder, A.Z., Vincent, J.L., Lustig, C., Head, D., Raichle, M.E., et al., 2007. Disruption of large-scale brain systems in advanced aging. Neuron 56 (5), 924–935. Available from: https://doi.org/10.1016/j.neuron.2007.10.038.

Assenza, G., Capone, F., di Biase, L., Ferreri, F., Florio, L., Guerra, A., et al., 2017. Corrigendum: Oscillatory activities in neurological disorders of elderly: biomarkers to target for neuromodulation. Front. Aging Neurosci. 9, 252. Available from: https://doi.org/10.3389/fnagi.2017.00252.

Babiloni, C., Binetti, G., Cassetta, E., Cerboneschi, D., Dal Forno, G., Del Percio, C., et al., 2004. Mapping distributed sources of cortical rhythms in mild Alzheimer's disease. A multicentric EEG study. Neuroimage 22 (1), 57–67. Available from: https://doi.org/10.1016/j.neuroimage.2003.09.028.

Babiloni, C., Lizio, R., Del Percio, C., Marzano, N., Soricelli, A., Salvatore, E., et al., 2013. Cortical sources of resting state EEG rhythms are sensitive to the progression of early stage Alzheimer's disease. J. Alzheimers Dis. 34 (4), 1015–1035. Available from: https://doi.org/10.3233/JAD-121750.

Baker, S.N., 2007. Oscillatory interactions between sensorimotor cortex and the periphery. Curr. Opin. Neurobiol. 17 (6), 649–655. Available from: https://doi.org/10.1016/j.conb.2008.01.007.

Basar, E., Schurmann, M., Basar-Eroglu, C., Karakas, S., 1997. Alpha oscillations in brain functioning: an integrative theory. Int. J. Psychophysiol. 26 (1–3), 5–29.

Basar, E., Basar-Eroglu, C., Guntekin, B., Yener, G.G., 2013. Brain's alpha, beta, gamma, delta, and theta oscillations in neuropsychiatric diseases: proposal for biomarker strategies. Suppl. Clin. Neurophysiol. 62, 19–54.

Baumeister, J., Barthel, T., Geiss, K.R., Weiss, M., 2008. Influence of phosphatidylserine on cognitive performance and cortical activity after induced stress. Nutr. Neurosci. 11 (3), 103–110. Available from: https://doi.org/10.1179/147683008X301478.

Beason-Held, L.L., 2011. Dementia and the default mode. Curr. Alzheimer Res. 8 (4), 361–365.

Beauchamp, M.S., Sun, P., Baum, S.H., Tolias, A.S., Yoshor, D., 2012. Electrocorticography links human temporoparietal junction to visual perception. Nat. Neurosci. 15 (7), 957–959. Available from: https://doi.org/10.1038/nn.3131.

Berman, M.H., Halper, J.P., Nichols, T.W., Jarrett, H., Lundy, A., Huang, J.H., 2017. Photobiomodulation with near infrared light helmet in a pilot, placebo controlled clinical trial in dementia patients testing memory and cognition. J. Neurol. Neurosci. 8 (1). Available from: https://doi.org/10.21767/2171-6625.1000176.

Berry, S.D., Thompson, R.F., 1978. Prediction of learning rate from the hippocampal electroencephalogram. Science 200 (4347), 1298–1300.

Bindoli, A., Rigobello, M.P., 2013. Principles in redox signaling: from chemistry to functional significance. Antioxid. Redox Signal. 18 (13), 1557–1593. Available from: https://doi.org/10.1089/ars.2012.4655.

Binnewijzend, M.A., Schoonheim, M.M., Sanz-Arigita, E., Wink, A.M., van der Flier, W.M., Tolboom, N., et al., 2012. Resting-state fMRI changes in Alzheimer's disease and mild cognitive impairment. Neurobiol. Aging 33 (9), 2018–2028. Available from: https://doi.org/10.1016/j.neurobiolaging.2011.07.003.

Blass, J.P., Baker, A.C., Ko, L., Black, R.S., 1990a. Induction of Alzheimer antigens by an uncoupler of oxidative phosphorylation. Arch. Neurol. 47 (8), 864–869.

Blass, J.P., Baker, A.C., Ko, L.W., 1990b. Alzheimer's disease: inborn error of metabolism of late onset? Adv. Neurol. 51, 199–200.

Brandenberger, G., 2003. [The ultradian rhythm of sleep: diverse relations with pituitary and adrenal hormones]. Rev. Neurol. (Paris) 159 (11 Suppl.), 6S5–10.

Brenner, R.P., Ulrich, R.F., Spiker, D.G., Sclabassi, R.J., Reynolds 3rd, C.F., Marin, R.S., et al., 1986. Computerized EEG spectral analysis in elderly normal, demented and depressed subjects. Electroencephalogr. Clin. Neurophysiol. 64 (6), 483–492.

Brion, J.P., 1998. Neurofibrillary tangles and Alzheimer's disease. Eur. Neurol. 40 (3), 130–140. Available from: https://doi.org/10.1159/000007969.

Buckner, R.L., Snyder, A.Z., Shannon, B.J., LaRossa, G., Sachs, R., Fotenos, A.F., et al., 2005. Molecular, structural, and functional characterization of Alzheimer's disease: evidence for a relationship between default activity, amyloid, and memory. J. Neurosci. 25 (34), 7709–7717. Available from: https://doi.org/10.1523/JNEUROSCI.2177-05.2005.

Buckner, R.L., Sepulcre, J., Talukdar, T., Krienen, F.M., Liu, H., Hedden, T., et al., 2009. Cortical hubs revealed by intrinsic functional connectivity: mapping, assessment of stability, and relation to Alzheimer's disease. J. Neurosci. 29 (6), 1860–1873. Available from: https://doi.org/10.1523/JNEUROSCI.5062-08.2009.

Castello, M.A., Soriano, S., 2014. On the origin of Alzheimer's disease. Trials and tribulations of the amyloid hypothesis. Ageing Res. Rev. 13, 10–12. Available from: https://doi.org/10.1016/j.arr.2013.10.001.

Chao, L., 2017. Photobiomodulation for improving brain function in dementia (PBM dementia). Retrieved from: https://clinicaltrials.gov/ct2/show/NCT03160027.

Coben, L.A., Danziger, W.L., Berg, L., 1983. Frequency analysis of the resting awake EEG in mild senile dementia of Alzheimer type. Electroencephalogr. Clin. Neurophysiol. 55 (4), 372–380.

Coben, L.A., Danziger, W., Storandt, M., 1985. A longitudinal EEG study of mild senile dementia of Alzheimer type: changes at 1 year and at 2.5 years. Electroencephalogr. Clin. Neurophysiol. 61 (2), 101–112.

Corder, E.H., Saunders, A.M., Strittmatter, W.J., Schmechel, D.E., Gaskell, P.C., Small, G.W., et al., 1993. Gene dose of apolipoprotein E type 4 allele and the risk of Alzheimer's disease in late onset families. Science 261 (5123), 921–923.

Damoiseaux, J.S., Prater, K.E., Miller, B.L., Greicius, M.D., 2012. Functional connectivity tracks clinical deterioration in Alzheimer's disease. Neurobiol. Aging 33 (4), . Available from: https://doi.org/10.1016/j.neurobiolaging.2011.06.024828 e819–830.

de Freitas, L.F., Hamblin, M.R., 2016. Proposed mechanisms of photobiomodulation or low-level light therapy. IEEE J. Sel. Top. Quantum Electron. 22 (3). Available from: https://doi.org/10.1109/JSTQE.2016.2561201.

De Taboada, L., Yu, J., El-Amouri, S., Gattoni-Celli, S., Richieri, S., McCarthy, T., et al., 2011. Transcranial laser therapy attenuates amyloid-beta peptide neuropathology in amyloid-beta protein precursor transgenic mice. J. Alzheimers Dis. 23 (3), 521–535. Available from: https://doi.org/10.3233/JAD-2010-100894.

Drachman, D.A., 2014. The amyloid hypothesis, time to move on: amyloid is the downstream result, not cause, of Alzheimer's disease. Alzheimers Dement. 10 (3), 372–380. Available from: https://doi.org/10.1016/j.jalz.2013.11.003.

Du, H., Guo, L., Yan, S.S., 2012. Synaptic mitochondrial pathology in Alzheimer's disease. Antioxid. Redox Signal. 16 (12), 1467–1475. Available from: https://doi.org/10.1089/ars.2011.4277.

Dubois, B., Padovani, A., Scheltens, P., Rossi, A., Dell'Agnello, G., 2016. Timely diagnosis for Alzheimer's disease: a literature review on benefits and challenges. J. Alzheimers Dis. 49 (3), 617–631. Available from: https://doi.org/10.3233/JAD-150692.

Dunkin, J.J., Leuchter, A.F., Newton, T.F., Cook, I.A., 1994. Reduced EEG coherence in dementia: state or trait marker? Biol. Psychiatry 35 (11), 870–879.

Farivar, S., Malekshahabi, T., Shiari, R., 2014. Biological effects of low level laser therapy. J. Lasers Med. Sci. 5 (2), 58–62.

Filippini, N., MacIntosh, B.J., Hough, M.G., Goodwin, G.M., Frisoni, G.B., Smith, S.M., et al., 2009. Distinct patterns of brain activity in young carriers of the APOE-epsilon4 allele. Proc. Natl. Acad. Sci. U.S.A. 106 (17), 7209–7214. Available from: https://doi.org/10.1073/pnas.0811879106.

Fleisher, A.S., Sherzai, A., Taylor, C., Langbaum, J.B., Chen, K., Buxton, R.B., 2009. Resting-state BOLD networks versus task-associated functional MRI for distinguishing Alzheimer's disease risk groups. Neuroimage 47 (4), 1678–1690. Available from: https://doi.org/10.1016/j.neuroimage.2009.06.021.

Fraser, S.P., Suh, Y.H., Djamgoz, M.B., 1997. Ionic effects of the Alzheimer's disease beta-amyloid precursor protein and its metabolic fragments. Trends Neurosci. 20 (2), 67–72.

Friedland-Leuner, K., Stockburger, C., Denzer, I., Eckert, G.P., Muller, W.E., 2014. Mitochondrial dysfunction: cause and consequence of Alzheimer's disease. Prog. Mol. Biol. Transl. Sci. 127, 183–210. Available from: https://doi.org/10.1016/B978-0-12-394625-6.00007-6.

Gabuzda, D., Busciglio, J., Chen, L.B., Matsudaira, P., Yankner, B.A., 1994. Inhibition of energy metabolism alters the processing of amyloid precursor protein and induces a potentially amyloidogenic derivative. J. Biol. Chem. 269 (18), 13623–13628.

Gardner, E., 2017. Alzheimer's drugs: will 2017 bring better news? Retrieved from: <https://www.pharmaceutical-technology.com/features/featurealzheimers-drugs-will-2017-bring-better-news-5775766/>.

Gasparini, L., Racchi, M., Benussi, L., Curti, D., Binetti, G., Bianchetti, A., et al., 1997. Effect of energy shortage and oxidative stress on amyloid precursor protein metabolism in COS cells. Neurosci. Lett. 231 (2), 113–117.

Giaquinto, S., Nolfe, G., 1986. The EEG in the normal elderly: a contribution to the interpretation of aging and dementia. Electroencephalogr. Clin. Neurophysiol. 63 (6), 540–546.

Gibson, G.E., Shi, Q., 2010. A mitocentric view of Alzheimer's disease suggests multi-faceted treatments. J. Alzheimers Dis. 20 (Suppl. 2), S591–607. Available from: https://doi.org/10.3233/JAD-2010-100336.

Goate, A., Chartier-Harlin, M.C., Mullan, M., Brown, J., Crawford, F., Fidani, L., et al., 1991. Segregation of a missense mutation in the amyloid precursor protein gene with familial Alzheimer's disease. Nature 349 (6311), 704–706. Available from: https://doi.org/10.1038/349704a0.

Greenstein, Y.J., Pavlides, C., Winson, J., 1988. Long-term potentiation in the dentate gyrus is preferentially induced at theta rhythm periodicity. Brain Res. 438 (1–2), 331–334.

Greicius, M.D., Srivastava, G., Reiss, A.L., Menon, V., 2004. Default-mode network activity distinguishes Alzheimer's disease from healthy aging: evidence from functional MRI. Proc. Natl. Acad. Sci. U.S.A. 101 (13), 4637–4642. Available from: https://doi.org/10.1073/pnas.0308627101.

Hamblin, M.R., 2016. Shining light on the head: photobiomodulation for brain disorders. BBA Clin. 6, 113–124. Available from: https://doi.org/10.1016/j.bbacli.2016.09.002.

Hashmi, J.T., Huang, Y.Y., Osmani, B.Z., Sharma, S.K., Naeser, M.A., Hamblin, M.R., 2010. Role of low-level laser therapy in neurorehabilitation. PM R 2 (12 Suppl. 2), S292–305. Available from: https://doi.org/10.1016/j.pmrj.2010.10.013.

Hasselmo, M.E., 2006. The role of acetylcholine in learning and memory. Curr. Opin. Neurobiol. 16 (6), 710–715. Available from: https://doi.org/10.1016/j.conb.2006.09.002.

Hellweg, R., Wirth, Y., Janetzky, W., Hartmann, S., 2012. Efficacy of memantine in delaying clinical worsening in Alzheimer's disease (AD): responder analyses of nine clinical trials with patients with moderate to severe AD. Int. J. Geriatr. Psychiatry 27 (6), 651–656. Available from: https://doi.org/10.1002/gps.2766.

Herrup, K., 2015. The case for rejecting the amyloid cascade hypothesis. Nat. Neurosci. 18 (6), 794–799. Available from: https://doi.org/10.1038/nn.4017.

Howard, M.W., Rizzuto, D.S., Caplan, J.B., Madsen, J.R., Lisman, J., Aschenbrenner-Scheibe, R., et al., 2003. Gamma oscillations correlate with working memory load in humans. Cereb. Cortex 13 (12), 1369–1374.

Hu, X., Das, B., Hou, H., He, W., Yan, R., 2018. BACE1 deletion in the adult mouse reverses preformed amyloid deposition and improves cognitive functions. J. Exp. Med. 215 (3), 927–940. Available from: https://doi.org/10.1084/jem.20171831.

Huang, C., Wahlund, L., Dierks, T., Julin, P., Winblad, B., Jelic, V., 2000. Discrimination of Alzheimer's disease and mild cognitive impairment by equivalent EEG sources: a cross-sectional and longitudinal study. Clin. Neurophysiol. 111 (11), 1961–1967.

Hyman, J.M., Wyble, B.P., Goyal, V., Rossi, C.A., Hasselmo, M.E., 2003. Stimulation in hippocampal region CA1 in behaving rats yields long-term potentiation when delivered to the peak of theta and long-term depression when delivered to the trough. J. Neurosci. 23 (37), 11725–11731.

Iaccarino, H.F., Singer, A.C., Martorell, A.J., Rudenko, A., Gao, F., Gillingham, T.Z., et al., 2016. Gamma frequency entrainment attenuates amyloid load and modifies microglia. Nature 540 (7632), 230–235. Available from: https://doi.org/10.1038/nature20587.

Iqbal, K., Liu, F., Gong, C.X., 2014. Alzheimer disease therapeutics: focus on the disease and not just plaques and tangles. Biochem. Pharmacol. 88 (4), 631–639. Available from: https://doi.org/10.1016/j.bcp.2014.01.002.

Jensen, O., Mazaheri, A., 2010. Shaping functional architecture by oscillatory alpha activity: gating by inhibition. Front. Hum. Neurosci. 4, 186. Available from: https://doi.org/10.3389/fnhum.2010.00186.

Jensen, O., Kaiser, J., Lachaux, J.P., 2007. Human gamma-frequency oscillations associated with attention and memory. Trends Neurosci. 30 (7), 317–324. Available from: https://doi.org/10.1016/j.tins.2007.05.001.

Jeong, J., 2004. EEG dynamics in patients with Alzheimer's disease. Clin. Neurophysiol. 115 (7), 1490–1505. Available from: https://doi.org/10.1016/j.clinph.2004.01.001.

Jones, D.T., Machulda, M.M., Vemuri, P., McDade, E.M., Zeng, G., Senjem, M.L., et al., 2011. Age-related changes in the default mode network are more advanced in Alzheimer disease. Neurology 77 (16), 1524–1531. Available from: https://doi.org/10.1212/WNL.0b013e318233b33d.

Kapogiannis, D., Mattson, M.P., 2011. Disrupted energy metabolism and neuronal circuit dysfunction in cognitive impairment and Alzheimer's disease. Lancet Neurol. 10 (2), 187–198. Available from: https://doi.org/10.1016/S1474-4422(10)70277-5.

Karran, E., Hardy, J., 2014. A critique of the drug discovery and phase 3 clinical programs targeting the amyloid hypothesis for Alzheimer disease. Ann. Neurol. 76 (2), 185–205. Available from: https://doi.org/10.1002/ana.24188.

Klimesch, W., 2012. alpha-band oscillations, attention, and controlled access to stored information. Trends Cogn. Sci. 16 (12), 606–617. Available from: https://doi.org/10.1016/j.tics.2012.10.007.

Klimesch, W., Sauseng, P., Hanslmayr, S., 2007. EEG alpha oscillations: the inhibition-timing hypothesis. Brain Res. Rev. 53 (1), 63–88. Available from: https://doi.org/10.1016/j.brainresrev.2006.06.003.

Klunk, W.E., Engler, H., Nordberg, A., Wang, Y., Blomqvist, G., Holt, D.P., et al., 2004. Imaging brain amyloid in Alzheimer's disease with Pittsburgh Compound-B. Ann. Neurol. 55 (3), 306–319. Available from: https://doi.org/10.1002/ana.20009.

Knyazev, G.G., 2007. Motivation, emotion, and their inhibitory control mirrored in brain oscillations. Neurosci. Biobehav. Rev. 31 (3), 377–395. Available from: https://doi.org/10.1016/j.neubiorev.2006.10.004.

Kowalski, J.W., Gawel, M., Pfeffer, A., Barcikowska, M., 2001. The diagnostic value of EEG in Alzheimer disease: correlation with the severity of mental impairment. J. Clin. Neurophysiol. 18 (6), 570–575.

Lane, N., 2006. Cell biology: power games. Nature 443 (7114), 901–903. Available from: https://doi.org/10.1038/443901a.

Larson, J., Wong, D., Lynch, G., 1986. Patterned stimulation at the theta frequency is optimal for the induction of hippocampal long-term potentiation. Brain Res. 368 (2), 347–350.

Leuchter, A.F., Spar, J.E., Walter, D.O., Weiner, H., 1987. Electroencephalographic spectra and coherence in the diagnosis of Alzheimer's-type and multi-infarct dementia. A pilot study. Arch. Gen. Psychiatry 44 (11), 993–998.

Leuner, K., Muller, W.E., Reichert, A.S., 2012. From mitochondrial dysfunction to amyloid beta formation: novel insights into the pathogenesis of Alzheimer's disease. Mol. Neurobiol. 46 (1), 186–193. Available from: https://doi.org/10.1007/s12035-012-8307-4.

Liebe, S., Hoerzer, G.M., Logothetis, N.K., Rainer, G., 2012. Theta coupling between V4 and prefrontal cortex predicts visual short-term memory performance. Nat. Neurosci. 15 (3), . Available from: https://doi.org/10.1038/nn.3038456–462, S451–452.

Lim, L., 2017. Vielight Neuro RX Gamma – Feasibility Pilot. Retrieved from: <https://vielight.com/references/>.

Lopez-Armada, M.J., Riveiro-Naveira, R.R., Vaamonde-Garcia, C., Valcarcel-Ares, M.N., 2013. Mitochondrial dysfunction and the inflammatory response. Mitochondrion 13 (2), 106–118. Available from: https://doi.org/10.1016/j.mito.2013.01.003.

Macrides, F., Eichenbaum, H.B., Forbes, W.B., 1982. Temporal relationship between sniffing and the limbic theta rhythm during odor discrimination reversal learning. J. Neurosci. 2 (12), 1705–1717.

Mandelkow, E.M., Mandelkow, E., 2012. Biochemistry and cell biology of tau protein in neurofibrillary degeneration. Cold Spring Harb. Perspect. Med. 2 (7), a006247. Available from: https://doi.org/10.1101/cshperspect.a006247.

Maquet, P., Degueldre, C., Delfiore, G., Aerts, J., Peters, J.M., Luxen, A., et al., 1997. Functional neuroanatomy of human slow wave sleep. J. Neurosci. 17 (8), 2807–2812.

Mevel, K., Chetelat, G., Eustache, F., Desgranges, B., 2011. The default mode network in healthy aging and Alzheimer's disease. Int. J. Alzheimers Dis. 2011, 535816. Available from: https://doi.org/10.4061/2011/535816.

Minoshima, S., Giordani, B., Berent, S., Frey, K.A., Foster, N.L., Kuhl, D.E., 1997. Metabolic reduction in the posterior cingulate cortex in very early Alzheimer's disease. Ann. Neurol. 42 (1), 85–94. Available from: https://doi.org/10.1002/ana.410420114.

Morris, G., Berk, M., 2015. The many roads to mitochondrial dysfunction in neuroimmune and neuropsychiatric disorders. BMC Med. 13, 68. Available from: https://doi.org/10.1186/s12916-015-0310-y.

Muller, W.E., Eckert, A., Kurz, C., Eckert, G.P., Leuner, K., 2010. Mitochondrial dysfunction: common final pathway in brain aging and Alzheimer's disease—therapeutic aspects. Mol. Neurobiol. 41 (2–3), 159–171. Available from: https://doi.org/10.1007/s12035-010-8141-5.

Neuper, C., Pfurtscheller, G., 2001. Evidence for distinct beta resonance frequencies in human EEG related to specific sensorimotor cortical areas. Clin. Neurophysiol. 112 (11), 2084–2097.

Palop, J.J., Chin, J., Roberson, E.D., Wang, J., Thwin, M.T., Bien-Ly, N., et al., 2007. Aberrant excitatory neuronal activity and compensatory remodeling of inhibitory hippocampal circuits in mouse models of Alzheimer's disease. Neuron 55 (5), 697–711. Available from: https://doi.org/10.1016/j.neuron.2007.07.025.

Passarella, S., Karu, T., 2014. Absorption of monochromatic and narrow band radiation in the visible and near IR by both mitochondrial and non-mitochondrial photoacceptors results in photobiomodulation. J. Photochem. Photobiol. B 140, 344–358. Available from: https://doi.org/10.1016/j.jphotobiol.2014.07.021.

Patel, K.T., Stevens, M.C., Pearlson, G.D., Winkler, A.M., Hawkins, K.A., Skudlarski, P., et al., 2013. Default mode network activity and white matter integrity in healthy middle-aged ApoE4 carriers. Brain Imaging Behav. 7 (1), 60–67. Available from: https://doi.org/10.1007/s11682-012-9187-y.

Pearson, H.A., Peers, C., 2006. Physiological roles for amyloid beta peptides. J. Physiol. 575 (Pt 1), 5–10. Available from: https://doi.org/10.1113/jphysiol.2006.111203.

Petrella, J.R., Sheldon, F.C., Prince, S.E., Calhoun, V.D., Doraiswamy, P.M., 2011. Default mode network connectivity in stable vs progressive mild cognitive impairment. Neurology 76 (6), 511–517. Available from: https://doi.org/10.1212/WNL.0b013e31820af94e.

Pfurtscheller, G., Stancak Jr., A., Neuper, C., 1996. Event-related synchronization (ERS) in the alpha band—an electrophysiological correlate of cortical idling: a review. Int. J. Psychophysiol. 24 (1–2), 39–46.

Poyton, R.O., Ball, K.A., 2011. Therapeutic photobiomodulation: nitric oxide and a novel function of mitochondrial cytochrome c oxidase. Discov. Med. 11 (57), 154–159.

Purushothuman, S., Johnstone, D.M., Nandasena, C., Mitrofanis, J., Stone, J., 2014. Photobiomodulation with near infrared light mitigates Alzheimer's disease-related pathology in cerebral cortex – evidence from two transgenic mouse models. Alzheimers Res. Ther. 6 (1), 2. Available from: https://doi.org/10.1186/alzrt232.

Rasch, B., Born, J., 2013. About sleep's role in memory. Physiol. Rev. 93 (2), 681–766. Available from: https://doi.org/10.1152/physrev.00032.2012.

Saltmarche, A.E., Naeser, M.A., Ho, K.F., Hamblin, M.R., Lim, L., 2017. Significant improvement in cognition in mild to moderately severe dementia cases treated with transcranial plus intranasal photobiomodulation: case series report. Photomed. Laser Surg. 35 (8), 432–441. Available from: https://doi.org/10.1089/pho.2016.4227.

Scahill, R.I., Schott, J.M., Stevens, J.M., Rossor, M.N., Fox, N.C., 2002. Mapping the evolution of regional atrophy in Alzheimer's disease: unbiased analysis of fluid-registered serial MRI. Proc. Natl. Acad. Sci. U.S.A. 99 (7), 4703–4707. Available from: https://doi.org/10.1073/pnas.052587399.

Selkoe, D.J., 1996. Amyloid beta-protein and the genetics of Alzheimer's disease. J. Biol. Chem. 271 (31), 18295–18298.

Sherrington, R., Rogaev, E.I., Liang, Y., Rogaeva, E.A., Levesque, G., Ikeda, M., et al., 1995. Cloning of a gene bearing missense mutations in early-onset familial Alzheimer's disease. Nature 375 (6534), 754–760. Available from: https://doi.org/10.1038/375754a0.

Sorg, C., Riedl, V., Muhlau, M., Calhoun, V.D., Eichele, T., Laer, L., et al., 2007. Selective changes of resting-state networks in individuals at risk for Alzheimer's disease. Proc. Natl. Acad. Sci. U.S.A. 104 (47), 18760–18765. Available from: https://doi.org/10.1073/pnas.0708803104.

Spires-Jones, T., Knafo, S., 2012. Spines, plasticity, and cognition in Alzheimer's model mice. Neural. Plast. 2012, 319836. Available from: https://doi.org/10.1155/2012/319836.

Swartz, B.E., Goldensohn, E.S., 1998. Timeline of the history of EEG and associated fields. Electroencephalogr. Clin. Neurophysiol. 106 (2), 173–176.

Swerdlow, R.H., Burns, J.M., Khan, S.M., 2014. The Alzheimer's disease mitochondrial cascade hypothesis: progress and perspectives. Biochim. Biophys. Acta 1842 (8), 1219–1231. Available from: https://doi.org/10.1016/j.bbadis.2013.09.010.

Szabados, T., Dul, C., Majtenyi, K., Hargitai, J., Penzes, Z., Urbanics, R., 2004. A chronic Alzheimer's model evoked by mitochondrial poison sodium azide for pharmacological investigations. Behav. Brain Res. 154 (1), 31–40. Available from: https://doi.org/10.1016/j.bbr.2004.01.016.

Thompson, P.M., Hayashi, K.M., de Zubicaray, G., Janke, A.L., Rose, S.E., Semple, J., et al., 2003. Dynamics of gray matter loss in Alzheimer's disease. J. Neurosci. 23 (3), 994–1005.

Uhlhaas, P.J., Singer, W., 2006. Neural synchrony in brain disorders: relevance for cognitive dysfunctions and pathophysiology. Neuron 52 (1), 155–168. Available from: https://doi.org/10.1016/j.neuron.2006.09.020.

Vanderwolf, C.H., 1969. Hippocampal electrical activity and voluntary movement in the rat. Electroencephalogr. Clin. Neurophysiol. 26 (4), 407–418.

Verret, L., Mann, E.O., Hang, G.B., Barth, A.M., Cobos, I., Ho, K., et al., 2012. Inhibitory interneuron deficit links altered network activity and cognitive dysfunction in Alzheimer model. Cell 149 (3), 708–721. Available from: https://doi.org/10.1016/j.cell.2012.02.046.

Waypa, G.B., Smith, K.A., Schumacker, P.T., 2016. O_2 sensing, mitochondria and ROS signaling: the fog is lifting. Mol. Aspects Med. 47-48, 76–89. Available from: https://doi.org/10.1016/j.mam.2016.01.002.

Webster, M.T., Pearce, B.R., Bowen, D.M., Francis, P.T., 1998. The effects of perturbed energy metabolism on the processing of amyloid precursor protein in PC12 cells. J. Neural. Transm. (Vienna) 105 (8–9), 839–853. Available from: https://doi.org/10.1007/s007020050098.

Westlye, E.T., Lundervold, A., Rootwelt, H., Lundervold, A.J., Westlye, L.T., 2011. Increased hippocampal default mode synchronization during rest in middle-aged and elderly APOE epsilon4 carriers: relationships with memory performance. J. Neurosci. 31 (21), 7775–7783. Available from: https://doi.org/10.1523/JNEUROSCI.1230-11.2011.

Wilson, R.S., Segawa, E., Boyle, P.A., Anagnos, S.E., Hizel, L.P., Bennett, D.A., 2012. The natural history of cognitive decline in Alzheimer's disease. Psychol. Aging 27 (4), 1008–1017. Available from: https://doi.org/10.1037/a0029857.

Winson, J., 1974. Patterns of hippocampal theta rhythm in the freely moving rat. Electroencephalogr. Clin. Neurophysiol. 36 (3), 291–301.

Wu, S., Zhou, F., Wei, Y., Chen, W.R., Chen, Q., Xing, D., 2014. Cancer phototherapy via selective photoinactivation of respiratory chain oxidase to trigger a fatal superoxide anion burst. Antioxid. Redox Signal. 20 (5), 733–746. Available from: https://doi.org/10.1089/ars.2013.5229.

Zomorrodi, R., Saltmarche, A.E., Loheswaran, G., Ho, K.F., Lim, L., 2017. Complementary EEG evidence for a significantly improved Alzheimer's disease case after photobiomodulation treatment. In: Paper Presented at the 26th Annual Scientific Conference, Canadian Academy of Geriatric Psychiatry Toronto.

Zomorrodi, R., Loheswaran, G., Pushparaj, A., Lim, L., 2018. Modulation of brain oscillatory activity to drive brain function via near-infra red stimulation (tNIRS). In: Paper Presented at the AAPB 49th Annual Scientific Meeting, Orlando, Florida.

Chapter 30

Electroencephalography as the diagnostic adjunct to transcranial photobiomodulation

Reza Zomorrodi[1], **Genane Loheswaran**[2] **and Lew Lim**[2]
[1]*Temerty Centre for Therapeutic Brain Intervention, Centre for Addiction and Mental Health, Toronto, ON, Canada,*
[2]*Vielight Inc., Toronto, ON, Canada*

30.1 Introduction

Over the past three decades, our knowledge of brain functions has been expanded by new technologies that allow us to modulate and assess neural activity in specific functional or anatomical regions of the brain. Noninvasive and safe techniques such as transcranial magnetic stimulation (TMS) or transcranial electrical stimulation are employed extensively in research and treatment of many neurodegenerative and neuropsychiatric disorders (Hoy and Fitzgerald, 2010; Dayan et al., 2013; Schulz et al., 2013; Downar et al., 2016). Measuring the brain response to any electrochemical intervention has been possible with several safe but indirect measurements such as functional magnetic resonance imaging (fMRI), functional near-infrared spectroscopy (fNIRS), magnetoencephalography and electroencephalography (EEG). These tools are classified based on spatial and temporal resolutions. For instance, fMRI provides information about brain functions with high spatial (about 5 mm) and low temporal resolution (about 5 seconds), but EEG provides low spatial (about 10 mm) and very high temporal resolution (about 0.001 seconds). However, compared to other tools, EEG is a much more economical and easier to use as a diagnostic tool, which can be employed in many research laboratory or clinics.

In the next sections, we briefly review the use of EEG in investigating photobiomodulation (PBM) as a new noninvasive method. In addition, we propose our hypothesis on PBM's mechanism of action in the brain and review current evidence on brain response to transcranial PBM.

30.2 Electroencephalography

EEG is a technique that measures the brain's electrical fields at the surface of the scalp and has been used over the years to study the brain's function. This electrical potential field yields a topographic distribution of voltage values, which mainly is the result of synchronous neuronal activity. Indeed, EEG reflects the temporal and spatial summation of dendritic postsynaptic potentials in large groups of pyramidal cells that are in parallel alignment to each other, and perpendicular to the surface of the scalp. The postsynaptic potentials make the biggest contribution toward EEG, but other processes such as nonsynaptic currents, glial cells activity, and calcium and sodium dendritic spikes play a role and modulate the measured signal (Murakami and Okada, 2006; Nunez and Srinivasan, 2006; Kirschstein and Kohling, 2009).

The main advantages of EEG are its high temporal resolution (~ 1 ms) and ability to directly measure cortical activity at the population-level. However, it has to be noted that our interpretation of the EEG signal is limited by low spatial resolution (> 1 cm), and the inability to assess subcortical or asynchronous neural activities (Cohen, 2017).

EEG exhibits oscillatory patterns that can be grouped into a few characteristic frequency bands. Each of these rhythmic activities tends to occur in networks (Nimmrich et al., 2015) and depends on connectivity pattern, type of excitatory or inhibitory synaptic connection, as well as intrinsic neuronal excitability. The EEG signal is characterized by its

Photobiomodulation in the Brain. DOI: https://doi.org/10.1016/B978-0-12-815305-5.00030-0

amplitude, phase, and its spatio-temporal patterns. Commonly used and new advanced analysis techniques expose different features of EEG, and successfully link these to certain cognitive and neural functions of the brain. Perception, memory, emotion, language, action, and other cognitive processes have been associated with spatio-spectro-temporal properties, cross-frequency coupling, functional connectivity, event-related potential response, entropy, and so on. Likewise, abnormal cortical oscillations have been observed in many neurodegenerative and neuropsychiatric disorders and are even correlated with the severity of symptoms.

It is worth mentioning that aside from all the valuable information provided by EEG features and even its potential as a biomarker of neuropathophysiology, considering EEG as an epiphenomenon or fundamental mechanism in cortical information processing is still under debate (Kirov et al., 2009; Lopes da Silva, 2013; Cohen, 2017).

30.3 Brain waves

As mentioned above, synchronized neural activities generate detectable rhythmic oscillations on the scalp surface. It has been hypothesized that synchronizations within the same frequency regulate communication between distant groups of neurons, and is the underlying mechanism of the cognitive functions (Onojima et al., 2018). Frequencies can either be higher, indicating faster oscillations occurring over smaller brain areas, or lower—indicating slower oscillations extending over larger brain areas (Nimmrich et al., 2015). Two properties determine the frequency of the rhythmic oscillation: intrinsic, such as type of ionic channels, type of synapse, connection to other brain region, and extrinsic type of external event or cognitive task (Wang, 2010).

Neural oscillations are classified into five canonical frequency bands to describe different brain functions: delta, theta, alpha, beta, and gamma oscillations.

30.3.1 Delta oscillations

Delta oscillations are usually in the 1–3 Hz range. These rhythms are found in the thalamus of the brain and are associated with deep sleep (Amin and Malik, 2013).

In contrast to the other oscillations, delta waves are enhanced in different neurodegenerative diseases where the function of interneurons is impaired (Nimmrich et al., 2015). Another distinct characteristic of delta oscillations is their much more distant, indirect effects on cognitive function as compared to the other oscillations. Several pieces of evidence infer to the role of delta waves in silencing interferences from other, external neural networks during mental tasks. During the go——no go task, a paradigm in which participants must only respond to one stimulus (most commonly, by pressing a button) and must ignore all other stimuli, increased delta power is found in the central, parietal, and temporal regions of the brain (Harmony, 2013). A crucial finding is that during the no go condition, there is also increased delta power, but in the frontal regions—which Harmony (2013) proposes is indicative of delta's inhibition of the reflexive, motor response to pressing the button when seeing the nontargeted stimulus. This further suggests that delta may be needed for inhibiting sensory afferents that interfere with internal concentration. Furthermore, a consistent finding is that delta activity is highest in states where interneurons and thalamocortical inputs are inactive, seemingly pointing to the role of delta in stopping interferences that disrupt concentration during mental tasks (Harmony, 2013).

30.3.2 Theta oscillations

Theta bands have a frequency of approximately 4–7 Hz (Lisman and Jensen, 2013), are found in the hippocampus, and are associated with dreaming sleep (Amin and Malik, 2013).

Theta bands are activated and increased, in general, during the memory processes of encoding, retention, and retrieval (Raghavachari et al., 2001; Amin and Malik, 2013). Some examples include theta synchronization during memory tasks, theta desynchronization during semantic retrieval, and increase in power, which varies with task difficulty, during maintenance and memory retrieval of working memory tasks (Amin and Malik, 2013). Evidence from other studies showed an increased in theta power prior to a stimulus predicted better memory performance (Lisman and Jensen, 2013). In addition to working memory, theta is associated with long-term memory (Lisman and Jensen, 2013).

30.3.3 Alpha oscillations

Alpha bands have frequencies ranging from 8 to 12 Hz and are typically found in the brain's posterior regions and central area (Amin and Malik, 2013; Nimmrich et al., 2015).

Alpha waves are seen in a relaxed but resting state, usually with the eyes closed (Groppe et al., 2013). Alpha oscillations are most likely involved in the temporal fluctuations of neural network inhibition. Certain neural mechanisms have been shown to produce alpha oscillations, such as thalamocortical loops, rhythmically firing pyramidal cells, local interneurons and finally, interactions of synaptic inputs with different time constraints. The fact that multiple, distinct neural mechanisms produce these frequencies suggests that there are many independent alpha generators in the brain (Klimesch, 2012).

Typical changes in alpha activity are seen during tasks measuring long-term and short-term memory, as well as during the three processes of memory: encoding, retrieval, and retention (Amin and Malik, 2013). Some specific examples of alpha changes during EEG measures include: increased alpha power during working memory maintenance, desynchronization during memory performance tasks, decreased amplitude during calculation tasks, increased activity during working memory encoding, and finally upper alpha activity during semantic retrieval (Amin and Malik, 2013). Moreover, features of the alpha band such as its amplitude and peak frequency depend on the cognitive task being performed, and the cognitive region and layer being measured (Cohen, 2017).

30.3.4 Beta oscillations

Beta oscillations lie somewhere in the middle range of neural oscillation frequencies bands, usually at around 13–30 Hz.

Beta waves are found in the brain's frontal regions and are associated with an alert and concentrated state. Out of the five types of oscillations, beta has traditionally been viewed as the least relevant to memory performance, with the exception of one type of task—beta power shows to decrease, along with an increase in theta power, during visual working memory tasks (Amin and Malik, 2013).

Beta oscillations have been thought to be exclusively associated with motor and sensory processing, with research indicating they are not very much involved with memory and cognitive functioning (Spitzer and Haegens, 2017). However, Spitzer and Haegens (2017) report on recent, accumulating findings that indicate beta waves may actually play a supplementary role. The authors proposed that beta oscillation mediated ensemble formation within and between cortical areas, in accordance to a current task's demands. Similarly to alpha, theta, and gamma waves, beta waves also seem to increase during working memory maintenance (Spitzer and Haegens, 2017). Beta synchronization seems to coincide with currently relevant task rules and stimulus categories. Specifically, typical changes in beta waves are seen late in the delay process of working memory maintenance—indicating that beta oscillations lead to the reactivation of cortical representations that are needed with the changing demands of a task. So rather than a persistent memory storage, beta changes are needed for the momentary updating (or reactivation) of working memory content according to changing demands of a task. Although the beta's role in memory is still under debate, some evidence suggests that beta facilitates network-level communication with its reactivation of cortical representations (Spitzer and Haegens, 2017).

30.3.5 Gamma oscillations

Gamma oscillations have much higher frequencies, lying in the range of 30–50 Hz. Gamma bands are found in the somatosensory cortex of the brain and associated brain states producing gamma bands, which are seen as sudden bursts of activity (Amin and Malik, 2013).

Evidence from studies indicates that gamma bands are associated with long-term memory (Lisman and Jensen, 2013). For example, gamma power as well as spike gamma coherence in the hippocampus are both shown to be higher during successful encoding of information as opposed to unsuccessful encoding in monkey studies. Similar effects have been seen in human subjects, in the hippocampus and also in the cortical regions (somatosensory cortex) (Lisman and Jensen, 2013). Gamma bands are also related to working memory, as studies have reported increased gamma power with increasing number of items being held in memory during working memory tasks (Lisman and Jensen, 2013).

30.4 Photobiomodulation as a new noninvasive brain stimulation method

Different forms of electrical stimulation (e.g., transcranial direct current stimulation, transcranial alternating current stimulation, transcranial rippled current stimulation) and magnetic stimulation (e.g., TMS, repetitive TMS, theta-burst) are outstanding tools to interact and modify brain functions through interaction with multiple neurotransmitters and neural networks. By implementing different protocols of these stimulation techniques, one may identify networks mediating cognitive or noncognitive function via suppressing or activating a certain neural population.

It has to be mentioned that most pharmacological interventions and electro/magneto-neuromodulations act through the neurotransmitter receptors or voltage-dependent ionic channels. However, the electrochemical properties are not the only features of the brain that may respond to external stimulation. Cytochrome c oxidase (CCO), a photo-receptor, which is located in the inner membrane of mitochondria responds to light, which provides another possible modulatory technique: PBM.

PBM was introduced by Endre Mester in 1967 and defined as the application of low levels of visible red or near-infrared (NIR) light to stimulate a biological system (Hamblin, 2016). The primary target of PBM is via CCO, found in the inner membrane of mitochondria which drives cellular metabolism (Karu, 2014). CCO is the last unit of the electron transport chain, transferring electrons from cytochrome c to molecular oxygen, and thus completing ATP synthesis in the mitochondria. By targeting CCO, PBM leads to oxygen consumption and increased energy production via mitochondrial oxidative phosphorylation (Wang et al., 2016). Since CCO is an inducible enzyme, effects are long-lasting.

The mechanism of PBM interaction can be classified into direct and indirect effects. The direct effects include increasing the activity of ion channels such as the Na^+/K^+ ATPase (Farivar et al., 2014; Giordano et al., 2017); and the indirect effects include regulating important secondary messengers such as calcium, cyclic adenosine monophosphate, and reactive oxygen species—all of which result in diverse biological cascades (Farivar et al., 2014; Passarella and Karu, 2014). These biological cascades lead to effects such as the maintenance of homeostasis and activating protective, antioxidant and proliferative gene factors, as well as the systematic responses, such as cerebral blood flow, which is deficient in neurocognitive disorders (Hashmi et al., 2010; Passarella and Karu, 2014; Hamblin, 2016; Giordano et al., 2017; Vargas et al., 2017).

30.5 The causal link between photobiomodulation and neural oscillations

Very few studies have looked at how PBM may interact with the brain, and explored the mechanism and possible cognitive outcomes. Here, we briefly review a few of the prevailing hypotheses and experimental evidence:

30.5.1 Maintaining homeostasis

One of the most prominent responses to PBM is the activation of sodium pumps and the Na^+/K^+ ATPase, which leads to greater membrane stability and resistance to depolarization (Konstantinovic et al., 2013). In a study combining transcranial laser stimulation (TLS), a PBM technique, and TMS, administering TLS before TMS was found to significantly reduce the size of the motor evoked potential (MEP) amplitude, contrary to typical TMS paradigms where the MEP amplitude is significantly increased. The increase in MEP is the result of an action potential in cortical pyramidal cells, representing increased cortical excitability; thus these findings indicated that TLS actually induced resistance to the depolarization normally caused by TMS (Konstantinovic et al., 2013). Maintaining stable membrane potential—in other words, maintaining homeostasis—depends on ATP levels, substantiating the role of PBM in regulating homeostasis.

30.5.2 Calcium signaling

Another possible association between brain response and PBM could be calcium signaling. PBM comes into play due to its indirect effects of activating secondary messengers, in particular Ca^{2+}. In an animal study (Moreau et al., 2018), the exposure of muse hippocampal neurons to infrared stimulation (IRS), increased intracellular release of Ca^{2+}, which led to the triggering of electrical spikes from these neurons. It has been proposed that IRS activates phospholipase C (PLC), which leads to phosphoinositide-signaling pathways, resulting in intracellular Ca^{2+} release from the endoplasmic reticulum. In addition, IRS leads to the activation of the Ca^{2+} class found in the G-protein coupled to PLC, allowing for the release of Ca^{2+} with the pulses of IRS (Moreau et al., 2018).

Another type of photo-receptor may be opsins (OPN), which are light-sensitive, G-protein coupled receptors that when activated, lead to extracellular release of Ca^{2+} (Hamblin, 2017). When activated, OPN leads to the opening of transient receptor channels (TRPs) which leads to an influx of extracellular Ca^{2+} into the cell. TRPs are altered by phosphoinositides and are activated by green as well as NIR lights (Hamblin, 2017).

It has to be noted that this pathway for IRS can be applicable to nonexcitable cell as well (Moreau et al., 2018). However, increased levels of Ca^{2+}, specifically, in interneurons and their vital role in neural network oscillation could be the missing piece in linking PBM to brain waves.

The synchronization of oscillatory networks depends on interneuron activity (Nimmrich et al., 2015). Interneurons have widespread axonal plexus, and it is this widespread aspect that enables synchronous inhibition of multiple neurons within their circuits. Interneurons are connected by synapses that can be either chemical or electrical, and these synapses lead to synaptic inhibition transmitting powerful and synchronous rhythmic signals throughout entire networks. One of the most prominent ways of categorizing interneurons and their functions is by their expression of calcium-binding proteins and peptides (Nimmrich et al., 2015). There are two main types Ca^{2+}-binding interneurons: parvalbumin (PV)- and cholecystokinin (CCK)-binding interneurons. PV interneurons are fast spiking and more consistent, involved in generating fast network oscillations whereas CCK interneurons are slower and much less reliable, with weaker contributions to network oscillations (Del Pino et al., 2017). Dysfunction of PV interneurons leads to reduced activity of gamma oscillations. On the other hand, the role of CCK interneurons in oscillations is less clearly defined but is hypothesized to be related to cell firing during theta oscillations (Del Pino et al., 2017).

However, more research is needed to deduce the exact mechanism by which PBM and EEG activity are connected to each other; the roles of calcium and resistance to depolarization (maintaining homeostasis) are not concrete but they could be the starting points to understand the relationship, if any, between PBM and brain stimulation.

30.6 Evidence for transcranial photobiomodulation influences on brain oscillations

Beside a large body ofempirical evidence at the cellular level (Hamblin et al., 2017), human studies have demonstrated the therapeutic effects of PBM in many psychiatric and neurological disorders expressed as cognitive enhancement. For instance, the effects of PBM have been reported in patients with (1) chronic traumatic brain injury (TBI) by improving the self-awareness, social functioning, sleep quality, cognitive, and mood states (Naeser et al., 2014; Morries et al., 2015); (2) major depression disorder or TBI with comorbid depression by the alleviation of depressive and anxiety symptoms (e.g., Hamilton scales) and PTSD scores (Schiffer et al., 2009; Naeser et al., 2011; Disner et al., 2016); (3) Parkinson's disease by improvement of motor and cognitive function (Johnstone et al., 2014; Reinhart et al., 2016); (4) Alzheimer's disease (AD) by improvement in memory, attention, sleep quality, mood states, and significant reduction in dementia scores (Maksimovich, 2015; Berman et al., 2017; Saltmarche et al., 2017).

In addition, the notable enhancements in cognitive function of healthy subjects have been reported (Barrett and Gonzalez-Lima, 2013; Blanco et al., 2017a,b). These studies showed improvement in short/long-term memory and executive functions.

Neurophysiological assessments, for revealing the impact of PBM on brain activity, are a missing research piece to quantify the action mechanisms and identify immediate brain response to PBM and specifically to the transcranial PBM. The main questions for all noninvasive transcranial stimulations methods (e.g., transcranial current or magnetic stimulation) are how to reach the region of interest, adjust the activity and evaluation of the brain responses. EEG and fMRI have been extensively employed to explore the effect of noninvasive brain stimulation. However, only a few studies demonstrated brain responses to the transcranial PBM. In a study on healthy, elderly subjects using EEG and fMRI, Vargas et al. (2017) showed a significant increase in resting-state EEG alpha, beta, and gamma power and more efficient prefrontal blood-oxygen-level dependent-fMRI response. In another study (Grover et al., 2017), the acute effect on an event-related brain response, indexed as P300, was reported for 31 healthy participants. In two other neurophysiological studies in healthy subjects, TMS was employed to investigate the impact of the transcranial PBM on cortical excitability. After irradiation of motor cortex (i.e., M1 for 8–10 minutes) by NIR light (810, 905 nm wavelength), a transient reduction of cortical excitability and increase of cortical inhibition have been observed (Konstantinovic et al., 2013; Chaieb et al., 2015). Using fNIRS, Tian et al. (2016) demonstrated a significant increase in hemoglobin oxygenation and a decrease in deoxygenation in both cerebral hemispheres over time during irradiation (10 minutes) and post-irradiation (6 minutes) after applying 1064-nm laser for 10 minutes to the forehead (Tian et al., 2016).

In a randomized, double-blind, controlled-placebo study, we investigated the brain response in 20 healthy subjects using the Vielight neuro gamma. This device delivers NIR light pulse (810 nm wavelength, 40 Hz) through five nonlaser light emitting diodes (e.g., frontal, left/right temporal, precuneus cortex, and intranasal) for a 20-minute session (Zomorrodi et al., 2019). The power spectrum and connectivity analysis were performed to assess the difference in the induced change between active and sham intervention. The analysis revealed that PBM delivered by the device produced frequency-dependent effects on endogenous brain activity. A reduction of low-frequency power (i.e., delta and theta) and an increase of high-frequency power (i.e., alpha, beta, and gamma) were observed following active stimulation, replicating the earlier case report for the moderately impaired AD patient (Zomorrodi et al., 2017). Comparison between active and sham groups demonstrated a significant difference in delta (1–4 Hz), theta (4–7 Hz), alpha (8–12 Hz), beta (12–30 Hz, and gamma (30–50 Hz).

The weighted phase lag index (wPLI) and graph theory measures revealed a significant change between active and sham stimulation, limited to the alpha and gamma frequencies. Findings from this study provide early evidence that

PBM modulates cortical oscillations that impact brain connectivity in a frequency-dependent manner. The study confirms that light can penetrate the brain sufficient enough to produce significant electrophysiological changes.

The wPLI and graph theory measures are employed to evaluate effective connectivity alteration. Results revealed significant change between active and sham stimulation, limited to alpha and gamma frequencies. Findings from this study provide early evidence that PBM modulates cortical oscillations that impact brain connectivity in a frequency-dependent manner. The study confirms that noncoherent NIR light at 810 nm penetrates the brain sufficiently enough to produce significant electrophysiological changes.

30.7 The potential use of electroencephalography with photobiomodulation for brain disorders

EEG has been used in the evaluation of a large variety of dementias and encephalopathies which include AD, dementia with Lewy bodies, Pick's disease, vascular dementia, Creutzfeldt–Jakob disease, and many others (Neiman, 2017). Our work with AD and dementia underlines the potential of using EEG as an adjunct tool to PBM. Numerous studies have reported increases in theta and delta activities and decreases in alpha and beta activities in patients with AD (Coben et al., 1983, 1985; Brenner et al., 1986; Giaquinto and Nolfe, 1986; Huang et al., 2000; Babiloni et al., 2004, 2013; Jeong, 2004). In a case study, we found that when delivering 810 nm pulsed at 40 Hz, we were able reverse these EEG signatures associated with AD and produce significant improvements in the cognition measures provided by the mini mental status examination and the Alzheimer's disease assessment scale—cognitive subscale (Zomorrodi et al., 2017).

The EEG measures presented in the healthy brains in the previous section (Zomorrodi et al., 2019) make this set of parameters a template to achieve desirable outcomes in AD and encourage the exploration of other parameters that may be helpful with other brain disorders. For example, a subset of attention deficit hyperactivity disorder is identified with elevated theta brain oscillation (Clarke et al., 2011). It suggests the opportunity to correct this through PBM by adjusting the parameters to attenuate theta power, reversing an unfavorable EEG pattern. These preliminary evidence calls for further investigations into identifying EEG signatures with specific brain disorders and to further observe the impact of modifying these signatures by varying PBM parameters.

30.8 Discussion and conclusion

To date EEG is a tool that has not been widely used as a measure of PBM effect on the brain. It is not as precise in identifying specific location of activity as fMRI or change in blood oxygenation as for fNIRS. However, its main advantage is the relative convenience of use, presentation of real-time brain response, and its measures of electrophysiological changes representing abnormal cortical brain function. This information can lead to developing effective interventions with PBM, based on the early evidence that varying selected parameters may modify the EEG patterns.

Abnormalities in brain waves can provide information on brain states, which can lead to clues on how certain disorders and insults can be addressed. Neurofeedback practices have been a way to correct these abnormalities. However, given the evidence of PBM effect on brain conditions, PBM could be a quicker intervention and appears to give more predictable outcomes, particularly in neurological cases. Hence EEG could be the ideal adjunct diagnostic tool for PBM.

We are in the pioneering stages of combining EEG with PBM. Much more investigative work is still needed but as more evidence is gathered, it becomes increasingly clear that EEG could make PBM a more effective medical intervention.

References

Amin, H., Malik, A.S., 2013. Human memory retention and recall processes. A review of EEG and fMRI studies. Neurosciences (Riyadh) 18 (4), 330–344.

Babiloni, C., Binetti, G., Cassetta, E., Cerboneschi, D., Dal Forno, G., Del Percio, C., et al., 2004. Mapping distributed sources of cortical rhythms in mild Alzheimer's disease. A multicentric EEG study. Neuroimage 22 (1), 57–67.

Babiloni, C., Lizio, R., Del Percio, C., Marzano, N., Soricelli, A., Salvatore, E., et al., 2013. Cortical sources of resting state EEG rhythms are sensitive to the progression of early stage Alzheimer's disease. J. Alzheimers Dis. 34 (4), 1015–1035.

Barrett, D.W., Gonzalez-Lima, F., 2013. Transcranial infrared laser stimulation produces beneficial cognitive and emotional effects in humans. Neuroscience 230, 13–23.

Berman, M.H., Halper, J.P., Nichols, T.W., Jarrett, H., Lundy, A., Huang, J.H., 2017. Photobiomodulation with near infrared light helmet in a pilot, placebo controlled clinical trial in dementia patients testing memory and cognition. J. Neurol. Neurosci. 8 (1).

Blanco, N.J., Maddox, W.T., Gonzalez-Lima, F., 2017a. Improving executive function using transcranial infrared laser stimulation. J. Neuropsychol. 11 (1), 14–25.

Blanco, N.J., Saucedo, C.L., Gonzalez-Lima, F., 2017b. Transcranial infrared laser stimulation improves rule-based, but not information-integration, category learning in humans. Neurobiol. Learn. Mem. 139, 69–75.

Brenner, R.P., Ulrich, R.F., Spiker, D.G., Sclabassi, R.J., Reynolds 3rd, C.F., et al., 1986. Computerized EEG spectral analysis in elderly normal, demented and depressed subjects. Electroencephalogr. Clin. Neurophysiol. 64 (6), 483–492.

Chaieb, L., Antal, A., Masurat, F., Paulus, W., 2015. Neuroplastic effects of transcranial near-infrared stimulation (tNIRS) on the motor cortex. Front. Behav. Neurosci. 9, 147.

Clarke, A.R., Barry, R.J., Dupuy, F.E., Heckel, L.D., McCarthy, R., Selikowitz, M., et al., 2011. Behavioural differences between EEG-defined subgroups of children with attention-deficit/hyperactivity disorder. Clin. Neurophysiol. 122 (7), 1333–1341.

Coben, L.A., Danziger, W.L., Berg, L., 1983. Frequency analysis of the resting awake EEG in mild senile dementia of Alzheimer type. Electroencephalogr. Clin. Neurophysiol. 55 (4), 372–380.

Coben, L.A., Danziger, W., Storandt, M., 1985. A longitudinal EEG study of mild senile dementia of Alzheimer type: changes at 1 year and at 2.5 years. Electroencephalogr. Clin. Neurophysiol. 61 (2), 101–112.

Cohen, M.X., 2017. Where does EEG come from and what does it mean? Trends Neurosci. 40 (4), 208–218.

Dayan, E., Censor, N., Buch, E.R., Sandrini, M., Cohen, L.G., 2013. Noninvasive brain stimulation: from physiology to network dynamics and back. Nat. Neurosci. 16 (7), 838–844.

Del Pino, I., Brotons-Mas, J.R., Marques-Smith, A., Marighetto, A., Frick, A., Marin, O., et al., 2017. Abnormal wiring of CCK(+) basket cells disrupts spatial information coding. Nat. Neurosci. 20 (6), 784–792.

Disner, S.G., Beevers, C.G., Gonzalez-Lima, F., 2016. Transcranial laser stimulation as neuroenhancement for attention bias modification in adults with elevated depression symptoms. Brain Stimul. 9 (5), 780–787.

Downar, J., Blumberger, D.M., Daskalakis, Z.J., 2016. The neural crossroads of psychiatric illness: an emerging target for brain stimulation. Trends Cogn. Sci. 20 (2), 107–120.

Farivar, S., Malekshahabi, T., Shiari, R., 2014. Biological effects of low level laser therapy. J. Lasers Med. Sci. 5 (2), 58–62.

Giaquinto, S., Nolfe, G., 1986. The EEG in the normal elderly: a contribution to the interpretation of aging and dementia. Electroencephalogr. Clin. Neurophysiol. 63 (6), 540–546.

Giordano, J., Bikson, M., Kappenman, E.S., Clark, V.P., Coslett, H.B., Hamblin, M.R., et al., 2017. Mechanisms and effects of transcranial direct current stimulation. Dose Response 15 (1), 1559325816685467.

Groppe, D.M., Bickel, S., Keller, C.J., Jain, S.K., Hwang, S.T., Harden, C., et al., 2013. Dominant frequencies of resting human brain activity as measured by the electrocorticogram. Neuroimage 79, 223–233.

Grover Jr., F., Weston, J., Weston, M., 2017. Acute effects of near infrared light therapy on brain state in healthy subjects as quantified by qEEG measures. Photomed. Laser Surg. 35 (3), 136–141.

Hamblin, M.R., 2016. Shining light on the head: photobiomodulation for brain disorders. BBA Clin. 6, 113–124.

Hamblin, M.R., 2017. Mechanisms and applications of the anti-inflammatory effects of photobiomodulation. AIMS Biophys. 4 (3), 337–361.

Hamblin, M.R., Pires de Sousa, M.V., Agrawal, T., 2017. Handbook of Low-Level Laser Therapy. Pan Stanford Publishing Pte. Ltd., Singapore, 1170 pp.

Harmony, T., 2013. The functional significance of delta oscillations in cognitive processing. Front. Integr. Neurosci. 7, 83.

Hashmi, J.T., Huang, Y.Y., Osmani, B.Z., Sharma, S.K., Naeser, M.A., Hamblin, M.R., 2010. Role of low-level laser therapy in neurorehabilitation. PMR 2 (12 Suppl. 2), S292–305.

Hoy, K.E., Fitzgerald, P.B., 2010. Brain stimulation in psychiatry and its effects on cognition. Nat. Rev. Neurol. 6 (5), 267–275.

Huang, C., Wahlund, L., Dierks, T., Julin, P., Winblad, B., Jelic, V., 2000. Discrimination of Alzheimer's disease and mild cognitive impairment by equivalent EEG sources: a cross-sectional and longitudinal study. Clin. Neurophysiol. 111 (11), 1961–1967.

Jeong, J., 2004. EEG dynamics in patients with Alzheimer's disease. Clin. Neurophysiol. 115 (7), 1490–1505.

Johnstone, D.M., el Massri, N., Moro, C., Spana, S., Wang, X.S., Torres, N., et al., 2014. Indirect application of near infrared light induces neuroprotection in a mouse model of parkinsonism – an abscopal neuroprotective effect. Neuroscience 274, 93–101.

Karu, T., 2014. Cellular and molecular mechanisms of photobiomodulation (low-power laser therapy). IEEE J. Sel. Top. Quant. Electr. 20 (2), 143–148.

Kirov, R., Weiss, C., Siebner, H.R., Born, J., Marshall, L., 2009. Slow oscillation electrical brain stimulation during waking promotes EEG theta activity and memory encoding. Proc. Natl. Acad. Sci. U.S.A. 106 (36), 15460–15465.

Kirschstein, T., Kohling, R., 2009. What is the source of the EEG? Clin. EEG Neurosci. 40 (3), 146–149.

Klimesch, W., 2012. alpha-band oscillations, attention, and controlled access to stored information. Trends Cogn. Sci. 16 (12), 606–617.

Konstantinovic, L.M., Jelic, M.B., Jeremic, A., Stevanovic, V.B., Milanovic, S.D., Filipovic, S.R., 2013. Transcranial application of near-infrared low-level laser can modulate cortical excitability. Lasers Surg. Med. 45 (10), 648–653.

Lisman, J.E., Jensen, O., 2013. The theta-gamma neural code. Neuron 77 (6), 1002–1016.

Lopes da Silva, F., 2013. EEG and MEG: relevance to neuroscience. Neuron 80 (5), 1112–1128.

Maksimovich, I., 2015. Dementia and cognitive impairment reduction after laser transcatheter treatment of Alzheimer's disease. World J. Neurosci. 5, 189–203.

Moreau, D., Lefort, C., Pas, J., Bardet, S.M., Leveque, P., O'Connor, R.P., 2018. Infrared neural stimulation induces intracellular Ca(2 +) release mediated by phospholipase C. J. Biophotonics 11 (2).

Morries, L.D., Cassano, P., Henderson, T.A., 2015. Treatments for traumatic brain injury with emphasis on transcranial near-infrared laser phototherapy. Neuropsychiatr. Dis. Treat. 11, 2159–2175.

Murakami, S., Okada, Y., 2006. Contributions of principal neocortical neurons to magnetoencephalography and electroencephalography signals. J. Physiol. 575 (Pt 3), 925–936.

Naeser, M.A., Saltmarche, A., Krengel, M.H., Hamblin, M.R., Knight, J.A., 2011. Improved cognitive function after transcranial, light-emitting diode treatments in chronic, traumatic brain injury: two case reports. Photomed. Laser Surg. 29 (5), 351–358.

Naeser, M.A., Zafonte, R., Krengel, M.H., Martin, P.I., Frazier, J., Hamblin, M.R., et al., 2014. Significant improvements in cognitive performance post-transcranial, red/near-infrared light-emitting diode treatments in chronic, mild traumatic brain injury: open-protocol study. J. Neurotrauma 31 (11), 1008–1017.

Neiman, E.S., 2017. EEG in Dementia and Encephalopathy. Retrived from: <https://emedicine.medscape.com/article/1138235-overview>.

Nimmrich, V., Draguhn, A., Axmacher, N., 2015. Neuronal network oscillations in neurodegenerative diseases. Neuromol. Med. 17 (3), 270–284.

Nunez, P.L., Srinivasan, R., 2006. A theoretical basis for standing and traveling brain waves measured with human EEG with implications for an integrated consciousness. Clin. Neurophysiol. 117 (11), 2424–2435.

Onojima, T., Goto, T., Mizuhara, H., Aoyagi, T., 2018. A dynamical systems approach for estimating phase interactions between rhythms of different frequencies from experimental data. PLoS Comput. Biol. 14 (1), e1005928.

Passarella, S., Karu, T., 2014. Absorption of monochromatic and narrow band radiation in the visible and near IR by both mitochondrial and non-mitochondrial photoacceptors results in photobiomodulation. J. Photochem. Photobiol. B 140, 344–358.

Raghavachari, S., Kahana, M.J., Rizzuto, D.S., Caplan, J.B., Kirschen, M.P., Bourgeois, B., et al., 2001. Gating of human theta oscillations by a working memory task. J. Neurosci. 21 (9), 3175–3183.

Reinhart, F., Massri, N.E., Chabrol, C., Cretallaz, C., Johnstone, D.M., Torres, N., et al., 2016. Intracranial application of near-infrared light in a hemi-parkinsonian rat model: the impact on behavior and cell survival. J. Neurosurg. 124 (6), 1829–1841.

Saltmarche, A.E., Naeser, M.A., Ho, K.F., Hamblin, M.R., Lim, L., 2017. Significant improvement in cognition in mild to moderately severe dementia cases treated with transcranial plus intranasal photobiomodulation: case series report. Photomed. Laser Surg. 35 (8), 432–441.

Schiffer, F., Johnston, A.L., Ravichandran, C., Polcari, A., Teicher, M.H., Webb, R.H., et al., 2009. Psychological benefits 2 and 4 weeks after a single treatment with near infrared light to the forehead: a pilot study of 10 patients with major depression and anxiety. Behav. Brain Funct. 5, 46.

Schulz, R., Gerloff, C., Hummel, F.C., 2013. Non-invasive brain stimulation in neurological diseases. Neuropharmacology 64, 579–587.

Spitzer, B., Haegens, S., 2017. Beyond the status quo: a role for beta oscillations in endogenous content (re)activation. eNeuro 4 (4).

Tian, F., Hase, S.N., Gonzalez-Lima, F., Liu, H., 2016. Transcranial laser stimulation improves human cerebral oxygenation. Lasers Surg. Med. 48 (4), 343–349.

Vargas, E., Barrett, D.W., Saucedo, C.L., Huang, L.D., Abraham, J.A., Tanaka, H., et al., 2017. Beneficial neurocognitive effects of transcranial laser in older adults. Lasers Med. Sci. 32 (5), 1153–1162.

Wang, X., Tian, F., Soni, S.S., Gonzalez-Lima, F., Liu, H., 2016. Interplay between up-regulation of cytochrome-c-oxidase and hemoglobin oxygenation induced by near-infrared laser. Sci. Rep. 6, 30540.

Wang, X.J., 2010. Neurophysiological and computational principles of cortical rhythms in cognition. Physiol. Rev. 90 (3), 1195–1268.

Zomorrodi R., Loheswaran G., Pushparaj A. and Lim L. 2019. Pulsed Near Infrared Transcranial and Intranasal Photobiomodulation Significantly Modulates Neural Oscillations: a pilot exploratory study. Sci. Rep. (Under Review).

Zomorrodi, R., Saltmarche, A.E., Loheswaran, G., Ho, K.F., Lim, L., 2017. Complementary EEG evidence for a significantly improved Alzheimer's disease case after photobiomodulation treatment. In: 26th Annual Scientific Conference, Canadian Academy of Geriatric Psychiatry Toronto.

Chapter 31

Can photobiomodulation enhance brain function in older adults?

Agnes S. Chan[1,2], Michael K. Yeung[1,3] and Tsz L. Lee[1]

[1]*Department of Psychology, The Chinese University of Hong Kong, Hong Kong, China,* [2]*Chanwuyi Research Center for Neuropsychological Well-Being, The Chinese University of Hong Kong, Hong Kong, China,* [3]*Department of Neurology and Neurosurgery, Montreal Neurological Institute, McGill University, Montreal, QC, Canada*

Epidemiological studies have shown that the world's elderly population continues to grow at an accelerating rate in the last few decades, and this growth rate exceeds that for the total population (Anderson and Hussey, 2000; World Health Organization, 2016). At the same time, the life expectancy of the elderly population, which is defined as people aged 60 years or above, has been rising steadily (World Health Organization, 2016). According to the United Nations (2017), the elderly population has doubled from 465 million in 1988 to 991 million in 2018, which refer to 9.0% and 13.0% of the total world population, respectively; it is projected that the elderly population will further double by 2048. In addition, it is well-documented that aging is associated with declines in mental abilities and functional activities (Royall et al., 2004; Schrack et al., 2013; West, 1996), which may be associated with increased healthcare and social expenditures (Dang et al., 2001; Lubitz et al., 2003; Reinhardt, 2003). For example, it has been estimated that healthcare expenditure for older adults aged 65 or above is three to five times more than that for younger individuals in any year per capita (Reinhardt, 2003). In addition, age-related social expenditures in 13 countries are projected to increase from an average of 19% of gross domestic product (GDP) in 2000 to an average of 26% of GDP by 2050, in which half of the increase will be caused by healthcare expenditure and old-age pension payments (Dang et al., 2001). Furthermore, a greater limitation in instrumental activity of daily living is associated with a larger healthcare expenditure per year (Lubitz et al., 2003). Because aging may add huge financial burdens to the elderly themselves, their family members, and the society (de Meijer et al., 2013), interventions that can improve the mental abilities of older adults to support their functional status are both socially and clinically important. Thus the objective of the present chapter is to discuss the possibility of using photobiomodulation (PBM) as a remediation for age-associated cognitive deterioration. To describe the cognitive deterioration associated with the aging brain, we will first review a neurophysiological model of aging in the following section.

31.1 Frontal lobe deterioration and normal human aging

31.1.1 Structural and functional deteriorations of the frontal lobe in normal human aging

It is well-documented that aging is associated with progressive brain deterioration, and converging evidence suggests that the frontal lobe of the cerebral cortex is more vulnerable to deterioration than other parts of the brain (West, 1996). There are various kinds of structural and functional changes that happen to the frontal lobe with advancing age (see Table 31.1); these neurobiological changes can be observed at both the macroscopic and microscopic levels.

With regard to macroscopic changes, several magnetic resonance imaging (MRI) studies have reported an age-related reduction in gray and white matter volume in different parts of the frontal lobe, including the dorsolateral prefrontal cortex (dlPFC) and orbitofrontal cortex (OFC) (Bartzokis et al., 2001; Grieve et al., 2005; Kalpouzos et al., 2009; Raz et al., 1997, 2005). More importantly, this age-related atrophy in brain tissues is more pronounced in the dlPFC and OFC than in other parts of the brain, such as the temporal and parietal cortices, posterior cortical and thalamic regions, and limbic and paralimbic structures (Grieve et al., 2005; Kalpouzos et al., 2009; Raz et al., 1997, 2005).

Photobiomodulation in the Brain. DOI: https://doi.org/10.1016/B978-0-12-815305-5.00031-2

TABLE 31.1 A summary of (A) structural and (B) functional deteriorations of the frontal lobe in healthy older adults.

Change	Characteristic	Study (Method)	Sample size (age)
(A) Structural			
Atrophy of neural tissue	Age-related reduction in gray and white matter volume in the frontal and temporal lobes	Bartzokis et al. (2001) (MRI)	70 (19–76 years)
	Accelerated gray matter loss in focal regions of the frontal and parietal lobes, including the dlPFC, but gray matter perseveration in limbic and paralimbic structures	Grieve et al. (2005) (MRI)	223 (8–79 years)
	Greater age-related reduction in gray matter volume in the frontal lobes (i.e., dlPFC, OFC) than in posterior cortical and thalamic regions	Kalpouzos et al. (2009) (MRI)	45 (20–83 years)
	Age-related reduction in gray matter volume in the prefrontal cortex (i.e., dlPFC, OFC) and relative sparing of the temporal and parietal cortices. Age-related reduction in white matter volume was smaller for prefrontal and superior parietal cortices	Raz et al. (1997) (MRI)	148 (18–77 years)
	Greater age-related atrophy in lateral prefrontal and orbitofrontal cortices than in posterior cortical regions	Raz et al. (2005) (MRI)	72 (20–77 years)
Degradation of white matter microstructure	Reduced fractional anisotropy in frontal but not posterior (i.e., temporal, parietal, occipital) systems	Head et al. (2004) (DTI)	25 (19–28 years) and 25 (66–88)
	Reduced fractional anisotropy in frontal but not posterior systems	Pfefferbaum et al. (2005) (DTI)	10 (22–37 years) and 10 (65–79 years)
	Greater age-related fractional anisotropy reduction in frontal white matter than in temporal and posterior white matter	Salat et al. (2005) (DTI)	38 (21–76 years)
Thinning of cortical layers	Greater cortical thinning in the superior, inferior, and medial frontal gyri and superior temporal cortex than in the inferior temporal, parietal, and medial temporal cortex	Fjell et al. (2009) (MRI)	883 (18–93 years)
	Thinning of the prefrontal and motor cortices and relative sparing of the temporal cortex	Salat et al. (2004) (MRI)	106 (18–93 years)
Retraction of dendritic branches	Age-related dendritic regression of layer V but not layer IIIc pyramidal cells in the prefrontal cortex	De Brabander et al. (1998) (Golgi-Cox)	8 (49–90 years)
	Age-related reduction in number of basal dendrites of Layer V pyramidal cells in the primary motor cortex	Nakamura et al., (1985) (Golgi)	8 (14–96 years)
Loss of synaptic connections	An average of 20% age-related reduction in presynaptic terminal counts in the frontal cortex	Masliah et al. (1993) (Immuno-labeling)	25 (16–98 years)
Decrease in the number of neurochemical receptors	Age-related reduction in dopamine D1 receptor binding potential in the dlPFC and caudate related to frontal and parietal activation during working memory processing	Backman et al. (2011) (PET)	20 (22–30 years) and 20 (65–75 years)
	Reduced densities of dopamine D1 receptors and their high-agonist affinity sites in the frontal cortex	de Keyser et al. (1990) (Golgi-Cox)	32 (19–88 years)
	Faster age-related declines in dopamine D2/D3 receptor subtypes in frontal and anterior cingulate cortices than in medial temporal and thalamic regions	Kaasinen et al. (2000) (PET)	24 (19–74 years)
	Reduced densities of dopamine D2 receptors in the frontal and anterior cingulate cortices associated with reduced glucose metabolism in frontal and cingulate cortices	Volkow et al. (2000) (PET)	37 (24–86 years)
	Age-related reduction in serotonin 5-HT_2 receptor availability with advance aging	Wang et al. (1995) (PET)	19 (21–49 years)

(Continued)

TABLE 31.1 (Continued)

Change	Characteristic	Study (Method)	Sample size (age)
(B) Functional			
Reduction in cerebral oxygenation and blood flow	Greatest age-related reduction in regional cerebral blood flow in frontal regions and lowest reduction in occipital regions	Bentourkia et al. (2000) (PET)	10 (21–36 years) and 10 (55–75 years)
	Reduced regional cerebral blood flow and cerebral metabolic rate of oxygen in gray matter in frontal, temporo-sylvian, and parieto-occipital cortex	Pantano et al. (1984) (PET)	18 (19–50 years) and 9 (55–76 years)
Reduction in glucose uptake	Greatest age-related reduction in regional cerebral blood flow in frontal and striatal regions and lowest reduction in occipital regions	Bentourkia et al. (2000) (PET)	10 (21–36 years) and 10 (55–75 years)
	Greater age-related reduction in glucose metabolism bilaterally in superior medial frontal (i.e., dlPFC, OFC), motor, and anterior cingulate cortices than in posterior cingulate, limbic, and thalamic regions and basal ganglia	Kalpouzos et al. (2009) (PET)	45 (20–83 years)
	Reduced glucose metabolism in frontal and cingulate cortices	Volkow et al. (2000) (PET)	37 (24–86 years)
Dysregulation of gene expression	Age-related downregulation of frontopolar genes involved in promoting synaptic transmission, intra-neuronal signaling, and neuronal survival; age-related upregulation of frontopolar genes involved in cellular defense to oxidative stress and inflammation and in DNA repair	Lu et al. (2004) (microarray)	30 (26–106 years)
	Age-related downregulation of genes involved in neuronal transmission and signal transduction in the lateral prefrontal neurons (BA 9 or 47); age-related upregulation of genes involved in inflammation and cellular defenses in lateral prefrontal glial cells	Erraji-Benchekroun et al. (2005) (microarray)	39 (13–79 years)
	Age-related changes in gene expression more evident in superior frontal gyrus and posterior cingulate cortex than in medial temporal lobe structures; age-related downregulation of genes that mediate energy production and intracellular signaling particularly in males; age-related upregulation of genes that mediate inflammatory responses in both sexes	Berchtold et al. (2008) (microarray)	55 (20–99 years)

Note: *dlPFC*, dorsolateral prefrontal cortex; *DTI*, diffusion tensor imaging; *OFC*, orbitofrontal cortex; *PET*, positron emission tomography; *MRI*, magnetic resonance imaging.

In addition, some studies have reported an age-related reduction in cortical thickness that is greater in the prefrontal and motor cortices than in the temporal and parietal lobes (Fjell et al., 2009; Salat et al., 2004). Furthermore, some diffusion tensor imaging studies have reported age-related changes in white matter microstructure, which are associated with less efficient neural transmission, in the prefrontal cortex (Head et al., 2004; Pfefferbaum et al., 2005; Salat et al., 2005). More importantly, these studies have reported a greater reduction in fractional anisotropy, which is a measure associated with myelination in white matter, in frontal regions than in posterior cortical regions in healthy older adults compared with younger adults.

In addition, aging is associated with functional changes that primarily occur in the frontal lobe. For example, positron emission tomography (PET) studies have reported a reduction in regional cerebral oxygenation and blood flow in the gray matter of the frontal cortex (Bentourkia et al., 2000; Pantano et al., 1984). Although this age-related reduction in oxygenation utilization does not seem to be specific to the frontal lobe (Bentourkia et al., 2000; Pantano et al., 1984), it is more pronounced in frontal cortical regions as compared to posterior regions, such as the occipital cortex (Bentourkia et al., 2000). In addition, some PET studies have reported a lower rate of glucose metabolism in the frontal cortex and anterior cingulate cortex in normal, older adults as compared with young adults (Bentourkia et al., 2000; Kalpouzos et al., 2009; Volkow et al., 2000). Consistent with the anterior-to-posterior gradient of neuroanatomical changes, these studies have reported a greater age-associated decline in glucose uptake rate in the frontal lobe (i.e., the dlPFC, OFC, anterior cingulate cortex) than in the posterior cortical regions and subcortical regions (Bentourkia et al., 2000; Kalpouzos et al., 2009).

The age-related structural changes in the frontal lobe can also be observed in individual neurons. For example, some studies have reported age-related dendritic regression of layer V pyramidal cells in the prefrontal cortex (De Brabander et al., 1998) and the primary motor cortex (Nakamura et al., 1985). This age-related dendritic shrinkage is accompanied with a reduction in presynaptic terminal counts (i.e., number of synapses) in the frontal cortex (Masliah et al., 1993). In addition, several studies have reported an age-related reduction in the density of dopamine D1 (Backman et al., 2011; de Keyser et al., 1990) and D2/D3 receptors (Kaasinen et al., 2000; Volkow et al., 2000) and in the availability of serotonin 5-HT_2 receptors (Wang et al., 1995) in neurons of the frontal cortex and anterior cingulate cortex. This age-related reduction in neurochemical receptors is greater in frontal regions than in other parts of the brain, such as the medial temporal cortex and the thalamus for dopamine D2/D3 receptors (Kaasinen et al., 2000) and the occipital cortex for serotonin 5-HT_2 receptors (Wang et al., 1995).

In addition, the age-associated functional changes extend to subcellular molecular processes in neurons and/or glial cells, which affect frontal lobe more than temporal lobe structures (Berchtold et al., 2008). Such changes include dysregulation of gene expression (Berchtold et al., 2008; Erraji-Benchekroun et al., 2005; Lu et al., 2004) and dysfunction of mitochondria (Lu et al., 2004; see Yankner et al., 2008, for review). For example, DNA microarray studies in humans have found an age-related downregulation of genes that promote synaptic plasticity, intra-neuronal signaling, neuronal survival or genesis, and/or energy (i.e., adenosine triphosphate, ATP) production in frontopolar (Lu et al., 2004), lateral prefrontal (BA 9 or 47) (Erraji-Benchekroun et al., 2005), and superior frontal neurons (Berchtold et al., 2008). These patterns suggest that aging is associated with impaired neural transmission and cellular maintenance or proliferation in prefrontal cortex neurons. In addition, research studies have found an age-related upregulation of genes that mediate cellular defense to oxidative stress and inflammatory responses or that mediate DNA repair in frontopolar neurons (Lu et al., 2004), superior frontal neurons (Berchtold et al., 2008), and lateral prefrontal glial cells (Erraji-Benchekroun et al., 2005). These patterns may reflect greater occurrences of oxidative stress and DNA damage in prefrontal neurons and glia cells with advancing age. Moreover, the alterations in gene expression may be mediated by mitochondrial dysfunction (Berchtold et al., 2008; Lu et al., 2004), which leads to increased levels of reactive oxygen species (ROS; i.e., oxidative stress) and to impaired ATP synthesis that is necessary for DNA repair. In summary, a diverse variety of studies have reported structural and functional deteriorations of the frontal lobe with advancing age. These age-related changes are evident across tissue, neuronal, and molecular levels.

31.1.2 Cognitive declines in frontal lobe functioning in normal human aging

Consistent with the relationships between aging and neurobiological changes in the frontal lobe, the cognitive deterioration associated with aging is more prominent on mental functions that are primarily mediated by the frontal lobe (West, 1996). Specifically, a substantial number of research studies have reported a decline in both the dorsolateral and ventromedial aspects of frontal lobe functioning in normal older adults.

With regard to dorsolateral frontal lobe functioning (see Table 31.2A), empirical studies have shown that older adults performed more poorly than younger adults did in standardized neuropsychological tests and experimental paradigms that measure shifting (Kramer et al., 1999, 1994; MacPherson et al., 2002), inhibition (Andrés and Van der Linden, 2000; Fjell et al., 2017; Hillman et al., 2006; Salthouse, 2010; Spieler et al., 1996; Van Der Elst et al., 2006; West and Alain, 2000), updating (Dobbs and Rule, 1989; Van der Linden et al., 1994), monitoring (MacPherson et al., 2002; Lamar and Resnick, 2004; West et al., 1998), verbal fluency (Chan and Poon, 1999; Brickman et al., 2005; Tombaugh et al., 1999; Troyer, 2000), and problem solving (Andrés and Van der Linden, 2000). In addition, some studies have reported age-related declines in other cognitive domains, such as sustained attention (Carriere et al., 2010; Chen et al., 1998; Mani et al., 2005) and free-recall or source memory (Perlmutter, 1979; Levine et al., 1997; Stuss et al., 1996). Importantly, these studies have shown that the performance of older adults is similar to that of patients with dorsolateral frontal lesion rather than to that of patients with temporal lesion (Levine et al., 1997; Stuss et al., 1996). Thus aging is associated with a decline in dorsolateral frontal lobe functioning in humans.

With regard to ventromedial frontal lobe functioning (see Table 31.2B), empirical studies have reported an age-related decline in the abilities to recognize emotions from facial expressions (Calder et al., 2003; Isaacowitz et al., 2007; Sullivan et al., 2007), voices (Ryan et al., 2010), and bodily gestures (Ruffman et al., 2009) and across

TABLE 31.2 Cognitive declines in (A) dorsolateral and (B) ventromedial frontal lobe functioning in normal older adults.

Function	Task	Performance in older adults	Study	Sample size (age)
(A) Dorsolateral				
Executive function				
Shifting	Task Switching	Slower reaction time on switching trials particularly	Kramer et al. (1999)	16 (18–30 years) and 16 (60–75 years)
	Wisconsin Card Sorting Test	More perseverative errors and fewer categories completed	Kramer et al. (1994)	32 (18–28 years) and 30 (60–74 years)
		Age-related increase in number of perseverative errors	MacPherson et al. (2002)	30 (20–38 years), 30 (40–59 years), and 30 (61–80 years)
Inhibition	Flanker	An aged-related slowing in responding on both congruent and incongruent conditions of the flanker task	Hillman et al. (2006)	241 (15–71 years)
		Slower and less accurate responses on both the congruent and incongruent conditions of the flanker task	Salthouse (2010)	62 (18–39 years), 89 (40–59 years), and 114 (60 years or above)
	Hayling Sentence Completion Test	A longer time to complete Part B (response inhibition) but not Part A (response initiation) and an increased number of errors not fully accountable by slower processing speed in Part B	Andrés and Van der Linden (2000)	47 (20–30 years) and 48 (60–70 years)
	Stroop	An age-related increase in Stroop time for all conditions	Fjell et al. (2017)	63 (23–52 years) and 56 (63–86 years)
		Age-related increases in Stroop time for all conditions and in the effect of interference (i.e., incongruent > neutral)	Spieler et al. (1996)	27 (17–26 years), 25 (58–79 years), and 25 (80–93 years)
		Age-related increases in Stroop time for all conditions and in the effect of interference [i.e., Incongruent color naming > (word reading + color naming)/2]	Van Der Elst et al. (2006)	1788 (24–81 years)
		Increases in Stroop time for all conditions and in the effect of interference	West and Alain (2000)	12 (M = 27.1 years) and 12 (M = 69.5 years)
Updating	*n*-Back	An age-related shift in occurrence of first error on the 1- and 2-back but not 0-back tasks	Dobbs and Rule (1989)	228 (30–99 years)
	Running memory	An increased number of errors as a function of list length	Van der Linden et al. (1994)	18 (19–27 years) and 18 (60–75 years)
Monitoring	Self-ordered Pointing Test	A larger number of errors	Lamar and Resnick (2004)	20 (20–40 years) and 20 (60–80 years)
		An age-related increase in total number of errors	MacPherson et al. (2002)	30 (20–38 years), 30 (40–59 years), and 30 (61–80 years)
		More perseverative and forgetting errors	West et al. (1998)	40 (18–40 years) and 40 (73–78 years)

(Continued)

TABLE 31.2 (Continued)

Function	Task	Performance in older adults	Study	Sample size (age)
Verbal fluency	Phonemic fluency	Age-related declines in number of words generated	Brickman et al. (2005)	471 (21–82 years)
		Age-related decrease in word production	Tombaugh et al. (1999)	1300 (16–95 years)
		Age-related declines in switching frequency, but not cluster size or number of words generated	Troyer (2000)	411 (18–91 years)
	Semantic fluency	Age-related declines in number of words generated	Brickman et al. (2005)	471 (21–82 years)
		Reduction in the number of words produced	Chan and Poon (1999)	316 (7–95 years)
		Reduction in the number of words produced	Tombaugh et al. (1999)	1300 (16–95 years)
		Age-related declines in number of words generated, switching frequency, but not cluster size	Troyer (2000)	411 (18–91 years)
Problem solving	Brixton Test	An increased number of errors not fully accounted by slower processing speed	Andrés and Van der Linden (2000)	47 (20–30 years) and 48 (60–70 years)
	Tower of London	A larger number of moves required and longer initiation and subsequent times to solve the problems	Andrés and Van der Linden (2000)	47 (20–30 years) and 48 (60–70 years)
Sustained attention	Continuous Performance Test	Age-related decreases in hit rate and sensitivity	Chen et al. (1998)	345 (20–65 years)
		An age-related increase in number of commission errors	Mani et al. (2005)	32 (19–82 years)
	Delayed response task	Increased errors only when distractors are present and at long but not short delays	Chao and Knight (1997)	12 (20–22 years) and 12 (57–71 years)
	Sustained Attention to Response Task	Aged-related decline in reaction time and increase in errors in adulthood. Age-related change in number of anticipations and omissions was relatively stable over time	Carriere et al. (2010)	638 (14–77 years)
Episodic memory	Conditional associative learning task	Like patients with dorsolateral prefrontal lesion, older adults had deficits in inhibiting interferences	Levine et al. (1997)	20 (18–39 years), 20 (63–83 years), and 14 patients with focal frontal lesion
	Verbal list learning test	Impaired organizational control processes; performance comparable to that in patients with frontal rather than limbic damage	Stuss et al. (1996)	20 (20–39 years), 20 (40–64 years), and 20 (65–79 years)
	Word list learning	Impaired free and cued verbal recall; free recall being more impaired; relatively intact recognition memory	Perlmutter (1979)	48 (18–29 years) and 48 (59–70 years)

(B) Ventromedial				
Emotion recognition				
Facial expressions	Emotion labeling task	Age-related decline in recognition of anger and fear but not of other basic facial emotions	Calder et al. (2003)	45 (18–30 years) in Study 1; 227 (17–70 years) in Study 2a; 125 (18–75 years) in Study 2b
		Age-related decline in recognition of anger, disgust, fear, and happiness, but not of sadness, surprise, or neutral expressions	Isaacowitz et al. (2007)	357 (18–85 years)
		Deficits in recognizing fear and anger in older adults in Study 1; deficits in recognizing anger in Study 2	Sullivan et al. (2007)	27 (20–37 years) and 27 (61–95 years) in Study 1; 30 (18–32 years) and 30 (60–87 years) in Study 2
Vocalizations	Emotion labeling task	Impaired recognition of sadness and anger in vocal expressions	Ryan et al. (2010)	40 (17–29 years) and 40 (60–84 years)
Body gestures	Emotion labeling task	Impaired recognition of anger, sadness, fear, and happiness in bodily expressions in older adults	Ruffman et al. (2009)	30 (21–34 years) and 30 (62–81 years)
Cross-modal expressions	Emotion matching tasks (voice/face; voice/body; voice/word)	Impaired anger recognition in the word task; Impaired sadness, anger, and disgust recognition in the body task; and impaired happiness, sadness, anger, disgust, and fear recognition in the face task; intact surprise recognition in any task	Ruffman et al. (2009)	26 (18–24 years) and 26 (64–84 years)
Decision making	Cambridge Gambling Task	An quadratic, age-related change in the ability to make optimal choices across lifespan	Deakin et al. (2004)	177 (17–73 years)
	Iowa gambling task	An quadratic, age-related change in the ability to decide advantageously across lifespan	Beitz et al. (2014)	1583 (5–89 years)
		Impaired ability to make advantageous decisions in older adults	Denburg et al. (2005)	80 (26–85 years)
		An quadratic, accelerated age-related decline in the ability to decide advantageously	Fein et al. (2007)	112 (18–55 years) and 52 (56–85 years)
Flexible reward-learning	Object discrimination reversal paradigm	Impaired acquisition and reversal learning of stimulus-reinforcement contingencies in the context of 80:20 probabilistic reinforcement in older adults	Mell et al. (2005)	20 (M = 23.2 years) and 20 (M = 67.6 years)
	Reward-based learning task	Impaired acquisition, reversal learning, and transfer learning of stimulus-reward associations in older adults	Weiler et al. (2008)	30 (19–33 years) and 30 (50–71 years)

modalities (Ruffman et al., 2009). These emotion recognition deficits seem to be more pronounced for negative emotions (e.g., anger, fear, sadness) than for positive (e.g., happiness, surprise) emotions (Calder et al., 2003; Ruffman et al., 2009; Ryan et al., 2010; Sullivan et al., 2007). In addition, some studies have reported poorer decision-making ability in older adults as compared with younger adults, such that older adults tend to make suboptimal choices and fail to anticipate the long-term consequences of their actions in general (Beitz et al., 2014; Deakin et al., 2004; Denburg et al., 2005; Fein et al., 2007). Furthermore, some studies have reported an age-related decline in the abilities to acquire and reverse stimulus-reward associations (Mell et al., 2005; Weiler et al., 2008), suggesting poorer behavioral flexibility in response to changing reinforcement contingencies over age. Thus aging is also associated with a decline in ventromedial frontal lobe functioning.

In summary, aging is associated with a decline in cognitive functions that are preferentially mediated by the frontal lobe of the brain. Many research studies have shown that a decline in frontal lobe functioning, especially inhibitory control and attentional shifting, are reliable risk factors for subsequent declines in global cognitive functioning (Clark et al., 2012) and for future falls and impairment in activities of daily living (Bell-McGinty et al., 2002; Herman et al., 2010; Kearney et al., 2013; Mirelman et al., 2012; Royall et al., 2004) in normal older adults. Thus interventions that can effectively maintain or enhance frontal lobe functioning in normal elderly population are clinically important.

31.1.3 Conventional interventions for improving frontal lobe functioning in normal older adults

To slow down the decline in frontal lobe functioning associated with human aging, much research has been devoted to developing interventions that target at maintaining or improving frontal lobe functioning of normal older adults. Conventional interventions can be broadly divided into two types: behavior-based interventions and lifestyle-based interventions.

Much research has examined the effects of behavioral-based interventions on frontal lobe functioning of normal older adults. Some studies have found that 4–8 weeks of training based on video game playing improved certain aspects of EF, such as task switching, updating, and sustained attention, of normal older adults (Anguera et al., 2013; Basak et al., 2008; Maillot et al., 2012). These improvements could be evident on far-transfer tasks (Anguera et al., 2013; Basak et al., 2008; Maillot et al., 2012) and mediated by increased frontal lobe activity during cognitive tasks (Anguera et al., 2013). Nevertheless, some older adults may have a relatively negative attitude toward video games, which leads to low compliance and thus few cognitive-enhancing effects (Boot et al., 2013). In addition, some studies have shown that cognitive training based on tasks that are well-structured or place little demand on multitasking, may only bring limited EF or memory improvement on far-transfer tasks for normal older adults (Ball et al., 2002; Dahlin et al., 2008). Notably, a meta-analytic study has found that cognitive training that focuses on practicing EF or WM tasks yields only a small improvement (i.e., mean Cohen's d of approximately 0.2) in EF/attention as assessed by far-transfer tasks (Karbach and Verhaeghen, 2014). Thus behavioral-based interventions seem to exert limited beneficial effects on frontal lobe functioning of normal older adults in general.

On the other hand, some studies have examined the effects of lifestyle-based interventions on frontal lobe functions in normal older adults. Although some studies have reported cognitive-enhancing effects of these interventions (Dustman et al., 1984; Prehn et al., 2016), many studies have reported insignificant or small effects of lifestyle-based interventions on different aspects of frontal lobe functioning, such as selective attention and free-recall memory, of normal older adults (Clark et al., 2011; Hassmén et al., 1992; Lautenschlager et al., 2008; Voss et al., 2010; Williamson et al., 2009). Specifically, aerobic exercise has been the most studied lifestyle-based approach to improve executive control in healthy, elderly people (Colcombe and Kramer, 2003; Smith et al., 2010; Young et al., 2015). However, many randomized controlled trials have reported insignificant or very small effects of aerobic exercise (e.g., 4–12 months of walking programs) relative to active (e.g., nonaerobic exercise or psychoeducation programs) and/or passive (i.e., waitlist) control groups on inhibitory control, verbal fluency, sustained attention, and/or reasoning of normal older adults (Blumenthal et al., 1989; Hassmén et al., 1992; Lautenschlager et al., 2008; van Uffelen et al., 2008; Voss et al., 2010; Williams and Lord, 1997; Williamson et al., 2009; see Young et al., 2015, for review). Thus lifestyle-based interventions, including aerobic exercise, may bring only limited improvements in frontal lobe function for normal older adults.

In summary, interventions that are based on cognitive training and lifestyle changes bring only limited improvements in frontal lobe functioning for normal older adults. Based on these limitations, alternative interventions that can effectively protect against the decline in frontal lobe functioning of normal, elderly humans are needed. In the next section, we will review the neurophysiological effects of PBM and discuss the possibility of using this technique as an intervention for the aging brain.

31.2 Photobiomodulation and neuroenhancement

31.2.1 Mechanisms of action of photobiomodulation

PBM is a noninvasive technique that delivers red to near-infrared light, with wavelengths ranging from 600 to 1100 nm, to targeted sites of the body (Hamblin, 2016). When the light is shone on the scalp and impacts the brain through the skull, this procedure is referred to as transcranial PBM. Although most of the light is absorbed or scattered by nonneural tissue as it penetrates through the scalp, skull, and cerebrospinal fluid, a small fraction of the delivered light may reach neural tissue (e.g., Jagdeo et al., 2012), because biological tissue is relatively transparent to red to near-infrared light (Jobsis, 1977; Villringer and Chance, 1997). The mechanisms underlying action of PBM are multifaceted and involve intracellular activities, extracellular adaptations, and morphological alterations. When applied to the brain, the neurobiological effects of PBM can be broadly classified into two types, which are outlined below.

First, PBM may increase energy production and oxygen supply to nerve cells and boost their metabolism through interaction between photon energy from near-infrared light and cytochrome c oxidase (CCO), which is the terminal enzyme of the mitochondrial respiratory chain (Avci et al., 2013; Hamblin, 2016; Karu, 2000; Wong-Riley et al., 2005). As CCO absorbs photon energy from the light, nitric oxide (NO) is released, thereby increasing the production of ATP (Hamblin, 2008; Lane, 2006; Sheppard et al., 2005) and providing extra metabolic energy for neural transduction (Tafur and Mills, 2008). These PBM-induced increases in CCO activity (Liang et al., 2008; Wong-Riley et al., 2001, 2005), NO (Sharma et al., 2011), oxygen consumption (Poyton and Ball, 2011), and ATP (Dong et al., 2015; Oron et al., 2007; Ying et al., 2008) have been demonstrated in numerous cell studies. In addition, based on the vasodilating effect of NO (Ignarro et al., 1999), both animal (Uozumi et al., 2010) and human (Salgado et al., 2015) studies have shown that PBM increases the diameter of blood vessels, thereby increasing regional cerebral blood flow.

In addition, PBM may slow down neuron death and dendritic atrophy (Huang et al., 2013; Meng et al., 2013; Yan et al., 2017; Yu et al., 2015; Yang et al., 2018) and promote morphogenesis and proliferation of dendrites and neurons (Fukuzaki et al., 2015; Meng et al., 2014; Yang et al., 2018). Such neuroprotective effects may be achieved in several ways. First, the increased CCO enzyme activity due to PBM may lead to a transient burst of ROS, which may activate signaling pathway that leads to antioxidant and antiapoptotic effects inside the cells (Waypa et al., 2016). In addition, PBM may suppress production of pro-inflammatory cytokines and increase production of antiinflammatory cytokines, changing inflammatory status in the neuronal environment (Yang et al., 2018). Furthermore, PBM increases ATP synthesis, which facilitates tissue development and repair (Karu, 2010; Khakh and Burnstock, 2009; Rathbone et al., 1992). Thus increased ATP production might promote neuron proliferation and differentiation (Fukuzaki et al., 2015; Yang et al., 2018). Moreover, PBM may upregulate expression of neurotrophic genes (Gomes et al., 2012; Yan et al., 2017), which is abundant throughout the (human) brain, particularly in the prefrontal cortex and hippocampus (Pezawas et al., 2004). Such upregulated gene expression may prevent dendritic atrophy and neuron death (Dong et al., 2015; Lee et al., 2017; Meng et al., 2013) and promote the growth and differentiation of new neurons and dendrites (Huang and Reichardt, 2001; Meng et al., 2014).

31.2.2 Photobiomodulation for enhancing brain functions in humans

As in vitro studies have provided evidence supporting the promising bioenergetic and cytoprotective effects of PBM (Hamblin, 2008; Oron et al., 2007; Sharma et al., 2011), some studies have evaluated the possibility of applying PBM as a neuropsychological intervention for enhancing brain functions in humans. Recently, there have been an increasing number of case studies and controlled experiments that examined the effects of PBM on brain functions in both healthy and clinical human populations. These studies in general have provided some empirical evidence for the potential utility of PBM for enhancing cognitive functions, including those that are primarily mediated by the frontal lobe, in both healthy and clinical populations (see Table 31.3).

31.2.2.1 Healthy humans

The controlled experiments have consistently reported beneficial effects of PBM on various aspects of frontal cognitive functions, including set shifting (Blanco et al., 2017a), rule-based category learning (Blanco et al., 2017b), sustained

TABLE 31.3 Effects of photobiomodulation on brain functions in (A) healthy humans and in patients with (B) Alzheimer disease, (C) stroke, (D) traumatic brain injury, and (E) depression.

Study (design)	Subject (year of age)	Light source	Treatment parameters	Treatment sites	Task/scale	Function improved
(A) Healthy						
Barrett and Gonzalez-Lima (2013) (Controlled experiment)	40 healthy young adults (range: 18–35)	Laser (1064 nm) Model CG-5000 (HD Laser Center, Dallas, USA)	Single session for 4 min; 60 J/cm^2, CW	2 sites: right frontal pole	Psychomotor vigilance task, delayed match-to-sample task, and the Positive and Negative Affect Schedule	Sustained attention, short-term memory, and affective state
Blanco et al. (2017a) (Controlled experiment)	30 healthy subjects ($M = 20.4$, $SD = 1.64$)	Laser (1064 nm) Model Cell Gen laser (HD Laser Center, Dallas, USA)	Single session for 8 min; 60 J/cm^2, CW	2 sites: F4 and Fp2	Wisconsin Card Sorting Test	Set shifting
Blanco et al. (2017b) (Controlled experiment)	118 healthy young adults ($M = 19$, $SD = 1.91$; range: 17–35)	Laser (1064 nm) Cell Gen laser (HD Laser Center, Dallas, USA)	Single session for 8 min; 60 J/cm^2, CW	2 sites: upper and lower portion of area in FP2, F4, and F8 sites	Category learning task	Rule-based category learning
Hwang et al. (2016) (Controlled experiment)	60 healthy subjects ($M = 23.47$, $SD = 3.82$)	Laser (1064 nm) model CG-5000 (HD Laser Center, Dallas, USA)	Single session for 8 min; 60 J/cm^2, CW	2 sites: medial and lateral to the right frontal polar (FP2) point	Psychomotor vigilance task and delayed match-to-sample task	Sustained attention and working memory
Moghadam et al. (2017) (Controlled experiment)	39 healthy subjects ($M = 21$, $SD = 1.83$, range: 18–24)	LED (Iranbargh, Tehran, Iran)	Single session for 2.5 min; 60 J/cm^2, CW	1 site: FP2	Parametric Go/No-Go task	Sustained attention
Vargas et al. (2017) (Controlled experiment)	12 older adults with subject memory complaints ($M = 62.27$, $SD = 14.21$, range: 49–90)	Laser (1064 nm) (HD Laser Center, Dallas, USA)	5 sessions in 5 weeks; 250 mW/cm^2, CW	2 sites: forehead at 4.2 cm diameter medial and lateral sites	Psychomotor vigilance task and delayed match-to-sample task	Sustained attention, short-term memory
(B) AD						
Berman et al. (2017) (Pilot study)	11 AD patients ($M = 81.8$, $SD = 6.18$, range: 74–95)	LED (1060–1080 nm)	28 sessions, 6 min each, in 28 days, 10 Hz	Full head	Mini Mental Status Exam, quantitative EEG, Alzheimer's disease Assessment Scale-Cognitive	Executive function (memory, attention, task switching)
Saltmarche et al. (2017) (Case study)	5 AD patients ($M = 77.6$, $SD = 7.23$, range: 72–90)	LED (810 nm) Model "810" and "Neuro" (Vielight Inc., Toronto, Canada)	14 clinic treatments or 84 home treatments in 12 weeks; 10.7 or 24.6 + 13.8 J/cm^2; 10 Hz	Sites targeting the default mode network: The mesial prefrontal cortex, precuneus, posterior cingulate cortex, inferior parietal lobe, and hippocampus	Mini Mental Status Exam, Alzheimer's Disease Assessment Scale-Cognitive	Global cognitive function

(C) Stroke						
Boonswang et al. (2012) (Case study)	1 chronic brainstem stroke patient (29)	LED (660 and 850 nm) Model XR-3T-1 (THOR, London, UK)	8 sessions, 32 min each, in 8 weeks; energy density = .95 J/cm^2; spot area = 0.20 cm^2	32 sites: the cerebral cortices, brainstem, cervical spine (8 sites); the core musculature and lymphatics (24 sites)	N/A	Mood symptoms, and sensory and motor functioning
Lampl et al. (2007) (Controlled experiment)	120 acute ischemic stroke patients ($M = 69.62$, range: 40–85)	Laser (808 nm) (Neurothera PhotoThera Inc., Carlsbad, CA)	Single session, each site for 2 min, for a total of 40 min; energy density = 1.2 J/cm^2 (cortex); frequency = CW	20 sites: full head	The National Institutes of Health Stroke Scale, Modified Rankin Scale, Barthel index, and Glasgow Outcome Scale	Neurological functioning
Naeser et al. (2012)	3 aphasic stroke patients (2 chronic nonfluent aphasia due to stroke, 1 logopenic variant, primary progressive aphasia)	LED (9 red, 52 near-infrared diodes)	18 sessions in 16 weeks; energy density = 13 J/cm^2; frequency = 146 Hz; spot diameter = 2.1 in.	3 sites: perisylvian and midline	Boston diagnostic aphasia examination and Picture Naming Task	Auditory comprehension (bilateral and dorsomedial); Picture naming (left only)
Zivin et al. (2009) (Controlled experiment)	660 acute ischemic stroke patients (Active: $M = 70.4$, $SD = 12.6$; Sham: $M = 70.0$, $SD = 11.9$)	Laser (808 nm) (Neurothera PhotoThera Inc., Carlsbad, CA)	Single session, each site for 2 min, for a total of 40 min; energy density = 1.2 J/cm^2 (cortex) frequency = CW	20 sites: full head	The National Institutes of Health Stroke Scale and Modified Rankin Scale	Neurological functioning (Moderate and moderate-severe but not severe stroke patients)
Zivin et al. (2014) (Controlled experiment: prematurely terminated)	1000 acute ischemic patients (planned sample)	Laser (808 nm) (Neurothera PhotoThera Inc., Carlsbad, CA)	Single session, each site for 2 min, for a total of 40 min; energy density = 1.2 J/cm^2 (cortex); frequency = CW	20 sites: full head	The National Institutes of Health Stroke Scale and Modified Rankin Scale	N/A
(D) TBI						
Hesse et al. (2015) (Case study)	5 TBI patients ($M = 47.2$, $SD = 20.86$)	Laser (785 nm) (Power Twin 21 by MKW Laser system)	30 sessions, 10 min each, in 6 weeks; 10 mW/cm^2, CW	5 locations: superior crest of the fossa sphenoidale on the forehead	Revised Coma Recovery Scale	Alertness and awareness
Naeser et al. (2011) (Case study)	2 TBI patients (59, 52)	LED (633 and 870 nm)	Single session per week for 6 years or single session per day for 4 months; 25.8 mW/cm^2 or 22.2 mW/cm^2, CW	2 sites: left and right forehead	Stroop test	Inhibition

(Continued)

TABLE 31.3 (Continued)

Study (design)	Subject (year of age)	Light source	Treatment parameters	Treatment sites	Task/scale	Function improved
Naeser et al. (2014) (Case study)	11 TBI patients ($M = 44.3$, $SD = 13.7$, range: 26–62)	LED (633 and 870 nm) Model 1100 (MedX Health, Toronto, Canada)	18 sessions, 20 min each, in 6 weeks; 13 J/cm^2, CW, 22.48 cm^2	11 sites: frontal, parietal, and temporal areas of the brain	Stroop test and California Verbal Learning Test	inhibition, task switching, verbal learning and memory; PTSD symptoms
(E) Depression						
Cassano et al. (2015) (Case study)	4 MDD patients ($M = 47$, $SD = 14$)	Laser (808 nm) (Neurothera PhotoThera Inc., Carlsbad, CA)	6 sessions, 8 min each, in 3 weeks; 84 J/cm^2, CW	4 sites: right and left forehead center at 20 and 40 mm from sagittal line	Hamilton Depression Rating Scale	Depressive symptoms
Disner et al. (2016) (Controlled experiment)	51 patients with elevated depression symptoms ($M = 19.37$, $SD = 3.05$)	Laser (1064 nm) Model CG-5000 (Cell Gen Therapeutics, Dallas, USA)	2 sessions, 8 min each; 60 J/cm^2, CW	2 sites: medial and lateral of left and right side of forehead	Dot-probe task, Center for Epidemiologic Studies—Depression Scale, and Negative Bias Assessment	Depressive symptoms and negative attention bias
Schiffer et al. (2009) (Case study)	10 MDD patients ($M = 35.1$, $SD = 7.14$, range: 25–46)	LED (810 nm) (Marubeni America Corp, Santa Clara, CA)	Single session for 8 min; 60 J/cm^2, CW	2 sites: F3 and F4	The Hamilton Depression Rating Scale, Hamilton Anxiety Rating Scale, Positive and Negative Affect Scale	Depressive and anxiety symptoms

Note: *AD*, Alzheimer disease; *CW*, continuous wave; *MDD*, major depressive disorder; *PTSD*, posttraumatic stress disorder; *TBI*, traumatic brain injury.

attention (Barrett and Gonzalez-Lima, 2013; Hwang et al., 2016; Moghadam et al., 2017; Vargas et al., 2017), and working memory (Barrett and Gonzalez-Lima, 2013; Hwang et al., 2016; Vargas et al., 2017) in healthy young and older adults. Specifically, individuals who underwent active PBM treatment reacted faster on the psychomotor vigilance and parametric go/no-go tasks as compared to placebo, suggesting a beneficial effect of PBM on sustained attention (Barrett and Gonzalez-Lima, 2013; Hwang et al., 2016; Moghadam et al., 2017; Vargas et al., 2017). In addition, individuals who received active PBM treatment had shorter memory retrieval latency and higher accuracy on the delayed match-to-sample task compared with those who did not, suggesting a beneficial effect of PBM on working memory (Barrett and Gonzalez-Lima 2013; Hwang et al., 2016; Vargas et al., 2017). Furthermore, individuals who had just completed a single session of stimulation were faster in learning category rules (Blanco et al., 2017b) and shifting sets (Blanco et al., 2017a) as compared to sham control. Thus transcranial PBM may lead to an immediate enhancement in cognitive functions that are primarily mediated by the frontal lobe in healthy humans.

31.2.2.2 Alzheimer disease

Several studies have examined the short-term and long-term effects of transcranial PBM on cognitive functions in Alzheimer disease (AD) patients (see Table 31.3B; Berman et al., 2017; Saltmarche et al., 2017). In addition, two longitudinal studies have reported significant improvements in global cognitive functioning, memory, visual attention, and/or task switching in AD patients who received 14–84 sessions of laser treatment (Berman et al., 2017; Saltmarche et al., 2017). Although there are only a few studies on the efficacy of PBM on dementia, the results so far are encouraging.

31.2.2.3 Stroke

Several studies have examined the effectiveness of transcranial PBM in alleviating the neurological and physical symptoms of acute (Lampl et al., 2007; Zivin et al., 2009, 2014) and chronic (Boonswang et al., 2012; Naeser et al., 2012) stroke patients (see Table 31.3C). Regarding the effects of PBM for acute stroke, three double-blind controlled trials (i.e., NEST-1, NEST-2, NEST-3) have been conducted. The initial two studies (i.e., NEST-1, NEST-2) have reported that acute stroke patients who received active PBM treatment had significantly greater improvements in neurological outcome at 90 days after stroke compared with those who did not (Lampl et al., 2007; Zivin et al., 2009; Stemer et al., 2010). Nevertheless, the third study of the series (i.e., NEST-3), which planned to enroll 1000 patients with moderate or moderate-severe acute ischemic stroke at up to 50 sites, was prematurely terminated due to an expected lack of success at interim analysis (Zivin et al., 2014).

In addition, some case studies have examined the effects of transcranial PBM for chronic stroke patients (Boonswang et al.; 2012; Naeser et al., 2012). Boonswang et al. (2012) found that a 29-year-old female who suffered a brainstem stroke 2 years ago had improved physical functioning after completing eight sessions of LED therapy. In addition, Naeser et al. (2012) found that a 16-week transcranial LED therapy that targeted the bilateral perisylvian and supplementary motor area led to improved auditory comprehension.

31.2.2.4 Traumatic brain injury

Several case studies have examined the cognitive-enhancing effects of PBM for traumatic brain injury (TBI) patients (see Table 31.3D; Hesse et al., 2015; Naeser et al., 2011, 2014). Two case studies have shown that TBI patients who received at least 18 sessions stimulation exhibited improvements in inhibition and task switching performance (Naeser et al., 2011, 2014) and in verbal learning and free-recall memory (Naeser et al., 2014). In addition, one case study has reported improved alertness and awareness in five patients with disorder of consciousness who received 30 sessions of laser therapy (Hesse et al., 2015). Therefore PBM might enhance inhibitory control, mental flexibility, verbal learning, and conscious awareness in TBI patients through stimulation of frontal brain regions.

31.2.2.5 Major depressive disorder

Some studies have examined the effects of PBM on mood in patients with major depressive disorder (MDD) or mild depressive symptoms (see Table 31.3E). Patients with MDD or mild depressive symptoms exhibited fewer depressive symptoms, as indicated by lower scores on standardized depression inventories, following one (Schiffer et al., 2009), two (Disner et al., 2016), or six sessions (Cassano et al., 2015) of PBM therapy. A reduction in anxiety symptoms has also found in MDD patients who received a single session of transcranial laser therapy (Schiffer et al., 2009). In addition, Disner et al. (2016) found that transcranial PBM to the right but not left prefrontal region led to a reduction in depressive symptoms in adults with elevated depression symptoms. Taken together, some preliminary evidence suggests that PBM may have positive effects on mood, depending on the location of stimulation site.

31.3 Photobiomodulation for normal older adults: a potential intervention for the aging brain

Converging evidence from structural neuroimaging, functional neuroimaging, and DNA studies suggest that normal human aging is associated with neurobiological changes that preferentially affect the frontal lobe of the cerebral cortex. These deteriorations are accompanied with a remarkable age-related decline in frontal lobe functions, that is, executive cognitive functions for the dorsolateral aspect and socio-emotional functions for the ventromedial aspect of frontal lobe function. Conventional interventions that target at enhancing frontal lobe functioning of normal older adults focus on cognitive training and lifestyle changes. However, these interventions often require intensive training and may bring only limited improvements in frontal cognitive functions for older adults.

On the other hand, empirical evidence from multidisciplinary research has come together to suggest that PBM may be an effective means to protect nerve cells and enhance brain functioning. With regard to neuroprotective effects, converging evidence from animal and human studies, both in vivo and in vitro, have shown that the delivery of a moderate amount of red or near-infrared light to nerve cells, as in transcranial PBM, can increase ATP synthesis and cerebral blood flow, thus increasing energy and oxygen supply to nerve cells for metabolic activity. In addition, PBM can activate cellular pathways and transcription factors that protect nerve cells from oxidative stress, inflammation, and programmed cell death and that encourage the formation of new neurons and synapses (e.g., dendritic branching). These physiological effects seem to have promising effects on preventing or slowing down the structural and functional deteriorations that are consistently observed in the aging human brain, especially the prefrontal cortex.

We recently conducted a controlled experiment to determine the effects of PBM treatment on frontal cognitive functions in normal older adults (Chan et al., 2019). We recruited 30 older adults, who were randomly assigned into experimental and control groups. In the experimental group, continuous wave irradiation with red and near-infrared light (i.e., 633 and 870 nm) to the Fp1, Fp2, and Pz sites of the scalp was delivered for 7.5 minutes. In the control group, the same procedure was applied, except the device was not turned on. A frontal cognitive assessment, which included tests for executive cognitive functions (i.e., the flanker and category fluency tasks), was administered before and after PBM treatment to assess the immediate effects of PBM. Our results showed that older adults who received the active PBM treatment had immediate improvements in inhibitory control and mental flexibility, as indicated by faster reaction time on the flanker task and a larger number of words generated on the category fluency task, compared with those who received the sham treatment.

Despite the potential beneficial effects of PBM on frontal cognitive functions of normal older adults, several challenges that concern treatment parameters have to be overcome to optimize the usefulness of this technique. That is, previous human studies on the effects of transcranial PBM of brain function have utilized different treatment parameters across studies, making it difficult to compare the effects of different parameters. The wavelength of light utilized in different studies has varied, from red to near-infrared wavelengths. In addition, the mode of light delivery has varied from pulsed mode at 10, 40, or 100 Hz to continuous wave, and the energy density delivered has varied from 2.95 to 84 J/cm^2. Furthermore, the number of sites stimulated has varied from 1 to 20 individual sites on the head (Lampl et al., 2007), and the spot area stimulated has varied from 0.20 to 22.48 cm^2 (Naeser et al., 2014). Given the large variability in treatment parameters across these different studies, further studies that compare different treatment parameters are needed to find out the set of parameters that works best for each population, including general elderly population.

In conclusion, we propose that PBM can be an effective treatment option to improve frontal lobe functions in normal older adults, because the world's population is aging at an accelerating rate (World Health Organization, 2016) and a decline in frontal lobe functioning may predict subsequent cognitive and functional declines in normal older adults. PBM may be promising in improving the quality of life of the general elderly population and in alleviating the financial burden of healthcare and social expenditures on the elderly themselves, their families, and society.

Acknowledgment

The authors would like to thank Henry Lee in the neuropsychology laboratory at the Department of Psychology of The Chinese University of Hong Kong for his efforts in preparing the tables and proofreading the manuscript.

Conflict of interest

The authors declare that the research was conducted in the absence of any commercial or financial relationships that could be construed as a potential conflict of interest.

References

Anderson, G.F., Hussey, P.S., 2000. Population aging: a comparison among industrialized countries. Health Aff. 19 (3), 191–203.

Andrés, P., Van der Linden, M., 2000. Age-related differences in supervisory attentional system functions. J. Gerontol. B Psychol. Sci. Soc. Sci. 55 (6), P373–P380.

Anguera, J.A., Boccanfuso, J., Rintoul, J.L., Al-Hashimi, O., Faraji, F., Janowich, J., et al., 2013. Video game training enhances cognitive control in older adults. Nature 501 (7465), 97.

Avci, P., Gupta, A., Sadasivam, M., Vecchio, D., Pam, Z., Pam, N., et al., 2013. Low-level laser (light) therapy (LLLT) in skin: stimulating, healing, restoring. In: Seminars in Cutaneous Medicine and Surgery. Frontline Medical Communications, p. 41.

Backman, L.J., Fong, G., Andersson, G., Scott, A., Danielson, P., 2011. Substance P is a mechanoresponsive, autocrine regulator of human tenocyte proliferation. PLoS One 6 (11), e27209.

Ball, K., Berch, D.B., Helmers, K.F., Jobe, J.B., Leveck, M.D., Marsiske, M., et al., 2002. Effects of cognitive training interventions with older adults: a randomized controlled trial. JAMA 288 (18), 2271–2281.

Barrett, D., Gonzalez-Lima, F., 2013. Transcranial infrared laser stimulation produces beneficial cognitive and emotional effects in humans. Neuroscience 230, 13–23.

Bartzokis, G., Beckson, M., Lu, P.H., Nuechterlein, K.H., Edwards, N., Mintz, J., 2001. Age-related changes in frontal and temporal lobe volumes in men: a magnetic resonance imaging study. Arch. Gen. Psychiatry 58 (5), 461–465.

Basak, C., Boot, W.R., Voss, M.W., Kramer, A.F., 2008. Can training in a real-time strategy video game attenuate cognitive decline in older adults? Psychol. Aging 23 (4), 765.

Beitz, K.M., Salthouse, T.A., Davis, H.P., 2014. Performance on the Iowa Gambling Task: from 5 to 89 years of age. J. Exp. Psychol. 143 (4), 1677.

Bell-McGinty, S., Podell, K., Franzen, M., Baird, A.D., Williams, M.J., 2002. Standard measures of executive function in predicting instrumental activities of daily living in older adults. Int. J. Geriatr. Psychiatry 17 (9), 828–834.

Bentourkia, M., Bol, A., Ivanoiu, A., Labar, D., Sibomana, M., Coppens, A., et al., 2000. Comparison of regional cerebral blood flow and glucose metabolism in the normal brain: effect of aging. J. Neurol. Sci. 181 (1-2), 19–28.

Berchtold, N.C., Cribbs, D.H., Coleman, P.D., Rogers, J., Head, E., Kim, R., et al., 2008. Gene expression changes in the course of normal brain aging are sexually dimorphic. Proc. Natl. Acad. Sci. U.S.A. 105 (40), 15605–15610.

Berman, M.H., Halper, J.P., Nichols, T.W., Jarrett, H., Lundy, A., Huang, J.H., 2017. Photobiomodulation with near infrared light helmet in a pilot, placebo controlled clinical trial in dementia patients testing memory and cognition. J. Neurol. Neurosci. 8 (1).

Blanco, N.J., Maddox, W.T., Gonzalez-Lima, F., 2017a. Improving executive function using transcranial infrared laser stimulation. J. Neuropsychol. 11 (1), 14–25.

Blanco, N.J., Saucedo, C.L., Gonzalez-Lima, F., 2017b. Transcranial infrared laser stimulation improves rule-based, but not information-integration, category learning in humans. Neurobiol. Learn. Mem. 139, 69–75.

Blumenthal, J.A., Emery, C.F., Madden, D.J., George, L.K., Coleman, R.E., Riddle, M.W., et al., 1989. Cardiovascular and behavioral effects of aerobic exercise training in healthy older men and women. J. Gerontol. 44 (5), M147–M157.

Boonswang, N.A., Chicchi, M., Lukachek, A., Curtiss, D., 2012. A new treatment protocol using photobiomodulation and muscle/bone/joint recovery techniques having a dramatic effect on a stroke patient's recovery: a new weapon for clinicians. BMJ Case Rep. bcr0820114689.

Boot, W.R., Champion, M., Blakely, D.P., Wright, T., Souders, D., Charness, N., 2013. Video games as a means to reduce age-related cognitive decline: attitudes, compliance, and effectiveness. Front. Psychol. 4, 31.

Brickman, A.M., Paul, R.H., Cohen, R.A., Williams, L.M., MacGregor, K.L., Jefferson, A.L., et al., 2005. Category and letter verbal fluency across the adult lifespan: relationship to EEG theta power. Arch. Clin. Neuropsychol. 20 (5), 561–573.

Calder, A.J., Keane, J., Manly, T., Sprengelmeyer, R., Scott, S., Nimmo-Smith, I., et al., 2003. Facial expression recognition across the adult life span. Neuropsychologia 41 (2), 195–202.

Carriere, J.S., Cheyne, J.A., Solman, G.J., Smilek, D., 2010. Age trends for failures of sustained attention. Psychol. Aging 25 (3), 569.

Cassano, P., Cusin, C., Mischoulon, D., Hamblin, M.R., De Taboada, L., Pisoni, A., et al., 2015. Near-infrared transcranial radiation for major depressive disorder: proof of concept study. Psychiatry J. 2015, 352979.

Chan, A.S., Poon, M.W., 1999. Performance of 7-to 95-year-old individuals in a Chinese version of the category fluency test. J. Int. Neuropsychol. Soc. 5 (6), 525–533.

Chan, A.S., Lee, T.L., Yeung, M.K., Hamblin, M.R., 2019. Photobiomodulation improves the frontal cognitive function of older adults. Int. J. Geriatr. Psychiatry 34, 369–377.

Chao, L.L., Knight, R.T., 1997. Prefrontal deficits in attention and inhibitory control with aging. Cereb. cortex 7 (1), 63–69.

Chen, W.J., Hsiao, C.K., Hsiao, L., Hwu, H., 1998. Performance of the Continuous Performance Test among community samples. Schizophr. Bull. 24 (1), 163.

Clark, P.G., Blissmer, B.J., Greene, G.W., Lees, F.D., Riebe, D.A., Stamm, K.E., 2011. Maintaining exercise and healthful eating in older adults: the SENIOR project II: study design and methodology. Contemp. Clin. Trials 32 (1), 129–139.

Clark, L.R., Schiehser, D.M., Weissberger, G.H., Salmon, D.P., Delis, D.C., Bondi, M.W., 2012. Specific measures of executive function predict cognitive decline in older adults. J. Int. Neuropsychol. Soc. 18 (1), 118–127.

Colcombe, S., Kramer, A.F., 2003. Fitness effects on the cognitive function of older adults: a meta-analytic study. Psychol. Sci. 14 (2), 125–130.

Dahlin, E., Nyberg, L., Bäckman, L., Neely, A.S., 2008. Plasticity of executive functioning in young and older adults: immediate training gains, transfer, and long-term maintenance. Psychol. Aging 23 (4), 720.

Dang, T., Antolin, P., Oxley, H., 2001. Fiscal Implication of Ageing: Projections of Age-Related Spending. OECD Economics Department Working Papers, no. 305, OECD Publishing, Paris.

Deakin, J., Aitken, M., Robbins, T., Sahakian, B.J., 2004. Risk taking during decision-making in normal volunteers changes with age. J. Int. Neuropsychol. Soc. 10 (4), 590–598.

De Brabander, J., Kramers, R., Uylings, H., 1998. Layer-specific dendritic regression of pyramidal cells with ageing in the human prefrontal cortex. Eur. J. Neurosci. 10 (4), 1261–1269.

de Keyser, J., De Backer, J., Vauquelin, G., Ebinger, G., 1990. The effect of aging on the D1 dopamine receptors in human frontal cortex. Brain Res. 528 (2), 308–310.

de Meijer, C., Wouterse, B., Polder, J., Koopmanschap, M., 2013. The effect of population aging on health expenditure growth: a critical review. Eur. J. Ageing 10 (4), 353–361.

Denburg, N.L., Tranel, D., Bechara, A., 2005. The ability to decide advantageously declines prematurely in some normal older persons. Neuropsychologia 43 (7), 1099–1106.

Disner, S.G., Beevers, C.G., Gonzalez-Lima, F., 2016. Transcranial laser stimulation as neuroenhancement for attention bias modification in adults with elevated depression symptoms. Brain Stimul. 9 (5), 780–787.

Dobbs, A.R., Rule, B.G., 1989. Adult age differences in working memory. Psychol. Aging 4 (4), 500.

Dong, T., Zhang, Q., Hamblin, M.R., Wu, M.X., 2015. Low-level light in combination with metabolic modulators for effective therapy of injured brain. J. Cereb. Blood Flow Metab. 35 (9), 1435–1444.

Dustman, R.E., Ruhling, R.O., Russell, E.M., Shearer, D.E., Bonekat, H.W., Shigeoka, J.W., et al., 1984. Aerobic exercise training and improved neuropsychological function of older individuals. Neurobiol. Aging 5 (1), 35–42.

Erraji-Benchekroun, L., Underwood, M.D., Arango, V., Galfalvy, H., Pavlidis, P., Smyrniotopoulos, P., et al., 2005. Molecular aging in human prefrontal cortex is selective and continuous throughout adult life. Biol. Psychiatry 57 (5), 549–558.

Fein, G., McGillivray, S., Finn, P., 2007. Older adults make less advantageous decisions than younger adults: cognitive and psychological correlates. J. Int. Neuropsychol. Soc. 13 (3), 480–489.

Fjell, A.M., Walhovd, K.B., Fennema-Notestine, C., McEvoy, L.K., Hagler, D.J., Holland, D., et al., 2009. One-year brain atrophy evident in healthy aging. J. Neurosci. 29 (48), 15223–15231.

Fjell, A.M., Sneve, M.H., Grydeland, H., Storsve, A.B., Walhovd, K.B., 2017. The disconnected brain and executive function decline in aging. Cereb. Cortex 27 (3), 2303–2317.

Fukuzaki, Y., Shin, H., Kawai, H.D., Yamanoha, B., Kogure, S., 2015. 532 nm Low-power laser irradiation facilitates the migration of GABAergic neural stem/progenitor cells in mouse neocortex. PLoS One 10 (4), e0123833.

Gomes, L.E.A., Dalmarco, E.M., André, E.S., 2012. The brain-derived neurotrophic factor, nerve growth factor, neurotrophin-3, and induced nitric oxide synthase expressions after low-level laser therapy in an axonotmesis experimental model. Photomed. Laser Surg. 30 (11), 642–647.

Grieve, S.M., Clark, C.R., Williams, L.M., Peduto, A.J., Gordon, E., 2005. Preservation of limbic and paralimbic structures in aging. Hum. Brain Mapp. 25 (4), 391–401.

Hamblin, M.R., 2008. The role of nitric oxide in low level light therapy. In: Mechanisms for Low-Light Therapy III. International Society for Optics and Photonics, p. 684602.

Hamblin, M.R., 2016. Shining light on the head: photobiomodulation for brain disorders. BBA Clin. 6, 113–124.

Hassmén, P., Ceci, R., Bäckman, L., 1992. Exercise for older women: a training method and its influences on physical and cognitive performance. Eur. J. Appl. Physiol. Occup. Physiol. 64 (5), 460–466.

Head, D., Buckner, R.L., Shimony, J.S., Williams, L.E., Akbudak, E., Conturo, T.E., et al., 2004. Differential vulnerability of anterior white matter in nondemented aging with minimal acceleration in dementia of the Alzheimer type: evidence from diffusion tensor imaging. Cereb. Cortex 14 (4), 410–423.

Herman, T., Mirelman, A., Giladi, N., Schweiger, A., Hausdorff, J.M., 2010. Executive control deficits as a prodrome to falls in healthy older adults: a prospective study linking thinking, walking, and falling. J. Gerontol. Series A 65 (10), 1086–1092.

Hesse, S., Werner, C., Byhahn, M., 2015. Transcranial low-level laser therapy may improve alertness and awareness in traumatic brain injured subjects with severe disorders of consciousness: a case series. J. Neurol. Neurosci. 6 (2).

Hillman, C.H., Motl, R.W., Pontifex, M.B., Posthuma, D., Stubbe, J.H., Boomsma, D.I., et al., 2006. Physical activity and cognitive function in a cross-section of younger and older community-dwelling individuals. Health Psychol. 25 (6), 678.

Huang, E.J., Reichardt, L.F., 2001. Neurotrophins: roles in neuronal development and function. Annu. Rev. Neurosci. 24 (1), 677–736.

Huang, Y., Nagata, K., Tedford, C.E., McCarthy, T., Hamblin, M.R., 2013. Low-level laser therapy (LLLT) reduces oxidative stress in primary cortical neurons in vitro. J. Biophotonics 6 (10), 829–838.

Hwang, J., Castelli, D.M., Gonzalez-Lima, F., 2016. Cognitive enhancement by transcranial laser stimulation and acute aerobic exercise. Lasers Med. Sci. 31 (6), 1151–1160.

Ignarro, L.J., Cirino, G., Casini, A., Napoli, C., 1999. Nitric oxide as a signaling molecule in the vascular system: an overview. J. Cardiovasc. Pharmacol. 34 (6), 879–886.

Isaacowitz, D.M., Löckenhoff, C.E., Lane, R.D., Wright, R., Sechrest, L., Riedel, R., et al., 2007. Age differences in recognition of emotion in lexical stimuli and facial expressions. Psychol. Aging 22 (1), 147.

Jagdeo, J.R., Adams, L.E., Brody, N.I., Siegel, D.M., 2012. Transcranial red and near infrared light transmission in a cadaveric model. PLoS One 7 (10), e47460.

Jobsis, F.F., 1977. Noninvasive, infrared monitoring of cerebral and myocardial oxygen sufficiency and circulatory parameters. Science (New York, N.Y.) 198 (4323), 1264–1267.

Kaasinen, V., Vilkman, H., Hietala, J., Nagren, K., Helenius, H., Olsson, H., et al., 2000. Age-related dopamine D2/D3 receptor loss in extrastriatal regions of the human brain. Neurobiol. Aging 21 (5), 683–688.

Kalpouzos, G., Chetelat, G., Baron, J.C., Landeau, B., Mevel, K., Godeau, C., et al., 2009. Voxel-based mapping of brain gray matter volume and glucose metabolism profiles in normal aging. Neurobiol. Aging 30 (1), 112–124.

Karbach, J., Verhaeghen, P., 2014. Making working memory work: a meta-analysis of executive-control and working memory training in older adults. Psychol. Sci. 25 (11), 2027–2037.

Karu, T. 2000. Mechanisms of low-power laser light action on cellular level. In: Effects of Low-Power Light on Biological Systems V. International Society for Optics and Photonics, p. 1.

Karu, T., 2010. Mitochondrial mechanisms of photobiomodulation in context of new data about multiple roles of ATP. Photomed. Laser Surg. 28 (2), 159–160.

Kearney, F.C., Harwood, R.H., Gladman, J.R., Lincoln, N., Masud, T., 2013. The relationship between executive function and falls and gait abnormalities in older adults: a systematic review. Dement. Geriatr. Cogn. Disord. 36 (1-2), 20–35.

Khakh, B.S., Burnstock, G., 2009. The double life of ATP. Sci. Am. 301 (6), 84–92.

Kramer, A.F., Humphrey, D.G., Larish, J.F., Logan, G.D., 1994. Aging and inhibition: beyond a unitary view of inhibitory processing in attention. Psychol. Aging 9 (4), 491.

Kramer, A.F., Hahn, S., Gopher, D., 1999. Task coordination and aging: explorations of executive control processes in the task switching paradigm. Acta Psychol. (Amst) 101 (2–3), 339–378.

Lamar, M., Resnick, S.M., 2004. Aging and prefrontal functions: dissociating orbitofrontal and dorsolateral abilities. Neurobiol. Aging 25 (4), 553–558.

Lampl, Y., Zivin, J.A., Fisher, M., Lew, R., Welin, L., Dahlof, B., et al., 2007. Infrared laser therapy for ischemic stroke: a new treatment strategy: results of the NeuroThera Effectiveness and Safety Trial-1 (NEST-1). Stroke 38 (6), 1843–1849.

Lane, N., 2006. Cell biology: power games. Nature 443, 901–903.

Lautenschlager, N.T., Cox, K.L., Flicker, L., Foster, J.K., van Bockxmeer, F.M., Xiao, J., et al., 2008. Effect of physical activity on cognitive function in older adults at risk for Alzheimer disease: a randomized trial. JAMA 300 (9), 1027–1037.

Lee, H.I., Lee, S., Kim, N.G., Park, K., Choi, B.T., Shin, Y., et al., 2017. Low-level light emitting diode (LED) therapy suppresses inflammasome-mediated brain damage in experimental ischemic stroke. J. Biophotonics 10 (11), 1502–1513.

Levine, B., Stuss, D.T., Milberg, W.P., 1997. Effects of aging on conditional associative learning: process analyses and comparison with focal frontal lesions. Neuropsychology 11 (3), 367.

Liang, H.L., Whelan, H.T., Eells, J.T., Wong-Riley, M.T., 2008. Near-infrared light via light-emitting diode treatment is therapeutic against rotenone- and 1-methyl-4-phenylpyridinium ion-induced neurotoxicity. Neuroscience 153 (4), 963–974.

Lu, T., Pan, Y., Kao, S., Li, C., Kohane, I., Chan, J., et al., 2004. Gene regulation and DNA damage in the ageing human brain. Nature 429 (6994), 883.

Lubitz, J., Cai, L., Kramarow, E., Lentzner, H., 2003. Health, life expectancy, and health care spending among the elderly. N. Engl. J. Med. 349 (11), 1048–1055.

MacPherson, S.E., Phillips, L.H., Della Sala, S., 2002. Age, executive function and social decision making: a dorsolateral prefrontal theory of cognitive aging. Psychol. Aging 17 (4), 598.

Maillot, P., Perrot, A., Hartley, A., 2012. Effects of interactive physical-activity video-game training on physical and cognitive function in older adults. Psychol. Aging 27 (3), 589.

Mani, T.M., Bedwell, J.S., Miller, L.S., 2005. Age-related decrements in performance on a brief continuous performance test. Arch. Clin. Neuropsychol. 20 (5), 575–586.

Masliah, E., Mallory, M., Hansen, L., DeTeresa, R., Terry, R.D., 1993. Quantitative synaptic alterations in the human neocortex during normal aging. Neurology 43 (1), 192–197.

Mell, T., Heekeren, H.R., Marschner, A., Wartenburger, I., Villringer, A., Reischies, F.M., 2005. Effect of aging on stimulus-reward association learning. Neuropsychologia 43 (4), 554–563.

Meng, C., He, Z., Xing, D., 2013. Low-level laser therapy rescues dendrite atrophy via upregulating BDNF expression: implications for Alzheimer's disease. J. Neurosci. 33 (33), 13505–13517.

Meng, C., He, Z., Xing, D., 2014. Low-level laser therapy promotes dendrite growth via upregulating brain-derived neurotrophic factor expression. In: Twelfth International Conference on Photonics and Imaging in Biology and Medicine (PIBM 2014). International Society for Optics and Photonics, p. 92301G.

Mirelman, A., Herman, T., Brozgol, M., Dorfman, M., Sprecher, E., Schweiger, A., et al., 2012. Executive function and falls in older adults: new findings from a five-year prospective study link fall risk to cognition. PLoS One 7 (6), e40297.

Moghadam, H.S., Nazari, M.A., Jahan, A., Mahmoudi, J., Salimi, M.M., 2017. Beneficial effects of transcranial light emitting diode (LED) therapy on attentional performance: an experimental design. Iran. Red Crescent Med. J. 19 (5).

Naeser, M.A., Saltmarche, A., Krengel, M.H., Hamblin, M.R., Knight, J.A., 2011. Improved cognitive function after transcranial, light-emitting diode treatments in chronic, traumatic brain injury: two case reports. Photomed. Laser Surg. 29 (5), 351–358.

Nakamura, S., Akiguchi, I., Kameyama, M., Mizuno, N., 1985. Age-related changes of pyramidal cell basal dendrites in layers III and V of human motor cortex: a quantitative Golgi study. Acta Neuropathol 65 (3–4), 281–284.

Naeser, M., Ho, M., Martin, P., Treglia, E., Krengel, M., Hamblin, M., et al., 2012. Improved language after scalp application of red/near-infrared light-emitting diodes: pilot study supporting a new, noninvasive treatment for chronic aphasia. Procedia-Soc. Behav. Sci. 61, 138–139.

Naeser, M.A., Zafonte, R., Krengel, M.H., Martin, P.I., Frazier, J., Hamblin, M.R., et al., 2014. Significant improvements in cognitive performance post-transcranial, red/near-infrared light-emitting diode treatments in chronic, mild traumatic brain injury: open-protocol study. J. Neurotrauma 31 (11), 1008–1017.

Oron, A., Oron, U., Streeter, J., Taboada, L.D., Alexandrovich, A., Trembovler, V., et al., 2007. Low-level laser therapy applied transcranially to mice following traumatic brain injury significantly reduces long-term neurological deficits. J. Neurotrauma 24 (4), 651–656.

Pantano, P., Baron, J.C., Lebrun-Grandie, P., Duquesnoy, N., Bousser, M.G., Comar, D., 1984. Regional cerebral blood flow and oxygen consumption in human aging. Stroke 15 (4), 635–641.

Perlmutter, M., 1979. Age differences in adults' free recall, cued recall, and recognition. J. Gerontol. 34 (4), 533–539.

Pezawas, L., Verchinski, B.A., Mattay, V.S., Callicott, J.H., Kolachana, B.S., Straub, R.E., et al., 2004. The brain-derived neurotrophic factor val66-met polymorphism and variation in human cortical morphology. J. Neurosci. 24 (45), 10099–10102.

Pfefferbaum, A., Adalsteinsson, E., Sullivan, E.V., 2005. Frontal circuitry degradation marks healthy adult aging: evidence from diffusion tensor imaging. Neuroimage 26 (3), 891–899.

Poyton, R.O., Ball, K.A., 2011. Therapeutic photobiomodulation: nitric oxide and a novel function of mitochondrial cytochrome c oxidase. Discov. Med. 11 (57), 154–159.

Prehn, K., Jumpertz von Schwartzenberg, R., Mai, K., Zeitz, U., Witte, A.V., Hampel, D., et al., 2016. Caloric restriction in older adults—differential effects of weight loss and reduced weight on brain structure and function. Cereb. Cortex 27 (3), 1765–1778.

Rathbone, M.P., Middlemiss, P.J., Kim, J., Gysbers, J.W., DeForge, S.P., Smith, R., et al., 1992. Adenosine and its nucleotides stimulate proliferation of chick astrocytes and human astrocytoma cells. Neurosci. Res. 13 (1), 1–17.

Raz, N., Gunning, F.M., Head, D., Dupuis, J.H., McQuain, J., Briggs, S.D., et al., 1997. Selective aging of the human cerebral cortex observed in vivo: differential vulnerability of the prefrontal gray matter. Cereb. Cortex 7 (3), 268–282.

Raz, N., Lindenberger, U., Rodrigue, K.M., Kennedy, K.M., Head, D., Williamson, A., et al., 2005. Regional brain changes in aging healthy adults: general trends, individual differences and modifiers. Cereb. Cortex 15 (11), 1676–1689.

Reinhardt, U.E., 2003. Does the aging of the population really drive the demand for health care? Health Aff. 22 (6), 27–39.

Royall, D.R., Palmer, R., Chiodo, L.K., Polk, M.J., 2004. Declining executive control in normal aging predicts change in functional status: the Freedom House Study. J. Am. Geriatr. Soc. 52 (3), 346–352.

Ruffman, T., Sullivan, S., Dittrich, W., 2009. Older adults' recognition of bodily and auditory expressions of emotion. Psychol. Aging 24 (3), 614.

Ryan, M., Murray, J., Ruffman, T., 2010. Aging and the perception of emotion: processing vocal expressions alone and with faces. Exp. Aging Res. 36 (1), 1–22.

Salat, D.H., Buckner, R.L., Snyder, A.Z., Greve, D.N., Desikan, R.S., Busa, E., et al., 2004. Thinning of the cerebral cortex in aging. Cereb. Cortex 14 (7), 721–730.

Salat, D.H., Tuch, D.S., Greve, D.N., van der Kouwe, A.J., Hevelone, N.D., Zaleta, A.K., et al., 2005. Age-related alterations in white matter microstructure measured by diffusion tensor imaging. Neurobiol. Aging 26 (8), 1215–1227.

Salgado, A.S., Zângaro, R.A., Parreira, R.B., Kerppers, I.I., 2015. The effects of transcranial LED therapy (TCLT) on cerebral blood flow in the elderly women. Lasers Med. Sci. 30 (1), 339–346.

Salthouse, T.A., 2010. Selective review of cognitive aging. J. Int. Neuropsychol. Soc. 16 (5), 754–760.

Saltmarche, A.E., Naeser, M.A., Ho, K.F., Hamblin, M.R., Lim, L., 2017. Significant improvement in cognition in mild to moderately severe dementia cases treated with transcranial plus intranasal photobiomodulation: case series report. Photomed. Laser Surg. 35 (8), 432–441.

Schiffer, F., Johnston, A.L., Ravichandran, C., Polcari, A., Teicher, M.H., Webb, R.H., et al., 2009. Psychological benefits 2 and 4 weeks after a single treatment with near infrared light to the forehead: a pilot study of 10 patients with major depression and anxiety. Behav. Brain Funct. 5 (1), 46.

Schrack, J.A., Zipunnikov, V., Goldsmith, J., Bai, J., Simonsick, E.M., Crainiceanu, C., et al., 2013. Assessing the "physical cliff": detailed quantification of age-related differences in daily patterns of physical activity. J. Gerontol. Series A 69 (8), 973–979.

Sharma, S.K., Kharkwal, G.B., Sajo, M., Huang, Y., De Taboada, L., McCarthy, T., et al., 2011. Dose response effects of 810 nm laser light on mouse primary cortical neurons. Lasers Surg. Med. 43 (8), 851–859.

Sheppard, F.R., Kelher, M.R., Moore, E.E., McLaughlin, N.J., Banerjee, A., Silliman, C.C., 2005. Structural organization of the neutrophil NADPH oxidase: phosphorylation and translocation during priming and activation. J. Leukoc. Biol. 78 (5), 1025–1042.

Smith, P.J., Blumenthal, J.A., Hoffman, B.M., Cooper, H., Strauman, T.A., Welsh-Bohmer, K., et al., 2010. Aerobic exercise and neurocognitive performance: a meta-analytic review of randomized controlled trials. Psychosom. Med. 72 (3), 239–252.

Spieler, D.H., Balota, D.A., Faust, M.E., 1996. Stroop performance in healthy younger and older adults and in individuals with dementia of the Alzheimer's type. J. Exp. Psychol. 22 (2), 461.

Stemer, A.B., Huisa, B.N., Zivin, J.A., 2010. The evolution of transcranial laser therapy for acute ischemic stroke, including a pooled analysis of NEST-1 and NEST-2. Curr. Cardiol. Rep. 12 (1), 29–33.

Stuss, D.T., Craik, F.I., Sayer, L., Franchi, D., Alexander, M.P., 1996. Comparison of older people and patients with frontal lesions: evidence from word list learning. Psychol. Aging 11 (3), 387.

Sullivan, S., Ruffman, T., Hutton, S.B., 2007. Age differences in emotion recognition skills and the visual scanning of emotion faces. J. Gerontol. B Psychol. Sci. Soc. Sci. 62 (1), P53–P60.

Tafur, J., Mills, P.J., 2008. Low-intensity light therapy: exploring the role of redox mechanisms. Photomed. Laser Surg. 26 (4), 323–328.

Tombaugh, T.N., Kozak, J., Rees, L., 1999. Normative data stratified by age and education for two measures of verbal fluency: FAS and animal naming. Arch. Clin. Neuropsychol. 14 (2), 167–177.

Troyer, A.K., 2000. Normative data for clustering and switching on verbal fluency tasks. J. Clin. Exp. Neuropsychol. 22 (3), 370–378.

United Nations, Department of Economic and Social Affairs, Population Division, 2017. World Population Prospects: The 2017 Revision, custom data acquired via website.

Uozumi, Y., Nawashiro, H., Sato, S., Kawauchi, S., Shima, K., Kikuchi, M., 2010. Targeted increase in cerebral blood flow by transcranial near-infrared laser irradiation. Lasers Surg. Med. 42 (6), 566–576.

Van Der Elst, W., Van Boxtel, M.P., Van Breukelen, G.J., Jolles, J., 2006. Normative data for the Animal, Profession and Letter M Naming verbal fluency tests for Dutch speaking participants and the effects of age, education, and sex. J. Int. Neuropsychol. Soc. 12 (1), 80–89.

Van der Linden, M., Brédart, S., Beerten, A., 1994. Age-related differences in updating working memory. Br. J. Psychol. 85 (1), 145–152.

van Uffelen, J.G., Chinapaw, M.J., van Mechelen, W., Hopman-Rock, M., 2008. Walking or vitamin B for cognition in older adults with mild cognitive impairment? A randomised controlled trial. Br. J. Sports Med. 42 (5), 344–351.

Vargas, E., Barrett, D.W., Saucedo, C.L., Huang, L., Abraham, J.A., Tanaka, H., et al., 2017. Beneficial neurocognitive effects of transcranial laser in older adults. Lasers Med. Sci. 32 (5), 1153–1162.

Villringer, A., Chance, B., 1997. Non-invasive optical spectroscopy and imaging of human brain function. Trends Neurosci. 20 (10), 435–442.

Volkow, N.D., Logan, J., Fowler, J.S., Wang, G., Gur, R.C., Wong, C., et al., 2000. Association between age-related decline in brain dopamine activity and impairment in frontal and cingulate metabolism. Am. J. Psychiatry 157 (1), 75–80.

Voss, M.W., Prakash, R.S., Erickson, K.I., Basak, C., Chaddock, L., Kim, J.S., et al., 2010. Plasticity of brain networks in a randomized intervention trial of exercise training in older adults. Front. Aging Neurosci. 2, 32.

Wang, G., Volkow, N.D., Logan, J., Fowler, J.S., Schlyer, D., MacGregor, R.R., et al., 1995. Evaluation of age-related changes in serotonin 5-HT2 and dopamine D2 receptor availability in healthy human subjects. Life. Sci. 56 (14), PL249–PL253.

Waypa, G.B., Smith, K.A., Schumacker, P.T., 2016. O_2 sensing, mitochondria and ROS signaling: the fog is lifting. Mol. Aspects Med. 47, 76–89.

Weiler, J.A., Bellebaum, C., Daum, I., 2008. Aging affects acquisition and reversal of reward-based associative learning. Learn. Mem. 15 (4), 190–197.

West, R.L., 1996. An application of prefrontal cortex function theory to cognitive aging. Psychol. Bull. 120 (2), 272–292.

West, R., Alain, C., 2000. Age-related decline in inhibitory control contributes to the increased Stroop effect observed in older adults. Psychophysiology 37 (2), 179–189.

West, R., Ergis, A., Winocur, G., Saint-Cyr, J., 1998. The contribution of impaired working memory monitoring to performance of the Self-Ordered Pointing Task in normal aging and Parkinson's disease. Neuropsychology 12 (4), 546.

Williams, P., Lord, S.R., 1997. Effects of group exercise on cognitive functioning and mood in older women. Aust. N. Z. J. Public Health 21 (1), 45–52.

Williamson, J.D., Espeland, M., Kritchevsky, S.B., Newman, A.B., King, A.C., Pahor, M., et al., 2009. Changes in cognitive function in a randomized trial of physical activity: results of the lifestyle interventions and independence for elders pilot study. J. Gerontol. Series A 64 (6), 688–694.

Wong-Riley, M.T., Bai, X., Buchmann, E., Whelan, H.T., 2001. Light-emitting diode treatment reverses the effect of TTX on cytochrome oxidase in neurons. Neuroreport 12 (14), 3033–3037.

Wong-Riley, M.T., Liang, H.L., Eells, J.T., Chance, B., Henry, M.M., Buchmann, E., et al., 2005. Photobiomodulation directly benefits primary neurons functionally inactivated by toxins: role of cytochrome c oxidase. J. Biol. Chem. 280 (6), 4761–4771.

World Health Organization, 2016. World Health Statistics 2016: Monitoring Health for the SDGs Sustainable Development Goals. World Health Organization.

Yan, X., Liu, J., Zhang, Z., Li, W., Sun, S., Zhao, J., et al., 2017. Low-level laser irradiation modulates brain-derived neurotrophic factor mRNA transcription through calcium-dependent activation of the ERK/CREB pathway. Lasers Med. Sci. 32 (1), 169–180.

Yang, L., Tucker, D., Dong, Y., Wu, C., Lu, Y., Li, Y., et al., 2018. Photobiomodulation therapy promotes neurogenesis by improving post-stroke local microenvironment and stimulating neuroprogenitor cells. Exp. Neurol. 299, 86–96.

Yankner, B.A., Lu, T., Loerch, P., 2008. The aging brain. Annu. Rev. Pathmechdis. Mech. Dis. 3, 41–66.

Ying, R., Liang, H.L., Whelan, H.T., Eells, J.T., Wong-Riley, M.T., 2008. Pretreatment with near-infrared light via light-emitting diode provides added benefit against rotenone-and MPP -induced neurotoxicity. Brain Res. 1243, 167–173.

Young, J., Angevaren, M., Rusted, J., Tabet, N., 2015. Aerobic exercise to improve cognitive function in older people without known cognitive impairment. Cochrane Database Syst. Rev. 4, CD005381.

Yu, Z., Li, Z., Liu, N., Jizhang, Y., McCarthy, T.J., Tedford, C.E., et al., 2015. Near infrared radiation protects against oxygen-glucose deprivation-induced neurotoxicity by down-regulating neuronal nitric oxide synthase (nNOS) activity in vitro. Metab. Brain Dis. 30 (3), 829–837.

Zivin, J.A., Albers, G.W., Bornstein, N., Chippendale, T., Dahlof, B., Devlin, T., et al., 2009. Effectiveness and safety of transcranial laser therapy for acute ischemic stroke. Stroke 40 (4), 1359–1364.

Zivin, J., Sehra, R., Shoshoo, A., Albers, G., Bornstein, N., Dahlof, B., et al., 2014. NeuroThera® Efficacy and Safety Trial–3 (NEST-3): a double-blind, randomized, sham-controlled, parallel group, multicenter, pivotal study to assess the safety and efficacy of transcranial laser therapy with the NeuroThera® laser system for the treatment of acute ischemic stroke within 24 h of stroke onset. Int. J. Stroke 9 (7), 950–955.

Further reading

Meadmore, K.L., Dror, I.E., Bucks, R.S., 2009. Lateralisation of spatial processing and age. Laterality 14 (1), 17–29.

Parkin, A.J., Walter, B.M., Hunkin, N.M., 1995. Relationships between normal aging, frontal lobe function, and memory for temporal and spatial information. Neuropsychology 9 (3), 304.

Chapter 32

Noninvasive neurotherapeutic treatment of neurodegeneration: integrating photobiomodulation and neurofeedback training

Marvin H. Berman[1], Trent Nichols[1], Jason Huang[2] and Damir Nizamutdinov[2]
[1]*Quietmind Foundation, Elkins Park, PA, United States,* [2]*Department of Neurosurgery, Baylor Scott & White Health, Dallas, TX, United States*

You never change things by fighting the existing reality. To change something, build a new model that makes the existing model obsolete.

Buckminster Fuller

32.1 Photobiomodulation and neurotherapy introduction

There is a rapidly growing appreciation for the therapeutic value of photobiomodulation (PBM) as a safe and potent antiinflammatory agent (Bradford et al., 2005) for treating functional neurodegenerative, inflammatory, and infection-based conditions (Readhead et al., 2018). Research efforts have expanded globally, and evidence regarding the efficacy of PBM continues to mount in support of its use in an ever-widening range of conditions. This chapter reviews how PBM is being deployed as a component in the multimodal treatment of neurodegenerative disorders, for example, dementia and Parkinson's disease. PBM is viewed as an effective tissue-level intervention to reduce cortical inflammation and increase cellular oxygenation and perfusion, which along with electroencephalogram (EEG) biofeedback training, is shown to support the renormalization of neural connectivity. Combining the two approaches is showing promise as a method for safely and effectively impeding neuronal damage and the formation of neurofibrillary plaques and tangles while retraining intra- and interhemispheric communication.

The historical background and mechanisms of action underlying the effects of PBM and EEG biofeedback (neurofeedback) training have previously been described (Berman et al., 2017). It has been shown that near-infrared (NIR) light can support healing and improved motor, cognitive, behavioral, and metabolic functioning (Hamblin, 2016). Combining the two approaches along with functional and integrative biomedical treatments (e.g., stem cell, gene therapy, and optogenetically delivered chemotherapy) would comprise a systemic intervention strategy that can add considerable support to the efficacy of what is now being called precision medicine.

The current functional medical approach to evaluating and treating dementia has been addressed by Bredesen and colleagues (Bredesen, 2017), and the focus here will be on current measurement techniques, for example, quantitative EEG (QEEG) and near-infrared spectroscopy (NIRS) and clinical applications employing photobiomodulation and brainwave biofeedback or neurofeedback training. There is a growing trend toward integrating neurophysiological with noninvasive neurotherapeutic methods to positively influence healthy functioning. Neurodegenerative disorders, being of a more systemic nature, are good candidates for such intervention strategies.

Photobiomodulation in the Brain. DOI: https://doi.org/10.1016/B978-0-12-815305-5.00032-4

Neurotherapeutics is a systemically-based noninvasive approach to treatment of neurodegenerative and neuropsychiatric conditions utilizing directed energy stimulation that promotes repair, regrowth, and concurrent renormalization of physiological functions and neural connectivity. The need for systemic interventions in treating neurodegenerative disorders has been illuminated by Bredesen's (2017) research efforts demonstrating the multidetermined roots of functional, neurocognitive, and behavioral challenges. The following list represents the main diagnostic components identified by Bredesen for evaluating and treating patients with dementia. Biomarkers with an X also respond positively to transcranial photobiomodulation (Table 32.1).

32.2 Pathophysiology of neurodegeneration

The etiology and pathogenesis of Alzheimer's disease (AD) is complex, with many viral, genetic, and environmental risk factors contributing including oxidative stress and insulin resistance (Readhead et al., 2018). The expression of many genes, and upregulation of multiple pathogenic pathways, results in amyloid β (Aβ) peptide deposition, tau hyperphosphorylation, inflammation, reactive oxygen species (ROS), mitochondrial disorders, insulin resistance, methylation defects, and down-regulation of neuroprotective factors as well as regional cerebral hypoperfusion. Antibody therapy directed against tau and amyloid beta, vaccines, and other methods to decrease tau and/or amyloid have not been successful despite considerable pharmaceutical and biotech efforts (Cummings et al., 2014).

The current pathophysiology of AD is characterized by the accumulation of neurofibrillary tangles (NFTs) of hyperphosphorylated tau protein and subsequently Aβ plaques. This accumulation of misfolded protein aggregates is a common feature of many neurodegenerative diseases including Alzheimer's. The ubiquitin proteasome pathway is responsible for most of protein degradation (Dantuma and Bott, 2014). Aβ accumulation impairs the ubiquitin-proteasome system (Nichols, 2014). Myeku (2014) demonstrated that tau-driven 26S proteasome impairment and cognitive dysfunction in a mouse model can be prevented early in the disease course by activating cAMP-PKA signaling (Myeku et al., 2016), and these represent key components that underlie PBM's mechanism of action. PBM results in increased mitochondrial ATP and proteasomal clearance of tau and β amyloid in mouse models that develop AD.

Eli Lilly in Indianapolis announced a major change to its closely watched clinical trial for the Alzheimer's drug solanezumab which failed to reach statistical significance (Underwood, 2016). "A major challenge of such trials is how to measure the drug's benefits," says Dennis Selkoe, a neurologist at Brigham and Women's Hospital in Boston, who was not involved in the Lilly trial. "Although people with early Alzheimer's may show mild memory impairment and problems with attention and focus, they can often follow recipes, make a cup of coffee, or drive a car," Selkoe reiterates. "Such abilities are unlikely to change much over the course of an 18-month clinical trial."

Animal trials often demonstrate improvement in a therapy long before human clinical trials show any benefit. Recently, a report on animal models has been presented in which NIR PBM was used to treat AD pathology in K369I tau transgenic model engineered to develop NFTs, and the APPs/PSEN1dE9 transgenic model (APP/PS1) to develop amyloid plaques (Purushothuman et al., 2014)

A kinase inhibitor, K252a, was able to prevent the typical motor deficits in the tau (P301L) transgenic mouse model (JNPL3) and markedly reduce soluble aggregated hyperphosphorylated tau (Le Corre et al., 2006; Kim et al., 2013). Le Corre and Kim reported the first NIR fluorescent ratiometric probe, CyDPA2, that targets tau aggregates. The specificity of CyPDA2 to aggregated tau was evaluated with in vitro hyperphosphorylated tau proteins (pTau), as well as ex vivo tau samples from AD human brain samples and the tauopathy transgenic mouse model, P301L. Purushothuman et al. (2014) also observed decreased levels of hyperphosphorylated tau NFTs after use of NIR LED therapy in K3 mice.

DNA methyltransferase 3A (DNMT3A) is one of two human de novo DNA methyltransferases essential for transcription regulation during cellular development and differentiation. There is increasing evidence that RNA plays a role in directing DNA methylation to specific genomic locations within mammalian cells. Here, two modes of RNA regulation of DNMT3A in vitro were described. A single-stranded RNA molecule that was antisense to the E-cadherin promoter binds tightly to the catalytic domain in a structurally dependent fashion causing potent inhibition of DNMT3A activity. Two other RNA molecules bind DNMT3A at an allosteric site outside the catalytic domain, causing no change in catalysis. A potent and specific in vitro modulation of DNMT3A activity by RNA supports in vivo data that RNA interacts with DNMT3A to regulate transcription (Schietinger, 2012). The observation that NIR light regulates transcription may occur through this mechanism (Fig. 32.1).

At the present time proof that NIR light can regulate site-specific subtelomeric DNA methylation and may affect DNA methylation has not been presented; however, blue light optogentically has been demonstrated to increase

TABLE 32.1 Bredesen diagnostic components evaluating and treating patients with dementia.

Biomarker/functional mechanisms	PBM positive effect (X)
Bredesen protocol and photobiomodulation	
1. Decrease Aβ production	X
2. Increase Aβ degradation and clearance	X
3. Decrease Aβ oligomerization	
4. Increase BDNF (brain derived nerve factor)	X
5. Increase NGF (nerve growth factor)	
6. Increase G-CSF	
7. Increase ADNP	
8. Decrease p-tau	X
9. Decrease homocysteine	
10. Build synapses	
11. Decrease 4/2	X
12. Increase A/G ratio (albumin/globulin)	
13. Decrease inflammation	X
14. Inhibit NF-kB	X
15. Increase GSH (glutathione)	
16. Increase antioxidants	
17. Decrease iron, copper, increase zinc–target of Zn	
18. Increase CBF	
19. Increase Ach	
20. Increase nAChR α 7 signaling	
21. Increase Aβ transport	
22. Decrease ApoE4 effect	
23. Increase GABA	
24. Decrease NMDA glutamate receptors	
25. Optimize hormones	
26. Increase vitamin D	
27. Decrease pro-NGF	
28. Decrease caspase-6	
29. Decrease N-APP	
31. Increase memory	X
32. Increase energy	X
33. Increase mitochondrial function	X
34. Increase mitochondrial protection	X
35. Increase chaperone expression	X

Functional and inflammatory biomarkers associated with neurodegenerative disorders with indicators for those that positively respond to transcranial near-infrared photobiomodulation.

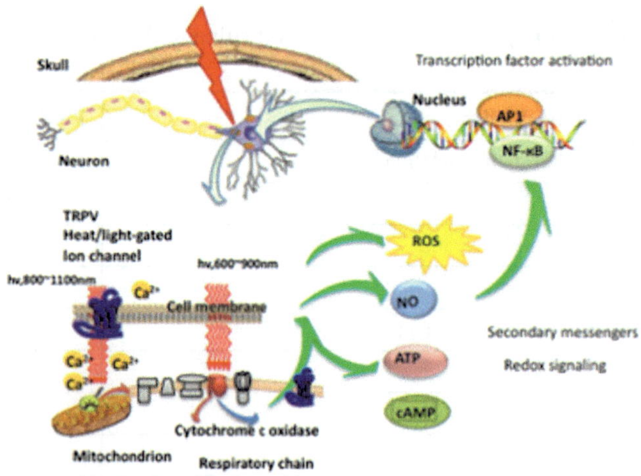

FIGURE 32.1 **Molecular and intracellular mechanisms of transcranial low level laser (light) or photobiomodulation on transcription factor activation.** *AP1*, Activator protein 1; *ATP*, adenosine triphosphate; Ca^{2+}, calcium ions; *cAMP*, cyclic adenosine monophosphate; *NF-kB*, nuclear factor kappa B; *NO*, nitric oxide; *ROS*, reactive oxygen species; *TRPV*, transient receptor potential vanilloid (Hamblin, 2016).

methylation selectively at subtelomeric CpG sites on the six examined chromosome ends as reported by SR Choudhury and associates at Purdue. This blue-light activation resulted in a progressive increase in telomere length over three generations of HeLa cell replication. They concluded that targeting of DNMT3A at the subtelometric DNA locations increases methylation at specific genomic locations in a HeLa cell model (Choudhury et al., 2016).

32.3 Photobiomodulation therapy

Low-level laser therapy was initially the term used to describe the therapeutic application of low intensity monochromatic light energy, with photobiomodulation having become the recommended comprehensive terminology encompassing both laser and LED light energy. There is no human evidence to suggest a clinical therapeutic advantage from the use of coherent light sources (as opposed to noncoherent light emitting diodes, LEDs), while there are substantially increased safety concerns, for example, tissue heating related damage. ANSI 2 135.3 (2011) and IEC 60825 document United States and international laser safety standards.

There are numerous instances of combination of laser and LED-based treatments, especially in the commercialization of PBM applications and devices. We are therefore exclusively advocating the development of LED-based technology for self-administered transcranial and intraocular applications. Our advocacy is also in response to the growing recognition that a significant proportion of modern healthcare assessment and intervention service delivery is shifting from face-to-face office-based structure to telemedicine and cloud-based applications (Delaney, 2017).

A search of the world's medical literature (250 + trials) of pharmacotherapeutic agents to treat Alzheimer's fails to provide any published evidence of long-term improvement (Cummings et al., 2014). PBM trials using low level NIR stimulation in treatment of traumatic brain injury (TBI) have been previously published in 2015 by Naesar and Hamblin. NIR light passes readily through the scalp and skull and arrives at the upper 1–5 cm of the brain. The primary photoreceptors for (600–950nm) red and NIR light are in the mitochondrial respiratory chain (Naeser and Hamblin, 2015). Cortical neurons are rich in mitochondria with increased biochemical pathways such as increased ATP and signaling pathways activated by ROS. PBM is based upon the ability of the light to alter cell metabolism as it is absorbed by hemoproteins and cytochrome c oxidase in particular (Kim, 2014). Regulation of gene expression and neurotransmitter behavior in the hippocampus and other brain regions typically associated with memory disorders are showing to be effective PBM targets. Most notable was the increase of brain derived nerve factor mRNA and enhancement of dendrite production and density in the hippocampus coupled with overall dendrite growth, density, and neuronal survival (Meng et al., 2013), further supported by Grillo et al. (2013), Ojha et al. (2011), Bradford (2007), and increasing production of molecular chaperones.

32.4 Near infrared photobiomodulation decreases synaptic vulnerability to Aβ

Comerta and researchers at University of Texas Galveston were working on synaptic dysfunction due to disruption of the binding of Aβ and tau oligomers, which are among the earliest impairments in AD. They had reported that a group of individuals referred to as nondemented with Alzheimer's neuropathology (NDAN) who had Aβ oligomer at the synapses but had retention of cognitive function differed from a group of demented AD subjects. They showed that these nondemented individuals displayed similar levels of soluble Aβ oligomers throughout their central nervous system (CNS), but their synapses were devoid of Aβ oligomers, suggesting that NDAN subjects are somehow resistant to Aβ oligomer production. They investigated the ability of NIR light to reduce synaptic susceptibility to Aβ oligomer binding, thereby increasing synaptic functions. They utilized wild type (Wt) mice to determine the impact of NIR light treatment on the binding of Aβ oligomers to isolated synaptosomes and long-term potentiation in the hippocampus of PBM-treated mice in the presence and absence of Aβ oligomers. Findings included significantly reduced Aβ1−42 at the synapses of the 6-month-old Tg2576 mice that overexpressed human amyloid precursor protein (APP). These changes coincided with a retention and increase in post PBM synaptic mitochondrial health in both Wt and Tg2576 and CD-1 mouse models. This study provides additional evidence that specific PBM protocols can effectively reduce synaptic vulnerability to damaging Aβ oligomers, thus furthering NIR light therapy as a viable treatment for AD (Comerota et al., 2017). These and other findings, especially work by Chazot and colleagues at University of Sunderland and Durham University helped further clarify the biochemistry underlying PBM's mechanism of action (see Fig. 32.2) (Berman et al., 2017).

Investigators at the University of Sunderland (Grillo et al., 2013) reported on an animal model of dementia (TASTPM mice) using noninvasive 1072 nm pulsed (10 Hz) stimulation. This was the first peer-reviewed publication describing this higher wavelength being employed as a potential treatment modality. Here we see the reduction in the number of small plaques placebo (no light) versus active 6-minute exposures over 2 consecutive days, twice weekly over 5 months (Fig. 32.3).

The original laboratory studies involved a cell line (lymphocytes exposed to UVA) and pretreated neurons in culture then exposed to varying concentrations of nitric oxide (Bradford et al., 2005). Subsequent media interest (Derbyshire, 2008) attracted the attention of researchers at Quietmind Foundation who were studying the effect of noninvasive brain-wave biofeedback (neurofeedback) training on cognitive and behavioral symptoms in subjects with early to mid-stage dementia. The present first author and colleagues at the Quietmind Foundation then began a collaboration in 2011 with the Cerebrolite's inventor, Gordon Dougal, MD, BSEE, a physician and electronic engineer, to conduct the first human clinical trials of PBM as a treatment for cognitive and behavioral symptoms of dementia.

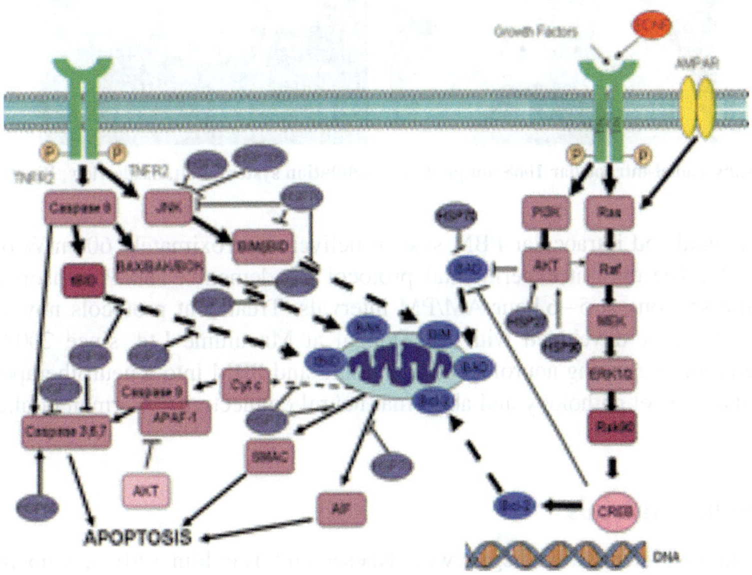

FIGURE 32.2 **Photobiomodulation mechanism of action.** A range of selective HSPs have been shown to be upregulated following in vivo treatment of Alzheimer's mice (HSP27, 60, 70, 90, 105) (Grillo et al., 2013); the role of these proteins in mitochondrial function, apoptosis, and chaperone-mediated protein folding are highlighted.

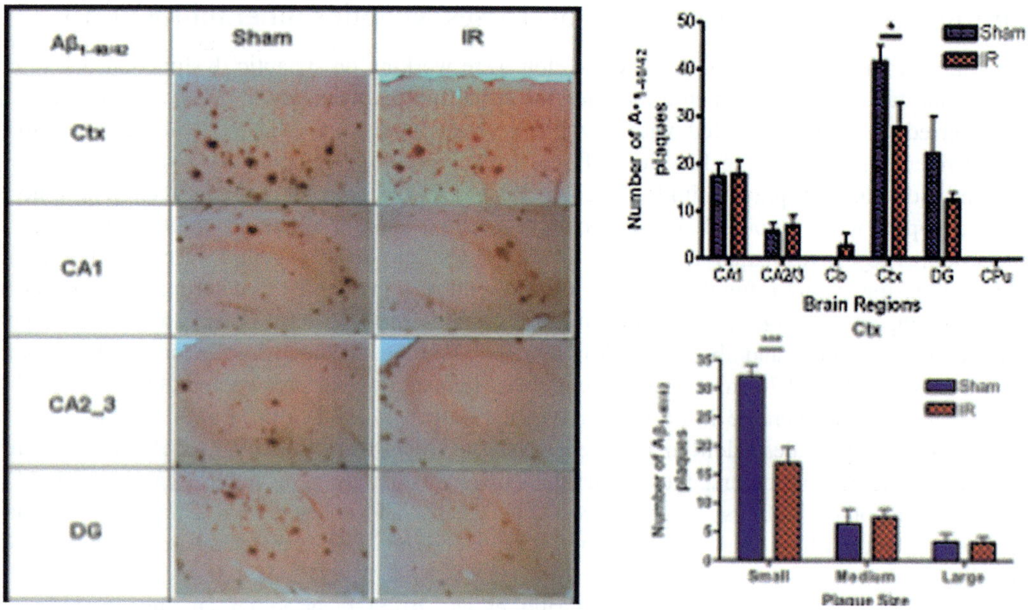

FIGURE 32.3 **Animal model of dementia treated with 1072 nm LLL.** Note tx group had a reduced number of small plaques. *CA1*, Cornus ammonis; *Ctx*, cortex; *DG*, dentate gyrus.

FIGURE 32.4 **Cerebrolite transcranial-intraocular 1068 nm photobiomodulation system.** Left, inside view; Right, top view.

The Cerebrolite transcranial and intraocular PBM system delivers approximately 600 mW of 1065–1075 nm photic stimulation pulsed at 10 Hz. The current experimental protocol for dementia and Parkinson's disease called for two daily, 5-minute stimulation sessions at 5–6 hour AM/PM intervals. Treatment protocols now employ the fifth generation device design, collaboratively developed with Dr. Dougal at Maculume Ltd. since 2008, tested by the research group at QMF with the goal of integrating neurofeedback training and PBM into a neurotherapeutic application that can simultaneously improve tissue-level pathology and abnormal neural connectivity (Berman et al., 2017) (Fig. 32.4).

32.5 Early human clinical trials

Human trials involving photobiomodulation began with Naeser and Hamblin (2015), who reported 11 chronic TBI patients whose cognition improved following treatment with red and NIR LEDs applied transcranially to forehead and scalp at 10 minutes per area and red light nasally with 18 outpatient sessions. Neuropsychological testing at 1, 2, and after 18 sessions of LED treatments demonstrated improvement in the Stroop test for executive function.

A pilot PBM trial involving early to moderate severity Alzheimer's was conducted, and we reported in 2017 that pilot double blind, placebo-controlled trial ($n = 11$; 6 active, 3 controls, and 2 dropouts) assessing the effect of 28 consecutive, 6-minute transcranial sessions of NIR stimulation. Patients were recruited from several local continuing care communities and using print and online media. All subjects were independently diagnosed with probable Alzheimer's dementia by a neurologist by means of the criteria of NIA-OA. Testing included mini mental status exam (MMSE), QEEG, and Alzheimer's disease assessment scale-cognitive (ADAS-Cog) that were administered the first day of treatment and within 3 days of completing the required 28 consecutive daily 6-minute exposure sessions. Surface cortical perfusion was measured before and after each transcranial and intraocular exposure using frontal (FP1 and FP2) NIRS wherein a 2-minute baseline was recorded using near surface infrared spectroscopy developed at the Biocomp Research Hemoencephalography and Bioexplorer software (Janow, 2002). Pre- and posttreatment QEEG changes were recorded after each session and are described in greater detail elsewhere (Berman et al., 2017).

The most significant changes in transcortical electrical activity were the normalization of central alpha (8–12 Hz) amplitudes and delta (0–4 Hz) and theta (4–8 Hz) hypocoherence and phase lag, that is, internodal correlation (Fonseca et al., 2013). Improved delta and theta are related to improved sleep architecture alertness and attention, and decreased alpha can result in reduced anxiety. John Nash, PhD, recently commented that, "Delta waves are wide geographically and raise wide regions of neurons closer to threshold, with the fast waves riding on the large oceanic swells of delta. Lack of delta causes lack of this widespread integration of frontal systems; also prevents effective sleep onset and restorative sleep" (Nash, 2018).

The substantial impact PBM that can elicit is quite apparent on both measures of EEG power and coherence activity and thus these might serve as noninvasively determined biomarkers to discriminate between different neurodegenerative disorders including Parkinson's and AD (Fonseca et al., 2013), and evaluating the efficacy of pharmacotherapeutic and directed energy-based interventions (Figs 32.5 and 32.6).

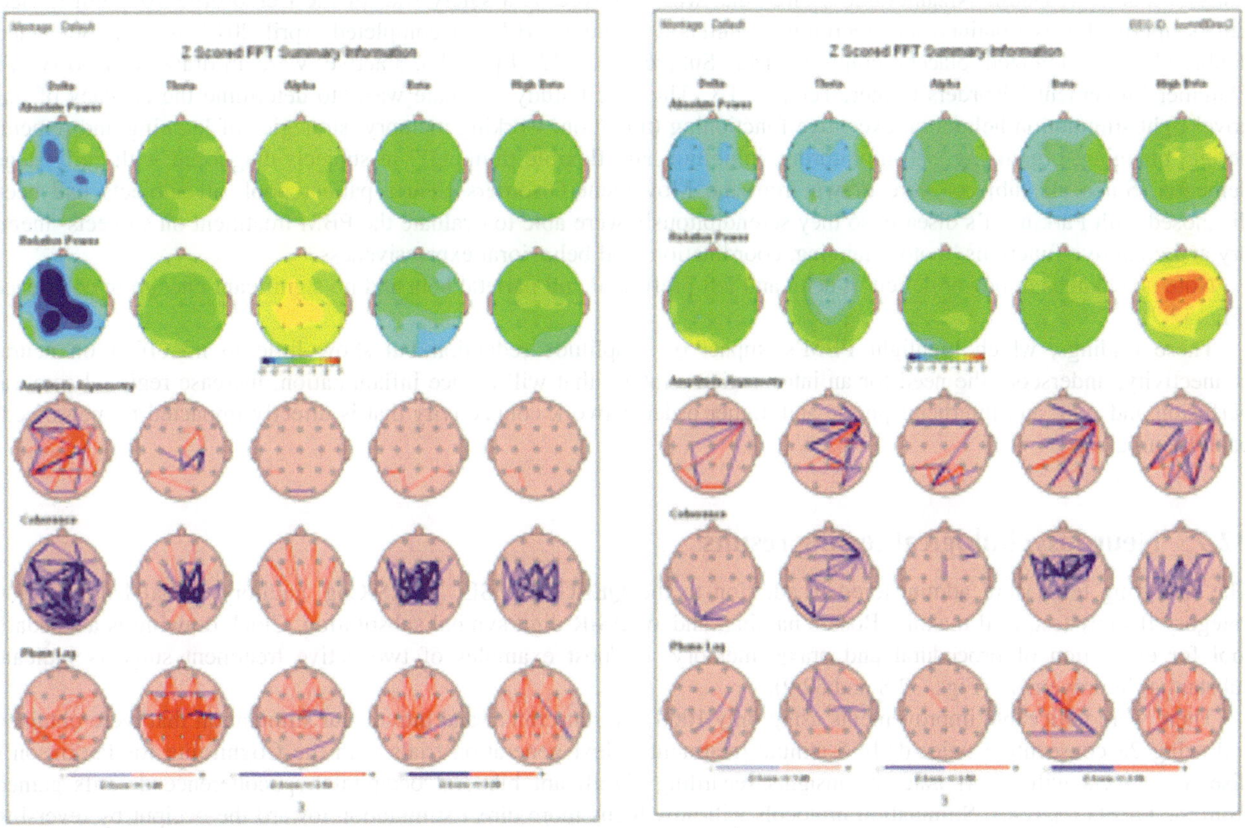

FIGURE 32.5 **Quantitative EEG changes following 1065–1075 nm transcranial and intraocular stimulation.** Left pre-PBM stimulation QEEG; Right post-PBM stimulation QEEG. Green color indicates regions where values of power were between 0 and 1 SD; yellow between 1 and 2 SD; red between 2 and 3 SD. Pink represents pre- and postcoherence summary maps.

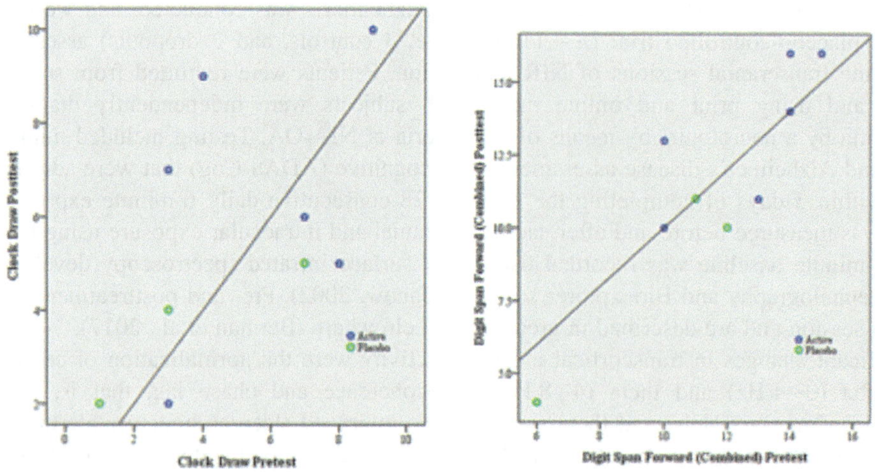

FIGURE 32.6 **Graphs comparing active and placebo clock drawing and digit span forward scores.** Results showed active treatment subjects tended to show greater improvement in executive functioning; clock drawing, immediate recalls, praxis memory, visual attention, and task switching (Trails A and B). Statistical significance could not be achieved due to low sample size. Improvements in clock drawing demonstrated moderate sensitivity and specificity for detecting executive cognitive dysfunction in people even with normal MMSE.

32.6 Digit span measures

A recently completed study was conducted by Dr. Jason Huang, Chairman, Dept. of Neurosurgery, Baylor Scott & White Health (BSWH), Temple, TX, an affiliate of Texas A&M Health Science Center and postdoctoral fellow and study coordinator, Damir Nizamutdinov, PhD, that was approved as a safety trial of the Cerebrolite 1068 nm device. Ethics approval was obtained and recruitment initiated in June 2017 and completed April 2018 as a single-center, double-blind, randomized, placebo-controlled trial. Subjects ($N = 12$, 4 placebo, 8 active) were all drawn from BSWH's Plummer Movement Disorders Center, Temple, TX. The stated study rationale was, "to determine the efficacy of this novel light stimulation helmet on executive functioning (attention, working memory, strategies of learning and remembering, planning, organizing, self-monitoring, inhibition and flexible thinking)" in subjects diagnosed with early-stage dementia. Since all subjects were drawn from the Movement Disorders Center patient pool, all subjects had been diagnosed with Parkinson's disease, so they serendipitously were able to evaluate the PBM treatment on subjects' memory and cognitive functions, motor planning, coordination, and behavioral expressiveness.

Note no changes in power deficit in delta and left prefrontal and parietal beta and no significant change in coherence (Fig. 32.7).

These findings, which highlight PBM's impact on amplitude reduction but show little to no effect on neural connectivity, underscore the need for an intervention strategy that will reduce inflammation, increase regional cerebral perfusion and ATP and interhemispheric and higher order network connectivity that is directly impacted by neurofeedback training.

32.7 Neuropsychological testing results

The following tests were administered in addition to the QEEG, MMSE, ADAS-Cog, auditory verbal learning test, category fluency test, trail making, Boston naming, and WAIS-R digit symbol substitution. Clock drawing is a standard tool for evaluation of procedural and praxis memory. Pre/post examples of two active treatment subjects indicate substantial improvement (Figs. 32.8 and 32.9).

There were noticeable improvements using only 1068 nm transcranial and intraocular stimulation that were obtained following 28 consecutive twice-daily 5-minute treatments. Next generation trials with the Cognitolite for Parkinson's disease subjects will incorporate the insights regarding significant bilateral occipital hypocoherence deficits gained from the QEEG analyses. Stimulation protocols will now focus more direct stimulation toward the occiput by reversing the ocular arrays of the ocular device to stimulate the foramen and thereby increase the level of PBM stimulation to the

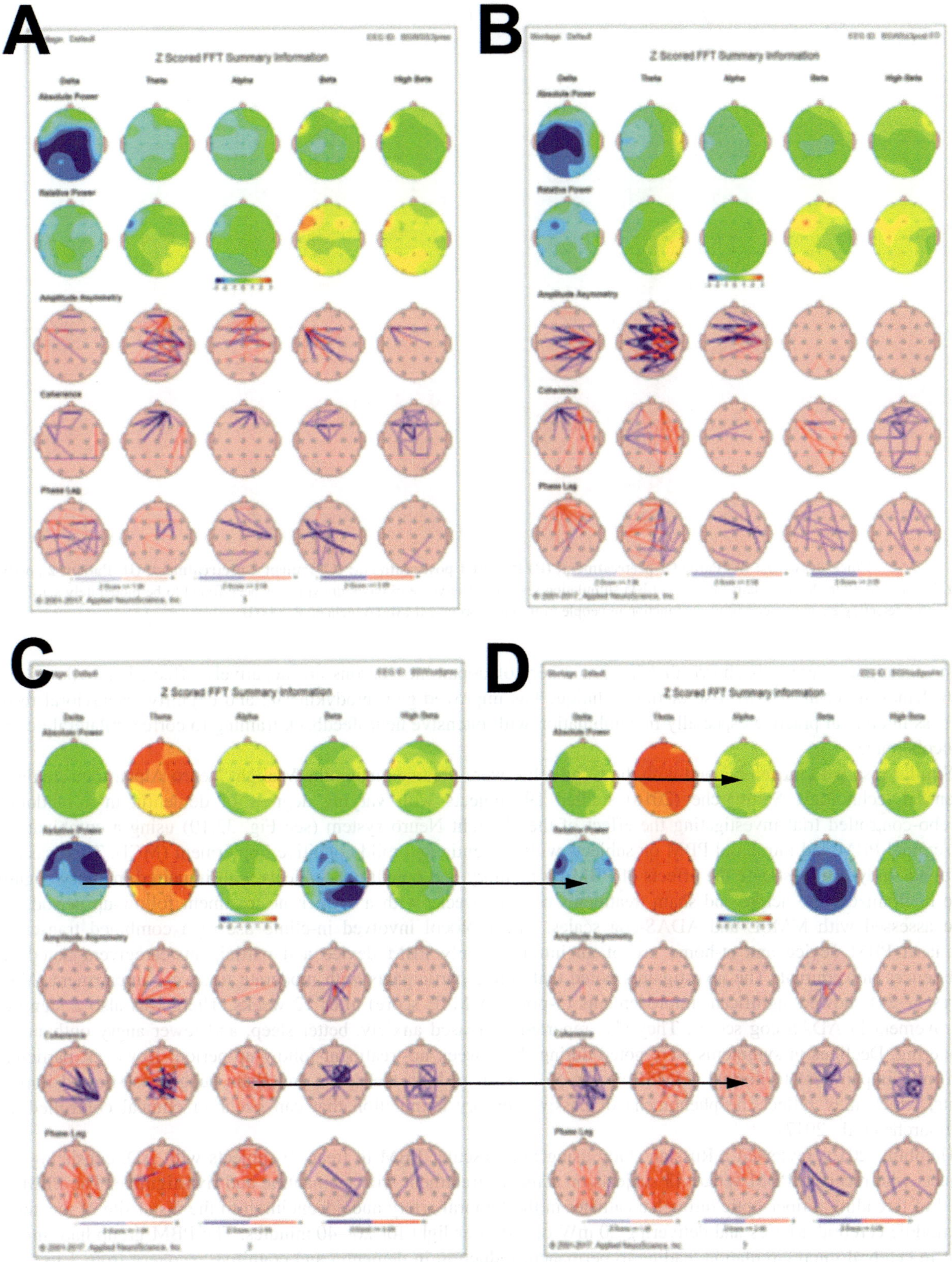

FIGURE 32.7 **Baylor Scott & White Pilot PBM study results: Pre/post active placebo QEEGs. (A) Preplacebo treatment; (B) postplacebo treatment; (C) preactive treatment; (D) postactive treatment.** Post active 28 consecutive twice daily, 5-minute treatment shows normalized delta relative power (*red circle*) and absolute alpha amplitude (*blue circle*) and decreasing alpha and beta coherence (2.5–2.0 SD).

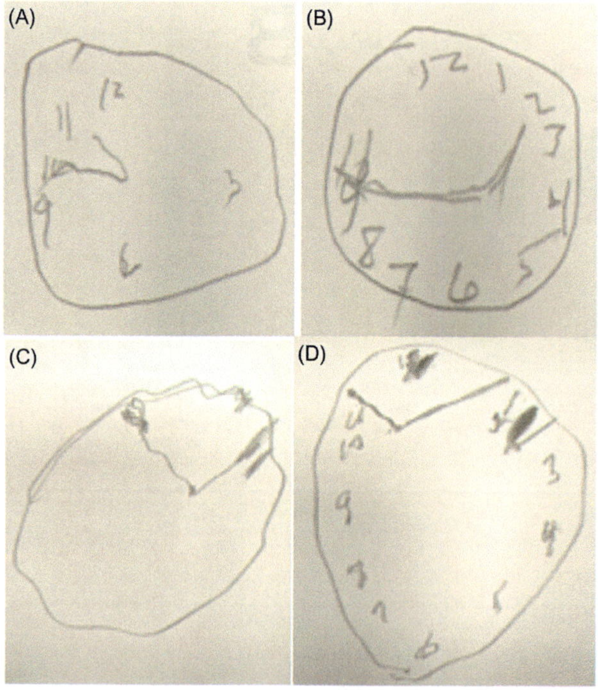

FIGURE 32.8 **Clock drawings: (A) Patient 1 pretreatment; (B) Patient 1 posttreatment; (C) Patient 2 pretreatment; (D) Patient 2 posttreatment.** Before and after 28 twice daily, 5-minute PBM stimulation sessions. The value of clock drawing has been found to be moderately sensitive and specific for detecting executive cognitive dysfunction in people even with normal MMSE (Angela et al., 2002).

substantia nigra where it is shown that large numbers of dopamine neurons are negatively affected (Novikova et al., 2006; Kinoshita et al., 2015). Use of this technique has improved gait, bradykinesia, and cognitive-behavioral responsiveness in clinical practice especially in combination with intensive neurofeedback training to correct bilateral occipital hypocoherence.

Similar results to those obtained with the Cognitolite were obtained using the Vielight Neuro Alpha transcranial and intranasal technology. Saltmarche (2016) studied 19 subjects with varying degrees of dementia in a randomized placebo-controlled trial investigating the effect of the Vielight Neuro system (see Fig. 32.10) using a combination of transcranial PBM and intranasal PBM on subjects with dementia and mild cognitive impairment (MCI). This was a single blind study to investigate the effects of PBM on memory and cognition. Subjects with impaired memory/cognition were randomized into active and sham treatments over 12 weeks with a 4-week no-treatment follow-up period. They were assessed with MMSE and ADAS-cog scales. The protocol involved in-clinic use of a combined transcranial-intranasal PBM device and at-home use of an intranasal-only PBM device and participants/caregivers noted daily experiences in a journal. The investigators noted that active participants with moderate to severe impairment (MMSE scores 5–24) showed significant improvements (5-points MMSE score) after 12 weeks. There was also a significant improvement in ADAS-cog scores. They also reported decreased anxiety, better sleep, and fewer angry outbursts and wandering. Declines in symptoms were noted during the 4-week no-treatment follow-up period. However, participants with mild impairment to normal (MMSE scores of 25–30) in both the active and sham subgroups showed improvements, which may reflect the phenomena of passive sensory stimulation that can occur in clinical controlled trials (Saltmarche et al., 2017).

Hamblin (2016) reviewed a Russian study using intravascular PBM to treat 89 subjects with AD, consisting of 46 subjects who received PBM versus 43 subjects on standard treatment with memantine and rivastigmine. The PBM consisted of threading a fiber-optic through a catheter in the femoral artery and advancing it to the distal site of the anterior and middle cerebral arteries and delivering 20 mW of red laser light for 20–40 minutes. The PBM group had improvement in cerebral microcirculation leading to permanent reduction in dementia and cognitive recovery from 1 to 7 years (Maksimovich, 2015).

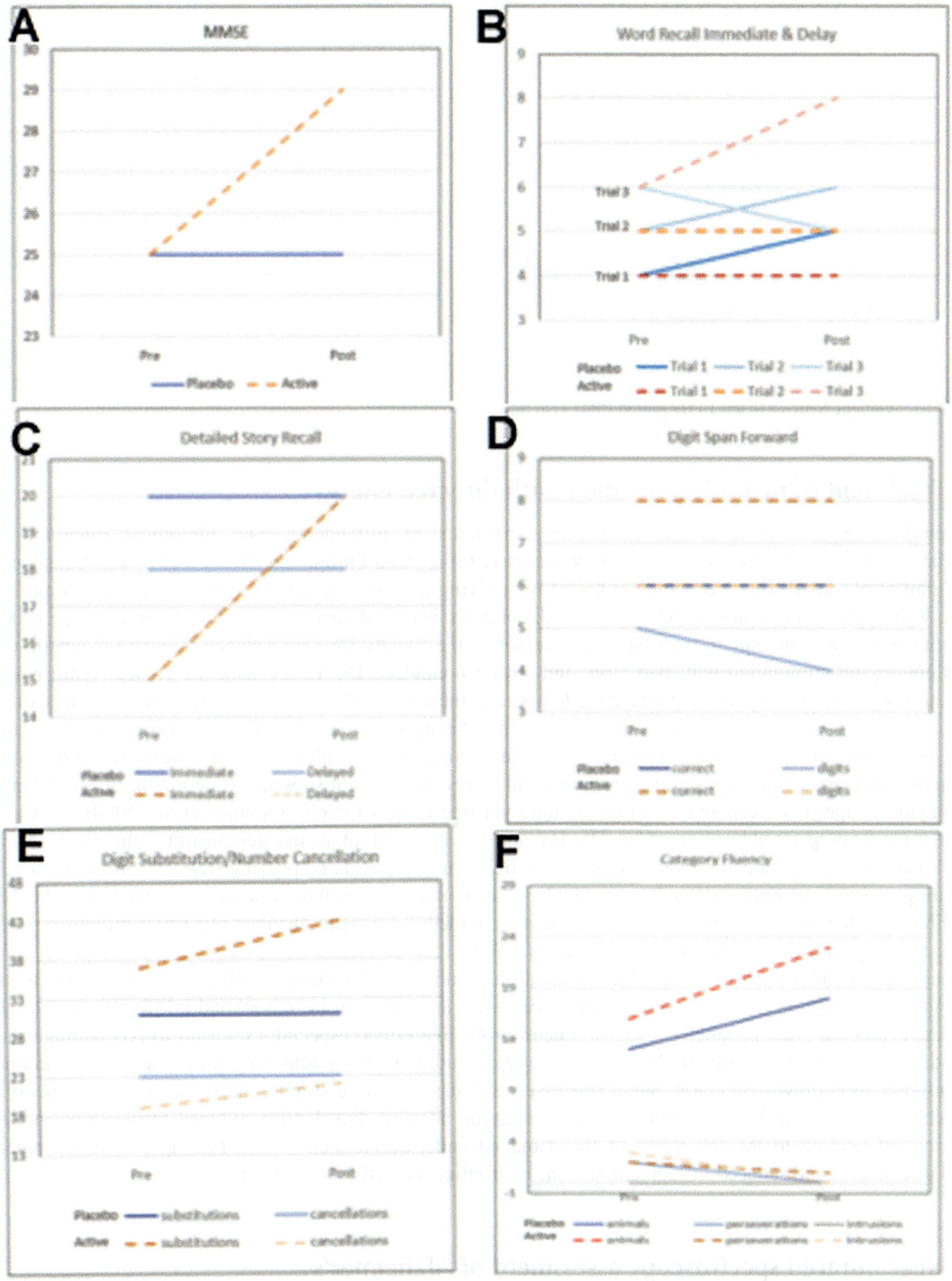

FIGURE 32.9 Placebo-active treatment comparisons. (A) MMSE; (B) word recall immediate and delayed; (C) delayed story recall; (D) digit span forward; (E) digit substitution number cancellation; (F) category fluency. Subjects with same pretreatment (MMSE = 25) score on ADAS-Cog neuropsychological measures.

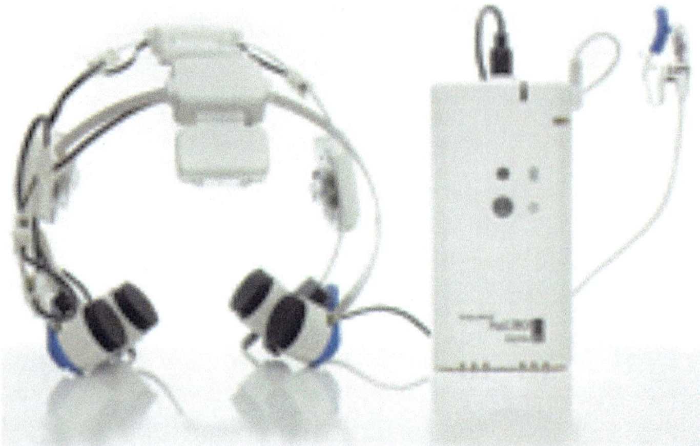

FIGURE 32.10 Vielight Neuro 810 nm 10 Hz pulsed PBM system.

32.8 Treatment of neurodegeneration with directed energy

Photobiomodulation is one of several noninvasive methods of influencing core biological functions including transcranial magnetic and pulsed electromagnetic fields (EMFs; Pena-Philippides et al., 2014) and transcranial ultrasound (Tufail et al., 2010) that are being employed in studying and treating a wide range of functional and metabolic disorders. The application of neuromodulation in Alzheimer's was first reported by Nichols and Pearce (2006) using moderate magnetic field therapy to treat AD. Their report in the society for neuroscience involved magnetic therapy of 0.5 Tesla employing stimulation with two large and strong nonpulsed DC EMFs with the subject lying between two large electromagnets. The treatment produced a temporary increase in the magnetic force on the atoms of the body resulting in a higher velocity and precession of certain orbiting electrons, thereby increasing electron transfer and chemical reactions. AD is a neurodegenerative disease secondary to oxidative stress, associated with genetic and environmental factors such as exposure to pesticides and heavy metals with subsequent depletion of mitochondrial protective enzymes, superoxide dismutase and glutathione via free radical toxicity (Nichols et al., 2000). This along with gene expression demonstrated by Wang, Che, Du, Ha, and Yarema at Hopkins has demonstrated that thousands of genes could be up and down regulated by moderate magnetic fields in two human embryonic stem cell lines by modulation of cell signaling and differentiation (Wang et al., 2009). NASA investigators have also demonstrated that picoTesla time varying EMFs on human neuronal cells results in similar molecular genetic changes regarding growth potential measured by gene chip analysis of 10,000 genes (Goodwin, 2003).

An innovative combination of static EMF stimulation combined with PBM by Blivet and colleagues as treatment against Aβ25–35 peptide induced toxicity in mice with an increase in mitochondrial function and adenosine triphosphate synthesis. Such parameters suggest a neuroprotective effect, especially relevant to AD pathogenesis where mitochondrial proteostasis is particularly affected. Several studies report astonishing properties of transcranial PBM, such as inflammation down-regulation, repair processes, and tissue healing stimulation for treatments in neurology and neuropsychiatry. They also hypothesized that β-nicotinamide adenine dinucleotide reduced nitric oxide, proteins, and ribonucleic acid leading to the self-repair of the damaged cells, accelerated wound healing and tissue regeneration, increased circulation, and reduced inflammation from microglial cells (Blivet et al., 2017).

32.9 Near infrared spectroscopy assessment of Alzheimer's

AD diagnosis has been handicapped by diagnosis only at autopsy. Recent advances in positron emission tomography imaging has been demonstrated to have a diagnostic edge over magnetic resonance imaging (MRI) and computed tomography, but is expensive and still limited in availability. NIR spectroscopy (NIRS) is now being utilized for diagnosis because the structural integrity of the cerebral vasculature, especially of cortical microvessels, is markedly compromised in patients with AD and in mouse models of AD. NIRS is a noninvasive neuroimaging tool used now to measure activation-induced changes in cerebral hemoglobin concentration. By this technique, changes in the optical absorption of light are recorded over time and are used to estimate the functionally evoked changes in cerebral oxyhemoglobin and deoxyhemoglobin concentrations that result from local cerebral vascular and oxygen metabolic effects during brain activities.

Recently, Van Beeka and associates have shown that subjects with early stage AD have a reduced cerebral blood flow velocity (CBFV) and an increased cerebrovascular resistance compared with age-matched healthy controls. Alzheimer subjects had smaller spontaneous cerebral blood flow oscillations, relative to blood pressure (BP) oscillations. They studied 21 subjects with mild to moderate AD and 20 age-matched controls, and investigated how oscillations in CBFV and O_2Hb are associated with spontaneous and induced oscillations in BP at the very low (VLF = 0.05 Hz) and low frequencies (LF = 0.1 Hz).They had discovered from Obrig's (2000) work that NIRS and functional MRI have revealed spontaneous oscillatory changes in cerebral cortical oxygenation. They then applied spectral and transfer function analysis to quantify dynamic cerebral autoregulation and brain tissue oxygenation (Van Beeka et al., 2010).

Because AD is a progressive neurodegenerative disease defined as a clinical condition with declining function in various cognitive domains, such as memory, reasoning, judgment, executive function, praxis, visuospatial abilities, and language. Since these cognitive disturbances commonly occur in AD, PBM may help to detect disease-specific alterations in the brain that may be of diagnostic value and/or useful for therapeutic monitoring. Hock et al. (1996), in a study of NIRS, showed that increase in cerebral Hb oxygenation in response to brain activation declines with physiological aging. They examined 29 individuals, including 12 healthy young volunteers and 17 healthy elderly volunteers. In comparison to the young group, the elderly participants showed a significantly lower mean increase in [HbO_2] and [HbT] levels in the frontal cortex during performing calculation tasks. Further regression analysis supports the hypothesis of an age-dependent decline in activation-induced regional increase of [HbO_2]. They concluded that this phenomenon is more pronounced in the parietal cortex than in the frontal cortex. The possible mechanism could have resulted from altered regional brain function, altered neurovascular coupling caused by neurodegeneration, or anatomical changes during aging and neurodegeneration (Hock et al., 1996).

In another study using NIRS, they found that the medial temporal lobe, including the entorhinal cortex, amygdala, and hippocampus, was one of the first regions of the brain to show AD-related neurodegeneration. The observed memory declines in the early and prodromal stages of AD are associated with degeneration of the medial temporal lobe, especially in the entorhinal cortex and hippocampus. These findings are reflected in a study of NIRS conducted by Fladby et al. of 13 subjects who revealed impaired response to olfactory stimulation in the temporal cortex with subjective memory complaints, MCI, or very mild AD. Because the pathway for the olfactory response is through the entorhinal cortex, the response may be diminished at an early stage of AD (Fladby et al., 2004) and may be remediated with the application of intranasal PBM, for example, the Vielight 810, and Neuro devices. Interventions with the Vielight Neuro with subjects diagnosed with dementia have produced positive results in small case series reports with significant improvements reported for both cognitive functions and behavior, for example, reduced aggressive behavior, anxiety, and wandering, along with improved sleep (Saltmarche et al., 2017).

32.10 Conclusion

PBM applications continue to evolve as evidence of its efficacy continues to appear in the peer-reviewed literature. The PubMed database shows there are between approximately 5300 and 6300 studies containing the search terms "photobiomodulation therapy," "low level laser therapy," or "low level light therapy." Clinical applications will continue to evolve from Quietmind Foundation's research focusing on at-home and day treatment program designs and development for treatment of neurodegenerative and neuropsychiatric disorders. Integrating tissue level treatment with PBM and neurofeedback for renormalization of neural connectivity is rooted in the understanding that, "in the final analysis, the only part of our being that holds a relationship with the external world is the nervous system". In so doing, we can leverage the benefits of a more efficient (robustly adaptive) functioning CNS interacting within a healthier biochemical ecosystem. It is hoped that by taking a systems-centered approach we are increasing our capacity to address the etiological complexities inherent in neurodegenerative disorders and more successfully design public health delivery systems that can support both neurorehabilitation treatment and prevention strategy development (Mabry and Kaplan, 2013).

Collective efforts within and among the relevant disciplines are now focusing on design, engineering and algorithm development to integrate neurophysiological (LORETA *z*-score neurofeedback), cardio-diagnostic (heart rate variability, HRV) and frequency-specific, directed energy and PBM to deliver scalable, safe, reliable, and effective home-based, affordable, neurotherapeutic treatment solutions. Such programs would increase tissue-level health, thereby improving neurophysiological sensitivity, that is, enhanced capacity to discriminate similarities and differences (Agazarian and Gantt, 2000) and functional health, resulting in improved neurocognitive functioning. HRV coherence and EEG coherence and phase related activity can then be used as a measure of broad systemic flexibility and, as such, a neuro marker that relates to our capacity for adaptive responsiveness.

Neurotherapeutic techniques measuring evoked EEG dominant frequency activity can be utilized to guide treatment, first by introducing pulsed electromagnetic stimulation below the level of conscious awareness, and monitoring changes in dominant frequency activity. This real time dominant frequency variability operationally understood as its approximation of randomness can then be evaluated and coupled with normalization of EEG amplitude, coherence, phase lock, and phase reset (Thatcher, 2012; Thatcher et al., 2014).

Continued studies are needed to construct algorithms that will combine the physiological (HRV) and neuroelectrical (EEG) inputs to dynamically analyze using age, gender, and handedness-based normative analytics to trigger delivery of appropriate directed energy stimulation to support normalized CNS functioning through the delivery of subconscious, contentless stimuli to facilitate more finely and adaptively differentiated structural and functional neurophysiological organization. This conceptualization of neurotherapeutic interventions aligns with the view of the CNS as a self-organizing and self-correcting system through its dynamic interaction at the boundary of internally and externally sourced experience. Technologies arising from integrating neurophysiological and functional medical paradigms are then likely to significantly improve treatment delivery systems and clinical outcomes especially for systemic neurodegenerative disorders. Future applications will integrate the delivery of noninvasive stimulation including full-body and transcranial colored (syntonic) and infrared light, ultrasound, microcurrent, pulsed EMFs, infra-low electromagnetic energy, and digitally transformed analog sound.

References

Agazarian, Y.M., Gantt, S., 2000. Autobiography of a Theory: Developing a Theory of Living Human Systems and Its Systems-Centered Practice (International Library of GroupAnalysis, 11), first ed. Jessica Kingsley Publisher, London.

Angela, J., Tench, S., Baker, V., 2002. The value of clock drawing in identifying executive cognitive dysfunction in people with normal Mini-Mental State Examination score. Can. Med. Assoc. J. 167, 859–864.

Berman, M.H., Halper, J.P., Nichols, T.W., et al., 2017. Photobiomodulation with near infrared light helmet in a pilot, placebo controlled clinical trial in dementia patients testing memory and cognition. J. Neurol. Neurosci. 8, 1. Available from: https://doi.org/10.21767/2171-6625.1000176.

Blivet, G., Meunier, J., Roman, F.J., Touchon, J., 2017. Neuroprotective effect of a new photobiomodulation technique against Aβ25–35 peptide–induced toxicity in mice: novel hypothesis for therapeutic approach of Alzheimer's disease suggested. Alzheimer's Dementia 4, 54–63. Available from: https://doi.org/10.1016/j.trci.2017.12.003.

Bradford, A., Barlow, P.L., Chazot, 2005. Probing the differential effects of infrared light sources IR1072 and IR880 on human lymphocytes: evidence of selective cytoprotection by IR1072. J. Photochem. Photobiol. B 81, 9–14.

Bredesen, D., 2017. The End of Alzheimer's: The First Program to Prevent and Reverse Cognitive Decline. Penguin, Random House, New York.

Comerota, M.M., Krishnan, B., Taglialatela, G., 2017. Near infrared light decreases synaptic vulnerability to amyloid beta oligomers. Sci. Rep. 7 (1), 15012. Available from: https://doi.org/10.1038/s41598-017-15357-.

Choudhury, S.R., Cui, Y., Narayanan, A., et al., 2016. Optogenetic regulation of site-specific subtelomeric DNA methylation. Oncotarget 7 (31), 50380–50391. Available from: https://doi.org/10.18632/oncotarget.10394.

Cummings, J.L., Morstorf, T., Zhong, K., 2014. Alzheimer's disease drug-development pipeline: few candidates, frequent failures. Alzheimers Res. Ther. 6 (37), 2–7. Available from: https://doi.org/10.1186/alzrt269.

Dantuma, N.P., Bott, L.C., 2014. The ubiquitin-proteasome system in neurodegenerative diseases: precipitating factor, yet part of the solution. Front. Mol. Neurosci. 31 (7), 70. Available from: https://doi.org/10.3389/fnmol.00070.

Delaney, D., 2017. Why Healthcare Is Continuing Its Shift to the Cloud. Health Data Management, April 27, 2017. <https://www.healthdatamanagement.com/opinion/why-healthcare-is-continuing-its-shift-to-the-cloud>.

Derbyshire, D., 2008. Dementia patient makes 'amazing' progress after using infra-red helmet. Mail Online, July 15. https://www.dailymail.co.uk/health/article-1034936/Dementia-patient-makes-amazing-progress-using-infrared-helmet.html. accessed Jun 21 2019.

Fladby, T., Bryhn, G., Halvorsen, O., et al., 2004. Olfactory response in the temporal cortex of the elderly measured with near-infrared spectroscopy: a preliminary feasibility study. J. Cereb. Blood Flow Metab. 24, 677–680.

Fonseca, L.C., Tedrus, G.M., Carvas, P.N., Machado, E.C., 2013. Comparison of quantitative EEG between patients with Alzheimer's disease and those with Parkinson's disease dementia. Clin. Neurophysiol. 124 (10), 1970–1974. Available from: https://doi.org/10.1016/j.clinph.2013.05.001. Epub 2013 Jun 5.

Goodwin, T.J., 2003. Physiological and Molecular Genetic Effects of Time-Varying Electromagnetic Fields on Human Neuronal Cells. ntrs.nasa.gov.

Grillo, S., Duggett, N.A., Ennaceur, A., Chazot, P.L., 2013. Non-invasive infra-red therapy (1072 nm) reduces β-amyloid protein levels in the brain of an Alzheimer's disease mouse model, TASTPM. J. Photochem. Photobiol. B 123, 13–22. Available from: https://doi.org/10.1016/j.jphotobiol.2013.02.015.

Hamblin, M.R., 2016. Shining light on the head: photobiomodulation for brain disorders. BBA Clin. 6, 113–124.

Hock, C., Villringer, K., Müller-Spahn, M., et al., 1996. Near infrared spectroscopy in the diagnosis of Alzheimer's disease. Ann. N.Y. Acad. Sci. 777, 22–29.

Janow, L., 2002. Bioexplorer Object-Oriented Neurofeedback Training Software. <http://cyberevolution.com/beoverview.htm>.

Kim, H.P., 2014. Lightening up light therapy: activation of retrograde signaling pathway by photobiomodulation. Biomol. Ther. 22 (6), 491–496.

Kim, H.Y., Sengupta, U., Shao, P., et al., 2013. Alzheimer's disease imaging with a novel Tau targeted near infrared ratiometric probe. Am. J. Nucl. Med. Mol. Imaging 3 (2), 102–117.

Kinoshita, K., Tada, Y., Muroi, Y., Unno, T., Ishii, T., 2015. Selective loss of dopaminergic neurons in the substantia nigra pars compacta after systemic administration of MPTP facilitates extinction learning. Life Sci. 137, 28–36. Available from: https://doi.org/10.1016/j.lfs.2015.07.017.

Le Corre, S., Klafki, H.W., Plesnila, N., et al., 2006. An inhibitor of tau hyperphosphorylation prevents severe motor impairments in tau transgenic mice. Proc. Natl. Acad. Sci. U.S.A. 103 (25), 9673–9678.

Maksimovich, I.V., 2015. Dementia and cognitive impairment reduction after laser transcatheter treatment of Alzheimer's disease. World J. Neurosci. 5, 188–202.

Mabry, P.L., Kaplan, R.M., 2013. Systems science: a good investment for the public's health. Health Educ. Behav. 40 (1 Suppl.), 9S–12S.

Meng, C., He, Z., Xing, D., 2013. Low-level laser therapy rescues dendrite atrophy via upregulating BDNF expression: implications for Alzheimer's disease. J. Neurosci. 33 (33), 13505–13517.

Myeku, N., Clelland, C.L., Emrani, S., Kukushkin, N.V., Yu, W.H., Goldberg, A.L., et al., 2016. Tau-driven 26S proteasome impairment and cognitive dysfunction can be prevented early in disease by activating cAMP-PKA signaling. Nat. Med. 22 (1), 46–53. Available from: https://doi.org/10.1038/nm.4011.

Naeser, M.A., Hamblin, M.R., 2015. Traumatic brain injury: a major medical problem that could be treated using transcranial, red/near-infrared LED photobiomodulation. Photomed. Laser Surg. 33 (9), 443–446. Available from: https://doi.org/10.1089/pho.2015.3986.

Nash, J., 2018. Personal Communication.

Nichols, T.W., 2014. Hyperphosphorylation of tau protein in Down's dementia and Alzheimer's disease; methylation and implications in prevention and therapy. J. Alzheimer's Dis. Parkinsonism 4, 159. Available from: https://doi.org/10.4172/2161-0460.1000159.

Nichols, T.W., Pearce, L.A., Stokesbary, D.I., Bonlie, D.R., 2000. Clinical observations in magnetic molecular energizing on Alzheimer's and Parkinson's patients and their possible; association with heavy metals. Soc. Neurosci. 26, A229.

Novikova, L., Garris, B.L., Garris, D.R., Lau, Y.S., 2006. Early signs of neuronal apoptosis in the substantia nigra pars compacta of the progressive neurodegenerative mouse 1-methyl-4-phenyl-1,2,3,6-tetrahydropyridine/probenecid model of Parkinson's disease. Neuroscience 140 (1), 67–76.

Obrig, H., Neufang, M., Wenzel, R., et al., 2000. Spontaneous low frequency oscillations of cerebral hemodynamics and metabolism in human adults. Neuroimage 12, 623–639.

Ojha, J., Karmegam, R.V., Masilamoni, J.G., Terry, A.V., Cashikar, A.G., 2011. Behavioral defects in chaperone-deficient Alzheimer's disease model mice. PLoS One 6 (2), e16550. Available from: https://doi.org/10.1371/journal.pone.0016550.

Pena-Philippides, J.C., Yang, Y., Bragina, O., Hagberg, S., Nemoto, E., Roitbak, T., 2014. Effect of pulsed electromagnetic field (PEMF) on infarct size and inflammation after cerebral ischemia in mice. Transl. Stroke Res. 5 (4), 491–500. Available from: https://doi.org/10.1007/s12975-014-0334-1.

Purushothuman, S., Johnstone, D.M., Nandasena, C., Mitrofanis, J., Stone, J., 2014. Photobiomodulation with near infrared light mitigates Alzheimer's disease-related pathology in cerebral cortex – evidence from two transgenic mouse models. Alzheimers Res. Ther. 6 (1), 2. Available from: https://doi.org/10.1186/alzrt232.

Readhead, B., et al., 2018. Multiscale analysis of independent Alzheimer's cohorts finds disruption of molecular, genetic, and clinical networks by human herpesvirus. Neuron. Available from: https://doi.org/10.1016/j.neuron.2018.05.023.

Saltmarche, A.E., Naeser, M.A., Ho, K.F., Hamblin, M.R., Lim, L., 2017. Significant improvement in cognition in mild to moderately severe dementia cases treated with transcranial plus Schietinger CH, Reich NO. (2012) RNA modulation of the human DNA methyltransferase 3A. Nucleic Acids Res. 40 (17), 8550–8557. Available from: https://doi.org/10.1093/nar/gks537.

Thatcher, R.W., 2012. Coherence, phase differences, phase shift, and phase lock in EEG/ERP analyses. Dev. Neuropsychol. 37 (6), 476–496. Available from: https://doi.org/10.1080/87565641.2011.619241.

Thatcher, R.W., North, D.M., Biver, C.J., 2014. LORETA EEG phase reset of the default mode network. Front. Hum. Neurosci. 8, 529. Available from: https://doi.org/10.3389/fnhum.2014.00529. eCollection 2014.

Tufail, Y., Matyushov, A., Baldwin, N., Tauchmann, M.L., Georges, J., Yoshihiro, A., et al., 2010. Transcranial pulsed ultrasound stimulates intact brain circuits. Neuron 66 (5), 681–694. Available from: https://doi.org/10.1016/j.neuron.2010.05.008.

Underwood, E., 2016. Why the big change to Lilly's Alzheimer's trial is not evidence its drug has failed again. Science. Available from: https://doi.org/10.1126/science.aaf9811.

Van Beeka, A., Lagroa, J., Olde-Rikkert, M., et al., 2010. Oscillations in cerebral blood flow and cortical oxygenation in Alzheimer's disease. Neurobiol. Aging 33, 428.e21–428.e31.

Wang, Z., Sarje, A., Che, P.L., Yarema, K., 2009. Moderate strength (0.23-0.28T) static magnetic fields (SMF) modulate signaling and differentiation in human embryonic cells. BMC Genomic 10, 356. Available from: http://www.biomedicalcentral.com/1471-2164/10/35.

Further reading

Cassano, P., Petrie, S., Mischoulon, D., Cusin, C., Katnani, H., Yeung, L., et al., 2018. Neuromodulation with near-infrared light for mood disorders: targeting mitochondria. In: Special Issue: 20th Annual Conference of the International Society for Bipolar Disorders, Mexico City, Mexico, 7–10 March 2018 Volume 20, Issue S1, March 2018, pp. 45–46.

Chan, C.S., Gertler, T.S., Surmeier, D.J., 2010. A molecular basis for the increased vulnerability of substantia nigra dopamine neurons in aging and Parkinson's disease. Mov. Disord. 25 (Suppl. 1), S63–S70. Available from: https://doi.org/10.1002/mds.22801.

Davidov, D.M., Steward, R., Ritchie, K., Chaudieu, I., 2010. Resilience and mental health. Clin. Psychol. Rev. 30, 479–495.

Dougal, G., 2016. Maculume, Ltd. Durham. Personal Communication.

Lim, L., 2018 Enhancing Brain Functions With Near Infrared Light. Integrative Practitioner. Sponsored white paper published online.

Naeser, M.A., Zafonte, R., Krengel, M.H., et al., 2014. Significant improvements in cognitive performance post-transcranial, red/near-infrared light-emitting diode treatments in chronic, mild traumatic brain injury: open-protocol study. J. Neurotrauma 31, 1008–1017.

Naoi, M., Maruyama, W., 1999. Cell death of dopamine neurons in aging and Parkinson's disease. Mech. Ageing Dev. 111 (2–3), 175–188.

Rango, M., Bresolin, N., 2018. Brain mitochondria, aging, and Parkinson's disease. Genes 9, 250. Available from: https://doi.org/10.3390/genes9050250. March 2018, 45–46.

Saltmarche, A.E., Naeser, M.A., 2017. Intranasal photobiomodulation: case series report. Photomed. Laser Surg 35 (8), 432–441. Available from: https://doi.org/10.1089/pho.2016.4227.

Toomim, H., 2011 Biocomp Research Institute. Los Angeles, CA. <https://biocompresearch.org/heg-instruments>.

University of Sunderland, January 26, 2008. New research could help reverse the biological clock for dementia patients. ScienceDaily. Retrieved June 29, 2018 from: <www.sciencedaily.com/releases/2008/01/080124104917.htm>.

Wang, R., Dong, Y., Lu, Y., Zhang, W., Brann, D.W., Zhang, Q., 2018. Photobiomodulation for global cerebral ischemia: targeting mitochondrial dynamics and functions. Mol. Neurobiol. Available from: https://doi.org/10.1007/s12035-018-1191-9. June 27, 2018.

Chapter 33

Transcranial photobiomodulation therapy: observations from four movement disorder patients

Catherine Hamilton[1], David Hamilton[1], Frank Nicklason[1,2] and John Mitrofanis[1]
[1]Department of Anatomy, University of Sydney, Sydney, NSW, Australia, [2]Geriatric Medicine, Royal Hobart Hospital, Hobart, TAS, Australia

33.1 Introduction

Transcranial red to near infrared light ($\lambda = 600-1070$ nm) therapy, also known as photobiomodulation, has been used to treat many neurological conditions in humans, from Alzheimer's disease (Saltmarche et al., 2017) to depression (Schiffer et al., 2009) and from traumatic brain injury (Naeser et al., 2011) to stroke (Lapchak et al., 2007) and lower back pain (Holanda et al., 2016; Chow and Armati, 2016). In each of these studies, as in all those involving experimental animals, photobiomodulation leads to beneficial outcomes, from improved locomotive behavior and/or cognition to a better survival of neurones and their synaptic terminations (Hamblin, 2016; Johnstone et al., 2016; Mitrofanis, 2017). Further, photobiomodulation has no reported side-effects and is not toxic to tissue. Indeed, the US Food and Drug Administration has approved many photobiomodulation devices for therapeutic use in humans.

From studies on animal models of Parkinson's and Alzheimer's disease, a key finding of photobiomodulation is that it can be neuroprotective, being able to slow the progression of neurodegeneration (Johnstone et al., 2016; Mitrofanis, 2017). This disease-modifying feature of the therapy makes it an attractive option for use in humans, mainly because all the current treatments for both neurological conditions are symptomatic, and do not slow or stop the disease progression.

In this chapter, we document four cases of patients with movement disorders, one with progressive supranuclear palsy and three with Parkinson's disease, that applied transcranial photobiomodulation therapy. Each patient used an in-house developed photobiomodulation helmet, lined with light emitting devices (LEDs) of various wavelengths across the red to near infrared light range (i.e., 670, 810, 850, 940 nm).

33.2 Case descriptions

In the section that follows, the clinical histories and changes in signs and symptoms, before and during photobiomodulation, will be described separately for each patient. The use of the photobiomodulation helmet was voluntary and progress was assessed by either the patient themselves, their spouse, or their attending medical practitioners. Each patient consented for their case to be included in this chapter.

33.2.1 Progressive supranuclear palsy: Patient FH

Patient FH, a 68 year old male was diagnosed with progressive supranuclear palsy 3 years ago. His major signs and symptoms at the time of diagnosis included: impaired vertical gaze, his vision became "blurry" when looking up and down, for example, when undertaking computer work or walking; impaired speech, particularly when trying to speak with volume; impaired fine motor skills, for example, his writing was difficult to read, with very small words and little separation of letters; difficulty in maintaining balance, he had fallen many times; a persistent cough, related to a difficulty in clearing saliva, eating or drinking; suffered emotional lability, with frequent bouts of anger and mood change;

Photobiomodulation in the Brain. DOI: https://doi.org/10.1016/B978-0-12-815305-5.00033-6

and impaired sleeping. FH had ceased reading books some 12 months earlier but he did not show any sign of cognitive impairment. Finally, FH had spells of pathological laughing, but these were rather infrequent and short-lasting.

Two years after diagnosis, FH started using a photobiomodulation helmet, constructed in-house using a lamp shade-like base lined internally with strips of 670 and 940 nm LEDs, both wired together (Fig. 33.1A,B). At a later date (~5 months after first use), the 940 nm LEDs were replaced by 810 nm LEDs. During this period, he used an intranasal light device (Bionase) with 660 nm LEDs (*arrow* Fig. 33.1B) as well. After exploring a range of different exposure times and frequency of use periods, the most favorable combination for FH was found to be: 10 minutes with 660 nm from the intranasal device and 670 nm from the helmet (applied at the same time), followed subsequently by 10–15 minutes with 940 or 810 nm from the helmet, undertaken twice daily.

From approximately a month after first using photobiomodulation, improvements were evident in many of FH's signs and symptoms (Table 33.1), from his speech becoming clearer and easier to understand, to his coughing not being

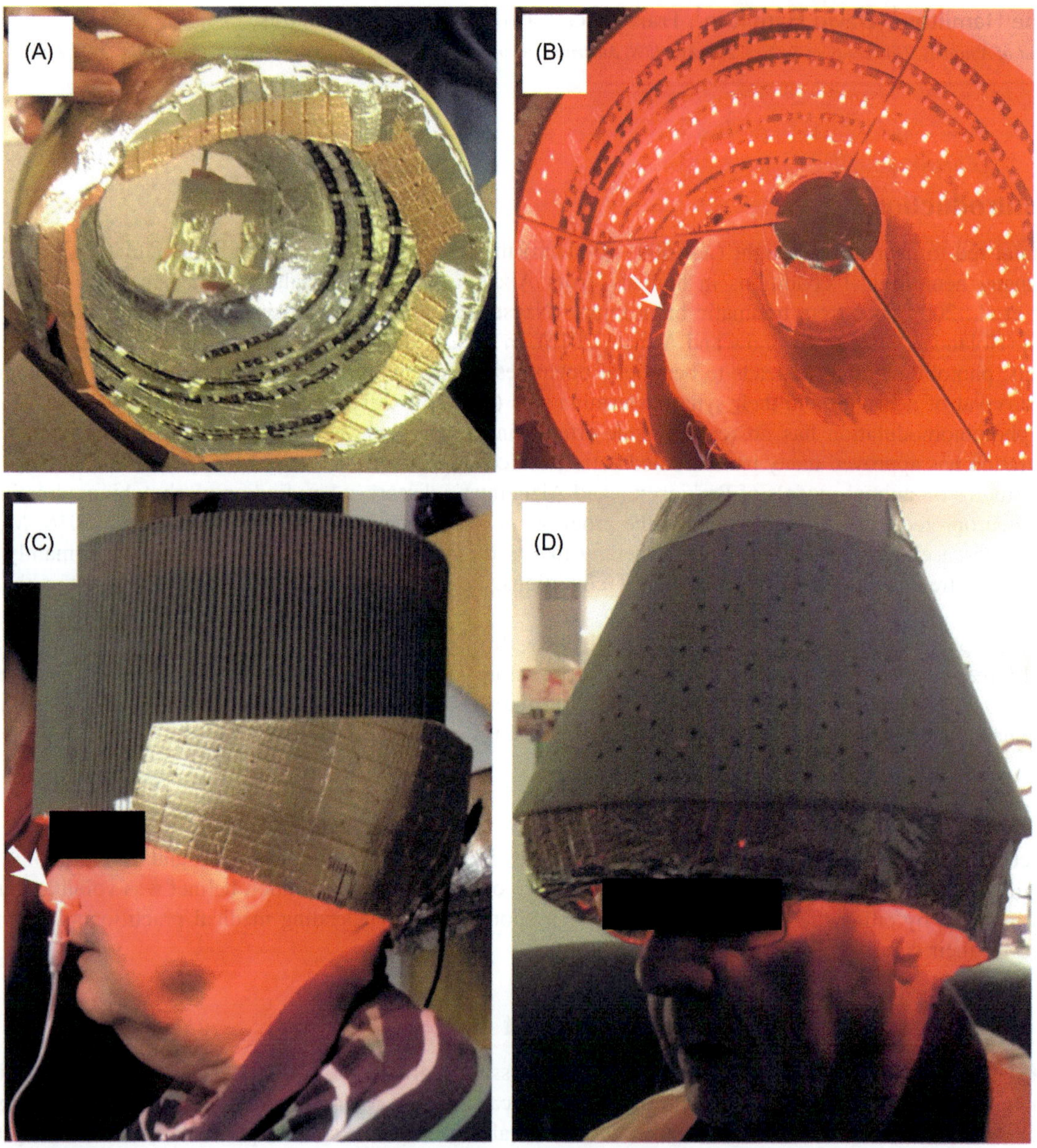

FIGURE 33.1 Photobiomodulation helmets (A) inside of helmet lined with LEDs (670 and 940 nm) not turned-on, (B) inside of helmet turned-on (670 and 940 nm), (C) PSP patient FH using helmet of 670 and 940 nm LEDs; FH used a photobiomodulation intranasal device of 660 nm LED (*arrow*), (D) Parkinson's patient BS using helmet of 670 and 850 nm LEDs.

TABLE 33.1 Patient FH.

Initial signs/symptoms	Improvement after photobiomodulation?
Impaired vertical gaze	–
Impaired speech	✓
Impaired fine motor skills	✓
Impaired balance	✓
Persistent cough	✓
Emotional lability	–
Impaired sleeping	–
Trouble reading	✓

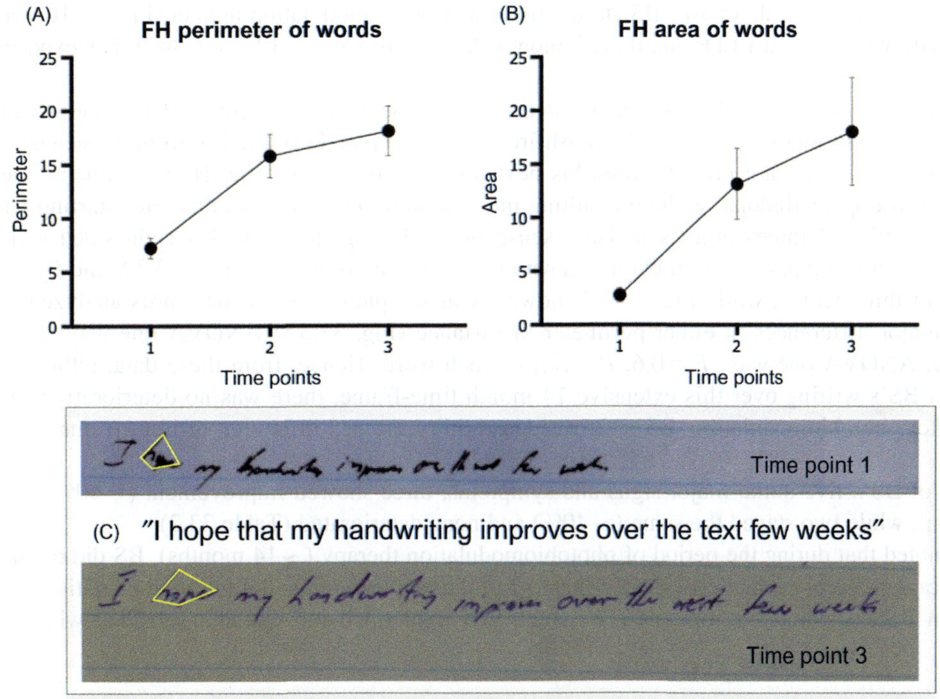

FIGURE 33.2 Analysis of writing from patient FH (PSP): analysis of from before photobiomodulation (time point 1) and during its course, 5 (time point 2) and 6 (time point 3) months after commencement of therapy. Graphs show means and SEMs: (A) shows changes in perimeter of words while (B) shows area of words. (C) shows a real sample of FH's writing at two time points (1 and 3). He wrote the sentence "I hope that my handwriting improves over the next few weeks." Each word was outlined (*yellow lines*; C) and the program calculated the area and perimeter of distance of the words.

as frequent and as explosive. Regarding FH's fine motor skills, we were able to analyze these further, by measuring changes in his writing. Fig. 33.2C shows a sample of a 10-word sentence that he wrote before photobiomodulation began (time point 1) and after approximately 6 months of use (time point 3). The differences between the two samples were striking, with the later one being much more legible, each word being larger and having more clearly separated letters (Fig. 33.2C). We were able to quantify these changes by measuring the perimeter of distance (Fig. 33.2A) and area (Fig. 33.2B) of each word in this sentence at the different stages using ImageJ software. Each word was outlined (*yellow lines*, Fig. 33.2C) and the program calculated the perimeter of distance and area of each word. Our analysis indicated an 80%–85% increase in perimeter of distance and a 60% increase in area of words in the sentence from before light therapy (time point 1) to approximately 5 (time point 2) and 6 months (time point 3) of use. These differences were significant for both perimeter of distance (ANOVA one-way: $F = 9.6$, $P < .0001$) and area (ANOVA one-way: $F = 5$, $P < .01$). His balance appeared to improve also, with him reporting fewer falls since photobiomodulation therapy began. In addition, he started reading novels again, something he enjoyed doing before onset of his debilitating condition. During this period, his cognition was as good as ever. His vertical gaze, emotional lability, and sleeping patterns, on the other hand, showed no improvement.

In summary, of FH's eight initial major signs and symptoms, five showed improvement (~65%) after photobiomodulation treatment, while three stayed the same (~35%). Moreover, none deteriorated (Table 33.1).

33.2.2 Parkinson's disease: Patient BS

Patient BS, a 75 year old male, was diagnosed with Parkinson's disease 5 years ago. His sense of smell was lost early and he developed right-sided signs of the disease, resting tremor, akinesia and rigidity, together with developing a difficulty in fine motor skills. The left side of his body showed no sign of impairment. A keen lawn bowler, he had, in the previous 12 months played with his left (nondominant) hand as he could no longer release the ball with his right hand. Otherwise well, he walked daily, did tai chi weekly, and had been active in starting and leading community groups. His daily medications included levodopa/carbidopa and rasagiline.

Approximately 3 years after diagnosis, BS started using a photobiomodulation helmet (Fig. 33.1C and D), similar to FH's, fitted initially with 670 nm LEDs and then, 3 months later, with 850 nm LEDs as well. His exposure was 30 minutes per day.

After 6 weeks of photobiomodulation, BS noted an improvement in his sense of smell, being able to smell his roast chicken dinner and smoke from nearby bushfires, for example. Further, his tremor became less prominent (Table 33.2), and was usually not present when his next dose of drugs were due. He subsequently ceased the early morning dose of levodopa/carbidopa, with no resulting increase in tremor. Five months after starting photobiomodulation, there were continued improvements in BS's sense of smell, together with his right-sided signs—tremor and akinesia—leading to a resumption of him playing lawn bowls with his right hand. Fig. 33.3A and B shows an analysis of BS's writing of three words, while Fig. 33.3C shows a real sample of one of the words analyzed, "Northbourne." There were no major differences in either perimeter of distance (Fig. 33.3A; ANOVA one-way: $F = 0.3$, $P = .9$) or area [Fig. 33.3B; ANOVA one-way: $F = 0.6$, $P = .7$)] of each word. Hence, from these data, although there were no improvements in BS's writing over this extensive 14 month time-frame, there was no deterioration. In a similar way to his writing after photobiomodulation, BS's right-sided rigidity showed little sign of improvement, albeit no sign of deterioration.

In summary, of BS's five initial major signs and symptoms, three showed improvement (~60%) after photobiomodulation treatment, while two stayed the same (~40%) and none deteriorated (Table 33.2).

It should be noted that during the period of photobiomodulation therapy (~14 months), BS did not develop any parkinsonian signs or symptoms on his left side. Further, all of the abovementioned photobiomodulation-induced improvements evident in BS's signs and symptoms (i.e., smell and right-sided movements) subsided when he was tired or

TABLE 33.2 Patient BS.

Initial signs/symptoms	Improvement after photobiomodulation?
Smell	✓
Tremor	✓
Akinesia	✓
Rigidity	–
Impaired fine motor skills	–

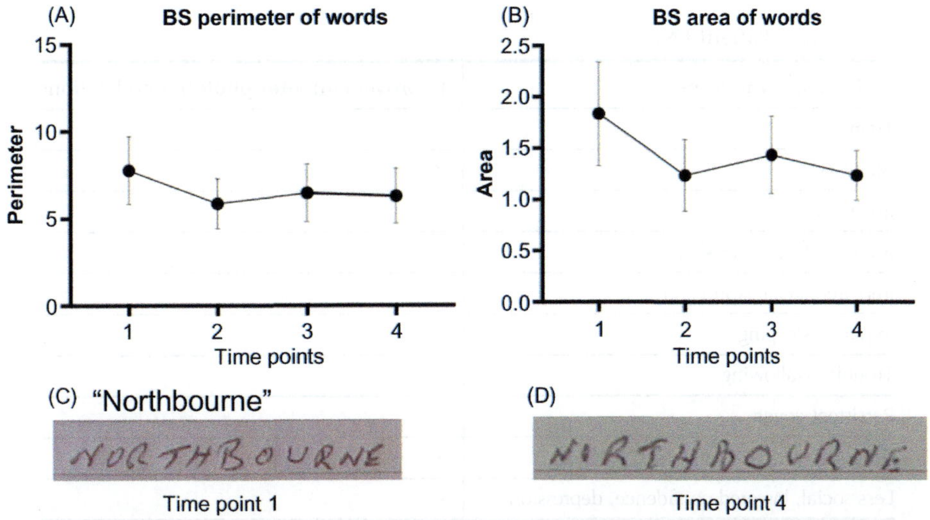

FIGURE 33.3 Analysis of writing from patient BS (Parkinson's disease): analysis of from before photobiomodulation (time point 1) and during its course, 6 (time point 2), 9 (time point 3), and 14 (time point 4) months after commencement of therapy; (C and D) Graphs show means and SEMs: (A) shows changes in perimeter of words, while (B) shows area of words. (C) Shows a real sample of BS's writing at two time points (1 and 4), using the word "Northbourne" as an example. (D) over Northbourne time point 4.

unwell. In fact, during a short, 2 week period where he developed a severe influenza, his signs (i.e., tremor, akinesia) returned in earnest, notwithstanding the continued photobiomodulation therapy. Once the influenza symptoms subsided, his parkinsonian signs and symptoms subsided also.

33.2.3 Parkinson's disease: Patient PN

Patient PN, a 63 year old male, was diagnosed with Parkinson's disease 2 and 1/2 years ago. His major signs and symptoms at the time of diagnosis included the following: episodes of resting tremor (worse when stressed); akinesia; a progressive change in gait; impaired fine motor skills; impaired facial movement with intermittent drooping of the left corner of the mouth; trouble sleeping; trouble swallowing; persistent coughing; fatigue; and finally, less socially interactive, depressed, and low in confidence—he used to be very talkative, but started feeling he was inside a shell. In general, he developed a "slowing down" in day-to-day activities. He once was a keen handyman around the house, but now, found it difficult to not only start a movement, but also to keep the movement going. A few months after diagnosis, PN had a bad fall and hit his head. While in hospital after the fall, he had a series of seizures (determined later to be drug-related). Thereafter, he had his driver's license canceled. His daily medications included levodopa/carbidopa and benztropine.

Nine months after diagnosis, PN started using a photobiomodulation helmet, similar to that described for the others (see Section 33.2.1), lined internally with strips of 670 and 810 nm LEDs, each with separate switches. His exposure was 10 minutes for each wavelength twice a day.

Approximately 4 weeks after first use of photobiomodulation, there was a noticeable reduction in his tremor (Table 33.3). At about this time, PN had his driver's licence restored. After about 7–8 weeks, the following improvements were evident: moving a lot better (less akinetic), walking better and faster (improved gait), sleeping better, more animation in his face, more energy and getting a lot more done (less fatigue), coughing much less, able to swallow easier (e.g., pills), and feels more confident and socially interactive and far less depressed (Table 33.3). Over the next few months, all these improvements stabilized and PN felt as though photobiomodulation had restored many features of his previous day-today activities and self-confidence. In terms of his fine motor skills, as measured by his writing, Fig. 33.4A and B shows an analysis of PN's writing of a seven word sentence using ImageJ software, while Fig. 33.4C shows real samples of the sentence analyzed. Although slight increases were evident over the 10 month time-frame (time point 4), these increases did not reach significance for either perimeter of distance (Fig. 33.4A; ANOVA one-way: $F = 0.1$, $P = .9$) or area (Fig. 33.4B; ANOVA one-way: $F = 0.1$, $P = .9$) of each word. Hence, although no major improvements were evident, there was no clear deterioration over this extensive 10 month time-frame.

In summary, of PN's 10 initial major signs and symptoms, nine showed improvement (90%) after photobiomodulation treatment, while one stayed the same (10%) and none deteriorated (Table 33.3).

TABLE 33.3 Patient PN.

Initial signs/symptoms	Improvement after photobiomodulation?
Tremor	✓
Akinesia	✓
Impaired gait	✓
Impaired fine motor skills	–
Impaired facial expression	✓
Impaired sleeping	✓
Trouble swallowing	✓
Persistent cough	✓
Fatigue	✓
Less social, lowered confidence, depression	✓

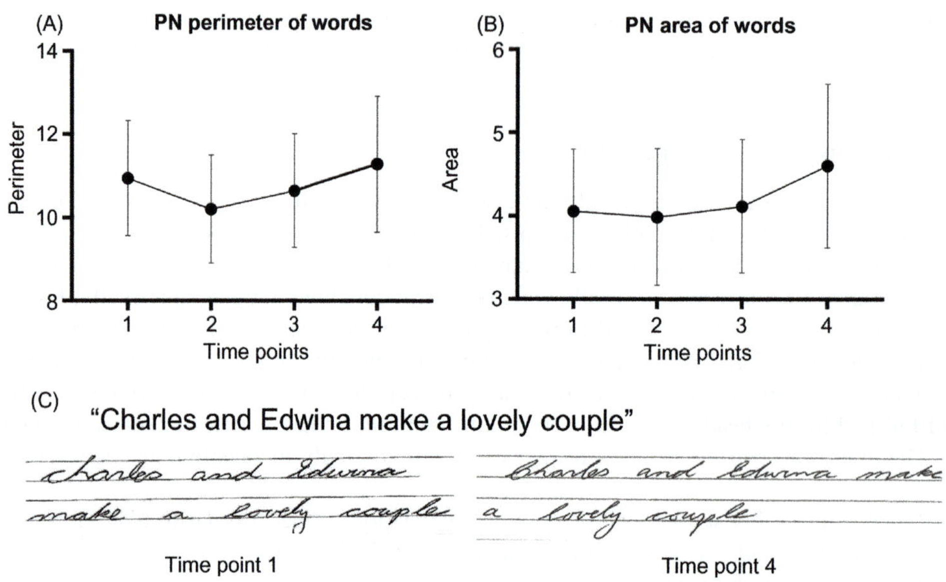

FIGURE 33.4 Analysis of writing from patient PN (Parkinson's disease): analysis of from before photobiomodulation (time point 1) and during its course, 2 (time point 2), 4 (time point 3) and 10 (time point 4) months after commencement of therapy; (C and D) Graphs show means and SEMs: (A) shows changes in perimeter of words, while (B) shows area of words. (C) Shows a real sample of PN's writing at two time points (1 and 4): "Charles and Edwina make a lovely couple."

33.2.4 Parkinson's disease: Patient MH

Patient MH, a 61 year old male, was diagnosed with Parkinson's disease 4 years ago. His major signs and symptoms included the following: difficulty using right hand (dominant), in particular impaired fine motor skills (he once was a keen trout fisherman, but became unable to tie a fly onto a line; further, he developed trouble writing, not being able to write cursive, but to print only and very slowly); less facial animation; changed gait with a reduced stride; fatigue; trouble making decisions and maintaining thoughts, lowered confidence, anxiety and social phobia; hesitant speech; difficulty sleeping and much apathy, developing a "could not be bothered with it all" attitude. His daily medications included levodopa/carbidopa.

TABLE 33.4 Patient MH.

Initial signs/symptoms	Improvement after photobiomodulation?
Impaired fine motor skills	✓
Impaired facial expression	✓
Impaired gait	✓
Fatigue	✓
Poor decision making	✓
Less social, lowered confidence	✓
Hesitant speech	✓
Impaired sleeping	–
Apathy	✓

Some 3 and 1/2 years after the diagnosis, MH started using a photobiomodulation helmet, similar to that described for the others (see above), lined internally with strips of 670 and 810 nm LEDs, each with separate switches. His exposure time was 10 minutes for each wavelength twice a day.

After about 10 days of photobiomodulation, MH reported that his parkinsonian signs and symptoms actually got worse, together with developing sore and watery eyes. He stopped using the helmet for a day or so—up until these signs and symptoms subsided—and then resumed use of the therapy. Within a month, MH reported that he was feeling better, he was more confident, socially interactive, and his thought process much clearer. Much to his delight, he started fishing again and had to be reminded that he had previously said he "couldn't be bothered." At about this time, there were a few initial technical issues with the helmet, namely the 670 nm lights did not work and MH was using the 810 nm lights only. The issue was resolved within a few days and MH resumed using both wavelengths. Over the next few months, all these improvements stabilized. In addition, there were improvements in his speech and gait, together with his face being more animated (Table 33.4). Further, MH started driving a lot more, for example, doing all the driving on an annual holiday of about 5 months duration; before using photobiomodulation, when his parkinsonian signs were becoming more marked, he did little of the driving during this annual road trip. His fine motor skills—as measured by his writing—showed improvement also. Fig. 33.5C shows a sample of a 13 word sentence that he wrote before photobiomodulation began (time point 1) and after approximately 2 months of use (time point 3). Our ImageJ analysis indicated a 30%–40% increase in perimeter of distance and a 55%–65% increase in area of words in the sentence from before light therapy (time point 1) to approximately 2 (time point 3) and 3 months (time point 4) of use. These differences were significant for both perimeter of distance (ANOVA one-way: $F = 3.7$, $P < .01$) and area (ANOVA one-way: $F = 7.8$, $P < .0001$). In contrast to these improvements in MH's sign and symptoms, his disturbed sleeping patterns showed little improvement after photobiomodulation.

In summary, of MH's nine initial major signs and symptoms, eight showed improvement ($\sim$90%) after photobiomodulation treatment, while one stayed the same ($\sim$10%) and none deteriorated (Table 33.4).

33.3 Discussion

Although there have been many previous experimental studies on the impact of photobiomodulation in animal models of Parkinson's disease (Hamblin, 2016; Johnstone et al., 2016; Mitrofanis, 2017), there have been few clinical case reports on this treatment in patients (Zhao et al., 2003; Maloney et al., 2010; Burchman, 2011). Each of these few previous clinical reports have indicated improvements in parkinsonian signs in patients after transcranial photobiomodulation, but they were very brief in scope and offered minimal details of treatment paradigms, together with the changes in signs and symptoms.

Our case report here is the first to describe outcomes in movement disorder patients, one with progressive supranuclear palsy and three with Parkinson's disease, with the use of a photobiomodulation helmet. Our report catalogs changes to a range of their particular signs and symptoms, including an analysis of their fine skilled movements (i.e., analysis of

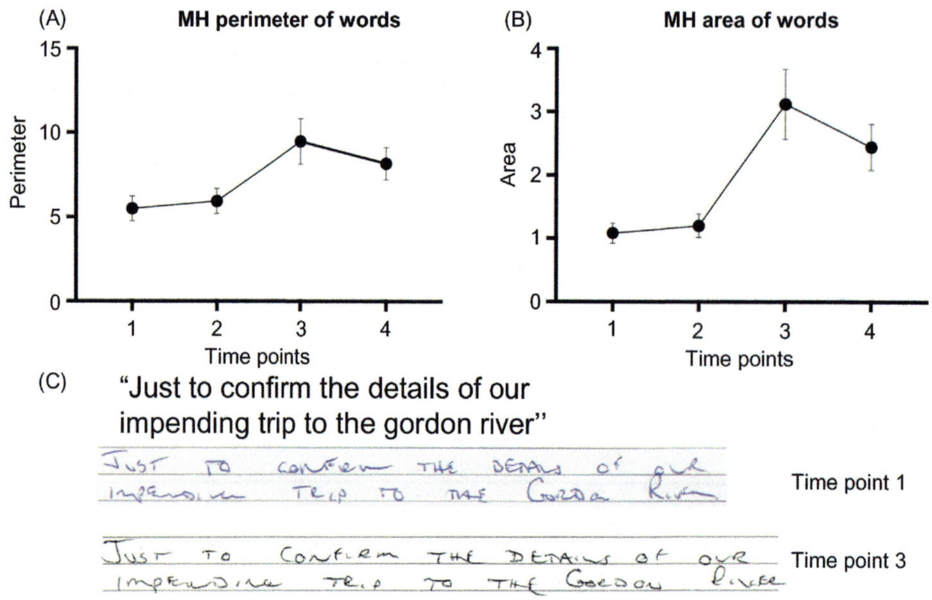

FIGURE 33.5 Analysis of writing from patient MH (Parkinson's disease): analysis of from before photobiomodulation (time point 1) and during its course, 1 (time point 2), 2 (time point 3) and 3 (time point 3) months after commencement of therapy; (C and D) Graphs show means and SEMs: (A) shows changes in perimeter of words, while (B) shows area of words. (C) Shows a real sample of MH's writing at two time points (1 and 3): "Just to confirm the details of our impending trip to the Gordon River."

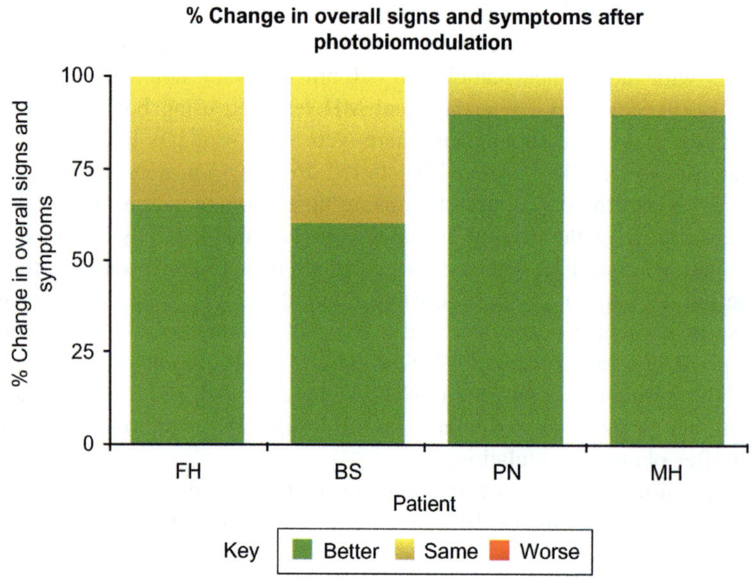

FIGURE 33.6 Graphs showing the percent change in overall signs and symptoms of each patient after photobiomodulation therapy. Of the initial signs and symptoms in each patient, the majority showed overall improvement (*green regions*), a minority stayed the same (*yellow*), and none got worse (*red*). The pattern of therapy outcome were similar in each patient.

writing). For each patient, transcranial photobiomodulation generated subtle but distinct improvements, enough to make a difference in their day-to-day lives. Of the initial signs and symptoms, ~75% showed overall improvement, ~25% stayed the same, and none got worse (see Fig. 33.6).

We are confident that each of these improvements were due to photobiomodulation and not to any other factor, in particular, a change in medication. Indeed, during the period of photobiomodulation, three patients had no increase in their dosage of medication, while one actually had a reduction (BS).

The improvements after photobiomodulation were evident in both the motor signs and the nonmotor symptoms of the four patients. For instance, two patients showed improvements in tremor (BS, PN), akinesia (BS, PN), and gait (PN, MH), while two others had improved social interactions and confidence levels (PN, MH), and one patient had a substantial improvement in their sense of smell (BS), and another with sleeping patterns (PN). Thus, rather than influence a specific sign or symptom, for example either tremor or sleeping, the benefits of photobiomodulation were widespread, impacting many different types of signs and symptoms, somewhat dependent on the individual.

It is of considerable importance that none of the four patients had any sign or symptom that got worse during photobiomodulation (see Fig. 33.6), nor did they develop any long-term and consistent side-effects. This aligns well with the findings of many previous reports indicating no safety concerns regarding this therapy in either humans or experimental animals (Hamblin, 2016; Johnstone et al., 2016; Mitrofanis, 2017). Further, each of the Parkinson's disease patients continued with their medications without complications while using photobiomodulation which suggested that this treatment was compliant with drug therapy. It remains to be determined whether photobiomodulation can be used with equal compliance in patients receiving deep brain stimulation.

It is tempting to speculate that several features of the photobiomodulation therapy had slowed or stopped the progression of the disease in the four patients. First, in terms of their fine motor skills, as measured by their writing, two patients showed no change (BS, PN), while the other two, quite remarkably, showed an improvement (FH, MH). This suggests that, for all four patients, there was no deterioration in these skills over this time period, and in the case of patients PN and BS, this was over extensive periods of 10 and 14 months respectively. Further, for two of these patients (FH, MH), photobiomodulation actually enhanced their fine motor movements. Second, patient BS had parkinsonian signs limited to his right-side, and during the course of his photobiomodulation treatment, his left-side remained free of disease. Notwithstanding, these two positive indications of a slowing of disease progression or neuroprotection by photobiomodulation in the four patients, requires much further investigation in humans with, for example, a set of controls and placebo helmet devices.

The mechanism that underlies the beneficial outcomes of photobiomodulation in the four patients is not known, although previous studies have indicated two possibilities. First, by direct stimulation, where photobiomodulation activates mitochondrial function directly by being absorbed by a photoacceptor, such as cytochrome c oxidase, that then increases ATP (adenosine triphosphate) energy levels and prompts the expression of stimulatory and protective genes within the cell (Hamblin, 2016; Johnstone et al., 2016; Mitrofanis, 2017). In essence, direct stimulation activates intrinsic protective mechanisms within the distressed cells to help them survive (Johnstone et al., 2016; Mitrofanis, 2017). Second, by indirect stimulation, where photobiomodulation triggers recruitment of a middle man, such as cells of the immune system (Liebert et al., 2014; Johnstone et al., 2016; Mitrofanis, 2017). Photobiomodulation may stimulate immune cells that then swarm to the region of distressed cells and help them survive, by increasing the expression of antiinflammatory cytokines while decreasing the pro-inflammatory ones (Johnstone et al., 2016; Mitrofanis, 2017). In the context of our findings here, it is likely that indirect stimulation is the main mechanism involved. Previous studies have reported that transcranial photobiomodulation in humans cannot penetrate to the very deep lying brainstem and basal ganglia (Hamblin, 2016; Johnstone et al., 2016; Mitrofanis, 2017), the main areas of lesions in both Parkinson's disease (Blandini et al., 2000; Bergman and Deuschl, 2002) and progressive supranuclear palsy (Boxer et al., 2017). However, transcranial photobiomodulation can reach the blood vessels in the skin over the cranium and within the meninges to activate immune cells; indeed, photobiomodulation has been shown to influence immune cell function in experimental animals (Muili et al., 2012). It is perhaps relevant that patient BS, despite his continued use of the photobiomodulation helmet, had all his parkinsonian signs and symptoms return after contracting a severe dose of influenza. After his influenza symptoms subsided, so did his parkinsonian signs and symptoms. This would suggest that the influenza virus had overwhelmed his immune system, so much so that photobiomodulation treatment was rendered ineffective. The other key point from this episode, and following on from the issue raised above regarding neuroprotection, is that light treatment is not "apply once and all is fixed" therapy; it requires consistent application in order to be effective. Overall, it is clear that further investigation is required as to the mechanism(s) involved in transcranial photobiomodulation, in particular to identify the particular type of immune cell (and other circulatory cell or molecule) that is activated and contributes to the beneficial outcomes of the treatment.

Finally, it is worth noting that in two of the patients (PN, MH), there were clear improvements in their overall depressive state, social interaction, level of anxiety, and self-confidence. Similar improvements have been noted in a previous study examining the impact of photobiomodulation on anxiety and depression in patients. In that study, patients were markedly less depressed and anxious for up to 4 weeks after a single treatment to each side of the head (Schiffer et al., 2009). In relation to our report here, it is possible that photobiomodulation was, not only beneficial to the neurodegenerative process of the patients, but also tbeneficial in improving the functional circuitry of their frontal lobe, resulting in an overall improvement in their psychological state (Schiffer et al., 2009).

33.4 Conclusion

Although the photobiomodulation-induced improvements in all the four movement disorder patients are encouraging (Fig. 33.6), these early observations need to be developed further by a double-blind clinical trial on a large cohort of patients. Such a trial could include strategies to determine whether the photobiomodulation treatment is symptomatic or indeed neuroprotective, as has been reported in many animal models of Parkinson's disease (Hamblin, 2016; Johnstone et al., 2016; Mitrofanis, 2017).

Acknowledgment

We thank Light Ahead Inc. for provision of the light devices, Ron Brown for assistance in the development of the light devices, and Julie Douglas for assistance with speech pathology records.

References

Bergman, H., Deuschl, G., 2002. Pathophysiology of Parkinson's disease: from clinical neurology to basic neuroscience and back. Mov. Disord. 17 (Suppl. 3), S28–S40.

Blandini, F., Nappi, G., Tassorelli, C., Martignoni, E., 2000. Functional changes of the basal ganglia circuitry in Parkinson's disease. Prog. Neurobiol. 62, 63–88.

Boxer, A.L., Yu, J.-T., Golbe, L.I., Litvan, I., Lang, A.E., Höglinger, G.U., 2017. Advances in progressive supranuclear palsy: new diagnostic criteria, biomarkers, and therapeutic approaches. Lancet Neurol. 16, 552–563. Available from: https://doi.org/10.1016/S1474-4422(17)30157-6.

Burchman, M., 2011. Using photobiomodulation on a severe Parkinson's patient to enable extractions, root canal treatment, and partial denture fabrication. J. Laser Dent. 19, 297–300.

Chow, R.T., Armati, P.J., 2016. Photobiomodulation: implications for anesthesia and pain relief. Photomed. Laser Surg. Available from: https://doi.org/10.1089/pho.2015.4048.

Hamblin, M.R., 2016. Shining light on the head: photobiomodulation for brain disorders. BBA Clin. 6, 113–124. Available from: https://doi.org/10.1016/j.bbacli.2016.09.002.

Holanda, V.M., Chavantes, M.C., Silva, D.F.T., de Holanda, C.V.M., de Oliveira, J.O., Wu, X., et al., 2016. Photobiomodulation of the dorsal root ganglion for the treatment of low back pain: a pilot study. Lasers Surg. Med. 48, 653–659. Available from: https://doi.org/10.1002/lsm.22522.

Johnstone, D.M., Moro, C., Stone, J., Benabid, A.-L., Mitrofanis, J., 2016. Turning on lights to stop neurodegeneration: the potential of near infrared light therapy in Alzheimer's and Parkinson's disease. Front. Neurosci. 9, 500. Available from: https://doi.org/10.3389/fnins.2015.00500.

Lapchak, P.A., Salgado, K.F., Chao, C.H., Zivin, J.A., 2007. Transcranial near-infrared light therapy improves motor function following embolic strokes in rabbits: an extended therapeutic window study using continuous and pulse frequency delivery modes. Neuroscience 148, 907–914. Available from: https://doi.org/10.1016/j.neuroscience.2007.07.002.

Liebert, A., Bicknell, B., Adams, R., 2014. Protein conformational modulation by photons: a mechanism for laser treatment effects. Med. Hypotheses YMEHY 7445. Available from: https://doi.org/10.1016/j.mehy.2013.12.009.

Maloney, R., Shanks, S., Maloney, J., 2010. The application of low-level laser therapy for the symptomatic care of late stage Parkinson's disease: a non-controlled, non-randomized study. In: Am Soc Laser Med Surg Abs 30th ASLMS Conference, Phoenix, p. 185.

Mitrofanis, J., 2017. Why and how does light therapy offer neuroprotection in Parkinson's disease? Neural Regen. Res. 12, 574–575.

Muili, K.A., Gopalakrishnan, S., Meyer, S.L., Eells, J.T., Lyons, J.-A., 2012. Amelioration of experimental autoimmune encephalomyelitis in C57BL/6 mice by photobiomodulation induced by 670 nm light. PLoS One 7, e30655. Available from: https://doi.org/10.1371/journal.pone.0030655.

Naeser, M.A., Saltmarche, A., Krengel, M.H., Hamblin, M.R., Knight, J.A., 2011. Improved cognitive function after transcranial, light-emitting diode treatments in chronic, traumatic brain injury: two case reports. Photomed. Laser Surg. 29, 351–358. Available from: https://doi.org/10.1089/pho.2010.2814.

Saltmarche, A.E., Naeser, M.A., Ho, K.F., Hamblin, M.R., Lim, L., 2017. Significant improvement in cognition in mild to moderately severe dementia cases treated with transcranial plus intranasal photobiomodulation: case series report. Photomed. Laser Surg. Available from: https://doi.org/10.1089/pho.2016.4227.

Schiffer, F., Johnston, A.L., Ravichandran, C., Polcari, A., Teicher, M.H., Webb, R.H., et al., 2009. Psychological benefits 2 and 4 weeks after a single treatment with near infrared light to the forehead: a pilot study of 10 patients with major depression and anxiety. Behav. Brain Funct. 5, 46. Available from: https://doi.org/10.1186/1744-9081-5-46.

Zhao, G., Guo, K., Dan, J., 2003. 36 case analysis of Parkinson's disease treated by endonasal low energy He-Ne laser. Acta Acad. Med. Qingdao Univ. (Chin.) 39, 398.

Chapter 34

Cerebral blood flow in the elderly: impact of photobiomodulation

Afonso Shiguemi Inoue Salgado[1,5], Francisco José Cidral-Filho[2,3], Daniel Fernandes Martins[2,3], Ivo I. Kerppers[4] and Rodolfo Borges Parreira[1,5]

[1]*Salgado Institute of Integrative Health, Londrina, Brazil,* [2]*Experimental Neuroscience Laboratory (LaNEx), University of Southern Santa Catarina, Palhocça, Santa Catarina, Brazil,* [3]*Postgraduate Program in Health Sciences, University of Southern Santa Catarina, Santa Catarina, Brazil,* [4]*Laboratory of Neuroanatomy and Neurophysiology, University of Centro-Oeste, Guarapuava, Brazil,* [5]*Residency Program in Integrative Physical Therapy at UNIFIL University, Londrina, Brazil*

34.1 Introduction

The world's population is aging at a faster pace than it has ever before. This demographic tendency is directly linked to lower fertility rates and a steep reduction in mortality, especially in the western world.

The aging process is progressive and dynamic; and involves biochemical, morphological, and functional alterations that progressively modify the organism, making it more susceptible to the deleterious effects of intrinsic and extrinsic factors, and insults that will eventually culminate in death (Federal Interagency Forum on Aging-Related Statistics: Older Americans – Key Indicators of Well-Being, 2016). After the sixth decade of life, one of the main alterations that occurs, is the degradation of the neurological system and the development of neurovascular diseases and cerebral vascular dysfunction due to decreased cerebral blood flow (CBF) and abnormal brain metabolism; which consequently elevate the risk of dementia related diseases (Morrisson and Hof, 1997; Gsell et al., 2000).

Continuous and sufficient CBF is vital to maintain good neuronal function, as is cerebral perfusion, assessed by blood flow rate per unit volume of tissue. CBF is an important indicator of brain health and, its disturbance may suggest the compromise of vascular function and/or its metabolism. Therefore, it is not surprising that the reduction or interruption of CBF is associated with many diseases such as hypertension, ischemic cerebrovascular accidents, and Alzheimer's disease (Gsell et al., 2000).

Understanding how CBF changes in cognitively healthy, elderly individuals, can be an important way to differentiate between what is normal from what is abnormal in neurophysiology.

34.2 Brain hemodynamics in the elderly

It is a well-known fact that CBF in the elderly individual is affected; nevertheless, whether the alterations derive from the normal aging process, or if they are related to cerebral tissue atrophy in the elderly, for example, is not yet fully understood (Chen et al., 2011). Nevertheless, it is known that blood flow velocity tends to decrease in the arteries with increasing age. A population study with elderly individuals free from any cerebrovascular accidents or dementia, reported a decrease in CBF velocity, mainly in the posterior cerebral and basilar arteries. An increase in arterial stiffness, that is, the medial and anterior cerebral arteries, was also detected (Yang et al., 2010). This association between CBF and the aging process could be linked to decreased cardiac output, diminished metabolic demands, abnormal hemodynamic values and alterations in blood vessel size, for example (Fu et al., 2006; Kusunoki et al., 1999).

Cerebrovascular homeostasis is maintained by vascular resistance which adapts itself to variations in blood flow thus preserving an adequate and stable CBF (Paulson et al., 1990).

Photobiomodulation in the Brain. DOI: https://doi.org/10.1016/B978-0-12-815305-5.00034-8

Blood vessel dynamics, known as vasomotricity, is established through contraction and relaxation of the endothelial cells and smooth muscle cells, and is modulated by rhythmic, spontaneous variations in the vessel lumen triggered by oscillations in endothelial cell membrane potentials.

It has been suggested that the vascular endothelium could initiate the synchronic vasomotricity induced by inositol 1,4,5-trisphosphate-mediated discharge of Ca^{2+} from its intracellular storage depots, which will activate chloride-dependent channels and depolarize vascular smooth muscle cells. This depolarization activates calcium channels, which synchronize adjacent cells through the inward flow of extracellular calcium. The subsequent calcium release activates potassium channels hyperpolarizing the cells and providing a negative feedback for when a new contraction is ready to be initiated. This rhythmic contraction and relaxation pattern helps maintain CBF (Haddock and Hill, 2002).

In vitro studies have shown that nitric oxide (NO) plays an important role in vasomotor regulation. NO is a gaseous molecule produced by NO synthase isoforms, neuronal and endothelial (eNOS). It is released by endothelial cells and nitrergic perivascular neurons, and is actively involved in vasomotor regulation by decreasing vascular resistance, promoting vasodilation, and increasing local blood flow, all of which contribute to adequate CBF. Endothelial NO functions as an antiplatelet, antiproliferative, antisclerotic, and antithrombotic agent regulating CBF. Consequently, endothelial dysfunction may negatively affect brain circulation and eventually lead to serious cerebral problems (Toda et al., 2009). On the other side, NO in excessive amounts may lead to certain brain pathologies such as stroke, multiple sclerosis and Alzheimer's disease. Excessive NO (iNOS induced NO) production may result from pathological conditions in which iNOS is upregulated, as in inflammatory processes such as hyperglycemia. Endothelial dysfunction may also increase oxidative stress which, in turn, will induce iNOS (Toda et al., 2009).

Evidence suggests that a neural component is very important to vasomotricity, that is, it has been demonstrated that astrocytes are implicated in regulation of vascular tonus (Filosa and Iddings, 2013), as they release potent vasoactive agents such as NO, glutamate, ATP, prostaglandins, and epoxyeicosatrienoic acids (Filosa and Iddings, 2013; Filosa et al., 2004; Carmignoto and Gomez-Gonzalo, 2010). Anatomically, astrocytes are situated in apposition to arterioles and capillaries, which facilitates the transmission of vasoactive signals/agents (Filosa and Iddings, 2013). Therefore, it has been suggested that the release of Ca^{2+} by perivascular astrocytes is one of the main drivers of vascular tonus (Iadecola, 2004).

Vascular alterations may lead to cerebral lesions, especially in the white matter, and as has been previously demonstrated (Tsao et al., 2013; Schmahmann et al., 2008), aging may be an important factor. A population study with elderly people has shown that hardening of the arteries may lead to white matter damage and cognitive decline (Tsao et al., 2013; Schmahmann et al., 2008).

The white matter blood supply is predominantly derived from penetrating branches of the subarachnoid artery, which runs through the perpendicular cortical layers of the brain surface and penetrates the white matter along these fibers (van den Bergh and van der Eecken, 1968).

Yang and collaborators (2016) have shown, by measuring resistivity (RI) and pulsatility (PI) indexes, that increased arterial stiffness may lead to large alterations in CBF. RI and PI are hemodynamic parameters commonly used to measure the blood vessel resistance that reflects the vessel's local extensibility (Staub et al., 2006; Roher et al., 2011a) and is associated with the speed of CBF (Franceschi et al., 1995; Vicenzini et al., 2007). RI is correlated with aging and cardiovascular risks (Frauchiger et al., 2001; Staub et al., 2006), whereas PI reflects distal cerebrovascular resistance (Lim et al., 2009). A transcranial Doppler ultrasound study has demonstrated that arterial stiffening increases cerebral vessel pulsatility (PI), and leads to small vessel damage through increased mechanical fatigue of the blood vessel smooth muscle cells (Hayashi et al., 2003).

Vicenzini and collaborators (2007) reported that vasomotor amplitude (assessing blood flow velocities during hypercapnia and hypocapnia) of the middle cerebral artery was reduced in patients with Alzheimer's disease and vascular dementia when compared to asymptomatic subjects. It is worth mentioning that even without a significant hemodynamic change of the blood vessels, diffusion and perfusion were abnormal in the white matter (Sam et al., 2016).

Decreased blood flow velocity and increased pulsatility are multifactorial and normally occur in the elderly population. These changes reflect structural and hemodynamic changes of blood vessels and have several repercussions, for example, increased arterial stiffness, decreased compliance, and microvascular congestion, which may be associated with dementia-related diseases, such as Alzheimer's (Roher et al., 2011b).

CBF changes generally occur due to dysregulation of cerebral vascular tone. Thus, altered cerebral hemodynamics may be aggravated by a cholinergic deficit resulting in depressed cerebrovascular and vasomotor responses (Marco et al., 2015), which will disrupt brain perfusion and may have important implications for associated vascular diseases, with an increased risk for Alzheimer's disease, stroke, and cognitive decline, for example. It is worth remembering that CBF disorders are less severe in elderly individuals who do not suffer from dementia (ten Dam et al., 2007).

34.3 Effect of photobiomodulation of the brain in the elderly

Some studies have demonstrated the positive effects of photobiomodulation of the brain in humans (Schiffer et al., 2009; Nawashiro et al., 2012). But in the elderly, research on cerebral circulation is still very scarce. Salgado and collaborators (2015) observed using Doppler ultrasonography that transcranial low level light therapy (LLLT) with an LED device (light-emitting diode—627 nm, 70 mW/cm^2, 10 J/cm^2, for a total of 2 minutes) applied to the frontal and parietal encephalic regions in elderly individuals, twice a week for 4 weeks, increased blood flow velocity in the middle and basilar cerebral arteries. Likewise, there was a reduction in PI and RI in the three arteries that were analyzed.

Other studies, conducted with populations other than the elderly, have also shown that PBM improves CBF. For example, Nawashiro and collaborators (2012) demonstrated that LLLT with LEDs (850 nm, 11.4 mW/cm^2, 20.5 J/cm^2, 30 minutes, twice a day) applied to the frontal region of the head, increased local CBF by 20% in comatose patients. Similarly, Schiffer and collaborators (2009) with transcranial LEDs (810 nm, 250 mW/cm^2; 60 J/cm^2; 4 minutes) applied to the frontal region of the head of subjects with depression and anxiety, increased local CBF. These studies suggest that, at least in part, LED irradiation can penetrate the skin and skull and reach the cerebral cortex and arteries in human subjects, producing measurable effects, such as increased CBF.

The fact that LED irradiation increases CBF is of great importance because chronic hypoperfusion of the brain impairs the supply of oxygen and nutrients to the brain and the brain-blood barrier; which leads to buildup of harmful substances and compromises venous return. Additional studies have shown that decreased CBF is associated with neuronal and synaptic impairment; the elderly population has higher cerebral vascular resistance, which indirectly leads to edema formation, dilatation of periarterial spaces, retention of interstitial fluids, and deposition of β-amyloid peptides in the vascular wall; all of which can be observed in many cases of Alzheimer's disease (Kalback et al., 2004; Henry-Feugeas, 2007; Okamoto et al., 2012).

It is worth remembering that Salgado and collaborators (2015) reported that LED irradiation reduced PI and RI in cerebral arteries, which suggests that this intervention could function, at least in part, as a protective factor against cardiovascular disease, as increased PI and RI is clinically associated to cardiovascular dysfunction (Staub et al., 2006; Giller et al., 1990; Lim et al., 2009). In this context, LED irradiation in the red and near infrared spectrum promotes vasodilation by increasing NO concentration through dissociation of cytochrome c oxidase-nitric oxide complexes in the mitochondrial cell membrane, which in turn elevates oxygen consumption and increases ATP production (Hamblin, 2008).

PBM of the vascular endothelium, promotes upregulation of phosphorylation of eNOS, increasing its activity (Lee et al., 2017), and therefore contributes to functional improvement in blood vessels. In rats with an ischemic lesion of the middle cerebral artery, LED irradiation performed prior to the ischemic injury induced better functional and neurological results compared to LED irradiation performed postischemia (Lee et al., 2017). This observation suggests that preemptive PBM exerted a protective factor against ischemic injury via eNOS. NO, synthesized by eNOS is an important regulator involved in vascular homeostasis (He et al., 2013). During PBM, a transient increase in NO metabolites was also seen as a by-product of increased NO bioavailability (Mitchell and Mack, 2013). NO has an important role in maintaining basal CBF, by acting directly upon the vascular endothelium and improving blood vessel function, that is, its extensibility and resistance, as well as, in neurotransmission, and signal transduction (Tanaka, 1996; Moncada et al., 1991). Given the current data, we suggest that PBM with transcranial LED irradiation could be used as a prophylactic therapy in the elderly population, who are commonly affected by vascular and cognitive alterations derived from the aging process.

References

Carmignoto, G., Gomez-Gonzalo, M., 2010. The contribution of astrocyte signaling to neurovascular coupling. Brain Res. Rev. 63, 138–148.

Chen, J.J., Rosas, H.D., Salat, D.H., 2011. Age-associated reductions in cerebral blood flow are independent from regional atrophy. Neuroimage 55, 468–478.

Federal Interagency Forum on Aging-Related Statistics: Older Americans – Key Indicators of well-being, 2016.

Filosa, J.A., Iddings, J.A., 2013. Astrocyte regulation of cerebral vascular tone. Am. J. Physiol. 305, H609–H619.

Filosa, J.A., Bonev, A.D., Nelson, M.T., 2004. Calcium dynamics in cortical astrocytes and arterioles during neurovascular coupling. Circ. Res. 95, e73–e81.

Franceschi, M., Alberoni, M., Bressi, S., et al., 1995. Correlations between cognitive impairment, middle cerebral artery flow velocity and cortical glucose metabolism in the early phase of Alzheimer's disease. Dementia 6, 32–38.

Frauchiger, B., Schmid, H.P., Roedel, C., et al., 2001. Comparison of carotid arterial resistive indices with intima-media thickness as sonographic markers of atherosclerosis. Stroke 32, 836–841.

Fu, C.H., Yang, C.C., Kuo, T.B., 2006. Age-related changes in cerebral hemodynamics and their correlations with cardiac autonomic functions. Neurol. Res. 28, 871–876.

Giller, C.A., Hodges, K., Batjer, H.H., 1990. Transcranial Doppler pulsatility in vasodilation and stenosis. J. Neurosurg. 72, 901–906.

Gsell, W., De Sadeleer, C., Marchalant, Y., et al., 2000. The use of cerebral blood flow as an index of neuronal activity in functional neuroimaging: experimental and pathophysiological considerations. J. Chem. Neuroanat. 20, 215–224.

Haddock, R.E., Hill, C.E., 2002. Differential activation of ion channels by inositol 1,4,5-trisphosphate (IP3) – and ryanodine-sensitive calcium stores in rat basilar artery vasomotion. J. Physiol. 545, 615–627.

Hamblin, M.R., 2008. The role of nitric oxide in low-level light therapy, Biomedical Optics (BiOS), 6846. International Society for Optics and Photonics, 684602-1.

Hayashi, K., Mori, K., Miyazaki, H., 2003. Biomechanical response of femoral vein to chronic elevation of blood pressure in rabbits. Am. J. Physiol. 284, H511–H518.

He, X., Zhao, M., Bi, X.Y., Yu, X.J., Zang, W.J., 2013. Delayed preconditioning prevents ischemia/reperfusion-induced endothelial injury in rats: role of ROS and eNOS. Lab. Invest. 93, 168–180.

Henry-Feugeas, M.C., 2007. MRI of the 'Alzheimer syndrome'. J. Neuroradiol. 34, 220–227.

Iadecola, C., 2004. Neurovascular regulation in the normal brain and in Alzheimer's disease. Nat. Rev. Neurosci. 5, 347–360.

Kalback, W., Esh, C., Castano, E.M., et al., 2004. Atherosclerosis, vascular amyloidosis and brain hypoperfusion in the pathogenesis of sporadic Alzheimer's disease. Neurol. Res. 26, 525–539.

Kusunoki, K., Oka, Y., Saito, M., et al., 1999. Changes in the visibility of intracranial arteries on MRA with normal aging. Neuroradiology 41, 813–819.

Lee, H.I., Lee, S.W., Kim, S.Y., Kim, N.G., et al., 2017. Pretreatment with light-emitting diode therapy reduces ischemic brain injury in mice through endothelial oxide synthase-dependent mechanisms. Biochem. Biophys. Res. Commun. 486, 945–950.

Lim, M.H., Cho, Y.I., Jeong, S.K., 2009. Homocysteine and pulsatility index of cerebral arteries. Stroke 40, 3216–3220.

Marco, L.Y., Farkas, E., Martin, C., Venneri, A., et al., 2015. Is vasomotion in cerebral arteries impaired in Alzheimer's disease? J. Alzheimer Dis. 46, 35–53.

Mitchell, U.H., Mack, G.L., 2013. Low-level laser treatment with near-infrared light increases venous nitric oxide levels acutely: a single-blind, randomized Trial of efficacy. Am. J. Phys. Med. Rehabil. 92, 151–156.

Moncada, S., Palmer, R., Higgs, E., 1991. Nitric oxide: physiology, pathophysiology, and pharmacology. Pharmacol. Rev. 43, 109Y42.

Morrisson, J.H., Hof, P.R., 1997. Life and death of neurons in the aging brain. Science 278, 412–419.

Nawashiro, H., Wada, K., Nakai, K., Sato, S., 2012. Focal increase in cerebral blood flow after treatment with near-infrared light to the forehead in a patient in a persistent vegetative state. Photomed. Laser Surg. 30, 231–233.

Okamoto, Y., Yamamoto, T., Kalaria, R.N., et al., 2012. Cerebral hypoperfusion accelerates cerebral amyloid angiopathy and promotes cortical microinfarcts. Acta Neuropathol. 123, 381–394.

Paulson, O.B., Strandgaard, S., Edvinsson, L., 1990. Cerebral autoregulation. Cerebrovasc. Brain Metab. Rev. 2, 161–192.

Roher, A.E., Garami, Z., Tyas, S.L., et al., 2011a. Transcranial Doppler ultrasound blood flow velocity and pulsatility index as systemic indicators for Alzheimer's disease. Alzheimers Dementia 7, 445–455.

Roher, A.E., Garami, Z., Tyas, S.L., Maarouf, C.L., et al., 2011b. Transcranial Doppler ultrasound blood flow velocity and pulsatility índex as systemic indicators for Alzheimer's disease. Alzheimers Dementia 7, 445–455.

Salgado, A.S., Zângaro, R.A., Parreira, R.B., Kerppers, I.I., 2015. The effects of transcranial LED therapy (TCLT) on cerebral blood flow in the elderly women. Lasers Med. Sci. 30, 339–346.

Sam, K., Peltenburg, B., Conklin, J., Sobczyk, O., et al., 2016. Cerebrovascular reactivity and white matter integrity. Neurology 87, 2333–2339.

Schiffer, F., Johnston, A.L., Ravichandran, C., et al., 2009. Psychological benefits 2 and 4 weeks after a single treatment with near-infrared light to the forehead: a pilot study of 10 patients with major depression and anxiety. Behav. Brain Funct. 5, 46.

Schmahmann, J.D., Smith, E.E., Eichler, F.S., Filley, C.M., 2008. Cerebral white matter: neuroanatomy, clinical neurology, and neurobehavioral correlates. Ann. N.Y. Acad. Sci. 1142, 266–309.

Staub, D., Meyerhans, A., Bundi, B., et al., 2006. Prediction of cardiovascular morbidity and mortality: comparison of the internal carotid artery resistive index with the common carotid artery intima-media thickness. Stroke 37, 800–805.

Tanaka, K., 1996. Is nitric oxide really important for regulation of the cerebral circulation? Yes or No? Keio J. Med. 45, 14–27.

ten Dam, V.H., van den Heuvel, D.M., de Craen, A.J., et al., 2007. Decline in total cerebral blood flow is linked with increase in periventricular but not deep white matter hyperintensities. Radiology 243, 198–203.

Toda, N., Ayajiki, K., Okamura, T., 2009. Cerebral blood flow regulation by nitric oxide: recent advances. Pharmacol. Rev. 61, 62–97.

Tsao, C.W., Seshadri, S., Beiser, A.S., et al., 2013. Relations of arterial stiffness and endothelial function to brain aging in the community. Neurology 81, 984–991.

van den Bergh, R., van der Eecken, H., 1968. Anatomy and embryology of cerebral circulation. Prog. Brain Res. 30, 1–26.

Vicenzini, E., Ricciardi, M.C., Altier, M., et al., 2007. Cerebrovascular reactivity in degenerative and vascular dementia: a transcranial Doppler study. Eur. Neurol. 58, 84–89.

Yang, Y., Lovett-Racke, A.E., Racke, M.K., 2010. Regulation of Immune responses and autoimmune encephalomyelitis by PPARs. PPAR Res. 104705.

Yang, D., Cabral, D., Gaspard, E.M., Lipton, R.B., Rundek, T., Derby, C.A., 2016. Cerebral hemodynamics in the elderly: a transcranial Doppler study in the Einstein Aging Study Group. J. Ultrasound Med. 35, 1907–1914.

Further reading

Elias, M.F., D'agostino, R.B., Elias, P.K., et al., 1995. Neuropsychological test performance, cognitive functioning, blood pressure, and age: the Framingham Heart Study. Exp. Aging Res. 21, 369–391.

Farmer, M.E., Kittner, S.J., Abbott, R.D., et al., 1990. Longitudinally measured blood pressure, antihypertensive medication use, and cognitive performance: the Framingham Study. J. Clin. Epidemiol. 43, 475–480.

Chapter 35

Transcranial photobiomodulation for major depressive and anxiety disorders and for posttraumatic stress disorder

Marco Antonio Caldieraro[1,2] and Paolo Cassano[3,4]

[1]*Universidade Federal do Rio Grande do Sul, Department of Psychiatry and Forensic Medicine, Porto Alegre, RS, Brazil,* [2]*Hospital de Clínicas de Porto Alegre, Department of Psychiatry, Porto Alegre, RS, Brazil,* [3]*Depression Clinical and Research Program, Department of Psychiatry, Massachusetts General Hospital, Boston, MA, United States,* [4]*Center for Anxiety and Traumatic Stress Disorders, Department of Psychiatry, Massachusetts General Hospital, Boston, MA, United States*

35.1 The potential of transcranial photobiomodulation for the anxious and depressed

Depression and anxiety are the most common mental health problems in the United States. Anxiety disorders affect 18.1% of adults in the general population in a given year (Kessler et al., 2005b), and 28.8% of adults at some point during their life-time (Kessler et al., 2005a). Major depressive disorder (MDD) is the single disorder with the highest lifetime prevalence among American adults (16.6%) (Kessler et al., 2005a), with a 12-month prevalence of 6.7% (Kessler et al., 2005b). Moreover, the comorbidity between MDD and an anxiety disorder is common, and results in worse outcomes for both disorders (Moscati et al., 2016). These disorders are also leading causes of disability. Depressive disorders are responsible for 40.5%, and anxiety disorders account for 14.6% of the global burden of years lost to mental and substance use disorders due to disability (Whiteford et al., 2013).

Current standard treatments for MDD are medications and psychotherapy (Olfson et al., 2016). However, up to one-third of patients do not achieve remission after several adequate antidepressant trials (Rush et al., 2009). Moreover, relapses are frequent after an initial remission (Sinyor et al., 2010), and pharmacological treatments present burdensome side effects (Cassano and Fava, 2004; Kennedy et al., 2016). Evidence-based psychotherapy requires frequent sessions and specialized professionals, therefore access is limited (Parikh et al., 2016).

For MDD patients who do not respond, tolerate, or accept medications or psychotherapy, device-based treatments can be offered. Those include electroconvulsive therapy (ECT), repetitive transcranial magnetic stimulation (rTMS), transcranial direct current stimulation, vagus nerve stimulation (VNS), magnetic seizure therapy, and deep brain stimulation. Consistent evidence supports the use of ECT, rTMS, and VNS for the treatment of MDD, while the other options are still experimental treatments (Milev et al., 2016). Despite evidence of efficacy, ECT, rTMS, and VNS have shortcomings. ECT is a complex procedure and requires anesthesia. Both ECT and rTMS are relatively expensive and require patients to frequently attend the treatment facility: up to two or three times a week for ECT (Bauer et al., 2013) and every weekday for rTMS (Perera et al., 2016). VNS is invasive since it requires the implant of a vagus nerve stimulator (Aaronson et al., 2013).

Anxiety disorders are a group of different conditions including generalized anxiety disorder (GAD), social anxiety disorder, panic disorder, agoraphobia, and specific phobias. Posttraumatic stress disorder (PTSD) is also characterized by significant anxiety symptoms but is classified by DSM-5 in a different category (trauma- and stressor-related disorders, (American Psychiatric Association, 2013)). Similar to MDD, standard treatments for anxiety disorders and PTSD are medication and psychotherapy, and most first line treatment medication for these disorders are antidepressants (Bandelow et al., 2008). Treatment resistance is a problem for a significant number of anxiety or PTSD patients receiving standard treatments (Bandelow et al., 2008), but differently from MDD, there are no device-based treatment approved for these disorders.

Photobiomodulation in the Brain. DOI: https://doi.org/10.1016/B978-0-12-815305-5.00035-X

The significant proportion of MDD, anxiety disorders and PTSD patients who do not respond to or tolerate standard treatments for their disorder indicate the need for new treatments. Although research on transcranial photobiomodulation (t-PBM) for mood and anxiety disorders is still preliminary—especially in regards of its efficacy—this modality is a promising new treatment. So far, t-PBM has not been associated with sexual side-effects, weight gain, or cognitive disturbances, which are instead common in the long-term use of antidepressant medications. Since self-administration at home of t-PBM is considered safe, its cost and time-demands are modest when compared to ECT, rTMS, VNS (and even to evidence-base psychotherapy). Therefore, if effective and well-tolerated, t-PBM could become a widely accessible intervention for the acute treatment, continuation, and maintenance of mood and anxiety disorders, scalable to the needs of the United States. The following sections present preliminary evidence on the efficacy and safety of t-PBM for MDD, anxiety disorders and PTSD.

35.2 Transcranial photobiomodulation for major depressive disorder

Different pathways relevant for MDD are affected by t-PBM. Modern pathophysiological models for MDD associate this disorder with brain hypometabolism (Mayberg et al., 2000; Videbech, 2000; Drevets et al., 2002; Kennedy et al., 2007) and mitochondrial dysfunction (Iosifescu et al., 2008; Hroudová et al., 2013; Morava and Kozicz, 2013; Karabatsiakis et al., 2014; Bansal and Kuhad, 2016), with increased oxidative stress (Ozcan et al., 2004; Eren et al., 2007; Sarandol et al., 2007; Shungu et al., 2012; Spanemberg et al., 2014), with inflammatory processes (Dowlati et al., 2010; Anisman and Hayley, 2012; Liu et al., 2012; Köhler et al., 2017) and with decreased neuroplasticity and neurogenesis (Autry and Monteggia, 2012; Duman, 2014).

Photobiomodulation (PBM) acts on all these pathways. Near-infrared (NIR) and red light deliver energy to the cytochrome C oxidase, and stimulate the mitochondrial respiratory chain leading to increased ATP production (Yu et al., 1997; Mochizuki-Oda et al., 2002; Oron et al., 2007b). This initial action triggers a cascade of other cellular mechanisms. NIR can induce short bursts of reactive oxygen species leading to the activation of antioxidant mechanisms resulting in reduction of the oxidative stress (de Freitas and Hamblin, 2016). Antiinflammatory effects were also demonstrated. NIR (810 nm) used as a treatment for pain in patients with rheumatoid arthritis, decreased production of TNF-α, IL-1β, and IL-8 (Yamaura et al., 2009). Light with the same wavelength (810 nm) decreased cellular infiltration in a rat model of spinal cord injury (Anders, 2009). Moreover, transcranial NIR (800 nm) reduced neuroinflammation in mice models of traumatic brain injury (TBI) (Khuman et al., 2012). PBM also stimulates neurogenesis and protects against cell death. Animal research has shown that NIR improves neurogenesis and synaptogenesis, via increase of brain-derived neurotrophic factor (Oron et al., 2007a; Ando et al., 2011; Wu et al., 2012a; Xuan et al., 2015). Other neurotrophic mechanisms have also been proposed for PBM such as the inhibition of GSK-3β and of pro-apoptotic molecules (de Freitas and Hamblin, 2016).

Animal studies indicate that t-PBM with NIR or red light produces an antidepressant effect on behavioral tests (Ando et al., 2011; Tanaka et al., 2011; Mohammed, 2016; Xu et al., 2016), which is comparable to the effect of fluoxetine (Wu et al., 2012b) and of citalopram (Salehpour et al., 2016). Transcranial stimulation with NIR was associated with increased ATP biosynthesis and with increased mitochondrial complex IV expression and activity in the prefrontal cortex (PFC) (Xu et al., 2016). Also, increased hippocampal neurogenesis (Tanaka et al., 2011) and a neuroprotective effect (Ando et al., 2011) were reported in animals treated with NIR t-PBM.

Consistent with the preclinical evidence, studies in humans reinforce the potential of t-PBM as a novel treatment for MDD. However, clinical investigations are still preliminary in nature. Most studies include small samples and have significant methodological limitations. Moreover, optimal stimulation parameters are still to be established.

An open trial treated 10 patients with treatment resistant depression with a single session of NIR t-PBM on the forehead, using an LED instrument (Marubeni America Corp.) (Schiffer et al., 2009). The NIR was delivered on EEG sites F3 and F4, which cover the dorsolateral prefrontal cortex (DLPFC) bilaterally. The stimulation parameters were: wavelength 810 nm, irradiance 250 mW/cm^2, fluence 60 J/cm^2, time 4 minutes per site. At weeks 2 and 4 posttreatment, a significant decrease was observed in depressive symptoms. The remission rate of MDD at week 2 [operationalized as a Hamilton Depression Rating Scale—21 items (HAM-D21) score <10] was 60%. Four weeks posttreatment, depressive symptoms were lower than pretreatment, but significantly higher than after two weeks, suggesting a limited duration of the effect of a single treatment.

Our group at MGH performed a pilot study of six sessions (two sessions-a-week for 3 weeks) of t-PBM with NIR (808 nm) in four participants with moderate to severe MDD (nontreatment resistant) (Cassano et al., 2015). At each treatment session, NIR was administered to the forehead bilaterally at four sites (2 minutes per site, 12.56 cm^2 window each), using a class IV laser (Photothera, Inc.) and the following treatment parameters: NIR irradiance

700 mW/cm^2 and fluence 84 J/cm^2, for a total NIR energy of 2.40 kJ per session. The depression severity assessed by the HAM-D—17-item (HAM-D17), decreased from 19.8 ± 4.35 at baseline, to 13.0 ± 5.35 at the endpoint (5 weeks after the end of the treatment). This difference was large enough to achieve statistical significance even in this small sample.

Attention bias modification (ABM) is a cognitive intervention designed to improve symptoms by decreasing negative attentional bias, but to this date its efficacy in depression could not be demonstrated (Mogoaşe et al., 2014). A randomized clinical trial (RCT) assessed if t-PBM could enhance the effects of ABM in individuals with elevated depressive symptoms (Disner et al., 2016). The participants ($n = 51$) were randomized to receive two sessions of right forehead, left forehead, or sham t-PBM. The interval between the treatments was 48 hours and all participants received one session of ABM before and another after each session of t-PBM. The light treatment used NIR (1064 nm) delivered by a laser device (Cell Gen Therapeutics) at two sites (13.6 cm^2 window each) for 4 minutes per site. The treatment parameters were: power 3.4 W, irradiance 250 mW/cm^2, and a fluence of 60 J/cm^2 per site, with a total energy of 1.63 kJ per session. In subjects who responded to the ABM, the improvement was enhanced by the right t-PBM, while no significant effect was observed for the left and sham t-PBM.

Our group at MGH also performed a RCT designed to assess the efficacy of t-PBM as a primary treatment for MDD (Cassano et al., 2018). Participants ($n = 21$) were randomized to receive bilateral stimulation on the prefrontal cortex (EEG sites F3 and F4) twice-a-week for 8 weeks, or sham treatment in the same sites and with the same frequency. Treatment was delivered by an LED device (Photomedex Inc.) emitting NIR (823 nm) (Fig. 35.1). The duration of the initial session was 20 minutes, and subsequent sessions could be extended up to 30 minutes based on clinical judgment. Treatment parameters were: power 1 W, CW, irradiance 33.2 mW/cm^2, and a fluence of 40–60 J/cm^2 per site, with a total energy of 3.4 kJ per session and 45.6 kJ for the entire treatment. The mean change in HAM-D17 total score in subjects receiving t-PBM in NIR-mode was significantly greater than in subjects receiving sham-mode with the baseline-observation carried forward approach [BOCF: NIR ($n = 10$) -10.8 ± 7.55 vs sham ($n = 11$) -4.4 ± 6.65; $P = .047$)] and with the completers approach [NIR ($n = 6$) -15.7 ± 4.41 vs sham ($n = 7$) -6.1 ± 7.86; $P = .031$)]. However, the threshold for significance was not reached with the last observation carried forward approach [NIR ($n = 10$) -10.8 ± 7.55 vs sham ($n = 9$) -5.33 ± 7.04; $P = .119$)]. Fig. 35.2 illustrates the mean HAM-D17 total scores over the course of the study for the two t-PBM groups (BOCF and completers). Remission (HAM-D17 ≤ 7) was observed in 50% of subjects who received the active treatment and in 18% of those in the sham group. These findings showed a trend toward significance ($P = .12$), however, due to limited power, did not detect group differences. The high remission rates are promising preliminary data suggesting the need for more well-powered studies.

Little is known about the long-term antidepressant effects of PBM. Our group reported the case of a patient receiving PBM for 31 months, as an add-on to antidepressant medication to treat depression with anxious distress (Caldieraro et al., 2018). The treatment was started with intranasal PBM, and t-PBM was added in the last 9 months. The treatment was well-tolerated, and a continuous improvement in the anxious symptoms was observed during the overall treatment follow-up, with a more pronounced improvement in depression observed after the addition of the t-PBM.

All clinical studies on t-PBM for MDD used NIR and aimed to stimulate the forebrain, which is consistent with dysfunction in the DLPFC associated with the disorder. Therefore, current evidence supports an antidepressant effect only for the stimulation at these sites. However, penetration of light is higher when light is delivered to other parts of the scalp, such as the occipital skull (Jagdeo et al., 2012). New studies must answer if stimulating other areas is associated with an antidepressant effect, and if this effect is similar of even greater than what is observed by stimulating the DLPFC. Current studies also suggest that multiple sessions may be necessary for a sustained antidepressant effect.

35.3 Transcranial photobiomodulation for anxiety disorders and for posttraumatic stress disorder

The effects of t-PBM on anxiety and fear were also studied in clinical and preclinical studies. Using a rat model to study anxiety and posttraumatic syndromes, Rojas et al. reported that t-PBM with red light (660 nm) was associated with improved fear extinction and lessened fear recall (Rojas et al., 2012).

In the study from Schiffer et al. (2009), described above, 9 of the 10 participants with MDD also had a comorbid anxiety disorder. The authors studied the effect of a single session of t-PBM on the anxiety symptoms, assessed by the Hamilton Anxiety Scale (HAM-A). Similar to what was observed with the depressive symptoms, the lowest symptom scores occurred 2 weeks posttreatment when mean HAM-A decreased 14.9 ± 9.6 points from baseline ($P = 0.002$).

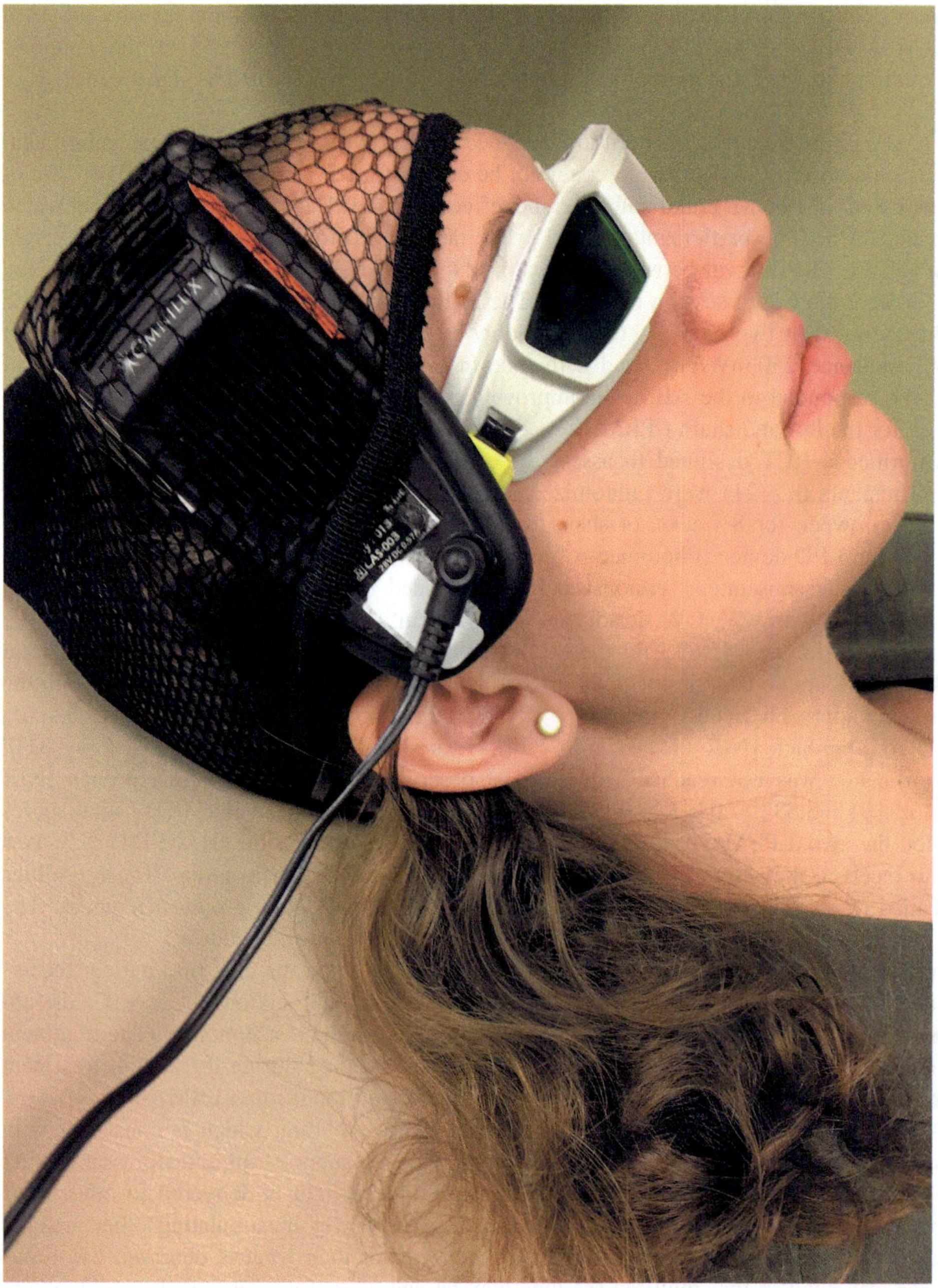

FIGURE 35.1 ELATED-2 study: the picture shows the handheld portions of the Omnilux New U devices, which are pressed against the forehead of the subject, bilaterally and simultaneously on the F3 and F4 sites (reference to EEG placement sites).

At week 4, mean HAM-A was still significantly lower than baseline scores ($P = 0.008$)—with a decrease of 9.0 ± 7.5 points from baseline—but significantly higher than at week 2 ($P = 0.004$) (Schiffer et al., 2009). This study indicates that, like for depressive symptoms, the improvement on anxiety after a single session of t-PBM is transient.

In an open study, 11 patients with mild chronic TBI were treated with 18 sessions of t-PBM. Four out of the 11 participants of the study met the criteria for PTSD (Naeser et al., 2014). During each session (three times a week for 6 weeks), the light was delivered to 11 sites on the scalp (22.5 cm^2 each) placed throughout the midline and bilaterally on frontal, parietal, and temporal areas. The treatment used a combination of NIR (870 nm) and red light (633 nm) emitted by a LED device (MedX Health Model 1100). The total duration of each treatment session was 20 minutes, with light being delivered to six sites for 10 minutes at one time and then to the other five sites for 10 minutes. The treatment parameters were: power 0.5 W, irradiance 22.2 mW/cm^2, and a fluence of 13.2 J/cm^2, with a total energy of 3.26 kJ per session. The four participants with comorbid PTSD presented a remarkable reduction on the PCL-C (PTSD checklist−civilian) scores.

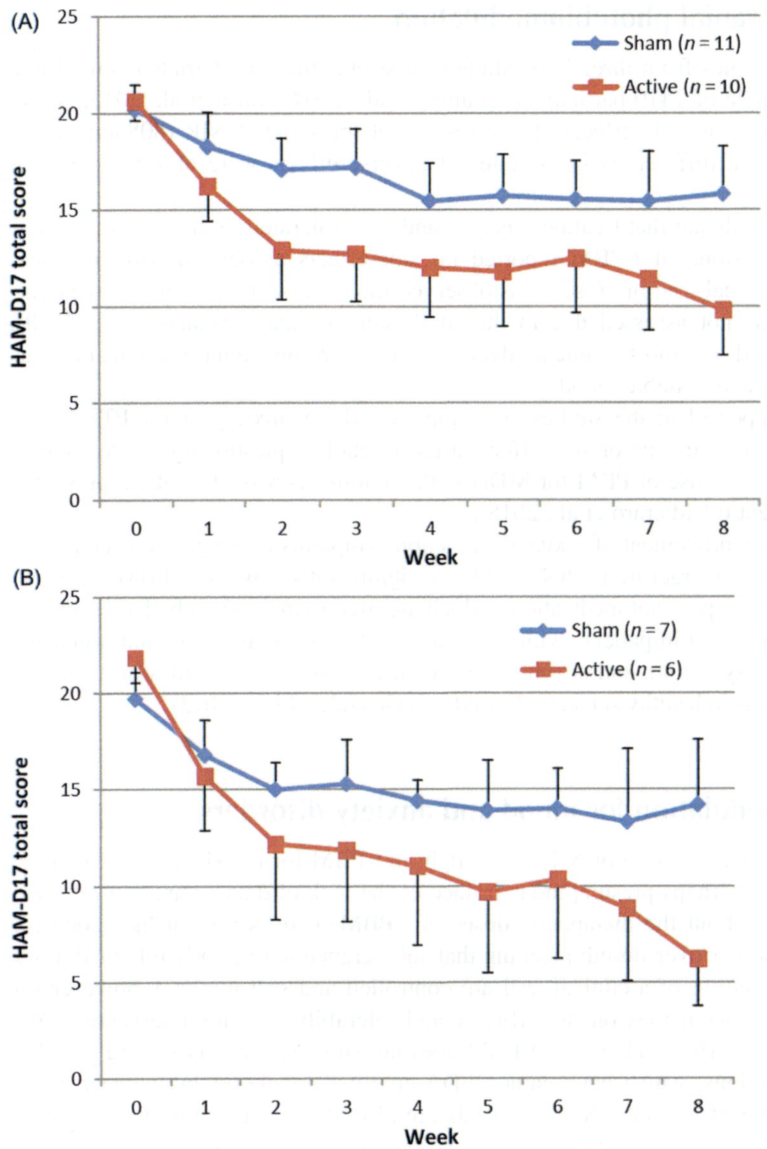

FIGURE 35.2 ELATED-2 study: graph of mean HAM-D17 total score in subjects treated with near-infrared light (NIR—*red line*) and in subjects treated with sham (*blue line*); (A) all study subjects included ($n = 21$); (B) only completers included ($n = 13$).
(A) All subjects (intent to treat—endpoint carried forward) (mean ± SE). Error bars on one side of each line; (B) completers only (imputed values for three missing data points from interim visits) (mean ± SE). Error bars on one side of each line.

A group providing t-PBM for patients with different neuropsychiatric conditions reported over 50 patients treated for PTSD. According to the authors, virtually all patients presented a remarkable improvement in their emotional stability and quality of life, as assessed by standardized instruments (Stephan et al., personal communication). As described above, in a case report of a patient receiving PBM for 31 months for MDD with anxious distress, a continuous improvement on the anxiety symptoms was observed during the overall treatment period (Caldieraro et al., 2018).

These studies on t-PBM for anxiety and PTSD have limitations due to their small sample sizes. Moreover, only a few studies enrolled patients with a primary diagnosis of an anxiety disorder; conversely, in most studies anxiety was not the primary outcome. These limitations temper our enthusiasm, despite the promising results reported so far. In terms of magnitude of the anxiolytic effect, these initial reports suggest that the therapeutic effects of t-PBM on anxiety and on posttraumatic stress could be even greater than on depression. This could be particularly relevant for patients suffering from PTSD, since in many cases available pharmacological treatments produce only modest improvements. Our group is currently conducting a two-site clinical trial to test the efficacy of t-PBM for the treatment of GAD, in collaboration with New York University. This study is a step forward toward adequately powered, controlled trials testing t-PBM in anxiety disorders.

35.4 Safety and tolerability of transcranial photobiomodulation

The strongest evidence on the safety of t-PBM comes from three large studies on stroke, the NEST trials (NeuroThera effectiveness and safety trial), with a pooled sample of 1410 participants (Lampl et al., 2007; Huisa et al., 2013; Hacke et al., 2014). These trials assessed the therapeutic and side effects of one session of transcranial NIR (808 nm) in the first 24 hours after a stroke. Overall, no significant differences in the rates of adverse effects were observed between groups: active versus sham (simulated) treatment.

The clinical studies on t-PBM for MDD also indicate that treatment is safe and well-tolerated by depressed patients. Two uncontrolled studies using one and six sessions of t-PBM reported no side-effects associated with treatment (Schiffer et al., 2009; Cassano et al., 2015). A clinical trial of 16 sessions observed more side-effects in the active treatment group, however statistical significance was not assessed due to the small sample size (Cassano et al., 2018). However, no serious adverse events were observed; the most frequent adverse effects were insomnia, illusions (such as seeing vivid colors or "tasting from an ashtray"), and irritable mood.

Similarly, no serious adverse events were reported in the studies assessing t-PBM for anxiety or for PTSD. The study reported by Schiffer et al., reported no adverse events or side effects after a detailed questioning of the patients (Schiffer et al., 2009). In the case report of long-term use of PBM for MDD with anxious distress, the patient presented headaches apparently associated with the treatment (Caldieraro et al., 2018).

Interestingly, the ELATED-2 trial reported an improvement of sexual dysfunction symptoms in the group receiving the active treatment (Cassano et al., 2019). If confirmed in larger trials, this would be a significant strength of t-PBM since sexual dysfunction is a common side effect of most antidepressant medications, which are used to treat not only depression but also anxiety. Moreover, cognition is frequently impaired in patients with MDD and with anxiety disorders; unfortunately, current treatments have little or no effects on this symptom, and at times they induce worsening of it. In turn, treatment with t-PBM was shown to exert pro-cognitive effects in healthy subjects (Barrett and Gonzalez-Lima, 2013).

35.5 Dosing transcranial photobiomodulation for mood and anxiety disorders

There is no consensus on the effective and well-tolerated dose of NIR or red light in t-PBM for mood and anxiety disorders. Writing a section on the dosing of t-PBM is perhaps presumptuous in face of the lack of data comparing different doses of t-PBM. Yet, because so little is known about the therapeutic doses of t-PBM there is a great likelihood for ongoing and future clinical trials to be under-dosed or over-dosed: meaning that subtherapeutic or poorly tolerated doses might be used. The risks for the patients in the setting of a clinical trial are controlled and still minimal, however for the field of t-PBM there is a risk of premature conclusions on the efficacy and tolerability of this intervention. We should instead be prepared for the likelihood of negative trials where t-PBM does not show superiority toward placebo. Both pharmacological and device-based interventions, which have gained FDA-approval for mood and anxiety disorders, have typically done so in spite of several negative trials. Negative trials are also at times resulting from exceedingly high placebo response, which is likely strong in patients who are invested in complementary and alternative therapies. A trial with high placebo response and negative results should be considered a failed study, where the procedures to contain placebo response of depressed and anxious participants to its expected levels (27%–38% reduction of symptoms on placebo) failed (Khan et al., 2017). Not surprisingly, given the challenges of dose finding and of placebo-response in psychiatry, less than 50% of placebo-controlled trials on antidepressant medications have demonstrated superiority of their studied intervention (Khan et al., 2000, 2002). In light of the challenges ahead for the field of t-PBM, we hereby offer some clinically informed and yet unproven observations on the dosing of t-PBM, which might inform treatment decisions in future studies:

- In the acute phase of treatment for mood and anxiety disorders, t-PBM sessions with FDA-cleared devices are typically started with once a week or twice a week frequency, and gradually increased to three or four times a week and possibly daily based on clinical response and as tolerated.
- In the continuation and maintenance phase of treatment, when t-PBM sessions are decreased to once every 2 weeks, mood and anxiety symptoms tend to reoccur. In some cases, even lowering to once a week sessions might precipitate a syndrome relapse.
- Defining the optimal dose is difficult because (1) penetration of red and NIR t-PBM hinges on multiple somatic characteristics; (2) individual differences are expected and; (3) antidepressants effects typically require weeks to manifest. A potential strategy to identify the therapeutic dose is to titrate the dose of each session or the number of weekly sessions up to the induction of mild side effects, such as transient headaches or mild irritability.

- Elderlies might need higher doses with daily sessions and even at times twice daily sessions, possibly due to larger subarachnoid space, due to lower hydration of tissues and to brain atrophy (decreased neuronal density).
- Children might respond to lower doses both in terms of energy delivered per session and in terms of total number of sessions. They might need as little as half the number of sessions compared to adults. In preteens (young children) a quarter of the adult dose per session is a reasonable starting dose. Children have less intervening tissues between skin and brain (in accordance to their stage of physical development) and higher neuronal density (Tsujimoto, 2008).
- While comparing different t-PBM devices, it is useful to refer to the total energy delivered per session and per week at skin level. Some devices with low irradiance (e.g., 15 mW/cm^2) are surprisingly still effective, if used on large surfaces of skin, for fairly long sessions (e.g., 40 minutes) and frequently during the week. This suggests that cumulative effects of t-PBM might be at play in determining a therapeutic response.
- Clinician should resist the temptation to conceptualize t-PBM as a single treatment modality. It is very much possible that based on chosen parameters the primary mechanism of action for neuromodulation changes. It is possible that at a very low dose t-PBM acts by inducing electromagnetic fields near to the brain rather then delivering photon energy to the brain.
- It is unclear if the chosen sites of the source of t-PBM on the scalp and on the forehead matter in the depressed and anxious patients. However, it is expected that depending on the localization on the scalp, different cortical areas will be preferentially irradiated. The F3 and F4 sites will preferentially serve to irradiate the DLPFC; Fp1-Fpz-Fp2 instead will allow preferential irradiation of the ventromedial prefrontal cortex.
- t-PBM is not deprived of risks of manic or hypomanic switches, similar to most antidepressant therapies. Even intranasal PBM (NIR) with trivial penetration to the brain has been associated with a hypomanic syndrome characterized by excessive productive activities and hyperness of mood.

35.6 Conclusion

The treatment with t-PBM has the potential to become an effective, safe, and widely available treatment for MDD, anxiety disorders, and PTSD. It has a novel mechanism of action, it is easy to administer, and inexpensive. However, clinical research on t-PBM is still preliminary and current evidence is not enough to consider it a standard treatment for these disorders. More studies are necessary to adequately test the efficacy of this treatment and to define the optimal stimulation parameters. Some of these studies are already being performed and will contribute to clarify the role of t-PBM in clinical practice.

References

Aaronson, S.T., Carpenter, L.L., Conway, C.R., Reimherr, F.W., Lisanby, S.H., Schwartz, T.L., et al., 2013. Vagus nerve stimulation therapy randomized to different amounts of electrical charge for treatment-resistant depression: acute and chronic effects. Brain Stimul. 6 (4), 631–640.

American Psychiatric Association, 2013. Diagnostic and Statistical Manual of Mental Disorders (DSM-5®). American Psychiatric Pub.

Anders, J.J., 2009. The potential of light therapy for central nervous system injury and disease. Photomed. Laser Surg. 27 (3), 379–380.

Ando, T., Xuan, W., Xu, T., Dai, T., Sharma, S.K., Kharkwal, G.B., et al., 2011. Comparison of therapeutic effects between pulsed and continuous wave 810-nm wavelength laser irradiation for traumatic brain injury in mice. PLoS One 6 (10), e26212.

Anisman, H., Hayley, S., 2012. Inflammatory Factors Contribute to Depression and Its Comorbid Conditions. Science Signaling 5 (244), 45.

Autry, A.E., Monteggia, L.M., 2012. Brain-derived neurotrophic factor and neuropsychiatric disorders. Pharmacol. Rev. 64 (2), 238–258.

Bandelow, B., Zohar, J., Möller, H.-J., WFSBP Task Force on Treatment Guidelines for Anxiety, Obsessive-Compulsive and Post-Traumatic Stress Disoders, Hollander, E., Kasper, S., et al., 2008. World Federation of Societies of Biological Psychiatry (WFSBP) guidelines for the pharmacological treatment of anxiety, obsessive-compulsive and post-traumatic stress disorders – first revision. World J. Biol. Psychiatry 248–312.

Bansal, Y., Kuhad, A., 2016. Mitochondrial dysfunction in depression. Curr. Neuropharmacol. 14 (6), 610–618.

Barrett, D.W., Gonzalez-Lima, F., 2013. Transcranial infrared laser stimulation produces beneficial cognitive and emotional effects in humans. Neuroscience 230, 13–23.

Bauer, M., Pfennig, A., Severus, E., Whybrow, P.C., Angst, J., Möller, H.-J., et al., 2013. World Federation of Societies of Biological Psychiatry (WFSBP) guidelines for biological treatment of unipolar depressive disorders, part 1: update 2013 on the acute and continuation treatment of unipolar depressive disorders. World J. Biol. Psychiatry 334–385.

Caldieraro, M.A., Sani, G., Bui, E., Cassano, P., 2018. Long-term near-infrared photobiomodulation for anxious depression complicated by Takotsubo Cardiomyopathy. J. Clin. Psychopharmacol. 38 (3), 268–270.

Cassano, P., Fava, M., 2004. Tolerability issues during long-term treatment with antidepressants. Ann. Clin. Psychiatry 16 (1), 15–25.

Cassano, P., Cusin, C., Mischoulon, D., Hamblin, M.R., De Taboada, L., Pisoni, A., et al., 2015. Near-infrared transcranial radiation for major depressive disorder: proof of concept study. Psychiatry J. 2015, 352979.

Cassano, P., Petrie, S.R., David, M., Cusin, C., Katnani, H., Yeung, A., et al., 2018. Transcranial photobiomodulation for the treatment of major depressive disorder. The ELATED-2 Pilot Trial. Photomed. Laser Surg. (accepted for publication).

Cassano, P., Dording, C., Thomas, G., Foster, S., Yeung, A., Uchida, M., et al., 2019. Effects of transcranial photobiomodulation with near-infrared light on sexual dysfunction. Lasers Surg. Med. 51, 127–135.

de Freitas, L.F., Hamblin, M.R., 2016. Proposed mechanisms of photobiomodulation or low-level light therapy. IEEE J. Sel. Top. Quantum Electron. 22 (3), 348–364.

Disner, S.G., Beevers, C.G., Gonzalez-Lima, F., 2016. Transcranial laser stimulation as neuroenhancement for attention bias modification in adults with elevated depression symptoms. Brain Stimul. 9 (5), 780–787.

Dowlati, Y., Herrmann, N., Swardfager, W., Liu, H., Sham, L., Reim, E.K., et al., 2010. A meta-analysis of cytokines in major depression. Biological Psychiatry. 67 (5), 446–457.

Drevets, W.C., Bogers, W., Raichle, M.E., 2002. Functional anatomical correlates of antidepressant drug treatment assessed using PET measures of regional glucose metabolism. Eur. Neuropsychopharmacol. 12 (6), 527–544.

Duman, R.S., 2014. Pathophysiology of depression and innovative treatments: remodeling glutamatergic synaptic connections. Dialogues Clin. Neurosci. 16 (1), 11–27.

Eren, I., Naziroğlu, M., Demirdaş, A., 2007. Protective effects of lamotrigine, aripiprazole and escitalopram on depression-induced oxidative stress in rat brain. Neurochem. Res. 32 (7), 1188–1195.

Hacke, W., Schellinger, P.D., Albers, G.W., Bornstein, N.M., Dahlof, B.L., Fulton, R., et al., 2014. Transcranial laser therapy in acute stroke treatment: results of neurothera effectiveness and safety trial 3, a phase III clinical end point device trial. Stroke 45 (11), 3187–3193.

Hroudová, J., Fišar, Z., Kitzlerová, E., Zvěřová, M., Raboch, J., 2013. Mitochondrial respiration in blood platelets of depressive patients. Mitochondrion 13 (6), 795–800.

Huisa, B.N., Stemer, A.B., Walker, M.G., Rapp, K., Meyer, B.C., Zivin, J.A., et al., 2013. Transcranial laser therapy for acute ischemic stroke: a pooled analysis of NEST-1 and NEST-2. Int. J. Stroke 8 (5), 315–320.

Iosifescu, D.V., Bolo, N.R., Nierenberg, A.A., Jensen, J.E., Fava, M., Renshaw, P.F., 2008. Brain bioenergetics and response to triiodothyronine augmentation in major depressive disorder. Biol. Psychiatry 63 (12), 1127–1134.

Jagdeo, J.R., Adams, L.E., Brody, N.I., Siegel, D.M., 2012. Transcranial red and near infrared light transmission in a cadaveric model. PLoS One 7 (10), e47460.

Karabatsiakis, A., Böck, C., Salinas-Manrique, J., Kolassa, S., Calzia, E., Dietrich, D.E., et al., 2014. Mitochondrial respiration in peripheral blood mononuclear cells correlates with depressive subsymptoms and severity of major depression. Transl. Psychiatry 4 (6), e397.

Kennedy, S.H., Konarski, J.Z., Segal, Z.V., Lau, M.A., Bieling, P.J., McIntyre, R.S., et al., 2007. Differences in brain glucose metabolism between responders to CBT and venlafaxine in a 16-week randomized controlled trial. AJP 164 (5), 778–788.

Kennedy, S.H., Lam, R.W., McIntyre, R.S., Tourjman, S.V., Bhat, V., Blier, P., et al., 2016. Canadian Network for Mood and Anxiety Treatments (CANMAT) 2016 clinical guidelines for the management of adults with major depressive disorder: Section 3. Pharmacological treatments. Can. J. Psychiatry 61 (9), 540–560.

Kessler, R.C., Berglund, P., Demler, O., Jin, R., Merikangas, K.R., Walters, E.E., 2005a. Lifetime prevalence and age-of-onset distributions of DSM-IV disorders in the National Comorbidity Survey Replication. Arch. Gen. Psychiatry 62 (6), 593–602.

Kessler, R.C., Chiu, W.T., Demler, O., Merikangas, K.R., Walters, E.E., 2005b. Prevalence, severity, and comorbidity of 12-month DSM-IV disorders in the National Comorbidity Survey Replication. Arch. Gen. Psychiatry 62 (6), 617–627.

Khan, A., Warner, H.A., Brown, W.A., 2000. Symptom reduction and suicide risk in patients treated with placebo in antidepressant clinical trials: an analysis of the Food and Drug Administration database. Arch. Gen. Psychiatry 57 (4), 311–317.

Khan, A., Khan, S., Brown, W.A., 2002. Are placebo controls necessary to test new antidepressants and anxiolytics? Int. J. Neuropsychopharmacol. 5 (3), 193–197.

Khan, A., Fahl Mar, K., Faucett, J., Khan Schilling, S., Brown, W.A., 2017. Has the rising placebo response impacted antidepressant clinical trial outcome? Data from the US Food and Drug Administration 1987-2013. World Psychiatry 16 (2), 181–192.

Khuman, J., Zhang, J., Park, J., Carroll, J.D., Donahue, C., Whalen, M.J., 2012. Low-level laser light therapy improves cognitive deficits and inhibits microglial activation after controlled cortical impact in mice. J. Neurotrauma 29 (2), 408–417.

Köhler, C.A., Freitas, T.H., Maes, M., de Andrade, N.Q., Liu, C.S., Fernandes, B.S., et al., 2017. Peripheral cytokine and chemokine alterations in depression: a meta-analysis of 82 studies. Acta Psychiatrica Scandinavica 135 (5), 373–387.

Lampl, Y., Zivin, J.A., Fisher, M., Lew, R., Welin, L., Dahlof, B., et al., 2007. Infrared laser therapy for ischemic stroke: a new treatment strategy: results of the NeuroThera Effectiveness and Safety Trial-1 (NEST-1). Stroke 38 (6), 1843–1849.

Liu, Y., Ho, R.C.-M., Mak, A., 2012. Interleukin (IL)-6, tumour necrosis factor alpha (TNF-α) and soluble interleukin-2 receptors (sIL-2R) are elevated in patients with major depressive disorder: a meta-analysis and meta-regression. J Affect Disord 139 (3), 230–239.

Mayberg, H.S., Brannan, S.K., Tekell, J.L., Silva, J.A., Mahurin, R.K., McGinnis, S., et al., 2000. Regional metabolic effects of fluoxetine in major depression: serial changes and relationship to clinical response. BPS 48 (8), 830–843.

Milev, R.V., Giacobbe, P., Kennedy, S.H., Blumberger, D.M., Daskalakis, Z.J., Downar, J., et al., 2016. Canadian Network for Mood and Anxiety Treatments (CANMAT) 2016 clinical guidelines for the management of adults with major depressive disorder: Section 4. Neurostimulation treatments. Can. J. Psychiatry 61 (9), 561–575.

Mochizuki-Oda, N., Kataoka, Y., Cui, Y., Yamada, H., Heya, M., Awazu, K., 2002. Effects of near-infra-red laser irradiation on adenosine triphosphate and adenosine diphosphate contents of rat brain tissue. Neurosci. Lett. 323 (3), 207–210.

Mogoaşe, C., David, D., Koster, E.H.W., 2014. Clinical efficacy of attentional bias modification procedures: an updated meta-analysis. J. Clin. Psychol. 70 (12), 1133–1157.

Mohammed, H.S., 2016. Transcranial low-level infrared laser irradiation ameliorates depression induced by reserpine in rats. Lasers Med. Sci. 31 (8), 1651–1656.

Morava, E., Kozicz, T., 2013. Mitochondria and the economy of stress (mal)adaptation. Neurosci. Biobehav. Rev. 37 (4), 668–680.

Moscati, A., Flint, J., Kendler, K.S., 2016. Classification of anxiety disorders comorbid with major depression: common or distinct influences on risk? Depress. Anxiety 33 (2), 120–127.

Naeser, M.A., Zafonte, R., Krengel, M.H., Martin, P.I., Frazier, J., Hamblin, M.R., et al., 2014. Significant improvements in cognitive performance post-transcranial, red/near-infrared light-emitting diode treatments in chronic, mild traumatic brain injury: open-protocol study. J. Neurotrauma 31 (11), 1008–1017.

Olfson, M., Blanco, C., Marcus, S.C., 2016. Treatment of adult depression in the United States. JAMA Intern. Med. 176, 1482–1491.

Oron, A., Oron, U., Streeter, J., De Taboada, L., Alexandrovich, A., Trembovler, V., et al., 2007a. low-level laser therapy applied transcranially to mice following traumatic brain injury significantly reduces long-term neurological deficits. J. Neurotrauma 24 (4), 651–656.

Oron, U., Ilic, S., De Taboada, L., Streeter, J., 2007b. Ga-As (808 nm) laser irradiation enhances ATP production in human neuronal cells in culture. Photomed. Laser Surg. 25 (3), 180–182.

Ozcan, M.E., Gulec, M., Ozerol, E., Polat, R., Akyol, O., 2004. Antioxidant enzyme activities and oxidative stress in affective disorders. Int. Clin. Psychopharmacol. 19 (2), 89–95.

Parikh, S.V., Quilty, L.C., Ravitz, P., Rosenbluth, M., Pavlova, B., Grigoriadis, S., et al., 2016. Canadian Network for Mood and Anxiety Treatments (CANMAT) 2016 clinical guidelines for the management of adults with major depressive disorder: Section 2. Psychological treatments. Can. J. Psychiatry 61 (9), 524–539.

Perera, T., George, M.S., Grammer, G., Janicak, P.G., Pascual-Leone, A., Wirecki, T.S., 2016. The Clinical TMS Society Consensus Review and Treatment Recommendations for TMS Therapy for Major Depressive Disorder. Brain Stimul. 9 (3), 336–346.

Rojas, J.C., Bruchey, A.K., Gonzalez-Lima, F., 2012. Low-level light therapy improves cortical metabolic capacity and memory retention. J. Alzheimers Dis. 32 (3), 741–752.

Rush, A.J., Warden, D., Wisniewski, S.R., Fava, M., Trivedi, M.H., Gaynes, B.N., et al., 2009. STAR*D: revising conventional wisdom. CNS Drugs 23 (8), 627–647.

Salehpour, F., Rasta, S.H., Mohaddes, G., Sadigh-Eteghad, S., Salarirad, S., 2016. Therapeutic effects of 10-Hz pulsed wave lasers in rat depression model: a comparison between near-infrared and red wavelengths. Lasers Surg. Med. 48 (7), 695–705.

Sarandol, A., Sarandol, E., Eker, S.S., Erdinc, S., Vatansever, E., Kirli, S., 2007. Major depressive disorder is accompanied with oxidative stress: short-term antidepressant treatment does not alter oxidative-antioxidative systems. Hum. Psychopharmacol. 22 (2), 67–73.

Schiffer, F., Johnston, A.L., Ravichandran, C., Polcari, A., Teicher, M.H., Webb, R.H., et al., 2009. Psychological benefits 2 and 4 weeks after a single treatment with near infrared light to the forehead: a pilot study of 10 patients with major depression and anxiety. Behav. Brain Funct. 5 (1), 46.

Shungu, D.C., Weiduschat, N., Murrough, J.W., Mao, X., Pillemer, S., Dyke, J.P., et al., 2012. Increased ventricular lactate in chronic fatigue syndrome. III. Relationships to cortical glutathione and clinical symptoms implicate oxidative stress in disorder pathophysiology. NMR Biomed. 25 (9), 1073–1087.

Sinyor, M., Schaffer, A., Levitt, A., 2010. The sequenced treatment alternatives to relieve depression (STAR*D) trial: a review. Can. J. Psychiatry 55 (3), 126–135.

Spanemberg, L., Caldieraro, M., Arrua Vares, E., Wollenhaupt de Aguiar, B., Yuri Kawamoto, S., Parker, G., et al., 2014. Biological differences between melancholic and nonmelancholic depression subtyped by the CORE measure. Neuropsychiatr. Dis. Treat. 10, 1523.

Tanaka, Y., Akiyoshi, J., Kawahara, Y., Ishitobi, Y., Hatano, K., Hoaki, N., et al., 2011. Infrared radiation has potential antidepressant and anxiolytic effects in animal model of depression and anxiety. Brain Stimul. 4 (2), 71–76.

Tsujimoto, S., 2008. The prefrontal cortex: functional neural development during early childhood. Neuroscientist 14 (4), 345–358.

Videbech, P., 2000. PET measurements of brain glucose metabolism and blood flow in major depressive disorder: a critical review. Acta Psychiatr. Scand. 101 (1), 11–20.

Whiteford, H.A., Degenhardt, L., Rehm, J., Baxter, A.J., Ferrari, A.J., Erskine, H.E., et al., 2013. Global burden of disease attributable to mental and substance use disorders: findings from the Global Burden of Disease Study 2010. Lancet 382 (9904), 1575–1586.

Wu, Q., Xuan, W., Ando, T., Xu, T., Huang, L., Huang, Y.-Y., et al., 2012a. Low-level laser therapy for closed-head traumatic brain injury in mice: effect of different wavelengths. Lasers Surg. Med. 44 (3), 218–226.

Wu, X., Alberico, S.L., Moges, H., De Taboada, L., Tedford, C.E., Anders, J.J., 2012b. Pulsed light irradiation improves behavioral outcome in a rat model of chronic mild stress. Lasers Surg. Med. 44 (3), 227–232.

Xu, Z., Guo, X., Yang, Y., Tucker, D., Lu, Y., Xin, N., et al., 2016. Low-level laser irradiation improves depression-like behaviors in mice. Mol. Neurobiol. 34 (1), 13.

Xuan, W., Agrawal, T., Huang, L., Gupta, G.K., Hamblin, M.R., 2015. Low-level laser therapy for traumatic brain injury in mice increases brain derived neurotrophic factor (BDNF) and synaptogenesis. J Biophotonics 8 (6), 502–511.

Yamaura, M., Yao, M., Yaroslavsky, I., Cohen, R., Smotrich, M., Kochevar, I.E., 2009. Low level light effects on inflammatory cytokine production by rheumatoid arthritis synoviocytes. Lasers Surg. Med. 41 (4), 282–290.

Yu, W., Naim, J.O., McGowan, M., Ippolito, K., Lanzafame, R.J., 1997. Photomodulation of oxidative metabolism and electron chain enzymes in rat liver mitochondria. Photochem. Photobiol. 66 (6), 866–871.

Chapter 36

Action at a distance: laser acupuncture and the brain

Nicholas Alexander Wise
Department of Physical Medicine and Rehabilitation, UNC Chapel Hill School of Medicine, Chapel Hill, NC, United States

36.1 Background

One of the more mysterious applications of the (already somewhat mysterious) field of photobiomodulation therapy (PBMT) is laser acupuncture (LA). Defined as the nonthermal stimulation of traditional acupuncture or reflex points with a low-level laser, LA differs from the broader field of PBMT in that the irradiated points are not necessarily the target of the intended effect but, instead, are chosen based on traditional Chinese medicine (TCM) theory or an associated bodily system, usually to achieve a nonlocal or systemic result. Despite its increasing popularity in the West (Cui et al., 2017), relative safety (Chan et al., 2017), and rapidly expanding body of research showing clinical efficacy for many conditions (Vickers et al., 2018; Xiang et al., 2017; Lin et al., 2017; Kung et al., 2017; Li et al., 2017), the lack of a clear mechanism of action through which acupuncture exerts its effects on human physiology is a significant hurdle, and for LA, that hurdle is even larger, for as many questions that have been answered, many more arise. While an exhaustive exploration of acupuncture theory and literature is beyond the scope of this chapter, a brief description will be helpful to provide context for the discussion of the effects of the modern development of LA and its unique ability to influence the function of the brain.

36.1.1 Acupuncture and meridian theory

Acupuncture is a nearly 3000-year-old system of healthcare that is still one of the most widely-used treatments in the world. As the centerpiece of TCM, it developed and evolved empirically over several millennia to include many diverse concepts and theories, such as the opposing forces of *yin* and *yang*, and the five elements (wood, fire, earth, metal, and water). These terms may sound outdated or out of place in a discussion of modern neuroscience, however they can be interpreted as early attempts to describe the concepts of homeostasis and human phenotypes. A consistent theme running throughout the history of TCM, is that animals and humans contain discrete pathways, called meridians, that originate on the extremities and terminate in the internal organs from which they derive their name (Zhou and Benharash, 2014a). Meridians are thought to exist close to the surface of the skin and act as biological ley lines, allowing the system-wide transfer of vital energy (*qi*) over a network that seemingly exists outside of recognized hardwired or neurological pathways. There are 12 principal meridians in classical acupuncture theory that are bilateral and symmetrical, plus 2 midline meridians, and 361 specific points (acupoints) that act as nodes through which one can interface with the *qi* network (Pacific, 1993); in addition, many other secondary points have been found to lie outside of the main meridian network which effectively doubles that number. It is thought that sufficient stimulation of the correct acupoint(s) can normalize either a local excess or deficiency of *qi*, and allow the meridian, and subsequently the body, to return to homeostasis and restore health.

36.1.2 Physical properties of meridians and acupoints

Despite the long history and empirical development of the methods used, the anatomical underpinnings of acupuncture have only recently begun to be explored. In fact, many questions remain about the physical composition of meridians

Photobiomodulation in the Brain. DOI: https://doi.org/10.1016/B978-0-12-815305-5.00036-1

and acupoints, not the least of which is, do they actually exist? The first systematic review to examine the anatomical nature of acupoints was performed in 1980 by Chan, and the conclusion was that anatomical structures at acupoints did not differ in any significant way from nonacupoints, and the effect must be primarily neurologically mediated (Chan, 1984). In addition, many acupoints are indeed located on meridians that overlie major neural pathways, such as the distal pericardium (PC) meridian over the median nerve, the bladder (BL) meridian following the sciatic nerve up the back of the leg, and points on the stomach meridian (ST36, 37) that follow the path of the deep peroneal nerve. Many other acupoints have no clear spatial relationship to large neural pathways, however, and numerous other anatomically-based theories have attempted to explain the presence of meridians. These include: a unique form of vessels called Bonghan corpuscles (Liu et al., 2013), fascial plane lines as guides for meridians (Langevin and Yandow, 2002; Bai et al., 2011), channels of polarized molecules (Lo, 2002), and most simply, acupoints as merely myofascial trigger points (Melzack et al., 1977). The mechanism for the specificity of acupoints has been attributed to nearby dense peripheral nerve bundles (Chan, 1984), supporting a neurological route for acupuncture effects; increased density of mast cells at acupoints, which provides a mechanism for the immunological influence seen clinically (Zhou and Benharash, 2014a; Cheng et al., 2009; Wang et al., 2014a); and areas of viscerally-referred cutaneous neurogenic inflammation (Kim et al., 2017), which suggests the existence of a somatotopic reflex system as an organizing principle of the human body.

The biophysical properties of acupoints have also been extensively explored. Acupoints have been repeatedly demonstrated to possess specific electrical properties, such as high conductance and potential, low impedance and resistance, increased capacitance, as well as increased power spectral density (Ahn and Martinsen, 2007; Zhou et al., 2014). Thermal differences (Yang et al., 2007a,b, 2017), channels of low hydraulic resistance (Zhang et al., 2008), enhanced acoustic/vibratory characteristics (Lee et al., 2004), and most intriguingly perhaps for the topic of LA, unique optical properties have also been demonstrated in human and animal meridians (Sang Min et al., 2001; Yang et al., 2007b; Yang et al., 2009; Jovanić et al., 2009; Zhong et al., 2010). Yang et al. (2009) showed that 633 nm light attenuates less along a 1.0 cm distance of the PC meridian ($78.8 \pm 6.4\%$) than along the same distance in nonmeridian directions ($87.1\% \pm 3.0\%$) ($P < .05$; $n = 20$). Employing an advanced biomedical diagnostic technique called optical coherence tomography (OCT), Zhong et al. (2010) successfully distinguished acupoints from nonacupoints. They used OCT to show a significant increase in the optical attenuation coefficients of acupoint PC8 after 10 minutes of irradiation with a 100 mW 808 nm laser probe compared with sham LA. In addition, there have been attempts to update acupuncture theory by incorporating modern scientific concepts such as quantum entanglement (DeSmul, 1996; Wang et al., 2017a), holographic theory (Dale, 1999; Curtis and Hurtak, 2004), and biophotonics (Wang et al., 2014b; Schlebusch et al., 2005; Pokorny et al., 2012). These theories are intriguing and perhaps someday will be able to be scientifically tested, but at present, a clear picture of the biological basis of meridians and acupoints has yet to emerge.

36.1.3 Microsystems

In addition to the full-body meridian macrosystem, there is evidence for numerous somatotopic microsystems located on circumscribed locations throughout the body (Dale, 1999). Beginning with Dr. Paul Nogier's discovery and introduction of the auriculotherapy (ear) microsystem to the West in the early 1950s, numerous other examples have been documented and used, such as foot and hand reflexology (Miura et al., 2013; Nakamaru et al., 2008), Korean hand acupuncture (Park and Cha, 2012; Litscher, 2002), Su Jok therapy, ECIWO (embryo containing the information of the whole organism) therapy (Zhang, 1987), Yamamoto new scalp acupuncture (YNSA), cranial reflex therapy (Wise, 2007), oral (dental/tongue/gum) acupuncture (Gleditsch, 1978 #7933) (Simma et al., 2009), and iridodiagnostics (Ma, 2015). While the true nature of microsystem activity is unclear, they are thought to be cutaneous projections of the nervous system and largely modulated through the autonomic nervous system. Laser stimulation has been reportedly used successfully on several of these microsystems, including auriculotherapy (Round et al., 2013), Su Jok (Nedeljkovic et al., 2008), YSNA (Yamamoto et al., 2007), and chiropractic cranial reflexes (Wise, 2010), indicating that they may work through whatever mechanism it is that governs LA.

36.1.4 Acupuncture methods

Stimulation is traditionally performed by inserting and manipulating thin needles into specific acupoints, either individually or, more commonly, in groups. The needles are usually inserted to a depth of 1/4 to 1/2 in. below the surface of the skin, or until a specific sensation called *deqi* is felt. *Deqi* is often described as a dull heaviness, numbness or a tingle, and it is thought by many to be an important component of the acupuncture experience, and perhaps the most important influence on therapeutic efficacy (Lundeberg, 2013; Hori et al., 2010). While actual puncture with needles is still the most common form of stimulation in TCM, effective noninvasive methods, like acupressure, which is the application of manual or physical stimulation of acupoints, and moxibustion, the application of heat though burning herbs (traditionally mugwort) on or near the acupoint, have been used for centuries (Chen and Wang, 2014; Mehta et al., 2017). The modern era brought the use of electroacupuncture (EA) in the 1950s and LA in the 1970s as effective alternatives to needling (Jun et al., 2015).

Many studies have validated the increasing use and efficacy of EA. EA can be performed either as an addition to traditional acupuncture, where the current is conducted through an inserted needle, or performed noninvasively via direct application of the current to the skin by an EA device. The electrical current used in EA may be pulsed at different frequencies, which may lead to different physiological effects via the release of different endogenous analgesic compounds (Mayor, 2013; Qiu et al., 2015; Lee and Kim, 2017). Guo et al. (1996) showed that low-frequency EA stimulation (2 Hz) caused the release of enkephalin precursor proteins, whereas high-frequency stimulation (100 Hz) increased expression of dynorphin precursors. Xiang et al. (2014) showed that 2 Hz but not 100 Hz EA evoked a significant increase in mu-opioid receptor binding potential in the anterior cingulate cortex (ACC), the caudate nucleus, the putamen, the temporal lobe, the somatosensory cortex, and the amygdala in rhesus monkeys. A 2018 systematic review and meta-analysis of EA for depression concluded that EA performed equally to antidepressants, but with a decreased risk of adverse events (Li et al., 2018).

36.2 Laser acupuncture

The advent of medical laser technology in the late 1960s inspired the use of low-level laser as an alternate method of stimulating acupuncture points. There was early evidence that LA was effective and could achieve similar results to needle acupuncture (NA), but with the significant advantages of being completely noninvasive, painless, and requiring shortened treatment times (Whittaker, 2004). LA is generally regarded to be clinically comparable to traditional NA. A systematic review by Baxter et al. (2008) and a follow-up by Law et al. (2015) concluded that LA is an effective treatment for musculoskeletal pain and dysfunction, but with two very important qualifications: an appropriate treatment dosage was necessary, and results were best at long-term follow-up. Unfortunately heterogeneity of the pooled data did not allow the authors of either review to determine an effective dose from the meta-analysis, however Baxter concluded that power outputs of at least 10 mW and dosages of at least 0.5 J point are advisable in LA (Baxter, 2009).

36.2.1 Potential mechanisms of laser acupuncture

The mechanism through which LA exerts its homeostatic effects is currently unknown. It is possible that a mechanism inherent in the PBM effect is amplified and works synergistically with some unique feature of acupoints. Perhaps the answer lies in the higher skin nitric oxide (NO) concentration and expression of neuronal nitric oxide synthesis (nNOS) in acupoints, compared to control points (Liang et al., 2008; Lundeberg, 2013; Ma, 2017). One of the known effects of PBM is a local increase in the release of NO, which acts as a potent vasodilator and cell signaling molecule (Hamblin, 2017). A recent study compared the dose-dependent release of NO from irradiated acupoints (PC4, PC5) and nonacupoints (Jiang et al., 2017b). LA was performed using 658 nm near-infrared laser with the power of 12, 24, 48 mW for 20, 40, 60 minutes. They found that the NO released at the nonacupoint following 40 minutes of laser stimulation only increased slightly, perhaps due to the normal effect of PBM on NO, which happens in any tissue; however the 24 and 48 mW LA more than doubled the NO production at PC4 and PC5 compared to the nonacupoint ($P < .05$), and peak efficiency in NO production was noted at 24 mW. This finding indicates that the LA-induced release of NO is specific to acupoints, and is dose-dependent. A follow-up study by the same authors found that LA stimulation of PC6 on one side increases NO production on the contralateral side as well, and that releases of NO from both LPC6 and RPC6 after LA at RPC6 are greater than those of after LA LPC6, which may indicate that the acupoint has lateralized specificity (Jiang et al., 2017a).

36.2.2 The deqi question

The imperceptibility of LA raises the question of the importance of *deqi* in regards to clinical efficacy. If indeed the *deqi* sensation is one of the most important components of NA, one might wonder how LA can be clinically effective without it? *Deqi* is thought to involve all nerve fiber types, from the fast-conducting myelinated Aβ fibers to the slow-conducting unmyelinated C fibers, although the slower conducting fibers in the musculotendinous layers are likely to be behind most of the sensation (Hui et al., 2007; Zhou and Benharash, 2014b). PBMT has numerous documented inhibitory and analgesic effects on peripheral pain nerves via disruption of microtubule arrays and fast axonal flow (Chow et al., 2011), however this effect may be dose-dependent and wavelength dependent. In a study with implications for LA, Chow et al. found that irradiating a single point overlying the rat sciatic nerve (and presumably the BL meridian) with 808 nm infrared laser for 120 seconds (54 J, 18 J/cm^2) increased somatosensory-evoked potential (SSEP) amplitudes when compared with delivery of the same total energy to four points, which caused decreased SSEP amplitudes and conduction block (Chow et al., 2012). Irradiation with red (650 nm) laser for 30 or 120 seconds caused no change in SSEPs. These findings are contrary to their own previous findings which showed significant neuronal inhibition with both red and infrared wavelengths (Yan et al., 2011 #7954), but consistent with the biphasic effect of hyperpolarization followed by depolarization. If the primary effect of LA is neurologically mediated, many questions still remain about the differing effects of dose, wavelength, frequency, and irradiance on neural tissue and acupoints.

36.3 Acupuncture and the brain

Modern brain research methods have delivered much information and insight into the neural correlates of the myriad clinical effects of acupuncture. This body of research has grown rapidly since the first study in the mid-1990s (Yoshida et al., 1995); for example, a PubMed search of "(acupuncture or EA) and (fMRI or PET or EEG)" produced more than 882 references as of May 2018. However, relatively few studies have specifically examined the brain effects of LA. A similar search for (LA and (fMRI or PET or EEG or NIRS)) (and excluding nonrelevant topics like laser-evoked pain) produced 37 results. In order to delineate the unique effects of LA on brain function, we will first briefly summarize some interesting issues and findings from what we know about the brain effects of NA and EA.

36.3.1 Functional magnetic resonance imaging

Functional magnetic resonance imaging (fMRI) is an important form of brain imaging that is used to detect changes in cerebral hemodynamics under certain task conditions. fMRI studies show correlative increases or decreases in a blood-oxygen-level-dependent (BOLD) signal, which can be interpreted as activation or deactivation of specific brain areas. fMRI has excellent spatial resolution (~1–2 mm) but poor temporal resolution: peak BOLD signal is seen ~6–9 seconds after onset of the neural activity (Fröhlich, 2016). fMRI studies have demonstrated robust effects of NA on brain activity and revealed that acupuncture's effects are largely (but not completely) explained as being mediated by the central nervous system. Although more recent studies have benefitted from more rigorous design and statistical analysis, it should be noted that the methodology and validity of some early fMRI acupuncture studies are in question (Cho et al., 1998; Qiu et al., 2016). Issues such as publication bias, unsuitable controls, neglected carryover effects of block design, overly generous interpretation of activation/deactivation, and questionable point specificity have resulted in a heterogenous body of lower-quality evidence about which it is often difficult to make conclusions (Beissner and Henke, 2011). With that caveat, let us take a look at what is out there.

A systematic review of NA and fMRI studies was conducted by He et al. (2015). They analyzed 82 studies from 2008 to 2014, comprising a total of 2263 subjects and a variety of designs and methods. They summarize that "acupuncture could not only evoke brain activation in sensorimotor brain areas and widespread deactivation in the limbic–paralimbic neocortical network, but also modulate the connectivity of several brain regions, including antinociceptive, memory and affective brain regions, within DMN, SMN and amygdale-associated brain network. These regions process information in circuits that could broadly be assumed to engage: the affective (amygdala, hippocampus), sensory (thalamus, primary and secondary somatosensory cortices), cognitive (ACC, anterior insula) and inhibitory (PAG, hypothalamus) processing during the experience of pain. Most fMRI studies of acupuncture show it may recruit distributed cortical and subcortical brain networks that are also implicated in both inhibitory and facilitating effects in the pain-modulation system for both sensation and affective pain perception. However, correlation between other therapeutic effects induced by acupuncture and the corresponding neuroimaging changes has not been well studied." In their summary, they describe a wide variety of responses in regard to point specificity in the included studies: some results were completely consistent with the idea that disorder-specific acupoints evoke hemodynamic changes in disorder-implicated brain areas, while others failed to replicate those results. They also emphasize that most of the studies were performed on a single point on healthy subjects, which is unlike real-world clinical practice, since (1) clinical acupuncture usually involves stimulation of multiple points simultaneously and (2) acupuncture is thought to play a homeostatic role, and the effects on brain activation may be quite different between healthy subjects and those with a pathological imbalance. They point out other factors such as small sample size, gender differences, psychological state, and carryover effects that increase the heterogeneity of the analysis and make the results difficult to interpret.

fMRI investigations of the neural correlates of the *deqi* sensation by Asghar et al. (2010) and Hui et al. (2010). Hui et al. (2010) found that *deqi* was associated with significant deactivation in the limbic–paralimbic–neocortical network, which is closely related to the default mode network (DMN). The DMN is a group of brain regions that are active when the brain is in a resting state, that is, not performing any specific task, and is mentioned frequently in the acupuncture literature and seems to play a significant role in the brain effects of acupuncture. A recent systematic review reported that verum acupuncture usually increases DMN and sensorimotor network connectivity with pain-, affective- and memory-related brain areas (Cai et al., 2018), making this an issue that requires further study. Future studies are needed to determine if LA modulates the DMN in a similar fashion.

The central tenet of point specificity appears to be supported by many later studies that have demonstrated the effective uniqueness of acupoints compared to the surrounding tissue and sham (nonacupoints) (Xing et al., 2013; Campbell, 2006; Wang et al., 2012; Qin et al., 2011; Li et al., 2012, 2016). This being acupuncture, however, the literature is not without its share of contradictory results. The first study to demonstrate point specificity with fMRI was eventually retracted, as several of the authors decided they no longer could support the concept after failure to replicate the original finding (Cho et al., 1998). This may be due to the use of NA on nonacupoints as a control. Brain imaging studies that have used real stimulation on sham acupoints as a control have demonstrated that there may not be such a thing as a totally inert point on the body, which makes using verum acupuncture on a true control point rather difficult (Quah-Smith et al., 2010; Dincer and Linde, 2003).

The choice of controls in fMRI studies is a very important consideration due to the fact that a needle inserted into the skin may trigger certain brain activity that may be misinterpreted as point-specific. A 2013 meta-analysis of 28 fMRI studies sought to explore the question of point specificity by analyzing the brain activation patterns that could be attributed solely to the insertion of a needle into the body (Chae et al., 2013). They showed that needle stimulation without regard to location evoked brain activation in the so-called pain matrix (insula, thalamus, ACC, somatosensory cortex, primary visual cortex, inferior frontal cortex, superior temporal cortex, superior temporal gyrus, and cerebellum), and caused significant deactivation in the DMN (medial prefrontal cortex, subgenual ACC, caudate, amygdala, posterior cingulate cortex, thalamus, parahippocampus, and cerebellum). Hui et al. (2010) also reported that when acupuncture induced sharp pain, the deactivation in the DMN was attenuated or reversed in direction. On its face, this does not seem very surprising: a painful sensation (however small) would theoretically cause the brain to wake up from its resting state. However, it does illustrate some of the issues present in interpreting brain activation patterns with NA and fMRI—issues that LA conveniently avoids completely.

36.4 Laser acupuncture and the brain

36.4.1 Animal studies

Relatively few animal studies examining the brain effects of LA have been published in English. Jittiwat showed that LA at GV20 significantly decreased the brain infarct volume in cortical and subcortical areas in cerebral ischemic rats (Jittiwat, 2017). In addition, LA increased the catalase, glutathione peroxidase, and superoxide-dismutase (SOD). A decreased infarct volume plus increased SOD are strongly associated with reduced neurological deficit after stroke. Sutalangka et al. examined LA at HT7 (a point associated with learning and memory) or sham in an animal model of Alzheimer's disease (Sutalangka et al., 2013). LA was performed once a day for 14 days with a violet laser (405 nm,100 mW, spot diameter of 500 mm, 10 minutes). The rats who received verum LA showed a cognitive enhancement effect compared to sham, as well as elevation of acetylcholine (ACh) in the hippocampus. ACh plays an important role in learning and memory and the authors suggest that LA at HT7 may improve cholinergic function in the hippocampus, which in turn gives rise to enhanced spatial memory. The same point (HT7) was examined for its effects on animal models of autism in two studies by Khongrum et al. (Khongrum and Wattanathorn, 2015, 2017). In both studies, LA was performed once daily for 10 minutes at HT7 on both the left and the right sides (405 nm, 100 mW (0.100 J/s), diameter of 500 mm). The results of the first study showed that LA at HT7 improved behavioral outcomes as well as markers of oxidative stress in the cerebral cortex, striatum, and hippocampus. In the second study (2017), they examined the histological changes in the cerebellum induced by the same LA at HT7 protocol. Compared to sham, the LA rats had significantly decreased oxidative stress, as well as decreased levels of the pro-inflammatory cytokine IL-6 in the cerebellum. In addition, they found enhanced Purkinje cell survival and increased GABAergic function as a result of the improvements in oxidative stress status and inflammation induced by LA at HT7. The behavioral and biomarker improvements evoked by LA in these animal models of neurodegenerative diseases make this an exciting area for future studies.

36.4.2 Laser acupuncture and functional magnetic resonance imaging

The first evidence of LA's effectiveness at modulating human brain activity (and point specificity) using fMRI came from a 2002 study ($n = 10$) that examined laser stimulation of a point on the left little toe; bladder 67 (BL67) which is traditionally used to treat a variety of eye disorders (Siedentopf et al., 2002). The laser parameters reported were 10 mW power output and a wavelength of 670 nm. Compared to the sham LA condition, they found significant activation of the visual cortex within Brodmann's area (BA) 18 (cuneus) and BA19 (occipital cortex) in the left hemisphere, areas which are involved in higher order visual processing, and no activation within corresponding areas of the right hemisphere. They conclude that the brain activity found "is not due to peripheral afferent input from the dermal mechanoreceptors because our brain activation map present only ipsilateral activity within BA 18 and 19. Due to these findings, we assume that the Merkel and Ruffini corpuscles, which are augmented at acupoints, are not involved in the underlying mechanism of acupuncture." One of the strengths of this early study, and LA in general, is the ability to use sham LA as a true control: there was no tactile stimulation from the laser and therefore no noise generated in the somatosensory cortex or pain matrix.

To investigate the potential difference in brain activation on fMRI between traditional NA and LA, Siedentopf et al. (2005) performed LA on gall bladder 43 (GB43) in 22 healthy male volunteers. The acupoint chosen for this study is a point traditionally used for deafness, dizziness, tinnitus, ear diseases, headache and migraine. GB43 is located on the foot between the fourth and fifth toe, which is far enough away from the MRI scanner to avoid creating artifacts—a significant concern for fMRI investigations of NA. Laser parameters were 670 nm (red) and 10 mW power output. The results indicated that LA of the left GB43 acupoint activated the left thalamus, left nucleus rubber, and the brainstem, with no activation in the right hemisphere. LA of the right GB43 point activated the central midbrain, extending paramedially to the right, and the placebo group (sham LA) showed no significant brain activations for either side. Of particular interest in both of these studies is (1) the point specificity of BL67 and GB43, activating areas associated with visual and auditory processing, respectively, and (2) the fact that LA of a single point activated ipsilateral brain regions only. If LA stimulation was solely transmitted to the brain via the neurological route from skin mechanoreceptors to the afferent somatosensory pathway, contralateral activation would be expected, as those tracts crossover at the brainstem. Together these findings both support the existence of an alternate informational path from the periphery to the brain that does not cross the midline (e.g., meridians).

36.4.3 The frequency question

Many laser devices used in LA and PBMT have the ability to modulate their output at different pulse frequencies, and the question of whether there are frequency-specific pulsing effects in LA (like EA) is an interesting and complex one. While it has been shown in vitro and in vivo, PBMT studies that are pulsing the laser at different frequencies can cause different physiological effects (Ando et al., 2011; Ilic et al., 2006; Hashmi et al., 2010), the exact mechanism of pulsing in LA is, yet again, unclear, as traditional PBMT does not cause neural membrane depolarization and firing the way that EA does. A fascinating study in 2010 by Hsieh et al. set out to examine whether different stimulation modes of LA on the same point would activate different brain areas in the similar fashion to EA. Using fMRI to look at BOLD signals, they compared the effects of a continuous wave (CW) laser stimulation and one that was modulated at 10 Hz with rest period and sham LA. The team demonstrated that even though the laser was the same (30 mW, 808 nm), and the acupoint was the same (left Kidney 1, K1), the 10 Hz modulation activated distinct brain areas when compared to CW. The CW LA showed significant activation of the inferior parietal lobule in the left parietal lobe, and the 10 Hz LA uniquely activated the left inferior parietal lobule and the left supramarginal gyrus, areas related to attention and memory. This study raises some obvious questions of the importance of frequency modulation in LA, such as: how does frequency modulation in LA actually work? Will different frequencies activate different brain regions? Can brainwaves be modulated for even entrained rom the periphery by pulsed LA? Clearly, more studies are needed to answer these frequency-related questions.

36.4.4 Laser acupuncture and depression

Acupuncture has long been used as a treatment for depression. The 2018 Cochrane review of acupuncture and depression states that "acupuncture may result in a moderate reduction in the severity of depression when compared with treatment as usual or no treatment. Use of acupuncture may lead to a small reduction in the severity of depression when compared with control acupuncture. Effects of acupuncture versus medication and psychological therapy are uncertain owing to the very low quality of evidence"(Smith et al., 2018). LA does not have an extensive body of evidence to support its use for depression, but there are three promising studies by Quah-Smith et al. that examined the acute brain effects of LA on a set of acupoints that are traditionally used for mood disorders: left liver 8 (LR8), right liver 14 (LR14), left heart 7 (HT7), and conception vessel 14 (CV14) on the midline, as well as a nonacupoint in the abdomen for a control. Their first study in 2010 performed LA on these points in healthy subjects ($n = 10$) with an 808 nm and 25 mW laser delivered though a fiber optic arm (Quah-Smith et al., 2010). A block design of 20 seconds was used in which the subject received either verum or sham (laser off) LA at a single acupoint, and this cycle was repeated four times per acupoint and once per control. Since the laser produces no thermal or other sensations, the subject was successfully blinded as to the treatment phase. Analysis of the BOLD signal showed a trend of point-specific patterns of ipsilateral activation of the frontal cortex, limbic cortex, and subcortical caudate, and deactivation on the contralateral side. Results for individual points included: LR8 activated the ipsilateral limbic cortex, and deactivated the middle frontal gyri bilaterally, as well as the contralateral temporal cortex and caudate. LR14 stimulation activated the contralateral frontal cortex and parietal cortex, and deactivated the contralateral occipital cortex and cerebellum. CV14 activated the left limbic cortex, with no significant deactivations. Stimulation of HT7 caused no significant activation or deactivation, but the control point (nonacupoint) activated the contralateral postcentral gyrus in the parietal cortex and deactivated the contralateral limbic cortex. Interestingly, the contralateral somatosensory cortex was only activated by LR14 and the control point, and not by any of the other points.

The same investigators repeated the same LA protocol in 2013, but this time in depressed patients ($n = 10$) (Quah-Smith et al., 2013b). They used the same laser parameters (808 nm, 25 mW, 4 J per point) on the same set of acupoints (LR8, LR14, HT7, CV14) plus the additional kidney 3 (KI3) to compare their results with the healthy subjects from their 2010 study. What they found, perhaps not unsurprisingly, was that LA activated a much more extensive network of brain regions in the depressed patients compared to the healthy subjects from their previous study. This difference may be due to the relative amount of activation at baseline of the two groups. While both the healthy and depressed subjects showed the most activation in the frontal and temporal lobes, depressed patients also showed significant activation in the inferior parietal lobule (LR14), modulation in the cerebellum with CV12 and LR8, and modulation in brain regions involved in the DMN. While the exact neuroanatomical basis for depression is unclear, abnormal function of the DMN is thought to play a role in depression (Ng et al., 2017).

The third trial performed by this team using the same LA protocols for depression was a double-blinded RCT in 2013 in which 47 participants were randomized to receive LA or sham at the same acupoints (LR14, CV14, LR8, HT7, and KI3) (Quah-Smith et al., 2013a). The LA was administered twice a week for 4 weeks and once a week for another 4 weeks, for a total of 12 sessions. The primary outcome was a change in severity of depression at 8 weeks using the Hamilton depression rating scale (HAM-D), and secondary outcomes were the change in severity of depression using the Quick Inventory for Depression-Self Reporting (QID-SR) and the Quick Inventory for Depression-Clinician (QIDS-CL). At 8 weeks, participants in the active laser group showed greater improvement on the primary and clinician-rated secondary outcome measures HAM-D (mean 9.28 (SD 6.55) vs mean 14.14 (SD 4.78) $P < .001$) and QIDS-CL (mean 8.12 (SD 6.61) vs mean 12.68 (SD 3.77) $P < .001$). The self-report QIDS-SR scores improved in both groups but did not differ significantly between the groups. At 3 months, the QIDS-SR scores of the active laser participants remained significantly lower than baseline.

36.4.5 Laser acupuncture and cerebral blood flow

Litscher et al. have performed many studies exploring the effect of different forms of LA on brain function. In 2000 they conducted a multiparametric examination of the effect of NA and LA on brain circulation and the bispectral index of EEG. They found that both NA and LA (19 mW, 685 nm) of six acupoints (LI4, ST30, BL60, BL65, BL66, BL67) increased the mean blood flow velocity in the posterior cerebral artery in specific brain areas. In 2004 Litscher performed LA on 18 healthy volunteers in a randomized controlled crossover trial using functional multidirectional transcranial ultrasound Doppler sonography (fTCD) ($n = 17$) and fMRI ($n = 1$) (Litscher et al., 2004). Reported laser parameters were: 30–40 mW power output, 685 nm wavelength; 20-minute irradiation time; beam diameter: 500 μm; energy density of about 4.6 kJ/cm^2 at each acupoint and a total sum of 36.8 kJ/cm^2 for all acupoints. Simultaneous LA on four acupoints (LI4, ST36, BL60, BL67) resulted in an increase of mean blood flow velocity in the posterior cerebral artery measured by fTCD [before stimulation (mean ± SE): 42.2 ± 2.5; during stimulation: 44.2 ± 2.6; after stimulation: 42.3 ± 2.4 cm/s, n.s.]. Mean blood flow velocity in the middle cerebral artery decreased insignificantly. The fMRI analysis of one subject showed that bilateral stimulation of the acupoints produced bilateral positive activation over the frontal cortex, as well as an increase of the BOLD signal in the left superior occipital gyrus (BA19). The activation of the visual cortex (presumably by the two vision-related acupoints, BL60 and BL67) agrees with the earlier findings by Siedentopf et al. (2002).

36.4.6 Laser acupuncture and brain oscillations

Another insight into brain state and function comes from electroencephalogram (EEG) recordings of cortical oscillations. EEG can detect the minute fluctuations in brain rhythms that give insight into spatial and temporal dynamics of brain activation. The analgesic effect of NA has been shown to be related to reducing the power of gamma oscillations (30–100 Hz) in bilateral prefrontal cortex, mid-cingulate cortex, primary somatosensory cortex, and insula (Hauck et al., 2017). A 2013 randomized trial comparing acupuncture with clonazepam in 80 patients with generalized anxiety disorder used EEG to explore changes in brainwaves at 6 weeks. The NA group showed superior reductions in the HAM-A compared to those in the clonazepam group ($P < .05$), plus increased alpha power and decreased theta power (Zhou et al., 2013).

Unfortunately to date there are few published explorations of LA in humans using EEG. While not strictly LA, one study from 2012 measured differences in brain oscillations before and after using laser stimulation on the palm of the hand (Laser parameters: 6 laser diodes, 830 nm, 7 mW each, and pulse frequency of 10 Hz) (Wu et al., 2012). They found significant activations in the alpha and theta bands, with deactivation of beta, very similar to changes in brain state induced by mindfulness meditation (Lomas et al., 2015). Even though acupoints in the hand were not specifically targets, it is likely that Pericardium 8 (PC8), was stimulated at the very least, which raises the question: what is the effect of incidental acupoint stimulation when performing traditional PBMT? Are the overall effects enhanced by "accidental" acupoint stimulation or is the lack of specificity an issue? EEG is an important method of quantifying brain activity and is sorely underutilized in LA research.

36.4.7 Laser acupuncture for stroke and neurorehabilitation

There is intriguing evidence from Naeser et al. that LA is effective in assisting the neurorehabilitation of stroke patients with paralysis, including case reports (Naeser, 1997; Naeser et al., 2011), as well as evidence of lesion-site improvement from a CT-scan study (Naeser et al., 1995). In the latter study, Naeser treated seven stroke patients (ages 48–71 years; 5 males), five with hemiplegia, including severely reduced or absent voluntary finger movement, and two patients with hand paresis. The reported laser parameters were: 20 mW, 780 nm NIR CW laser, with a 1-mm-diameter aperture. LA stimulation was performed for 20 seconds per point (51 J/cm^2) on the shallower points on the hands and face, and deeper acupuncture points (on the arms and legs) received LA for 40 seconds per point (103 J/cm^2). The patients were treated 2–3 times per week for 3–4 months. Overall, five of seven of the patients (71.4%) showed improvement on physical examination by a blinded assessor after conclusion of treatment, including increases in range of motion, grip strength, and hand dexterity tests. Based on the CT scans, they were able to determine that the size of the lesion in the motor pathway areas correlated to whether or not they responded to LA treatment, with the cutoff being at >50% of the brain area (severe paralysis). These results warrant further investigation of LA as a mono- and adjuvant therapy in neurorehabilitation.

36.4.8 The wavelength question

The vast majority of PBMT research studies have used laser devices that fall within the range of red to near infrared (632–1064 nm), and even within that range, there is evidence that physiological effects are not only dose-dependent, but wavelength-specific as well (Albuquerque-Pontes et al., 2015; Ang et al., 2012; Pereira et al., 2014; Santos et al., 2014; Usumez et al., 2014). However, there is some evidence to show that lasers using shorter wavelengths may also evoke specific physiological results on tissue (Wang et al., 2017b; Hwang et al., 2015). Litscher's research group in Austria has shown the effectiveness of nontraditional wavelengths of light for use in LA. In 2010 they found that violet laser stimulation at GV14 on the upper back increased the blood flow velocity in the basilar artery significantly ($P < .001$) compared with the reference interval before LA (Litscher et al., 2010). The reported laser parameters were: 405 nm, 110 mW, diameter 500 μm, treatment time of 10 minutes. They also found that the same violet-LA protocol increased peripheral microcirculation (Wang et al., 2011) and skin temperature distribution (Litscher et al., 2011). Litscher has also reported positive results in modulating neurological responses using yellow and green laser light (Litscher et al., 2015, 2018). These initial explorations raise the interesting question of the potential therapeutic aspects of different wavelengths that can be used in LA. Since the depth of penetration of blue, green, and yellow light is thought to not be more than a few millimeters deep, how deep are acupoints exactly? We know needles are usually inserted up to ½ in. deep, but how deep does laser light need to penetrate in order to get its therapeutic effect? Future LA studies are needed with direct comparison between wavelengths to determine the optimal color for the job in brain stimulation.

36.5 Conclusion

The recent explosion in neuroscience research and advances in brain imaging technology have settled the question of whether or not acupuncture influences brain function. The answer is clearly in the affirmative. It is now clear that acupoint stimulation, whether by needle, laser, or electricity, can modulate brain function and thanks to fMRI technology, we know a lot about where. We just don't yet know exactly how or why. It is safe to say that the effects of acupuncture cannot be explained away by the placebo effect or an expectation of benefit anymore. However, we also know that the real answer is, of course, complicated, as acupuncture and the placebo response do actually share some similar brain pathways for triggering the release of endogenous opiates. The historical problems that acupuncture has had with hard evidence are partly due to questionable research methods and heterogeneity of results, however it may also be due to the fact that acupuncture truly is a nonlinear approach to a nonlinear system (human physiology), and it needs to be studied as such by improved methodology. The advantages of incorporating modern laser technology in this complex field, such as the ability to use an inactivated laser as a credible true sham in double-blinded RCTs, will go a long way to help solve some of these research problems for the future, and would seem to be clearly required.

References

Ahn, A.C., Martinsen, O.G., 2007. Electrical characterization of acupuncture points: technical issues and challenges. J. Altern. Complement. Med. 13, 817–824.

Albuquerque-Pontes, G.M., Vieira Rde, P., Tomazoni, S.S., Caires, C.O., Nemeth, V., Vanin, A.A., et al., 2015. Effect of pre-irradiation with different doses, wavelengths, and application intervals of low-level laser therapy on cytochrome c oxidase activity in intact skeletal muscle of rats. Lasers Med. Sci. 30, 59–66.

Ando, T., Xuan, W., Xu, T., Dai, T., Sharma, S.K., Kharkwal, G.B., et al., 2011. Comparison of therapeutic effects between pulsed and continuous wave 810-nm wavelength laser irradiation for traumatic brain injury in mice. PLoS One 6, e26212.

Ang, F.Y., Fukuzaki, Y., Yamanoha, B., Kogure, S., 2012. Immunocytochemical studies on the effect of 405-nm low-power laser irradiation on human-derived A-172 glioblastoma cells. Lasers Med. Sci. 27, 935–942.

Asghar, A.U., Green, G., Lythgoe, M.F., Lewith, G., Macpherson, H., 2010. Acupuncture needling sensation: the neural correlates of deqi using fMRI. Brain Res. 1315, 111–118.

Bai, Y., Wang, J., Wu, J.P., Dai, J.X., Sha, O., Tai Wai Yew, D., et al., 2011. Review of evidence suggesting that the fascia network could be the anatomical basis for acupoints and meridians in the human body. Evidence-based complementary and alternative medicine: eCAM 2011, 260510.

Baxter, G.D, 2009. Laser acupuncture: effectiveness depends upon dosage. Acupunct. Med. 27, 92.

Baxter, G.D., Bleakley, C., Mcdonough, S., 2008. Clinical effectiveness of laser acupuncture: a systematic review. J. Acupunct. Meridian Stud. 1, 65–82.

Beissner, F., Henke, C., 2011. Methodological problems in FMRI studies on acupuncture: a critical review with special emphasis on visual and auditory cortex activations. Evid. Based Complement. Alternat. Med. 2011, 607637.

Cai, R.L., Shen, G.M., Wang, H., Guan, Y.Y., 2018. Brain functional connectivity network studies of acupuncture: a systematic review on resting-state fMRI. J. Integr. Med. 16, 26–33.

Campbell, A., 2006. Point specificity of acupuncture in the light of recent clinical and imaging studies. Acupunct. Med. 24, 118–122.

Chae, Y., Chang, D.-S., Lee, S.-H., Jung, W.-M., Lee, I.-S., Jackson, S., et al., 2013. Inserting needles into the body: a meta-analysis of brain activity associated with acupuncture needle stimulation. J. Pain 14, 215–222.

Chan, M.W.C., Wu, X.Y., Wu, J.C.Y., Wong, S.Y.S., Chung, V.C.H., 2017. Safety of acupuncture: overview of systematic reviews. Sci. Rep. 7, 3369.

Chan, S.H., 1984. What is being stimulated in acupuncture: evaluation of the existence of a specific substrate. Neurosci. Biobehav. Rev. 8, 25–33.

Chen, Y.W., Wang, H.H., 2014. The effectiveness of acupressure on relieving pain: a systematic review. Pain Manag. Nurs. 15, 539–550.

Cheng, K., Shen, X.Y., Ding, G.H., Wu, F., 2009. Relationship between laser acupuncture analgesia and the function of mast cells. Zhongguo Zhen Jiu 29, 478–483.

Cho, Z.H., Chung, S.C., Jones, J.P., Park, J.B., Park, H.J., Lee, H.J., et al., 1998. New findings of the correlation between acupoints and corresponding brain cortices using functional MRI. Proc. Natl. Acad. Sci. U.S.A. 95, 2670–2673.

Chow, R., Armati, P., Laakso, E.L., Bjordal, J.M., Baxter, G.D., 2011. Inhibitory effects of laser irradiation on peripheral mammalian nerves and relevance to analgesic effects: a systematic review. Photomed. Laser Surg. 29, 365–381.

Chow, R., Yan, W., Armati, P., 2012. Electrophysiological effects of single point transcutaneous 650 and 808 nm laser irradiation of rat sciatic nerve: a study of relevance for low-level laser therapy and laser acupuncture. Photomed. Laser Surg. 30, 530–535.

Cui, J., Wang, S., Ren, J., Zhang, J., Jing, J., 2017. Use of acupuncture in the USA: changes over a decade (2002-2012). Acupunct. Med. 35, 200–207.

Curtis, B.D., Hurtak, J.J., 2004. Consciousness and quantum information processing: uncovering the foundation for a medicine of light. J. Altern. Complement. Med. 10, 27–39.

Dale, R.A., 1999. The systems, holograms and theory of micro-acupuncture. Am. J. Acupunct. 27, 207–242.

DeSmul, A., 1996. Very new waves in very old meridians: quantum medical physics of the living. Acupunct. Electrother. Res. 21, 15–20.

Dincer, F., Linde, K., 2003. Sham interventions in randomized clinical trials of acupuncture – a review. Complement. Ther. Med. 11, 235–242.

Fröhlich, F., 2016. Network Neuroscience. Elsevier Science.

Gleditsch, J., 1978. Mouth acupuncture. Zahnarzt 22 (5), 259–264.

Guo, H.F., Tian, J., Wang, X., Fang, Y., Hou, Y., Han, J., 1996. Brain substrates activated by electroacupuncture of different frequencies (I): comparative study on the expression of oncogene c-fos and genes coding for three opioid peptides. Brain Res. Mol. Brain Res. 43, 157–166.

Hamblin, M.R., 2017. Mechanisms and applications of the anti-inflammatory effects of photobiomodulation. AIMS Biophys. 4, 337–361.

Hashmi, J.T., Huang, Y.Y., Sharma, S.K., Kurup, D.B., De Taboada, L., Carroll, J.D., et al., 2010. Effect of pulsing in low-level light therapy. Lasers Surg. Med. 42, 450–466.

Hauck, M., Schroder, S., Meyer-Hamme, G., Lorenz, J., Friedrichs, S., Nolte, G., et al., 2017. Acupuncture analgesia involves modulation of pain-induced gamma oscillations and cortical network connectivity. Sci. Rep. 7, 16307.

He, T., Zhu, W., Du, S.-Q., Yang, J.-W., Li, F., Yang, B.-F., et al., 2015. Neural mechanisms of acupuncture as revealed by fMRI studies. Auton. Neurosci. 190, 1–9.

Hori, E., Takamoto, K., Urakawa, S., Ono, T., Nishijo, H., 2010. Effects of acupuncture on the brain hemodynamics. Auton. Neurosci. 157, 74–80.

Hui, K.K.S., Nixon, E.E., Vangel, M.G., Liu, J., Marina, O., Napadow, V., et al., 2007. Characterization of the "deqi" response in acupuncture. BMC Complement. Altern. Med. 7, 33.

Hui, K.K., Marina, O., Liu, J., Rosen, B.R., Kwong, K.K., 2010. Acupuncture, the limbic system, and the anticorrelated networks of the brain. Auton. Neurosci. 157, 81–90.

Hwang, M.H., Shin, J.H., Kim, K.S., Yoo, C.M., Jo, G.E., Kim, J.H., et al., 2015. Low level light therapy modulates inflammatory mediators secreted by human annulus fibrosus cells during intervertebral disc degeneration in vitro. Photochem. Photobiol. 91, 403–410.

Ilic, S., Leichliter, S., Streeter, J., Oron, A., Detaboada, L., Oron, U., 2006. Effects of power densities, continuous and pulse frequencies, and number of sessions of low-level laser therapy on intact rat brain. Photomed. Laser Surg. 24, 458–466.

Jiang, W.-L., Wei, H.-J., Guo, Z.-Y., Ni, Y.-R., Yang, H.-Q., Xie, S.-S., 2017a. Changes in nitric oxide releases of the contralateral acupoint during and after laser acupuncture at bilateral same-name acupoints in human. Evid. Based Complement. Alternat. Med. 2017, 5763458.

Jiang, W.L., Wei, H.J., Guo, Z.Y., Ni, Y.R., Yang, H.Q., Xie, S.S., 2017b. Effects of different-intensity laser acupuncture at two adjacent same-meridian acupoints on nitric oxide and soluble guanylate cyclase releases in human. Microcirculation 24.

Jittiwat, J., 2017. Laser acupuncture at GV20 improves brain damage and oxidative stress in animal model of focal ischemic stroke. J. Acupunct. Meridian Stud. 10, 324–330.

Jovanić, B., Nikolovski, D., Radenković, B., Despotović, M., 2009. Optical properties of acupunctural points as diagnostic method. Acta Phys. Polon. A 116, 693–696.

Jun, M.H., Kim, Y.M., Kim, J.U., 2015. Modern acupuncture-like stimulation methods: a literature review. Integr. Med. Res. 4, 195–219.

Khongrum, J., Wattanathorn, J., 2015. Laser acupuncture improves behavioral disorders and brain oxidative stress status in the valproic acid rat model of autism. J. Acupunct. Meridian Stud. 8, 183–191.

Khongrum, J., Wattanathorn, J., 2017. Laser acupuncture at HT7 improves the cerebellar disorders in valproic acid-rat model of autism. J. Acupunct. Meridian Stud. 10, 231–239.

Kim, D.H., Ryu, Y., Hahm, D.H., Sohn, B.Y., Shim, I., Kwon, O.S., et al., 2017. Acupuncture points can be identified as cutaneous neurogenic inflammatory spots. Sci. Rep. 7, 15214.

Kung, Y.Y., Hwang, S.J., Li, T.F., Ko, S.G., Huang, C.W., Chen, F.P., 2017. Trends in global acupuncture publications: an analysis of the Web of Science database from 1988 to 2015. J. Chin. Med. Assoc. 80, 521–525.

Langevin, H.M., Yandow, J.A., 2002. Relationship of acupuncture points and meridians to connective tissue planes. Anat. Rec. 269, 257–265.

Law, D., Mcdonough, S., Bleakley, C., Baxter, G.D., Tumilty, S., 2015. Laser acupuncture for treating musculoskeletal pain: a systematic review with meta-analysis. J. Acupunct. Meridian Stud. 8, 2–16.

Lee, G., Kim, W., 2017. The modulatory effect of acupuncture on the activity of locus coeruleus neuronal cells: a review. Evid. Based Complement. Alternat. Med. 2017, 9785345.

Lee, M.S., Kim, Y.C., Moon, S.R., Shin, B.C., Jeong, D.M., 2004. Hydrodynamic analysis of waveforms induced by vibrational stimuli at meridian and non-meridian points. Am. J. Chin. Med. 32, 977–984.

Li, A., Li, X.L., Zhang, F., Yue, J.H., Yuan, C.S., Li, K., et al., 2016. A functional magnetic resonance imaging study of the neuronal specificity of an acupoint: acupuncture at Rangu (KI 2) and its sham point. Intern. Med. J. 46, 973–977.

Li, J., Wang, Q., Liang, H., Dong, H., Li, Y., Ng, E.H., et al., 2012. Biophysical characteristics of meridians and acupoints: a systematic review. Evid. Based Complement. Alternat. Med. 2012, 793841.

Li, X., Wang, R., Xing, X., Shi, X., Tian, J., Zhang, J., et al., 2017. Acupuncture for myofascial pain syndrome: a network meta-analysis of 33 randomized controlled trials. Pain Physician 20, E883-e902.

Li, X.B., Wang, J., Xu, A.D., Huang, J.M., Meng, L.Q., Huang, R.Y., et al., 2018. Clinical effects and safety of electroacupuncture for the treatment of post-stroke depression: a systematic review and meta-analysis of randomised controlled trials. Acupunct. Med. 36, 284–293.

Liang, Y., Ma, S.X., Chen, J.X., 2008. Effect of nitric oxide on noradrenergic function and skin electric resistance of acupoints and meridians. Zhen Ci Yan Jiu 33, 213–216.

Lin, Y.C., Wan, L., Jamison, R.N., 2017. Using integrative medicine in pain management: an evaluation of current evidence. Anesth. Analg. 125, 2081–2093.

Litscher, D., Wang, G., Gaischek, I., Wang, L., Wallner-Liebmann, S., Petek, E., 2015. Yellow laser acupuncture – a new option for prevention and early intervention of lifestyle-related diseases: a randomized, placebo-controlled trial in volunteers. Laser Ther. 24, 53–61.

Litscher, D., Wang, J., Litscher, G., Li, G., Bosch, P., Van Den Noort, M., et al., 2018. Gender differences in laser acupuncture-results of a crossover study with green and yellow laser at the ear point Shenmen. Medicines (Basel) 5, 24.

Litscher, G., 2002. Computer-based quantification of traditional chinese-, ear- and Korean hand acupuncture: needle-induced changes of regional cerebral blood flow velocity. Neurol. Res. 24, 377–380.

Litscher, G., Rachbauer, D., Ropele, S., Wang, L., Schikora, D., Fazekas, F., et al., 2004. Acupuncture using laser needles modulates brain function: first evidence from functional transcranial Doppler sonography and functional magnetic resonance imaging. Lasers Med. Sci. 19, 6–11.

Litscher, G., Huang, T., Wang, L., Zhang, W., 2010. Violet laser acupuncture – part 1: Effects on brain circulation. J. Acupunct. Meridian Stud. 3, 255–259.

Litscher, G., Wang, L., Huang, T., Zhang, W., 2011. Violet laser acupuncture – part 3: pilot study of potential effects on temperature distribution. J. Acupunct. Meridian Stud. 4, 164–167.

Liu, J.L., Jing, X.H., Shi, H., Chen, S.P., He, W., Bai, W.Z., et al., 2013. Historical review about research on "Bonghan System" in China. Evid. Based Complement. Alternat. Med. 2013, 636081.

Lo, S.Y., 2002. Meridians in acupuncture and infrared imaging. Med. Hypotheses 58, 72–76.

Lomas, T., Ivtzan, I., Fu, C.H., 2015. A systematic review of the neurophysiology of mindfulness on EEG oscillations. Neurosci. Biobehav. Rev. 57, 401–410.

Lundeberg, T., 2013. Acupuncture mechanisms in tissue healing: contribution of NO and CGRP. Acupunct. Med. 31, 7–8.

Ma, S.X., 2017. Nitric oxide signaling molecules in acupoints: toward mechanisms of acupuncture. Chin. J. Integr. Med. 23, 812–815.

Ma, Y., 2015. Neuropsychological mechanism underlying antidepressant effect: a systematic meta-analysis. Mol. Psychiatry 20, 311–319.

Mayor, D., 2013. An exploratory review of the electroacupuncture literature: clinical applications and endorphin mechanisms. Acupunct. Med. 31, 409–415.

Mehta, P., Dhapte, V., Kadam, S., Dhapte, V., 2017. Contemporary acupressure therapy: adroit cure for painless recovery of therapeutic ailments. J. Tradit. Complement. Med. 7, 251–263.

Melzack, R., Stillwell, D.M., Fox, E.J., 1977. Trigger points and acupuncture points for pain: correlations and implications. Pain 3, 3–23.

Miura, N., Akitsuki, Y., Sekiguchi, A., Kawashima, R., 2013. Activity in the primary somatosensory cortex induced by reflexological stimulation is unaffected by pseudo-information: a functional magnetic resonance imaging study. BMC Complement. Altern. Med. 13, 114.

Naeser, M.A., 1997. Neurological rehabilitation: acupuncture and laser acupuncture to treat paralysis in stroke, other paralytic conditions, and pain in carpal tunnel syndrome. J. Altern. Complement. Med. 3, 425–428.

Naeser, M.A., Stiassny-Eder, D., Galler, V., Hobbs, J., Bachman, D., Lannin, L.N., 1995. Laser acupuncture in the treatment of paralysis in stroke patients: a CT scan lesion site study. Am. J. Acupunct. 23, 13–28.

Naeser, M.A., Saltmarche, A., Krengel, M.H., Hamblin, M.R., Knight, J.A., 2011. Improved cognitive function after transcranial, light-emitting diode treatments in chronic, traumatic brain injury: two case reports. Photomed. Laser Surg. 29, 351–358.

Nakamaru, T., Miura, N., Fukushima, A., Kawashima, R., 2008. Somatotopical relationships between cortical activity and reflex areas in reflexology: a functional magnetic resonance imaging study. Neurosci. Lett. 448, 6–9.

Nedeljkovic, M., Ljustina-Pribic, R., Savic, K., 2008. Innovative approach to laser acupuncture therapy of acute obstruction in asthmatic children. Med. Pregl. 61, 123–130.

Ng, S.K., Urquhart, D.M., Fitzgerald, P.B., Cicuttini, F.M., Hussain, S.M., Fitzgibbon, B.M., 2017. The relationship between structural and functional brain changes and altered emotion and cognition in chronic low back pain: a systematic review of MRI and fMRI Studies. Clin. J. Pain 34, 237–261.

Pacific, W.H.O.R.O.F.T.W., 1993. Standard Acupuncture Nomenclature: A Brief Explanation of 361 Classical Acupuncture Point Names and their Multilingual Comparative List, WHO Regional Office for the Western Pacific, Manila.

Park, H.J., Cha, C., 2012. The effect of Korean hand acupuncture on young, single Korean students with irritable bowel syndrome. Gastroenterol. Nurs. 35, 403–414.

Pereira, T.S., Flecha, O.D., Guimaraes, R.C., De Oliveira, D., Botelho, A.M., Ramos Gloria, J.C., et al., 2014. Efficacy of red and infrared lasers in treatment of temporomandibular disorders – a double-blind, randomized, parallel clinical trial. Cranio 32, 51–56.

Pokorny, J., Martan, T., Foletti, A., 2012. High capacity optical channels for bioinformation transfer: acupuncture meridians. J. Acupunct. Meridian Stud. 5, 34–41.

Qin, W., Bai, L., Dai, J., Liu, P., Dong, M., Liu, J., et al., 2011. The temporal-spatial encoding of acupuncture effects in the brain. Mol. Pain 7, 19.

Qiu, K., Jing, M., Sun, R., Yang, J., Liu, X., He, Z., et al., 2016. The status of the quality control in acupuncture-neuroimaging studies. Evid. Based Complement. Alternat. Med. 2016, 3685785.

Qiu, Z.Y., Ding, Y., Cui, L.Y., Hu, M.L., Ding, M.X., 2015. The expression patterns of c-Fos and c-Jun induced by different frequencies of electroacupuncture in the brain. Evid. Based Complement. Alternat. Med. 2015, 343682.

Quah-Smith, I., Sachdev, P.S., Wen, W., Chen, X., Williams, M.A., 2010. The brain effects of laser acupuncture in healthy individuals: an FMRI investigation. PLoS One 5, e12619.

Quah-Smith, I., Smith, C., Crawford, J.D., Russell, J., 2013a. Laser acupuncture for depression: a randomised double blind controlled trial using low intensity laser intervention. J. Affect Disord. 148, 179–187.

Quah-Smith, I., Suo, C., Williams, M.A., Sachdev, P.S., 2013b. The antidepressant effect of laser acupuncture: a comparison of the resting brain's default mode network in healthy and depressed subjects during functional magnetic resonance imaging. Med. Acupunct. 25, 124–133.

Round, R., Litscher, G., Bahr, F., 2013. Auricular acupuncture with laser. Evid. Based Complement. Alternat. Med. 2013, 984763.

Sang Min, L., Chunho, C., Kwang-Sup, S., Gilwon, Y., 2001. Measurement and analysis of optical properties of acupuncture points and channels. In: ISIE 2001. 2001 IEEE International Symposium on Industrial Electronics Proceedings (Cat. No.01TH8570), vol. 1, pp. 655–656.

Santos, L.A., Marcos, R.L., Tomazoni, S.S., Vanin, A.A., Antonialli, F.C., Grandinetti Vdos, S., et al., 2014. Effects of pre-irradiation of low-level laser therapy with different doses and wavelengths in skeletal muscle performance, fatigue, and skeletal muscle damage induced by tetanic contractions in rats. Lasers Med. Sci. 29, 1617–1626.

Schlebusch, K.P., Maric-Oehler, W., Popp, F.A., 2005. Biophotonics in the infrared spectral range reveal acupuncture meridian structure of the body. J. Altern. Complement. Med. 11, 171–173.

Siedentopf, C.M., Golaszewski, S.M., Mottaghy, F.M., Ruff, C.C., Felber, S., Schlager, A., 2002. Functional magnetic resonance imaging detects activation of the visual association cortex during laser acupuncture of the foot in humans. Neurosci. Lett. 327, 53–56.

Siedentopf, C.M., Koppelstaetter, F., Haala, I.A., Haid, V., Rhomberg, P., Ischebeck, A., et al., 2005. Laser acupuncture induced specific cerebral cortical and subcortical activations in humans. Lasers Med. Sci. 20, 68–73.

Simma, I., Gleditsch, J.M., Simma, L., Piehslinger, E., 2009. Immediate effects of microsystem acupuncture in patients with oromyofacial pain and craniomandibular disorders (CMD): a double-blind, placebo-controlled trial. Br. Dent. J. 207, E26.

Smith, C.A., Armour, M., Lee, M.S., Wang, L.Q., Hay, P.J., 2018. Acupuncture for depression. Cochrane Database Syst. Rev. 3, Cd004046.

Sutalangka, C., Wattanathorn, J., Muchimapura, S., Thukham-Mee, W., Wannanon, P., Tong-Un, T., 2013. Laser acupuncture improves memory impairment in an animal model of Alzheimer's disease. J. Acupunct. Meridian Stud. 6, 247–251.

Usumez, A., Cengiz, B., Oztuzcu, S., Demir, T., Aras, M.H., Gutknecht, N., 2014. Effects of laser irradiation at different wavelengths (660, 810, 980, and 1,064 nm) on mucositis in an animal model of wound healing. Lasers Med. Sci. 29, 1807–1813.

Vickers, A.J., Vertosick, E.A., Lewith, G., Macpherson, H., Foster, N.E., Sherman, K.J., et al., 2018. Acupuncture for chronic pain: update of an individual patient data meta-analysis. J. Pain 19, 455–474.

Wang, D.J., Chang, X.R., Yan, J., Wang, X.J., Wang, T., Liu, J., et al., 2012. Comparative study on therapeutic effect between acupuncture at special acupoints and non-specific acupoints in foot yangming meridian for functional dyspepsia. Zhongguo Zhen Jiu 32, 703–708.

Wang, G., Litscher, D., Tian, Y., Gaischek, I., Jia, S., Wang, L., et al., 2014a. Acupoint activation: response in microcirculation and the role of mast cells. Medicines 1, 56–63.

Wang, X., Huang, J., Han, J., Yang, M., Pang, J., Zhao, X., 2014b. Recent progress of traditional Chinese medical science based on theory of biophoton. Front. Optoelectron. 7, 28–36.

Wang, J., Wu, B., Chen, S., 2017a. Discussion on quantum entanglement theory and acupuncture. Zhongguo Zhen Jiu 37, 1233–1237.

Wang, Y., Huang, Y.Y., Wang, Y., Lyu, P., Hamblin, M.R., 2017b. Red (660 nm) or near-infrared (810 nm) photobiomodulation stimulates, while blue (415 nm), green (540 nm) light inhibits proliferation in human adipose-derived stem cells. Sci. Rep. 7, 7781.

Wang, L., Huang, T., Zhang, W., Litscher, G., 2011. Violet laser acupuncture--part 2: Effects on peripheral microcirculation. J. Acupunct. Meridian Stud. 4, 24–28.

Whittaker, P., 2004. Laser acupuncture: past, present, and future. Lasers Med. Sci. 19, 69–80.

Wise, N.A., 2007. Cranial Laser Reflex Technique for post-stroke muscle spasm: a case report. In: North American Association of Laser Therapy (NAALT) Annual Conference. Tucson, AZ.

Wise, N.A., 2010. Reduction of chronic musculoskeletal pain with cranial laser reflex technique (CLRT): a randomized controlled trial using pressure algometry. AIP Conf. Proc. 1226, 184–191.

Wu, J.-H., Chang, W.-D., Hsieh, C.-W., Jiang, J.-A., Fang, W., Shan, Y.-C., et al., 2012. Effect of low-level laser stimulation on EEG. Evid. Based Complement. Alternat. Med. 2012, 11.

Xiang, A., Cheng, K., Shen, X., Xu, P., Liu, S., 2017. The immediate analgesic effect of acupuncture for pain: a systematic review and meta-analysis. Evid. Based Complement. Alternat. Med. 2017, 3837194.

Xiang, X.H., Chen, Y.M., Zhang, J.M., Tian, J.H., Han, J.S., Cui, C.L., 2014. Low- and high-frequency transcutaneous electrical acupoint stimulation induces different effects on cerebral mu-opioid receptor availability in rhesus monkeys. J. Neurosci. Res. 92, 555–563.

Xing, J.J., Zeng, B.Y., Li, J., Zhuang, Y., Liang, F.R., 2013. Acupuncture point specificity. Int. Rev. Neurobiol. 111, 49–65.

Yamamoto, T., Schockert, T., Boroojerdi, B., 2007. Treatment of juvenile stroke using Yamamoto New Scalp Acupuncture (YNSA) – a case report. Acupunct. Med. 25, 200–202.

Yan, W., Chow, R., Armati, P.J., 2011. Inhibitory effects of visible 650-nm and infrared 808-nm laser irradiation on somatosensory and compound muscle action potentials in rat sciatic nerve: implications for laser-induced analgesia. J. Peripher. Nerv. Syst. 16, 130–135.

Yang, H., Xie, S., Li, H., Wang, Y., 2009. On optics of human meridians. Sci. China Series G: Phys. Mech. Astron. 52, 502–507.

Yang, H.Q., Xie, S.S., Hu, X.L., Chen, L., Li, H., 2007a. Appearance of human meridian-like structure and acupoints and its time correlation by infrared thermal imaging. Am. J. Chin. Med. 35, 231–240.

Yang, H.Q., Xie, S.S., Liu, S.H., Li, H., Guo, Z.Y., 2007b. Differences in optical transport properties between human meridian and non-meridian. Am. J. Chin. Med. 35, 743–752.

Yang, Z., Zhou, M., Wang, X., Zhao, Y., Chen, Z., Lan, Y., et al., 2017. Review on skin temperature of acupoints. Zhongguo Zhen Jiu 37, 109–114.

Yoshida, T., Tanaka, C., Umeda, M., Higuchi, T., Fukunaga, M., Naruse, S., 1995. Non-invasive measurement of brain activity using functional MRI: toward the study of brain response to acupuncture stimulation. Am. J. Chin. Med. 23, 319–325.

Zhang, W.B., Tian, Y.Y., Li, H., Tian, J.H., Luo, M.F., Xu, F.L., et al., 2008. A discovery of low hydraulic resistance channel along meridians. J. Acupunct. Meridian Stud. 1, 20–28.

Zhang, Y., 1987. ECIWO Biology and Medicine: A New Theory of Conquering Cancer and Completely New Acupuncture Therapy. Neimenggu People Press, Beijing.

Zhong, H.Q., Zhang, Z.D., Guo, Z.Y., Wei, H.J., Yang, H.Q., He, Y.H., et al., 2010. Using OCT image to distinguish human acupoint from non-acupoint tissues after irradiation with laser in vivo: a pilot study. Chin. Opt. Lett. 8, 418–420.

Zhou, Q., Gai, S., Lin, N., Zhang, J., Zhang, L., Yu, R., et al., 2014. Power spectral differences of electrophysiological signals detected at acupuncture points and non-acupuncture points. Acupunct. Electrother. Res. 39, 169–181.

Zhou, W., Benharash, P., 2014a. Effects and mechanisms of acupuncture based on the principle of meridians. J. Acupunct. Meridian Stud. 7, 190–193.

Zhou, W., Benharash, P., 2014b. Significance of deqi response in acupuncture treatment: myth or reality. J. Acupunct. Meridian Stud. 7, 186–189.

Zhou, X.F., Li, Y., Zhu, H., Chen, L.L., 2013. Impacts of acupuncture at twelve meridians acupoints on brain waves of patients with general anxiety disorder. Zhongguo Zhen Jiu 33, 395–398.

Chapter 37

Signature wounds of war: a case study

George Louis Lindenfeld[1,2] and George Rozelle[3]

[1]RESET Therapy Professional Training Institute LLC, Sarasota, FL, United States, [2]RESET Therapy Professional Training Institute LLC, Hendersonville, NC, United States, [3]MindSpa Integrative Wellness Center, Sarasota, FL, United States

37.1 Introduction

Much progress in therapeutic applications has been made over the course of my 48 years as a clinical psychologist, however rarely in my career, have I encountered such a rapid and lasting transformative change. That is, until approximately 6 years ago when I was introduced to the bioacoustical utilization device (BAUD) patented in 2006 by Dr. Frank Lawlis. Delving into this new intervention and utilizing it fully in my practice, I found that amazingly, many of my patients recovered from a myriad of difficulties that plagued them. Furthermore, accompanying conditions such as: depression, sleep disorder, chronic pain, etc., improved as well. Looking deeper into the mechanisms of action of this tool, I came to understand that the binaural sound utilized by the BAUD, was able to quickly and permanently alter the emotional components of long-term stored memory.

In contrast, it has only been a short time ago, that I was introduced to the world of infrared light, particularly in regards to its ability to improve the consequences of traumatic brain injury (TBI). Given that by 2007, it was noted in a RAND Corporation Report that: "1.64 million service members who had been deployed for OEF/OIF as of October 2007, we estimate that approximately 300,000 individuals currently suffer from PTSD or major depression and that 320,000 individuals experienced a probable TBI during deployment... Some specific groups, previously understudied—including the Reserve Components and those who have left military service—may be at higher risk of suffering from these conditions" (Invisible Wounds of War: Psychological and Cognitive Injuries, Their Consequences, and Services to Assist Recovery—RAND_MG720. pdf, 2008).

Given my relative newness to the field of photobiomodulation, some background information may be helpful in appreciating how I came to be invited to write a chapter in a book about this topic. In 2016, I was working on my latest book called: *The Treatment of PTSD Comorbid Conditions.* Two of the chapters (Dementia and Traumatic Brain Injury), focused on conditions that appear to be responsive to photobiomodulation. I took a giant leap and contacted Dr. Michael Hamblin, asking him to consider writing the forward to my newest undertaking. Graciously, he responded to my request, writing material pertinent to the entire book as well as to the subject matter of the two specified chapters. I thought that a good start to this chapter would be to share a portion of those comments as a way of appreciating Dr. Hamblin's contribution to neuroplasticity's transformative potential in the broadest sense.

> *One of the most important concepts that has arisen in this emerging field of neuroscience is called 'synaptic plasticity.' In this sense, acoustical neuromodulation and Photobiomodulation share a common mechanistic foundation resting on neuroplasticity. As I read about the applications of RESET Therapy to the PTSD comorbid conditions ... I was struck with an awareness of the inherent potential in each and every cell in our body to heal and restore itself ... it is clear that ... acoustical neuromodulation interrupts the memory reconsolidation process within the retention network of the brain thereby permitting it to reset to pre-trauma levels.*
>
> *... He (Lindenfeld), proposes that when we combine the treatment he calls RESET Therapy with Photobiomodulation, we emerge with a perfectly balanced, "peanut-butter and jelly sandwich." The concept here is that intervention with RESET blocks the insidious effects of unresolved trauma from perpetuating damage to the body as a whole. On the other hand, PBMT triggers: an increase in regional cerebral blood flow; improves brain oxygenation; increases ATP levels that are crucial for effective brain function. Apparently, one shared function of both the light that induces photobiomodulation, and the sound that*

Photobiomodulation in the Brain. DOI: https://doi.org/10.1016/B978-0-12-815305-5.00037-3

returns the cells to their inherent growth potential, is their apparent ability to diminish neuro-inflammation. This phenomenon is especially damaging to the brain if excessively prolonged.

Dr. Michael Hamblin

This synthesis of two different interventions truly complement each other in the healing of body, mind, and soul. Through sound, we facilitate the erasure of the insidious effects of unresolved trauma with its accompanying neutralization of the neuroinflammatory consequences on the body and mind. Through light, we accelerate an increase in cerebral oxygenation and blood flow as well as increased ATP production. These two interventions truly supplement each other within the context of a transformative process made possible through, synaptic plasticity.

As one who appreciates the magic of a peanut butter and jelly sandwich, particularly when the jelly is pineapple, I am thrilled to contribute to Dr. Hamblin's newest effort. To be clear, let us consider the reconsolidation enhancement by stimulation of emotional triggers (RESET Therapy), aspect to be equivalent to the pineapple jelly part. My challenge was to conceptualize how to best integrate the joint utilization of these transformative treatments. After giving this much thought, I decided to first apply RESET Therapy to remediate the posttraumatic stress disorder (PTSD) symptoms in a combat veteran and then follow up with photobiomodulation for the residual, underlying TBI component.

In regards to the peanut butter, I have earlier acknowledged that I am a newcomer to the field of photobiomodulation. I have recently celebrated my 79th birthday and decided about two years ago, to explore the potential that infrared light might offer me in sustaining my own cognitive abilities during aging. Simultaneously, my awareness of the signature wounds of our engaged military personnel, was becoming increasingly evident within the recent population of returning combat veterans with PTSD. This awareness emerged as my colleagues and I were engaged in a pilot project in the Sarasota, Florida region that to date has involved six combat veterans from three eras (Vietnam; Gulf Wars; Iraq/Afghanistan). Without going into too much detail at this point, the results of this intervention were extremely impressive with each of the volunteer veterans receiving four treatment sessions of RESET Therapy. Their posttest results will be shown later indicating a significant reduction in their PTSD symptomatology to the point that they no longer meet diagnostic criteria necessary for the condition.

We obtained approval from the QuietMind Foundation, an institutional review board, for our research noting that we would exclude veterans with a history of TBI or active substance abuse. We did this initially with the thought that we would avoid compounding our primary focus of remediating symptoms associated only with PTSD. With veterans from earlier eras, this was not difficult to do. However, with veterans from Iraq/Afghanistan, we found a much higher concurrence between the two conditions. While performing preliminary screening of veterans interested in participating in the study, it quickly became apparent that many of them had experienced mild to moderate levels of TBI. Therefore, with the approval of the QuietMind Foundation, study inclusion criteria were modified to allow veterans with mild to moderate cases of TBI to participate.

Our inquiry into this matter revealed that the National Center for PTSD had noted an increase in TBI in the following 2017 update: "The conflicts in Iraq and Afghanistan have resulted in increased numbers of Veterans who have experienced traumatic brain injuries (TBI). The Department of Defense and the Defense and Veteran's Brain Injury Center estimate that 22% of all combat casualties from these conflicts are brain injuries, compared to 12% of Vietnam related combat casualties. 60% to 80% of soldiers who have other blast injuries may also have traumatic brain injuries" (Traumatic Brain Injury and PTSD, 2017).

Since we were now including veteran volunteers with this condition, I thought it would be prudent to explore what current remediative treatments were available for TBI. My basic impression was that these were primarily supportive or palliative rather than restorative interventions as noted in the following perspective: Effective treatment for postconcussive symptoms (PCS) immediately following mild traumatic brain injury (mTBI) includes reassurance, support, education about mTBI, and symptom management. However, effective treatments for chronic postconcussive-like symptoms, particularly with mental health comorbidities, remain unclear (Vanderploeg et al., 2019).

Another investigator focusing on treatment for TBI found that: "Overall, counseling for adults with TBI has largely involved the use of behavioral, CBT, and other directive methods in the group and/or individual modalities. It tends to focus on skill development, symptoms management and reduction, social and interpersonal competence, and mood regulation. The counseling tends to be part of a larger treatment plan for the clients, as they struggle to re-attain proficiency in multiple areas of their lives, involving relationships, work, and self-efficacy (maucieri-2012).

Other, more recent investigators, have found that interventions for the treatment of TBI are rather limited. They note that: "While there have been some encouraging reports of improvement with the use of hyperbaric oxygen, published military studies show no benefit. Pharmacology has been limited to symptomatic amelioration" (Morries et al., 2015). "Recent multi-site studies of pharmacological agents to directly treat concussion and brain injury have failed"

(Willyard, 2015). "The overlap in symptoms between TBI and post-traumatic stress disorder has contributed to the use of medications which have been shown to be potentially harmful to the injured brain in those with a TBI". (Henderson, 2016) "For example, anti-psychotic drugs have been shown to be neurotoxic and can potentially impede any neurological regeneration which might otherwise occur" (Phelps et al., 2015).

The only treatment interventions noted in a 2016 article by Santos et al., that offer promise of TBI cortical remediation include: repetitive transcranial magnetic stimulation, transcranial direct current stimulation, and transcranial led therapy. As my focus in this chapter is on the use of low-level laser therapy, I concentrated only on this particular intervention. The study authors noted that: "In humans the use of laser therapy has also started to be used with different focuses. A low-power laser therapy was used to verify the biostimulant effects of this therapy in tissue repair. It might be noted that after completion of the dental procedure patients undergoing therapy had vascular hyperemia increased local blood circulation, increased metabolism, collagen synthesis, stimulation produce endorphins, inhibiting nociceptor signals, which brought analgesic, anti-inflammation, tissue edema and reduction in healing effect. On the other hand, when specifically applied to the CNS, they have also had their related results to those found in animal models and in clinical use" (Santos et al., 2016).

Researchers in a 2014 study explored the use of 4 weeks of light therapy to treat fatigue following TBI. They reported that: "After controlling age, gender, and baseline depression, treatment with high-intensity blue light therapy resulted in reduced fatigue and daytime sleepiness during the treatment phase, with evidence of a trend toward baseline levels 4 weeks after treatment cessation. These changes were not observed with lower-intensity yellow light therapy or no treatment control conditions. There was also no significant treatment effect observed for self-reported depression or psychomotor vigilance performance. Blue light therapy appears to be effective in alleviating fatigue and daytime sleepiness following TBI and may offer a noninvasive, safe, and nonpharmacological alternative to current treatments" (Sinclair et al., 2014).

Returning to issues with our recently returning veterans, improvised explosive devices have become the: "weapons of choice for the insurgent enemy in Iraq and Afghanistan. More soldiers are surviving these blast injuries due to improved torso protection yet are sustaining head and neck injuriess in numbers that exceed those from previous wars. Although moderate and severe traumatic head injuries are easily identified and aggressively treated, mild traumatic brain injuries (m-TBIs), or concussions, had previously been deemed inconsequential and often overlooked. Recently, however, the U.S. Department of Defense and Veterans Health Administration have placed emphasis on identifying service members at risk for m-TBI because a select number continue to have disabling symptoms that can negatively affect quality of life. Research regarding the effects and treatment of blast injury are gaining momentum, but further work needs to be accomplished" (Snell and Halter, 2010).

A 2016 study investigated mild TBI deficits in recently deployed veterans with mild traumatic brain injury (mTBI). "Veterans discharged from 2007 to 2012 were recruited from Veterans Affairs clinics... The mTBI group performed significantly worse on all of the executive and nonexecutive measurements with the exception of Category Fluency, after controlling for age, depression effort, and combat exposure. Depression and combat exposure were greater for the mTBI group" (Gaines et al., 2016). Furthermore, "The long-term effects of mTBI can leave service members with insomnia, anxiety, emotional distress, and impaired cognitive functioning. As of March 2015, there have been more than 320,344 diagnosed cases of traumatic brain injury (TBI) (in theater and in garrison) in the U.S. military since 2000" (Dickey, 2016).

As noted earlier, since deciding to include those with mild to moderate TBI in our study, an opportunity arose to research into photobiomodulation devices when my colleague, Dr. George Roselle and I copresented in September 2016 at a Society for Neurofeedback and Research (ISNR) workshop in Orlando, Florida. While there, I had the occasion to meet with Dr. Lew Lim, representing VieLight, a firm which produces photobiomodulation devices. Dr. Lim and his colleagues previously researched and ultimately developed an intranasal photobiomodulation device as early as 1995. They consequently made it commercially available around 10 years later as a noninvasive method of introducing therapeutic photonic energy into the human body.

After engaging in discussion related to our mutual interests and my personally trying out a VieLight device, I asked Dr. Lim for advice related to my work with combat veterans. He suggested consideration of the VieLight 810 which is a nonlaser, intranasal, brain photobiomodulation device. This instrument enables photonic transmission past the intranasal channel to the deep ventral brain areas. The device is pulsed at 10 Hz, 50% duty cycle. It shuts off automatically after 25 minutes of treatment time, operating on a single AA battery. The LED beam footprint spans the underside of the brain up into the midbrain area. With these specifications, the power density is 13 mW/cm^2. In a 2014 study, "near infrared energy stimulation produced beneficial effects on frontal cortex functions including working memory and sustained attention" (Gonzalez-Lima and Barrett, 2014).

Following the conference, I looked into Dr. Lim's 2015, inventor's notes focusing primarily on the following comments: "it is unlikely that photons from traditional transcranial positions on the head can reach the important primal regions of the brain that are located on the underside of the brain. In reality, these regions are much closer to the nasal areas than to the scalp. Amongst other functions, these areas govern memory, behavior and emotions. In a nutshell, they determine the "essence of the person" in everyone. The nuclei are located in these ventral areas (including the hippocampus, entorhinal cortex, and the ventral medial prefrontal cortex) and they form the important subdivisions of the key network of the brain called the default mode network (DMN).

"Since light penetrates significantly more deeply in to the brain from an intranasal position (in the nose) than from transcranial (on the head) locations, it is logical that a light source located in the intranasal position is crucial if one is looking for a comprehensive PBM treatment of the brain. On its own merit, researchers have already found that Intranasal Light Therapy has positive outcomes with neurologic conditions in humans, such as insomnia, mild cognitive impairment... Unlike the transcranial method, photons from the nasal cavity area can be efficiently directed to the brain tissues, as there are no scalp and hair to act as barriers" (Microsoft Word—Neuro inventor notes draft 1 Hil 2—Neuro-inventor-notes.pdf, 2015).

37.2 RESET Therapy

Shifting now to our RESET Therapy study entitled: *A Paradigm Shift in The Treatment of Post-Traumatic Stress Disorder: A Pilot Study*, our volunteer combat veterans were initially screened in a diagnostic intake process by the primary investigator (Lindenfeld), wherein they acknowledged experiencing a preponderance of PTSD symptoms. The volunteers denied having a major psychiatric disorder; complex PTSD; psychotic manifestations; active suicidal or homicidal ideation; severe brain injury interfering with speech, writing and purposeful action; ringing in the ears (tinnitus); or current addictive difficulties including the use of recreational agents. They provided their US Department of Defense Certificate of Release or Discharge from active duty (DD Form 214) documenting their service period and honorable discharge.

Following their diagnostic intake, selected volunteers were seen by a single, independent doctoral-level psychometrician (Rosenfield). The exception was QEEG, brain mapping data acquired by another of the study's principal investigators (Rozelle). Psychometric intervention began with the CAPS-5, which is considered to be the gold standard in PTSD assessment. (Weathers et al., 2001). It is used to assess the 20 DSM-5 PTSD symptoms on a 0–4 scale with a possible total of 80 points.

Included in the CAPS-5 are questions that target the onset and duration of symptoms, subjective distress, impact of symptoms on social and occupational functioning, improvement in symptoms since a previous administration, overall response validity, overall PTSD severity, and specifications for the dissociative subtype (depersonalization and derealization). This was followed by a neuropsychological computer-based assessment called the CNS vital signs (CNSVS). This neuropsychological computer-based assessment is designed to evaluate the neurocognitive status of patients.

A neurocognitive report is produced which represents the raw test scores, domain scores with a comparison of age-adjusted norms, and an overall neurocognition index. Specifically, the instrument measures eleven basic brain functions including composite memory, verbal memory, visual memory, executive function, processing speed, psychomotor speed, reaction time, complex attention, cognitive flexibility, simple visual attention, and motor speed. The psychometric characteristics of the tests in the CNSVS battery are very similar to the characteristics of the conventional neuropsychological tests upon which they are based. "CNSVS is suitable for use as a screening instrument, or as a serial assessment measure" (Gualtieri and Johnson, 2006). Finally, the RESET Therapy treatment was delivered by one PhD level clinical psychologist, (Lindenfeld) who had extensive experience in the treatment process.

RESET Therapy is the treatment process utilizing the BAUD and its protocols developed by Dr. Frank Lawlis in 2006 that interfere with targeted trauma memories. Through the use of binaural sound, it blocks repeated restoration of trauma material through the reconsolidation process after it is selectively lit up in the emotional part of the brain through the patient's intentional focus. I use the term target in RESET Therapy and suggest to my patient that we are going to turn off the switch in the brain that produces the PTSD symptoms (Lindenfeld and Bruursema, 2015). For those individuals whose trauma is specific rather than developmental such as that which occurs in adverse childhood experiences, remediation tends to be rapid and permanent.

Two dials of the BAUD volume controller were initially adjusted by the patient with eyes closed to assure a balanced binaural sound effect. Then the counselor/therapist slowly adjusted the frequency knob, to a level where the patient noticed a resonance effect with the targeted trauma. Subsequently, the disrupter knob was adjusted across a beat

TABLE 37.1 Pre and posttreatment (4 sessions) RESET Therapy CAPS-5 scores.

ID	Pretest	Posttest	Difference (%)
V-2	52	05	90
V-4	49	22	55
GW-1	47	04	91
GW-2	77	01	98
I/A-1	69	24	65
I/A-3	74	37	50

frequency in the 0–20 Hz range to a point where the veteran noticed a dropping out of the emotional component of the activated emotional material. The desired effect of the binaural sound was to disrupt the memory reconsolidation process.

Because it was necessary to identify the trauma frequency, it was explained in detail to the subject that the treatment should light up the target in the emotional region (limbic system) of the brain. To accomplish this, it was necessary to use sensory aspects of a selected trauma rather than merely thinking about it. Thus, the subject was informed that the best way to create this experience was to imagine being in it fully and completely by bringing in all of the involved senses and thoughts that were present at the time of the experience, including sight, sound, smell, skin sensation, etc.

An explanation was provided about the subjective units of distress scale (SUDS), which ranges from a level of 0–100 and has been used to determine the subjective level of intensity of the targeted trauma (Kaplan and Smith, 1995). An intensity score was determined by the veteran at the most intensive point during the period of imagery and then again following the treatment intervention. The therapist observed and monitored physiological reactivity including facial expression, respiration rate, muscle tension, etc. The veteran was instructed to hand signal at the point of self-activation of a selected trauma at which time the therapist would begin to slowly adjust the frequency knob of the BAUD.

The veteran was also instructed to hand-signal again when the discomfort level increased to a maximum level. His/her observed physiological indicators and sense of self-awareness were also used to help verify the most sensitive frequency level. While, alternative means of tuning in for those who were dissociated from their bodily sensation were available through hands-on training opportunities leading to certification in this procedure, this intervention was not necessary. When there was certainty that the frequency setting resonated with the trauma target, the final stage of the tuning-in procedure took place. The final step involved the use of the disrupter knob of the BAUD, which was adjusted to a beat frequency at which the veteran perceived that the target was beginning to weaken in intensity. At this point, a 5-minute tuning in trial was begun. SUDS levels were acquired both prior to and following this 5-minute trial.

Following four, 1-hour treatment sessions wherein the average administration of the binaural sound (RESET) was for a 15- to 20-minute period, we found that the largest effect of the treatment with this small sample came from the CAPS-5 data. All six subjects showed sharp reductions in their CAPS-5 score with the mean CAPS-5 score falling from 61.3 (SD 13.5) to 15.5 (SD 14.3). Sample median scores agreed well with the mean scores.

Note the pretreatment versus the posttreatment changes on the CAPS-5 in Table 37.1 of our six volunteers following four RESET Therapy treatments. Typically, scores of 50 and above are confirmation of the presence of PTSD although indicators need to be present in all categories. The following represent the era the Veteran served in: V = Vietnam; GW = Gulf Wars; IA = Iraq/Afghanistan. Positive changes forthcoming from the treatment range from an amazing high of 98% to a low of 50% following the posttest. All veterans in the group no longer qualify for the diagnosis of PTSD according to the CAPS-5 criteria. The level of *t*-test significance was: $P = .00025$. Our volunteer veteran (GW-2), on his third CAPS-5 administration, 1 year from the first administration, indicated a score of 0.

Two of our volunteers out of the six (GW-2, I/A-3), reported having the dual diagnoses of PTSD and TBI during their initial intake interview. When the treatment phase ended, only GW-2 indicated interest in later utilization of photobiomodulation treatment for his TBI condition. Paradoxically, his treatment results revealed the greatest degree of change from his pretreatment PTSD level (98%). Similar results were evident on his pre- and postbrain mapping material while his TBI indicators remained constant.

Based on these results, we were able to assume that PTSD symptomatology were cleared from the current functioning level and consequently, difficulties related to memory, fatigue, and the sustaining of attention were likely associated with his remaining TBI condition. We now shift to the case study aspect of this chapter.

37.3 Case study

At his diagnostic intake, this veteran (GW-2) reported extremely high levels of intrusive memories, psychological distress, and emotional physiological reactivity. He also reported a great deal of anger toward his family and those around him. He reported an instance of threatening a driver and his passengers at a traffic stop with a handgun. His wife was extremely concerned about his behavior in the home.

Historically, he reported a brief loss of consciousness in a service-connected vehicle accident where the vehicle he was in, rolled over. He recalls being nauseous and dizzy for a few days following this incident. In his younger years, he also engaged in a great deal of contact sports remembering being dinged a number of times. He denied any specific cognitive problems prior to treatment.

While in Bosnia in 2002, as a member of the NATO peacekeeping force, this US army sergeant was involved in the discovery of as many as 10 holes in the ground where rampaging Serbs attempted to conceal their atrocities since 1995. One such site contained 417 corpses. The sergeant watched grieving survivors flock to the excavation sites, scouring the remains for clues to the identity of the victims. Even after 7 years in the ground, the handiwork of an "ethnic cleansing" massacre outside Srebrenica emitted a stench that made him gag.

He took a toothbrush and tried scrubbing it from his nostrils until they bled. "But it was all up here," he said as he pointed a forefinger to his head. "You can never forget seeing these mangled, dismembered bodies, especially the children," recalls this 43-year-old father. "You could pull out a little shirt and you could shake bones from it. Or there'd be a little skull, and you could see where the machetes had hit them. Or a shirt might have a missing sleeve, and you knew what happened."

What this veteran saw, left an imprint that followed him home. His internalized anger evolved into road-rage. Most of those around him never saw a full portrait of his Jekyll-Hyde personality that: kept losing it in traffic; became hypervigilant for tailgaters; was ready to bolt from his SUV at the next intersection to physically confront the perceived offenders. They didn't see the guns he planted around his home hidden beneath the kitchen sink, below the dining room table, behind the shower head, behind the refrigerator, some within reach of his young children. Nor were they around for the nightmares filled with imagery of Bosnian corpses—"it was like there was somebody pulling me down into that hole"—from which he awoke in panic, ready to swing.

"Ultimately, I was Baker Acted, (involuntarily hospitalized) and found myself in the mental health ward at a VA hospital in footed pajamas and they're checking me for scars, abrasions, needle marks. My psychiatrist was convinced I was an avid cocaine user because that's how he said I was able to stay up at all hours to complete all my projects."

Following his completed RESET Therapy treatment sequence, this veteran reported that almost all of his distressing symptoms had abated. He added that certain stimuli could still elicit some distress, but that this effect was manageable. On post-RESET therapy testing, he no longer met the criteria for a diagnosis of PTSD although no cognitive measure was found to be lower on posttest results. He compared his computer-generated electroencephalogram (EEG) brain images, that were recorded as he revisited his own overseas trauma, to that of an electrical storm.

"You know how the weather radar has looked in a storm with angry reds and orange? My initial brain map looked like I was in the middle of an intensive downpour. It was like six inches of rain and flooding over the next hour in every adjacent direction. After my treatment experience, my angry reds and oranges had dissolved into balmy greens. It was all clear skies—debilitating symptoms obliterated."

Most remarkable were the anecdotal reports of the man following the post-RESET therapy assessment. He reported that he went to the VFW for the first time, and has felt sufficiently comfortable and safe to resume using his nighttime CPAP. He reported that visiting family members who were unaware of the man's participation in a PTSD treatment remarked that a behavioral change was evident. He quoted visiting family members as saying, "He seemed to be more eager to participate in group activities, such as going to the beach."

He described going on an "Easter egg hunt," not for Easter eggs, but for the weapons he had secreted around his home. He reported removing six guns from hiding places in his house, in places where he felt vulnerable. He reported, "I didn't know I had that many" in the shower, behind the toilet, in his shaving drawer, underneath the dining room table, and mounted in the cupboard behind the dishes. Reminded that he has children in the house who would have had access to these weapons, the veteran indicated that he now felt that he had narrowly escaped tragedy. "Now I sleep great. I went from driving this big-ass V-8 Tundra with god knows how many guns I had in it, praying for a confrontation on the road, to driving a four-cylinder Honda Hybrid getting 49 miles a gallon in the right lane. How about that?"

A meeting was scheduled with the veteran in order to discuss the next phase of his treatment. He understood that he would be provided with a VieLight 810 intranasal, brain photobiomodulation device. The use of this instrument was fully explained and he agreed to use it at least one to two times per day over the course of a threemonth period in order to maximize the effect. He also understood that he would be reevaluated after approximately three months of use in order to compare EEG brain maps and CNSVS changes.

Following his 3-month trial, subjectively, he reported that: "medically, my cholesterol levels have reduced by around 20% as measured against my prior blood test at the VA so that's one less medication (Simvastatin), that I'm required to take daily. My doctor's first response was to pat me on the shoulder for being a good soldier for taking my medication as I was supposed to. I told him that I was taking part in this intranasal light study for soldiers with TBI. He thought that I was involved with some 'do it yourself' kit. I'm also off the blood pressure medication" (Lisinopril).

"Typically, around this time of year, I experience a nasal infection and in fact, I was contemplating some kind of nasal surgery. I have not experienced the kind of nasal problem I have had in recent years. With my sleep apnea, typically I used it for seven days a week and currently, I'm using it three times a week. It started off with a 'No Mask Monday' and then proceeded to a 'Without It Wednesday' to a 'Take It Off Thursday'. The cool thing about this treatment is not knowing what to expect and then being able to tell you what I got".

"My fatigue experience has reduced from a full two-hour movie to a one-half hour sitcom. My prior experience with headaches have minimized to at least 40% below when I started using the light. This makes me much more productive with favorable household chores. I've also experienced an increased interest and productivity in intimacy matters."

"With short-term memory matters, previously I had four little units that, when I pressed them, they would activate some location that would be revealed on my phone through an app, such as locating my keys. Within the last three to four weeks, I haven't had to use the memory device at all. With long-term memory, it seems like I can retain information for a longer period of time. It hasn't been necessary for others to remind me of things like dental appointments that are scheduled in the future. Additionally, my patience seems to have increased particularly with my kids. My driving has improved. I've come to realize that I don't always need to be driving at around 90 mph. I can't think of any particular recent incident where someone cut me off and me becoming emotional about it like I did in the past. Overall, this light stuff is amazing."

Treatment results: The CNSVS is a computerized neuropsychological test designed to evaluate the neurocognitive status of patients. It covers a range of mental processes. A neurocognitive report presents the raw test scores, domain scores with a comparison of age-adjusted norms, and an overall neurocognition index. Specifically, the instrument measures eleven basic brain functions including composite memory, verbal memory, visual memory, executive function, processing speed, psychomotor speed, reaction time, complex attention, cognitive flexibility, simple visual attention, and motor speed. The psychometric characteristics of the tests in the CNSVS battery are very similar to the characteristics of the conventional neuropsychological tests upon which they are based.

Our volunteer's scores were all determined to be valid in all three treatment administrations. In looking at the data, the neurocognition index appears to have improved between the first and second test remaining in the average range across all administrations. Tests that measure higher levels of cognitive functioning reveal an upward trend including the following: verbal memory; cognitive flexibility; executive functioning; complex attention. The scores in this particular cluster increased from the average range to above average. Our speculation is that the volunteer's use of the VieLight 810 at least, partially contributed to this change.

Weakening tendencies were also evident in a number of scores including: composite memory and simple attention. Slippage is also present in reaction time (Low) as well as visual memory (low average). Certainly, some of findings may be due to natural variation in scoring. As noted with the higher cognitive functioning scores, the overall trend is encouraging particularly we only used the intranasal unit. All standard scores on the CNSVS for the three evaluation experiences are provided in Table 37.2.

Neuropsych questionnaire: The NPQ-short form (NPQ-45) is used to scores patient responses in terms of 20 symptom clusters: inattention, hyperactivity-impulsivity, learning problems, memory, anxiety, panic, agoraphobia, obsessions and compulsions, social anxiety, depression, mood instability, mania, aggression, psychosis, somatization, fatigue, sleep, suicide, pain, and substance abuse. The NPQ is not meant to be a diagnostic indicator but it is reliable (patients tested twice, patient-observer pairs, two observers) and discriminates patients with different diagnoses. Scores generated by the NPQ correlate reasonably well with commonly used rating scales, and the test is sensitive to the effects of treatment (Gualtieri, 2007). Scores are based on a scale of 0 (not a problem) to 300 (severe). As a rule, scores above

TABLE 37.2 CNS-vital signs.

Domain scores	Pre-RESET	Post-RESET	VieLight 810
	Standard score 3/01/17	Standard score 3/27/17	Standard score 3/01/18
Neurocognition index	94	103	101
Composite memory	104	114	100
Verbal memory	95	109	112
Visual memory	110	113	89
Psychomotor speed	95	104	107
Reaction time	76	88	74
Complex attention	103	108	111
Cognitive flexibility	92	103	112
Processing speed	90	101	101
Executive function	92	103	112
Simple attention	106	94	94
Motor speed	102	105	109

TABLE 37.3 Neuropsych questionnaire (NPQ) SF-45.

	Pre-RESET 3/01/17	Post-RESET 3/27/17	VieLight 810 3/01/18
Attention	280 (Severe)	140 (Mild)	0 (Not a problem)
Memory	250 (Moderate)	200 (Moderate)	0 (Not a problem)
Panic	267 (Severe)	0 (Not a problem)	0 (Not a problem)
Mood stability	300 (Severe)	100 (Mild)	0 (Not a problem)
Fatigue	300 (Severe)	100 (Mild)	0 (Not a problem)
Suicide	233 (Moderate)	33 (Not a problem)	0 (Not a problem)
Impulsive	240 (Moderate)	120 (Mild)	40 (Not a problem)
Anxiety	267 (Severe)	133 (Mild)	0 (Not a problem)
Depression	260 (Severe)	100 (Mild)	0 (Not a problem)
Aggression	275 (Severe)	75 (Mild)	0 (Not a problem)
Sleep	300 (Severe)	50 (Not a problem)	0 (Not a problem)
Pain	225 (Moderate)	100 (Mild)	0 (Not a problem)

225 indicate a severe problem; scores from 150 to 224 indicate a moderate problem; scores from 75 to 145 indicate a mild problem.

Our volunteer's results on this instrument parallel his other comorbid measures. His primary complaint pertaining to aggression is dramatically reduced. Other measures appear to relate to his assumed TBI condition such as his memory score, which is the only remaining item still in the moderate range following his treatment experience. His score on the suicide item is noteworthy in that he no longer views this as a primary problem for him. The same may be said for his prior sleep issues (Table. 37.3).

Photobiomodulation: In our study, prior to the use of intranasal photobiomodulation, (VieLight 810) and after this treatment, nineteen channels of raw EEG data were collected from our volunteer utilizing a BrainMaster Discovery amplifier using an ElectroCap with standard international 10–20 electrode placements. Baseline recording conditions were eyes closed resting for 5 minutes. Our volunteer then was asked to trigger a traumatic memory (same for each recording) for another 5 minutes by experiencing it on a sensory level similar to when the event actually occurred.

The collected data was imported into Neuroguide for visual inspection, selection of artifact free EEG, and statistical analysis (LSNDBweb.pdf, n.d.) The data were then compared to the Neuroguide reference normative data base matched for age, gender, and handedness (Thatcher, 2011).

We found that prior to the provision of photobiomodulation, the RESET therapy intervention produced significant changes in the volunteer's PTSD trigger conditions resulting in the remediation of his trauma-related symptoms. However, the resting QEEG baseline remained stable, thus serving as a pretreatment baseline for the current investigation of the effect of near-infrared (nIR) neuromodulation on his TBI condition.

The QEEG test is used to evaluate the nature and severity of deregulation in the brain such as in mild to moderate traumatic brain injury (MTBI). It provides a quantitative assessment of areas of brain dysfunction and information on impaired conduction and connectivity between different regional neural networks in the brain. The assessment of impaired connectivity is based on abnormal measurements of coherence and phase. Furthermore, the TBI discriminant and concussion index provides information on the presence of a pattern in the EEG that is often found in patients with a history of mTBI. The TBI discriminant and concussion index also provides information about connectivity and excitability of brain regions.

The initial prephotobiomodulation treatment summary analysis revealed significant slowing with elevated bitemporal delta (1–4 Hz) and bilateral elevated theta (4–8 Hz), especially in parietal and temporal areas. Asymmetry measures pointed to right temporal dysregulation. Hypo-coherence was evident in delta. As noted earlier, the post-PTSD intervention resting baseline showed no discernible difference.

As evident in Fig. 37.1, the TBI discriminant places the subject solidly in the mild brain injury group with a moderate level of severity. Ramifications for daily living include reduced executive functioning, sleep impairment, short-term memory impairment, emotional dysregulation, and poor impulse control. He reported particular difficulties with higher level cognitive functioning. As also reported in his narrative, he specified particular problems with short and long-term memory.

As evident in Fig. 37.2, posttreatment nIR application revealed impressive diminishment in the volunteer's TBI discriminant and concussion index. Also, post-nIR brain mapping revealed significant reduction in delta and a slight reduction in theta. Amplitude asymmetry improved. There was an increase in left posterior temporal hypo coherence. Furthermore, the TBI discriminant analysis was significant at the 97.5 probability level with moderate level of severity before nIR home treatment. After nIR treatment the TBI analysis showed improvement with an 85% probability and mild level of severity (Fig. 37.2). This means that statistical markers commonly found in mild brain injury have moved toward normalization.

Discussion: The central theme of this chapter is the utilization of dual, sensory based forms of clinical intervention that appear to enhance synaptic plasticity in a complementary manner. The choice of beginning treatment intervention with RESET therapy first derived from the author's clinical experience and candidly, comfort level with the procedure. This leads to the question of: would similar results have been forthcoming if the photobiomodulation procedure were used initially or solely? Furthermore, one might ask, does the apparent slippage in some of the scores on the CNSVS derive from the sole use of an intranasal infrared light procedure.

A follow-up question would focus on the role that the DMN has within overall cognitive functioning and how it might contribute within the context of the remediation process. This DMN is perceived to be a system of highly correlated interacting brain regions that are distinct from other networks in the brain. The network has been found to be most active when the brain is awake but, in a resting state. It has also been found to be active: "when the individual is thinking about others, thinking about themselves, remembering the past, and planning for the future" (Buckner et al., 2008). "The network activates by default when a person is not involved in a task... The DMN has been shown to be negatively correlated with other networks in the brain such as attention networks" (Broyd et al., 2009).

A VieLight Neuro Gamma unit designed to enhance the DMN has been personally utilized by one of the authors (Lindenfeld). This cranial placed unit delivers a 40 Hz pulse that correlates with gamma brainwave oscillations. It is intended to enhance the functioning of the cortical nodes of the DMN (bilateral mesial prefrontal cortex, precuneus/posterior cingulate cortex, angular gyrus, and hippocampus). The device is known to enhance memory function and cognition. Specifically, we wonder if the VieLight 810 intranasal unit and Neuro Gamma unit were to have been used in concert, might we have seen improvement rather than slippage in visual memory, reaction time, and simple attention on the CNSVS.

Montage: LinkEars EEG ID: 1.001.01_EC

Traumatic brain injury discriminant analysis*

TBI discriminant score = 0.18 TBI probability index = 97.5%

The TBI Probability Index is the subject's probability of membership in the mild traumatic brain injury population. (see Thatcher et al, EEG and Clin. Neurophysiol., 73: 93-106, 1989.)

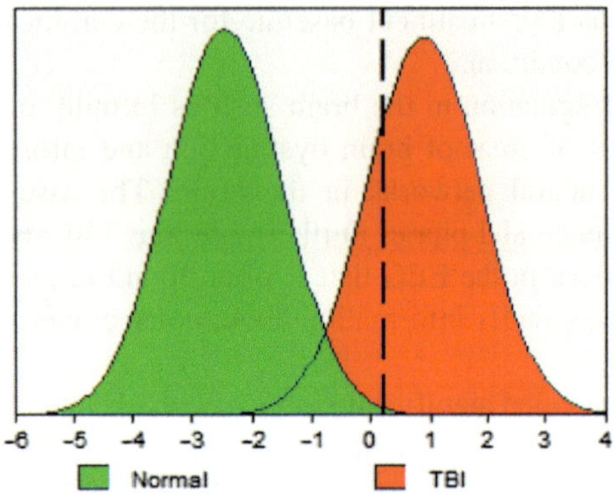

			RAW	Z
FP1-F3	COH	Theta	84.53	0.20
T3-T5	COH	Beta	66.70	1.11
C3-P3	COH	Beta	70.94	-1.57
FP2-F4	PHA	Beta	0.26	-0.66
F3-F4	PHA	Beta	-0.23	-0.63
F4-T6	AMP	Alpha	-5.69	0.50
F8-T6	AMP	Alpha	-51.67	0.62
F4-T6	AMP	Beta	56.36	1.66
F8-T6	AMP	Beta	14.69	1.90
F3-O1	AMP	Alpha	-49.80	0.33
F4-O2	AMP	Alpha	-40.69	0.43
F7-O1	AMP	Alpha	97.95	-0.24
F4-O2	AMP	Beta	33.62	1.26
P3	RP	Alpha	39.35	-0.96
P4	RP	Alpha	40.37	-0.95
O1	RP	Alpha	43.37	-0.99
O2	RP	Alpha	41.62	-1.29
T4	RP	Alpha	20.75	-1.55
T5	RP	Alpha	37.25	-1.16
T6	RP	Alpha	34.41	-1.50

TBI severity index = 5.29

This severity score places the patient in the MODERATE range of severity.

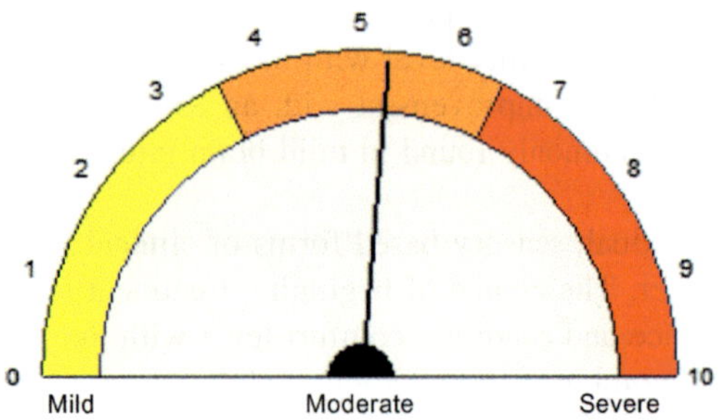

			RAW	Z
FP1-C3	COH	Delta	62.23	0.72
FP1-FP2	COH	Theta	77.39	-1.41
O1-F7	COH	Alpha	20.82	-0.65
O2-T6	COH	Alpha	85.13	0.17
P3-O1	COH	Beta	79.33	0.55
FP1-T3	PHA	Theta	-0.01	-1.68
T3-T4	PHA	Theta	-5.34	-0.52
O1-F7	PHA	Alpha	-41.98	1.29
F7-F8	PHA	Alpha	-1.35	-0.25
T5-T6	PHA	Beta	2.35	0.28
C3-F7	AMP	Delta	21.34	-0.61
FP2-F4	AMP	Delta	51.73	1.41
C4-F8	AMP	Delta	-20.69	-2.31
O1-O2	AMP	Theta	-2.15	-0.20
P3-F7	AMP	Alpha	106.92	0.17
FP2-P4	AMP	Alpha	-71.93	0.03

The TBI Severity Index is an estimate of the neurological severity of injury. (see Thatcher et al, J. Neuropsychiatry and Clinical Neuroscience, 13(1): 77-87, 2001.)

FIGURE 37.1 Traumatic brain injury pre-nIR.

Included in this chapter was reference to a pilot study wherein six combat-veterans from three different eras experienced rather dramatic remission of their PTSD symptoms including comorbid features. We are in the process of submitting our findings to peer reviewed journals in order to make this intervention more widely available to the professional community. With 20 veterans and active military personnel continuing to take their lives on a daily basis, RESET

Montage: LinkEars EEG ID: 3.001.01_EC

Traumatic brain injury discriminant analysis*

TBI discriminant score = –0.37 TBI probability index = 85.0%

The TBI Probability Index is the subject's probability of membership in the mild traumatic brain injury population. (see Thatcher et al, EEG and Clin. Neurophysiol., 73: 93-106, 1989.)

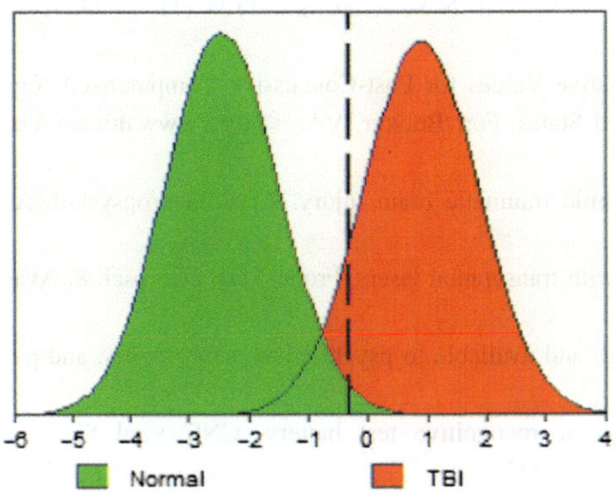

			RAW	Z
FP1-F3	COH	Theta	75.37	-1.08
T3-T5	COH	Beta	69.49	1.29
C3-P3	COH	Beta	75.76	-0.75
FP2-F4	PHA	Beta	-0.08	-1.16
F3-F4	PHA	Beta	0.06	-1.16
F4-T6	AMP	Alpha	-41.46	-0.30
F8-T6	AMP	Alpha	-79.97	-0.03
F4-T6	AMP	Beta	38.26	1.06
F8-T6	AMP	Beta	-2.42	1.36
F3-O1	AMP	Alpha	-40.57	0.50
F4-O2	AMP	Alpha	-43.85	0.38
F7-O1	AMP	Alpha	79.40	-0.62
F4-O2	AMP	Beta	13.39	0.85
P3	RP	Alpha	39.19	-0.98
P4	RP	Alpha	48.18	-0.43
O1	RP	Alpha	48.98	-0.69
O2	RP	Alpha	48.32	-0.90
T4	RP	Alpha	36.64	-0.29
T5	RP	Alpha	42.33	-0.86
T6	RP	Alpha	48.18	-0.60

TBI severity index = 1.79

This severity score places the patient in the MILD range of severity.

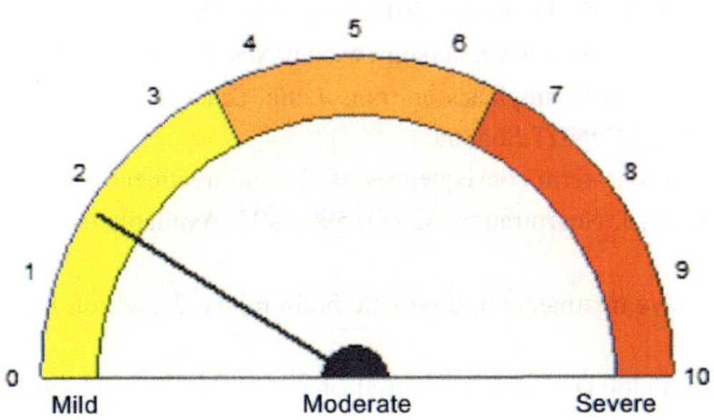

			RAW	Z
FP1-C3	COH	Delta	41.33	-1.02
FP1-FP2	COH	Theta	75.05	-1.79
O1-F7	COH	Alpha	34.64	0.28
O2-T6	COH	Alpha	86.07	0.32
P3-O1	COH	Beta	77.38	0.29
FP1-T3	PHA	Theta	4.65	0.63
T3-T4	PHA	Theta	34.81	0.96
O1-F7	PHA	Alpha	-35.03	1.12
F7-F8	PHA	Alpha	0.54	-0.96
T5-T6	PHA	Beta	1.92	0.07
C3-F7	AMP	Delta	-24.05	-2.04
FP2-F4	AMP	Delta	-56.93	-1.37
C4-F8	AMP	Delta	2.13	-1.45
O1-O2	AMP	Theta	-10.02	-0.71
P3-F7	AMP	Alpha	70.10	-0.70
FP2-P4	AMP	Alpha	-93.51	-0.41

The TBI Severity Index is an estimate of the neurological severity of injury. (see Thatcher et al, J. Neuropsychiatry and Clinical Neuroscience, 13(1): 77-87, 2001.)

FIGURE 37.2 Traumatic brain Injury Post nIR.

therapy may potentially begin to alter this stark reality. In a similar spirit, our first responders experience cumulative stress within the context of their service to our communities. We are unable to capture their incidence of suicide due to the way in which data is collected for each of the professions involved in the first responder system.

Shifting now to TBI, current estimates exceed 20% among returning veterans. While infrared light offers promise, more intensive, well-funded research is clearly necessary to establish that this approach is able to ameliorate the signature wounds of war. As earlier described, we have found in our studies of combat veterans that symptoms of both PTSD and MTBI are commonly present. In this single case study, we have shown that a noninvasive simple home training option using nIR light can demonstrate change in a treated TBI combat veteran. It goes without question that based on this tentative finding that additional research with more subjects is needed to further this line of investigation.

References

Broyd, S.J., Demanuele, C., Debener, S., Helps, S.K., James, C.J., Sonuga-Barke, E.J.S., 2009. Default-mode brain dysfunction in mental disorders: a systematic review. Neurosci. Biobehav. Rev. 33 (3), 279–296. Available from: https://doi.org/10.1016/j.neubiorev.2008.09.002.

Buckner, R.L., Andrews-Hanna, J.R., Schacter, D.L., 2008. The brain's default network. Ann. N.Y. Acad. Sci. 1124 (1), 1–38. Available from: https://doi.org/10.1196/annals.1440.011.

Dickey, N.W., 2016. Review of the Scientific Evidence of Using Population Normative Values for Post-Concussive Computerized Neurocognitive Assessments. Defense Health Agency/Defense Health Board Falls Church United States, Fort Belvoir, VA. <http://www.dtic.mil/docs/citations/AD1017708>.

Gaines, K.D., Soper, H.V., Berenji, G.R., 2016. Executive functioning of combat mild traumatic brain injury. Appl. Neuropsychol. Adult 23 (2), 115–124. Available from: https://doi.org/10.1080/23279095.2015.1012762.

Gonzalez-Lima, F., Barrett, D.W., 2014. Augmentation of cognitive brain functions with transcranial lasers. Front. Syst. Neurosci. 8. Available from: https://doi.org/10.3389/fnsys.2014.00036.

Gualtieri, C.T., 2007. An internet-based symptom questionnaire that is reliable, valid, and available to psychiatrists, neurologists, and psychologists. Medscape Gen. Med. 9 (4), 3.

Gualtieri, C.T., Johnson, L.G., 2006. Reliability and validity of a computerized neurocognitive test battery, CNS Vital Signs. Arch. Clin. Neuropsychol. 21 (7), 623–643. Available from: https://doi.org/10.1016/j.acn.2006.05.007.

Henderson, T.A., 2016. Multi-watt near-infrared light therapy as a neuror.pdf. (n.d.). Retrieved from http://www.sjzsyj.org/CN/article/openArticlePDF.jsp?id = 1799

Invisible Wounds of War: Psychological and Cognitive Injuries, Their Consequences, and Services to Assist Recovery—RAND_MG720.pdf, 2008. <https://www.rand.org/content/dam/rand/pubs/monographs/2008/RAND_MG720.pdf>.

Lindenfeld, G., Bruursema, L.R., 2015. Resetting the Fear Switch in PTSD: A Novel Treatment Using Acoustical Neuromodulation to Modify Memory Reconsolidation. <http://www.academia.edu/12683048/Resetting_the_Fear_Switch_in_PTSD_A_Novel_Treatment_Using_Acoustical_Neuromodulation_to_Modify_Memory_Reconsolidation>.

LSNDBweb.pdf, n.d. <http://appliedneuroscience.com/LSNDBweb.pdf>.

maucieri-2012-tbi-article.pdf, n.d. <https://www.psychologytoday.com/files/attachments/139539/maucieri-2012-tbi-article.pdf>.

Microsoft Word—Neuro inventor notes draft 1 Hil 2—Neuro-inventor-notes.pdf, 2015. <http://www.emersonww.com/Neuro-inventor-notes.pdf>.

Morries, L.D., Cassano, P., Henderson, T.A., 2015. Treatments for traumatic brain injury with emphasis on transcranial near-infrared laser phototherapy. Neuropsychiatr. Dis. Treat 11, 2159–2175. Available from: https://doi.org/10.2147/NDT.S65809.

Phelps, T.I., Bondi, C.O., Ahmed, R.H., Olugbade, Y.T., Kline, A.E., 2015. Divergent long-term consequences of chronic treatment with haloperidol, risperidone, and bromocriptine on traumatic brain injury–induced cognitive deficits. J. Neurotrauma 32 (8), 590–597. Available from: https://doi.org/10.1089/neu.2014.3711.

Santos, J.G.R.P., dos, Paiva, W.S., Teixeira, M.J., 2016. The current role of non-invasive treatments in traumatic brain injury. J. Neurol. Disord. 4 (5). Available from: https://doi.org/10.4172/2329-6895.1000294.

Sinclair, K.L., Ponsford, J.L., Taffe, J., Lockley, S.W., Rajaratnam, S.M.W., 2014. Randomized controlled trial of light therapy for fatigue following traumatic brain injury. Neurorehabil. Neural Repair 28 (4), 303–313. Available from: https://doi.org/10.1177/1545968313508472.

Snell, F.I., Halter, M.J., 2010. A signature wound of war. J. Psychosoc. Nurs. Mental Health Serv. 1–7. Available from: https://doi.org/10.3928/02793695-20100108-02.

Thatcher, R.W., 2011. Neuropsychiatry and quantitative EEG in the 21st Century. Neuropsychiatry 1 (5), 495–514. Available from: https://www.yellowbrickprogram.com/ArticlePDF/Neuroscience-and-qEEG-21st-Century.pdf.

Traumatic Brain Injury and PTSD: Focus on Veterans—PTSD: National Center for PTSD, 2017. [General Information]. <https://www.ptsd.va.gov/professional/co-occurring/traumatic-brain-injury-ptsd.asp> (last updated 06.11.17.).

Vanderploeg, R.D., Belanger, H.G., Curtiss, G., Bowles, A.O., Cooper, D.B., 2019. Reconceptualizing rehabilitation of individuals with chronic symptoms following mild traumatic brain injury. Rehabil. Psychol. 64 (1), 1–12. Available from: https://doi.org/10.1037/rep0000255.

Weathers, F.W., Blake, D.D., Schnurr, P.P., Kaloupek, D.G., Marx, B.P., & Keane, T.M., 2013. *The Clinician-Administered PTSD Scale for DSM-5 (CAPS-5)*. - PTSD: National Center for PTSD. [General Information]. Retrieved from https://www.ptsd.va.gov/professional/assessment/adult-int/caps.asp

Willyard, C., 2015. Why is there no pill to treat concussions? Retrieved from https://www.statnews.com/2015/12/28/concussion-pill/

Chapter 38

Transcatheter intracerebral photobiomodulation in degenerative brain disorders: clinical studies (Part 1)

Ivan V. Maksimovich

Clinic of Cardiovascular Diseases Named After Most Holy John Tobolsky, Moscow, Russia

38.1 Introduction

Alzheimer's disease (AD) is the most common neurodegenerative disease (Alzheimer's Association Report, 2017). Despite many years of research, the etiology and pathogenesis of AD are not fully understood. Considering the development of AD, it is necessary to take into account that this disease arises not only due to the disorders in amyloid beta and tau metabolism in the cerebral tissue and vascular walls, but also due to disorders in cerebral angioarchitectonics, decreased cerebral microcirculation, development of hypoperfusion and hypoxia (Waldemar et al., 2007; Burton et al., 2009; Weiner et al., 2015; Maksimovich, 2008b, 2012c; Zlokovic, 2010, 2011; Iadecola, 2010; Baloiannis and Baloiannis, 2012; Grammas et al., 2014; Kimbrough et al., 2015). In this regard, cerebral small vessel disease specific for AD is recognized as one of the causes of the development of the disease (De la Torre, 2016; Baloyannis, 2015; Pantoni, 2010; Cai et al., 2015).

For the first time, back in the 1930s of the last century, the vascular contribution to the research of AD was proposed by Morel, who described the disease as cerebral dysoric or druzoidal angiopathy (Morel, 1950).

Vascular and microcirculatory changes with AD have a specific complex character and begin to develop with the reduction of cerebral capillaries (Maksimovich, 2008b, 2011; Zlokovic, 2011; Baloiannis and Baloiannis, 2012; De la Torre, 1997; Kalaria, 2002; Brown and Thore, 2011). In the tissue of the hippocampus, the number of capillaries decreases, they become thinner, their branching decreases, and as a result, distal arterial blood flow is diminished (Maksimovich, 2008b, 2012c; Zlokovic, 2011; Baloiannis and Baloiannis, 2012; Grammas et al., 2014).

The developing process leads to the development of hypovascular zones, first in the temporal area and then in the frontoparietal area (Maksimovich, 2011, 2012c, 2015; Baloiannis and Baloiannis, 2012; De la Torre, 1997; Kalaria, 2002).

Hemodynamically, the blood flowing along the arterial branches cannot pass through the contracted arterioles and capillaries, which consequently leads to the opening of arteriovenous shunts in the temporal and frontoparietal areas (Maksimovich, 2011, 2012c). This is a natural defensive reaction of the body to the disorders in blood flow along the distal arterioles and capillaries (Maksimovich, 2008b, 2012c). These arteriovenous shunts are characterized by an active dumping of arterial blood into the venous bed, which causes its overflow, stagnation, and disorders in venous outflow. This restructuring of the blood supply disturbs cerebral hemodynamics even more (Maksimovich, 2011).

Changes in the microcirculatory bed lead to hypoxia, cause death of mitochondria in the cells, damages the smooth endoplasmic reticulum and Golgi apparatus, loss of synapses, degeneration, and death of neurons (Baloiannis and Baloiannis, 2012; De la Torre, 1997, 2016; Kalaria, 2002; Maksimovich, 2011; Brown and Thore, 2011). The combination of these cerebral microcirculatory, hemodynamic changes with tissue damage determines the lesion of the neurovascular unit (Zlokovic, 2010, 2011; Iadecola, 2010).

Capillary and hemodynamic disorders affect the balance of amyloid beta causing a decrease in its excretion and an increase in its accumulation (Zlokovic, 2010, 2011; Baloyannis, 2015). This process leads to the deposition of amyloid

Photobiomodulation in the Brain. DOI: https://doi.org/10.1016/B978-0-12-815305-5.00038-5

beta in the cerebral tissue and the vascular wall, which in turn causes a decrease in the elasticity of the microvessels, and it narrows their lumen, thereby further reducing the intracerebral blood flow (Kimbrough et al., 2015).

The combination of all these pathological processes disturbs cerebral microcirculation even more and promotes the development of intracerebral hypoxia (Kimbrough et al., 2015; Cai et al., 2015; De la Torre and Stefano, 2000; Love and Miners, 2016) simultaneously causing the dysfunction of the blood–brain barrier (Bell and Zlokovic, 2009; Montagne et al., 2015).

While AD progresses, natural physiological intracerebral angiogenesis decreases (Maksimovich, 2011; Brown and Thore, 2011).

Growing cerebrovascular changes lead to increased tortuosity of the distal arterial branches and capillaries (Maksimovich, 2008b, 2012c).

The combination of these changes in cerebral angioarchitectonics and microcirculation is very specific, it can be found only in patients with AD and is combined into the concept of "dyscirculatory angiopathy of Alzheimer's type (DAAT)" (Maksimovich, 2008b, 2011, 2012a,c, 2015). Patients with other cerebrovascular and neurodegenerative diseases do not have the combination of similar changes (Maksimovich, 2008b, 2011, 2012a).

The more these disorders become pronounced, the faster the cerebral hypoperfusion and hypoxia that contribute to the development of AD are aggravated (Maksimovich, 2012a,c; De la Torre, 2016; Nelson et al., 2016). As a result, mechanisms causing the progress of the disease and its clinical symptoms work for a long time (Zlokovic, 2010; Kimbrough et al., 2015; Maksimovich, 2015).

In connection with the recently obtained data concerning hemodynamic disorders in AD, an increasing number of researchers indicates the need to develop new methods of treating this disease. These methods should be firstly directed at the restoration of cerebral microcirculatory blood supply, secondly, at normalization of the exchange of amyloid beta in cerebral tissues, and thirdly, at the development of regenerative processes in the cerebral tissue (Maksimovich, 2008b, 2011; Grammas et al., 2014; Love and Miners, 2016; Naeser and Hamblin, 2011).

It has been found out in experimental and clinical studies that natural compensatory stimulation of neurogenesis occurs along with the development of hypoxic and ischemic lesions in the brain, with neuronal regeneration taking place predominantly near the functioning vessels, especially in the hippocampal area (Jin et al., 2006; Galvan and Jin, 2007; Oron et al., 2006; Naeser and Hamblin, 2011).

One of the promising directions of research in revascularization and regeneration of cerebral tissue is the use of low-energy laser technologies also known as photobiomodulation (PBM) (Maksimovich, 2012a,c, 2015; De la Torre, 2016; Naeser and Hamblin, 2011; Hamblin, 2017).

Even in the initial period of the development of laser technology, which was connected with studies on experimental animals, it was discovered that laser energy of low power density (1–5 mW/cm^2) does not damage healthy tissues of the body, while increasing the energy of mitochondria and stimulating cellular metabolism that accelerates the regeneration of various tissues (Deviatkov, 1993).

Studies aimed at understanding the mechanisms of the effect of low-energy laser radiation on nerve tissues show high receptivity of the cerebral tissue. It is revealed that low-energy lasers restore the exchange of ATP in the mitochondria of neurons, prevent the death of neurons, and also stimulate neurogenesis, thereby causing regenerative processes in the cerebral tissue (Oron et al., 2006; Naeser and Hamblin, 2011; Hamblin, 2017).

These studies have made it possible to develop and successfully apply the method of transcranial laser PBM or low level laser therapy (LLLT) for the treatment of ischemic, traumatic cerebral lesions, dementia, and AD (Hashmi et al., 2010; Hamblin, 2018).

Since cerebral ischemic and neurodegenerative diseases develop against the background and because of blood supply disorders, the establishment of methods for treating AD should necessarily be connected with correcting the cause too, by restoring blood supply to damaged tissues (Maksimovich, 2011, 2012c; De la Torre, 2016).

Earlier, in carrying out experimental studies of the effect of laser energy on different layers of the vascular wall, it was found that in the range from 600 to 700 nm, laser wavelength is moderately absorbed by atherosclerotic tissues, intima, and the muscle layer of cerebral vessels (1.5–2 absorption units) (Fig. 38.1). The depth of penetration of laser energy into cerebral tissues is about 20–40 mm. It was also found that as the wavelength is decreased, the absorption of laser energy increases significantly, and as the wavelength is increased, the absorption decreases allowing a large part of it to pass into the surrounding tissues (Aleinikov et al., 1987a, 1987b; Maksimovich et al., 1988).

Studies on experimental animals show that with transluminal (transcatheter) application, laser energy with a wavelength of 633 nm (helium-neon laser) actively stimulates physiological angiogenesis causing rapid opening of arterial and capillary collateral vessels (Aleinikov et al., 1987a; Maksimovich, 2004).

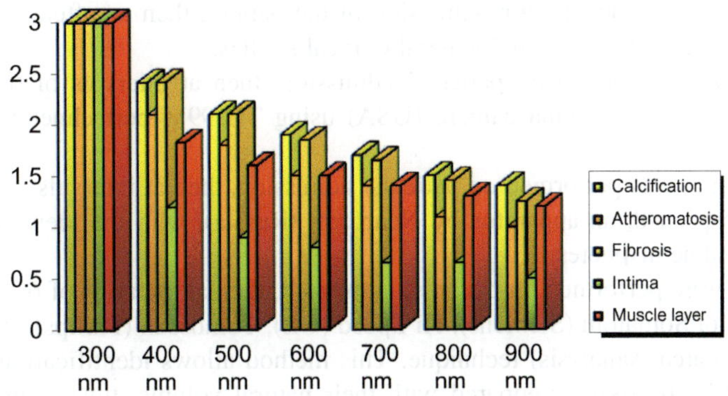

FIGURE 38.1 The graph of absorption of the vascular wall of laser energy in the wavelength range of 300–900 nm.

These studies allowed the development of the "method of transluminal (transcatheter) laser revascularization of cerebral blood vessels" (Maksimovich, 2006, 2008a). One of the aspects of this method was the restoration of cerebral distal arterial and capillary blood supply by stimulating angiogenesis by transcatheter intracerebral action of low-energy lasers, which is essentially transcatheter intracerebral PBM or transcatheter intracerebral LLLT.

Considering the question of the mechanism of low-energy laser effect on cerebral tissues, it should be noted that the effect is complex in nature and includes:

- stimulation of natural, physiological angiogenesis, which causes pronounced collateral and capillary revascularization of cerebral tissues;
- restoration of the exchange processes of ATP in the mitochondria of neurons;
- stimulation of neurogenesis in cerebral tissues.

The present work is devoted to the clinical application of transcatheter intracerebral laser (PBM) in the treatment of AD to stimulate angiogenesis, restore the cerebral distal arterial and capillary blood supply, and stimulate metabolic and regenerative processes in the cerebral tissue.

38.2 Materials and methods

All examination and transcatheter intracerebral interventions in this study were carried out with the approval of the ethics committee and with the consent of patients and their relatives.

38.2.1 Patient selection criteria

1. the consent of patients and their relatives for the examination and treatment;
2. the patients' somatic state allowing to carry out the necessary examination and treatment;
3. signs of developing dementia and cognitive disorders in the patients examined;
4. involutive changes in the temporal and frontal parietal areas of the brain.

200 people with different stages of AD were examined. Of these, 93 patients aged from 34 to 80 (mean age 67.5) were selected, 32 (34.40%) men and 61 (65.59%) women.

For the study, we selected patients with AD of varying severity but without pronounced severe comorbidities.

38.2.2 Patient examination plan

The patients were examined in accordance with the following parameters:

- Clinically, the severity of dementia was determined using the clinical dementia rating (CDR) scale (Morris, 1993). Primary testing was carried out on patients' admission, repeated at discharge, then at intervals of 6–12 months.
- Cognitive functions were assessed using the minimental state examination (MMSE) (Folstein et al., 1975). Primary testing was carried out on patients' admission, repeated at discharge, then at intervals of 6–12 months.

- Laboratory examination was conducted upon admission of the patient, then as often as it was necessary. The diagnostics included coagulatory, biochemical and general clinical studies.
- Brain scintigraphy (SG) was performed on patients' admission, then at intervals of 6–12 months. The procedure was performed with Ohio nuclear gamma camera (USA) using TC 99M pertechnetate 555 in dynamic and static modes.
- Rheoencephalography (REG) was performed on patients' admission, then at intervals of 6–12 months. The research was carried out with Reospectr-8, an apparatus by Neurosoft (Russia), with the identification of disorders in pulse blood filling in the cerebral hemispheres.
- CT and MRI of the brain were performed on patients' admission, then at intervals of 6–12 months. The examination was carried out by means of Somatom (Siemens), Hi Speed (GE), Tomoscan (Philips), Apetro Eterna (Hitachi) using the ATAA (advance tomo area Aanalysis) technique. This method allows identification of the percentage decrease in the volume of temporal lobe tissue compared with their natural volume, thus showing involutional changes in temporal lobes (Maksimovich, 2011). Using the tomography dementia rating scale (TDR), morphometric assessment of the severity of dementia and AD stages was made (Maksimovich, 2012b, 2017). The method allows to morphometrically determine the stage of dementia in AD by the severity of atrophic changes in cerebral temporal lobes shown during CT or MRI.
- Cerebral multigated angiography (MUGA) was performed according to the classical technique with transfemoral access using Advantx (GE) devices. Capillary blood flow was recorded using the angio vision computer program (Maksimovich, 2011, 2012c). This program allows you to identify the changes in the density and number of black pixels in the examined area of the cerebral image during the passage of the radiopaque substance through the microvessels (Omnipack 350) (Maksimovich, 2011, 2012a). In the subsequent period, Philips proposed a similar program supplemented with 2D effects (De Lin and Jackson, 2012).

38.2.2.1 Patient examination results

- CDR testing detected dementia of varying severity in 83 (89.25%) patients;
- MMSE testing showed cognitive disorders of varying severity in all 93 (100%) patients;
- Laboratory examination revealed no significant abnormalities;
- SG showed a decrease in blood flow in the cerebral hemispheres in all 93 (100%) patients;
- REG showed a decrease in the volume of pulse blood filling in carotid basins in all 93 (100%) patients;
- CT and MRI demonstrated involuntary changes in the brain manifested by atrophy of temporal lobes of varying severity in all 93 (100%) patients. Morphometric determination of dementia stages was performed in all 93 (100%) patients.
- MUGA revealed the phenomena of DAAT in all 93 (100%) patients (Table 38.1 and Fig. 38.2).

TABLE 38.1 DAAT signs in the patients examined.

Cerebral vascular changes	Total
Reduction of the number of arterioles and capillaries in the temporal and frontoparietal areas with the formation of hypovascular zones	93 (100%)
Development of multiple arteriovenous shunts in the basins of arteries which supply blood to the temporal and frontoparietal areas	93 (100%)
Early venous dumping of arterial blood through arteriovenous shunts into the venous channel	93 (100%)
Abnormal widening of the lateral venous branches receiving blood from the arteriovenous shunts	84 (90.32%)
Congestion of venous blood at the border of the frontal and parietal areas caused by excessively increased blood flow from the arteriovenous shunts	85 (91.40%)
Increased looping of distal intracranial arterial branches	74 (79.57%)

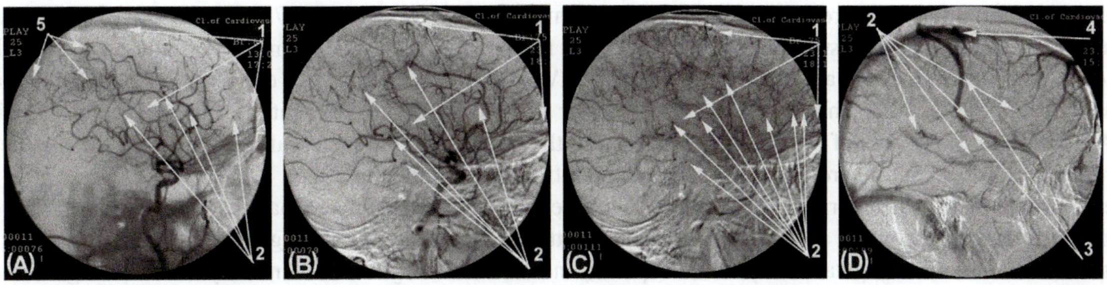

FIGURE 38.2 DAAT signs according to cerebral MUGA data. (A) Left internal carotid artery angiogram, arterial phase; (B) left internal carotid artery angiogram, arterial phase; (C) left internal carotid artery angiogram, capillary phase; (D) left internal carotid artery angiogram, venous phase. (1) Reduction of the number of arterioles and capillaries in the temporal and frontal parietal areas with the formation of hypovascular zones; (2) multiple arteriovenous shunts; (3) abnormally widened lateral venous branch; (4) stagnation of venous blood.

38.2.2.2 Patient selection

According to the morphometry of atrophic changes in the temporal lobes, severity of dementia and cognitive impairment, the examined patients were divided into the following groups:

- Ten (10.75%) people had preclinical AD stage of TDR-0. These patients showed no dementia but they had increasing memory disorders and demonstrated a decrease in cognitive function to 26–28 MMSE points. Cerebral involutional changes were manifested by atrophy of the temporal lobes with a decrease in tissue volume by 4%–8%. Each of these patients had direct relatives suffering from AD.
- Twenty-six (27.96%) people had early AD stage of TDR-1. These patients had mild dementia with a 2-year history of the disease. Clinically, dementia corresponded to the level of CDR-1, cognitive functions were reduced to 20–25 MMSE points. Cerebral involutional changes were manifested by atrophy of the temporal lobes with a decrease in tissue volume by 9%–18%;
- Forty (43.01%) people had middle AD stage of TDR-2. These patients had moderate dementia with an anamnesis of 2–6 years. Clinically, dementia corresponded to the level of CDR-2, cognitive functions were reduced to 12–19 MMSE points. Cerebral involutional changes were manifested by atrophy of the temporal lobes with a decrease in tissue volume by 19%–32%;
- Seventeen (18.28%) people had severe AD stage of TDR-3. These patients had severe dementia with an anamnesis of 7–12 years. Clinically, dementia corresponded to the level of CDR-3, cognitive functions were reduced to 7–11 MMSE points. Cerebral involutional changes were manifested by atrophy of the temporal lobes with a decrease in tissue volume by 33%–62%;

38.2.3 Treatment methods

38.2.3.1 Test group

Forty-eight (51.61%) patients, 17 (35.42%) men and 31 (54.58%) women, underwent transcatheter intracerebral PBM.

These patients had the following severity of dementia: TDR-0—4 (8.33%) patients; TDR-1—16 (33.33%) patients; TDR-2—21 (43.76%) patients; TDR-3—7 (14.58%) patients.

In the group of patients with TDR-0, the preclinical stage of the disease, transcatheter intracerebral PBM was performed against the background of increasing memory impairment.

In the groups of patients with TDR-1, TDR-2, TDR-3 with AD, transcatheter intracerebral PBM was performed in the period from 12 months to 12 years after the onset of the symptoms of the disease.

38.2.3.1.1 The method of transcatheter intracerebral photobiomodulation

Under local anesthesia, using Seldinger technique, the common femoral artery was catheterized with an introducer of 6–7 F in diameter. Through this introducer, guiding catheters were conducted and installed in the corresponding branches of the internal carotid artery. Then, a thin, flexible, fiber-optic light guide tool (of a diameter from 25 to 100 μm) connected to the laser is introduced. The fiber-optic instrument is taken to the distal sections of the anterior and middle cerebral arteries where laser radiation is performed (Aleinikov et al., 1987a). If necessary, for X-ray television inspection, periodically, in small doses, a solution of radiopaque substance is injected (Omnipack 350). PBM is carried out using a helium-neon laser with a power of 25 mW. The total time of laser exposure is about 20–40 minutes (Aleinikov et al., 1987b; Maksimovich et al., 1988). After the intracerebral intervention, repeated cerebral MUGA is made. According to the results of the study, we determine the severity of intracerebral angiogenesis, the degree of collateral revascularization, and the restoration of microcirculation (Maksimovich, 2015).

After tracateter and itracerebral PBM, the patients received disaggregant, anticoagulant, antioxidant, vasodilating and nootropic therapy according to generally accepted patterns. The patients were given: Aspirin, depending on the parameters of the blood coagulation system, Heparin, indirect anticoagulants, Pentoxifylline 100 mg, Complamin 150 mg, Inosin 200 mg, Nootropil (Piracetam) 1200 mg (or Gliatilin 1000 mg) intravenously, with a drop counter, No. 10–15, and then were given pills. In the period that followed, the courses of pills were repeated two times a year. The patients did not receive any specific therapy aimed at treating AD.

38.2.3.2 Control group

Forty-five (48.39%) patients, 16 (35.36%) men and 39 (64.44%) women, received conservative treatment.

These patients had the following severity of dementia: TDR-0—6 (13.33%) patients; TDR-1—13 (28.89%) patients; TDR-2—15 (33.34%) patients; TDR-3—11 (24.44%) patients.

Conservative treatment was performed according to conventional schemes (Moskvin, 2008; Maksimovich, 2013). Patients from TDR-0 group received: Nootropil (Piracetam) 2400 mg per day (courses of 3–4 months), or Gliatilin 1200 mg per day (courses of 4–6 months). Patients from groups TDR-1, TDR-2, TDR-3 received Memantine 5–20 mg per day or Rivastigmine 3–12 mg per day. Simultaneously, patients of all groups received vasoactive drugs: Pentoxifylline 800 mg per day for 3 months and Complamin 450 mg per day for 2–3 months, which was repeated two times a year.

38.3 Results

38.3.1 Test group

38.3.1.1 Immediate result after transcatheter intracerebral photobiomodulation

In the period of transcatheter intracerebral PBM, as well as in the early and distant period, there were no complications in any case.

According to cerebral MUGA carried out after the transcatheter intracerebral PBM, all 48 (100%) patients demonstrated a good immediate result manifested in stimulation of angiogenesis, improvement and restoration of collateral and capillary blood supply, reduction of arteriovenous shunts, as well as improvement of venous outflow (Figs. 38.3 and 38.4).

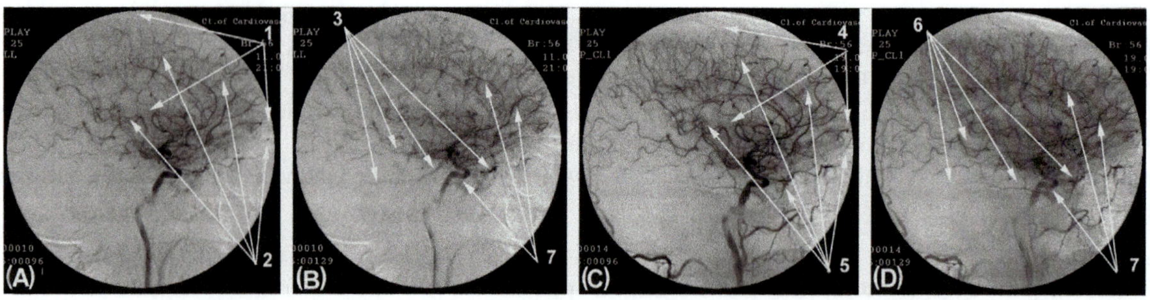

FIGURE 38.3 **Patient A**, 56 years old, male (TDR-2) Left internal carotid artery angiogram before and transcatheter intracerebral laser PBM. (A) Arterial phase before transcatheter intracerebral laser PBM; (B) late arterial phase before transcatheter intracerebral laser PBM; (C) arterial phase after transcatheter intracerebral laser PBM; (D) late arterial phase after transcatheter intracerebral laser PBM; (1) hypovascular areas in temporal and frontoparietal regions; (2) multiple arteriovenous shunts in temporal and frontoparietal regions; (3) dumping of arterial blood through arteriovenous shunts into venous channel, simultaneous contrast of arteries and veins; (4) stimulation of atheiogenesis, collateral and capillary bed recovery in temporal and frontoparietal region; (5) closing of arteriovenous shunts; (6) reduction of arteriovenous shunts in temporal and frontoparietal region, no simultaneous contrasting of arteries and veins; (7) intracerebral arterial branches.

38.3.1.2 *Early period (1–6 months) after transcatheter intracerebral photobiomodulation*

In all four (100%) patients with preclinical AD (TDR-0), memory improvement was clinically noted, as well as recovery of cognitive functions to 28–30 MMSE points.

According to CT and MRI data, all four (100%) patients showed an increase in the temporal lobes volume along with narrowing of Sylvian fissures and restoration of the level of the subarachnoid space.

According to SG and REG, all four (100%) patients demonstrated normalization of the rate of blood flow and pulse blood flow in the cerebral hemispheres.

All 16 (100%) patients with early AD (TDR-1) clinically showed a decline in dementia. Six (37.50%) of them featured improvement of cognitive functions up to 25–26 points and 10 (62.50%) of them to 27–28 MMSE points.

According to CT and MRI data, all 16 (100%) patients had an increase in the volume of cerebral temporal lobes along with narrowing of Sylvian fissures and restoration of the level of the subarachnoid space.

According to SG and REG data, all 16 (100%) patients showed normalization of blood flow and pulse blood filling in the cerebral hemispheres.

In all 21 (100%) patients with middle AD stage (TDR-2), a decrease in the level of dementia was clinically noted. Twelve (57.14%) of them showed improvement in cognitive functions to the level of 19–20 MMSE points and nine (42.86%) patients—to the level of 21–22 point by MMSE.

According to CT and MRI data, all 21 (100%) patients had an increase in the volume of cerebral temporal lobes, accompanied by a narrowing of Sylvian fissures and a decrease in the subarachnoid space.

According to SG and REG, all 21 (100%) patients had signs of restoring the rate of blood flow and pulse blood flow in the cerebral hemispheres.

In all seven (100%) patients with late AD stage (TDR-3), a decrease in the level of dementia was clinically noted, accompanied by improvement in cognitive functions to 11–12 MMSE point.

According to CT and MRI data, all seven (100%) patients had a tendency to an increase in the volume of cerebral temporal lobes accompanied by narrowing of Sylvian fissures and a decrease in the subarachnoid space.

According to SG and REG data, all seven (100%) patients showed signs of restoration in the rate of blood flow and pulse blood filling in the cerebral hemispheres.

38.3.1.3 *Long-term (1–7 years) results after transcatheter intracerebral photobiomodulation*

In 12 months after transcatheter intracerebral PBM, all four(100%) patients with preclinical AD (TDR-0) stage showed restoration of memory and cognitive functions up to 28–30 MMSE points (Table 38.2).

According to CT and MRI data, all four (100%) patients demonstrated recovery of cerebral temporal lobes to the age norm, along with narrowing of Sylvian fissures and restoration of the subarachnoid space. Consequently, all four (100%) patients had normal indicators of healthy people in accordance with the criteria listed above.

SG and REG demonstrated that all four (100%) patients showed normalization of the rate of blood flow and pulse blood filling in the cerebral hemispheres.

The observed positive dynamics lasted during the entire period of supervision.

In 12 months after the treatment, all 16 (100%) patients with early AD stage (TDR-1) had no evidence of dementia and showed persistent recovery of cognitive functions up to 27–28 MMSE points (Table 38.2).

According to CT and MRI data, all 16 (100%) patients demonstrated almost complete recovery of the cerebral temporal lobes to the age norm, along with narrowing of Sylvian fissures and restoration of the subarachnoid space. Consequently, all 16 (100%) patients were transferred to TDR-0 group according to the criteria listed above.

According to SG and REG data, all 16 (100%) patients showed normalization of the rates of blood flow and pulse blood filling in the cerebral hemispheres.

The observed positive dynamics lasted during the entire period of supervision

In 12 months after the treatment, all 21 (100%) patients with middle level AD stage (TDR-2) had a decrease in the level of dementia and improvement of cognitive functions up to 21–22 MMSE points. In a more distant period, 12 (57.14%) patients showed a further decrease in the level of dementia and further restoration of cognitive functions up to 23–25 MMSE points. In nine (42.86%) patients, cognitive functions were maintained at the same level of 21–22 MMSE points (Table 38.2).

The condition of all 21 (100%) patients remained stable for 4 years, after which there was a tendency toward reduction in cognitive functions down to 20–21 MMSE points.

According to CT and MRI data, all 21 (100%) patients had an increase in the volume of cerebral temporal lobes, along with narrowing of Sylvian fissures and subarachnoid space, in 12 months after the treatment. In a more distant period, 12 (57.14%) patients showed further growing involutional changes of temporal lobes to 15%–18% tissue volume decrease. In nine (42.86%) patients, further decrease in involutional changes was not detected (Fig. 38.5). Consequently, 21 (100%) patients were transferred to TDR-1 group according to the criteria listed above.

According to SG and REG data, all 21 (100%) patients retained positive dynamics of blood flow rate and pulse blood filling in the cerebral hemispheres.

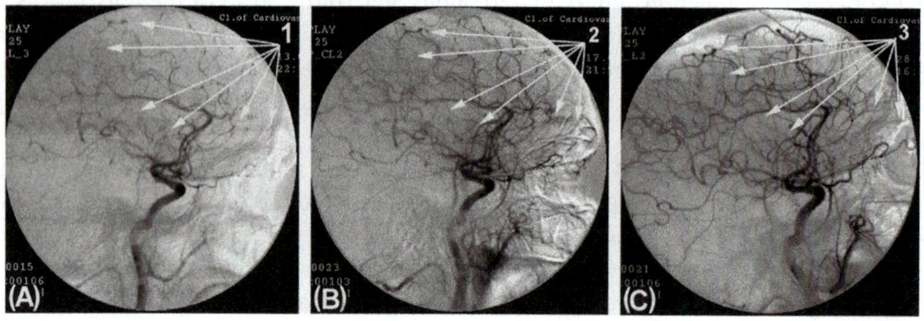

FIGURE 38.4 **Patient T**, 61 years old, male (TDR-2) Left internal carotid artery angiogram before and after transcatheter intracerebral laser PBM. (A) Arterial phase, before transcatheter intracerebral laser PBM; (B) arterial phase, after transcatheter intracerebral laser PBM; (C) arterial phase, 5 years after transcatheter intracerebral laser PBM. (1) hypovascular areas in temporal and frontoparietal regions; (2) stimulation of atheiogenesis, collateral and capillary bed recovery in temporal and frontoparietal region; (3) further progression of atheiogenesis, enhancement of collateral and capillary bed in temporal and frontoparietal region.

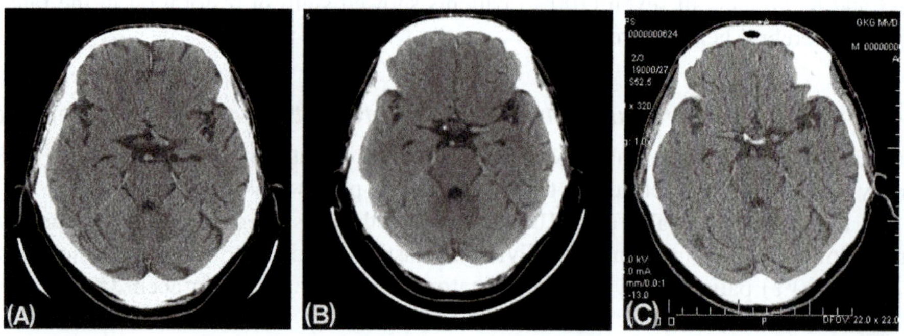

FIGURE 38.5 **Same patient T**, 61 years old, male (TDR-2) Brain CT before and after transcatheter intracerebral laser PBM. (A) Before transcatheter intracerebral laser PBM. Decrease in temporal lobes volume: left by 22%, right by 28%; (B) after transcatheter intracerebral laser PBM. Recovery of temporal lobes volume: left by10%, right by 12%; (C) 5 years after transcatheter intracerebral laser PBM. Recovery of temporal lobes volume to age norm.

In 12 months after the treatment, all seven (100%) patients with severe AD stage (TDR-3) showed a decrease in the level of dementia and improvement of cognitive functions. Four (57.14%) of them had an improvement in cognitive functions up to 11–14 MMSE points, three (42.86%) of them—up to 15–19 MMSE points (Table 38.2).

According to CT and MRI data, all seven (100%) patients showed a decrease in the atrophy of cerebral temporal lobes to 28%–34%, along with narrowing of Sylvian fissures and the subarachnoid space, 12 months after the treatment. Consequently, five (71.43%) patients were transferred to TDR-2 group, two (28.57%) patients remained in TDR-3 group.

The stable condition lasted for about 2–2.5 years, after which cognitive functions dropped to 11–14 MMSE points.

According to SG and REG data, all seven (100%) patients retained positive dynamics of the rate of the blood flow and pulse blood flow in the cerebral hemispheres.

Between 2 and 6 years after transcatheter intracerebral PBM, repeated cerebral MUGA was carried out for nine (21.74%) patients. In all nine (21.74%) patients, further progression of angiogenesis was observed, accompanied by collateral and capillary revascularization (Fig. 38.4).

TABLE 38.2 Clinical results of treatment of patients of test and control groups in the long-term period.

Group	Summary frequency table						
	Severity degree before treatment	Severity degree after treatment healthy	Severity degree after treatment TDR-0	Severity degree after treatment TDR-1	Severity degree after treatment TDR-2	Severity degree after treatment TDR-3	Totals
Test group	TDR-0	4	0	0	0	0	4
Column percentage		100.00%	0.00%	0.00%	0.00%	0.00%	
Row percentage		100.00%	0.00%	0.00%	0.00%	0.00%	
Sample proportion		4.30%	0.00%	0.00%	0.00%	0.00%	4.30%
Test group	TDR-1	0	16	0	0	0	16
Column percentage		0.00%	100.00%	0.00%	0.00%		
Row percentage		0.00%	100.00%	0.00%	0.00%	0.00%	
Sample proportion		0.00%	17.20%	0.00%	0.00%		17.20%
Test group	TDR-2	0	0	21	0	0	21
Column percentage		0.00%	0.00%	100.00%	0.00%		
Row percentage		0.00%	0.00%	100.00%	0.00%	0.00%	
Sample proportion		0.00%	0.00%	22.58%	0.00%	0.00%	22.58%
Test group	TDR-3	0	0	0	5	2	7
Column percentage		0.00%	0.00%	0.00%	100.00%		
Row percentage		0.00%	0.00%	0.00%	71.43%	0.00%	
Sample proportion		0.00%	0.00%	0.00%	5.38%	0.00%	7.53%
TOTALS of test group		**4**	**16**	**21**	**5**	**2**	**48**
Sample proportion		8.33%	33.33%	43.75%	10.42%	4.17%	100.00%
Control group	TDR-0	0	3	3	0	0	6
Column percentage			100.00%	60.00%	0.00%	0.00%	
Row percentage		0.00%	50.00%	50.00%	0.00%	0.00%	
Sample proportion		0.00%	3.23%	3.23%	0.00%	0.00%	6.45%
Control group	TDR-1	0	0	2	11	0	13
Column percentage		0.00%	0.00%	40.00%	100.00%	0.00%	
Row percentage		0.00%	0.00%	15.38%	84.62%	0.00%	
Sample proportion		0.00%	0.00%	2.25%	12.36%	0.00%	13.98%
Control group	TDR-2	0	0	0	0	15	15
Column percentage			0.00%	0.00%	0.00%	57.69%	
Row percentage		0.00%	0.00%	0.00%	0.00%	100.00%	
Sample proportion		0.00%	0.00%	0.00%	0.00%	16.13%	16.13%
Control group	TDR-3	0	0	0	0	11	11
Column percentage			0.00%	0.00%	0.00%	42.31%	
Row percentage		0.00%	0.00%	0.00%	0.00%	100.00%	
Sample proportion		0.00%	0.00%	0.00%	0.00%	11.83%	11.83%
TOTALS of control group		**0**	**3**	**5**	**11**	**26**	**45**
Sample proportion		0.00%	6.67%	11.11%	24.44%	57.78%	100.00%
Column total		**4**	**19**	**26**	**16**	**28**	**93**
Total percentage		4.30%	20.43%	27.96%	17.20%	30.11%	100.00%

The original data, the distribution of which is given in the summary table, were analyzed using the nonparametric Mann-Whitney test. The statistical analysis showed that at the beginning of the therapy there were no significant differences between the Test and Control groups ($P > .05$), while after the treatment these differences became significant ($P < .01$). The analysis of the indicator dynamics by means of the Mann-Whitney test also revealed significant differences between the groups: the effect of the therapy in the Test group was significantly higher than in the Control Group ($P < .01$).

38.3.2 Control group

38.3.2.1 Early period (1–6 months) after the beginning of conservative treatment

Patients with preclinical AD stage (TDR-0) had a tendency to improve their memory and restore cognitive functions to the level of 27–28 MMSE points against the background of conservative treatment.

Patients from TDR-1 group had a tendency to moderate stabilization of their condition.

Patients from TDR-2 and TDR-3 groups showed a tendency for a further increase in dementia and a decrease in cognitive functions.

According to CT and MRI, no structural cerebral changes were detected in all patients of the control group.

SG and REG showed a weak tendency to restore the rates of blood flow and pulse blood filling in the cerebral hemispheres in all 45 (100%) patients of the control group.

38.3.2.2 The long-term period (1–5 years) after the beginning of conservative treatment

In the period of 2 years after the onset of conservative treatment, all six (100%) patients with preclinical AD stage (TDR-0) experienced improvement in memory and stabilization of cognitive functions at the level of 27–29 MMSE points. In a more distant period, three (50.00%) of them showed no signs of dementia, another three (50.00%) of them noted the appearance of initial signs of dementia accompanied by a decrease in cognitive functions to the level of 24–25 points (Table 38.2).

According to CT and MRI, one (16.67%) patient showed no signs of growing involutional changes, which are manifested by a decrease in tissue volume of the temporal lobes. Three (50.00%) patients showed a tendency to a decrease in the tissue volume of temporal lobes, two (33.33%) showed a marked decrease in the tissue volume of the temporal lobes to 14%–18% (Fig. 38.6A and B).

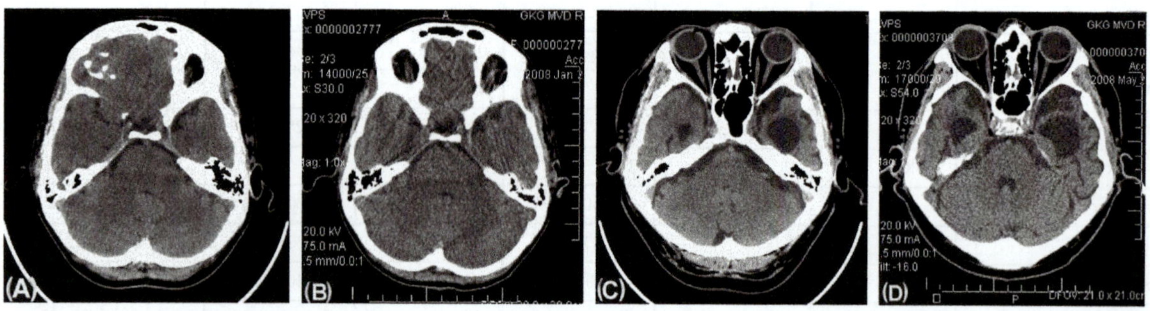

FIGURE 38.6 Brain CT of control group patients before and after 2 years after the beginning of conservative treatment. (A) **Patient Z,** 34 years old, female preclinical stage (TDR-0) before conservative treatment. Decrease in temporal lobes volume: left by 4%, right by 8%; (B) **same patient Z,** 36 years old, female (TDR-1) after 2 years of conservative treatment. Increased atrophy. Decrease in temporal lobes volume: left by 14%, right by 18%; (C) **Patient S,** 67 years old, male (TDR-3) before conservative treatment. Decrease in temporal lobes volume: Total atrophy of the temporal lobes 58%; (D) **same patient S,** 69 years old, female (TDR-3) after 2 years of conservative treatment. Increased atrophy. Increase in the atrophy up to 64%.

Consequently, three (50.00%) of the patients, in accordance with the criteria listed above, were transferred to the group of patients with AD TDR-1, and three (50.00%) patients remained in TDR-0 group (Table 38.2).

SG and REG showed that all six (100%) patients had a tendency to restore the rate of blood flow and pulse blood filling in the cerebral hemispheres.

For 2–3 years after the beginning of conservative treatment, all 13 (100%) patients with early AD stage (TDR-1) showed stabilization of the level of dementia and the level of cognitive functions. In another 3 years and further on, all 13 (100%) patients in this group experienced an increase in the signs of dementia and a decline in cognitive functions. In two (15.38%) patients, cognitive functions decreased to 20–21 MMSE points, in 11 (84.62%) patients to 18–19 MMSE points.

According to CT and MRI data, in the period of 2–3 years after the beginning of conservative treatment, all 13 (100%) patients showed an increase in cerebral involutional changes with growing cerebral atrophy and with a decrease in the volume of temporal lobes to 12%–24%.

Consequently, 11 (84.62%) patients were transferred to TDR-2 group in accordance with the criteria listed above. Two (15.38%) patients remained in TDR-1 group (Table 38.1).

SG and REG demonstrated that all 13 (100%) patients showed weak dynamics in restoring the speed of blood flow and pulse blood filling.

All 15 (100%) patients with middle AD stage (TDR-2) featured growing dementia and growing cognitive impairment to 11–12 MMSE points in a period of more than 12 months after the beginning of conservative treatment. In a more distant period, there was a further increase in dementia and a decrease in cognitive functions to the level of 9–11 MMSE points.

According to CT and MRI data, in the period of more than 1.5–2 years after the onset of conservative treatment, all 15 (100%) patients had an increase in cerebral involutional changes with growing atrophy and a decrease in the volume of temporal lobes to 33%–40%.

Consequently, all 15 (100%) patients were transferred to TDR-3 group according to the criteria listed above (Table 38.2).

SG and REG showed that nine (60.00%) patients had a decrease in the rates of cerebral blood flow and pulse blood filling, and six (40.00%) had no changes in the indicators.

All 11 (100%) patients with severe AD stage (TDR-3) had growing dementia and a decrease in cognitive functions to 7–8 MMSE points 12 months after the beginning of conservative treatment.

According to CT and MRI data, all 11 patients showed an increase in cerebral involutional changes with growing atrophy and with a decrease in the volume of temporal lobes to 40%–55% (Fig. 38.6C and D) (Table 38.2).

SG and REG showed that three (27.27%) patients had a decrease in the rates of cerebral blood flow and pulse blood filling, and eight (72.73%) patients featured no pronounced dynamics of the indicators.

38.4 Discussion

The etiology and pathogenesis of AD are complex. On the one hand, the deposits of TAU and amyloid beta in the cerebral tissue and vascular wall lead to neurodegeneration. On the other hand, specific vascular, microvascular and venous changes manifested in DAAT lead to disorders in blood supply, cause hypoxia, and disrupt metabolic processes in the cerebral tissue (Maksimovich, 2008b, 2011, 2012a, 2013). These two processes are inextricably linked; they exacerbate each other, lead to cerebral dysfunction, neurodegeneration, and AD development (De la Torre, 2016; Maksimovich, 2011, 2015; Iadecola, 2004).

Since DAAT is detected not only in patients with clinical AD stages (TDR-1, TDR-2, TDR-3), but also in people with the preclinical AD stage of TDR-0, as well as in descendants of patients with AD (Maksimovich, 2012c), it is obvious that these changes are primary and are possibly inherited while AD develops.

These factors must be taken into account when working out the ways of treating this disease.

The use of Memantine or Rivastigmine is aimed at improving metabolic processes in the cerebral tissue, at stimulating the transmission of nerve impulses and at inhibiting the formation of amyloid beta. These drugs are not effective enough and do not always give the desired therapeutic effect (Matsunaga et al., 2014; Grossberg et al., 2015). The use of Pentoxifylline and Complamin to improve cerebral microcirculation is effective during early, unexpressed lesions of the cerebral microcirculatory bed (Maksimovich, 2015).

As a result, the treatment proved effective only for those control group patients who had early AD stages. Moreover, the effect was temporary and resulted in stabilizing the present state, not in improving it.

In test group patients, the mechanism of action of the transcatheter intracerebral laser PBM is complex (De la Torre, 2016; Maksimovich, 2015; Naeser and Hamblin, 2011; Hamblin, 2017; Rojas et al., 2012).

As shown by the present studies, the laser energy in the red area of the spectrum stimulates the process of physiological angiogenesis, causes collateral and capillary revascularization, reduces hypoxia, and restores the nutrition of the cerebral tissues. The obtained data are confirmed by studies carried out earlier (Maksimovich, 2004, 2015). Cerebral revascularization contributes to the recovery of amyloid beta release process, and it also contributes to the normalization of its metabolism in tissues, which is confirmed by studies conducted by other authors (Zlokovic, 2011; Maksimovich, 2006, 2015; Bell and Zlokovic, 2009).

During transcatheter intracerebral application, laser energy penetrates into cerebral tissues for 2–4 cm affecting the mitochondrial apparatus in neurons and triggering the activation and recovery of cerebral energy and metabolic processes. At the same time, laser action stimulates neurogenesis and causes regenerative processes in the cerebral tissues. The data obtained are supported by studies of other authors performed with transcranial PBM, which also show that, on the one hand, laser energy acting on mitochondria stimulates the exchange of ATP thereby restoring the energy resource of cells, and on the other hand, stimulates neurogenesis, which contributes to the recovery of cerebral tissue (Naeser and Hamblin, 2011; Hamblin, 2017, 2018; Hashmi et al., 2010; Song et al., 2012; Rojas et al., 2012; Konstantinović et al., 2013; Purushothuman et al., 2014).

It is interesting to note that in some works of other authors it was reported that when transcranial PBM is performed, direct destruction of amyloid beta by laser energy cannot be excluded (Yang et al., 2010).

For test group patients, transcatheter intracerebral PBM led to reduction in cerebral involutional changes and an increase in the tissue volume of temporal lobes, which indicates the development of regenerative processes in the cerebral tissue.

The resulting neuroprotective effect was observed for a long time and was accompanied by a decrease in the level of dementia in CDR testing and TDR determination, as well as by improvement or restoration of cognitive functions during MMSE testing.

As a result, patients with preclinical AD TDR-0 stage were transferred to the category of practically healthy people. Patients with more severe mental disorders and more severe stages of the disease (TDR-1, TDR-2, TDR-3) were transferred to groups with an earlier stage of AD.

In patients from TDR-0 and TDR-1 groups with early stages of the disease, the effect was observed throughout the observation period in the period of more than 10 years.

In patients from groups TDR-2, TDR-3 with later stages of the disease and more severe neurodegenerative and involutional changes in the cerebral tissue, a pronounced positive effect was noted for 2.5–4 years. In the subsequent period, there was a decrease in cognitive functions.

The difference in the results obtained in test group patients with different stages of AD is due to the fact that patients with advanced stages of the disease develop severe, widespread changes in cellular and tissue structures with widespread deposition of amyloid beta. Severe neurodegeneration and marked deposition of amyloid beta in the tissues do not allow full restoration of structural changes in the cerebral tissue.

38.5 Conclusion

Transcatheter intracerebral application of low-energy lasers in the red area of the spectrum during PBM is a pathogenetically substantiated, physiological and effective method of treating AD. Stimulation of angiogenesis leads to the restoration of collateral and capillary blood supply thereby producing cerebral revascularization and restoring the exchange of amyloid beta in the tissue. Stimulation of ATP metabolism in the mitochondria of neurons improves cellular and tissue metabolism. All these mechanisms lead to stimulation of neurogenesis, which causes regenerative processes in the cerebral tissue.

The use of transcatheter intracerebral laser PBM allows to reduce the level of dementia and improve cognitive and mental functions for a long time.

It cannot be excluded that the use of this method for people with preclinical AD stage of TDR-0, as well as in patients with early clinical stage of TDR-1, will make it possible to avoid the appearance or further progression of the disease.

The application of the method for patients with advanced AD stages of TDR-2, TDR-3 allows to return the patients to a more active, full life for a long time.

38.6 Conflict of interest

The authors declared no potential conflicts of interest with respect to the research, authorship, and/or publication of this article.

38.7 Funding

The authors received no financial support for the research, authorship, and/or publication of this article.

References

Aleinikov, V.S., Maksimovich, I.V., Masychev, V.I., et al., 1987a. Investigation of the regimes of action and the mechanism of destruction of atherosclerotic plaques in laser angioplasty. Electr. Ind. 1, 9–15.

Aleinikov, V.S., Maksimovich, I.V., Masychev, V.I., 1987b. The use of laser radiation in total atherosclerotic lesions of the vascular bed. Electr. Ind. 8, 15–18.

Alzheimer's Association Report, 2017. Alzheimer's disease facts and figures. 2017. Alzheimer's Dementia 13 (4), 325–373.

Baloiannis, S.J., Baloiannis, I.S., 2012. The vascular factor in Alzheimer's disease: a study in Golgi technique and electron microscopy. J. Neurol. Sci. 322, 117–121.

Baloyannis, S.J., 2015. Brain capillaries in Alzheimer's disease. Hell J. Nucl. Med. 18 (Suppl. 1), 152.

Bell, R.D., Zlokovic, B.V., 2009. Neurovascular mechanisms and blood-brain barrier disorder in Alzheimer's disease. Acta Neuropathol. 118, 103–113.

Brown, W.R., Thore, C.R., 2011. Review: Cerebral microvascular pathology in ageing and neurodegeneration. Neuropathol. Appl. Neurobiol. 37 (1), 56–74.

Burton, E.J., Barber, R., Mukaetova-Ladinska, E.B., Robson, J., Perry, R.H., Jaros, E., et al., 2009. Medial temporal lobe atrophy on MRI differentiates Alzheimer's disease from dementia with Lewy bodies and vascular cognitive impairment: a prospective study with pathological verification of diagnosis. Brain 132 (Pt1), 195–203.

Cai, Z., Wang, C., He, W., Tu, H., Tang, Z., Xiao, M., et al., 2015. Cerebral small vessel disease and Alzheimer's disease. Clin. Interv. Aging 23 (10), 1695–1704.

De la Torre, J.C., 1997. Hemodynamic consequences of deformed microvessels in the brain in Alzheimer's disease. Ann. N.Y. Acad. Sci. 26, 75–91.

De la Torre, J.C., 2016. Cerebral perfusion enhancing interventions: a new strategy for the prevention of Alzheimer dementia. Brain Pathol. 26 (5), 618–631.

De Lin, M., Jackson, E.F., 2012. Applications of imaging technology in radiation research. Radiat. Res. 177 (4), 387–397.

De la Torre, J.C., Stefano, G.B., 2000. Evidence that Alzheimer's disease is a microvascular disorder: the role of constitutive nitric oxide. Brain Res. Brain Res. Rev. 34 (3), 119–136.

Deviatkov, N.D., 1993. Application of electronics in medicine and biology. Electr. Equip.: Microwave Equip. 1, 66–76.

Folstein, M.F., Folstein, S.E., McHugh, P.R., 1975. "Mini-mental state". A practical method for grading the cognitive state of patients for the clinician. J. Psychiatr. Res. 12 (3), 189–198.

Galvan, V., Jin, K., 2007. Neurogenesis in the aging brain. Clin. Interv. Aging 2, 605–610.

Grammas, P., Martinez, J., Sanchez, A., Sanchez, A., Yin, X., Riley, J., et al., 2014. A new paradigm for the treatment of Alzheimer's disease: targeting vascular activation. J. Alzheimers Dis. 40 (3), 619–630.

Grossberg, G.T., Farlow, M.R., Meng, X., Velting, D.M., 2015. Evaluating high-dose rivastigmine patch in severe Alzheimer's disease: analyses with concomitant memantine usage as a factor. Curr. Alzheimer Res. 12 (1), 53–60.

Hamblin, M.R., 2017. Mechanisms and mitochondrial redox signaling in photobiomodulation. Photochem. Photobiol. 94, 199–212.

Hamblin, M.R., 2018. Photobiomodulation for traumatic brain injury and stroke. J. Neurosci. Res. 96, 731–743.

Hashmi, J.T., Huang, Y.Y., Osmani, B.Z., Sharma, S.K., Naeser, M.A., Hamblin, M.R., 2010. Role of low-level laser therapy in neurorehabilitation. Arch. Phys. Med. Rehabil. 2 (12 Suppl. 2), S292–S305.

Iadecola, C., 2004. Neurovascular regulation in the normal brain and in Alzheimer's disease. Nat. Rev. Neurosci. 5 (5), 347–360.

Iadecola, C., 2010. The overlap between neurodegenerative and vascular factors in the pathogenesis of dementia. Acta Neuropathol. 120 (3), 287–396.

Jin, K., Wang, X., Xie, L., Mao, X.O., Zhu, W., Wang, Y., et al., 2006. Evidence for stroke-induced neurogenesis in the human brain. Proc. Natl. Acad. Sci. U.S.A. 103, 13,198–13,202.

Kalaria, R., 2002. Small vessel disease and Alzheimer's dementia: pathological considerations. Cerebrovasc. Dis. 13, 48–52.

Kimbrough, I.F., Robel, S., Roberson, E.D., Sontheimer, H., 2015. Vascular amyloidosis impairs the gliovascular unit in a mouse model of Alzheimer's disease. Brain 138 (Pt 12), 3716–3733.

Konstantinović, L.M., Jelić, M.B., Jeremić, A., Stevanović, V.B., Milanović, S.D., Filipović, S.R., 2013. Transcranial application of near-infrared low-level laser can modulate cortical excitability. Lasers Surg. Med. 45 (10), 648–653.

Love, S., Miners, J.S., 2016. Cerebral hypoperfusion and the energy deficit in Alzheimer's Disease. Brain Pathol. 26 (5), 607–617.

Maksimovich, I.V., 2004. Transluminal Laser Angioplasty in the Treatment of Ischemic Brain Leasons (M.D. Ph.D. dissertation), Russian University of Friendship of the People, Moscow.

Maksimovich, I.V., 2006. Method for Endovascular Treatment of Alzheimer's Disease. Patent, No.2297860 Russian.

Maksimovich, I.V., 2008a. Method and Device for Endovascular Treatment of Alzheimer's Disease. Patent No. 7389776.USA.

Maksimovich, I.V., 2008b. Radiodiagnostics of Alzheimer's disease. Diagn. Interv. Radiol. 4, 27–38.

Maksimovich, I.V., 2011. Dyscirculatory angiopathy of Alzheimer's type. J. Behav. Brain Sci. 1 (2), 57–68.

Maksimovich, I.V., 2012a. Vascular factors in Alzheimer's disease. Health 4 (9A), 735–742.

Maksimovich, I.V., 2012b. The tomography dementia rating scale (TDR) – The rating scale of Alzheimer's disease stages. Health 4 (9A), 712–719.

Maksimovich, I.V., 2012c. Certain new aspects of etiology and pathogenesis of Alzheimer's disease. Adv. Alzheimer's Dis. 1 (3), 68–76.

Maksimovich, I.V., 2013. Disorders of cerebrovascular angioarchitectonics and microcirculation in the etiology and pathogenesis of Alzheimer's disease. Adv. Alzheimer's Dis. 2 (4), 171–181.

Maksimovich, I.V., 2015. Dementia and cognitive impairment reduction after laser transcatheter treatment of Alzheimer's disease. World J. Neurosci. 5 (3), 189–203.

Maksimovich, I.V., 2017. Morphometric definition of Alzheimer's disease stages by means of The Tomography Dementia Rating Scale (TDR). Brain Disord. Ther. 6, 1–4.

Maksimovich, I.V., Lesnoy, M.I., Zubov, V.V., 1988. Transluminal laser angioplasty with low-intensity laser radiation. Application of Lasers in Surgery and Medicine. 21–25, Samarkand.

Matsunaga, S., Kishi, T., Iwata, N., 2014. Combination therapy with cholinesterase inhibitors and memantine for Alzheimer's disease: a systematic review and meta-analysis. Int. J. Neuropsychopharmacol. 18 (5), pii: pyu 115.

Montagne, A., Barnes, S.R., Sweeney, M.D., Halliday, M.R., Sagare, A.P., Zhao, Z., et al., 2015. Blood-brain barrier breakdown in the aging human hippocampus. Neuron 85 (2), 296–302.

Morel, F., 1950. An apparently dyshoric and topical angiopathy. Monatsschr. Psychiatr. Neurol. 120 (5–6), 352–357.

Morris, J.C., 1993. The Clinical Dementia Rating (CDR): current version and scoring rules. Neurology 11 (43), 2412–2414.

Naeser, M.A., Hamblin, M.R., 2011. Potential for transcraniallaser or LED therapy to treatstroke, traumatic brain injury, and neurodegenerative disease. Photomed. Laser Surg. 29 (7), 443–446.

Nelson, A.R., Sweeney, M.D., Sagare, A.P., Zlokovic, B.V., 2016. Neurovascular dysfunction and neurodegeneration in dementia and Alzheimer's disease. Biochim. Biophys. Acta 1862 (5), 887–900.

Oron, A., Oron, U., Chen, J., Eilam, A., Zhang, C., Sadeh, M., et al., 2006. Low-level laser therapy applied transcranially to rats after induction of stroke significantly reduces long-term neurological deficits. Stroke 37, 2620–2624.

Pantoni, L., 2010. Cerebral small vessel disease: from pathogenesis and clinical characteristics to therapeutic challenges. Lancet Neurol. 9 (7), 689–701.

Purushothuman, S., Johnstone, D.M., Nandasena, C., Mitrofanis, J., Stone, J., 2014. Photobiomodulation with near infrared light mitigates Alzheimer's disease-related pathology in cerebral cortex-evidence from two transgenic mouse models. Alzheimers Res. Ther. 6 (1), 2.

Rojas, J.C., Bruchey, A.K., Gonzalez-Lima, F., 2012. Low-level light therapy improves cortical metabolic capacity and memory retention. J. Alzheimers Dis. 32 (3), 741–752.

Song, S., Zhou, F., Chen, W.R., 2012. Low-level laser therapy regulates microglial function through Src-mediated signaling pathways: implications for neurodegenerative diseases. J. Neuroinflammation 18 (9), 219.

Waldemar, G., Dubois, B., Emre, M., Georges, J., McKeith, I.G., Rossor, M., et al., 2007. Recommendations for the diagnosis and management of Alzheimer's disease and other disorders associated with dementia: EFNS guideline. Eur. J. Neurol. 14 (1), e1–e26.

Weiner, W.W., Veitch, D.P., Aisen, P.S., Beckett, L.A., Cairns, N.J., Cedarbaum, J., et al., 2015. 2014 Update of the Alzheimer's Disease Neuroimaging Initiative: a review of papers published since its inception. J. Alzheimer's Dementia 11 (6), e1–e120.

Yang, X., Askarova, S., Sheng, W., Chen, J.K., Sun, A.Y., Sun, G.Y., et al., 2010. Low energy laser light (632.8 nm) suppresses amyloid-β peptide-induced oxidative and inflammatory responses in astrocytes. Neuroscience 171 (3), 859–868.

Zlokovic, B.V., 2010. Neurodegeneration and the neurovascular unit. Nat. Med. 16 (12), 1370–1371.

Zlokovic, B.V., 2011. Neurovascular pathways to neurodegeneration in Alzheimer's disease and other disorders. Nat. Rev. Neurosci. 3, 723–738.

Chapter 39

Transcatheter intracerebral photobiomodulation in ischemic brain disorders: clinical studies (Part 2)

Ivan V. Maksimovich
Clinic of Cardiovascular Diseases Named after Most Holy John Tobolsky, Moscow, Russia

39.1 Introduction

Ischemic lesions of the brain are increasingly observed among the population of different countries (Gillum et al., 2011; Abou-Chebl, 2013). The most common cause of cerebral ischemia is atherosclerosis, which affects various vascular pools (Pasi et al., 2012; Maksimovich, 2012).

The brain has a high variability in its angioarchitectonics and microcirculation. Being the most highly vascularized human organ, it is extremely sensitive to hypoxia and needs full constant blood supply more than any other organ (Maksimovich, 2012). The most important feature of cerebral blood supply is that one cubic centimeter of cerebral tissue contains about 3–4 thousand capillaries, which is much more than in any other organs and tissues (Gjulev et al., 2002). This need of the cerebral tissue for high capillary blood flow leads to the fact that the slightest disorders in the blood supply cause the development of hypoxia and subsequent ischemia (Caplan et al., 2014; Maksimovich, 2016; Crozier, 2012).

The development of cerebral atherosclerosis is a rather complex and slow process (Gjulev et al., 2002). Atherosclerosis simultaneously affects both the main and distal arterial branches and capillaries leading to cerebral small vessel disease (Maksimovich, 2012; Schmidtke and Hull, 2005; Brown and Thore, 2011). One type of development of cerebral atherosclerosis may cause disseminated lesions in small arterial branches and capillaries (Maksimovich, 2016; Brown and Thore, 2011; Caplan, 2016). Gradually developing hypoxia disrupts the metabolism of ATP in the mitochondria of neurons leading to the death of individual cells and their conglomerates (Maksimovich, 2016; Schmidtke and Hull, 2005). This process is accompanied by transient ischemic attacks and lacunar microstrokes, which at an early stage may repeatedly occur without significant clinical symptoms (Maksimovich, 2012, 2016; Gjulev et al., 2002; Schmidtke and Hull, 2005; Brown and Thore, 2011; Caplan, 2016). As a result, multiple foci of gliosis develop in different parts of the brain. This course of the disease causes slow growing neurodestruction and neurodegeneration, which gradually leads to the development of cognitive impairment, dementia, and decrease in the activity of daily life (Pasi et al., 2012; Maksimovich, 2012; Brown and Thore, 2011; Pantoni, 2010; Qureshi and Caplan, 2014; Pendlebury et al., 2011).

If the process predominates at the subcortical level, depending on the localization of lesions in some parts of the white matter, the patient develops a tendency toward Binswanger's disease (BD) or vascular parkinsonism (VP) (Tomimoto, 2011; Rosenberg et al., 2016). At an early, preclinical stage of the disease, the foci of cerebral gliosis have small dimensions and are disseminated (Rosenberg et al., 2016; Akiguchi et al., 2014). With further development, the ischemic foci progressively merge into larger demyelination sites gradually causing leukoaraiosis (Akiguchi et al., 2014; Thompson and Marsden, 1987). The widespread subcortical demyelination leads to severe neurodegenerative changes and development of BD (Maksimovich, 2017c; Ramos-Estebanez et al., 2011). With the development of VP, the atherosclerotic process proceeds in a similar manner, however, in this case the foci of gliosis and microstrokes predominate in the thalamus, basal ganglia, and bridge (Brown and Thore, 2011; Maksimovich, 2017b).

Photobiomodulation in the Brain. DOI: https://doi.org/10.1016/B978-0-12-815305-5.00039-7

In another way of cerebral atherosclerosis development, the process goes on more rapidly in more proximal parts of the arterial bed, which leads to stenotic and occlusive lesions of larger arteries (Frölich et al., 2012; Maksimovich, 2017a). This course of atherosclerosis leads to more extensive ischemia and the development of larger focal ischemic strokes that spread in the white and gray matter (Abou-Chebl, 2013; Maksimovich, 2012, 2017a; Crozier, 2012; Qureshi and Caplan, 2014).

It should be noted that if cerebral hypoxia and ischemia develop for many years, they stimulate slow natural regenerative processes in the cerebral tissue. Angiogenesis is stimulated in the cerebral tissue. Angiogenesis, as a protective reaction, causes the development of the collateral arterial and capillary beds. Collateral vessels allow the delivery of blood to ischemic sites from other vascular pools (Maksimovich, 2012, 2016; Gjulev et al., 2002). As a result, the severity of cerebral ischemic lesions depends not only on the localization and spreading of the atherosclerotic process, but also on the level of development of the collateral blood supply (Gjulev et al., 2002; Maksimovich, 2016). Consequently, ischemic strokes in older age, with developed natural collateral blood supply, proceed more easily than those in a younger age (Maksimovich, 2016; Crozier, 2012). Simultaneously, hypoxia and ischemia compensatively stimulate neurogenesis leading to regenerative processes in the cerebral tissue. In this case, regeneration of neurons takes place more actively near functioning vessels (Jin et al., 2006; Galvan and Jin, 2007; Naeser and Hamblin, 2011).

Treatment of ischemic lesions of the brain is a rather difficult task. Conservative treatment methods for various types of cerebral ischemia have limited effectiveness. The effect of drugs, on the one hand, is directed at the vasodilatory effect and the improvement of microcirculation, and on the other hand, at the restoration of metabolic processes in the cerebral tissue (Maksimovich, 2016). However, the effectiveness of the drugs commonly used is rather limited (Qureshi and Caplan, 2014; Maksimovich, 2017c).

In this regard, conservative treatment is effective in the early stages of ischemic diseases (Maksimovich, 2012; Caplan et al., 2014). During this period, ischemic lesions are not overly widespread and do not lead to diffuse hippoperfusion, hypoxia, and neurodegeneration (Maksimovich, 2016; Tomimoto, 2011; Ramos-Estebanez et al., 2011).

At later stages of the disease characterized by widespread vascular and microvascular lesions, along with severe ischemic, neurodestructive and neurodegenerative changes, conservative treatment becomes ineffective (Caplan et al., 2014; Maksimovich, 2016, 2017a; Brown and Thore, 2011).

Commonly reconstructive and interventional surgery is performed on large extracerebral arterial branches. These methods have proved effective in surgical interventions on brachiocephalic arteries (Takaiwa et al., 2013; Matsumaru et al., 2004). In the case of distal intracerebral atherosclerosis, this kind of surgery is difficult to perform due to the intracranial location of the vessels and their small diameter (Caplan et al., 2014; Maksimovich, 2016; Derdeyn and Chimowitz, 2007; Altinbas et al., 2014).

This state of affairs requires the development of new, more effective methods of treating cerebral ischemic lesions. One of the promising directions in this area is the use of laser energy, especially since brain tissue has high tropism for laser action (Maksimovich, 2012, 2016; Naeser and Hamblin, 2011).

As was noted in Chapter 38, Transcatheter intracerebral photobiomodulation in degenerative brain defects: clinical studies (Part 1), a new perspective direction in the treatment of ischemic, traumatic, and neurodegenerative brain lesions has been developed as a result of experimental and clinical studies. This method is called transcranial photobiomodulation (PBM), or low-level laser (light) therapy (LLLT) and has demonstrated its high efficiency (Naeser and Hamblin, 2011; Hashmi et al., 2010; Song et al., 2012; Konstantinović et al., 2013; Hamblin, 2018a, 2018b).

Another direction of laser treatment of cerebral ischemia lesions is transcatheter intracerebral laser treatment—the method of transluminal (transcatheter) laser revascularization of cerebral blood vessels (Maksimovich, 2004, 2006, 2008).

This method has two applications:

- direct transcatheter cerebral revascularization aimed at restoring the lumen and patency of large cerebral vessels by means of high-energy lasers;
- indirect transcatheter cerebral revascularization aimed at stimulating angiogenesis and at restoring blood supply through intracerebral collateral and capillary revascularization by means of low-energy lasers. This method is one of the types of PBM—transcatheter intracerebral photobiomodulation, or transcatheter intracerebral LLLT. This direction is the subject of this study.

The effect of low-power laser energy on cerebral tissues has three main mechanisms: stimulation of angiogenesis, which causes pronounced collateral and capillary revascularization, restoration of ATP exchange in the mitochondria of neurons, stimulation of neurogenesis and regenerative processes in cerebral tissues.

The present research is devoted to the clinical application of transcatheter intracerebral PBM in the treatment of ischemic brain lesions for stimulation of angiogenesis, restoration of cerebral distal arterial and capillary blood supply, stimulation of metabolic and regenerative processes in the cerebral tissue.

39.2 Materials and methods

All the examination and transcatheter intracerebral interventions were performed with the approval of the ethics committee, as well as with the consent of the patients examined and treated and their relatives.

39.2.1 Patient selection criteria

1. The consent of patients, as well as their relatives, to conduct the necessary examination and treatment;
2. The somatic state of patients allowing to carry out the necessary examination and treatment;
3. The presence of cerebral ischemic disorders, cerebral involutional changes, dementia, cognitive impairment and decrease in daily life activity.

1872 patients aged from 29 to 81 (mean age 74.5) suffering from various types of ischemic brain lesions were examined, of whom 1365 (72.92%) were men and 507 (27.08%) were women.

Of these, 1708 (91.24%) patients with distal intracerebral ischemic lesions were selected and were divided into the following groups:

- 911 patients with cerebral arteriosclerosis and with chronic cerebrovascular insufficiency, aged from 29 to 81 (mean age 74.5): 622 (73.87%) men, 220 (26.13%) women. Of these, 23 (2.52%) patients had BD, 46 (5.05%) had VP;
- 797 patients with cerebral atherosclerosis and previous ischemic stroke of varying severity aged from 30 to 81 (mean age 74.5): 598 (73.03%) men, 199 (24.97%) women. (The study involves patients who suffered a stroke in the period of 6 months to 6 years before the treatment).

39.2.2 Patient screening plan

- Computed tomography and magnetic resonance imaging of the brain were carried out according to the conventional schemes using Somatom (Siemens), Hi Speed(GE), Tomoscan (Philips), Apetro Eterna (Hitachi);
- Assessment of the clinical severity of dementia was accomplished in accordance with the clinical dementia rating scale (CDR) (Morris, 1993);
- Evaluation of cognitive functions was done using minimental state examination (MMSE) (Folstein et al., 1975);
- Evaluation of daily activities was carried out using index Bartels functional evaluation (IB) (Mahoney and Barthel, 1965);
- Laboratory diagnostics included coagulatory, biochemical and general clinical tests;
- Brain scintigraphy (SG) (we used TC 99M pertechnetate 555) was done on Ohio Nuclear, US gamma camera;
- Rheoencephalography (REG) was done on the apparatus Reospectr-8, Neurosoft, Russia (disorders in pulse blood flow in the cerebral hemispheres were identified);
- Cerebral multigated angiography (MUGA) was carried out on the Advantx (GE) devices using the conventional method of transfemoral access. Radiopaque substance (Omnipack 350) was automatically administered in the dose of 10–12 mL intracarotid artery and of 7–8 mL intravertebrally. The examination was carried out in direct and lateral projections in the mode of constant subtraction with a recording rate of 25 frames per second (Maksimovich, 2012, 2016). Peripheral capillary blood flow was determined with the computer program angio vision (Maksimovich, 2004). This program allows recording changes in the number and density of black pixels in the required area of the cerebral angiographic image as the radiopaque substance passes through the distal arterioles and capillaries (Maksimovich, 2006, 2008). In subsequent years, the Philips Company offered a similar program supplemented with 2D effects (De Lin and Jackson, 2012).

39.2.3 Analysis of patients

The results of the examination of patients with ischemic brain lesions are presented in Table 39.1.

TABLE 39.1 The results of the examination of patients.

Identified changes	Group 1 N-911	Group 2 N-797
Multiple deposits of calcium salts in intracerebral vessels	758 (83.21%)	662 (83.06%)
General involutive changes in the brain with expansion of the subarachnoid space	735 (80.68%)	668 (83.81%)
Expansion of Sylvian fissures	612 (67.18%)	673 (84.44%)
Nonocclusive hydrocephalus symptoms	422 (46.32%)	496 (62.23%)
Leukoaraiosis symptoms	203 (22.28%)	0
Disseminated foci of gliosis in the white matter	136 (14.93%)	119 (14.93%)
Single or multiple microcysts (2-5 mm)	141 (15.48%)	45 (5.65%)
Microfocal postischemic cysts	0	527 (66.12%)
Midfocal postischemic cysts	0	195 (24.47%)
Macrofocal postischemic cysts	0	75 (9.41%)
Foci of gliosis in the basal ganglia of the white matter	52 (5.71%)	0
Clinical dementia identification		
CDR-1	284 (31.17%)	483 (60.60%)
CDR-2	18 (1.98%)	209 (26.22%)
CDR-3	0	45 (5.65%)
Cognitive impairment		
Decrease to 20-25 MMSE points	272 (29.86%)	487 (61.10%)
Decrease to 12-19 MMSE points	18 (1.98%)	263 (33.00%)
Decrease to 7-11 MMSE points	0	47 (5.90%)
Everyday life activities		
Decrease of Barthel Index (IB) below 100 points	67 (7.35%)	733 (91.97%)
Laboratory tests		
Increased level of lipids in the blood	729 (80.02%)	574 (72.02%)
Signs of hypercoagulation	683 (74.97%)	517 (64.87%)
SG		
Decrease in blood flow in cerebral hemispheres	911 (100%)	797 (100%)
REG		
Decrease in the volume pulse blood flow in the carotid system	911 (100%)	797 (100%)
MUGA		
Stenotic lesions in distal basins of intracerebral branches	911 (100%)	635 (79.67%)
Occlusive lesions in distal basins of intracerebral branches	85 (9.33%)	797 (100%)
Depletion of capillary contrasting (lesion of the capillary bed)	838 (91.99%)	797 (100%)
Two-sided atherosclerotic lesions	869 (95.39%)	773 (96.99%)

39.2.4 Selection of patients

39.2.4.1 Group 1

Patients with distal intracerebral atherosclerosis and with chronic cerebrovascular insufficiency, among them:

-Test group 1

579 (63.56%) patients who underwent transcatheter intracerebral PBM.

-Control group 1

332 (36.44%) patients who received conservative treatment.

39.2.4.2 Group 2

Patients with distal intracerebral atherosclerosis who previously had ischemic stroke, among them:

-Test group 2

496 (62.23%) patients who underwent transcatheter intracerebral PBM. According to the size of the ischemic lesion, the patients were divided into the following groups: microfocal strokes—285 (57.46%), midfocal strokes—153 (30.85%), macrofocal strokes—58 (11.69%).

-Control group 2

301 (37.77%) patients received conservative treatment. According to the size of the ischemic lesion, the patients were divided into the following groups: microfocal strokes—242 (80.40%), midfocal strokes—42 (13.95%), macrofocal strokes—17 (5.47%);

The changes identified during the examination in groups of patients are presented in Table 39.2.

39.2.5 Methods of treating patients

39.2.5.1 Test group 1, Test group 2—Transcatheter intracerebral photobiomodulation was carried out

Under local anesthesia, following Seldinger technique, the common femoral artery is punctured and catheterized with a setting of an introducer of 6–7 F in diameter. Through this introducer, guiding catheters are drawn coaxially being first installed in the internal carotid artery and then conducted into its intracerebral branches. Through these catheters a thin, flexible, fiber-optic light guide instrument, which is connected to the laser and is 25–100 μm in diameter, is introduced. This fiber-optic instrument is guided to the distal sections of the anterior and middle cerebral arteries, where laser exposure is then performed (Maksimovich, 2006). During the intervention, under X-ray television control, Omnipac 350 is injected in small doses to position the light guide instrument. PBM is carried out using helium-neon laser with a power of 25 miles Wt. The total laser exposure time is about 20–40 minutes (Maksimovich, 2008). After the transcatheter intracerebral intervention, a second cerebral MUGA is done. Based on the results of the MUGA, the severity of intracerebral angiogenesis, the extent of collateral and capillary revascularization and restoration of microcirculation are assessed (Maksimovich, 2012, 2016).

After transcatheter and intracerebral PBM, the patients received disaggregant, anticoagulant, antioxidant, vasodilating and nootropic therapy. The patients were given: Aspirin, depending on the parameters of the blood coagulation system; Heparin, indirect anticoagulants; Pentoxifylline 100 mg, Complamin 150 mg, Inosin 200 mg, Nootropil (Piracetam) 1200 mg (or Gliatilin 1000 mg) intravenously, with a drop counter, numbers 10–15, and then they were given pills. In the period that followed, the courses of pills were repeated two times a year.

39.2.5.2 Control group 1, Control group 2—Conservative treatment was received

Conservative treatment was given to patients who, for one reason or another, could not undergo transcatheter intracerebral PBM, or they and their relatives did not want to have it conducted.

These patients underwent therapeutic treatment with similar schemes and doses of drugs used in the postoperative period by patients of the test groups. The patients were given: Aspirin, depending on the parameters of the blood coagulation system; Heparin, indirect anticoagulants; Pentoxifylline 100 mg, Complamin 150 mg, Inosin 200 mg, Nootropil (Piracetam) 1200 mg (or Gliatilin 1000 mg) intravenously, with a drop counter, numbers 10–15, and then they were given pills. In the following period, the patients also underwent repeated courses of infusions and pills for one or two months twice a year.

TABLE 39.2 Identified changes in the groups of examined patients.

Identified changes	Group 1 (N-911)		Group 2 (N-797)	
	Test (N-579)	Control (N-332)	Test (N-496)	Control (N-301)
CT and MRI				
General involutive changes in the brain with expansion of subarachnoid space	459 (79.27%)	276 (83.13%)	416 (83.87%)	252 (83.72%)
Multiple deposits of calcium salts in intracerebral vessels	481 (83.07%)	277 (83.43%)	411 (82.86%)	251 (83.39%)
Expansion of Sylvian fissures	390 (67.36%)	222 (66.87%)	419 (84.48%)	254 (84.39%)
Non-occlusive hydrocephalus symptoms	269 (46.46%)	153 (46.08%)	308 (62.10%)	188 (62.46%)
Leukoaraiosis symptoms	135(23.32%)	68 (20.48%)	0	0
Disseminated foci of gliosis in the white matter	93 (16.06%)	43 (12.95%)	74 (14.92%)	45 (14.95%)
Single or multiple microcysts (2–5 mm)	90 (15.54%)	51 (15.36%)	28 (5.65%)	17 (5.65%)
Microfocal postischemic cysts	0	0	285 (57.46%)	242 (80,40%)
Midfocal postischemic cysts	0	0	153 (30.85%)	42 (13.95%)
Macrofocal postischemic cysts	0	0	58 (11.69%)	17 (5.65%)
Foci of gliosis in the basal ganglia of the white matter	41 (7.08%)	11 (3.31%)	0	0
Clinical dementia identification				
CDR-1	185 (31.95%)	99 (29.82%)	300 (60.48%)	183 (60.80%)
CDR-2	13 (2.25%)	5 (1.51%)	137 (27.62%)	72 (23.92%)
CDR-3	0	0	29 (5.85%)	16 (5.31%)
Cognitive impairment				
Decrease to 20-25 MMSE points	175 (30.22%)	97 (29.22%)	303 (61.09%)	184 (61.13%)
Decrease to 12-19 MMSE points	13 (2.25%)	5 (1.51%)	164 (33,06%)	99 (32.89%)
Decrease to 7-11 MMSE points	0	0	29 (5.85%)	18 (5.98%)
Everyday life activities				
Decrease of Barthel Index (IB) below 100 points	41 (7.08%)	26 (7.83%)	456 (91.94%)	277 (92.03%)
Laboratory tests				
Increased level of lipids in the blood	463 (79.97%)	266 (80.12%)	357 (71.98%)	217 (72.09%)
Signs of hypercoagulation	435 (75.13%)	248 (74.70%)	322 (64.92%)	195 (64.78%)
SG				
Reduction of blood flow in cerebral hemispheres	579 (100%)	332 (100%)	496 (100%)	301 (100%)
REG				
Reduction of volume pulse blood flow in the carotid system	579 (100%)	332 (100%)	496 (100%)	301 (100%)
MUGA				
Stenotic lesions in distal basins of intracerebral branches	579 (100%)	332 (100%)	395 (79.64%)	240 (79.73%)
Occlusive lesions in distal basins of intracerebral branches	54 (9.33%)	31 (9.34%)	496 (100%)	301 (100%)
Depletion of capillary contrasting (lesion of the capillary bed)	532 (91.88%)	306 (92.17%)	496 (100%)	301 (100%)
Two-sided atherosclerotic lesions	553 (95.51%)	316 (95.18%)	481(96.98%)	292 (97.01%)

39.2.6 Evaluation of results

The results obtained after the treatment were evaluated by means of CDR, MMSE, IB:

- Good clinical result is an almost complete restoration of mental and motor functions and daily life activity.
- Satisfactory clinical result is an incomplete restoration of mental and motor functions and daily life activity.
- Relatively satisfactory clinical result is a partial restoration of mental and motor functions and daily life activity.
- Relatively positive clinical result is an absence of negative dynamics with an insignificant restoration of mental and motor functions and daily life activity.

39.3 Results

39.3.1 Test group 1—Patients with intracerebral atherosclerosis and chronic cerebrovascular insufficiency

39.3.1.1 Immediate results

According to MUGA, a good direct angiographic result manifested in pronounced angiogenesis and collateral and capillary revascularization, was obtained in 556 (96.03%) patients (Fig. 39.1 (B, C), (Fig. 39.2)).

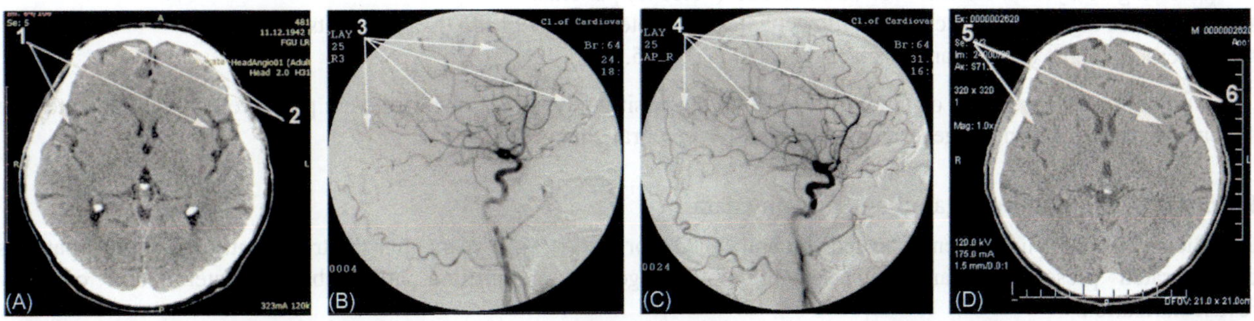

FIGURE 39.1 Patient M, 64 years old, male. Atherosclerosis of cerebral vessels, distal form, chronic cerebrovascular insufficiency. Before and after transcatheter intracerebral PBM. (A) CT of the brain, before the intervention: (1) moderate expansion of the subarachnoid space; (2) expansion of Sylvian fissures; (B) right internal carotid artery angiogram, arterial phase, before the intervention: (3) atherosclerotic lesion of distal branches of the right middle and front cerebral artery; (C) right internal carotid artery angiogram, arterial phase, after the intervention: stimulation of angiogenesis, collateral arterial and capillary revascularization of the right hemisphere; (D) CT of the brain. 12 months after the intervention: (5) restoration of Sylvian fissures; (6) reduction of subarachnoid space.

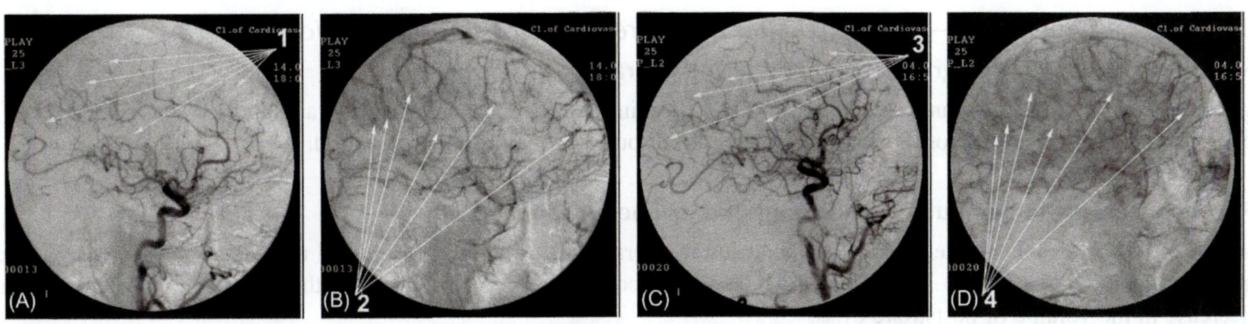

FIGURE 39.2 Patient S, female, 68 years old, distal atherosclerotic lesions of cerebral vessels, BD, (CDR-1), before and after the conducted transcatheter intracerebral PBM. (A) Left internal carotid artery angiogram, arterial phase, before the intervention: (1) depletion of the capillary bed of the brain white matter; (B) left internal carotid artery angiogram, venous phase: (2) multiple arteriovenous shunts; (C) left internal carotid artery angiogram, arterial phase, after the Intervention: (3) angiogenesis stimulation, arterial collateral and capillary bed restoration; (D) left internal carotid artery angiogram, venous phase: (4) closure of multiple arteriovenous shunts.

39.3.1.2 Early period (1–6 months) after transcatheter intracerebral photobiomodulation

Results according to scintigraphy and rheoencephalography

Improvement of blood flow and volume pulse blood flow in the cerebral hemispheres was obtained in all 579 (100%) patients. The received positive dynamics persisted throughout the observation period.

Results according to computed tomography and magnetic resonance imaging

Within 6 months after transcatheter intracerebral PBM, all 579 (100%) patients showed a tendency to decrease in the general involutional changes in the brain, which was accompanied by narrowing of the subarachnoid space.

Results of assessment of mental and motor functions

All 198 (100%) patients who previously had mental and motor function disoreders showed pronounced positive dynamics.

39.3.1.3 Long-term (1–10 years) after transcatheter intracerebral photobiomodulation

Results according to computed tomography and magnetic resonance imaging

In the following period of 12–24 months:

- general involutional changes in the cerebral cortex decreased in 434 (94.55%) patients (Fig. 39.1 (B, C)), (Fig. 39.2);
- narrowing of Sylvian fissures was noted in 355 (91.03%) patients;
- decrease in signs of nonocclusive hydrocephalus was detected in 124 (46.09%) patients;
- decrease in signs of leukoaraiosis was detected in 35 (25.93%).

In the more distant period of time (2–10 years), the positive dynamics obtained were preserved, which indicates the appearance of signs of neurogenesis and regenerative cerebral changes.

Results according to multigated angiography data

69 patients (11.91%) underwent repeated cerebral MUGA or MRA 2–10 years after the treatment. 67 (97.10%) patients demonstrated preservation and enhancement of angiogenesis manifested in cerebral collateral and capillary revascularization.

39.3.2 Test group 2—patients with intracerebral atherosclerosis and previous ischemic stroke

39.3.2.1 Immediate results

According to MUGA, a good immediate angiographic result, manifested in pronounced angiogenesis, collateral and capillary revascularization, was obtained in 471 (94.96%) patients (Fig. 39.3 (B, C)), (Fig. 39.4 (B, C)).

39.3.2.2 Early period (1–6 months) after intracerebral photobiomodulation

Results according to scintigraphy and rheoencephalography

Improvement of blood flow and volume pulse blood flow in the cerebral hemispheres was observed in all 496 (100%) patients, while the received positive dynamics persisted throughout the observation period.

Results according to computed tomography and magnetic resonance imaging

In the period of 6 months after the transcatheter intracerebral PBM, the patients showed a tendency to decrease in the general involutional changes in the brain, which was accompanied by narrowing of the subarachnoid space and a decrease in the volume of poststroke cysts.

Results of assessment of mental and motor functions

All 496 (100%) patients who previously mental and motor function disorders showed marked positive dynamics with an increase in IB.

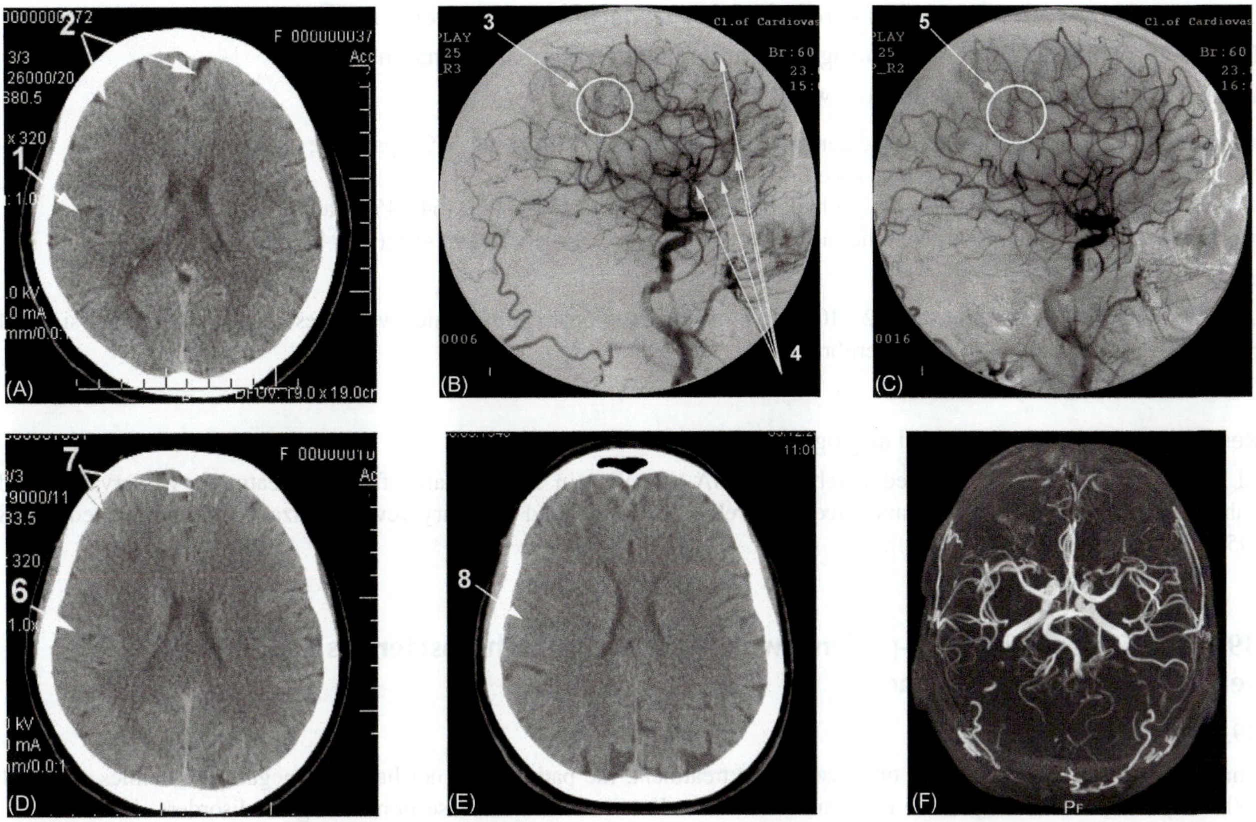

FIGURE 39.3 Patient H, female, 60 years old. Atherosclerosis of cerebral vessels, chronic cerebrovascular insufficiency, ischemic stroke in the basin of the middle cerebral artery to the right. Before and after transcatheter intracerebral PBM. (A) CT of the brain, before the intervention: (1) moderate heterogeneous postischemic cyst in the right middle cerebral artery region. (2) Moderate expansion of the subarachnoid space; (B) right internal carotid artery angiogram, arterial phase, before the intervention: (3) occlusion of distal branches of the right middle cerebral artery; (4) multiple stenosis of intracranial branches; (C) right internal carotid artery angiogram, arterial phase, after the intervention: (5) stimulation of angiogenesis, collateral arterial and capillary revascularization of the right hemisphere; (D) CT of the brain, 12 Months after the Intervention: (6) reduction in the size of the postischemic cyst with signs of cerebral tissue structure recovery; (7) subarachnoid space restoration; (E) CT of the brain, 6 years after the Intervention: (8) the structure of the right hemisphere cerebral tissue is restored, no signs of residual effects of the postischemic cyst; (F) MRA of the brain. 6 years after the Intervention: the patency and lumen of the distal branches of the right internal carotid artery are completely preserved, there is further progression of collateral revascularization.

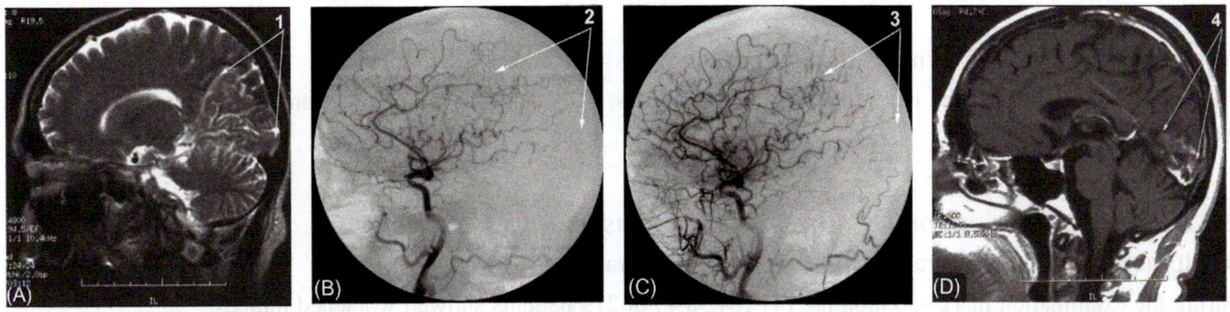

FIGURE 39.4 Patient A, female, 58 years old (CDR-2). Atherosclerosis of cerebral vessels, chronic cerebrovascular insufficiency, extensive ischemic stroke in the occipital parietal region of the right hemisphere. Before and after transcatheter intracerebral PBM. (A) MRI of the brain, before the intervention: (1) a huge postishemic cyst in the right occipital parietal region of the right hemisphere. (B) Right internal carotid artery angiogram, arterial phase, before the intervention: (2) multiple occlusions of the distal branches of the right medial cerebral artery; (C) right internal carotid artery angiogram, arterial phase, after the intervention: (3) stimulation of angiogenesis, marked collateral arterial and capillary revascularization of the right hemisphere; (D) MRI of the brain, 10 months after the Intervention: (4) significant reduction in the size of the postischemic cyst with signs of cerebral tissue structure recovery.

39.3.2.3 *Long-term (1–10 years) results after transcatheter intracerebral photobiomodulation*

Results according to computed tomography and magnetic resonance imaging

In the period of 12–24 months that followed:

- general involutional changes in the cerebral cortex decreased in 476 (95.97%) patients;
- narrowing of Sylvian fissures was noted in 393 (93.79%) patients;
- decrease in the signs of nonocclusive hydrocephalus was detected in 200 (64.94%) patients;
- decrease in the volume of postischemic cysts after strokes was detected in 461 (92.94%) patients (Fig. 39.3 (A, D)), (Fig. 39.4 (A, D)).

In a more distant period of time (2–10 years), the obtained positive dynamics were preserved indicating the signs of neurogenesis and restoration of the cerebral tissue (Fig. 39.3 (A, E)).

Results according to multigated angiography data

71 (13.45%) patients had repeated cerebral MUGA or MRA in 2–10 years after the treatment. Preservation and enhancement of angiogenesis, manifested in cerebral collateral and capillary revascularization, was obtained in 68 (95.77%) patients (Fig. 39.3 (B, F)).

39.3.3 Control group 1—patients with intracerebral atherosclerosis and chronic cerebrovascular insufficiency

39.3.3.1 *Immediate results*

Immediately after the first course of conservative treatment, the patients did not have any negative dynamics, and 201 (60.54%) of them showed positive dynamics manifested in a moderate decrease in neurological disorders.

39.3.3.2 *Early period (1–6 months) after conservative treatment:*

Results according to scintigraphy and rheoencephalography

Partial improvement in blood flow and pulse blood flow in the cerebral hemispheres was obtained in 219 (65.96%) patients. The improvement was not stable and did not last.

Results according to computed tomography and magnetic resonance imaging

In the period of 6 months after the first course of the conservative treatment, there was no significant reduction in involutional changes in the brain tissue.

Results of assessment of mental and motor functions

All 104 (100%) patients, who had previously had disorders in mental and motor functions, showed moderate positive dynamics.

39.3.3.3 *Long-term period (1–10 years) after conservative treatment:*

Results according to scintigraphy and rheoencephalography

During the examination in 12–36 months, 141 (64.38%) of 219 patients showed worsened indices.

Results according to computed tomography and magnetic resonance imaging

In the period of 12–24 months after the treatment, there was no significant decrease in involutive changes in the brain tissue. In a more distant period, 235 (70.78%) patients experienced an increase in general involutional changes, expansion of Sylvian fissures, increasing signs of nonocclusive hydrocephalus and of leukoaraiosis (Fig. 39.5 (A-E)).

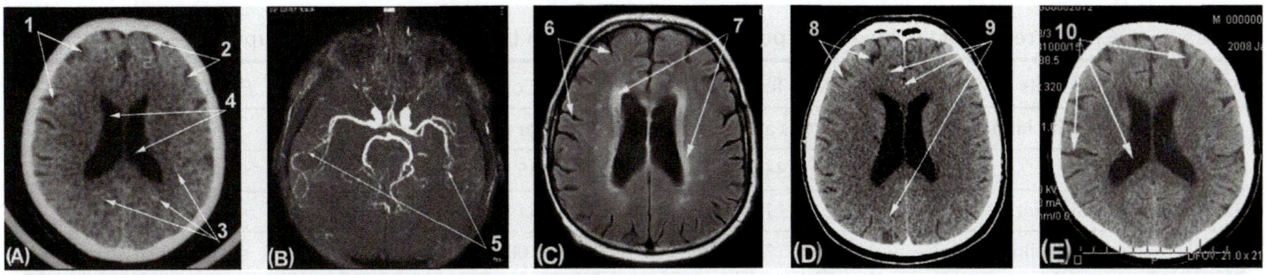

FIGURE 39.5 Patient K, 61 years old, male. Atherosclerosis of cerebral vessels, distal form, chronic cerebrovascular insufficiency. Before and after the start of conservative treatment. (A) CT of the brain, before the treatment: (1) convexital sulci expansion; (2) subarachnoid space expansion; (3) multiple deposits of calcium salts in intracerebral vessels; (4) ventricles enlargement; (B) MRA of the brain, before the treatment: (5) multiple stenosis of intracranial branches; (C) MRI of the brain, 1 years after the beginning of the treatment: (6) further progressing of convexital sulci expansion; (7) signs of developing leukoaraiosis; (D) CT of the brain. 3 years after the beginning of the treatment: (8) further progressing of convexital sulci expansion; (9) increased deposition of calcium salts in intracerebral vessels; (E) CT of the brain, 6 years after the beginning of the treatment: (10) further progression of atrophic changes in the brain, an increase in the expansion of convexital sulci, expansion of the ventricles and increasing signs of nonocclusive hydrocephalus.

39.3.4 Control Group 2—patients with intracerebral atherosclerosis and previous ischemic stroke

39.3.4.1 Immediate results

Immediately after the first course of the conservative treatment, there was no negative dynamics in the patients; 119 (39.53%) patients had a stable condition and demonstrated positive dynamics manifested in a moderate decrease in neurological, mental, and motor disorders.

39.3.4.2 Early period (1–6 months) after conservative treatment

Results according to scintigraphy and rheoencephalography

According to SG and REG, a partial improvement in blood flow and pulse blood flow in the cerebral hemispheres was obtained in 192 (63.79%) patients. The improvement was not stable and did not last.

Results according to computed tomography and magnetic resonance imaging

In the period of 6 months after the first course of the conservative treatment, there was no significant decrease in the involutional changes in the brain tissue.

Results of assessment of mental and motor functions

All 277 (100%) patients, who had previously had abnormalities of mental and motor functions, showed some positive dynamics with a moderate increase in IB.

39.3.4.3 Long-term period (1–10 years) after the conservative treatment

Results according to scintigraphy and rheoencephalography

During the examination in 12–36 months, 107 (55.73%) out of 192 patients demonstrated worsened indices.

Results according to computed tomography and magnetic resonance imaging

In the period of 12–24 months after the conservative treatment, there was no significant decrease in involutional changes in the brain tissue. The volume of postischemic cysts after the stroke did not decrease. In the more distant period, 216 (71.76%) patients showed an increase in general involutional changes, expansion of Sylvian fissures, and growing signs of nonocclusive hydrocephalus and of leukoaraiosis.

39.3.5 Clinical results in the long-term period

Clinical results after the treatment in the test and control groups are presented in Table 39.3.

TABLE 39.3 Clinical results in the long-term period after treatment in the test and control groups.

After-treatment results	Test group	Control group	*P* (chi-square)
Group 1. Patients with intracerebral atherosclerosis and chronic cerebrovascular insufficiency.			
Good clinical result	444	35	<.001
Satisfactory clinical result	116	53	<.001
Relatively satisfactory clinical result	17	108	<.001
Relatively positive clinical result	2	136	<.001
Total	579	332	
Group 2. Patients with intracerebral atherosclerosis and progression of microfocal stroke.			
Good clinical result	251	51	<.001
Satisfactory clinical result	29	60	<.001
Relatively satisfactory clinical result	5	102	<.001
Relatively positive clinical result	0	29	<.001
Total	285	242	
Group 2. Patients with intracerebral atherosclerosis and progression of midfocal stroke.			
Good clinical result	92	0	<.001
Satisfactory clinical result	38	8	<.001
Relatively satisfactory clinical result	23	10	<.001
Relatively positive clinical result	0	24	<.001
Total	153	42	
Group 2. Patients with intracerebral atherosclerosis and progression of macrofocal stroke.			
Good clinical result	12	0	<.001
Satisfactory clinical result	22	0	<.001
Relatively satisfactory clinical result	24	3	<.001
Relatively positive clinical result	0	14	<.001
Total	58	17	

To define the differences of the symptoms under study, we conducted an analysis of contingency tables with Chi-square criterion. All the signs revealed significant differences ($P < .01$).

39.4 Discussion

The development of ischemic cerebral lesions is a complex, multicomponent process (Maksimovich, 2012). Under natural conditions, the body reacts to the development of cerebral ischemia with a complex of restorative processes. Reduction of cerebral blood flow stimulates angiogenesis, which manifests itself by opening collateral channels (Maksimovich, 2016). Developed hypoxia stimulates metabolic processes and neurogenesis in the cerebral tissue (Jin et al., 2006; Galvan and Jin, 2007; Naeser and Hamblin, 2011). However, unfortunately, these processes proceed slowly, and do not have sufficient time to act during the development of ischemic lesions.

Treatment of cerebral ischemic lesions should also be multicomponent and directed at various processes occurring during the development of ischemia of cerebral tissues (Maksimovich, 2016; Naeser and Hamblin, 2011).

Conservative treatment may seem attractive due to its simplicity, but it is limited by the effectiveness of existing drugs (Maksimovich, 2012). The drugs used in modern clinical practice do not allow achieving pronounced, persistent cerebral revascularization and restoration of metabolic and regenerative processes in the cerebral tissue. Consequently, the resulting therapeutic effect of the conservative treatment is mainly associated with partial improvement of the cerebral blood supply and metabolic processes, but this is often not enough for the functional recovery of ischemic brain tissues. As a result, in control groups 1 and 2, good and satisfactory clinical results were obtained only in a limited number of patients with chronic cerebrovascular insufficiency, micro and midfocal ischemic strokes.

The effect of laser energy on the brain is complex and multicomponent. With transcatheter intracerebral PBM, laser light penetrates deeply into the cerebral tissues. The brain tissue is an optically active environment; the depth of penetration of coherent radiation at a wavelength of 633 nm (helium neon laser) with transcatheter intracerebral application is 20–40 mm (Maksimovich, 2004; Maksimovich et al., 1988). As a result, not only the vascular wall, but also large volumes of surrounding cerebral tissues are exposed to laser action. According to the data of other authors, with the direct action of lasers with a wavelength of 808 nm, the penetration depth in the tissue of the cadaveric brain is 40–50 mm (Tedford et al., 2015), which coincides with our previous studies (Maksimovich, 2004; Maksimovich et al., 1988).

Atherosclerosis is a systemic disease; it affects the intracerebral vessels of both hemispheres. The degree of the development of atherosclerosis may be different, and in one hemisphere, ischemic lesions may be more pronounced than in the other. However, in the hemisphere with less pronounced atherosclerotic changes, the signs of hypoxia and ischemia still exist. In this connection, with ischemic lesions, reconstructive and transcatheter operative interventions, both on brachiocephalic and intracerebral vessels, are desirable to be carried out from two sides (Maksimovich, 2012, 2016, 2017a; Caplan et al., 2014; Takaiwa et al., 2013; Matsumaru et al., 2004). In this regard, to improve the blood supply of the entire brain, transcatheter intracerebral PBM was conducted from two sides in all cases, both in the right and left hemispheres.

By affecting the vascular wall and surrounding cerebral tissues, laser energy stimulates angiogenesis, causing collateral and capillary revascularization of the brain (Maksimovich, 2004, 2012, 2017a,c). As a result, in the patients of test groups 1 and 2, transcatheter intracerebral PBM improved the blood supply of the entire brain, which naturally contributed to its greater functional recovery.

According to many authors, affecting ischemic cerebral tissues, laser energy stimulates and restores metabolic processes in neurons, increases the level of and restores the exchange of adenosine triphosphate in mitochondria (Naeser and Hamblin, 2011; Hamblin, 2018b; Hennessy and Hamblin, 2017; Cassano et al., 2016; Karu, 2010), which is confirmed by this work. In all patients of test groups 1 and 2, after a two-sided transcatheter intracerebral PBM, rehabilitation period is rather short, which indicates the restoration of metabolic processes in the cerebral tissue.

As many authors point out, when affecting cerebral tissues, laser energy stimulates neurogenesis and the formation of new neurons, which in turn causes cerebral regenerative processes (Naeser and Hamblin, 2011; Hashmi et al., 2010; Hamblin, 2018a; Oron et al., 2006; Moskvin and Khadartsev, 2017). These data are also confirmed by this study. After transcatheter intracerebral PBM, all patients with chronic cerebrovascular insufficiency from test group 1 noted the narrowing and restoration of Sylvian fissures, reduction of the subarachnoid space, and an increase in the mass of the cerebral tissue, which indicates the development of regenerative processes. In 10–12 months after the transcatheter intracerebral PBM, all test group 2 patients with intracerebral atherosclerosis and previous ischemic stroke showed decreased cerebral involutional changes, narrowing of Sylvian fissures, narrowing of the subarachnoid space, and moreover, a reduction in the size of the postischemic cyst, which also indicates the development of regenerative processes in the cerebral tissue. It should be noted that in a more distant period, regenerative changes in the brain continued to progress.

As a result, in the early period after the treatment, patients in the test groups experienced a decrease in the signs of dementia (CDR), restoration of cognitive (MMSE) and mental functions, as well as restoration of daily life activity (IB). Consequently, in test group 1, in patients with chronic cerebrovascular insufficiency, a good and satisfactory clinical result after transcatheter intracerebral PBM was obtained in 560 (96.72%) cases. In test group 2, 444 (89.52%) patients with cerebral atherosclerosis and ischemic stroke showed good and satisfactory clinical results after transcatheter intracerebral PBM.

39.5 Conclusion

Transcatheter intracerebral PBM is pathogenetically grounded and is an effective method of treatment of ischemic lesions of the brain. Stimulating angiogenesis, laser energy promotes cerebral revascularization and restores collateral and capillary blood supply to ischemic tissues. By influencing the exchange of ATP in mitochondria of neurons, laser energy improves cellular and tissue metabolism. Stimulating neurogenesis, laser energy causes regenerative processes in the cerebral tissue. This complex laser action during transcatheter intracerebral PBM allows to significantly improve daily life activity, restore cognitive and mental functions, and reduce the level of dementia in patients with different severity of cerebral ischemic lesions, thus allowing them to return to active daily life. The obtained positive clinical result is observed for a long time.

In general, conservative treatment has a positive clinical effect only in the early, lighter stages of cerebral ischemic lesions.

Conflict of interest

The authors declared no potential conflicts of interest with respect to the research, authorship, and/or publication of this article.

Funding

The authors received no financial support for the research, authorship, and/or publication of this article.

References

Abou-Chebl, A., 2013. Management of acute ischemic stroke. Curr. Cardiol. Rep. 15 (4), 348–354.

Akiguchi, I., Budka, H., Shirakashi, Y., Woehrer, A., Watanabe, T., Shiino, A., et al., 2014. MRI features of Binswanger's disease predict prognosis and associated pathology. Ann. Clin. Transl. Neurol. 1 (10), 813–821.

Altinbas, A., Algra, A., Martin, M., Brown, M.M., Featherstone, R.L., Kappelle, L.J., et al., 2014. Effects of carotid endarterectomy or stenting on hemodynamic complications in the International Carotid Stenting Study: a randomized comparison. Int. J. Stroke 9 (3), 284–290.

Brown, W.R., Thore, C.R., 2011. Review: cerebral microvascular pathology in ageing and neurodegeneration. Neuropathol. Appl. Neurobiol. 37 (1), 56–74.

Caplan, L.R., 2016. The effect of small artery disease on the occurrence and management of large artery disease. JAMA Neurol. 73 (1), 19–20.

Caplan, L.R., Thomas, A.J., Inoa, V., 2014. Interventional treatment of brain ischemia related to intracranial cerebrovascular occlusive lesions. Curr. Opin. Neurol. 27 (1), 1–7.

Cassano, P., Petrie, S.R., Hamblin, M.R., Henderson, T.A., Iosifescu, D.V., 2016. Review of transcranial photobiomodulation for major depressive disorder: targeting brain metabolism, inflammation, oxidative stress, and neurogenesis. Neurophotonics 3, 031404.

Crozier, S., 2012. Withdrawing and withholding treatments in acute severe stroke patients in the elderly. Rev. Pract. 62 (9), 1243–1245.

De Lin, M., Jackson, E.F., 2012. Applications of imaging technology in radiation research. Radiat. Res. 177 (4), 387–397.

Derdeyn, C.P., Chimowitz, M.I., 2007. Angioplasty and stenting for atherosclerotic intracranial stenosis: rationale for a randomized clinical trial. Neuroimaging Clin. N. Am. 17 (3), 355–ix.

Folstein, M.F., Folstein, S.E., McHugh, P.R., 1975. "Mini-mental state". A practical method for grading the cognitive state of patients for the clinician. J. Psychiatr. Res. 12 (3), 189–198.

Frölich, A.M., Psychogios, N.M., Klotz, E., Schramm, R., Knauth, M., Schramm, P., 2012. Angiographic reconstructions from whole-brain perfusion CT for the detection of large vessel occlusion in acute stroke. Stroke 43 (1), 97–102.

Galvan, V., Jin, K., 2007. Neurogenesis in the aging brain. Clin. Interv. Aging 2, 605–610.

Gillum, R.F., Kwagyan, J., Obisesan, Th.O., 2011. Ethnic and geographic variation in stroke mortality trends. Stroke 42 (2), 3294–3296.

Gjulev, N.M., Pustozertsev, V.G., Gjulev, S.N., 2002. Cerebrovascular Diseases. BINOM, Moscow.

Hamblin, M.R., 2018a. Photobiomodulation for traumatic brain injury and stroke. J. Neurosci. Res. 96, 731–743.

Hamblin, M.R., 2018b. Mechanisms and mitochondrial redox signaling in photobiomodulation. Photochem. Photobiol. 94 (2), 199–212.

Hashmi, J.T., Huang, Y.Y., Osmani, B.Z., Sharma, S.K., Naeser, M.A., Hamblin, M.R., 2010. Role of low-level laser therapy in neurorehabilitation. Arch. Phys. Med. Rehabil. Suppl. 2, S292–S305.

Hennessy, M., Hamblin, M.R., 2017. Photobiomodulation and the brain: a new paradigm. J. Opt. 19 (1), 013003.

Jin, K., Wang, X., Xie, L., Mao, X.O., Zhu, W., Wang, Y., et al., 2006. Evidence for stroke-induced neurogenesis in the human brain. Proc. Natl. Acad. Sci. U.S.A. 103, 13,198–13,202.

Karu, T.I., 2010. Multiple roles of cytochrome c oxidase in mammalian cells under action of red and IR-A radiation. IUBMB Life 62, 607–610.

Konstantinović, L.M., Jelić, M.B., Jeremić, A., Stevanović, V.B., Milanović, S.D., Filipović, S.R., 2013. Transcranial application of near-infrared low-level laser can modulate cortical excitability. Lasers Surg. Med. 45 (10), 648–653.

Mahoney, F.I., Barthel, D.M., 1965. Functional evaluation: the barthel index. Maryland State Med. J. 14, 61–65.

Maksimovich, I.V., 2004. Transluminal Laser Angioplasty in Treatment of Ischemic Lesions of a Brain (Ph.D. Dissertation), Russian University of Friendship of the People, Moscow.

Maksimovich, I.V., 2006. Method for Carrying out Transluminal Laser-Induced Brain Revascularization in Atherosclerotic Injury Cases. Patent RU, 2297861.

Maksimovich, I.V., 2008. Method of Transluminal Laser Revascularization of Cerebral Blood Vessels Having Atherosclerotic Lesions. Patent US, 7490612.

Maksimovich, I.V., 2012. Transcatheter treatment of atherosclerotic lesions of the brain complicated by vascular dementia development. World J. Neurosci. 2 (4), 200–209.

Maksimovich, I.V., 2016. Transcatheter cerebral revascularization in the treatment of atherosclerotic lesions of the brain. Brain Disord. Ther. 5 (1), 1–8.

Maksimovich, I.V., 2017a. Results of brain transcatheter laser revascularization in the treatment of the consequences of ischemic stroke. J. Vas. Dis. Treat. 1 (1), 2–5.

Maksimovich, I.V., 2017b. Transcatheter laser revascularization of the brain in atherosclerotic parkinsonism treatment. J. Neurol. Neurophysiol. 8 (1), 86 (Suppl.).

Maksimovich, I.V., 2017c. Possibilities of application of transcatheter treatment of vascular dementia with Binswanger's disease. Glob. J. Health Sci. 9 (6), 13–21.

Maksimovich, I.V., Lesnoy, M.I., Zubov, V.V., 1988. Transluminal laser angioplasty with low-intensity laser radiation. Application of Lasers in Surgery and Medicine. pp. 21–25, Samarkand.

Matsumaru, Y., Tsuruta, W., Takigawa, T., Hyodo, A., Sato, H., Matsumura, A., 2004. Percutaneous transluminal angioplasty for atherosclerotic stenoses of intracranial vessels. Interv. Neuroradiol. 10 (Suppl. 2), 17–20.

Morris, J.C., 1993. The Clinical Dementia Rating (CDR): current version and scoring rules. Neurology 11 (43), 2412–2414.

Moskvin, S.V., Khadartsev, A.A., 2017. Basic Techniques of Low Level Laser Therapy. Triada, Tver, 144 pp.

Naeser, M.A., Hamblin, M.R., 2011. Potential for transcraniallaser or LED therapy to treatstroke, traumatic brain injury, and neurodegenerative disease. Photomed. Laser Surg. 29 (7), 443–446.

Oron, A., Oron, U., Chen, J., Eilam, A., Zhang, C., Sadeh, M., et al., 2006. Low-level laser therapy applied transcranially to rats after induction of stroke significantly reduces long-term neurological deficits. Stroke 37, 2620–2624.

Pantoni, L., 2010. Cerebral small vessel disease: from pathogenesis and clinical characteristics to therapeutic challenges. Lancet Neurol. 9 (7), 689–701.

Pasi, M., Poggesi, A., Salvadori, E., Pantoni, L., 2012. Post-stroke dementia and cognitive impairment. Front Neurol. Neurosci. 30, 65–69.

Pendlebury, S.T., Wadling, S., Silver, L.E., Rothwell, P.M., 2011. Transient cognitive impairment in TIA and minor stroke. Stroke 42, 3116–3121.

Qureshi, A.I., Caplan, L.R., 2014. Intracranial atherosclerosis. Lancet 383 (9921), 984–998.

Ramos-Estebanez, C., Moral-Arce, I., Gonzalez-Mandly, A., Dhagubatti, V., Gonzalez-Macias, J., Munoz, R., et al., 2011. Vascular cognitive impairment in small vessel disease: clinical and neuropsychological features of lacunar state and Binswanger's disease. Age Ageing 40 (2), 175–180.

Rosenberg, G.A., Wallin, A., Wardlaw, J.M., Markus, H.S., Montaner, J., Wolfson, L., et al., 2016. Consensus statement for diagnosis of subcortical small vessel disease. J. Cereb. Blood Flow Metab. 36 (1), 6–25.

Schmidtke, K., Hull, M., 2005. Cerebral small vessel disease: how does it progress? J. Neurol. Sci. 229–230, 13–20.

Song, S., Zhou, F., Chen, W.R., 2012. Low-level laser therapy regulates microglial function through Src-mediated signaling pathways: implications for neurodegenerative diseases. J. Neuroinflammation 18 (9), 219.

Takaiwa, A., Kuwayama, N., Akioka, N., Kurosaki, K., Hayashi, N., Endo, S., et al., 2013. Effect of carotid endarterectomy on cognitive function in patients with asymptomatic carotid artery stenosis. Acta Neurochirurgica 155, 627–633.

Tedford, C.E., DeLapp, S., Jacques, S., Anders, J., 2015. Quantitative analysis of transcranial and intraparenchymal light penetration in human cadaver brain tissue. Lasers Surg. Med. 47, 312–322.

Thompson, P.,D., Marsden, C.,D., 1987. Gait disorder of subcortical arteriosclerotic encephalopathy: Binswanger's disease. Mov. Disord. 2, 1–8.

Tomimoto, H., 2011. Subcortical vascular dementia. Neurosci. Res. 71 (3), 193–199.

Chapter 40

Russian low level laser therapy techniques for brain disorders

Sergey V. Moskvin[1] and Andrey V. Kochetkov[2]

[1]*The Federal State-Financed Institution "O.K. Skobelkin State Scientific Center of Laser Medicine under the Federal Medical Biological Agency" of Russia, Moscow, Russia,* [2]*Federal State-Funded Educational Institution of Continuing Professional Education "Institute of Advanced Training of FMBA of Russia", Moscow, Russia*

40.1 Introduction

In recent years, there has been a significant growth in the prevalence of vascular diseases, including brain disorders, leading to the increased frequency of acute cerebrovascular accidents (CVAs). More than 6 million people in the world have CVA annually, more than 450,000 cases per year are registered in Russia (Murav'yeva and Karpova, 2014). The problem of the prevention of CVA remains an important and high priority task in both economic and social terms. It is increasingly evident that it is the primary prevention of vascular catastrophes that contributes to the most effective reduction in the number of fatal cases after strokes, and the prevention of severe consequences. Therefore, further research and study of methods that contribute to the prevention of development of CVA of different etiology is necessary.

One of the branches of medicine, where low level laser therapy (LLLT) demonstrates very good results, is the treatment and rehabilitation of patients with chronic cerebral ischemia (cerebral atherosclerosis, hypertensive encephalopathy, sequelae of cerebral infarction). Since 1994 we have constantly worked with this category of patients in two large scientific medical centers; we have significant amount of experience of clinical application of LLLT (more than 10,000 patients). During this time, the effectiveness of many LLLT techniques was tested and their parameters were optimized. It was shown that intravenous laser blood illumination (ILBI) and noninvasive laser blood illumination (NLBI) are the two main methods that should be used in cerebrovascular diseases.

Our data are confirmed by numerous studies of specialists from other medical centers in Russia. We gathered a lot of clinical experience, which allows us to state that in many respects, LLLT has currently no alternative, first of all, in the effectiveness of preventing the development of CVA.

LLLT is one of the methods of physiotherapy which gained its popularity in the USSR first and then in Russia. In English-speaking publications devoted to the subject, it is stated that Hungarian researchers were the first to suggest the method (Mester et al., 1968). However, dozens of studies on the therapeutic application of low-intensity laser light (LILL) were carried out in a number of former Soviet Republics at that time and hundreds of articles and even monographs were published, but as they were all written in the Russian language, they were ignored by the global professional community. It is undeniable that at the moment, Russia is the leader in this field, as it is Russian specialists who create the most efficient techniques of laser therapy and its effects.

The absolute safety of LLLT and its efficiency were proven long ago (Kapustina et al., 1996; Moskvin, 1997); the mechanisms of the therapeutic (biological) effect of low-intensity laser illumination have been studied thoroughly (Moskvin, 2003, 2008), which makes it possible to develop the trend more actively, in general and in particular, together with its different techniques.

Photobiomodulation in the Brain. DOI: https://doi.org/10.1016/B978-0-12-815305-5.00040-3

40.2 Protocol requirements of low level laser therapy procedures in Russia, low level laser therapy techniques

Protocol requirements are strictly obligatory as the necessity to set all the parameters of the method listed below have been clearly proven. It will be impossible to get a predictable and appropriate response to the impact of laser light and to achieve the desired therapeutic effect if even one of the parameters is implemented incorrectly.

Let us take into account the fact that in most cases, minimum energy LILL, which is used to influence the human body for therapeutic purposes, is required for the successful implementation of LLLT techniques. However, there are techniques that require power density limits, but there are not many of them.

All LLLT techniques must contain the following information (Moskvin, 2014, 2016; Gerasimenko et al., 2015):

1. Laser light wavelength is measured in nanometers (nm). Here are the wavelengths which are the most common in laser therapy:
   - 365–405 nm—ultraviolet (UV) spectrum,
   - 440–445 nm—blue spectrum,
   - 520–525 nm—green spectrum,
   - 635 nm—red spectrum,
   - 780–785 nm—infrared (IR) spectrum,
   - 890–904 nm—IR spectrum.

   It is useless to illuminate the same area with laser and/or incoherent light sources with a different wavelength simultaneously because of inhibitory interaction. It is possible that some of these findings will cause disbelief or surprise for some, despite the facts being proved. We have made detailed analyses of the work of several authors (more than 20), and also carried out our own research, confirming the complete ineffectuality of using LILL of two wavelengths simultaneously, 635 and 904 nm. On the other hand, a combination, that is, consecutive application of LILL with different wavelengths is exceptionally efficient. But it must only be done in the following manner: first, the red laser 635 nm—exposure time of 1.5 minutes, then a 1.5 minute break, followed by an IR-laser 904 nm for 1.5 minutes. These time intervals are fixed and cannot be varied, since they are determined by the periods of wave propagation of an increased concentration of Ca^{2+}, initiated by the action of LILL, which are exactly 100 and 300 seconds (maximum concentration), no more, no less. A slight difference by 10 seconds (1.5 minutes = 90 seconds) in the methodology from the ideal variant (100 + 100 + 100 seconds) is due to the need to switch from one mode to another, since the exposure must be completed before the stimulation phase and must begin after the inhibitory phase of this biological rhythm (Moskvin, 2014).

   Laser operation mode: continuous, modulated, pulse (super-pulsed).
2. LILI power.

   The average power of continuous lasers, operating in continuous or modulated modes is measured in milliwatts (mW), and the pulse (peak) power of pulsed lasers is measured in watts (W).
3. Modulation frequency or pulse for a pulsed mode is a quantity of fluctuations (pulses) per a unit of time (second). It is measured in Hertz (Hz, 1/s).
4. The duration of light pulse is a very important parameter for pulsed lasers, it is constant (most often 100–150 ns). Average power of pulsed lasers ($P_{av.}$) is directly proportional to pulse power (P_p), pulse duration (τ_p), and frequency (F_p): $P_{av.} = P_p \times \tau_p \times F_p$.
5. Illumination area. It is measured in square centimeters (cm^2).

   The required area is almost always defined by the technique itself without unnecessary measurements, for example, for a contact-mirror technique, the area is supposed to be 1 cm^2. Laser diodes in matrix emitters must be arranged so that their impact area is multiplied by power density. For example, eight (most often) pulsed laser diodes, each with the power of 10 W, are placed on an area of 8 cm^2, and upon the contact with the skin the power density will be 10 W/cm^2 respectively. For laser acupuncture or ILBI the area is not indicated, because the impact area is too small, and dispersion and absorption of the laser light energy in the volume of biological tissues are of primary importance.
6. Power density (PD). It is measured in watts and milliwatts per square centimeter (W/cm^2 or mW/cm^2).
7. Exposure (exposure time) on one area (zone) and total duration of the procedure are measured in seconds or minutes. This is an extremely important parameter that can hardly ever be changed. Total duration of the LLLT procedure (consistent effect on all the areas) should not exceed 20, or 5 minutes for one area (except for ILBI).
8. Exposure localization (technique).
9. The number of procedures per course of treatment and their frequency.

Calculation of energy is measured in joules (J or W · s), and energy density (J/cm^2 or W · s/cm^2) is not carried out as there is no need for this information for efficient laser therapy.

It is expedient to include one of the general systemic methods into a LLLT protocol (laserpuncture and/or ILBI) and combined with the direct impact on the affected area (local, transdermal or abdominal technique, and a combined method—laser phoresis).

Local low-intensity laser illumination is applied directly on the affected area located close to the surface of the body or in contact through a mirror head, or remote at a small distance from the surface (1–2 cm) in a stable manner.

The following types of LILL are most often used for local laser illumination:

- continuous LILL of red spectrum (635 nm), CW PD—10–15 mW/cm^2;
- pulsed LILL of red spectrum (635 nm), peak PD—4–5 W/cm^2, pulse duration 100–150 ns, frequency 80–10,000 Hz;
- pulsed IR LILL (890–904 nm), peak PD—8–10 W/cm^2, pulse duration 100–150 ns, frequency 80–10,000 Hz.

For pulsed lasers, the frequency varies according to the desired effect: regeneration and an antiinflammatory effect—80–150 Hz, anesthesia—3000–10,000 Hz. There are up to two to three local zones for one area, exposure to each is 2–5 minutes. A more than 5-minute exposure to a single zone is not recommended.

40.3 Intravenous laser blood illumination

A continuous mode LILI is used; the exposure is implemented intravenously through special disposable sterile light guides with a special puncture needle, most often in the cubital vein (Figs. 40.1 and 40.2, zone 1) (Geynits et al., 2012).

Different techniques with the use of laser light of different spectrum (Table 40.1) are used at the present time for ILBI implementation:

- ILBI-635 (wavelength of 635 nm, red spectrum, power of 1.5–2 mW, exposure 10–20 minutes) has a universal effect, and a positive effect on the immune system as well as on the function of the tissues.
- ILBI-525 (wavelength of 525 nm, green spectrum, power of 1.5–2 mW, exposure 7–8 minutes) is recommended for maximum enhancement of the function of tissues.
- ILBI-365 and ILBI-405—laser ultraviolet blood illumination (LUVBI, wavelength of 365–405 nm, power of 1.5–2 mW, exposure 3–5 minutes) is preferable for the correction of immune disorders as a result of a disease or injury.

There are many options for the technique and parameter variation which must not be broken.

Power (1.5–2 mW) is not changed, but can be increased up to 20–25 mW in some cases with the use of special laser emitting heads or is changed from one procedure to another. But it is necessary to be extremely careful with this parameter and use it only on purpose, and for some defined applications.

Exposure. "Standard" time for ILBI-635 procedure can be increased, sometimes to 25–30 minutes, but not more (Meshalkin and Sergievsky, 1989). It is necessary to know the peculiarities of ILBI-635 application in the older age group (decrease exposure by two times) (Davydenko, 2006). There is a rule in pediatrics that says: "the younger the age the less the exposure" (Moskvin et al., 2009, 2010), for ILBI-635 the exposure is decreased to 5–7 minutes, though we

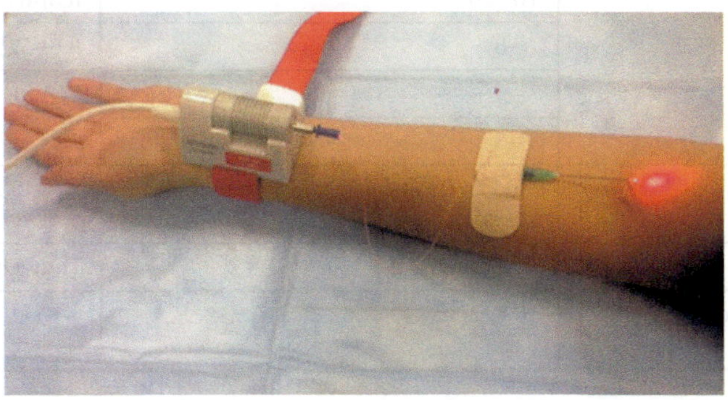

FIGURE 40.1 Intravenous laser blood illumination procedure.

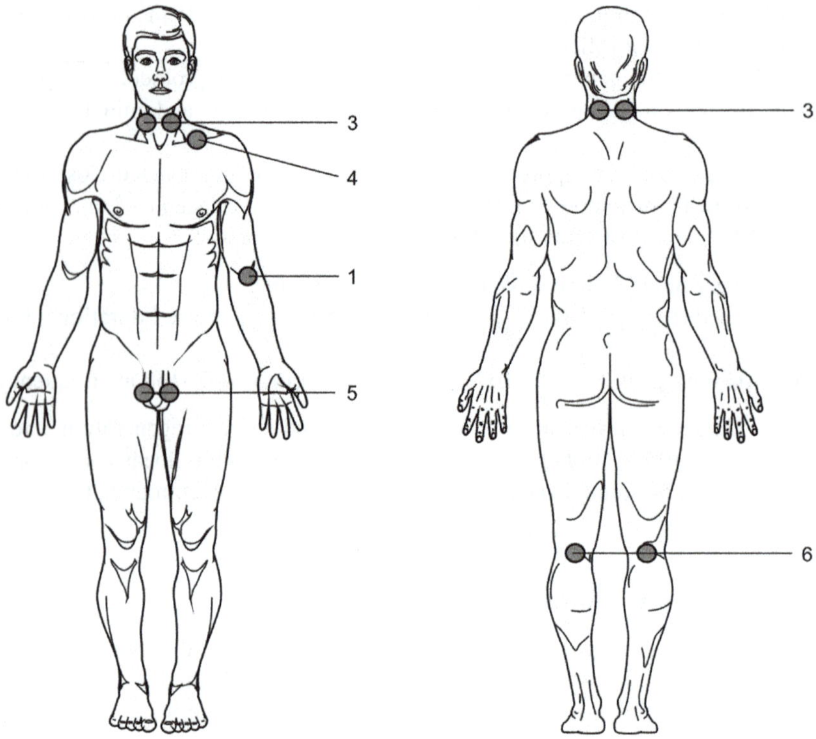

FIGURE 40.2 Main areas of laser blood illumination.

TABLE 40.1 ILBI-635 («classical», basic), ILBI-525, ILBI-365, ILBI-405 (LUVBI) techniques.

Parameter	Value	Note
Laser light wavelength, nm (spectrum)	635 (red)	ILBI-635
	525 (green)	ILBI-525
	ILBI-365 (ultraviolet)	LUVBI
	ILBI-405 (violet)	
Laser operation mode	Continuous	–
Illumination power*, mW	1.5–2	At the output of a disposable light guide
Exposure, min	10–20	ILBI-635
	7–8	ILBI-525
	3–5	LUVBI
	3–5	
Localization	Median cubital vein (v. *mediana cubiti)*	Fig. 40.2, zone 1 (in the left or right arm)
Technique	Intravenous	Through a disposable sterile light guide
Number of procedures per course of treatment	10–12	–

TABLE 40.2 ILBI-525 + LUVBI (basic) technique.

Parameter	Value	Note
Laser light wavelength, nm (spectrum)	365–405 (UV)	LUVBI
	520–525 (green)	ILBI-525
Laser operation mode	Continuous	–
Illumination power*, mW	1.5–2	At the output of a disposable light guide
Exposure, min	3–5	LUVBI
	7–8	ILBI-525
Localization	Median cubital vein (*v. mediana cubiti*)	Fig. 40.2, zone 1 (in the left or right arm)
Technique	Intravenously	Through a disposable sterile light guide
Number of procedures per a course of treatment	10–12	Daily, alternating ILBI-525 and LUVBI every other day

TABLE 40.3 ILBI-635 + LUVBI technique.

Parameter	Value	Note
Laser light wavelength, nm (spectrum)	365–405 (UV)	LUVBI
	635 (red)	ILBI-635
Laser operation mode	Continuous	–
Illumination power*, mW	1.5–2	At the output of a disposable light guide
Exposure, min	3–5	LUVBI
	10–20	ILBI-635
Localization	Median cubital vein (*v. mediana cubiti*)	Fig. 40.2, zone 1 (in the left or right arm)
Technique	Intravenous	Through a disposable sterile light guide
Number of procedures per a course of treatment	10–12	Daily, alternating ILBI-635 and LUVBI every other day

are sure, that for children it is almost always possible to substitute the intravenous technique with the external illumination of the supraclavicular area.

Combined techniques ILBI-525 + LUVBI (Table 40.2) and ILBI-635 + LUVBI (Table 40.3) are gaining more and more popularity. We emphasize the fact that illumination is implemented every other day; it is strictly prohibited to implement ILBI with a different wavelength for the same patient on the same day, especially simultaneously.

Alternating the procedures allows the optimization of the effect on the immune system on the days when LUVBI is implemented, as well as the function of tissues on the days when ILBI-635 or ILBI-525 are implemented (more efficient option).

40.4 Noninvasive laser blood illumination

It is implemented over large blood vessels (arteries and veins) close to the injured area. Pulsed lasers of red (635 nm) or infrared (890–904 nm) spectrum and matrix (eight laser diodes) emitters with illumination areas of 8 cm^2 (Fig. 40.3),

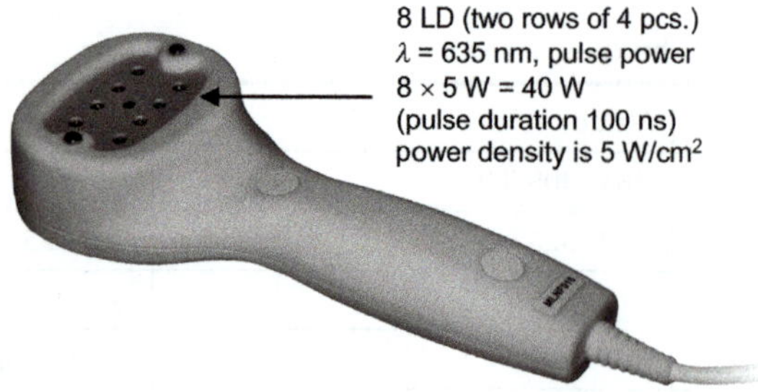

FIGURE 40.3 Matrix laser emitting head ML-635-40 (Laser therapy device "LASMIK") for efficient noninvasive laser blood illumination.

TABLE 40.4 NLBI technique.

Parameter	Value	Note
Laser light wavelength, nm (spectrum)	635 (red)	NLBI-635
	904 (infrared)	NLBI-904
Laser operation mode	Pulsed	–
Light pulse duration, ns	100–150	–
Illumination peak power, W	30–40	Matrix emitting head, NLBI-635
	60–80	Matrix emitting head, NLBI-904
Peak power density, W/cm^2 (surface area of 10cm^2)	3–4	NLBI-635
	6–8	NLBI-904
Frequency, Hz	80–150	–
Exposure on 1 zone, min	2–5	–
Number of zones	2–4	Symmetrically
Localization	On the projection of large blood vessels close to the lesion area	See text
Technique	Contact	Through a transparent nozzle
Number of procedures per a course of treatment	10–12	Daily

or as an option, with a single laser with a mirror head with illumination area of 1 cm^2 are used for NLBI. The power density is identical in any case (Moskvin, 2017) (Table 40.4):

- NLBI-635, the most effective option, pulsed LILL of red spectrum (635 nm), peak PD—4–5 W/cm^2, pulse duration 100–150 ns, frequency 80 Hz;
- NLBI-904, pulsed IR LILL (890–904 nm), peak PD—8–10 W/cm^2, pulse duration 100–150 ns, frequency 80 Hz.

The following illumination localizations are used for NLBI (Fig. 40.2):

- the projection of the common carotid artery (carotid sinus area) symmetrically (zone 2),
- the projection of the vertebral artery symmetrically (zone 3),
- supraclavicular area on the left (zone 4),
- vascular bundles in the groin area symmetrically (zone 5),
- popliteal symmetrically (zone 6).

The pulse frequency is fixed (80–150 Hz);the question of the possibility and permissibility of frequency increase (the average power for pulsed lasers) has not been studied yet. It is recommended to illuminate symmetrical zones with an exposure time of 2–5 minutes on each zone. It is prohibited to illuminate one zone for more than 5 minutes.

The analysis of the publications devoted to the research of the mechanisms of the therapeutic effect of one of the most well-known LLLT techniques—laser blood illumination, as well as the analysis of its long experience of application allows us to speak with confidence about the prospects of this trend. Moreover, both methods: ILBI and NLBI are developing independently, since each method has its own advantages and disadvantages.

The replacement of UV blood illumination with UV lamps by laser ultraviolet blood illumination (LUVBI) has significantly simplified the technique and increased its efficiency. The most efficient options for ILBI are combined techniques: ILBI-635 + LUVBI and ILBI-525 + LUVBI. The most efficient technique for NLBI is the use of low-intensity pulsed laser light with a wavelength of 635 nm and peak power of up to 40 W.

NLBI is always about the impact over large blood vessels. Illumination of peripheral vessels in any localization, like a "laser watch" on the wrist (Litscher and Litscher, 2016) or endonasal (Liu et al., 2010) (Chinese versions) is a suboptimal variation of the method (Moskvin, 2014). In addition, endonasal exposure can be extremely dangerous for women, as it is accompanied by reflex excitation of hypothalamic formations that control the secretion of biologically active substances involved in many processes: stimulation of uterine contraction, regulation of circulatory and reproductive systems, control of the production of various hormones (estrogens, follicle stimulating hormone, etc.) (Serov et al., 1988).

We also do not yet recommend the use of a transcranial technique, which is the subject of a lot of publications, since the method has not been worked out. For example, all the processes that can occur in this case have not been fully studied. It is known that pulsed IR LILL penetrates deeply, covering almost the entire brain, causing activation of microcirculation in the affected area; but light is also absorbed by neurons, and accordingly there is a potential possibility of activating some undesirable processes that we do not know yet. This issue needs to be additionally and very carefully studied.

40.5 The analysis of the literature on the use of low level laser therapy in patients with various cerebrovascular disorders

The first successful experience was obtained by Russian clinicians as early as the 1970s. During these years several thousand scientific studies have been published in Russian (Kochetkov and Moskvin, 2004a,b; Kochetkov et al., 2012). Even a brief analysis of these publications can take more than 100 pages, so we will give several examples from our studies, namely, how laser exposure by a matrix pulse red laser light (wavelength 635 nm, pulse power 40 W on an area of 8 cm^2, pulse duration 100 ns, exposure 5 minutes) impacts on microcirculation (Fig. 40.4) and central hemodynamics (Fig. 40.5). Below we will present the results of some studies, recommendations based on these scientific data, as well as evidence of extensive clinical experience for the application of laser therapy with different types of chronic and acute cerebral ischemia.

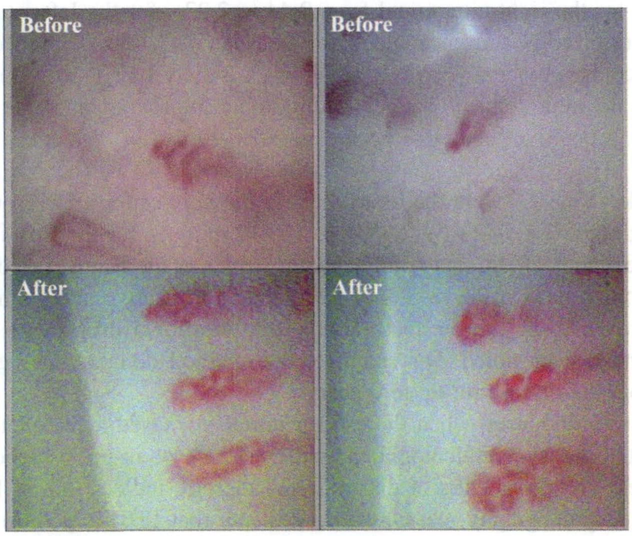

FIGURE 40.4 At the top. A patient suffering from type 2 diabetes (about 4 years) before treatment (depleted capillary network, reduced blood flow, capillary deformation typical of diabetic microangiopathy). At the bottom. The same patient after ten sessions of LLLT—increase of capillary network density, blood flow velocity, form of capillaries (study material—prof. Duvanskiy V.A.).

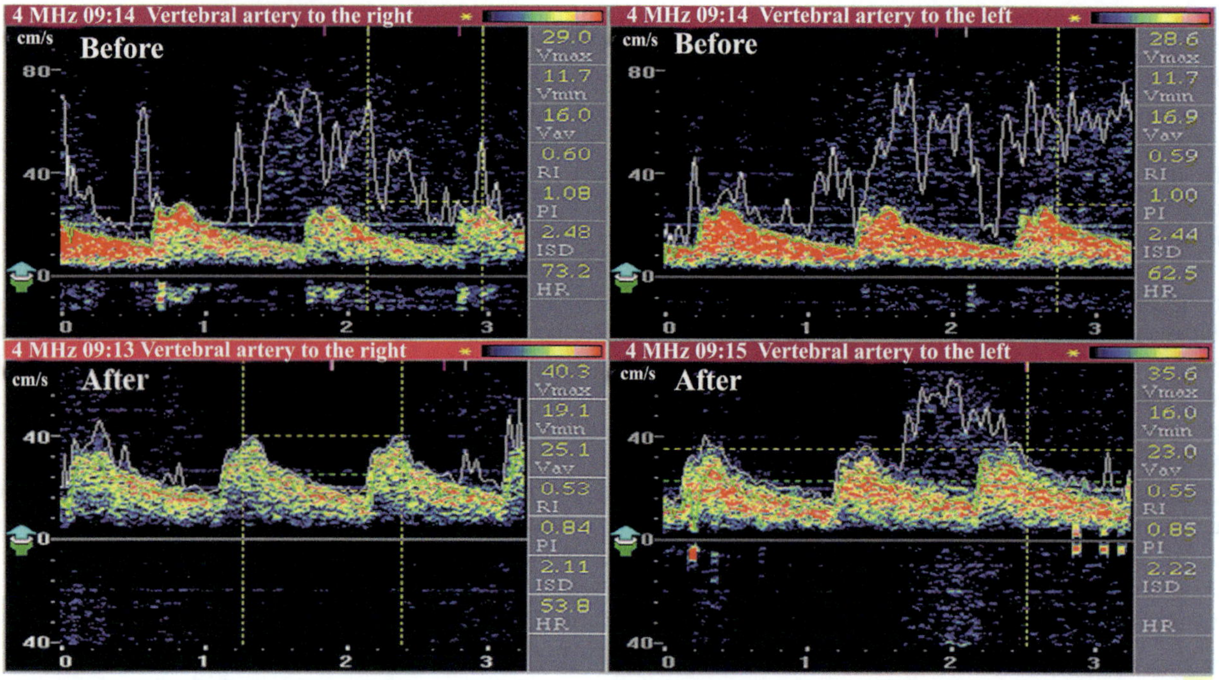

FIGURE 40.5 Ultrasound dopplerography dynamics after a course of matrix laser therapy. Dynamics of changes in vertebral arteries according to ultrasound dopplerography after one procedure of noninvasive laser blood illumination (matrix laser emitting head ML-635-40, wavelength 635 nm, pulse power 40 W, power density 10 W/cm^2, frequency 80 Hz, exposure 5 min symmetrically to the right and left vertebral arteries). Arteriodilatate effect.

Fig. 40.4 shows how the disturbed microcirculation on the arm of a woman with type 2 diabetes mellitus is restored after 10 daily procedures. The achieved effect persists up to 3 months, and the technique is widely used in the treatment of patients with metabolic disorders. This study, performed at the Institute of Laser Medicine of the FMBA of Russia, was chosen as an example only because of its visibility. The same result was obtained with the disturbance of microcirculation of other localization and etiology.

Fig. 40.5 clearly shows a pronounced arterio-dilatation effect after a single illumination of vertebral arteries in a patient with dyscirculatory encephalopathy. Fifteen minutes after the end of the procedure, the pulsation index (PI) decreased from 2.48 to 2.11—for the right artery, and from 2.44 to 2.22—for the left artery. A typical course of LLLT in patients with cerebral atherosclerosis and encephalopathy consists of 10 NLBI procedures daily. It is necessary to conduct preventive treatment courses twice a year, consisting of 2–3 daily LLLT procedures.

If we consider the clinical and neurological changes leading to an improvement in the structure of the "syndrome of stem-cerebellar insufficiency," positive changes occur in 95%–96% of patients according to our long-term observations.

Skupchenko and Makhovskaya (1993) were the first to justify the application of ILBI-635 (wavelength 635 nm, output power at the end of the light guide 1.5–2 mW, exposure 10–12 minutes for the first session and 15–20 minutes for subsequent sessions) in the treatment of patients with ischemic stroke in the acute period, dyscirculatory encephalopathy, and multiple sclerosis. Comparison of the results of treatment in the active and control groups made it possible to establish that laser therapy proved to be more effective than any traditional drug or medicinal methods of treatment. Some peculiarities of LLLT have been noted. For example, this method demonstrates good treatment results only in patients who have suffered from the disease for up to 1 year.

The effect of LLLT is more pronounced when there is an initial increased sympathetic tone of the autonomic nervous system. Patients with initial parasympathicotonia had either a decrease in pulse blood filling, or else there were no statistically significant changes. The parameters of diastolic and dicrotic indices during ILBI were significantly lower than normal. Data from rheoencephalography measurements indicate that ILBI is able to reduce the initially increased tone of the cerebral arteries, increase and normalize the pulse blood filling of the brain, reduce peripheral vascular resistance, and improve venous outflow. LLLT is more effective in patients with a predominance of the sympathetic tone of the autonomic nervous system. Patients with baseline parasympathicotonia, according to rheoencephalography, are resistant to LILL action. The results of rheoencephalographic studies suggest that ILBI improves the functioning of the microcirculation system and the restoration of symmetry between the carotid and vertebrobasilar arterial basins by increasing blood filling in the affected basin (Makhovskaya, 1993; Skupchenko and Makhovskaya, 1993). The normalizing effect of ILBI has been confirmed by other researchers (Solovyeva, 1993).

In acute cerebrovascular disorders (stroke), acute disorders of cerebral circulation (ADCC) in hyperacute and acute phases, ILBI-635 accelerates the regression of cerebral symptoms and focal neurological manifestations. In patients with residual effects from a previous stroke, as well as a history of chronic cerebrovascular insufficiency, ILBI-635 significantly improves the psycho-affective sphere and regresses the local neurological deficit, promotes cerebral blood flow, which is manifested in increasing its speed, enhancing the functioning of natural anastomoses and reducing interhemispheric asymmetry. ILBI-635 significantly inhibits platelet aggregation (two times more intense and two and a half times longer than acetylsalicylic acid), has a significant activating effect on the fibrinolytic system, contributes to a moderate decrease in activity of the hemocoagulation system in patients with cerebral vascular lesions of ischemic nature, raises the arterio-venous difference by shifting the oxygen-hemoglobin equilibrium curve to the right by 20%, which contributes to the increased utilization of oxygen by the brain tissue. Under the influence of LLLT, due to the optimization of metabolic processes in the brain tissue, changes in the parameters of bioelectrical activity are observed in the form of electrophysiological signs: increasing the amplitude and intensity of the alpha rhythm, reducing interhemispheric asymmetry and the expression of slow rhythms. The combined use of laser light and acetylsalicylic acid contributes to the strengthening and stabilization of the antiaggregation effect of the latter (Steblyukova, 1989).

Gorbunov et al. (2003, 2005) presented their experience with another, noninvasive LLLT method in the early rehabilitation of patients after acute disorders of cerebral circulation (ADCC). Pulsed IR LILL, wavelength 890–904 nm, power 4–6 W, frequency 80–150 Hz, exposure 5 mins were used, applied to the surface of the common carotid arteries (CCA) or on the surface of vertebral arteries (VA). A pronounced therapeutic and prophylactic effect of pulsed IR LILL was demonstrated using this technique for various forms of cerebrovascular insufficiency, including after an acute CVA of ischemic and hemorrhagic nature. The beneficial effect of pulsed IR LILL manifests itself in the development of collateral circulation, cerebral hemodynamic and vascular reserve, normalization of blood viscosity, regression of intravascular microcoagulation and aggregation with a resulting decrease in microcirculatory disorders; which increases the rate and maturity of neuroplastic processes in the central nervous system, hence increasing the positive results of the rehabilitation of patients (Churaleva, 2006).

For the last 15–20 years, neurologists in Russia have been divided according to their preferences. Some believe that the best method is ILBI for the treatment of patients with cerebrovascular diseases, while others choose NLBI on the surface of VA or CCA, depending on the localization of the area of the vascular lesion.

We will present the data of only a few studies, but there are supporters of both points of view. There are a lot of publications, but the quality of many of them raises a few questions, while others should be considered very critically. Nevertheless, it is absolutely certain that the possibilities of optimizing laser therapy methods are far from exhausted, and the vast majority of works have been performed using far from optimal laser exposure regimens, which can explain the suboptimal results achieved in some of the studies. Our 25year clinical experience in laser therapy and research has made it possible to confidently determine the most effective parameters of laser techniques, based on theoretical developments and the understanding of the mechanisms of biomodulatory and therapeutic effects of LILL. So far, there is not much scientific data available regarding the combined methods of treatment and rehabilitation after ADCC, especially physiotherapeutic methods.

Karneev (2007) proved that light exposure with the standard ILBI-635 method (wavelength 635 nm, output power at the end of the light guide 1.5–2 mW) in treatment of patients with chronic cerebral ischemia should not exceed 15–20 minutes, while maximum stimulation of the antioxidant system (AOS) is observed, which is an indirect sign of the activation of metabolic processes. Exceeding this exposure (illuminating for up to 30–45 minutes), as well as increasing the laser illumination power (by more than 3–4 mW), leads to a sharp decrease in all clinical indices, and can be especially dangerous for patients with the severe form of the disease. With optimal parameters, ILBI-635 exerts a pronounced positive effect on all AOS parameters (Fig. 40.6). Also, the main indices of the central blood flow are normalized, and a correlation ($rs = 0.73$, $P < .002$) is found between the relative increase in oxygen saturation of hemoglobin and the change in the blood flow linear velocity along the middle cerebral artery with laser therapy. These data are consistent with the results of earlier studies (Perminova, 1994; Rassomakhin, 1996; Solovyeva, 1993), and other researchers, who used laser light with a wavelength of 670 nm and power of 3–4 mW for ILBI (Nechipurenko et al., 2009).

Let us dwell on the issues of dosage, power limitation, and LILL exposure during ILBI, since a violation of the rules known to us can lead to negative and even tragic consequences. For example, Weber et al. (2007), having reviewed the literature and having conducted independent studies in which the correct parameters for the ILBI procedure with different

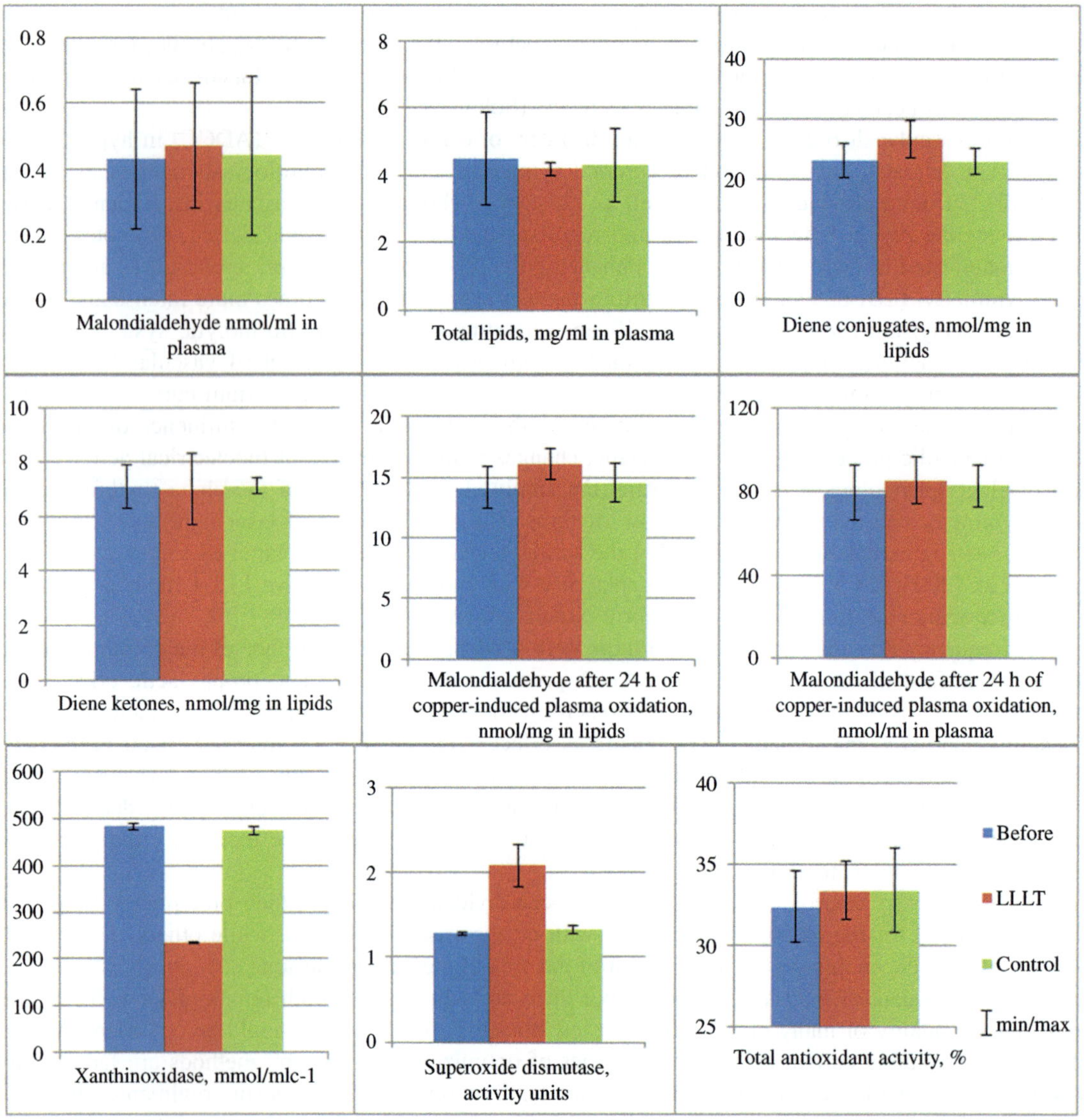

FIGURE 40.6 The dynamics of parameters of lipid peroxidation and antioxidant system in patients with chronic brain ischemia of the first stage (based on data of Karneev, 2007).

wavelengths are specified, had nevertheless, produced a device with dangerously high laser illumination power: IR 810 nm, 100 mW; red 635 nm, 100 mW;green 532 nm, 50 mW; blue 447 nm, 100 mW; UV 375 nm, 5–10 mW (http://www.webermedical.com). Such parameters in any case cannot be used for any category of patients, especially with cerebrovascular diseases. The thing that saved these patients who were being treated by this device, is that the other parameters of the techniques that were recommended were completely wrong. This is good news for the patients. The influence of different laser light sources on the human body were mutually inhibited by using simultaneous illumination by LILL with different wavelengths, and therefore did not endanger the patients. This is what happens when unprofessional people who do not know the basics of laser therapy begin to make a business "treating" their patients.

Kandyba (2002) used ILBI-635 (wavelength 633 nm, continuous mode, 1.5 mW power, 20 minutes exposure, seven courses per day every second day) for ischemic cerebral circulation disorders in the pathology of extracranial arteries in elderly and senile patients. In the main subgroup ($n = 36$), patients underwent ILBI while the control group ($n = 43$) were treated according to the standard scheme (reperfusion and cerebroprotective therapy). The greatest improvement in dynamics after a course of ILBI was observed due to a change in the content of glucose-6-phosphate dehydrogenase, glutathione reductase and glutathione peroxidase, which indicates a significant increase in antioxidant defense mechanisms that compensate for the negative effects of hypoxia as a result of prolonged cerebral ischemia (Table 40.5).

TABLE 40.5 The dynamics of parameters of antioxidant system and lipid peroxidation after ILBI-635.

Parameter	Subgroup	Before treatment	After treatment	Normal range
Concentration of reduced glutathione (μmol/g of hemoglobin)	Main (n = 26)	3.65 ± 0.54[a]	5.10 ± 0.19[a]	Male 5.18–6.38 Female 7.46–8.66
	Control (n = 18)	3.42 ± 0.42	3.21 ± 0.20	
Sulfhydryl groups of proteins (μmol/g of hemoglobin)	Main (n = 26)	8.01 ± 0.56	5.36 ± 0.38	Male 4.95–6.25 Female 5.58–6.88
	Control (n = 18)	7.08 ± 0.38	6.84 ± 0.74	
Malondialdehyde (nmol/g of hemoglobin)	Main (n = 26)	8.67 ± 0.35	7.90 ± 0.91	Male 5.16–6.38 Female 3.04–4.06
	Control (n = 18)	8.21 ± 0.23	10.60 ± 0.94	
Glucose-6-phosphate dehydrogenase (μmol/min g of hemoglobin)	Main (n = 26)	7.35 ± 0.54[a]	13.1 ± 1.13[a]	Male 4.93–6.17 Female 5.78–6.88
	Control (n = 18)	5.23 ± 0.45	5.41 ± 0.59	
Glutathione reductase (μmol/min g of hemoglobin)	Main (n = 26)	235.3 ± 6.5[a]	347.8 ± 13.9[a]	Male 185–233 Female 184–242
	Control (n = 18)	219.0 ± 11.8	225.8 ± 10.7	
Glutathione peroxidase (μmol/min g of hemoglobin)	Main (n = 26)	2.76 ± 0.028[a]	4.22 ± 0.34[a]	Male 4.90–5.38 Female 5.96–7.28
	Control (n = 18)	2.64 ± 0.023	3.32 ± 0.07	
Catalase (μmol/min g of hemoglobin)	Main (n = 26)	17.70 ± 1.21	18.71 ± 0.27	Male 27.0–29.2 Female 36.5–41.3
	Control (n = 18)	16.70 ± 1.09	18.84 ± 0.32	

[a]*$P < .05$ the reliability of differences in antioxidant system data before and after treatment on the basis of linear correlation and nonparametric methods.*

Anatskaya et al. (2015) studied the effect of LLLT on the level of circulating endothelial progenitor cells (EPC) in the acute period of lacunar cerebral infarctions (LACI) and concluded that ILBI in the first 48 hours of LACI in cerebral microangiopathy, significantly increased the level of circulating EPC of CD34+ with the phenotype of the earliest cells CD309+ and CD133+; population CD31+ cells with the phenotype CD309+; endothelial cells expressing CD117+ and pluripotent cells with the phenotypes CD34CD45+ and CD45+ CD117+. Stabilization of the blood–brain barrier (BBB) can be facilitated by an increase in the level of circulating CD31+ CD71+ cells, which increase the tightness of the endotheliocyte junctions of small cerebral arteries and arterioles, preventing the pathological permeability of the BBB and promoting the formation of capillaries. The obtained results make it possible to recommend ILBI to diminish endothelial dysfunction, by reducing the level of circulating endothelial cells, induction of poststroke angiogenesis of the cerebral microcirculatory bed, and stabilization of the BBB in the acute LACI period.

In an experimental model (nonbreeding dogs), it was shown that ILBI-635 (wavelength 633 nm, power 10 mW, exposure 20 minutes) had a pronounced antihypoxic effect and increased survival after 20 minute acute hypoxia of the brain by 25% (Galeeva, 1992).

The question of the advantages of intravenous or surface (noninvasive) methods of blood laser illumination in patients with chronic cerebral ischemia remains open, there are supporters for both methods. Altman and Penzina (2009) came to the conclusion that ILBI-635 (635 nm wavelength, 1.5 mW power) is the most effective, more than NLBI. However, NLBI was performed with suboptimal parameters (wavelength 890 nm, pulse mode, one laser diode power 4 W, frequency of 1500 Hz). The method of determining exposure individually (Penzina, 2008) also raises questions.

In contrast to this point of view, Eltsova (2000) believes that NLBI is more successful, with more acceptable parameters (wavelength 890 nm, pulse mode, one laser diode power 5–7 W, frequency 80 Hz, 2 minutes per sine-carotid zone symmetrically) and is not only comparable in efficiency to ILBI-635 (wavelength 633 nm, power 1 mW, exposure 20 minutes) in the treatment of patients with atherosclerotic dyscirculatory encephalopathy, but it allows the elimination of a number of problems inherent in the intravenous method, such as invasiveness, trauma, higher cost, and the need to use consumables (disposable light guides and needles). The comparison based on the changes in the clinical picture, the state of microcirculation, and the parameters of the lipid transport system.

I would also like to add to some comments to this research—the NLBI must absolutely be carried out using only the matrix laser head in pulsed mode, but not by one that illuminates only a small single point. This has been clearly proven, both by relevant comparative studies, and confirmed by many years of practice in medical centers around the world.

It is clear that patients suffering from cerebrovascular diseases can also be successfully treated by other means of therapy, not only with laser blood illumination (although this is the most common method of treatment). For example, Brodovskaya (1989) showed that consecutive exposure to continuous LILL in the red spectrum (wavelength 633 nm, power 25 mW, exposure 100 seconds) in the region of the unpaired suboccipital zone (C1-C3 and C7-T1) paravertebrally and in synocarotid zones symmetrically, after a course of 10 sessions daily, had a positive effect on the state of cerebral hemodynamics, the bioelectrical activity of the brain, lipidocoagulogram, and the sympathetic-adrenal system. This allows one to consider LLLT pathogenetically justified and an adequate method of treatment and prophylaxis of cerebral haemodynamic insufficiency.

Some authors report the rather high efficiency of laser acupuncture in the treatment of patients with various disorders of the cerebral circulation (Nevmerzhitskaya, 2007; Reukov, 2010), but we do not consider this option to be the first choice basic approach.

Neurocirculatory dystonia, which includes cervical-cranial syndrome, can significantly and negatively affect the blood supply to the brain. LLLT allows for the normalization of the vegetative tone in an average of 57% of cases, cerebral hemodynamics (volume velocity of blood flow in vertebral arteries is increased by 1.3 times), and psychophysiological status (severity of pain and values of the index of functional impairment due to pain in the neck are reduced by 3–5.9 times) (Ljutkevich, 2008). A course of 10–12 daily procedures is performed and repeated after 6 months. The LLLT technique (LILL wavelength 890 nm, pulse mode, one laser diode, peak power 5–7 W), included a combined (sequential) action:

- NLBI on the synocarotid region symmetrically for 1–2 mins (frequency 80 Hz);
- for cervical-thoracic paravertebral zones (C1-T4) for 1 minute per zone (frequency 1500 Hz);
- at acupuncture points (laser acupuncture) with a sympatholytic and sedative effect (frequency 80 Hz, exposure 30 seconds per point): P7 (lecue), GI4 (hegu), GI11 (qi chi), C3 (shao hai), C7 (shen men), C9 (shao chun), E36 (tsu san li), VB20 (feng chi), VG14 (yes zhui).

Laser acupuncture in vertebrobasilar insufficiency is recommended by other authors, albeit with the use of a continuous LILL of the red spectrum (wavelength 633 nm, power 2–3 mW) (Choi, 1990).

The course of supra-arterial applications of pulsed red LILL with a wavelength of 635 nm, projected on to the common carotid artery (CCA) in the synocarotid zone and the vertebral artery (VA) in the suboccipital zone is accompanied by a clinically significant reduction in the most prominent symptoms of cerebrovascular insufficiency—cephalgia, dizziness, cerebrostenic cognitive and psychovegetative disorders, and to a lesser extent, focal symptoms are regressed. A Doppler ultrasound confirms the development of collateral circulation, due to the improvement of the functional state and/or inclusion of extra-intracranial and intracranial blood flow mechanisms. Compression and laser tests for the diagnosis of the cerebral hemodynamic reserve (CHR) showed its development in 85% of patients after 10–12 procedures, but only in the basins of the main arteries of the head that were not affected by hemodynamically significant stenosis. The obtained data testify to a higher hemodynamic efficiency of laser therapy with the use of pulsed LILL with a wavelength of 635 nm, in comparison to 904 nm, since at the last where positive dynamics in the form of development of cerebral CHR and collateral circulation are achieved only in 50% of patients (Kochetkov, 1998).

Taking into account the variations in the development of CHR with the supra-arterial combined use of a pulsed LILL with a wavelength of 635 and 904 nm for the treatment of patients with hemodynamically significant stenosis (HSS), especially in the carotid system, important conclusions have been made with regard to the different application methods of LLLT for different degrees of stenotic lesions of common carotid arteries (CCA) or projecting on to the surface of vertebral arteries (VA). The data obtained by us testify to the undoubted benefits of such combined treatment for various cerebrovascular pathologies, and more importantly, for the treatment of patients with cerebral stroke (Kochetkov and Moskvin, 2004a,b).

It has been proven by our researchers that it is exceptionally important during the NLBI procedure in the treatment of patients with chronic disorders of the cerebral circulation to adhere to optimal time of 2 and 5 minute treatment, both on the projection of the VA and the CCA (Kosmynin, 2005; Kochetkov and Moskvin, 2004a,b).

Using IR pulsed LILL (wavelength 890 nm, frequency 80 Hz, 10 W pulsed peak power, one laser diode) over the VA in the suboccipital zone, symmetrically for 5 minutes per zone on patients with different degrees of vetebrobasillary insufficiency (VBI) is promoting positive enhancement of the linear velocity of the blood flow in the main

TABLE 40.6 The average blood flow velocity (cm/s) in the main arteries of the head after the use of laser therapy ($M \pm m$).

Artery	LLLT ($n = 33$)		Placebo ($n = 15$)	
	Before treatment	After treatment	Before treatment	After treatment
Internal carotid	23.4 ± 1.4	26.1 ± 1.1	22.9 ± 1.4	23.7 ± 1.5
Vertebral	10.2 ± 1.0	13.9 ± 1.5[a]	10.8 ± 1.6	11.4 ± 0.9

[a]*P* < .05 in comparison with initial values.

TABLE 40.7 The state of the hemostasis system after the use of laser therapy ($M \pm m$).

Index	LLLT ($n = 33$)		Placebo ($n = 15$)	
	Before Treatment	After Treatment	Before treatment	After treatment
Time of plasma coagulation, s	110.6 ± 5.0	124.6 ± 3.1[a]	111.2 ± 4.1	113.4 ± 5.2
Plasma tolerance to heparin, s	514 ± 11	520 ± 26	512 ± 29	525 ± 17
Concentration of fibrinogen, g/l	4.9 ± 0.6	4.0 ± 0.5	4.8 ± 0.4	4.7 ± 0.5
Fibrinolytic activity, %	14.7 ± 0.9	14.1 ± 0.8	14.9 ± 0.9	15.5 ± 0.8
Aggregation of platelets induced by ADP,%	35.2 ± 1.4	30.1 ± 1.9[a]	36.3 ± 3.1	35.5 ± 2.4
Aggregation of platelets induced by epinephrine,%	44.5 ± 2.4	37.1 ± 2.1[a]	45.6 ± 4.3	43.7 ± 4.1

[a]*P* < .05 in comparison with initial values.

arteries of the head (Table 40.6), an increase in the blood filling of the vessels of the vertebral-basilar system alongside a decrease in their tone and improvement in the state of microcirculation and hemostasis is observed (Table 40.7); which is accompanied by a regression of neurological deficit, cognitive and psychoemotional violations at different degrees of the development of the pathological process. The application of LLLT contributes to a decrease in the severity of neurological manifestations, depressive and hypochondriacal disorders, and accompanies a reliable increase of the vertebral arterial blood filling and a decrease in their tone (Lapochkin, 2004).

After a LLLT course, there was a tendency for hypocoagulation in the form of a decrease in fibrinogen concentration, an increase in the plasma coagulation time ($P < .05$), and a decrease in platelet aggregation ($P < .05$). Clinical improvement in such patients was associated with a positive effect of LT not only on the state of cerebral hemodynamics, but also on the hemostasis and microcirculation system (Lapochkin, 2004).

Excellent treatment results can be obtained by combining laser blood illumination (ILBI or NLBI) with other methods of treatment. For example, Bendlin (2015) combined transcranial electrostimulation, and every second day NLBI (904 nm wavelength, pulsed mode, ML-904-80 Matrix emitting head Lasmik device, 8 laser diodes, total peak power 40 W, frequency 80 Hz) for 5 minutes symmetrically on the VA, for a course treatment of 8–10 procedures. Dynamics of changes in neurological symptoms in patients with vertebrobasilar insufficiency as a result of treatment are presented in Fig. 40.7.

Elchaninov (2002) showed a high efficiency by combining plasmapheresis and laser blood illumination (wavelength 635 nm, continuous mode, 2.5 mW power, 15 minutes exposure) in the treatment of young patients with chronic cerebral ischemia.

We adhere to the point of view that, according to the totality of the indications, the most effective technique is the NLBI pulsed LILL of the red spectrum (wavelength 635 nm, matrix illuminating head consisting of 8 laser diodes of

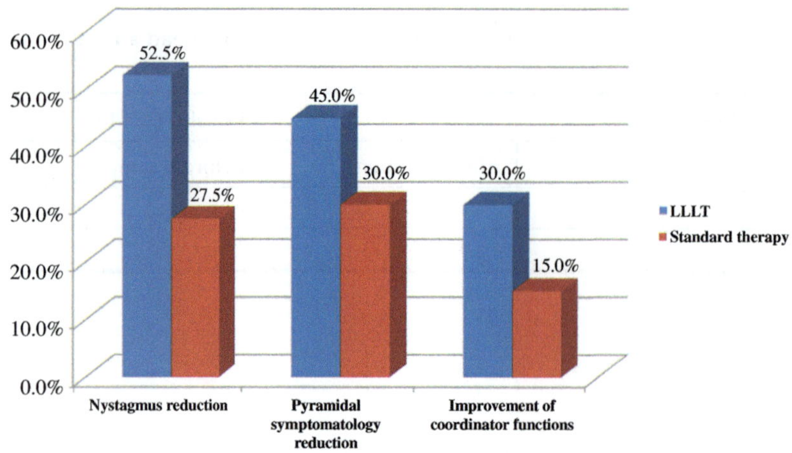

FIGURE 40.7 The dynamics of neurologic symptoms in patients with vertebral-basilar insufficiency percent (Bendlin, 2015).

5 W each, peak power density 3–4 W/cm^2) for 2 or 5 minutes symmetrically on the VA with vertebrobasilar insufficiency and, projecting on to the CCA with more pronounced violations of the cerebral circulation in the sinocarotid area. A course of 10–15 procedures of laser therapy is performed daily, and preventive courses that are 3–5 procedures should be repeated every 5–6 months. (Kochetkov et al., 2009; Leiderman, 2010; Leiderman and Schekina, 2010; Leiderman et al., 2009a,b,c,d, 2010a,b,c, 2011). Of course, other methods of treatment should also be included.

In the outpatient setting of one of Moscow's polyclinics, 102 patients aged from 49 to 70 years old, (mean age 61.5 years—33% men, 67% women) with comprehensive clinical data confirming the presence of CCI patients stages I–II. The patients excluded from the study were those over the age of 70; with a long history (more than 10 years) or severe arterial hypertension (AH)—arterial pressure (AP) systolic higher than 160 mmHg, AD diastolic higher than 95 mmHg.; circulatory insufficiency above Stage I; ischemic heart disease (IHD) with angina pectoris above functional class II (FC); with clinically pronounced signs of respiratory, renal and hepatic insufficiency; uncontrolled diabetes mellitus (DM); with diseases of the thyroid gland with hyperthyroid function; with marked cognitive disorders (less than 24 points on the MMSE scale); with stenosis of the main arteries of the head, the presence of unstable atherosclerotic plaques, bilateral and/or associated hemodynamically significant stenosis (HSS) in the carotid system (CS) and vertebral-basilar system (VBS).

All observed patients received basal therapy: drug treatment (vasoactive, nootropic, disaggregate agents) and were randomly assigned to two groups comparable in age, sex, clinical symptomatology: the base group (52 people) and control group (50 people). The first group (52 people) were treated with basic LLLT performed with the pulsed matrix illuminating heads (wavelength 635 nm) according to the method we developed (see above) every second day, three times a week. A course consists of 9–11 procedures. In the second, control group (50 people), sham therapy was performed by a nonfunctional laser light (placebo procedure), including using the apparatus and applying the laser emitting head to the impact area, without illumination, which was provided by special technical means. The medical staff who conducted the procedures were not informed of the exposure option (double blind control). The duration of treatment in both groups was 21–22 days.

Laser illumination was carried out sequentially on posterolateral surfaces of the neck and projected on to the surface of vertebral arteries (VA) and cervical sympathetic plexuses (two zones), using the stable contact technique. The treatment was performed using the matrix illuminating head ML-635-40 (laser therapy device Lasmik, manufacturer—the Research center, Matrix, Moscow, Russia). The LILL parameters were as follows: wavelength 635 nm; impulse response frequency 80 Hz; pulse peak power 40 W (8 laser diodes); exposure per one zone: first procedure—2 minutes, all subsequent exposure times—5 minutes. Assessment of the patients' condition was carried out before and after the treatment using a single diagnostic algorithm, including a clinical and neurological examination, as well as additional research methods: a neuroimaging survey was carried out to verify the diagnosis (according to indications, computer or magnetic resonance imaging of the brain); To assess the degree of cognitive deficits and exclude dementia, the mini-mental state examination (MMSE) scale was used (Folstein et al., 1975).

An assessment of the current mental state was carried out using scales: a scale for assessing the psychological general well-being (PGWB) index (Dupuy, 1984); the Beck depression inventory (Beck et al., 1961); and the scale of reactive and personal anxiety of Spielberger–Khanin (Khanin, 1976; Spielberger et al., 1970). To study the state of the autonomic nervous system, a questionnaire was used to identify autonomic dysfunction (Wein, 2003); Kerdo's vegetative index was determined (Kérdö, 1966); and the method of cardiointervalography (CIG) was used (Bayevsky et al., 1984).

The state of microcirculation was evaluated by method of conjunctival bulb angioscopy performed by a slit lamp (Inami, Japan). To assess the state of the cerebral circulation, ultrasound was used: ultrasound dopplerography (USD) and duplex scanning (DS) of the extracranial departments of the main arteries of the head, transcranial Doppler sonography (TDS) using the Philips SD-800. The blood flow linear velocity was measured in the main arteries of the head to assess the cerebral hemodynamic reserve (CHR) using the Gosling's pulsatility index (PI), with a functional hypercapnic test with determination of the coefficient of reactivity to hypercapnia (Nikitin, 1998).

The statistical processing of results was carried out by the methods of analysis of variance statistics with the verification of the significance of the results by the student's criterion with a given probability equal to 95%, as well as the nonparametric Mann-Whitney U-test, chi-square test, and Fisher test. Calculations were carried out using the BIOSTAT program. Differences were estimated as significant starting at $P < .05$. All indicators are presented as the mean $\pm$ standard deviation ($M \pm \sigma$).

The duration of the course of the disease (CCI) ranged from 3 to 12 years. Etiological factors of development of chronic cerebrovascular insufficiency in 79 (77.4%) patients were atherosclerotic damage lesions of the main arteries of the head, and in 23 (22.5%)—there was a combination of atherosclerosis and high blood pressure (HPB). Analysis of the severity of clinical-neurological symptoms showed that 35 (34.3%) patients had stage I chronic cerebral ischemia (CCI) and 67 patients (65.6%) had stage II.

Clinical and instrumental signs of concomitant ischemic heart disease and angina pectoris were noted in 29 (28.4%) patients, of which 14 (13.7%) suffered a small-focal myocardial infarction. Clinical angiological and ultrasound signs of one- or two-sided atherosclerotic lesion damage of arteries of the lower extremities were revealed in 15 (14.7%) patients. Clinical and laboratory signs of compensated type two diabetes were observed in nine (8.8%) patients. Other vascular risk factors were also identified: a strongly aggravated family history of cardiovascular diseases (78%); hypercholesterolemia (69%); smoking (28%); and 22 (21.5%) patients underwent transient ischemic attacks (TIA) without a gross residual neurologic deficit. The time elapsed since the TIA episode was at least 3 weeks.

At the time of screening and distribution in groups, all patients had cognitive and psychoemotional abnormalities as leading signs in the clinico-neurological picture. Patients complained of headaches (92%), dizziness (74%), memory loss (84%), irritability, anxiety (83%), sleep disturbance (76%), fatigue (82%), violation of perception and reproduction of new information (78%). The main characteristics of changes in the emotional-volitional sphere were low moods, suspiciousness, quick temper, difficulty in changing mental settings, which had a negative impact on the quality of life of patients.

In patients with chronic cerebral ischemia stage I, disorders in the neurological status were of a subsyndromic nature. Multiple diffuse focal neurological symptoms were revealed in the form of a lack of convergence (five patients), mild asymmetry of nasolabial folds (eight), deviation of the tongue (three), anisoreflexia (nine), coordination disorders (six).

In patients with CCI stage II, vertebral-basilar insufficiency (VBI) syndrome was diagnosed in 52 (78.8%) cases. Carotid insufficiency syndrome (CIS) shows in 12 (16.7%) patients. Combined lesions were detected in three (4.5%) patients with stage II. Clinically, VBI syndrome was manifested by vestibular-atactic disorders in 48 patients, lack of oculomotor innervation in 36, cranial and bulbar innervation in 8, conductive pyramidal motor and sensory disorders in 10. CCI was associated with the presence of slight pyramidal insufficiency or hemi-type sensory disorders—in three patients, the initial signs of amyostatic syndrome—in four, pseudobulbar syndrome—in four. Fifty-eight percent of patients diagnosed with vertebrogenic syndrome of the cervical and upper chest level, in 6%—compression radiculopathy C5-C7. The dynamics of the severity of pain were assessed by a visual analog scale (VAS). In the motor sphere, there was a restriction of the range of movements in the cervical spine, the flattening of the cervical lordosis, the asymmetry of the height of the shoulder girdle, the tension of the individual muscular groups of the neck and extremities. The vertebrogenic nature of the lesion was confirmed by functional radiography, CT or MRI. In 48 patients, according to the MRI of the brain, focal changes of vascular genesis, periventricular and/or subcortical leukoareosis were detected. According to the MMSE test, the degree of cognitive impairment in 58% of the examined category of patients corresponded to the syndrome of moderate cognitive dysfunction (24–27 points on the scale). When assessing the patients' background of depression using the Beck questionnaire: light depression was detected in 32% of patients and moderate depression in 68% of patients. When assessing the severity of anxiety manifestations on the Spielberger–Khanin scale, they were found to increase practically in all subjects before treatment. The average level of reactive and personal anxiety was observed in 53% of cases, and in the remaining 47% of cases, a high level of anxiety was detected. When assessing the quality of life with the index of general psychological well-being (IGPW) before treatment, a significant decrease in the level of normal indices ($P < .05$) by 46% was noted. The domains that showed the most significant decreases included general health, vital energy, and emotional well-being. In the examined group of CCI patients, signs of autonomic dysfunction were revealed. According to blood pressure meters, the sum of points, normally equal to 15, averaged at 42.8 ± 5.5 points before the treatment. Evaluation of the vegetative tone on the Kerdo's vegetative index showed the presence of vegetative imbalance, an increase in the functional activity of the sympathetic-adrenal system in patients with CCI. The average value of the Kerdo's vegetative index was 15.7 ± 1.9. According to the cardio-intervalography, sympathetic influences also prevailed prior to treatment. Hypersympathicotonia was observed in 52% of patients, sympathicotonia—in 36% of patients, vagotonia in 7%, and normotonia in 5% of cases, respectively. The predominance of increased tone of the sympathetic nervous system indicated the tension of adaptive reactions of the organism in patients with CCI.

Patients initially examined with conjunctival biomicroscopy were found to have manifested microcirculation disturbances by an increase in the tortuosity of the microvessels, the granularity of the blood flow, and the presence of aggregations of the type of the "sludge" phenomenon dotted blood flow, the formation of focal venous stasis and areas where the blood thickens. According to USD, DS and TCD, the following signs of cerebral hemodynamic disturbances were revealed: the presence of hemodynamically significant stenosis (HSS) (>60%) in eight patients in the common carotid artery (CCA) bifurcation zone or in the internal carotid artery (ICA), 16 in extracranial departments of the VA; decrease in blood flow linear velocity for CCA/ICA and VA—in 84% of cases; the presence of significant asymmetry of blood flow linear velocity between the sides of CS and VBS in 49% of cases; increase in PI values for the main arteries of the head—in 85% of cases; reduction of cerebrovascular reactivity (CVR) (according to reactivity coefficient to hypercapnia)—in 53% of cases; the presence of signs of venous dyscirculation—in 57% of patients.

A good toleration of LLLT procedures in all patients was noted, as well as pronounced compliance (90%). Overall, 49 patients (94%) in the main group reported improvement in their overall condition, while only 27 (54%) in the control group reported the same thing—which was significantly lower than the main group (chi-square, $P < .01$).

Negative effects (patient condition worsening) were not noted. In a number of cases, a single procedure of laser therapy was accompanied by the development of a sedative effect, a reduction or disappearance of headaches, and also an antihypertensive effect with a concomitant mild form of hypertension, a decrease in systolic (by 8–10 mmHg) and diastolic blood pressure (by 5–7 mmHg).

During treatment, subjective symptoms in the main group of patients regressed significantly ($P < .01$, chi-square test) faster than in the control group (Fig. 40.8).

The greatest clinical effect after the course of LLLT was manifested in the reduction of headaches, the severity of asthenic syndrome and sleep disorders. There was also an improvement in the state of cerebral function, primarily in the form of regression of mental fatigue, memory improvement, and concentration of attention. A subjective clinical effect of therapy in most patients manifested itself by the fifth to seventh procedure and was stabilized by the end of treatment. Patients of the treatment group noted an increase in energy and improved overall physical condition. The degree of fixation on negative emotions and poor health also decreased.

In contrast to the control group, the treatment group showed a much more pronounced regression of pain symptoms in the region of the cervical spine (Fig. 40.9, Table 40.8). This was accompanied by a reduction in the stiffness of the

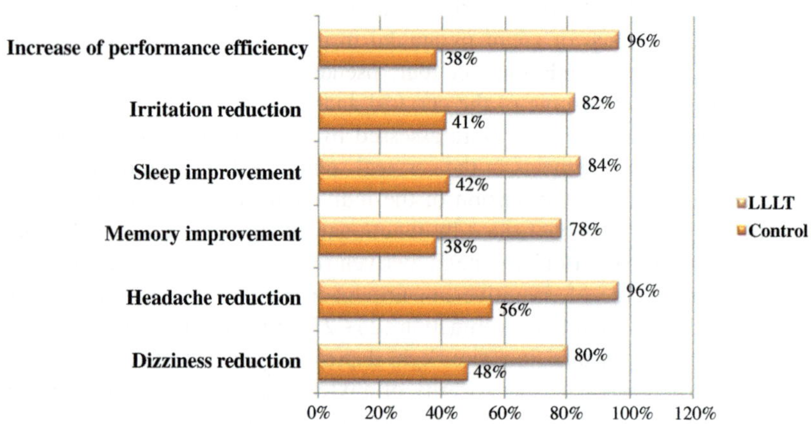

FIGURE 40.8 The dynamics of the main neurologic symptoms (Leiderman, 2010).

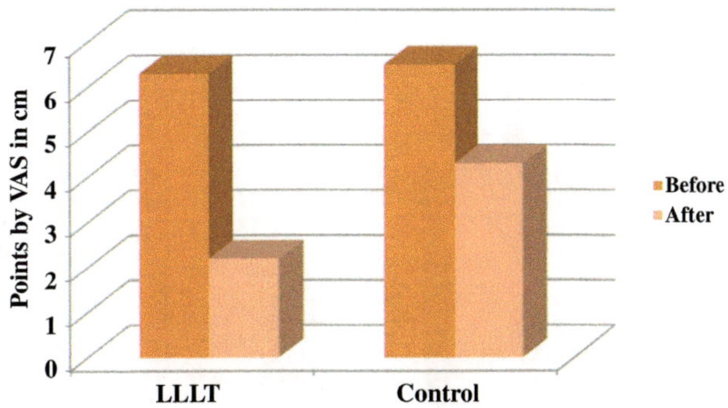

FIGURE 40.9 The dynamics of vertebrogenic pain syndrome by VAS scale (Leiderman, 2010).

TABLE 40.8 Dynamics of the vertebrogenic pain syndrome after a course of LLLT in patients with CCI according to VAS.

Parameters	Main group		Control group	
	Before treatment	After treatment	Before treatment	After treatment
Points according to VAS	6.3 ± 2.5	2.2 ± 0.9[a,b]	6.5 ± 1.9	4.3 ± 0.8[a]

[a]Significant differences with respect to the baseline results ($P < .05$).
[b]Significant differences between groups ($P < .05$).

muscles in the shoulder girdle, the elimination of muscle hypertonia, paresthesia of the upper limbs and an increase in the volume of movements in the cervical spine.

By the end of the course of treatment, the patients in the main group also had more pronounced dynamics of focal neurological symptoms compared to the control group. The greatest regression was found in oculomotor disturbances, deficiency of VII and XII pairs of cranial nerves, impaired coordination, cochlear and atactic disorders. Out of 28 patients, a decrease in the severity of the leading cerebrovascular symptoms with VBI syndrome in the main group was noted in 26 patients (92.8%), and in the control group—out 27 patients, 14 were observed to have a decrease (51.8%), which was significantly lower in comparison with the main group (chi-square, $P < .05$). The dynamics of extrapyramidal and pyramidal pseudobulbar symptoms were nonsignificant and unreliable.

Overall, objective improved dynamics of neurological status were observed in 43 (82%) patients in the main group and in 19 (38%) in the control group, which was significantly lower in comparison to the main group (chi-square, $P < .05$). Improvement of the psychoemotional status of patients after treatment was expressed as a decrease in emotional lability and an improvement in mood.

The patients of the main group reliably normalized their emotional state, according to the data of the Beck depression inventory. In the control group, the manifestations of depression also decreased, but these changes were unreliable (Fig. 40.10, Table 40.9).

After the end of the treatment course, the indices of personal anxiety in both groups remained practically unchanged, however, the reactive anxiety in the main group, in contrast to the control group, significantly decreased (Fig. 40.11).

When analyzing the effect of treatment on the quality of life of patients, reliable, positive dynamics in PGWB index were established. In the group of patients who received a course of LLLT, almost all indicators of PGWB were higher than in the placebo group. The most significant shifts were revealed on the scales characterizing emotional well-being (by 46%, $P < .01$), general health (by 35%, $P < .01$), and vital energy (by 40%, $P < .01$). Mean PGWB index after treatment was 82.5 ± 6.5 points in the main group, and 65.5 ± 7.5 points in the control group, which is significantly lower than in the main group ($P < .05$). Thus, the conducted studies reflect the high efficiency of the application of the pulsed laser matrixes of red spectrum in the correction of emotional and affective disorders, which, undoubtedly, is also significant for the stabilization of cognitive function. At the end of treatment course, the treatment group were noted to have a significant decrease in signs of autonomic dystonia using the questionnaire of vegetative dysfunction (by 55.3%), and in the control group—only by 24.7%. A statistically significant decrease in the absolute value of the Kerdo's vegetative index in the main group was revealed, which reflected the tendency to restore the vegetative balance (Fig. 40.12).

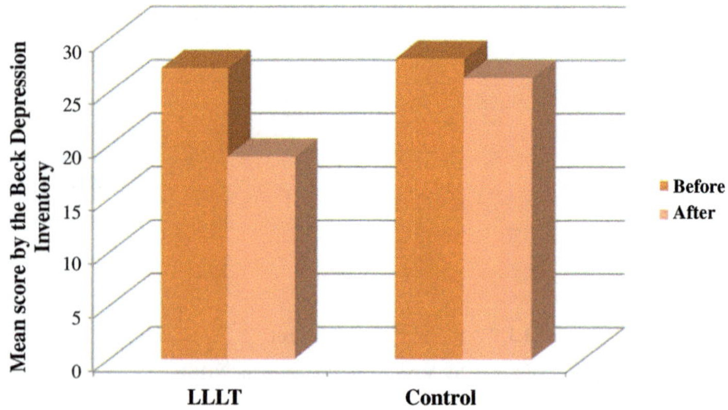

FIGURE 40.10 The dynamics of depression symptoms by the Beck depression inventory in patients with dyscirculatory encephalopathy (Leiderman, 2010).

TABLE 40.9 Dynamics of depression symptoms according to the Beck Depression Inventory in patients with CCI after a course of LLLT.

Group	Average score	
	Before treatment	After treatment
Main	27.2 ± 2.5	18.9[a] ± 2.9
Control	28.1 ± 3.1	26.3 ± 2.9

[a]Significant differences with respect to baseline results ($P < .05$).

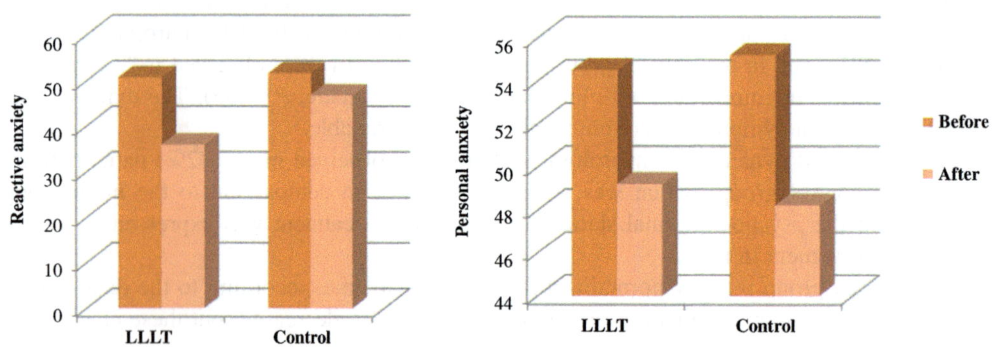

FIGURE 40.11 The dynamics of anxiety level indicators by the Spielberger–Khanin scale (Leiderman, 2010).

According to the cardio-intervalography (CIG) data, the main group saw a decrease in the number of patients with hypersympathicotonia and an increase in the number of patients with normotonia (Fig. 40.13).

Thus, segmental exposure to pulsed matrix laser illumination with a wavelength of 635 nm causes a neurodynamic response of a systemic nature, manifested in a change in the functional state of the nervous system and in the correction of the vegetative status. The normalizing effect of LLLT on the sympathoadrenal system is shown: vegetative homeostasis changed toward a decrease in sympathetic activity, which contributes to the inhibition of vasoconstrictor effects of spinal stimulation.

During the treatment, the main group showed positive changes in the microcirculation system according to the data of the bulboangioscopy. The effects of a single application of LILL were accompanied by an increase in the intensity of capillary blood flow and the dilatation of arterioles. By the end of the treatment course, a significant positive effect

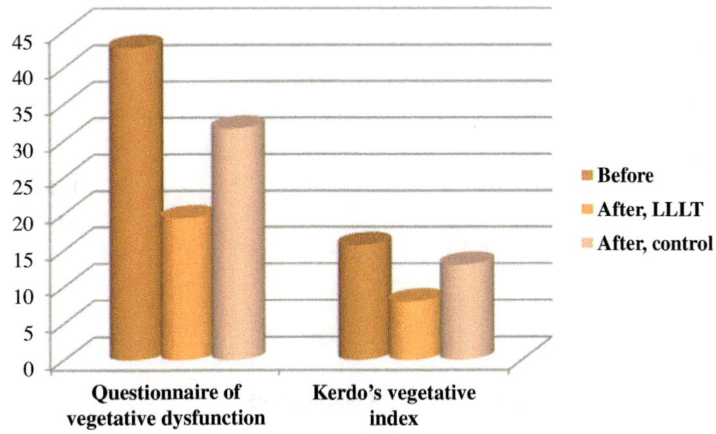

FIGURE 40.12 The dynamics of the values of vegetative indicators (Leiderman, 2010).

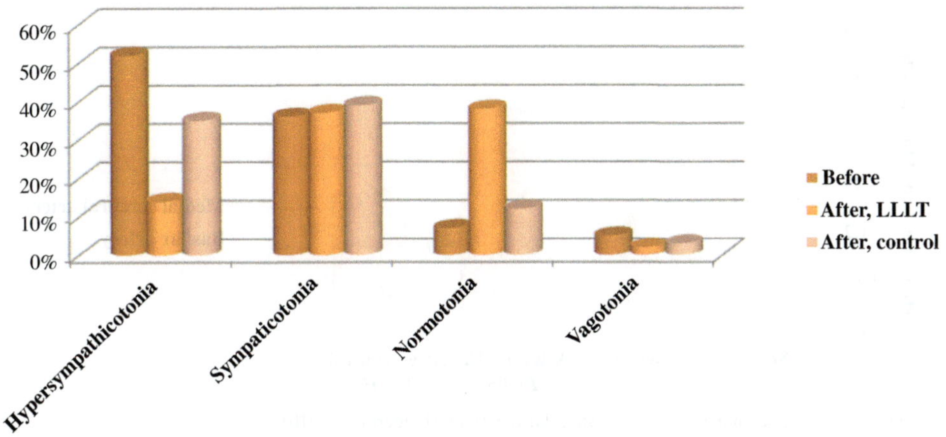

FIGURE 40.13 Cardiointervalography data (percentage of patients) Leiderman, 2010).

from LLLT on the microcirculatory system was noted. The main components of microcirculatory disturbances were reduced: intravascular—marked regression of erythrocyte aggregates, a decrease in the granularity of blood flow; extravasal—in the form of a decrease in perivascular edema, stagnant phenomena; and also vascular, manifested by the development of the capillary bed in the avascular zones, that is, neovasculogenesis. As a result, a decrease in the conjunctival index (CI), reflecting the degree of microcirculation impairment, in the main group occurred on average by 35.4% ($P < .05$). In the control group, there was a tendency to improve microcirculation, but the dynamics of CI were unreliable (Fig. 40.14).

A USD study after the treatment course revealed more significant dynamics of the cerebral hemodynamic (CH) values in the main group compared with the control group. Improvement of blood flow velocity in the CCA/ICA system was noted in 54% of cases (in control—34%, $P < .05$) and VA—in 82% (in control - 35%, $P < .05$).

The disappearance or decrease in the asymmetry of blood flow linear velocity between the carotid system and vertebralbasilar system was observed in 59% of cases in the main group and in 21% in the control group ($P < .05$). LLLT produced an arteriodilating effect. The decrease in the initially elevated PI in the basin of the middle cerebral artery (MCA) in the main group occurred on average by 10.5% ($P < .05$, by the U criterion), and in the control—by 1.3% ($P > .05$); in the basin of basilar artery (BA)—by 9.4% in the main group ($P < .05$) and 1.1%– in the control ($P > .05$). In 62% of cases in the main group and in 28% of cases in the control there was a decrease in venous circulation ($P < .05$, by the chi-square test).

Comparison of the indices of the central nervous system between the beginning of the course and after laser treatment made it possible to quantify the cerebrovascular reactivity (CVR), reflecting the adaptive capabilities of the cerebral circulation system. In patients with CCI stages I-II, the reactivity of the cerebral vessels to laser therapy was sufficient to increase the CHR (Fig. 40.15).

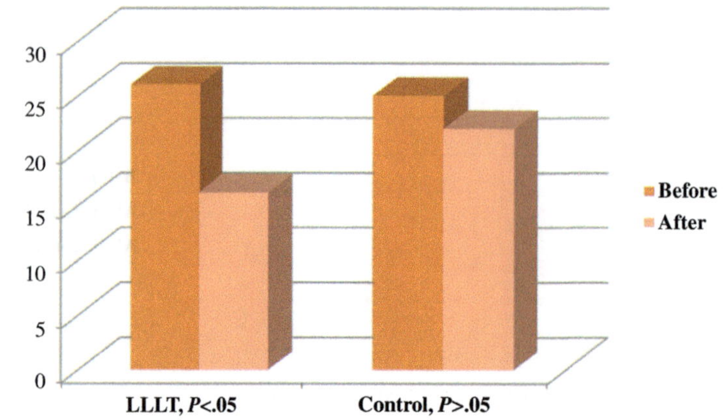

FIGURE 40.14 The dynamics of the conjunctival index (Leiderman, 2010).

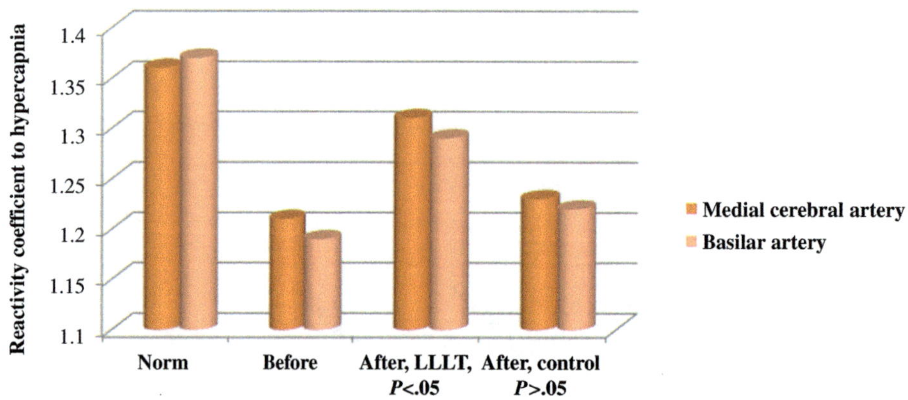

FIGURE 40.15 The dynamics of the parameters of cerebrovascular reactivity (Leiderman, 2010).

In the main group of patients, a statistically significant increase in reactivity coefficient to hypercapnia was observed, which indicated an increase in the reserve of vasodilation of resistive vessels and an improvement in autoregulatory capabilities of cerebral blood flow. At the end of the treatment course, reliable signs of the development of collateral circulation were revealed: activation of the flow in the anterior and posterior connective arteries, eye anastomosis in 29 (56%) patients in the main group and only 7 (14%) in the control ($P < .05$).

The obtained results testify to the stimulating effect on the state of the cerebral vascular reserve that the LILL method that we have developed exerts, as well as an increase in its dilatation component and an increase in the tolerance of the brain to ischemia and hypoxia. The main features of the methodology, which determine its high efficiency are:

- unique, redundant, red laser diodes operating in pulsed mode with a peak power of 5 W at a pulse duration of 10 ns and a pulse repetition rate of 10,000 Hz;
- type of emitting heads—MATRIX **ONLY**;
- optimum frequency—80 Hz;
- optimum exposure at 2 and 5 minutes;
- selection of the localization of application.

A few words about the pulsed red laser diodes and their uniqueness, which, so far are only produced by our company (Research Center "Matrix"). Mitsubishi announced in 2015, that it is the only company that will start production of similar laser diodes (wavelength 638 nm), but no more than 2.5 W (https://www.photonicsonline.com/doc/mitsubishi-electric-launching-w-nm-red-laser-diode-0001), which is not sufficient for effective treatment.

Subjective assessment of the results of treatment by patients is also important, the patients gladly agreed to receive laser therapy procedures—simple, painless, nontime consuming and which allows the patient to gain almost immediate relief (Fig. 40.16).

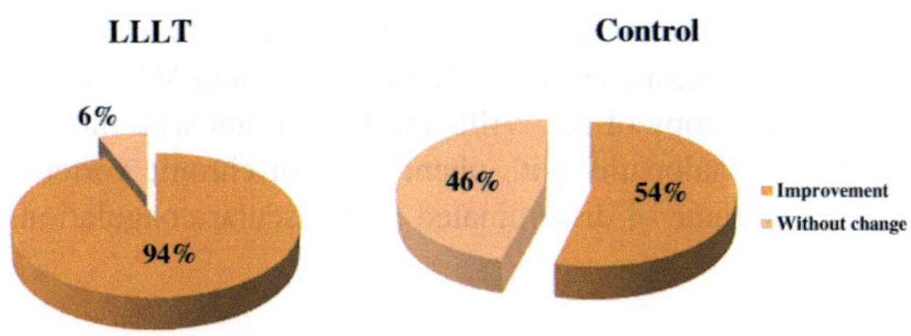

FIGURE 40.16 Subjective evaluation of treatment outcomes (Leiderman, 2010).

Studies show a general homeostatic orientation in the implementation of therapeutic effects of red pulsed laser illumination. Clinical changes of the segmental effects of the laser matrix on the paravertebral zones of the cervical spine in patients with early forms of cerebral circulatory disturbances are accompanied by a decrease in afferent excitability, a change in interneuron conduction, which leads to a decrease in the tone of the vascular wall. Influencing the zone of cervical sympathetic plexus and VA, LILL can affect the nervous mechanism of regulation of cerebral circulation, its vascular mechanism (photoinduced dilatation of arterioles, stimulation of endothelial relaxing factors production) and mechanisms of long-term adaptation in the form of neovasculogenesis. The observed significant regression of pain and muscular-tonic cervical syndrome contributes to reducing the vertebrogenic effect on VA, improving venous outflow and eliminating brain dyscirculation.

According to the follow-up data, the following year after the treatment course, the clinical results were most stable in the treatment group and were observed to remain so over 6–8 months, and in the control group only within 3–4 months. ($P < .05$ by the Fisher criterion). Remote clinical results (observation for 24 months) showed that the combination of drug therapy with repeated courses of treatment according to the developed scheme of laser therapy allowed to reduce the frequency of the development of acute disorders of cerebral circulation (stroke) in the main group by 2.5 times in comparison with the control ($P < .05$).

It has been a relatively long time since the indications and contraindications for the use of laser therapy in acute disorders of cerebral circulation (stroke) had been developed (Gorbunov et al., 2003), and it is imperative that they are followed.

40.6 Indications

1. In ischemic types (atherothrombotic, cardioembolic, and lacunar subtypes) of strokes with a residual deficiency of mild and moderate severity, including repeated strokes; at hemispheric localization—in one of the basins of the carotid system or zones of adjacent blood circulation, developed alongside atherosclerotic lesions of the main arteries of the head, hypertension and with their combination; beginning with the third to fourth week (transient ischemic attack and minor stroke) and fourth to fifth week with cerebral stroke (CS) of the disease.
2. In the case of hemorrhagic type of stroke with volume of bleeding up to 15–20 mL, including hemorrhagic transformation of the primary ischemic focus, developed alongside AH or its combination with the defeat of the main arteries of the head; at hemispheric localization; beginning with fourth to fifth weeks (nontraumatic subarachnoid hemorrhage) and fifth to sixth weeks (parenchymal and parenchymal-ventricular hemorrhage) of the disease.
3. When the patient age does not exceed 70 years; a neurologic deficit of mild and moderate severity in the absence of symptoms requiring specialized neurosurgical intervention or observation (intracranial hypertension, meningeal syndrome), as well as paroxysmal disorders of consciousness, gross disorders of orientation and perception; a satisfactory somatic state (compensation of cardiac and coronary, respiratory, endocrine, renal and hepatic insufficiency).
4. Atherosclerotic dyscirculatory encephalopathy, including combination with stroke (see above) with functional and organic blood flow disorders or stenosis, hemodynamically significant stenosis (HSS > 60%) of one of the carotid arteries (internal or total), or occlusion of one from the internal carotid arteries (in the presence of a collateral overflow through the ipsilateral orbital anastomosis), but not with HSS or occlusion of both carotid arteries.
5. Various forms of dyscirculatory encephalopathy, including in combination with strokes with the phenomena of latent syndrome of disseminated intravascular coagulation, thrombophilia or hypercoagulable state.
6. With hemorrhages in the occipital part of the hemisphere with the volume of the outflow of blood up to 10–15 mL, including with hemorrhagic transformation of the focus of primary ischemic lesion, developed in combination with arterial hypertension or its combination with the defeat of the main arteries of the head, starting at the fifth to sixth weeks of illness.

7. Atherosclerotic dyscirculatory encephalopathy, including in combination with CVA (see above) with functional and organic disturbances of blood flow—stenosis, HSS > 60% of one or both VA, or occlusion of one VA (in the presence of collateral overflow in the anisotropy of the Willis circle), but not with the occlusion of both VAs.
8. Various forms of dyscirculatory encephalopathy with phenomena of chronic vertebral-basilar insufficiency, as well as with the phenomena of latent syndrome of disseminated intravascular coagulation, thrombophilia or hypercoagulable state.

40.7 Contradictions

1. The presence of an active atherosclerotic plaque (heterogeneous structure, uneven contours, etc.) in the illumination zone, expressed and/or widespread atherosclerotic or other in nature (aortoarteritis, etc.), damage to the arteries of the brain and heart, especially in the development of cerebrocardial syndrome, acute coronary insufficiency or hemodynamically significant cardiac rhythm disturbances in the acute period of a stroke.
2. Suspicion or presence (according to an angiography) of an aneurysm or arteriovenous malformation not completely cut off from the bloodstream.
3. Infratenorial (posterior cranial fossa) localization of the hemorrhage.
4. Hemorrhagic syndrome (hematuria, etc.).

The development of laser therapy techniques does not stand still. On the contrary, it is actively developing, but the general principles of application of the method should always be strictly observed. Of course, we all understand that complex treatment, in which laser therapy is only a component, will be much more effective than just using a single method. Examples of this multimodal approach, especially those often used in the rehabilitation of stroke patients, are commonly applied when traditional pharmacotherapy and medical-physical therapy are being used along with various methods of physiotherapy and balneology (Amro Ismail, 2008; Karimova, 2004; Kovalyov, 2014; Sidyakina, 2012; Khatkova, 2013; Cherevaschenko et al., 2013). It is necessary to take into account the age of patients in preparation for the treatment regimen. For example, it is shown that laser therapy in the complex treatment of gerontological patients with cerebrovascular pathology exerts a more pronounced and stable hypotensive effect compared to other physiotherapeutic methods (Karimova, 2004).

Let us cite as an example several patents for demonstrating the variety of existing LLLT methods used in Russia for the treatment of patients with cerebrovascular diseases (Table 40.10).

This chapter has not covered yet another very important aspect of the issue—the prevention of chronic cerebrovascular disorders, and as a consequence, the development of stroke. We are certainly convinced that if everyone after 30

TABLE 40.10 Russian patents—LLLT methods for treating patients with cerebrovascular diseases.

Pathology	Technique	Result	Reference
Acute ischemic stroke, locomotor disorders	Paravertebrally, localization depends on the involved area, pulsed IR LILL (890 nm, 10 W), 4 min on each side alternately, comprising acupuncture and therapeutic exercises, 12–15 daily procedures per course.	Higher clinical effectiveness, reduced length of staying in hospital and the disability rate in the patients.	Patent 2487739 RU
Cerebral ischemic insult	ILBI, 633 nm, 2 mW, no more than 30 min, combined with medicamentous therapy and laser acupuncture, 10 daily procedures per course.	Reduced treatment duration, restored as much as possible disturbed functions of extremities and eliminated speech disturbances.	Patent 2145895 RU
Cerebrocranial trauma	Pulsed IR LILL (890 nm, 80–1500 Hz), on anterior temporal areas, carotid arteries, in paravertebral application to cervical region of the vertebral column for 2–4 min, 10 daily procedures per course.	Accelerated treatment course; reduced drug doses; adjusted immune responses; alleviated paresis and paralysis symptoms.	Patent 2156149 RU
Cerebro-vascular insufficiency	NLBI, pulsed IR LILL (890 nm, 5–7 W, 80–150 Hz), alternately on the projection area of common carotid and vertebral arteries for 4–6 min, 10–12 daily procedures per course.	Clinical improvement, increased duration of remission.	Patent 1780770 SU

(Continued)

TABLE 40.10 (Continued)

Pathology	Technique	Result	Reference
Chronic cerebral ischemia	Continuous LILL (540 nm, 0.9 mW), on skin projections of postganglionary sympathetic long carotid artery, as well as its wall, for 3–5 min per field, 10 daily procedures per course.	Higher efficiency of treatment.	Patent 2347594 RU
Cognitive disorders accompanying chronic cerebral ischemia	Pulsed IR LILL, 890 nm, 70–100 W (matrix of laser diodes, illumination area of 20–30 cm^2), 1–5 mW/cm^2, 1000–3000 Hz, on internal carotid arteries and vertebral arteries, projections of frontal and temporal lobes of both cerebral hemispheres, as well as upper cervical ganglion. Duration of exposure of each zone is 120 current pulse strikes of the patient, recorded during the session. 10–15 daily sessions before noon.	Improvement of cognitive functions due to biosynchronization of laser exposure with phases of tissue blood filling increasing rhythms.	Patent 2607942 RU
Craniocerebral injuries in patients of elder and senile age	Pulsed IR LILL (890 nm, to carotid arteries from both sides—daily beginning from the third to fourteenth day after the injury during 10 days, 1–3 min on each, 1500–3000 Hz, 5–10 W; to spinal arteries from two sides,—daily beginning from the third to fourteenth day after the injury during 10 days, 1.5–3 min on each, 80–150 Hz, 5–10 W; on the right and left suprarenal gland projection—daily beginning from the fifth to seventh day after the injury during 5–7 days, 1–2 min on each, 80–150 Hz, 3–7 W; on thymus projection—daily beginning from the fifth to seventh day after the injury during 5–7 days, 1–4 min, 80–600 Hz, 3–7 W.	Enhanced effectiveness and reliability of treatment of craniocerebral injuries exactly in patients of elder and senile age.	Patent 2207172 RU
Discirculatory encephalopathy	NLBI, pulsed IR (890 nm, 5–7 W, 5 Hz) for 0.5–2 min on ulnar vascular fascicles and sinocarotid zones every second day. 10–12 daily procedures per course.	Stabilization of cerebral hemodynamics, lipid profile, arterial pressure values, improvement of clinical neurological semiology, functional condition of vegetative nervous system, cognitive and emotional-volitional spheres.	Patent 2449822 RU
Disseminated sclerosis	Laser acupuncture, 690 nm, 15–60 s per point (P7; P2; P1; GI17; GI18; RP20; C6; C7; V2; V11–V35; MC8; TR20; VB19; VB8; VG1–VG14). 10–15 daily procedures per course.	The disappearance of such symptoms as nystagmus, impaired coordination, muscle twitching, euphoria, increased duration of remission	Patent 2005457 RU
Disseminated sclerosis	Transcutaneously, continuous IR LILL (700–1000 nm, 20–150 mW) sequentially during one session: on injured spinal cord segments projections from the back side—in scanning mode during 10–120 min, on brain—through occipital and temporal foramens during 5–60 min per each foramen, on axillary and inguinal lymph nodes are treatment during 1–10 min per each node. Daily sessions with pauses on holidays during 3–6 months.	The prevention of disease progressive development; restored work of injured organs and systems.	Patent 2200041 RU

(*Continued*)

TABLE 40.10 (Continued)

Pathology	Technique	Result	Reference
Encephalopathy	Continuous IR LILL (860 nm, 30 mW, modulation frequency 70 Hz), on projections of vertebral arteries along the entire length from the formation to the cranial opening, four fields for 3 min starting within a cerebral affection, then it is symmetric. Thereafter, continuous LILL (633 nm, 10 mW, modulation frequency 70 Hz) on projections of internal carotid arteries at the orbital level within a cerebral affection then symmetrically, two fields for 1 min. 5–10 daily procedures per course.	Higher clinical effectiveness due to the balanced improvement of blood flow and brain metabolism.	Patent 2392985 RU
Epilepsy	ILBI, 633 nm, 5 mW, 30 min, seven daily procedures per course, repeated course every 6 months.	The method makes it possible to give up medicamental therapy and reduce number of attacks. Stable remission.	Patent 2131754 RU
Epilepsy	ILBI, continuous red LILL (633 nm, 2 mW, 20 min), pulsed IR LILL (890 nm, 8 W, 80–1500 Hz), 1 min on each zone symmetrically: temporal, parietal and paravertebral (C1–C6), laser acupuncture, continuous IR LILL (1300 nm, 2 mW), on corporal points symmetrically during 1 min: VG24, VG23, VG21, VG20, VG19, VG18, VG16, VB20, temporal, mastoid and paravertebral zones (C1–C6). LILL (1300 nm, 2 mW) 20 s on corporal points: VC12 and E13—daily, every other day—GI4, 2 points and E36, 2 points—the first day, MC6, 2 points and RP6, 2 points—the second day. 10–16 procedures per course, the first 5 procedures—daily, the rest—every other day. The course is repeated in 3–3.5 months at least three times.	Enhanced effectiveness of treatment and reduced treatment period.	Patent 2149655 RU
Epilepsy	NLBI, pulsed IR LILL (890 nm, 5–7 W, 80–150 Hz), 15 daily procedures per course.	Improved mental measurements, relieved proneness to convulsive activity, suppressed epileptogenic zones, improved metabolic and energy cerebral processes, normalized cerebral blood flow regulation, recovered cortical neurodynamic disorders.	Patent 2353411 RU
Focal cerebral injuries in the cases of craniocerebral traumas	After surgically removing of injured brain lesion foci a catheter is set in the brain zone injured most of all. LILL (633 nm, 5 mW) is applied via the catheter during 5 min in the nearest postoperative period under clinical electrophysiological control during 2–5 days.	Enhanced effectiveness of treatment; reduced neuroloric manifestations; increased consciousness level due to the prevention of inflammatory process and secondary necrosis of brain tissue.	Patent 2194549 RU
Inflammatory brain diseases	Plasmapheresis and laser blood illumination (633 nm, continuous, 2.2 mW), exposure time of 10 min, seven daily procedures per course.	Enhanced effectiveness of treatment; retained antimicrobial properties with increased adsorption on leukocytes, strengthening the membrane of leukocytes, protecting it from the damaging effect of the antimicrobial drug.	Patent 2286154 RU

(*Continued*)

TABLE 40.10 (Continued)

Pathology	Technique	Result	Reference
Nervous system pathology	ILBI (633 nm, continuous, 1.5–2.5 мВт, 20–40 min) or NLBI (890 nm, pulsed, 20 W, 1.2–4.8 Hz, 10 min), or ILBI/NLBI alternating every second procedure. This is alternated with laser acupuncture and transcranial scanning of brain areas using pulsating IR LILL.	Lessening the negative influence of drugs on the organism.	Patent 2169021 RU
Parkinson's disease	ILBI (633 nm, 1 mW, 20 min) and drug therapy, five daily procedures per course, repeating in a year.	Higher efficiency of therapy due to stimulating the system of antioxidant tissue protection, increased protective body reserves and correction of immune disorders.	Patent 2255775 RU
Parkinson's disease	Laser acupuncture, 690 nm, 10–15 daily procedures per course.	Increased duration of remission.	Patent 1785440 SU
Vertebrobasilar insufficiency	Continuous red (633 nm, 2.5 mW) on facial reflexogenic zones and NLBI pulsed IR laser (890 nm, 3 W, 150 Hz), six to eight procedures per course.	A persistent positive clinical effect, the method allows accelerating medical and labor rehabilitation of patients after an ischemic attack.	Patent 2180207 RU
Vertebrobasilar insufficiency	NLBI, pulsed IR laser (890 nm, 3 W, 600 Hz), 600 Hz, 2–3 min per elbow and popliteal folds, laser acupuncture (VB20 and VB21, 1 min per each point), The course is 10 days twice every second month.	Reduced treatment time and the amount of medication.	Patent 2356527 RU
Vertebrobasilar insufficiency	Continuous red LILL (633 nm, 2 mW) combined with the drug solutions.	Prolonged remission and recovered work capacity ensured by improving rheological blood properties, reinforced vascular wall of the vertebral zone, fast, effective, stable improvement of cerebral blood circulation in the vertebrobasilar artery at the level of microvasculature, improved venous blood outflow with no side effects of the conducted therapy.	Patent 2464052 RU
Vertebrobasilar insufficiency	Pulsed IR LILL (890 nm, 8 W), paravertebral zones on cervical and upper thoracic segments on both sides, additionally transcranial electrostimulation, 12–14 procedures per course.	Reduced complaints and irritability; improved general health state, cerebral blood circulation and vibration sensitivity.	Patent 2315636 RU

years of age conducted preventive laser therapy every 4–5 months, illuminating daily two to three times with a matrix pulsed laser emitting head (wavelength 635 nm, pulse mode, peak power 40 W per 8 laser diodes, frequency 80 Hz, exposure 2 minutes) symmetrically applied to the sinocarotid zone, could either prevent a stroke completely, or reduce the likelihood of developing a stroke by as much as ten times. Such a simple procedure could save the lives of millions of people if it becomes more wide-spread in modern healthcare.

References

Altman, D.A., Penzina, E.B., 2009. Initial manifestations of cerebral blood supply insufficiency: peculiarities of cerebral hemodynamics and effects of differentiated laser therapy // Traditional and new directions of vascular surgery and angiology. - Chelyabinsk, pp. 4–6. (Russian)

Anatskaya, L.N., Goncharova, N.V., Severin, I.N., et al., 2015. Influence of intravenous laser blood illumination on the level of circulating endothelial progenitor cells in the acute period of lacunar cerebral infarctions/News NANB, 3:24–29. (Russian)

Bayevsky, R.M., Kirillov, O.I., Kletskin, S.Z., 1984. Mathematical Analysis of Heart Rate Measurements Under Stress. Study, Moscow, p. 221. (Russian).

Beck, AT, Ward, C.H., Mendelson, M., et al., 1961. An inventory for measuring depression. Arch. Gen. Psych. 4, 561–571.

Bendlin, I.D., 2015. Transcranial electrostimulation and magnetolaser therapy of patients with vertebrobasilar insufficiency. Author's abstract. PhD of medical sciences. St. Petersburg, p. 22. (Russian)

Brodovskaya, A.M., 1989. Low-intensity laser light in the complex treatment of patients with initial manifestations of cerebral blood supply deficiency. Author's abstract. PhD of medical sciences. Odessa, p. 15. (Russian)

Cherevaschenko, L.A., Ledovskaya, T.P., Berezhnaya, E.V., Kulikov, N.N., Cherevaschenko, I.A., 2013. Dynamics of cerebral blood flow parameters under the influence of radon baths and laser craniopuncture in patients with organic lesions of the central nervous system. Spa Med. 2, 42–46. (Russian).

Churaleva, O.V., 2006. Features of neurofunctional and immunological disorders in the early recovery period of ischemic stroke and their correction. PhD of medical sciences. Novosibirsk, p. 18. (Russian)

Davydenko, T.E., 2006. Intravascular laser blood illumination in holistic therapy of the widespread atherosclerosis of elderly and senile age patients (Thesis). Saint Petersburg. (Russian).

Dupuy, H.J., 1984. The psychological general well-being (PGWB) index. In: Werner, N.K., Mattson, M.E., Furberg, C.F., Elinson, J. (Eds.), Assessment of Quality of Life in Clinical Trials of Cardiovascular Therapies. Le Jacp, Atlanta, pp. 170–183.

Elchaninov, A.P., 2002. Hereditary and acquired factors of thrombophilia and therapy of chronic cerebral ischemia in young people. Author's abstract. PhD of medical sciences. St. Petersburg, p. 32. (Russian)

Eltsova, G.N., 2000. Comparative efficacy of cutaneous and intravenous laser therapy in patients with atherosclerotic dyscirculatory encephalopathy. Abstract. PhD of medical sciences, p. 18. (Russian)

Folstein, M.F., Folstein, S.E., McHugh, P.R., 1975. «Mini-mental state». A practical method for grading the cognitive state of patients for the clinician. J. Psych. Res. 12 (3), 189–198.

Galeeva, O.P., 1992. Cranio-cerebral infusion and endovascular laser action in acute cerebral hypoxia. Author's abstract. PhD of medical sciences, p. 24. (Russian)

Gerasimenko, M.U., Geynits, A.V., Moskvin, S.V. (Eds), 2015. Laser therapy in treatment and rehabilitation, and preventive programs: clinical guidelines. Triada, Moscow. (Russian).

Geynits, A.V., Moskvin, S.V., Achilov, A.A., 2012. Intravenous laser blood illumination.: Triada, Moscow–Tver. (Russian).

Gorbunov, F.E., Kochetkov, A.V., Minenkov, A.A., et al., 2003. The Use of Low-Intensity Laser Illumination in the Infrared Range in the Early Rehabilitation of Patients After Acute Disorders of Cerebral Circulation. Technics, Moscow, p. 17. (Russian).

Ismail, A., 2008. Physiotherapy in complex rehabilitation of patients with cerebrovascular pathology. Pedagog. Psychol. Med. Biol. Probl. Phys. Educat. Sports 9, 7–12.

Kandyba, D.V., 2002. Ischemic disorders of cerebral circulation in the pathology of extracranial arteries. Author's abstract. Doct. med. sciences. St. Petersburg, p. 45. (Russian)

Kapustina, G.M., Moskvin, S.V., Titov, M.N., 1996. Intravenous laser blood illumination (ILBI). Med. Marketing Media 24, 20–21 [Russian].

Karimova, E.A., 2004. Medico-social aspects of the choice of the rehabilitation program for elderly and senile patients with the consequences of ischemic stroke. Author's abstract. PhD of medical sciences. Saratov, p. 24. (Russian)

Karneev, A.N., 2007. Cerebral resistance to oxidative stress in patients with chronic cerebral ischemia. Author's abstract. Doct. honey. sciences, p. 50. (Russian)

Kérdö, I., 1966. Ein aus Daten der Blutzirkulation kalkulierter index zur beurteilung der vegetativen tonuslage von I. Kérdö. Acta neurovegetativa 29 (2), 250–268.

Khanin, Y.L., 1976. A short guide to the Ch.D. Spielberger scale of reactive and personal anxiety p. 18. (Russian)

Khatkova, S.E., 2013. New approaches to the system of complex rehabilitation of patients with post-spastic spasticity. Author's abstract. PhD of medical sciences, p. 48. (Russian)

Kochetkov, A.V., 1998. Medical physical factors at the stage of early rehabilitation of patients with cerebral stroke. Abstract. PhD of medical sciences, p. 47. (Russian)

Kochetkov, A.V., Moskvin, S.V., 2004a. Laser therapy of patients with cerebral stroke. Triada, Tver. (Russian).

Kochetkov, A.V., Moskvin, S.V., 2004b. Laser therapy of patients with cerebral stroke. Triada, Tver, p. 51. (Russian).

Kochetkov, A.V., Leiderman, N.E., Moskvin, S.V., 2009. Matrix pulse lasers of the red spectrum of illumination in the complex treatment of cerebrovascular disease. Collection of reports and theses of the forum "World of Health", pp. 107–108. (Russian)

Kochetkov, A.V., Moskvin, S.V., Karneev, A.N., 2012. Laser therapy in neurology. Triada, Moscow–Tver. (Russian).

Kosmynin, A.G., 2005. Application of laser therapeutic "Matrix" in atherosclerotic discirculatory encephalopathy. Author's abstract. dis. cand. med. sciences, p. 27. (Russian)

Kovalyov, G.F., 2014. Correction of hemodynamic disorders in the main arteries of the head at an extracranial level in patients with chronic cerebral ischemia in a Spa. Abstract. dis. cand. med. sciences. Pyatigorsk, p. 22. (Russian)

Lapochkin, O.L., 2004. Magnetic-laser therapy of patients with circulatory failure in the vertebral-basilar system. Author's abstract. PhD of medical sciences, 21 p. (Russian)

Leiderman, N.E., 2010. The use of pulsed matrix lasers of 0.63μm in the complex treatment of patients with discirculatory encephalopathy. Author's abstract. PhD of medical sciences, p. 26. (Russian)

Leiderman, N.E., Kochetkov, A.V., Moskvin, S.V., 2009a. Experience of application of matrix pulsed lasers of the red spectrum of illumination in the complex treatment of cerebrovascular disease. Materials of scientific-practical. Conf. with Int. participation "Laser medicine of the XXI century," p. 74. (Russian)

Leiderman, N.E., Moskvin, S.V., Kochetkov, A.V., 2009b. Application of matrix lasers of the red spectrum of radiation in complex regenerative treatment of patients with cerebrovascular pathology. Collection of theses of the IY scientific-practical conference "New technologies of regenerative medicine of FMBA of Russia". - M. (1), 55–56. (Russian).

Leiderman, N.E., Moskvin, S.V., Kochetkov, A.V., 2009c. Technology LASMIK in complex restorative treatment of patients with chronic brain ischemia. Vestnik New Med. Technol. 2 (4), 104–106. TXYI. (Russian).

Leiderman, N.E., Moskvin, S.V., Kochetkov, A.V., 2009d. Efficiency of 0.63-micron matrix pulsed lasers in the treatment of patients with chronic cerebral ischemia. Collected theses of Int. scientific and practical work. Conf. "The use of lasers in medicine and biology. - Kharkov (3), 21–22. (Russian).

Leiderman, N.E., Kochetkov, A.V., Moskvin, S.V., 2010a. Influence of laser action by pulsed matrix illuminators of 0.63 microns on the clinical and functional state of patients with discirculant encephalopathy. Materials of XXXIII Intl. scientific and practical work. Conf. "The use of lasers in medicine and biology." Uzhgorod, pp. 21–23. (Russian)

Leiderman, N.E., Kochetkov, A.V., Moskvin, S.V., 2010b. New technology of laser therapy in the rehabilitation treatment of patients with chronic cerebrovascular insufficiency. Clin. Neurol. 1 (2), 20–25. (Russian).

Leiderman, N.E., Shchekina, R.V., Partigulova, E.V., 2010c. Efficiency of pulsed red matrix lasers in correction of cognitive and psychovegetative disorders in patients with discirculatory encephalopathy. Med. Bull.e Ministry Internal Affairs 1 (3), 27–32. (Russian).

Leiderman, N.E., Shchekina, R.V., Zayonchkovskaya, I.M., 2011. Dyscirculatory encephalopathy - the possibilities of physiotherapy in a complex of restorative treatment. Physiotherapist 3, 21–26. (Russian).

Leiderman, N.E., Schekina, R.V., 2010. The technology of laser therapy in the complex treatment of patients with discirculatory encephalopathy. Hospital. 2 (2), 34–37. (Russian).

Litscher, G., Litscher, D., 2016. A laser watch for simultaneous laser blood illumination and laser acupuncture at the wrist. Integr. Med. Int. 3, 75–81.

Liu, T.C.Y., Wu, D.F., Gu, Z.Q., Wu, M., 2010. Applications of intranasal low intensity laser therapy in sports medicine. J. Innovation Opt. Health Sci. 3 (1), 1–16.

Ljutkevich, A.A., 2008. Changes in autonomic regulation and cerebral hemodynamics in cervical-cranial syndrome, correction methods. Author's abstract. PhD of medical sciences. Novosibirsk, p. 21. (Russian)

Makhovskaya, T.G., 1993. Intravascular laser therapy with ischemic disorders of cerebral circulation. Abstract. PhD of medical sciences. Perm, p. 24. (Russian)

Meshalkin, E.N., Sergievsky, V.S., 1989. Application of low energy helium-neon laser in cardiology and cardiac surgery. In: Skobelkin, O.K. (Ed.), Lasers in surgery. Medicine, Moscow, pp. 238–243. (Russian).

Mester, E., Ludani, G., Selyer, M., Szende, B., Total, G.J., 1968. The stimulating effect of low power laser rays on biological systems. Laser Rev. 1, 3–8.

Moskvin, S.V., 1997. Laser therapy like a modern stage of heliotherapy (historical aspect). Lasernaya meditsina 1 (1), 44–49. (Russian).

Moskvin, S.V., 2003. Laser therapy in dermatology: vitiligo. Tekhnika, Moscow. (Russian).

Moskvin, S.V., 2008. About mechanism of therapeutic influence of low-frequency laser radiation. Bull. New Med. Technol. 15 (1), 166–172 (Russian).

Moskvin, S.V., 2014. The effectiveness of laser therapy. Series "Effective laser therapy" 2, Triada, Moscow-Tver. (Russian).

Moskvin, S.V., 2016. Basics of laser therapy. Series "Effective laser therapy" 1, Triada, Moscow-Tver. (Russian).

Moskvin, S.V., 2017. Low-level laser therapy in Russia: history, science and practice. J. Lasers Med. Sci. 8 (2), 56–65.

Moskvin, S.V., Nasedkin, A.N., Osin, A.Y., Khan, M.A., 2009. Laser therapy in pediatrics. Triada, Moscow-Tver. (Russian).

Moskvin, S.V., Nasedkin, A.N., Osin, A.Y., Khan, M.A., 2010. Laser therapy in pediatrics. EKSMO, Moscow. (Russian).

Murav'yeva, V.N., Karpova, E.N., 2014. Modern concepts of risk factors and prevention of stroke (literature review). Int. J. Exp. Ed. 3, 59-45. (Russian).

Nechipurenko, N.I., Likhachev, S.A., Griboyedova, TV., 2009. Influence of laser hemotherapy on the state of lipid peroxidation, the content of trace elements in the blood and cutaneous microhemodynamics in patients with chronic cerebral ischemia. J. Grodno State Med. Univ. 2, 79–81 (Russian).

Nevmerzhitskaya, I.V., 2007. Application of frequency-modulated system magnetolaser laser for restorative correction of initial manifestations of cerebral circulation disorders. Author's abstract. PhD of medical sciences. Moscow, p. 25. (Russian)

Nikitin, Y.M., 1998. Ultrasonic dopplerographic diagnosis of vascular diseases. Y.M. Nikitin, A.I. Trukhanov (Eds.), pp. 64–114. (Russian)

Penzina, E.B., 2008. Neuro-immune disorders and their correction with low-intensity laser radiation in men with initial manifestations of cerebral blood supply deficiency. Author's abstract. PhD of medical sciences. Ekaterinburg, p. 29. (Russian)

Perminova, L.G., 1994. Clinico-physiological characteristics of patients with discirculatory encephalopathy in the process of intravenous laser therapy. Author's abstract. PhD of medical sciences. Nizhny Novgorod, p. 19. (Russian)

Rassomakhin, A.A., 1996. Clinico-biochemical and clinical-immunological parallels in endovascular laser therapy in patients with dyscirculatory encephalopathy. Abstract. PhD of medical sciences. Saratov, p. 23. (Russian)

Reukov, A., 2010. Correction of post-stroke motor disorders by combined use of magnetic laser therapy and acupuncture. Military Med. J. 7, 56–57 (Russian).

Serov, V.N., Kozhin, A.A., Zhukov, V.V., Khusainova, I.S., 1988. Laser therapy in endocrinological gynecology. Publishing house of Rostov State University, Rostov-on-Don. (Russian).

Sidyakina, I.V., 2012. Perfection of the system of rehabilitation programs in acute and acute periods of severe cerebral ischemic stroke. Author's abstract. PhD of medical sciences, p. 42. (Russian)

Solovyeva, E.Y., 1993. Application of intravenous laser therapy in patients with atherosclerotic encephalopathy. Author's abstract. PhD of medical sciences, p. 16. (Russian)

Spielberger, C.D., Gorsuch, R.L., Lushene, R.E., 1970. STAI manual for the state–trait anxiety inventory consulting. Psychologist. Palo Alto Press.

Spupchenko, V.V., Makhovskaya, T.G., 1993. Laser therapy in neurology. Samara-Khabarovsk, p. 81. (Russian)

Steblyukova, I.A., 1989. Clinical efficacy of intravenous laser therapy and its combination with drugs in cerebral vascular lesions of an ischemic nature. Author's abstract. PhD of medical sciences, p. 21. (Russian)

Weber, M.H., Fußgänger-May, T., Wolf, T., 2007. The intravenous laser blood illumination - introduction of a new therapy. Dt. Ztschr. F. Akup 50 (3), 12–23.

Wein, A.M., 2003. Vegetative Disorders. Clinic, Diagnosis, Treatment. MIA, Moscow, p. 752. (Russian).

Patent 1780770 SU, Int.Cl. A61N 5/06. Method for treating cerebrovascular insufficiency / F.E. Gorbunov, I. N. Danilova, N.I. Strelkova et al. – № 4855072/14, Date of filing: 27.02.1990; Publ. 15.12.1992, Bul. № 46.

Patent 1785440SU, Int. Cl. A61H 39/00, A61N 5/06. Method for treating parkinsonism / L.A. Chernyak. – № 4398085/14; Date of filing: 13.01.1988; Publ. 30.12.1992, Bul. № 48.

Patent 2005457RU, Int. Cl. A61H39/00, A61N5/06. Method for treating disseminated sclerosis / L.A. Chernyak. – № 4380432/14; Date of filing: 18.02.1988; Publ. 15.01.1994.

Patent 2131754 RU, Int.Cl. A61N5/06. Method of epilepsy monotherapy / O. A. Eremin, M. A. Makhotin, M. V. Fatkhullin et al. – № 98117473/14; Date of filing: 21.09.1998; Publ. 20.06.1999.

Patent 2145895 RU, Int.Cl. A61N5/06, A61H39/08, A61K31/07, A61K31/16, A61K31/195, A61K31/355, A61K35/30, A61K35/14. Method of cerebral ischemic insult treatment / V. I. Shmatov, A. M. Litvinov. – № 96124033/14; Date of filing: 24.12.1996; Publ. 27.02.2000.

Patent 2149655 RU, Int.Cl. A61N 5/06, A61H 39/06. Method for treating the cases of epilepsy/ V. I. Gernets, S. V. Moskvin. – № 97116791/14; Date of filing: 16.10.97; Publ. 27.08.2000, Bul. № 15, Priority 16.10.97.

Patent 2156149 RU, Int.Cl. A61N5/067. Method for applying magnetic laser therapy to treat the cases of cerebrocranial trauma / V. I. Kuprijanov, I. A. Izotov, A. R. Evstigneev, V. N. Madykin. – № 99113726/14; Date of filing: 28.06.1999; Publ. 20.09.2000.

Patent 2169021 RU, Int.Cl. A61N5/06, 5/067, A61H39/00, 39/06. Method for treating central nervous system pathology / A. V. Gorjuchko. – № 2000110529/14; Date of filing: 27.04.2000; Publ. 20.06.2001.

Patent 2180207 RU, Int.Cl. A61H39/00, A61N5/06, A61N5/067. Method for treating the cases of insufficiency / N.A. Jakovlev. – № 2000125482/14; Date of filing: 11.10.2000; Publ. 10.03.2002.

Patent 2194549 RU, Int.Cl. A61N5/067. Method for treating focal cerebral injuries in the cases of craniocerebral traumas / R. D. Kasumov, Ju.V. Zotov, L.B. Dzhabarova et al. – № 99115058/14; Date of filing: 12.07.1999; Publ. 20.12.2002.

Patent 2200041 RU, Int.Cl. A61N5/067. Method for treating patients for sclerosus mul tiplexus by means of laser radiation / V.A. Ovsjannikov, I.M. Eliseeva, A.P. El'chaninov, M.V. Burmistrova. – № 2000101036/14; Date of filing: 18.01.2000; Publ. 10.03.2003.

Patent 2207172 RU, Int.Cl. A61N5/067. Laser therapy method for treating the cases of craniocerebral injuries in patients of elder and senile age / D. B. Sumnaja, A.I. Kozel', G.K. Popov, D.B. Giller. – № 2002115857/14; Date of filing: 13.06.2002; Publ. 27.06.2003.

Patent 2255775 RU, Int.Cl. A61N 5/067, G01N 33/50. Method for treating parkinson's disease / V.V. Poleshchuk, T.V. Vitreshchak, L.V. Komel'kova et al. – № 2003137107/14; Date of filing: 25.12.2003; Publ. 10.07.2005, Bul. № 19.

Patent 2286154 RU, Int.Cl. A61K 31/545, A61P 31/04, A61M 1/34, A61N 5/067. Method for introducing for treating inflammatory brain disease cases / V.A. Shatrov, S.V. Mozhaev, O.V. Ostrejko. – № 2005104368/14; Date of filing: 17.02.2005; Publ. 27.10.2006, Bul. № 30.

Patent 2315636 RU, Int.Cl. A61N 1/36, A61N 5/06. Method for treating patients for vibration disease aggravated with vertebrobasilar insufficiency syndrome / A.V. Zarubin, A.G. Shiman, G.I. Shvartsman и др. – № 2006114229/14; Date of filing: 25.04.2006; Publ. 27.01.2008, Bul. № 3.

Patent 2347594 RU, Int.Cl. A61N 5/06. Method for treatment of chronic cerebral ischemia / Ja.N. Mashkovskaja, N.E. Ivanova, V.V. Kir'janova, A. M. Malova. – № 2007137763/14; Date of filing: 11.10.2007; Publ. 27.02.2009, Bul. № 6.

Patent 2353411 RU, Int.Cl. A61N 5/06, A61N 5/067, A61N 2/04. Epilepsy management / V.V. Butukhanov, V.A. Sorokovikov. – № 2007115259/14; Date of filing: 23.04.2007. Publ. 27.04.2009, Bul. № 12.

Patent 2356527 RU, Int.Cl. A61H 39/00, A61N 5/067. Therapy of vertebrobasilar insufficiency / E.V. Denisova. – № 2007145247/14; Date of filing: 05.12.2007; Publ. 27.05.2009, Bul. № 15.

Patent 2392985 RU, Int.Cl. A61N 5/067. Method of treating encephalopathy / T.G. Tyshkevich. – № 2009118075/14; Date of filing: 12.05.2009; Publ.27.06.2010, Bul. № 18.

Patent 2449822 RU, Int.Cl. A61N 5/067. Method of integrated treatment of patients with discirculatory encephalopathy / L.T. Gil'mutdinova, D.R. Iseeva, E.M. Nazarova et al. – № 2011106724/14; Date of filing: 22.02.2011; Publ. 10.05.2012, Bul. № 13.

Patent 2464052 RU, Int.Cl. A61N 5/067, A61K 31/375, A61K 31/455, A61K 33/14, A61P 9/10, A61M 5/158. Method of treating vertebrobasilar insufficiency / A.G. Pakhomov, I.A. Pakhomov. – № 2011116627/14; Date of filing: 26.04.2011; Publ. 20.10.2012, Bul. № 29.

Patent 2487739 RU, Int.Cl. A61N 2/04, A61N 5/067, A61H 39/08, A61P 1/00. Method of treating locomotor disorders accompanying acute ischemic stroke / A.S. Reukov, V.V. Kir'janova, P.I. Guzalov. – № 2012118122/14; Date of filing: 03.05.2012; Publ. 20.07.2013, Bul. № 20.

Patent 2607942 RU. Int.Cl. A61N5/00, A61N5/067. Method for treating cognitive disorders accompanying chronic cerebral ischemia / S.L. Zaguskin, D.V. Bakuzova. – № 2015118860; Date of filing: 19.05.2015; Publ. 10.12.2016, Bul. № 34.

Chapter 41

Laser treatment of central nervous system injuries: an update and prospects

L. Longo
Institute for Laser Medicine, International Academy for Laser Medicine and Surgery, Florence, Italy

41.1 Introduction

Laser treatments of the central and peripheral nervous systems date from the late 1960s. One of the earliest articles by Fork (1971) in *Science* on argon laser treatment of nerves, came shortly after Mester et al. (1968) published their work on tissue stimulation with ruby and other wavelength lasers.

Then, Walker (1983) treated amputation neuromas of the lower limbs with ruby laser, and over the years, many other authors have dealt with the topic. There are many cited references in the *Proceedings of the International Congress of Laser Medicine* which is held in Florence and is now in its 30th edition. The *Proceedings of Laser Florence* 1999, 2000, 2001, 2002, 2003, and 2004, were published by SPIE (Longo et al., 1999); *Laser Florence* 2007 by Linux, Prague; *Laser Florence* 2008, 2009, 2010, 2011, and 2012 by the American Institute of Physics (Longo, 2008–2012); *Laser Florence* 2013 and 2015 by Medimond, Bologna (Longo, 2012-2013-2015); and *Laser Florence* 2017, once again by SPIE (Longo, 2017).

There are many other collections of bibliographies on the subject (Rochkind, 2008, 2009a,b; Longo, 2010; Hamblin et al., 2017), as well as approximately 100 other articles concerning laser treatment of the central nervous system (CNS) in international journals. My first book (Longo, 1986), and especially my latest *Laser Manual of Medical Technology*, describe laser's mechanisms of action in the diagnosis, treatment and surgery of all human tissues (Longo, 2015) (Fig. 41.1)

We began using this type of treatment at the end of 2003, selecting our patients on the basis of the inclusion/exclusion criteria listed in Table 41.1 (Longo, 2010). In the beginning the only patients who came to us had spinal cord damage with complete lesions, ranging from 6 months to many years, with complete lack of motor and sensory function classified AIS A, without any hope for improvement, and they brought medical reports written by eminent physicians who recommended against any type of treatment because they considered it useless.

However, things turned out differently, and we also began treating patients with incomplete spinal cord lesions, classified as AIS B, that is those who still had some sensation below the level of the lesion.

In all of our patients we gradually try to eliminate all their pharmacological medications on the basis of the results obtained—or which can be obtained—with laser treatment and associated physiotherapy. The patients always continued with personalized program of physiotherapy which is not the conservative type that is generally prescribed by the protocols of spinal units and centers of rehabilitation.

As to the exclusions, we do not even start treating patients who had undergone inadvisable surgeries such as muscle transposition, or had ruptured and/or strained tendons since these lesions cannot be remedied with regenerative laser treatment.

If there is no measurable and objective improvement after the first cycle of laser treatment we immediately interrupt the treatment. We also interrupt the treatment if the patient uses drugs, especially cocaine or excessive amounts of cannabis. We also stop treatment if the patient continues using nonprescribed drugs or if he or she does not comply with the recommended treatment intervals, or stops physiotherapy, or does not follow the prescribed program.

Photobiomodulation in the Brain. DOI: https://doi.org/10.1016/B978-0-12-815305-5.00041-5

R. Fork, 1971

J. Walker, 1981

S. Rockhind et al.

J.J. Anders et al.

Y. Asagai et al.

K. Atsumi et al.

J. Basford et al.

K. Byrnes et al.

M. Dyson et al.

T. Oshiro et al.

LASER FLORENCE Proceedings–1999-2000-2001-2002-2003-2004-2017 (SPIE Publisher)

2007 (LibruxPublisher)–2008-2009-2010-2011 (American Institute of Physics Publisher)

Bibliographic review J. Tuner-L. Hode, *The Laser Therapy Handbook*, Prima Pub, 2003

Bibliographic personal review, L. Longo, in *Laser Physics Letter,* Wiley Publ, 2010

L. Longo-*Laser Manual of Medical Technology,* OEO Publisher, Firenze, 2014

M. R. Hamblin et al.–*Hand book of Low Level Laser Therapy*, Pan Stanford Publisher, Singapore, 2017

Many Others 1972–2017

FIGURE 41.1 References on the effects of nonsurgical laser therapy on central nervous tissue.

TABLE 41.1 Selection criteria of patients with traumatic spinal cord injuries.

Inclusion	Exclusion
• Both sexes, 15–60 years old • Lesion occurred 1 year or longer ago • NMR or CT: total and subtotal lesion of CNS • AIS A and B • No therapy possible • Stop drugs, progressively • Continue obligatory physical therapy	• Inadvisable surgery • No improvement after first cycle of laser treatment • Use of drugs not prescribed • Interruption of physical therapy

Source: *Longo, L., 2010. Nonsurgical laser and light in the treatment of chronic diseases. Laser Phys. Lett. 7(11), 771–786.*

41.2 Clinical experience

From 2004 to 2016, we treated a total of 301 patients with traumatic lesions of the CNS (Table 41.2). Seventy-two dropped out for various reasons before completing the entire course of treatment, 10 had traumatic cerebral lesions. Therefore, our observations on posttrauma spinal lesions are based on 219 subjects.

Table 41.3 shows the different types of lasers we used. The numbers stand for the mean dose necessary for each patient. We always begin with this standard dosage and then gradually adjust it according to the results obtained.

We try to use at least three types of lasers (Longo, 2010; Longo et al., 1997; Yonezu and Kogure, 2013; Orchardson et al., 1997; McCaughey et al., 2010; Paula et al., 2014), of varying wavelengths: 800 nm diode, CO_2 10,600 nm and Nd:YAG 1064 nm, each in different doses to try to obtain different effects: antiinflammatory (Longo, 2010; Hamblin et al., 2017), antiedema, nerve regeneration (McCaughey et al., 2010; Paula et al., 2014; Wu et al., 2009; Yoshimi, 2007), and functional recovery.

TABLE 41.2 Patients treated from year 2004 until 2016.

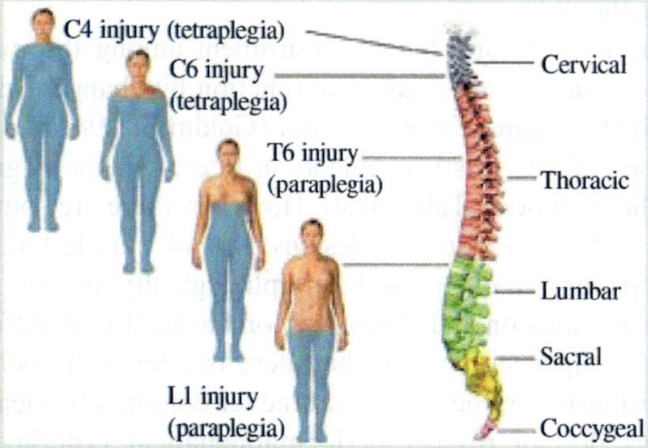

301 in total

72 Patients stopped the treatment

10 Patients had posttraumatic Injuries of the Brain—BI

219 had posttraumatic Injuries of the Spine Cord—SCI

Source: *www.longolaser.it.*

TABLE 41.3 Laser therapy of the SCI—parameters.

	Antiinflammation and edema reabsorption	Nervous regeneration	Muscle tone regulation	Antiinflammation and peripheral regulation of muscle tone
Laser	Diode 808, 915 nm wavelength	Diode 808, 915 nm wavelength	CO_2 10,600 nm wavelength	Nd:YAG, diode 1064 nm wavelength
Output power (W)	10	10	15	5
Spot size	5 cm	5 cm	10 cm	6 mm
Fluence	12 J/cm^2	4 J/cm^2	36 J/cm^2	35 J/cm^2/passage
Total energy	720 J	240 J	Variable	Variable
Wave form	1000 Hz	10 Hz	Continuous wave	1 Hz
Tissue target	Lesion	Nerve trigger points and coherence domains	Around the lesion	Lesion's area and muscle-tendon joints
Sessions a day	4	4	4	4
Sessions a cycle	At first 3 cycles, 20 irradiations each, with interval of 1 month Further cycles, 8–12 irradiations each, in average with interval of 1 month			

Source: *Longo, L., 2010. Nonsurgical laser and light in the treatment of chronic diseases. Laser Phys. Lett. 7(11), 771–786; Longo, L., Giubilo, F., Romanelli, C., Longo, D., 2017. PT laser and physical therapy applied to traumatic central nervous system injuries: update. In: Hamblin, M., et al. (Eds.), Handbook of Low Level Laser Therapy. Pan Stanford Publisher, Singapore.*

Muscle tone can be affected in both modes (Yoshimi, 2007; Asagai et al., 2005; Ushigome et al., 2008; Ohkuni et al., 2008), that is, we can trigger hypertonia, contractions and spasms as well as hypotonia and flaccidity simply by modifying the dose.

The laser is the only known instrument among the many existing physical treatments that can have an opposite effect on the same tissue and same function by changing the dose. Schawlow said that the laser is the only instrument that can light a cigarette and put it out (Goldman, 1982).

In these patients it is very important to establish a diagnosis and then follow them up over time to monitor the outcome of the treatment (Table 41.4). However, there are some factors that must always be kept in mind. First of all, there are no two identical cases: the lesions, loss of muscle tone and function, and response to the treatments all vary from patient to patient. In other words, morphologically and topographically similar lesions of the CNS always involve a different loss of function and always respond to treatment differently.

Another important point is that there is often a dissociation between the anatomical and functional aspects of the lesion. Furthermore, the lesions are never completely measurable, reproducible, or repeatable. Since the patients and the lesions do not respect basic mathematical criteria, and given all the variables, statistical analyses are not applicable.

And yet, all over the world, statistical analyses are applied to these lesions just the same. It is on the basis of those analyses that diagnostic and treatment protocols are established, and decisions made as to whether the patient is treatable or untreatable—that is, whether he has any hope of functional and anatomical recovery. Unfortunately the patient is frequently given the verdict: "Dear Sir (or Madam), you are in a wheelchair and will remain in it for the rest of your life" or, "No, don't worry, you will walk again." It would be wiser not to issue these verdicts, be they positive or negative because of all the reasons listed above. It is the protocol that must be adapted to the patient, and not the patient to the protocol: This technical error is made all over the world and almost always in Italy, with the result of often giving a wrong prognosis and incorrect treatment causing enormous psychological and physical harm (Table 41.4).

To assess treatment we have the physiatrist's clinical neurological and instrumental examinations (Table 41.5).

On the clinical level we follow more or less standardized international scales for assessing sensory and motor sensitivity below the lesion level such as the American Spinal Injuries Association (ASIA), the Asworth for muscle tone, the Franklin scale for motor function, and many other scales for each function.

However, the patient must always be the focus of the assessment. Therefore we must take into account what he and all the people around him have to say (Table 41.6).

Among the instrumental examinations, diagnostic imaging gives us an enormous amount of information, as does complete muscular assessment from the simple electromyogram to surface electromyography, to somato-motor and

TABLE 41.4 Diagnosis and therapy of CNS traumatic lesions.

- Similar CNS lesions always give different function loss and different answer to the therapy
- Anatomy and function often are dissociated
- Lesion is never totally measurable, reproducible, repeatable
- *Statistical criteria: Not valid, because there are many variable factors*
- *Protocols*
- Always relative, never absolute!!

TABLE 41.5 Treatment evaluation.

- *Specialistic clinic evaluation*: AIS score, Modified Ashworth Scale, Franklin Scale, others
- *Auto-evaluation* of the patient and his family
- NMR, CT
- EMG, sEMG
- ESSP, ESMP (somato-sensor and motor evoked potentials)
- HHD (hand-held dynamometers)

TABLE 41.6 **Specialistic clinical evaluation.**

AIS score, Ashworth Scale modified, Franklin, others
• *Nontotally objective* • Not totally measurable • Easy reproducibility • Repeatable

TABLE 41.7 NMR, CT, Rx.

• Objective • Lesion not totally measurable ("blind" solid angle of about 25 degree) • Reproducible • Repeatable • Often *dissociation* between anatomical and clinical signs

somatosensory evoked potentials as well as dynamometry, that is the measurement of muscle strength. However, each of these diagnostic tools has its limits. Specialized clinical examinations are never completely objective because any international assessment scale is subject to a certain degree of subjectivity both on the part of the patient and the examining physician.

Diagnostic imaging is clearly a more objective method, but it has the drawback of a blind spot of about 25–30 degrees in which we see absolutely nothing. Therefore the diagnosis of a complete or incomplete lesion is merely presumed and not absolute. Although imaging techniques offer good reproducibility and repeatability there is often poor correlation between anatomical and clinical signs (Table 41.7).

For example, one patient had a T11-L1 lesion with dislocation of the stumps, rupture of the thoracic aorta and, without a doubt, a complete lesion (already in 2004) with a series of statements from the Italian, Austrian, Swiss, and North American "Gotha" of neurosurgery, neurology, and physiatry who recommended against any type of treatment. We treated him in 2006, he stood up in 2007, has been walking with braces since the end of 2007, and from ASIA A he is now ASIA D. Each year the patient had a CAT at the lesion level and the lesion was always evident and unchanged except for the total absence of edema and signs of phlogosis, which had disappeared after about 60 applications. The lesion remained unchanged until 2011, when, at last a column of tissue appeared which suggested that the lesion had begun to experience healing. Therefore there was an evident dissociation between the clinical signs, which clearly began improving since 2007, and the anatomical imaging signs which only improved in 2011 (Figs. 41.2–41.3).

Muscle assessment is objective, repeatable, and reproducible, but it is subject to certain variables that may include errors in posture and errors in selecting the measuring points, the condition of the skin surface and other morphological and logistical factors (Table 41.8).

For example, Fig. 41.4 shows a 43-year-old male with a complete T4 lesion. After a cycle of laser treatment, the left side began showing a response to surface electromyography, and with 70 treatments his right side also began normalizing.

Tables 41.9–41.11 summarize the patients treated according to the topography of the lesions: the patients who interrupted treatment are shown in yellow. There were 7 patients out of 219 who remained unchanged. However, several patients interrupted the treatment, not for lack of results, but for the slowness of seeing results; they may have had expectations that were too high having been poorly advised, or they may have had "poor behavior" that is, not following the schedules we give them, and sometimes also because of the costs. We had a total of three relapses: one due to an excess of physiotherapy and one for having totally cut-out the physiotherapy. In two cases, albeit through opposite causes, an edema probably reformed on the site of the lesion and began compressing the perilesional tissue and interrupting any connection between the two damaged stumps. A third patient relapsed because a fall from the wheelchair caused another traumatic injury on the site of the original lesion.

Table 41.12 summarizes the results obtained.

May 2006

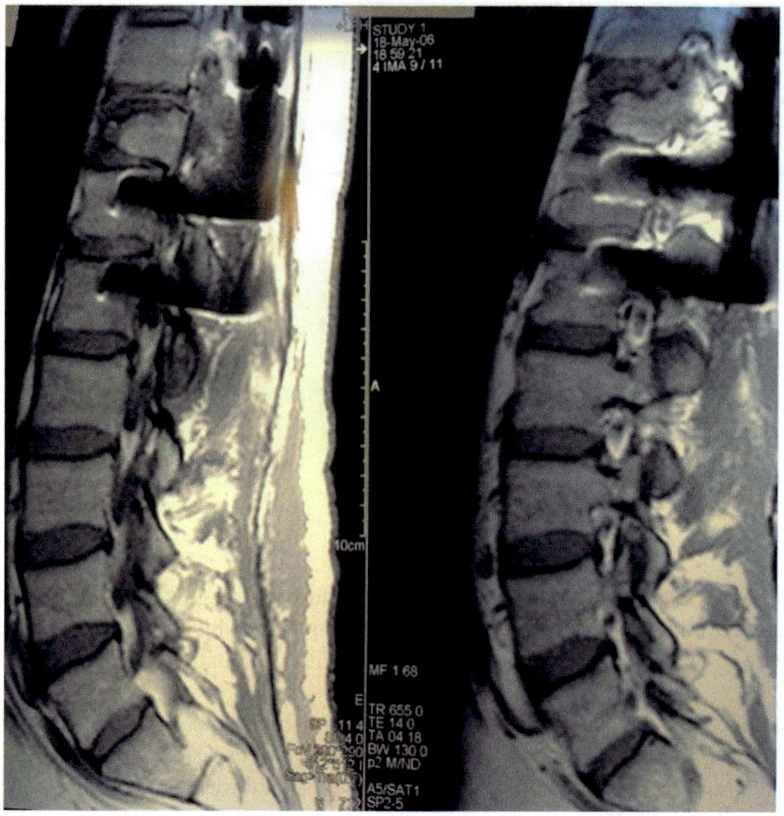

FIGURE 41.2 Patient—31-year-old male, complete lesion T11-L1, ASIA A, from the year 2004. Laser therapy started in June 2006, he started to walk with tutors and walking aids from the end of the year 2007, after 134 irradiations, ASIA D. *www.longolaser.it.*

Examining the patient before and after each laser treatment cycle, we consider the results positive if, after 20 applications, superficial and/or deep tactile, and/or pain, and/or heat and/or pressure sensitivity falls, at least two dermatomes below the lesion. Smooth muscle must show at least 1 degree of improvement per cycle, while voluntary muscle responses depend on the quality and amount of physiotherapy during and in the intervals between cycles. Anal sphincter control and everything related to bowel function and reflexology normalized in 90% of the patients, on the average after 120 applications. In the past 5 years, 52 out of 56 patients with complete lesions had normal rectal exploration and bowel function. We read everywhere that in patients with complete lesions, these functions cannot be restored once they have been lost. Perhaps it is time to correct the old books. Bladder/urethral sphincter control was almost completely restored in the women, but never in the men.

Nearly all the patients, both men and women, resumed sexual activity including erectile, ejaculatory, and sensory function, in fact some of the patients have produced children.

One hundred sixty-one patients recovered sufficiently to be able to stand after an average of 120 applications and 42 patients can walk with braces.

We carry out clinical assessments on the average of once every 6 months; instrumental tests every 2 months for neuromuscular function and X-ray assessments every year. In cases of traumatic cerebral lesions, electroencephalograms and psychological evaluations are done every 3 months.

The cause/effect relationship between applied laser treatment and results obtained cannot be scientifically controlled or proven since we cannot conduct proper double-blind or single-blind controls on homogeneous groups analyzing one variable at a time. However, we can make some scientific observations (Tables 41.13–41.14).

For example, due to personal reasons, 39 patients interrupted laser treatment for more than 2 years. When they returned their conditions were unchanged with the same level of sensitivity below the lesion as when they had they left. The condition was not the same as at the start of treatment, but as at the end, which means that they all maintained the positive results they had obtained, without however, gaining further improvement. Resuming laser treatment after an interval of years, the patients began to improve again with further positive results. If these improvements were not due to the laser treatment (rather than being spontaneous recovery), how come the patients did not continue to improve

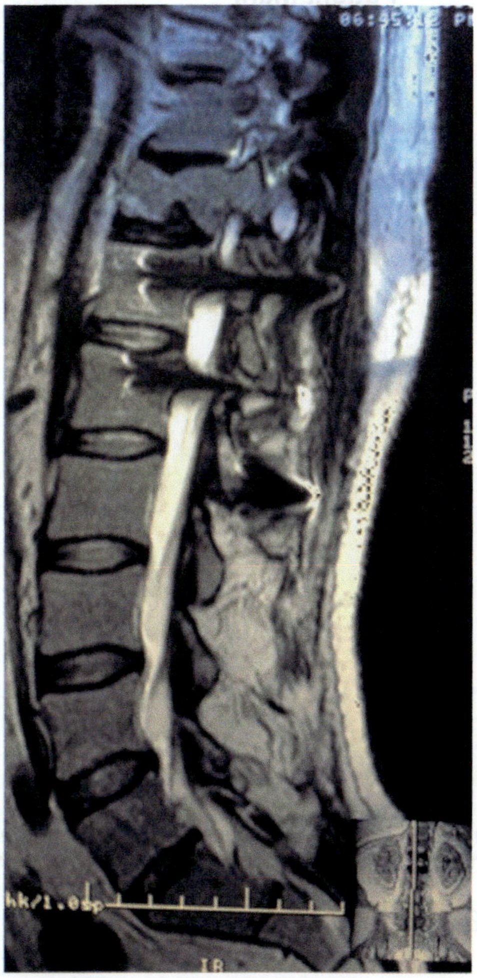

FIGURE 41.3 October 2011: column of new nervous tissue can partially repair the lesion. *www.longolaser.it.*

TABLE 41.8 sEMG, EMG, ESSP, ESMP, HHD.

- – Objective
- – Reproducible
- – Repeatable
- • Artefacts caused from different condition of the skin and of different sites of measuring
- • Only on/off activity
- • Discomfort for the patient (needles)

during the period that they had suspended the laser treatments? And, how come they began to improve again after they resumed the treatments?

There are practically no adverse reactions except for the potential danger of a first degree burn measuring a few centimeters (Table 41.15). Another potential effect is a photobiomodulation overdose which can cause opposite, albeit, transient effects: inflammation and anaesthetization, instead of reducing the local inflammation and facilitating functional recovery below the lesion. We have never seen this, but theoretically it could happen if the lasers are used by nonexpert physicians or physiotherapists (Longo, 2010). Complications: for the patient(s) none, for the physician, many.

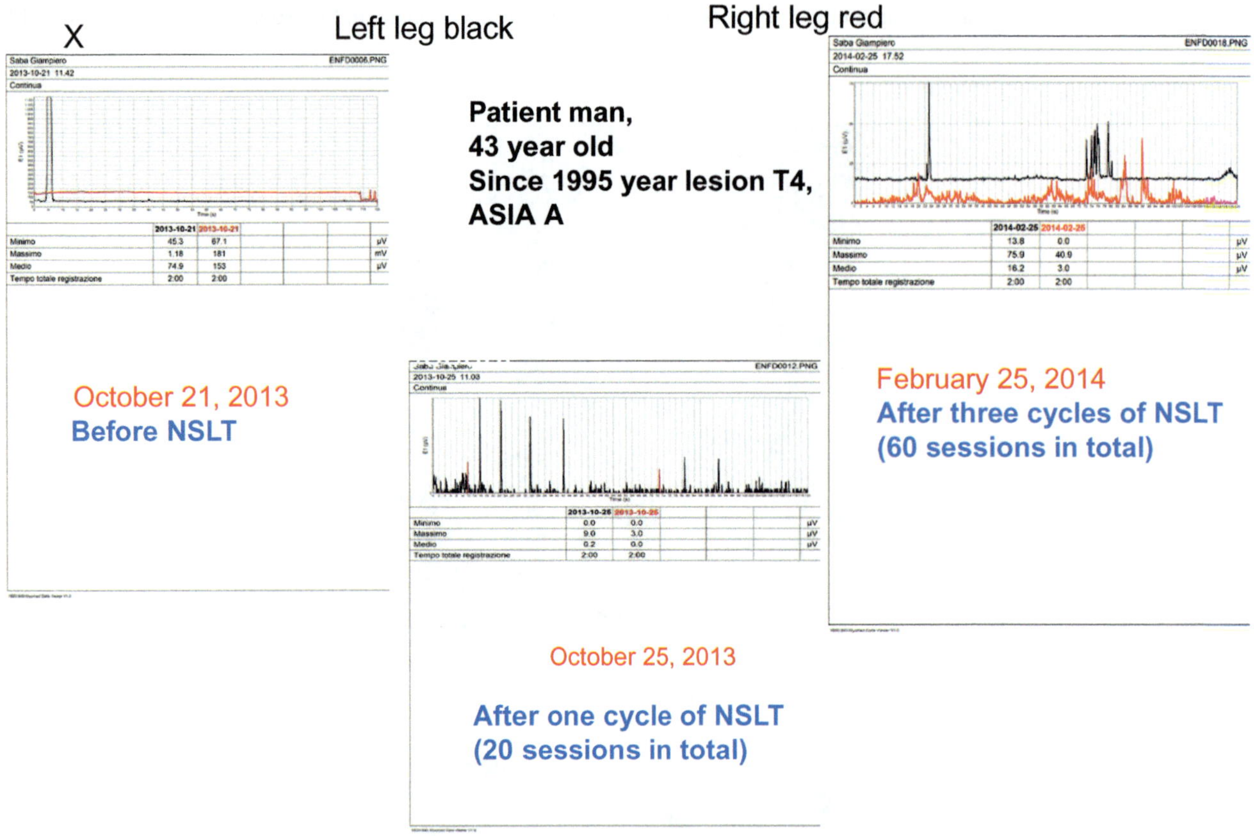

FIGURE 41.4 sEMG of quadriceps muscle.

TABLE 41.9 Patients treated from year 2004 until 2016 in blue: patients which interrupted the therapy.

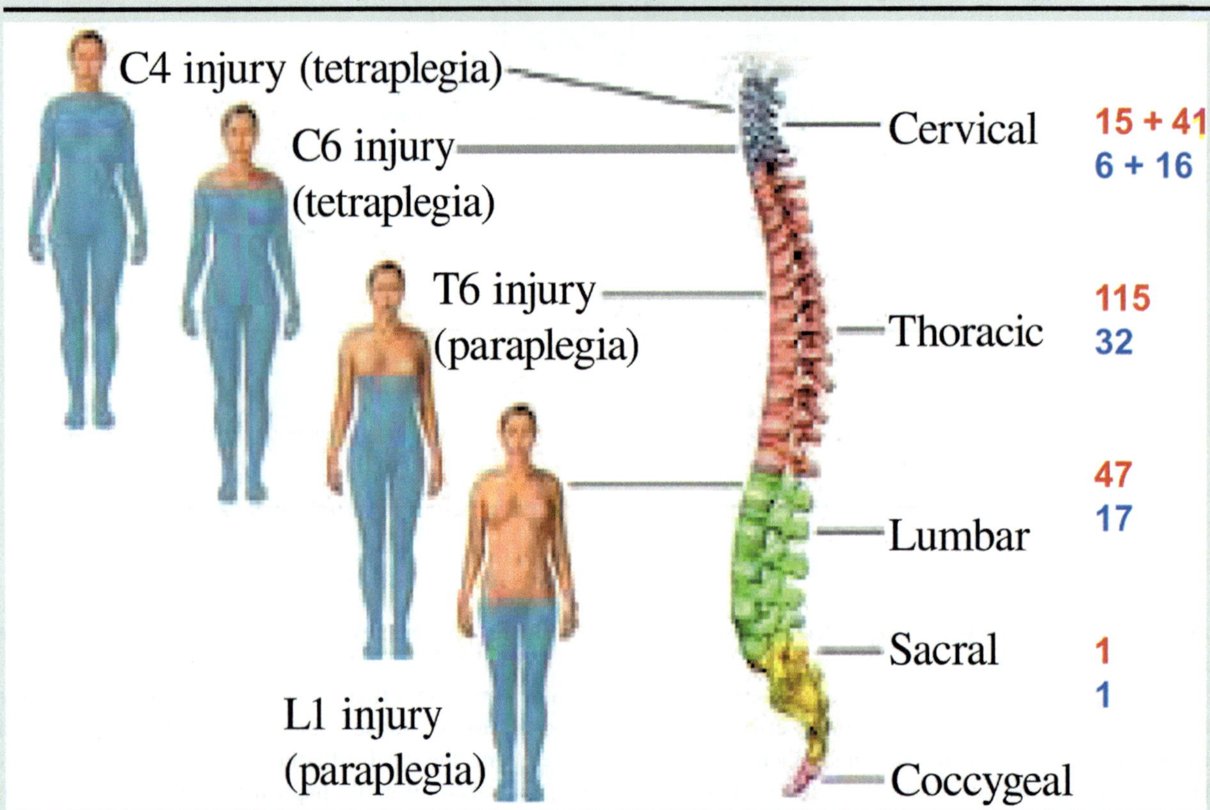

Source: *www.longolaser.it.*

TABLE 41.10 Patients without result after 20 sessions.

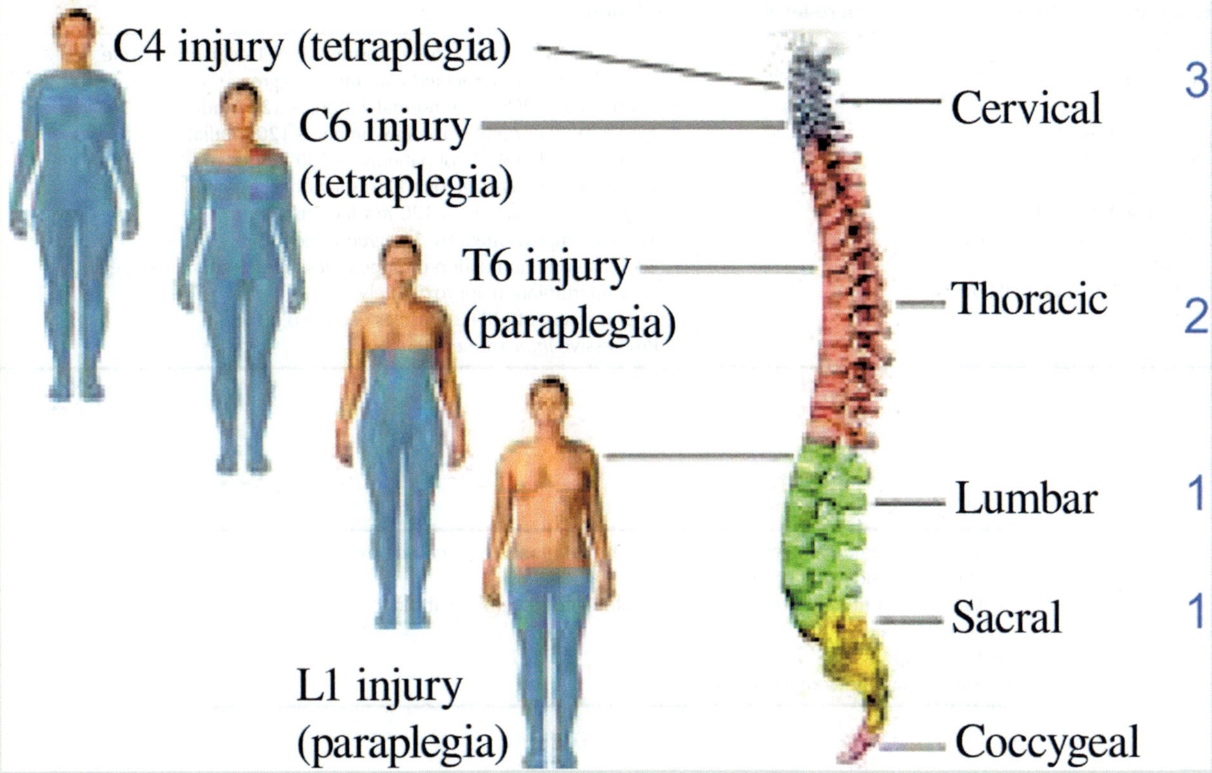

Source: *www.longolaser.it.*

TABLE 41.11 Interruption of laser treatment 72 patients on 301 treated in total.

No results after first cycle	7 patients
Less and slow results	28
Incorrect rhythm of the cycles	13
Expensive therapy	21
Temporary results	3

For example, after about 100 applications per patient normalizing muscle tone, we managed to have six muscle relaxant pumps removed, this means a considerable loss for the pharmaceutical industry because the pumps are included in the protocols of nearly every spinal unit and are proposed immediately after a lesion occurs, especially in young patients subject to acute spastic syndromes immediately following an injury. These pumps cost €5–6, inserting them the same, and there is the cost of the drug that must be administered for the rest of the patient's life; in addition he or she will likely be psychologically impaired as well. With laser treatment, as muscle tone and sensory function are recovered, the pumps can be removed, and, except for a few cases, it may even be better not to insert them at all. And then there is jealousy among colleagues, which unfortunately exists in every professional field, not just in medicine.

TABLE 41.12 Evaluation of the results until today on 301 CNSI patients.

Sensation Min returned to two metamere levels below the lesion	
• Involuntary motor	• Improvement in muscle tone, posture, MIN 1 degree per cycle
• Voluntary motor	• Variable, strictly connected with fitness degree
• Anal sphincter	• Improvement 90% until normalization, ~120 irradiations
• Urethral sphincter	• *No* for men, normalization in women, ~120 irradiations
• Sexual activity	• Quite normal in 99 % of patients, ~100 irradiations
• Stand up	• *171 p*, after an average of 100 irradiations
• Walking(ASIA C–D)	• *42 p*, after an average of 120 irradiations
• ASIA and other classifications	• Minimal improvement by *1 degree/6 months*
• NMR, CT	• Edema and inflammation disappear, lesion of medulla reduced after 60 irradiation approximately
• SSEP, SMEP, sEMG, HHD	• Improvement of MIN 2 muscles each two cycles
• EEG	• Progressive *improvement*

TABLE 41.13 Questions.

- The improvement of these patients could be spontaneous or caused by other therapies, such as physical therapy alone.
- Cause/effect ratio between laser and results is not controlled and not scientifically demonstrated.
- Follow-up?
- Side effects and complications?

TABLE 41.14 Answers.

- Thirty-nine patients stopped laser treatment for 2 years and more, for personal reasons.
- When they come back to continue NSLT, the situation of these patients was the same as the moment they stopped the treatment: follow-up was positive, as always.
- Once they restarted the therapy, the improvement restarted a second time.

TABLE 41.15 Side effects and complications.

- *Side effect*—Skin burn first degree, exceptional
- *Complications*: not for patient
- *Only* for physician: stop myorelaxant and other drugs, Jealousy of colleagues and more…!

41.3 Mechanisms of action

There are many hypothesis on the mechanisms of action (Table 41.16):

- There is an important antiinflammatory and antiedema effect (Longo, 2010).
- There is stimulation of neural stem cell regeneration, because lasers can make the neural stem cells multiply, regenerate, differentiate, and migrate (Anders et al., 2005, 2008).
- There is direct stimulation of neuron function, and in cultures there have been transformations of glial cells into functional neurons capable of transmitting nervous electric stimuli, a transformation made possible by 36 genetic mutations (Anders, 2009; Oron et al., 2007).

TABLE 41.16 **Hypothesis of mechanism of action.**

1. Active hyperemic, antiinflammatory, antiedema effects
2. Stimulation of nerve and progenitor cells regeneration[a]: *new tissue welding or bypass*
3. Stimulation of neuron function: transformation of the neurons from glial cells to functional cells, between 36 genic mutations
4. Influence on the bioplasma: energy around the body from 300 until 2000 nm (Inyushin and Chekurov, 1975)
5. Influence human energy field (Brennan et al., 1987)
6. All mechanisms listed below

[a]*Copyright 2006 by Anders, J., Longo, L., Waynant, R., Ilev, I., Romanczyk, T. USHUS Bethesda, FDA, USA; ILM Firenze, Italy.*

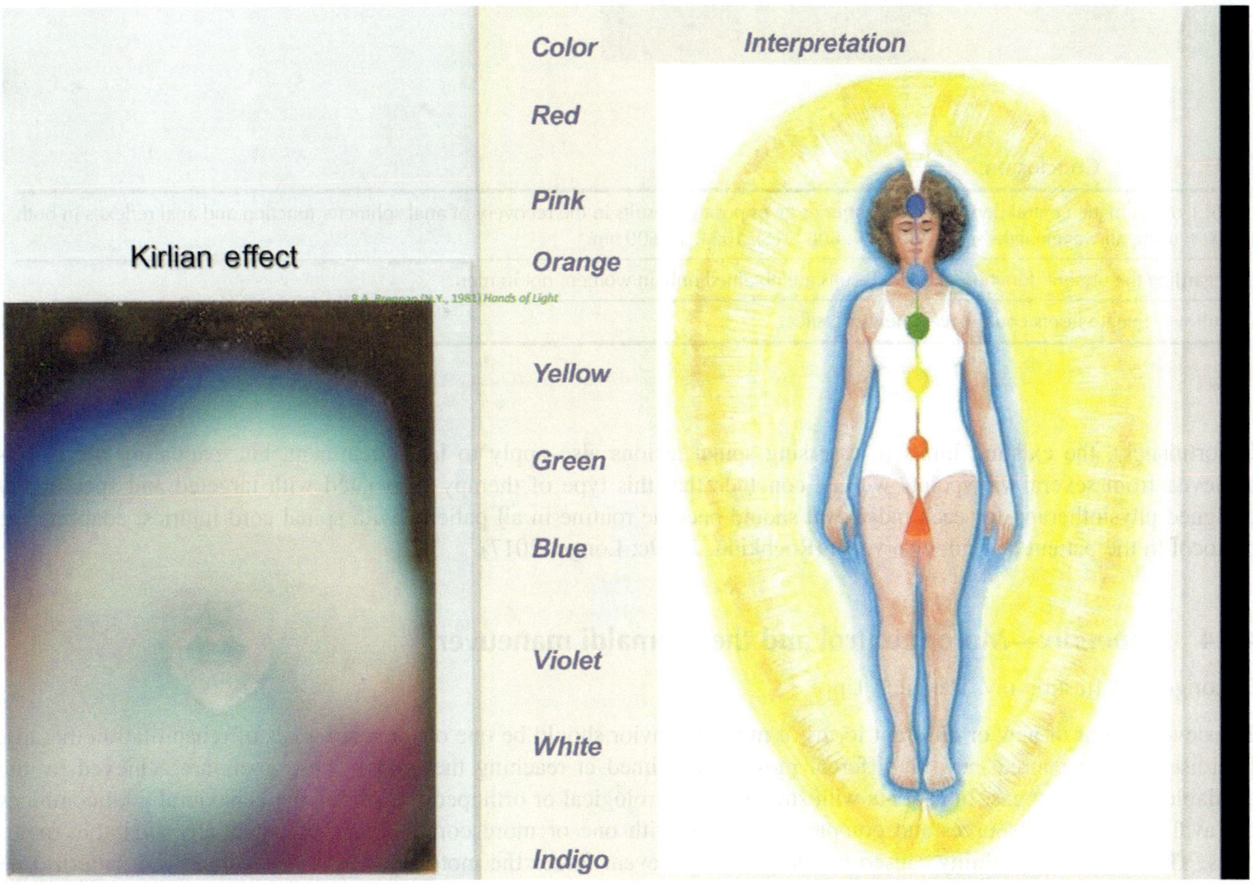

FIGURE 41.5 Body energy anatomy.

Then, there is the influence on the cell metabolism. We know that we produce energy that we exchange with the environment continuously, and these energies are harmoniously distributed throughout the body and around it, in a sort of cap that surrounds the body and is called bioplasma. Russian scientists have identified and measured that light ranges from 200 to 2000 nm (Inyushin and Chekurov, 1975). Lasers are light and probably fit into this mechanism as well.

All of these mechanisms can interact with each other. Fig. 41.5 shows the anatomical energy of the human body (Brennan, 1987), on the left is my energy anatomy in 1996, it will be very different now, it was measured in Orlando (Florida), and this is the energetic anatomy of a flower (Clarke) (Fig. 41.6). All living beings have an energetic anatomy. And the energy/matter interaction is indispensable for life.

In conclusion (Table 41.17), on the basis of our findings, we can say that by using different wavelengths at the same time, nonsurgical laser treatment of traumatic spinal lesions gives positive objective results. These results can be measured with different clinical and instrumental methods which are synergic but the values are not absolute.

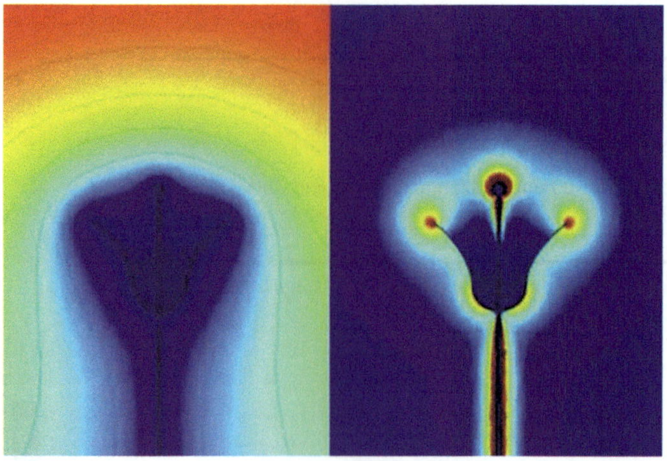

FIGURE 41.6 Electrical potential of flower J. Clarke et al., *Science* (2012). *www.longolaser.it.*

TABLE 41.17 Conclusion.

NSLT of traumatic central nervous system injuries gives positive results in the recovery of anal sphincter function and anal reflexes in both sexes, using all wavelengths simultaneously: 808 (915), 1064, 10,600 nm.
Regarding the urethral function, positive results are obtained only in women, not in men.
Multiwavelength lasers could have synergistic effect.

Unfortunately, the existing limits in assessing spinal lesions also apply to laser treatment, but evaluating the results achieved from several viewpoints we can conclude that this type of therapy associated with targeted and specifically designed physiotherapy for each individual should become routine in all patients with spinal cord injuries, adapting the protocol to the patient and not vice versa (Rochkind, 2009c; Longo, 2017).

41.4 Appendix—Motor control and the Grimaldi maneuver

D. Longo, G. Cherubini, V. Mangè, P. Lippi

The development of new or different forms of motor behavior should be one of the main goals of rehabilitation therapy. Exercise should generate new or different movements aimed at reaching the goals. Those goals are achieved by the available motor resources. In patients with strokes, or neurological or orthopedic diseases, the behavioral goal combines the available motor resources and completes the task with one or more combinations of potentially available motor units. The reduced availability caused by the pathologic event forces the motor system to use stereotyped patterns that have little flexibility and little redundancy, in other words, they are inadequate. The traumatic event can produce a drastic reduction in motor resources needed to achieve the goal: one example is the onset of spasticity in neurological diseases. Physiologically spasticity is defined as a motor disorder characterized by a velocity-dependent increase in the tonic stretch reflexes (muscle tone) with exaggerated tendon jerks, resulting from hyperexcitability of the stretch reflexes as one component of the upper motor neuron (UMN) syndrome (Lance, 1980).

The therapeutic exercise that uses the principle of shortening and solicitation of stress intention can lead to a reduction in spasticity, the recovery of strength and coordination, and lastly, when possible, all perfectly coherent with the equilibrium point hypothesis and threshold control. The neural control of motor actions involves changes in parameters of the system. A well established form of parametric control—threshold control—is briefly reviewed with a major focus on how it helps to solve several motor problems, in particular, the problem of the relationship between posture and movement and redundancy problems in the control of multiple muscles and joints (Feldman, 2006).

It is widely known that the emergence of new and/or different behaviors can take place via two processes: facilitation and learning. Facilitation does not necessarily involve a stable modification of the motor components, since the removal of the facilitation generally leads the motor system back to the previous level of competency. Therefore, the only road to take is that of directing the process back to learning, in the sense of developing new and/or different behaviors. For the ecological school, the learning process is identified by the acquisition of an order of superior or different systemic complexities. In this case, the complex dynamic system is identified as the motor perception apparatus of living beings.

Learning, from this standpoint is an emerging phenomenon, since it is takes place as a spontaneous phenomenon. However, it can only occur in the presence of specific physical conditions (critical instability) that become evident with the onset of a new and/or different motor behavior which could not be expressed before. Emerging phenomena can only occur spontaneously in the presence of complex dynamic systems with many degrees of freedom. It has been shown that central control levels are able to change a component (λ) of the threshold length value, at which the activity of muscle is initiated. By shifting the thresholds of appropriate muscles, the nervous system produces movement or, if movement is blocked, isometric torques (Feldman, 2006).

The neuromuscular system behaves similarly to a mass-spring system; it is a specific type of oscillatory system. According to Latash, the group, or assembly of many oscillatory systems controls the dynamics of the attractors. The principle of motor abundance suggests that CNS facilitates solutions that are equally capable of completing the task. It does so by imposing restrictions on the body-environment systems and allowing the solution to emerge at a given real state of the system (which is never repeated in subsequent trials) (Latash Mark, 2012). The restrictions represent a guide for the dynamic processes so that the actions are not caused by restrictions, but rather some actions are excluded by them (Scott Kelso et al., 1980)

The actions are characterized by self-organization, lacking a controller and are stable, flexible, redundant, and competitive. In some ways they are similar to motor learning processes. Spinal mechanisms adjust independently with respect to variable contextual conditions (Grimaldi et al., 1986).

According to Gibson and Turvey's description, motor behavior is guided by information. Information understood ecologically is a special kind of resource—a resource about the nature and whereabouts of other resources understood as opportunities for action (i.e., affordances) that further biological processes (Turvey and Carello, 2011).

The information is direct and structured with respect to the isomorphic action/perception. In terms of direct perception, tissue deformations acquire informational characteristics: the stimuli that reach the receptor surfaces are unique and specific to the action, and have properties which traverse the different states of matter, that is, physical, physiological, and psychological states. In transformation processes, the properties remain unchanged (isomorphism). Perhaps in this way we can better understand the self-descriptive character of information in complex dynamic systems since the real properties (stimulation patterns) do not have to be compared with an internal model or with the representation in order to be able to describe reality.

We suggest that when it is applied to a motor system, Grimaldi's muscle shortening maneuver (MSM) produces instability in the information field, and informational catastrophe that leads to instability of the neuronal flow needed for perceptive and motor control. The instability of the information field can be mediated by any receptive organ (tactile, visual, proprioreceptive) when faced with such a disturbance or disruption the nervous system is physically forced to make and/or generate different relations, abandoning stereotyped behaviors, and releasing degrees of freedom. It is possible that during the instability there is a loss of isomorphism, which is restored with the change in the order of systemic complexity (learning). This process produces an informational catastrophe in the neuromuscular spindles forcing them to set new muscle thresholds. This kind of maneuver is an active but involuntary training for the patient (Longo et al., 2017).

The MSM produces a shortening of the target muscle when it should create tension in stretching. The energy absorbed by stretching, which should transform into electrical activity and, therefore, into muscle strength, can or should find a new channel that should lead to the evocation of new and/or different motor components, configuring a learning process. The emerging motion, as clinically manifested, consists of setting new thresholds, the respective characteristics remain unchanged, and is probably allowed by the activating latent synapses, or by significant changes in the sensitivity to stretching. The consequence may be the appearance of previously absent voluntary movements, with increases in the strength, range, and the speed of executing the movement. Clinical observations could also detect increased strength on repetition of the motion, the development of new trajectories often accompanied by a reduction of any muscle hypertonia in neurologically impaired patients or the development of motion due to the action of the antagonist muscle and/or other synergy. Reduced pain is a frequent result in orthopedic diseases treated with MSM exercise therapy. Some of the method's protocols offer a useful alternative or complementary resource to existing treatments. Once a coordinative structure is set up, it is redundant, stable, precisely flexible, and it maintains a behavior characteristic of some nonlinear oscillatory systems. The new coordinative structure can protect the system against disturbances (clearly within limits) and therefore provide a certain amount of stability to motor performance. The authors support processes of self-organization since "groups of neurons that share similar thresholds and similar range pairing together give rise to the neuronal firing" and are able to produce a coordinated pattern to achieve the motor goal. When the process is evoked for the first time after the lesion-causing event, it can lead to the appearance of new motor components which will complement the existing motor functions. The new motor component is automatically and immediately available. The features of the MSM maneuver can be summarized as follows: (1) segmentary and overall assessment of the functional system, to detect the principal motor dysfunction; (2) identification and preparation of the exercise setting (the exercise does not take long, an average of 15 minutes can be sufficient); (3) segmentary and overall assessment a few minutes after completion of the exercise, and an assessment a few days later.

Some manual maneuvers can be done with the patient in different positions. The most important variables of the exercise are the therapeutic range or range of joint excursion—which may vary—and the speed of execution. The absence of structured muscle-tendon retractions is essential for the range of motion required by the exercise.

References

Anders, J., 2009. The potential of light therapy for central nervous system injury and disease. Photomed. Laser Surg. 27 (3), 379–380.

Anders, J., Longo, L., Waynant, R., Romanczick, T. Light as a Replacement for Mitogenic Factors on Progenitor Cells. Docket No: 9551-024 Pct U.S. Provisional Application Serial No. 60/666,582 filed March 31, 2005.

Anders, J.J., Romanczyk, T.B., Ilev, I.K., Moges, H., Longo, L., Wu, X., et al., 2008. Light support neurite outgrowth of human neural progenitor cells in vitro: the role of P2Y receptors. IEEE J. Sel. Top. Quant. Electron. 14 (1), 118–125.

Asagai, Y., Watanabe, Y., Oshiro, T., Yamamoto, K., 2005. Suppression of myotonia in cerebral palsy and adjunctive effect of low level laser therapy on intensive functional training. Laser Ther. 14 (4), 171–178.

Brennan, B., 1987. Hands of Light. Bantam Dell Publishing Group, New York.

Clarke J. Electrical potential of the flowers, Science, 2010.

Feldman, A.G., 2006. The nature of voluntary control of motor actions. In: Latash, M., Lestienne, F. (Eds.), Motor Control and Learning. Springer Science, New York.

Fork, R.L., 1971. Laser stimulation of nerve cells in aplasia. Science 171, 907–908.

Goldman, L., 1982. The Biomedical Laser. Springer Verlag Publ, New York.

Grimaldi, L.P., Fantozzi, M., Marri, P., Lippi, P., Catelani, G., 1986. Evocazioni di componenti motorie assenti nelle lesioni del sistema nervoso centrale II - Criteri di organizzazione degli atti motori. Giardini Editori e Stampatori in Pisa, Pisa.

Hamblin, M., Pires De Sousa, M.V., Agrawal, T. (Eds.), 2017. Handbook of Low Level Laser Therapy. Pan Stanford Publishing.

Inyushin, V.M., Chekurov, P.R., 1975. Biostimulation through Laser Radiation of Bioplasma. Kazakh State University, USSR Hill and Ghosak, Copenhagen University.

Lance, J.W., 1980. Symposium. In: Feldman, R.G., Young, R.R., Koella, W.P. (Eds.), Spasticity: Disordered Motor Control. Year Book Medical Pubs, Chicago, pp. 485–495.

Latash Mark, L., 2012. The bliss (not the problem) of motor abundance (not redundancy). Exp. Brain Res. 1, 1–5.

Longo, D., Longo, L., Lippi, P., Cherubini, G., Mangè, V., 2017. Effects of laser therapy and Grimaldi's muscle shortening manoeuvre on motor control of subjects with incomplete spinal cord injuries. Laser Ther. 26 (3), 203–209. Available from: https://doi.org/10.5978/islsm.17-OR-16.

Longo, L., 1986. Terapia Laser. USES Editore, Florence.

Longo, L., 2008–2012. Advances in laserology – Selected papers of Laser Florence. In: Proceedings of American Institute of Physics Publ., Melville, NY, vol. 1142, 2008, 2009; vol. 1226, 2010; vol. 1486, 2011, 2012.

Longo, L., 2010. Non surgical laser and light in the treatment of chronic diseases. Laser Phys. Lett. 7 (11), 771–786.

Longo, L., 2012–2013–2015. Advances in Laserology – Selected papers of Laser Florence. Medimond Publisher, Bologna.

Longo, L., 2015. Laser Manual of Medical Technology. OEO, Florence.

Longo, L., 2017. Non surgical laser and light in the treatment of central and peripheral nervous system injuries. Photomed. Laser Surg. 35 (4), 181–183 (Guest Editorial).

Longo, L., in press. Laser Florence 2017, Proceedings of SPIE, vol. 2018, ISSN 1605-7422.

Longo, L., Simunovic, Z., Postiglione, M., Postiglione, M., 1997. Laser therapy for fibromyositic rheumatism. J. Clin. Laser Med. Surg. 15 (5), 217–220.

Longo, L., Hofstetter, A., Pascu, M.L., Waidelich, W., 1999. Progress in Biomedical Optics and Imaging – Laser Florence, vol. 1, No 37, Proceedings of SPIE, vol. 4166, 1999; Laser Florence 2000, vol. 2, No 35, Proceedings of SPIE, vol. 4606, 2000; Laser Florence 2001, vol. 3, No 28, Proceedings of SPIE, vol. 4903, 2001; Laser Florence 2002, vol. 4, No 35, Proceedings of SPIE, vol. 5287, 2002; Laser Florence 2003, vol. 5, No 28, Proceedings of SPIE, vol. 5610, 2003; Laser Florence 2004, vol. 5 No 28 Proceedings of SPIE, vol. 5610, 2004, ISSN 1605-7422.

McCaughey, R.G., Chlebicki, C., Wong, B.J.F., 2010. Novel wavelengths for laser nerve stimulation. Lasers Surg. Med. 42, 69–75.

Mester, E., Ludany, G., Selyei, M., Szende, B., Tota, G.J., 1968. The stimulating effects of low power laser rays on biological systems. Laser Rev. 1, 3–9.

Ohkuni, I., Ushigome, N., Harada, T., Oshiro, T., Musya, Y., Mizutani, K., et al., 2008. Low level laser therapy for cerebral palsy. Laser Ther. 17 (1), 29–33.

Orchardson, R., Peacock, J.M., Whitters, C.J., 1997. Effect of pulsed Nd:Yag laser radiation on action potential conduction in isolated mammalian spinal nerve. Lasers Surg. Med. 21, 142–148.

Oron, U., Ilic, S., De Taboada, L., Streeter, J., 2007. Ga-As (808 nm) laser irradiation enhances ATP production in human neuronal cells in culture. Photomed. Laser Surg. 25 (3), 180–182.

Paula, A.A., Nicolau, R.A., de O Lima, M., Selgado, M.A.C., Cogo, J.C., 2014. Low-intensity laser therapy effect on the recovery of traumatic spinal cord injury. Lasers Med. Sci. 29, 1849–1859.

Rochkind, S., 2008. Phototherapy and nerve tissue repair Lecture notes in Electrical Engineering 12. In: Waynant, R., Tata, D.B. (Eds.), Proceedings of Light-Activated Tissue Regeneration and Therapy Conference. Springer, pp. 408–412.

Rochkind, S., 2009a. Review of 30-years experience: laser phototherapy in neuroscience and neurosurgery: Part 1. Muscle and nerve. Laser Ther. 18 (1), 27–38.

Rochkind, S., 2009b. Review of 30-years experience: laser phototherapy in neuroscience and neurosurgery: Part 2. Nerve cells, brain and spinal cord. Laser Ther. 18 (3), 127–136.

Rochkind, S., 2009c. Phototherapy in peripheral nerve injury for muscle preservation and nerve regeneration. Photomed. Laser Surg. 27 (2), 219–220.

Scott Kelso, J.A., Holt, K.G., Kugler, P.N., Turvey, M.T., 1980. 2 On the concept of coordinative structures as dissipative structures: II. Empirical lines of convergence. Adv. Psychol. 1, 49–70.

Turvey, M.T., Carello, C., 2011. Obtaining information by dynamic (effortful) touching. Philos. Trans. Soc. Lond. B: Biol. Sci. 366 (1581), 3123–3132.

Ushigome, N., Harada, T., Okuni, I., Ohshiro, T., Musya, Y., Mizutani, K., et al., 2008. Effects of low level laser therapy (LLLT) on spasticity caused by cerebral vascular accidents (CVAS). Laser Ther. 17 (2), 95–99.

Walker, J., 1983. Laser therapy of amputation neuroma. Neurosci. Lett. 43, 339.

Wu, X., Dmitriev, A.E., Cardoso, M.J., Viers-Costello, A.G., Borke, R.C., Streeter, J., et al., 2009. 810 nm wavelength light: an effective therapy for transected or contused rat spinal cord. Lasers Surg. Med. 41, 36–41.

Yonezu, T., Kogure, S., 2013. The effect of low-level laser irradiation on muscle tension and hardness compared among three wavelengths. Laser Ther. 22 (3), 201–207.

Yoshimi, A., 2007. Application of LLLT in patients with cerebral palsy of the adult tension athetosis type. Nippon Laser Igakkaishi 28, 74–76.

Chapter 42

Photobiomodulation treatment for brain disorders: posttraumatic stress disorder (PTSD) and dementia

Randy Lamartiniere[1], Rhett Bergeron[2], Ronald Aung-Din[3], Matthew Bennett[4], William Stephan[5] and Louis Banas[6]

[1]Photo medicine Clinic, Baton Rouge, LA, United States, [2]Real Health Medical, Roswell, GA, United States, [3]Sarasota, FL, United States, [4]Patterson, CA, United States, [5]Buffalo, New York, United States, [6]Laser Innovations, Amherst, New York, United States

42.1 Introduction (clinical team)

Dr. Randy Lamartiniere is an internal medicine physician with 30 years of experience. He has long been interested in different types of therapy although skeptical of unproven or unscientific treatments. He met Mr. Lou Banas after being introduced by one of his patients. Lou Banas employed cold lasers. After meeting with Mr. Banas he became interested in cold laser therapy and was impressed by the science behind the therapy which included mainly accelerating the natural healing processes of tissue in general. They began treating patients with various conditions with quite impressive results. Since that time many patients suffering from several bothersome and painful conditions have had dramatic and rapid resolution of these problems. Two of the more interesting cases have been an adult male with long-term posttraumatic stress disorder (PTSD) from a childhood emotional trauma who had impressive results after only three treatments,and a woman who suffered a recent concussion as well as PTSD from a motor vehicle accident with similar results despite a poor prognosis related by her neurologist. The results can be seen as almost too good to believe, which is one of the hurdles in marketing this therapy.

Rhett Bergeron is an alternative medicine physician trained as an internist with a very busy practice in Atlanta specializing in hormone replacement, cancer therapy, weight loss, and now thanks to Mr. Banas, pain management and neuro-rehabilitation. Since meeting Mr. Banas he has found the results of photobiomodulaltion therapy surprising and significant. Two cases come to mind that the reader will find interesting. The 42 year old mail suffering from severe PTSD and other brain damage due to meningitis. He served as a detective at a local community station. He had a very supportive wife because he was nonfunctional and could not drive a car without getting lost, and he scored very high on a PTSD interview. After five treatments he was totally functional, his PTSD score came down, and can drive his car with no problem. The second case involved a long-term patient who suffered from Lyme disease. This is a very intelligent, young man who again scored very high on a PTSD score. He felt strongly that anyone suffering from a major disease like Lyme would have a PTSD element. Again after just a few treatments he was essentially much better and his PTSD score showed he significantly improved.

Lou Banas, is a certified laser therapist. This describes a new breed of therapists that are bringing medicine of the future here and now. Laser systems are a revolutionary approach for sports injuries that have shown significant success at seven universities and with two NFL teams. Several years ago, Dr. Richard M. Restak published *The Brain, The Last Frontier* emphasizing the fact it is the one organ we still want to learn more about and how to treat it (Restak, 1980). It is well-documented that many scientific discoveries are the result of a serendipitous event (compare Isaac Newton). In 2009, we were the catalyst for such a discovery. The opportunity to use this new technology, originally called low level laser therapy now referred to as photobiomodulation first arose in Dr. Stephan's private medical practice in Buffalo, New York. Dr. Stephan, a well-respected, primary care physician, placed heavy emphasis on preventive and alternative medicine. He had been using MB for over 11 years and as a primary care physician sees all types of injuries. On several

Photobiomodulation in the Brain. DOI: https://doi.org/10.1016/B978-0-12-815305-5.00042-7

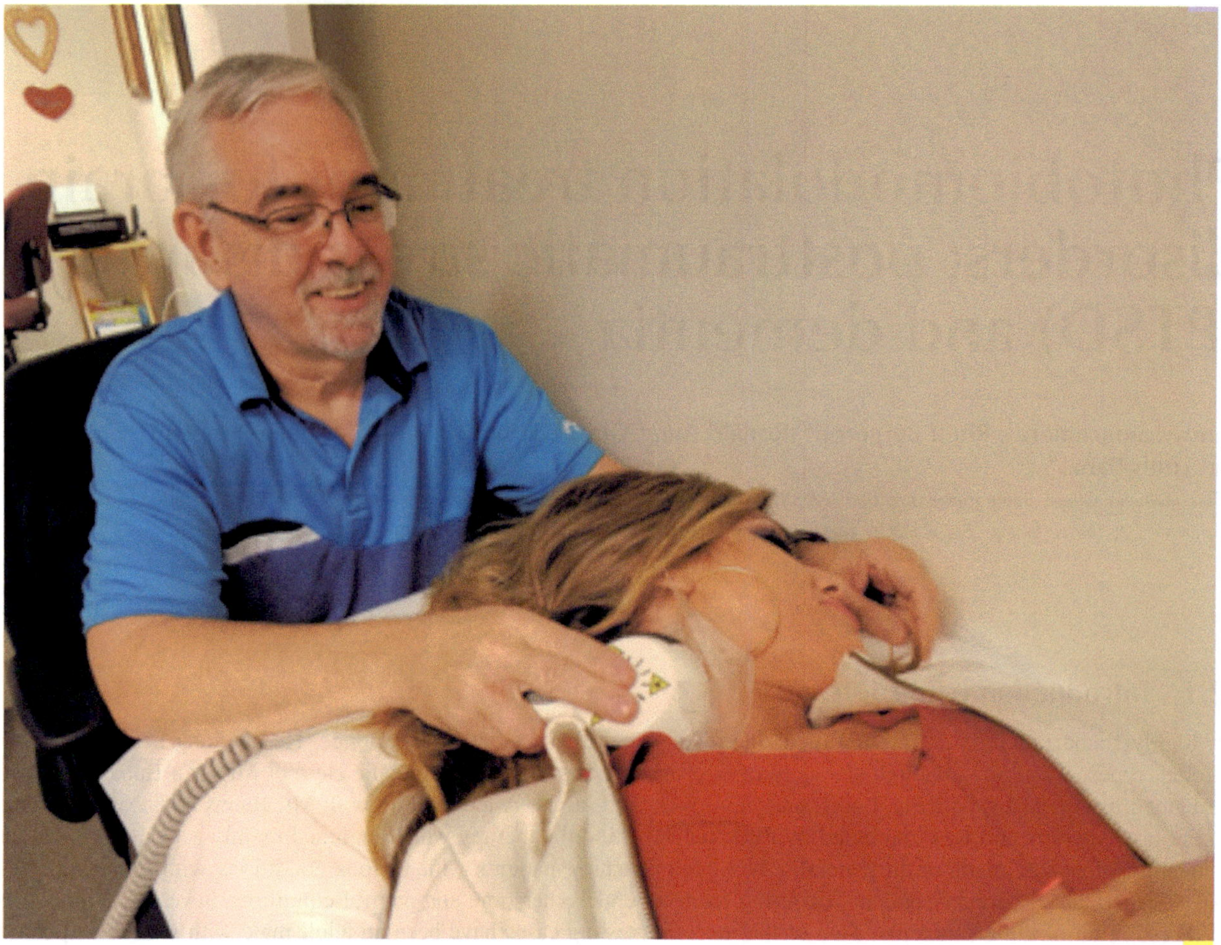

FIGURE 42.1 Randy Lamartiniere M.D. demonstrating cold laser treatment.

occasions, I was able to assist him with the treatment of patients. Although we had a great success treating many different types of injuries, the most remarkable successes were for migraine, concussion, and dementia. Patients who had suffered a concussion or migraine had significant and sometimes total recovery with these treatments. We knew this success was remarkable but we assumed we were only treating superficially, eventually realizing the skull is translucent! Fig. 42.1 shows Dr. Lamartinere applying photobiomodulation to a patient.

42.2 Original concussion case

Although this chapter is dedicated to dementia and PTSD, this concussion case is worth noting. Six years ago, we treated a young man who was hit with a lead pipe two to three times and had intractable migraines for 2 years; he was healed with just four treatments and the case was published in November 2012 (Stephan et al., 2012). As a result of this discovery, we started addressing the issue of dementia and PTSD which has reached epidemic proportions in this country. The following patients were treated utilizing the Theralase TLC 900 series. The cluster head with a treatment area of 20 cm^2 includes 5×905 nm near infrared superpulsed laser diodes each with 100 mW average power and a peak power of 50 W, at a pulse duration of 200 nanoseconds, and 4 CW-red 660 nm laser diodes each with a power of 25 mW (Kester et al., 2014). All patients received three treatments over a 1-week period. Five sites were treated including four treatments on the prefrontal cortex and one at the circle of Willis each for a duration of 2 minutes. See the Theralase website for full parameters of the laser device (Theralase, 2018). Figs. 42.2 and 42.3 illustrate the Theralase laser system.

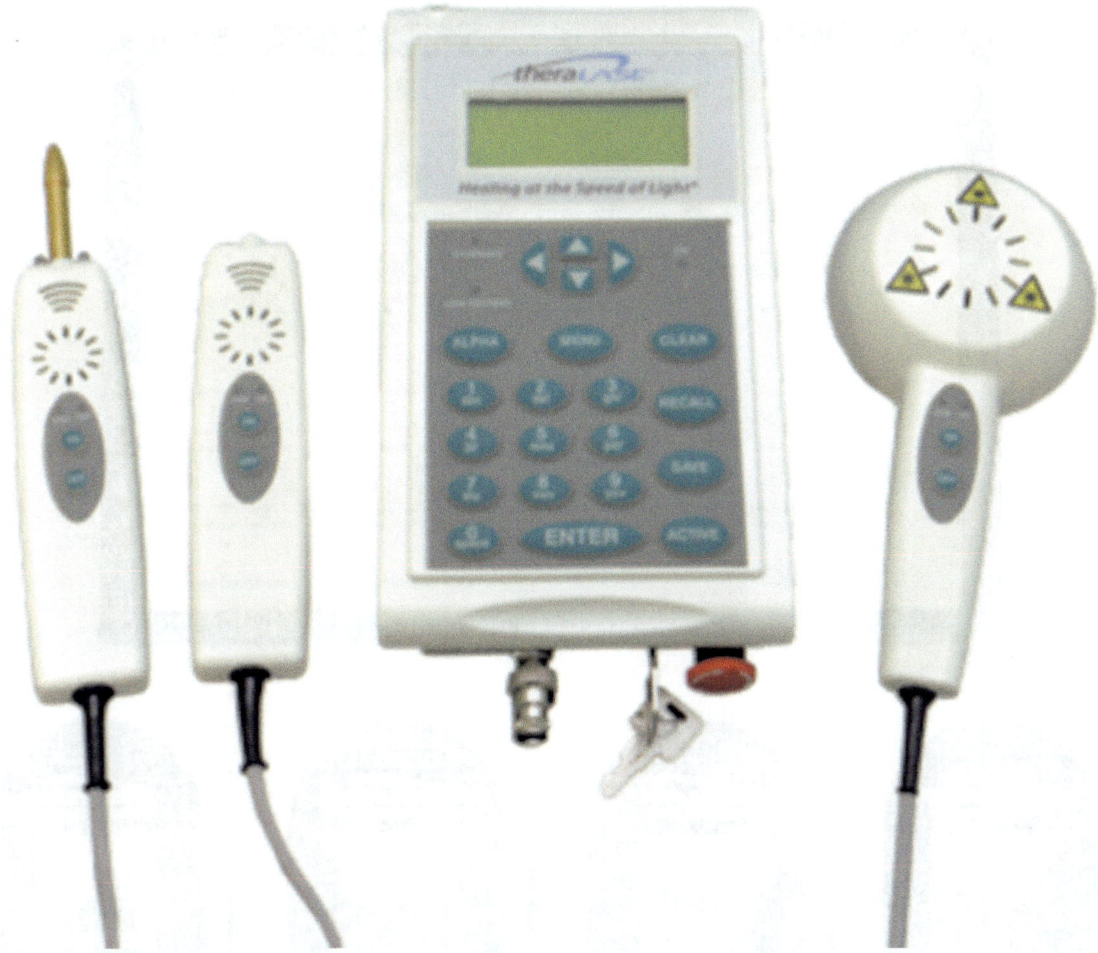

FIGURE 42.2 Theralase system showing different applicator heads.

42.3 Posttraumatic stress disorder evaluation

The cooccurring disorders program screening assessment questionnaire is universally accepted as a valid assessment of an individual's degree of emotional distress due to a traumatic event(s). A score of 35 indicates a mild case and a score of 55 or above indicates a severe disorder. The individuals all presented here had scores over 55 and unless indicated; scores are presented before the first and after the fourth session. Over 50 individuals were treated over a 1-year period at five different locations which included: Phoenix, AZ; Buffalo, NY; Lafayette, IN; Bossier City, LA; and Sarasota, FL. Virtually all patients exhibited and proclaimed remarkable improvement in their emotional stability and quality of life which is indicated in their PTSD scores.

Below is a copy of the interview questionnaire:

Instructions: This questionnaire asks about problems you may have had after a very stressful experience involving actual or threatened death, serious injury, or sexual violence. It could be something that happened to you directly, something you witnessed, or something you learned happened to a close family member or close friend. Some examples are a serious accident; fire; disaster such as a hurricane, tornado, or earthquake; physical or sexual attack or abuse; war; homicide; or suicide.

First, please answer a few questions about your worst event, which for this questionnaire means the event that currently bothers you the most. This could be one of the examples above or some other very stressful experience. Also, it could be a single event (for example, a car crash) or multiple similar events (for example, multiple stressful events in a war-zone or repeated sexual abuse). How long ago did it happen? ______________________________ (please estimate if you are not sure)

FIGURE 42.3 Theralase system showing different treatment protocols.

Did it involve actual or threatened death, serious injury, or sexual violence?

Yes No

How did you experience it?

It happened to me directly.

I witnessed it.

I learned about it happening to a close family member or close friend.

I was repeatedly exposed to details about it as part of my job (for example, paramedic, police, military, or other first responder).

Other, please describe.

If the event involved the death of a close family member or close friend, was it due to accident or violence, or was it due to natural causes?

Below is a list of problems that people sometimes have in response to a very stressful experience. Keeping your worst event in mind, please read each problem carefully and then circle one of the numbers to the right to indicate how much you have been bothered by that problem in the past month. Page 1 of 2 **PCL-5 with Criterion A** (14 August 2013) National Center for PTSD	**Not at all**	**A little bit**	**Moderate problem**	**Quite a bit**	**Extreme problem**
In the past month, how much were you bothered by:					
1. Repeated, disturbing, and unwanted memories of the stressful experience?	0	1	2	3	4
2. Repeated, disturbing dreams of the stressful experience?	0	1	2	3	4
3. Suddenly feeling or acting as if the stressful experience were actually happening again (as if you were actually back there reliving it)?	0	1	2	3	4
4. Feeling very upset when something reminded you of the stressful experience?	0	1	2	3	4
5. Having strong physical reactions when something reminded you of the stressful experience (for example, heart pounding, trouble breathing, sweating)?	0	1	2	3	4
6. Avoiding memories, thoughts, or feelings related to the stressful experience?	0	1	2	3	4
7. Avoiding external reminders of the stressful experience (for example, people, places, conversations, activities, objects, or situations)?	0	1	2	3	4
8. Trouble remembering important parts of the stressful experience?	0	1	2	3	4
9. Having strong negative beliefs about yourself, other people, or the world (for example, having thoughts such as: I am bad, there is something seriously wrong with me, no one can be trusted, the world is completely dangerous)?	0	1	2	3	4
10. Blaming yourself or someone else for the stressful experience or what happened after it?	0	1	2	3	4
11. Having strong negative feelings such as fear, horror, anger, guilt, or shame?	0	1	2	3	4
12. Loss of interest in activities that you used to enjoy?	0	1	2	3	4
13. Feeling distant or cut off from other people?	0	1	2	3	4
14. Trouble experiencing positive feelings (for example, being unable to feel happiness or have loving feelings for people close to you)?	0	1	2	3	4
15. Irritable behavior, angry outbursts, or acting aggressively?	0	1	2	3	4
16. Taking too many risks or doing things that could cause you harm?	0	1	2	3	4
17. Being "superalert" or watchful or on guard?	0	1	2	3	4
18. Feeling jumpy or easily startled?	0	1	2	3	4
19. Having difficulty concentrating?	0	1	2	3	4
20. Trouble falling or staying asleep?	0	1	2	3	4

42.4 Case studies for posttraumatic stress disorder

The following case studies for PTSD were treated by the above-mentioned providers in the cities indicated.

Case 1: This patient initially presented with depression, but I realized later it was PTSD (questionnaire not utilized). A 34 year old business owner in Buffalo, NY claimed he had fired at least 32 people in his small company in the last 2 1/2 years due to his inability to cope with stressful situations. He claimed to be suffering from depression. I decided to try photobiomodulation therapy to increase blood flow as this might improve brain function. After the third session he told me how much better he was, and that the real problem that had been haunting him was that he had witnessed the murder of his mother when he was only 9 years old and the murderer was never convicted; imagine the psychological damage! He returned for his fourth 12 minute treatment. His life was totally turned around and he was forever grateful.

Case 2: A 69-year-old Vietnam Vet from Rochester, NY after 50 years experienced emotional problems, and had just coincidently contacted the VA the week before becoming aware of the treatments. He holds an executive management position but knew he had psychological issues. After completing a series of four treatments over an 8-day period he felt significant relief. He did not follow up with the VA. PTSD scores: pretreatment 47; posttreatment 27.

Case 3: A 54-year-old woman from Rochester, NY was on SSI due to a diagnosis of PTSD as the result of an abusive spouse. After six treatments, her score indicated remarkable improvement in her emotional well-being. She was given three more maintenance treatments over a 4-month period and her testimonial video can be viewed at www.paintherapyusa.com. PTSD scores: pretreatment 62; posttreatment 25.

Case 4: A 75-year-old woman from Sarasota, FL was forced into prostitution at the age of 19 in New York City. After the second treatment, she had remembered she had an abortion at the age of 21. This event was "blanked out" in her memory and she cried in the office on subsequent occasions stating she had "killed" a baby, something she felt was deeply against her convictions. After the fifth treatment among other improvements she now could enjoy dining, driving her car by herself, and her handwriting was now legible. Questionnaire not utilized.

Case 5: A 45-year-old man in Phoenix, AZ endured multiple beatings as a child and subsequently was accidently electrocuted at work and thrown 30 ft. to the ground. The doctors felt the fall restarted his heart. Because of the accident, he underwent stress management, temper management, and had severe muscular skeletal pain for several months. The PTSD questionnaire was administered and he scored very high. After five treatments over a 10-day period he no longer suffered from PTSD, was virtually pain free, was relaxed, no longer short tempered, and was no longer prone to sudden outbursts of anger. PTSD scores: pretreatment 67; posttreatment 32. Additional data:

Event	Gender/ age	Scores: 35 = minimal PTSD; 45 = moderate; 55 above severe	
Mother murdered	m/34	None	First client, mother confronted babysitter after son reported abuse, babysitter murdered her and not convicted, reported major improvement in life
Sexual abuse 4/2015	f/75	57-37-27-20	
Spousal abuse	f/45	58-38-33 = 34	Caucasian women divorced for 10 years on SSI for PTSD, claims total healing and goes back to work
Military incident	m/82	62-47-25	Korea vet worked as MP, witnessed terrible crime also lost wife 6 months previous
Iraqi vet	m/34	69-45-65-39	Night sweats, nightmare
Physical abuse	f/42	60-56-56-43-37	
Spousal abuse	f/23	44-38-38-35	
Rape 14 years old	f/22	44-38-38-35	Living in half way house became heroin addict
Abandoned as child	f/24	55-42-29	Living at half way house for drug abuse, lived with various relative's or 12 years
Attacked at holding center	f/2	64-49-35	Drug issues very attractive Caucasian attacked at Erie Co holding, lawsuit pending
Physical abuse by boyfriend	f/42	53-32	Pictures provided of physical abuse
Spousal abuse	f/44	58-44-25	
Anxiety dysfunctional family	m/45	50-31	
Spousal abuse	f/35	48-29	
Vietnam vet	m/68	52-57-47-37	Working manager of large corp, after 50 years decided to seek VA help, canceled appointment after 4th treatment
Iraqi vet sniper	m/42	66-54-49-40	Detective in local police force has "parade of faces" then acquires viral infection of brain, major brain "fog" cannot drive, very confused, Total healing 5 TX
Dysfunctional family	m/45	65-49-46-40	
Daughter suicide	f/54	45-34	Three treatments all that were needed
Lyme disease, emotional	m/47	52-43	Very bright young man informed me that anyone suffering a major health problem will most likely have PTSD
Witnessed horrific accident	m/42	44-37	Extreme anxiety driving car or any threatening situation. While treating him he saw my hand move near him and told me to stop. Called me on the way home to tell me of major improvement
Spousal abuse	f/52	55-30-17	
Abused by priests	m/52	56-37-31	Successful attorney never told anyone but me, healed after 3 TX
Sudden death of father	f/33	57-38-31	Sudden death of beloved father who just best leukemia 3 months earlier, suddenly dies
Sudden death of father	f/31	55-39	Sudden death of beloved father who just best leukemia 3 months earlier, suddenly dies
Physical abuse by husband 35 years ago	f/72	47-30-27	Now happily married 15 years old issue healed
Physical abuse by boyfriend (1 year ago)	f.41	52-37-27	

Physical abuse by spouse 29 years ago	f/71	55-39-29	Divorced for 25 years old issue resolved
Afghan vet	m/34	47-34	Professional retired officer M.D. wife insisted he see me. "I did not realize I had such problems" healed 2 TX
Daughter has stroke	f/72	55-34-24	Family history of strokes, daughter needs 24 hour care
Car accident 10 years ago	f/34	54-34-35	Only issues unresolved is riding in car as passenger
Published psychologist	f/72	47-32	On website looked for solution for 25 years
Afghan vet saw much combat	m/29	54-39-31	Retired veteran healed as denoted, however after working on firing range after smelling smoke came back for two more treatments and was again healed
Iraq veteran saw severe combat	m/32	45-40	Had only two treatments
Medical assistant college incident	f/32	47-31-27	Wanted third treatment

In conclusion, due to the remarkable success treating PTSD in both the civilian and military population, I have set up a foundation called "The Orthogenesis Project."

42.4.1 Case studies for dementia

There are more than 5 million Americans currently living with Alzheimer's disease (AD) and according to the American Alzheimer's Association every 67 seconds someone in the United States develops the disease. Estimates of costs to American society are $214 billion including Medicare and Medicaid. These costs are staggering, yet if we do not find a cure the cost in today's costs would be $1.2 trillion by 2050. Extensive research is being conducted worldwide and recent studies show that AD could be more closely related to small vessel disease (SVD) than previously thought. However, regarding photobiomodulation therapy, two controlled dementia studies are currently underway in San Francisco and Boston using similar technology. The technology used here (described above) which is similar, is now in the midst of conducting a PTSD study in the Atlanta area. As recently as May of this year *JAMA Network Journals* reported that cerebral SVD and AD disease pathology appear to be associated, new research indicates (Kester et al., 2014). Our study supports the hypothesis that the pathways of SVD and AD are inter-connected. SVD could provoke amyloid pathology while AD-associated cerebral amyloid pathology may lead to auxiliary vascular damage researchers conclude (Attems and Jellinger, 2014). Dr. Sandra Black (a prominent Alzheimer's researcher in Toronto, Canada) during a recent radio talk show theorized that exercise is one area for prevention that could be a solution in mitigating the problem (Middleton et al., 2018). The inference here is that increased blood flow during cardio vascular exercise might mitigate the buildup of plaque and protein aggregates. Specialists concede that vascular dementia is directly related to SVD which by estimates accounts for 10%–20% of all dementia cases. Alzheimer's is more insidious in which the vascular system is hampered by the buildup of amyloid plaque and tau protein and as a result of this build up, much attention is being given to prevention as well as a cure.

The following are a few dementia case studies from those which we have treated. Although PTSD cases described above were carefully documented with the PTSD questionnaire, this was not the case with dementia, which although having several questionnaires and scores available, it was too time consuming to ask the clients to cooperate. However we did use the following questionnaire on a few occasions.

Name Age is patient alert? Level of education

1. What day of the week is it?
2. What is the year?
3. What state are we in?
4. Please remember these five objects. I will ask you what they are later. Apple; Pen; Tie; House; Car.
5. You have $100 and you go to the store and buy a dozen apples for $3 and a tricycle for $20. How much did you spend? How much do you have left?
6. Please name as many animals as you can in 1 minute. 0–4 animals; 5–9 animals; 10–14 animals; 15 + animals
7. What were the five objects I asked you to remember? 1 point for each one correct.
8. I am going to give you a series of numbers and I would like you to give them to me backwards. For example, if I say 42, you would say 24.

   87 649 8537

9. This is a clock face. Please put in the hour and minute markers if the time is 10 minutes to 11 o'clock. Hour markers okay Time correct
10. Please place an X in the triangle.
    Which of the above figures is largest?
11. I am going to tell you a story. Please listen carefully because afterwards, I'm going to ask you some questions about it. Jill was a very successful stockbroker. She made a lot of money on the stock market. She then met Jack, a devastatingly handsome man. She married him and had three children. They lived in Chicago. She then stopped work and stayed at home to bring up her children. When the children were teenagers, she went back to work. She and Jack lived happily ever after. What was the female's name? What work did she do? When did she go back to work? What state did she live in?

Case 1: An 88 year old woman who was wheel chair bound started having simple memory problems (dementia) 8 years previously. She recently was forced to move in with her son due to aberrant behavior in the skilled nursing facility. She would have violent outbursts toward others when confused, or if more than two or three people were in her presence. Her energy level was very low. Six treatments were administered over a 12-day period. Her son reported that she was much more relaxed and outbursts were nonexistent. In addition, in the evening she maintained a much higher energy level and was much more alert. The son was very pleased with the change in her entire demeanor and after a 5-week period noticed a small decline in her condition. Therefore we have now decided on a maintenance regimen of monthly treatments.

Case 2: A 38 year old female, under the care of this primary care physician for 3 months had a history of stroke and possible TBI caused by spousal abuse. She sought treatment for severe migraines of various degrees three to four times per week over a 5-year period. Her previous primary care physician had her on a very high dosage of OxyContin and I was asked by her current primary care physician if I could help wean her off. After the first treatment her headache pain was almost totally eliminated, however due to cutting back on her medication the migraines continued. A course of eight treatments (sometimes on an emergency basis) were administered over a 14-day period and she received significant relief every time. However, after the seventh treatment the mother reported her daughter began speaking in full sentences and her memory had returned. In addition she no longer needed to walk with a cane. The client was followed up intermittently over a 2-year period and no regression was apparent.

Case 3: The practice manager of an orthopedic practice asked if I could possibly help her after observing my work with injured patients in her office. (The initial use of this type of modality was for muscular-skeletal injury.) Her 77 year old father was first diagnosed with dementia 12 years previously, and although he was taking several medications for memory problems, still had to be admitted to a nursing home. His memory loss was severe and he recognized his daughter but did not know her name even though she visited him three to four times per week. After the fourth treatment, the daughter called and reported to me that his brother stopped by to see him for the first time in months and he not only called him by name but they cried and had a conversation. He was treated the next day and upon our arrival while sitting in the dining room he recognized his daughter, waved and called out to her by name. This indeed was a startling moment. A fifth treatment was administered however the daughter then left for a 10-day vacation. Upon her return, he had reverted back to his old self and after several treatments administered later, there was no change.

Case 4: A 67 year old postal worker who was still working part-time was getting concerned about memory lapses and loss of the sense of smell. Both his parents died of AD in their 80s. We proposed a series of six treatments to start as a preventive measure to which he readily agreed. He reports that he has regained his sense of smell. In addition, he reported he had three small, fleeting headaches something that he never experiences. We conjecture this may be the result plaque or tau protein moving through his system. We will continue to give him treatments monthly.

Case 5: A 61 year old retired housewife experienced a very rapid decline over a 9 month period. They had planned on moving to another city, where she would have more family support. However we encouraged her daughter who acted as caregiver, to allow the patient to undergo a least a series of six treatments. After the second treatment, the receptionist was shocked that the patient asked to sign her own credit card receipt. After the sixth treatment, the daughter reported that she now dresses herself without help, and if she makes a mistake the daughter points it out and she corrects the problem herself. In addition, she can now get into the car and fasten her seat belt without any assistance. Also, she had a problem eating a sandwich whereas she would pick up the top slice of bread and the daughter would need to correct her, which was no longer a problem.

Case 6: An 86 year old male with a 10 year history of dementia has been treated continuously on a regular basis since November 2014. The subject is a retired medical researcher and his wife is a retired school teacher. Careful documentation is being kept by his wife and the clinician. At the beginning, within 2 weeks of commencing treatment, the major improvement reported by the wife was the fact that he slept throughout the night, giving the family much relief from anxiety and sleep deprivation; previous to this time he would wander around the house at all hours of the night. His verbal echoing also was minimized in the first few months, and conversation was improved. A memory test was administered twice a week and he showed stabilization. Although he has declined significantly since 2018, which may be attributed to less frequent treatments due to his physical decline and his inability to walk, there is a new therapeutic outcome to relate. He was taken to the hospital with pneumonia and could no longer swallow. Aspiration was a major fear and he was sent home and was told he might have a week left to live. Intensive laser treatments after 1 week brought him back to his previous state. Previously only his prefrontal cortex was treated (on occasion the circle of Willis) However, we now treated the brainstem and along with dally doses of atropine he now eats normally and on occasion will respond with the nodding of his head or mouthing a thank you. His caretaker wife gives us much credit for his survival. Much credit must also be given to the wife for the many activities she engages him in, including puzzles and math games, but much credit is also given to the ongoing photobiomodulation treatments.

42.5 Conclusion and future directions

The anecdotal case studies cited in this chapter have provided us with the foundation to conduct further formal controlled studies. Currently, such a study concerning dementia is on-going in San Francisco and Boston using similar technology. Harvard has completed a Phase I study and Phase II is being planned a controlled, formal study regarding PTSD to be held in Atlanta in the fall of 2018. The providers who contributed to this chapter all are currently providing this service to all their patients at minimal cost, and completely free to the military veterans. Photobiomodulation technology may be a breakthrough approach to the treatment of brain disease and brain trauma.

References

Attems, J., Jellinger, K.A., 2014. The overlap between vascular disease and Alzheimer's disease – lessons from pathology. BMC Med. 12, 206.

Kester, M.I., Goos, J.D., Teunissen, C.E., Benedictus, M.R., Bouwman, F.H., Wattjes, M.P., et al., 2014. Associations between cerebral small-vessel disease and Alzheimer disease pathology as measured by cerebrospinal fluid biomarkers. JAMA Neurol. 71, 855–862.

Middleton, L.E., Black, S.E., Herrmann, N., Oh, P.I., Regan, K., Lanctot, K.L., 2018. Centre- versus home-based exercise among people with mci and mild dementia: study protocol for a randomized parallel-group trial. BMC Geriatr. 18, 27.

Restak, R.M., 1980. ISBN-10: 9780446969758 The Brain, The Last Frontier. Warner Books, Inc.

Stephan, W., Banas, L.J., Bennett, M., Tunceroglu, H., 2012. Efficacy of super-pulsed 905 nm. low level laser therapy (LLLT) in the management of traumatic brain injury (TBI): a case study. World J. Neurosci. 2 (4), 231–233.

Theralase, I., 2018. <http://theralase.com/tlc-1000/>,Toronto, Canada.

Chapter 43

What we don't know and what the future holds

Michael R. Hamblin
Wellman Center for Photomedicine, Massachusetts General Hospital, Boston, MA, United States

43.1 Questions, or what we don't know

When we started planning this volume, we had no idea it would grow into the substantial tome that it has become. The final chapter will seek to answer some of the pressing questions that have been raised in preceding chapters contained within this book.

43.2 What are the best diseases and conditions to be treated?

This is one of the most compelling questions to arise from this volume. People are rightly suspicious of any treatment that claims to cure everything (or almost everything). "Anything that sounds too good to be true, almost certainly is" is another mantra that is frequently repeated. So should photobiomodulation (PBM) to the brain be chiefly directed toward traumatic brain injuries resulting from strokes and head injuries for which the most preclinical data and even a considerable number of clinical studies already exist.

Other disorders that come under this heading include brain damage resulting from birth trauma (cerebral palsy), global ischemia occurring after a heart attack, or other major trauma that interrupts the blood supply to the brain. If these are indeed the most beneficial conditions to be treated, then another intriguing question arises. Should PBM be applied in the acute phase soon after the injury has occurred? Or perhaps PBM would be more effective in patients with chronic brain damage. Many studies in animal models from the Oron laboratory (Oron et al., 2006, 2007, 2012) and from the Hamblin laboratory (Xuan et al., 2014a,b; Ando et al., 2011; Wu et al., 2012) suggest that in animal models, the greatest benefit is obtained when light is delivered to the head shortly after the brain injury has occurred (a few hours). However, the reports from Naeser et al. (2011, 2014) suggest that PBM has remarkable effects in human patients who have suffered from chronic traumatic brain injury (TBI). A recent report in an animal model suggests that transcranial photobiomodulation (tPBM) could be effective in global cerebral ischemia such as that suffered when the heart is restarted after a coronary arrest (Wang et al., 2019).

On the other hand, perhaps the major benefits might be more in the area of neurodegenerative diseases, of which there are now a host of different manifestations recognized. Alzheimer's disease and age-related dementia is an ever-growing epidemic that is predicted to affect a large proportion of the population, as the average age continues to increase because other diseases are gradually being conquered. Preliminary studies suggest that PBM is very effective in patients with Alzheimer's (Saltmarche et al., 2017; Berman et al., 2017). Workers in Australia led by Mitrofanis have devoted considerable effort to establishing PBM as a treatment for Parkinson's disease (PD; Mitrofanis, 2017).

Or even yet, perhaps the best applications might be in the area of psychiatric disorders where intriguing preliminary evidence suggests remarkable benefits in the treatment of depression (Cassano et al., 2015), anxiety (Schiffer et al., 2009), posttraumatic stress disorder (Naeser et al., 2011), insomnia (Henderson and Morries, 2015a,b), and addiction (Zalewska-Kaszubska and Obzejta, 2004).

Photobiomodulation in the Brain. DOI: https://doi.org/10.1016/B978-0-12-815305-5.00043-9

Then we have neurodevelopmental disorders such as autism spectrum disorder (Leisman et al., 2018), attention deficit hyperactivity disorder, and intellectual disability. As yet, only preliminary anecdotal reports exist in this area. Finally we have the area of cognitive enhancement where there have actually been quite a few published reports (de la Torre, 2017).

43.3 How important is light penetration to the brain?

Due to the considerable success of PBM for the brain, the question of how important is light penetration to the brain has become essential. Many tissue optics studies have explored the penetration of various wavelengths of light through the scalp and skull (Pitzschke et al., 2015), or even just through the skull alone in animal models (Lapchak et al., 2015; Jagdeo et al., 2012) or human cadaver heads (Tedford et al., 2015; Hart and Fitzgerald, 2016; Wan et al., 1981), and attempted to estimate how much light actually reaches the cortex. Certainly it is true, that many investigators remain firmly convinced that high power is more effective than lower power. One of the major reasons put forward for the failure of the NEST-3 clinical trial for acute stroke, was the lack of sufficient light penetration to the brain, despite the fact that the surface power of the laser ($\sim$750 mW/cm^2) required active cooling of the laser head, in order to avoid thermal damage to the shaved scalp (Lapchak and Boitano, 2016). Moreover, Henderson and Morries have stated they are firmly convinced that high power densities are absolutely necessary to achieve sufficient light penetration to the brain (Henderson and Morries, 2015a,b).

On the other hand, we have the undoubted success of Vielight and the Neuro-alpha and Neuro-gamma devices. By most accounts these devices would be considered considerably under-powered. The total power is only $<$200 mW (combined transcranial and intranasal) and the power density is 41 mW/cm^2 (Saltmarche et al., 2017).

43.4 What about systemic effects?

There is evidence for systemic effects of PBM for brain disorders, both in animal models and in humans. One of the most convincing examples of the systemic effects of PBM was demonstrated by Johnstone et al. (2014). These workers studied a mouse model of PD caused by MPTP injection. They compared 670-nm light directed specifically at either the head or body. Indirect application (remote PBM or abscopal PBM) produced a significant rescue of tyrosine hydroxylase-positive cells in the SNc at the low MPTP dose, but this protection was not as robust as that achieved by direct irradiation of the head. In another study (Kim et al., 2018) they showed that remote preconditioning with PBM (670 nm light to the mouse back) protected mice against a subsequent challenge with MPTP. Because remote PBM produced a similar degree of neuroprotection to ischemic preconditioning of the leg, they suggested that the two approaches shared a common mechanism.

The systemic effects in humans are evidenced by the use of the Vielight transcranial light emitting diode (LED) device simultaneously combined with intranasal LEDs (Saltmarche et al., 2017). Fig. 43.1 shows the Vielight Neuro device. It also appears from studies carried out in China that intranasal LEDs used alone, can have many different positive effects on brain function (Wang, 2006; Xu et al., 2001, 2002a,b,c, 2003).

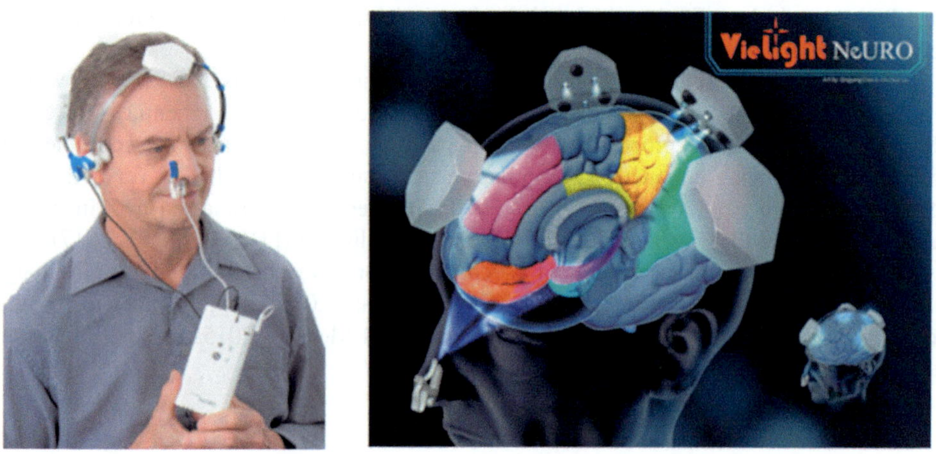

FIGURE 43.1 Vielight Neuro combined transcranial and intranasal device.

43.5 What is the best way to deliver light?

This is another highly important and hotly debated question. In many clinical studies the forehead has been the site of choice for tPBM. It is likely there are two convincing reasons for this choice. Firstly, the forehead is without hair, and hair does act as a barrier to light penetration. Hair blocks light by both absorption and scattering, and the balance between these two depends on the exact wavelength of light. Despite the fact that the amount of hair on the human head varies widely among individuals of both genders in color, amount, and thickness, there are few (if any) studies, looking at the effect of hair on light penetration to the brain.

In other studies, a LED helmet that covers all or most of the surface of the head has been utilized. There is yet a third model of light delivery that has been used, typified by the approach of Vielight. This approach uses specific locations on the head in order to deliver the light to very localized areas of the cortex. For instance, the nodes of the default mode network consist of the following discrete brain areas: medial prefrontal cortex, posterior cingulate cortex, inferior parietal lobule, lateral temporal cortex, hippocampus, and the precuneus.

43.6 How important is pulsing?

How does pulsed light differ from CW at the cellular and molecular level, and how is the outcome of PBM affected? If pulsing is more efficacious, then at what pulse parameters is the optimal outcome achieved? In particular, what is the ideal pulse repetition rate or frequency to use?

43.6.1 Pulse parameters and light sources

There are five parameters that could be specified for pulsed light sources. The pulse width or duration or ON time (PD) and the pulse interval or OFF time (PI) are measured in seconds. Pulse repetition rate or frequency (F) is measured in Hz. The duty cycle (DC) is a unitless fractional number or percent. The peak power and average power are measured in watts.

Pulse duration, pulse repetition rate, and DC are related by the simple equation;

$$\mathrm{DC} = F \times \mathrm{PD}$$

Peak power is a measure of light intensity during the pulse duration, and related to the average power (measured in watts) by:

$$\text{Average power} = \text{Peak power} \times F \times \mathrm{PD}$$

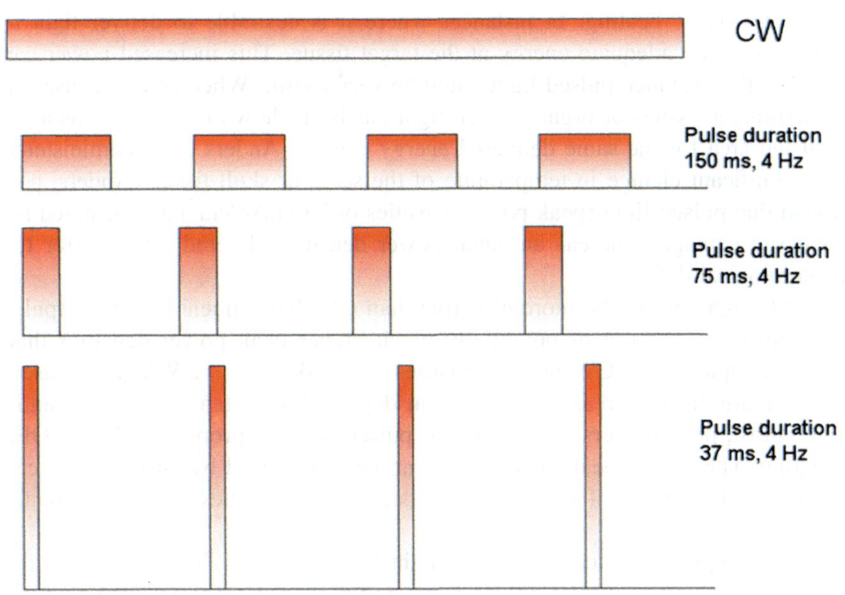

FIGURE 43.2 Diagram comparing the structure of CW light with pulsed light of various pulse durations (37, 75, 150 ms) and the same pulse frequency (4 Hz).

Alternatively,

$$\text{Peak power} = \frac{\text{Average power}}{\text{DC}}$$

In all cases, it is necessary to specify any two out of three of: PD, *F*, and DC, and either the peak or average power for the pulse parameters to be fully defined.

Fig. 43.2 graphically shows the relationship between peak power and pulse duration.

43.6.2 Types of pulsed light sources

Five major types of pulsed lasers (or other light sources) are commonly utilized: (1) Q-switched, (2) gain-switched, (3) mode-locked, (4) superpulsed, and (5) chopped or gated. Each utilizes a different mechanism to generate light in a pulsed as opposed to continuous manner, and vary in terms of pulse repetition rates, energies, and durations. However the first three classes of truly pulsed lasers mentioned above are in general not used for PBM; instead super-pulsed or gated lasers are mainly used. The concept of super-pulsing was originally developed for the carbon dioxide laser used in high power tissue ablative procedures. The idea was that by generating relatively short pulses (μs) the laser media could be excited to higher levels than those normally allowed in CW mode where heat dissipation constraints limit the maximum amounts of energy that can be used to excite the lasing media. With the original carbon dioxide superpulsed lasers, the short pulses would confine the thermal energy in the tissue (by making the pulse duration less than the thermal diffusion time) reducing collateral thermal damage to normal tissue.

Another type of laser that particularly benefited from super-pulsing is the gallium-arsenide (GaAs) diode laser. This laser has a wavelength in the region of 904 nm and pulse duration usually in the range of 100–200 ns. Another semiconductor laser amenable to super-pulsing is the indium-gallium-arsenide (In-Ga-As) diode laser. It emits light at a similar wavelength (904–905 nm) as the GaAs diode laser, producing very brief pulses (200 ns) of high frequencies (in the range of kilohertz). These pulses are of very high peak powers (1–50 W) and an average power of 60 mW. Theoretically, the super-pulsed GaAs and In-Ga-As lasers allow for deep penetration without the unwelcome effects of CW (such as thermal damage), as well as allowing for shorter treatment times.

The other major class of pulsed light sources used in PBM are simply CW lasers (usually diode lasers) that have a pulsed power supply generated by a laser driver containing a pulse generator. This technology is described as chopped or gated. It is also equally feasible to use pulse generator technology to pulse LEDs or LED arrays (Valchinov and Pallikarakis, 2005).

43.6.3 Why could pulsing be important in photobiomodulation?

Pulsed light offers numerous potential benefits. Because there are "quench periods" (pulse OFF times) following the pulse ON times, pulsed lasers can generate less tissue heating. In instances where it is desirable to deliver light to deeper tissues increased powers are needed to provide adequate energy at the target tissue. This increased power can cause tissue heating at the surface layers and in this instance pulsed light could be very useful. Whereas CW causes an increase in temperature of the intervening and target tissues or organ, pulsed light has been shown to cause no measurable change in the temperature of the irradiated area for the same delivered energy density. Anders et al. administered pulsed light to pig craniums, and found no significant change in temperature of the scalp or skull tissue (Anders, personal communication). Ilic et al. (2006) found that pulsed light (peak power densities of 750 mW/cm^2) administered for 120 seconds produced no neurological or tissue damage, whereas an equal power density delivered by CW (for the same number of seconds) caused marked neurological deficits.

Aside from safety advantages, pulsed light might simply be more effective than CW. The "quench period" (pulse OFF times) reduces tissue heating, thereby allowing the use of potentially much higher peak power densities than those that could be safely used in CW. For example, when CW power densities at the skin of ≥ 2 W/cm^2 are used, doubling the CW power density would only marginally increase the treatment depth while potentially significantly increasing the risk of thermal damage; in contrast, peak powers of ≥ 5 W/cm^2 pulsed using appropriate ON and OFF times might produce little, or no tissue heating. The higher peak powers that can be safely used by pulsing light can overcome tissue heating problems and improve the ability of the laser to penetrate deep tissues achieving greater treatment depths.

There may be other biological reasons for the improved efficacy of pulsed light (PW) over CW. The majority of the pulsed light sources used for PBM have frequencies in the 2.5–10,000 Hz range and pulse durations are commonly in the range of a few milliseconds. This observation suggests that if there is a biological explanation of the improved effects of pulsed light it is either due to some fundamental frequency that exists in biological systems in the range of tens to hundreds of Hz, or alternatively due to some biological process that has a time scale of a few msec. Two possibilities for what these biological processes could actually be occur to us. Firstly, it is known that mammalian brains have waves that have specific frequencies. Electroencephalography studies have identified four distinct classes of brain waves (Westmoreland et al., 1990; Lopes da Silva, 1991). Alpha waves (8–13 Hz) occur in adults who have their eyes closed or who are relaxed (Kirschfeld, 2005). Beta waves (14–40 Hz) mainly occur in adults who are awake, alert, or focused (Thuroczy et al., 1994). Delta waves (1–3 Hz) occur mainly in infants, adults in deep sleep, or adults with brain tumors (Picchioni et al., 2009). Theta waves (4–7 Hz) occur mainly in children ages 2–5 years old and in adults in the twilight state between sleeping and waking or in meditation (Baijal and Srinivasan, 2010). The possibility of resonance occurring between the frequency of the light pulses and the frequency of the brain waves may explain some of the results with tPBM using pulsed light.

Secondly, there are several lines of evidence that ion channels are involved in the subcellular effects of PBM. Some channels permit the passage of ions based solely on their charge of positive (cationic) or negative (anionic) while others are selective for specific species of ion, such as sodium or potassium. These ions move through the channel pore single file nearly as quickly as the ions move through free fluid. In some ion channels, passage through the pore is governed by a "gate," which may be opened or closed by chemical or electrical signals, temperature, or mechanical force, depending on the variety of channel. Ion channels are especially prominent components of the nervous system. Voltage-activated ion channels underlie the nerve impulse and while transmitter-activated or ligand-gated channels mediate conduction across the synapses.

There is a lot of literature on the kinetics of various classes of ion channels but in broad summary it can be claimed that the time scale or kinetics for opening and closing of ion channels are of the order of a few milliseconds. For instance Gilboa et al. (2005) used pulses having a width 10 ms and a period of 40 ms (25 Hz). Other reports on diverse types of ion channels have given kinetics with timescales of 160 ms (Priestley and Kemp, 1994), 3 ms (Schneggenburger and Neher, 2000); and one paper giving three values of 0.1, 4, and 100 ms (Kampa et al., 2004). Potassium and calcium ion channels in the mitochondria and the sarcolemma may be involved in the cellular response to PBM (Chow et al., 2007; Karu, 2008; Karu et al., 2004).

Thirdly there is the possibility that one mechanism of action of PBM on a cellular level is the photodissociation of nitric oxide from a protein binding site (heme or copper center) such as those found in cyctochrome c oxidase (Lane, 2006). If this process occurs it is likely that the NO would rebind to the same site even in the presence of continuous light. Therefore, if the light was pulsed multiple photodissociation events could occur, while in CW mode the number of dissociations may be much smaller.

43.6.4 Effect of pulsing photobiomodulation for the brain

The Hamblin Laboratory published a study that compared the same dose of 810 nm PBM (36 J/cm^2 at 50 mW/cm^2) delivered either by CW, pulsed at 10 Hz, or pulsed at 100 Hz in a mouse TBI model (Ando et al., 2011). A single PBM dose was delivered 4 hours post-TBI; and the beneficial effect on cognitive function was statistically better with 10 Hz than it was with either CW and 100 Hz, which were both equally effective, but in turn were statistically better than TBI controls (see Fig. 43.3).

Another interesting animal study came from a group at MIT led by Li-Huei Tsai (Iaccarino et al., 2016). This report originated in an optogenetics study where light-sensitive channelrhodopsin ion channels were transfected into the brains of genetically-engineered Alzheimer's mice using a viral vector. They found that optogenetically driving fast-spiking parvalbumin-positive-interneurons at gamma (40 Hz), but not other frequencies, reduced the levels of amyloid-β (Aβ) peptides in the brain. Gene expression profiling revealed induction of genes associated with morphological transformation of microglia, and histological analysis confirmed increased microglia colocalization with Aβ. Subsequently, they designed a noninvasive 40 Hz light-flickering regime that reduced Aβ levels in the visual cortex and mitigated plaque load in aged mice. A Danish group (Ismail et al., 2018) carried out a pilot clinical study in which human AD patients were exposed to 40 Hz pulsed light for 2 hours each day (morning and evening) for 10 days. PET scans specific for amyloid plaque showed no significant reduction in plaque levels. However this study was based upon the beneficial effects of pulsed light being mediated through the eyes, whereas in small animals such as mice the light may have directly affected the brain. Moreover 10 days was a relatively short time to expect to see a significant benefit.

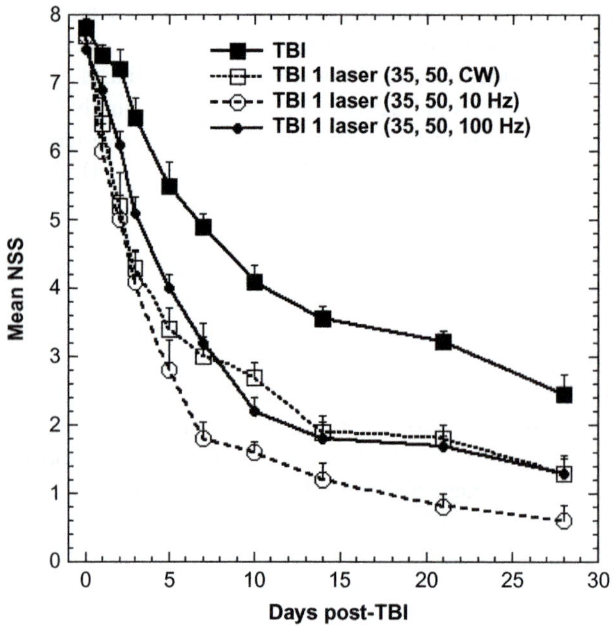

FIGURE 43.3 Time course of neurological severity score (NSS) of mice with TBI receiving either control (no laser-treatment), or 810-nm laser (36 J/cm^2 delivered at 50 mW/cm^2 with a spot size of 0.78 cm^2) in either CW, PW 10 Hz or PW 100 Hz modes. Results are expressed as mean ± S.E.M. ($n = 10$). *Taken from Ando, T., Xuan, W., Xu, T., Dai, T., Sharma, S.K., Kharkwal, G.B., et al., 2011. Comparison of therapeutic effects between pulsed and continuous wave 810-nm wavelength laser irradiation for traumatic brain injury in mice. PLoS One 6 (10), e26212 no permission required.*

43.7 How important is the location on the head?

It is noteworthy that in the initial tLED PBM studies carried out in the Naeser lab (Karbe et al., 1998; Saur et al., 2006). to treat chronic stroke patients with left hemisphere stroke and aphasia, placing the LED cluster head at the vertex, stimulating both the R supplementary motor area (SMA) and the L SMA simultaneously, was not beneficial—in fact it worsened naming (a measure used in aphasia research). This bilateral tLED placement protocol had placed the LED cluster heads on the left and the right sides of the head (left and right perisylvian areas). When the LEDs were placed only on the left side of the head (left perisylvian areas), improved naming was observed. Thus, the results from the Naeser study (Naeser et al., 2012) with tLEDs supported the MRI findings that document recovery in aphasia—that is, better recovery is associated with increased activation primarily in the left hemisphere perilesional areas only (including the L SMA).

43.8 How important is the biphasic dose response?

Our animal study in a mouse model of TBI was one of the few studies to have clearly shown that the well-known biphasic dose response curve applied in tPBM. We found that NIR laser (810 nm, 18 J/cm^2) applied to the head daily for 3 days was superior to the same dose either applied once, or applied once a day for 14 consecutive days (Xuan et al., 2013). This result was sufficiently intriguing for us to explore it further. We found that the 14 daily laser treatments did not permanently damage the brain or the cognitive function of the mice, but merely delayed the onset of the PBM-induced benefit for some time (2 months). In other words, the cognitive scores of the mice that had received a TBI followed by 14 PBM sessions were lower than TBI controls at 4 weeks, but changed to being higher than TBI controls at 8 weeks post-TBI (Xuan et al., 2016). By contrast, the mice that received three daily PBM sessions had increasingly better cognitive scores not only up to 4 weeks, but up to 8 weeks as well. The reasons for these interesting findings were proposed to be due to the excessive number PBM sessions causing neuroinflammation (as indicated by raised expression of GFAP) which delayed the regenerative effect of the PBM, until the resulting neuroinflammation had subsided.

43.9 What about cognitive enhancement and preconditioning?

A few studies have investigated the effect of PBM on brain function in normal laboratory animals. El-Massri et al. (2018) treated mice with PBM (670 nm) for 20 minutes per day, commencing at 5 months old and continued for 8 months. They found a clear reduction in glial cell numbers (astrocytes and microglia) in the striatum after PBM in aged mice. By contrast, the number of two types of striatal interneurons (parvalbumin + and encephalopsin +), together with the density of striatal dopaminergic terminals (and their midbrain cell bodies), remained unchanged after such treatment. The reduction of gliosis may have implications for prevention of cognitive decline and reversal of brain aging. Michalkova et al. (2008) showed that treatment with 1072 nm LEDs (6 minutes daily for 10 days) in middle-aged (12 months) female CD-1 mice improved deficits in working memory in a 3D-maze.

There have been several studies reporting enhancement of cognitive functions in normal, healthy adults. These studies have often involved student volunteers who have been recruited as experimental subjects. tPBM (1064 nm) produced significant improvements in sustained attention and short-term memory retrieval (Barrett and Gonzalez-Lima, 2013). These effects of PBM were found to be comparable to the cognitive enhancing effects of vigorous aerobic exercise. In a follow-on study they combined tPBM with aerobic exercise (EX, treadmill running) (Hwang et al., 2016). Both tPBM and EX reduced reaction time in the psychomotor vigilance task (sustained attention) and increased the number of correct responses in the delayed match-to-sample task (working memory), but the combination was not significantly better than either treatment alone.

In another study, the same group found improved performance on the Wisconsin Card Sorting Task, a gold standard neuropsychological measure of executive function, in healthy, young participants (Blanco et al., 2017a,b). They then showed that a prefrontal cortex mediated reflective system that learns categories using explicit rules was improved by tPBM, while a striatally mediated reflexive learning system that forms gradual stimulus-response associations was not (Blanco et al., 2017a,b). They then proceeded to test tPBM (to the right or left forehead) combined with attention bias modification (ABM) in 51 adults diagnosed with depression (Disner et al., 2016). One and two weeks later, right tPBM led to greater depression symptom improvement among participants who responded to ABM (i.e., attention was directed away from negative stimuli). Minimal changes were observed in the left and sham tPBM

Johnstone's laboratory (Kim et al., 2018) demonstrated a preconditioning approach in a mouse model of PD caused by 1-methyl-4-phenyl-1,2,3,6-tetrahydropyridine (MPTP) injection. They showed that remote preconditioning with PBM (670 nm light to the mouse back) protected mice against a subsequent challenge with MPTP. Because remote PBM produced a similar degree of neuroprotection to ischemic preconditioning of the leg, they suggested the two approaches shared a common mechanism. Another (as yet unpublished) preclinical study of PBM for the brain used preconditioning to the head. This study used light delivered to the head in neonatal mice, 6 hours before the animals were subjected to a cerebral hypoxia protocol. This involved surgical ligation of the left carotid artery, followed by 1 hour of breathing of a gas mixture containing only 6% oxygen. Mice that received PBM preconditioning performed better on mazes that measure cognitive performance and memory and learning. The 6-hour time point was determined by a series of experiments that measured ATP content in the brains of normal mice at different times after PBM. Six hours was the highest, and the investigators argued that this should be the optimum preconditioning time.

43.10 How does photobiomodulation compare with other noninvasive brain stimulation techniques?

Two other major types of noninvasive brain stimulation therapies are transcranial magnetic brain stimulation (TMS) and transcranial direct current stimulation (tDCS). Transcranial electrical stimulation of the cortex was first demonstrated in 1980 (Merton and Morton, 1980), but was never used as a medical therapy because of the undesirable pain it caused. A much more convenient method was invented by Barker and colleagues only 5 years later, namely TMS (Barker et al., 1985). Each of these noninvasive treatment techniques are reviewed briefly here. Fig. 43.4 schematically illustrates these noninvasive brain stimulation techniques (as well as DBS discussed in Section 43.11).

43.10.1 Transcranial magnetic brain stimulation

TMS relies on the induction of small electrical currents within the brain by a magnetic field, which passes through the skull. Thus, the application of TMS is painless and therefore widely used for noninvasive stimulation of the human brain. The stimulation works by passing a large (5 kA or more), brief (<1 ms) current through the wired and insulated coil placed on the subject's scalp. The brief current flowing through the coil generates a magnetic pulse that penetrates

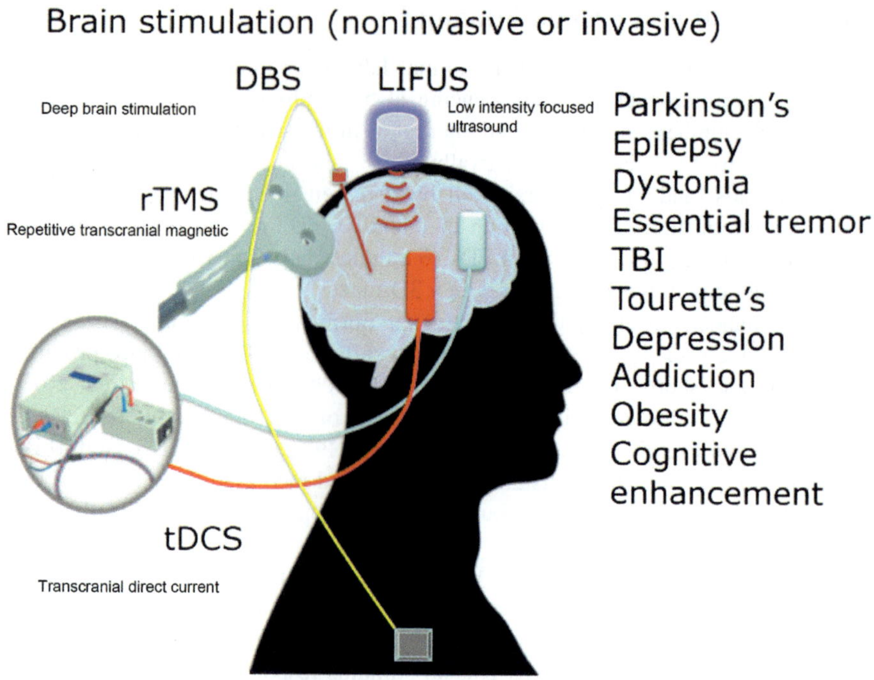

FIGURE 43.4 Noninvasive brain stimulation techniques (rTMS, tDCS, LIPUS) plus deep brain stimulation (DBS).

the skull and in turn induces small eddy currents in electrically conductive regions—that is, in the underlying brain tissue (Sparing and Mottaghy, 2008). Relatively focal stimulation can be achieved by combining two circular coils to form a figure-of-eight (or butterfly-shaped) coil. The magnetic fields sum up at the point of intersection of both coils. Using spherical model approximation, it has been estimated that the spatial resolution of TMS is in the centimeter range (10–20 mm).

Repetitive TMS of appropriate frequency, intensity, and duration can lead to transient increases or decreases in excitability of the targeted cortex that last beyond the duration of the rTMS train itself (Pascual-Leone et al., 1998). Slow (1 Hz) rTMS has been shown to decrease cortical excitability in humans (Maeda et al., 2000). Conversely, fast rTMS (5, 10, or 20 Hz) can induce a transient increase in cortical excitability (Pascual-Leone et al., 1994).

In humans, there is a dominant hemisphere for language, which in most cases is the left hemisphere. After a left hemisphere stroke when aphasia is present, it has been observed that there is much brain reorganization for language that takes place over the following weeks and 2–3 months. Indeed, during the first few months, there is increased activation in the right hemisphere. However, after a year or more, in those cases who have better language recovery, the increased right activation becomes less, and there is a shift to increased left hemisphere activation, mostly in the left perilesional areas (Saur et al., 2006; Szaflarski et al., 2013). The increased, overactivation that remains in the right hemisphere language homologues is considered by many to be maladaptive, or only "...partially compensatory, and that optimal neuroplastic changes eventually involve recruitment of perilesional areas" in the left hemisphere (Shah et al., 2013).

There have been several studies in the past decade that used slow, 1 Hz, rTMS to inhibit overactivation in right frontal cortex to improve language in chronic stroke. With rTMS the motor threshold is first established for the first dorsal interosseous muscle; and the rTMS is delivered at 90% of that motor threshold for any given patient. The optimum area to suppress with rTMS in these chronic aphasia cases has been the right pars triangularis, located in the right inferior frontal gyrus (a right hemisphere homologue of Broca's language area in the left hemisphere) (Martin et al., 2004; Naeser et al., 2005; Hamilton et al., 2010; Barwood et al., 2011; Weiduschat et al., 2011). These rTMS treatments are usually given for 20 minutes (10 days, 5 days per week). Follow-up fMRI studies have shown that for post-rTMS, there was suppression of the right frontal areas, and this was associated with improved ability to name pictures and increased number of words per phrase length (Martin et al., 2009), and overall language improvement (Barwood et al., 2011; Weiduschat et al., 2011).

The rTMS studies with aphasia patients have been based on the notion of interhemispheric inhibition, where the overactivation observed in the right hemisphere following left hemisphere stroke was associated with less inhibitory dominance from the injured, dominant left hemisphere. Therefore, parts of the right hemisphere became disinhibited, and more overactive. Thus, suppressing a part of the right hemisphere was effective to improve language, beginning with the earliest studies (Martin et al., 2004; Naeser et al., 2005). The role of the right hemisphere, however, is still unclear (Fama and Turkeltaub, 2014).

A new type of rTMS, intermittent theta burst (iTBS) with a short, excitatory stimulation (Huang et al., 2005) has recently been used to excite the left hemisphere, and improve language in chronic aphasia patients who have had left hemisphere stroke (Szaflarski et al., 2011). The iTBS treatments consisted of bursts of three pulses at 50 Hz given every 200 ms, in 2-second trains, repeated every 10 seconds over 200 seconds, for a total of 600 pulses. Theta burst was administered at 80% of motor threshold (10 days, 5 days per week). An improvement in the white matter integrity (increase in fractional anisotropy values on DTI, MRI scans) was noted near the stimulation sites, as well as distal areas (Allendorfer et al., 2012). Many stimulation sites were perilesional in the left inferior or superior frontal areas; areas that had shown peak activation on fMRI scans during language tasks (Szaflarski et al., 2011). Significant improvement was observed in a timed, verbal fluency task (number of words recalled in specific category—e.g., animals), post the iTBS treatment series.

Repetitive TMS has also been used to treat visual neglect and paralysis following stroke (Lefaucheur, 2006; Liew et al., 2014). The role of rTMS and tDCS on rehabilitation of upper limb and hand paresis has been reviewed (Di Pino et al., 2014) and these authors suggested "a bimodal balance-recovery model that links interhemispheric balancing and functional recovery to the structural reserve spared by the lesion." The noninvasive brain stimulation would be tailored to the needs of individual patients. The use of tDCS for stroke patients has also been reviewed in the Cochrane database (Elsner et al., 2013).

43.10.2 Transcranial direct current stimulation

During application of tDCS, electrical currents are applied constantly over a longer period of time than with rTMS, and usually in the order of minutes, at much lower intensities to achieve changes in cortical excitability that persist even after stimulation has ceased. In comparison to TMS, tDCS requires inexpensive hardware and the procedure is simple. The most important component is a current generator, which is capable of delivering a constant electrical current flow of up to 2 mA. The electrical current is delivered through two sponge electrodes soaked in saline solution. Typically, these electrodes have a relatively large surface of 20–35 cm^2 that limits the focality of stimulation. However, the large size keeps current densities low (one of the critical safety parameters). Nevertheless, subjects may feel a mild tingling or itching sensation on the scalp beneath the electrodes. tDCS relies on the assumption that a weak constant direct current (DC) polarizes tissue. Stimulation is usually applied for a few minutes (up to 30 minutes). Depending on the direction of current flow, that is, polarity, tDCS can be delivered either as cathodal, inhibitory, or anodal, excitatory.

There are several studies that have used tDCS to improve language in chronic stroke patients with aphasia. Most applied excitatory, anodal tDCS to the left hemisphere (Baker et al., 2010; Fridriksson et al., 2011; Marangolo et al., 2013); or inhibitory, cathodal tDCS to the right hemisphere (Kang and Paik, 2011). The first rDCS study with chronic aphasia patients, however, applied cathodal tDCS to the left hemisphere, and reported improvement in naming (Monti et al., 2008).

Some studies have suggested that the relative level of left-hemispheric dominance for language, in an individual person prior to stroke, would also have relevance to variability in recovery patterns (Heiss and Thiel, 2006). These authors have suggested that for long-term recovery, RH recruitment may be less efficient than restoring the LH network. Patients with better recovery have been observed to have higher activation in specific areas such as the L superior temporal gyrus and L SMA; the results from the (Naeser, 2012) study support the findings of the rTMS and tDCS studies reviewed above with aphasia patients, suggesting that these other types of noninvasive brain stimulation are also more effective, when the left hemisphere is stimulated with excitatory stimulation, and/or the right hemisphere is suppressed with inhibitory stimulation.

While the noninvasive transcranial treatment methods reviewed in this chapter (tPBM including tLEDs, rTMS and tDCS) vary in terms of methodology, they all seek to increase brain function by stimulating neural activity, are thought to subsequently stimulate *neurogenesis* (the formation of new neurons, or brain cells) and *synaptogenesis* (the formation of new synapses or connections between neurons) in the rehabilitation of TBI and stroke. Neurogenesis is the generation of neuronal precursors and birth of new neural cells. Two key sites for adult human neurogenesis include the subventricular zone of the lateral ventricles, and the subgranular layer of the dentate gyrus in the hippocampus (Eriksson et al., 1998). "Neurogenesis persists in the adult mammalian brain, where it can be stimulated by physiological factors, such as growth factors and environmental enrichment, and by pathological processes, including ischemia..." (Jin et al., 2006).

Increased vascular endothelial growth factor (VEGF) has been associated with enhanced postischemic neurogenesis and migration of newly formed neurons, in VEGF-over-expressing transgenic mice, consistent with a possible role in ischemic brain tissue repair (Wang et al., 2007). VEGF could provide a therapeutic target for more chronic brain repair. NIR laser therapy has been observed to increase VEGF and angiogenesis following myocardial infarction in animal studies (Tuby et al., 2006). Scalp application of red/NIR LED could stimulate secretion of VEGF, thus promoting angiogenesis and neurogenesis.

Brain plasticity has been defined as "the brain's ability to undergo functional and structural alterations in response to internal and external environmental changes" (May et al., 2007). An example of brain plasticity in normal adult humans was observed after only five rTMS sessions to the left superior temporal gyrus, when macroscopic cortical changes were then observed in Heschl's gyrus using MRI. The rTMS was applied at 1 Hz, with 2000 pulses per session, at 110% of motor threshold. In another study by this group, it was observed that after 3 months of learning a new motor task (juggling), there was transient gray matter increase in extrastriate motion specific visual area hMT/V5 bilaterally, and in the left inferior parietal sulcus. These changes degraded nearly to baseline 3 months after the juggling stopped (May et al., 2007; Draganski et al., 2004; Draganski and May, 2008).

It is believed that increased neuroplasticity is a foundation for stroke rehabilitation and recovery (Di Pino et al., 2014). Future research with tPBM, rTMS, and tDCS should establish the ideal treatment parameters for any given modality. These modalities would not be administered at the same time in a stroke patient, but it is possible that optimal results could be obtained if they were used sequentially, or in alternating periods of treatment.

43.10.3 Low intensity pulsed ultrasound

Ultrasound (US) is a term used for sound waves at higher frequencies than the audible range for human hearing. Ultrasound has been extensively used in medicine and industry since World War II. In addition to numerous imaging applications, therapeutic US has been employed or investigated for lithotripsy, sonophoresis, gene therapy, bone healing, drug delivery, peripheral nerve blocking, and localized tumor ablation (Sundaram et al., 2017). While high intensity focused US (HIFU) has been quite widely studied by tissue ablation and for cancer therapy due its particular ability to stimulate the immune system (van den Bijgaart et al., 2017), low intensity focused ultrasound (LIFU) does not produce and heating or damage of tissue. Tyler et al. (2008) reported that LIFU could stimulate membrane ion channels and enhance synaptic transmissions. They also proposed a protocol to use LIFU to stimulate the motor cortex in the mouse brain (Tufail et al., 2011). It was subsequently realized that LIFU had a distinct advantage over rTMS and tDCS in that the spatial resolution within the brain was much better (Rezayat and Toostani, 2016). Moreover LIFU is ideally suited to be used within a functional MRI scanner (Yuan et al., 2017). Legon et al. worked with human volunteers and found that the lateral and axial spatial resolution of the LIFU beam was 4.9 and 18 mm, respectively (Legon et al., 2014). Electroencephalographic recordings showed that LIFU significantly attenuated the amplitudes of somatosensory evoked potentials elicited by median nerve stimulation. Monto et al. (2016) published a first-in-man report describing the use of LIFU to stimulate the thalamus in a 25-year-old man who was in a minimally conscious state after a road traffic accident. The patient progressed from having only a fleeting awareness of the outside world to being able to answer questions and attempt to walk.

43.11 Could an invasive approach be considered?

The success of deep brain stimulation (DBS) especially for PD over many years has led many to question, whether PBM would be better delivered by some invasive procedure such as inserting an optical fiber into the brain. This would be similar to the way an electrode is inserted into the brain in DBS. In 1983 a chance clinical observation related to drug abuse (Langston et al., 1983) led to the creation of a monkey model for PD by intravenous injection of MPTP (*N*-methyl-4-phenyl-1,2,3,6-tetrahydropyridine) (Burns et al., 1983). This new model was characterized by selective destruction of dopaminergic neurons in the pars compacta of the substantia nigra. Investigation of this model led to the discovery of basal ganglia-thalamocortical loops, and showed over-activity in a part of the basal ganglia called the subthalamic nucleus (STN) in Parkinson's. Bergman et al. (1990) showed that surgical lesions created in the STN could lead to reversal of experimental parkinsonism.

A neurosurgeon in Grenoble, France called Benabid (Benabid et al., 1989) then developed a technique to implant electrodes on both sides of the brain in three patients with disabling akinetic-rigid PD and severe motor fluctuations (Limousin et al., 1995). FDA approval was first received in 1997, and DBS has now been studied for essential tremor, dystonia, epilepsy TBI, depression, and chronic pain (Chen et al., 2013).

The first reports of DBS being used to treat PD at Professor Benabid's clinic were published in 1993 (Benabid et al., 1993), although the team had first performed the procedure in 1987. Today the treatment of PD using DBS is carried out worldwide. Current estimates are that approximately 20,000 patients with movement disorders have been treated with DBS, including dystonia, a much rarer disease than PD, but one that can affect children (Vidailhet et al., 2013).

The Mitrofanis laboratory in Australia has experimented with implanting of optical fibers into the brains of experimental animals. This laboratory has generally studied models of PD and they have explored whole head light delivery and remote light delivery to the back, and have shown major improvements in many measures of PD severity (Johnstone et al., 2015). They have also inserted optical fibers into the brains of these experimental animals including rats (Reinhart et al., 2016), mice (Moro et al., 2013), and nonhuman primates (El-Massri et al., 2017).

43.12 What does the future hold?

It appears that the field of PBM for brain disorders in on the cusp of success. As this book shows, it is now being widely explored in many different countries of the world for a vast range of different indications. Perhaps what has been holding back the field the most, is the relative uncertainty of what is the best business model for companies in PBM to pursue. Pharmaceutical companies have a well-defined business model of selling drugs administered as oral tablets to be taken daily and generally sold as 30- or 90-day prescriptions. These drugs may even be taken by patients for years or decades (Dierks et al., 2016). In a similar manner, companies engaged in the fight against cancer can develop expensive biologics that require IV infusions, but are unlikely to be administered over the long term. In the days when lasers were accepted as the most likely light sources for PBM, at least expensive laser systems could be considered to be important pieces of medical equipment. One consideration that drove the progression of Photothera Inc. through the NEST-1–3 trials (Lapchak and Boitano, 2016) was the realization that if it had been successful in obtaining FDA approval, then laser systems costing in excess of $100,000 would have been required equipment in emergency rooms in every major hospital throughout the country.

The question now however is, how are companies expected to make a satisfactory return on investment when LED devices costing mere hundreds of dollars (or at the most a few thousand dollars) may be all that is required? The excellent safety and almost total lack of side-effects of LED devices, suggests that mass market home use devices may be the way to go. However, how are the major advertising and marketing efforts that will undoubtedly be required before take-up of this somewhat unconventional method of treating brain disorders penetrates to a sizable fraction of the potential market, to be paid for? If doctors could prescribe LED devices for their patients, then at least the barrier of funding widespread public education campaigns might be avoided.

The amazing growth of the internet has led to word of mouth (spread of information) about all aspects of PBM, and PBM for the brain in particular. Numerous small companies selling PBM devices have sprung up, and even home-made LED devices are becoming talked about. PBM for the brain has become of some interest to the biohacking community (Yetisen, 2018).

References

Allendorfer, J.B., Storrs, J.M., Szaflarski, J.P., 2012. Changes in white matter integrity follow excitatory rTMS treatment of post-stroke aphasia. Restor. Neurol. Neurosci. 30 (2), 103–113.

Ando, T., Xuan, W., Xu, T., Dai, T., Sharma, S.K., Kharkwal, G.B., et al., 2011. Comparison of therapeutic effects between pulsed and continuous wave 810-nm wavelength laser irradiation for traumatic brain injury in mice. PLoS One 6 (10), e26212.

Baijal, S., Srinivasan, N., 2010. Theta activity and meditative states: spectral changes during concentrative meditation. Cogn. Process. 11 (1), 31–38.

Baker, J.M., Rorden, C., Fridriksson, J., 2010. Using transcranial direct-current stimulation to treat stroke patients with aphasia. Stroke 41 (6), 1229–1236.

Barker, A.T., Jalinous, R., Freeston, I.L., 1985. Non-invasive magnetic stimulation of human motor cortex. Lancet 1 (8437), 1106–1107.

Barrett, D.W., Gonzalez-Lima, F., 2013. Transcranial infrared laser stimulation produces beneficial cognitive and emotional effects in humans. Neuroscience 230, 13–23.

Barwood, C.H., Murdoch, B.E., Whelan, B.M., Lloyd, D., Riek, S., O'Sullivan, J.D., et al., 2011. Improved language performance subsequent to low-frequency rTMS in patients with chronic non-fluent aphasia post-stroke. Eur. J. Neurol. 18 (7), 935–943.

Benabid, A.L., Pollak, P., Hommel, M., Gaio, J.M., de Rougemont, J., Perret, J., 1989. [Treatment of Parkinson tremor by chronic stimulation of the ventral intermediate nucleus of the thalamus]. Rev. Neurol. (Paris) 145 (4), 320–323.

Benabid, A.L., Pollak, P., Seigneuret, E., Hoffmann, D., Gay, E., Perret, J., 1993. Chronic VIM thalamic stimulation in Parkinson's disease, essential tremor and extra-pyramidal dyskinesias. Acta Neurochir. Suppl. (Wien) 58, 39–44.

Bergman, H., Wichmann, T., DeLong, M.R., 1990. Reversal of experimental parkinsonism by lesions of the subthalamic nucleus. Science 249 (4975), 1436–1438.

Berman, M.H., Halper, J.P., Nichols, T.W., Jarrett, H., Lundy, A., Huang, J.H., 2017. Photobiomodulation with near infrared light helmet in a pilot, placebo controlled clinical trial in dementia patients testing memory and cognition. J. Neurol. Neurosci. 8 (1).

Blanco, N.J., Saucedo, C.L., Gonzalez-Lima, F., 2017a. Transcranial infrared laser stimulation improves rule-based, but not information-integration, category learning in humans. Neurobiol. Learn. Mem. 139, 69–75.

Blanco, N.J., Maddox, W.T., Gonzalez-Lima, F., 2017b. Improving executive function using transcranial infrared laser stimulation. J. Neuropsychol. 11 (1), 14–25.

Burns, R.S., Chiueh, C.C., Markey, S.P., Ebert, M.H., Jacobowitz, D.M., Kopin, I.J., 1983. A primate model of parkinsonism: selective destruction of dopaminergic neurons in the pars compacta of the substantia nigra by N-methyl-4-phenyl-1,2,3,6-tetrahydropyridine. Proc. Natl. Acad. Sci. U.S.A. 80 (14), 4546–4550.

Cassano, P., Cusin, C., Mischoulon, D., Hamblin, M.R., De Taboada, L., Pisoni, A., et al., 2015. Near-infrared transcranial radiation for major depressive disorder: proof of concept study. Psychiatry J. 2015, 352979.

Chen, X.L., Xiong, Y.Y., Xu, G.L., Liu, X.F., 2013. Deep brain stimulation. Interv. Neurol. 1 (3–4), 200–212.

Chow, R.T., David, M.A., Armati, P.J., 2007. 830 nm laser irradiation induces varicosity formation, reduces mitochondrial membrane potential and blocks fast axonal flow in small and medium diameter rat dorsal root ganglion neurons: implications for the analgesic effects of 830 nm laser. J. Peripher. Nerv. Syst. 12 (1), 28–39.

de la Torre, J.C., 2017. Treating cognitive impairment with transcranial low level laser therapy. J. Photochem. Photobiol. B 168, 149–155.

Dierks, R.M., Bruyere, O., Reginster, J.Y., Richy, F.F., 2016. Macro-economic factors influencing the architectural business model shift in the pharmaceutical industry. Expert Rev. Pharmacoecon. Outcomes Res. 16 (5), 571–578.

Di Pino, G., Pellegrino, G., Assenza, G., Capone, F., Ferreri, F., Formica, D., et al., 2014. Modulation of brain plasticity in stroke: a novel model for neurorehabilitation. Nat. Rev. Neurol. 10 (10), 597–608.

Disner, S.G., Beevers, C.G., Gonzalez-Lima, F., 2016. Transcranial laser stimulation as neuroenhancement for attention bias modification in adults with elevated depression symptoms. Brain Stimul. 9 (5), 780–787.

Draganski, B., May, A., 2008. Training-induced structural changes in the adult human brain. Behav. Brain Res. 192 (1), 137–142.

Draganski, B., Gaser, C., Busch, V., Schuierer, G., Bogdahn, U., May, A., 2004. Neuroplasticity: changes in grey matter induced by training. Nature 427 (6972), 311–312.

El Massri, N., Lemgruber, A.P., Rowe, I.J., Moro, C., Torres, N., Reinhart, F., et al., 2017. Photobiomodulation-induced changes in a monkey model of Parkinson's disease: changes in tyrosine hydroxylase cells and GDNF expression in the striatum. Exp. Brain Res. 235 (6), 1861–1874.

El Massri, N., Weinrich, T.W., Kam, J.H., Jeffery, G., Mitrofanis, J., 2018. Photobiomodulation reduces gliosis in the basal ganglia of aged mice. Neurobiol. Aging. 66, 131–137.

Elsner, B., Kugler, J., Pohl, M., Mehrholz, J., 2013. Transcranial direct current stimulation (tDCS) for improving function and activities of daily living in patients after stroke. Cochrane Database Syst. Rev. 11, CD009645.

Eriksson, P.S., Perfilieva, E., Bjork-Eriksson, T., Alborn, A.M., Nordborg, C., Peterson, D.A., et al., 1998. Neurogenesis in the adult human hippocampus. Nat. Med. 4 (11), 1313–1317.

Fama, M.E., Turkeltaub, P.E., 2014. Treatment of poststroke aphasia: current practice and new directions. Semin. Neurol. 34 (5), 504–513.

Fridriksson, J., Richardson, J.D., Baker, J.M., Rorden, C., 2011. Transcranial direct current stimulation improves naming reaction time in fluent aphasia: a double-blind, sham-controlled study. Stroke 42 (3), 819–821.

Gilboa, G., Chen, R., Brenner, N., 2005. History-dependent multiple-time-scale dynamics in a single-neuron model. J. Neurosci. 25 (28), 6479–6489.

Hamilton, R.H., Sanders, L., Benson, J., Faseyitan, O., Norise, C., Naeser, M., et al., 2010. Stimulating conversation: enhancement of elicited propositional speech in a patient with chronic non-fluent aphasia following transcranial magnetic stimulation. Brain Lang. 113 (1), 45–50.

Hart, N.S., Fitzgerald, M., 2016. A new perspective on delivery of red-near-infrared light therapy for disorders of the brain. Discov. Med. 22 (120), 147–156.

Heiss, W.D., Thiel, A., 2006. A proposed regional hierarchy in recovery of post-stroke aphasia. Brain Lang. 98 (1), 118–123.

Henderson, T.A., Morries, L.D., 2015a. SPECT perfusion imaging demonstrates improvement of traumatic brain injury with transcranial near-infrared laser phototherapy. Adv. Mind Body Med. 29 (4), 27–33.

Henderson, T.A., Morries, L.D., 2015b. Near-infrared photonic energy penetration: can infrared phototherapy effectively reach the human brain? Neuropsychiatr. Dis. Treat. 11, 2191–2208.

Huang, Y.Z., Edwards, M.J., Rounis, E., Bhatia, K.P., Rothwell, J.C., 2005. Theta burst stimulation of the human motor cortex. Neuron 45 (2), 201–206.

Hwang, J., Castelli, D.M., Gonzalez-Lima, F., 2016. Cognitive enhancement by transcranial laser stimulation and acute aerobic exercise. Lasers Med. Sci. 31 (6), 1151–1160.

Iaccarino, H.F., Singer, A.C., Martorell, A.J., Rudenko, A., Gao, F., Gillingham, T.Z., et al., 2016. Gamma frequency entrainment attenuates amyloid load and modifies microglia. Nature 540 (7632), 230–235.

Ilic, S., Leichliter, S., Streeter, J., Oron, A., DeTaboada, L., Oron, U., 2006. Effects of power densities, continuous and pulse frequencies, and number of sessions of low-level laser therapy on intact rat brain. Photomed. Laser Surg. 24 (4), 458–466.

Ismail, R., Hansen, A.K., Parbo, P., Braendgaard, H., Gottrup, H., Brooks, D.J., et al., 2018. The effect of 40-Hz light therapy on amyloid load in patients with prodromal and clinical Alzheimer's disease. Int. J. Alzheimers Dis. 2018, 6852303.

Jagdeo, J.R., Adams, L.E., Brody, N.I., Siegel, D.M., 2012. Transcranial red and near infrared light transmission in a cadaveric model. PLoS One 7 (10), e47460.

Jin, K., Wang, X., Xie, L., Mao, X.O., Zhu, W., Wang, Y., et al., 2006. Evidence for stroke-induced neurogenesis in the human brain. Proc. Natl. Acad. Sci. U.S.A. 103 (35), 13198–13202.

Johnstone, D.M., el Massri, N., Moro, C., Spana, S., Wang, X.S., Torres, N., et al., 2014. Indirect application of near infrared light induces neuroprotection in a mouse model of parkinsonism—an abscopal neuroprotective effect. Neuroscience 274, 93–101.

Johnstone, D.M., Moro, C., Stone, J., Benabid, A.L., Mitrofanis, J., 2015. Turning on lights to stop neurodegeneration: the potential of near infrared light therapy in Alzheimer's and Parkinson's disease. Front. Neurosci. 9, 500.

Kampa, B.M., Clements, J., Jonas, P., Stuart, G.J., 2004. Kinetics of Mg^{2+} unblock of NMDA receptors: implications for spike-timing dependent synaptic plasticity. J. Physiol. 556 (Pt 2), 337–345.

Kang, E.K., Paik, N.J., 2011. Effect of a tDCS electrode montage on implicit motor sequence learning in healthy subjects. Exp. Transl. Stroke Med. 3 (1), 4.

Karbe, H., Thiel, A., Weber-Luxenburger, G., Herholz, K., Kessler, J., Heiss, W.D., 1998. Brain plasticity in poststroke aphasia: what is the contribution of the right hemisphere? Brain Lang. 64 (2), 215–230.

Karu, T.I., 2008. Mitochondrial signaling in mammalian cells activated by red and near-IR radiation. Photochem. Photobiol. 84 (5), 1091–1099.

Karu, T.I., Pyatibrat, L.V., Afanasyeva, N.I., 2004. A novel mitochondrial signaling pathway activated by visible-to-near infrared radiation. Photochem. Photobiol. 80 (2), 366–372.

Kim, B., Mitrofanis, J., Stone, J., Johnstone, D.M., 2018. Remote tissue conditioning is neuroprotective against MPTP insult in mice. IBRO Rep. 4, 14–17.

Kirschfeld, K., 2005. The physical basis of alpha waves in the electroencephalogram and the origin of the "Berger effect". Biol. Cybern. 92 (3), 177–185.

Lane, N., 2006. Cell biology: power games. Nature 443 (7114), 901–903.

Langston, J.W., Ballard, P., Tetrud, J.W., Irwin, I., 1983. Chronic Parkinsonism in humans due to a product of meperidine-analog synthesis. Science 219 (4587), 979–980.

Lapchak, P.A., Boitano, P.D., 2016. Transcranial near-infrared laser therapy for stroke: how to recover from futility in the NEST-3 clinical trial. Acta Neurochir. Suppl. 121, 7–12.

Lapchak, P.A., Boitano, P.D., Butte, P.V., Fisher, D.J., Holscher, T., Ley, E.J., et al., 2015. Transcranial Near-Infrared Laser Transmission (NILT) profiles (800 nm): systematic comparison in four common research species. PLoS One 10 (6), e0127580.

Lefaucheur, J.P., 2006. Stroke recovery can be enhanced by using repetitive transcranial magnetic stimulation (rTMS). Neurophysiol. Clin. 36 (3), 105–115.

Legon, W., Sato, T.F., Opitz, A., Mueller, J., Barbour, A., Williams, A., et al., 2014. Transcranial focused ultrasound modulates the activity of primary somatosensory cortex in humans. Nat. Neurosci. 17 (2), 322–329.

Leisman, G., Machado, C., Machado, Y., Chinchilla-Acosta, M., 2018. Effects of low-level laser therapy in autism spectrum disorder. Adv. Exp. Med. Biol. 1116, 111–130.

Liew, S.L., Santarnecchi, E., Buch, E.R., Cohen, L.G., 2014. Non-invasive brain stimulation in neurorehabilitation: local and distant effects for motor recovery. Front. Hum. Neurosci. 8, 378.

Limousin, P., Pollak, P., Benazzouz, A., Hoffmann, D., Le Bas, J.F., Broussolle, E., et al., 1995. Effect of parkinsonian signs and symptoms of bilateral subthalamic nucleus stimulation. Lancet 345 (8942), 91–95.

Lopes da Silva, F., 1991. Neural mechanisms underlying brain waves: from neural membranes to networks. Electroencephalogr. Clin. Neurophysiol. 79 (2), 81–93.

Maeda, F., Keenan, J.P., Tormos, J.M., Topka, H., Pascual-Leone, A., 2000. Modulation of corticospinal excitability by repetitive transcranial magnetic stimulation. Clin. Neurophysiol. 111 (5), 800–805.

Marangolo, P., Fiori, V., Calpagnano, M.A., Campana, S., Razzano, C., Caltagirone, C., et al., 2013. tDCS over the left inferior frontal cortex improves speech production in aphasia. Front. Hum. Neurosci. 7, 539.

Martin, P.I., Naeser, M.A., Theoret, H., Tormos, J.M., Nicholas, M., Kurland, J., et al., 2004. Transcranial magnetic stimulation as a complementary treatment for aphasia. Semin. Speech Lang. 25 (2), 181–191.

Martin, P.I., Naeser, M.A., Ho, M., Doron, K.W., Kurland, J., Kaplan, J., et al., 2009. Overt naming fMRI pre- and post-TMS: two nonfluent aphasia patients, with and without improved naming post-TMS. Brain Lang. 111 (1), 20–35.

May, A., Hajak, G., Ganssbauer, S., Steffens, T., Langguth, B., Kleinjung, T., et al., 2007. Structural brain alterations following 5 days of intervention: dynamic aspects of neuroplasticity. Cereb. Cortex 17 (1), 205–210.

Merton, P.A., Morton, H.B., 1980. Stimulation of the cerebral cortex in the intact human subject. Nature 285 (5762), 227.

Michalikova, S., Ennaceur, A., van Rensburg, R., Chazot, P.L., 2008. Emotional responses and memory performance of middle-aged CD1 mice in a 3D maze: effects of low infrared light. Neurobiol. Learn. Mem. 89 (4), 480–488.

Mitrofanis, J., 2017. Why and how does light therapy offer neuroprotection in Parkinson's disease? Neural Regen. Res. 12 (4), 574–575.

Monti, A., Cogiamanian, F., Marceglia, S., Ferrucci, R., Mameli, F., Mrakic-Sposta, S., et al., 2008. Improved naming after transcranial direct current stimulation in aphasia. J. Neurol. Neurosurg. Psychiatry 79 (4), 451–453.

Monti, M.M., Schnakers, C., Korb, A.S., Bystritsky, A., Vespa, P.M., 2016. Non-invasive ultrasonic thalamic stimulation in disorders of consciousness after severe brain injury: a first-in-man report. Brain Stimul. 9 (6), 940–941.

Moro, C., Torres, N., El Massri, N., Ratel, D., Johnstone, D.M., Stone, J., et al., 2013. Photobiomodulation preserves behaviour and midbrain dopaminergic cells from MPTP toxicity: evidence from two mouse strains. BMC Neurosci. 14, 40.

Naeser, M., 2012. Improved language after scalp application of red/near-infrared light-emitting diodes: pilot study supporting a new, noninvasive treatment for chronic aphasia. Procedia. Soc. Behav. Sci. 61, 138–139.

Naeser, M.A., Martin, P.I., Nicholas, M., Baker, E.H., Seekins, H., Kobayashi, M., et al., 2005. Improved picture naming in chronic aphasia after TMS to part of right Broca's area: an open-protocol study. Brain Lang. 93 (1), 95–105.

Naeser, M.A., Saltmarche, A., Krengel, M.H., Hamblin, M.R., Knight, J.A., 2011. Improved cognitive function after transcranial, light-emitting diode treatments in chronic, traumatic brain injury: two case reports. Photomed. Laser Surg. 29 (5), 351–358.

Naeser, M.A., Ho, M., Martin, P.I., Treglia, E., Krengel, M.H., Hamblin, M.R., et al., 2012. Improved language after scalp application of red/near-infrared light-emitting diodes: pilot study supporting a new, noninvasive treatment for chronic aphasia. Procedia Soc. Behav. Sci. 61, 138–139.

Naeser, M.A., Zafonte, R., Krengel, M.H., Martin, P.I., Frazier, J., Hamblin, M.R., et al., 2014. Significant improvements in cognitive performance post-transcranial, red/near-infrared light-emitting diode treatments in chronic, mild traumatic brain injury: open-protocol study. J. Neurotrauma 31 (11), 1008–1017.

Oron, A., Oron, U., Chen, J., Eilam, A., Zhang, C., Sadeh, M., et al., 2006. Low-level laser therapy applied transcranially to rats after induction of stroke significantly reduces long-term neurological deficits. Stroke 37 (10), 2620–2624.

Oron, A., Oron, U., Streeter, J., de Taboada, L., Alexandrovich, A., Trembovler, V., et al., 2007. low-level laser therapy applied transcranially to mice following traumatic brain injury significantly reduces long-term neurological deficits. J. Neurotrauma 24 (4), 651–656.

Oron, A., Oron, U., Streeter, J., De Taboada, L., Alexandrovich, A., Trembovler, V., et al., 2012. Near infrared transcranial laser therapy applied at various modes to mice following traumatic brain injury significantly reduces long-term neurological deficits. J. Neurotrauma 29 (2), 401–407.

Pascual-Leone, A., Valls-Sole, J., Wassermann, E.M., Hallett, M., 1994. Responses to rapid-rate transcranial magnetic stimulation of the human motor cortex. Brain 117 (Pt 4), 847–858.

Pascual-Leone, A., Tormos, J.M., Keenan, J., Tarazona, F., Canete, C., Catala, M.D., 1998. Study and modulation of human cortical excitability with transcranial magnetic stimulation. J. Clin. Neurophysiol. 15 (4), 333–343.

Picchioni, D., Killgore, W.D., Balkin, T.J., Braun, A.R., 2009. Positron emission tomography correlates of visually-scored electroencephalographic waveforms during non-Rapid Eye Movement sleep. Int. J. Neurosci. 119 (11), 2074–2099.

Pitzschke, A., Lovisa, B., Seydoux, O., Zellweger, M., Pfleiderer, M., Tardy, Y., et al., 2015. Red and NIR light dosimetry in the human deep brain. Phys. Med. Biol. 60 (7), 2921–2937.

Priestley, T., Kemp, J.A., 1994. Kinetic study of the interactions between the glutamate and glycine recognition sites on the *N*-methyl-D-aspartic acid receptor complex. Mol. Pharmacol. 46 (6), 1191–1196.

Reinhart, F., Massri, N.E., Chabrol, C., Cretallaz, C., Johnstone, D.M., Torres, N., et al., 2016. Intracranial application of near-infrared light in a hemi-parkinsonian rat model: the impact on behavior and cell survival. J. Neurosurg. 124 (6), 1829–1841.

Rezayat, E., Toostani, I.G., 2016. A review on brain stimulation using low intensity focused ultrasound. Basic Clin. Neurosci. 7 (3), 187–194.

Saltmarche, A.E., Naeser, M.A., Ho, K.F., Hamblin, M.R., Lim, L., 2017. Significant improvement in cognition in mild to moderately severe dementia cases treated with transcranial plus intranasal photobiomodulation: case series report. Photomed. Laser Surg. 35 (8), 432–441.

Saur, D., Lange, R., Baumgaertner, A., Schraknepper, V., Willmes, K., Rijntjes, M., et al., 2006. Dynamics of language reorganization after stroke. Brain 129 (Pt 6), 1371–1384.

Schiffer, F., Johnston, A.L., Ravichandran, C., Polcari, A., Teicher, M.H., Webb, R.H., et al., 2009. Psychological benefits 2 and 4 weeks after a single treatment with near infrared light to the forehead: a pilot study of 10 patients with major depression and anxiety. Behav. Brain Funct. 5, 46.

Schneggenburger, R., Neher, E., 2000. Intracellular calcium dependence of transmitter release rates at a fast central synapse. Nature 406 (6798), 889–893.

Shah, P.P., Szaflarski, J.P., Allendorfer, J., Hamilton, R.H., 2013. Induction of neuroplasticity and recovery in post-stroke aphasia by non-invasive brain stimulation. Front. Hum. Neurosci. 7, 888.

Sparing, R., Mottaghy, F.M., 2008. Noninvasive brain stimulation with transcranial magnetic or direct current stimulation (TMS/tDCS)-From insights into human memory to therapy of its dysfunction. Methods 44 (4), 329–337.

Sundaram, K.M., Chang, S.S., Penson, D.F., Arora, S., 2017. Therapeutic ultrasound and prostate cancer. Semin. Intervent. Radiol. 34 (2), 187–200.

Szaflarski, J.P., Vannest, J., Wu, S.W., DiFrancesco, M.W., Banks, C., Gilbert, D.L., 2011. Excitatory repetitive transcranial magnetic stimulation induces improvements in chronic post-stroke aphasia. Med. Sci. Monit. 17 (3), CR132–CR139.

Szaflarski, J.P., Allendorfer, J.B., Banks, C., Vannest, J., Holland, S.K., 2013. Recovered vs. not-recovered from post-stroke aphasia: the contributions from the dominant and non-dominant hemispheres. Restor. Neurol. Neurosci. 31 (4), 347–360.

Tedford, C.E., DeLapp, S., Jacques, S., Anders, J., 2015. Quantitative analysis of transcranial and intraparenchymal light penetration in human cadaver brain tissue. Lasers Surg. Med. 47 (4), 312–322.

Thuroczy, G., Kubinyi, G., Bodo, M., Bakos, J., Szabo, L.D., 1994. Simultaneous response of brain electrical activity (EEG) and cerebral circulation (REG) to microwave exposure in rats. Rev. Environ. Health 10 (2), 135–148.

Tuby, H., Maltz, L., Oron, U., 2006. Modulations of VEGF and iNOS in the rat heart by low level laser therapy are associated with cardioprotection and enhanced angiogenesis. Lasers Surg. Med. 38 (7), 682–688.

Tufail, Y., Yoshihiro, A., Pati, S., Li, M.M., Tyler, W.J., 2011. Ultrasonic neuromodulation by brain stimulation with transcranial ultrasound. Nat. Protoc. 6 (9), 1453–1470.

Tyler, W.J., Tufail, Y., Finsterwald, M., Tauchmann, M.L., Olson, E.J., Majestic, C., 2008. Remote excitation of neuronal circuits using low-intensity, low-frequency ultrasound. PLoS One 3 (10), e3511.

Valchinov, E.S., Pallikarakis, N.E., 2005. Design and testing of low intensity laser biostimulator. Biomed. Eng. Online 4 (1), 5.

van den Bijgaart, R.J., Eikelenboom, D.C., Hoogenboom, M., Futterer, J.J., den Brok, M.H., Adema, G.J., 2017. Thermal and mechanical high-intensity focused ultrasound: perspectives on tumor ablation, immune effects and combination strategies. Cancer Immunol. Immunother. 66 (2), 247–258.

Vidailhet, M., Jutras, M.F., Grabli, D., Roze, E., 2013. Deep brain stimulation for dystonia. J. Neurol. Neurosurg. Psychiatry 84 (9), 1029–1042.

Wan, S., Parrish, J.A., Anderson, R.R., Madden, M., 1981. Transmittance of nonionizing radiation in human tissues. Photochem. Photobiol. 34 (6), 679–681.

Wang, F., 2006. Therapeutic effect observation and nurse of intranasal low intensity laser therapy on insomnia. J. Commun. Med. 4, 58.

Wang, R., Dong, Y., Lu, Y., Zhang, W., Brann, D.W., Zhang, Q., 2019. Photobiomodulation for global cerebral ischemia: targeting mitochondrial dynamics and functions. Mol. Neurobiol. 56, 1852–1869.

Wang, Y., Jin, K., Mao, X.O., Xie, L., Banwait, S., Marti, H.H., et al., 2007. VEGF-overexpressing transgenic mice show enhanced post-ischemic neurogenesis and neuromigration. J. Neurosci. Res. 85 (4), 740–747.

Weiduschat, N., Thiel, A., Rubi-Fessen, I., Hartmann, A., Kessler, J., Merl, P., et al., 2011. Effects of repetitive transcranial magnetic stimulation in aphasic stroke: a randomized controlled pilot study. Stroke 42 (2), 409–415.

Westmoreland, B.F., Klass, D.W., 1990. Unusual EEG patterns. J. Clin. Neurophysiol. 7 (2), 209–228.

Wu, Q., Xuan, W., Ando, T., Xu, T., Huang, L., Huang, Y.Y., et al., 2012. Low-level laser therapy for closed-head traumatic brain injury in mice: effect of different wavelengths. Lasers Surg. Med. 44 (3), 218–226.

Xu, C., Wang, L., Liu, J.H., Tan, Y., Li, Q., 2001. Endonasal low energy He-Ne laser treatment of insomnia. Qian Wei J. Med. Pharm. 18, 337–338.

Xu, C., Wang, L., Shang, X., Li, Q., 2002. The treatment of Alzheimer's disease with hypoenergy He-Ne laser. Prac. J. Med. Pharm. 19, 647–648.

Xu, C., Wang, L., Lu, C., 2002. Endonasal low energy He-Ne laser treatment of poststroke depression. Prac. J. Med. Pharm. 19, 893.

Xu, C., Wu, Z., Wang, L., Shang, X., Li, Q., 2002. The effects of endonasal low energy He-Ne laser treatment of insomniaon on Sleep EEG. Prac. J. Med. Pharm. 19, 407–408.

Xu, C., Lu, C., Wang, L., Li, Q., 2003. The effects of endonasal low energy He-Ne laser therapy on antioxydation of Parkinson's disease. Prac. J. Med. Pharm. 20, 816–817.

Xuan, W., Vatansever, F., Huang, L., Wu, Q., Xuan, Y., Dai, T., et al., 2013. Transcranial low-level laser therapy improves neurological performance in traumatic brain injury in mice: effect of treatment repetition regimen. PLoS One 8 (1), e53454.

Xuan, W., Vatansever, F., Huang, L., Hamblin, M.R., 2014a. Transcranial low-level laser therapy enhances learning, memory, and neuroprogenitor cells after traumatic brain injury in mice. J. Biomed. Opt. 19 (10), 108003.

Xuan, W., Agrawal, T., Huang, L., Gupta, G.K., Hamblin, M.R., 2014b. Low-level laser therapy for traumatic brain injury in mice increases brain derived neurotrophic factor (BDNF) and synaptogenesis. J. Biophotonics 9999 (9999).

Xuan, W., Huang, L., Hamblin, M.R., 2016. Repeated transcranial low-level laser therapy for traumatic brain injury in mice: biphasic dose response and long-term treatment outcome. J. Biophotonics 9 (11–12), 1263–1272.

Yetisen, A.K., 2018. Biohacking. Trends Biotechnol. 36 (8), 744–747.

Yuan, Y., Dong, Y., Hu, S., Zheng, T., Du, D., Du, J., et al., 2017. Reduced apparent diffusion coefficient in various brain areas following low-intensity transcranial ultrasound stimulation. Front. Neurosci. 11, 562.

Zalewska-Kaszubska, J., Obzejta, D., 2004. Use of low-energy laser as adjunct treatment of alcohol addiction. Lasers Med. Sci. 19 (2), 100–104.

Index

Note: Page numbers followed by "*f*" and "*t*" refer to figures and tables, respectively.

B

C

D

J

K

L

N

T

9780128153055